P9-AEZ-601

DRYING '85

DRYING '85

Edited by

Ryozo Toei

Kyoto University, Japan

and

Arun S. Mujumdar

McGill University, Canada

HEMISPHERE PUBLISHING CORPORATION

Washington New York London

DISTRIBUTION OUTSIDE NORTH AMERICA

SPRINGER-VERLAG

Berlin Heidelberg New York Tokyo

DRYING '85

1 2 3 4 5 6 7 8 9 0 BCBC 8 9 8 7 6 5

Library of Congress Cataloging in Publication Data

Main entry under title:

Drying '85.

Selection of papers from the 4th International Drying Symposium held in Kyoto, Japan, Jul. 9-12, 1984, under the sponsorship of the Society of Chemical Engineers, Japan.
Includes bibliographies and indexes.
1. Drying—Congresses. I. Tōei, Ryōzo, date- II. Mujumdar, A. S. III. International Symposium on Drying (4th : 1984 : Kyoto, Japan) IV. Kagaku Kōgaku Kyōkai (Japan)
TP363.D7925 1985 660.2′8426 85-746
ISBN 0-89116-444-8 Hemisphere Publishing Corporation

DISTRIBUTION OUTSIDE NORTH AMERICA:
ISBN 3-540-15710-7 Springer-Verlag Berlin

CONTENTS

SECTION V VACUUM AND FREEZE DRYING

SECTION VI DRYING OF FOODSTUFFS

SECTION X INDUSTRIAL DRYING SYSTEMS

SECTION XI PHYSICAL PROPERTIES RELATED TO DRYING

SECTION XII MISCELLANEOUS TOPICS

PREFACE

It is a pleasure to present this volume consisting of a selection of papers presented at the 4th International Drying Symposium (IDS) held in Kyoto, July 9-12, 1984, under the sponsorship of the Society of Chemical Engineers, Japan. This book includes full texts of the ten keynote lectures and several papers which were presented in the technical sessions in the 4th IDS.

As editors we had the arduous and unpleasant task of selecting papers for this volume. We were delighted with the high quality manuscripts submitted. We are grateful to the authors for their efforts in meeting the manuscript guidelines and also the deadlines. Aside from quality, we have attempted to maintain in this volume a balance of coverage in terms of topics as well as geography. Space restrictions prevented us from presenting all the high quality papers for publication in DRYING 85.

Numerous papers of archival interest have been released for general publication in international journals. Several will no doubt appear in Drying Technology - An International Journal. Many others will find their way into other standard engineering journals as well. Over 120 papers from 30 countries were included in the Proceedings volumes distributed to all registrants in Kyoto.

Numerous individuals, professional societies and organizations have contributed, directly or indirectly, to the success of the 4th IDS and hence to DRYING 85 as well. We wish to acknowledge in particular the monumental efforts of the Secretary of the 4th IDS, Professor M. Okazaki. We are indebted to the Society of Chemical Engineers, Japan, as well as the several co-sponsoring professional societies around the world for their hearty blessings to the IDS series.

We wish to record our sincere appreciation of the staff of Hemisphere and Springer Verlag for their efficient production and handling of this publication. Also, our thanks are due to Purnima Mujumdar for her services as our Editorial Assistant at the final stages of this work.

Chemical Engineering Department
Kyoto University
Kyoto, Japan

Ryozo Toei

Chemical Engineering Department
McGill University
Montreal, Canada

Arun S. Mujumdar

THE FOURTH INTERNATIONAL DRYING SYMPOSIUM

Kyoto International Conference Hall, Kyoto, Japan

July 9 - 12, 1984

Editors

R. Toei, Department of Chemical Engineering, Kyoto University, Kyoto, Japan
A.S. Mujumdar, Department of Chemical Engineering, McGill University, Montreal, Canada

Sponsor

The Society of Chemical Engineers, Japan

Co-Sponsors

American Institute of Chemical Engineers
Australian National Committee of the Institution of Chemical Engineers
The Institution of Engineers, Australia
Canadian Society of Chemical Engineers
Deutsche Vereinigung für Chemie- und Verfahrenstechnik, Federal Republic Germany
Indian Institute of Chemical Engineers
The Korean Institute of Chemical Engineers
The Institution of Chemical Engineers, London

Advisory Members

V.A. Borodulya, U.S.S.R.; E.J. Crosby, U.S.A.; W.J.M. Douglas, Canada;
I. Filkova, Czechoslovakia; W.H. Gauvin, Canada C.W. Hall, U.S.A.;
H. Hayashi, Japan; L. Imre, Hungary; M. Karel, U.S.A.;
R.B. Keey, New Zealand; J.J. Kelly, Ireland; H.G. Kessler, F.R. Germany;
C.J. King, U.S.A.; B. Kisakürek, Turkey; A.I. Morgan, U.S.A.;
S. Ohtani, Japan; M. Okazaki, Japan; O.E. Potter, Australia;
J.R. Puiggali, France; D. Reay, U.K.; E. Rotstein, Argentina;
E.U. Schlünder, F.R. Germany; R.M. Shah, India; A. Sorensen, Denmark;
C. Strumiłło, Poland; W. Tanthapanichakoon, Thailand;
H.A.C. Thijssen, The Netherlands; S. Whitaker, U.S.A.

NOMENCLATURE

The Symbols listed here are recommended for use in all papers contained in the Proceedings. Subscripts and/or superscripts should be used to distinguish symbols recomended for more than one quantity.

Symbol	Quantity	SI Unit
A	area	m^2
a	thermal diffusivity	m^2/s
a_t	turbulent thermal diffusivity (eddy thermal Diffusivity)	-
C_B	molecular concentration of B	mol/m^3
c_p	specific heat capacity at constant pressure	J/(K kg)
c_v	specific heat capacity at constant volume	J/(K kg)
D	diffusion coefficient	m^2/s
d	diameter	m
d_{equ}	equivalent (hydraulic) diameter	m
E	energy	J
F	force	N
f	specific free energy	J/kg
G	weight	N
g	specific Gibbs function	J/kg
g_n	acceleration of gravity = 9.80665	m/s^2
H	enthalpy	J
h	specific enthalpy	J/kg
Δh_v	specific latent heat of vaporization	J/kg
K	equilibrium constant	-
l	length	m
M	molar mass	kg/mol
m	mass	kg
$\dot{m}$	mass flux density	$kg/(m^2 s)$
N	Avogadro number	mol^{-1}

Symbol	Quantity	SI Unit
n	amount of substance	mol
P	power	W
p	pressure	N/m^2
Q	heat, quantity of heat	J
$\dot{q}$	heat flux density	W/m^2
R	universal gas constant = 8.3144	J/(mol K)
r	radius	m
S	entropy	J/K
s	specific entropy	J/(kg K)
T	thermodynamic temperature	K
t	time	s
u	velocity	m/s
V	volume	m^3
V_m	molar volume	m^3/mol
v	specific volume	m^3/kg
$\dot{v}$	volumetric flux density	$m^3/(m^2 s)$
W	work	J
w	water content	kg/kg
Greek letters		
α	heat transfer coefficient	$W/(K m^2)$
α_r	absorptance for radiation	-
α,β,γ	plane angles	rad
β	mass transfer coefficient	m/s
γ	cubic expansion coefficient	K^{-1}
Δ	difference	-
δ	thickness	m
ε	emissivity	-
η	dynamic viscosity	kg/(s m)
θ	Celsius temperature = $T - T_0$, T_0 = 273.15 K	°C
λ	thermal conductivity	W/(K m)
μ	chemical potential	J/mol
ν	kinematic viscosity	m^2/s
ρ	mass density	kg/m^3
ρ_B	mass concentration of B	kg/m^3
ρ_l	density of liquid	kg/m^3

Symbol	Quantity	SI Unit
Greek letters (Continued)		
ρ_v	density of vapor	kg/m^3
σ	Stefan-Boltzmann constant = $5.67032 \cdot 10^{-8}$	$W/(K^4 m^2)$
σ	surface tension	N/m
τ_s	shear stress	N/m^2
ϕ	relative humidity	-
ψ	void fraction	-
Coordinates		
x,y,z	Cartesian coordinates	
r,ϕ,z	cylindrical coordinates	
r,ϕ,ψ	spherical coordinates	

Symbol and definition	Name
Dimensionless parameters	
$Ar = \frac{g_n l^3 \Delta\rho}{\nu^2 \rho}$	Archimedes number
$B_i = \frac{\alpha l}{\lambda_{solid}}$	Biot number
$F_o = \frac{a t}{l^2}$	Fourier number
$Fr = \frac{u}{\sqrt{g_n l}}$	Froude number
$Gr = \frac{g_n l^3 \gamma \Delta T}{\nu^2}$	Grashof number
$Ja = \frac{c_p \rho_l \Delta T}{\rho_v \Delta h_v}$	Jakob number
$Le = \frac{a}{D}$	Lewis number
$Nu = \frac{\alpha l}{\lambda}$	Nusselt number

Symbol and definition	Name
Dimensionless parameters (Continued)	
$Pe = \frac{u\,l}{a} = Re \cdot Pr$	Peclet number
$Pe^* = \frac{u\,l}{D} = Re \cdot Sc$	Peclet number for mass transfer
$Pr = \frac{\nu}{a} = \frac{c_p\,\eta}{\lambda}$	Prandtl number
$Re = \frac{u\,l}{\nu}$	Reynolds number
$Sc = \frac{\nu}{D}$	Schmidt number
$Sh = \frac{\beta\,l}{D}$	Sherwood number
$St = \frac{\alpha}{\rho\,u\,c_p} = \frac{Nu}{Re \cdot Pr}$	Stanton number
$St^* = \frac{\beta}{u} = \frac{Sh}{Re \cdot Sc}$	Stanton number for mass transfer
$We = \frac{u^2\,\rho\,l}{\sigma}$	Weber number

DRYING '85

SECTION I: TEXTS OF KEYNOTE LECTURES AT IDS '86, KYOTO, JAPAN

OPPORTUNITIES FOR FUNDAMENTAL RESEARCH IN DRYING

Dr. Carl W. Hall

National Science Foundation
Washington, D.C. 20550

ABSTRACT

Physical phenomena at the molecular level, such as van der Walls interactions, can be used effectively for understanding and modeling drying relationships. Drying can be modeled utilizing fundamental phenomena, in contrast to most of the previous work in which these phenomena were used to explain empirical relationships.

New instruments and techniques provide an opportunity for advancing the knowledge of fundamental research in drying. The computer provides an opportunity for obtaining, storing, and analyzing a large quantity of data. Previously rather limited amounts of data were available from which analyses were made. New statistical and mathematical modeling approaches, a basis of much new fundamental research, need to be developed.

The two major developments make possible representing the drying relationship utilizing molecular building blocks to construct the product: newly-developed or about to be available instruments to measure and "see" the water molecule for a substrate (advanced electron microscopes), and high-speed, high-capacity computers (supercomputers).

The water molecule has a diameter of 2 Å and a bond length of 0.9 Å. Water molecules are on a substrate which has a wide range of dimensions but much larger than water. As an example, hemoglobin protein with a molecular weight of 68,000 daltons has dimensions of 38 Å by 150 Å. Spectroscopy has been used to determine the presence of elements or molecules, such as water. During drying we are interested not only in the presence of water molecules but also the amount and movement of water. Hopefully, the observations and measurements during drying can be done without destroying the relationships between water and the substrate. A challenge for researchers is to utilize these techniques for nondestructive evaluation (NDE). The implications of the water molecule as a dipole and drying have not been explored--possibly another opportunity for fundamental research.

The scanning electron microscope (SEM) has a resolution of 100 to 200 Å; the transmission electron microscope (TEM), as low as 1-5 Å; and a scanning transmission electron microscope (STEM), combining the above with additional features, is expected to have a resolution of 0.5 Å. The photoelectron microscope (PEM) is presently used to survey the walls of cells. Further, the possibility of looking in three dimensions, to observe and characterize the movement, change of state, vaporization and condensation of water, offers new opportunities for research. The flow of water (liquid and gas) will be modeled as it moves through the products (fats, sugars, cellulose, proteins, minerals, etc.) in much the same way that the flow of electrons or ions can be represented as they move through metals. Likewise, the effect on elemental units of the product by browning, checking, splitting, or as represented by destruction of vitamins or stabilization of enzymes, can be related. These relationships can be utilized further to develop systems and equipment for drying. Admittedly, these measurement techniques have not been extensively developed for drying of biological products but offer considerable opportunities.

The supercomputer can be used to represent and model the drying phenomena from the most elemental component (electron, chemical, radical, atom, molecule) for which information is available and applied to the entire drying system. The interaction of the water and substrate, energy involved, flow of water either as gas or liquid, and escape and measurement of water moving from the product (transcutaneous) can be used as parameters for the design of drying equipment. The components of the system can be modeled and the system can be represented with many parameters on the computer. The results of this approach using fundamental relationships can be used for several products and applications, providing results of a generic nature, while minimizing but not eliminating many of the time-consuming repetitive laboratory tests and empirical results from field tests.

1. INTRODUCTION

There have been few major breakthroughs in drying for the past 20 years. Improvements have been made in equipment and efficiency primarily by applying new technology rather than new fundamentals. Fundamental engineering research in drying offers the possibility of a breakthrough in understanding and carrying out drying operations. Fundamental engineering research is that research which delineates the mechanisms of an engineering process and provides information which is generic, and underpins the technology. Such an activity is generally long-range and transferable to potential users. Research in drying is often cross-interdisciplinary or multidisciplinary in approach on complex phenomena which do not have unique solutions. Fundamental engineering research includes understanding phenomena which generally involves interactions between energy and matter. Although I will speak primarily of fundamental research from the standpoint of engineering, participation by people in many other fields will be required to identify the basic phenomena involved in drying. Engineers will have to strengthen and broaden their background; and their interaction with biological scientists will have to increase in order to make major advances in understanding of biological materials.

The search for fundamental information may be technology-driven, science-driven, curiosity-driven, or market-driven. My presentation will focus on the potential for fundamental research through technology-driven developments of recent years, and these are the electron microscope, instrumentation and computer developments. It goes without saying that to these developments must be added the necessary ingredients of a strong science base and investigators who are very inquisitive.

2. OBJECTIVE

The objective of fundamental research in drying is to determine, represent, and evaluate moisture flow under dynamic conditions with different substrates and energy sources under varying environmental conditions.

3. PROCEDURES

Fundamental research in drying can be arrived at by (1) using science-based models to represent the molecular basis of action and reactions, and (2) using technology-based advances for measurements, computations, and modeling.

4. SCIENCE-BASED MODELS

These procedures can be used to study the product and its contained moisture and will be based on the physical, chemical and biological characteristics of the product. The characteristics of water molecules, the forces between water molecules, and the forces between the water molecules and the substrates (often biological) are important phenomena to be represented.

The general approach is to build the molecules on the substrates, using an inert material like sand, or a biological material such as proteins, fat, fiber, or combinations thereof. A potentially fruitful research approach is to consider the distribution of molecules within the product and represent the movement of those molecules during drying. This approach is in contrast to the conventional procedure of making most of the measurements external to the product, then making predictions on the internal mechanism of moisture movement. With the fantastic capabilities of the supercomputer now available, and even greater capabilities in the future, molecules or groups of molecules and their interactions can be represented and tracked. (See Table 1.)

To give an idea of the numbers to be handled, an ounce of water contains about a million, million, million, million molecules. It would have been unreasonable to think in terms of tracing even a small portion of those molecules prior to the present capabilities in computing.

The expanded calculations capability afforded by the supercomputer also leads to new opportunities in the development of macroscopic drying models. This approach is especially relevant in engineering research and awaits immediate exploitation by the research community.

Several molecular models are available based on previous work in surface chemistry and related areas. Many standard books cover these relationships; examples are recent books by Kihara, et al (1976), Levine, et al (1976), March (1974), and Speakman (1966), which direct attention to molecular energy and molecular and intermolecular forces.

Some aspects of the molecular approach include:

1. The intermolecular binding energy is one to 10 percent of the typical chemical bond energy (March, 1974).
2. Future work should aim at interpreting the first order process which is unimolecular fragmentation and radiationless

Table 1. Some Main Features of U.S. and Japanese Supercomputers

Machine Names	NEC SX-2	CRAY X-MP	CYBER 205 SERIES 600	FUJITSU FACOM VP-200	HITACHI S-810
Manufacturers	NEC	Cray Research	Control Data	Fujitsu	Hitachi
Peak Capacity (MFLOPS)	1300	630	400	533	800
Machine Cycle (ns)	6	9.5	20	7.5	15
Number of Pipe Lines	2 piplines ×4 sectors	6 pipelines (×2)	2 pipelines ×4 sectors	3 pipelines ×2 sectors	6 pipelines ×2 sectors
Vector Register Capacity (Byte)	80 K	4 K (×2)	None	64 K	64 K
Main Memory (Byte) Maximum Capacity	256 M	32 M	64 M	256 M	256 M
Number of Gates/ Logical Chips	1000 gates [.25ns]	16 gates	168 gates [0.6–0.8ns]	400 gates [0.35n] 1300 gates [35ns]	550 gates [0.35ns] 1500 gates [0.45ns]
Vector Register Chips (Access Time)	1 k bits [3.5ns]	N/A	None	1 K bits [5.5ns]	1 K bits [4.5ns]
Main Memory Chips (Access Time)	64 K bits [40ms]	4 K bits	16 K bits [80ms]	64 K bits [55ns]	16 K bits [40ms]
Cooling Method	Water Cool	Freon Cool	Freon Cool	Forced Air Cool	Forced Air and Water Cool

From SIAM News, v. 17, n. 1, p. 5, January 1984

transition, and the second order process which is bimolecular collisional force (Levine, 1976).

3. Free molecular flow occurs when the smallest cross section of a path is larger than the mean free path of the vapor molecules.
4. For gas at normal temperature and pressure, the mean free path is in the order of 1,000 Angstroms (Å).
5. Various models of molecular forces and positions and states are identified in Table 2.

Table 2. Some Useful Models Related to Drying

1. van der Waals' interactions
2. Arrhenius Equation
3. Coulomb's Law on attraction between charges
4. Force between magnetic or electric dipoles
5. Molecular flow through holes (Clausing's Equation)
6. Molecular theory of viscosity
   Gases - Maxwell
   Liquids - Andrade; Eyring
*7. Viscous flow
   Hagen-Poiseuille Equation
   Navier - Stokes (capillary) Equation
8. Langmuir, water surface tension
9. Freundlich Equation (adsorption on solids)
10. Eyring
11. Onsager's
*12. Gibb's (adsorption on solids)

*macroscopic modeling

Basic properties of water may be useful and helpful in fundamental studies involving the advanced instrumentation and computer capabilities in fundamental research on drying. Some examples are:

1. Water is made up of polar molecules.
2. H-O-H atoms are at an angle of 105°.
3. O-H distance is 0.95 Å. The water molecule is approximately 2 Å in diameter.
4. Moisture greatly affects electronic properties of dielectrics, and of cellulose materials and derivatives.
5. The force between two magnetic or electric dipoles can be represented by Coulomb's law.
6. The molecular vibration of liquid water is about 10^{-14} to 10^{-13} sec. (suggests possibility of laser to influence reactions).
7. Water has the property of self-ionization into OH and H ions.
8. In some of its electronically-excited states, the water molecule becomes linear.
9. Water is a good solvent for ionic materials and a poor solvent for large molecular molecules which do not ionize. (Thewlis, 1962)

5. MEASURING INSTRUMENTS AND METHODS

The name "superscope" is used to represent a whole new field of instruments to "see" objects previously unseen. The trend to develop instruments to see smaller and smaller objects will continue. Even the classical principles of physics, which we took for granted for many years are being challenged. There the notion that measuring instruments could not be used to measure a dimension substantially shorter than the wavelength of light is now being re-evaluated. The electron microscope, developed in the 1930's, has been the basis for many new developments permitting measurements to less than 100 Angstroms. Two Angstroms is presently the limit of resolution of the electron microscope; this limit is being challenged by present research (Crewe), in which the STEM (scanning transmission electron microscope) under development is expected to have a resolution of 0.5 Angstroms. Such a development would make it possible to see water molecules and the arrangements and spacing of the atoms within the water molecule. An entire field of electron microscopy has developed with magnification going up at least 200,000 times, and now with the most advanced HVEM (high-voltage electron microscope), up to 2,600,000 times. (Table 3)

In comparison, the best light microscopes provided magnification of 2,000 times and the early electron microscopes provided magnification of 12,000 times. With the new capabilities of electron microscopes, one will be able to "see" and study spacing of atoms at about 1 Angstrom (Å), or a water molecule at 2 Å, or protein cells, which at their smallest dimension would be about 40 Å.

Table 3. Representative Magnification of Viewed Objects

Best Light Microscopes	2.000 x
Early Electron Microscopes	12,000 x
1984 Electron Microscopes	200,000 x
Most Advanced Enhanced Electron Microscopes (HVEM)	2,600,000 x

If one can tag or identify molecules or groups of molecules and follow those from their original site to the outside of the product, representations previously postulated become proven and, therefore, more believable and useful.

A summary of the type of electron microscopes is given in Table 5.

Table 5. Types of Electron Microscopy

SEM	Scanning Electron Microscope
TEM	Transmission Electron Microscope
STEM	Scanning Transmission Electron Microscope
PEM	Photoelectron Microscopy
IETS	Inelastic Electron Tunneling Microscopy
STM	Scanning Tunneling Microscopy
SAM	Scanning Acoustic Microscopy
ARM	Atomic Resolution Microscope
HVEM	High Voltage Electron Microscope
APS	Analytical Positron Spectroscopy
SLAM	Scanning Laser Acoustic Microscope
FESTEM	Field Emission Scanning Transmission Electron Microscope
PIES	Penning Ionization Electron Spectroscopy

The features of each device are beyond the scope of this paper. It should be noted that computers are critical to the design and control of these devices.

The application of the principles of physical chemistry to surface phenomena, that is being done so productively with metals, offers great promise for understanding surface reactions of water on biological material. Using different energy sources--electrons, lasers, molecular beams, positrons--to bombard

Table. 4. Representative Sizes

Spacing of atoms in molecule	1 Å
Radius of oxygen molecule	1.4 Å
Water molecule	2 Å
Atom, size of simple	5 Å
DNA fiber diameter	20 Å
Protein cells, usually more than	40 Å
Hemoglobin	150 x 40 Å
Bacteria	2 microns
Red blood cells	8 microns
Starch molecule	20 microns
Hair, diameter	75 microns

Note:
Particle of 80 Å visible at 90,000 x magnification
1 Å = 10^{-8} cm = 10^{-10} m = 10^{-4} microns
1 micron = 10^4 Å = 10^{-3} mm = 10^{-6} m
Ionic range, 1-100 Å
Macromolecular range, 10^2 to 10^4 Å

the surface with different angles and intensities, various surface phenomena can be determined. It must be recognized that drying is a surface plus a below-surface phenomenon. To obtain a better understanding, the energy distribution could be studied as related to the product, moisture content, and temperature. In metal surface work, a gas such as nitrogen is studied. Here I would propose a study for the desorbing of water molecules from the surface using thermal energy from pulsed laser. This should be done in real time to provide dynamic response data (Anon., 1983, C&EN).

Electron microscopy is destructive of biological materials. By using higher voltage and, hence higher energy, there is usually less damage because the time of treatment is greatly reduced. Less total energy is used at higher voltages and the effect on the product reduced. Although the thermal desorption of molecules from a metal surface still needs much study to describe and predict, considerably more effort is needed to describe desorption (adsorption) from a biological surface. The Arrenhius equation that is used in drying and is applied on a gross basis, can be applied to adsorbed molecules on the surface on a molecular basis. Measuring instruments and methods would include the following:

1. Rate and nature of gas formation might be studied in biological material with biosensors now available or to be developed.
2. Use of appropriate laser energy to move the water molecule and an electron microscope to measure the movement of those water molecules. Vibrational spectroscopy of adsorbed molecules is a developing technique in which the reaction of a water molecule can be studied with selective vibrational excitation.
3. Thermal effects on products may be determined by spin resonance spectroscopy. (Hillman, et al, 1983)
4. The major challenge is to adapt these new instruments and techniques to biological materials during the drying process in such a way that the moisture and material relationships will not be greatly changed through the testing process.

Many activities are now being reported on designing molecules by computer. These might be done in three dimensions (3-D) and in color. But the molecules cannot be built and represented without basic data. A tool is available to us to develop information from which water molecules can be built on different proteins and starches. The water molecules could be removed with various energy inputs with the actions and results appearing on a monitor.

Before the computer model can be built, considerable research must be devoted to see what is happening--structure, energy, change of state, and change in location. A data base is developed for later modification and/or use for computer graphics. This will be done for drying in which the water molecules on appropriate substrates will be modeled. As that molecule changes state and moves (or moves and changes state) with various energy input, fundamental aspects in drying will be defined. These data can only be obtained, interpreted and integrated through fundamental research in drying.

Interestingly, a recent report (Marx, 1983) states that "cellular water, which occupies roughly 70 percent of the cell volume, may be organized, although this point is still subject to dispute." Statements such as these suggest many opportunities and possibilities for fundamental research in drying. The answer to these and similar statements will provide opportunities for improving the efficiency and effectiveness of moving energy to the molecule and of the flow of moisture from molecular sites.

An important instrument for determining the molecular structure is NMR (nuclear magnetic resonance spectroscopy). Using NMR, the composition, bonding, and three-dimensional arrangement of the atoms and molecules can be investigated. Two major advantages of

NMR are that it provides (1) a nondestructive technique and (2) a means of studying complicated and large molecules.

5. MACROSCOPIC MOISTURE TRANSPORT STUDIES

Macroscopic studies of drying are typically directed at specific systems used in engineering. Among the many topics of concern, one of the most important is heat and mass transfer, including, of course, moisture transport. Current interest within this topic deals with drying in porous media as represented by fibrous materials used in building insulation and by soils in the ground of our earth which is sometimes employed to store solar energy. The general description of the drying process in these materials is very complicated. In addition to the difficulty in specifying the solid structure in which moisture transport and heat transfer take place, there is also a multitude of physical phenomena that must be accounted for. These phenomena include (Eckert and Pfender, 1978; Ogniewicz and Tien, 1981): heat transport by conduction, convection and radiation; vapor transport by diffusion and convection; flow of liquid due to capillary action and gravity; and condensation or freezing with the attendant release of latent heat. Recent studies (Ogniewicz and Tien, 1981) suggest that a relatively small rate of condensation can have a significant effect on the thermal performance of porous insulation. Furthermore, porous properties, such as thermal diffusivity, vapor diffusivity and thermal conductivity also have a strong influence on energy and moisture transfer in the materials. The detailed modeling of these transport mechanisms and the understanding of their precise effects have been limited by the computer capability on inadequate thermophysical property information. The availability of supercomputers will remove the first of these shortcomings. The current state-of-the-art is based on simplified approaches, including the use of a one-dimensional model. While such a model provides an understanding of gross features, the opportunity now exists to acquire a more accurate quantification. Additional experimental studies, such as the recent efforts to measure heat and mass transfer coefficient as a function of moisture content in moist soils (Shah, et al, 1984), are also needed.

As an illustration of the macroscopic approach to the moisture transport or migration problem, consider the specific situations in which the drying medium is subjected to a periodic temperature fluctuation. In some cases, the drying medium may be treated as a semi-infinite, porous, unsaturated solid whose surface is exposed to the thermal fluctuation. With the assumption that the thermophysical properties are constant, and that the porous material does not change dimensions while being dried, the one-dimensional conservation equations describing the energy and moisture transport may be solved relatively easily using the computer. Such a study has been completed (Faghri and Eckert, 1980) by assuming further that the gas in the voids is stagnant and that the transport of sensible heat is negligible relative to the transport of heat of evaporation. The results from such a relatively simple analysis provide a method for predicting the amplitude of the moisture fluctuation and also its penetration depth. Typical results plotted using dimensionless parameters are shown in Figure 1. This figure indicates the fluctuations of the moisture content, W, and the fluctuations of the temperature, T, as a function of distance, X, and time, τ. The other parameters that are important to this problem are evident in the figure, and these are:

- K = moisture diffusion coefficient
- α = thermal diffusivity of the porous material
- $\frac{K}{\alpha}$ = Luikov number
- W = moisture content at time $\tau = 0$
- D^* = moisture transport parameters related to rate of change of vapor mass fraction with temperature
- τ_0 = period of the temperature fluctuation

imposed at the surface

T_A = temperature amplitude

From a comparison of the thermal and moisture fields, it is evident that there is a drop in the moisture content on the surface (x=0) as the temperature increases, and this trend is consistent with the physics of the problem. It is also clear that a great disparity exists in the penetration depths. The figures shown are applicable to soils, and for these it may be shown by a simple calculation using information contained in the dimensionless plate that the penetration depth for moisture is roughly 20 percent of that for heat.

The above example is indicative of the physical insight that is made possible through macroscopic modeling of the heat and moisture migration problem. But, as pointed out earlier, the conclusions that may be derived from current modeling efforts can at best be treated as qualitative illustrations of trends. More accurate results can be provided by considering two- or three-dimensional analysis that incorporates variable thermophysical properties. Such an analysis will become more easily implemented when supercomputers are more generally applied to these problems. Also, that variable thermophysical property is an important consideration in drying is illustrated by recent data (Shah, et al, 1984) which show that the moisture diffusion coefficient changes by approximately two orders of magnitude when the soil condition changes from nearly dry to within 15 percent of saturation. Accurate determination of many similar properties important in drying analysis is now possible, and with it we can expect an increase in advanced modeling studies. And since drying research involves many fundamental disciplines, we can look forward to more interdisciplinary efforts as more difficult problems are tackled, and it will not be surprising to see that many of these efforts will involve international cooperation.

A framework for approaching separation in generic sense is provided by Giddings (1978).

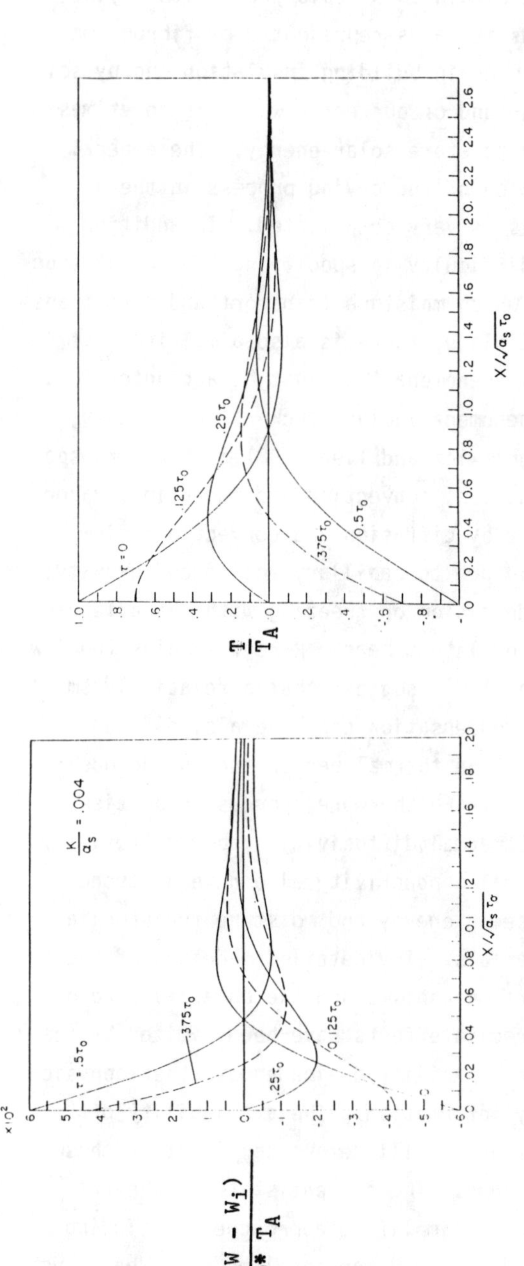

Figure 1. Dimensionless moisture content and temperature fluctuations as a function of dimensionless distance and time [after Faghri and Eckert (1980)]

He presents the view that "separation is a spatial disengagement among components of a mixture." He describes the six basic categories for separation methods: static (non-flow); flow; continuous force; discontinuous force; parallel flow; and perpendicular flow. The equations for these categories provide a basis for a common thread for separation, including drying, on a macroscopic basis.

SUMMARY

The field of drying (and wetting) will benefit from recent developments in instrumentation and computers. Instruments developed over the past 25 years will permit an opportunity for investigators to look closely at macroscopic events. Instruments now being developed will provide an opportunity to look at microscopic and submicroscopic phenomena. Angstroms will replace microns as units of comparison. Mathematical models based on physical and biological theories and relationships will evolve from experimental work. The supercomputer will be used to control and record and analyze data obtained from measuring instruments. Further, the computer will be used increasingly to simulate molecular events as part of the drying process. A new era of research is before us.

REFERENCES

Adams, Tom, "Acoustic Microscope Probes Materials," High Technology, 2(5):78-81, Sept./Oct. 1982).

Anon., "Theory Predicts Thermal Decomposition Rates," Chemical & Engineering News, 61(8): 24 (Feb. 21, 1983).

Anon., "Center Will Advance Electron Microscopy," Industrial Research and Development, 25(11): 41 (Nov. 1983).

Anon., "Electron Microscopy Goes 3-D," Science News, 123(25):396 (June 18, 1983).

Anon., "Microscope Sees with Sound," Science Digest, 91(6):27 (June 1983).

Anon., "New Methods Shed Light on Surface Chemistry," Chemical & Engineering News, 61(37):30-32 (Sept. 12, 1983).

Anon., "New Scope Brings Atoms into Focus," Science Digest, 91(10):35 (Oct. 1983).

Anon., "Positrons Useful in Probing Pores," Chemical & Engineering News, 61(37):48 (Sept. 12, 1983).

Bash, Paul A., et al., "Van der Waals Surfaces in Molecular Modeling," Science, 222:1325-1327 (Dec. 23, 1983).

Baum, Rudy, "Photoelectron Applied to Cells," Chemical & Engineering News, 60(50):34 (Dec. 13, 1982).

Baum, Rudy, "Zero-Field NMR Advances Molecular Structure Determinations," Chemical & Engineering News, 61(50):23-24 (Dec. 12, 1983).

Chandler, D., Weeks, J. D., and Andersen, H. C., "Van der Waals Picture of Liquids, Solids, and Phase Transformation," Science, 220:787-794 (May 20, 1983).

Crewe, A. V., "High-Resolution Scanning Transmission Electron Microscopy," Science, 221:325-330 (July 23, 1983).

Cromie, W. J., "High-Resolution Electron Microscopy Offers a View of the Atomic World," Mosaic, 14(4):15-19 (July/August, 1983).

Derra, S., "Crewe Plans STEM to Resolve 0.5 Å," Industrial Research and Development, 25(3):51-52 (March 1983).

DeYoung, H. G., "Biosensors, the Meeting of Biology and Electronics," High Technology, 3(11):41-49 (Nov. 1983).

Eckert, E. R. G., and Pfender, E., "Heat and Mass Transfer in Porous Media," Proceedings of Sixth International Heat Transfer Conference, Toronto, Canada, 6:1-12 (1978)

Eisenberg, D., and Kauzman, W., The Structure and Properties of Water, Oxford University Press, Oxford, England (1969).

Faghri, M., and Eckert, E. R. G., "Moisture Migration Caused by Periodic Temperature Fluctuations in an Unsaturated Porous Medium," Warme-und Stoffubertragung, 14:217-223 (1980).

Fendler, J. H., "Membrane Mimetic Chemistry," Chemical & Engineering News, 62(1):25-38 (Jan. 2, 1984)

Franks, Felix (editor), Water, A Comprehensive Treatise, 6 volumes, Vol. 1, The Physics and Physical Chemistry of Water, Plenum Press, New York, N.Y., 596 p. (1972)

Giddings, J. Calvin, "Basic Approaches to Separation, Analysis and Classification of Methods According to Underlying Transport Characteristics," Separation Science and Technology, 13(1):3-24 (1978).

Gregg, S. J., and Singh, K. S. W., Adsorption Surface Area and Porosity, Academic Press, Inc., New York, N.Y., 340 p. (1982).

Griffith, J. D., Electron Microscopy in Biology, Wiley-Interscience, New York, N.Y., 296 p. (1981).

Halle, J. M., and Mansoori, G. A., Molecular-Based Study of Fluids, American Chemical Society, Washington, D.C., 524 p. (1983).

Harris, J. E., Electron Microscopy of Proteins, Academic Press, Inc., New York, N.Y., 362 p. (1981).

Hasted, J. B., Aqueous Dielectrics, Chapman and Hall, London, England, 302 p. (1973).

Hillman, G. C., et al., "Determination of Thermal Histories of Archeological Cereal Grains with Spin Resonance Spectroscopy," Science, 222:1235-1236 (Dec. 16, 1983).

Johnson, Roger S., "It Makes One Dandy Package - Even if It's Only Skin Deep," Smithsonian, 13(12):151-154 (March 1983).

Jorgenson, J. W., and Lukacs, K. D., "Capillary Zone Electrophoresis," Science, 222:266-272 (1983).

Jortner, J., and Pullman, B., Intramolecular Dynamics, D. Reidel Publishing Co., New York, N.Y. (1982).

Khanna, S. K., and Lambe, J., "Inelastic Electron Tunneling Spectroscopy," Science, 220:1345-1353 (June 24, 1983).

Kihara, T., and Ichimaru, S., Intermolecular Forces, J. Wiley and Sons, New York, N.Y., 182 p. (1976).

Levine, R. D., and Bernstein, R. B., Molecular Reaction Dynamics, Oxford University Press, Oxford, England, 205 p. (1974).

Levine, R. D., and Jortner, J. (editors), Molecular Energy Transfer, J. Wiley and Sons, New York, N.Y., 310 p. (1976).

March, N. H. (editor), Orbital Theories of Molecules and Solids, Clarendon Press, Oxford, England (1974).

Marx, J. L., "Organizing the Cytoplasm," Science, 222:1109-1111 (Dec. 9, 1983).

Ogniewicz, Y., and Tien, C. L., "Analysis of Condensation in Porous Insulation," International Journal of Heat and Mass Transfer, 24(3):421-429 (1981).

Rawls, R., "Penning Spectroscopy Studies Molecules," Chemical & Engineering News, 61(31):20 (Aug. 1, 1983).

Reese, K. M., "The Coming of Age of NMR," Mosaic, 14(5):14-19 (1983).

Robinson, A. L., "IBM Images Surfaces by Electron Tunneling," Science, 220:43 April 1, 1983).

Shah, D. J., Ramsey, J. W., and Wang, M., "An Experimental Determination of the Heat and Mass Transfer Coefficients in Moist, Unsaturated Soils," International Journal of Heat and Mass Transfer, to appear in 1984.

Speakman, J. C., Molecules, McGraw Hill Book Co., New York, N.Y., 158 p. (1966).

Tucker, J. B., "Designing Molecules by Computer," High Technology, 4(1):52-59 (Jan. 1984).

Wolfenden, R., "Waterlogged Molecules," Science, 222:1087-1093 (Dec. 9, 1983).

SHORT-CUT CALCULATION FOR NON-ISOTHERMAL DRYING OF SHRINKING AND NON-SHRINKING PARTICLES AND OF HOLLOW SPHERES CONTAINING AN EXPANDING CENTRAL GAS CORE

H.A.C.Thijssen and W.J.Coumans

Gist Brocades, Delft,
and Department of Chemical Engineering,
Eindhoven University of Technology, Netherlands

ABSTRACT

A procedure is presented for the calculation of non-isothermal drying rates of shrinking and non-shrinking particles of different geometries including rectangular bodies, cylinders, spheres and hollow cylinders and spheres. The implication on drying rate of changes in volume of the gas bubble in hollow spheres is taken into account.

Starting point are the experimentally observed isothermal drying rates versus fraction water removed of a slab of the material of interest at two temperature levels.

With the aid of a short cut calculation method based on the exact numerical computer solutions of the diffusion equation valid for concentration dependent diffusion coefficients and complex boundary conditions the information obtained from the isothermal slab drying experiments is transformed to other geometries and non-isothermal conditions.

The transformations are derived for the four main drying phases, including the constant surface water activity period, the period with decreasing surface water activity, the penetration period where the surface water concentration is in equilibrium with the surrounding gas phase and the central water concentration has still hardly changed and finally the regular regime period where the surface remains in equilibrium with the gas phase and the drying rate is no longer affected by the initial conditions.

The calculation is executed in a step-to-step procedure where the conditions for each new step are obtained from the experimental drying curves, the transformation equations valid for the relevant drying phase and heat and mass balances.

The calculation procedure is performed easily and the results do, notwithstanding the various simplifications, deviate no more than a few percent from the actual drying rates and total drying time.

1. INTRODUCTION

In drying non-shrinking or shrinking particles of the main geometries and hollow spheres containing and expanding or shrinking gas phase, which generally is the case in e.g. spray-drying, the drying rate can be calculated as a function of both the fraction of water removed and the temperature by solving numerically the non-linear diffusion problem. This approach is very complex due to the non-linearity of the sorption isotherms, the strongly water concentration dependency of the water diffusion coefficient and the varying particle temperature during drying. The numerical calculation requires highly sophisticated programming-skill and is only possible with the aid of high speed large memory computers. The numerical calculation becomes even much more complicated in case of more complex geometries such as rectangular bodies, finit length cylinders, etc.

Liou and Bruin [1] presented for single phase isotropic particles of the main geometries including spheres, cylinders and slabs an approximate, but yet very accurate, short-cut method for the calculation of concentration profiles and drying times for non-linear diffusion problems with a power relation between diffusion coefficient and water concentration. It has been developed for drying at constant particle temperature, uniform initial water concentration and Dirichlet boundary conditions. The condition of a constant surface water concentration according to their boundary conditions implies a Biot number approaching infinity and consequently the absence of a constant drying rate period (in fact a constant surface water activity period). Their method requires only a few minutes on a pocket calculator.

In the present paper their method is further simplified and generalized to variable surface water concentration, variable temperature and other geometries, such as rectangular bodies, finite length cylinders and spheres containing an expanding or shrinking gas core. The drying rates of spherical particles containing one central gas void can be calculated both for an expanding gas bubble and a non-shrinking and shrinking liquid shell. It is assumed that the liquid phase surrounding the disperse gas bubble remains fully fluid until a certain critical water concentration is reached and that surface tension effects can be neglected. Consequently the pressure inside the disperse gas phase stays equal to the outside pressure at water concentrations higher than this mean critical water concentration. It is moreover assumed that at water concentrations below this critical value the liquid phase becomes more or less rigid and prevents changes in volume of the central gas bubble.

2. THE EQUATIONS DESCRIBING THE DEHYDRATION OF PARTICLES

2.1. The general diffusion equation

A generalized formulation of the diffusion equation [2] is expressed by the following equation:

$$\frac{\delta m}{\delta \tau} = \frac{\delta}{\delta \Phi}\left(D_r X^2 \frac{\delta m}{\delta \Phi}\right) \tag{1}$$

which holds for concentration dependent diffusion in non-shrinking and shrinking particles of any geometry.

Before treating the short-cut calculation procedures, we shall first define the dimensionless variables of equation (1) for non-shrinking and shrinking homogeneous systems (liquid or solid) respectively.

Non-shrinking systems. For non-shrinking homogeneous systems the dimensionless concentration m is defined as:

$$m = \frac{\rho_w}{\rho_{w,o}} \tag{2}$$

with ρ_w the water concentration at time t and $\rho_{w,o}$ the initial water concentration.

The dimensionless time τ is defined as:

$$\tau = \frac{D_o t}{R^2} \tag{3}$$

Where D_o represents the value of the diffusion coefficient at $\rho_{w,o}$ and R is the radius or half-thickness of the specimen.

The dimensionless space coordinate Φ is defined as:

$$\Phi = \left(\frac{r}{R}\right)^{\nu+1} \tag{4}$$

with r the space coordinate and ν a geometry parameter as will be defined in eqs. (7), (13) and (14).

The dimensionless diffusion coefficient D_r is defined as:

$$D_r = \frac{D}{D_o} \tag{5}$$

with D the diffusion coefficient.

The dimensionless variable X is defined as:

$$X = (\nu+1)\Phi^{\nu/(\nu+1)} \tag{6}$$

The geometry parameter ν is for solid slabs, cylinders of infinite length and spheres defined as

$$\nu+1 = \frac{\text{surface body}}{\text{volume body}} R \tag{7}$$

and for these three main geometries $(\nu+1)$ amounts 1, 2 and 3 respectively.

It is obvious that equation (7) cannot be used in equation (1) for calculating the drying behaviour of other geometries. For example a cube would also have a $\nu+1$ value of 3, but dries at the same value of R much more slowly than a solid sphere. In order to avoid confusion the product of specific surface and R will for other geometries be indicated by $\nu'+1$.

For the three main geometries (solid slabs, cylinders and spheres) the values of the product of the Sherwood numbers in the regular regime with the corresponding values of $\nu+1$ are plotted versus $\nu+1$ in figure 1. The curve connecting the three points is called the main geometry relation. For a concentration independent diffusion coefficient the main geometry relation is described by:

$$Sh_d(\nu+1) = 4.93 + 5.86\nu + 0.77\nu^2 \tag{8}$$

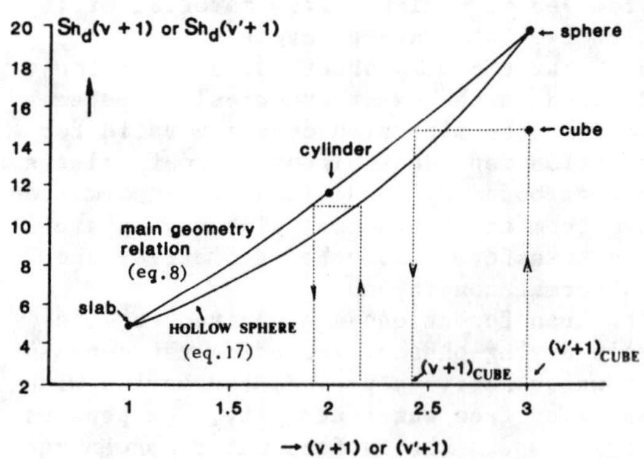

Fig.1: Construction of the geometry parameter $(\nu+1)$ of a fictive body from given values of $(\nu'+1)$ and Sh_d belonging to an arbitrary body e.g. cube, hollow sphere

It can be analytically proven that two bodies 1 and 2 of different geometries having a constant water concentration at the surface (valid during penetration period and regular regime) show an identical drying behaviour if:

$$Sh_{d,1}(\nu_1'+1) = Sh_{d,2}(\nu_2'+1) \tag{9}$$

where

$$Sh_d = \frac{2F}{(\bar{m}-m_i)\bar{D}_r} \tag{10}$$

with

$$F = \frac{n_{w,i} R}{\rho_{w,o} D_o} \qquad (11)$$

in which $n_{w,i}$ is the evaporation flux and F is a dimensionless flux parameter and

$$\bar{D}_r = \frac{1}{\bar{m}-m_i} \int_{m_i}^{\bar{m}} D_r \, dm \qquad (12)$$

where $\bar{m}$ and m_i indicate the mean and the interface concentration respectively.

For most geometries such as rectangular bodies, finite cylinders, hollow spheres and hollow cylinders (tubes) the Sherwood number in the regular regime can easily be calculated for a concentration independent diffusion coefficient. From equation (9) it will be clear that a fictive body 1 with geometry parameter ν_1 belonging to the main geometry relation (eq. (8)) will have the same drying behaviour during the penetration period and regular regime as a body 2 of an arbitrary geometry if

$$Sh_{d,1}(\nu_1+1) = Sh_{d,2}(\nu_2'+1) \qquad (13)$$

Consequently through the analytical calculation of Sh_d in the regular regime for D is constant the corresponding value of the geometry parameter $\nu+1$ of the fictive body is obtained with equations (8) and (13). In figure 1 the point $(\nu'+1,\ Sh_d(\nu'+1))$ calculated for a cube and the corresponding geometry parameter $\nu+1$ are indicated. From the exact numerical solutions of equation (1) it could surprisingly be concluded that the calculation of the geometry parameter as described above stays also valid for strong concentration dependent diffusion coefficients, if the value of $\nu_{a=0}$, valid for a constant diffusion coefficient and obtained from eqs.(8) and (13), is multiplied with a simple correction factor. The value of ν_a of the fictive body to be used in eq. (1) now becomes:

$$\nu_a = \nu_{a=0} \left[1 + 0.13(2-\nu_{a=0}) \frac{a}{a+1}\right] \qquad (14)$$

In this equation a expresses the concentration dependency of the diffusion coefficient for a power law dependence of the diffusion coefficient on the water concentration as defined by eq. (32).

Thus, first the value of ν is simply calculated for the situation of a concentration independent diffusion coefficient and subsequently corrected with eq.(14) for concentration dependent diffusion coefficients.

Hollow sphere, non-shrinking shell. For a hollow sphere equation (7) becomes

$$\nu'+1 = \frac{3(1-\lambda)}{1-\lambda^3} \qquad (15a)$$

in which

$$\lambda = \frac{R_c}{R_c+R} \qquad (15b)$$

with R_c the radius of the central gas bubble and R_c+R the outside radius of the sphere, as indicated by figure 2.

For non-shrinking hollow spheres the relation between the radius R_o of the gas bubble free sphere, the bubble radius R_c and λ reads:

$$\frac{1}{\lambda^3} - 1 = \left(\frac{R_o}{R_c}\right)^3 \qquad (16)$$

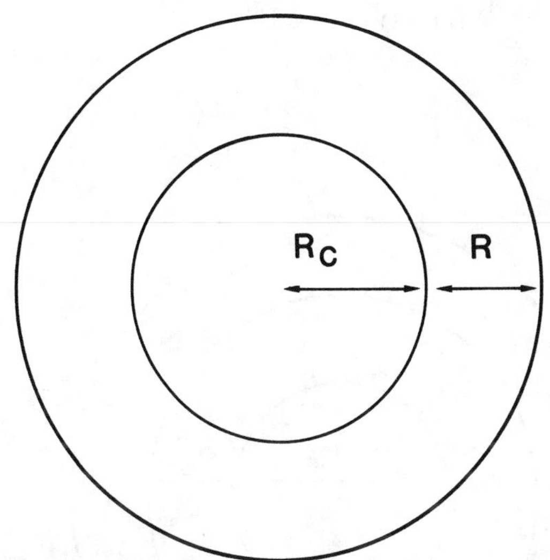

Fig.2: Non-shrinking spherical particle with central gas core

The analytical solution of the diffusion equation for hollow spheres with constant diffusion coefficient reads:

$$Sh_d(\nu'+1) = [6.44 - 2.29(1+\nu') + 0.78(1+\nu')^2](1+\nu') \qquad (17)$$

The lower curve in figure 1 represents equation (17) and from equation (8) and (17) follows

$$\frac{(\nu+1)}{(\nu'+1)} = 1.33 - 0.44(\nu'+1) + 0.11(\nu'+1)^2 \qquad (18)$$

where $\nu+1$ is the geometry parameter of a fictive solid body which corresponds with the hollow sphere.

Shrinking systems. For shrinking systems the application of a reference component mass centered coordinate system avoids the complication of a moving boundary [3]. In this coordinate system the dimensionless concentration m is now defined as:

$$m = \left(\frac{\rho_w}{\rho_s}\right) \Big/ \left(\frac{\rho_{w,o}}{\rho_{s,o}}\right) \qquad (19)$$

with ρ_s the mass concentration of the non-volatile component(s).

The dimensionless time τ is defined as:

$$\tau = \frac{D_o \rho_{s,o}^2}{d_s^2 R_s^2} t \qquad (20)$$

with $\rho_{s,o}$ the initial mass concentration of the non-volatile component(s), d_s the density of the non-volatile component(s) and R_s the radius or half-thickness of the system in case the water concentration would be zero. This so-called "dry solids radius" R_s is related to the initial radius R_o of the system by:

$$R_s = \left(\frac{\rho_{s,o}}{d_s}\right)^{1/(\nu+1)} \cdot R_o \qquad (21)$$

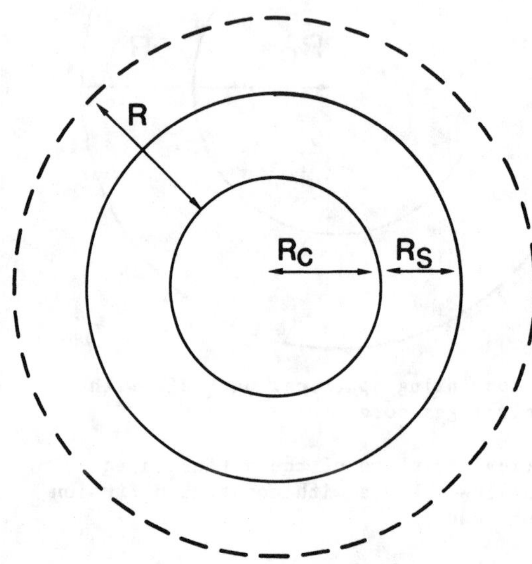

Fig.3: Shrinking hollow sphere with dry shell of thickness R_s

$(\nu'+1)$ for shrinking hollow spheres is defined by:

$$\nu'+1 = \frac{3(1-\lambda^*)}{1-\lambda^{*3}} \qquad (22)$$

where

$$\lambda^* = \frac{R_c}{R_c + R_s} \qquad (23)$$

in which $R_c + R_s$ is the outside radius of the dry sphere ($\bar{m}$=0) with a gas core R_c, see figure 3 and if $R_{s,o}$ is the "dry solids radius" of the gas bubble free sphere:

$$\left(\frac{1}{\lambda^{*3}} - 1\right) = \left(\frac{R_{s,o}}{R_c}\right)^3 \qquad (24)$$

The value of the geometry parameter for shrinking systems $(\nu+1)$ follows according to equation (18) again from the corresponding value of $(\nu'+1)$ calculated with equation (22).

The dimensionless space coordinate Φ is now defined as:

$$\Phi = \int_0^r \rho_s r^\nu dr \Big/ \int_0^R \rho_s r^\nu dr = \frac{\nu+1}{\rho_{s,o} R_o^{\nu+1}} \int_0^r \rho_s r^\nu dr \qquad (25)$$

The dimensionless diffusion coefficient is defined as:

$$D_r = \frac{D \rho_s^2}{D_o \rho_{s,o}^2} \qquad (26)$$

and the dimensionless variable X as:

$$X = (\nu+1) \left[\int_0^\Phi \left(1 + \frac{d_s \rho_{w,o}}{d_w \rho_{s,o}} m\right) d\Phi \right]^{\nu/(\nu+1)} \qquad (27)$$

Equation (27) only holds if the density of the non-volatile component(s) does not change upon mixing.

<u>Initial and boundary conditions for non-shrinking and shrinking systems</u>. The initial and boundary conditions are:

$$\tau = 0 \quad 0 \le \Phi \le 1 \qquad m = 1 \qquad (28)$$

$$\tau > 0 \quad \Phi = 0 \qquad X\frac{\partial m}{\partial \Phi} = 0 \qquad (29)$$

$$\Phi = 1 \qquad m = m_i \qquad (30)$$

$$\text{if } Bi \to \infty \qquad m = m_i = 0$$

Where the Biot number Bi is defined by:

$$Bi = \frac{k_g R}{D_o} \qquad \text{for non-shrinking systems} \qquad (31a)$$

$$Bi = \frac{k_g \rho'_{wi,sat} d_s R_s}{D_o \rho_{so}^2} \qquad \text{for shrinking systems} \qquad (31b)$$

in which k_g is the mass transfer coefficient in gas phase and $\rho'_{wi,sat}$ is the saturation water concentration at the interface in the gas phase.

During drying four drying phases can be distinguished viz.:

1. the drying period with constant surface water activity
2. the drying period with decreasing surface water activity
3. the penetration period where $m_{\Phi=1} = 0$ and $m_{\Phi=0} = 1$

4. the regular regime curve where $m_{\Phi=1} = 0$ and $m_{\Phi=0} << 0.8$

For Bi = 0	phase 3, the penetration period, is fully absent
For Bi >> 0	phase 1 is very short
For Bi → ∞	phases 1 and 2 are absent
For $M_s \geq 2000$	phase 2 is absent (M_s is mean mol.wt.solids)
For $M_s \geq 2000$	and for F* or $F \leq (1 + \nu)$ phases 2 and 3 are absent.

3. AN APPROXIMATE METHOD FOR THE COMPUTATION OF THE WATER FLUX AND DRYING TIME

3.1. Non-shrinking systems

The concentration dependency of the diffusion coefficient is approximated by:

$$D_r = m^a \quad (32)$$

in which the exponent a generally is a function of m.

The relation between D_r and m is schematically indicated by figure 4.

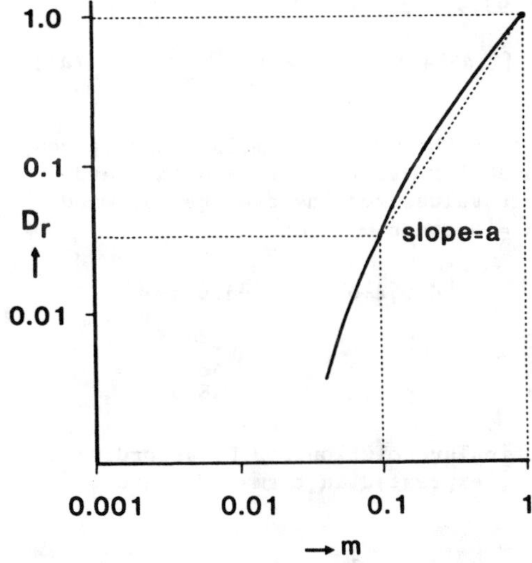

Fig.4: Schematical effect of m on D_r

The water flux follows from the mass balance:

$$\frac{d\bar{m}}{d\tau} = -FX_i \quad (33)$$

where $\bar{m}$ is the average value of the dimensionless concentration in the particle, F the flux parameter and X_i the value of X at $\Phi = 1$; it appears that X_i is equal to $\nu + 1$.
Further holds:

$$F = D_r X_i \left(\frac{\partial m}{\partial \Phi}\right)_i \quad (34)$$

Liou [1] introduced two auxiliary variables, i.e. the drying efficiency U and the flux function G.

$$U = 1 - \bar{m} \quad (35)$$

$$G = \frac{UF}{\nu' + 1} \quad (36)$$

One should be aware that in eq. (36) not $(\nu+1)$ but $(\nu'+1)$, expressing the specific surface area times R, has to be used.

<u>Phase 1 and 2, the drying periods with almost constant surface water activity and decreasing surface water activity</u>. During the constant water activity period the relation between G and U yields a straight line and is independent of the geometry parameter. It can be analytically shown that for high drying rates $G_{max,1}$ at the end of that linear relationship is constant and independent of the mass transfer coefficient in the gas phase k_g and also independent of the geometry parameter. The relationship between G and U for the four drying phases is schematically indicated in figure 5. For lower drying rates the end of the linear relationship between G and U ends approximately at the curve in figure 5 which is parallel to the regular regime curve

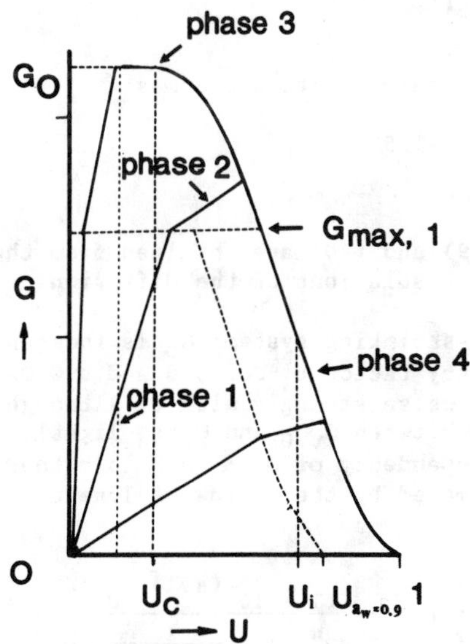

Fig.5: Schematical presentation of G function of a slab

and starts at G=0 for an U value corresponding with a water activity of about 0.9.
Phase 2: because during this period the averaged evaporation flux is about half that of the constant water activity period the mean slope of the relation between G and U will also be about half of that during the constant water activity period.

During the constant water activity period the relation of G versus U can be calculated for any other value of the mass transfer coefficient by multiplying a known relation of G versus U with the ratio of the mass transfer coefficients according to:

$$G_2 = G_1 * \frac{k_{g,2}}{k_{g,1}} \tag{37}$$

Phase 2 will end at the intersection of its G-function with either the G-function of the penetration period (phase 3) or with the G-function of the regular regime (phase 4).

<u>Phase 3, the penetration period with $m_i = 0$ and $m_{\Phi=0} = 1$</u>. It can be derived for a slab that during the penetration period for $m_i = 0$ ($Bi \to \infty$) and $m_{\Phi=0} = m_c = 1$ strictly holds:

$$G_0 = \frac{2}{\pi} \qquad \text{for } a = 0 \tag{38}$$

and that for $0 \leq a < \infty$ and $m_i = 0$, $m_{\Phi=0} = 1$ G stays approximately constant, the deviation being less than a few percent.

The values of G_o for $0 \leq a < \infty$ and $0 \leq \nu \leq 2$ can be calculated from the relation:

$$G_o = \frac{\pi^2 + e^2 a}{2(a+2)^3} \left(\frac{a+1}{a+2}\right)^a + \left(\frac{2}{\pi} - \frac{\pi^2}{16}\right) \frac{1}{2^a} \tag{39}$$

for $U \to 0$ and $Bi \to \infty$

For $0 \leq a \leq 1$ this equation becomes:

$$G_o = \frac{2}{\pi} (a + 1)^{-1.5} \tag{40}$$

equations (39) and (40) are obtained from the exact computer solutions of the diffusion equation.

For non-shrinking systems G_o is independent of the geometry factor ν. For $U>0$ and $\nu>0$ $G_{a,\nu}$ starts to decrease strongly with U. Although the relationship between $G_{\nu>0}$ and U is slightly curved the dependency of $G_{\nu>0}$ on U is rather well approximated by the following linear relationship:

$$G_{a,U,\nu} = G_o \left[1 - \frac{(1 + 10^{-(a+1)})\, U}{0.82^a \left(1 + \frac{1}{\nu(\nu + 1)}\right)} \right] \tag{41}$$

Equation (41) is a good approximation of the numerically calculated values of G valid for the penetration curve. The equations for the penetration period stay strictly valid as long as the centre concentration $m_{\Phi=0}$ remains equal to unity. From numerical computer calculations the following relation between U_c, being the value of U at which $m_{\Phi=0}$ starts to deviate from unity, and the geometry factor ν could be derived:

$$U_c = \frac{2}{a + 2} \left[(\nu + 1)^{0.26} - 0.625 \right] \tag{42}$$

Because for $U>U_c$ the relationship between $m_{\Phi=0}$ and U is linear, we can write:

$$m_{\Phi=0} = 1 \qquad \text{for } 0 \leq U \leq U_c \tag{43}$$

and

$$m_{\Phi=0} = \frac{1 - U}{1 - U_c} \qquad \text{for } U_c \leq U \leq 1 \tag{44}$$

<u>Phase 4, the regular regime period with $m_i = 0$ and $m_{\Phi=0} < 1$</u>. Schoeber [2,5] defined an average Sherwood number Sh_d for the disperse phase (equation (10)).

Equation (10) can be rewritten as

$$F = \frac{Sh_{d,a,\nu}\, \bar{m}^{a+1}}{2(a + 1)} \tag{45}$$

In the regular regime these average Sherwood numbers are constant and independent of U. For $D_r = m^a$, Sh_d is a function of a and ν only. From the numerical solutions of Schoeber the following relation could be obtained:

$$Sh_d(\nu+1) = 4.93 + 5.86\nu + 0.77\,\nu^2 + [2.45 + 4.96\nu + 2.77\nu^2] \frac{a}{a+2} \tag{46}$$

Equation (46) is the so-called main geometry relation extended for values of a between 0 and ∞.

The limit values for the average Sherwood numbers at $a = 0$ and $a \to \infty$ are:

geometry		$Sh_{d,\nu,a=0}$	$Sh_{d,\nu,a\to\infty}$
slab	$\nu=0$	$\frac{1}{2}\pi^2$	e^2
cylinder	$\nu=1$	5.783	4e
sphere	$\nu=2$	$2/3\pi^2$	$e^{8/3}$

The particular solution for F, according to equation (45), expressed in terms of U and G becomes:

$$G = \frac{Sh_{d,a,\nu}\, U(1 - U)^{a + 1}}{2(a + 1)(\nu' + 1)} \tag{47}$$

Because of the definition of the G function eq.(47) contains $(\nu'+1)$ in the denominator. The function of G pocesses an inflection point i for:

$$U_i = \frac{2}{a + 2} \tag{48}$$

Equation (47) holds for $U \geq U_i$.
For values of G at $U \leq U_i$ the G function in the regular regime can be approximated by a straight relationship between G and U. The negative slope equals the ratio of the slope of the tangent to the G function in $U=U_i$ and $(1+10^{-(a+1)})$. Thus:

$$\frac{G-G_{U_i}}{U-U_i} = -\frac{Sh_d}{2(\nu+1)(a+1)} \cdot \left(\frac{a}{a+2}\right)^a \cdot \frac{1}{1+10^{-(a+1)}} \quad (49)$$

The regular regime, its inflection point U_i and the approximative linear G function for U values smaller than U_i is schematically depicted in figure 6. From eq. (47), valid for $U_i \leq U \leq 1$, the values of a can be calculated as a function of U

$$a = \frac{\ln\left[1 + \frac{0.01}{n_w} \cdot \frac{dn_w}{dU}\right]}{\ln\left[1 - \frac{0.01}{1-U}\right]} - 1 \quad \text{at U} \quad (50)$$

The value of D_o can for one value of a in the regular regime and corresponding value of U be calculated with the following equation as obtained from eq. (11) and (47)

$$D_o = -\frac{2R_o}{\rho_{w,o}\,\bar{Sh}_d} \cdot (1-U)^{-a} \frac{dn_w}{dU} \quad \text{for a at U} \quad (51)$$

Consequently a can be obtained from the experimentally observed relation between n_w and ρ_w or U in the regular regime. In principle the dependence of a on U can be extrapolated to U=0.

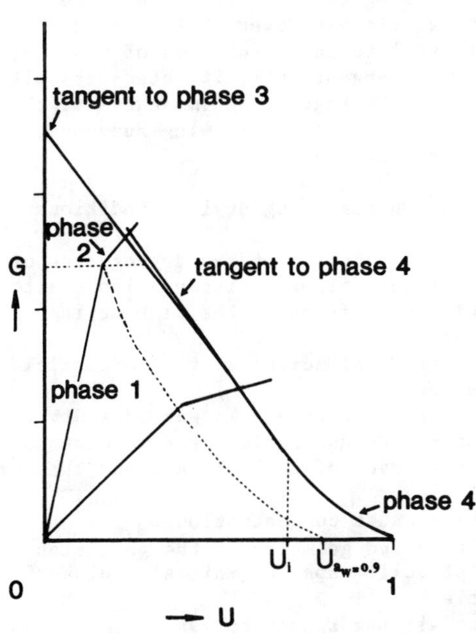

Fig.6: Schematical presentation of G function of a sphere

3.2. Shrinking systems

When shrinkage of the system occurs during the desorption process, the mass flux has to be corrected for this phenomenon. Because the mass flux $n_{w,i}$ in a reference component mass centered coordinate system is related to the flux parameter F* for shrinking systems by:

$$F^* = \frac{n_{w,i}\, d_s R_s}{\rho_{w,o} D_o \rho_{s,o}} \quad (52)$$

the correction for shrinkage can directly be applied to F*.

The mass balance for shrinking systems reads:

$$\frac{dm}{d\tau} = -X_i F^* \quad (53)$$

$$F^* = -X_i\, D_r \left(\frac{\delta m}{\delta \Phi}\right)_{\Phi=1} \quad (54)$$

The shrinking factor H is defined by:

$$H = \frac{F^*}{F} \quad (55)$$

It can be derived that

$$H^o = \lim_{U \to 0} H = \left[1 + \frac{d_s \rho_{w,o}}{d_w \rho_{s,o}}\right]^{\nu/(\nu+1)} \quad (56)$$

Equation (56) can also be used by good approximation for U > 0 during phase 1, 2 and 3.

Finally, the shrinkage factor H in the regular regime, expressed in terms of the efficiency U becomes:

$$H = 1 + \frac{Sh_{d,\infty,\nu}}{Sh_{d,a,\nu}} \left[\left(1 + \frac{d_s \rho_{w,o}}{d_w \rho_{s,o}} (1-U)\right)^{\nu/(\nu+1)} - 1\right] \quad (57)$$

The determination of the drying time now follows straightforward from the approximate and partly linearised G* function.

For an arbitrary geometry, including shrinking spheres with a expanding central gas core, the dimensional drying time is calculated by integrating the mass balance (eq. (53)):

$$t = \int_0^U \frac{d_s^2 R_s^2 U}{(\nu+1) D_o \rho_{so}^2 X_i G^*} \cdot dU \quad (58a)$$

where according to equation (27) for Φ=1:

$$X_i = (\nu+1)\left[1 + \frac{d_s \rho_{wo}}{d_w \rho_{so}} (1-U)\right]^{\nu/(\nu+1)} \quad (58b)$$

This integral can be solved either graphically or numerically.

4. CALCULATION OF THE NON-ISOTHERMAL DRYING RATE OF SHRINKING AND NONSHRINKING PARTICLES OF THE VARIOUS GEOMETRIES AND OF SPHERES WITH AN EXPANDING GAS CORE

4.1. Slab drying experiments

For the material of interest two isothermal slab drying curves in the range $0 \leq U \leq 0.95$ have to be obtained experimentally at two temperature levels. Moreover the equilibrium value of U corresponding with a water activity 0.9 has to be experimentally determined.

In order to obtain the information quickly and accurately the drying is preferably done with a thin layer in the range $0.4 \leq U \leq 0.95$ ($Bi \simeq 10^5$) and with a thick layer in the range $0 \leq U \leq 0.5$ ($Bi \simeq 10^7$). Information regarding the mass transfer coefficient k_g in the gas phase is obtained under identical experimental conditions with a layer of gelled water, see eq. (59):

$$n_{w,i} = k_g(\rho'_{w,i} - \rho'_{w,\infty}) \tag{59}$$

in which $\rho'_{w,i}$ denotes the water concentration in the gas phase in equilibrium with the surface water concentration $\rho_{w,i}$ and $\rho'_{w,\infty}$ is the water concentration in the bulk of the gas phase.

These drying experiments yield $n_{w,i}$ as a function of U at a constant value k_g of the mass transfer coefficient in the gas phase. Both $n_{w,i}$ functions should preferably show a region where $n_{w,i}$ stays constant. The values of G can be calculated with eqs. (11), (35) and (36) if D_o would be known. The value of D_o at temperature T can be calculated from the regular regime with eqs. (50) and (51).

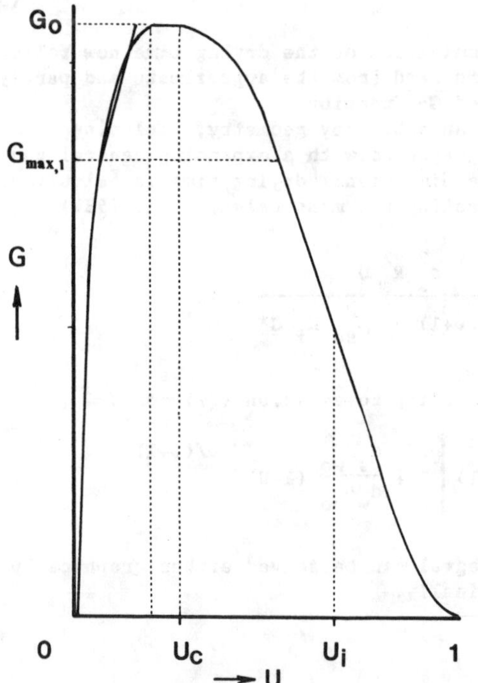

Fig.7: Experimentally observed G function of a slab

An experimentally obtained G versus U function is schematically presented in figure 7 for one temperature. If in the drying experiment the Bi number has been chosen sufficiently high, the relation between the experimentally observed value of G and U will stay at a maximum value over a certain trajectory of U.

For $Bi \to \infty$ this maximum value will be equal to G_o at $U = 0$. Consequently the value G_o, being the first point of the penetration curve, is easily obtained.

From the experimental G_o value the value of a in $U = 0$ is calculated with eq. (39). If the maximum in G function is not clearly present the value of a is obtained by extrapolation from the regular regime (eqs. (50) and (51)).

The experimentally obtained values of G as a function of U can be inter- or extrapolated to other temperatures with the Arrhenius relation:

$$\ln \frac{G_1}{G_2} = \frac{E_G}{R}\left(\frac{1}{T_2} - \frac{1}{T_1}\right) \tag{60}$$

The value of the activation energy is obtained from the slab drying experiments at two temperature levels.

<u>The calculation of G as function of U for</u> $0 \leq U \leq 1$ <u>and</u> $0 \leq \nu \leq 2$ <u>from the experimental slab drying curve</u>. All relations calculated from the experimental G function for $\nu = 0$, the complete picture is obtained for any value of ν between 0 and 2 by following the G function valid for phase 1 and 2, stepping over at the intersection with the tangent to the G function of phase 3, following this tangent until it intersects with the tangent to the regular regime curve of phase 4 until $U_i = a/(a+2)$ and following subsequently the regular regime curve.

4.2. Information regarding drying conditions

For the calculation of the drying rate of the particles the following information has to be available in addition to the slab drying experiments:

- mass transfer coefficient k_g to the particle in the dryer
- entrance air humidity y' (kg water/kg dry air)
- degree of mixing and residence time distribution in the dryer of both the air and the drying particles
- the initial water concentration $\rho_{w,o}$
- initial size and geometry of the particles
- in case of hollow spheres initial radius of gas bubble
- kg of dry air per kg dry solids, x'
- minimum residence time of particles in dryer
- entrance air temperature T'_{enter}
- maximum allowable relative humidity exit air
- density of the non-volatile (solids) fraction

4.3. Heat and mass balances and mass transfer coefficient in the gas phase

Next to the properties of the G and G* functions heat and mass balances and information regarding the mass transfer coefficient k_g in the gas phase are needed for the calculation of the evaporation rate and particle temperature as a function of the fraction of water evaporated.

Heat and mass balances. In a non-isothermal drying process the particle temperature T_U is calculated from the psychrometric properties of the surrounding air including the water content of the air y'_U and air temperature T'_U and from the ratio of the evaporation fluxes $n_{w,i,U}/n_{wb,U}$ of the actual particle and of the same particle at wet bulb conditions (pure water) respectively. The relation is given by:

$$\frac{n_{w,i,U}}{n_{wb,U}} \simeq \frac{T'_U - T_U}{T'_U - T'_{wb,U}} \qquad 0 \leq U \leq 1 \qquad (61)$$

The temperatures of the air T'_U, of the particle T_U and of the wet bulb $T_{wb,U}$ are connected by the straight wet-bulb line in the Mollier diagram. In this diagram y'_U and T'_U are points of the wet-bulb line.

The values y'_U and T'_U corresponding with the fraction water removed U are obtained from heat and mass balances and information regarding the way and degree of mixing in both phases. In case of co-current flow without residence time distribution like in very tall spray-dryers, the following balances hold if changes in enthalpy of the particles, due to temperature changes, are neglected:

enthalpy balance

$$x'(H'_{enter} - H'_U) = \Delta H_v U \frac{\rho_{w,o}}{\rho_{s,o}} \qquad (62)$$

mass balance

$$x'(y'_U - y'_{enter}) = U \frac{\rho_{w,o}}{\rho_{s,o}} \qquad (63)$$

where H' and ΔH_v are the enthalpy of the air and the heat of vaporization of water respectively.

For any value of U the psychrometric properties H_U and y'_U are calculated with eqs. (62) and (63). These values yield in the Mollier diagram the local air temperature T'_U and wet bulb temperature $T_{wb,U}$. Subsequently the local particle temperature T_U is calculated with eq. (61).

Mass transfer coefficient in gas phase. The value of k_g generally is not only dependent on flow conditions but also on U. In a spray-dryer at not a too short distance from the point of atomization holds:

$$Sh' = \frac{k_g 2R}{D'} = 2 \qquad (64)$$

where Sh' is the Sherwood number in the gas phase, D' is the diffusion coefficient of water in the gas phase and 2R the external diameter of the droplet at U.

4.4. Calculation procedure

Firstly the $G_{\nu=0}$ functions for $0 \leq U \leq 1$ are calculated from the slab drying experiments at the two corresponding temperatures. From the $G_{\nu=0}$ function the value of a is obtained as function of the averaged water concentration.

Secondly the activation energy E_G is calculated for the drying periods 1,2 and 3,4.

Thirdly the value of G* is calculated for the actual geometry in phase 1 or 2 for $U = \Delta U$ being a small increment in U. To this end the particle temperature is chosen equal to the wet bulb temperature of the entering air. With eq. (59) the evaporation flux is calculated and subsequently for the given geometry the value $G^*_{U=\Delta U}$ is calculated with eqs. (36) and (52).

Fourthly in a step-to-step procedure with stepsize ΔU the G* function for phase 1 and 2 is calculated from the $G_{\nu=0}$ function, the transformation equations, the mass and heat balances, eqs. (61), (62) and (63) and the Mollier diagram. In each step the temperature of the particle is calculated, the effect of temperature on G* is taken into account with the Arrhenius equation (60) and if present the increase in internal bubble radius is calculated from the equilibrium vapour pressure in the bubble. Consequently in every step the geometry factor ν is adapted. The G* function of phase 1 and 2 is continued until its G* value exceeds the G* value which would be calculated with the aid of the penetration curve.

Fifthly the G* function in the penetration region is calculated. If the central bubble radius starts to increase the value of G* belonging to the tangent of the actual value of the geometry factor should be chosen. The penetration and regular regime tangents are followed until the G* value becomes higher than the G* value calculated with the regular regime function (eq. (47) and (57)).

Sixthly the G* function is calculated in the regular regime.

Finally from the non-isothermal and non-isogeometry G* function the drying time is calculated with eqs. (58).

NOMENCLATURE

a	power in concentration dependence of diffusion coefficient	
Bi	Biot number	
d	density	kg/m^3
D	diffusion coefficient	m^2/s
E	activation energy	J/mol
F	flux parameter	
G	flux function	
H	shrinkage factor	
H'	enthalpy of gas phase	J/kg
ΔH_v	latent heat of vaporization	J/kg
i	inflection point	
j	mass flux	kg/m^2s
k_g	mass transfer coefficient	m/s
m	dimensionless concentration	

M	molecular weight	kg/mol
n	mass flux	kg/m^2s
r	space coordinate	m
R	radius of half-thickness of the system	m
Sh_d	average Sherwood number of the dispersed phase	
Sh	Sherwood number in the gas phase	
t	time	s
T	temperature	°K
U	efficiency drying process, $U = 1 - \bar{m}$	
x	kg dry air per kg dry solids	
X	dimensionless variable	
y	air humidity, kg water/kg dry air	

Greek symbols

λ	ratio of internal bubble radius and total radius	
ν	geometry parameter	
Φ	dimensionless space coordinate	
ρ	mass concentration	kg/m^3
τ	dimensionless time	

Subscripts

c	centre
g	gas phase
G	referring to flux function
i	interface; or referring to inflection point
o	initial
r	reduced
s	solid
sat	saturation
w	water

Superscripts

$\bar{}$	average value
*	referring to shrinking systems
'	referring to gas phase; or referring to an arbitrary geometry

REFERENCES

1. Liou, J.K. and Bruin S., Int.J.Heat Mass Transfer, vol.25, 1221 (1982);
2. Schoeber, W.J.A.H., Regular regimes in sorption processes. Ph.D. Thesis, Eindhoven University of Technology, The Netherlands (1976);
3. Lyn, v.d.J., Simulation of heat and mass transfer in spray drying. Ph. D.Thesis, Agricultural University Wageningen, The Netherlands (1976);
4. Verhey, J.G.P., Neth.Milk Dairy, vol.24, 96 (1970), vol.25, 246 (1971), vol.26, 186 (1972), vol.26, 203 (1972), vol.27, 3 (1973), vol.27, 19 (1973);
5. Schoeber, W.J.A.H. and Thijssen, H.A.C., Paper presented at the AIChE meeting on dehydration and concentration of foods, Los Angeles, U.S.A. (1975);

MOISTURE TRANSPORT MECHANISMS DURING THE DRYING OF GRANULAR POROUS MEDIA

Stephen Whitaker

Department of Chemical Engineering, University of California
Davis, CA 95616, U.S.A.

ABSTRACT

The theoretical foundations of heat, mass and momentum transport in granular porous media are reviewed, and the process of drying a layer of wet sand with warm, dry air is considered. The gas-phase momentum transport problem is examined from the point of view of the species momentum equation, and the analysis supports prior approximations of the momentum transport process. The liquid-phase moisture transport process is then re-examined in the light of past theoretical and experimental studies, and we conclude that the traditional concepts of relative permeability and capillary pressure saturation relations must be modified in order to obtain agreement between theory and experiment.

1. INTRODUCTION

When a layer of wet sand is dried by passing warm dry air over the sand, as illustrated in Fig. 1, one encounters the basic components of most drying processes. During the early stages of drying, the gas-phase mass transfer resistance in the surrounding air dominates the process and a constant drying rate occurs. While the behavior of the drying rate is especially simple, the physical processes that are taking place are not. The controlling factors of heat and mass transfer at the interface between the wet porous medium and the surrounding air are quite complex, and only a few papers have dealt with the details of the transition from volume-averaged transport equations within a porous medium to the point equations in the surrounding homogeneous phase. The most recent analysis of this type is by Ross [1], and the approach needs to be extended to the heat and mass transfer processes that occur at the porous medium — air interface illustrated in Fig. 1. In the analysis of momentum transfer by Ross [1] a "slip velocity" appears in the boundary condition, and in a more qualitative analysis of the heat transfer process at a phase interface, the

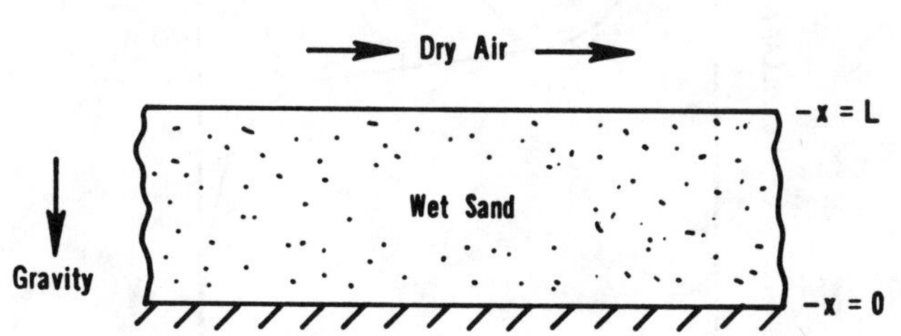

FIGURE 1
Basic Drying Process

author [2] provides evidence of the existence of a "temperature jump" at the interface. The need for a solid, theoretical analysis of the interfacial transport process will become apparent when the heat transfer process is examined in detail.

While the drying rate may be constant during the early stages of drying, the moisture distribution within the porous medium can take on a variety of forms depending on the nature of the capillary pressure-saturation curve. The internal moisture distribution during the constant rate period has been studied extensively by Chen [3] who found that the liquid-phase motion owing to capillary action is always quasi-steady. For a capillary pressure-saturation curve of the type shown in Fig. 2, Chen's work indicates that the saturation distribution can be calculated by the laws of hydrostatics provided the saturation is everywhere greater than the irreducible saturation, S_o. The fractional saturation used here is defined by

$$S = \frac{\psi_\beta \rho_\beta + \psi_\gamma \langle \rho_1 \rangle^\gamma}{\rho_\beta (1-\psi_\sigma)} \tag{1}$$

in which the β, γ and σ-phase are identified in Fig. 3. The liquid-phase flow in a saturated or partially saturated porous medium can be represented by

$$\langle \underset{\sim}{u}_\beta \rangle = -\frac{1}{\eta_\beta} \underset{\approx}{K}_\beta \cdot (\underset{\sim}{\nabla} \langle p_\beta \rangle^\beta - \rho_\beta \underset{\sim}{g}) \tag{2}$$

in which $\langle \underset{\sim}{u}_\beta \rangle$ represents the <u>phase average</u> velocity and $\langle p_\beta \rangle^\beta$ represents the <u>intrinsic phase average</u> pressure in the liquid phase. If one assumes that the capillary pressure, defined by

$$\langle p_c \rangle = \langle p_\gamma \rangle^\gamma - \langle p_\beta \rangle^\beta \tag{3}$$

can be expressed as a function of the saturation and the temperature, Eq. 2 can be expressed as

$$\langle \underset{\sim}{u}_\beta \rangle = \frac{1}{\eta_\beta} \underset{\approx}{K}_\beta \cdot \left[\left(\frac{\partial \langle p_c \rangle}{\partial S} \right) \underset{\sim}{\nabla} S + \left(\frac{\partial \langle p_c \rangle}{\partial \langle T \rangle} \right) \underset{\sim}{\nabla} \langle T \rangle - \underset{\sim}{\nabla} \langle p_\gamma \rangle^\gamma + \rho_\beta \underset{\sim}{g} \right] \tag{4}$$

Chen's studies [3] have shown that when $S_o \leq S \leq 1$ the saturation distribution can be determined by a solution of the quasi-steady form of the moisture transport equation [4, Sec. 5] which takes the form

$$\underset{\sim}{\nabla} S = -\rho_\beta \underset{\sim}{g} / (\partial \langle p_c \rangle / \partial S) \tag{5}$$

This form results from Eq. 4 when the moisture flux owing to temperature gradients and gas-phase pressure gradients can be ignored.

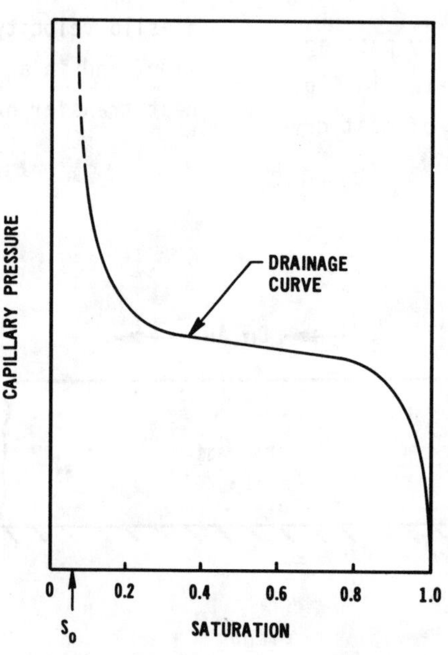

FIGURE 2
Capillary Pressure - Saturation Curve

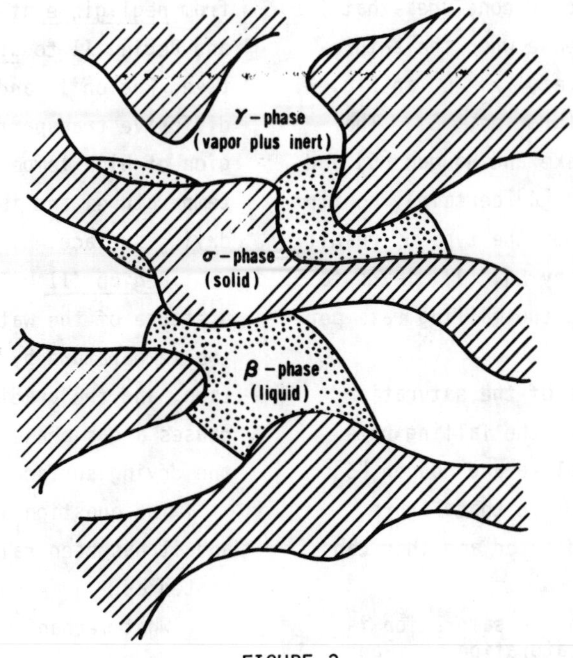

FIGURE 3
Wet Granular Porous Medium

The permeability tensor in Eq. (4) can be expressed in terms of the single phase permeability $\underset{\approx}{K}$ and the relative permeability k_r to obtain

$$\langle \underset{\sim}{u}_\beta \rangle = \frac{k_r}{\eta_\beta} \underset{\approx}{K} \cdot \left[\left(\frac{\partial \langle p_c \rangle}{\partial S} \right) \underset{\sim}{\nabla} S + \left(\frac{\partial \langle p_c \rangle}{\partial \langle T \rangle} \right) \underset{\sim}{\nabla} \langle T \rangle - \underset{\sim}{\nabla} \langle p_\gamma \rangle^\gamma + \rho_\beta \underset{\sim}{g} \right] \quad (6)$$

Experimental measurements of the relative permeability indicate the type of functional dependence shown in Fig. 4, and this can be approximated by

$$k_r = \left(\frac{S - S_o}{1 - S_o} \right)^3 \quad (7)$$

If one accepts this representation of the liquid-

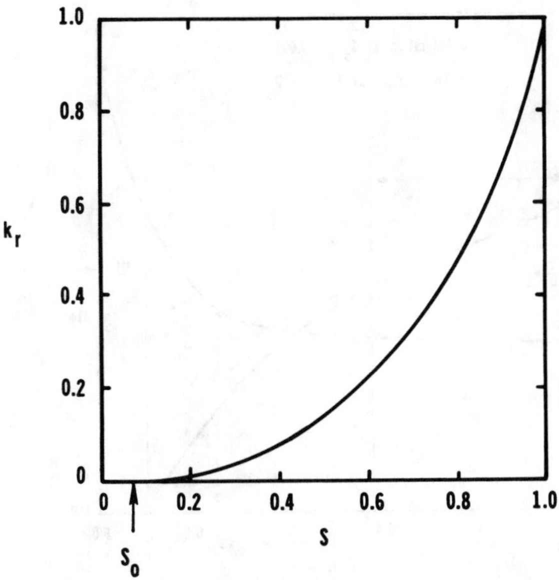

FIGURE 4
Relative Permeability for the Wetting Phase

phase flow phenomena, one quickly concludes that the constant rate period should cease when the saturation at the drying surface, i.e., $x = L$ in Fig. 1, approaches the irreducible saturation, S_o. The experimental results of Ceaglske and Hougen [5] and the calculations of the author [4] certainly confirm this; however, the acceptance of the type of result presented in Fig. 4, or given by Eq. (7), leads to a paradoxical situation during the falling rate period of drying.

Experimental measurements of the saturation and temperature profiles during the falling rate period are available from Ceaglske and Hougen [5] and are illustrated in Fig. 5. In that figure three distinct regions are indicated and they can be described as follows:

Region I: In this region the saturation is greater than the irreducible saturation S_o and the liquid phase is continuous. Capillary action causes the liquid to move toward the drying surface and the temperature in this region is nearly uniform.

Region II: In this region the saturation is below the irreducible saturation and according to Eqs. (6) and (7) there is no liquid-phase transport of moisture. Significant evaporation occurs in this region and the temperature gradient ranges from negligible at the interface between Region I and Region II to significant at the interface between Region II and Region III. The gas-phase diffusive transport of moisture is from the region of high temperature to the region of low temperature, and is thus directed away from the drying surface.

Region III: In this region the partial pressure of the water vapor is no longer tied to the temperature by the Clausius-Clapeyron equation, and the gradient in the partial pressure causes a gas-phase diffusive transport toward the drying surface.

The question at this point should be clear and it has been raised before [6]. Here it is stated as:

> What mechanism of moisture transport in Region II is responsible for maintaining continuity of moisture flux between Region I and Region III?

If there is no liquid-phase moisture transport in Region II, there must be a mechanism of moisture transport in the gas phase that is large enough to overcome the diffusive transport and match the liquid-phase transport at the edge of Region I and the gas-phase transport in Region III. If gas-phase convective transport is significant in

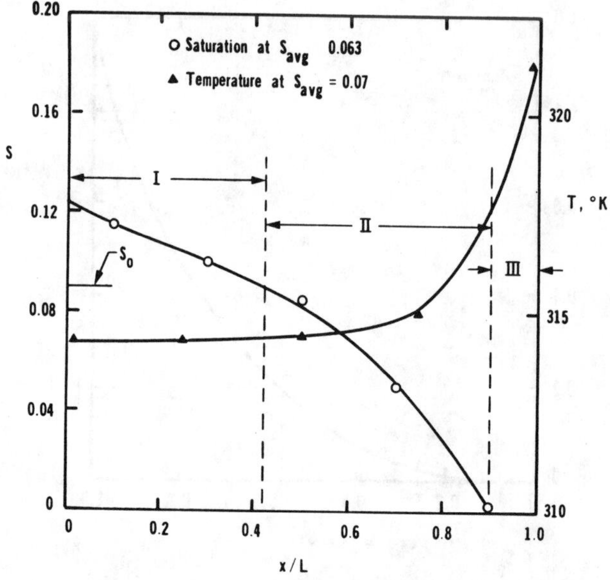

FIGURE 5
Experimental Saturation and Temperature Profiles During the Falling-Rate Period

Region II it may also be significant in Region III and any predictive theory of drying must take into account this mechanism of moisture transport. This means that gas-phase pressure gradients in Regions II and III must be determined and used with Darcy's law to compute the convective transport. In an earlier study [6] the use of the gas-phase momentum equation was avoided by means of a plausible argument that is common in the solution of what are known as <u>diffusion problems</u> [7, Sec. 17.2]. In the next section we examine, in greater detail than in previous studies, the problem of mass transport in the gas phase. We conclude that gas-phase convective transport can be ignored in Regions II and III, and we then turn our attention to the problem of liquid-phase transport in Region II.

2. GAS-PHASE CONVECTIVE TRANSPORT

In earlier studies of the process of drying granular porous media, the starting point for the development of a predictive theory of drying was based on the following set of transport equations [4, Sec. 1]

$$\frac{\partial \rho_A}{\partial t} + \nabla \cdot (\rho_A \mathbf{v}_A) = 0, \quad A = 1, 2 \tag{8}$$

$$0 = -\nabla p + \rho \mathbf{g} + \eta \nabla^2 \mathbf{v} \tag{9}$$

$$\rho c_p \left(\frac{\partial T}{\partial t} + \mathbf{u} \cdot \nabla T\right) = \lambda \nabla^2 T \tag{10}$$

In addition, Fick's law of diffusion was used in the analysis of the gas-phase transport of moisture. At this time, it would appear that the physics of the drying process can be better understood if the analysis is initiated in terms of the <u>species</u> continuity equations given by Eq. (8), the <u>overall</u> thermal energy equation [8, Sec. II] given by Eq. (10) and the <u>species momentum equations</u>. The role of the species momentum equation in the analysis of diffusional processes has been discussed at great length elsewhere [9, Sec. 1], and for a non-reacting dilute gas mixture we can express the governing equations as

$$\frac{\partial \rho_A}{\partial t} + \nabla \cdot (\rho_A \mathbf{v}_A) = 0, \quad A = 1, 2, \ldots N \tag{11}$$

$$\rho_A \left(\frac{\partial \mathbf{v}_A}{\partial t} + \mathbf{v}_A \cdot \nabla \mathbf{v}_A\right) = -\nabla p_A + \rho_A \mathbf{g} + \nabla \cdot \boldsymbol{\tau}_A + p \sum_{\substack{B=1 \\ B \neq A}}^{B=N} \frac{x_A x_B (\mathbf{v}_B - \mathbf{v}_A)}{\mathcal{D}_{AB}}, \quad A = 1, 2, \ldots N \tag{12}$$

$$\rho c_p \left(\frac{\partial T}{\partial t} + \mathbf{u} \cdot \nabla T\right) = \lambda \nabla^2 T - \nabla \cdot \sum_{A=1}^{A=N} \mathbf{j}_A h_A \tag{13}$$

In general the diffusive flux of thermal energy in Eq. (13) is neglected to obtain Eq. (10), and in a comprehensive analysis of the problem of mass transport in porous catalysts [9] it is shown that the gas-phase species momentum equations can be simplified to

$$0 = -\nabla p_A + \nabla \cdot \boldsymbol{\tau}_A + p \sum_{\substack{B=1 \\ B \neq A}}^{B=N} \frac{x_A \mathbf{N}_B - x_B \mathbf{N}_A}{c \mathcal{D}_{AB}}, \quad A = 1, 2 \ldots N \tag{14}$$

Here $\mathbf{N}_A$ represents the molar flux given explicitly by

$$\mathbf{N}_A = c_A \mathbf{v}_A \tag{15}$$

One should think of Eq. (11) as the governing differential equation for ρ_A (or c_A in the molar form) and Eq. (12) or Eq. (14) as the governing differential equation for the species velocity $\mathbf{v}_A$. The volume-averaged form of Eq. (14) requires considerable effort for the result must reduce to the Graham relation [10, Chap 6] under certain circumstances and Darcy's law [11] under other conditions. For isothermal conditions and a binary system in which $|\mathbf{N}_B| \ll |\mathbf{N}_A|$, the volume-averaged form of Eq. 14 is given by [9, Sec. 4]

$$0 = \psi_\gamma \mathbf{B} \cdot \nabla \langle c_A \rangle^\gamma + \psi_\gamma \langle x_B \rangle^\gamma \langle \mathbf{N}_A \rangle^\gamma / \mathcal{D}_{AB} + \frac{\psi_\gamma \eta_\gamma \mathbf{B} \cdot \mathbf{K}_\gamma^{-1}}{RT} \cdot \left[\frac{\langle c_A \rangle^\gamma (M_A M_B^{1/2} + M_B M_A^{1/2}) + \langle c_A \rangle^\gamma M_A^{3/2}}{(\langle c_A \rangle^\gamma M_A + \langle c_B \rangle^\gamma M_B)(\langle c_A \rangle^\gamma M_A^{1/2} + \langle c_B \rangle^\gamma M_B^{1/2})}\right] \cdot \langle \mathbf{N}_A \rangle^\gamma \tag{16}$$

Here $\mathbf{B}$ is a tensor of order one and $\mathbf{K}_\gamma$ represents the gas-phase permeability tensor to be used in conjunction with Darby's law. It is important to note that Eq. (16) is valid only for the case of bulk flow and diffusion; however, the analysis for the transition regime and Knudsen flow is also available [9, Sec. 4]. The last term in Eq. 16 represents the momentum exchange between species A and the solid phase, and when this term is important convective transport makes a significant contribution to the gas-phase moisture transport. Conversely, when the last term is negligible we can ignore convective transport and the diffusive representation of gas-phase moisture transport used by Whitaker and Chou [6] becomes valid. In order to estimate the magnitude of the last term we use

$$\mathbf{B} = \mathbf{O}(1) \quad , \quad \mathbf{K} = \mathbf{O}(\ell_\gamma^2) \tag{17}$$

in which ℓ_γ represents the characteristic length for the γ-phase. When the molecular weights of species A and B are the same order of magnitude we also have

$$\left[\frac{\langle c_A\rangle^\gamma (M_A M_B^{1/2} + M_B M_A^{1/2}) + \langle c_A\rangle^\gamma M_A^{3/2}}{(\langle c_A\rangle^\gamma M_A + \langle c_A\rangle^\gamma M_B)(\langle c_A\rangle^\gamma M_A^{1/2} \quad \langle c_B\rangle^\gamma M_B^{1/2})}\right] = \mathbf{O}\left(\frac{1}{c}\right) \tag{18}$$

and from this we conclude that convective transport can be ignored when the following constraint is satisfied

$$\frac{\psi_\gamma \eta_\gamma}{RT\ell_\gamma^2} \frac{1}{c} \ll \frac{\psi_\gamma \langle x_B\rangle^\gamma}{\mathcal{D}_{AB}} \tag{19}$$

A more convenient form is given by

$$\langle x_B\rangle^\gamma \gg \frac{\eta_\gamma \mathcal{D}_{AB}}{p\ell_\gamma^2} \tag{20}$$

and for the conditions under investigation, the term on the right hand side of Eq. (20) is on the order of 10^{-8}. This means that the last term in Eq. (16) can be discarded and the molar flux is given by

$$\langle \mathbf{N}_A\rangle^\gamma = -\frac{\mathbf{D}_{eff}\cdot\nabla\langle c_A\rangle^\gamma}{1 - \langle x_A\rangle^\gamma} \tag{21}$$

Here the effective diffusivity tensor is given by

$$\mathbf{D}_{eff} = \mathbf{B}\mathcal{D}_{AB} \tag{22}$$

and Eq. (21) represents the molar form of the flux expression used by Whitaker and Chou [6] in their analysis of the drying problem illustrated in Fig. 1. The type of analysis leading to Eq. (16) has also been explored by Hadley [12] and that development also supports the idea that convective transport can be neglected when Eq. (20) is valid.

In the analysis presented by Whitaker and Chou [6] the existence of Region II is prohibited because the <u>absence</u> of a liquid-phase flux <u>toward</u> the drying surface and the <u>presence</u> of a gas-phase diffusive flux <u>away</u> from the drying surface presents an impossible situation. Their comparison between theory and experiment is illustrated in Fig. 6, and there one can see that Region II is replaced with a <u>jump</u> in the saturation on the order of S_o. Clearly the experimental data illustrate an entirely different type of behavior and the resolution of this paradox seems to reside with the existence of a liquid-phase moisture flux below the irreducible saturation, S_o.

3. LIQUID-PHASE TRANSPORT

The liquid-phase transport of moisture has already been discussed in Sec. 1 and the pertinent expressions for the liquid phase velocity are given by Eqs. (6) and (7). For the problem under consideration, there is no significant contribution of either the temperature gradient or the gas-phase pressure gradient to the velocity of the liquid and Eq. 6 can be simplified to

$$\langle \mathbf{u}_\beta\rangle = \left(\frac{k_r}{\eta_\beta}\right)\mathbf{K}\cdot\left[\left(\frac{\partial\langle p_c\rangle}{\partial S}\right)\nabla S + \rho_\beta \mathbf{g}\right] \tag{23}$$

On the basis of the experimental data of Ceaglske and Hougen [5], the work of Whitaker and Chou [6] and the analysis described in Sec. 2, it seems clear that some liquid-phase mass transport <u>must</u> occur in Region II. In order to permit this transport to take place within the framework of the current theoretical analysis, we propose the following two <u>deviations</u> from the traditional thought concerning two phase flow in porous media:

I. The capillary pressure tends toward <u>infinity</u> at a value of the saturation, S_{oo}, that is significantly lower than

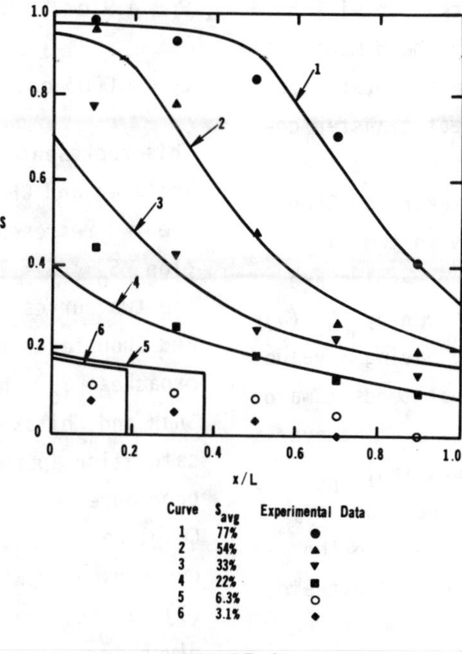

FIGURE 6

Saturation Profiles of Ceaglske and Hougen Compared with the Theory of Whitaker and Chou

S_o and difficult to distinguish from zero on an experimental basis.

II. The relative permeability tends toward zero at a saturation, S_{oo}, and the relative permeability for saturations lower than S_o is difficult to distinguish from zero on an experimental basis.

It is important to note that zero and infinity have no precise meaning and when we use these terms we really mean "small relative to something" and "large relative to something". The above statements can be made somewhat more precise in terms of the restrictions:

I. $\langle p_c \rangle \gg \sigma/\ell$, when $0 \leq S_{oo} < S_o$

II. $k_r \ll 1$ when $S_{oo} \leq S \leq S_o$

Here σ represents the surface tension and ℓ represents the characteristic particle diameter or pore diameter for the porous medium. Clearly this means that we are considering variations of $\langle p_c \rangle$ and k_r with respect to S that are very difficult to determine by traditional experiments.

At this point we have chosen to depart from our earlier objective of comparison between theory and experiment in the absence of adjustable parameters and we are now willing to explore the following adjustments:

1) The capillary pressure-saturation curve is required to fit experimental data in the range for which data is available, but may be finite for saturations less than S_o and approaching S_{oo}.
2) The relative permeability-saturation curve is required to fit experimental data in the range, $S_o \leq S \leq 1$, but may be non-zero for saturations approaching S_{oo}.
3) The film mass transfer coefficient is fixed by the experimental data, but the film heat transfer coefficient can be adjusted at will.

With this type of flexibility available in a theoretical framework as complex as that encountered in the drying process, one should expect that theory and experiment would match rather nicely - and they do. That they do is irrelevant. What is relevant is the nature of the capillary pressure - saturation curve, the relative permeability - saturation curve, and the relation of the film heat transfer coefficient to that predicted by the traditional heat and mass transfer analogy.

4. THEORY AND EXPERIMENT

The theoretical foundations have been presented earlier [4,6,8] and the details of the new results presented here will be available elsewhere

[13]. In this presentation the discussion will be confined to the adjustments that have been made in the capillary pressure-saturation relation, the relative permeability, and the film heat transfer coefficient. As a point of reference, we note that the theoretical calculations of Whitaker and Chou [6] give rise to the discontinuities in the saturation profiles shown in Fig. 6. Those results are based on the relative permeability given by Eq. (7) with the irreducible saturation, S_o, having a value of 0.09, and a capillary pressure that tends toward infinity as the saturation approaches S_o. However, if we propose a new irreducible saturation, S_{oo}, with a value of 0.009 we can adjust the capillary pressure and relative permeability to obtain the agreement between theory and experiment illustrated in Fig. 7. This type of agreement must surely be characterized as phenomenal and it is important to discuss the adjustments that must be made in order to achieve this type of agreement. To begin with, we note that the dimensionless capillary pressure was represented by

$$\frac{\langle p_c \rangle}{\rho_\beta g L} = \left\{ P\left[1-e^{-Q(1-S)}\right] + R(1-S) + U(S-S_{oo})^{-F} \right\} \Big/ L \tag{24}$$

in which

$P = 4.9$ cm, $Q = 40.0$, $R = 1.8$ cm, $F = 4.4$

$U = 0.0003$ cm, $S_{oo} = 0.009$

This representation is compared with that used by Whitaker and Chou in Fig. 8 and there we see that the two representations are quite close in the region $S_o < S \leq 1$. The crucial difference between the two curves is that the one used by Whitaker and Chou tends to infinity as the saturation approaches S_o while the representation used by Shah and Whitaker [13] tends to infinity as the saturation approaches S_{oo}. Both curves are in good agreement with the experimental data of Ceaglske and Hougen. The relative permeability curve used by Whitaker and Chou is given by Eq. (7), and the relation used by Shah and Whitaker is given by

$$k_r = \left(\frac{S-S_o}{1-S_o}\right)^3, \quad S \geq S_o + \varepsilon \tag{25}$$

$$k_r = A\left(\frac{S-S_{oo}}{1-S_{oo}}\right)^B, \quad S \leq S_o + \varepsilon \tag{26}$$

in which

$A = 552.6$, $B = 7.42$, $\varepsilon = 0.055$, $S_o = 0.09$

$S_{oo} = 0.009$

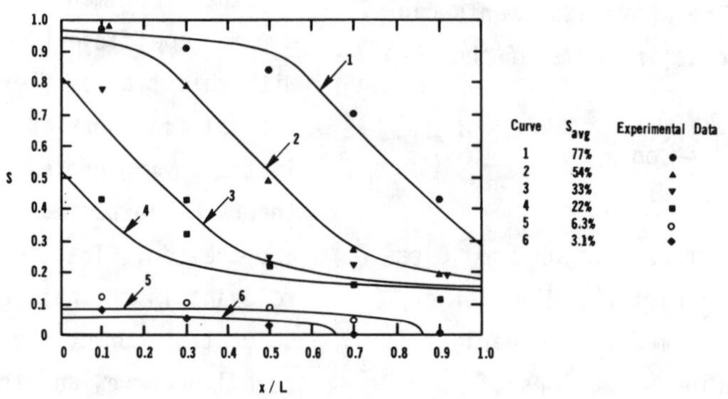

FIGURE 7

Comparison of Theory and Experiment

Using S_{oo} as the Irreducible Saturation

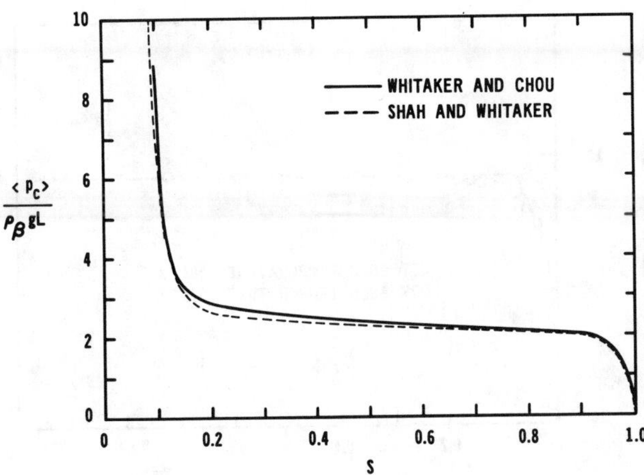

FIGURE 8
Capillary Pressure - Saturation Curves

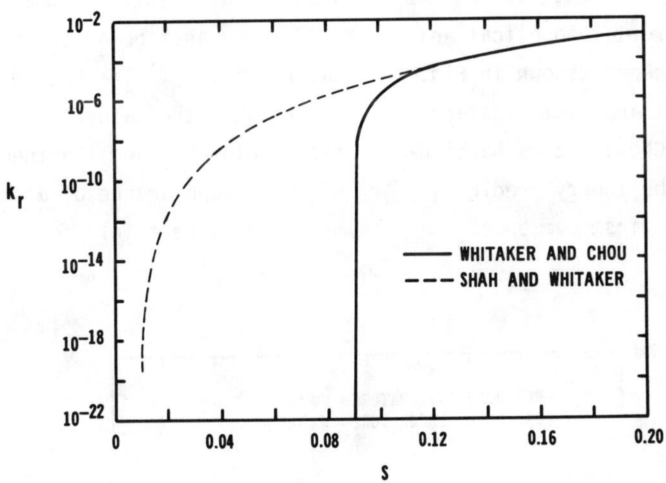

FIGURE 9
Modified Relative Permeability

A graphical comparison is shown in Fig. 9 and there one can see that deviations from Eq. (7) occur only for values of the relative permeability less than or equal to 10^{-4}. It should now be apparent that we have adjusted the representations for $\langle p_c \rangle$ and k_r in a manner that produces no disagreement with the experimental values and, in fact, these functions have been adjusted in regions where experimental data are difficult, if not impossible, to obtain.

In order to compare theory and experiment for the rate of drying, a mass transfer coefficient has

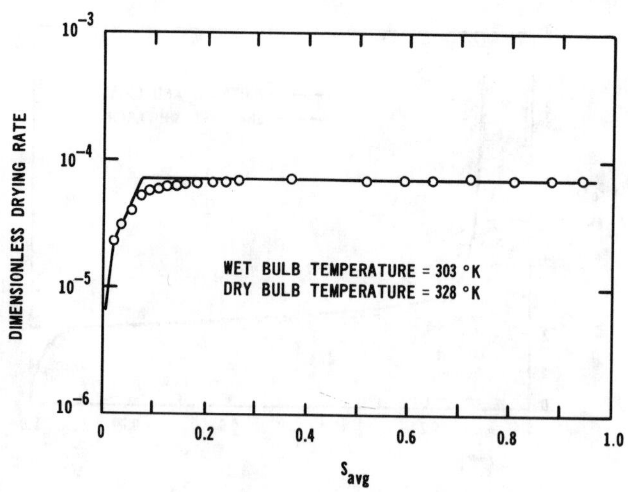

FIGURE 10
Comparison of Theoretical and Experimental
Drying Rates

been chosen to fit the experimental data during the constant-rate period. The current theoretical analysis leads to the type of agreement shown in Fig. 10, and this represents a great improvement over that obtained by Whitaker and Chou. It is particularly important to note that the theory predicts two falling-rate periods. The first commences when the saturation at the drying surface is about 0.07, i.e. just below the traditional irreducible saturation, S_o. The second falling-rate period begins when the saturation at the drying surface falls below S_{oo} and the theoretical model predicts the appearance of a dry region. The experimental data certainly follow the trend of the

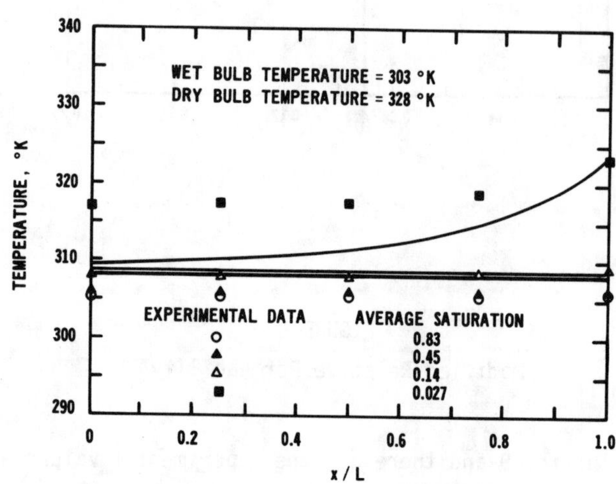

FIGURE 11
Comparison of Theoretical and Experimental
Temperature Profiles

theoretical predictions, but the transition from the constant-rate period to the second falling-rate period occurs over a wider range of the average saturation.

In order to compare predicted and measured temperature profiles, one requires a film heat transfer coefficient in addition to a knowledge of the thermal properties of the wet porous medium. The latter are relatively easy to estimate and a heat and mass transfer analogy can be used to predict the film heat transfer coefficient. Since the Prandtl and Schmidt numbers are near unity, the analogy consists of equating the Sherwood and Nusselt numbers. When this is done, the agreement between theory and experiment is very poor. However, if we allow the film heat transfer coefficient to be an adjustable parameter we obtain the type of agreement illustrated in Fig. 11. The theoretical results shown in Fig. 11 were obtained using a film heat transfer coefficient that is about two times larger than the value predicted from the mass transfer coefficient. Clearly this is not a minor change and the matter of heat and mass transfer coefficients at the surface of a porous medium need to be considered in greater detail.

5. CONCLUSIONS

By treating the relative permeability and capillary pressure as adjustable functions we are able to obtain excellent agreement between measured and predicted saturation profiles during the drying of a granular porous medium. The adjustments made to these functions are not in disagreement with the experimental values for the capillary pressure and relative permeability; however, they are contrary to the traditional thought concerning these functions. The predicted drying rate illustrates a constant-rate period, a first falling-rate period and a second falling-rate period. The two falling-rate periods are generated by the nature of the capillary pressure and relative permeability functions, thus providing some support for the manner in which these functions have been adjusted. In order to obtain reasonable agreement between measured and predicted temperature profiles, a film heat transfer coefficient was used that was about five times larger than that predicted by analogy with the mass transfer coefficient.

ACKNOWLEDGEMENT

This work was supported by NSF Grant CPE-81165 28.

NOMENCLATURE

Roman Letters

c	total molar concentration, moles/m^3
c_A	molar concentration of species A, moles/m^3
c_p	constant-pressure heat capacity, kcal/kgK
$\underset{\approx}{D}_{eff}$	effective diffusivity tensor, m^2/s
$\mathcal{D}_{AB}$	molecular diffusivity, m^2/s
$\underset{\sim}{g}$	gravity vector, m/s^2
h_A	partial mass enthalpy, kcal/kg
$\underset{\sim}{j}_A$	mass diffusive flux, kg/m^{2}s
$\underset{\approx}{K}$	Darcy's law permeability, m^2
$\underset{\approx}{K}_\beta$	liquid-phase permeability, m^2
k_r	relative permeability for the liquid phase
ℓ	characteristic length for the void space, m
L	Depth of the porous medium, m
M_A	molecular weight of species A, kg/kg-mole
$\underset{\sim}{N}_A$	molar flux vector, moles/m^{2}s
p	total pressure, N/m^2
$\langle p_\gamma \rangle^\gamma$	intrinsic phase average pressure for the gas phase, N/m^2
$\langle p_\beta \rangle^\beta$	intrinsic phase average pressure for the liquid phase, N/m^2
$\langle p_c \rangle$	capillary pressure, N/m^2
$\mathcal{R}$	universal gas constant, Nm/mole K
S	saturation
S_o	traditional irreducible saturation
S_{oo}	modified irreducible saturation
T	temperature, K
$\langle T \rangle$	spatial average temperature, K
$\underset{\sim}{u}$	mass average velocity, m/s
$\langle \underset{\sim}{u}_\beta \rangle$	phase average velocity for the liquid phase, m/s
$\underset{\sim}{v}_A$	species velocity, m/s
x	distance, m

Greek Letters

η_β	viscosity of the liquid, Ns/m^2
λ	thermal conductivity, kcal/s m K

ρ_β	density of the liquid, kg/m^3
$\langle\rho_1\rangle^\gamma$	phase average density of the water in the gas phase, kg/m^3
σ	surface tension, N/m
$\underset{\approx}{\tau}_A$	viscous stress tensor for species A, N/m^2
ψ_β	volume fraction of the liquid phase
ψ_γ	volume fraction of the gas phase
ψ_σ	volume fraction of the solid phase

REFERENCES

1. Ross, S. M., "Theoretical Model of the Boundary Condition at a Fluid-Porous Interface," AIChE Journal, vol. 29, 840 (1983).
2. Whitaker, S., "Radiant Energy Transport in Porous Media," Ind. Engng. Chem. Fundls., vol. 19, 210 (1980).
3. Chen, S., "Some Investigations on Drying of Granular Porous Materials," M.S. Thesis, Dept. of Chemical Engineering, University of California at Davis, 1982.
4. Whitaker, S., "Heat and Mass Transfer in Granular Porous Media," Advances in Drying, Vol. 1, edited by A. S. Mujumdar, Hemisphere Pub. Corp., New York (1980).
5. Ceaglske, N. H. and Hougen, O. A., "Drying Granular Solids," Ind. Eng. Chem., vol. 29, 805 (1937).
6. Whitaker, S. and Chou, W. T-H., "Drying Granular Porous Media - Theory and Experiment," Drying Tech., vol. 1, 3 (1983).
7. Bird, R. B., Stewart, W. E. and Lightfoot, E. N., Transport Phenomena, John Wiley & Sons, New York (1960).
8. Whitaker, S., "Simultaneous Heat, Mass and Momentum Transfer in Porous Media: A Theory of Drying," Advances in Heat Transfer, edited by J. P. Harnett and T. F. Irvine, Jr., Academic Press (1977).
9. Whitaker, S., "Transport Process with Heterogeneous Reaction," 25th Conicet Anniversary Reactor Design Conference, Santa Fe, Argentina (1983).
10. Jackson, R., Transport in Porous Catalysts, Elsevier Publishing Co., New York (1977).
11. Whitaker, S., "Advances in the Theory of Fluid Motion in Porous Media," Ind. Eng. Chem., vol. 61, 14 (1969).
12. Hadley, G. R., "Theoretical Treatment of Evaporation Front Drying," Int. J. Heat Mass Transfer, vol. 25, 1511 (1982).
13. Shah, G. and Whitaker, S., "Liquid Phase Transport in Porous Media During Drying - An Alternative View of the Concept of Irreducible Saturation," to be submitted to Ind. Engng. Chem. Fundls. (1984).

ASPECTS OF SOLAR DRYING

L. Imre

Technical University Budapest,
H-ISES, UNESCO E.N.C.S.E
1111 Budapest, HUNGARY

ABSTRACT

The oil crisis and the increasing danger of pollution of the environment compelled mankind to a renewed stock-taking of possible energy sources and reconsider the viewpoints of energy economy.

In the course of these considerations the renewable energy sources came into the focus of interest and the possibilities in solar energy have also been rediscovered.

Setting out from the incontestable advantages of solar energy there was an enthusiastic upswing at the initial stage, which was followed by a disheartened period and at present a realistic way of looking at the possibilities is existing.

The prerequisite of utilizing solar energy for drying is rentability, i.e. an acceptable payback time.

The realization of the importance of economy aspects followed by detailed analysis lead to the knowledge of the main influencing factors, their role and mechanizm. Difficulties arise from the time-dependent and diluted character of solar radiation. This involves problems requiring different means for their solution /heat storer, auxiliary energy source, large collector surfaces, control system/ and so the investment cost are considerable.

For the economical utilization of solar energy for drying it is necessary to coordinate the drying purposes with the specific characteristic of solar radiation /e.g. small power demand; products requiring low temperature drying, fitting the peak requirements to the highest radiation period/. Several agricultural and forestry products correspond to these requirements.

The other possibility for increasing rentability is the reduction of investment costs. In this respect there are two possible ways: to use simple, cheap, plain constructions or to develop multi-purposed and integrated systems used throughout the whole year.

In any case the task has to be handled as an optimum problem and can be solved by the use of computer simulation.

Appreciating the advantages of using solar energy for drying considerable research and developing work is in progress all over the world and economical solutions have already been found. However, the period of searching for new ways has not yet come to an end.

1. INTRODUCTION

The utilization of solar energy for drying was most probably one of the first purposeful technological activities of man.

The natural methods of drying employed from ancient times have been kept until recently; however, in the last few decades they have been ousted from most fields of application. The requirements of modern cultivation technology in agriculture have incited to the introduction of drying methods, which can guarantee reliable product quality, high efficiency, need little manpower, and are independent of the weather. Considerable part of the drying plants have been installed with oil heating and have come into very difficult situation because of the oil crisis. Quantities of certain types of dried products had to be reduced considerably and some oil heated drying plants had to be stopped.

Experts are considering several possibilities for the elimination of the difficulties. For instance in the field of preserving fodder methods, which do not need drying, came in the foreground. At the same time the possible energy sources had to be reconsidered and the renewable energy sources, among them the possibilities in using solar energy had to be revaluated. In recent years an impetuous work for developing the modes of application and equipment has got a start, and several solar dryers have been put into operation. Experts opinion, however, differs concerning the importance and prospects of solar drying. Those supporting solar drying emphasize its possibilities, while other experts dispute its importance in satisfying the technical and economical requirements. There are also experts who are non-committal as they think to find contradictions in the application of solar drying.

2. VIEWPOINTS OF CONSIDERATION

2.1. Characteristic Features

The most important features of solar energy can be summarized as follows [1].

a/ The radiated energy of the Sun is at disposal, free of charge,

b/ it is also available at regions of the Earth where no other energy source can be found,

c/ it cannot be monopolized,

d/ it has not any polluting effect.

e/ The radiation is periodical and has breaks,

f/ the intensity of radiation is time-dependent, changes with seasons and parts of the day,

g/ radiation is "deluted", the energy flux density is low,

h/ the intensity of radiation is weather-dependent.

Features a/...d/ offer significant advantages, while those of e/...h/ originate certain difficulties in realization.

2.2. Criteria of Valuation

For using solar energy an appropriate equipment is needed. According to the energy criterion the utilization of solar energy is justified, when in the service life of the equipment more energy can be gained than is necessary for the manufacturing. However, the realization of this criterion is not sufficient for the economical application of solar drying. There are several criteria used for economic valuation of solar energy systems [2, 3].

Solar dryers with very simple construction /tent-, cabinet-type dryers, etc./ for replacing natural drying, are often compared to results obtained by way of natural drying /drying time, protection against weather and parasites, etc.[7]/.

It is possible to use a kind of valuation based on the comparison between the first and maintenance costs of drying equipment using traditional energy source and, the same costs of a solar dryer, if drying output is the same [8] or, all costs of producing 1 kg dried material may be compared [9]. These latter methods can generally be applied, i.e. in cases, too, when the traditional and solar dryers differ not only in the energy source used but also in the buildup of the actual drier /material forwarding system, control, etc./.

A further characteristic of the valuation is the payback time which gives a good possibility for surveying the economic correlations of the technical solutions of solar dryers. The analysis given below is based on the supposition that every advantageous and disadvantageous property can be characterized by an economic parameter. It is also supposed that the "dryer" itself is identical in the solar and traditional systems.

2.3. Payback Time

Payback time can also be defined in many ways [2]. The following definition will be used here [4]: payback time is the time at which the sum of the investment /C/ and the annual expenses /mC/ with compounded interest is equal to the total savings /S/ with compounded interest gained by the use of solar energy instead of fuel. In a dynamic economic environment, according to Böer [4]

$$C/1+r/^n + \frac{/1+r/^n - /1-i/^n}{r - i} mC = \frac{/1+r/^n - /1-e/^n}{r - e} S . \qquad /1.a/$$

The payback time for the condition i = e

$$n = \frac{\lg /1 - \frac{r - i}{\frac{S}{C} - m}/}{\lg /\frac{1 + i}{1 + r}/} . \qquad /1.b/$$

Payback time depends on the quotient of the yearly attainable savings S and the first cost C and on the value of m. The economy of a solar dryer can be enhanced by increasing S and decreasing C. /E.g. for a return within 5 years $\frac{S}{C} \geq 0{,}24$, when r = 0,055; m = 0,05; i = 0,05; e = 0,09 [4]/.

The values of S and C are interdependent with the technical solution. This latter, at the same time, determines the technological possibilities.

3. CORRELATIONS BETWEEN SAVINGS AND TECHNOLOGY

3.1. Components of the Savings

The yearly savings S consist of two main parts

$$S = S_1 + S_2 \qquad /2/$$

where S_1 is the annually displaced energy cost,

S_2 is further savings by the use of solar energy.

3.2. The Anually Displaced Energy Cost

The cost of energy carrier displaced yearly by solar energy

$$S_1 = E \cdot a \, . \qquad /3/$$

The value of E is the function of several factors

$$E = E_o - \sum_i L_i \, , \qquad /4/$$

where E_o is the energy incident on the surface of the collector during the operation time t_o:

$$E_o = A \int_0^{t_o} I/t/ \, dt \, , \qquad /5/$$

the sum of the energy losses

$$\sum_{i=1}^{3} L_i = L_1 + L_2 + L_3 \, , \qquad /6/$$

where L_1 energy loss of the collector,
L_2 parasitic losses,
L_3 exit loss of the drier.

In equation /5/, A is the surface of the collectors necessary for a given drier. The global irradiance I/t/ is the function of the geographical position; its influence on the rentability is very significant /i.e. the possible economic efficiency is dependent on the geographical position/. If the operation time of the dryer t_o is small /e.g. yearly 1-2 months only/, then E_o and therewith S are also small. It is reasonable therefore, to make the yearly exploitation as high as possible.

The energy loss of the collector L_1 is interdependent with the construction /the number of coverings the thickness of heat insulation, selectivity of adsorber surface etc./ and the overtemperature of the absorber: $\Delta T_a = T_a - T_w$. Using the long-term efficiency η_{1t}

$$L_1 = E_o [1 - \eta_{1t} / \Delta T_a /] \, . \qquad /7/$$

The necessary collector surface A for solving a drying problem with a given energy demand depends on η_{1t}. Fig.1/a shows a construction of a liquid type flat-plate collector. The long-term efficiency of the collector without covering, with 1 and 2 coverings, as a function of ΔT_a is presented in Fig.1/b /for May, Hungary/. Fig.1/c gives the formation of relationship $\frac{S}{C}$ of the collector as a function of ΔT_a, with different number of coverings. It is obvious that more economical solutions can be attained with smaller ΔT_a; thus solar dryers with flat-plate collectors can be applied economically for materials requiring expressedly low temperature drying. Another possible way may be: to use flat-plate collectors for preheating the working medium only.

Energy losses L_2 and L_3 will be referred later.

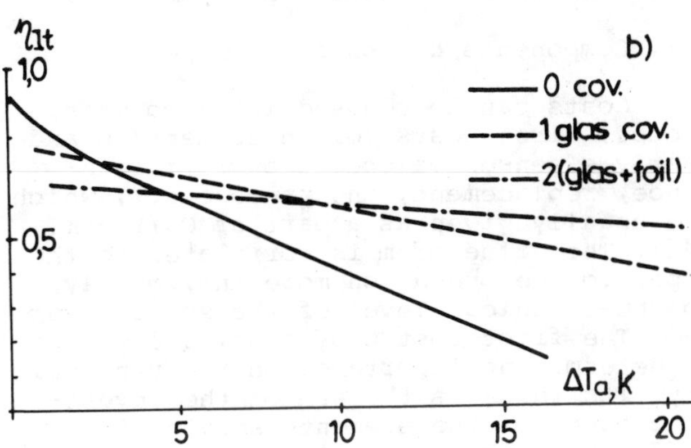

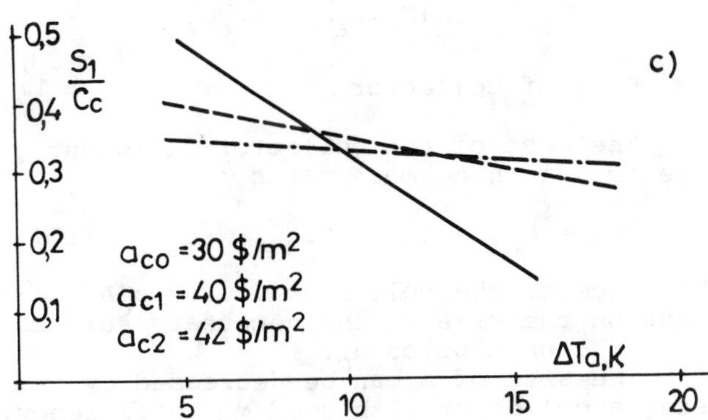

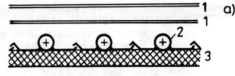

Fig. 1 Construction of a liquid-type flat-plate collector /a/, the long-term efficiency /b/ and the $\frac{S}{C}$ as the function of absorber's overtemperature ΔT_a

3.3. Savings from Other Reasons

Savings S_2 obtained as a result of using solar energy

$$S_2 = S_{21} + S_{22} \, , \qquad /8/$$

where S_{21} is the yearly savings gained by avoiding environment pollution, i.e. the costs which should be spent on neutralizing the pollution effects of driers using fuels to avoid greater damage in the environment.

S_{22} is the yearly savings obtained by the use of solar energy in regions without other energy sources /transportation costs of other energy carriers, or savings through the improvement of the quality of dried products/.

4. CORRELATIONS OF COSTS AND TECHNOLOGY

4.1. Components of Costs

Costs can be classed into two main groups: first costs /of investment/ C and yearly expenses of operation Y /maintenance, replacement, insurance etc./, which are usually given as a part of C /Y = mC, [4]/. The value of m is correlated to the type, to the operation mode and, mostly, to the technical level of the solar dryer.

The first cost C of a solar dryer is of determining importance on the rentability. Its value is the sum of the investment costs of the elements serving for difference purposes:

$$C = C_C+C_S+C_p+C_H+C_A+C_{ah}+C_D+C_o \qquad /9/$$

4.2. Cost of Collector

The cost of the collector C_C is put together of three main parts

$$C_C = C_{C1} + C_{C2} + C_{C3} \qquad /10/$$

The price of the collector /$C_{C1} = a_C A$/ depends on the size of the necessary surface A and the unit price a_C.

The size of A can be decreased by using a collector with good η_{lt} efficiency. There is a relation partly between the long-term efficiency and the construction, partly between the construction and the unit price a_C; thus design and dimensioning of flat-plate collectors is an optimum problem when a_C should be minimized [2,10]. Unit price a_C can be reduced by good construction and careful choice of materials. For a given task the value of A can also be determined by optimisation [20].

The cost of the supporting structure of the collectors, C_{C2} is sometimes quite significant. In the case of high performance driers, by the use of collector construction integrated into the roof structure can save not only the costs of the supporting structure, but also the traditional covering of the roof [5, 6, 11]. In such cases C_{C2} may become even of negative value. With roof collectors the maintenance may be more difficult, and have an increasing effect on the value of m.

C_{C3} is the instalment cost of the collector. With great surface panel-type collectors the volume of on-the-spot mounting work and C_{C3} is usually slightly greater than that of the prefabricated smaller collector units.

4.3. Choice of the Working Medium of the Collector

The choice of the working medium of the collector has also some influcence on a_C. For drying purposes collectors with air as working medium have to be considered in the first place /direct systems, Fig.2./.

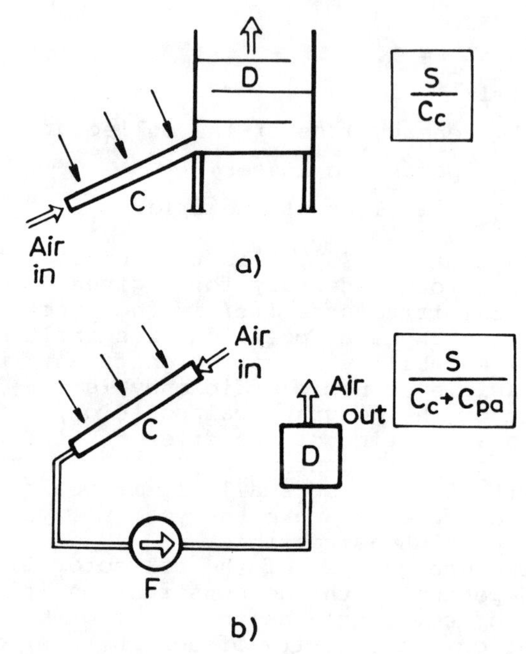

Fig. 2 Direct solar dryers and actual composition of $\frac{S}{C}$ ratio, with natural convection /a/, and forced convection /b/

The application of air-type collectors has significant advantages, yet disadvantages and limitations as well. Advantages lie in their simplicity, operation safety /no danger of leaking, no pressure problems in idle run/; disadvantages are: in large collectors the even distribution of air is hard to realize, space demand and cost of air ducts /C_{pa}/ may be very high. Because of that, the application of indirect-system /liquid--type collectors and air-water heat exchangers/ come into view especially with high performances. The reduction of C_p costs may be somewhat counter-balanced by the surplus cost of the liquid-air heat exchanger C_H and by the parasitic energy requirement of pump P /see L_2 in Equ./6//. It must also be kept in mind that the liquid-type collector has to be operated on a higher temperature level than the air-type one to ensure the same inlet air temperature.

The application of liquid-type collectors may also be supported by another circumstance: the solar energy gained by the collectors can be utilised even out of the drying time. It could be seen from relationship /5/ that the actual operation time t_o has a determining influence on E_o and therewith on S, too, since the primary source of saving is the utilized solar energy.

The possible maximum value of t_o can be taken as equalling the number of sunny hours per year. The striving to have as high an operation time as possible appears to be justified. In some cases drying can be carried on through a long period of the year /secondary coal recovery from slurry ponds [12], drying of lumber/. Drying in agriculture is adjusted to the ripening of the crop. In the case of multi-purpose dryers different materials can be dried to increase t_o, but a full yearly utilization can be hoped for only in the case of satisfying other heat consumers, e.g. technological hot water supply in farms, heating, air conditioning or bioenergetic processes. The utilization of the dryer's collector system for satisfying other heat demands puts forward the application of liquid working medium and indirect systems /Fig.3; ΔS/.

There are two ways of attaining the purpose: the application of liquid-type collector + liquid-air heat exchanger, or the application of a two-medium: water-air /hybrid/ collector /Fig.3/d/. Its absorber is made up of finned tubes. The liquid working medium flows in the pipes when there is no demand for drying. /The liquid side can also be connected to a heat storage/. The hybrid collector works as a water-air heat exchanger in the radiation breaks [19]. The cost of the hybrid collector is not significantly higher than that of the liquid-type one. However, the cost of the air-side pipe system /C_{pa}/ remains thus, the application of a hybrid system can be justified only for solar dryers of small and medium performance.

4.4. Cost Factors of Heat Storage

In the breaks of solar radiation the solar dryer can be operated in the following modes:
a/ out of operation,
b/ operates on heat storage,
c/ operates on auxiliary energy source.

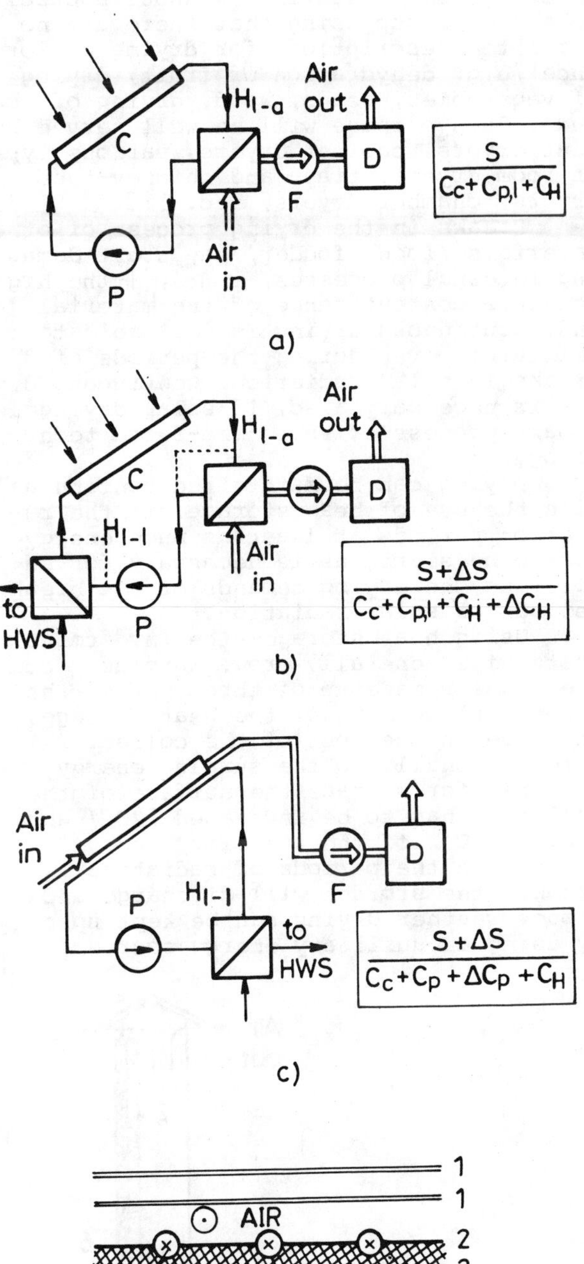

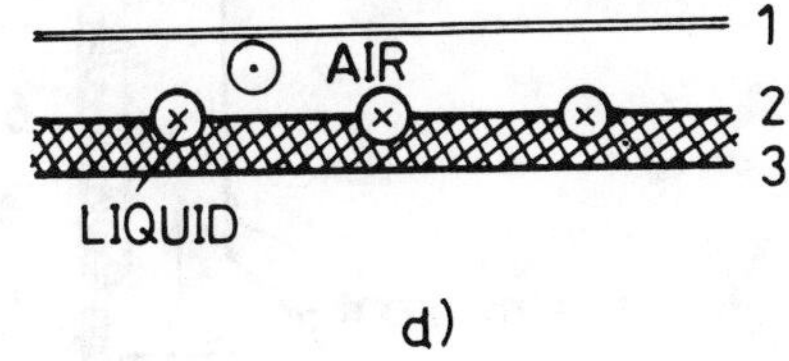

Fig. 3 Schemes of indirect solar dryers and actual composition of $\frac{S}{C}$ ratio,
a/ simple indirect system;
b/ connected to HWS;
c/ hybrid system;
d/ construction of a hybrid collector /covering 1, absorber 2, heat insulation 3/

Drying of certain materials can be interrupted temporarily, without appreciable damage supposing that there are no definite prescriptions for drying performance /e.g. dehydration of fruit, drying of vegetables, seeds, wood, drying of coal mud/. Such purpose will be well served by tent dryers, cabinet dryers, various types of room dryers, tray- and chimney-type dryers, chamber dryers, etc. [1].

Breaks in the drying process of other materials /food, fodder/ may start damaging internal processes, at least in the high moisture content range of the material and, continuous drying is desirable to be maintained even during the periods of breaks in solar radiation. Continuous drying is necessary also, if solar drying is a part process of some time-rated technology.

Drying can be maintained continuously with the use of heat storage. In the radiation periods at least as much energy has to be stored as is necessary for satisfying the drying demands of the break periods of solar radiation.

Using heat storage, the investment costs will generally grow. Surplus costs are usually made up of three parts: the investment cost C_S of the heat storage, increase in the cost of the collector ΔC_C /proportionally to the surplus energy destined for storage the surface of the collector has to be increased [20]/ and, the cost C_{aH} of the auxiliary energy source. In the periods of radiation breaks, the storage will discharge and in cloudy weather drying can be kept up only by using an auxiliary energy source.

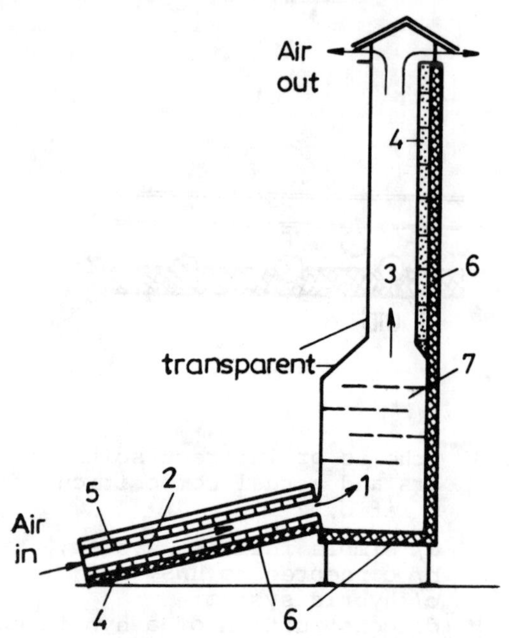

Fig. 4 Construction of a country dryer with natural convection and latent heat storage

The investment cost of heat storage depends on the type of storage employed. In case of air as working medium, according to Duffie and Close [8], the absorbent bed store gives lower costs than the rock bed store.

With air as working medium latent heat stores can also be used. Their advantage is a high specific heat storing capacity; further, a well defined phase change temperature, which is practically constant during the time of phase change [13, 14, 15].

Heat storers under direct radiation can be charged up to higher temperature than heated by the working medium. Fig.4. shows the schematic drawing of a chimney--type country dryer with directly irradiated latent heat storage integrated into the collector /1: drying space, 2: collector, 3: chimney, 4: fixed latent heat storage material as absorber, 5: latent heat storage plates can be pushed out on both sides, 6: heat insulation, 7: trays/.

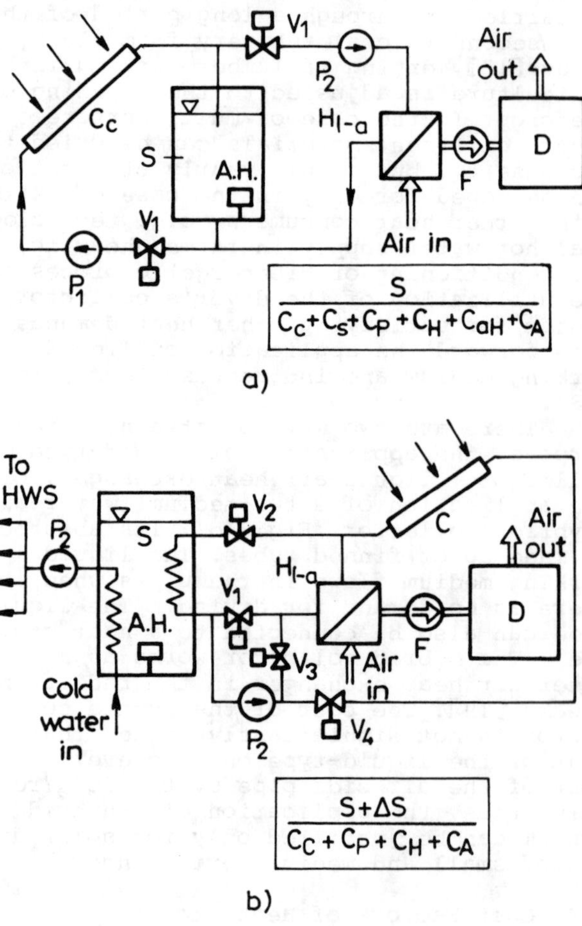

Fig. 5 Scheme of indirect solar dryers and actual composition of $\frac{S}{C}$ ratio, a/: with water storing tank /S/; b/: integrated to the hot water supply system of a farm

Latent heat storage plates can also be made of materials with different phase change temperature.

Low performance country dryers with water bags heat storage give cheap solution [16].

With liquid working medium the obvious solution is to employ a stratified or well-mixed heat storage tank /Fig.5/a/. The cost C_S of the heat storer and of its pipeline system generally has an increasing effect on the payback time of the solar dryer causing an increase /ΔC/ in the denominator of quotient $\frac{S}{C}$ and having no effect on the energy gained.

To ensure a full yearly utilization of the collector other energy consumers have to be served. For instance, if connected to a technological hot water supply /HWS/ system of a farm, there is a way of using the existing water tank of the HWS system as heat storage of the solar system /Fig.5/b; C_S = O/. In this case the original energy source of the HWS system can also be used as the auxiliary energy source of the solar system /C_{ah} can be saved/ and the costs that appear will be only those of the pumps and connecting pipes /multipurpose utilisation of equipment, integrated systems [5, 6]/.

4.5. Technical and Rentability Correlations of the Control System

The task of the control system is to ensure the selection of the operation mode of the solar system and to control the input characteristics of the drying air.

The operation of solar dryers without heat storage and auxiliary energy source usually do not need controlling.

With heat storage controlling elements are needed at least for changing of operation mode /operation of dryer from collector and/or heat storage/.

Integrated, complex solar systems have several possible operation modes. The selection of the working modes can be controlled from the collector outlet temperature of the working medium.

The temperature and/or the mass flow rate of the air entering the dryer can be controlled as the drying process proceeds.

Primary purpose of using control is the energy saving /i.e. to decrease L_3 in Equ./6/ / by, for instance, using recirculation [17], or by using intermittent drying operation [18]. In the latter case the period of break in drying is practically used according to the extent of internal moisture content equalization.

The costs of the control system /C_A/ may have a counter part in energy savings /ΔS/ partly through economic utilization of energy taken into the drying process, and further by the fact that the energy gained by the collector grows by increasing t_o when using an integrated complex solar system with control devices.

4.6. Other Connected Systems

The humid air leaving the solar dryer practically contains the full quantity of the solar energy getting into the dryer with the air, though a part of it in the form of evaporation heat. The energy of the departing air can partly be utilized, e.g. by the help of a heat pump [19].

Heat pumps or other supplementary energy converters are generally very costly, though cost C_o may have a considerable energy counter part /ΔS/.

Complex and connected energy systems are complicated; their operation and maintenance costs are higher /in relationship /1/ the value of m is greater than that of the simple systems/. Because of the simultaneous effect for S and C, the design of complex solar dryers with connected heat pump is also an optimum problem.

CONCLUSIONS

Criterium of the application of solar drying is rentability, which can be characterized by payback time.

Shortest payback time /0,5...3 years/ can be reached with simple-structure, cheap solar dryers, especially

in the case when the material to be dried needs mild, low temperature drying,

does not demand continuous operation and

the task of drying ensures the utilization of the dryer for a long period of the year.

Cheap construction in itself is not sufficient as a precondition of the rentability because the price may be interdependent with the life time of the drier and with the costs of re-establishment and maintenance.

In the case of materials requiring continuous drying, and of high performances the chances of rentability are reduced by the seasonal character of drying, the necessity of heat storage and of an auxiliary energy source. For such tasks rentability can be ensured for integrated complex solar systems, which can be exploited all the year round, and certain components of the system /heat storage, auxiliary energy source/ can be utilized for several purposes and thus their costs do not burden solely the solar dryer. For such solar drying systems a payback time less than 10 years can be attained.

The development of solar technology has lead so far to the elaboration of various means and methods offering many possibilities. In the present period the requirements of rentability come into the foreground.

When calculating the rentability of solar systems no emphasis has been laid so far on the cleanness of solar energy:

it involves no kind of environment pollution in contrast with drying technologies based on fuels. The reason for that may be in the fact that there are no methods accepted for the calculation of the benefit /S_{21}/ of this circumstance.

However, there is an increasing number of signs indicating pollution of the atmosphere which cause financially expressible damages /e.g. at some geographical spots acid rains cause contamination of the soil/.

Without appropriate counter measures atmosphere pollution may go on deteriorating and an artificial regenerating process may become essential.

Taking into consideration these facts it is really justified that solar drying has to be introduced in all fields where rentability and technological applicability make it possible. It is a common interest of mankind that coordinated actions supported by the governments of all nations be taken for the further development of solar technology.

NOMENCLATURE

a	unit price of substituted energy,	$/J
A	collector surface,	m^2
C	investment cost,	$
E	yearly utilized energy	J/a
E_o	incident radiation energy,	J/a
HWS	hot water supply,	
I	intensity of incident radiation,	W/m^2
L	yearly energy loss,	J/a
m	yearly maintenance cost factor,	
n	payback time,	a
r	rate of interest,	
S	yearly savings,	$/a
Y	yearly maintenance cost,	Y=mC
t	time,	s
t_o	yearly operation time of the collector,	s
η	efficiency,	

Indices

a	air
ah	auxiliary heater
A	controlling
C	collector
D	dryer
H	heat exchanger
lt	long term
l	liquid
p	pipe
o	other
S	store

Editorial Note - The word "rentability" is used synonimously with profitability in this paper.

REFERENCES

1. Imre, L., Solar drying, in Handbook of Industrial Drying, ed. A.S.Mujumdar, Marcel Decker, New York /in prep./
2. Duffie, J.A. and Beckman, W.A., Solar Engineering of Thermal Processes, Wiley, New York /1980/.
3. Lunde, P.J., Solar Thermal Engineering, Wiley, New York, /1980/.
4. Böer, K.W., Solar Energy, 20. 225 /1978/.
5. Imre, L., Farkas, I., Kiss, I.L., Molnár, K., Numerical Methods in Thermal Problems, vol. III. ed. R.W. Lewis, J.A. Johnson, W.R.Smith, Pineridge Press, Swansea /1983/.
6. Imre, L., Farkas, I., Fábri, L., Hecker, G., Proc. Int. Meeting on Energy Savings in Drying Processes, 88.1. Liege /1983/.
7. Eissen, W., Mühlbacher, W., Proc.Int. Conference on Solar Drying and Rural Development, Bordeaux, 263 /1983/.
8. Duffie, N.A., Close, D.J., Solar Energy 20. 405 /1978/.
9. Quinette, J.Y., Pruvot, J.M.,Daguenet, M., Proc. Int. Conference on Solar Drying and Rural Development, Bordeaux, 145 /1983/.
10. Vaishya, J.S., Subrahmaniyam, S., Bhide, V.G., Solar Energy, 26. 367 /1981/.
11. Peck, M.K., Proctor, D., Solar Energy, 28. 183 /1983/.
12. Helmer, W.A., Powder Technology spec. issue /in preparation/
13. Telkes, M., Mozzer, R.P., Proc. Annual Meeting Am. Section ISES, Denver, /1978/.
14. Marks, S.B., Solar Energy, 30. 45 /1983/.
15. Abhat, A., Solar Energy, 30. 313 /1983/.
16. Puiggali, J.R., Lara, M.A., Proc.3rd Int. Drying Symposium, Birmingham, vol. 1. 390 /1982/.
17. Auer, W.W., DRYING'80, ed. A.S. Mujumdar, Hemisphere P.C. New York, 292 /1980/.
18. Imre, L., Molnár, K., Farkas, I., Proc. Int. Conference of Solar Drying and Rural Development, Bordeaux, 93 /1983/.
19. Imre, L., Kiss, I.L., Molnár, K., Proc. 3rd Int. Drying Symposium, Birmingham, vol. 1. 370 /1982/.
20. Gordon, J.M., Rabl, A., Solar Energy, 28. 519 /1982/.

STEAM DRYING

Colin Beeby and Owen E. Potter

Department of Chemical Engineering, Monash University
Clayton, 3168, Vic., AUSTRALIA.

ABSTRACT

The possibility of drying in superheated steam or superheated solvent vapour has been realized for a long time but practiced very little. The advantages and disadvantages of such drying are discussed. Knowledge of equilibrium moisture content and drying rates is necessary when designing equipment for drying in superheated vapour, and some aspects of these are presented. A short discussion on large scale application of steam drying is included.

INTRODUCTION

The concept of drying by use of superheated steam has been known at least since early this century. Hausbrand [1] in 1908, discussed the merits of such a system and presented Fig. 1 illustrating the reheating of superheated steam and its return to the drying chamber. The lack of suitable equipment hindered early attempts to realise the potential advantages of superheated steam drying. Batch dryers, such as were first utilised were not energy-efficient and in any case were not very suitable for a condensible vapour because of the losses during start-up and shutdown. Operation under pressure compounded such problems and increased equipment costs with little in the way of compensating savings. Low-cost energy and lax environmental controls meant that there was little pressure to move in the direction of more energy-efficient drying and reduced emissions to atmosphere.

One process named after Fleissner [2] aimed to take advantage of certain properties of the material to be dried. A number of materials, including some brown coals, on being heated in water or steam, lose liquid water from the particles by a physico-chemical process. If the water is drained off, substantial dewatering is possible without the need for evaporation. A plant operated commercially, using this process in Austria till the mid-70's. A small section is included here on Fleissner drying but as will be seen later, the same physico-chemical processes also occur in the absence of steam. Nevertheless the use of steam as the drying medium in the commercial operation justifies its inclusion here.

Dryers developed along two main lines, *indirect* where heat is transferred to the drying material through a hot surface, and *direct* where heat is supplied by direct contact of hot gas with moist material. In the case of indirect dryers, even if there is a flow of non-condensible gas such as air through the dryer, the evolution of steam or solvent vapour in the vicinity of the hot surface means that the heat transfer is occurring as though only steam or solvent vapour is present. By the very nature of direct drying, the gas contacting the moist material can be any of superheated steam or solvent, air, flue gas or other gases.

Indirectly heated dryers operating with the atmosphere surrounding the moist material being the evaporated moisture, have been widely used since the turn of the century. Early chemical engineering texts Badger and McCabe, 1936 [3], Walker et al. 1937 [4], Perry 1941 [5] abound with illustrations and descriptions of pan or agitator dryers, drum dryers, tray dryers and rotary dryers, all of which were in common use.

Some would be and were, operated under vacuum. When drying heat sensitive materials, the lowering of operating pressure reduces the temperature at which the vapour "boils" off. These dryers were also operated at atmospheric pressure with a small stream of inert gas to carry away the moisture evolved or with a small exhaust blower to remove vapours. The same type of equipment is still in use today, nearly 50 years later. Perry, 1973 [6] suggests the use of various pan and rotary dryers when it is desirable to recover the evaporated solvent, and mentions operation under vacuum for heat-sensitive or readily oxidized materials. More recently the use of fluidized beds with immersed heating surfaces has been suggested for drying high moisture coal in an atmosphere of superheated steam, Potter and Keogh [7]. The much higher heat transfer coefficients typical of fluidized beds (200-500 W/m^2K), compared with agitated pans (20-50 W/m^2K) and similar devices allows much more efficient drying.

Although drying in a superheated steam or solvent atmosphere was quite common for indirect dryers, there had been very little direct drying in superheated vapours. Karrer, 1920 [8] reports the "novel" commercial batch drying of foundry sand with superheated steam in a specially converted drying oven. In 1937, Walker et al. [4] comment "This method of drying (using superheated vapour) seems capable of wide application ..", indicating that in spite of the known advantages, direct drying in superheated vapours was not widely used. In principle any type of conventional direct dryer can be converted to eliminate air and use superheated vapour as the heat source. Examples in commercial use include batch ovens, continuous tray dryers, fluidized beds and pneumatic conveyor dryers and aspects of these applications are discussed later.

Additionally we should include drying by means of radiation and dielectric heating since these could easily take place in an atmosphere of the evaporated liquid, but we shall restrict this paper to the more conventional thermal drying methods.

The important design parameters vary with the system. Equilibrium between superheated steam and solid particles is of importance. For direct heating superheated vapour to particle heat transfer coefficients are required. For indirect heating the heat-transfer coefficient between the surface and bed of particles or film of material is important in design. In both cases it needs to be considered whether any special factors arise when superheated steam is employed.

Gas-cleaning is another aspect needing study. Provided condensation is avoided, cyclones should prove satisfactory for denser materials and no new considerations should enter into their design. Again provided condensation is avoided, and the right materials chosen, bag-filters could be employed, even in pressurized systems, although as yet there has been little use of bag-filters in such applications. The volume of gas to be handled is greatly reduced so that gas-cleaning in principle becomes much less costly.

The advantages of superheated vapour drying become more manifest as the scale of operation is increased and it becomes possible to put to good use the vapour product from the drying operation. Energy economy on the large scale can only be achieved by such drying unless there is some unusual property to be exploited.

Conventional drying is a very energy intensive process, Keey[9], and in this era of high energy costs, the savings obtained by employing superheated vapour drying show it to be a desirable alternative, Svensson [10]. The tightening of environmental controls and the advantages of eliminating air may offer further incentives, Lane and Stern [11].

FLEISSNER DRYING

The Fleissner [2] process was developed at Koeflach, Austria in order to provide a lumpy dry fuel from a coal which would not briquette. In the Fleissner process, the raw coal lumps are loaded into a drying vessel which is closed and filled with saturated steam to a pressure of 12.5-25 bar. After thorough heating the pressure is reduced, sometimes to below atmospheric, the moisture in the coal flashes to steam, a considerable quantity of water having previously left the coal lump during the heating stage as a liquid. Early discussions of Fleissner drying may be found in Rosin [12] and Klein [13], [14]; Cooley and Lavine [15] and Harrington et al.[16] have studied the lignites of Dakota, USA. Brown coal in Victoria, Australia has been extensively studied for Fleissner drying applications but no industrial application has followed. Fig. 2, from Evans and Siemon [17] shows dewatering due to heating in water under pressure after separating coal from water before cooling. Data for heating in nitrogen are also presented, and are very similar. These authors envisaged that a 350 MW station using brown coal with 66.7% w/w moisture would require 30 dryers each 30 cm dia. and 20 m long operating at 5 MN/m^2. Aspects of the bonding of water to brown coal have been studied by Allardice and Evans [18] and by Stewart and Evans [19]. Figure 3 from Allardice and Evans [18] shows the weight of sorbed water as a function of relative vapour pressure for data from 30°C - 60°C. There is an hysteresis phenomenon on re-adsorption.

A comprehensive test was carried out between July 1963 and April 1964 on a range of low-rank coals by Bull [20]. Coals studied included Mae Moh from Thailand and Victorian brown coals from Anglesea, Bacchus Marsh and Morwell. Two different steam-pressures 400 and 570 p.s.i.g. were employed. While the system was still pressurized it was noted that 54-75 per cent of water in the coal had been expelled as liquid. (However, this does not mean that a clear separation of expelled water from the coal is easily possible in an engineering sense.) A beneficial aspect of the expulsion of water as liquid is that leaching accompanies the dewatering operation. The only mineral leached out in large amounts was sodium chloride. For example, relative to dry weight of coal, chloride ion was reduced from 0.47% w/w to one-third to half that quantity. The removal of large amounts of sodium chloride from the Morwell coal greatly reduced its boiler-fouling propensities when burnt in a test-rig.

ADVANTAGES AND DISADVANTAGES

Advantages and disadvantages are now briefly considered.

Advantages

a.(i) *Elimination of air or gas through the dryer produces energy savings* as argued by Lane and Stern, 1956 [11] and Thompson, 1955, [21] who describe the application of superheated steam to drying a fine particulate material in a continuous tray dryer. If the energy for drying is drawn from the available sensible heat of a hot gas entering and, after being cooled in the drying process, leaving the system, the ratio of gas to water will be significant and the energy economy poor. When drying in recirculating superheated steam the only material leaving the system is water-vapour equal in quantity to that dried from the particle. Accordingly a higher energy efficiency is possible.

Karrer, 1920 [8] describes some early experiments in which large batches of sand (7-9 tonne) were dried batchwise using superheated steam, and reported a large increase of thermal efficiency over conventional hot air drying. The result of these trials was the commercial application of superheated steam drying by a local foundry. Thermal efficiency was limited by the batchwise operation but nevertheless was found to be double that obtained with hot air. Further trials of large-scale batch drying of peat emphasized the lower energy consumption of steam drying.

a.(ii) *Multiple-effect operation* has been considered by Potter and co-authors (22),(23),(24), with operation in counter-current, co-current or cross-flow, the last being favoured where there is doubt about the fluidization behaviour of partially-dried solids. In considering a cross-flow situation, as in Fig. 4, fresh raw coal is added to, and dry coal removed from each effect, with the effects are at different pressures. The highest pressure effect is heated by the steam supply and vapour produced within the effect is cleaned and condenses in the heating tubes of the next effect, and so on. In principle, depending on the number of effects, it is possible to obtain one, two, three or more kilograms water evaporated per kg steam supplied. Hence energy economy of considerable magnitude is possible in this system. Mechanical vapour recompression is also a possibility, although the available compressors are generally designed for smaller pressure rises than might be desired for drying purposes. For instance, when mechanical vapour recompression is applied to evaporation of a liquid, the temperature-rise in converting saturated steam at atmospheric pressure to saturated steam at the higher pressure may be as small as 8 Cdeg. In a fluidized-bed steam-dryer the heat-transfer coefficient may be only 10 per cent of that for evaporation of clean liquids, so that a larger temperature (or pressure) rise will be sought for drying than for evaporation of clean liquids e.g. 50-100 Cdeg. Hence a pressure rise to 5 bar, at the least, might be indicated. Commercial operation is required to refine the requirements and possibilities.

a.(iii) Under certain circumstances *drying rates are increased*. This aspect is considered in the following sections where drying rates are discussed.

a.(iv) *Recovery of solvent is simplified*. This is fairly self-evident. Walker et al.[4], give an example of the use of superheated naphtha vapour in removing naphtha from wool or leather which had been degreased with naphtha. Basel and Gray [25] report the use of superheated vapour of hexane and isopropyl alcohol to separate solvent and alcohol adhering to polypropylene. These authors pointed out the ease of solvent recovery compared with conventional drying systems. A full-scale dryer operated from 1960 with superheating of solvent vapour which was then blown through a fluidized bed of polypropylene. In 1963, Yoshida and Hyodo [26] also emphasized the ease of solvent recovery with superheated solvent drying, removal of acetone from cellulose acetate fibre being considered.

a.(v) *Smaller gas volume aids solids recovery and gas clean-up*. Basel and Gray [25] pointed to higher volumetric heat capacities of organic vapours relative to air, which helped reduce equipment size and cost so far as superheated vapour drying was concerned. Lane and Stern, 1956 [11] and Thompson in 1955 [21], in discussing the application of superheated steam to a continuous tray dryer claimed as an advantage a much-reduced dust problem due to the reduction in gas-volume to be handled and argued that, if necessary, the exhaust vapours can be condensed and the small amount of dust filtered from the condensate (provided the dust is insoluble). If some method of gas-cleaning is employed the large reduction in gas-volume is of clear advantage e.g. for a bag-filter there will be a proportional reduction in filtration surface area.

a.(viii) Use of superheated steam with coal or other combustible solids *reduces risk of explosion and fire-hazards*. Flue-gas dryers may catch fire but clearly this cannot eventuate in a steam environment.

a.(vii) In certain cases, *product quality is enhanced*. Karrer in 1920 [8] reported that drying of cabbage and hay in superheated steam enabled better colour retention. In 1948, Blaw-Knox [27] argued product quality as a main reason for introducing their "Vapor Desolventizers". Superheated solvent vapour was used as the heat source for removing residual solvent from extracted soybean flakes and the benefit arose from the elimination of localized overheating of flakes on hot metallic heat transfer surfaces. A similar process was described in 1953 by Chu et al. [28]. Yoshida and Hyodo in 1963 [26], concluded that cellulose acetate fibre was much stronger as a result of extrusion and drying in the vapour atmosphere. The batch drying of timber in superheated steam was described in 1961, by Kollmann [29] referring to German practice in the 1950's. He reported that timber quality was more consistent. Product uniformity and quality assurance were benefits of a commercial superheated steam explosion-puffing and drying process described by Akao and Aonuma [30]. Drying of puffed soybean meal in a superheated steam pneumatic conveyor dryer gave a sterile product of low moisture with good long-term storage qualities. In the superheated steam pneumatic conveyor drying of paper-pulp as described by Svensson [10], consistent product quality was obtained.

a.(viii) *Control is simpler* because drying rates and product final moisture can be easily controlled by monitoring the vapour temperature, Kollmann [29], Svensson [9]. More complex humidity control is necessary on conventional air dryers.

Disadvantages

b.(i) Generally higher temperatures create *problems with temperature-sensitive materials* while operation under vacuum, to reduce temperature, adds to complexity. The temperatures are higher since the boiling point of the vapour is equalled or exceeded with superheated steam or vapour, whereas in conventional drying the temperature approaches the (lower) wet-bulb temperature. The problem is important for some polymers e.g. Basel and Gray [25], who conclude that maximum bed and vapour inlet temperatures must be specified to avoid product degradation and to prevent fusion of the particles. Operation at lower pressure, they point out, adds considerably to equipment costs. Kollmann [29], notes that timber dried at too high a temperature is subject to an undesirable discoloration.

b.(ii) *Harder to achieve a low-moisture level.* This involves the equilibrium between the drying solid and superheated steam and is discussed in a separate section.

b.(iii) *Purging at start-up and shut-down with an inert gas if solids are combustible or if a combustible solvent vapour is employed.* The extent to which purging at start-up and/or shut-down is necessary depends very much on the dryer-type and the material. In a fluidized bed dryer, if indirectly heated, working with coal and steam, it would be most convenient to start-up and shut-down while fluidizing with an inert gas e.g. washed flue gas. By this means it is easily possible to raise the temperature on starting to above the boiling point of water before steam is introduced, and on cooling to introduce an inert gas to sweep out the steam and reduce the temperature. The requirement for some inert gas to be available does increase the cost.

b.(iv) *Some difficulties in feeding and discharge.* Feeding moist solids can be a source of problems in any system where it is required, but perhaps there is in the case of steam-drying an additional factor arising from the condensation which occurs when cold solids and steam are contacted. Each case must be studied in its proper context. In large-scale steam-drying it should be possible to preheat the coal in steam-heated rotary dryers using waste steam. This will be advantageous in lock-hopper systems feeding pressurised units since the higher partial pressure of water will serve to exclude air. Also condensation will be minimised.

b.(v) *Need to avoid condensation on walls and in lines.* Liquid water in particulate systems can be troublesome and in steam-dryers is best avoided by thermal insulation, or, in extreme cases as in very cold countries, steam-tracing might be required. The design requirement would be that the wall temperature should exceed the temperature of condensation by, say, 5 C deg. This can be achieved with a thin layer of insulation. The heat losses from a very large dryer, say 8 m x 7 m x 3.5 m, would be substantially less than 1 percent of the heat load.

OPERATION UNDER PRESSURE

Operation under pressure has some desirable aspects. As an example the volume of gas to be handled is greatly reduced. On the other hand there are disadvantages - maintenance is more demanding, vessels may require approval as pressure-vessels, feeding is a more complex operation, possibly requiring lock-hopper systems.

Heat-transfer. In convective flow situations as between a pipe-wall and steam flowing inside the tube, **pressure of operation** has an effect as shown in Table 1 which relates to a constant steam velocity at different pressures.

Potter et al. [23] presented Fig. 5 showing the maximum heat-transfer coefficient predicted for steam fluidized beds of dry coal by three predictions, Grewal and Saxena [31], Gelperin et al. [32] and Heyde and Klocke [33]. The predictions indicate a beneficial effect of pressure, due largely to the higher temperatures required as pressure rises. The effect of pressure is not as large as in convective flow situations. For an explanation see the section on heat-transfer to packed beds.

TABLE 1

Ratio of heat-transfer coefficient between wall and gas for barely superheated steam in pipeflow at a given temperature to that for 100°C, 1 bar steam, when actual gas-velocity is maintained constant.

PRESSURE (bar)	TEMPERATURE (°C)	$\frac{h}{h_{100}}$
1	100	1
4.7	150	3.45
15.5	200	9.93

Reduced gas volumes. The volumes to be handled are approximately inversely proportional to the pressure, so significant reduction in gas-volume occurs and this must have a beneficial effect on overall equipment size. In particular it may be anticipated that where cyclones are used for gas-cleaning the number required will be inversely proportional to the pressure. So far as bag-filters are concerned, if the approach velocity is kept constant despite operation at higher pressures, the bag-filter area is inversely proportional to the pressure. If it is assumed that the approach velocity to the filter-cloth is proportional to the particle's terminal velocity then the advantage is not so well marked, but still very considerable. Table 2 is reproduced from Potter et al. [23] for a steam-fluidized bed dryer. In this Table a 4.3 fold increase in pressure suggests a bag-filter area reduction of about 2.2. If the gas-cleaning were to be accomplished in some wash-tower device then the cross-sectional area would be reduced approximately inversely to pressure.

TABLE 2

Load Factor and Bag-filter Area at Different Pressures

PRESSURE MN/m^2	$U_{above\ bed}$ m/s	Load Factor	Bag-Filter Area Factor
1.3	0.24	3.84	0.26
0.43	0.34	2.25	0.45
0.101	0.50	1.0	1.0
0.007	0.83	0.15	6.7

Load factor = ratio of moisture removal per unit plan area at a particular pressure, if enough heat transfer area is provided.

Bag-filter area factor = ratio of bag-filter area at a particular pressure to that at atmospheric pressure, assuming the design velocity is proportional to the terminal velocity.

Fluidization quality improves at higher

pressure. It is generally considered that when fluidized beds are employed the quality of fluidization is improved by operation under pressure. Rowe et al. [34] observed by X-rays two alumina powders (262, 450 μm) and observed a transition to fine-particle behaviour at 2 bar and more so at higher pressures. They measured the minimum velocity at which bubbling occurs, U_{mb} and U_{mf} and noted that U_{mb}/U_{mf} increased from unity at 1 bar to 1.1 at 10 bar. Such behaviour would generally be considered to indicate an improvement in the quality of fluidization. An improvement in behaviour might also be indicated for agitated pan dryers and rotary steam-tube dryers operated under pressure.

Large vessels a disadvantage? When handling very large throughputs the vessels employed will eventually rise in size to the limit of manufacture, transportation and maintenance. A cylindrical vessel 6 m diameter operated at 5 bar, such as might be required for large scale steam drying of brown coal, would require a wall thickness of about 2.5 cm which is unremarkable. Tube lenths up to 17 m are readily obtainable. Hence the pressures being employed or being considered do not impose undue restraints on design and manufacture.

Feeding pressurized systems. Some form of lock-hopper arrangement will presumably be required. Preheating the feed will reduce the amount of air or other gas carried into the system. Akao and Aonuma [30] and Anon. [35] describe special equipment for introducing and removing material from pressurized dryers.

EQUILIBRIUM

The moisture in wet material is present in many different forms ranging from free moisture, where the vapour pressure exerted is the same as the vapour pressure of the liquid solvent, through capillary and colloidal moisture, to physically adsorbed and chemically attached moisture. The stronger the bonding between moisture and material is, the lower is the partial pressure exerted by the moisture. Wet material on exposure to an atmosphere of fixed partial pressure of vapour, will lose moisture until remaining is that which exerts the same partial pressure as in the surrounding atmosphere.

In superheated vapour drying, since there is only one gaseous component the partial pressure of vapour in the dryer is the same as the total pressure. To remove solvent which is in any way bonded and thereby exerting a vapour pressure lower than the normal vapour pressure of the free solvent, the temperature must be raised above the solvent boiling point at the pressure of operation. Removal of water in superheated steam at a pressure of one atmosphere for instance, would require a temperature of at least 100°C.

For non-porous solids or where the moisture is held in large pores or voids, so long as the temperature is maintained above the normal boiling point of the solvent, material will dry completely. Chu et al. [36] noted that drying sand in superheated steam gave an equilibrium moisture content of zero just as for air drying. Lane and Stern [11] point out that the void spaces of the dried product are filled with superheated vapour which should be considered as residual moisture. Obviously this "equilibrium" moisture cannot be removed by superheated vapour drying. Simple calculation shows for example for sand dried in steam the residual moisture is only of the order 0.02% by weight and in most cases could be safely neglected.

Potter and Keogh [7] report the pilot plant drying of a mineral material (non-porous aluminium hydroxide crystals) with superheated steam, resulting in a final moisture content of less than 0.1%. This material contained very few pores less than 200 Å and moisture associated with the feed material was essentially "free".

In capillary porous solids there is a lowering of the vapour pressure of the liquid due to the energy associated with the curvature of the moisture interfaces within the substance. The bond energy is given by the Kelvin equation:

$$-\Delta G = -RT\,\ln(P/P^{o}) = (2\sigma V_L \cos\theta)/r \quad ..(1)$$

The removal of moisture from these fine pores by superheated vapour drying requires the temperature to be raised above the normal boiling point. For steam at a pressure of 1 atmosphere, the relationship between the size of the largest pores which remain filled with water and the degree of superheat is shown in Fig. 6. Obviously unless a material contains a significant proportion of very fine pores (< 100 Å), the material will be effectively dry at quite small values of superheat. Toei et al. [37] report on the drying of small spheres of porous biscuit-ware (pore radius about 100 Å) in superheated steam. Drying at temperatures as low as 114°C still removed 95% of the moisture from the pores. Fig. 6 suggests that 102°C would have been sufficient to remove water from 100 Å pores and that at 114°C all pores larger than about 14 Å would be emptied.

Fig. 7 shows the equilibrium moisture content of fresh F.C.C. in superheated steam at one atmosphere pressure. This material contains a considerable volume of fine pores less than 50 Å (during mercury porosimetry such pores are only just beginning to be filled under an applied pressure of 150 MN/m^2). Considerable degree of superheat is required to remove the moisture from such pores.

Equilibrium moisture contents in superheated steam are presented for several types of complex materials - various timbers, Kollmann [29] paper-pulp, Svennson [10] and brown coal, Ho [38]. All of these exhibit a similar relationship between moisture content and temperature with a rapid reduction in moisture for small increase of temperature above saturation. As the moisture level decreases much higher superheat is required to remove further moisture. The data for the various materials are replotted in Fig. 8 on logarithmic co-ordinates. Even though the materials have very different structure their curves are remarkably similar with moisture content varying with superheat raised to a negative power between half and two-thirds.

This applies over the range of superheats of practical interest; about 3 to 50°C. Obviously this would not apply at very low values of superheat where the moisture content approaches the finite value for the saturation temperature. For instance brown coals exhibit a saturation moisture content of about 67% (w/w dry basis) at 100°C, Stewart and Evans [19] and timbers a value of about 25 to 30%, Kollmann [29].

The preceding has referred to examples where the volatile material is water, but other solvents would be expected to behave in similar fashion. Basel and Gray [25] report very low levels of volatile solvent in polymers and organic acids dried in the superheated solvent.

DIRECT DRYING

Any conventional drying process involves the transfer of heat and mass between the drying medium and the wet solid and is often characterized by two or more distinct periods. The first is an initial heat up period during which time the wet solid material absorbs heat from its surroundings and is heated up from its initial temperature to a temperature where moisture begins to evaporate from the solid. Following this is the period generally referred to as the constant rate period. Under constant drying conditions, so long as the solid surface remains wetted with water, the removal of moisture continues at a constant rate. When the moisture level is so low that the surface is no longer wetted the drying rate decreases further as more moisture is removed. This is known as the falling rate period.

Wenzel and White [39] note that drying in a medium of superheated steam rather than air does not alter the general characteristics of the drying process. Figures 9 and 10 taken from Luikov [40] show the drying of single pellets of cellulose in air and steam and show the heat-up, constant rate and falling rate periods to be similar in nature for the two drying mediums.

Heat-up period. In both conventional and superheated vapour drying, the heat-up or induction period involves heat transfer from the hot drying gas to the cold moist solid. When drying with air or other non-condensable gas, such heating results in a sensible heat change of the gas as the temperature of the moist solid is raised from its initial value to the temperature at which drying begins (normally close to the wet bulb temperature). When the drying gas is a condensable vapour, drying normally occurs at the boiling temperature. The heat-up period therefore involves transfer of a larger quantity of heat from the vapour, and, particularly if the degree of superheat is not high, some vapour will condense on the cold surface of the moist material during the heat-up period. The quantity of condensed vapour depends on various factors such as the thermal diffusivity of the material being heated, its moisture content and the degree of superheat of the drying vapour. Trommelen and Crosby [41] calculate that drops of water initially at 30°C experience an increase in weight of about 12.5% when exposed to atmospheric pressure superheated steam of 150°C. Obviously the increase in moisture level means that more drying needs to be accomplished, in that the condensed water has to be re-vaporised.

The drying time for material which is predominantly liquid, such as solution droplets, will be little affected by this increase in moisture during heat-up. However for materials initially containing low moisture levels the increase may represent a major portion of the moisture to be removed and may extend the time required for drying. For instance sand initially at 30°C and containing 10% moisture will increase in moisture level to 14% during heat-up in superheated steam at 150°C, representing a 40% increase in the amount of water to be removed during drying.

The increase in moisture during heating of materials in superheated vapours has been reported by various researchers e.g. Luikov [40] reports drying cellulose pellets in steam, (see Fig. 10), Yoshida and Hyodo [42] when drying food in steam, Lee and Ryley [43] and Trommelen and Crosby [41] on drying drops of water and drops of various solutions and suspensions, again in steam. For drying in superheated organic vapours, the increase during induction will be larger due to the typically lower latent heat of vaporization compared with water/steam. Moyers [44] reports increase in solvent weight of 20% during heat-up of drops of ethyl acetate solution in the superheated vapour.

The amount of moisture condensing is reduced if the fresh material absorbs heat by means other than conduction through the superheated vapour. In continuous directly heated fluidized bed dryers heat transferred by contact between cold wet particles and the bed of hot dried particles will reduce the amount of condensation on the feed material. It is important to be aware of the changes which occur during the heat-up stage and to make allowance during design. The changes caused by surface wetting need to be considered. Special start-up procedures may be necessary to avoid agglomeration in particulate systems such as spouted and fluidized beds where for correct operation it is necessary that particles remain separate, Basel and Gray [25].

Constant rate drying period. As mentioned previously, under constant drying conditions, the constant rate period extends from the end of the heat-up period to a time when the internal movement of moisture is no longer fast enough to maintain a wet surface. For drying in air or other non-condensable gas the relevant rate processes are the conduction of heat from the hot bulk gas to the cooler surface through a gas film, and the diffusion of moisture from the surface through this film to the bulk. The driving forces for these transfer processes are respectively, the temperature difference between bulk and surface and the vapour partial pressure difference between surface and bulk. The surface temperature approaches the wet-bulb temperature. In drying in superheated vapour, heat is conducted similarly through a vapour film to the drying surface under the influence of a temperature difference driving force. However since the surrounding gas is composed soley of the vapour, removal of moisture from the surface occurs not by means of diffusional mass transfer but by bulk flow with pressure difference as the driving force. Chu et al. [28] indicate that an increase of temperature at the surface of only a few hundredths of a degree provides a sufficient pressure difference to account for the flow of evaporated vapour. For instance, a 1 mm diameter drop of water drying in superheated steam at 150°C requires a pressure only 10^{-6} N/m^2 above atmospheric pressure to provide sufficient driving force for vapour removal. For practical purposes then, the resistance to moisture removal from the surface can be neglected and the temperature of the surface during the constant rate period is maintained at the boiling point of the solvent corresponding to the pressure at which the dryer is operating, Wenzel and White [39], Chu et al. [36], Lane and Stern [11], Toei et al. [45]. Trommelen and Crosby [41] present moisture and temperature histories of drops of water and various solutions and suspensions dried in superheated steam. For all materials there is a period during which the temperature of the drop remained at, or just slightly above, the boiling point. Since the size diminishes as drying proceeds there is no constant rate period, but the period during which the temperature is constant corresponds to that where the surface is maintained wet. Luikov [40] when drying cellulose pellets in superheated steam, shows the surface to remain at the boiling point during the constant rate period. (see Fig. 10).

Toei et al. [37] showed that the constant rate period of superheated steam drying continued longer than for air drying even though the constant rates were the same for both media (Fig. 11) which suggests the moisture is more mobile under conditions of superheated steam drying. Luikov [40] suggests that the larger moisture diffusion coefficient during steam drying is due to the higher temperature at which the material dries.

Since the constant drying rate is unrestricted by removal of vapour from the surface, the prevailing temperature difference between bulk gas and the particle surface and the appropriate heat transfer coefficient determine the drying rate. Chu et al. [36] investigated the heat transfer coefficients for drying of pans of sand wet with water in streams of air and superheated steam and a mixture of both, and concluded that the data for all gases could be correlated by a single equation of the form:

$$\left(\frac{h}{C_{pg}\dot{m}}\right)\left(\frac{C_{pg}\mu}{k_g}\right)^{2/3} = a\left(\frac{L\dot{m}}{\mu}\right)^n \qquad ..(2)$$

where a and n were dependent on the regime of operation whether laminar, transitional or turbulent.

Toei et al. [45] determined that the heat transfer to drops of water again in air, superheated steam and mixtures of both could be correlated by the single equation :

$$\left(\frac{hd_p}{k_g}\right) = 2.0 + 0.65\left(\frac{d_p\dot{m}}{\mu}\right)^{1/2}\left(\frac{C_{pg}\mu}{k_g}\right)^{1/3} \qquad ..(3)$$

showing no significant difference with gas composition.

The heat transfer coefficients determined during the constant rate drying of single porous ceramic spheres in similar gas mixtures, Toei et al [46], could also be correlated by the above equation. The correlation of heat transfer coefficients obtained for the through-flow drying of a fixed bed of similar porous spheres in superheated steam, was indistinguishable from that for air drying.

Lee and Ryley [43] studied the drying of drops of water in air and superheated steam and found a difference in correlation of results for the different gases. Although of the same form as the above equation, the constant was found to be 0.6 for air and 0.74 for superheated steam rather than the value of 0.65 for the gas mixtures tested by Toei et al. [45].

Lee and Ryley [43] point out the modification to the heat transfer rate by the mass flow of vapour from the surface. The heat transfer coefficient for single spheres drying in stationary vapour is reduced from the value for non-drying spheres as shown below:

$$\frac{\text{Nu drying}}{\text{Nu non-drying}} = \frac{Z}{e^z - 1} \quad \text{where } z = \frac{W\,C_{pg}}{h} \qquad ..(4)$$

This correction, although strictly for stagnant vapour, was found to differ little from unity for the conditions normally encountered in vapour drying, Lee and Ryley [43], and can be safely neglected. Montlucon [47] presents a more complete theoretical analysis of the heat and mass transfer associated with a drop evaporating into its vapour.

Confusion exists as to the relative drying rates of air and superheated steam, Yoshida and Hyodo [48]. Sjenitzer [49] suggests comparison on the basis of equal temperature difference driving force. Trommelen and Crosby [41] compared behaviour at the same gas temperature and found almost invariably that air drying exhibited more rapid drying although as the temperature of the drying medium increased, rates in air and steam became similar. Luikov [40] showed the higher drying rates obtained in air reflect the larger temperature difference available for heat transfer since the material dries at the wet bulb temperature of the air rather than the boiling point as in steam drying. Yoshida and Hyodo [48] measured the evaporation rates of water into air, humid air and superheated steam in a wetted wall column. When compared at the same gas mass velocity and temperature an inversion temperature was found, Fig. 12. For gas temperatures less than this, drying proceeded faster in air than superheated steam, and above this temperature drying in steam was more rapid. This reflects the larger temperature difference available with air drying especially at gas temperatures just above the liquid boiling temperatures. The smaller driving force for steam drying is offset by the increased heat transfer coefficient. Under the conditions employed, this inversion temperature was in the region of 160°C to 180°C, decreasing as the mass velocity of the drying gas was increased. Trommelen and Crosby [41] found the drying rates to become the same at a higher temperature (400°C.)

Falling rate period. Once the surface of the particle is no longer moist the drying rate begins to decrease and the process enters the falling rate period. During this period the transfer of moisture and heat within the drying material become rate limiting, the drying rate being determined by the nature of the material and not just by the liquid and vapour properties. The increased mobility of moisture within the material at the higher temperatures prevailing during steam drying has already been mentioned. Yoshida and Hyodo [26], [42] during the drying of fibres in superheated acetone and foodstuff in steam, found the skin or film that forms when the surface dries out is much more porous and permeable to the flow of vapour from within the material, enhancing both the rate of drying and the product quality compared with air drying. Trommelen and Crosby [41] found that drops of film-forming materials such as food products and detergent, dried faster in steam than air during the falling rate period and pointed out the lower resistance to vapour transfer through the more pliable film that results on drying in steam.

Overall rates of drying. The general characteristics of the drying process are not altered when the drying medium is superheated vapour rather than air. The major differences are the condensation on the particle during the heat-up period and the elevated surface temperature during the constant rate period. The overall rates of drying can be faster or slower in superheated vapours than air depending on many factors - the amount of condensation occurring during the induction period, whether the gas temperature is above or below the inversion temperature and the nature of the material being dried being just some of them. Heat transfer rates can be readily predicted using the correlations developed for air as the bulk medium, whilst the mass transfer resistance for vapour transport into the bulk gas can be neglected. Resistance to both vapour and moisture movement within the drying material appear to be reduced for superheated vapour drying, increasing drying rates during the falling rate periods.

INDIRECT DRYING

In indirect dryers the heat required for evaporating moisture from wet material is transferred to the drying material from a hot surface separating the heating medium from the material to be dried. The heating medium is usually steam but can be hot flue gases or other hot fluids depending on the temperatures required for drying. The heat source need not necessarily be a hot fluid, electric heating could be used to maintain the hot surface. The material processed in indirect dryers is often of a granular nature but such diverse forms as sheets, pastes, sludges and solutions are handled by various equipment.

Since the heat for drying is not supplied by a flow of hot air contacting the particles and which constantly removes evaporated moisture from the vicinity of the drying material, the region around the drying material is filled with evaporated moisture, the evolution of vapour sweeping away non-condensables. Any free moisture is removed from the material at the boiling point corresponding to the prevailing pressure just as when considering direct drying in superheated vapours. Similarly removal of any moisture bound to the solid requires elevation of the temperature above the boiling point.

The type of equipment used for indirect drying depends mainly on the form of the moist material, its sensitivity to temperature and the quantity of material to be dried. Early texts such as Badger and McCabe [3] provide descriptions of the many types of dryers and indicate the appropriate applications.

When the wet material or dried product is conveniently handled on trays, shelf or tray dryers can be used. The trays are placed on hollow steam or hot water heated shelves which gradually heat the material to a temperature such that the moisture evaporates under the pressure existing in the dryer. Because of the batchwise operation and the amount of labour involved in filling and emptying the trays, the use of such dryers is restricted to small amounts of relatively valuable material. If the material to be dried is temperature sensitive and/or if more rapid drying is required the tray dryer can be operated under vacuum, the vapour produced being condensed in a condenser placed between the dryer and vacuum source.

Material which is granular or crystalline and must be handled in large quantities, can be dried in rotary dryers provided it is not so sticky as to build up on the walls or heating tubes. A rotary dryer consists of a cylinder rotated on suitable bearings and usually is slightly inclined to the horizontal. Feed is introduced at one end and dried product discharged at the other. Heating can be supplied by steam tubes fastened in concentric rows inside the cylinder rotating with it, Perry [6], or by hot combustion gases passing around the shell and through ducts within the dryer, Badger and McCabe [3]. With suitable design these dryers are available not just for drying of water from materials but also removal and subsequent recovery of solvents and drying of material which cannot be exposed to atmospheric or combustion gases, Perry [6].

Agitated dryers are used to process similar materials as rotary dryers and can also handle solids of a more sticky nature. Rotating scrapers keep stirring the drying material from the jacketed heating surface. Agitated dryers can be used for continuous operation but batch-wise operation is more usual. Drying under vacuum is common for the reasons already explained.

Screw or paddle dryers transfer heat by rotating internally heated screws or paddles in the drying material and are suitable for solvent recovery if used in closed cycle operation, Anon. [50].

For drying slurries or liquid solutions, drum dryers can be used. These usually consist of two steam heated cylindrical drums which rotate toward each other. The liquid to be dried is fed into the V-shaped space between the drums. Doctor knives are used to remove dried product from the drum. For vacuum operation an evacuated casing may be used, Badger and McCabe [3]

In the manufacture of paper and textiles the continuous sheet which is formed is carried in zig zag fashion over a series of steam-heated rotating cylinders, the speed of rotation of the cylinders being matched to the speed of the sheet produced. A modification of this type of dryer is one in which pasty material is carried on an endless wire mesh belt passing between the drying rolls.

Fluidized beds have long been used for direct drying of granular material. The addition of heating surfaces within the bed enables higher efficiencies by reducing the air flow through the system, Huthwaite [41], Herron and Hummel [52]. The catalogues from manufacturers such as Escher Wyss and Niro Atomizer illustrate the various types and applications of indirectly heated fluidized beds for drying. Liborius [53] mentions the possibility of completely eliminating air from the dryer using heating panels to supply the necessary heat for drying and superheated vapour for fluidizing.

Rates of Drying

In all the previously mentioned forms of indirect dryer, heat for drying is transferred from the hot surface through a layer of dried or partially-dried material to the region where evaporation is taking place. As drying proceeds the evaporation region moves further from the hot surface and the increasing resistance to heat transfer through the dried material reduces the drying rate. Periodic replenishment of undried material adjacent to the surface by mixing or stirring as in the various rotary or mechanically agitated dryers enhances the average heat transfer from the surface and allows faster drying. The very rapid mixing typical of fluidized beds is responsible for the very high (in comparison) heat transfer coefficients at immersed surfaces. Typical heat transfer coefficients for some indirect dryers are given in Perry [6] and vary from about 20 W/m^2 K for vacuum shelf tray dryers and 30 to 90 W/m^2 K for rotary or agitated types, to values of 150 to 500 $W.m^2/K$ at immersed surfaces in fluidized beds, indicating the importance of replacement of dried material at the heating surface. The drying of sheet material is, in some ways, analogous to the tray drying of granular material, where during drying the evaporation region moves away from the heated surface and heat for drying must be transferred through a porous layer of dried material.

Since heat transfer between packed, agitated or fluidized beds and submerged surfaces is the rate limiting step in indirect drying, it is necessary to understand the processes by which heat is transferred and the relative importance of these processes in the particular types of equipment.

Heat transfer to a bed of particles occurs in two sequential stages, Schlünder [54], Xavier and Davidson [55], Baskakov [56]. In the first stage heat is transferred from the heating surface of the adjacent layer of particles. A wall to particle heat transfer coefficient, h_{wp}, is ascribed to this process. The second stage is the conduction of this heat into the bed of particles. If the bed of particles is maintained isothermal for instance, by perfect mixing, rapid replacement of particles at the surface or latent heat absorption, the resistance to heat transfer by conduction into the bed becomes negligible and the wall to particle heat transfer coefficient becomes limiting. In the other extreme heat conduction becomes dominant if the particles are immobile, residence times at the surface are sufficiently long, or if there is only sensible heat consumption within the bed. In most types of dryers neither resistance can be completely neglected although one or other may be predominant. The relative importance of each can vary through the drying process within one piece of equipment, Schlünder [54].

Wall Heat Transfer Coefficient

Various equations are available for the prediction of the heat transfer coefficient for heat transfer from the surface to the first layer of particles. Xavier and Davidson [55] in their discussion of fluidized bed heat transfer suggest

$$h_{wp} = \frac{k_g}{\delta} \quad \text{with } d_p/4 < \delta < d_p/10 \qquad \ldots(5)$$

Obviously the coefficient increases as the gas conductivity increases and/or particle size decreases. The solid properties are assumed to have no effect on the coefficient.

Schlünder [54] has analysed in a semi-rigorous way, the heat conduction through the gaseous gap between the first layer of particles and the surface. By allowing for the reduction of gas thermal conductivity in the region close to the point of contact where the separation distance is comparable to the mean free path length of the gas molecule, the following expression for the heat transfer coefficient is obtained.

$$h_{wp} = \psi \frac{4k_g}{d_p} \left\{ (1 + \frac{2s}{d_p}) \ln (1 + \frac{d_p}{2s}) - 1 \right\} \ldots(6)$$

Where ψ is 0.75 to 0.85.

This is again independent of solid properties and is suitable for use in any of packed, agitated and fluidized beds.

Olbrich [57] for stagnant packed beds shows the wall coefficient to be given by:

$$h_{wp} > 2.12 \frac{k^o_e}{d_p} \quad ..(7)$$

where k^o_e is the effective thermal conductivity of the stagnant packed bed, which will be discussed in the next Section. Olbrich [57] notes that when gas is flowing through the packed bed the wall bed coefficient must be modified.

Decker and Glicksman [58] have considered the effect of the surface roughness of the particle and have derived a series of expressions which could be used to determine the wall-particle heat transfer coefficient. The surface roughness was not found to have a strong effect on the coefficient and they suggest using:

$$h_{wp} = 16 \frac{k_g}{d_p} \text{ for metal particles and}$$

$$h_{wp} = 12 \frac{k_g}{d_p} \text{ for ceramic materials.} \quad ..(8)$$

For beds of sand particles in air at ambient temperatures, predictions of wall-particle heat transfer coefficients tend to vary as between different authors, predicted Nusselt numbers, $h_{wp} d_p/k_g$, falling in range 4-10, Xavier et al. [55], range 10-30, Schlünder [54], 12, Decker et al. [58] or > 13, Olbrich [57].

Heat Conduction Within the Bed

It is convenient to consider the bed of granular material as a homogeneous medium to which Fourier's theory of heat conduction can be applied. For a semi infinite packed bed contacting a plane surface the expression for the instantaneous heat transfer coefficient is:

$$h_{bed} = \frac{1}{\sqrt{\pi}} \sqrt{k_e \rho_e C_{pe}} \frac{1}{\sqrt{t}} \quad ..(9)$$

The time-averaged coefficient is :

$$\bar{h}_{bed} = \frac{2}{\sqrt{\pi}} \sqrt{k_e \rho_e C_{pe}} \frac{1}{\sqrt{t_R}} \quad ..(10)$$

where t_R is the time that the material is held at the heating surface.

The effective bed conductivity k_e can be determined using the relationships developed for steady-state heat conduction through packed beds, which consist of a component pertaining to the stagnant bed of solid particles and a flow dependent component. The stagnant bed effective conductivity can be determined using Kunii and Smith [59] or the more convenient expression of Krupiczka [60] below.

$$\frac{k^o_e}{k_g} = \left(\frac{k_s}{k_g}\right)^N \quad ..(11)$$

where $N = 0.280 - 0.757 \log_{10}\varepsilon - 0.057 \log_{10}(k_s/k_g)$.

Xavier and Davidson [55] add the flow dependent component to the stagnant component to give the following expression for the effective conductivity:

$$k_e = k^o_e + 0.1 \rho_g C_{pg} d_p U \quad ...(12)$$

Olbrich [57] suggests a similar form but with a value of 1/7 replacing the constant 0.1. The other effective properties ρ_e Cp_e of the quasi homogeneous material are simply determined from below:

$$\rho_e C_{pe} = (1-\varepsilon)\rho_s C_{ps} + \varepsilon\rho_g C_{pg} \quad ..(13)$$

Since the gas density is normally several orders of magnitude less than the solid density the second term can be neglected.

Overall Heat Transfer Coefficient.

When the conditions are such that neither heat transfer from wall to particles nor heat conduction within the bed solely describe the process, the overall resistance is taken as the sum of the resistances of each stage. Xavier and Davidson [55] simply add the resistance of the time averaged conduction within the bed to the wall-particle resistance i.e.

$$\frac{1}{h} = \frac{1}{h_{wp}} + \frac{1}{\bar{h}_{bed}} \quad ..(14)$$

Schlünder [54] determines the time-averaged instantaneous overall heat transfer coefficient and determines:

$$h = \bar{h}_{bed}\left\{1 + \frac{\bar{h}_{bed}}{2h_{wp}} \ell n\left(\frac{1}{1+\frac{2h_{wp}}{\bar{h}_{bed}}}\right)\right\} \quad ..(15)$$

Although quite different in form, both expressions reduce to the same limits if either the conduction in the bed or heat transfer from wall to particle is the rate determining stage. If the two stages are comparable the overall coefficients vary only by about 10%

The difficulty in using these expressions is the determination of the fictive contact time t_R. For non-agitated systems such as tray shelf dryers or heaters, t is simply the time after commencement of the process and t_R, the total process time. For mechanically stirred systems such as agitated or rotary dryers t_R is not necessarily the same as the time constant of the mixing device i.e. rotational speed, but is probably strongly related to it. Schlünder [54] reports that in a laboratory scale stirred bed up to 15 revolutions of the stirrer were necessary to achieve a complete replacement of solids at the surface. For fluidized beds the contact time is especially difficult to estimate and is compounded by the surface not being entirely covered with solids at any given instant due to the passage of gas bubbles through the bed. Various authors, Xavier and Davidson [55], Martin [61], Bock [62] present methods to determine both the fraction of surface not covered and the effective contact time of solids at the surface.

In fluidized beds, particularly for beds of large particles (greater than about 1 mm), heat transfer from the surface by gas convection becomes another important mechanism for heat transfer. Gas convection heat transfer is considered to be in parallel with the series combination of surface to particle transfer and conduction into the bed. A method for estimation is given by Gabor [63].

Application to Drying Processes.

During drying processes, the heat transferred through the heating surface is absorbed by the drying material as both sensible and latent heat. The nature of the material changes during drying particularly if the original moisture content is very high, and this is reflected by the thermal properties of the material. To allow for latent heat consumption Schlünder [54] proposes replacing the effective specific heat C_{pe} by the total enthalpy change of the particulate material, i.e. $(dH/dT)_e$ where

$$\left(\frac{dH}{dT}\right)_e = C_{pe} + \left(\frac{\partial H}{\partial X}\right)_e \cdot \frac{\partial H}{\partial T} \quad ..(16)$$

If the heat transferred is absorbed as latent heat there is no resistance to conduction within the bed and the transfer rate is limited by the wall to particle coefficient as mentioned earlier. Obviously as the material dries out the heat transfer will be limited by both the wall particle moisture and the resistance due to a layer of dried material. The instantaneous heat transfer coefficient will be given by:

$$\frac{1}{h} = \frac{1}{h_{wp}} + \frac{x(t)}{k_e} \quad ..(17)$$

where $x(t)$ is the distance of the evaporating front from the heating surface and k_e is the effective conductivity of the dried material.

The estimation of effective conductivity for packed beds of solid granular material is straightforward as mentioned earlier. However since many drying operations involve porous materials which may be filled to a greater or lesser extent with moisture, estimation of the effective solid conductivity and hence the effective bed conductivity if difficult. Luikov [40] discusses the effect of structure of the particle on the apparent particle conductivity. For particles where the void spaces are unconnected the apparent conductivity is approximately given by the sum of the conductivities of solid material and void gas weighted on a volume basis, Kunii and Smith [59] i.e.

$$k_e = (1-\varepsilon_s)k_s + \varepsilon_s k_g \quad ..(18)$$

For particles where the structure more closely represents a packing of solid spheres the methods of Kunii and Smith [59] or Krupiczka [60] could be used. The apparent conductivity of material which is completely wet could be estimated by a similar method, replacing the gas conductivity with the conductivity of the liquid phase, Luikov [40]. For particles in which only some of the void spaces are filled with water estimation of apparent conductivity is difficult. Allowance must be made for internal transfer of heat by vaporization of moisture and condensation in cooler regions. This mechanism can cause the apparent conductivity of a partially wet particle to be greater than that for fully wet, Luikov [40].

Okazaki et al. [64] present a method for estimation of the effective thermal conductivity of wet granular beds which allows for the effect of enhanced conductivity through the liquid film at the points of contact of the packing. Leyers [65] also discusses the effect on conductivity of area rather than point contact.

For fluidized beds, the flow dependent part of the effective conductivity is estimated by using the incipient fluidizing velocity as the superficial gas velocity through the packing, and is particularly important for large particles. For agitated and pan or tray dryers the gas passing through the granular bed is the evaporated vapour. The drying rates for these types of equipment are sufficiently low for the flow dependent effect on effective conductivity to be neglected, although the evaporation rate is sufficiently high to maintain an atmosphere of the vapour in the vicinity of the solids and so the thermal properties of the vapour should be used.

Operation at pressures above or below atmospheric pressure affects the heat transfer coefficient through the vapour thermal conductivity. Although thermal conductivity is unaffected by pressure at normal pressures, it is temperature-dependent and, as pointed out earlier, since the drying temperature is close to the saturation temperature at the prevailing pressure, a change in pressure will affect the vapour conductivity. Schlünder also notes that the effect of operation at vacuum is to increase the mean free path length of the vapour molecule which causes a further reduction in the gas conductivity in the region of the points of contact of particles and surface. Heat transfer coefficients for indirectly heated vapour dryers could be expected to increase with increase in pressure.

Very little has been published on the heat transfer coefficients pertaining to solids fluidized by vapours close to the condensing temperature. Kondo et al. [66] measured tube to bed coefficients obtained using superheated steam to fluidize micro-spherical catalyst particles. The data obtained correlated well with data for fluidization with air indicating the same mechanisms apply to heat transfer in both media. Mickley and Fairbanks [67] in heating a bed of microspheres fluidized with ammonia found coefficients about 50 per cent higher with ammonia than with air, whereas using a bed of glass beads the coefficients for air and ammonia are very similar. They explain the observation in terms of an additional heat-capacity of the solids because of adsorbed ammonia. Baskakov et al. [68] sprayed water onto porous charcoal particles and found that spraying led to an increase in heat-transfer coefficient from 190 to 240 W/(m^2)(C), the particles absorbing 13 per cent w/w moisture. They argued that the increase in heat-transfer coefficient was brought about through increase in the particle heat capacity alone.

Fig. 12 shows the maximum heat transfer coefficient (with respect to fluidizing gas velocity) for various materials - non-porous sands and ballotini and porous F.C.C. - fluidized in air and steam at various temperatures. The effect of gas temperature is taken into account by plotting the maximum coefficient against the gas thermal conductivity.

For F.C.C. a much higher coefficient is evident in steam as against air, when compared at the same gas conductivity. Even the non-porous sands and ballotini exhibit a coefficient consistently 20 to 25% higher in steam than in air. This behaviour is contrary to the predictions of all correlations and indicates that an extra mechanism of heat transfer may be operating when using steam. From the evolution of heat which occurs on admission of steam to a bed of air-dried F.C.C. it is evident that moisture is adsorbed into the pores of the particles and the desorption of this moisture at the hot tube surface and consequent readsorption in the bulk bed could account for the higher heat transfer coefficient in steam over that in air. No explanation is given for the very consistent 20 to 25% increase in steam for the non-porous sands and ballotini, although capillary adsorption at points of contact may be a factor.

Where there is the possibility of desorption at the hot surface followed by re-adsorption in the (relatively) cold bulk, then there is the possibility of an additional mechanism of heat-transfer to the bulk bed. When a material such as brown coal is considered, allowance has to be made for the required moisture level e.g. a typical as-mined brown coal contains 2 kg water per kg dry coal and exerts the vapour pressure of water until the moisture is reduced to about 0.8 kg water per kg dry coal, there-after drying is from capillaries, the partial pressure of water vapour gradually decreasing as the moisture-level is reduced. At very low moisture levels, desorption becomes a factor. The behaviour is hysteretic as the dry coal is exposed to the water vapour and picks up moisture initially by adsorption and then by capillary condensation. Typically the coal picks up only about 0.2 kg water/kg dry coal when the partial pressure of water vapour is restored to one bar. It is clear that if superheated steam is used to fluidize an adsorbent solid, the heat-transfer coefficient will be increased by adsorption-desorption processes, an equilibrium situation being reached so far as adsorbed water is concerned.

LARGE SCALE SYSTEMS

Although the principles of drying with super-heated vapour have been understood for many years, there have been few commercial applications, particularly on the large scale, making use of the inherent advantages. One process which has been developed specifically to make use of the possible energy efficiencies is the *superheated steam drying of paper pulp* practised by Mo Do Chemetics AB in Sweden, Svennson [10], Anon. [35]. Developed in collaboration with the Chalmers University of Technology, Gothenburg, the dryer uses low-pressure steam at 2 to 5 bar pressure to pneumatically transport disintegrated paper pulp through the tube side of a series of tubular heat exchangers. Steam at higher pressure, 8-15 bar, is used as a heat source in this indirect dryer. During processing of up to 150 tonne of dry pulp per 24 hours, the moisture of the pulp is reduced from about 1.2 kg per kg dry pulp to about 0.15 kg per kg dry pulp.

The steam evaporated from the pulp is bled from the system and used in place of process steam for lumber drying and combustion air preheating enabling a very low net energy consumption for the drying process. Values of 0.4 to 0.5 GJ per tonne pulp are quoted, Svennson [10], for steam drying in contrast to 3.0 to 3.5 GJ per tonne pulp typical of conventional oil-fired flash dryers. Savings in the cost of energy are quoted as about $10 per tonne dry pulp on an air dried cost of $19 per tonne (1980 US dollars).

The first industrial dryer began operation in 1978 and has demonstrated the distinct advantages of drying pulp in superheated steam, particularly when the system is integrated with the mill.

Utilization of abundant deposits of *brown coal* is restricted by the high moisture content in its as-mined state, typically up to 2 kg/kg dry coal for Victorian brown coals. Large scale drying of such material would enable its efficient use for power generation or as an alternative industrial fuel, but an efficient process for drying must be used. The steam-heated steam-fluidized bed drying system has been under development at Monash University and some results have been reported, [7], [22], [23], [24], [69]. Considering electric power production, the first system proposed the application to a 350 MW power station fired with raw brown coal. In this case the drying was to be conducted in superheated steam at atmospheric pressure using a portion of the steam exhausting the power station high pressure turbine as a heat source for drying. The condensed heating steam was returned to the boiler and the steam evolved from the coal was used to generate clean steam which could be let down through a low pressure turbine and produce more power. The dried coal was burned in the boiler raising its thermal efficiency from 65% to an estimated 82%. (Actually the figure employed should have been 88-92% since this range would apply for a new boiler designed for dry brown coal). The savings in brown coal used was about 8%. If a triple effect dryer was used the coal savings approach 20%.

It has been estimated that for a 2000 MW station in mid-1978 values the savings were as follows:

- Initial capital *saving* $(A) 42 million
- Net Present Value of Oil used at start-up or shut-down and in firing difficulties. $(A) 42 million or
- Or if briquettes are used instead of oil $(A) 16 million.

This integrated drying scheme has virtues in a situation where the power-station must meet, wholly or partially, the diurnal variations of power demand. A relatively small amount of storage capacity for dry brown coal makes possible the following scenario:

Maximum electric power production:
Using dry brown coal from storage. No drying.

85% maximum:
Dryers operating to dry all coal used for combustion.

50% maximum:
Small boiler turndown, dryers dry more coal than needed for combustion. Excess dried coal is stored.

A clear advantage arises due to the fact that as much drying as possible is undertaken while the power demands are low. By this means it is never necessary to have a high turndown on the boiler.

The cheapest drying system would take advantage of the diurnal electric power turn down by maintaining boiler operation to dry coal. The drying cost of excess coal produced would then be relatively small since the power station capital cost is assignable to electric power production. The overall situation in Victoria may thus be considered as a choice between, either :-

(a) Constructing a new power station of conventional design burning raw brown coal and producing electricity only, or :

(b) Establishing a new power station fed with dry coal to be produced in dryers bought from the savings of about 30% in power station capital cost. Such a station will produce both the required electricity and a dry coal product.

For dry coal production in conjunction with electric power generation, Keogh and Potter [69] have considered the case where a power station is operated to produce in the steady state as much dry coal as possible. In this case steam from the high-pressure turbines is wholly used for drying and steam from the dryers is cleaned and utilised in low-pressure turbines, there being no intermediate pressure turbines. Their calculations indicated a very favourable cost of drying. Such a system has been costed by Davy McKee Pacific Ltd., indicating for a production level of 5 million tons per annum dry coal from 15 million tons per annum raw coal production costs of $(A)22 per ton dry coal could possibly be realised [70], [71]. The system involved the simultaneous production of 250 MW electricity. However, if dry brown coal production is tied in with power stations subject to diurnal variation of electricity demand, the cost could be substantially lower.

Another application which has been considered is the drying of *alumina hydrate* before calcination. A paper discussing this question at length has been prepared and will be published elsewhere.

CONCLUSION

Drying in superheated steam, particulaly for large-scale applications, should have a comparatively bright future. Much however depends on the price of energy which is presently declining. The factors ensuring increasing energy demands in the world as a whole continue to be significant and the upward pressure on energy prices will develop again as the world economy recovers.

The position for smaller-scale applications is not so clear-cut but for these applications there will probably be a small but steady increase.

Much needs to be done to ensure the development of economical designs for the large-scale. For this scale of operation the best gas-cleaning devices are not clearly established and may in any case differ for differing materials. Experience of pressurized units needs to be extended. Still requiring much attention is the matter of solids feeding which is common to all dryers and to other equipment as well. Here again the best methods for large plant have not been thoroughly established and developmental expenditure is required.

NOMENCLATURE

Symbols used are those recommended for use in papers at this Symposium except for the following:

Symbol	Quantity	Units
a	constant in Eqn.(2)	-
d_p	particle diameter	m
ΔG	Gibbs free energy	J/mol
H	specific enthalpy	J/kg
k	thermal conductivity	W/(Km)
L	length of pan or tray	m
N	exponent in Eqn.(11)	-
Nu	Nusselt number ($h_{wp} d_p/kg$)	-
n	pressure	N/m^2
r	pore radius	m
s	function of mean free path length [54]	m
t_R	fictive contact time	s
U	superficial gas velocity	m/s
V_L	specific molar volume of liquid	m^3/mol
W	evaporation rate	$kg/(m^2s)$
X	water content	kg/kg
Z	variable in Eqn.(4)	-
Greek letters		
ε	voidage	-
θ	contact angle	radius
μ	gas viscosity	kg/(ms)
ψ	constant in Eqn.(6)	-

Subscripts

e	effective property of granular bed
g	gas or vapour
s	solid
bed	relating to granular bed
wp	wall to particle

Superscripts

o	stagnant bed property
o	referring to free liquid
-	time-averaged mean value

REFERENCES

1. Hausbrand, E., Drying by Means of Air and steam, 3rd revised English edn., Scott, Greenwood & Son, London, (1924).
2. Fleissner, H., Sonderdruck aus der Sparwirtschaft, vols. 10 & 11, (1927).
3. Badger, W.L. and McCabe, W.L., Elements of Chemical Engineering, McGraw Hill, New York, (1936).
4. Walker, W.H., Lewis, W.K,, McAdams, W.H. and Gilliland, E.R., Principles of Chemical Engineering, 3rd Edn., McGraw-Hill, New York, (1937).
5. Perry, R.H., Chemical Engineers' Handbook, 2nd edn., McGraw-Hill, New York, (1941).
6. Perry, R.H., Chemical Engineers' Handbook. 5th edn., McGraw-Hill, Kogakusha, Tokyo, (1973).
7. Potter, O.E. and A.J. Keogh, J. Inst. Energy, vol. 52, No. 412, pp. 143-149, (1979).
8. Karrer, J., reported in Engineering, vol.110, pp. 821-822, (1920).
9. Keey, R.B., Introduction to Industrial Drying Operations, Pergamon Press, Oxford (1978).
10. Svensson, C., Proc. 2nd Int. Symp. on Drying, Montreal, Canada, vol. 2, pp. 301-307, (1980).
11. Lane, A.M. and Stern, S., Mech Eng.,Vol. 78, No. 5, pp. 423-426, (1956).
12. Rosin, P.O., Braunkohle, vol. 28, pp.649-658. (1929).
13. Klein, H., Braunkohle, vol. 29, pp. 1-10 and 21-30, (1930).
14. Klein, H., Braunkohle, vol. 31, pp. 138-143, (1932).
15. Cooley, A.M. Jr. and Lavine, I., Ind. & Eng. Chem., vol. 25, pp. 221-224, (1933).
16. Harrington, L.C., Parry, V.F. and Koth, A., U.S. Bur. Mines Tech., Paper 633, (1942).
17. Evans, D. and Siemon, S.R., J. Inst. Fuel, pp. 413-419 (1970).
18. Allardice, D.J. and Evans, D.G., Fuel, vol.50, pp. 201-210, (1970) and ibid, pp. 236-249, (1970).
19. Stewart, R. and Evans, D., Fuel, vol. 46, pp. 263-274, (1967).
20. Bull, F.A., Drying of Brown Coals by the Fleissner Process, University of Melbourne, Brown Coal Laboratory, (1964).
21. Thompson, E.T., Chemical Engineering, vol.62, pp. 104-105, Sept. (1955).
22. Potter, O.E., Proc. 2nd Int. Drying Symp., Montreal, Canada, vol. 2, pp. 396-400 (1980).
23. Potter, O.E., Beeby, C.J., Fernando, W.J.N. and Ho, P., Proc. 3rd Int. Drying Symp., Birmingham, England, vol. 2, pp.115-123, (1982).
24. Potter, O.E., Beeby, C.J. and Ho, H., Specialists Meeting on Coal Fired MHD Power Generation, Sydney, 4-6 November, pp. 3.3.1-3.3.6, (1981).
25. Basel, L. and Gray, E., Chem. Eng. Prog., vol. 58, No. 6, pp. 67-70, (1962).
26. Yoshida, T. and Hyodo, T., Ind. & Eng. Chem. Proc. Des. Dev., vol. 2, No. 1, pp. 52-56, (1963).
27. Blaw-Knox Co., Chem. Eng. Prog., vol. 44, No. 3, 25, (1948).
28. Chu, C.J. Lane, A.M. and Conklin, D., Ind. & Eng. Chem.,vol. 45, No. 7, p.1586-1591, (1953).
29. Kollmann, F.P.P., Forest Products Journal, pp. 505-515, Nov., (1961).
30. Akao, T. and Aonuma, T., Proc. 1st Int. Symp. on Drying, Montreal, Canada, pp. 117-121, (1978).
31. Grewal , N.S. and Saxena, S.C., Ind. & Eng. Chem. Proc. Des. Dev., vol. 20, pp. 108-116, (1981).
32. Gel'perin, N.I., Einshtein, V.G. and Korotyanska, L.A., Int. Chem. Eng., vol. 9, No. 1, pp. 137-142, (1969).
33. Heyde, M. and Klocke, H.J., Int. Chem. Eng. vol. 20, No. 4, pp. 283-599, (1980).
34. Rowe, P.N., Foscolo, P.U., Hoffman, A.C. and Yates, J.G., Proc. 4th Int. Conf. of Fluidization, Kashikojima, Japan, pp. 1-7-1 to 1-7-8, (1983).
35. Anon., Process and Chem. Eng., Feb., 33-36, (1980).
36. Chu, J.C., Finelt, S., Hoerrner, W. and Lin, M., Ind. & Eng. Chem., vol.53, No.3, pp. 275-280, (1959).
37. Toei, R., Okazaki, M., Kimura, M., and Kubota, K., Kagaku Kogaku (abridged version), vol. 5, No. 1, pp. 140-141, (1967).
38. Ho, P., Moisture Equilibria of Victorian Brown Coals in Superheated Steam, M. Eng. Sci. thesis, Monash University, Melbourne, Victoria, (1984).
39. Wenzel, L. and White, R.R., Ind. & Eng. Chem. vol. 43, No. 6, pp. 1829-1937.(1951).
40. Luikov, A.V., Heat and Mass Transfer in Capillary-Porous Bodies, Pergamon Press, Oxford, (1966).
41. Trommelen, A.M. and Crosby, E.J., A.I.Ch.E. J., vol. 16, No. 5, pp. 857-867, (1970).
42. Yoshida, T. and Hyodo, T., Food Engineering, vol. 38, no. 4, pp. 86-87, (1966).
43. Lee, K. and Ryley, D.J., Trans. ASME J. Heat Transfer, pp. 445-451, Nov., (1968).
44. Moyers, C.G. Jr., Proc. 1st Int. Symp. on Drying, Montreal. Canada, pp. 224-229, (1978).
45. Toei, R., Okazaki, M., Kubota, K., Ohashi, K., Katoaka, K. and Mizuta, K., Kagaku Kogaku (Abridged edition), vol. 4. No. 2, pp. 220-223, (1966).
46. Toei, R., Okazaki, M., Kimura, M. and Vieda, H., Kagaku Kogaku (abridged edition), vol. 5, No. 1, pp. 139-140, (1967).
47. Montlucon, J., Int. J. Multiphase Flow, vol. 2, pp. 171-182, (1975).
48. Yoshida, T. and Hyodo, T., Ind. & Eng. Chem. Proc. Des. Dev., vol. 9, No. 2, pp.207-214, (1970).

49. Sjenitzer, F., Chem. Eng. Sci., vol. 1, No. 3, pp. 101-117, (1952).
50. Anon., Chem. Eng., Vol. 86, No. 23, pp. 101-102, (1979).
51. Huthwaite, J.A., The Chemical Engineer, No. 285, pp. 295-303, (1974).
52. Herron, D. and Hummel, C., Chem. Eng. Prog. vol. 76, no. 1, pp. 44-52, (1980).
53. Liborius, E., paper presented at Chemeca '82 Conference, Sydney, (1982).
54. Schlünder, E.U., Chem. Eng. Comm., vol. 9., pp. 273-302, (1981).
55. Xavier, A.M. and Davidson, J.F., A.I.Ch.E. Symp. Ser., vol. 77, No. 208, pp. 368-373, (1981).
56. Baskakov, A.P., Int. Chem. Eng., vol. 4, No. 2, pp. 320-324, (1964).
57. Olbrich, W.E., Proceedings of Chemeca '70 Conference, Melbourne and Sydney, Australia, Session 6A, pp. 101-119, (1970).
58. Decker, N.A. and L.R. Glicksman, A.I.Ch.E. Symp. Ser, vol. 77, No. 208, pp. 341-349, (1981).
59. Kunii, D., and Smith, J.M., A.I.Ch.E.J., vol. 6. No. 1, pp. 71-78, (1960).
60. Krupiczka, R., Int. Chem. Eng., vol. 7, No. 1, pp. 122-144, (1967).
61. Martin, H., Chem. Eng. Comm., vol. 13, No. 1, pp. 1-16, (1981).
62. Bock, H.J., Proc. 4th Int. Conf. on Fluidization, Kashikojima, Japan, pp. 5-2-1 to 5-2-8, (1983).
63. Gabor, J.D., Chem. Eng. Sci., vol. 25, No.6, pp. 979-984, (1970).
64. Okazaki, M., Ito, I., and Toei, R., A.I.Ch.E. Symp. Ser., vol. 73, No. 163, pp. 164-176, (1977).
65. Leyers, H.J., Chemie-Ing.-Techn., vol. 44, No. 19, pp. 1109-1115, (1972).
66. Kondo, T., Yoshida, K., and Kunii, D., Kagaku Kogaku (abridged edn.), vol. 4, No.1, pp. 200-201, (1966).
67. Mickley, H.S. and Fairbanks, D.F., A.I.Ch.E.J., vol. 1, No. 3, pp. 374-384, (1955).
68. Baskakov, A.P., Berg, B.V., Vitt, O.K., Filippovsky, N.F., Kirakosyan, V.A., Goldobin, J.M. and Maskaev, V.K., Pow. Tech., vol. 8, pp. 273-282, (1973).
69. Keogh, A.J., and Potter, O.E., Fuel Processing Technology, vol. 4, pp. 217-227, (1981).
70. O'Beirne, R.J. and Yamaoka, Y., Drying and Curing, Australia Institute of Energy, Melbourne Group, July, 1983.
71. Owens, L.W., O'Beirne, R.J., Truman, A.H., and Yamaoka, Y., Coaltech. Australia, pp. 75-92, Oct. (1983).

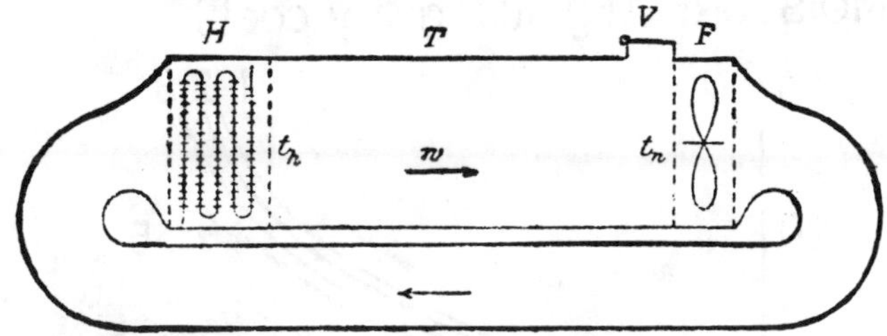

Fig. 1 Superheated steam dryer after Hausbrand [1].

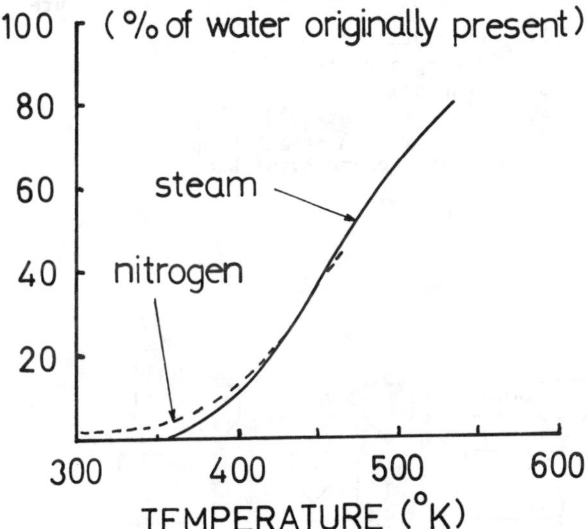

Fig. 2 Dewatering of Yallourn brown coal by heating in steam or nitrogen with coal removed from the water before cooling.

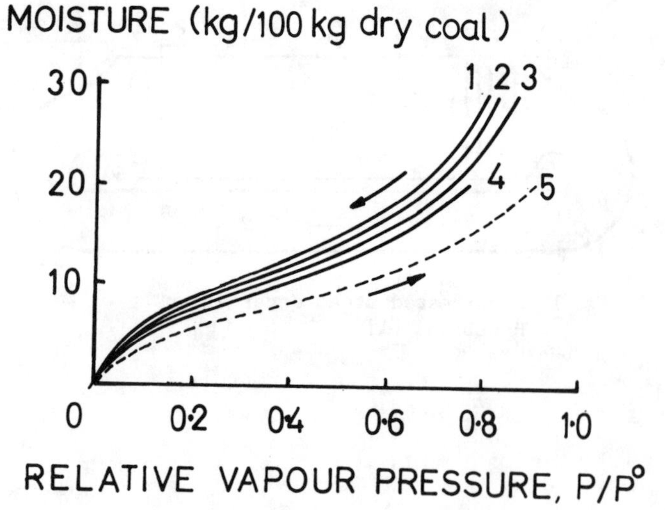

MOISTURE (kg/100 kg dry coal)

30

20

10

0 0·2 0·4 0·6 0·8 1·0

1 2 3 4 5

RELATIVE VAPOUR PRESSURE, P/P°

Fig. 3 Isotherms of Victorian brown coal in air at various temperatures.

1. desorption at 30°C
2. desorption at 40°C
3. desorption at 49°C
4. desorption at 60°C
5. readsorption at all temperatures

from Allardice and Evans [18]

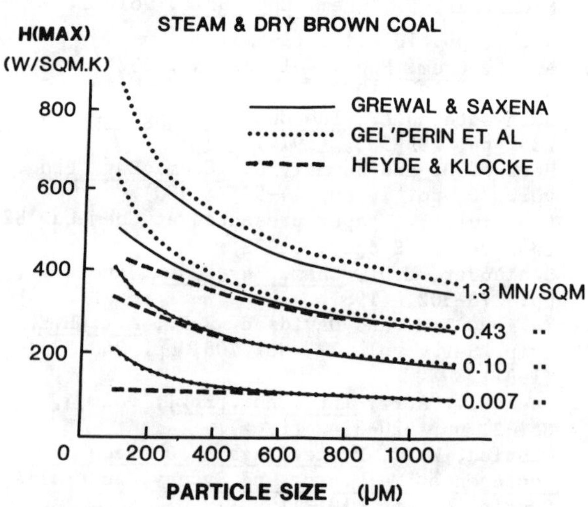

Fig. 5 Heat transfer coefficients for dry brown coal based on predictions of [31], [32], [33].

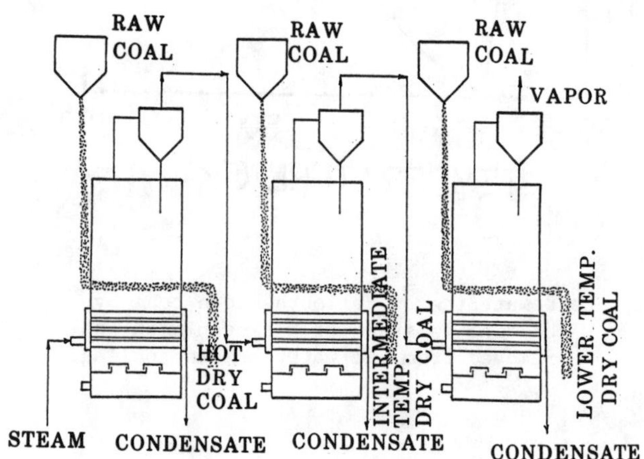

Fig. 4 Multiple effect superheated steam dryer for brown coal from Potter et al. [23].

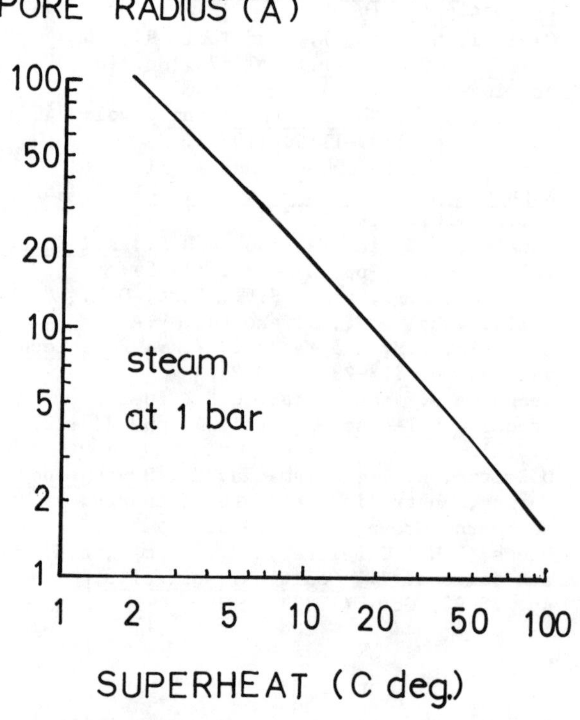

Fig. 6 The largest pore remaining at a given steam superheat.

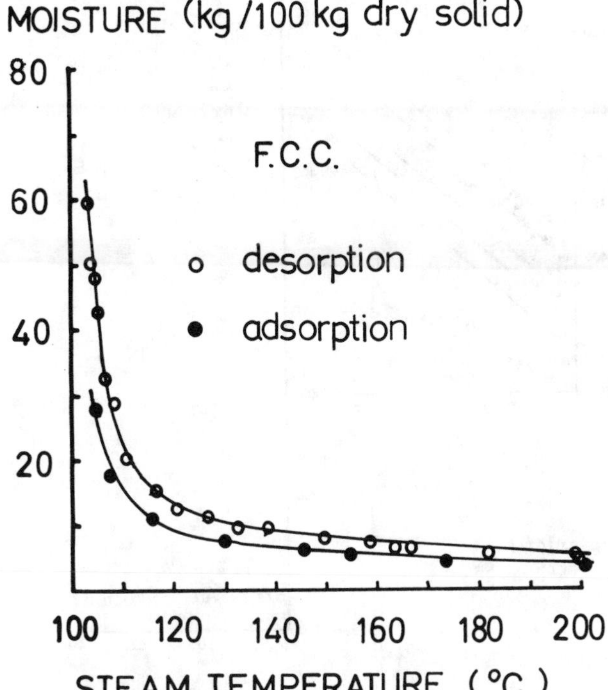

Fig. 7 Moisture equilibrium of F.C.C. in superheated steam.

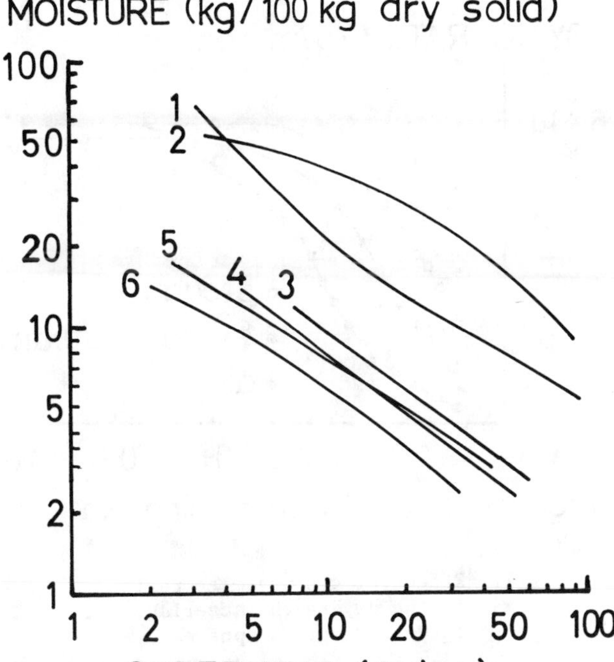

Fig. 8 Moisture equilibrium in superheated steam for various material.

1. F.C.C.
2. Brown coal [38]
3. paper pulp at 3 bar [10]
4. paper pulp at 1 bar [10]
5. timber (eucalyptus) [29]
6. timber (Spruce and Beech) [29]

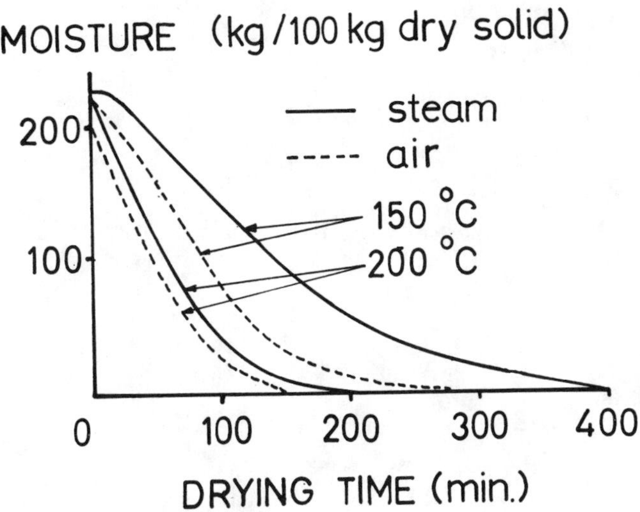

Fig. 9 Drying of a cellulose pellet in air and steam at 150°C and 200°C from Luikov [40].

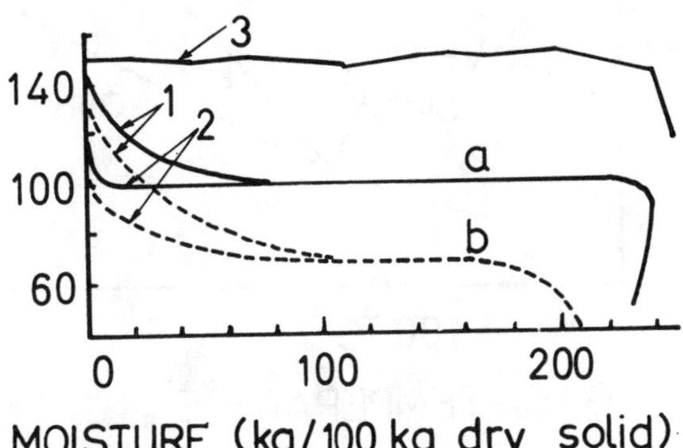

Fig.10 Drying of cellulose pellet in (a) superheated steam (b) air showing
(1) surface temperature of pellet
(2) centre temperature of pellet
(3) gas temperature
from Luikov [40].

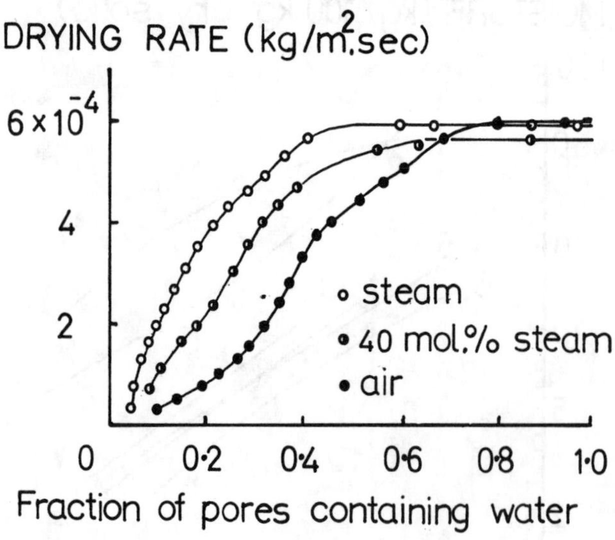

Fig.11 Effect of gas phase composition on the drying characteristic curve for porous spheres.

from Toei et al. [37].

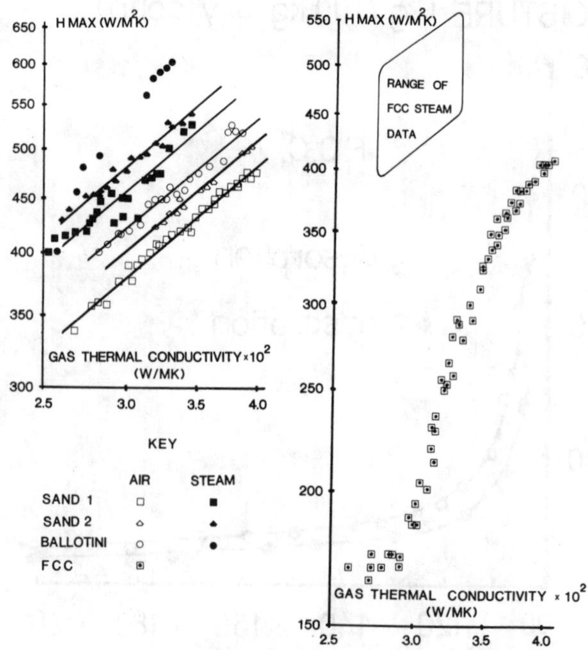

Fig.13 Maximum tube-to-bed heat transfer coefficients for various materials fluidized in air and superheated steam.

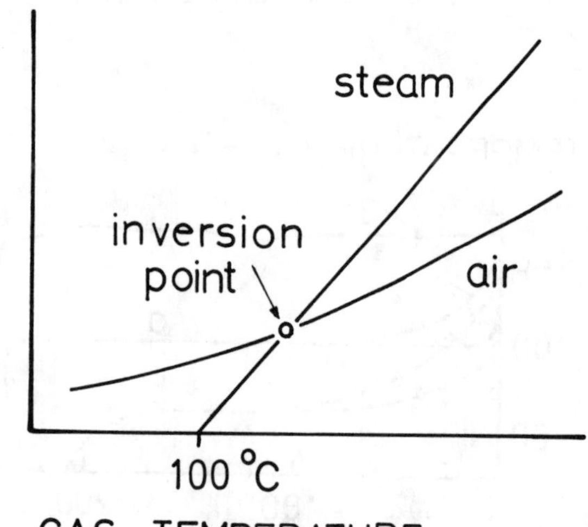

Fig.12 The effect of change of gas temperature on evaporation rate in air and superheated steam.

from Yoshida and Hyodo [48].

CONTROL OF FOOD-QUALITY FACTORS IN SPRAY DRYING

C. Judson King

Department of Chemical Engineering
University of California
Berkeley, CA 94720, U.S.A.

ABSTRACT

Research over the past 15 to 20 years has provided considerable fundamental insight into mechanisms of phenomena governing quality of spray-dried food products. Important advances have been made in understanding rates of drying, retention of volatile flavor and aroma substances, changes in particle size and shape (morphology), stickiness and agglomeration, and extents of thermal degradation reactions. Recent research results in each of these areas, obtained at Berkeley and elsewhere, are described and interpreted. From these results conclusions are drawn regarding effective means of improving and controlling product-quality factors.

Losses of volatile components are governed by a ternary diffusion process with the drops. The loss rates are greatest in the immediate vicinity of the atomizer, before sufficient drying of the drop surfaces occurs for a selective-diffusion mechanism to become dominant. Foaming and incorporation of an emulsified oil phase or surfactants all have substantial effects on volatiles loss. These effects can be interpreted through transport analyses.

Particle inflation during drying requires entrainment or nucleation of an air bubble as a first step, and is then governed by evaporation of water vapor into the bubble. Tendencies to form surface folds on particles are governed by the viscosity of the concentrated solution, and are described by a dimensionless group involving viscosity, surface tension, particle size, and exposure time.

Bulk mixing, convection patterns, and drop trajectories may be interpreted through the configuration of the dryer, as well as the ratio of the air supply rate to the rate of air entrainment into the spray by momentum exchange as the drops decelerate. These patterns determine the local temperature and humidity and thus directly influence rates of drying and loss of volatile components.

1. INTRODUCTION

Spray drying is used extensively for liquid and slurry foods, as well as a large number of other substances. For foods, there are a number of quality factors which are at least as important as the various economic factors associated with spray drying. Dried foods have a tendency to be lacking in flavor and aroma compared with the original material. This translates directly into lower consumer appeal. Another important property is bulk density, which is a critical control parameter. The bulk density depends directly upon particle size and shape. Some food substances--notably fruit juices--are very difficult to spray dry because of tendencies toward particle stickiness. In other cases, for example milk and coffee, controlled stickiness of particles is sought, so as to enable agglomeration of the product, which improves handling properties and redispersability.

Each of the properties described above is controlled by transport processes occurring within a spray dryer. These include momentum transfer, heat transfer, and mass transfer. The purpose of this paper is to explore the ways in which these properties depend upon transport processes, and to point out the logic through which a deeper understanding of the mechanisms of these processes can lead to process improvement.

The presentation here builds upon earlier reviews of spray drying, notably those of King, et al.[1], Bruin and Luyben [2], Thijssen [3], and Kerkhof and Schoeber [4].

2. DRYING RATE

Perhaps the most obvious transport processes are those governing drying rate--the rate of evaporation of water. This is a simultaneous heat and mass transfer process. In the early stages of drying, the rate is determined by resistances to heat and mass transfer external to the drops, with the drops reaching the wet-bulb temperature [5,6]. This situation has been well described and has been incorporated into many drying models [e.g., 7,8]. However, for food materials, mass-transfer resistances within the particles rapidly become important and dominant. For simple spherical particles the internal mass-transfer resistance can be described from a knowledge of diffusion coefficients [9], but when particles undergo changes in size and shape (expansion, shrivelling, blowholes, etc.) the situation becomes much more complicated to describe. Thus there should be a close connection between changes in particle morphology and the resultant drying rate.

3. RETENTION OF VOLATILE FLAVOR AND AROMA SUBSTANCES

Important flavor and aroma substances tend to have very high activity coefficients in the aqueous phase [10], and are therefore highly volatile. If they are to be retained in the product, it is necessary that a substantial mass-transfer barrier to their loss be imposed within the particles. This is accomplished by the selective-diffusion mechanism, described by Thijssen and co-workers [3,11], whereby the surface water content is reduced sufficiently so that the diffusion coefficients of volatile substances become substantially less than that of water.

In reality, volatiles loss is determined by a multicomponent diffusion process, in which water, the main bulk solute(s), and the volatile compound(s) must be considered as separate components. This follows from the fact that the water will have a substantial outward flux, whereas the main bulk solute(s) do not. The three-component diffusion process has been considered by Chandrasekaran and King [12], adopting a reference plane of no net volume flow. A more convenient frame of reference is probably the (moving) surface of no net flux of the main bulk solute(s), as used by Kerkhof [13].

Losses of volatile acetates in sucrose solutions have been monitored by Kieckbusch and King [14], as a function of location in the region near the atomizer. This was done by means of a chilled, continuous sample collector, with analysis of samples by flame-ionization gas chromatography. Quite large losses are observed in the atomizer region, before the selective-diffusion mechanism comes into play. The rates of loss have been interpreted semi-quantitatively in terms of mass-transfer models [14,15,16]. Volatiles loss is most effectively correlated and interpreted as a function of % water evaporation, because of similarity considerations [4,15]. These measurements have been extended to higher nozzle pressures, higher gas temperatures, and greater distances from the atomizer by Etzel and King [17], who used a sampler involving direct collection in chilled methanol. A typical relationship for the region near the atomizer is shown in Figure 1. [17].

Further research with sampling at various locations within a dryer has demonstrated the influence of an emulsified oil phase [15], which serves to alter the break-up mechanism and the drop-size distribution, and to extract and sequester those volatile compounds which distribute heavily into the oil phase. Experiments with pressure atomizers run under high (7 MPa) pressures have demonstrated the approach to the asymptotic retention predicted by the selective-diffusion theory [17] (see Figure 2), while experiments with and without surfactants added to sucrose-solution feeds have indicated that surfactants are effective in reducing volatiles loss, by virtue of suppression of droplet circulation and oscillation [16]. Results showing this effect for spraying water containing propyl acetate are shown in Figure 3. Natural surfactants are present in plenteous proportions in many food products. Experiments have also been carried out under conditions which can lead to foaming of sprayed drops [18]. Dissolution of a highly soluble gas under pressure has little effect, because desorption of bubbles does not occur within a time frame comparable with that over which most of the volatiles loss occurs. Mechanical admixing of gas does lead to foamed drops, and causes additional losses of volatiles, as has been observed experimentally and also interpreted theoretically in terms of diffusion mechanisms [18].

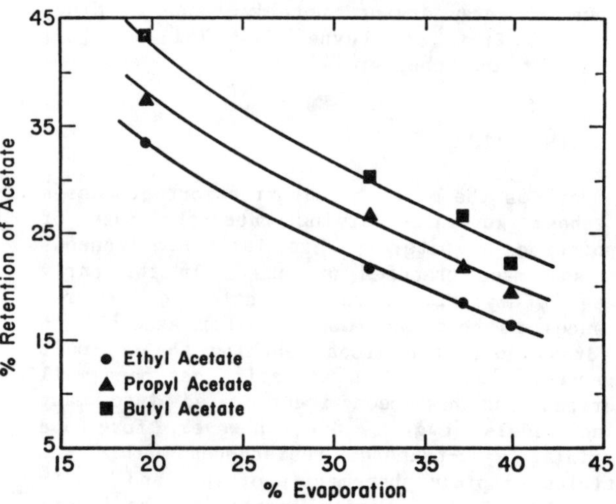

Figure 1. Retention of Acetates for 50% w/w Sucrose Solution, Fan-spray Atomizer, 7.00 MPa Atomization Pressure, 200°C Column Air Feed Temperature, and 49.5°C Liquid Feed Temperature [17].

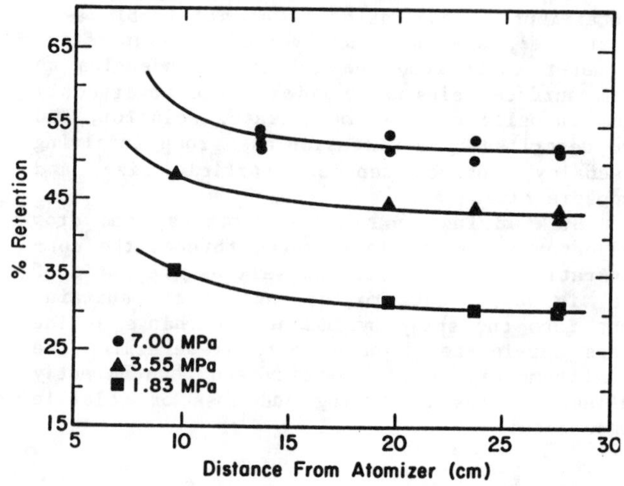

Figure 2. Retention of Propyl Acetate vs. Axial Distance for Different Atomizer Pressures; Other Conditions the Same as Those for Figure 1 [17].

Furuta, et al [19] have measured retentions of ethanol during drying of single drops of maltodextrin solutions, also finding general agreement with the directional predictions of the selective-diffusion model.

Some of the process implications which may be drawn from this knowledge are the following:

- Control of conditions near the atomizer is important for effective volatiles retention.
- Through the selective-diffusion theory one can reason that it is desirable to obtain very rapid initial evaporation rates, again to enhance volatiles retention.
- An emulsified oil phase can be effective for sequestering volatiles against loss, especially if the degree of dilution of the reconstituted product is sufficient to cause the volatiles to be released from the oil phase.
- Foaming of the feed will in most cases increase the degree of volatiles loss.
- In cases where the feed is low enough in surfactant content to allow droplet circulation, oscillation and/or turbulence, volatiles retention can be improved by adding appropriate surfactants.

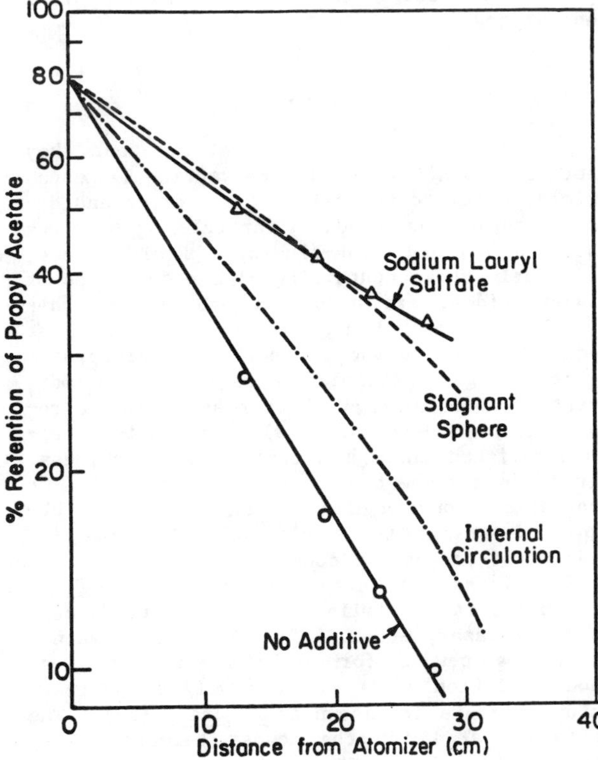

Figure 3. Calculated and Experimental Retentions of Propyl Acetate for Spraying Water With and Without Added Sodium Lauryl Sulfate. Atomizer Pressure = 7.0 MPa; Liquid Feed Temperature = 17°C; Air Temperature = 25°C [16].

- Higher dissolved-solids contents of the feed liquid will lead to higher volatiles retention.

4. CHANGES IN PARTICLE MORPHOLOGY

Microscopic examination of spray-dried particles has shown that numerous different particle morphologies result. Shrunken spherical particles are formed for some substances, such as sucrose and lactose, but this is the exception rather than the rule. Particles formed from other substances tend to exhibit features such as particle expansion through formation of internal voids, development of folds and crinkles on particle surfaces, and development of blowholes. In some cases, particles appear to have undergone one or more cycles of expansion, blowhole development, and particle shrivelling.

The morphology of spray-dried particles is an interesting curiosity in itself, but it is also important practically. Particle morphology directly affects bulk density. It affects drying rates and may also influence rehydration characteristics and volatiles loss.

Marshall and co-workers [20,21,22] carried out several of the earlier important phenomenological studies on particle morphologies achieved during spray drying. Buma and Henstra [23] analyzed morphological changes in milk and milk constituents during spray drying, deriving conclusions regarding the contributions of different constituents to morphological changes. This work was extended by Verhey [24], who also carried out experiments which revealed the fact that internal voidage tends to be under vacuum, communicates slowly with the environment through diffusion, and is filled with gas with a composition characteristic of the atomizer region. An important application of this knowledge was the conclusion that it should be possible to eliminate internal voidage and produce particles of unusually high bulk density by blanketing the atomizer region with steam. This has been verified experimentally [24,25].

Several different experimental approaches have been used to study particle morphologies. The simplest is merely to collect dried powder and examine it by either optical or scanning-electron microscopy (SEM). Charlesworth and Marshall [22] and subsequently Crosby and co-workers [26,27] have suspended drops from fine fibers, which hold them in a stationary position. Toei and co-workers hold drops in place by means of an ultrasonic field [28,29]. Both these techniques yield valuable information but do require using relatively large drops and monitoring changes over time scales of at least a few minutes, or more. A modified approach uses an ultrasonic field to hold a drop in place in an upflowing, warm-air stream [19,30]. In this way, times for the appearance of solid crystals on the surfaces of drops of aqueous solutions of inorganic substances have been measured and interpreted through mass-transfer models [30].

At Berkeley, we have recently developed a device which forms a single stream of droplets of uniform size and then allows them to fall through a heated column, giving a uniform and controllable temperature-time history during drying. The dry particles are then observed by SEM, and transmission optical microscopy is used to provide supplemental observations on voidage and crystallization in dry or still-liquid particles [31,32,33].

Some of the commonly encountered types of particle morphology are shown in Figures 4a-4f [33]. The particle in Figure 4a is a commercial spray-dried coffee product, unagglomerated. It exhibits a combination of internal voidage, surface shrivelling, and a blowhole. The remaining photographs are of particles from the device with the single falling stream of uniform-size drops. The particle in Figure 4b is formed from 30% skim milk; it exhibits a blowhole and is expanded, but has a smooth surface. The particles in Figure 4c are formed from uniform-size drops of 30% maltodextrin solution, dried under identical conditions. One has expanded and is smooth; the other is shrivelled. Figure 4d shows a particle formed from 10% coffee extract and subsequently broken open. It exhibits multiple internal voids. The particles in Figures 4e and 4f are formed from mixtures of varying proportions of maltodextrin and a 7:1 sucrose:fructose mixture. The particle in Figure 4e (50% maltodextrin) has a smooth surface, while that in Figure 4f (75% maltodextrin) displays surface folds.

Greenwald and King [32] considered the mechanism(s) by which internal voids form and particles inflate during spray drying. Nucleation of bubbles of water vapor, as such, seems not to occur, even for particle temperatures well above 100°C. A more likely mechanism starts with formation of a bubble of air or whatever other inert gas may be present. This can come from mechanical entrainment of bubbles during atomization [24], or can come from bubble formation through desorption of dissolved air or gas. A driving force for desorption will occur as water is evaporated and as the drop temperature rises, lowering the solubility of air. Even a deaerated liquid feed can absorb air very efficiently in the atomizer region, where the drops are at relatively low temperature. The mass-transfer mechanisms for absorption are the same that cause very rapid losses of volatiles in the atomizer region.

Following formation, the bubble of air or other inert gas grows by vaporization of water to reach the equilibrium partial pressure of water, which increases as the drop temperature increases. As the mole fraction of air in the bubble consequently decreases, the bubble, and hence the drop, grow in size.

By this mechanism, the morphology can be strongly affected by the initial nucleation step for air bubbles. The occurrence of nucleation requires the presence of a heterogeneous nucleation site within the drop, which is a statistical phenomenon. Presumably, particles such as that shown in Figure 4d contained several nucleation sites, resulting in multiple voids. The two particles in Figure 4c differ either because the one on the left contained a nucleation site and that on the right did not, or because the particle on the right expanded, developed a blowhole, and then shrivelled.

Alexander [33] has considered the mechanisms that cause folds and shrivelling ridges to be present or absent on particle surfaces. It appears that folds, once formed by uneven shrinkage forces, will tend to flow under a surface-tension driving force. The flow is resisted by the viscosity of the concentrate. By virtue of this mechanism, the tendency to develop folds or not is described by a dimensionless group in which viscosity is by far the most sensitive variable. This group is

$$\frac{\sigma}{\eta d} \left| \frac{dr}{dt} \right|$$

where σ is surface tension, η is viscosity, d (outside the derivative) is diameter, and $|dr/dt|$ is the (absolute) rate of change of particle radius due to drying. Thus the tendency to develop folds or not is related to a critical value of the viscosity of the concentrate at the time when folding can start to occur. In Figures 4e and 4f, the feed containing a greater proportion of maltodextrin (Figure 4f) develops a higher viscosity at a given degree of drying and therefore develops folds. The feed solution containing more of the mono/disaccharide mixture (Figure 4e) is less viscous, and flow is able to smooth out folds rather than letting them develop. The critical-viscosity concept at play here is very similar to the analysis used earlier [34] to interpret the occurrence of "collapse" during freeze-drying.

5. STICKINESS AND AGGLOMERATION

Some substances become so sticky that they cannot be spray dried by conventional means. Particles stick to the walls of the dryer and degrade, and/or the product particles clump together and are not free-flowing. Fruit juices, particularly those containing high proportions of monosaccharides, are notorious problems in this way. On the other hand, many products, such as dried milk and instant coffee, are agglomerated to form larger particle aggregates with more attractive appearance and better handling and reconstitution properties. Agglomeration is usually accomplished through a controlled stickiness, created through heat and/or humidity [25]. A mechanistic understanding of stickiness is obviously desirable to enable selection and improvement of processing conditions.

Stickiness is often assessed through a sticky-point test, similar to that described originally by Lazar, et al. [35]. Results of such a test for a powder formed by freeze-drying an aqueous solution of a 7:1 (by weight) sucrose: fructose mixture are shown in Figure 5 [36]. The temperature at which the powder becomes sticky decreases as the moisture content increases. Similar results have been obtained for a number of other products [35, etc.].

Downton, et al. [36] have postulated a mechanism of sticking and agglomeration through viscous flow, driven by surface tension and forming bridges between particles. The mechanism, the analysis, and the concept of a critical viscosity therefore become similar to the aforementioned

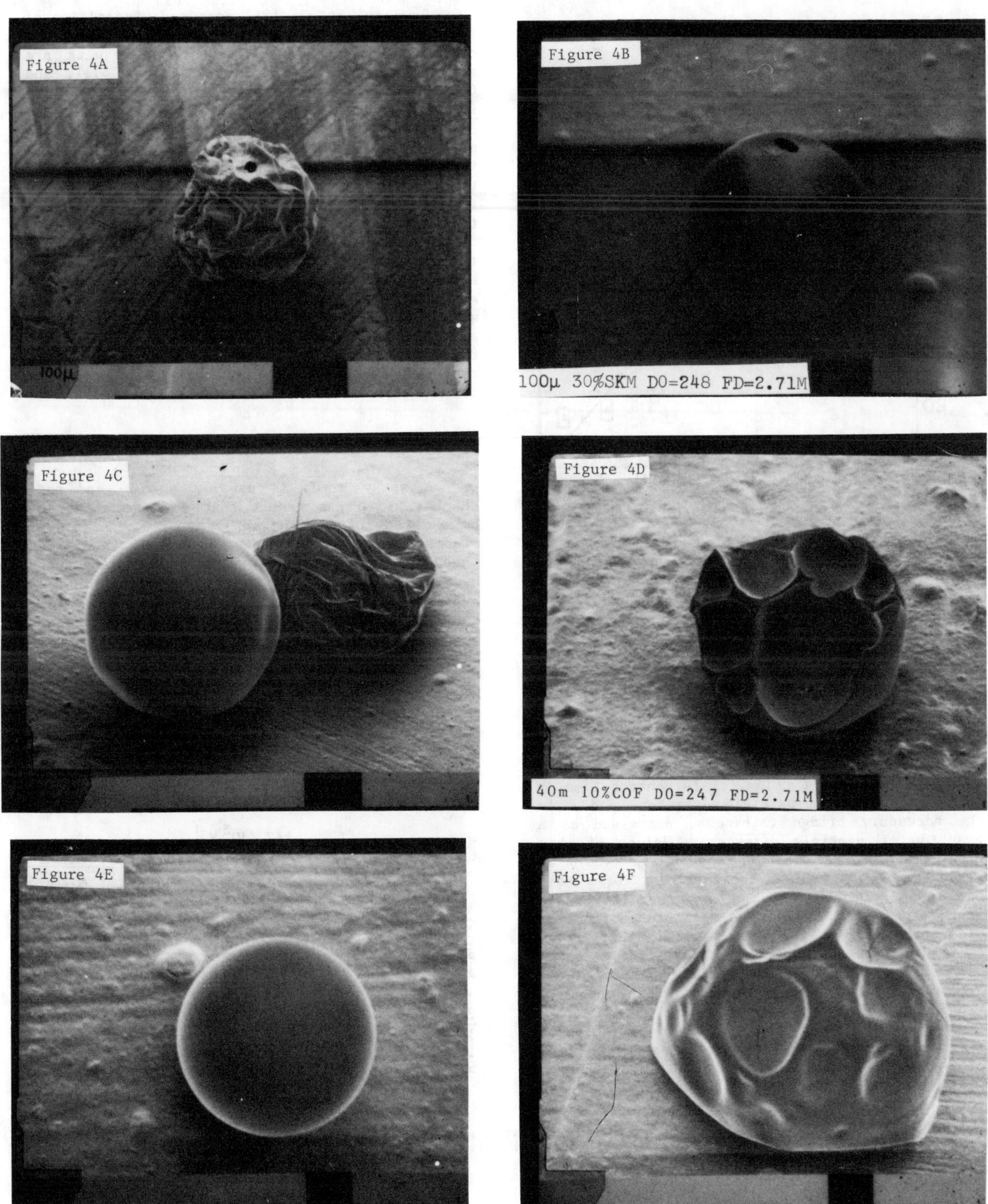

Figure 4. Examples of Different Particle Morphologies.

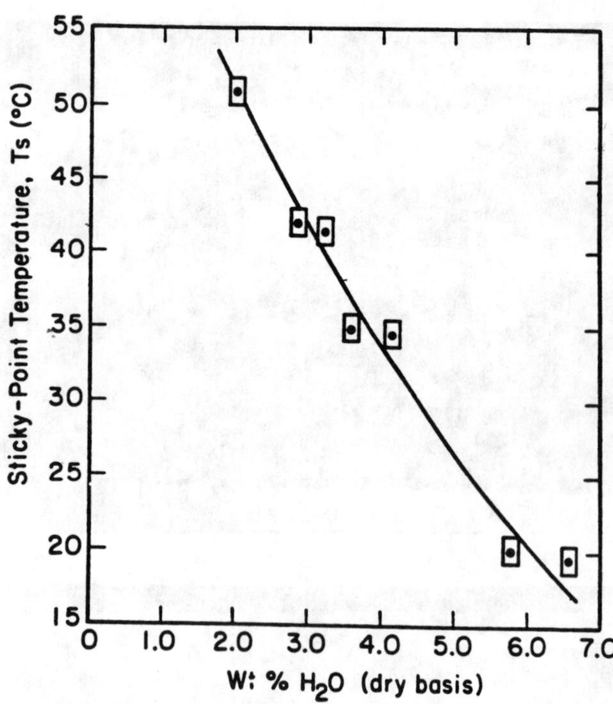

Figure 5. Experimental Sticky-Point Measurements for a Powder Composed of a Homogeneous 7:1 (w/w) Sucrose:Fructose Mixture [36].

analyses of collapse during freeze drying [34] and of the development of surface folds during spray drying [33]. The applicable dimensionless group is

$$\frac{\sigma t_c}{\eta f d}$$

where σ is surface tension, t_c is contact time, η is viscosity, d is particle diameter, and f is the necessary bridge thickness, expressed as a fraction of particle diameter. Values of this group much greater than one cause stickiness or agglomeration. Values substantially less than one do not. Because of the large sensitivity of viscosity to temperature and moisture content, an order-of-magnitude analysis is sufficient. The similarity between experimental moisture content-temperature relationships describing stickiness and structural collapse has also been noted by Tsouroflis, et al. [37].

The size of a bridge needed to cause sticking of particles may be estimated as 0.1 to 1.0% of the particle diameter, on the basis of the investigations by Schubert [38] and Rumpf [39] on the strength of bridges between particles. Building that factor into the flow model (f = 0.001 to 0.01) yields a predicted critical viscosity in the range of 10^9 to 10^{11} mPa s. For amorphous particles with concentrate viscosities greater than this value, there will be insufficient flow to cause sticking during short (1 to 10s) contacts. For viscosities below this critical value, flow and stickiness will occur.

Downton, et al. [36] measured viscosities of bulk quantities of highly concentrated solutions of 7:1 sucrose:fructose mixture. Results are shown by the curves for 2.37, 3.64, 6.11, and 6.99% water (w/w) in solution, given in Figure 6. Also shown in Figure 6 are the experimentally determined sticky points (from Figure 5) for the same mixture. It can be seen that the critical viscosities for stickiness do indeed fall centrally within the predicted range and are relatively constant. This lends strong confirmation to the postulated mechanism.

From this mechanism of stickiness one can rationalize why certain processing approaches help the stickiness problem. These include the addition of small amounts of high-molecular-weight material and the introduction of cool air along the walls of the dryer, both of which can raise viscosities by orders of magnitude. A more rational design of agglomerators should also be possible with this mechanistic insight.

6. MIXING AND FLOW PATTERNS OF AIR AND SPRAY

A spray dryer is usually a wide and open chamber, with the result that a large amount of bulk mixing of air and spray occur. A number of studies [e.g. 40,41] have been made of mixing phenomena in the entire drying chamber, and techniques for modelling these phenomena have been put forward by Crowe [42] and others. The mixing and flow patterns are important for food-product quality for many reasons--among them the facts that they determine particle residence times and the driving forces (temperature, humidity) for the mass and heat transfer phenomena which govern drying rates and rates of volatiles loss.

From the standpoint of volatiles retention and the initial steps which affect ultimate particle morphology, the most critical region is that

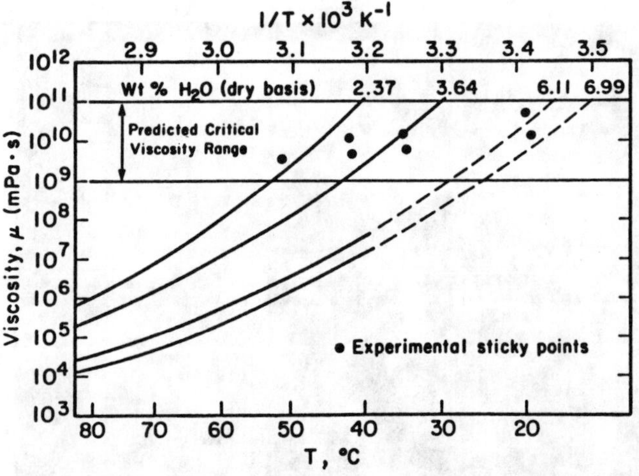

Figure 6. Experimentally Determined Viscosities and Sticky Points for Concentrates Composed of a Homogeneous 7:1 (w/w) Sucrose:Fructose Mixture [36].

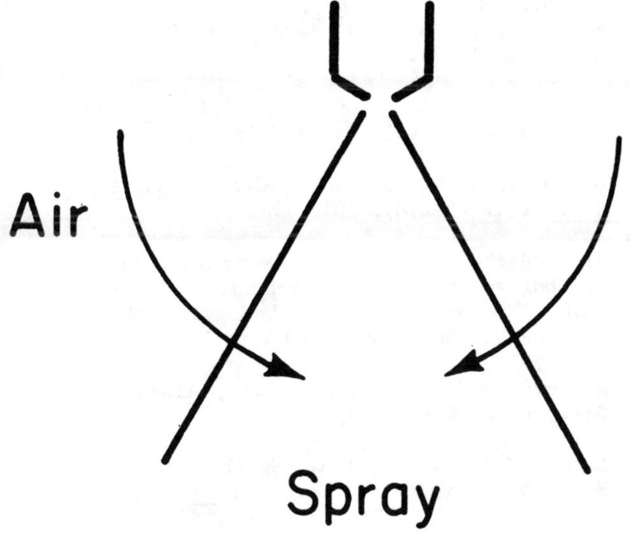

Figure 7. Entrainment of Air Into the Spray, Caused by Drop Deceleration in the Region Near the Atomizer.

very near the atomizer. When a pressure atomizer is used, this is a zone where the drops slow from the initial velocity at which they emanate from the atomizer, decelerating toward the terminal velocity. The deceleration effect brings large quantities of air into the spray region, through a momentum-exchange process [43] (See Figure 7). This phenomenon has been modelled by Rothe and Block, on the bases of a number of assumptions [44]. Zakarian [45] has modified their model by allowing for the distribution of drop sizes. However, these models still do not allow for the effect of turbulent mixing in the atomizer region, or for the fractionation which occurs among drops of different sizes because the drag of the entrained air draws the smaller particles preferentially inward.

Etzel [46] has observed the importance of comparing the rate of entrainment of air into the spray, on the one hand, with the feed rate of air, on the other hand. When the rate of air entrainment becomes comparable to or larger than the air feed rate, it will be necessary for there to be a large recycle of air toward the nozzle zone from locations lower in the dryer. The recycle flow is driven by the pumping action of the slowing spray, and serves to bring cooler and more humid air into the atomizer region. This should have a substantial influence on the driving forces for heat and mass transfer, and should thereby affect volatiles loss and other phenomena discussed in this paper.

ACKNOWLEDGMENT

The research from the Berkeley group described herein has been supported financially by the Chemical and Biochemical Processes Program, Division of Chemical and Process Engineering, National Science Foundation, U.S.A. Supplemental support was received through fellowships from General Foods Corporation, Conselho Nacional de Desenvolvimento Cientifico e Tecnologico (Brasil), and the Graduate Minority Program of the University of California.

REFERENCES

1. King, C.J., Kieckbusch, T.G., and Greenwald, C.G., Food-Quality Factors in Spray Drying, in Advances in Drying, ed. A.S. Mujumdar, vol. 3, Hemisphere Publ. Co., New York (1984).

2. Bruin, S., and Luyben, K.Ch.A.M., Drying of Food Materials: A Review of Recent Developments, in Advances in Drying, ed. A.S. Mujumdar, vol. 1, Hemisphere Publ. Co., New York (1980).

3. Thijssen, H.A.C., Lebensm. Wiss. u. Technol., vol. 12, 308 (1979).

4. Kerkhof, P.J.A.M., and Schoeber, W.J.A.H., Theoretical Modelling of the Drying Behaviour of Droplets in Spray Drying, in Advances in Preconcentration and Dehydration of Foods, ed. A. Spicer, pp. 349-397, Appl. Sci. Publs., London (1974).

5. Ranz, W.E., and Marshall, W.R., Jr., Chem. Eng. Prog., vol. 48, 141, 173 (1952).

6. Miura, K., Miura, T., and Ohtani, S., AIChE Symp. Ser., vol. 73 (163), 95 (1977).

7. Katta, S., and Gauvin, W.H., AIChE Jour., vol. 21, 143 (1975).

8. Pham, Q.T. and Keey, R.B., Trans. Instn. Chem. Engrs., vol. 55, 114 (1977).

9. Sano, Y. and Keey, R.B., Chem. Eng. Sci., vol. 37, 881 (1982).

10. Bomben, J.L., Bruin, S., Thijssen, H.A.C., and Merson, R.L., Aroma Recovery and Retention in Concentration and Drying of Foods, in Advances in Food Research, eds. G.F. Stewart, et al., vol. 20, Academic Press, New York (1973).

11. Thijssen, H.A.C., and Rulkens, W.H., De Ingenieur, (The Hague), vol. 80, Ch 45 (1968).

12. Chandrasekaran, S.K., and King, C.J., AIChE Jour., vol. 18, 513 (1972).

13. Kerkhof, P.J.A.M., A Quantitative Study of the Effect of Process Variables on the Retention of Volatile Trace Components in Drying, Ph.D. dissert., Tech. Univ. Eindhoven, Netherlands (1975).

14. Kieckbusch, T.G., and King, C.J., AIChE Jour., vol. 26, 718 (1980).

15. Zakarian, J.A., and King, C.J., Ind. Eng. Chem. Process Des. & Devel., vol. 21, 107 (1982).

16. Frey, D.D., and King, C.J., The Effects of Surfactants on Mass Transfer During Spray Drying of Aqueous Sucrose Solutions, A.I.Ch.E. Natl. Mtg., Denver, CO, August (1983).

17. Etzel, M.R., and King, C.J., Loss of Volatile Trace Organics During Spray Drying, Ind. Eng. Chem. Process Des. & Devel., in press (1984).

18. Frey, D.D., Experimental and Theoretical Investigations of Foam Spray Drying, Ph.D. dissert. in Chemical Engg., Univ. of California, Berkeley (1984).

19. Furuta, T., Okazaki, M., Toei, R., and Tsujmoto, S., Proc. 3rd Intl. Cong. on Engg. & Food, Paper 4-6, Dublin, Ireland, Intl. Union of Food Sci. & Technol., September 1983.

20. Duffie, J.A., and Marshall, W.R., Jr., Chem. Eng. Prog., vol. 49, 417, 480 (1953).

21. Crosby, E.J., and Marshall, W.R., Jr., Chem. Eng. Prog., vol. 54(7), 56 (1958).

22. Charlesworth, D.H., and Marshall, W.R., Jr., AIChE. Jour., vol. 6, 9 (1960).

23. Buma, T.J., and Henstra, S., Netherlands Milk & Dairy J., vol. 25, 278 (1971).

24. Verhey, J.G.P., Netherlands Milk and Dairy J., vol. 25, 246 (1971); ibid., vol. 26, 186, 203 (1972); ibid., vol. 27, 3 (1973).

25. Jensen, J.D., Food Technol., vol. 29(6), 60 (1975).

26. Abdul-Rahman, Y.A.K., Crosby, E.J., and Bradley, R.L., Jr., J. Dairy Sci., vol. 54, 1111 (1971).

27. Crosby, E.J., and Weyl, R.W., AIChE Symp. Ser., vol. 73(163), 82 (1977).

28. Toei, R., Okazaki, M., and Furuta, T., Drying Mechanism of a Non-Supported Droplet, in Proc. First Intl. Symp. on Drying, ed. A.S. Mujumdar, pp. 53-58, Science Press, Princeton, NJ, (1978).

29. Toei, R., and Furuta, T., AIChE Symp. Ser., vol. 78(218), 111 (1982).

30. Furuta, T., Okazaki, M., Toei, R., and Crosby, E.J., in Drying '82, ed. A.S. Mujumdar, pp. 157-164, Hemisphere Publ. Co., New York (1983).

31. Greenwald, C.G., and King, C.J., J. Food Process. Eng., vol. 4, 171 (1981).

32. Greenwald, C.G., and King, C.J., AIChE Symp. Ser., vol. 78(218), 101 (1982).

33. Alexander, K., Factors Governing Surface Morphology in the Spray-Drying of Foods, Ph.D. dissert. in Chemical Engg., Univ. of California, Berkeley (1983).

34. Bellows, R.J., and King, C.J., AIChE Symp. Ser., vol. 69(132), 33 (1973).

35. Lazar, M.E., Brown, A.H., Smith, G.S., Wong, F.F., and Lindquist, F.E., Food Technol., vol. 10(3), 129 (1956).

36. Downton, G.E., Flores-Luna, J.L., and King, C.J., Ind. Eng. Chem. Fundam., vol. 21, 447 (1982).

37. Tsouroflis, S., Flink, J.M., and Karel, M., J. Sci. Food Agric., vol. 27, 509 (1976).

38. Schubert, H., Chem. Ing. Tech., vol. 45(6), 396 (1973).

39. Rumpf, H., Chem. Ing. Tech., vol. 46(1), 1 (1974).

40. Keey, R.B., and Pham, Q.T., Chem. Eng. Sci., vol. 32, 1219 (1977).

41. Paris, J.R., Ross, P.N., Dastur, S.P., and Morris, R.L., Ind. Eng. Chem. Process Des. & Devel., vol. 10, 157 (1971).

42. Crowe, C.T., Modeling Spray-Air Contact in Spray-Drying Systems, in Advances in Drying, ed. A.S. Mujumdar, vol. 1, Hemisphere Publ. Co., New York, (1980).

43. Briffa, F.E.J., and Dombrowski, N., AIChE Jour., vol. 12, 708 (1966).

44. Rothe, P.H., and Block, J.A., Int. J. Multi-Phase Flow, vol. 3, 263 (1977).

45. Zakarian, J.A., Volatiles Loss in the Nozzle Zone During Spray Drying of Emulsions, Ph.D. dissert. in Chemical Engg., Univ. of California, Berkeley (1979).

46. Etzel, M.R., Volatiles Loss During Spray Drying of Liquid Foods, Ph.D. dissert. in Chemical Engg., Univ. of California, Berkeley (1983).

MODELLING CONTINUOUS CONVECTION DRYERS FOR PARTICULATE SOLIDS - PROGRESS AND PROBLEMS

David Reay

Engineering Sciences Division, A.E.R.E. Harwell,
Oxfordshire OX11 ORA, ENGLAND.

ABSTRACT

This paper first outlines the potential value of fundamental models of drying operations and identifies the information which they must contain. The latter is divided into feedstock characteristics and equipment characteristics. Our present ability to formulate reliable feedstock models and equipment models for fluid bed dryers, pneumatic conveying dryers, rotary dryers and spray dryers is then assessed.

1. TYPES OF MODEL

A model of a drying operation is a theoretical device which permits prediction of how a particular piece of hardware will perform on a particular feedstock with specified operating conditions. Three types of model may be identified:

(a) scale-up factor models;
(b) input/output models;
(c) fundamental models.

In practice, most dryers are designed today using simple scale-up factor models and operated using simple input/output models. Fundamental models are usually much more complicated and are only just beginning to have an impact on practical dryer design and operation. To understand the driving force for developing them it is essential to appreciate the limitations of the simpler types of model.

The dryer manufacturer uses a scale-up factor model when he conducts pilot plant tests, varying operating conditions until the product specification is met, and then estimates the dimension of the full-scale dryer using empirical scale-up rules based on his past experience. This generally results in a workable design, but there are some drawbacks. Firstly, the region within which he knows his scale-up rules work is circumscribed by the range of his past experience with regard to both feedstocks and size of the full-scale plant; going outside this range is a gamble for both the vendor and his client, and if the gamble fails the consequences are costly for both. Secondly, the customer has no technical basis for assessing the adequacy of a proposed design because normally the vendor will not reveal his jealously guarded scale-up factors; the divergence in equipment sizes proposed by different vendors for a specified duty is evidence of the diversity of scale-up rules in use for nominally similar equipment. Thirdly, when the customer has his dryer he has no basis for assessing the effects of future changes in feedstock or operating conditions.

In practice, the customer usually learns about the impact of changes in feedstock and operating conditions through a posteriori correlation of inputs and outputs. This is the input/output model. The model may be an intuitive understanding gained by plant operators after long experience, or occasionally it may be formalised in an operations research model after a series of planned variations. Again, the model cannot safely be extrapolated outside the range of past experience, and this limits its usefulness. There is also the danger of producing off-specification product during the learning phase.

A fundamental model is one which is based on a mathematical representation of the physical processes occurring in the dryer. It is grounded in theory, and so may be contrasted with the empirical correlation of experience which is at the heart of the other two types of model. It requires some experimental data, particularly on the characteristics of the feedstock, but in principle these can be obtained from bench-scale or pilot plant tests. It ought to be a more powerful tool than the other two types of model, since in theory its range of application is not circumscribed by past experience. However, for many drying operations we are not yet able to construct satisfactory fundamental models, either because we do not know enough about the physical processes in the equipment or because we have not yet devised suitable techniques for acquiring data on the feedstock. The purpose of this paper is to review progress and outstanding problems in this area.

2. STRUCTURE OF A FUNDAMENTAL MODEL

It is important to realise at the outset that in any continuous drying situation a specific feedstock is being dried in a specific type of dryer, and that to model the operation in a fundamental manner we need information on the characteristics of both the feedstock and the dryer. In fact, we need two models which may be called the feedstock model and the equipment model respectively. The relationship between them is shown in Fig. 1.

The feedstock model describes in mathematical terms the drying characteristics of the feedstock. It will be concerned with drying kinetics and equilibria, and with any morphological, structural

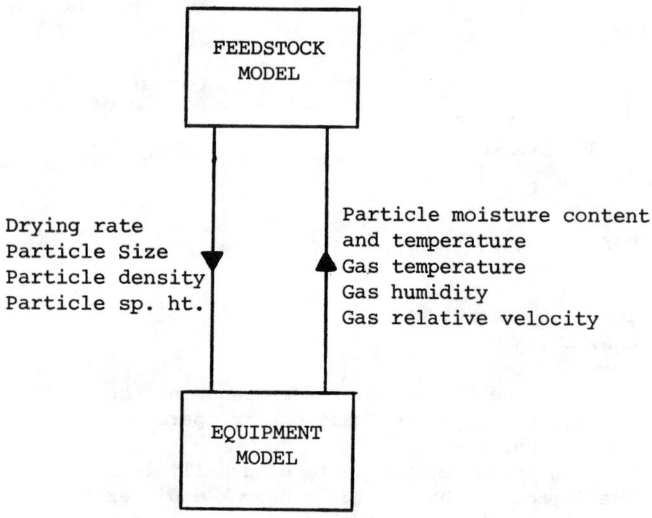

Fig. 1 Components of a fundamental model of a drying operation

and composition changes in the particles during drying. Its main function is to calculate the drying rate which will be observed at any particle moisture content and temperature and local conditions of gas temperature, humidity and relative velocity. A secondary function is to generate new values of the particle size, density and specific heat if these have changed as a result of changes in the particle moisture content and temperature. Ideally, the feedstock model should be independent of the type of dryer and its parameters should be determinable from bench-scale tests on a small quantity of feedstock.

The equipment model describes in mathematical terms the dryer in which the feedstock is being processed. It will contain information on gas-to-particle heat transfer, aerodynamics and particle transport and will state explicitly how these are related to equipment design variables. Its function is to calculate the change in particle moisture content in a given dryer volume (or element of volume), given estimates of the drying rate and particle properties generated by the feedstock model; alternatively, it may calculate the dryer volume needed to accomplish a given change (or incremental change) in particle moisture content. In either case it generates concurrently the changes in particle temperature, gas temperature, gas humidity and relative velocity to be fed back into the feedstock model for the next stage in the simulation. The basic set of equations for calculating these changes depends on whether the particles and gas are flowing co-currently, counter-currently or in cross-flow. The equipment model calls on the feedstock model for numerical values of the drying rate and particle properties, but its form is independent of the specific feedstock. If its parameters cannot be calculated from existing theory they will usually have to be estimated from measurements on a pilot plant or a full-scale plant and limits placed on extrapolation of the data.

A good fundamental model of a drying operation can only be constructed if we can devise both a good feedstock model and a good equipment model. If there are deficiencies in either, the overall model will be of strictly limited value. We now consider the problems of formulating a feedstock model, and after that we look at modelling of some common particulate solids dryers.

3. FEEDSTOCK MODELS

3.1 Introduction

A complete feedstock model would contain the following elements:

(a) one or more batch drying curves obtained on a small quantity of feedstock under controlled conditions of temperature, humidity and gas velocity;

(b) an algorithm for predicting the drying rate at any operating condition and moisture content, based on the experimental data obtained in (a);

(c) if the feedstock is hygroscopic, data on its equilibrium moisture content as a function of temperature and relative humidity;

(d) information on changes in particle size, density and specific heat with declining moisture content.

The first two elements will now be considered in more detail.

3.2 Measurement of Batch Drying Curves

There are several experimental techniques which have been developed:

(a) Pass air across a tray containing a sample of the material, with periodic or continuous weighing of the tray.

(b) Suspend a drop of the material in an air stream and weigh the drop periodically or continuously.

(c) Pass air through a fluidised or spouted bed of the material and periodically withdraw a sample for off-line determination of its moisture content.

(d) Pass air either upwards through a fluidised bed or downwards through a layer of the material on a mesh, with continuous measurement of the humidity of the exhaust air.

If we wish to construct a model of band drying, fluid bed drying or spouted bed drying there is clearly no difficulty in choosing a batch drying curve measurement technique which simulates the gas-solids contacting pattern in the dryer and which measures drying kinetics over a time-scale of the same order of magnitude as particle residence times in such dryers. Hence, extrapolation of the experimental data is minimised, which is desirable for confidence in the feedstock model. Unfortunately, the same cannot be said of some other particulate solids dryers.

Technique (b) simulates the gas-solids contacting pattern in a spray dryer, but at present it needs further development to be practicable for drop sizes below 1 to 2 mm and drying times below a few minutes. The latter restriction means that drying conditions have to be much milder than in commercial spray dryers. Hence, the experimental data have to be extrapolated considerably with regard to both particle size and operating conditions and this may introduce uncertainty into the feedstock model.

None of the above techniques simulates the gas-solids contacting pattern in a pneumatic conveying dryer. Furthermore, the particle residence time of 1 to 2 seconds is nearly two orders of magnitude lower than the time scale over which the fastest of the above techniques (technique (d) with a fast response hygrometer) operates. Hence, much milder drying conditions have to be used and considerable extrapolation is again required.

Also, none of the above techniques simulates the gas-solids contacting pattern in a cascading rotary dryer, although workers at the University of Lund in Sweden are attempting to move closer to it by using technique (d) with a pulsed air supply to simulate the periodic cascading of particles through air (1). Particle residence times in rotary dryers are several minutes, so there is no problem of time scale.

In summary, there may be considerable uncertainty at present in feedstock models using data gathered by present-day experimental techniques when these models are applied to several important particulate solids dryers. The uncertainty can be removed either by developing more appropriate experimental techniques or by demonstrating that the rather large extrapolations needed at present can be made successfully. The latter requires:

(a) reliable models or correlations for gas-solids heat transfer coefficients in commercial dryers to permit transformation of data from one gas-solids contacting pattern to another;

(b) reliable models for moisture migration in particles to permit transformation of data from mild to much more severe drying conditions.

Item (a) is reviewed in the sections on dryer models, while item (b) is discussed below.

3.3 Extrapolation of a Batch Drying Curve to other Drying Conditions

The ability to extrapolate measured drying kinetics data to conditions not covered experimentally is required not only for the purpose outlined above, but also to permit estimation of drying rates at the wide variety of conditions encountered by the particles as they travel through a co-current or counter-current dryer.

The simplest extrapolation technique is to assume that the characteristic drying curve hypothesis holids (2,3). This hypothesis postulates that at any time the drying rate is given by:

$$-\frac{dw}{dt} = \left(-\frac{dw}{dt}\right)_c f\left(\frac{w - w_e}{w_{cr} - w_e}\right) \qquad (1)$$

where w_{cr} = critical moisture content, w_e = equilibrium moisture content and $(-dw/dt)_c$ is the drying rate which would be observed at the prevailing conditions if the particles were still wet enough to exhibit constant rate drying. The function $f = 1$ for $w > w_{cr}$ and $1 < f < 0$ for $w_{cr} < w < w_e$. Essentially, the hypothesis states that a change in external conditions in the falling rate period will have the same proportional effect on the drying rate as in the constant rate period.

This hypothesis has been used frequently in recent drying literature, but is it correct? The meagre experimental evidence currently available suggests that it holds approximately for some materials but not for others. Ashworth and Carter (4) found it to work reasonably well for silica gel granules dried at various air temperatures (but always the same gas velocity) in a tray dryer. Schlünder (5) also observed that it successfully predicted the effect of air temperature changes on the drying of molecular sieve particles in a fluid bed; however, it failed for aluminium silicate particles dried in the same bed over the same range of temperatures. To date, no systematic experimental programme has been carried out on a wide range of materials and operating conditions to provide guidelines on when this simple extrapolation hypothesis may be used; such a programme is urgently needed. Until we understand when and why it works, so-called 'fundamental' models based on it will still contain a substantial empirical element.

The alternative is to base extrapolation on a mechanistic model of the drying process. There are three basic types of model:

(a) wetted surface models;

(b) receding evaporative interface models;

(c) diffusion models.

These have been reviewed critically by van Brakel (6).

The wetted surface model assumes that evaporation takes place only at the surface, and that liquid reaches the surface by capillary action. The internal resistance to liquid movement is assumed to be negligible. Basing extrapolation on this model has the same result as basing it on the characteristic drying curve hypothesis.

The receding evaporative interface model assumes that evaporation occurs only at the surface of a shrinking wet core which is surrounded by a dry shell of material. The internal resistance to liquid movement is assumed to be so large that the liquid is effectively stationary. The vapour diffusivity in the dry shell is determined by fitting the model to an experimental batch drying curve. If this model is a good description of what is happening inside the particles we would not expect the characteristic drying curve hypothesis to work for gas velocity changes, except when the dry shell is very thin. On the other hand, if the evaporative interface stays near the wet bulb temperature the hypothesis may give quite a good prediction of the effects of temperature and humidity changes in the gas stream.

The diffusion model assumes that moisture transport is by diffusion through the solid and that at every point the diffusion rate is proportional to the moisture concentration gradient. The moisture diffusivity is found by fitting the model to experimental batch drying data. It is often reported that to fit this model to a batch drying curve it must be assumed that the diffusivity declines exponentially with decreasing moisture content; this may be due to the material shrinking as it dries, or it may be due to an incorrect choice of model. If this model is a good description of what is happening inside the particles we would not expect the characteric drying curve hypothesis to work for changes in gas velocity, humidity or temperature.

Hence, by measuring several batch drying curves at various gas velocities and temperatures we can both check the validity of extrapolation using the characteristic drying curve hypothesis and, by noting where it works and where it does not, form an idea of the moisture removal mechanism and select an appropriate model. Caution should be excerised, however, if the batch drying curves were measured at much milder conditions than those which will be used in the commercial dryer; the moisture removal mechanism may be different at the more severe conditions. Suppose, for example, that the characteristic drying curve hypothesis holds well for both velocity and temperature changes at the mild conditions. This may be because the wetted surface model is valid at these conditions, i.e. liquid transport to the surface is not a rate-limiting step. At the more severe conditions the evaporation rate may be much faster than the liquid transport rate and a receding evaporative interface model may be more applicable; then the effect of gas velocity changes on drying rate will be quite different.

4. FLUID BED DRYERS

Developing a fundamental model of a fluid bed drying operation is relatively straightforward since the gas-solids contacting pattern can be duplicated quite well on the bench scale, the time-scale of the drying kinetics is no problem and particle transport through the dryer follows simple laws.

In a plug flow fluid bed dryer the particles flow along a fairly long and narrow channel and the bed is normally quite shallow. If the particles could be assumed to be in perfect plug flow the variation of their moisture content w with time t in the bed would be the same as the variation of w with t in a batch drying test conducted at the same bed depth, gas velocity and air inlet temperature. Hence, the required bed area for a given feed rate could be computed easily from such a test. In fact, there is always some axial dispersion of the particles. The extent of the axial dispersion is characterised by the axial dispersion number:

$$B = \frac{D_p \bar{t}}{L^2}$$

where D_p is the particle diffusivity and L is the bed length and $\bar{t}$ is the mean particle residence time. D_p may be estimated by the following correlation (7):

$$D_p = 3.71 \times 10^{-4} \frac{(u_G - u_{mf})}{u_{mf}^{1/3}} \ m^2/s \qquad (2)$$

where u_G = superficial gas velocity (m/s) and u_{mf} = minimum fluidisation velocity (m/s). The particle residence time distribution function $E(t)$ is then given by:

$$E(t) = \frac{1}{2\sqrt{\pi B}} \ exp \left[- \frac{(1-t/\bar{t})^2}{4B}\right] \qquad (3)$$

which is derived by analogy with chemical reactor theory. The mean product moisture content $\bar{w}_o$ is given by:

$$\bar{w}_o = \int_0^\infty w(t) \ E(t) \ dt \qquad (4)$$

where $w(t)$ is the afore mentioned batch drying curve.

It remains to predict how $w(t)$ will vary if the bed depth and gas velocity differ from those used in the batch test. Reay and Allen (8) studied the fluid bed drying of several materials and concluded that as a rough approximation they could be divied into two classes:

(a) Materials such as iron ore and ion exchange resin which lose their water comparatively easily; most of the drying occurs within a short distance of the distributor and the gas leaving the dense phase is probably close to equilibrium with the surface of the solid; the time needed to reach a given moisture content is proportional to the bed weight per unit area and inversely proportional to the gas velocity.

(b) Materials with a high internal resistance to mass transfer such as wheat, which dry very slowly; the gas leaving the dense phase is far from equilibrium with the solids and $w(t)$ is independent of the bed weight per unit area and the gas velocity.

Of course, it is conceivable that some type (a) materials may exhibit type (b) behaviour towards the end of drying if the last portion of moisture is difficult to remove.

The inlet air temperature is not usually a design variable with plug flow fluid bed dryers because it is almost always cheapest to set it at the highest value possible consistent with not damaging either the product or the dryer.

The other main type of fluid bed dryer is the well-mixed variety, in which the particle residence time distribution follows the perfect mixing law:

$$E(t) = \frac{1}{\bar{t}} \ exp \left(- \frac{t}{\bar{t}}\right) \qquad (5)$$

In this type of dryer the bed temperature is uniform, so the batch drying curve $w(t)$ must be measured at constant bed temperature rather than constant air inlet temperature. In practice, it is very difficult to keep the bed temperature constant in the early stages of measuring a batch drying curve. To overcome this problem Reay and Allen (9) have devised, and verified by experiment, a technique for transforming a measured constant air inlet temperature batch drying curve into a constant bed temperature batch drying curve at any desired bed temperature. The same authors (8) have also confirmed, by experiment with several feed-stocks in a pilot plant, that when $w(t)$ curve obtained in this manner is combined in equation (4) with the perfect mixing $E(t)$ defined by equation (5) a reasonably good prediction of the observed mean product moisture content $\bar{w}_o$ is obtained.

5. PNEUMATIC CONVEYING DRYERS

The chief requirements of an equipment model for pneumatic conveying dryers are methods for calculating the gas-to-particle heat transfer coefficient and the particle velocity up the tube.

All models published to date have estimated the gas-to-particle heat transfer coefficient by means of the Ranz-Marshall equation:

$$Nu = 2.0 + 0.6\, Re_p^{1/2} Pr^{1/3} \qquad (6)$$

However, this correlation was obtained from experiments on convective heat transfer to an isolated stationary spherical particle in a flowing gas (10). There are comparatively few studies of heat transfer from gas to solids in a flowing suspension. These few studies have been reviewed by Bandrowski and Kaczmarzyk (11), who also contributed measurements of their own. There is approximate concurrence that $Nu \propto Re_p^{0.8}$, which contrasts sharply with the relation between Nu and Re_p predicted by the Ranz-Marshall equation and must raise doubts about the validity of the latter for predicting gas-to-particle heat transfer rates in pneumatic conveying dryers. The proportionality constant was found by Bandrowski and Kaczmarzyk to decline with increasing particle concentration, which may be due to increased damping of the free-stream turbulence by the particles.

Particle transport velocities in the steady particle velocity region can now be predicted to within 20% for small particles. The work of Reay and Bahu presented at this symposium (12) shows conclusively that solids-wall friction should be included in the particle transport model for all but the largest dryers, and certainly if the overall model is to be used for scale-up from pilot plant data. Their solids-wall friction factor correlation provides a suitable basis for taking this effect into account. Further work is needed on particle transport in the acceleration zone, which is where much of the heat and mass transfer is likely to occur.

As mentioned in section 3.2, the feedstock model for a pneumatic conveying drying operation requires a great deal of extrapolation from currently-feasible bench-scale kinetics measurements. There is no reason in principle why this cannot be done, given reliable correlations for the gas-to-particle heat transfer coefficient in the two situations, but it has not yet been demonstrated successfully.

A particular problem arises with those feedstocks which enter the dryer as agglomerates of unpredictable size and which then break up in the dryer in a manner which cannot yet be predicted. At present we cannot predict heat transfer rates, drying rates or particle velocities for such materials. This situation is quite common, for example in the drying of filter cake where the wet cake is mixed with recycled dry product to make it dispersable in the dryer. A research study on this topic would be desirable.

Despite the uncertainties still requiring resolution, encouraging results have been reported in two recent attempts to simulate pilot plant pneumatic conveying dryers using fundamental models (13,14). Martin and Saleh (13) studied the drying of chalk and PVC and based their feedstock model on batch drying curves obtained in a bench-scale fluid bed dryer; this has the disadvantage that with fine particles the gas leaving the fluid bed is close to equilibrium, so it is difficult to determine the kinetics accurately. Andrieu and Bressat (14) also studied PVC drying, but for their feedstock model they assumed constant rate drying only, arguing that in such a short residence time only moisture on or near the surface would be removed.

Martin and Saleh predicted the general shape of their measured gas temperature, humidity and solids moisture content profiles correctly but did not get good quantitative agreement. Andrieu and Bressat obtained better quantitative agreement with their measurements, but they only began their simulation at 1 metre above the feed point, the measured values at that location being used to start the calculation. Hence, they did not attempt to simulate the crucial acceleration zone. Nevertheless, the results obtained by both sets of authors are sufficiently encouraging to raise hopes that a successful fundamental model for pneumatic conveying drying operations may not be too far away.

6. ROTARY DRYERS

Baker (15) has provided a very thorough critical review of current scientific understanding of the physical phenomena occurring in rotary dryers. The present section is of necessity very much briefer, and readers requiring more details should consult his work.

As with pneumatic conveying dryers, the chief problems in constructing a fundamental equipment model for rotary drying operations lie in obtaining reliable prediction of the gas-to-particle heat transfer coefficient and the particle transport velocity along the rotating drum. However, the motion of particles in a rotary dryer and their interaction with the air stream are more complex than in a pneumatic conveying dryer and current predictive models are very unreliable.

Most correlations for gas-to-particle heat transfer in rotary dryers take the form:

$$\alpha_v = \frac{K\, u_G^{\,n}}{d_d} \qquad (7)$$

where α_v is a heat transfer coefficient per unit dryer volume, u_G is the gas velocity and d_d the drum diameter. K and n are empirical constants. The values of n derived by various authors by fitting equation (7) to experimental data (often on rotary coolers) vary from zero to 0.8. Baker (15) has shown that the resulting predictions of α_v differ by an order of magnitude when the various correlations are applied to an identical situation.

Schofield and Glikin (16) used a more fundamental approach. From a model of flight lifting and cascading they estimated the surface area of particles in contact with the gas at any time. This was combined with an area-based single particle heat transfer coefficient to calculate the heat transfer rate, basing the Reynolds number term on the average gas velocity u_G in the drum.

Heat transfer rates calculated in this way were stated to be "much larger" than those observed in practice on an industrial rotary cooler, although no numbers are quoted. The most likely reason for the discrepancy was thought to be particle shielding, i.e. particles in the centre of the cascades not experiencing the full air velocity u_G.

This lack of understanding of the aerodynamics of gas flow through and around cascades of particles falling perpendicular to the gas stream is also a severe problem when attempting to predict particle transport rates in rotary dryers. Forward particle motion by simple cascading, i.e. in the absence of air flow, can be analysed reasonably well (16,17), but estimating the additional component of particle motion caused by air drag is difficult. Schofield and Glikin (16) used the standard drag curve for an isolated particle to estimate this effect, but did not test their predictions against experimental data. Kelly and O'Donnell (18) tried to allow for particle shielding by using a momentum balance, equating the pressure drop experienced by air flowing through the cascades to that of air flowing between them in order to estimate the true air velocity through the cascades; they also allowed for kiln action (important in an overloaded drum) and particle bouncing (important in an underloaded drum) but nevertheless their predictions of particle residence time were still 15 to 30% higher than experimental values measured in a pilot plant.

Thus, we may conclude that at present there is no satisfactory general equipment model for rotary dryers. The lack of a reliable general method of predicting heat transfer coefficients also creates uncertainty in the use of bench-scale drying kinetics data. Hence, it is not possible at present to construct reliable fundamental models of rotary drying operations.

7. SPRAY DRYERS

Spray drying operations are very difficult to model on a fundamental basis. There are severe problems in formulating both feedstock models and equipment models.

The first problem in feedstock models is the large extent of extrapolation, with regard to both drop size and drying time, needed from bench-scale tests using current techniques for acquiring drying kinetics data. Improved bench-scale techiques are being developed, for example laser observation of a levitated drop as reported by Toei, Okazaki and Furuta (19), but there is still a long way to go. Furthermore, these techniques generally use single droplets, whereas in a real spray dryer the drying kinetics may be influenced strongly by statistical events such as air bubble incorporation and shell rupturing which require averaging over many droplets to obtain an adequate picture.

Nevertheless, useful progress has been made in recent years in developing more sophisticated models of moisture migration in drying droplets which are helping to bridge the gap between bench-scale tests and spray drying practice. These are diffusion models with an exponential relation between moisture diffusivity and moisture content, which is a physically reasonable approach for colloidal food and dairy feedstocks, but may be less appropriate for slurries and salt solutions. Particular mention may be made of the regular regime approach developed by Schoeber and Thijssen (20), the short-cut method proposed by Schoeber (21) for estimating moisture diffusivity and the diffusion models accounting for particle inflation developed by van der Lijn (22) and Sano and Keey (23). The latter authors also developed from their model a proposed relation between particle size and drying time.

Even if we have an adequate moisture migration model, however, there still remains the problem of estimating the initial drop size and size distribution with which to start the simulation. Existing atomization correlations can only give very approximate predictions, although the situation is improving as more research is done on the atomization of non-Newtonian materials representative of spray dryer feedstocks (24,25). Furthermore, we are not yet able to predict the changes in particle size and density due to puffing, shrinking, cracking, shrivelling or rupturing during drying.

With regard to the equipment model, there seems little reason to doubt the validity of the Ranz-Marshall equation for predicting gas-to-particle heat transfer coefficients. The main problem is that gas flow patterns in industrial spray dryers are quite complicated. There is substantial turbulence, swirl and recirculation, particularly near the atomizer where a large fraction of the evaporation occurs. Knowledge of the gas flow patterns is important for calculating particle trajectories and residence times and also for estimating the particle Reynolds number to be used in the Ranz-Marshall heat transfer equation.

Two basic approaches have been used for simulating gas and particle flows in spray dryers:

(a) A simplified analytical representation of the gas flow pattern is employed, based on experimental measurements or correlations for the specific type of chamber. The droplet flow is superposed on this gas flow pattern with the assumption that the gas velocity vectors are not affected by the presence of the droplets, and droplet trajectories are calculated numerically by integrating the equation of motion through this flow field.

(b) The gas flow pattern for an empty chamber is derived by dividing it into a grid and solving the Navier-Stokes equation for every cell, the chamber dimensions and boundary conditions being specified. The spray is then introduced with an initial guess of the droplet trajectory. Momentum, heat and mass exchange between the spray and the gas are then calculated for each cell, and by a process of iteration gas and spray flow patterns are found which satisfy the continuity, momentum and energy equations at every cell boundary.

The first approach is exemplified by the models of Katta and Gauvin (26) and Keey and Pham (27), both of which simulated laboratory spray dryers with a pneumatic atomizer. Computer programs of only modest complexity and short run times are required. However, this approach is dependent on having experimental data on the gas flow pattern in the type of chamber concerned and this will not always be readily available. Furthermore, quite small changes in the air inlet arrangement can produce very significant changes in the gas flow pattern, so these models cannot be used for exploring potential changes in geometry or in air distribution systems. Finally, they ignore the effect which the droplets have on the gas flow pattern; while momentum transfer from droplets to gas may not be very significant (the feed rate is typically less than 10% of the gas mass flow rate), the droplets have a severe quenching effect on the hot gas which is likely to generate substantial convection effects.

The second approach to aerodynamic modelling (approach (b) above) has the potential to overcome all these disadvantages at the expense of greatly increased computational complexity and program running time. Its main use is likely to be for exploring potential changes in chamber geometry and air inlet arrangements prior to building hardware. Such models have been under development for some time for simulating gas and droplet flows in combustion chambers. The first recorded application to spray dryers was made by Crowe (28), who simulated a very simple laboratory spray dryer with a pneumatic atomizer; the predicted axial gas and spray temperature profiles and the predicted axial evaporation profile agreed well with experimental data. Recently, O'Rourke and Wadt (29) have used a similar basic approach (albeit with important differences in detail) to simulate a Niro co-current spray dryer with a rotary atomizer; their predictions have not yet received experimental verification, but they do not appear unrealistic.

In summary, we may conclude that much progress has been made over the past decade in developing improved feedstock models and improved equipment models for spray drying. However, the problems are difficult and much more remains to be done before satisfactory overall fundamental models can be constructed for spray drying operations.

8. CONCLUSIONS

Compared to some other branches of chemical science, the development of fundamental models for particulate drying operations is still in its infancy. Of the types of drying operations considered in this paper, only for the comparatively simple case of drying in a fluid bed is it possible at present to construct a satisfactory overall fundamental model. Reliable fundamental simulations of other types require improvements in our ability to characterise both feedstocks and equipment.

Improved feedstock characterisation requires developments in:

(a) experimental techniques for bench-scale measurement of drying kinetics, in order to reduce the extrapolation required from bench-scale to industrial drying conditions;

(b) models of moisture migration in particles, so that extrapolation and interpolation can be made with greater confidence.

These developments are needed particularly for simulation of pneumatic conveying drying and spray drying operations. In spray drying we also need better predictions of initial drop size and particle size and density changes during drying.

Improved equipment characterisation requires improvements in the following areas:

(a) prediction of gas-to-particle heat transfer coefficients and particle transport velocities in rotary dryers;

(b) prediction of gas flow patterns and particle trajectories in spray dryers.

Only when these problems have been solved will it be possible to realise the full potential benefits of fundamental models of these latter types of drying operations. Meanwhile, the rather inadequate fundamental models which we carr currently build may still give useful insights and at least qualitative pointers for solving practical drying problems, and they may give quantitative guidance on the effects of limited extrapolation from existing plant operations.

NOMENCLATURE

a	thermal diffusivity of gas	m²/s
Dp	particle diffusivity	m²/s
d_d	drum diameter	m
d_p	particle diameter	m
$E(t)$	particle residence time distribution function	-
K	constant (equation (7))	w_s^n/Km^{2+n}
L	bed length	m
n	constant (equation (7))	-
t	time	s
$\overline{t}$	mean particle residence time	s
u_G	gas velocity	m/s
u_{mf}	mimimum fluidisation velocity	m/s
u_p	particle velocity	m/s
w	particle moisture content	kg/kg
w_{cr}	critical moisture content	kg/kg
w_e	equilibrium moisture content	kg/kg
$\overline{w}_o$	mean product moisture content	kg/kg

Greek

α	gas-to-particle heat transfer coefficient	w/m²K
α_v	gas-to-particle heat transfer coefficient based on dryer volume	w/m³K
λ	thermal conductivity of gas	w/mK
ν	kinematic viscosity of gas	m²/s

Dimensionless Groups

$B = \frac{D_p\overline{t}}{L^2}$ axial dispersion number

$Nu = \frac{\alpha d_p}{\lambda}$ Nusselt number

$Pr = \frac{\nu}{a}$ Prandth number

$Re_p = \frac{(u_G - u_p)d_p}{\nu}$ Reynolds number

REFERENCES

1. Hallstrom, A. and Wimmerstadt, R., paper presented at meeting of the Europeon Federation of Chemical Engineering Working Party on Drying, Liege, Belgium (1983).
2. Keey, R.B., A.I.Ch.E. Symp. Series, Vol. 73, No. 163, p1, (1977).
3. Suzuki, M., Keey, R.B. and Maeda, S., A.I.Ch.E. Symp. Series, Vol. 73, No. 163, p47, (1977).
4. Ashworth, J.C. and Carter, J.W., Drying '80; Volume 1; Developments in Drying, p151-159, Hemisphere Publishing Corp., New York, (1980).

5. Schlünder, E.V., Chem.-Ing.-Techn., Vol. 48, 190 (1976).
6. van Brakel, J., Mass Transfer in Convective Drying, in Advances in Drying, ed. A.S. Mujumdar, Vol. 1, Chapter 7, Hemisphere Publishing Corp., New York (1980).
7. Reay, D., Proc. 1st Int. Drying Symposium, Montreal, p136 (1978).
8. Reay, D. and Allen, R.W.K., Proc. 3rd Int. Drying Symposium, Birmingham, Vol. 2, p130 (1982).
9. Reay, D. and Allen, R.W.K., J. Sep. Proc. Technol., Vol. 2, No. 4, p11 (1983).
10. Ranz, W.E. and Marshall, W.R., Chem. Eng. Progr., Vol. 48(3), 141 and Vol. 48(4), 173 (1952).
11. Bandrowski, J. and Kaczmarzyk, G., Chem. Eng. Sci., Vol. 33, 1303 (1978).
12. Reay, D. and Bahu, R.E., Proc. 4th Int. Drying Symposium, Kyoto, (1984).
13. Martin, H. and Saleh, A.H., Verfahrenstechnik, Vol. 16(3), p162 (1982).
14. Andrieu, J. and Bressat, R., Proc. 3rd Int. Drying Symposium, Birmingham, Vol. 2, p10, (1982).
15. Baker, C.G.J., Cascading Rotary Dryers, in Advances in Drying, Volume 2, ed. A.S. Mujumdar, Chapter 1, Hemisphere Publishing Corp., New York (1983).
16. Schofield, F.R. and Glikin, P.G., Trans. Inst. Chem. Engrs., Vol. 40, 183 (1962).
17. Saeman, W.C. and Mitchell, T.R., Chem. Eng. Progr., Vol. 50, 467 (1954).
18. Kelly, J.J. and O'Donnell, P., Trans. Inst. Chem. Engrs., Vol. 55, 243 (1977).
19. Toei, R., Okazaki, M. and Furuta, T., Proc. 1st Int. Drying Symposium, Montreal, p53 (1978).
20. Schoeber, W.J.A.H. and Thijssen, H.A.C., A.I.Ch.E. Symp. Series, Vol. 73, No. 163, p12 (1977).
21. Schoeber, W.J.A.H., Proc. 1st Int. Drying Symposium, Montreal, p1 (1978).
22. van der Lijn, J., Simulation of heat and mass transfer in spray drying, Ph.D. thesis, Agriculatural University, Wageningen, Netherlands (1976).
23. Sano, Y. and Keey, R.B., Chem. Eng. Sci., Vol. 37, 881 (1982).
24. Filkova, I. and Weberschinke, J., Proc. CHISA Congress, Prague (1981).
25. Filkova, I. and Cedik, C., Proc. 3rd Int. Drying Symposium, Birmingham, Vol. 1, p516, (1982).
26. Katta, S. and Gauvin, W.H., A.I.Ch.E. Journal, Vol. 21, 143 (1975).
27. Keey, R.B. and Pham, Q.T., The Chemical Engineer, p516 (1976).
28. Crowe, C.T., Modelling spray-air contact in spray drying systems, in Advances in Drying, Volume 1, ed. A.S. Mujumdar, Chapter 3, Hemisphere Publishing Corp., New York (1980).
29. O'Rourke, P.J. and Wadt, W.R., A Two-Dimensional, Two-Phase Flow Model for Spray Dryers, Los Alamos National Laboratory Report LA-9423-MS, U.S.A. (1982).

VACUUM CONTACT DRYING OF FREE FLOWING MECHANICALLY AGITATED PARTICULATE MATERIAL

Ernst Ulrich Schlünder

Institut für Thermische Verfahrenstechnik, University of Karlsruhe
7500 Karlsruhe, FRG

ABSTRACT

In vacuum contact drying of particulate material heat is supplied from a hot surface to adjacent layers of mechanically agitated material. The drying rate curves can be predicted from physical properties alone provided that the actual contact time of the material at the hot surface is known. For mechanically agitated particulate material the actual contact time has been correlated to the time scale of the agitator by the "penetration model". Only one empirical parameter, the so called mixing number N_{mix} is necessary in order to predict drying rates for bench scale as well as for industrial dryers.

1. INTRODUCTION

In vacuum contact drying of particulate material heat is supplied from a hot surface to adjacent layers of mechanically agitated particulate material. The generated vapor is removed by a vacuum pump. Thus, contact dryers can be run without an inert gas atmosphere. In order to achieve a uniform drying of the particulate material the particle layers are mechanically agitated. The mixing process can be carried out by stirrers, scrapers, paddles or rotating drums with or without flyers inside. Fig. 1 shows schematically two disc dryers with stirrers, a rotary dryer with a rotating drum and rotating scrapers inside and a drum dryer with a static drum and rotating paddles. The performance of these dryers will be analyzed in this paper.

2. PRESENT KNOWLEDGE

The drying rate $\dot{m}$ is affected by both heat supply and vapor removal. Fig. 2 illustrates the flow of heat into the product due to temperature gradients and the flow of vapor out of the product due to pressure gradients for a non-agitated packed bed of particulate material heated from below. The individual resistances are defined as follows:

$$\left.\begin{array}{rl} \text{contact:} & 1/\alpha_{ws} = (T_w - T_o)/\dot{q}_o \quad (1) \\ \text{bulk penetration:} & 1/\alpha_{sb} = (T_o - T_b)/\dot{q}_o \quad (2) \\ \text{particle penetration:} & 1/\alpha_p = (T_b - T_s)/\dot{q}_{z_T} \quad (3) \end{array}\right\} \text{heat in}$$

$$\left.\begin{array}{rl} \text{particle permeation:} & 1/\beta_p = (p_s - p_b)/\dot{m} \quad (4) \\ \text{bulk permeation:} & 1/\beta_b = (p_b - p\ )/\dot{m} \quad (5) \end{array}\right\} \text{mass out}$$

I. Disc dryer

D_1 = 240 mm
D_2 = 214 mm
α_{ext} = 70 000 W/m²K
electrically heated

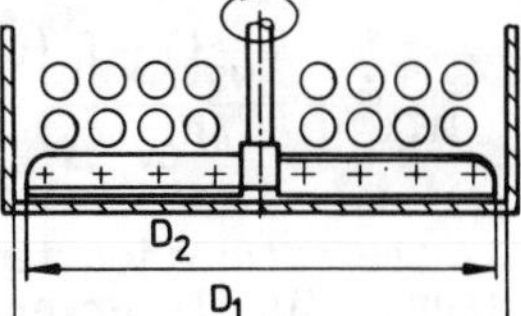

II. Disc dryer

D_1 = 400 mm
D_2 = 170 mm
α_{ext} = 675 W/m²K
oil heated

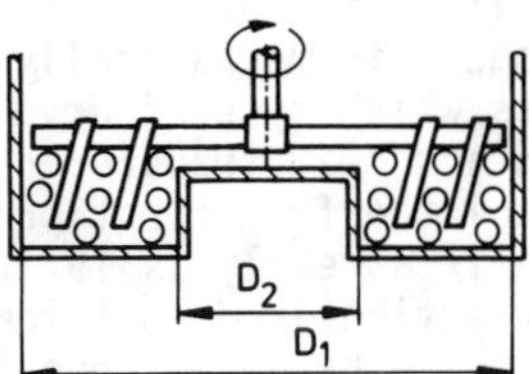

III. Rotary drum dryer

D = 340 mm
α_{ext} = 425 W/m²K
oil heated

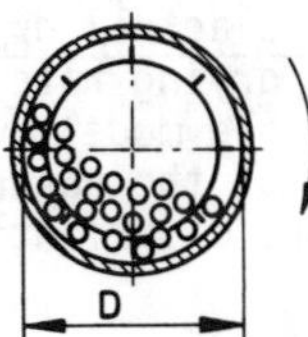

Paddle dryers

IV. Volume 5 l
D = 134 mm
α_{ext} = 3 500 W/m²K
steam heated

V. Volume 160 l
D = 506 mm
α_{ext} = 2 700 W/m²K
steam heated

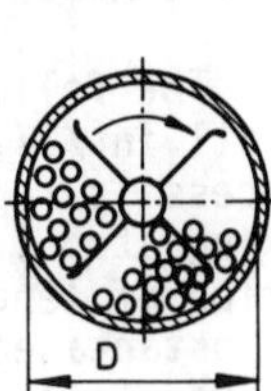

Fig. 1: Analyzed contact driers

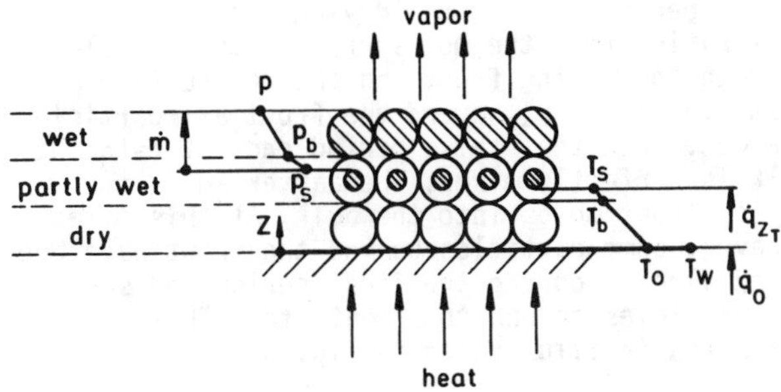

Fig. 2: Heat and mass transfer resistances in contact drying.

All these resistances ly in series. However, if the particulate material is mechanically agitated, we expect a random distribution of dry, partly wet and wet material, thus forming a random distribution of transport resistances partly in series and partly parallel. Since this situation is rather complicated and therefore is also difficult to be described by a suitable model, it is of great interest to find out whether - and if yes - which one of these five resistances might become rate controlling. This question leads to the recognition that a maximum drying rate $_{max}\dot{m}$ exists, which by no means can be superceeded.

The maximum drying rate $_{max}\dot{m}$. The contact resistance $1/\alpha_{ws}$ is the one which controls the maximum drying rate:

$$_{max}\dot{m} = \frac{\alpha_{ws}(T_w - T_s(p))}{\Delta h_{ev}(T_s)} \tag{6}$$

T_s is the saturation temperatur pertaining to the pressure p in the dryer while $\Delta h_{ev}(T_s)$ is the latent heat of evaporation pertaining to the saturation temperatur (T_s). Eq.(6) gives the upper limit of the drying rate, regardless whether the product is mechanically agitated or not, regardless which type of dryer is considered. The contact heat transfer coefficient α_{ws} can be predicted from first principles as has been shown in /1/ as early as 1971. In the meantime slight corrections and modifications of the respective formulae have been published. The latest version is given in /2/.

The actual drying rate. The prediction of the actual drying rate requires the establishment of a suitable model, which adequately describes the effect of the random particle motion and the moisture distribution on the drying rate.

3. THE MODEL

The following concept, which is described in detail in /3/, has been adopted. The steady mixing process is replaced by a sequence of unsteady mixing steps. During a certain (fictitious) period of length t_R the bulk is static. Thereafter an instantaneous perfect macromixing of the bulk occurs, followed by the same static period again. This is the concept which is otherwise known as the so called 'penetration theory'. During the static period a distinct 'drying front' is penetrating from the hot surface into the bulk. Between the moving front and the hot surface all particles are dry, beyond the front all particles are wet. When the static period ends at $t=t_R$ the bulk is perfectly mixed. Thereafter again the drying front moves into the bulk. At this time, however, some particles behind the front have been dried already during the first period and some of the particles beyond the front, too. This situation is illustrated in fig. 3.

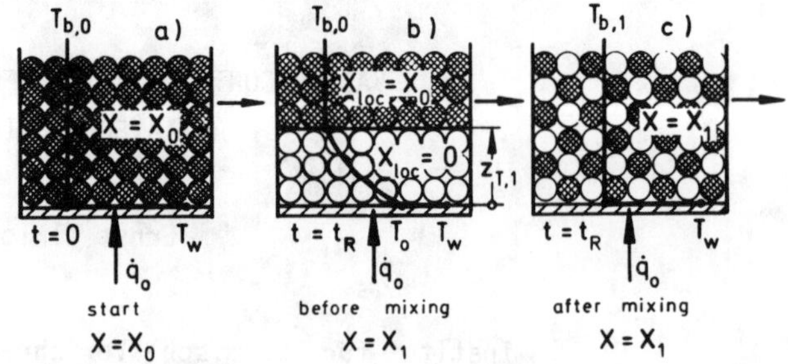

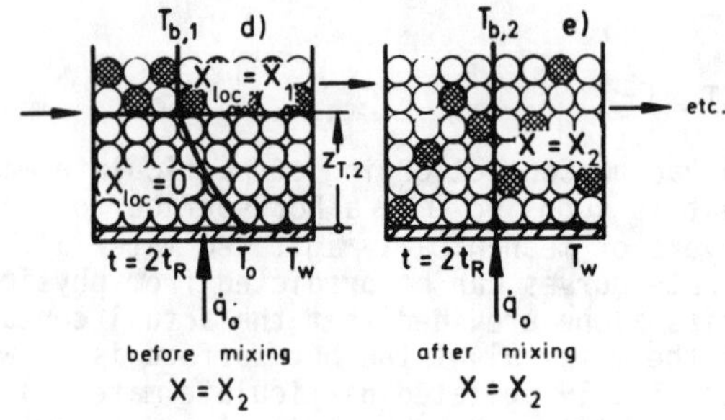

Fig. 3 Illustration of the new contact drying model

Provided that the packed bed may be considered as a quasi continuum, the bulk heat penetration resistance $1/\alpha_{sb}$ can be predicted by application of Fourier's theory. Therefore we need the over all heat conductivity of the dry packed bed λ_{bed}. This quantity can be calculated from standard equations as e.g. summarized in /5/.

The transient temperature profiles between the hot surface and the moving drying front during each (fictitious) static period ($0 \leq t \leq t_R$) follows from Neumann's solution /6/ of Fourier's equation for constant surface temperature T_o (see fig. 3):

$$\frac{T - T_b}{T_o - T_b} = 1 - \frac{\operatorname{erf}(z/2\sqrt{\kappa_{bed} t})}{\operatorname{erf} \zeta}, \tag{7}$$

where ζ is the reduced instantaneous position of the drying front

$$\zeta = \frac{z_T}{2\sqrt{\kappa_{bed}\, t}} \tag{8}$$

as given by eq. (11) obtained from Neumann's solution as well. $\kappa_{bed} = \lambda_{bed}/(\rho c_p)_{bed}$ is the over all thermal diffusivity of the dry packed bed. The time averaged heat penetration coefficient for the partly wet packed bed follows by definition from

$$\alpha_{sb,wet} = \frac{1}{t_R} \int_0^{t_R} \frac{-\lambda_{bed} \left(\frac{\partial T}{\partial z}\right)_{z=0}}{T_o - T_b}\, dt \,. \tag{9}$$

Substitution of $(\partial T/\partial z)_{z=0}$ yields

$$\alpha_{sb,wet} = \frac{2}{\sqrt{\pi}} \frac{\sqrt{(\lambda\rho c_p)_{bed}}}{\sqrt{t_R}} \cdot \frac{1}{\text{erf}\,\zeta} \tag{10}$$

with

$$\sqrt{\pi}\,\zeta\,\exp(\zeta^2)\text{erf}\,\zeta = \frac{c_{p,bed}(T_o-T_b)}{X\,\Delta h_{ev}} \,. \tag{11}$$

As X goes to zero, ζ goes to infinity and erf ζ goes to unity. In this case eq. (10) yields α_{sb} for the entirely dry packed bed:

$$\alpha_{sb,dry} = \frac{2}{\sqrt{\pi}} \frac{\sqrt{(\lambda\rho c_p)_{bed}}}{\sqrt{t_R}} \,. \tag{12}$$

Actually T_o is not constant. Disregarding this and following the engineering type of approach we eliminate T_o by simply putting into series the two individual resistances $1/\alpha_{ws}$ and $1/\alpha_{sb}$ in order to form an over all resistance:

$$\frac{1}{\alpha_{dry}} = \frac{1}{\alpha_{ws}} + \frac{1}{\alpha_{sb,dry}} \tag{13}$$

$$\frac{1}{\alpha_{wet}} = \frac{1}{\alpha_{ws}} + \frac{1}{\alpha_{sb,wet}} \tag{14}$$

Then we reach at

$$\frac{\alpha_{dry}}{\alpha_{ws}} = \frac{1}{1+\frac{\sqrt{\pi}}{2}\cdot\sqrt{\tau_R}} \tag{15a}$$

$$\tau_R = \frac{\alpha_{ws}^2}{(\lambda\rho c_p)_{bed}}\, t_R \tag{15b}$$

and

$$\frac{\alpha_{wet}}{\alpha_{ws}} = \frac{1}{1+(\alpha_{ws}/\alpha_{dry}-1)\text{erf}\,\zeta} \tag{16}$$

with

$$\sqrt{\pi}\zeta\,\exp(\zeta^2)\left\{1+\left(\frac{\alpha_{ws}}{\alpha_{dry}}-1\right)\text{erf}\,\zeta\right\} = \left(\frac{\alpha_{ws}}{\alpha_{dry}}-1\right)\frac{1}{\xi} \tag{17}$$

where

$$\xi = \frac{X\,\Delta h_{ev}}{c_{p,bed}(T_w-T_b)} \,. \tag{18}$$

Eq. (16) in connection with eq. (17) yields the reduced heat transfer coefficient of the wet bed α_{wet}/α_{ws} as a function of the reduced average moisture content of the bulk ξ with the reduced heat transfer coefficient of the dry bed α_{dry}/α_{ws} as parameter. Fig. 4 shows that for large moisture contents α_{wet} goes to α_{ws} because latent heat consumption, which keeps the bed isothermal, becomes predominant.

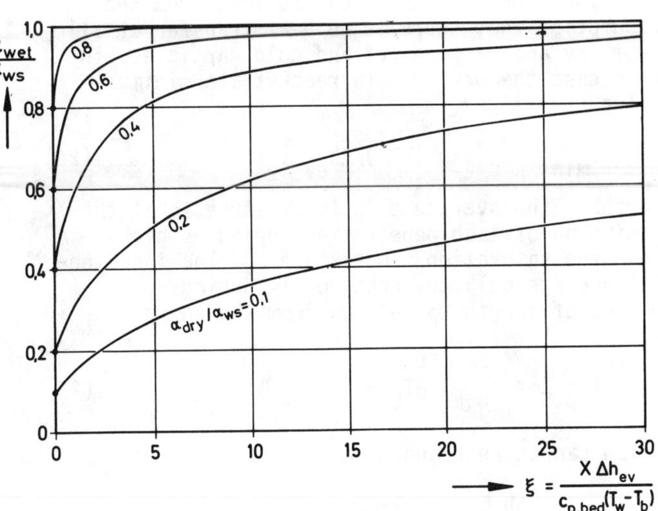

Fig. 4 Reduced heat transfer coefficient α_{wet}/α_{ws} of wet product as a function of reduced moisture content ξ with reduced heat transfer coefficient α_{dry}/α_{ws} of dry product as parameter.

Knowing α_{wet} we know the heat flux at the hot surface (z=0):

$$\dot{q}_o = \alpha_{wet}\,(T_w - T_b) \tag{19}$$

and the drying front $(z = z_T)$:

$$\dot{q}_{z_T} = \dot{q}_o \exp(-\zeta^2) \,. \tag{20}$$

The difference between these two heat fluxes warms up the dry material in between the hot surface and the drying front.

There are two limiting cases as to the prediction of the drying rate $\dot{m}$. Assume that the dry and warm particles are always cooled down again by direct contact with still wet particles after being mixed back into the bulk at $t=t_R$, then drying rate reaches its (relative) maximum value

$$\dot{m}_{max} = \dot{q}_o/\Delta h_{ev} \tag{21}$$

or introducing a normalized drying rate

$$\dot{\mu} = \dot{m}/_{max}\dot{m} \tag{22}$$

we obtain

$$\dot{\mu}_{max} = \frac{\dot{m}_{max}}{_{max}\dot{m}} = \frac{\alpha_{wet}}{\alpha_{ws}} \tag{23}$$

In this case the normalized drying rate curve of the bulk is identical to the ratio α_{wet}/α_{ws} as shown in fig. 4. The bulk temperature T_b at the beginning of each penetration period (t=0) is always equal to the saturation temperature T_s.

The drying rate $\dot{m}$ at zero bulk moisture content X is finite.

The other limiting case follows from the assumption that there is no heat transfer at all from dry and warm to wet and cold particles. In this case the drying rate reaches its minimum value

$$\dot{m}_{min} = \dot{q}_o \exp(-\zeta^2)/\Delta h_{ev} \, . \tag{24}$$

Further, the average bulk temperature T_b at the beginning of each penetration period is higher than the saturation temperature T_s. The incremental average bulk temperature ΔT_b during each period of length t_R follows from the energy balance

$$M_{dry}(c_{p,S}+X\, c_{p,L})\, \Delta T_b = (\dot{q}_o-\dot{q}_{z_T})\, A\, t_R \tag{25}$$

which can be rearranged to

$$\Delta T_b = \frac{\Delta h_{ev}}{c_{p,bed}+X\, c_{p,L}} \, \frac{1-\exp(-\zeta^2)}{\exp(-\zeta^2)} \, \Delta X \tag{26}$$

thus connecting the increment of the average bulk temperature ΔT_b with the decrement of the average bulk moisture content. Eq. (26) allows a stepwise calculation of the average bulk temperature T_b with decreasing moisture content X. The normalized minimum drying rate follows from eq. (24)

$$\frac{\dot{m}_{min}}{\max\dot{m}} = \frac{\alpha_{wet}}{\alpha_{ws}} \, \frac{T_w-T_b}{T_w-T_s} \exp(-\zeta^2) \, . \tag{27}$$

Fig. 5 shows both $\dot{\mu}_{max}$ and $\dot{\mu}_{min}$ as well as the normalized temperature difference $\Theta = (T_w-T_b)/(T_w-T_s)$ versus the average bulk moisture content X and ξ resp. for a medium grained product as specified in the legend. One finds that at large ξ there is only little difference between $\dot{\mu}_{max}$ and $\dot{\mu}_{min}$, while at low ξ in contrast to $\dot{\mu}_{max}$ the minimum drying rate tends towards zero. Since at low average bulk moisture contents also the probability of warm dry particles to collide with cold wet ones goes to zero, the minimum drying rate $\dot{\mu}_{min}$ is expected to be much closer to the actual one than $\dot{\mu}_{max}$. Consequently $\dot{\mu}_{min}$ is recommended for practical usage.

Summarizing so far, the drying rate $\dot{m}(X)$ can be predicted provided α_{dry} is known. α_{dry} can be calculated by eq. (15), provided that the length t_R of the fictitious static period is known. This is the remaining problem. We postulate that t_R should be a mechanical property of the system, which is formed by the mixing device in connection with the mechanical properties of the product. Then t_R should be a function of the time scale of the mixing device (reciprocal stirrer speed n r.p.m., drum revolutions per unit time etc.), introduced as t_{mix}. The reduced penetration time τ_R can be split into two dimensionless groups:

$$\tau_R = N_{therm} \cdot N_{mix} \, , \tag{28}$$

where

$$N_{therm} = \frac{\alpha_{ws}^2 \, t_{mix}}{(\rho c\lambda)_{bed}} \tag{29}$$

and

$$N_{mix} = \frac{t_R}{t_{mix}} \, . \tag{30}$$

The quantity N_{therm} follows from predictable quantities such as α_{ws} and $(\rho c\lambda)_{bed}$ and from the given time scale of the mixing device t_{mix}. The quantity N_{mix}, the so called 'mixing number', must be fitted to experimental results, since so far there does not exist a theory of the random particle motion. N_{mix} simply says how often the mixing device must have turned around until the product has been ideally mixed once - within the scope of the penetration model. Inserting eq.(28) into eq. (15) yields

$$\frac{\alpha_{dry}}{\alpha_{ws}} = \frac{1}{1+\frac{\sqrt{\pi}}{2}\sqrt{N_{therm}\, N_{mix}}} \tag{31}$$

For any given N_{mix} the drying rate $\dot{m}$ (X) can be predicted together with the bulk temperature T_b by a straight foreward stepwise calculation procedure starting from the initial moisture content X_{in} and the initial bulk temperatur T_b-T_s.

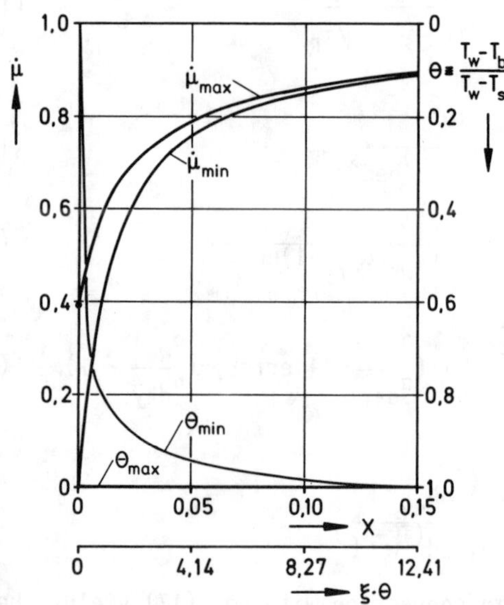

Fig. 5
Normalized drying rate curves and normalized temperature difference Θ. d = 1 mm; δ = 10 µm; p = 1 mbar ; T = 280 K ; λ_g = 0.0184 W/mK ; γ = 0.8 ; ε = 0.9 ; φ_A = 0.8 ; ψ = 0.4 ; λ_s= 1.1 W/mK ; φ_K = 0.008 ; $(\rho c_P)_{bed}$=10^6 J/m^{3}K; n = 0.75 rpm ; N_{mix} = 15 ; $(T_w - T_s)$ = 30 K .

4. EXPERIMENTAL RESULTS

Drying rates were measured by Günes /7/, Mollekopf /4/, Fritz /8/ and Vosteen /9/ in various contact dryers as shown in fig. 1. Günes and Mollekopf investigated the disc dryer No. I while Fritz and Vosteen took measurements with the dryers No. II, III, IV and V.

In the following figures 6 to 11 experimental drying rates are shown for various materials dried in the dryer I, resp. at constant pressure p, constant hot surface temperature T_W and constant speed of the stirrer n.

All individual experimental data are listed in /4/. The mixing number N_{mix}, as listed in the legends, was chosen to give the best data fit. The dryer No. I was heated electrically and the temperature of the hot surface T_W was measured directly by inserted thermocouples. The dryers No. II and III were heated with oil while the dryers No. IV and V were steamheated. In these cases the temperature T_W was eliminated by the temperature T_h, which is either the oil- or the steamtemperature by adding the external heat transfer resistance $1/\alpha_{ext}$ to the contact resistance

$$1/\alpha^*_{WS} = 1/\alpha_{WS} + 1/\alpha_{ext} \tag{32}$$

Günes /7/ and Mollekopf /4/ dried magnesium and aluminiumsilicate particles in the disc dryer No. I. Vosteen /9/ dried hygroscopic 'Perlkontakt' with an equilibrium moisture content of X_{eq} = 0,005 kg H_2O/kg dry prod. in the disc dryer No. II, which had two stirrer wings. He also dried aluminium silicate in the smaller paddle drier No. IV and PVC in the larger one No. V. The driers III, IV and V were filled half with the product, which means that the actual hot surface area is half of the total one.

Fig. 6 and 7 show the effect of heating temperature T_W on the drying rate curve. Fig. 6 shows results for coarse (d = 6 mm) and fig. 7 those for fine (d = 0.59 mm) ceramic material, both dried in the disc dryer No. I.

As a result of these runs the mixing number N_{mix} turned out to be independent of the hot surface temperature T_W, and the moisture content X thus indicating that the introduction of N_{mix} as a purely mechanical property of the system may be justified. Further, N_{mix} is in the same order of magnitude for coarse as well as for fine material. Eventually, the results confirm the effect, predicted by the theory, that the drying rate $\dot{m}$ is proportional to the temperature difference T_W-T_S only for coarse material, while for fine powders the effect of the temperature difference on the drying rate is much less. The reason is that the actual drying rate $\dot{m}$ is much closer to the maximum drying rate $_{max}\dot{m}$ for coarse material than for fine ones, which means that in the latter case the bulk resistance $1/\alpha_{sb,wet}$ becomes rate controlling. The bulk resistance, however, increases with increasing temperature difference as can be seen from eq.(16) and (17).

	d	T_w	p	n	λ_{bed}	α_{ws}	$_{max}\dot{m}$	N_{mix}
	mm	°C	mbar	min^{-1}	W/mK	W/m^2K	kg/m^2h	-
⊡	6.00	50.5	15.9	15.4	.170	44.5	2.36	15.50
⊙	6.00	61.1	16.0	15.4	.173	45.4	3.10	15.50
▷	6.00	70.6	16.9	15.4	.176	46.4	3.75	15.50
◇	6.00	80.9	16.3	15.4	.178	47.0	4.55	15.50
⋈	6.00	85.8	15.4	15.4	.179	47.1	4.95	15.50

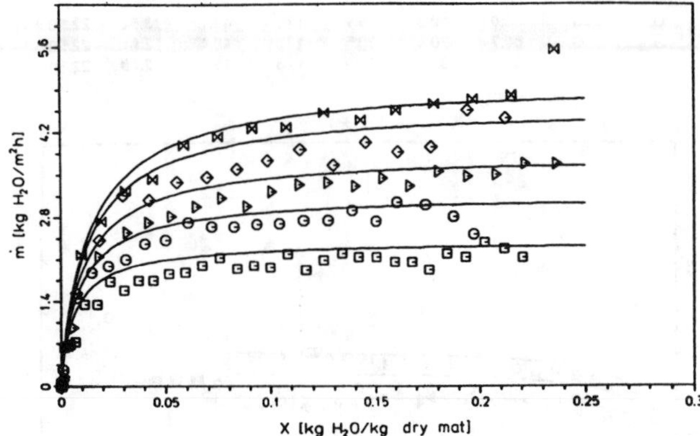

Fig. 6 Influence of hot surface temperature T_W on the drying rate curve. MgSi , ρ_{bed} = 980 kg/m^3 , $c_{p,bed}$ = 800 J/kgK, δ = 20 µm, d = 6.0 mm, p = 16 mbar, n = 15.4 rpm, N_{mix} = 15.5 . Dryer I.

	d	T_w	p	n	λ_{bed}	α_{ws}	$_{max}\dot{m}$	N_{mix}
	mm	°C	mbar	min^{-1}	W/mK	W/m^2K	kg/m^2h	-
⊡	.59	60.0	30.0	7.5	.137	312.2	16.35	6.00
⊙	.59	80.2	30.0	7.5	.140	318.5	26.10	6.00
▷	.59	100.8	30.0	7.5	.143	324.8	36,43	6.00

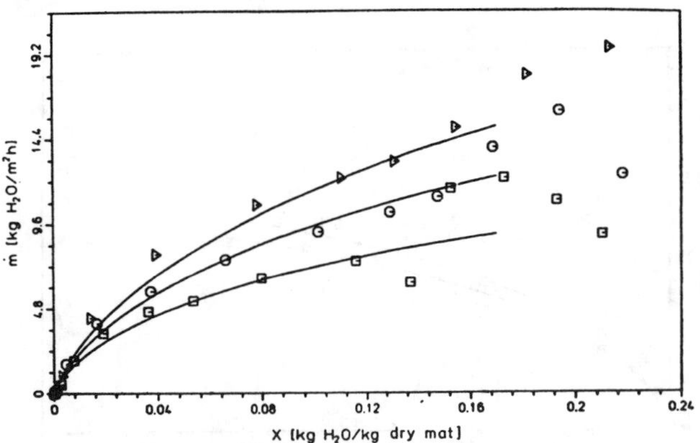

Fig. 7 Influence of hot surface temperature T_W on the drying rate curve. AlSi, ρ_{bed} = 1030 kg/m^3, $c_{p,bed}$ = 800 J/kgK, δ= 2.5 µm, d = 0.59 mm, p = 30 mbar, n = 7.5 rpm, N_{mix} = 6. Dryer I.

Fig. 8 and 9 exhibit the effect of pressure on the drying rate for coarse and fine material dried in the disc cryer No. I at pressures ranging from 10 to 30 mbar. As a result the fitted mixing number N_{mix} does not vary with pressure and has almost the same values as in the previous runs.

	d	T_w	p	n	λ_{bed}	α_{ws}	$_{max}\dot{m}$	N_{mix}
	mm	°C	mbar	min^{-1}	W/mK	W/m^2K	kg/m^2h	-
⊡	6.00	60.9	10.3	29.9	.170	43.1	3.32	22.50
⊖	6.00	60.7	20.4	30.5	.175	46.4	2.91	22.50
▷	6.00	60.9	33.6	29.9	.178	48.5	2.48	22.50

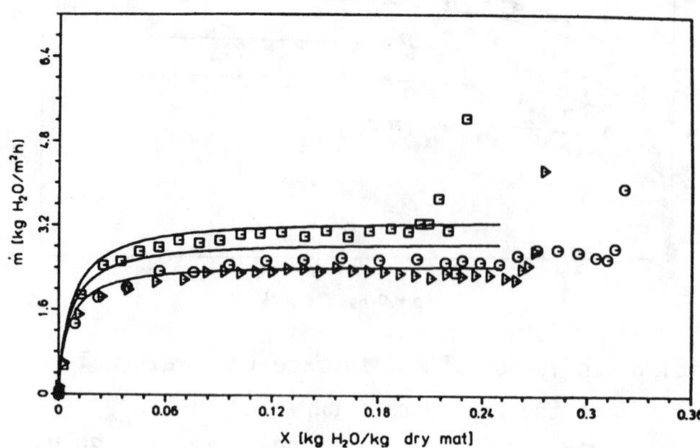

Fig. 8 Influence of the pressure p on the drying rate curve. MgSi, ρ_{bed} = 980 kg/m^3, $c_{p,bed}$ = 800 J/kgK, δ = 20 µm, d = 6.0 mm, T_W = 61°C, n = 30 rpm, N_{mix} = 22.5. Dryer I.

	d	T_w	p	n	λ_{bed}	α_{ws}	$_{max}\dot{m}$	N_{mix}
	mm	°C	mbar	min^{-1}	W/mK	W/m^2K	kg/m^2h	-
⊡	.83	60.0	10.0	7.5	.124	183.8	14.01	6.00
⊖	.83	60.0	30.0	7.5	.143	247.9	12.99	6.00

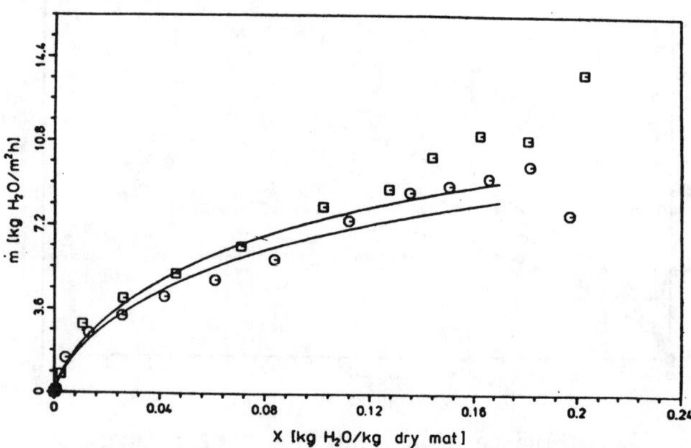

Fig. 9 Influence of the pressure p on the drying rate curve. AlSi, ρ_{bed} = 1030 kg/m^3, $c_{p,bed}$ = 800 J/kgK, δ= 2.5 µm, d=0.83 mm, T_W = 60°C, n = 7.5 rpm, N_{mix} = 6. Dryer I.

The runs depicted in the figures 10 to 11 have been selected to show how intensification of product mixing affects the drying rate. Parameter in these figures is the number of revolutions per unit time n of the stirrers.

	d	T_w	p	n	λ_{bed}	α_{ws}	$_{max}\dot{m}$	N_{mix}
	mm	°C	mbar	min^{-1}	W/mK	W/m^2K	kg/m^2h	-
⊡	6.00	70.9	1.9	.0	.151	31.8	3.35	-
⊖	6.00	71.7	1.9	.2	.151	31.9	3.39	2.00
▷	6.00	70.7	1.9	1.0	.151	31.8	3.34	5.00
◇	6.00	71.1	1.9	45.0	.151	31.8	3.36	22.50

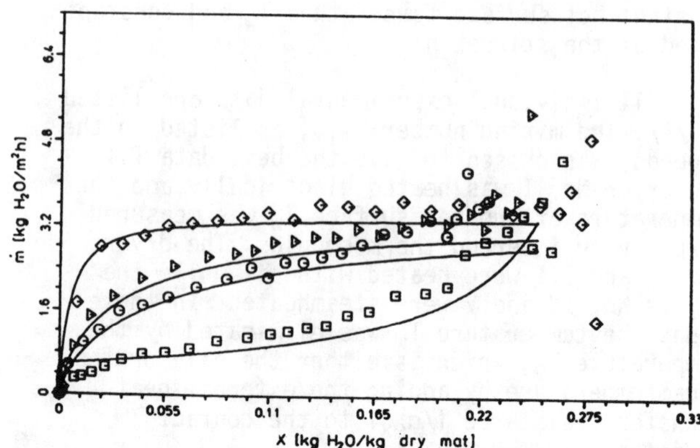

Fig. 10 Influence of the number n of revolutions on the drying rate curve. MgSi, ρ_{bed} = 980 kg/m^3, $c_{p,bed}$ = 800 J/kgK, δ = 20 µm, d = 6.0 mm, T_W = 71°C, p = 1.9 mbar. Dryer I.

	d	T_w	p	n	λ_{bed}	α_{ws}	$_{max}\dot{m}$	N_{mix}
	mm	°C	mbar	min^{-1}	W/mK	W/m^2K	kg/m^2h	-
⊡	.83	44.4	16.5	.0	.132	210.2	9.08	-
⊖	.83	44.4	16.5	1.0	.132	210.2	9.08	7.50
▷	.83	44.4	16.5	7.5	.132	210.2	9.08	18.75
◇	.83	44.4	16.5	22.5	.132	210.2	9.08	22.50
⋈	.83	44.4	16.5	45.0	.132	210.2	9.08	22.50

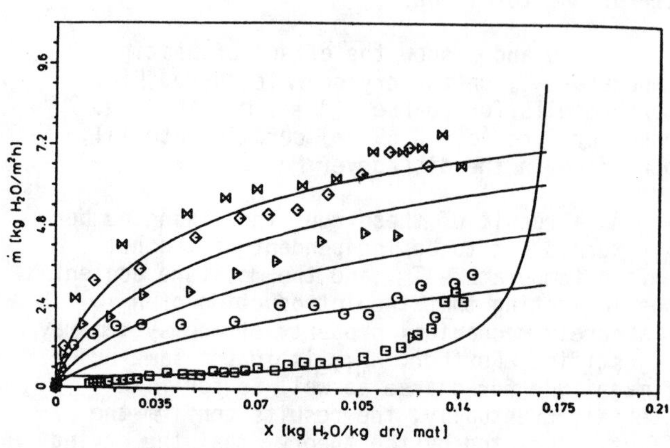

Fig. 11 Influence of the number n of revolutions on the drying rate curve. AlSi, ρ_{bed} = 1030 kg/m^3, $c_{p,bed}$ = 800 J/kgK, δ = 2.5 µm, d = 0.83 mm, T_W = 44.4°C, p = 16.5 mbar, Dryer I

Fig. 10 and 11 show drying rate curves for coarse and fine ceramic material dried in the disc and fine ceramic material dried in the disc dryer No. I. As a result, the fitted mixing numbers

N_{mix} are in between 2 and 25 for all types of dryers and for coarse as well as for fine grained material. The lower values pertain to low values of n and vice versa. The effect of N_{mix} on the drying rate is low as the actual drying rate comes close to the maximum drying rate which is true for coarse material. In this case the drying rate becomes contact resistance controlled. On the other hand is the drying rate curve sensitive to N_{mix} if the product is fine grained. In this case the drying rate becomes bulk resistance controlled. However, even in this case the strongest effect of N_{mix} on the drying rate $\dot{m}$ is given by the limiting proportionality

$$\dot{m} \sim 1/\sqrt{N_{mix}} \ .$$

Both the contact resistance $1/\alpha_{ws}$ and the reduced moisture content $\xi = X \Delta h_{ev}/c_{p,bed}(T_w - T_s)$ act as damping factors so that

$$\dot{m} \sim 1/N_{mix}^{0.2 \div 0.3}$$

may be quite a good guess for practical estimations. Since N_{mix} was found to be within 2 and 25 at least for the dryers and products investigated in this paper a rough guess of a constant value

$$N_{mix} = 7$$

would have caused no greater errors as to the drying rate than about $\pm$ 50 %.

5. THE MIXING NUMBER N_{mix}

The mixing number N_{mix} was determined by curve fitting when adapting the theory (eq.(27)) to experimental drying rate curves. It is the only free parameter in the theory. The number simply says, how many revolutions of the mixing device are necessary in order to achieve one perfect mixing of the product within the scope of the unterlying penetration model. Consequently, large values of N_{mix} indicate a less effective mixing of the product and vice versa. Since large values always appear at large numbers of revolutions per unit time, one may conclude, that intensifying the mixing power is always accompanied by a loss of mixing effectiveness. This is probably the only general result of this analysis. Quantitative evaluation of N_{mix} leads to the establishment of empirical power laws for each dryer or group of dryer. As a first attempt N_{mix} was correlated with the Froude number Fr, which at least for the drum driers makes some sense. For the disc dryers gravity forces are also involved, however, they do not fall in line with the centrifugal forces. Nevertheless, the Froude number was also used to correlate N_{mix} for disc dryers. The resulting power law correlations write

$$N_{mix} = C\ Fr^{x} \qquad (33)$$

with

$$Fr = \frac{(2\pi n)^2\ D}{2\ g} \ , \qquad (34)$$

where D is the diameter of the disc or of the drum, resp. Evaluation of the available data yielded numbers for C and x as follows:

	C	x
Disc dryers No. I and II	25	0.20
Drum dryer No. III	16	0.20
Paddle dryers No.IV and V	9	0.05

6. SUMMERY AND CONCLUSIONS

Drying rate curves $\dot{m}(X)$ for free flowing particulate material in direct contact with a hot surface and in pure steam atmosphere can be predicted from physical properties alone without any empirical (data fitted) parameter, provided that the actual contact time of the material at the hot surface is known. This presupposition is fulfilled in case that the particulate material forms a static packed bed resting on a hot surface. Evidence is given in the figures 10 and 11 (zero stirrer speed, n = 0). The drying rate curves can be predicted by a simple straight foreward stepwise calculation procedure (HP 41 C) using only physical properties in order to calculate the contact heat transfer resistance $1/\alpha_{ws}$, the dry bulk heat penetration resistance $1/\alpha_{dry}$ and eventually the wet bulk heat penetration resistance $1/\alpha_{wet}$.

If the particulate material is mechanically agitated the actual contact time is unknown. Two steps were necessary to overcome this problem. First a model, the so called penetration model, was introduced, which (arbitraryly) defines the actual contact time. Secondly, empirical correlations have been established, which connect the actual contact time with the time scale of the mixing device. These correlations reflect the fact that on the average the actual contact time is roughly ten times the time scale of the mixing device.

With this concept it was possible to obtain very good agreement between theory and experiments for five different direct contact dryers (disc, drum, paddle) of various size (11 up to 160 l) operated at various pressures (1 mbar up to 200 mbar), various temperature differences (10^oC up to 200^oC) and various time scales of the mixing motion ($.2 \leq n \leq 130$ rpm) and filled with different (monosized) free flowing fine and coarse material (170 micron up to 6.6 mm particle diameter). No particle resistance was introduced at all, which means that the liquid even at low average bulk moisture content X evaporates at the surface of the individual particles.

Therefore, the drying rate curve of the bulk $\dot{m}(X)$ is nothing but a bulk property rather than a particle property. This may be explained by the following reasons. Firstly, for each single particle the drying rate varies from a maximum value when it contacts the hot surface to almost zero when it stays in the bulk. During the latter period moisture profiles within the particles can relax, which favors surface evaporation. Secondly, in case that the moving drying front has passed over a certain number of particle layers, which is true for fine grained material, the bulk resistance is always greater than any heat conduction resistance formed by a dry shell of a single particle. Thirdly also the blockage of the hot surface by already dry particles reduces the average actual drying rate especially at low average bulk moisture contents X much more than any internal particle resistance.

This altogether may explain why obviously no internal particle resistances must be taken into account in direct contact drying. This result should be accepted at least until new facts demand revision.

NOMENCLATURE

A	m^2	hot surface area
A_{requ}	m^2	required hot surface area
c	-	constant, eq.(33)
$c_{p,bed}$	J/kg K	specific heat of dry bed
d	m	particle diameter
D	m	dryer diameter
Fr	-	Froude number, eq. (34)
g	m/s^2	gravity constant
Δh_{ev}	J/kg	latent heat of evaporation
$\dot{m}$	kg/m^2s	drying rate
$_{max}\dot{m}$	kg/m^2s	drying rate of perfectly mixed bed
$\dot{m}_{max}$	kg/m^2s	maximum drying rate of non-perfectly mixed bed
$\dot{m}_{min}$	kg/m^2s	minimum drying rate of non-perfectly mixed bed
M_{dry}	kg	mass of dry material
n	1/s	number of revolutions of mixing device
N_{therm}	-	dimensionless group, eq. (29)
N_{mix}	-	mixing number, eq. (30)
p	bar	pressure in dryer
p_b	bar	bulk pressure
p_s	bar	saturation pressure at drying front
$\dot{q}$	W/m^2	heat flux
$\dot{q}_o$	W/m^2	heat flux into bed
$\dot{q}_{z_T}$	W/m^2	heat flux into drying front
t	s	time
t_{mix}	s	time constant of mixing device ($\equiv \frac{1}{n}$)
t_R	s	contact time
t_{requ}	s	required drying time
T	K	temperature
T_b	K	bulk temperature
T_h	K	temperature of heating oil or vapour
T_o	K	interface temperature
T_s	K	saturation temperature
T_w	K	hot surface temperature
ΔT_b	k	increment of bulk temperature
x	-	constant in eq. (33)
X	kg/kg	moisture content
X_{eq}	kg/kg	equilibrium moisture content
X_{in}	kg/kg	initial moisture content
X_{out}	kg/kg	moisture content at end of drying process
ΔX	kg/kg	decrement of moisture content
z	m	length normal to hot surface
z_T	m	penetration depth
α	W/m^2K	heat transfer coefficient
α_{dry}	W/m^2K	overall heat transfer coefficient of dry bed
α_p	W/m^2K	particle heat transfer coefficient
α_{sb}	W/m^2K	bulk heat transfer coefficient
$\alpha_{sb,dry}$	W/m^2K	heat transfer coefficient of dry bulk
$\alpha_{sb,wet}$	W/m^2K	heat transfer coefficient of wet bulk
α_{wet}	W/m^2K	overall heat transfer coefficient of wet bed
α_{ws}	W/m^2K	contact heat transfer coefficient
α^*_{ws}	W/m^2K	modified contact heat transfer coefficient
β_b	kg/m^2sPa	bulk permeation coefficient
β_p	kg/m^2sPa	particle permeation coefficient
ζ	-	reduced penetration depth
Θ	-	normalized temperature difference
κ_{bed}	m^2/s	thermal diffusivity of bed
λ_{bed}	W/mK	thermal conductivity of bed
$\dot{\mu}$	-	normalized drying rate
ξ	-	reduced moisture content
ρ_{bed}	kg/m^2	density of bed
τ_R	-	reduced contact time

REFERENCES

1. Schlünder, E.U., Chem.-Ing.-Techn. 43, 651/654 (1971).
2. Mollekopf, N., and Martin, H., vt-verfahrenstechnik, 16, 701/706 (1982).
3. Schlünder, E.U., Chemie-Ing.-Techn. 53, 925/941 (1981).
4. Mollekopf, N., Wärmeübertragung an mechanisch durchmischtes Schüttgut mit Wärmesenken in Kontaktapparaten, Dissertation, Institut für Thermische Verfahrenstechnik, Universität Karlsruhe, Karlsruhe, (1983).
5. Schlünder, E.U., vt-verfahrenstechnik 18 (1984).
6. Carlslaw, H.S., Jaeger, J.C., Conduction of Heat in Solids, 2nd. ed., pp. 283, Oxford University Press (1959).
7. Günes, S., Schlünder, E.U. and Gnielinski, V., vt-verfahrenstechnik 14, 31/39 (1980).
8. Fritz, W., Untersuchungen zur Wärmeübertragung bei der Kontakttrocknung von körnigen und formlosen Produkten in Vakuumschaufeltrocknern, GVC-Fachausschußsitzung 'Trocknungstechnik', Graz (1982).
9. Vosteen, B., Wärmetransport in Kontaktkühlern und Kontakttrocknern, GVC-Fachausschußsitzung 'Trocknungstechnik', Graz (1982).
10. Zehner, P., Experimentelle und theoretische Bestimmung der effektiven Wärmeleitfähigkeit durchströmter Kugelschüttungen bei mäßigen und hohen Temperaturen, VDI-Forschungsheft, No. 588, Verein Deutscher Ingenieure, Düsseldorf (1973).
11. Bauer, R., Effektive radiale Wärmeleitfähigkeit gasdurchströmter Schüttungen aus Partikeln unterschiedlicher Form und Größenverteilung, VDI-Forschungsheft No. 582 (1977) Verein Deutscher Ingenieure, Düsseldorf, see also Int. Chem. Eng. 18, 181/203 (1978).
12. Schlünder, E.U., Proc. 7th Int. Heat Transfer Conf., München, vol. 1, RK 10, 195/212 (1982).

HEAT AND MASS TRANSPORT PROPERTIES OF HETEROGENEOUS MATERIALS

Morio Okazaki

Department of Chemical Engineering, Kyoto University
Kyoto 606, JAPAN

ABSTRACT

Relating to the drying mechanism of the wet porous materials, correlating (or predicting) methods for effective thermal conductivity of the wet porous solid and the apparent mass transfer coefficient of the adsorbed molecule in the adsorptive fine-porous solid are described.

1. INTRODUCTION

One of the ultimate purposes of the fundamental research on drying mechanism must be to establish a reliable predicting procedure of the relation between the drying rate and the moisture content, that is the drying characteristic curve, under any arbitrary drying condition. The phenomena of "drying is one of the typical simultaneous transport phenomena of heat and mass. Therefore, if we can obtain the solution of the simultaneous differential equations on heat and mass (moisture) transfer in the wet material, it is possible to predict the drying characteristic curve by using this solution.

Generally there exist two problems against solving the governing differential equations.

The first problem is that it is not so easy to make reasonable evaluations of the apparent thermal conductivity and moisture transfer coefficient (moisture diffusivity) in the wet material which exist in the above differential equations, because of those strong dependencies on the moisture and the temperature. The industrial raw materials to be dried are generally so-called "Porous Solids" which have very complicated structure character of pore network.

The second problem concerns with the mathematical procedure to obtain the solution of the simultaneous differential equations. Those equations are usually non-linear since the transport properties such as the thermal conductivity and the moisture transfer coefficient are of function of moisture content and/or the temperature. Even today, it might be inpractical to get the rigorous numerical solution by means of large computer. By this reason, developments on the practical method to obtain the approximate solutions are very required.

The purpose of this paper is to describe the recent developments concerning with the problem especially on the correlation and/or prediction method of the effective thermal conductivity of the wet porous solid and the apparent mass transfer coefficient in the adsorptive porous solid.

2. EFFECTIVE THERMAL CONDUCTIVITY OF WET POROUS MATERIAL

Effective thermal conductivities of wet porous material, such as packed bed of particles, bricks, thermal-insulation materials or sintered metals, are often used for various engineering calculations as well as for studies of solid drying.

In dried-up or fully saturated state, the porous material consists of two phases, solid and fluid (gas or liquid)phase. The wet porous material below saturation consists of three phases, solid, gas and liquid. In the following, correlation or prediction procedures of effective thermal conductivites of wet porous materials in which the thermal conductivity of the solid phase, λ_s, is greater than the one of the liquid phase, λ_l are mentioned first. Then those for case of $\lambda_s<\lambda_l$ are mentioned.

2.1. Characterization of Porous Solids

As a basis of the following analysis, we regard the porous solid as the bed of uniform spherical particles consolidated at the contact point between particles as shown in Fig. 1, where the degree of consolidation is specified by angle α, β or γ.

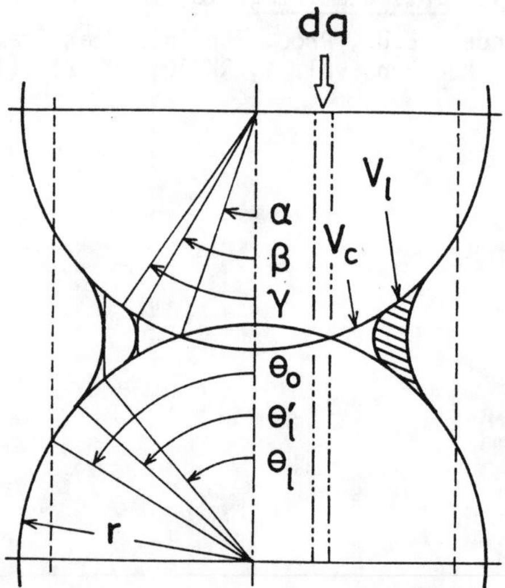

Fig. 1 Wedge water around a contact point

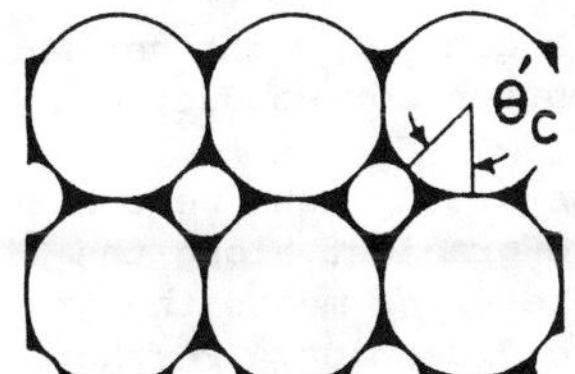

Fig. 2 Coalescing angle

The packed bed of particles in unconsolidated state is given by $\beta = 0$ for example.

The number of contact points Nc on one particle made by the sorrounding particles is of importance to characterize the porous solid. When one evaluates the effective thermal conductivity, it plays a particulary predominant role. The following Ridgway and Tarbuck's equation [1] can be used to approximately evaluate the number of contact points in the consolidated granular bed from the void fraction ψ.

$$N_c = 13.8 - \sqrt{232\psi - 57.2} \quad : \psi \geq 0.249 \tag{1}$$

2.2. Water Configuration in Bed

The configuration of water retained in bed must be considered as a stochastic phenomena and the rigorous treatment seems to be impossible. Thus we introduce the following assumptions.

An angle θ_l made at the center of a particle by a wedge water held around a contact point as shown in Fig. 1, is defined as a measure of the amount of the wedge water. If all of the wedge water in the bed uniformly grow up, they begin to coalesce with the neighbors just at the critical angle θ_c which is dependent only on the packing mode, provided that the particles are fine (cf. Fig. 2). After θ_l reaches θ_c, the voids can be thought to be suddenly filled up successively with water, as the water content increases further. Based on these assumptions, the state of retained water can be expressed as follows. As the apparent water content of the bed, Φ, gradually increases from zero, the wedge waters uniformly grow up and finally they reach θ_c, and thereafter they do not grow anymore, but voids saturated with water begin to appear, and the number of them increases until all of voids are saturated out.

Providing that the cross-sectional area exclusively occupied by a critical wedge water of θ_c, $\pi r^2 sin^2\theta_c$, is approximately equal to one $N_c/2$ th of the cross-sectional area of the particle, πr^2, the critical angle θ_c is given by

$$\theta_c = Sin^{-1}\sqrt{2/N_c} \tag{2}$$

On the other hand, the next relation exists between θ_l and θ'_l.

$$\theta'_l = Sin^{-1}\left\{\frac{cos^2\alpha - (1 - sin\theta_l)^2}{cos^2\alpha + (1 - sin\theta_l)^2}\right\} \tag{3}$$

or

$$\theta_l = Sin^{-1}(cos\alpha \cdot tan\theta'_l + 1 - cos\alpha/cos\theta'_l) \tag{4}$$

From Eqs. (2) and (3), θ_c can be evaluated. While the theoretical θ'_c is 45° for the cubic and 30° for the orthorhombic packing, the calculated values by Eq.(2) are 44.5° and 29.3° respectively.

2.3. Water Content of Bed

When the peripheries of the cross sections of the solid part, which is forced out around a contact point by sintering of the contact point, and the wedge water are assumed to be circular, these relative volumes to the volume of a particle are given by

$$V_c = (2 - 3cos\alpha + cos^3\alpha)/2 \tag{5}$$

$$= (3/2)cos\alpha\{(cos\alpha/cos\gamma) - 1\}^2 \times [1 - \{(\pi/2) - \gamma\}\ tan\gamma] \tag{6}$$

$$V_l = (3/2)cos\alpha\{(cos\alpha/cos\theta'_l) - 1\}^2 \times [1 - \{(\pi/2) - \theta'_l\}tan\theta'_l] - V_c \tag{7}$$

There is the next relation among α, β and γ.

$$sin\beta = (1/cos\gamma)(cos\alpha \cdot sin\gamma - cos\alpha + cos\gamma) \tag{8}$$

The values α and γ can be calculated from β by using Eqs.(5),(6) and (8), and then V_c and V_l can be calculated from β and θ_l by using Eqs.(3) and (5)~(7).

The "local" water content, ϕ, referred to a contact point, which is the degree of saturation of water to the void volume allotted to one contact point, is given by

$$\phi = N_c V_l (1 - \psi)/(2\psi) \tag{9}$$

According to the above simplifications on the water configuration, the relationship between the (apparent) water content of the bed, Φ, and the local water contents is represented as follows;

$$\text{(i)}\quad 0 < \Phi \leq \Phi_c\ (= \phi_c) \quad : \quad \Phi = \phi \tag{10}$$

$$\text{(ii)}\quad \Phi_c < \Phi < 1 \quad : \quad \Phi = a\phi_c + (1 - a) \tag{11}$$

,where

$$a = \frac{\text{number of wedge water of critical state}}{\text{total number of contact points}} \tag{12}$$

From Eq.(11),

$$a = (1 - \Phi)/(1 - \phi_c) \tag{13}$$

where ϕ_c can be calculated from Eq.(7) by substituting θ'_c to θ'_l.

2.4. Unit Cell of Bed

When the conductivity of the solid phase is greater than those of gas and liquid phase, a unit cell model as shown in Fig. 3 can be taken in order to evaluate the effective thermal conductivity. The

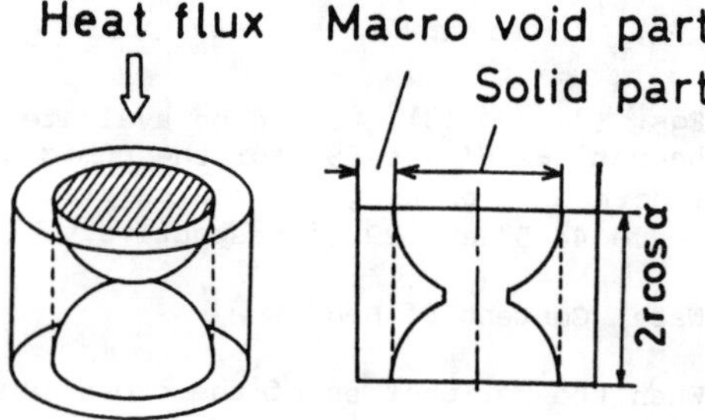

Fig. 3 Unit cell model

void fraction of the "solid part" in the cell, ψ_s, is given by

$$\psi_s = 1 - 2/(3\cos\alpha) \tag{14}$$

Therefore the fractional areas of the solid and the macro void part, across which heat flows, are given as follows;

$$\begin{array}{lll} & \psi \geq \psi_s & \psi < \psi_s \\ A_s : & (3/2)(1-\psi)\cos\alpha & 1 \\ A_v : & 1-(3/2)(1-\psi)\cos\alpha & 0 \end{array} \tag{15}$$

2.5. Apparent Conductivity of Solid Part

Effective number of contact points. Let us consider that how many contact points locating just on the macroscopic direction ($\delta=0$ in Fig. 4) of the heat transfer the real contact points, whose number is given by Eq.(1), randomly locating on the surface of the under hemi-sphere of a particle are equivalent to, from the contribution on conductive heat transfer.

The following assumptions are adopted relating to Fig.4.

(1) The temperature is uniform on the plane perpendicular to the direction of the macroscopic heat transfer.

(2) The heat transfer from a sphere to an adjacent one through the contact point can be substantially replaced with that from the center of the sphere to the one of the adjacent sphere along the line connecting them.

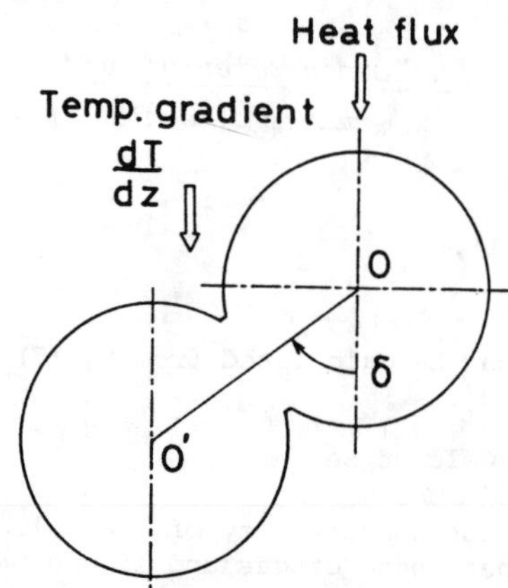

Fig. 4 Configuration of contact point

(3) The each contact point has an independent contribution for conduction.

From these assumptions we can assume that the temperature gradient along the line OO' becomes $\cos\delta \cdot dT/dz$ where dT/dz means the gradient along the macroscopic direction. Further, the displacement from the center O to the center O' is equivalent to the displacement of $2r \cdot \cos\delta$ along the macroscopic direction. Therefore finally we can assume that the heat flux through a contact point having an angle δ corresponds to $\cos^2\delta$ times of that having the angle $\delta=0$. The number of contact points per unit surface area of a sphere is given as $N_c/4\pi r^2$. The surface element corresponding to an angle element $d\delta$ is $2\pi r^2 \cdot \sin\delta \cdot d\delta$. Consequently the effective number of contact points on a hemisphere based on contribution of heat transfer, n_c, is given by Eq.(16).

$$n_c = \int_0^{\pi/2} \cos^2\delta \, \frac{N_c}{4\pi r^2} \, 2\pi r^2 \sin\delta \, d\delta = \frac{N_c}{6}, \; (\text{for } N_c \geq 6) \tag{16}$$

$$= 1, (N_c < 6) \tag{17}$$

When N_c is less than 6, we assume $n_c=1$ in order to satisfy that $\lambda_e = \lambda_s = \lambda_l$ in the case of $\lambda_s = \lambda_l$ for any viod fraction of bed.

Heat flux through a contact point. Let us consider the mechanisms of heat transfer at a moderate low temperature in the bed where there is no convection of fluid and no radiative transfer. It is necessary, for the present case, to take into account not only the conductive heat transfer through solid, liquid and gas, but also the enthalpy transfer accompanied by the water vapor diffusion, because that the bed is wet. In Fig. 1 illustrating the cross section of two particles in contact with each other, the heat transfer occurs in the vertical direction by the following four paths, assuming them to be unidirectional (Kunii et al.[2]).

a. Conduction through solid(particle), solid (sintered contact point) and solid(particle).

$$Q_1 = \int_0^{\beta} q_1(\theta)\, d\theta \quad ;[S - S - S] \tag{18}$$

b. Conduction through solid(particle), liquid (wedge water) and solid(particle).

$$Q_2 = \int_{\beta}^{\theta_l} q_2(\theta)\, d\theta \quad ;[S - L - S] \tag{19}$$

c. Conduction through solid(particle), gas(with enthalpy transfer) and solid(particle).

$$Q_3 = \int_{\theta_l}^{\theta_l'} q_3(\theta)\, d\theta \quad ;[S -(G + D) - S] \tag{20}$$

d. Conduction through solid(particle), gas(without enthalpy transfer) and solid(particle).

$$Q_4 = \int_{\theta_l'}^{\theta_0} q_4(\theta)\, d\theta \quad ;[S - G - S] \tag{21}$$

, where θ_0, which is an angle prescribing the part of the particle cross-sectional area exclusively possessed by one contact point, is given as follows, providing that this area is to be one-n_cth of the cross-sectional area of particle.

$$\pi(r \cdot sin\theta_0)^2/\pi r^2 = 1/n_c \tag{22}$$

$$\theta_0 = Sin^{-1}\sqrt{1/n_c} \tag{23}$$

The heat flux through one contact point, Q, is given by Eq.(24).

$$Q = Q_1 + Q_2 + Q_3 + Q_4 \tag{24}$$

As the total flux through one particle, that is through the solid part of "unit cell" is $n_c Q$, the apparent conductivity of the solid part is given by

$$\lambda_{es} = (\frac{n_c Q}{\pi r^2})/(\frac{dT}{dz}) = 2n_c K \tag{25}$$

$$K = (\lambda_s sin^2\beta)/2 + \lambda_l \kappa_l\{\kappa_l ln(\frac{\kappa_l - cos\theta_l}{\kappa_l - cos\beta}) + cos\theta_l - cos\beta\}$$

$$+(\lambda_d + \lambda g)\kappa_d\{\kappa_d ln(\frac{\kappa_d - cos\theta_l'}{\kappa_d - cos\theta_l}) + cos\theta_l' - cos\theta_l\}$$

$$+\lambda_g \kappa_g\{\kappa_g ln(\frac{\kappa_g - cos\theta_0}{\kappa_g - cos\theta_l'}) + cos\theta_0 - cos\theta_l'\} \tag{26}$$

where

$$\kappa_l = \lambda_s cos\alpha/(\lambda_s - \lambda_l)$$

$$\kappa_d = \lambda_s cos\alpha/(\lambda_s - \lambda_d - \lambda_g) \tag{27}$$

$$\kappa_g = \lambda_s cos\alpha/(\lambda_s - \lambda_g)$$

In the above equations, λ_d is the apparent conductivity of the enthalpy transfer contribution caused by so called "the evaporation-diffusion-condensation" of liquid(water) given by Eq.(28) [5].

$$\lambda_d = [M_l D_v p_T \Delta h_v/\{RT(p_T - p_s)\}](dp_s/dT) \tag{28}$$

2.6. Effective Thermal Conductivity

The apparent heat flux through the bed is the sum of the seperate fluxes through the solid part and the marco void part in the unit cell and the enthalpy flux through the void in case of wet state below saturation. Then the effective thermal conductivity is given as follows:

$$\Phi = 0 \quad : \quad \lambda_e = A_v\lambda_g + A_s\lambda_{es} \tag{29}$$

$$0 < \Phi \leqq \Phi_c \quad : \quad \lambda_e = A_v\lambda_g + (\psi/\mu_{ld})\lambda_d + A_s\lambda_{es} \tag{30}$$

$$\lambda_e|_{\Phi=\Phi_c} = (\lambda_e)_{crit.} \tag{31}$$

$$\Phi_c < \Phi < 1 : \quad \lambda_e = a(\lambda_e)_{crit.} + (1 - a)(\lambda_e)_{sat.} \tag{32}$$

$$\Phi = 1 \quad : \quad \lambda_e = A_v\lambda_l + A_s\lambda_{es} \tag{33}$$

$$= (\lambda_e)_{sat.} \tag{34}$$

where (λ_e)crit. and (λ_e)sat. are the conductivity at $\Phi=\Phi_c$ and $\Phi=1$ respectively.

As to the tortuosity factor, μ_{ld} of the vapor diffusion through the void in the bed, the following equation generally gives a good prediction for the typcal consolidated granular bed such as the glass filter or the sintered metal.

$$1 / \mu_{ld} = \sqrt{\psi} \quad : consolidated\ packed\ bed \tag{35}$$

, which is a special case of Bruggemann's equation [6] on the effective conductivity of emulsion.

On the other hand, the predicted μ_{ld} by Eq.(35) is usually too small as to the porous solid which has complicated void structure, such as the commercial brick. For this type of porous solid, the next empirical formula which was obtained by correlation of the observed μ_{ld} of Krischer et.al. [7] (cf. Fig. 5) is recommendable.

$$1 / \mu_{ed} = 0.39\ \psi^{1.2} \quad : commercial\ brick \tag{36}$$

2.7. Effective Thermal Conductivity in Case of $\lambda_s < \lambda_f$

The prediction procedure mentioned above is as for the wet porous solid in which the conductivity of the solid phase is larger than that of the fluid phase. Although the gas thermal conductivity never exceeds the solid conductivity, the liquid conductivity sometimes exceeds the solid conductivity. In the latter case, the above prediction is apt to give low effective conductivity in fully saturated state, (λ_e)sat. For this case, the following Bruggemann's Equation[6], which has been proposed to evaluate the effective conductivity of the emulsion, can be adopted.

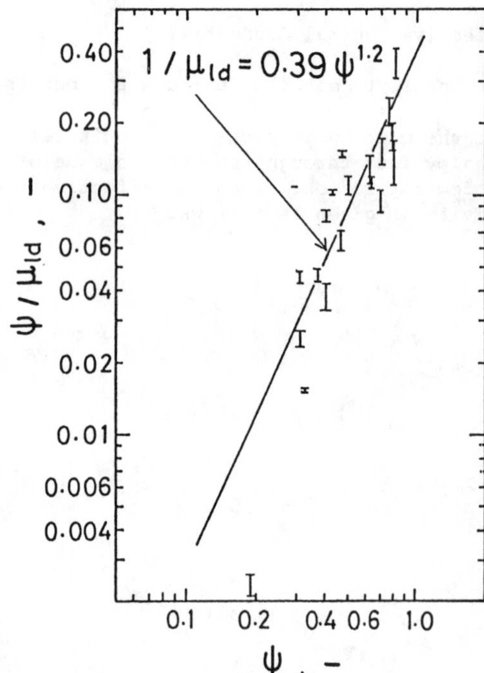

Fig. 5 Correlation of the tortuosity factors observed by Krischer [7]

$$\frac{\lambda_e - \lambda_s}{\lambda_l - \lambda_s} \cdot \left(\frac{\lambda_l}{\lambda_e}\right)^{1/3} = \psi \qquad (37)$$

(By substituting $\lambda_s=0$ into Eq.(37), Eq.(35) can be obtained.)
The predicted value by Eq.(37) is to be (λ_e)sat.

No.	System (14°C)	ψ	λ_s
1	Crushed glass - Castor oil	0.453	0.955
2	Crushed glass - Water	0.530	0.955
3	Quartz sand - Castor oil	0.454	5.23
4	Quartz sand - water	0.501	5.23

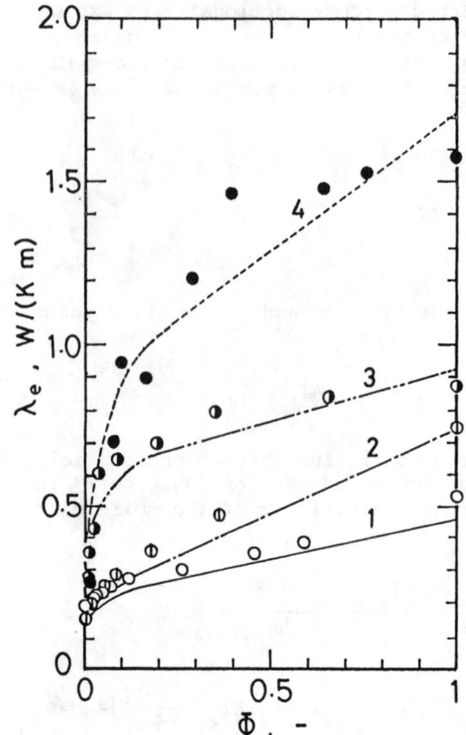

Fig. 6 Effective thermal conductivities (granular bed, no enthalpy effect) : Kimura [8]

2.8. Comparison of Model with Data

Fig. 6 shows the comparison of the predicted and the observed thermal conductivities of unconsolidated bed by Kimura[8] at relatively low temperatures where the effect of enthalpy transfer caused by vapor diffusion may be negligible. The data are presented in terms of an effective thermal conductivity and water content. The predicted results seem to be in good agreement with the observed over the whole range of water content. The values of λ_s, λ_g and λ_l used for prediction were taken from the paper [8].

The comparisons in the case of relatively high temperature where the effect of enthalpy transfer must be taken into account are shown in Figs.7 and 8 [3]. The value of λ_s of "Tottori Sand" was determined so as to fit the predicted conductivity with the observed at ϕ=0. That of "Glass Beads" is the observed value by a seperate measurement. The arrow heading right points to the predicted conductivity in the bone-dried state, and the one heading left to the extrapolated conductivity from the wet state to the dried-up state. The difference between them comes from the contribution of the enthalpy transfer at ϕ=0.

Though there are some larger discrepancies between the observed and the predicted than in Fig.6 , the observed results are seen to follow the predicted trend within a deviation of ±25%.

Figs. 9 ~ 13 concern with consolidated porous solids. Figs.9 and 10 present data of three kinds of glass filter, GF-1, 2 and 3. The values of β were determined by photographic method to the cross section of test piece. Further, the value of λ_s is also observed value. Therefore the predicted curves are from the present model without any arbitrary parameters. In spite of that the predicted at $\Phi=0$ are somewhat smaller than the observed in all cases, it would be said, from a practical point of view, that the agreement between the observed and the predicted over the whole range of water content favors the approach taken in the present prediction.

Figs.11, 12 and 13 present data from four kinds of brick, HB-18, B-4, A-1 and S. It was impossible to obtain the observed data of λ_s and β for them by separate experiments. Thus these two paraments were determined so that the observed and the predicted conductivity coincide with each

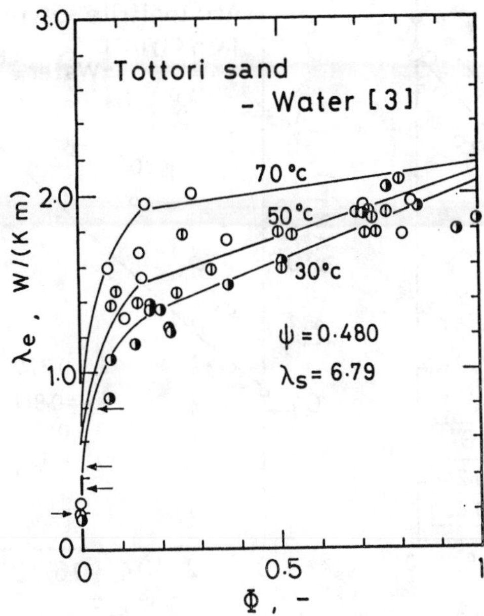

Fig. 7 Effective thermal conductivities (granular bed, enthalpy effect) [3]

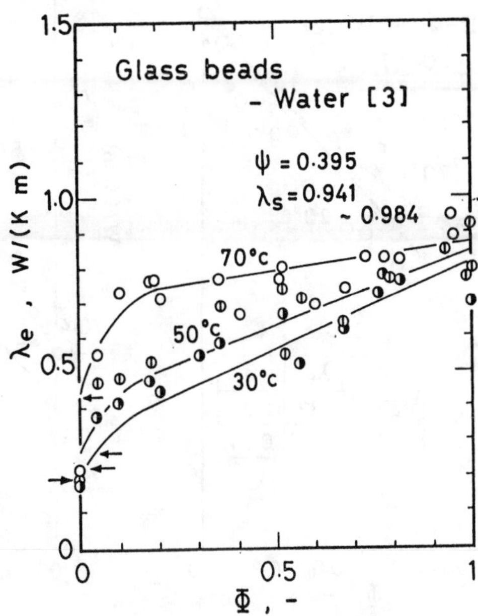

Fig. 8 Effective thermal conductivities (granular bed, enthalpy effect) [3]

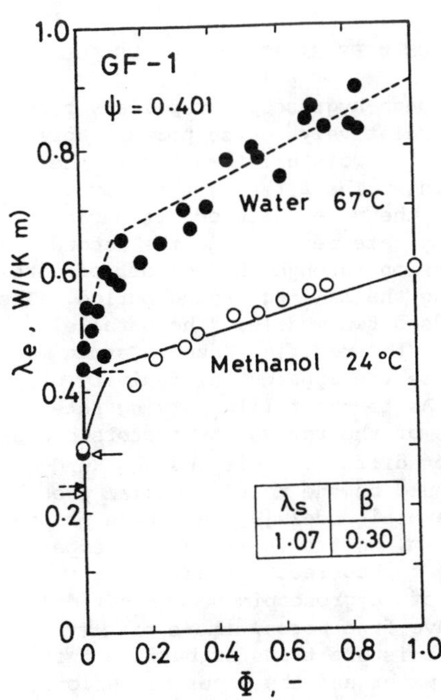

Fig. 9 Effective thermal conductivities (Glass filter : GF-1) [4]

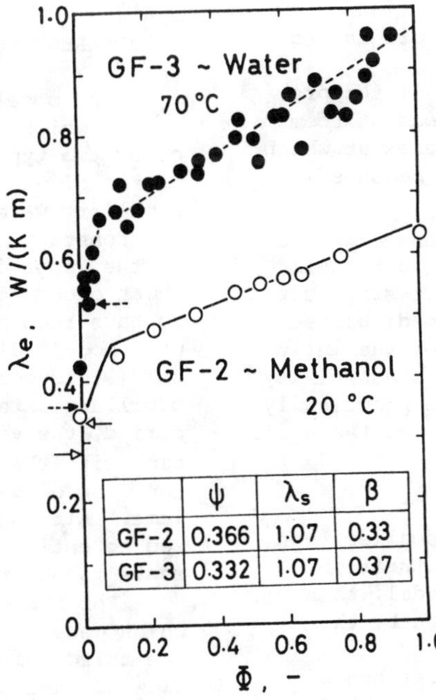

Fig. 10 Effective thermal conductivities (Glass filter : GF-2 and GF-3) [4]

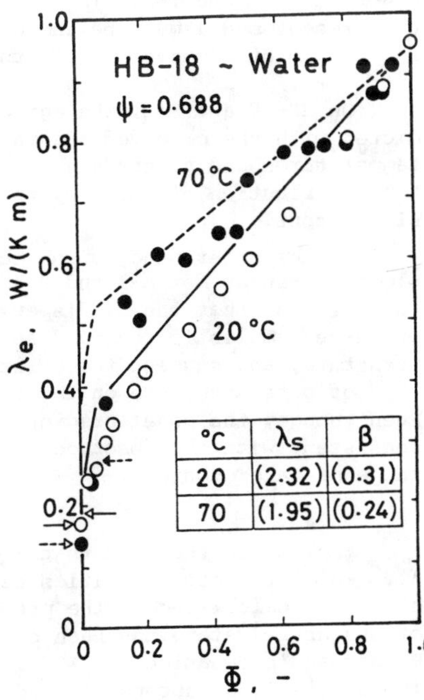

Fig. 11 Effective thermal conductivities (Brick : HB-18) [4]

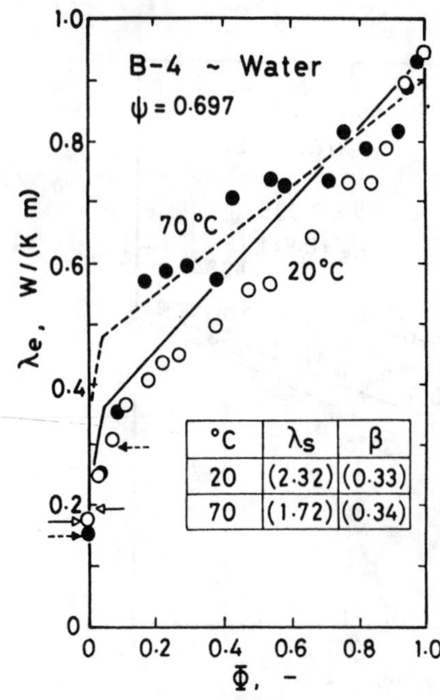

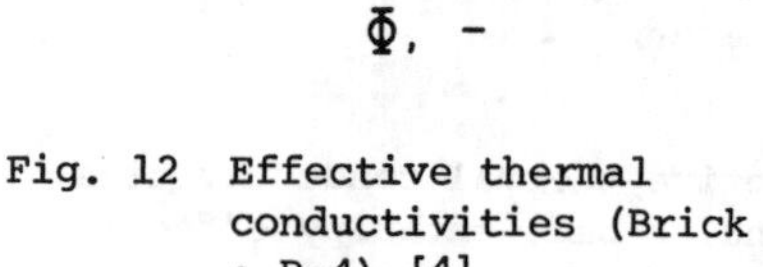

Fig. 12 Effective thermal conductivities (Brick : B-4) [4]

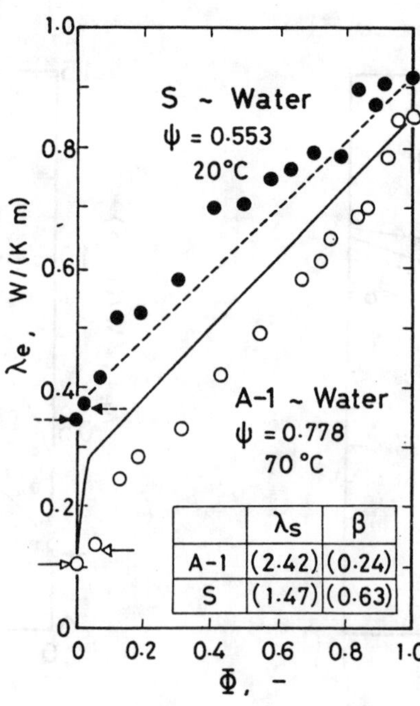

Fig. 13 Effective thermal conductivities (Brick : A-1 and S) [4]

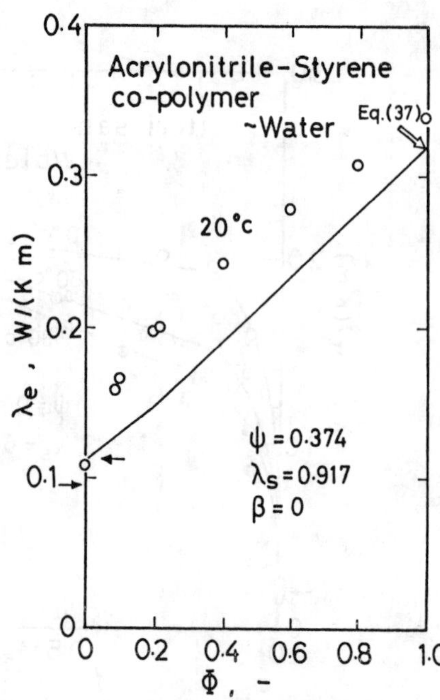

Fig. 14 Effective thermal conductivities (granular bed : ASP) [4]

other at $\Phi=0$ and $\Phi=1$. In these figures, therefore the present model must be thought as a kind of an interpolating model rather than a predicting one.

On HB-18 and B-4, the agreements of the predicted with the observed are fairly good at room temperatures and pretty high temperatures at which the contribution of enthalpy transfer becomes significant.

On the other hand, there occur some appreciable differences for A-1 and S. These facts seem to come from that the bi-dispersed pore structure of these samples is regarded as a mono-dispersed structure, and suggest the existence of the influence of pore structure on the thermal conductivity. Even though, the model predictions are practically consistent with the observed results over the whole range of water content.

Fig. 14 is the case that λ_s is smaller than λ_l. While the conductivities below saturation have been calculated by the present model, that of staturated state has been calculated by the Bruggemann's equation, Eq.(37). The sample is an unconsolidated granular bed of Acrylonitrile-Styrene co-polymer, of which the conductivity is about one third of that of water. Even though the predicted conductivity, which has been obtained without any arbitary parameters, are appreciably smaller than the observed especially in the unsaturated range expect the bone-dried and the saturated state, still it might be a kind of rough estimation. But it is certainly necessary to make some modification on the present model.

3. APPARENT MOISTURE DIFFUSIVITY IN POROUS SOLID

In so called "non-hygroscopic capillary porous solid" which have relatively coarse pores, there occur two types of the moisture transfer in the course of the drying. The first is the flow of capillary water in the pore which occurs during the constant drying rate period, and the second is the vapor diffusion through the network of pores which occurs during the falling drying period. There have been proposed two models, "the parallel pore model"[7] and "the modified Kozeny-Carman model"[9] to correlate the apparent diffusivity of capillary water. As to the falling drying rate period, the values of the tortuosity factors concerning with the vapor diffusion, viscous and Kundsen flow of gas would be the most essential problems. Although several models[9] have been proposed to estimate them, so far it would need experiments to determine the correct values.

For the case of "hygroscopic porous solids" which generally have fine pores, there exists the third type of moisture transfer between the flow of capillary water and the vapor diffusion. Because of the diverse characteristics of hygroscopic materials, it is impossible to give a general understanding on this type of mass transfer. Among various kinds of such material, the adsorptive material where the physical adsorption of water occurs is the most simple and typical one. This type of mass transfer in the adsorptive material, means the surface flow (or surface diffusion) of adsorbed gas molecules on the interior wall of pores. This moisture transfer becomes of importance in the relatively mild drying and moistening of dry product. The purpose of this chapter is to describe the latest developments in the understanding of the surface flow in the adsorptive material.

In the range of low partial pressure of the adsorbable gas, monolayer adsorption predominates, while multilayer adsorption gradually increases with increase of the pressure. Further, when the pressure is approaching the saturated pressure of the adsorbate, capillary condensation begins to occur in the pores of adsorbent. The adsorbed phase, which means the molecules adsorbed in both monolayer and multilayer, can be considered to be gradually replaced by the capillary condensed phase. Therefore, the apparent adsorbed phase should be divided into the actual adsorbed phase and the capillary condensed phase to interpret the mechanism of surface flow.

Consider the case that a plug of porous adsorbent contacts the gases having different partial pressure of adsorbate on the both sides. In the adsorbed phase the transport of adsorbed molecules is caused by the gradient of the amount adsorbed, or more directly, the gradient of the number density of the hopping molecules. On the other hand, the transport of capillary condensate can be regarded as Poiseuille flow of viscous liquid filling the pores of the porous media, caused by the gradient of capillary force. Hence, the above two transport mechanisms should be considered separately.

3.1. Surface Flow in Adsorbed Phase

Modified hopping molecule model. Hill[10] and Higashi et al.[11] regarded the surface flow phenomenon as a random walk process of the so-called "hopping molecules" on the surface. We chose the case that $E_{ao}>E_{so}>RT$ as the basis for the present model, where E_{so}, E_{ao} and RT are the potential barrier among adsorption sites, the differential heat of adsorption and the thermal motion energy of molecules respectively. In the case that $E_{ao}<E_{so}$, the surface flow does not take place. If $E_{so}<RT$, an adsorbed molecule would be expected to behave as a molecule in a two-dimensional gas.

When an adsorbed molecule gains energy E between E_{so} and E_{ao}, this molecule hops from site to site in the adsorption state. Before the molecule is desorbed, several "hoppings" occur and therefore surface flow ensues. On the other hand, an adsorbed molecule remains on the adsorption site when $E<E_{so}$ and is desorbed in the case that $E > E_{a0}$.

Consider the migration of "hopping molecules" on the surface. Fig. 15 shows the hopping of adsorbed molecules. These molecules randomly hop to one of the neighboring sites after various holding times as shown in Fig. 15. Here, it is assumed that only the adsorbed molecules exposed to the gas phase contribute to the surface flow and the adsorbed molecules in the lower layer do not make any contribution to the surface flow. Further, we divide the molecules exposed to the gas phase into two groups. One group is the molecules adsorbed on the vacant sites, and the other is the molecules forming a multilayer. All molecules of latter group make same contribution to the surface flow.

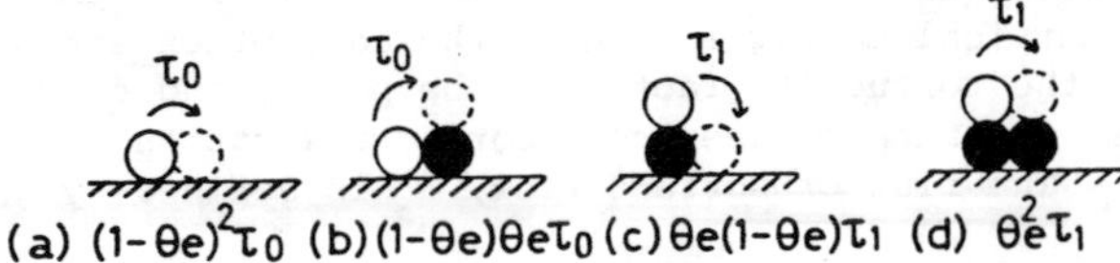

Fig. 15 Hopping modes of adsorbed molecules

Surface flow coefficient. The mean holding time of the adsorbed molecule in the first layer, τ_0, is given as follows.

$$\tau_0 = \tau_{s0}\int_0^{E_{a0}} f_{s0}(E)dE\Big/\int_{E_{s0}}^{E_{a0}} f_{s0}(E)dE \tag{38}$$

$$= \tau_{s0}(1-e^{-E_{a0}/RT})/(e^{-E_{s0}/RT}-e^{-E_{a0}/RT}) \tag{39}$$

,where f_{s0} is the distribution function of energy. Similary, for the adsorbed molecule in all layer above the first layer,

$$\tau_1 = \tau_{s1}(1-e^{-E_{a1}/RT})/(e^{-E_{s1}/RT}-e^{-E_{a1}/RT}) \tag{40}$$

where E_{a1} is equal to the heat of vaporization of the adsorbate by making the similar assumption as in the B.E.T. theory. E_{s1} is the activation energy for migration in the all layers above the first layer.

Fig. 15 illustrates the four kinds of hopping modes of the adsorbed molecule. Providing the energetically homogeneous surface, in Case (a) of Fig. 15 for example, the hopping probability is $(1-\theta_e)^2$ at the surface coverage θ_e, because the probabilities that an adsorbed molecule locates at a vacant site and it hops to another vacant site are both $(1-\theta_e)$, and its holding time is τ_0. Setting forward the similar considerations, the expected value of the holding time of an adsorbed molecule in the first layer, τ, is given by

$$\tau = (1-\theta_e)^2\tau_0+\theta_e(1-\theta_e)\tau_0+(1-\theta_e)\theta_e\tau_1+\theta_e^2\tau_1$$

$$= (1-\theta_e)\tau_0 + \theta_e\tau_1 \tag{41}$$

Regarding the surface flow as a random walk process, the surface flow coefficient D_s' based on the surface coverage θ_e can be expressed by the Einstein's equation:

$$D_s' = C'\delta^2/\tau \tag{42}$$

$$= C'\delta^2/[\tau_0\{1-\theta_e(1-\tau_1/\tau_0)\}] \tag{43}$$

Where C' is a constant to be determined from the geometrical configuration of the adsorption site and the tortuosity factor of the pores, and δ is the distance between neighboring adsorption sites.

The flux of surface flow, N_s is defined by

$$N_s = -\rho_{app} D'_s A dq_e/dz \quad (44)$$

$$= -\rho_{app} D_s A dq/dz \quad (45)$$

Then,

$$D_s = D_{s0}(\theta_e/\theta)(e^{-aE_{a0}/RT} - e^{-E_{a0}/RT}) / [(1-e^{-E_{a0}/RT})\{1-\theta_e(1-\tau_1/\tau_0)\}] \quad (46)$$

Here, we take the following assumptions.

$$E_{s0} = a \cdot E_{a0} \quad (47)$$

$$\tau_{s0} \cong \tau_{s1} \quad (48)$$

Then,

$$\tau_1/\tau_0 = (e^{-aE_{a0}/RT} - e^{-E_{a0}/RT})(1-e^{-E_{a1}/RT}) / \{(e^{-E_{s1}/RT} - e^{-E_{a1}/RT})(1-e^{-E_{a0}/RT})\} \quad (49)$$

When the surface is energetically heterogeneous, Eqs.(46) and (49) should be rewritten as follows.

$$D_s = \frac{D_{s0}(\theta_e/\theta)\int_{E^{\circ}_{a0}}^{E_{a0}} \frac{e^{-aE/RT} - e^{-E/RT}}{(1-e^{-E/RT})\{1-\theta_e(1-\tau_1/\tau_0)\}} g(E)dE}{\int_{E^{\circ}_{a0}}^{E_{a0}} g(E)dE} \quad (50)$$

$$\tau_1/\tau_0 = (e^{-aE/RT} - e^{-E/RT})(1-e^{-E_{a1}/RT}) / (e^{-E_{s1}/RT} - e^{-E_{a1}/RT})(1-e^{-E/RT}) \quad (51)$$

Here adsorbed molecules have various heat of adsorption, E. The function $g(E)$ is the number density of the adsorbed molecule which has the heat of adsorption between E and $(E + dE)$, and E_{a0} and E^0_{a0} are heats of adsorption at an arbitrary θ_e and $\theta_e=0$, respectively. The parameters except a and D_{s0} can be evaluate as follows[12], providing the adsorption isotherm obeys the B.E.T. equation.

$$\theta_e: \quad \theta = Cx/\{(1-x)(1-x+Cx)\} \quad (52)$$

$$\theta_e = \theta(1-x) \quad (53)$$

$$E_{s1}: \quad \mu_L = K_\mu T e^{E_{s1}/RT} \quad (54)$$

$$g(E): \quad E_{a0} = (E_{st} - RT - xE_{a1})/(1-x) \quad (55)$$

$g(E)$ is evaluated from the relation between E_{a0} and the amount adsorbed.

In summary, the surface flow coefficient, D'_s which concerns to the adsorbed phase is given by Eqs.(46) or (50), which contains two parameters, a and D_{s0}. These equations are just quantitative expressions of the dependences of the surface flow coefficient on the amount adsorbed and the temperature.

<u>Surface flow of adsorbed molecules under coexistence of capillary condensed phase.</u> When capillary condensation coexists, the surface area which contributes to the surface flow decreases because of occupation of capillary condensed phase in fine pores. Therefore, the surface flow rate in the adsorbed phase is given by Eq.(56).

$$N_d = -\rho_{app} D_s A (dq_{BET}/dz)\{(S_T - S_C)/S_T\} \quad (56)$$

$$= -P_d A dp/dz \quad (57)$$

$$P_d = (D_s \rho_{app})(dq_{BET}/dq)\{(S_T - S_C)/S_T\} \quad (58)$$

where $(S_T - S_C)/S_T$ is a correcting term which accounts for the decrease of the surface area contributing to the surface flow.

3.2. Flow of Capillary Condensate

The critical radii at which capillary condensation occurs can be given by the following Kelvin equation.

$$ln(p/p_s) = -2\sigma V_L cos\alpha/(r_c RT) \quad (59)$$

Further, the capillary suction pressure is given by

$$p_c = -2\sigma cos\alpha/r_c \quad (60)$$

According to the Kelvin equation the capillary condensation successively proceeds from fine pores to the larger ones as the gas-phase pressure of adsorbate increases. Hence the high pressure side of the porous plug is filled with capillary condensate more readily than the low pressure side. Once the capillary condensed phase is formed, the capillary force due to surface tension (Eq.(60)), is higher on the low pressure side than on the high pressure side.

Accordingly, if the gradient of the gas-phase total pressure exists, then both the gradients of the capillary force and the gas-phase total pressure act as driving forces in the transfer of the capillary condensate, and viscous flow of capillary condensate ensues.

The Kozeny-Carman eqation is given by:

$$\bar{u} = -(\psi^3 dp_f/dz)/(\mu_L K_C S^2 \rho_{app}^2) \tag{61}$$

where $\bar{u}$ is the mean flow velocity of condensate. Eq.(61), which was derived for the flow of fluid through porous media, can be assumed to be valid for the case that capillary condensation occurs in all pores. Providing that the pores ,where the capillary condensation does not take place, does not contribute to the flow of condensate, the relative volume of the capillary condensed phase $\psi(V_c/V_T)$ and the surface area of that phase might be used instead of ψ and S_c in Eq.(61), respectively.

Neglecting the gravity force and the gradient of the gas-phase total pressure because the solid under consideration is micro-porous, the flux of capillary condensate, N_c, is given by

$$\begin{aligned} N_c &= (\rho_L/M)\bar{u}A \\ &= -P_c A dp/dz \\ &= -\rho_L \psi^3 (V_c/V_T)^3 A (dp_c/dz)/(\mu_L K_C S_C^2 \rho_{app}^2 M) \end{aligned} \tag{62}$$

From Eqs.(59) and (60),

$$dp_c/dz = \{RT/(V_L p)\} dp/dz \tag{63}$$

Therefore

$$P_c = \rho_L^2 \psi^3 (V_C/V_T)^3 RT/(\mu_L K_C S_C^2 \rho_{app}^2 M^2 p) \tag{64}$$

$$= (S_T/S_C)^2 (V_C/V_T)^3 (p_s/p) P_c^0 \tag{65}$$

where P_c^0 is the permeability of capillary condensate in the case that all the pores in the porous solid are filled up with the capillary condensate.

Apparent surface flow coefficient. The apparent surface flow rate Ns is usually defined by

$$N_s = -P_{sapp} A (dp/dz) = -\rho_{app} D_{sapp} A (dq/dz) \tag{66}$$

Providing that the transport paths in the capillary condensed phase and in the adsorbed phase are approximately independent each other, the apparent surface permeability P_{sapp} is given as $(P_c + P_d)$. Consequently the apparent surface flow coefficient D_{sapp} is expressed by Eq.(67).

$$D_{sapp} = (P_c + P_d)/\{(dq/dp)\rho_{app}\} \tag{67}$$

where P_d and P_c are determined from Eqs.(58) and (65), respectively.

The volume V_c and the surface area S_c of the capillary condensed phase should be estimated to determine the apparent surface flow coefficient using Eqs.(58), (65) and (67). If an adsorption equilibrium in the phase without capillary condensation obeys the B.E.T. equation,

$$q_{BET} = q_m Cx/\{(1 - x)(1 - x + Cx)\} \tag{68}$$

The thickness of the adsorbed layer t can be evaluated by Eq.(69).

$$t = q_{BET} V_L / S_T \tag{69}$$

As the Kelvin radius r_c can be calculated from Eq.(59) by assuming that $\cos\alpha = 1$, capillary condensation occurs in the pores whose radii are smaller than $(r_c + t)$. Accordingly, V_C and S_C can be estimated from the pore-size distribution of the adsorbent. Providing that the B.E.T. type adsorption takes place uniformly on the surface of adsorbent, the amount adsorbed q_d in the phase where capillary condensation does not occur is given as Eq.(70).

$$q_d = (S_T - S_C) q_{BET} / S_T \tag{70}$$

The amount adsorbed q can be evaluated as the sum of that adsorbed in the adsorbed phase q_d and that in the capillary condensed phase q_c [13].

$$q = q_c + q_d \tag{71}$$

where q_c can be calculated from V_C.

3.3. Correlation of Surface Flow Coefficients

Fig. 16 shows apparent surface flow coefficients for ethylene, propyrene, iso-butane and sulfurdioxide gas in porous Vycor glass[12] at low amounts adsorbed where the capillary condensation does not take place. The solid lines in the figure are the correlated ones by the present model. According to the fact that the observed data including other published experimental ones[14 - 18] are well correlated, it can be said the correlating equations, Eqs.(46) and (50) satisfactorily explain the dependence of surface flow coefficient on the amount adsorbed and the temperature in the wide range from less than monolayer to multilayer adsorption [13, 19 - 22].

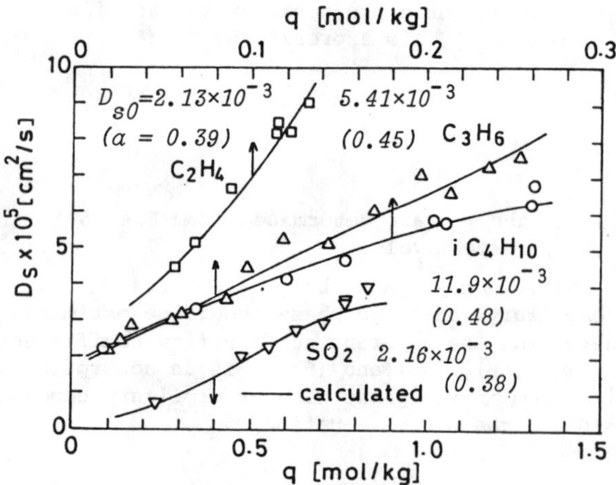

Fig. 16 Surface flow coefficients (Vycor glass ; 30°C) [12]

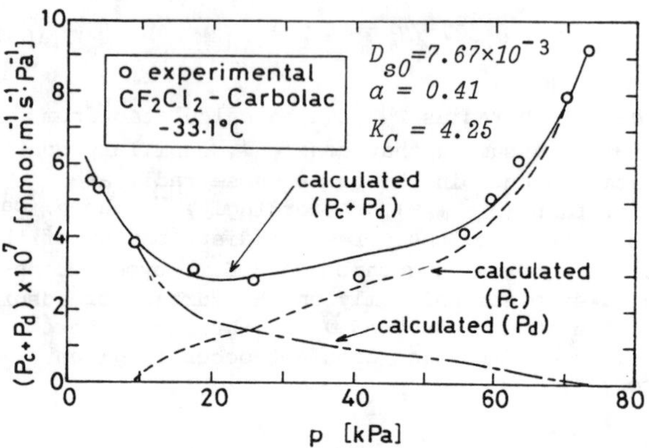

Fig. 17 Permeabilities of CF_2Cl_2 through Carbolac : Carman & Raal [14]

As an example of the surface flow under coexistence of the capillary condedsation, the calculated P_d, P_c and $(P_c + P_d)$ for the surface flow of CF_2Cl_2 through Carbolac [14] are shown in Fig.17. P_d decreases with increasing partial pressure of adsorbate, p, while P_c increases. The calcutated $(P_d + P_c)$ well agrees to the observed. The permeability of the adsorbed phase, Pd, can be calculated from Eqs.(50) and (58), where S_c and q_{BET} can be evaluated by Eqs.(59), (68) and (69) and the pore-size distribution determined by the method proposed by Dollimore and Heal [23] providing of B.E.T. adsorption. For the capillary condensed phase, the permeability of the condensate, P_c can be calculated by Eq.(65), where V_C can be evaluated by Eqs. (59), (68) and (69) and the pore-size distribution. The value of P_c^0 can be determined by the following two methods. It is determined from the observed permeability in the case where all pores are filled with condensate or from the parameter-fitting so as to coincide with observed apparent surface flow coefficients. Also the value can be roughly estimate by Eq.(64) where $K_C = 5$. In the present correlations, the values searched by parameter-fitting method were adopted.

Some examples of the correlation of the apparent surface flow coefficient are shown in Figs.18 to 22. Through those figures, the change of D_{sapp} with the amount adsorbed displays a typical inverse S-shaped curve. This change can be interpreted by the present model as follows. The surface flow coefficient increases with the amount adsorbed in the range of monolayer adsorption based on Eq. (50). When multilayer adsorption begins in the adsorbed phase and capillary condensation is formed in finer pores, the number of adsorbed molecules contributing to surface flow and the surface area $(S_T - S_C)$ begins to decrease. Hence, P_d decreases according to Eq.(58). On the other hand, the increase of P_c only makes up for the decrease of P_d, because the flow resistance of the capillary condensate is

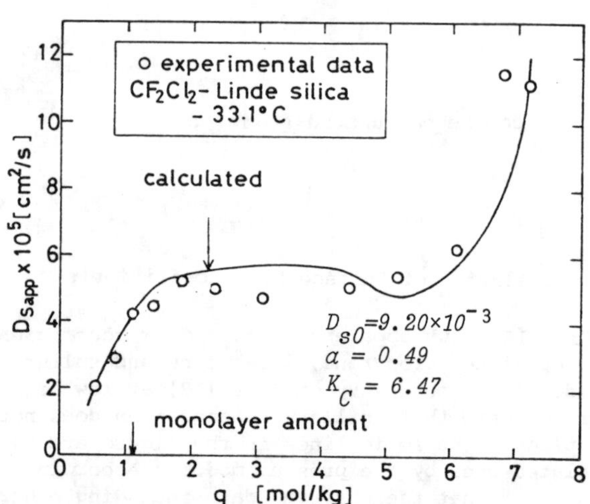

Fig. 18 Apparent surface flow coefficients of CF_2Cl_2 on Linde silica : Carman & Raal [14]

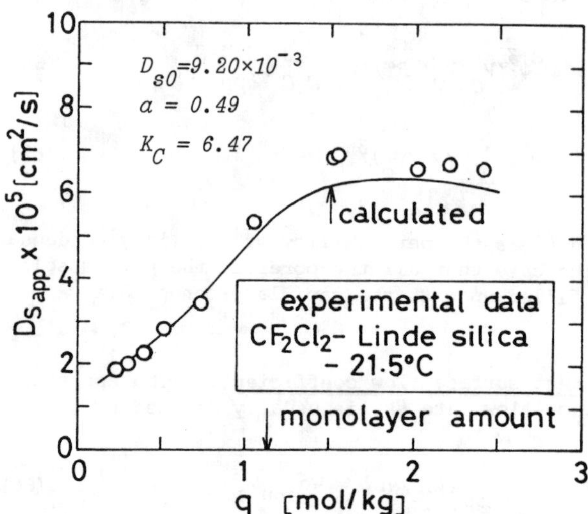

Fig. 19 Apparent surface flow coefficients of CF_2Cl_2 on Linde silica : Carman & Raal [14]

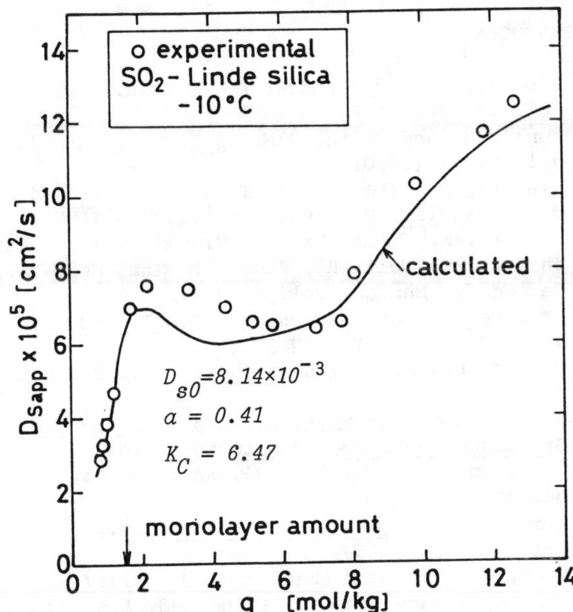

Fig. 20 Apparent surface flow coefficients of SO_2 on Linde silica : Carman & Raal [14]

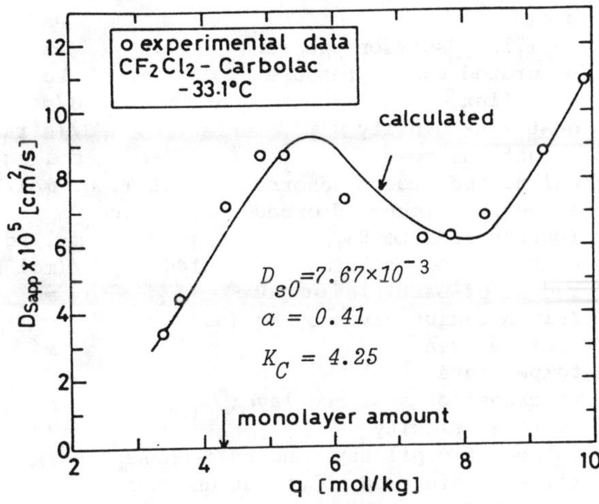

Fig. 21 Apparent surface flow coefficients of CF_2Cl_2 on Linde silica : Carman & Raal [14]

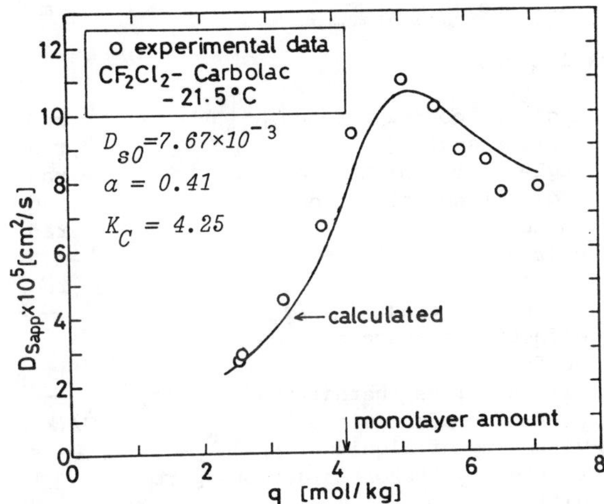

Fig. 22 Apparent surface flow coefficients of CF_2Cl_2 on Carbolac : Carman & Raal [14]

large in the finer pores. So, $(P_d + P_c)$ dose not show a large increase, while the slope of adsorption isotherm (dq/dp) steeply increase with the gas-phase pressure. Accordingly, the apparent surface flow coefficient D_{sapp} changes to show a maximum value due to Eq.(67). As the capillary condensation proceeds further, P_c increases and D_{sapp}-rises again, and gives a minimum value of D_{sapp}. From the above consideration, the apparent surface flow coefficient follows the inverse S-shaped dependence of the amount adsorbed.

Even though still there are three adjusting parameters which are difficult to be estimated without observed data, it might be said that the present model offers a pretty good perspective to the surface flow phenomena in the adsorptive fine-porous material, taking into account that the correlation curves simulate well the S-shaped observed data. The problems to be solved in the future work are to establish reliable evaluating methods for those parameters, and so as to do, it is necessary to accumulate more plenty of observed data to various kinds of adsorptive materials and adsorbates.

4. CONCLUSION

From a fundemental aspect on the drying mechanism of porous materials, the effective thermal conductivity of the wet porous solid and the apparent surface flow coefficient of the adsorbate in the adsorptive solid were described to give views about correlating or predicting methods.

NOMENCLATURE

A	fractional area of unit cell	—
	cross-sectional area	m^2
C	constant	—
C'	constant	m^2 kg/(s mol)
D	diffusion coefficient	m^2/s
E	energy, heat of adsorption	J/mol
E_{st}	isosteric heat of adsorption	J/mol
$g(E)$	number density of molecules adsorbed which have the heat of adsorption between E and $(E + dE)$	J/mol
Δh_v	specific latent heat of vaporization	J/kg
K	constant	—, N s/(K m^2)
M	molecular weight	kg/mol
N_c	number of contact point	—
N_s	rate of surface flow	mol/s
n_c	effective number of contact points	—
P	permeability of mass	mol/(m s Pa)

p	pressure	Pa
p_c	capillary suction pressure	Pa
p_s	saturated vapor pressure	Pa
Q	heat flux	J/s
q	heat flux density	J/(s rad)
	amount adsorbed	mol/kg
q_{BET}	calculated amount adsorbed by B.E.T.eq.	mol/kg
q_e	effective amount adsorbed for surface (corresponds to θ_e)	mol/kg
R	universal gas constant (= 8.3144)	J/(mol K)
r	radius of particle or pore	m
r_e	Kelvin radius given by Eq.(59)	m
S	surface area	m^2
T	temperature	K
t	thickness of adsorbed layer	m
$\overline{u}$	average velocity	m/s
V_C	volume of capillary condensed phase	m^3/kg
V_c	relative sintered volume around consolidated contact point to volume of a particle	—
V_L	molar volume of liquid (= M/ρ_L)	m^3/mol
V_l	relative volume of wedge water to volume of a particle	—
x	relative pressure (= p/p_s)	—
z	distance	m

Greek letters

α	angle of consolidation	rad
	contact angle	rad
β	angle of consolidation	rad
γ	angle of consolidation	rad
δ	angle	rad
	unit hopping distance	m
θ	angle	rad
	surface coverage (= q/q_m)	—
θ_e	effective surface coverage for surface flow	—
κ	dimensionless thermal conductivity	—
λ	thermal conductivity	W/(k m)
μ_L	viscosity of liquid	Pa s
μ_{ld}	tortuosity factor to gaseous pore diffusion	—
ρ	density	kg/m^3
σ	surface tension	N/m
τ	mean holding time of adsorbed molecule	s
Φ	macroscopic water content of wet porous solid	—
ϕ	local water content around contact point	—
ψ	void fraction	—

Suffix

a	adsorption
C	capillary condensed phase
c	capillary condensed phase
d	adsorbed phase
e	effective value
f	fluid
g	gas
l	liquid
m	monolayer adsorption
s	solid, solid part of unit cell
T	total
v	vapor, macro void part of unit cell
0	adsorbed molecule in monolayer
1	adsorbed molecule in multilayer

REFERENCES

1. Ridgway, K. and Tarbuck, K.J., Brit. Chem. Eng., vol. 12, 384 (1967)
2. Kunii, D. and Smith, J.M., A.I.Ch.E.Journal, vol. 6, 71 (1960)
3. Okazaki, M., Ito, I. and Toei, R., A.I.Ch.E.Symposium Series, No. 163, vol. 73, 164 (1977)
4. Okazaki, M., Yamasaki, T., Ninomiya, T., Nakauchi, H. and Toei, R., Proc. 7th Int. Heat Transfer Conf., Munich, vol. 6, 93 (1982)
5. Krischer, o. and Esdorn, H., V.D.I.Forsch., vol. 22, 1 (1956)
6. Bruggemann, D.A.G., Ann. Phys., vol. 24, 636 (1935)
7. Krischer, O. and Kast, W., Die wissenschaftlichen Grundlagen der Trocknungstechnik, Bd. 1, 3. Auflage, pp. 186-188, 209-236, Springer-Verlag, Berlin (1978)
8. Kimura, M., Kagaku Kogaku (Chem. Eng., Japan), vol. 23, 502 (1959)
9. Toei, R., Drying Mechanism of Capillary Prous Bodies, in Advances in Drying, ed. A.S.Mujumdar, vol. 2, pp. 268-297, Hemisphere Publishing Corp., Washington (1983)
10. Hill, T.L., J.Chem. Phys., vol. 25, 730 (1956)
11. Higashi, K., Ito, H. and Oishi, J., J. Japan Atom. Energy Soc., vol. 5, 846 (1963)
12. Okazaki, M., Tamon, H. and Toei, R., A.I.Ch.E.J., vol. 27, 262 (1981)
13. Tamon, H., Okazaki, M. and Toei, R., A.I.Ch.E.J., vol. 27, 271 (1981)
14. Carman, P.C. and Real, F.A., Proc. Roy. Soc., vol. 201A, 38 (1951)
15. Gilliland, E.R., Baddour, R.F., George, G.P. and Sladek, K.J., Ind. Eng. Chem. Fundam., vol. 13, 95 (1974)
16. Horiguchi, Y., Hudgins, R.R. and Silveston, P.L., Can. J. Cham. Eng., vol. 49, 76 (1971)
17. Ponzi, M., Papa, J., Rivalola, J.B.P. and Zgrablich, G., A.I.Ch.E.J., vol. 23, 347 (1977)
18. Ross, J.M., and Good, R.J., J.Phys. Chem., vol. 60, 1167 (1956)
19. Okazaki, M., Tamon, H., Hyodo, T. and Toei, R., A.I.Ch.E.J., vol. 27, 1035 (1981)
20. Tamon, H., Kyotani, S., Wada, H., Okazaki, M. and Toei, R., J. Chem. Eng. Japan, vol. 14, 136 (1981)
21. Toei, R., Imakoma, H., Tamon, H. and Okazaki, M., J. Chem. Eng. Japan, vol. 16, 364 (1983)
22. Toei, R., Imakoma, H., Tamon, H. and Okazaki, M., J. Chem. Eng. Japan, vol. 16, 431 (1983)
23. Dollimore, D. and Heal, G.R., J. Appl. Chem., vol. 14, 109 (1964)

INFLUENCE OF SURFACE ACTIVE AGENTS ON THE DRYING RATE OF SOLIDS

Marcel Loncin, Thomas Roth and Christian Bornhardt

Department of Food Engineering of the University
Kaiserstr. 12, D-7500 Karlsruhe, West Germany

1. INTRODUCTION

The drying rate $\dot{m}$ of moist solids in a turbulent stream of air may be expressed by the equation:

$$\dot{m} = \beta_p (p_s - p_{air}) \quad . \qquad (1)$$

The vapor pressure of water at the interface (ps) is influenced by

- the presence of membranes or waxes,
- the transport of solutes (due to the migration of water towards the interface) and their counterdiffusion,
- the modification of structure of the solid due to the dehydration at the surface,
- the presence of surface active agents.

The capillary pressure is proportional to the surface tension; therefore, for porous solids, an increase of drying rate due to the presence of surface active agents could be expected. This action is however mostly negligible compared with the influence of other factors.

It is also reported in the literature that the evaporation rate of pure water could be increased by the presence of surface active agents /1/. Although some theoretical explanations has been published /2,3/ this influence, if any, is slight and could be due to experimental errors /4,5/.

On the other hand the decrease of evaporation rate of water from ponds by use of small amount of cetyl alcohol (hexadecanol) is well known and used on a large scale in practice /6/. It is generally assumed that this retardation effect is very specific; some authors however /7,8/ found similar retardations with products as different as mineral oil, sodium stearate and cetyl alcohol.

It is also known that ethyl oleate possibly mixed with potassium carbonate increases the drying rate of grapes used for the manufacture of raisins /9,10/.

2. DECREASE OF DRYING RATE OF PURE WATER

The activity of water (a_w) initially defined as a ratio of fugacities is accurately approximated /11/ by the ratio

$$a_w = \frac{\text{vapor pressure of water in the product}}{\text{vapor pressure of pure flat water at the same temperature}} \quad . \qquad (2)$$

The vapor pressure of water at the interface (p_s) is therefore equal to the vapor pressure of pure water at the temperature of the product multiplied by the activity of water at the interface (a_{wo}).

The activity of water at the surface of meat has been investigated by Krispien /12/ mainly in relation with the growth of microorganisms. His method consists of measuring the relative humidity of air at various distances from the solid and extrapolating to distance zero. Although obviously inaccurate, because the extrapolation takes place into the boundary layer, this method shows clearly the influence of various factors, especially of the shape and of the presence of fat and membranes, on the growth of microorganisms.

Actually a_{wo} is a function not only of the solid but also of the relative humidity and of the velocity of the gas phase; it also strongly depends on time. It is clear that if the velocity of the air is equal to zero (closed system) and if this air remains a long time in contact with the solid, the relative humidity of the air will finally be equal to the a_w inside the solid. If the velocity of the air increases and if its relative humidity is below the a_w inside the solid, the activity of water at the interface decreases. If the diffusion phenomena in the solid are slow enough, compared with the rate of exchange between solid and air, a_{wo} becomes equal to the relative humidity of the air (Fig.1).

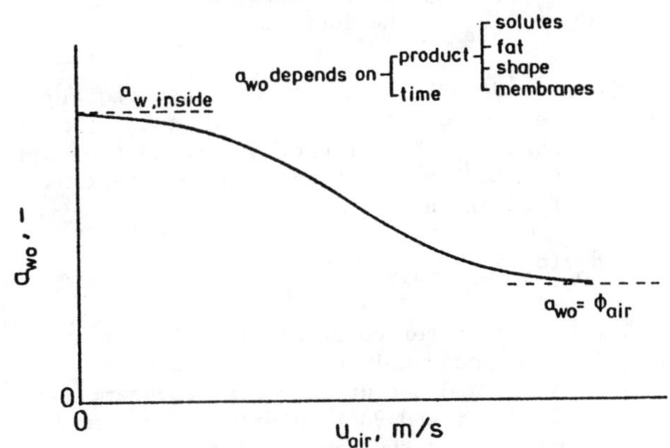

Fig.1 Influence of air velocity on the water activity at the surface (qualitatively)

Three methods have been developed /13-15/ based on the fact that for small samples (length or diameter 1 to $3 \cdot 10^{-2}$ m) the rate of heat conduction in the solid is considerable faster than the rate of the other heat and mass transfer. The temperature can therefore be assumed to be completely homogeneous throughout the sample.

The first method consists of simply measuring the temperature of the sample in a stream of air. If this temperature is equal to the wet bulb temperature, a_{wo} is equal to 1. If the temperature of the sample is higher, it is assumed that the air at the interface is in equilibrium with the sample and that its enthalpy per unit of mass of dry air is equal to the enthalpy in the bulk of air (The decrease in temperature is compensated by the increase of water vapor content).

These assumptions are exactly the same as for the routine determination of the relative humidity of air by use of wet and dry bulb temperatures. A simple graphical method using the enthalpy diagram for moist air allows to determine the relative humidity of the air at the surface which is equal to a_{wo} (Fig.2).

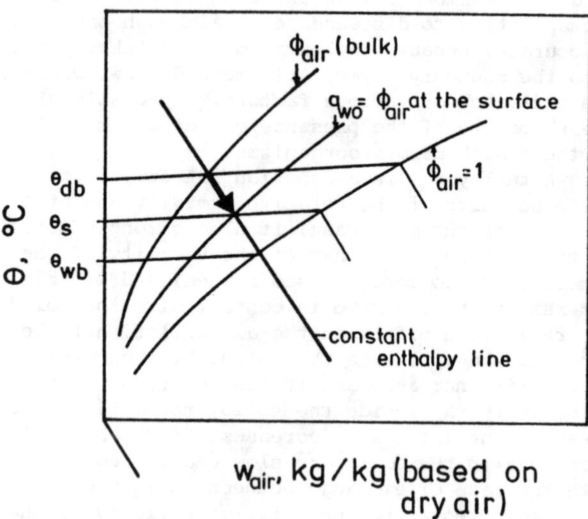

Fig.2 First method to determine a_{wo} from an enthalpy-diagramm for moist air : $a_{wo} = \Phi_{air}$ at the surface

The second method consists of measuring in addition the drying rate $\dot{m}$. As shown in Eq. (3a) $\dot{m}$ is proportional to the difference between the vapor pressure of water at the surface and the partial pressure of water in air:

$$\dot{m} = \beta_p (p_o a_{wo} - p_{air}) \tag{3a}$$

The mass transfer coefficient β_p in the air can easily and accurately be determined by measuring the temperature and the drying rate of a sample of agar gel (2 % agar and 98 % water) of the same shape and size under the same conditions. The experience has shown that the activity of water of such a gel is always equal to 1 even if 50 or 60 % of water are lost by drying.

The third method is based on the assumption that the change of enthalpy of the sample can be neglected. This means that the rate of heat transfer from the air to the sample is exactly equivalent to the rate of vaporization of water if drying is steady state. (This assumption is also made for the first method.) In this case the simple measurement of the drying rate of the sample compared with an agar gel allows to calculate a_{wo}.

The drying rate of the sample can be calculated from Eq.(3a)

or - if pseudo-steady state - from

$$\dot{m} = \frac{\alpha}{\Delta h_v} (\Theta_{air} - \Theta_s) \tag{3b}$$

The corresponding drying rate for free water is

$$\dot{m}_{free} = \beta_p (p_{wb} - p_{air}) \tag{4a}$$

or

$$\dot{m}_{free} = \frac{\alpha}{\Delta h_v} (\Theta_{air} - \Theta_{wb}) \tag{4b}$$

$$\text{with } \frac{\alpha}{\beta_p \Delta h_v} = 64.7 \quad /16/, \tag{5}$$

$$\frac{\dot{m}}{\dot{m}_{free}} = \frac{\Theta_{air} - \Theta_s}{\Theta_{air} - \Theta_{wb}} = \frac{p_o a_{wo} - p_{air}}{p_o - p_{air}} \tag{6}$$

Θ_{wb} and Θ_s can be calculated and from these, finally a_{wo}.

It should be pointed out that methods 1 and 3 are based on the measurement of the temperature and of the drying rate of the sample, resp. Method 2 is based on both. Although based on different principles the results are in excellent agreement as shown in Table 1 for samples maintained in a stream of air with a velocity of 2 m/s.

Table 1 Results of a_{wo}-calculation by the three methods

sample	drying conditions u_{air} = 1.5 m/s		θ_s (°C)	$\dot{m}$ (kg/m²s)	a_{wo} calculated by method:		
	θ_{air} (°C)	ϕ_{air}			1	2	3
peeled apples (after 1 h)	32	0.30	20.4	$2.53 \cdot 10^{-4}$	0.90	0.88	0.85
peeled apples (after 8 h)	32	0.30	26.08	$1.07 \cdot 10^{-4}$	0.49	0.49	0.47
peeled carrots (after 1 h)	32	0.30	20.0	$2.68 \cdot 10^{-4}$	0.93	0.92	0.90
agar gel	40	0.28	24.2	$3.85 \cdot 10^{-4}$	1.0	1.0	1.0
agar gel dipped in 1 % palmitic acid/ isopropanol	40	0.28	29.0	$2.79 \cdot 10^{-4}$	0.69	0.70	0.73
agar gel dipped in 1 % glyc. mono-stearate/water	40	0.28	32.5	$2.03 \cdot 10^{-4}$	0.52	0.53	0.56
agar gel dipped in 1 % hexadecanol/ isopropanol	40	0.28	32.5	$1.87 \cdot 10^{-4}$	0.52	0.52	0.56
agar gel dipped in 1 % soy oil-based monoglyceride/water	32	0.17	26.7	$1.00 \cdot 10^{-4}$	0.30	0.31	0.32
agar gel dipped in 1 % soy oil-based monoglyceride/water	12.3	0.15	11.3	$1.66 \cdot 10^{-5}$	0.20	0.19	0.20

Numerous experiments have shown /17/ that the described drying inhibition can only be achieved with suitable surface active agents. Surface activity is a necessary but not sufficient condition for effectivity. The following four classes of surfactants inhibit drying:

- fatty acids,
- fatty acid monoglycerids,
- n-alcohols (fatty alcohols) and
- fatty acid sucrose esters.

The studies show that the following factors play a role in influencing the relative reduction in drying rate:

- typ of surfactant,
- concentration of surfactant in the dipping bath up to a certain level,
- film pressure if application takes place from a film balance,
- type of solvent or suspension medium,
- air temperature and other drying conditions,
- character of the coated surface, and - in some cases -
- procedure of preparation and "age" of the dipping bath.

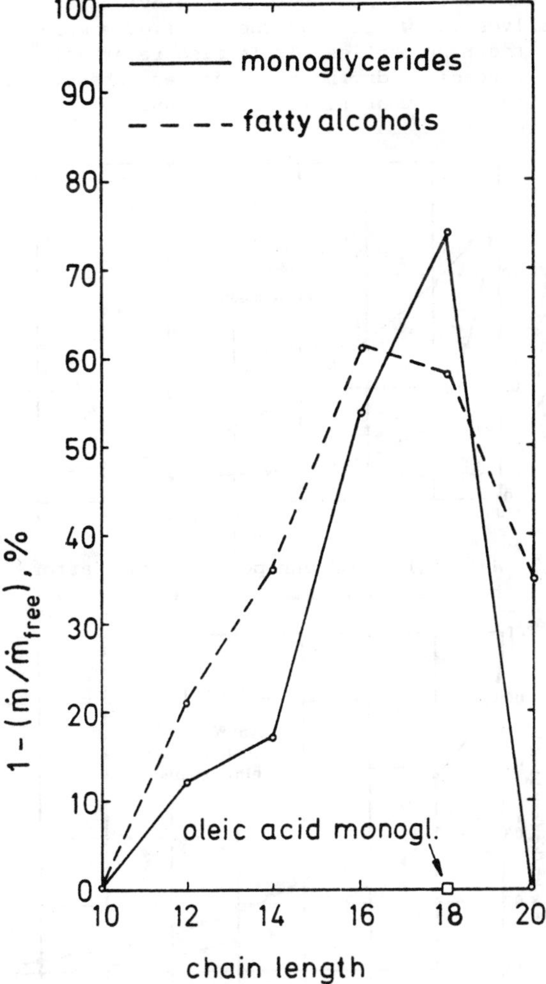

Fig.3 Reduction of drying rate of agar gel samples depending on kind of surfactant (1 % w/w in the dipping bath, Θ_{air} = 32 °C, Φ_{air} = 0.3, u_{air} = 1.5 m/s)

Fig.3 shows that the effect of the type of surfactant is very marked. Even for saturated and unsaturated hydrocarbon chains with the same number of carbon atoms there are great differences. This shows that not only polarity and size of the polar group, but also the chain length and the presence of a angle of tilt in the non-polar group is important for layer formation and retardation of drying.

The tests confirm that above a certain film thickness, which is caused by using higher dipping concentrations of the surfactants, no further inhibition of drying occurs. Previous explanations of such membrane effects, based on models of porous or liquid membranes, break down at this point, as in these cases the rate of mass transfer through the film should continously decrease with increasing membrane thickness. These results lead to the conclusion that an distinction must be drawn between effective and ineffective layers of molecules in the surfactant film, corresponding to different "degrees of order".

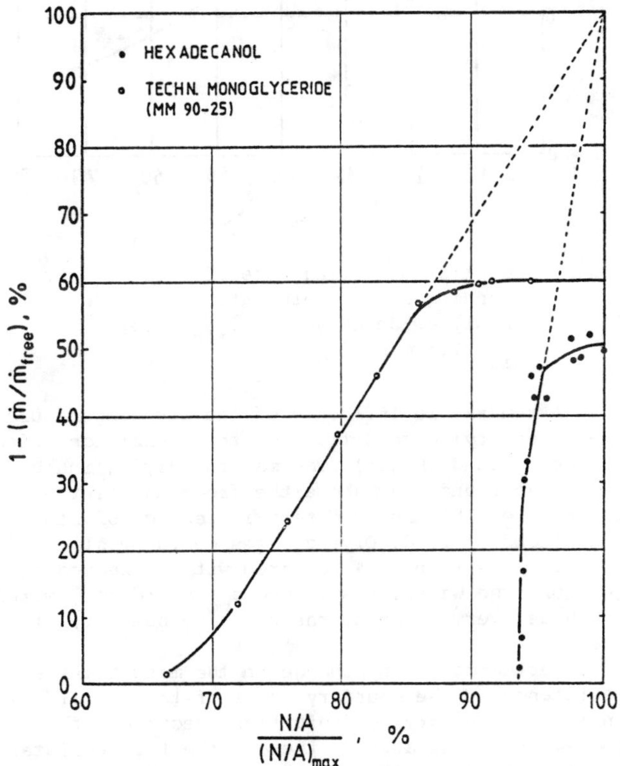

Fig.4 Reduction of drying rate of agar gel samples after transplantation of surfactant films with various relative film densities from a film balance (Θ_{air} = 32 °C, Φ_{air} = 0.1, u_{air} = 1.5 m/s)

It could be shown by a special technique of film transplantation to the samples, that it is only a monomolecular highly ordered film of surfactant molecules, which is responsible for the drying inhibition. Such a film may be prepared with a Langmuir film-balance and carefully transplantated to the sample to be investigated. The measure of the effectivity of drying inhibition is the film pressure or the molecular density in the surfactant film. It can be seen from Fig.4 that 65 % (for monoglyceride) or 93 % (for hexadecanol) of the maximal possible film density have to be reached to find an inhibition effect at all.

The relation between relative film density and inhibition of drying is a linear one in a wide range (Fig.4). But it is not possible to achieve a complete inhibition of drying with a monomolecular film of maximal density (interrupted line). This may be explained by irregularities (impurities) of the prepared film or by partially destruction of high-density films during transplantation.

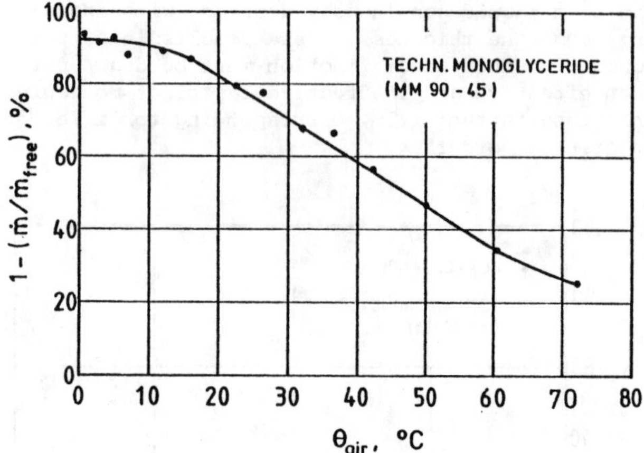

Fig.5 Reduction of drying rate of agar gel samples depending on air temperature (1 % w/w techn. monoglyceride MM 90-45, Φ_{air} = 0.3, u_{air} = 1.5 m/s)

A remarkable influence of the air temperature and other drying conditions on the inhibition effect has been found (Fig.5). The maximal drying inhibition was found just above the freezing point of the sample. This is important in respect of storing conditions of food. Drying rates could be slowed down to less then 10 % compared with untreated agar gel (or free water, resp.) and a_{wo} could be lowered to values very close to the relative humidity of the air.

Furthermore, any change in the mass transfer resistance in the boundary layer of the air ($1/\beta_p$) influences the drying inhibition, because with increasing resistance in the air the film resistance is becoming "less important".

That means, the inhibition of drying is most efficient for high film density, low temperature and strong drying conditions.

3. INCREASE OF DRYING RATE

As already mentioned a significant increase of drying rate can only be obtained by removal of an inhibition of mass transfer due to membranes or waxes. This is the case when ethyl oleate and possibly potassium carbonate are used for the manufacture of raisins.

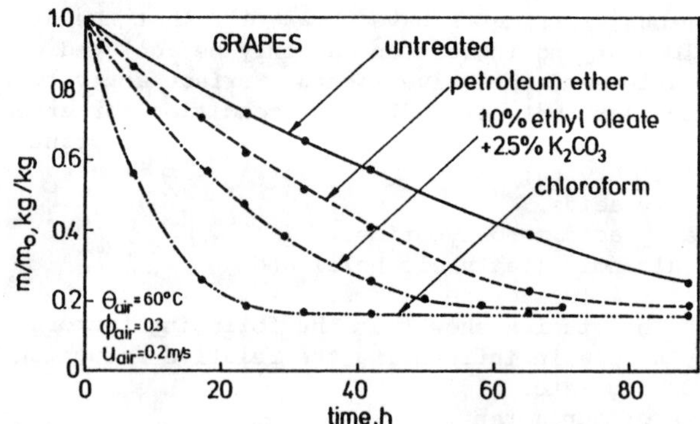

Fig.6 Drying rate of grapes after dipping/washing with different solvents/emulsions

Fig.6 shows the influence of various treatments on the drying rate of grapes at 60 °C in a stream of air with a relative humidity of 0.3 and a linear velocity of 0.2 m/s /18/. A drying rate even higher than with K_2CO_3 and ethyl oleate can be obtained by washing the grapes with chloroform. This is certainly due to its excellent solvent properties for waxes. It is not obvious if ethyl oleate only dissolves the waxes from the peels or simply modifies their properties. It is also remarkable that no increase in drying rate has been observed after treatment of other fruits like prunes with ethyl oleate.

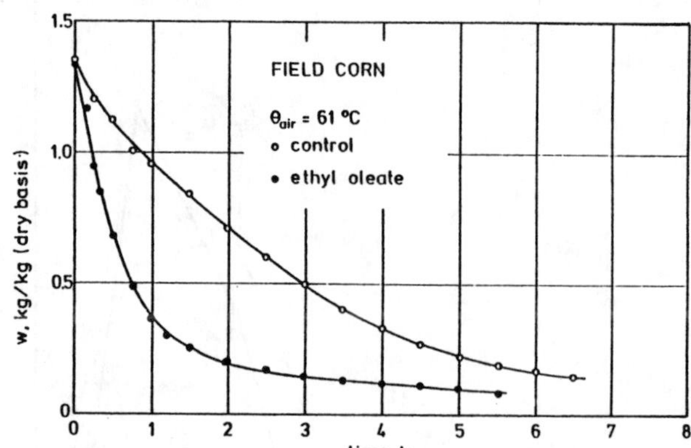

Fig.7 Weight loss of unripe field corn after treatment with ethyl oleate

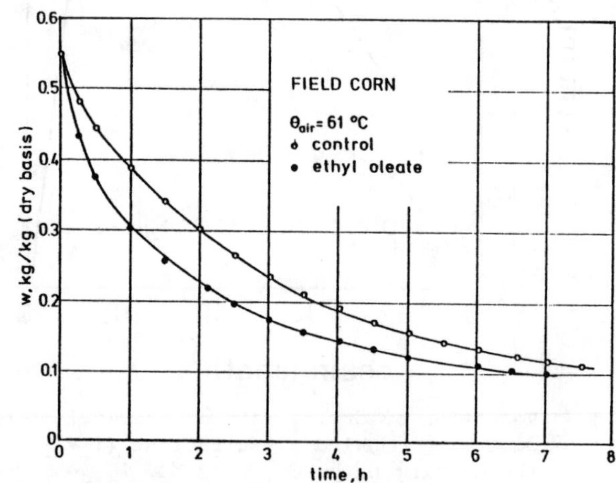

Fig.8 Weight loss of ripe field corn after treatment with ethyl oleate

The behavior of corn is extremely interesting. As shown on Fig.7 and Fig.8, a treatment with ethyl oleate increases strongly the drying rate if field corn is harvested unripe and contains 50 % moisture or more; the effect for normal ripe corn is much less important /19/. This limits the interest of ethyl oleate for this particular application.

Ethyl oleate is not miscible with water and various surface active agents can be used in order to stabilize the emulsion without influencing the increase of drying rate. An interesting application of such emulsions is the treatment of grapes on the vineyard in order to increase the sugar content and therefore the concentration of alcohol after fermentation in wine /18/.

4. CONCLUSIONS

Although sometimes mentioned in the literature, the drying rate of free water in a stream of air can not be significantly increased by addition of surface active agents.

Some lipophilic surface active agents like cetyl alcohol and particularly glycerol monostearate can considerably decrease the drying rate of free water. This action is extremely specific and due to a single oriented monolayer; the reduction of drying rate is much more important at low temperature (just above the freezing point) than at higher temperatures.

An increase of drying rate can only be ovserved when the surface active agent modifies the permeability of membranes or peels. This is the case for ethyl oleate which can strongly increase the drying rate of grapes or of unripe corn.

NOMENCLATURE

Symbol	Meaning	Unit
a_w	water activity	-
a_{wo}	water activity at the interface	-
$\dot{m}_{free}$	drying rate of free (pure, flat) water	$kg/(m^2s)$
m_o	total mass before start of drying	kg
N/A	film density	number of molecules/m^2
$(N/A)_{max}$	film density at the point of film collaps	number of molecules/m^2
p_{air}	partial pressure of water in the air outside the boundary layer (bulk)	Pa
p_o	vapor pressure of free (pure, flat) water	Pa
p_s	vapor pressure of water at the interface	Pa
p_{wb}	vapor pressure of free (pure, flat) water at wet bulb temperature	Pa
β_p	mass transfer coefficient taking into account a difference in water pressures as driving force (pressure-standardized)	s/m
Θ_{db}	dry bulb temperature	^oC
Θ_s	sample temperature	^oC
Θ_{wb}	wet bulb temperature	^oC

REFERENCES

1. Bull, H.B., J. Biol. Chem., vol. 123, 17 (1938).
2. Derjaguin, B.V., and Churaev, N.V., Structure of the Boundary Layers of Liquids and its Influence on the Mass Transfer in Fine Pores, in Progress in Surface and Membrane Science, eds. D.A. Cadenhead and J.F. Danielli, vol.14, pp. 69-130, Academic Press, New York (1981).
3. Derjaguin, B.V., Bakanov, S.P., and Kurghin, I.S., Disc. Farr. Soc., vol. 30, 96 (1960).
4. Blank, M., Monolayer Permeability, in Progress in Surface and Membrane Science, eds. D.A. Cadenhead and J.F. Danielli, vol. 13, pp. 89-139, Academic Press, New York (1979).
5. Blank, M., and Mussellwhite, P.R., J. Coll. Interf. Sc., vol. 27, 188 (1968).
6. La Mer, V., The Retardation of Evaporation by Monolayer, in Gas/Liquid and Liquid/Liquid Interface, Proc. 2. Int. Congr. of Surface Activity, ed. J.H. Schulman, pp. 259-261, Butterworths Scientific Publications, London (1957).
7. Cammenga, H.K., Klinge, H., Rudolph, B.-E., and Schulze, F.-W., Tenside Detergents, vol. 12, 19 (1975).
8. Wolf, E., Schulze, F.-W., Petrick, H.-J., Cammenga, H.K., and Barnes, G.T., Tenside Detergents, vol. 16, 58 (1979).
9. Barth, J., Grundlagenuntersuchungen über den Einfluß verschiedener Trocknungsverfahren auf das Trocknungsverhalten und die Qualität von Rosinen, Diplomarbeit, Universität Hohenheim, W. Germany (1982).
10. Eissen, W., Trocknung von Trauben mit Solarenergie, Ph. D. thesis, Universität Hohenheim, W. Germany (1983).
11. Loncin, M., and Merson, L., Food Engineering, Academic Press, New York (1979).
12. Krispien, K., Versuche zur Ermittlung der Oberflächenwasseraktivität (a_w-Wert) von Schlachttierkörpern und Fleisch zur Beeinflussung der a_w- und pH-Werte der Fleischoberfläche, Ph. D. thesis, Bundesanstalt für Fleischforschung, Kulmbach, and Freie Universität Berlin, W. Germany (1978).
13. Roth, T., and Loncin, M., ISOPOW3, Beaune, France, in press.
14. Roth, T., and Loncin, M., ICEF3, Dublin, Ireland, in press.
15. Roth, T., and Loncin, M., J. Coll. Interf. Sc., in press.
16. Loncin, M., Die Grundlagen der Verfahrenstechnik in der Lebensmittelindustrie, Verlag Sauerländer, Aarau (1969).
17. Roth, T., Zeitschr. f. Lebensmittel-Technologie u. -Verfahrenstechnik, vol. 33, 497 (1982).
18. Bornhardt, Chr., Beschleunigung der Trocknung von Trauben, Studienarbeit, Univ. Karlsruhe, W. Germany (1983).
19. Suarez, C., Loncin, M., and Chirife, J., J. Fd. Sc., in press.

THEORY OF CONSTANT-RATE EXPRESSION AND SUBSEQUENT RELAXATION

P.J. Banks

Division of Energy Technology,
Commonwealth Scientific and Industrial Research Organization
Highett, Victoria, 3190 AUSTRALIA

ABSTRACT

The removal of liquid from porous materials by expression, that is compression with drainage, is carried out using presses in various industries. Such mechanical dewatering uses much less energy than evaporative drying, and is being investigated at the CSIRO Division of Energy Technology initially as a means for dewatering Victorian brown coal (lignite) which is 2/3 water as mined.

The expression process depends on the permeability, compressibility and rheological properties of the porous material. The theory of one-dimensional constant-rate expression and subsequent relaxation of a porous material saturated with liquid is described, and solutions are presented for two linear models of material properties, namely a spring and a spring plus dashpot in series. Results from the theory are shown to explain the main features of the observed behaviour of brown coal and other materials.

The paper provides a basis for determining the properties of materials from which liquid needs to be expressed, and facilitating the design and operation of suitable presses.

1. INTRODUCTION

The removal of liquid from porous materials by expression, that is compression with drainage, occurs in both natural and technological systems. Examples are the consolidation of soil under load, and the use of presses for the separation of oil or juice from biological materials and in the dewatering of coal or mineral filter cakes. Such mechanical dewatering uses much less energy than evaporative drying [1,2]. Therefore, the study of the expression process is important for energy resource conservation and advance in biotechnology. Expression is being investigated at the CSIRO Division of Energy Technology initially as a means for dewatering Victorian brown coal (lignite) which is 2/3 water as mined [2].

This paper is directed to the interpretation of the results of constant-rate expression tests, in order to determine material properties and facilitate the design and operation of presses. Previous expression studies have considered mainly constant-pressure tests [3], while constant-rate testing has found favour for the measurement of the consolidation properties of soils [4].

The expression process depends on the permeability, compressibility and rheological properties of the porous material. In this paper, the theory of one-dimensional constant-rate expression and subsequent relaxation of a porous material saturated with a liquid is described, and solutions are presented for two linear models of material properties. Results from the theory are shown to explain the main features of the observed behaviour of apple chunks [1] and brown coal [2]. Schwartzberg et al. [1] also presented similar results for coffee grounds, chopped alfalfa and polyfoam sponge. The results for brown coal were obtained by the author's colleague D.R. Burton.

The theory given here complements previous studies of constant-rate expression and consolidation [5 to 8], and the explanation given here for the observed relaxation process clarifies and extends that given previously [1].

2. TEST RESULTS

Basic expression or consolidation tests are traditionally carried out as shown in Fig. 1. The material is contained in a cylinder, compressed by a piston on one face, and liquid expressed from the material passes through a porous membrane at the other face.

Tests results for apple chunks [1] are shown in Fig. 2, and similar results for a cylinder of as-mined brown coal [2] in Fig. 3. Piston force versus piston displacement is shown for a constant rate of piston movement until the piston is stopped. The subsequent decrease in piston force as the material relaxes is then shown versus

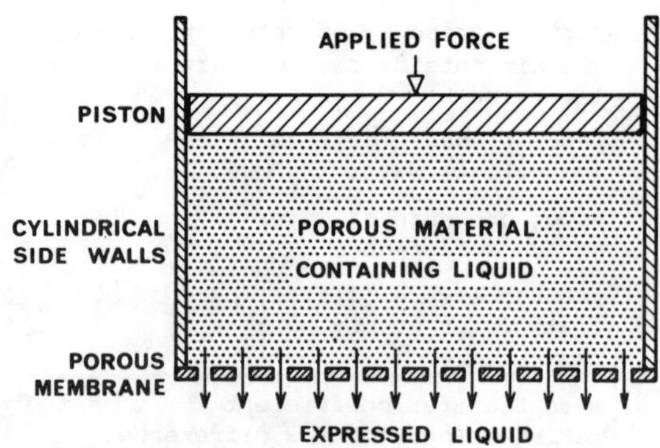

Fig. 1 Basic expression test

time. The test results show two main features:

Nonlinear compression. The compressibility of the material appears to decrease as compression proceeds.

Two-stage relaxation. An initial rapid decrease in piston force is followed by a gradual decrease.

Theoretical analysis is now presented to provide explanation of the test results. One-dimensional behaviour is assumed, which means that friction between the material and the cylindrical side walls is assumed negligible. The initial material length/diameter ratio was 1/4 for the brown coal test [2], justifying the assumption [9]. The value of this ratio for the apple test [1] may be inferred from information given to have been about 3, indicating a significant effect of side-wall friction [9].

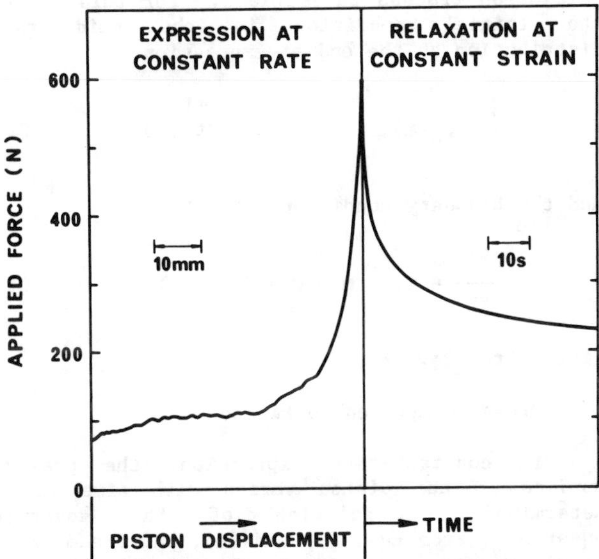

Fig. 2 Expression test record for delicious apple chunks, from [1, Fig. 4]

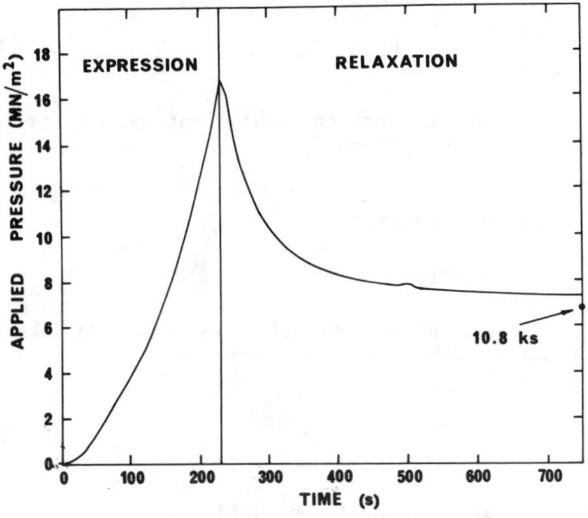

Fig. 3 Expression test record for brown coal [2]: Expression at constant rate, $d\bar{u}/dt = 1.67 \times 10^{-3}\ s^{-1}$ Relaxation at constant strain, $\bar{u} = 0.383$

3. GENERAL ANALYSIS

3.1 Governing equation

The mathematical equation governing expression is now derived using a model and analysis based on those of Smiles [10] and Shirato et al. [6]. Smiles' parameters are different but equivalent [8].

Consider a liquid-saturated porous material undergoing one-dimensional expression, as shown in Fig. 1. The material is contained in a cylinder with frictionless walls, compressed by a piston on one face, and liquid expressed from the material passes through a porous membrane at the other face. The distance between the faces decreases as liquid is removed, so that a position between the faces is specified by a material distance m, defined as the volume of material (excluding pores) per unit face area between the membrane and the position. The material is assumed incompressible so that its volume excluding pores does not change during compression. Therefore, the material distance is related to actual distance x and the void ratio of the material e (the ratio of pore volume to non-pore volume) by

$$\frac{dm}{dx} = \frac{1}{(1 + e)} \tag{1a}$$

and

$$m = \int_0^x \frac{dx}{(1 + e)} \tag{1b}$$

with x = m = 0 at the porous membrane face of the material. At the piston face, x = X which decreases during expression, while m = M which remains constant.

During expression, both the porous material and the contained liquid are in motion. The use of material distance enables the consideration of liquid motion alone in the analysis. Consider a control volume with its two end faces δm apart, each fixed relative to the porous material, and assume the pores to be always full of liquid and the liquid incompressible. Then, conservation of liquid gives

$$\frac{\partial e}{\partial t} = \frac{\partial}{\partial m}\left(k \frac{\partial p_1}{\partial m}\right) \tag{2}$$

with the volume flow rate of liquid across a face in the material per unit face area, relative to the material and towards m = 0, given by

$$k \frac{\partial p_1}{\partial m} = \frac{K}{\mu} \frac{\partial p_1}{\partial x} \tag{3}$$

where K is the permeability of the porous material, μ the dynamic viscosity of the liquid and p_1 the pressure of the liquid in the pores. From Eqs. (1a) and (3), the permeability parameter k is given by

$$k = \frac{K}{\mu\,(1 + e)} \tag{4}$$

and so varies with void ratio directly and via the permeability.

The pressure p_c acting on the liquid-saturated porous material at a position is balanced by two components, the liquid pressure p_l and the pressure p_s resulting from the compression of the porous structure of the material, since inertia effects may be neglected. Thus

$$p_c = p_l + p_s \tag{5}$$

Shear stresses due to friction between the porous material and the cylindrical side walls have been assumed negligible, and the effect of gravity may be neglected, so that

$$\frac{\partial p_c}{\partial x} = 0 \tag{6}$$

Using Eqs. (1a), (5) and (6), Eq. (2) may be written

$$\frac{\partial e}{\partial t} = - \frac{\partial}{\partial m} (k \frac{\partial p_s}{\partial m}) \tag{7}$$

Together with a constitutive equation for the porous material,

$$e = e (p_s, \frac{\partial}{\partial t}, t) \tag{8}$$

Eq. (7) becomes the equation governing expression.

3.2 Initial and Boundary Conditions

Prior to expression, the porous material is assumed to be uncompressed and of uniform void ratio, giving the initial condition

$$e = e_o \text{ and } p_s = 0, \quad 0 < m < M, \ t = 0 \tag{9}$$

with M the material distance between piston and membrane faces given by

$$M = \frac{X_o}{(1 + e_o)} \tag{10}$$

where X is the actual distance between faces and subscript o specifies values at t = 0.

In constant-rate expression, the piston is displaced at a constant velocity q, and for no liquid removal at the piston face, the rate of liquid removal through the porous membrane per unit face area is q. Thus, the boundary conditions are, using Eqs. (3), (5) and (6),

$$k \frac{\partial p_s}{\partial m} = - q, \quad m = 0, \ t > 0 \tag{11a}$$

$$\frac{\partial p_s}{\partial m} = 0, \quad m = M, \ t > 0 \tag{11b}$$

Integration of Eq. (7) and use of these boundary conditions gives

$$M \frac{d\bar{e}}{dt} = - q \tag{12}$$

where

$$M \bar{e} = \int_0^M e \, dm \tag{13}$$

It follows from Eq. (1a) that

$$\frac{dX}{dt} = - q \tag{14}$$

When the piston is stopped after a period of expression, the boundary conditions change and a relaxation process takes place. For this process the initial condition is the void ratio distribution at the end of expression,

$$e = e_e(m), \quad 0 < m < M, \ t = 0 \tag{15}$$

and the boundary conditions are

$$\frac{\partial p_s}{\partial m} = 0, \quad m = 0 \text{ and } M, \ t > 0 \tag{16}$$

giving from Eq. (7), $d\bar{e}/dt = 0$.

3.3 Pressure Applied to Piston

In constant-rate expression, the pressure applied to the piston varies with time and is determined by solution of the governing equation. From Eqs. (5) and (6), and assuming the liquid pressure drop across the porous membrane to be negligible, the applied pressure is given relative to that downstream of the membrane by

$$p_a = (p_s)_{m = 0} \tag{17}$$

In the relaxation subsequent to expression, Eq. (17) also applies.

4. ELASTIC MATERIAL

4.1 Mathematical Model

If the porous material is a purely elastic structure with compressibility

$$a = - \frac{de}{dp_s} \tag{18}$$

in general dependent on void ratio e, Eq. (7) becomes

$$\frac{\partial e}{\partial t} = \frac{\partial}{\partial m} (E \frac{\partial e}{\partial m}) \tag{19}$$

with the expression coefficient $E = k/a$ in general dependent on e.

It is convenient to put Eq. (19) in dimensionless form by defining,

$$\text{strain, } u = \frac{(e_o - e)}{(1 + e_o)} \qquad (20)$$

$$\text{dimensionless time, } \tau = \frac{E_o t}{M^2} = \frac{E_o (1 + e_o)^2 t}{X_o^2} \qquad (21)$$

$$\text{dimensionless material distance, } \xi = \frac{m}{M}$$

$$\text{and dimensionless expression coefficient, } U = \frac{E}{E_o}$$

giving

$$\frac{\partial u}{\partial \tau} = \frac{\partial}{\partial \xi} (U \frac{\partial u}{\partial \xi}) \qquad (22)$$

Dimensionless pressures are defined by

$$\sigma = \frac{a_o p}{(1 + e_o)} \qquad (23)$$

In dimensionless form, the initial and boundary conditions become:

For expression:

$$u = \sigma_s = 0, \quad 0 < \xi < 1, \ \tau = 0 \qquad (24)$$

$$U \frac{\partial u}{\partial \xi} = - P, \quad \xi = 0, \ \tau > 0 \qquad (25a)$$

$$\frac{\partial u}{\partial \xi} = 0, \quad \xi = 1, \ \tau > 0 \qquad (25b)$$

$$\text{giving } \frac{d\bar{u}}{d\tau} = P \quad \text{and} \quad \bar{u} = P\tau \qquad (26a,b)$$

$$\text{where } \bar{u} = \int_0^1 u \, d\xi = \frac{(X_o - X)}{X_o} \qquad (27)$$

$$\text{and } P = \frac{q X_o}{E_o (1 + e_o)^2} \qquad (28)$$

$\bar{u}$ is the overall strain of the porous material between piston and membrane faces, and P the dimensionless liquid pressure drop across the porous material for the expressed liquid flow rate right through the material.

For relaxation:

$$u = u_e(\xi), \quad 0 < \xi < 1, \ \tau = 0 \qquad (29)$$

$$\frac{\partial u}{\partial \xi} = 0, \quad \xi = 0 \text{ and } 1, \ \tau > 0 \qquad (30)$$

$$\text{giving } \frac{d\bar{u}}{d\tau} = 0 \quad \text{and} \quad \bar{u} = \bar{u}_e \qquad (31a,b)$$

In dimensionless form, Eq. (17) for the pressure applied to the piston becomes

$$\sigma_a = (\sigma_s)_{\xi = 0} \qquad (32)$$

and, from Eq. (18),

$$\sigma_s = \int_0^u (\frac{a_o}{a}) \, du \qquad (33)$$

4.2 Linear Model

For constant compressibility a and permeability parameter k, the expression coefficient E is constant, and Eqs. (22) and (33) become

$$\frac{\partial u}{\partial \tau} = \frac{\partial^2 u}{\partial \xi^2} \qquad (34)$$

and

$$\sigma_s = u \qquad (35)$$

The solution of Eq. (34) for the initial and boundary conditions, and the dimensionless applied pressure from Eqs. (32) and (35), are:

For expression:

$$u = \bar{u} + \frac{P}{3} [(1 - 3\xi(1 - \frac{\xi}{2})) - 6 \sum_{n=1}^{\infty} \frac{1}{(n\pi)^2} \exp (- (n\pi)^2 \tau) \cos (n\pi\xi)] \qquad (36)$$

$$\sigma_a = \bar{u} + \frac{P}{3} [1 - 6 \sum_{n=1}^{\infty} \frac{1}{(n\pi)^2} \exp (- (n\pi)^2 \tau)] \qquad (37)$$

For relaxation:

$$u = \bar{u}_e + \frac{P}{3} [6 \sum_{n=1}^{\infty} \frac{1}{(n\pi)^2} \exp (- (n\pi)^2 \tau) \cos (n\pi\xi)] \qquad (38)$$

$$\sigma_a = \bar{u}_e + \frac{P}{3} [6 \sum_{n=1}^{\infty} \frac{1}{(n\pi)^2} \exp (- (n\pi)^2 \tau)] \qquad (39)$$

The solution for the variation with time of the applied pressure is presented in Fig. 4. During expression at constant rate, dimensionless applied pressure σ_a is plotted versus overall strain $\bar{u}$, which is proportional to time. During subsequent relaxation, σ_a is plotted versus dimensionless time τ. During expression, σ_a is seen to increase in two stages; an initial transient asymptotes to a steady rate of increase. In the steady stage, there is a

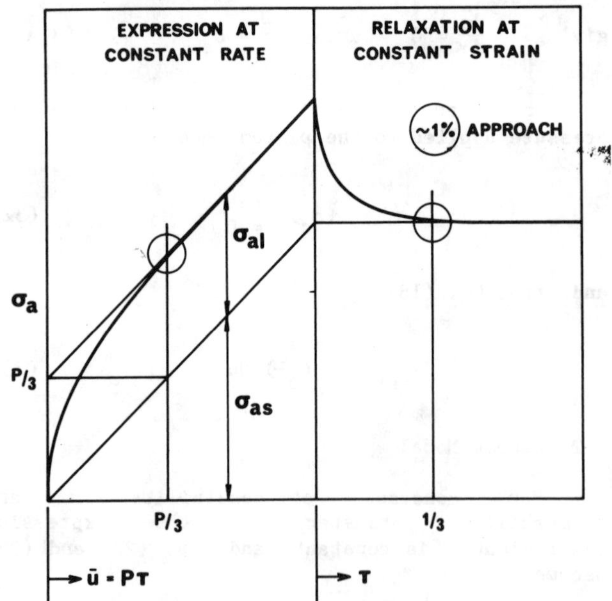

Fig. 4 Applied pressure versus time for linear elastic material model

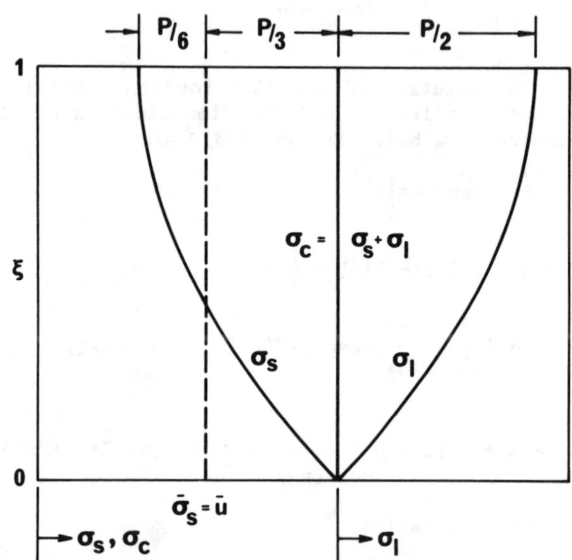

Fig. 5 Pressure profiles in steady stage of expression in Fig. 4

constant contribution from liquid pressure drop in the porous material, $\sigma_{al} = P/3$, which results from a constant profile of liquid pressure through the material. This profile develops in the transient stage progressively from the porous membrane face of the material [7, Fig. 2].

The profiles of liquid and solid pressure, σ_l and σ_s, in the steady stage are shown in Fig.5. These profiles are parabolic and their sum σ_c is constant through the material. The material is most compressed adjacent to the porous membrane, and liquid flows down its pressure gradient towards the membrane. When the expression process is halted by stopping the piston, liquid continues to flow towards the membrane because liquid pressure gradients exist, but liquid cannot pass through the membrane because the volume of material and contained liquid is not changing. Therefore, there is a redistribution of liquid over the length of the material until liquid pressure gradients diminish to zero, and the solid pressure is uniform with its mean value at the end of expression.

This redistribution of liquid causes the fall off of applied pressure during relaxation subsequent to expression shown in Fig. 4.

Relaxation and the first stage of expression involve changing pressure profiles through the material, and both occupy a dimensionless time of about 1/3, Fig. 4.

This theory does not explain the second stage of relaxation observed in the test results, Section 2, so that rheological properties for the porous material are now considered.

5. RHEOLOGICAL MATERIAL

5.1 Constitutive Equation

The constitutive equation relating stress and strain in a material generally includes time derivatives and time dependence because of the rheological or flow properties of the material. As a result, the governing equation for expression is more complex than for a purely elastic material.

There are two approaches to obtaining a constitutive equation, mechanistic and phenomenological. The first considers actual structures and deformation mechanisms on a molecular or microscopic scale, and the latter constructs models exhibiting the same behaviour as the material.

The structure of brown coal is most complex and largely unknown. Hence models representing observed behaviour are sought. The structure of apple material is better known and a possible mechanism for a simple model is mentioned in Section 6.2.

5.2 Maxwell Body

The Maxwell body consists of a spring and dashpot in series, giving the constitutive equation for compression

$$-\frac{de}{dt} = a\frac{dp_s}{dt} + r\,p_s \tag{40}$$

where r is the dashpot coefficient. Hence Eq. (7) becomes

$$a\frac{\partial p_s}{\partial t} + r\,p_s = \frac{\partial}{\partial m}\left(k\frac{\partial p_s}{\partial m}\right) \tag{41}$$

For a linear model, in which a, r and k are constant, Eq. (41) becomes, in dimensionless form,

$$\frac{\partial \sigma_s}{\partial \tau} + \frac{1}{\tau_T}\,\sigma_s = \frac{\partial^2 \sigma_s}{\partial \xi^2} \tag{42}$$

where τ_T is the dimensionless time for t = T, with T = a/f the relaxation time of the Maxwell body. This time is usually large for actual materials.

With $\tau_T >> 1/3$, the first stage in expression or subsequent relaxation is as for a linear purely elastic body, and the solution in the second stage for the Maxwell body follows from Eq. (42) with Eqs. (35) and (36) or (38) for $\tau = \infty$. Thus:

For expression:

$$\frac{d\sigma_s}{d\tau} + \frac{1}{\tau_T}\sigma_s = P \tag{43}$$

giving, with also Eq. (32),

$$\sigma_a = \frac{P}{3} + \bar{u}\,[(1 - \exp(-\tau/\tau_T))/(\tau/\tau_T)] \tag{44}$$

For relaxation:

$$\frac{d\sigma_s}{d\tau} + \frac{1}{\tau_T}\sigma_s = 0 \tag{45}$$

giving

$$\sigma_a = \sigma_{ase}\exp(-\tau/\tau_T) \tag{46}$$

with σ_{ase} given by the second term on the RHS of Eq. (44) at $\tau = \tau_e$.

Eqs. (44) and (46) are plotted in Fig.6 to indicate the effect on the second stages in expression and subsequent relaxation of replacing the purely elastic material by a Maxwell body. This replacement corresponds to introducing a dashpot in series into the material model, and is seen to result in a gradual reduction in applied pressure throughout expression and relaxation.

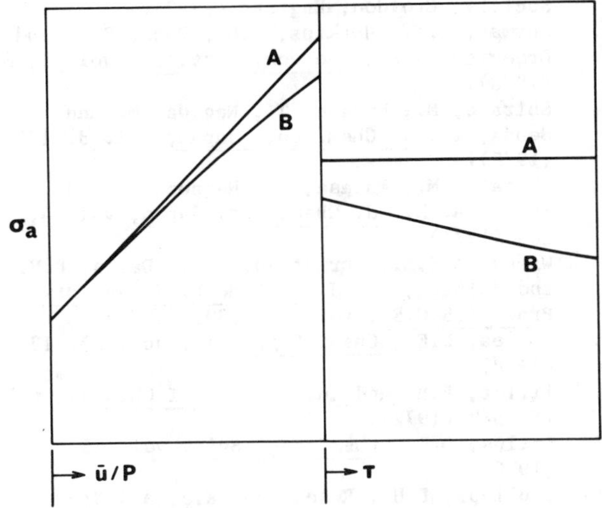

Fig. 6 Effect on second stages in Fig. 4 of introducing a dashpot in series into the material model (A to B), for $\tau_e = 0.3\,\tau_T$ and $\tau_T >> 1/3$

6. INTERPRETATION OF TEST RESULTS

6.1 Nonlinear Compression

Decrease of compressibility with increasing compression occurs in most materials, and explains the difference between the tests and linear theory in applied pressure versus time during expression, Figs. 2 to 4. For the brown coal test, the time for the transient stage of expression is estimated to be negligible for the coal properties given [11, 2] from other measurements.

For a rate of expression slow enough for liquid pressures to be negligible, $u = \bar{u}$ and the compression curve for a nonlinear elastic material may be obtained from Eqs. (32) and (33). In general, a solution to Eq. (22) with U a function of u is needed, and such a solution is being investigated by the author.

6.2 Two-stage Relaxation

The two stages of relaxation seen in the test results, Figs. 2 and 3, may be explained by liquid redistribution over the length of the porous material, Section 4.2, and rheological properties of the porous material, Section 5.2.

This explanation of relaxation clarifies and extends that given by Schwartzberg et al. [1, p.187] for their test results, including Fig. 2 above. These authors reported that "Two possible relaxation mechanisms have been qualitatively evaluated : 1) the decay of pressure inside the particles as fluid exudes from the particle interiors through pores in the particle walls; and 2) the equalization of internal stresses and strains in the press cake due to cake movement inside the test cylinder". The latter mechanism corresponds to liquid redistribution, and the former is equivalent in effect to the dashpot in the Maxwell body model of rheological properties. Schwartzberg et al. considered the two mechanisms as alternatives, and concluded that their test results did not support either mechanism. This conclusion is to be expected in view of the significant effect of side-wall friction in the tests, Section 2 above. Schwartzberg et al. recognised this effect and suggested that possibly "relaxation is due to a gradual reduction in wall friction".

For the brown coal test (Fig. 3), the magnitude and duration of the first stage of relaxation are consistent with the liquid redistribution mechanism for the properties of coal given [11, 2] from other measurements. Precise prediction is prevented by variation in coal properties between samples, and the partial irreversibility of compressive deformation of the coal, Section 6.3.

Rheological properties for brown coal have been inferred by previous researchers [11, 13] making this explanation for the second stage of relaxation apparent in Fig. 3 inevitable.

6.3 Parameters for Theoretical Models

Test results like Figs. 2 and 3 can be used together with the theoretical models explaining them to obtain material properties for the models, however there are complicating effects that require consideration:

Irreversibility. The deformation of the structure of the porous material may be describable by an elastic model, but this deformation may not be wholly reversible. This effect may be measured by allowing re-expansion after relaxation. However, the effect makes difficult the modelling of liquid redistribution during relaxation, since a portion of the material continues to contract while the rest re-expands. Evidence for the effect has been reported for alfalfa [12, p.680] and brown coal [11, 2].

Side-wall friction. Shear stresses due to friction between the porous material and the cylindrical side walls, Fig. 1, are assumed negligible in the theory, Section 3. A correction for the effect of these stresses can be estimated, but the stresses cause departure from one-dimensional behaviour, [9]. Therefore, the effect of side-wall friction needs to be minimised, by reduction of the initial material length/diameter ratio and/or coating the side walls.

Membrane resistance. The liquid pressure drop across the porous membrane has been assumed negligible in the theory. This assumption needs to be checked by measurement for the test membrane, porous material and expression rate. The possibility of an increase in membrane resistance during expression needs to be considered. Such an increase may be caused by clogging of the membrane with particles detached from the porous material or by compression of the membrane.

The inclusion of membrane resistance in the theory of constant-rate expression simply requires the addition of the membrane pressure drop for the given expression rate to the derived applied pressure. Therefore, the presence of membrane resistance in constant-rate tests is shown by a sudden increase in applied pressure when the piston is started, and a sudden decrease when it is stopped. Such sudden changes are not evident in the test results for brown coal (Fig. 3), but could be present in the apple test results (Fig. 2).

7. CONCLUSION

The theoretical analysis and discussion of test results given in this paper provide a basis for determining the properties of materials from which liquid needs to be expressed, and facilitating the design and operation of suitable presses.

ACKNOWLEDGEMENTS

The author is grateful for the stimulus of collaboration with D.R. Burton, and for helpful discussions with D.E. Smiles and B.G. Richards of the CSIRO Division of Soils, I.B. Donald and A.K. Parkin of the Department of Civil Engineering, Monash University, and A.H. Truman of Davy McKee Pacific Pty. Ltd.

NOMENCLATURE

a	porous-material compressibility, Eq.(18)	m^2/N
e	material pore volume/non-pore volume	-
$\bar{e}$	overall average void ratio, Eq. (13)	-
E	k/a, expression coefficient	m^2/s
k	permeability parameter, Eq. (4)	$m^4/N\ s$
K	porous-material permeability, Eq. (3)	m^2
m	material distance, Eqs. (1)	m
M	total material distance, Eq. (10)	m
p	pressure	N/m^2
P	dimensionless pressure drop, Eq. (28)	-
q	piston velocity	m/s
r	dashpot coefficient	$m^2/N\ s$
t	time	s
T	a/r, relaxation time of Maxwell body	s
u	strain, Eq. (20)	-
$\bar{u}$	overall strain, Eqs. (27)	-
U	E/E_o	-
x	distance	m
X	distance between piston and membrane faces	m
μ	liquid dynamic viscosity	$N\ s/m^2$
ξ	m/M	-
σ	dimensionless pressure, Eq. (23)	-
τ	dimensionless time, Eqs. (21)	-
τ_T	τ for t = T	-

Subscripts

a	pressure applied to piston
c	pressure acting on liquid-saturated porous material
e	at end of expression
l	of or due to liquid
o	at t = 0
s	due to compression of porous material structure

REFERENCES

1. Schwartzberg, H.G., Rosenau, J.R. and Richardson, G., A.I.Ch.E. Symp. Ser., vol. 73, no. 163, 177 (1977).
2. Burton, D.R., Proc. 12th Aust. Chem. Eng. Conf., Melbourne (1984).
3. Shirato, M., Murase, T. and Hayashi, N., Proc. World Filtration Cong. III, Downingtown, PA, USA, vol. 1, 280, Filtration Society, Croydon, England (1982).
4. Gorman, C.T., Hopkins, T.C., Deen, R.C. and Drnevich, V.P., Geotech. Test. J., vol. 1, 3 (1978).
5. Shirato, M., Murase, T., Negawa, M. and Senda, T., J. Chem. Eng. Japan, vol. 3, 105 (1970).
6. Shirato, M., Murase, T., Negawa, M. and Moridera, H., J. Chem. Eng. Japan, vol. 4, 263 (1971).
7. Wissa, A.E.Z., Christian, J.T., Davis, E.H. and Heiberg, S., J. Soil Mech. Found. Div., Proc. A.S.C.E., vol. 97, 1393 (1971).
8. Smiles, D.E., Chem. Eng. Sci., vol. 33, 1355 (1978).
9. Tiller, F.M. and Lu, W.-M., A.I.Ch.E.J., vol. 18, 569 (1972).
10. Smiles, D.E., Chem. Eng. Sci., vol. 25, 985 (1970).
11. Trollope, D.H., Rosengren, K.J. and Brown, E.T., Geotechnique, vol. 15, 363 (1965).
12. Schwartzberg, H.G., Proc. 2nd Pacific Chem. Eng. Cong., Denver, CO, USA, vol. 1, 675, A.I.Ch.E., New York (1977).
13. Covey, G.H. and Stanmore, B.R., Fuel, vol. 59, 123 (1980).

IMPORTANCE OF GAS PHASE MOMENTUM EQUATION IN DRYING ABOVE THE BOILING POINT OF WATER

Christian MOYNE and Alain DEGIOVANNI

Laboratoire d'Energétique et de Mécanique Théorique et Appliquée
Ecole Nationale Supérieure de la Métallurgie
et de l'Industrie des Mines, Parc de Saurupt, 54042 NANCY CEDEX
FRANCE

ABSTRACT

Classical models of drying take generally in account only two equations : an energy equation and a mass transfer equation. The derivation of such a system describing the conjugate heat and mass transfer inside an unsaturated porous body makes indispensable to neglect the gas phase momentum equation. Such a hypothesis leads from a theoretical point of view to obvious difficulties and as it has been already pointed out makes the physical mechanisms of mass transfer very unclear in some situations even below the boiling point of water. When drying takes place above this point the moisture vapor transport under the influence of a total pressure gradient is in pendular state one of the most important phenomena.

If we assume that the solid can be considered as continuous, homogeneous, isotropic and is locally in thermodynamic equilibrium, the mathematical solution of the problem involves the resolution of a system of three partial derivatives equations which are non-linear due to the variation of the phenomenological coefficients with moisture content.

Such a model is resolved in a one-dimensional case and allows us to check the validity of the theoretical assumptions. The computed values for temperature, moisture content and pressure are compared with experimental results obtained for drying in superheated steam of light concrete slabs.

1. INTRODUCTION

Classical models of drying take generally in account only two equations : an energy equation and a mass transfer equation. Such an approach which has been used extensively since the earlier modern works about drying (1 - 5) neglects the convective movement induced by a total pressure gradient in gaseous phase. Only LUIKOV (4, 5) notes the possibility to take it in account by adding to the two classical equations a third one about pressure.

The hypothesis of a constant gaseous pressure inside the porous media is often formulated with the other assumption that the inert component (air filling the voids) is stagnant (6 - 8). In particular, the theory of equivalent thermal conductivity as developed by KRISCHER (1) is based on such assumptions which appear a little contradictory. Nevertheless this approximation seems to be justified at least from a practical point of view in so far as drying takes place at temperatures well below the boiling point of water.

Two theoretical and completely different approachs highlight the role of total gaseous pressure for describing the simultaneous heat and mass transfer in porous media. The first one is derived on the basis of the thermodynamic of irreversible processes (9). The second one is a mechanical approach (10, 12). More in the last mentioned study WHITAKER et al. demonstrates very clearly the necessity to take in account the gas phase momentum equation even below the boiling point of water if we do not want a model with a good fit of the parameters but a physical description of the phenomena encountered in drying.

Numerous studies in the literature (not necessarily in the drying research field) consider the moisture transport induced by a total pressure gradient but with very simplified assumptions or other aim than a phenomenological description of the transport. In particular CROSS et al. (13, 14), DAYAN et al. (15, 16) make use of the receding plane model. HUANG et al. (17, 18), HARMATHY (19) are interested by the hygrothermal behaviour of concrete slabs with external conditions very close of those of the atmospheric surrounding. In contrary extrem conditions are investigated with respect to the safety of nuclear reactors or to the behaviour of material subject to fires (20 - 22).

In our laboratory we have been principally concerned with wood drying with particular reference with two types of dryer : vacuum dryer and high temperature convective drying. In the former the external pressure of the surrounding is lowered down to 7 kPa. The heat necessary to water evaporation is supplied to the wood probe by contact with a hot plate at temperature up to 80°C. We clearly demonstrate the drying rate acceleration related to experimental conditions which allow that the vapor pressure inside the probe may be above the external pressure (23). In the latter the drying fluid (superheated steam or moist air) at temperature above the boiling point of water up to 180°C flows above the board disposed in the test section of a wind-tunnel at atmospheric pressure. In this case for the most favourable experimental tested conditions we have measured an overpression of 1 atmosphere at the center of a board of 1 m long (24, 25). So we conclude that over the boiling point of water the total pressure gradient in gaseous phase is a major driving force which acts essentially on the gaseous transport inside the porous media as we shall demonstrate it latter.

Unfortunately wood is a very complex material The physical properties are strongly anisotropic. In particular the movement of capillary water is essentially longitudinal whereas hygroscopic water moves rather transversely. So a mathematical one dimensional model cannot permit us to analyse very accurately the physics of the process.

On the other end convective dryings above the boiling of water are today of great practical interest especially with superheated steam. Indeed we can combine (for no thermosensitive material) the advantage of a quick operation and of energy savings : steam can be always recirculated and more the heat supplied for vaporization can be recovered by vapor recompression.

In conclusion of this rapid survey of the literature and the previous results obtained by our own laboratory, the following items may be outlined :

- there is something to learn on the physics of drying by taking in account the gas phase momentum equation even below the boiling point of water,
- above this point, the total pressure gradient in gaseous phase is a very important driving force,
- from a practical point of view drying in superheated steam is a major research field in drying technology.

2. THEORY

Two general hypothesis are made in the mathematical modelling of drying :

- the porous medium is assumed to be continuous,
- the media is assumed to be in local thermodynamic equilibrium : the temperature of the three phases is identical and the vapor pressure is equal to the equilibrium vapor pressure (which is the saturated vapor pressure only outside the hygroscopic range).

2.1. Mass flux density expression

2.1.1. Liquid

For the capillary water, the liquid mass flux density is expressed as :

$$\dot{m}_l = -\frac{K_l}{\nu_l}\nabla\left(P - \frac{2\sigma}{r^*}\right) \qquad (1)$$

where P is the total pressure in gaseous phase.
K_l the liquid phase permeability
r^* the equivalent pore radius

Whithout specifying more Eq(1) may be written in the form :

$$\dot{m}_l = -a_{ml}(\nabla w + \delta_l \nabla T) - \frac{K_l}{\nu_l}\nabla P \qquad (2)$$

The "diffusion" coefficient a_{ml} and the "thermodiffusion" one δ_l take in account the capillary transport of water. The "diffusion" coefficient a_{ml} falls down for a critical moisture content which distinguishes between the often refered as "funicular" and "pendular" states of water. In the funicular period there are continuous threads of liquid in the material. The moisture transport takes place predominantly in liquid phase with capillary pressure as driving force. In the "pendular" state liquid phase is no more continuous. So the mobility of water decreases sharply. Other mechanisms in gaseous phase must be accounted to ensure the moisture transport.

The second term in equation (2) corresponds to a possible convective liquid movement under the influence of a pressure gradient in gaseous phase. In fact in the "funicular" state the variations of the gaseous total pressure are negligible compared to the capillary pressure variations. In the "pendular" state the liquid phase permeability K_l decreases in the same way as a_{ml}. So this term appears to be of little importance on drying kinetics.

In the hygroscopic range for the material under investigation we may suppose that the liquide phase transport is negligible. This would not be necessary the same for agricultural products for example. Nevertheless the formalism of equation (2) is sufficiently general to describe hygroscopic migration of water in these terms.

2.1.2. Gaseous phase

The mass flux density for water vapor q_v is given by :

$$\dot{m}_v = \omega_v(\dot{m}_a + \dot{m}_v) - \rho_g\, D f\, \nabla\omega_v \qquad (3)$$

Where ω_v is the mass fraction of vapor
D the diffusion coefficient for water vapor in air
f a factor which takes in account the flux reduction due to the reduction of the cross section, the tortuosity and the constrictivity of the pores.

The second term of Eq (3) is a diffusion one whereas the first one is a convective one resulting from the bulk motion of the fluid derived from the NAVIER-STOKES equations. How can we relate this velocity to the DARCY law? If we remember that (under somewhat drastic assumptions) DARCY law can be obtained from NAVIER-STOKES equations and if we know that these equations remain unaltered for a binary mixture if the same external force per unit mass acts on each species, the term $(\dot{m}_a + \dot{m}_v)$ may be written :

$$\dot{m}_a + \dot{m}_v = -\frac{K_g}{\nu_g}\nabla P \qquad (4)$$

by reproducing the same arguments just as for

a single fluid filling porous media. The only difference lies in the fact that we assume a velocity equal to 0 at the gas-liquid interface, which is not exactly true due to the evaporation. The combination of Eqs (3) et (4) yields :

$$\dot{m}_v = - \omega_v \frac{K_g}{\nu_g} \nabla P - \rho_g \, Df \, \nabla \omega_v \qquad (5)$$

In the same way for the inert component (air) we have :

$$\dot{m}_a = - \omega_a \frac{K_g}{\nu_g} \nabla P_g + \rho_g \, Df \, \nabla \omega_v \qquad (6)$$

2.2. Expression of the conservation laws

We must now express the conservation laws of energy, and the continuity equation for liquid phase, water vapor and air species.

. Energy : neglecting the convective terms and the pressure work yields :

$$\rho_o c_p \frac{\partial T}{\partial t} = \nabla(\lambda \nabla T) - \dot{q} \, \Delta h_v \qquad (7)$$

where $\dot{q}$ is the mass rate of evaporated water per volum unit

$c_p = c_{po} + c_{pl} \, w_l$ is the specific heat of the medium per dry solid mass unit

λ is the thermal conductivity of the solid which must be calculated supposing that the liquid and gaseous phases are quiescent.

. Liquid phase :

$$\rho_o \frac{w_l}{\partial t} = - \nabla \dot{m}_l - \dot{q} \qquad (8)$$

. Water vapor :

$$\rho_o \frac{\partial w_v}{\partial t} = - \nabla \dot{m}_v + \dot{q} \qquad (9)$$

. Air :

$$\rho_o \frac{\partial w_a}{\partial t} = - \nabla \dot{m}_a \qquad (10)$$

Now we define the moisture content w as :

$$w = w_l + w_v \neq w_l \qquad (11)$$

By adding Eq (8) and Eq (9), we form the continuity equation for w. On the other hand we made the assumption that in Eq (9) the accumulation term $\rho_o \frac{\partial \omega_v}{\partial t}$ is negligible compared to both the transport one $- \nabla \dot{m}_v$ and the evaporation one $\dot{q}$. Thus we express $\dot{q}$ as :

$$\dot{q} = \nabla \dot{m}_v \qquad (12)$$

We obtain finally the system of three partial derivatives equations :

$$\rho_o c_p \frac{\partial T}{\partial t} = \nabla(\lambda \nabla T) - \Delta h_v \nabla \dot{m}_v \qquad (13)$$

$$\rho_o \frac{\partial w}{\partial t} = - \nabla(\dot{m}_l + \dot{m}_v) \qquad (14)$$

$$\rho_o \frac{\partial w_a}{\partial t} = - \nabla (\dot{m}_a) \qquad (15)$$

If we use the expression of the mass flux given in the preceding section and if we choose the temperature T, the moisture content w, the total pressure in gaseous phase P as independent parameters this system becomes (the gas mixture is supposed to follow the perfect gas law) :

$$M \frac{\partial X}{\partial t} = \nabla (K \nabla X) \qquad (16)$$

where X is the vector column $\begin{Bmatrix} T \\ w \\ P \end{Bmatrix}$

M is a "capacity" matrix

$$M = \begin{bmatrix} \rho_o c_p & 0 & 0 \\ 0 & 1 & 0 \\ m_{31} & m_{32} & m_{33} \end{bmatrix}$$

$$m_{31} = - \frac{\psi M_a b}{\rho_o RT} \left(\frac{\partial p_v}{\partial T} + \frac{P - p_v}{T} \right) \qquad (17)$$

$$m_{32} = - \frac{\psi M_a b}{\rho_o RT} \left[\frac{\partial p_v}{\partial w} + \frac{\rho_o}{\rho_l b \psi} (P - p_v) \right]$$

$$m_{33} = + \frac{\psi M_a b}{\rho_o RT}$$

where $b = 1 - \frac{\rho_o w}{\rho_l \psi}$ is the saturation of the pores.

K is a "conductivity" matrix

$$K = k_{ij}$$

$$k_{11} = \lambda + \rho_o \Delta h_v D^* \frac{\partial \omega_v}{\partial T}$$

$$k_{12} = \rho_o \Delta h_v D^* \frac{\partial \omega_v}{\partial w}$$

$$k_{13} = \Delta h_v \left(\rho_o D^* \frac{\partial \omega_v}{\partial P} + \frac{K_g}{\nu_g} \omega_v \right)$$

$$k_{21} = a_{ml} \delta_l + D^* \frac{\partial \omega_v}{\partial T} \qquad (18)$$

$$k_{22} = a_{ml} + D^* \frac{\partial \omega_v}{\partial w}$$

$$k_{23} = \frac{1}{\rho_o} \left(\frac{K_g}{\nu_g} \omega_v + \frac{K_l}{\nu_l} \right) + D^* \frac{\partial \omega_v}{\partial P}$$

$$k_{31} = - D^* \frac{\partial \omega_v}{\partial T}$$

$$k_{32} = - D^* \frac{\partial \omega_v}{\partial w}$$

$$k_{33} = \frac{1}{\rho_o} \frac{K_g}{\nu_g} (1 - \omega_v) - D^* \frac{\partial \omega_v}{\partial P}$$

where $D^* = D \frac{\rho_g}{\rho_o} f$ is a vapor diffusion coefficient in porous media

$\omega_v = \frac{p_v M_v}{(P - p_v) M_a + p_v M_v}$ is the mass fraction of vapor.

To the three equations (16) is added the thermodynamic equation which gives us the value of the vapor pressure in the body :

$$p_v = p_{vs}(T)\ \phi(u, T) \qquad (19)$$

where $p_{vs}(T)$ is the saturated vapor pressure $\phi(u, T)$ is the relative humidity given by sorption isotherms.

3. NUMERICAL RESOLUTION

The preceding model was numerically resolved in a one dimensional case with symmetrical convective boundaries conditions (Fig 1) and uniform initial values :

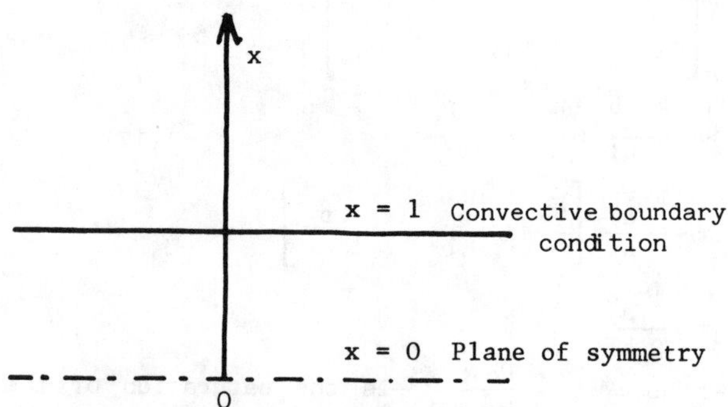

Figure 1 : Boundaries conditions

3.1. Boundary condition

. At x = 0 :

$$\frac{\partial T}{\partial x} = \frac{\partial w}{\partial x} = \frac{\partial P}{\partial x} = 0 \qquad (20)$$

At x = 1, two cases are distinguished. First we define a boiling temperature for adsorbed water T_b at the interface by the relation :

$$p_v(T_b)\ \phi(T_b, w_{x=1}) = P_\infty \qquad (21)$$

* If $T_{x=1} < T_b$: we can define two heat and mass transfer coefficients α, β which are related by the principle of analogy between the two transferts. So :

$$\alpha(T_{x=1}-T_\infty) = -\lambda\left(\frac{\partial T}{\partial x}\right)_{x=1} - \Delta h_v\ \dot{m}_1 \qquad (22)$$

$$\dot{m} = \dot{m}_1 + \dot{m}_v = \beta\ \rho_g\ .\left(\frac{Pv_{x=1}}{P_\infty} - \frac{P_{v\infty}}{P_\infty}\right) \qquad (23)$$

$$P_{x=1} = P_\infty \qquad (24)$$

* In the other case the total pressure at the interface cannot exceed the atmospheric pressure if we admit the boundary layer hypothesis and we cannot define a mass transfer coefficient for such situations. The conditions (23) is then remplaced by the relation:

$$T_{x=1} = T_b \qquad (25)$$

3.2. Numerical method

Due to the sharp variation of the phenomenological coefficients and to the boundaries conditions the problem is of course non-linear and we need a numerical method which is appropriate to such a case.

The discretization in space of the problem is made by application of the GALERKIN's method in m finite elements :

$$X(x, t) = \sum_j N_j(x)\ X_j(t) \qquad (26)$$

where $X_j(t)$ is the value of X at node $j(1 \leqslant j \leqslant m+1)$. $N_j(x)$ is the interpolation function which is taken as linear (P1).

The equation (16) premultiplied by each function N_i is integrated over the entire domain to yield :

$$A(Z)\frac{dZ}{dt} + B(Z)\ Z - J(Z) = 0 \qquad (27)$$

where Z is the column vector $\begin{Bmatrix} X_1 \\ \cdot \\ \cdot \\ \cdot \\ X_{m+1} \end{Bmatrix}$

A a "mass" matrix (to ensure the stability of the method A is diagonalized)
B a "stiffness" matrix
J a column vector which takes in account the boundaries conditions.

Due to the non-linearity of the problem A, B and J are (strong) functions of Z. Then the discretization in time uses a fully implicit method :

$$F(Z^{n+1}) = A(Z^{n+1})\frac{Z^{n+1} - Z^n}{\Delta t} - B(Z^{n+1})(Z^{n+1}) = 0 \qquad (28)$$

The resolution of this non-linear equation at time (n + 1) utilizes the NEWTON's method.

4. NUMERICAL AND EXPERIMENTAL RESULTS

To test the validity of our assumptions, simultaneously experimental and "numerical" drying of light concrete "Ytong" are investigated. The choice of such a material is made because drying at low temperature and some thermophysical properties of such a material were studied thoroughly by KRISCHER (1). So we can compare with some confidence the two results although at that time it seems not reasonable to put experimental points on numerical curves. Our purpose is to analyse the moisture convective transport under the influence of a total pressure gradient on drying.

4.1. Experimental results

A Ytong slab (0,05 m thick ; 0,20 m large, 1 m long) is dried in an aerodynamic return flow wind tunnel as shown on Figure 8. The pressure is the atmospheric one. The fluid is superheated steam at temperature T_∞ = 186°C and incipient flow velocity u_∞ = 13 m/s.

The slab placed in the horizontal plane of symmetry of the test section lies on three rods fixed on the plate of an electronic balance which allows a continuous weighing during drying.

The temperatures in the slab are obtained by means of thermocouples implanted 30 mm deep in the thickness of the slab.

The pressure inside the slab is measured by the following method : a hollow steel needle penetrates in a 50 mm drilled hole in the thickness of the slab. An epoxy resin makes the imperviousness. A very short flexible pipe connects the needle to a pressure transducer with a chamber capacity equal to 1 cm^3. Rather than an absolute measurement, this device allows us to follow the pressure evolution inside the material.

The recordings of the average moisture content $\overline{w}$, temperature at various positions in the thickness, pressure at the half thickness (measured at the half length of the board) are shown on figure 2.

The analysis of the experimental results shows a first period with a rather linear decrease of the average moisture content. During this period some negative observed values of measured pressure may be related to capillary pressures. The surface temperature of the slab is close to the boiling point of water.

At the beginning of the falling rate period the gaseous pressure increases. The outer part of the slab temperatures is above the boiling point of water whereas a steady temperature slightly under this point close to 95°C is observed in the inner part.

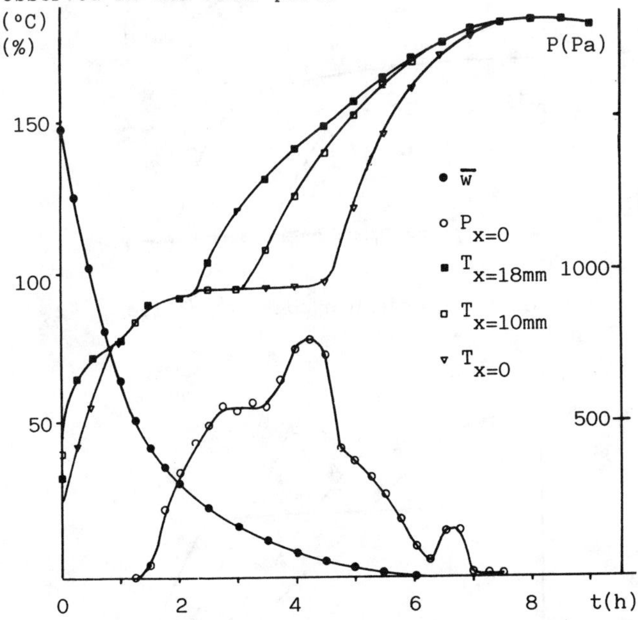

Figure 2: Experimental results

A more precise description of the transport mechanisms will be presented with the comparison between theoretical and experimental results.

4.2. Mathematical model

The values used in the mathematical model which are listed in Appendix are adapted from KRISCHER'S values (1). Only a few remarks will be made here :

- it must be stated again that the value of the thermal conductivity λ takes only in account the variation with moisture content w due to the change of the "geometry" of the medium and is not the "equivalent conductivity". These phenomena are described by the change phase terme of Eq(8),

- the liquide phase "diffusion" coefficient is shown on Fig. 9. The critical moisture content as defined in & 2.1.1. lies in the range(30%, 40%),

- the thermodiffusion coefficient δ_1 is evaluated on the basis of Eq (1) by means of the relation :

$$\delta_1 = -\frac{1}{\sigma}\frac{d\sigma}{dT}\frac{r^*}{dr^*/dw} \qquad (29)$$

The equivalent pore radius r* is determined experimentally by mercury porosimetry. The highest observed value is less than 6.10^{-4} K^{-1}. So this term can be neglected.

- In the case under investigation of drying at high temperature with superheated steam the diffusion transport of air filling initially the pores has been neglected compared to the convective movement. So we put $D^*_v = 0$.

- The intrinsic permeability of the material is supposed to be in the range 10^{-13} - 10^{-15} m^2.

The properties of this light concrete are assumed to be isotropic. Due to the length and width (thickness ratio, the process is supposed to be one-dimensional in the thickness (Ox direction as indicated on Fig. 8). We assume also a constant average heat transfert value. As a matter of fact due to boundary layers development in the flow direction from the leading edge, this coefficient varies along the slab.

The Figure 3 shows the time evolutions of average moisture content $\overline{w}$, temperatures inside the slab and pressure at its "center" given by the numerical method for a saturation permeability $K_{sat} = 10^{-13}$ m^2.

For the same conditions the temperature, moisture content and pressure profiles at different times are drawn respectively on Figures 4, 5 et 6.

4.3. Comparison between theory and experiment

The comparison of Fig (2) and (3) shows a good (at least qualitative) agreement between theory and experiment which allows us a more precise description of the phenomena.

During the first drying period the essential driving force is the capillary pressure. The moisture profile is rather parabolic. The surface temperature of the slab is the boiling point of water. The gaseous pressure is equal to the atmospheric one.

The appearance of the second drying period can be examined using the computed values. When the surface moisture content falls down the critical value, capillary forces do not act sufficiently to bring water to the surface. So it becomes in the hygroscopic range and to satisfy Eq (24) the surface temperature increases.

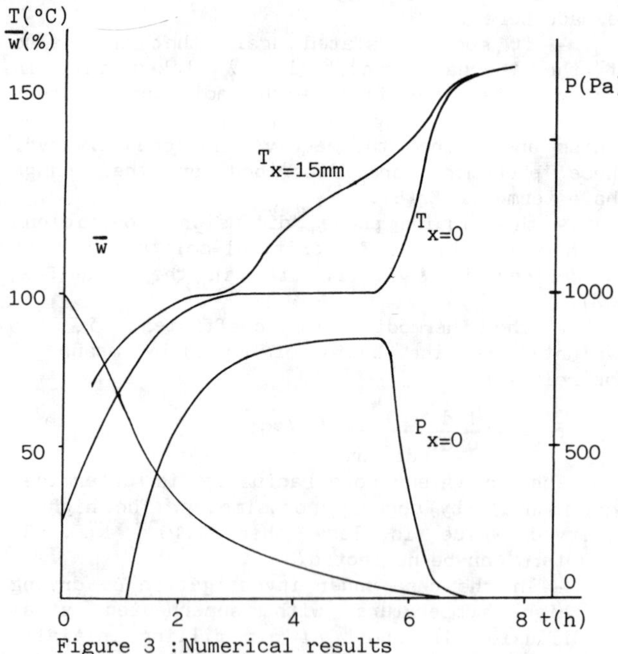

Figure 3 : Numerical results

Then the moisture content profiles show a very accentuated S shaped form which allows to distinguish three zones inside the body:

- a inner zone : moisture content is above the critical point, thermal conductivity is high enough. So temperature, moisture and pressure are rather constant,
- a outer zone : moisture content lies in the hygroscopic range. Temperature gradient and pressure gradient are rather constant. In this region heat transfer by conduction and mass transfer in vapor form under the action of pressure gradient take place in quasi-steady state,
- a transition region which penetrates inside the body where water evaporates. If the thickness of this transition region is neglected, the problem that we treat as a continuous one tends towards a classical description of receding plane type.

Finally we examine on Fig (7) the influence of permeability value on drying. We compare numerically the average moisture content evolution and the product $K_{sat}(P_{x=0} - P_{\infty})$ for the preceding value $K_{sat} = 10^{-13}\ m^2$ and a hundred times as low one $K_{sat} = 10^{-15}\ m^2$. In accordance with the capillary pressure mechanism, at high moisture content the drying rates are the same. At lower moisture content the body of greater permeability dries only slightly faster in the investigated range. If permeability is not a very sensitive parameter for drying kinetics, it strongly influences the pressure level inside the slab. So from an experimental point of view the non-observance of a total pressure gradient for a high permeability porous material can hide the importance of convective vapor transport.

On the other hand we have outlined that during the falling rate period the two major phenomena limiting drying kinetics are heat conduction through the "dried" zone and water vapor mass transport which act as two in series resistances. The weak influence of the permeability indicates that heat transfer resistance is much greater than the corresponding mass transfer one. So drying kinetics are heat transfer controlled.

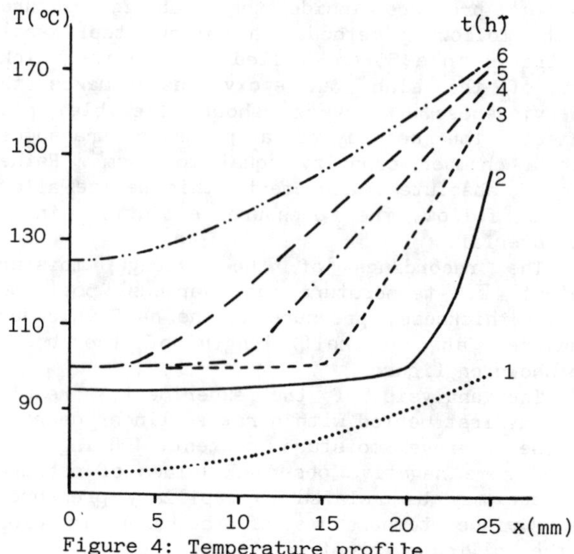

Figure 4: Temperature profile

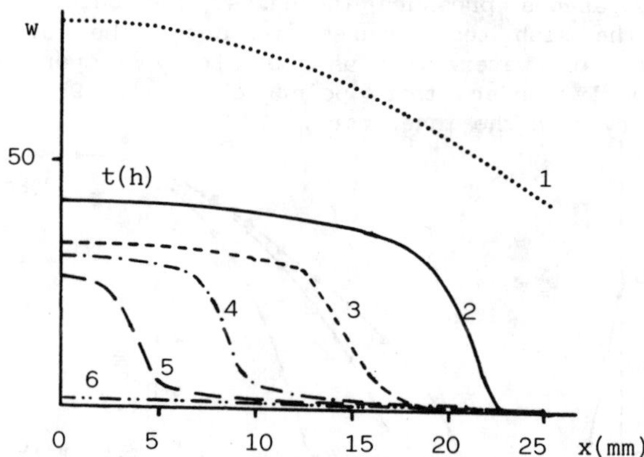

Figure 5 : Moisture content profile

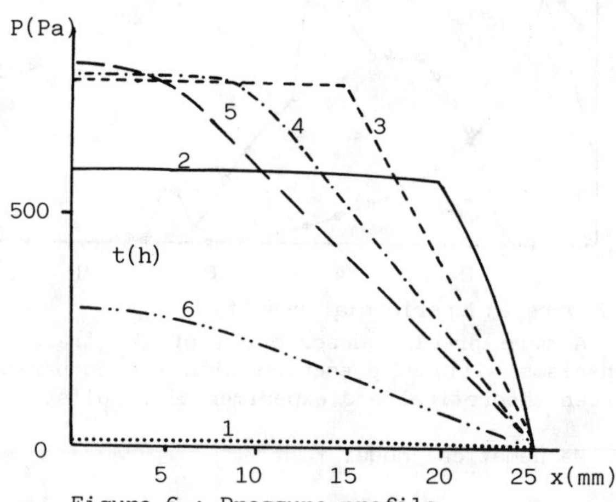

Figure 6 : Pressure profile

CONCLUSION

By comparison between numerical and experimental results we show the importance of moisture transport in gaseous phase under the influence of a total pressure gradient in the case of high temperature convective drying.

The foregoing analysis of the physical phenomena which leads to the numerical solution foreshadows that even under the boiling of water this mechanism can and must be considered.

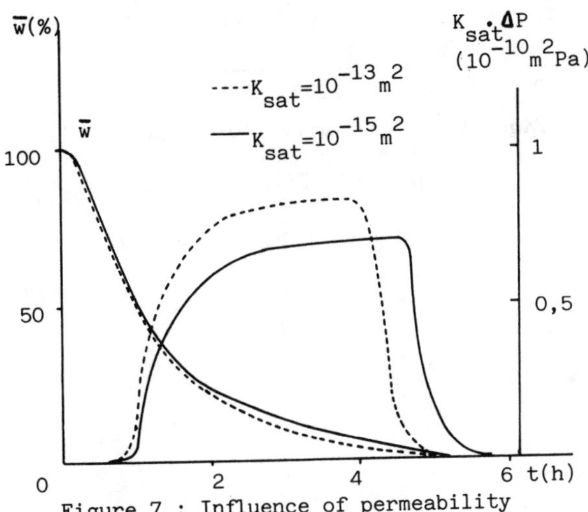

Figure 7 : Influence of permeability

REFERENCES

1. Krischer, O. and Kroll E., Springer Verlag (1956)
2. Philip, J.R. and De Vries, D.A., Trans. Amer. Geophys. Union, Vol. 38, pp.222-232 (1957)
3. De Vries, D.A., Trans. Amer. Geophs. Union, Vol. 39, pp. 909-916 (1958)
4. Luikov, A.V., Pergamon Press, Oxford, (1966)
5. Luikov, A.V., Int. J. of Heat and Mass Transfer Vol. 18, pp. 1-14 (1975)
6. Roques, M.A. and Cornish, A.R.H., The Second Pacific Chemical Engineering Congress, Denver, pp. 516-521 (1977)
7. Fortes, M. and Okos, M.R., Proceedings of the First Int. Symp. on Drying, Montreal, pp. 100-109 (1978)
8. Crausse, P., Bacon, G. and Bories, S., Int. J. of Heat and Mass Transfer, Vol. 24, pp. 991-1004 (1981)
9. Benet, J.C. and Jouanna, P., Int. J. of Heat and Mass Transfer, Vol. 26, pp. 1585-1595 (1983)
10. Whitaker S., Advances in Heat Transfer, Vol. 13, pp. 119-203 (1977)
11. Whitaker S., Advances in drying, MC Gill University, Vol. 1, pp. 23-61 (1980)
12. Chou W.T.H., Whitaker S., 3rd Int. Drying Symposium, Vol. 1, pp. 135-148 (1982)
13. Cross MM. M., Gibson R.D., Young R.W., Int. J. of Heat and Mass Transfer, Vol. 24, pp. 991-1004 (1981)
14. Cross MM. M., Gibson R.D., Young R.W., Int. J. of Heat and Mass Transfer, Vol. 22, pp. 827-830 (1979)
15. Dayan A., Glueker E.L., Int. J. of Heat and Mass Transfer, Vol. 25, pp. 1461-1467 (1982)
16. Dayan A., Int. J. of Heat and Mass Transfer, Vol. 25, pp. 1469-1476 (1982)
17. Huang C.L.D., Int. J. of Heat and Mass Transfer, Vol. 22, pp. 1295-1307 (1979)
18. Huang C.L.D., Siang H.H., Best C.H., Int. J. of Heat and Mass Tranfer, Vol. 22, pp. 257-266 (1979)
19. Harmathy T.Z., I & Ec Fundamentals, Vol.8, pp. 92-103 (1969)
20. England G.L., Sharp T.J., Trans. 1st Int. Conf. on Structural Mechanics in Reactor Technology, H2|4, PP. 129-143 (1971)
21. Morrison F.A.JR., Int. J. of Heat and Mass Transfer, Vol. 16, pp. 2331-2342 (1973)
22. Sahota M.S., Pagni P.J., Int. J. of Heat and Mass Transfer, Vol. 22, pp. 1069-1081 (1979)
23. Moyne C. et Martin M., Int.J. of Heat and Mass Transfer, Vol. 25, n° 12, p. 1839 (1982)
24. Basilico C. et Martin M., (to be published in l'Int. J. of Heat and Mass Transfer)
25. Basilico C., Moyne M. et Martin M., 3th Int. Drying symposium, Birmingham, England (1982)

APPENDIX

Values used in the numerical model

. ρ_o = 540 kg/m^3
. l = 0,025 m
. c_p = 837 + 4185 * w J/kg K
. ψ = 0,79
. $\Delta h_{vs}(T)$ = 2 500 800-2441 (T-273.15) J/kg
. $p_{vs}(T)$ = exp (25.270 - 5123.25/T) Pa

and in the hygroscopic range :

w < 0,07 $\quad \emptyset = \frac{w}{w_{hygr}} \left(2 - \frac{w}{w_{hygr}}\right)$

. $\lambda = \frac{0.339+0.308*w}{2.776+w*(0.525-1.174*w)}$ W/mK

. a_{ml} = 10 ** (-7. - 100 * (w - 0,4)^2) m^2/s if w > 0,4 and 10^{-7} if w < 0,4
. δ_1 = 0
. $w^* = \frac{w - w_{hygr}}{w_{sat} - w_{hygr}}$

$\frac{K_1}{K_{sat}} = w^{*2} (2 - w^{*2})$

$\frac{K_1}{K_{sat}} = 1 + w^{*2} (2 w^* - 3)$

. 10^{-13} m^2 < K_{sat} < 10^{-15} m^2

. α = 50 W/m^2 K $\quad \beta$ = 0,06 m/s

. T_i = 293 K $\quad w_i$ = 1 $\quad P_i$ = 101 325 Pa

. T_∞ = 453 K (superheated steam) P_∞ = 101325Pa

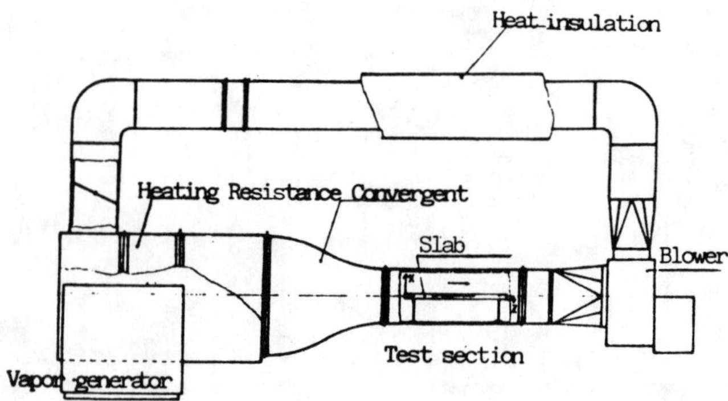

Figure 8 : Aerodynamic wind- tunnel

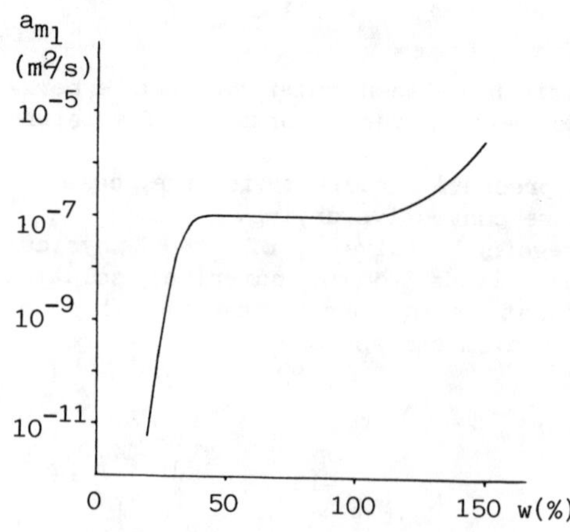

Figure 9 : Diffusivity of water in liquid phase

CONVECTIVE DRYING OF POROUS MATERIALS CONTAINING BINARY MIXTURES

Franz Thurner and Ernst-Ulrich Schlünder

Institut für Thermische Verfahrenstechnik, Universität Karlsruhe (TH)
7500 Karlsruhe, FRG

ABSTRACT

The estimation of the size of a dryer requires the knowledge of the drying rate as a function of the moisture content, called drying curve. In addition to that, for the drying of materials containing mixtures it is important to know the composition of the moisture as a function of the moisture content, called composition curve.

In the theoretical part of this work the drying curve and the composition curve for the drying of materials containing binary mixtures are defined. The influence of thermodynamic equilibrium, gas and liquid side mass transfer on the composition curve are estimated by solving the differential equation for the unsteady state diffusion equation in the liquid phase.

In the experimental part samples wetted with a mixture of isopropyl alcohol and water were dried in a drying channel. The drying curve was obtained by weighing the sample while the composition curve was determined by analyzing the outlet air with an infrared gas analyzer.

The investigation showed that the drying curves of materials containing binary mixtures have the same characteristical shape as the drying curves of pure compounds. Further it was found that the composition curve is influenced by thermodynamic equilibrium, gas and liquid side mass transfer depending on the drying conditions. If the dimensions of the sample are large the selectivity is mainly influenced by the liquid side mass transfer, thus the drying is mainly nonselective. Only at the beginning of the drying a selectivity can be achieved for a few minutes. If the dimensions of the sample are small the selectivity is mainly influenced by the thermodynamic equilibrium as well as the gas side mass transfer.

1. INTRODUCTION

The estimation of the size of a dryer requires the knowledge of the drying rate as a function of the moisture content, called drying curve. In addition to that, for the drying of materials containing mixtures it is important to know the composition of the moisture as a function of the moisture content, called composition curve. The composition curve is influenced by the thermodynamic equilibrium as well as by the velocity of the mass transfer. Knowing these influences on the composition curve it is possible to choose favourable drying conditions.

The expression "favourable drying conditions" can have different meanings /1/. Essentially one can distinguish between selective and nonselective drying. The favourable removal of one component is called selective drying. For example for the drying of foodstuff it is important to remove only the water whereas the aroma shall remain inside the material. The selective drying can also be used for a more efficient recycling of solvents. Most organic solvents are expensive and, therefore, it is not allowed to release them. If it is possible to achieve already a seperation of the various solvents during the drying the following separation effort can be reduced. In some cases the moisture is toxic or inflammable. Then it is not allowed to exceed certain concentrations in the drying air.

If none of the components is removed favourably, it is called nonselective drying. For example for the drying of varnish coatings it is important that none of the solvents is removed favourably in order to get a uniform coating.

For the design of a dryer for materials containing mixtures it is impossible to predict the drying curve and the composition curve due to the complicated heat and mass transfer. The existing models are restricted to the drying of materials containing pure compounds, in most cases water /2/. Therefore for the design of dryers it is necessary to rely on know-how from existing plants or to carry out experiments. This investigation has been undertaken in order to reduce the experimental effort and to get a better understanding of the drying behavior of materials containing mixtures.

2. THEORETICAL PART

2.1 Definitions

The drying curve. The drying curve of materials containing pure compounds is obtained by plotting the drying rate versus the moisture content. The drying curve of materials containing binary mixtures shall be defined in the same way.

The drying rate is the weight loss of the material to be dried per unit time and per unit surface area:

$$\dot{m}(t) = -\frac{1}{A}\frac{dM(t)}{dt} = -\frac{1}{A}\frac{d[M_1(t) + M_2(t)]}{dt} \qquad (1)$$

In this equation $M(t)$ is the overall mass of the wet material and $M_1(t)$ and $M_2(t)$ are the masses of component 1 and 2 in the liquid mixture.

The moisture content is the ratio of the mass of moisture $M_1(t)$ to the mass of bone-dry material M_S:

$$X(t) = \frac{M_1(t)}{M_S} = \frac{M_1(t) + M_2(t)}{M_S} \qquad (2)$$

A typical drying curve is shown in Fig. 1.

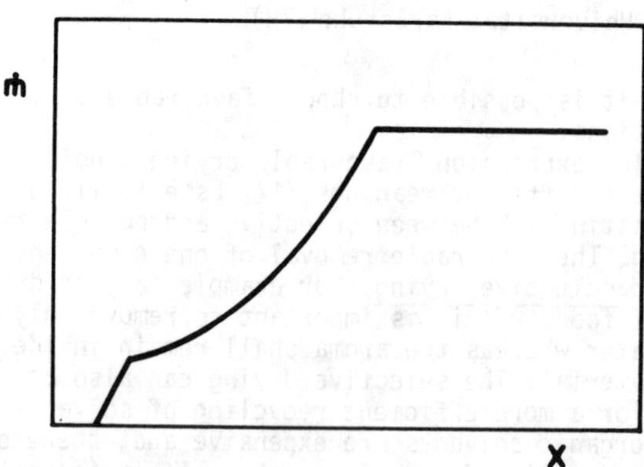

Fig. 1 Drying curve

The composition curve. During the drying of materials containing binary mixtures the composition of the moisture can be changed. In the following we consider the drying of a sample wetted with a mixture of component 1 and 2. This sample is dried convective with the drying agent 3. Component 1 is the more volatile component and component 2 the less volatile component. The molar evaporation stream $\dot{N}(t)$ is composed of the evaporation streams of both components $\dot{N}_1(t)$ and $\dot{N}_2(t)$. The selectivity S_1 is defined according to Schlünder /3/ as the difference between the relative evaporation stream of component 1

$$\dot{r}_1(t) = \frac{\dot{N}_1(t)}{\dot{N}_1(t) + \dot{N}_2(t)} \qquad (3)$$

and the mol fraction of component 1 in the liquid phase

$$\tilde{x}_1(t) = \frac{N_1(t)}{N_1(t) + N_2(t)} \,. \qquad (4)$$

Therefore the selectivity can be expressed by

$$S_1(t) = \dot{r}_1(t) - \tilde{x}_1(t) \,. \qquad (5)$$

Based on this definition one can distinguish between the following cases:

$S_1 > 0$: Component 1 is removed favorably; the concentration of component 1 in the moisture decreases.

$S_1 = 0$: None of the components is removed favorably; the composition of the moisture remains unchanged.

$S_1 < 0$: Component 2 is removed favorably; the concentration of component 1 in the moisture increases.

By plotting the molfraction of component 1 in the moisture $\tilde{x}_1$ versus the relative molar moisture content η the composition curve is obtained (Fig. 2). The relative molar moisture content is the ratio of the actual molar moisture N to the initial molar moisture N_0.

In Fig. 2 the composition curves for the different cases of selectivity are depicted.

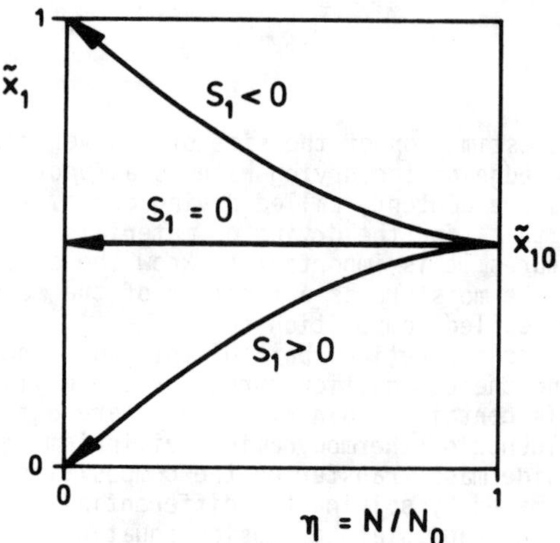

Fig. 2 Composition curve

Which of these cases does occur depends on the influence of thermodynamic equilibrium, gas and liquid side mass transfer. Their influence on the selectivity will be estimated in the next chapter by solving the differential equation for the diffusion in the liquid phase.

2.2 Estimation of the Selectivity

The estimation is based on the situation illustrated in Fig. 3. We consider a capillary perpendicular to the sample surface filled with liquid. The sample surface is exposed to the drying air. Due to the temperature gradient heat is transferred to the sample surface resulting in an evaporation of the liquid. The evaporation stream is swept off by the drying agent.

The liquid is a binary mixture. At the beginning of the drying it is distributed uniformly in the capillary. During the progress of the drying the concentration profiles will be developed, as shown in Fig. 3.

For the calculation the following assumptions are made:

- The evaporation of the liquid takes place at the sample surface; the liquid is transferred to the sample surface by capillary motion.
- The concentration in the bulk phase of the liquid remains constant, that means, that the capacity of the bulk phase is much larger than the capacity of the boundary layer.

- The temperature inside the sample is constant and is equal to the measured temperature for the constant rate period.
- Thermodynamic equilibrium is assumed at the sample surface.
- The contact time between the sample and the drying agent is short.

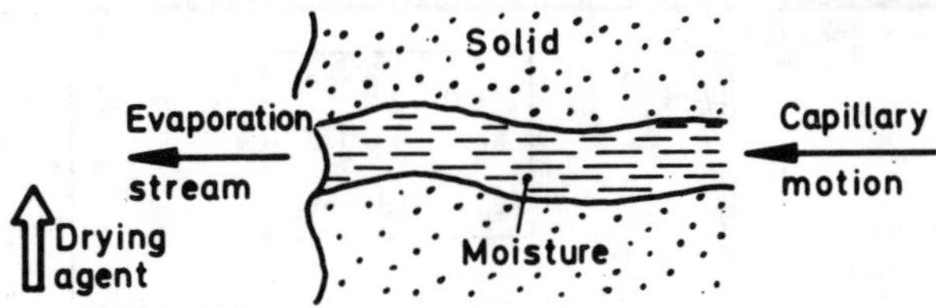

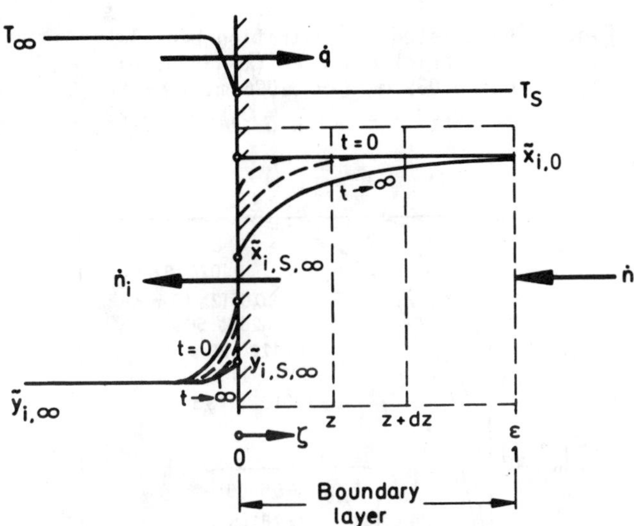

Fig. 3 Drying of materials containing binary mixtures

The aim of the calculation is to get the concentration profiles in the liquid as a function of space and time. Knowing this it is possible to calculate the selectivity as a function of the time.

The governing equations for the calculation are
- mass balances,
- kinetic equations for mass transfer
- equilibrium relationship.

The overall mass balance around the boundary layer of thickness ε (Fig. 3) is given by

$$\dot{N}_{in} = \dot{N}_{out} + \frac{dN}{dt}$$

$$\tilde{\rho}_1 A v = (\dot{n}_1 + \dot{n}_2) A$$

$$v = \frac{1}{\tilde{\rho}_1} (\dot{n}_1 + \dot{n}_2) \ . \tag{6}$$

The mass balance for component i around a differential element of the boundary layer is given by

$$\dot{N}_{i,in} = \dot{N}_{i,out} + \frac{dN}{dt}$$

$$\tilde{\rho}_1 \frac{\delta_1}{\mu} A \left.\frac{\partial \tilde{x}_i}{\partial z}\right|_{z+dz} + \tilde{\rho}_1 A v \tilde{x}_{i,z+dz} =$$

$$\tilde{\rho}_1 \frac{\delta_1}{\mu} A \left.\frac{\partial \tilde{x}_i}{\partial z}\right|_{z} + \tilde{\rho}_1 A v \tilde{x}_{i,z} + \tilde{\rho}_1 A \, dz \frac{\partial \tilde{x}_i}{\partial t}$$

$$\boxed{\frac{\delta_1}{\mu} \frac{\partial^2 \tilde{x}_i}{\partial z^2} + v \frac{\partial \tilde{x}_i}{\partial z} = \frac{\partial \tilde{x}_i}{\partial t}} \ . \tag{7}$$

For the mass transfer from the sample surface to the bulk of the air stream the linear kinetic equations can be used, because of the low concentrations:

$$\dot{n}_i = \tilde{\rho}_g \beta_{g,i} (\tilde{y}_{i,S} - \tilde{y}_{i,\infty}) \ . \tag{8}$$

The mass transfer coefficient in this equation is calculated with the correlations for forced convection around immersed bodies /4/.

At the sample surface thermodynamic equilibrium is assumed

$$\tilde{y}_{i,S} = \tilde{y}_i^*(T_S) = \frac{\gamma_i(\tilde{x}_{i,S}, T_S)\, \tilde{x}_{i,S}\, p_i^*(T_S)}{p_{ges}} \ . \tag{9}$$

In this equation the activity coefficient is calculated with the NRTL-method /5/ and the vapor pressure with the Antoine-equation. The parameters for these equations are taken from /6/.

This set of equations is sufficient for the calculation of the concentration profiles as a function of space and time. This can be done by solving the differential equation (Eq.(7)) for unsteady state diffusion in the liquid phase with the initial condition

$$\boxed{t = 0 : \tilde{x}_i(z,0) = \tilde{x}_{i,0}} \tag{10}$$

and the boundary conditions

$$z = 0 : \tilde{\rho}_g \beta_{g,i} (\tilde{y}_{i,S} - \tilde{y}_{i,\infty}) = \tilde{\rho}_1 \frac{\delta_1}{\mu} \left.\frac{\partial \tilde{x}_i}{\partial z}\right|_{z=0} + \dot{n}\, \tilde{x}_{i,S} \tag{11}$$

$$z = \varepsilon : \tilde{x}_{i,\varepsilon} = \tilde{x}_{i,0} \tag{12}$$

This differential equation was solved numerical using the explicit method of differences /7/ under the assumption that the velocity of the capillary motion v is constant. This assumption is justified, because for the drying of materials containing binary mixtures a constant rate period is obtained, too (see chapter 3.3). Further the thickness ε of the boundary layer has to be known.

The velocity of the capillary motion v can be calculated from a mass balance around the boundary layer for the steady state, which is achieved for t against infinity

$$\dot{n}_{1,\infty} = \dot{n}_\infty \tilde{x}_{1,0} \tag{13}$$

or with $\dot{n}_\infty = \dot{n}_{1,\infty} + \dot{n}_{2,\infty}$

$$\dot{n}_{1,\infty} = \dot{n}_{2,\infty} \frac{\tilde{x}_{1,0}}{1 - \tilde{x}_{1,0}} \tag{14}$$

By replacing the molar flux densities by Eq.(8) and using Eq.(9) for the thermodynamic equilibrium one obtains

$$\frac{1}{\tilde{x}_{1,S,\infty}} = 1 + \frac{\beta_{g,1}}{\beta_{g,2}} \frac{\gamma_1(\tilde{x}_{1,S,\infty},T_S)}{\gamma_2(\tilde{x}_{1,S,\infty},T_S)} \frac{p_1^*(T_S)}{p_2^*(T_S)} \frac{1-\tilde{x}_{1,0}}{\tilde{x}_{1,0}} \quad . \tag{15}$$

From this equation the interfacial concentration $\tilde{x}_{1,S,\infty}$ can be calculated by iteration. Knowing this the velocity of the capillary motion can be obtained from Eq.(6) by replacing the molar flux densities by Eq.(8).

The thickness of the boundary layer ε can be estimated from the concentration profile for the steady state. This can be obtained from Eq.(7) with $\partial\tilde{x}_i/\partial t = 0$:

$$\frac{\delta_1}{\mu} \frac{\partial^2 \tilde{x}_i}{\partial z^2} + v \frac{\partial \tilde{x}_i}{\partial z} = 0 \quad . \tag{16}$$

Solving this differential equation for the boundary conditions

$$z = 0 : \quad \tilde{x}_i = \tilde{x}_{i,S,\infty} \tag{17}$$

$$z \to \infty : \quad \tilde{x}_i = \tilde{x}_{i,0} \tag{18}$$

one obtains

$$\frac{\tilde{x}_{i,\infty}(z) - \tilde{x}_{i,S,\infty}}{\tilde{x}_{i,0} - \tilde{x}_{i,S,\infty}} = 1 - \exp\left[- \frac{v\mu}{\delta_1} z \right] . \tag{19}$$

The thickness of the boundary layer is arbitary defined as the spacial coordinate z, where

$$\frac{\tilde{x}_{i,\infty}(\varepsilon) - \tilde{x}_{i,S,\infty}}{\tilde{x}_{i,0} - \tilde{x}_{i,S,\infty}} = 0.999 \; . \tag{20}$$

With this definition ε can be calculated from Eq.(20).

With the calculated values of v and ε the differential equation (Eq.(10)) was solved. Figs. 4 and 5 show the obtained concentration profiles of a brick cylinder wetted with isopropyl alcohol(1)-water(2) for $\tilde{x}_{1,0} = 0.30$ and $\tilde{x}_{1,0} = 0,60$. In these figures the dimensionless concentration

$$\xi = \frac{\tilde{x}_1(z,t) - \tilde{x}_{1,S,\infty}}{\tilde{x}_{1,0} - \tilde{x}_{1,S,\infty}} \tag{21}$$

is plotted versus the dimensionless spacial coordinate

$$\zeta = \frac{z}{\varepsilon} \quad . \tag{22}$$

The dimensionless time is given by

$$\tau = \frac{\delta_1 \, t}{\mu \, \varepsilon^2} \quad . \tag{23}$$

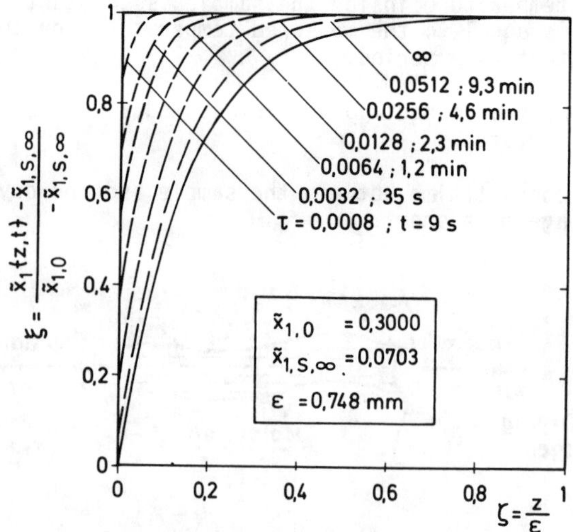

Fig. 4 Calculated concentration profiles of a brick cylinder ($\tilde{x}_{1,0} = 0.30$) d = 0.039 m, l = 0.096 m, T_∞ = 60 °C u_∞ = 0.20 m/s, $\tilde{y}_{i,\infty} = 0$, T_S = 20 °C

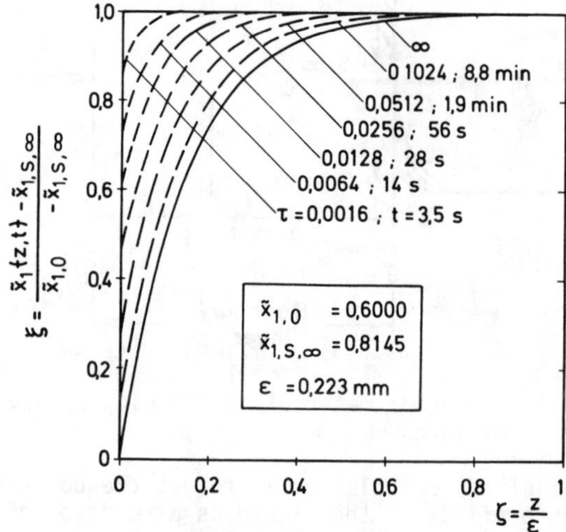

Fig. 5 Calculated concentration profiles of a brick cylinder ($\tilde{x}_{1,0} = 0.60$) d = 0.039 m, l = 0.096 m, T_∞ = 60 °C, u_∞ = 0.20 m/s, $\tilde{y}_{i,\infty} = 0$, T_S = 20 °C

From the figures can be seen that the order of magnitude of the time required to develop the concentration profile is about 10 minutes. Further it can be seen that the thickness of the boundary layer is about 0.5 mm.

Knowing the interfacial concentration as a function of time it is also possible to calculate the evaporation streams and thus the selectivity as a function of time:

$$S_1(t) = \frac{\dot{n}_1(t)}{\dot{n}_1(t) + \dot{n}_2(t)} - \tilde{x}_{1,0} \tag{24}$$

with

$$\dot{n}_i(t) = \tilde{\rho}_g \beta_{g,i} \, (\tilde{y}_{i,S}(t) - \tilde{y}_{i,\infty}) \; . \tag{25}$$

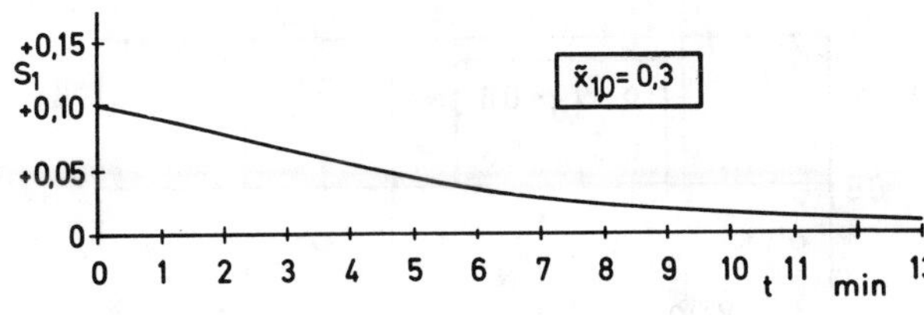

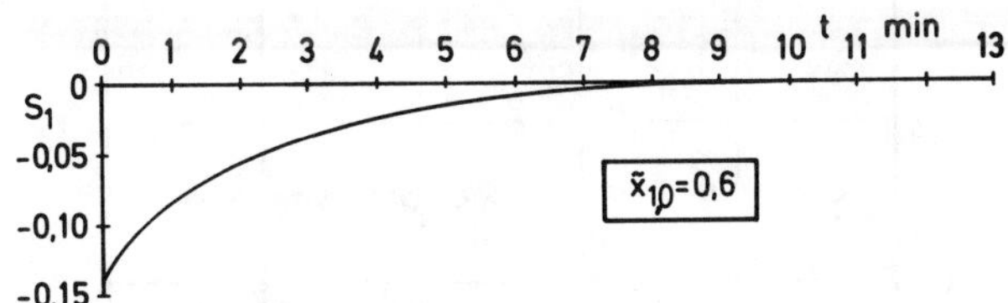

Fig. 6 Calculated selectivity of a brick cylinder
$d = 0.039$ m, $l = 0.096$ m,
$T_\infty = 60$ °C, $u_\infty = 0.20$ m/s, $\tilde{y}_{i,\infty} = 0$

Fig. 6 shows the selectivity as a function of time for the same conditions. It can be seen that selectivity is achieved only during the first minutes while the concentration profile is developed. During this period the selectivity depends on thermodynamic equilibrium as well as gas side mass transfer. After then the selectivity is zero. From the foregoing calculations for the drying of materials containing binary mixtures can be concluded:

- If the dimensions of the sample are larger than the thickness of the boundary layer the drying is nonselective, except beginning of the drying.
- If the dimensions of the sample are smaller than the thickness of the boundary layer the drying is selective. The selectivity is mainly influenced by thermodynamic equilibrium as well as gas side mass transfer.
- If the dimensions of the sample are larger than the thickness of the boundary layer the selectivity at the beginning of the drying is influenced by thermodynamic equilibrium as well as gas side mass transfer.

3. EXPERIMENTAL PART

3.1 Measuring Technique

In the experimental part of this work single porous bodies wetted with either isopropyl alcohol or water or a mixture of both were dried in a drying channel using dry air (Fig. 7).

During the drying of the samples the following measurements were carried out:

- The weight of the sample was continuously measured with a balance, thus giving the drying curve;
- the isopropyl alcohol concentration in the outled air was determined by means of an infrared gas analyzer, thus giving the composition curve.

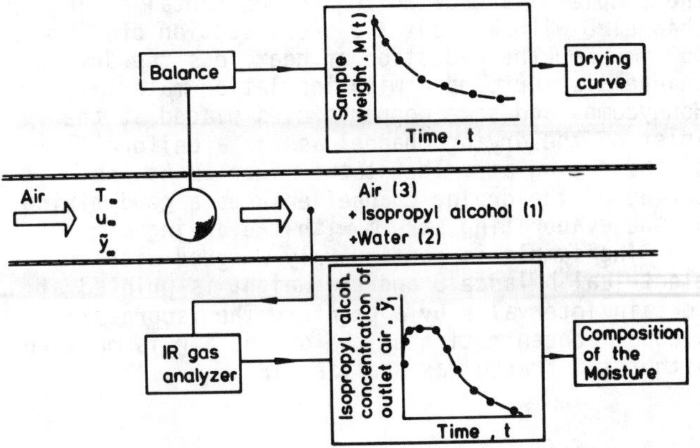

Fig. 7 Measuring technique

3.2 Experimental Equipment

Fig. 8 shows a flow diagram of the experimental equipment.

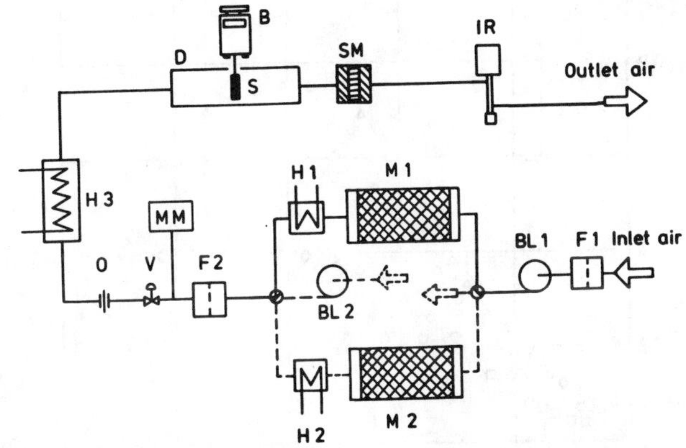

B	Balance	M1,2	Molecular sieve
BL1,2	Blower	MM	Moisture meter
D	Drying channel	O	Orifice plate flow meter
F1,2	Filter		
H1,2,3	Resistance heater	S	Sample
IR	Infrared gas analyser	V	Valve

Fig. 8 Flow diagram of the experimental equipment

The flowrate of air is produced by the blower BL1. Passing the filter F1 and the molecular sieve M1 or M2 dust and moisture are removed from the air. The filter F2 keeps back fines from the molecular sieve. The moisture meter MM is used to check the residual moisture of the drying air. If the capacity of one of the molecular sieves is exhausted, it can be regenerated by passing air with a temperature of 250 to 300 °C in countercurrent flow through it. The valve V is used to adjust the flow rate, which is measured with an orifice plate flow meter O. Then the air stream is heated up to the desired temperature with the electrical resistance heater H3. The air temperature is controlled via a thermocouple at the inlet of the drying channel using a PID-controller.

The samples S are dried in the horicontal drying channel D with a quadratic cross section of 150 x 150 mm. For the reduction of heat loss the drying channel is surrounded with insulation material. Honeycombs and some copper sieves placed at the inlet of the drying channel assure a uniform flow of the drying air. The static mixer SM at the outlet of the drying channel ensures a good mixing of the evaporating stream with the drying air.

The sample is continuously weighed with an electrical balance B and the weight is printed at certain intervalls by a printer. The isopropyl alcohol concentration of the outlet air is measured with the infrared gas analyzer IR.

3.3 Discussion of the Experimental Results

In this chapter some of the experimental results will be presented.

Drying of cylinders

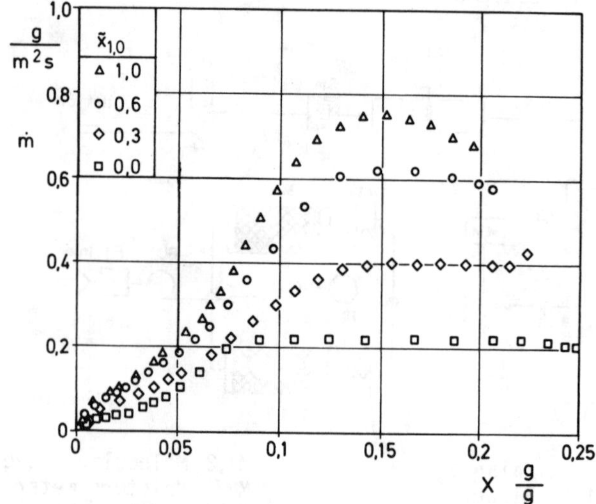

Fig. 9 Drying curves of a brick cylinder
$d = 0.0392$ m, $l = 0.0957$ m,
$T_\infty = 60$ °C, $u_\infty = 0.20$ m/s, $\tilde{y}_{i,\infty} = 0$

Fig. 9 shows the drying curves of a brick cylinder for constant drying conditions. In the figure the drying rate $\dot{m}$ is plotted versus the moisture content X. Depicted are the drying curves of the pure compounds isopropyl alcohol and water and two curves of mixtures of them. Brick has a high porosity resulting in a distinct constant rate period. This period is terminated at the critical moisture content and the drying rate decreases then due to the additional heat and mass transfer resistance of the already dried material. For the same drying conditions the drying rate of isopropyl alcohol is about three times higher than the one of water. The drying curves of the mixtures have the same characteristical shape than the ones of the pure compounds. The fact, that the drying curves of the mixtures do not approach either one of the pure compounds can be interpreted as a first indication for nonselective drying.

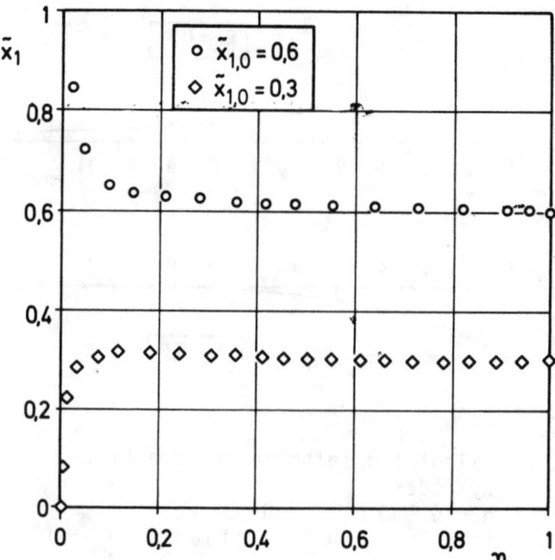

Fig. 10 Composition curves of a brick cylinder
$d = 0.0392$ m, $l = 0.0957$ m,
$T_\infty = 60$ °C, $u_\infty = 0.20$ m/s, $\tilde{y}_{i,\infty} = 0$

In Fig. 10 the composition curves of the same runs are shown. In the figure the molfraction of isopropyl alcohol in the moisture $\tilde{x}_1$ is plotted versus the relative moisture η. Depicted are two runs with different initial compositions $\tilde{x}_{1,0}$. It can be seen that under the conditions investigated no selectivity was achieved in a wide range.

From these results can be concluded that under the conditions investigated the selectivity is mainly influenced by the liquid side mass transfer (see chapter 2.2). The boundary layer ε was estimated to be 0.75 mm for $\tilde{x}_{1,0} = 0.30$ and 0.22 mm for $\tilde{x}_{1,0} = 0.60$. These are much smaller compared to the radius of the sample, which is 20 mm. Thus the liquid side mass transfer is the controlling step, which means no selectivity.

Drying of hollow cylinders. From chapter 2.2 can be seen that it is necessary to reduce the length of the diffusional path in the liquid in order to get a selectivity. This was done by using hollow cylinders of sintered bronze with a small wall thickness.

Fig. 11 shows the drying curves of a hollow sintered bronze cylinder for constant drying conditions. Depicted are the drying curves of the pure compounds isopropyl alcohol and water and two curves of mixtures of them. The initial moisture contents are much lower compared to brick, because of the higher solid density of bronze. The drying rates of the constant rate period are in agreement with the ones for the brick cylinder (Fig. 9), because the drying conditions are the same. The constant rate period is followed by the falling rate period.

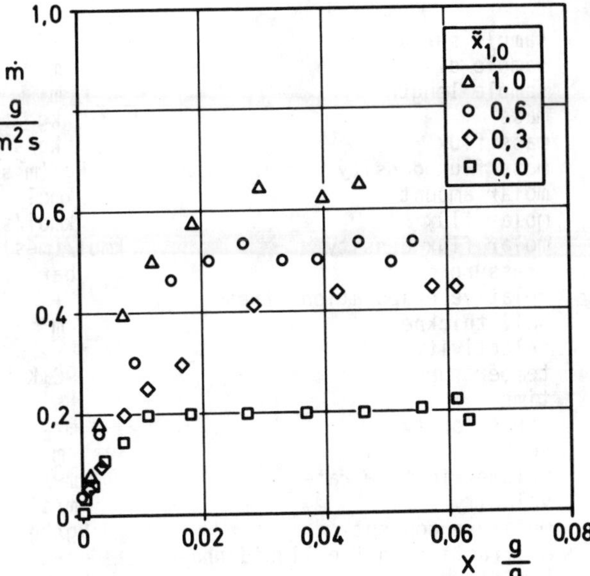

Fig. 11 Drying curves of a hollow sintered bronze cylinder
$d = 0.0310$ m, $l = 0.096$ m,
$s = 0.0005$ m, $T_\infty = 60\ ^{\circ}C$,
$u_\infty = 0.20$ m/s, $\tilde{y}_{i,\infty} = 0$

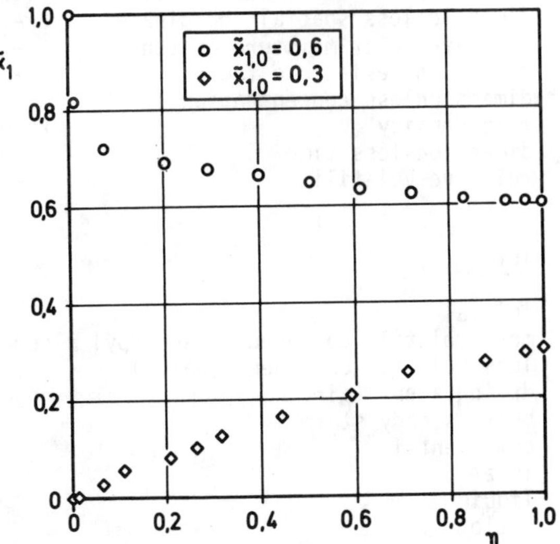

Fig. 12 Composition curves of a hollow sintered bronze cylinder
$d = 0.0310$ m, $l = 0.0960$ m,
$s = 0.0005$ m, $T_\infty = 60\ ^{\circ}C$,
$u_\infty = 0.20$ m/s, $\tilde{y}_{i,\infty} = 0$

Fig. 12 shows the composition curves for the same runs. Depicted are the two runs with different initial compositions. It can be seen, that selectivity was achieved, because the liquid side mass transfer is not any more the controlling step. The selectivity is mainly influenced by the thermodynamic equilibrium and the gas side mass transfer. That means that for an initial composition of $\tilde{x}_{10} = 0.60$, which is close to the azeotropic point, the faster diffusing component is removed favorably. In our case water is the faster diffusing component and thus the molfraction of isopropyl alcohol in the moisture increases. For an initial composition of $\tilde{x}_{10} = 0.30$ the higher relative volatility over-compensates this effect and the more volatile isopropyl alcohol is removed favorably. Thus the molfraction of isopropyl alcohol in the moisture decreases.

Selectivity at the beginning. In chapter 2.2 has been shown, that it is possible to achieve a selectivity at the beginning of the drying in any case. This selectivity is only influenced by the thermodynamic equilibrium as well as the gas side mass transfer. In order to check whether a selectivity at the beginning of the drying can be obtained the first runs with the brick cylinder (Figs. 9 and 10) were repeated measuring the concentration of isopropyl alcohol and water in the outlet air using two infrared gas analyzers. The results are shown in Fig. 13.

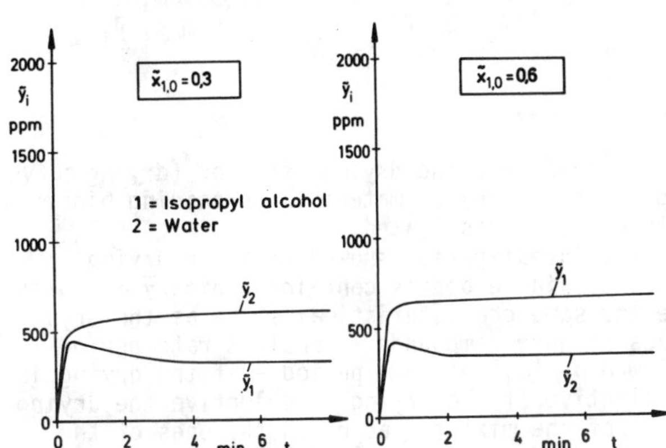

Fig. 13 Selectivity at the beginning of the drying
Sample: brick cylinder
$d = 0.0392$ m, $l = 0.0957$ m,
$T_\infty = 60\ ^{\circ}C$, $u_\infty = 0.20$ m/s, $\tilde{y}_{i,\infty} = 0$

From this figure can be seen, that at the first few minutes selectivity was achieved, indeed. For $\tilde{x}_{1,0} = 0.30$ isopropyl alcohol and for $\tilde{x}_{1,0} = 0.60$ water is removed favorably like expected.

For comparison the selectivity at the beginning of the drying was determined from the measurements and compared with the calculated selectivity (see chapter 2.2). From Fig. 14 can be seen that within the experimental accuracy a good agreement was achieved. From these results can be concluded, that it should be possible to achieve selectivity throughout the whole drying process by intermittent drying.

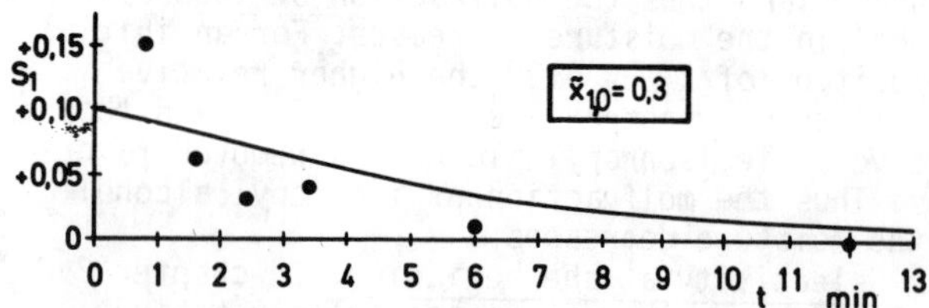

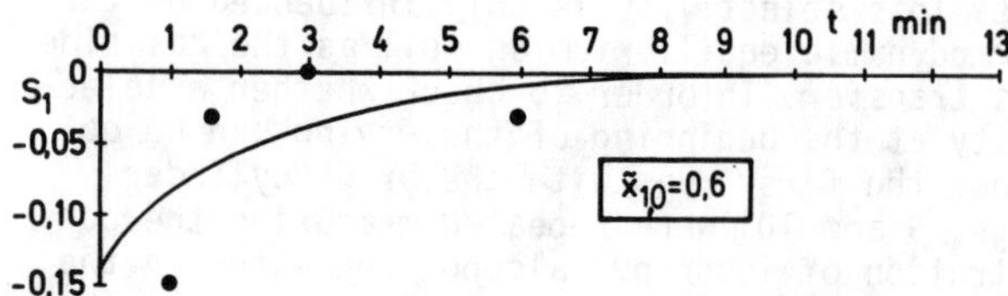

Fig. 14 Calculated and measured selectivity
Sample: brick cylinder
$d = 0.0392$ m, $l = 0.0957$ m,
$T_\infty = 60$ °C, $u_\infty = 0.20$ m/s, $\tilde{y}_{i,\infty} = 0$

4. CONCLUSIONS

In this work the drying behavior (drying curve, composition curve) of materials containing binary mixtures was investigated.

The investigation showed that the drying curves of single bodies containing binary mixtures have the same characteristical shape as the drying curves of pure compounds - constant rate period followed by falling rate period - if the drying is nonselective. If the drying is selective the drying curves of the mixtures approach the ones of the pure compounds.

Further it was found that the composition curves are influenced by the thermodynamic equilibrium, gas and liquid side mass transfer. If the dimensions of the sample are large compared to the thickness of the boundary layer the selectivity is mainly controlled by the liquid side mass transfer, which means no selectivity. Only at the very beginning of the drying a selectivity was obtained, which is controlled by thermodynamic equilibrium as well as by the gas side mass transfer. If the dimensions of the sample are small compared to the thickness of the boundary layer the liquid side mass transfer can be neglected. In this case the selectivity is mainly influenced by the thermodynamic equilibrium as well as by the gas side mass transfer.

ACKNOWLEDGMENT

The financial support of the Forschungsgemeinschaft für Luft- und Trocknungstechnik e.V., Frankfurt/Main, which made this investigation possible, is gratefully acknowledged.

NOMENCLATURE

A	sample surface	m^2
d	sample diameter	m
l	sample length	m
M	mass	kg
$\dot{M}$	mass flux	kg/s
$\dot{m}$	mass flux density, drying rate	kg/(m²s)
N	molar amount	kmol
$\dot{N}$	molar flux	kmol/s
$\dot{n}$	molar flux density	kmol/(m²s)
p	pressure	bar
$\dot{r}$	relative evaporation stream	-
s	wall thickness	m
S	selectivity	-
T	temperature	°C,K
t	time	s
u	air velocity	m/s
V	volume	m^3
$\dot{V}$	volumetric flow rate	m³/s
v	velocity	m/s
X	moisture content	kg/kg
$\tilde{x}$	molfraction in the liquid phase	-
Y	humidity	kg/kg
$\tilde{y}$	molfraction in the gas phase	-

Greek letters

β	mass transfer coefficient	m/s
γ	activity coefficient	-
δ	diffusion coefficient	m²/s
ε	thickness of the boundary layer	mm
ζ	dimensionless spacial coordinate	-
η	relative molar moisture content	-
μ	diffusion resistance factor	-
ξ	dimensionless concentration	-
ρ	mass density	kg/m³
τ	dimensionless time	-
ω_{12}	relative volatility	-

Subscripts

0	initial
1	more volatile component (isopropyl alcohol)
2	less volatile component (water)
3	drying agent (air)
∞	bulk, steady state
i	component i
in	inlet
l	liquid
out	outlet
S	surface
s	solid

Superscripts

*	equilibrium
~	molar

REFERENCES

1. Thurner, F., and Schlünder, E.U., Proc. 3d Int. Drying Symp., Birmingham, vol.2, 326 (1982).
2. Härtling, M., Messung und Analyse von Trocknungsverlaufskurven, Ph. D. thesis, Universität Karlsruhe, Karlsruhe (1978).

3. Schlünder, E.U., Chem.-Ing.-Tech., vol. 50, 749 (1978).
4. Schlünder, E.U., et. al., Heat Exchanger Design Handbook, Hemisphere, Washington (1983)
5. Renon, H. and Prausnitz, J.M., AIChE J., vol. 14, 135 (1968).
6. Gmehling, J., Onken, U., Vapor-Liquid-Equilibrium Data Collection, Chemistry Data Series, vol. 1, part 1, DECHEMA, Frankfurt/M. (1977).
7 Hornbeck, R.W., Numerical Methods, Quantum Publishers Inc., New York (1975).

THE MATHEMATICAL MODELLING OF DRYING PROCESS BASED ON MOISTURE TRANSFER MECHANISM

Zbigniew Przesmycki
Czesław Strumiłło

Research Centre, ZUT PL UNIPROT, Obywatelska 128/152
94-104 Łódź, POLAND
Łódź Technical University, Wólczańska 175, 90-924 Łódź, POLAND

ABSTRACT

A mathematical model of drying of hygroscopic-porous materials based on Krischer's approach to moisture transfer in dried materials has been proposed. The model exposes the phenomenon of receding of the evaporation zone into the dried body, and takes into account the dried body shrinkage.

The presented model as well as Krischer's and Berger and Pei's models have been solved numerically and the solutions have been compared with experimental results.

1. INTRODUCTION

The parameters necessary for dryers design and for carrying out the process of drying under optimum conditions can be found in both experimental and theoretical ways. The latter seems to be more convenient and economical in the present state of computational technics.

A theoretical treatment, however, needs the knowledge of the mechanism of moisture transfer in a dried body together with an appropriate mathematical description that appears to be still a problem requiring solution.

Over sixty years ago, Lewis [1] first published the paper dealing with moisture transfer during drying of solid materials and proposing also a simple mathematical description. Since then hundreds of various papers have been issued bringing a great number of different, very often partly contradictory views and opinions. The authors rarely give comprehensive, critical review of state of the art thereby making full understanding of the problem more difficult.

That is the reason why some authors attempt at presenting a retrospective review of the works done up to now. The papers by Keey [2,3], Van Brakel [4], Fortes and Okos [5], Toei [6] and Kisakürek [7] can be quoted among others. Also in our previous study [9] the mechanisms of moisture transfer in dried materials have been discussed. The hitherto existing approaches have been classified on the mathematical and phenomenological basis into six following groups, called theories: liquid diffusion, capillary flow, gas transfer, internal pressure, osmotic and simultaneous heat and mass transfer. The last theory appears to be the most general and useful in theoretical considerations of moisture transfer in dried body.

There exists a great deal of papers on simultaneous heat and mass transfer theory. Unfortunately so far no works can be recommended as useful in design and processing practice. It seems that a few of them, by Krischer [10], Philip and De Vries [11], Luikov [12,13], and Whitaker [14,15] deserve more attention.

Three, historically first studies [10-13] can be treated as the classical ones and in the field of possible applications each of them is somehow deficient. No clear superiority of any approach could be spotted and so, further evidence is needed. Whitaker's study [14,15], on the other hand, requires experimental verification.

Because Whitaker follows his line of attack in a consistent way (the next paper has been just published [17]) we intend to examine a few aspects of the drying process modelling basing on a classical treatment. For this purpose we formed the model described in Section 3. First, however, the mathematical description of heat and mass transfer fluxes is given with reference to the above mentioned theories and studies.

2. THE MATHEMATICAL DESCRIPTION OF HEAT AND MASS TRANSFER FLUXES IN A DRIED BODY

According to capillary flow theory, the following liquid moisture transfer equation can be written

$$\dot{\vec{m}}_l = -\mathbb{k}_l \rho_l \; \vec{\nabla}(\varphi + \varphi_G) \tag{1}$$

Limiting the number of parameters that affect capillary potential φ to moisture content S and temperature T

$$\varphi = \varphi(S,T) \tag{2}$$

the following equation is obtained

$$\dot{\vec{m}}_l = -k_l \rho_l \left(\frac{\partial \varphi}{\partial S} \vec{\nabla} S + \frac{\partial \varphi}{\partial T} \vec{\nabla} T + \vec{\nabla} \varphi_G \right) \qquad (3)$$

It is worth emphasizing that the functional dependence (2) is close to reality for homogeneous rigid media. Defining k_S and k_T as

$$k_S = k_l \frac{\partial \varphi}{\partial S} ; \qquad k_T = k_l \frac{\partial \varphi}{\partial T} \qquad (4,5)$$

the equation

$$\dot{\vec{m}}_l = -k_S \rho_l \vec{\nabla} S - k_T \rho_l \vec{\nabla} T - k_l \rho_l \vec{\nabla} \varphi_G \qquad (6)$$

is obtained.

The form of the above equation is similar to that obtained by Philip and De Vries [11]. If both gravitational and comprising $\vec{\nabla} T$ terms are neglected, equation (6) reduces to the form suitable for liquid diffusion theory

$$\dot{\vec{m}}_l = -k_S \rho_l \vec{\nabla} S \qquad (7)$$

Having in mind the way of its derivation, the equation of liquid diffusion often used (among others by Krischer [10])

$$\dot{\vec{m}}_l = -k \vec{\nabla} \omega \qquad (8)$$

where

$$k = k(\omega, T) \qquad (9)$$

appears to be rather controversial from a phenomenological point of view. However, for hygroscopic shrinking materials due to a large number of parameters affecting the capillary potential φ, the employment of eq.(8) can be treated as justified. Moreover, eq.(8) in the case of presence of the surface diffusion appears to be a phenomenological one and gives additionally a possibility for describing the moisture transfer due to osmotic or internal pressure.

The vapor transfer in a general case consists of two constituents, the molecular and convective ones. The appropriate transfer equations can be obtained by the application of Fick's and Darcy's laws, respectively. The importance of convective transport increases with the intensity of drying process. Taking into account the molecular species of transport only, the following equation holds

$$\dot{\vec{m}}_v = -M_v D_{eff} \vec{\nabla} C_v \qquad (10)$$

In a general case

$$C_v = C_v(S, T) \qquad (11)$$

and eq. (10) can be written in the form

$$\dot{\vec{m}}_v = -M_v D_S \vec{\nabla} S - M_v D_T \vec{\nabla} T \qquad (12)$$

where coefficients D_S and D_T are defined as follows

$$D_S = D_{eff} \frac{\partial C_v}{\partial S} ; \quad D_T = D_{eff} \frac{\partial C_v}{\partial T} \qquad (13,14)$$

For a practical application of that equation one can use the specific expression instead of eq.(11) as Philip and De Vries [11] did. Unfortunately, such treatment suffers from one considerable drawback, for there is no expression that would cover the whole period of the drying process. Therefore, the application range of the resultant equation describing the vapor transfer would be limited.

Note that eq.(12) expresses the vapor transfer by means of the same driving forces as for liquid transfer. It enables to add the appropriate forms of eqs.(6) and (12) to obtain the equation describing the total moisture transfer in dried body. Such an addition leads to Philip and De Vries form of expression [11]. It is evident that its usefulness is limited due to the same reason as in the case of eq.(12). If the gravitational term is neglected the equation describing the total moisture transfer in a dried body reduces to the form presented by Luikov [12,13]. One should realize, however, that from Luikov's derivation it follows that the term comprising $\vec{\nabla} T$ is related to the Soret effect rather than to temperature dependence of the capillary potential.

The total energy flux in a dried body consists of two constituents. First of them, conduction one, can be described by Fourier's equation

$$\dot{\vec{q}} = -\lambda_{eff} \vec{\nabla} T \qquad (15)$$

The second constituent denotes convective energy flux due to moisture movement. It is understood that the importance of each species of the heat transfer depends on the character of both drying conditions and dried media, and in any case has to be considered.

3. THE MATHEMATICAL MODEL

3.1. The Phenomenon of Receding Evaporation Front

Sherwood's investigations [18] showed the existence of the evaporation front which, when the drying process proceeds, retreats into the dried body.

The character of the moisture distribution in a dried body obtained by Luikov and Kolesnikov [19] confirms Sherwood's results and leads to the conclusion that one cannot establish the sharp evaporation plane dividing the body into two parts. The evaporation process takes place in some region called thus the evaporation zone. Then, in the material subjected to drying in a general case three zones can be distinguished (Fig. 1): dry, evaporation and wet. During the constant drying rate period the wet zone fills the whole body and evaporation process takes place from the drying medium surface. The other zones appear subsequently during the falling drying rate period.

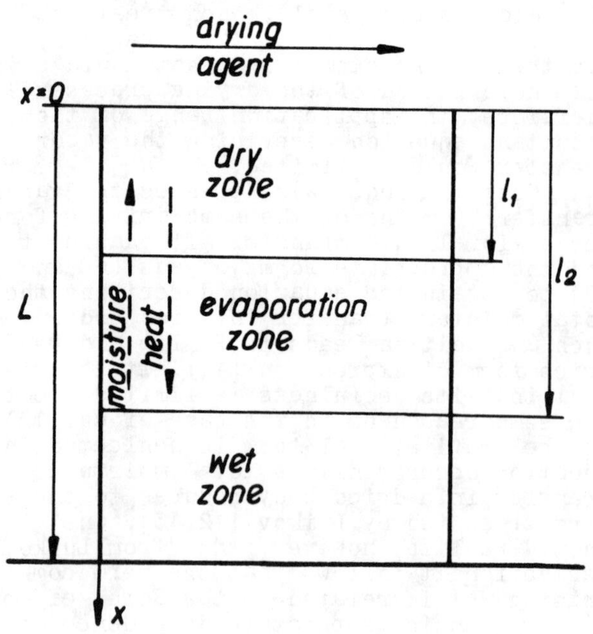

Fig. 1 The zones of dried material

3.2. General Assumptions

As we mentioned at the end of Section 1, the basic aim of the present paper is to give some contribution to the mathematical modelling based on the mechanism of moisture transfer. Having in mind the present state of knowledge in the area of drying modelling we will deal with a relatively simple drying process. Therefore, we limit our considerations to the following conditions.

- The drying process is of a convective character in the system of air-water being drying agent and moisture, respectively.
- The drying media are in contact with drying agent at one flat surface. The moisture and heat transfer is one-dimensional.
- The drying media are homogeneous and hygroscopic-porous according to the classification given in [4].
- The drying agent temperatures are below the water boiling point.

The last two assumptions justify the next ones.

- The participation of moisture transfer due to convection and gravity in the total moisture transfer is negligibly small.
- The heat transfer is due to conduction.
- Liquid diffusion equation is applicable in the description of liquid transfer in a dried body.

3.3. Heat and Mass Balance Equations

According to the general assumption, transfer equations (8), (10) and (15) were used to produce a set of differential balance equations describing the combined heat and mass transfer in each of the above mentioned zones of a dried body. The governing equations are given in the form which enables the shrinking phenomenon to be considered.

Wet zone. Assuming that mass transfer takes place only in a continuous liquid phase the following equation can be written

$$\frac{\partial(w\varrho_s)}{\partial t} = \frac{\partial}{\partial x}\left(k\,\frac{\partial(w\varrho_s)}{\partial x}\right) \qquad (16)$$

The unsteady heat transfer in this zone can be described as follows, assuming c_{s+l} to be constant

$$c_{s+l}\,\frac{\partial}{\partial t}\left(\varrho_{s+l}T\right) = \frac{\partial}{\partial x}\left(\lambda_{eff}\,\frac{\partial T}{\partial x}\right) \qquad (17)$$

In the constant drying rate period the body surface temperature is nearly constant.

Evaporation zone. It is assumed that in this region the moisture is transferred in both gaseous and liquid phases. For the hygroscopic-porous media this is a plausible hypothesis since, even when the free liquid continuity does not exist, water molecules adsorbed on the inner surfaces of solid form bridges between places of higher moisture content. Thus, even in the region of moisture content close to minimum hygroscopic moisture content, the liquid transfer due to surface diffusion is possible.

Differential equations of unsteady transfer of vapor and liquid, assuming that Ψ = const are as follows, respectively

$$M_v\Psi\,\frac{\partial c_v}{\partial t} = M_v\Psi\,\frac{\partial}{\partial x}\left(D_{eff}\,\frac{\partial c_v}{\partial x}\right) + m_{ph} \qquad (18)$$

$$\frac{\partial(w\varrho_s)}{\partial t} = \frac{\partial}{\partial x}\left(k\,\frac{\partial(w\varrho_s)}{\partial x}\right) - m_{ph} \qquad (19)$$

where m_{ph} is connected with evaporation or condensation processes in first stages of drying and, which is often neglected, with sorption or desorption processes in the last stages. Equations (18) and (19) lead to the expression describing the total moisture transfer in this zone

$$M_v\Psi\,\frac{\partial c_v}{\partial t} + \frac{\partial(w\varrho_s)}{\partial t} =$$

$$= M_v\Psi\,\frac{\partial}{\partial x}\left(D_{eff}\,\frac{\partial c_v}{\partial x}\right) + \frac{\partial}{\partial x}\left(k\,\frac{\partial(w\varrho_s)}{\partial x}\right) \qquad (20)$$

Heat transfer governing differential equation can be written in the form

$$c_{s+l}\,\frac{\partial}{\partial t}\left(\varrho_{s+l}T\right) = \frac{\partial}{\partial x}\left(\lambda_{eff}\,\frac{\partial T}{\partial x}\right) - \Delta h_{ph}\, m_{ph} \qquad (21)$$

provided that c_{s+1} = const.

The mass rate of phase change can be expressed as

$$m_{ph} = \frac{\partial}{\partial x}\left(k\,\frac{\partial(w\rho_s)}{\partial x}\right) - \frac{\partial(w\rho_s)}{\partial t} \qquad (22)$$

or

$$m_{ph} = M_v\,\psi\left(\frac{\partial C_v}{\partial t} - \frac{\partial}{\partial x}\left(D_{eff}\,\frac{\partial C_v}{\partial x}\right)\right) \qquad (23)$$

Dry zone. Through this zone moisture is transferred due to vapor diffusion only. Thus, the appropriate governing differential equations take the form

$$\frac{\partial C_v}{\partial t} = \frac{\partial}{\partial x}\left(D_{eff}\,\frac{\partial C_v}{\partial x}\right) \qquad (24)$$

and

$$c_{s+1}\,\frac{\partial}{\partial t}\left(\rho_{s+1}T\right) = \frac{\partial}{\partial x}\left(\lambda_{eff}\,\frac{\partial T}{\partial x}\right) \qquad (25)$$

3.4. Boundary Conditions

For the drying process under investigation the boundary conditions imposed at the interfaces between drying media and the drying agent as well as between each zone of the body can be formulated as follows (in accordance with Fig. 1)

Mass transfer

$$w = w_0 \quad \text{for } t = 0,\ L \geq x \geq 0 \qquad (26)$$

$$\beta M_v(\Delta C_v)_{l_1} = k\,\frac{\partial(w\rho_s)}{\partial x} + \psi D_{eff} M_v\,\frac{\partial C_v}{\partial x} \quad \text{for } t \geq 0,\ x = l_1 \qquad (27)$$

$$w = w_{hmin};\ c_v = c_v(w,T) \quad \text{for } t \geq t_{ss},\ x = l_1 \qquad (28)$$

$$w = w_{hmax};\ C_v = C_v(T) \quad \text{for } t \geq t_{st},\ x = l_2 \qquad (29)$$

$$\frac{\partial(w\rho_s)}{\partial x} = 0 \quad \frac{\partial C_v}{\partial x} = 0 \quad \text{for } t \geq 0,\ x = L \qquad (30)$$

Heat transfer

$$T = T_w \quad \text{for } t_{st} > t \geq 0 \quad L \geq x \geq 0 \qquad (31)$$

$$\alpha(T_a - T_i) = -\lambda_{eff}\,\frac{\partial T}{\partial x} \quad \text{for } t \geq t_{st},\ x = 0 \qquad (32)$$

$$\frac{\partial T}{\partial x} = 0 \quad \text{for } t \geq 0,\ x = L \qquad (33)$$

The boundary conditions given in the general form are modified according to the stage of the drying process.

For a constant drying rate period they take the following form

$$w = w_0 \quad \text{for } t=0 \quad L \geq x \geq 0 \qquad (26)$$

$$\beta M_v(C_{vi} - C_{va}) = k\,\frac{\partial(w\rho_s)}{\partial x} \quad \text{for } t_{st} > t \geq 0\ ,\ x = 0 \qquad (27a)$$

$$\frac{\partial(w\rho_s)}{\partial x} = 0\ ;\ \frac{\partial C_v}{\partial x} = 0 \quad \text{for } t \geq 0,\ x = L \qquad (28)$$

For the falling drying rate period when the dry zone does not occur still, it can be written

$$\beta M_v(C_{vi}-c_{va}) = k\,\frac{\partial(w\rho_s)}{\partial x} + \psi D_{eff} M_v\,\frac{\partial C_v}{\partial x} \quad \text{for } t_{ss} > t \geq t_{st},\ x = l_1 = 0 \qquad (27b)$$

$$w = w_{h\,max};\ C_v = C_v(T) \quad \text{for } t \geq t_{st},\ x = l_2 \qquad (29)$$

$$\frac{\partial(w\rho_s)}{\partial x} = 0;\ \frac{\partial C_v}{\partial x} = 0 \quad \text{for } t \geq 0,\ x = L \qquad (30)$$

$$\alpha(T_a - T_i) = -\lambda_{eff}\,\frac{\partial T}{\partial x} \quad \text{for } t \geq t_{st},\ x = 0 \qquad (32)$$

$$\frac{\partial T}{\partial x} = 0 \quad \text{for } t \geq 0,\ x = L \qquad (33)$$

For last stages of the falling drying rate period ($t \geq t_{ss}$) when all zones of the dried body occur, the general form of the boundary conditions is valid with self-evident absence of initial conditions (26) and (31).

3.5. Information Concerned with the Model Solution

The first supplementary equation necessary to solve heat and mass transfer equations is connected with a thermodynamic relation between vapor and liquid. If we assume that the liquid-vapor equilibrium exists inside the dried body, it implies that the partial vapor pressure is

- equal to its saturation value in the region of liquid content larger than the maximum hygroscopic moisture content w_{hmax}
- controlled by sorption isotherm in the hygroscopic region, i.e. in the region of the liquid content smaller than $w_{h\,max}$

This approach is presented by Berger and Pei [16]. In our study the well-known Nernst relation is employed in the range of $w > w_{h\,max}$, and for $w < w_{h\,max}$ sorption isotherms found experimentally by BET method are used.

The second supplementary equation joining the solid density ρ_s with moisture content w is as follows

$$\rho_s = \frac{\rho_0}{1 + \beta_v w} \qquad (34)$$

The coefficient β_v was determined experimentally in simple laboratory tests.

Two different drying media were investigated. These were clay and brick. Both belong to the group of the hygroscopic-porous materials, though only clay shrinks during the drying process.

Boundary conditions (27) and (32) contain the heat and mass transfer coefficient α and β, respectively. The proper choice of their evaluation method is one of the most important, unfortunately often not enough appreciated points of modelling procedure. It is relatively easy to estimate the coefficients α , and β especially for the constant drying rate period and to some extent for the falling drying rate period when dry zone does nor occur. Though, even for the constant drying rate period one can point the factors hindering the analysis (see [4]). For our purpose, the dimensionless equations given in [20] were employed. Whereas, for evaporation from submerged front we proposed the below reported procedure of the β estimation. Namely, the coefficient β was evaluated at the moment of the appearance of dry zone by means of eq.(27) using liquid and vapor content profiles, which have been found in the preceeding step. Taking into account the discrete character of the zones translation connected with the below mentioned solution procedure, one assumes that for the given location of the dry zone, mass transfer coefficient β was constant. The example of the evaluated dependence $\beta = \beta(l_1)$ were given in Fig. 2.

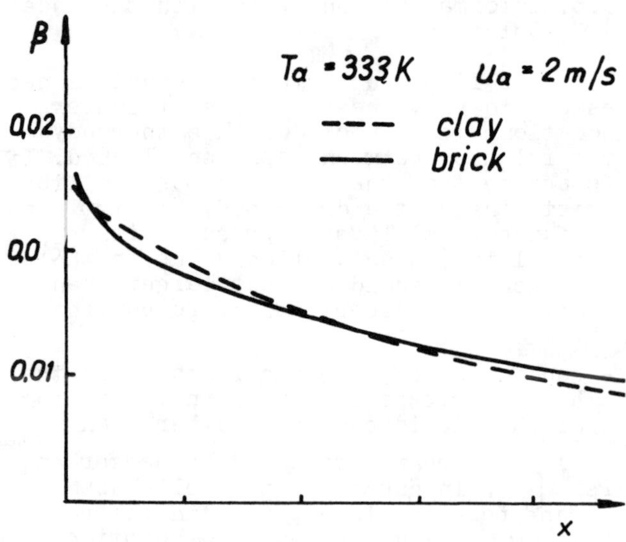

Fig. 2 Dependence of mass transfer coefficient β on position of the surface separating dry and evaporation zones

Liquid diffusivity k for clay was found in capillary rise tests with reference to Macey's data [21], and for brick we used Haertling's data [22], treating the solid density ρ_s as an identifying parameter. It is worth stressing that Heartling's values of liquid diffusivity k were found on the basis of the drying process. The effective diffusion coefficient D_{eff} was evaluated from the dependence $D_{eff} = D_{eff}(\psi)$ obtained on the basis of Currie's considerations [23]. The effective thermal conductivity λ_{eff} was found using the expression $\lambda_{eff} = \lambda_{eff}(\rho_s, w)$ given by Kaufman (Ref. in [24]) who investigated heat transfer in building materials.

Latent heat of phase change Δh_{ph} was estimated from the dependence $\Delta h_{ph} - \Delta h_v = \Delta h(\phi, T)$ given in [25].

In addition, there were several parameters associated with numerical solution of the non-linear, partial differential equations describing heat and mass transfer in the dried body.

4. EXPERIMENTAL WORK AND COMPUTER CALCULATIONS

A number of drying experiments were performed using a typical laboratory convective dryer [8]. The surface for heat and mass transfer works out for both experimental media at 20 sq.cm. The initial moisture content w_0 and the depth of the samples L amount for clay to 0.2 kg/kg, 0.03 m, and for brick 0.168 kg/kg, 0.05 m, respectively. For each material nine experiments were performed using the following air parameters: T_a = 353, 333, 318 K and u_a = 5, 3.65, 2 m/s. The duration of one experiment was 31 hrs. All experimental results were given in [8].

The proposed model and both Krischer's [10] and Berger and Pei's [16] models were solved numerically by means of the finite differences method. When material shrinkage occurs the second order partial derivates were approximated using the following expression

$$\left.\frac{\partial^2 w}{\partial x^2}\right|_i = \frac{2(w_{i-1,k}h_2 - w_{i,k}(h_1+h_2) + w_{i+1,k}h_1)}{(h_1+h_2)\, h_1 h_2} \tag{35}$$

where i and k are the indexes attributed to the space and time axes. The details of the numerical procedure were given elsewhere [8].

The simulation and experimental results were shown in the form of typical drying curves (due to difficulties in measuring the moisture profiles), and graphs of the temperature schedules. The examples of the results are presented in Figures 3 through 6.

5. DISCUSSION

The above presented model of drying of hygroscopic-porous media is based on Krischer's approach to moisture transfer in dried materials with Berger and Pei's correction concerning liquid-vapor equilibrium.

The model was built on the basis of the assumption that material density and transfer coefficients are the function of time and coordinate. The value of void fraction ψ was assumed to be constant. If the parameters ρ_s, k, λ_{eff} and D_{eff} are taken as constant the proposed model is reduced to the form presented by Krischer.

On the other hand, if we assume that ρ_s, k, λ_{eff} and D_{eff} are constant and ψ is a function of both location and time, Berger and Pei's model form is obtained.

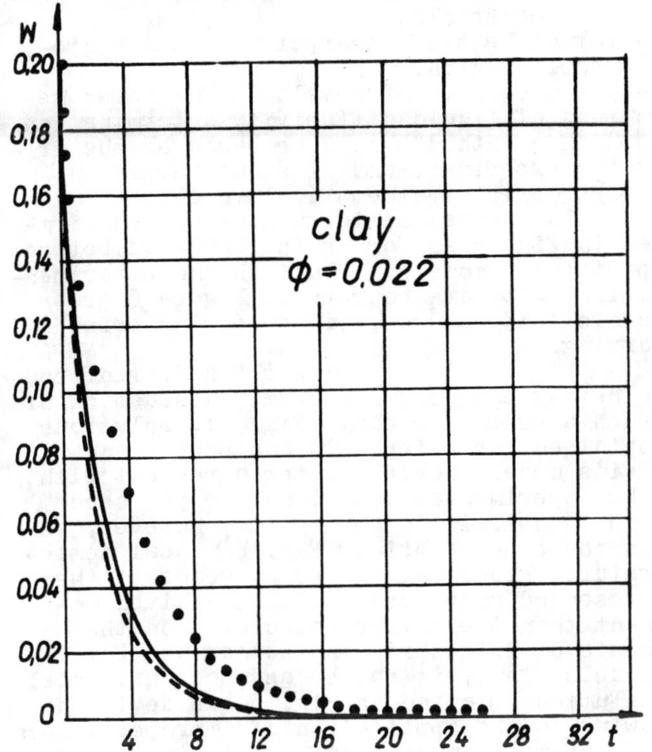

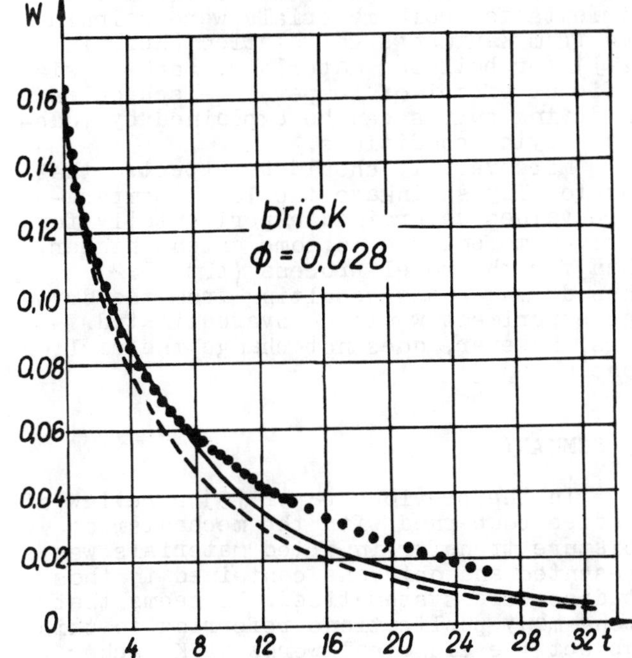

Fig. 3, 4 Drying curves; T_a= 353K; u_a= 5m/s
• experiment
—— presented model
--- Krischer's [10] and Berger and Pei's [16] models

It should be stressed that

- Krischer's model has been solved applying Berger and Pei's correction concerning the liquid-vapor equilibrium in the material being dried.
- Material density, transfer coefficients and void fraction were taken as constant only in differentiation procedure when the forms of particular models were derived. They were varying in calculations.

As follows from Figs. 3-6 differences between solutions of Krischer's and Berger and Pei's models are negligible. When this fact is taken into account and the equations describing total mass transfer in Krischer's and Berger and Pei's models in each zone of the dried body are considered, it may be concluded that vapor transfer does not control the total moisture transfer in the evaporation zone.

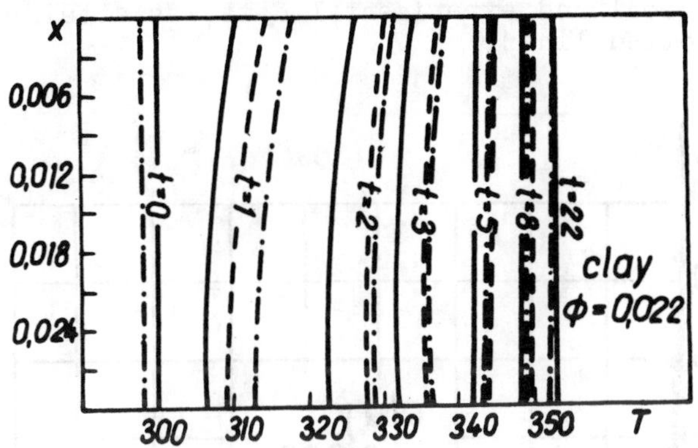

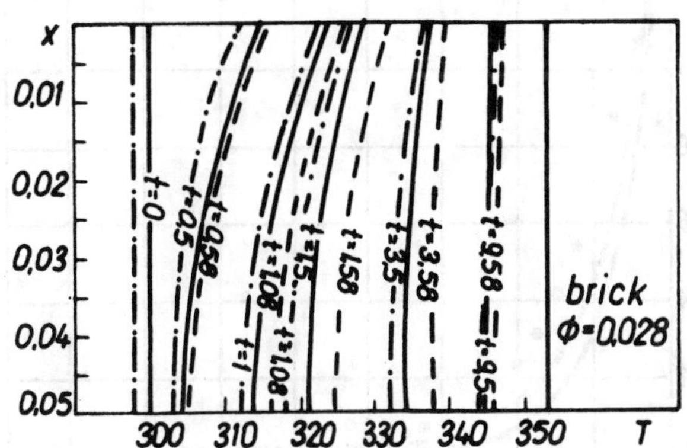

Fig. 5, 6 Temperature distributions; T_a = 353K; u_a = 5 m/s
—·— experiment
—— presented model
---- Krischer's [10] and Berger and Pei's [16] models

From Graphs 3-4 it follows that for brick the agreement of experimental and theoretical results is slightly better than for clay. In order to appreciate the deviations between theoretical and experimental results the following relative error has been defined.

$$E = \left| \frac{w_{exp} - w_{theor}}{w_{exp}} \right|_t \tag{36}$$

Its values change during the drying process from zero through some maximum to zero. The maximum value of the relative error can be treated as a measure of deviations between both kinds of the results. For clay it oscillates from 0.65 to 0.81 according to drying conditions and the model employed. And for brick the maximum values of the relative error fluctuate between 0.24 and 0.47. In both cases the best results were obtained using our model. Additionally the proposed model has been solved for clay with the values of liquid diffusivity obtained on the basis of a drying experiment [22]. So obtained results are presented in Fig. 7. Better consistency of theoretical and experimental data was attained (dotted line).

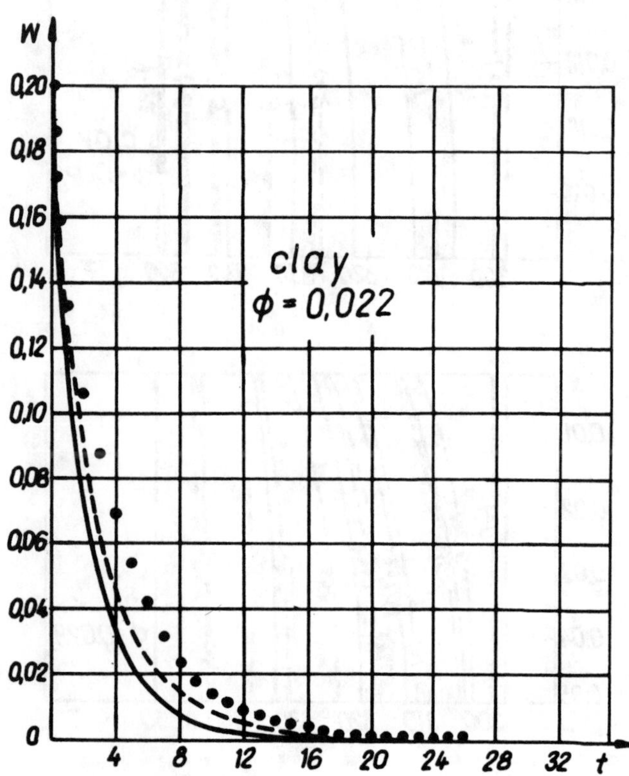

Fig. 7 Drying curves; T_a = 353K, u_a= 5m/s
- • experiment
- —— presented model; k - without drying experiment; E = 0.78
- ---- presented model; k - basing on drying experiment; E =0.35

The above discussion indicates that Krischer's model can be applied to the description of moisture transfer in dried hygroscopic-porous materials, taking into account

- Berger's and Pei's modification concerning liquid-vapor equilibrium in the material being dried;
- variation of transfer coefficients and changes of density in the case of shrinking materials;
- latent heat of desorption in the hygroscopic region;
- the necessity of developing a proper method of liquid diffusivity determination. Such a method should be based on the drying experiment and it should apply an appropriate mathematical model.

The above considerations are confirmed in Fig. 6 as far as the agreement between particular models' solutions and experimental results are concerned. Figure 6 presents temperature profiles during brick drying.

Slightly different conclusions can be drawn from a discussion of such a graph for clay (Fig. 5). Solutions obtained for Krischer's and Berger and Pei's models reveal better agreement with the experimental data than the proposed model does. At the same time, however, as in the case of brick (Fig. 6) local temperature values calculated by means of the presented model are lower in a given moment than the values calculated on the basis of Krischer's and Berger and Pei's models. Thus, it can be assumed that model solutions plotted in Fig. 5 are deviated owing to the application of improper values of thermal conductivity coefficients λ_{eff} for clay in the calculations. These coefficients for both materials were calculated from Kaufman's correlation (Ref. in [24]) for building materials. Lack of significant effect of temperature schedules on drying curves can be explained by moderate drying conditions.

Moreover, it should be expected that due to clay shrinkage the local temperature values determined experimentally for a given moment are to some extent higher than for the model process (p.p. 3.2). Thus drying rates resulting from the drying experiment would be overestimated. This, however, does not change the earlier conclusions.

6. SUMMARY

In the previous paper [9] a review of studies concerned with the mechanism of moisture transfer in dried materials was presented and opinions contained in those studies were systematized. It seems that among many publications referring to this subject the classical works by Krischer [10], Philip and De Vries [11], Luikov [12, 13] completed by Whitaker's studies [14, 15] cover generally the whole region connected with moisture and heat transfer in capillary-porous and hygroscopic-porous media (cf. the classification given in [4] p.219).

Since none of the above mentioned classical works is superior to the other and each of them has some deficiency, it seems to be justified that some aspects of these studies should be examined.

A mathematical model of moisture transfer in hygroscopic-porous materials has been proposed. It is based on Krischer's approach to moisture transfer [10] and takes into account Berger and Pei's modification [16] concerning liquid-vapor equilibrium in the dried material.

The model has the following characteristics.

- Emphasis is given to the phenomenon of evaporation zone retreat inside the dried material. It was assumed that in the general case the dried body could be divided into three zones: dry, evaporation and wet. In the evaporation zone the most controversial from the point of view of transfer phenomena, surface diffusion and vapor diffusion take place.
- The model takes into account material shrinkage during drying and latent heat of moisture desorption in the hygroscopic region.
- The variation of moisture and heat transfer coefficients during drying are also considered.

The presented model as well as Krischer's [10] and Berger and Pei's model [16] have been solved numerically under the following conditions.

- Krischer's model has been solved using the same relations describing liquid-vapor equilibrium in the dried material as in the case of proposed model and Berger and Pei's model.
- All models for a given material were solved using the same parameters.
- As testing materials clay and brick were used.
- The values of liquid diffusivity for brick and clay were determined by various methods.

The comparison of model solutions with our experimental data indicates that it is possible to apply the discussed and modified Krischer's approach presented in this paper in mathematical modelling of hygroscopic-porous bodies. However, it is necessary to use a proper method of liquid diffusivity determination. It seems that from the point of view of practical applications such a method should be based on a drying experiment jointly with the proper mathematical model of drying.

The procedure of model solution points to the necessity of paying more attention to boundary conditions especially in the falling drying rate period. Van Brakel [4] has started the discussion on the subject for the constant drying rate period.

NOMENCLATURE

c	- specific heat capacity	J/(kgK)
E	- relative error defined by eq. 36	
h_1, h_2	- integration steps	m
Δh_{ph}	- specific latent heat of phase change	J/kg
k	- liquid diffusivity	m^2/s
k_L	- hydraulic conductivity	m/s
m_{ph}	- mass rate of phase change	$kg/(m^3s)$
S	- moisture content	m^3/m^3
t	- time	hours
t_{ss}	- time after which the dry zone appears	hours
t_s	- constant drying rate period duration	hours
ρ_o	- constant, $\rho_o = \lvert \rho_s (1+\beta_v w) \rvert_{hmax}$	kg/m^3
φ	- capillary potential	m
φ_G	- gravitational potential	m
ψ_T	- porosity	-
ω	- moisture content	kg/m^3

Indexes

a	- air
eff	- transfer inside dried body
hmax	- maximum hygroscopic
hmin	- minimum hygroscopic
i	- interface
l	- liquid
s	- bone dry body
v	- vapor
0	- initial

REFERENCES

1. Lewis, W.K., J. Ind. Eng. Chem., vol.13, 427 (1921)
2. Keey, R.B., Theoretical Foundations of Drying Technology, in Advances in Drying ed. A.S. Mujumdar, vol. 1, Hemisphere Publ. Corp. (1980)
3. Keey, R.B., Proc. 3rd IDS, Birmingham UK vol. 1 (1982)
4. Van Brakel, J., Mass Transfer in Convective Drying, in Advances in Drying, ed. A.S. Mujumdar, vol. 1, Hemisphere Publ. Corp. (1980)
5. Fortes, M. and Okos, M.R., Drying Theories: Their Bases and Limitations as Applied to Foods and Grains, in Advances in Drying, ed. A.S. Mujumdar, vol. 1, Hemisphere Publ. Corp. (1980)
6. Toei, R., Drying Mechanism of Capillary Porous Bodies, in Advances in Drying, ed. A.S. Mujumdar, vol. 2, Hemisphere Publ. Corp. (1983)
7. Kisakürek, B., Proc. 3rd IDS, Birmingham UK, vol. 1 (1982)
8. Przesmycki, Z., Mathematical Modelling of Drying Process of Capillary-Porous Materials Taking into Account Mechanism of Moisture Transfer, Ph.D. thesis, Łódź Technical University (1980),(in Polish)
9. Przesmycki, Z., Strumiłło, C., Inż. Chem. i Proc., vol 4, 365 (1983),(in Polish)
10. Krischer, O., Die wissenschaftlichen Grundlagen der Trocknungstechnik, Springer-Verlag, Berlin (1956)
11. Philip, J.R. and De Vries, D.A., Trans. Am. Geophys. Union, vol.38, 222 (1957)

12. Luikov, A.V., Teoria Sushki, Energia, Moscow (1968),(in Russian)
13. Luikov, A.V., Int. J. Heat Mass Transfer, vol. 18, 1 (1975)
14. Whitaker, S., Simultaneous Heat,Mass and Momentum Transfer in Porous Media: A Theory of Drying, in Advances in Heat Transfer, ed. A.S. Mujumdar, vol. 13, Academic Press, N.Y. (1977)
15. Whitaker, S., Heat and Mass Transfer in Granular Porous Media, in Advances in Drying, ed. A.S. Mujumdar, Vol. 1, Hemisphere Publ. Corp. (1980)
16. Berger, D., Pei, D.T.C., Int. J. Heat Mass Transfer, vol. 16, 293 (1973)
17. Whitaker, S. and Chou, W.F.H., Drying Technology, vol. 1, 3 (1983-84)
18. Sherwood, T.K., Sushka tverdykh tel, Gostechizdat, Sverdlovsk-Moscow (1936)
19. Luikov, A.V. and Kolesnikov, A.G., Zhurn. Tech.Fiz., vol. 2, 708 (1932)(in Russian)
20. Smolsky, B.M. and Sergeev, B.M., Int. J. Heat Mass Transfer, vol. 5, 1011 (1962)
21. Macey, A.H., Trans. Brit. Cer. Soc., vol. 41, 73 (1942)
22. Haertling, M., Prediction of Drying Rates, in Developments in Drying, ed. A.S. Mujumdar, vol. 1, Hemisphere Publ. Corp. (1980)
23. Currie, J.A., Brit. J. Appl. Phys.,vol. 12, 275 (1961)
24. Luikov, A.V., Teoreticheskye osnovy stroitelnoy teplofizyki, Izd. Ak. Nauk BSSR, Minsk (1961),(in Russian)
25. Nikitina, L.M., Termodynamicheskye parametry i koefficienty massoteploperenosa vo vlazhnykh telakh, Energia, Moscow (1968) (in Russian)

NUMERICAL MODELING OF THE DRYING OF POROUS MATERIALS

G. Ronald Hadley

Fluid and Thermal Sciences Department
Sandia National Laboratories
Albuquerque, New Mexico, USA

ABSTRACT

The drying of porous materials represents a challenging class of problems involving several nonlinear transport mechanisms operative in both the liquid and vapor phases. The difficulties encountered in attempting to solve such problems have led most researchers to simplify their models in one of two ways: either 1) moisture migration is assumed to proceed only in the liquid phase towards a drying surface where it then evaporates, or 2) evaporation occurs at a stationary liquid-vapor interface and is transported to the drying surface as a vapor (evaporation front model). Although there are some situations in which one of these models may be adequate, most drying problems involve the transport of both phases simultaneously. The disposal of nuclear waste canisters in partially saturated geological formations offers a good example of this class of problems, a class for which presently existing solution methods are inadequate.

This paper presents a one-dimensional numerical solution technique for the transport of water, water vapor, and an inert gas through a porous medium. The fundamental equations follow from the application of volume averaging theory. These transient equations are finite differenced and solved using a predictor-type time integration scheme especially formulated to provide stable solutions of the resulting strongly coupled nonlinear equations. Solutions are presented for a simple problem involving the drying of a bed of sand, for which experimental data is available. The inclusion of vapor phase transport for this problem is shown to lead naturally to a prediction of the constant rate and falling rate periods of drying. Results are seen to be especially sensitive to the choice of a function representing the relative permeability of the bed for low saturations.

1. INTRODUCTION

When a partially saturated porous material dries, the coupled transport of heat, liquid water, water vapor, and air is often involved. Although this fact has been known for some time, a complete solution of drying problems has been severely hampered, due primarily to two factors: the coupling of the transport equations, and the strong nonlinearities in each equation. The equation describing the transport of the liquid phase contains nonlinear terms arising primarily from the shape of the capillary pressure curve. The gas transport equations are non-linear due to the inclusion of Darcy's law, and the heat equation due to temperature-dependent transport properties. The equations are coupled as a result of evaporation/condensation, the influence of gas pressure gradients on liquid motion, and convective cooling by all three fluid phases. These complications have rendered analytic solutions impossible for all but the simplest drying problems. Although the advent of the high-speed computer has opened up new possiblilties for numerical solution of the full set of equations, difficulties still persist due to the disparate time scales involved. This situation often results in numerical instabilities together with their associated time step restrictions.

Because of these difficulties, early work in the area of drying has concentrated on the solution of a single transport equation, usually with properties chosen so as to linearize the equation and enable the attainment of analytic solutions. Thus the transport of liquid water was treated in the early 1930's by Sherwood [1] as a linear diffusion process, for which analytic solutions were easily obtained. This treatment was recognized as deficient by Ceaglske and Hougen [2], and the linear diffusion equation later modified by Van Arsdel [3] to allow a variable diffusion coefficient. The resulting non-linear diffusion equation had been presented earlier by Richards [4], and has come to be known as Richard's equation. It has received widespread usage in the calculation of groundwater flow in partially saturated media, and is reasonably complete except for its neglect of gas phase pressure gradients. Thus, Richard's equation describes the movement of liquid due to capillary forces under conditions where the gas phase(s) may be neglected.

By the late 1950's, the desire to treat non-isothermal problems led to the inclusion of vapor transport by Philip and de Vries [5] in the form of an "extended" diffusion model. Modern technology has since produced a variety of situations in which drying is dominated by vapor transport. Cross, Gibson and Young [6] examined pressures generated during the drying of iron ore pellets at high temperatures. In this case, the vapor transport was modeled using Darcy's law, with the assumption that the vapor source was confined to a receding plane -- the so-called "evaporation front" model. This model represents an opposite extreme from Richard's equation since all liquid is assumed immobile and capillary forces are unable to compete with high gas pressure gradients. The evaporation front model was recently extended by Hadley [7], who considered the combined diffusion and flow of a binary gas system through a porous medium. This latter work also considered Knudsen diffusion, an effect not previously included in drying theories. Evaporation front models have been usually accompanied by the solution of the heat equation, sometimes with the assumption of an "evaporation temperature" at the surface of the front [8].

Many current problems in the area of drying and related fields may not be solved by considering either liquid or vapor transport alone. One example is the drying which takes place near a canister of nuclear waste which has been implaced deep in a partially saturated geological formation. In this situation, capillary forces and pressure gradients generated by canister heating are of comparable magnitude, so that both kinds of transport must be included. The "porous heat pipe" effect, in which latent heat is transported by liquid and vapor phases simultaneously, affords another example. In both these problems, high evaporation rates further complicate the solution by causing strong coupling between the heat and mass transfer equations.

The above class of problems has resisted solution until recently, due both to uncertainties in the formulation of the equations, as well as the numerical techniques necessary for their solution. Although the microscopic transport equations valid within a given phase have been known for some time, the formulation of macroscopic equations has until recently lacked a firm basis of development. This situation changed with the advent of volume averaging theory [9,10], in which known point equations were systematically averaged to produce equations valid on a macroscopic length scale. Whitaker [11] has used volume averaging to derive a general theory of drying that includes heat and mass transport by both the liquid and vapor phases. Whitaker's equations, when certain effects such as dispersion are omitted, are identical to the equations solved in this work, except for gas phase transport. There Whitaker included only Darcy's law, whereas the present treatment solves equations previously derived by Hadley [7], which also include binary diffusion and Knudsen diffusion.

The emphasis of the present work is on the numerical techniques employed to solve the coupled equations of heat and mass transfer which describe partially saturated flow. These techniques have been incorporated into the computer program PETROS, which solves the four transport equations (heat plus three fluids) in one dimension using finite difference methods. Careful attention has been paid to the time integration scheme, which must control the time step internally so as to avoid numerical instabilities. Although PETROS has so far been applied to a variety of interesting problems, the discussion in this paper will be limited to a study of the drying of a granular material. This problem is one of great interest, for which data is available for comparison, and serves to illustrate the advantages to be gained by the inclusion of vapor transport. Using this additional mechanism, the mass flux boundary condition at the drying surface may be properly modeled as a surface evaporation process. The result is a clear prediction of the constant rate and falling rate periods of drying. Comparison with the data of Ceaglske and Hougen [2] is for the first time extended to saturations below the "residual" saturation, at which point vapor transport dominates the problem. Drying profiles in this regime are shown to depend strongly on the way the relative permeability curve is modeled.

2. MODELING ASSUMPTIONS

The equations to be presented in the next section describe the transport of heat, and the masses of three fluid constituents (water, water vapor, and an inert gas) through a porous medium. These equations have been derived on the basis of a physical model, together with certain simplifying assumptions. Both the model and simplifications are described below:

2.1 Geometry

All modeling is one-dimensional, in planar, cylindrical, or spherical geometry. The porous medium is thus composed of slabs, cylindrical or spherical shells, each of which may be a different homogeneous and isotropic material. In each case the porous matrix is assumed to be non-deformable.

2.2 Liquid Transport

Liquid motion is transient and assumed to occur as a noninertial (Darcy) flow due to a pressure gradient or the force of gravity. The pressure gradient is made up of the following components:

Surface Tension Effects. Differences in pressure between the gas and liquid phases are assumed to be representable in terms of a single function of the form

$$p_c = p_c(s)f(T) \qquad (1)$$

where s is the saturation (water volume per unit void volume), and the function f(T) is assumed linear in the temperature. No hysteresis effects are included, so that for a drying problem, a capillary pressure curve appropriate to drying should be used.

Gas Pressure Gradients. Unlike the derivation of Richard's equation, the gas pressure is not assumed to be constant. Thus liquid motion may be driven by gradients in the gas pressure. Change of phase is allowed and included in the liquid mass balance. The Darcy's law employs a relative permeability which depends only upon s.

2.3 Gas Phase Transport

The motion of both gas phases is treated as noninertial and steady. The latter restriction follows from the assumption that gas velocities are much higher than liquid velocities, and equilibration times correspondingly shorter. The two gas species (water vapor + inert gas) are assumed ideal, and their motion includes the following effects:

a. Darcy flow of the combined mixture using a relative permeability which depends only upon s.

b. Knudsen diffusion of each component relative to the porous medium, assuming a single average pore size for each material.

c. Binary gaseous diffusion of one gas relative to the other.

A constant tortuosity factor may be inserted if desired for both types of diffusion. Change of phase is included in the water vapor mass balance.

2.4 Phase Change

The rate of phase change may be computed in either of two different ways. With the equilibrium model, the water vapor pressure is set equal to the equilibrium value when liquid is present above some threshold saturation; i. e.,

$$p_v = \begin{cases} p_{sat}(T) & \text{if } s \geq s_1 \\ \text{det. from mom. eq.} & \text{if } s < s_1 \end{cases} \tag{2}$$

For this case, evaporation rates may be determined by calculating the divergence of the water vapor flux. The other option is the non-equilibrium model, in which the evaporation rate is assumed to be given by

$$e = cs\left(p_{sat}(T) - p_v\right) \quad , \tag{3}$$

with c constant. For this case, (the method used for the drying calculations to be presented later), p_v is determined entirely from the solution of the pressure equations.

2.5 Heat Transfer

Heat transfer is assumed to be due to conduction and convection only, with dispersion effects neglected. The mixture is considered to be in thermal equilibrium, and is assigned a single thermal conductivity, which may depend on position and time either directly or indirectly through other problem variables. Convection of all fluid phases is included. The heat equation is solved transiently and includes latent heat effects. The latent heat of vaporization is allowed to depend linearly on temperature.

3. GOVERNING EQUATIONS

Equations consistent with the modeling assumptions listed above have been previously derived and will only be listed here. The heat and liquid mass transport equations were derived using volume averaging theory [12]. The gas transport equations were developed [7] using the dusty gas model introduced by Evans et. al. [13]. We present below equations appropriate for heat and mass transport through a porous medium of porosity ψ.

The continuity equation for liquid water is

$$\psi\rho_\ell \frac{\partial s}{\partial t} + \frac{1}{r^n}\frac{\partial}{\partial r}\left(r^n J_\ell\right) = -e \tag{4}$$

where the geometry indicator n takes on the value

$$n = \begin{cases} 0 & \text{planar geometry} \\ 1 & \text{cylindrical geometry} \\ 2 & \text{spherical geometry} \end{cases} \tag{5}$$

In accordance with Darcy's law, the liquid mass flux is given by

$$J_\ell = -\frac{\rho_\ell \kappa k_\ell}{\eta_\ell}\left[\frac{\partial p_\ell}{\partial r} - \rho_\ell g\right] \quad , \tag{6}$$

where κ is the saturated permeability of the matrix, and the force of gravity is in the direction of positive r. We may eliminate p_ℓ from equation (6) using the definition of capillary pressure:

$$p_c(s,T) = p - p_\ell \quad , \tag{7}$$

where p is the total gas pressure

$$p = p_v + p_a \quad . \tag{8}$$

Equation (6) thus becomes

$$J_\ell = -\frac{\rho_\ell \kappa k_\ell}{\eta_\ell}\left[\frac{\partial p}{\partial r} - \frac{\partial p_c}{\partial s}\frac{\partial s}{\partial r} - \frac{\partial p_c}{\partial T}\frac{\partial T}{\partial r} - \rho_\ell g\right]. \tag{9}$$

We express the dependence of capillary pressure on temperature via the relation

$$p_c(s,T) = p_c(s)\left[1 + \frac{d\ln\sigma}{dT}\left(T - T_0\right)\right] \tag{10}$$

where for convenience we take T_0 = 300K. Then equation (9) may be further expanded to read

$$J_\ell = -\frac{\rho_\ell \kappa k_\ell}{\eta_\ell}\left[\frac{\partial p}{\partial r} - f(T)\frac{dp_c}{ds}\frac{\partial s}{\partial r} - p_c(s)\frac{d\ln\sigma}{dT}\frac{\partial T}{\partial r} - \rho_\ell g\right] \quad , \tag{11}$$

where

$$f(T) \equiv 1 + \frac{d\ln\sigma}{dT}\left(T - T_0\right) \quad . \tag{12}$$

Equations (3),(4),(5),and (11) together with the functions $k_\ell(s)$, $\eta_\ell(T)$ and $p_c(s)$ thus specify the liquid transport.

The heat transport equation is

$$\rho c_p\Big|_{mix}\frac{\partial T}{\partial t} + \underline{c}_p \cdot \underline{J}\frac{\partial T}{\partial r} = \frac{1}{r^n}\frac{\partial}{\partial r}\left(r^n \lambda \frac{\partial T}{\partial r}\right) - e\Delta h_v(T) \quad , \tag{13}$$

where λ is the mixture conductivity, $\rho c_p\big|_{mix}$ is defined by

$$\rho c_p\Big|_{mix} \equiv (1 - \psi)\rho_m c_{pm} + \psi s \rho_\ell c_{p\ell} \quad , \tag{14}$$

and $\underline{c}_p$ and $\underline{J}$ are

$$\underline{c}_p = \begin{pmatrix} c_{p\ell} \\ c_{pa} \\ c_{pv} \end{pmatrix} \tag{15}$$

$$\underline{J} = \begin{pmatrix} J_\ell \\ J_a \\ J_v \end{pmatrix} \tag{16}$$

The two gas components (water vapor and an inert gas which we will call "air" and denote by the subscript "a") satisfy the steady state continuity equations

$$\frac{1}{r^n}\frac{\partial}{\partial r}\left(r^n J_v\right) = e \tag{17}$$

$$\frac{1}{r^n}\frac{\partial}{\partial r}\left(r^n J_a\right) = 0 \tag{18}$$

The mass fluxes are given by [7]

$$J_v = -\frac{\psi(1-s)m_v D_{vk}}{kT}\frac{\partial p_v}{\partial r} - \frac{p_v m_v \kappa k_v}{\eta_v kT}\frac{\partial p}{\partial r} + \frac{D_{vk}}{pD_{va}}\left(\frac{p_v J_a}{S^2} - p_a J_v\right) \tag{19}$$

$$J_a = -\frac{\psi(1-s)m_v S D_{vk}}{kT}\frac{\partial p_a}{\partial r} - \frac{p_v S^2 m_v \kappa k_v}{\eta_v kT}\frac{\partial p}{\partial r} + \frac{D_{vk}}{pD_{va}}\left(p_a S J_v - \frac{p_v J_a}{S}\right) \tag{20}$$

with

$$S \equiv \sqrt{\frac{m_a}{m_v}} \quad . \tag{21}$$

Equations (19) and (20) may be solved for the fluxes:

$$J_v = -\frac{\psi(1-s)m_v D_{vk}}{kT}\left[\frac{\frac{\partial p_a}{\partial r} + \left(1 + \frac{pSD_{va}}{p_v D_{vk}}\right)\frac{\partial p_v}{\partial r}}{1 + \frac{p_a S}{p_v} + \frac{pSD_{va}}{p_v D_{vk}}}\right] - \frac{\kappa k_v m_v p_v}{\eta_v kT}\frac{\partial p}{\partial r} \tag{22}$$

$$J_a = -\frac{\psi(1-s)m_v S D_{vk}}{kT}\left[\frac{\frac{\partial p_v}{\partial r} + \left(1 + \frac{pD_{va}}{p_a D_{vk}}\right)\frac{\partial p_a}{\partial r}}{1 + \frac{p_v S}{p_a} + \frac{pD_{va}}{p_a D_{vk}}}\right] - \frac{\kappa k_v m_v S^2 p_a}{\eta_v kT}\frac{\partial p}{\partial r} \tag{23}$$

Equations (17),(18),(22), and (23) together comprise a set of coupled nonlinear differential equations for the two pressures p_v, p_a.

The Knudsen diffusion coefficient is given by [7]

$$D_{vk} = \frac{2}{3}\frac{\mathbf{R}}{\tau}\sqrt{\frac{8kT}{\pi m_v}} \tag{24}$$

where $\mathbf{R}$ is the average pore radius and τ the tortuosity factor. The binary diffusion coefficient D_{va} is given by [14]

$$D_{va} = \frac{2.3 \times 10^{-5}}{\tau}\left(\frac{p_0}{p}\right)\left(\frac{T}{T_1}\right)^{1.81} \quad \frac{m^2}{s} \quad , \tag{25}$$

where $p_0 = 9.8 \times 10^4$ Pa and $T_1 = 256K$.

4. NUMERICAL METHODS

4.1 Differencing Method

We first construct the mesh shown in Figure 1. All primary variables (p_v, p_a, s, T) are defined at the node points, and all fluxes (J_v, J_a, J_ℓ) are defined at the midnode points indicated by dashed lines.

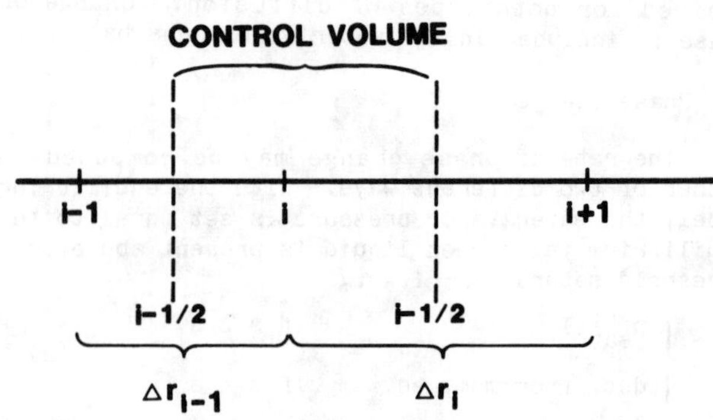

Fig. 1 Mesh used for finite differencing the transport equations.

The difference equations are formed by considering the region between dashed lines as a control volume, and integrating the continuity equations over this region. As an example, consider the saturation equation (4). We multiply both sides by r^n and integrate, giving

$$\rho_\ell \frac{\partial}{\partial t}\int_{i-1/2}^{i+1/2} \psi r^n s dr + r^n J_\ell \Big|_{i-1/2}^{i+1/2} = -\int_{i-1/2}^{i+1/2} r^n e dr \quad . \tag{26}$$

If we approximate s, e as being piecewise linear and ψ constant between mesh points, then the integrals in equation (26) are

$$\int_{i-1/2}^{i+1/2} \psi r^n s dr \approx r_i^n s_i \frac{\psi_i \Delta r_i + \psi_{i-1}\Delta r_{i-1}}{2} \tag{27}$$

$$\int_{i-1/2}^{i+1/2} r^n e dr \approx r_i^n e_i \frac{\Delta r_i + \Delta r_{i-1}}{2} \tag{28}$$

The correction terms to (27) and (28) are of order $n/4(\Delta r/r)$. Thus equations (27) and (28) are quite accurate except for locations very near the origin in curvilinear coordinates. The fluxes at the control volume boundaries are expressed using averaged properties and centered differences. For example,

$$J_\ell\Big|_{i+1/2} = -\frac{\rho_\ell \kappa k_\ell}{\eta_\ell}\Bigg|_{i+1/2}\left[\frac{p_{i+1} - p_i}{\Delta r_i} - \frac{dp_c}{ds}\Bigg|_{i+1/2} f\left(T_{i+1/2}\right)\frac{s_{i+1} - s_i}{\Delta r_i} - p_c\Big|_{i+1/2}\frac{d\ln\sigma}{dT}\frac{T_{i+1} - T_i}{\Delta r_i} - \rho_\ell g\right] \tag{29}$$

When equation (29) and its counterpart $J_\ell\Big|_{i-1/2}$, together with equations (27) and (28) are substituted into equation (26), the final difference equation for s is obtained. The values of s appearing in the flux expressions are taken to be advanced in time, so that a fully implicit difference equation is produced. This choice was made in order to produce the most stable possible numerical scheme. Notice that this difference method is fully conservative; that is, for zero evaporation rate, mass is conserved exactly [15] because the flux entering one control volume is the same as that leaving the adjacent control volume.

A difference equation for the temperature is obtained in a similar manner by first multiplying equation (13) through by r^n and integrating over a control volume. The difference equations for temperature and saturation are thus both fully implicit and are each solved using a simple tridiagonal algorithm [15]. The gas transport equations are differenced in a fashion similar to what has just been described. In this case, however, the resulting difference equations form a pair of coupled equations which require a block-tridiagonal algorithm. Since some nonlinear terms are present, a self-consistent solution is obtained by iteration until convergence is achieved. Such convergence is made easier by an initial linearization of the equations.

4.2 Time Integration Scheme

The difference equations presented above could be used to advance the problem in time by simply advancing s, then T, followed by a self-consistent pressure solution. However, the strong coupling between equations, together with possible nonlinear boundary conditions, makes this simple procedure inadequate to prevent numerical oscillations for some problems, even with a carefully controlled time step. Consequently, an elaborate time step procedure has been employed, which, for many problems of interest, will allow well-behaved solutions. In general, problems which produce high evaporation/condensation rates cause the greatest difficulty due to the coupling between the gas momentum and heat transfer equations. The time integration sequence employs an internally computed time step which is chosen to keep changes in s and T between time steps of a reasonable size. The time step is chosen using in part an algorithm discussed by Gresho et. al. [16]. Upon completing the nth time step, the norm of an effective relative error is first determined by

$$y_n = \max_i [s_i^n - s_i^{n-1}] + \frac{1}{300} \max_i [T_i^n - T_i^{n-1}] \qquad (30)$$

where s_i^n refers to the saturation at time step n, mesh point i. Then, for a required time step truncation error ε,

$$\Delta t_{n+1} = \Delta t_n \left[\frac{3\varepsilon\left(1 + \frac{\Delta t_{n-1}}{\Delta t_n}\right)}{y_n} \right]^{1/3} \qquad (31)$$

The ratio $\Delta t_{n+1}/\Delta t_n$ is, however, restricted to be ≤ 1.25 , and if a value below 0.8 is obtained, the entire time step is repeated using the new value Δt_{n+1}. Using the computed time step Δt_{n+1}, the $n+1^{st}$ time step advancement proceeds by first dividing Δt_{n+1} by the factor 2^m, where m is nominally four. Intermediate solutions are then obtained for a set of reduced time steps which are successively increased by a factor of two until the original time step Δt_{n+1} is recovered. All nonlinear terms are reevaluated prior to each intermediate solution so as to avoid sudden changes in coefficients. This procedure is cumbersome and time consuming, but appears to be necessary for problems involving high evaporation rates.

5. APPLICATION TO DRYING

5.1 Problem Description

The numerical solution technique described in the last section will now be applied to a fundamental problem of great interest, namely the drying of a granular porous material. We have chosen to simulate the experiment described by Ceaglske and Hougen [2] so a comparison with data may be made. This experiment is shown schematically in figure 2. Samples of a coarse, well-

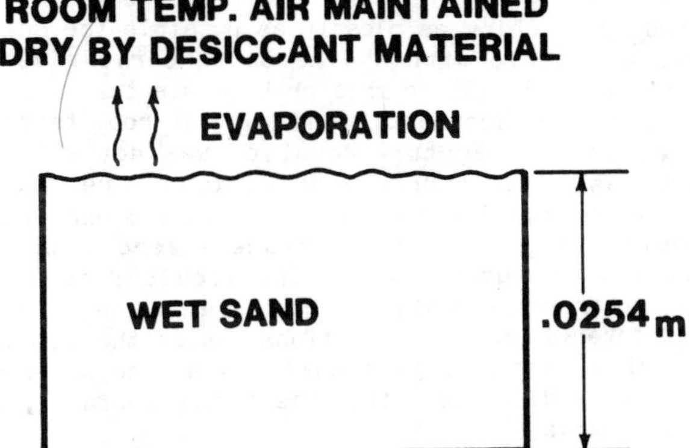

Fig. 2 Schematic of Ceaglske and Hougen's drying experiment.

characterized sand were placed in several identical containers and saturated with water. The containers were then all placed in an enclosed chamber at room temperature containing desiccant material, so that drying commenced due to molecular diffusion of water vapor from the surface of each sample toward the desiccant material. At certain unspecified intervals a single container was removed, sectioned, and the saturation of each section determined by weighing, drying, and reweighing. The result was a series of saturation profiles, each labeled by the value of the total average saturation, as shown by the data points in figure 3.

A numerical simulation of this experiment was attempted using the property values shown in Table 1.

Table 1. Property values used for the simulation of Ceaglske and Hougen's experiment.

κ	$1.0 \times 10^{-9}\ m^2$
T	300K
ψ	0.41
p_c	1225.

The permeability used was scaled from the average particle size using the empirical relation [17]

$$\kappa = 6.17 \times 10^{-4}\ d^2 \quad , \tag{32}$$

where d is the average particle diameter in meters. The relative permeability for this first simulation was given by the standard formula

$$k_\ell = \left(\frac{s - s_0}{1 - s_0}\right)^3 \quad , \tag{33}$$

with $s_0 = 0.085$ and k_ℓ zero for $s < s_0$. The Leverett curve [18] for drainage was used for the capillary pressure curve, scaled by the pressure given in Table 1. This scale pressure was determined by matching as closely as possible the scaled Leverett curve with the measured curve given in reference [2]. Since this problem was taken to be dominated by isothermal diffusion at room temperature, the temperature equation was not solved. Zero mass flux boundary conditions were imposed upon both liquid and vapor at the bottom end of the sample. At the drying surface a zero mass flux condition was used also for the liquid flow. For the vapor component, the mass flux was equated to that determined from conditions inside the chamber. Assuming zero relative humidity at the desiccant surface (a distance x from the drying surface), the mass flux is [7]

$$J_v = -D_{va}\frac{m_v}{kT}\frac{p_{v1}}{x} \quad , \tag{34}$$

where p_{v1} is the partial pressure of water vapor at the drying surface. The vapor mass flux diffusing across the first control volume boundary is

$$J_v = -\frac{\psi(1-s_1)m_v D_{va}}{kT\tau}\left(\frac{p_{v2} - p_{v1}}{\Delta r_1}\right) . \tag{35}$$

Equating the two expressions for mass flux then produces the boundary condition

$$p_{v1} = p_{v2}\left[\frac{1}{1 + \frac{\tau\Delta r_1}{x\psi(1-s_1)}}\right] . \tag{36}$$

These boundary conditions are considerably more realistic than those employed by previous simulations in that they describe water being drawn towards the surface by capillary forces where it then evaporates and diffuses away. In the simulation of this experiment, water moving into the first control volume (centered about mesh point 2) evaporated and diffused across the control volume boundary at location i=1/2 as a vapor. For the fine zoning employed, these processes essentially took place " at the surface". As the surface dried, the drop in mass loss rate due to the greater distance taken for vapor to reach the surface was automatically accounted for. A value of x equal to 0.002 meters was chosen to give a reasonable initial mass loss rate of $\sim 3.5 \times 10^{-4}$ kg/m^2-s, although no measured value was given in reference [2].

5.2 Saturation Profiles

A comparison between the computed and measured saturation profiles is shown in figure 3. The small discrepancies which are apparent prior to

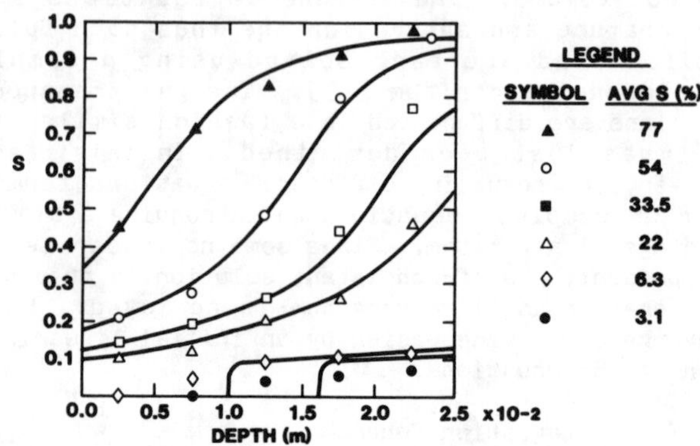

Fig. 3 Comparison between experimental data (symbols) and computer simulation (curves) using standard relative permeability. Curves are for same average saturation as corresponding symbols.

surface dryout are probably due to uncertainties in property curves, particularly the relative permeability, which was not measured for this sand. Certainly, the agreement in this regime is qualitatively very good. As a layer near the surface dries, however, a qualitative disagreement becomes evident. The computed curves show a sharp front occurring at a saturation just above the value of s_0 used for the liquid relative permeability cutoff. This results from the requirement that k_ℓ be essentially zero below $s_0 = 0.085$. This model was proposed early in the history of drying theory because of the belief that "pendular" rings of water at low saturations were unconnected and that consequently no liquid flow was possible. However, film flow can still take place along a wetted surface and thus provide a connecting path, albeit one of very low effective permeability. Although no specific model for film flow will be introduced in this paper, it is nonetheless instructive to observe the effect of a non-zero relative permeability on the calculation described above. (This idea was suggested to the author by S. Whitaker.)

Figure 4 shows a plot of the logarithm of the relative permeability described by equation (33) and used in the previous simulation. Although k_ℓ should be zero below $s_0 = 0.085$, it was actually set to 1×10^{-9} for numerical reasons. If, however, some phenomenon such as film flow is occurring, one would expect the curve to approach zero in a smoother manner, perhaps as shown by the dotted line. The drying simulation was then rerun with this modified k_ℓ function, to see what, if any, changes might occur in the solutions. The resulting saturation profiles from this second simulation are shown in figure 5. Since k_ℓ was not changed for saturations above 0.1, the top four

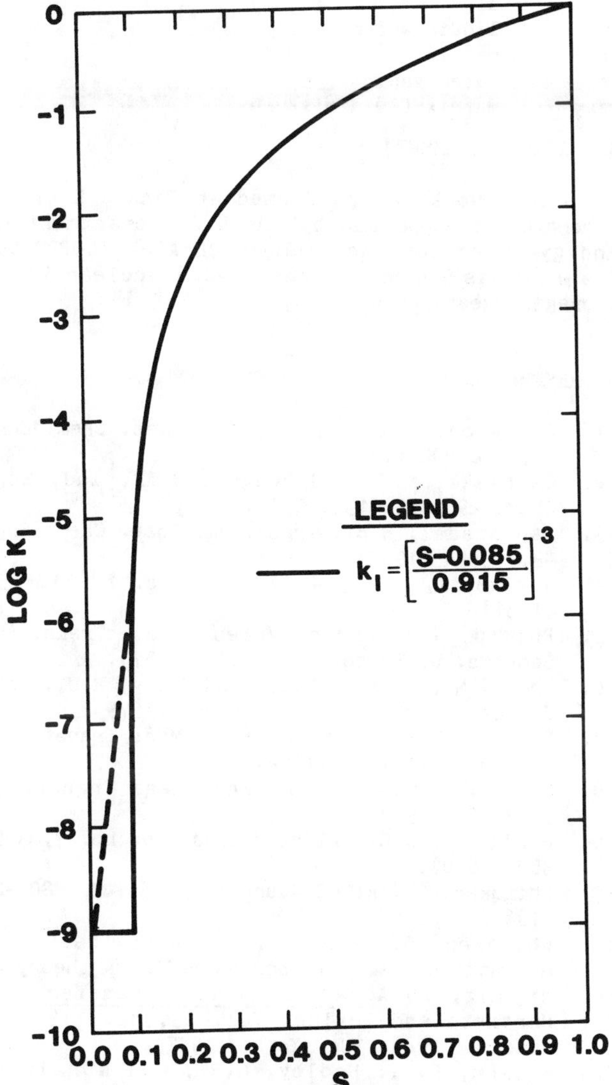

Fig. 4 Standard liquid relative permeability curve (solid line) described by equation (33). The function is modified below s = 0.1 as shown by the dashed curve.

curves are unchanged except for the narrow dry region near the surface displayed by curve four (s_{av} = 0.22). The last two curves show a marked improvement, however, in two repects. First, they display a less steep front than do the previous set of curves. Second, the plateau now drops well below s_0, in good agreement with the data.

No claim should be made at this point that the "modified" k_ℓ curve is altogether correct, because very little is known about the behavior of the capillary pressure curve in this region. However, these calculations clearly show the sensitivity of the solutions at low saturations to the functional form of k_ℓ. This fact is particularly impressive when one considers that k_ℓ was only changed in a region where it was already below 5×10^{-6} in magnitude. It is thus clear that more work is needed to clarify the forces involved at small saturations if the relationship between vapor and liquid transport is to be understood.

5.3 Drying Rates

Although a measurement of drying rates did not accompany the saturation profiles presented for the experiment already described, Ceaglske and Hougen did perform other experiments on the same sand

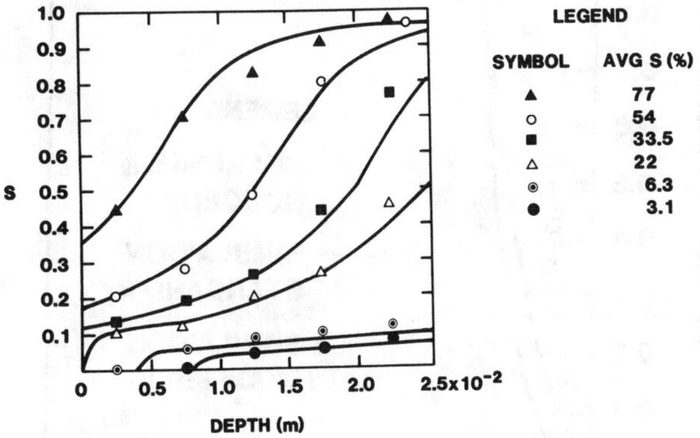

Fig. 5 Comparison between experimental data (symbols) and computer simulation (curves) using modified relative permeability. Curves are for same average saturation as corresponding symbols.

samples in which drying rates were measured. These experiments differed only in that drying resulted from the forced flow of dry heated air over the sample surface. Due to latent heat effects, however, the sample remained isothermal for the majority of the drying period. In figure 6 we compare computed drying rates for the previous isothermal simulations with Ceaglske and Hougen's measurements. The simulations clearly predict a constant rate drying period which persists until dryout occurs at the surface. (The falloff at high saturations is of numerical origin and could be removed by the use of finer zoning.) The predicted "falling rate" drying period shown by the solid line has a quite different shape than shown by the data. This effect results from the steep penetrating front diplayed in figure 3, which forces the vapor to diffuse through a greater path length than would be the case for a diffuse front. The modification in the relative permeability which smooths out that front is also seen to improve the shape of the drying rate curve in figure 6. Once again, the value of the relative permeability at low saturations is seen to have an appreciable effect upon the overall drying problem.

6. CONCLUSION

A numerical solution technique has been presented which solves in one dimension the coupled heat and mass flow equations for water, water vapor, and an inert gas moving through a porous medium. The physical effects included in the resulting computer program PETROS are more complete than for previous simulations, particularly for the gas phase transport. The latter includes effects due to binary gaseous diffusion and Knudsen diffusion as well as Darcy flow. PETROS is thus capable of simulating two-phase flow through porous media

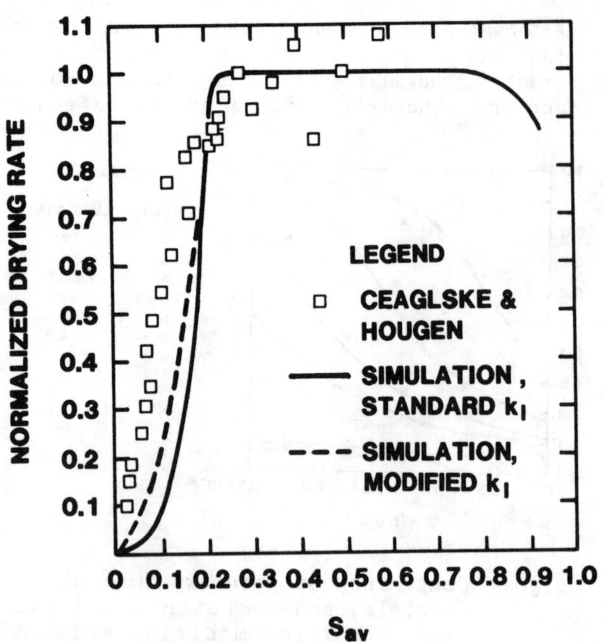

Fig. 6 Normalized drying rates plotted versus average saturation.

over a wide range of conditions. For drying problems, this means the inclusion of mass transport in both the liquid and vapor phases simultaneously.

Simulations of drying experiments performed by Ceaglske and Hougen on small samples of sand were presented. These simulations were carried out with realistic boundary conditions at the drying surface, whereby liquid transported by capillary action to the surface, then evaporated and diffused away. This same boundary condition could be employed for drying by a moving air stream, simply by adjusting the length of the effective boundary layer thickness over which diffusion takes place. The resulting simulation showed good agreement with the measured saturation profiles, even for saturations less than the residual value, provided the relative permeability was altered. This result implies that significant liquid phase transport may be occurring even when the remaining water is in the so-called "pendular" state.

The boundary conditions employed at the drying surface were found to lead in a natural way to a "constant rate" period of drying as previously noted by experimenters. However, when dryout occurred at the drying surface, the mass loss rate dropped in approximate agreement with experiment. As was suggested by Ceaglske and Hougen, the drop in drying rate is caused by the recession of the evaporation source into the porous medium and the associated greater path length to the drying surface.

7. NOMENCLATURE

k	(unsubscripted) Boltzmann's constant
k_v	gas phase relative permeability
m	molecular mass
Subscripts	
a	air
ℓ	liquid water
m	matrix
v	water vapor
mix	matrix-fluid mixture

8. ACKNOWLEDGMENT

This work was performed at Sandia National Laboratories Supported by the U.S. Department of Energy under contract number DE-AC04-76DP00789. The work was funded by the Nevada Nuclear Waste Storage Investigations project.

9. REFERENCES

1. Sherwood, T.K., Trans. Am. Inst. Chem. Eng. 27, 190-202 (1931).
2. Ceaglske, N.H. and Hougen, O.A., Ind. Eng. Chem. 29, 805-813 (1937).
3. Van Arsdel, W.B., Trans. Am. Inst. Chem. Eng. 43, 13-24 (1947).
4. Richards, L.A., J. Appl. Phys. 1, 318-333 (1931).
5. Philip, J.R. and de Vries, D.A., Trans. Am. Geophys. U. 38 no.2, 222-232 (1957).
6. Cross, M., Gibson, R.D., and Young, R.W., Int. J. Heat Mass Transf. 22, 47-50 (1979).
7. Hadley, G.R., Int. J. Heat Mass Transf. 25, no. 10, 1511-1522 (1982).
8. Gupta, L.N., Int. J. Heat Mass Transf. 17 ,313-321 (1974).
9. Slattery, J.C., AIChE Journal 16, no. 3, 345-352 (1970).
10. Whitaker, S., AIChE Journal 13, no. 3, 420-427 (1967).
11. Whitaker, S., Simultaneous heat, mass, and momentum transfer in porous media: A theory of drying, in Advances in Heat Transfer, ed. Hartnett and Irvine, Vol. 13, pp. 119-200, Academic Press, New York, (1977).
12. Wilson, R.K., Hadley, G.R., and Nunziato, J.W., Sandia National Laboratories Report SAND84-0746J, Sandia National Laboratories, Albuquerque, NM, (1984).
13. Evans, R.B. III, Watson, G.M., and Mason, E.A., J. Chem. Phys. 35, 2076-2083 (1961).
14. Eckert, E.R.G. and Drake, R.M.Jr., Analysis of Heat and Mass Transfer, p. 787, Mcgraw-Hill, New York, (1972).
15. Roache, P.J., Computational Fluid Dynamics, p. 28, Hermosa Publishers, Albuquerque, N.M., p. 28 (1972).
16. Gresho, P.M., Lee, R.L., and Sani, R.L., On the time-dependent solution of the incompressible Navier-Stokes equations in two- and three-dimensions, in Recent Advances in Numerical Methods in Fluids, Vol 1, Pineridge Press, Swansea, U. K. (1979).
17. Bear, J., Dynamics of Fluids in Porous Media, p. 133, American Elsevier, New York, (1972).
18. Leverett, M.C., Trans. A.I.M.E., 142, 341-358, (1941).

THEORETICAL PREDICTION OF MOISTURE TRANSFER BY CONVECTIVE DRYING FOR A FLAT PLATE

Vladimir Strongin and Irene Borde

Institutes for Applied Research, Energy Laboratory, Ben-Gurion University of the Negev, Beer-Sheva 84110, ISRAEL

ABSTRACT

A system of differential equations for moisture diffusion in a porous flat plate during convective drying was solved analytically. Relationships for the prediction of the moisture content at any point of the flat plate and at any moment of time for periods of constant and falling rates of drying were obtained. From these equations, the total moisture flux, the amount of liquid removed from the plate, and the values of the drying rates during the falling rate period were derived. The differential equations for moisture diffusion were based on the mathematical model of heat and mass transfer in capillary porous media. It was assumed that the liquid and its vapor were in thermodynamic equilibrium corresponding to the desorption isotherm, that the moisture transfer caused by the total pressure gradient was negligible in comparison with the moisture transfer caused by the diffusion gradient, and that the volume of the dried material did not change. For integration of the differential equations, the moisture transfer was considered a quasi-steady-state process. A mathematical criterion for estimating the relative thickness of a flat plate was introduced. Our theoretical predictions were shown to be in satisfactory agreement with experimental data.

1. INTRODUCTION

For a capillary-porous body undergoing convective drying we write the equations for moisture transport of the liquid moisture and its vapor respectively, as follows:

$$\rho \frac{\partial w}{\partial t} = \rho\nabla(D\nabla w) - I \quad (1)$$

$$\frac{\partial}{\partial t}(\psi_v \rho_v) = \nabla(D_v \nabla \rho_v) + I \quad (2)$$

These two equations follow from a general system of equations of mass and heat transfer [1] on the basis of the assumptions listed below:

i) the volume of the body does not change appreciably and can be assumed to be constant;

ii) the total pressure gradient within the body is so small that the molecular (filtration) transfer of moisture induced by it is significantly lower than the diffusion transfer;

iii) the transfer of moisture is calculated for mean values of the temperature, and then terms representing the thermodiffusion of moisture vanish from expressions for diffusion fluxes of liquid moisture and its vapor.

Once the system of Eqs.(1)-(2) has been obtained, we then make the following assumptions:

a) liquid moisture is in equilibrium with its vapor in accordance with the desorption isotherm, i.e., $\phi=\phi(w)$;

b) the liquid-moisture content w, which is virtually equal to the measured moisture content of the body, does not fall below the region of convexity of the desorption isotherm i.e., generally a curve with concave and convex regions (we do not consider the narrow range of concavity for low moisture contents of the monomolecular layer);

c) any convex segment ($w_m \geq w \geq w^*$) of the desorption isotherm can be described by a broken curve, consisting of two line segments ϕ_1 and ϕ_2 with different slopes k_1 and k_2 (Fig. 1), and $\phi(w)$ can be described analytically [including the range $w \geq w_m$ where $\phi(w)=1$] as follows:

$$\phi = \phi(w) = \begin{cases} \phi_s = 1 & (w \geq w_m) \\ \phi_1 = \phi_\zeta + k_1(w - w_\zeta) & (w_\zeta \leq w \leq w_m) \\ \phi_2 = \phi_\zeta - k_2(w_\zeta - w) & (w^* \leq w \leq w_\zeta) \end{cases} \quad (3)$$

where

$$k_1 = (1-\phi_\zeta)/(w_m - w_\zeta), \quad k_2 = (\phi_\zeta - \phi^*)/(w_\zeta - w^*) \quad (4)$$

d) D and D_v are equal to some average values in each of the above-described ranges of $\phi(w)$.

Since

$$\psi_v = \psi - (\rho/\rho_L)w, \quad \rho_v = \rho_s\phi \quad (5)$$

then, employing assumptions a) through d), we find from Eqs.(1)-(2) for $w \geq w_m$ ($\phi = \phi_s = 1$):

$$[1-(\rho_s/\rho_L)]\,\rho\frac{\partial w}{\partial t} = D\rho\nabla^2 w \quad (w \geq w_m) \quad (6)$$

where

$$I = -(\rho_s/\rho_L)\,\rho\frac{\partial w}{\partial t} \quad (w \geq w_m) \quad (7)$$

For $w^* \leq w \leq w_m$ ($\phi = \phi_1, \phi_2$) we obtain:

$$[\rho + \rho_s k_1 \psi_{v1} - (\rho/\rho_L)\rho_s\phi_1]\frac{\partial w_1}{\partial t} = \overline{D}_1\rho\nabla^2 w_1 \quad (w_m \geq w_1 \geq w_\zeta) \quad (8)$$

$$[\rho + \rho_s k_2 \psi_{v2} - (\rho/\rho_L)\rho_s\phi_2]\frac{\partial w_2}{\partial t} = \overline{D}_2\rho\nabla^2 w_2 \quad (w_\zeta \geq w_2 \geq w^*) \quad (9)$$

$$I_{1,2} = -\frac{(\rho/\rho_L)\phi_{1,2}D_{1,2} + k_{1,2}(D_{v1,2} - \psi_{v1,2}D_{1,2})}{(\rho/\rho_s)D_{1,2} + k_{1,2}D_{v1,2}} \quad (10)$$

In Eqs. (8)-(9)

$$\overline{D}_{1,2} = D_{1,2} + (\rho_s/\rho)k_{1,2}D_{v1,2} \quad (11)$$

Eq. (6) describes the transfer of moisture during the first stage of drying. It follows from the right-hand side of Eq. (6) that during the first stage only liquid moisture is transferred, despite the fact that, according to Eq. (7), it vaporizes within the body. This vaporization does not produce a vapor density gradient since the vapor remains saturated throughout the entire body, $\rho_v=\rho_s=$ const. During this stage the vaporization rate comprises a very small part of the change in the liquid - moisture content, $\rho(\partial w/\partial t)$, which is clear from Eq. (7) where $(\rho_s/\rho_L) \ll 1$, since $\rho_L \simeq 1000$ kg/$m^3 \gg \rho_s$, and $\rho(\partial w/\partial t)$ is virtually equal to the change in the total moisture content of the body (left-hand side of Eq. (6)).

The system of Eqs.(8)-(9) describes fully the transport of moisture during the second drying period if the mean moisture content of the material as of the start of this period, i.e., the critical moisture content w_{cr}, can be assumed to be equal to the maximum hygroscopic moisture content w_m. However, very frequently $w_{cr} \gg w_m$. In such a case, k_1 as described by Eq.(4) should be replaced by k_1 throughout Eqs.(8)-(11) such that:

$$k_1 = (1-\phi_\zeta)/(w_{cr}-w_\zeta) \tag{12}$$

and w_m should be replaced by w_{cr}

As can be seen from the right-hand sides of Eqs.(8)-(9), during the second drying period both the liquid moisture and its vapor move within the material. Since $D_V \gg D$, it is possible that in the "dry" region $w^* \le w \le w_\zeta$, where $k_2 \gg 1$, almost the entire moisture content may be transported in vapor form, despite the fact that the vapor content w_v is considerably less than the content of liquid moisture, w. The vaporization rate I_2 in this region may thus constitute a considerable part of $\rho(\partial w_2/\partial t)$.

Thus, during the second drying period a significant amount of moisture may be transported in vapor form. This fact notwithstanding the density of the total moisture content $\rho(w_v+w)$ and its change with time [left-hand sides, Eqs.(8) and (9), respectively] are virtually solely determined by the density of the liquid moisture content $\rho w_1, \rho w_2)$ and its change with time $\rho(\partial w_1/\partial t), \rho(\partial w_2/\partial t)$.

Let us now consider the convective drying of a plate. If both the plate surfaces are in contact with the drying gas, then we place the coordinate origin at the center of the plate and, by virtue of symmetry, we perform the calculations only for $0 \le x \le \delta$, where δ is half the thickness of the plate. If one of the surfaces is isolated from the drying gas, then δ is the total plate thickness.

We assume the initial moisture content of the plate (its liquid moisture content) to be the same throughout the plate, i.e., $w° =$ const. If, in addition $w° > w_m$, then the first and second drying periods, as described by Eq. (6) and Eqs.(7)-(8), respectively, will be observed.

We now integrate Eqs. (6) through (8) by the method of replacement of steady states. This means that we assume, as a first approximation, that the moisture transport is quasi steady, when during each time t the spatial variation in the overall flux of the liquid moisture and its vapor at any point in the plate is equal to zero.

In addition, we introduce during the initial stages of the two periods the concept of the "domain of influence", below which there is still no motion of moisture, and hence no change in the moisture content. The concept may be applied when we consider drying from the time that the moisture content is the same throughout the material in question. Then, the change in moisture content on the material's surface does not immediately induce a change in the internal moisture content. The domain of influence thus has a moving boundary, which moves into the material (toward the center of the plate in two-sided drying or toward its isolated surface when drying occurs only on one side).

2. THE FIRST DRYING PERIOD

Moisture transport during the first period is described by Eq. (6), where for a plate $\nabla^2 w = \partial^2 w/\partial x^2$. The initial and boundary conditions here are $(w)_{t=0}=w°$, $D\rho(\partial w/\partial x)_{x=0}= 0$,

$$D\rho(\partial w/\partial x)_{x=\delta} = \dot{m} = \text{const.} \tag{14}$$

The duration of the first period is defined by time t_m over which the moisture content w_δ on the plate's surface falls from an initial $w°$ to the the maximum hygroscopic value w_m.

As a first approximation, we assume the moisture transport to be quasi steady, and we consider it in sequence for the two time intervals: $0 \le t \le t_I$ and $t_I < t < t_m$. Over the interval $0 \le t \le t_I$ moisture transport occurs with formation of the domain of influence $\delta - x°(t)$, where $x°(t)$ is the moving boundary of this domain. At the boundary itself and further towards the center of the plate, the moisture content remains equal to its initial value $w°$. At $t \to t_I$ $x°(t) \to 0$, i.e., t_I is the time from which moisture w_o at the center of the plate (x=0) begins to change (Figs. 2 and 3).

Thus, in accordance with the method of solution suggested by us, we will obtain, instead of Eqs.(6) and (14), for the interval $0 \le t \le t_I$:

$$w = w° \qquad (0 \le x \le x°) \tag{15}$$

$$D\rho \frac{\partial^2 w}{\partial x^2} = 0 \qquad (x° \le x \le \delta) \tag{16}$$

With conditions

$$(\partial w/\partial x)_{x=\delta} = -\frac{\dot{m}}{\rho D}, \quad (w)_{x=x°}=w°, \quad (x°)_{t=0} = \delta \tag{17}$$

By integration of Eq.(16) for the first two conditions of (17) we obtain:

$$w = \begin{cases} w° & (0 \le x \le x°) \quad (18) \\ w° - \dfrac{\dot{m}}{\rho D}(x-x°) & (x° \le x \le \delta) \quad (19) \end{cases}$$

The quantity m of moisture which is removed (from the start of the first period) is then equal to:

$$m = \int_{x°}^{\delta} \rho(w°-w)\,dx = \frac{\dot{m}}{D}\int_{x°}^{\delta} (x-x°)\,dx = \frac{\dot{m}}{2D}(\delta-x°)^2 \tag{20}$$

On the other hand

$$\frac{dm}{dt} = \dot{m} = \text{const} \tag{21}$$

From Eqs. (20) and (21) we find, using the third condition of (17), $x°=x°(t)$, i.e.

$$x° = \delta - \sqrt{2Dt} \qquad (22)$$

Substituting Eq. (22) for x° into Eqs. (18)-(19), we obtain the dynamics of the moisture-content profile for the time $0 \le t \le t_I$.

Since $x°(t_I)=0$, we find from Eq.(22) that

$$t_I = \delta^2/2D \qquad (23)$$

As is clear from Eq.(20) the amount of moisture removed from the plate as of time t_I will be:

$$m_I = \dot{m}\,\delta^2/2D \qquad (24)$$

and the moisture-content profile, as follows from Eqs.(18)-(19), will become:

$$w_I = w° - (\dot{m}/\rho D)\,x \qquad (0 \le x \le \delta) \qquad (25)$$

Starting with time t_I the moisture content w_o at the center of the plate starts decreasing (from w°), and hence during the concluding interval $(t_I \le t \le t_m)$ of the first drying period it becomes a function of time, i.e., $w_o = w_o(t)$. Then, Eqs. (15)-(16) reduce to:

$$\rho D \frac{\partial^2 w}{\partial x^2} = 0 \qquad (0 \le x \le \delta) \qquad (26)$$

and the conditions described by (17) become:

$$(\partial w/\partial x)_{x=\delta} = -\frac{\dot{m}}{\rho D},\ (w)_{x=0} = w_o,\ (w_o)_{t=t_I} = w° \qquad (27)$$

Integrating Eq. (26) for the first two conditions of (27), we obtain from the moisture-content profile

$$w = w_o - (\dot{m}/\rho D)\,x \quad (t \ge t_I,\ 0 \le x \le \delta) \qquad (28)$$

The quantity of moisture removed from the plate toward time $t \ge t_I$ of the first period is

$$m = m_I + \int_0^{\delta} \rho(w_I - w)\,dx = \rho(w° - w_o)\,\delta + \dot{m}\delta^2/2D \qquad (29)$$

From Eqs.(29) and (21), we find, using the third condition of (27), the moisture content w_o at the center of the plate as a function of time

$$w_o = w° - \frac{\dot{m}}{\rho\delta}\left(t - \frac{\delta^2}{2D}\right) \qquad (t \ge t_I) \qquad (30)$$

Substituting Eq.(30) into Eq. (28), we obtain an expression for the moisture content at any point within the plate at any instant of time $t \ge t_I$:

$$w = w° - \frac{\dot{m}}{\rho\delta}\left(t - \frac{\delta^2}{2D} + \frac{\delta}{D}x\right) \quad (t \ge t_I,\ 0 \le x \le \delta) \qquad (31)$$

In particular, from Eq.(31) we obtain the following expression for the moisture content w_δ on the plate's surface $(x=\delta)$:

$$w_\delta = w° - \frac{\dot{m}}{\rho\delta}\left(t + \frac{\delta^2}{2D}\right) \quad (t \ge t_I) \qquad (32)$$

2.1 Parallelism of the Moisture Profiles

It follows from the solutions obtained by us above for the first drying period that the moisture content profile (in the region in which it varies) during this period decreases in parallel to itself. In fact, either Eq. (19) or Eq.(28) yields

$$w - w_\delta = \frac{\dot{m}}{\rho D}(\delta - x) \qquad (33)$$

i.e., the moisture content at any point of the profile is higher than the moisture content on the plate's surface by an amount which is independent of time and is determined solely by the coordinate of the point.

The conclusion as to the parallelism of moisture profiles can also be arrived at by finding $\partial w/\partial t$, the rates of lowering of the points of the moisture-content profile for times $t \le t_I$ and $t \ge t_I$. We find

$$\frac{\partial w(x,t)}{\partial t} = \begin{cases} 0 & (t \le t_I,\ 0 \le x \le x°) \quad (34) \\ -\dfrac{\dot{m}}{\rho\sqrt{Dt}} = \dfrac{\dot{m}}{\rho(\delta - x°)} & (t \le t_I,\ x° \le x \le \delta) \quad (35) \\ -\dot{m}/\rho\delta & (t \ge t_I,\ 0 \le x \le \delta) \quad (36) \end{cases}$$

From these equations, we can see that $\partial w/\partial t$ is independent of x, i.e., that the rates of lowering of all the points on the profile of w(x,t) in the corresponding ranges of t and x are the same, and hence the profile as a whole moves parallel to itself.

8 We can also see that up to time t_I, i.e., up to the time at which the moisture content at the center of the plate starts changing, the rate of lowering of the moisture-content profile (in the region in which it changes) decreases with time to a value $\dot{m}/\rho\delta$ (equal to the rate of drying of the plate, kg/kg sec) and then remains unchanged.

2.2 Relative Thickness of the Plate

The moisture content w_δ at the plate's surface $(x=\delta)$ starts decreasing from the start of the first period. We thus obtain from Eqs.(19) and (22)

$$w_\delta = w° - \frac{\dot{m}}{\rho D}(\delta - x°) = w° - \frac{\dot{m}}{\rho}\sqrt{2t/D} \qquad (37)$$

It is possible that w_δ will become equal to the maximum hygroscopic moisture content w_m at the end of the first drying period, while at this time the corresponding value of $x° = x°_m$ will still be greater than zero, $x°_m > 0$, and thus the moisture content at the center of the plate will still be equal to the initial moisture content. Obviously, this will occur, as is clear from Eq. (37), when

$$\delta \ge (w° - w_m)\,\rho D/\dot{m} \qquad (38)$$

The above described situation means that if condition (38) is satisfied, it will be impossible to achieve reduction in the moisture content at the center of the plate during the entire first drying period. Such a plate will be termed "thick".

On the other hand, if

$$\delta < (w° - w_m)\,\rho D/\dot{m} \qquad (39)$$

then the moisture content at the center of the plate will already start decreasing during the first drying period (starting at time t_I). Such a plate will be termed "thin."

If we introduce the concept of relative thickness δ

$$\delta = \dot{m}\,\delta/(w^o - w_m)\,\rho D \qquad (40)$$

then the plate is thick when $\delta>1$ and thin when $\delta>1$

Let us now find the mean moisture content in the thick and thin plates at the time of termination of the first drying period. This moisture content w_{cr} is usually termed the critical moisture content. From Eqs. (18)-(19) and (28) we obtain, respectively:

$$w_{cr} = \begin{cases} w^\circ - \dot{m}\delta/2\rho D\hat{\delta}^2 & (\hat{\delta} \geq 1) \quad (41) \\ w^\circ - \dot{m}\delta/2\rho D = w_m + \dot{m}\delta/2\rho D & (\hat{\delta} \leq 1) \quad (42) \end{cases}$$

3. THE SECOND DRYING PERIOD (FIG. 4)

From the time at which $w_\delta = w_m$, the condition $\dot{m}$ = const no longer holds true, since there is no longer free moisture on the surface, the entire mositure control being hygroscopic, i.e., bound. This time represents the start of the second drying period.

During the second drying period, the surface moisture content w_δ often decreases to the equilibrium moisture content w^* very much faster than the moisture content in the center of the material, and we can thus assume that $w_\delta = w^*$. The time in the second period will be measured from the period's start.

For the second drying period the initial moisture content profile is assumed to be equal to w_{cr} according to Eq. (41) or Eq. (42) depending on the relative thickness of the plate. Then, in accordance with the method of solution, for the time $0 \leq t \leq t_I$ (during which the moisture content at the center of the plate still does not drop below w_{cr}) we can replace Eqs. (8)-(9) with the following system equations:

$$w = w_{cr} \quad (0 \leq x \leq x^\circ), \quad \rho\bar{D}_1 \frac{\partial^2 w_1}{\partial x^2} = 0 \quad (x^\circ \leq x \leq \zeta) \tag{43}$$

$$\rho\bar{D}_2 \frac{\partial w2}{\partial x^2} = 0 \quad (\zeta \leq x \leq \delta) \tag{44}$$

with the boundary and initial conditions:

$$(w_1)_{x=x^\circ} = w_{cr}, \ (w_1)_{x=\zeta} = w_\zeta, \ (w_2)_{x=\zeta} = w_\zeta, \ (w_2)_{x=\delta} = w^\circ$$
$$[(\partial w_1/\partial x)/(\partial w_2/\partial x)]_{x=\zeta} = \bar{D}_2/\bar{D}_1, \ (x^\circ)_{t=0} = \delta \tag{45}$$

For the specific problem w_{cr} = const, w_ζ = const [see Eq. (3)], w^*=const (w^* corresponds to ϕ^* that is equal to the relative humidity of the drying gas, the fifth condition of (45) follows from the continuity of flow at the point ζ. In addition, $x^\circ = x^\circ(t)$ and $\zeta = \zeta(t)$.

Integrating Eqs. (43)-(44) for the first five conditions of (45), we find the following relationships for the time interval $0 \leq t \leq t_I$ of the second period:

Moisture content profiles

$$w_1 = w^* + (w_\zeta - w^*)\,[1 - \hat{D} + (\hat{D} + \hat{w})\,\frac{\delta - x}{\delta - x^\circ}] \quad (x^\circ \leq x \leq \zeta) \tag{46}$$

$$w_2 = w^* + (w_\zeta - w^*)\,(1 + \hat{w}/\hat{D})\,\frac{\delta - x}{\delta - x^\circ} \quad (\zeta \leq x \leq \delta) \tag{47}$$

Relationship between ζ and x° and also $\dot{m}(x^\circ)$

$$\delta - \zeta = (\delta - x^\circ)/(1 + \hat{w}/\hat{D}) \tag{48}$$

$$\dot{m} = \rho\bar{D}_2(w_\zeta - w^*)\,(1 + \hat{w}/\hat{D})\,\frac{1}{\delta - x^\circ} \tag{49}$$

Removed moisture

$$m = \frac{1}{2}\,\rho\,(w_\zeta - w^*)\,(\hat{w} + \frac{1 + \hat{w}}{1 + \hat{w}/\hat{D}})\,(\delta - x^\circ) \tag{50}$$

Further, since

$$\dot{m} = dm/dt \tag{51}$$

using Eqs. (49) and (50), we obtain

$$\frac{2\bar{D}_2}{\delta - x^\circ} = [\hat{D} - \frac{\hat{D} - 1}{(1 + \hat{w}/\hat{D})^2}]\,\frac{d(\delta - x^\circ)}{dt} \tag{52}$$

After integrating Eq. (52) for the last condition (45), we find an expression for $x^\circ(t)$

$$\delta - x^\circ = 2\sqrt{\bar{D}_2 t}\Big/[\hat{D} - \frac{\hat{D} - 1}{(1 + \hat{w}/\hat{D})^2}]^{1/2} \tag{53}$$

Setting $x^\circ = 0$ in Eq. (53), we obtain an expression for calculating t_I, the time at which the moisture content w_o at the center of the plate starts becoming smaller than w_{cr}

$$t_I = \frac{\delta^2}{4\bar{D}_2}\,[\hat{D} - \frac{\hat{D} - 1}{(1 + \hat{w}/\hat{D})^2}] \tag{54}$$

The mean moisture content of the plate at time t_I can be expressed as:

$$\bar{w}_I = w_{cr} - \frac{1}{2}\,(w_\zeta - w^*)\,(\hat{w} + \frac{1 + \hat{w}}{1 + \hat{w}/\hat{D}}) \tag{55}$$

In Eqs. (46)-(55)

$$\hat{D} = \bar{D}_2/\bar{D}_1, \quad \hat{w} = (w_{cr} - w_\zeta)/(w_\zeta - w^*) \tag{56}$$

Starting with time t_I, which corresponds to $x^\circ(t_I) = 0$, the moisture content w_o at the center of the plate falls below w_{cr} and thus becomes a function of time: $w_o = w_o(t)$. Then Eqs. (43)-(44) transform to

$$\rho\bar{D}_1 \frac{\partial^2 w_1}{\partial x^2} = 0 \quad (0 \leq x \leq \zeta) \tag{57}$$

$$\rho\bar{D}_2 \frac{\partial w2}{\partial x^2} = 0 \quad (\zeta \leq x \leq \delta) \tag{58}$$

and boundary and initial conditions (45) become

$$(w_1)_{x=0} = w_o, \ (w_1)_{x=\zeta} = w_\zeta, \ (w_2)_{x=\zeta} = w_\zeta, \ (w_2)_{x=\delta} = w^*,$$
$$[(\partial w_1/\partial x)/(\partial w_2/\partial x)]_{x=\zeta} = \bar{D}_2/\bar{D}_1, \ (w_o)_{t=t_I} = w_{cr} \tag{59}$$

Eqs. (57)-(58) with conditions (59) are valid up to some time t_{II} when $w_o(t_{II}) = w_\zeta$.

Integrating Eqs. (57)-(58) for the first five conditions of (59), we obtain over the time interval $t_I \leq t \leq t_{II}$, the following expressions for calculating the second drying period:

Moisture-content profiles

$$w_1 = w^* + (w_\zeta - w^*)\,[1 - \hat{D} + (\hat{D} + \hat{w}_o)\,\frac{\delta - x}{\delta}] \quad (0 \leq x \leq \zeta) \tag{60}$$

$$w_2 = w^* + (w_\zeta - w^*)\,(1 + \hat{w}_o/\hat{D})\,\frac{\delta - x}{\delta} \quad (\zeta \leq x \leq \delta) \tag{61}$$

Relationship between ζ and w_o

$$\delta - \zeta = \delta/(1 + \hat{w}_o/\hat{D}) \tag{62}$$

Flux of moisture from the plate

$$\dot{m} = \rho\bar{D}_2(w_\zeta - w^*)\,(1 + \hat{w}_o/\hat{D})/\delta \tag{63}$$

Quantity of moisture removed

$$m = \frac{1}{2}\,\rho\delta\,(w_\zeta - w^*)\,(2\hat{w} + \frac{\hat{D} - \hat{w}_o{}^2}{\hat{D} + \hat{w}_o}) \tag{64}$$

Further, using Eqs. (51), (63) and (64), we obtain a differential equation for w_o and t_o. Integration of that equation for the last condition of (59) yields the following relationship between $\hat{w}_o$ and t,

$$\frac{4\bar{D}_2}{\delta^2} t = \hat{D} - \frac{\hat{D}-1}{(1+\hat{w}_o/\hat{D})^2} + 2\hat{D}\ln\frac{1+\hat{w}/\hat{D}}{1+\hat{w}_o/\hat{D}} \quad (65)$$

In Eqs.(60)-(65) $\hat{D}$ and $\hat{w}$ are defined by Eq.(56), whereas

$$\hat{w}_o=(w_o-w_\zeta)/(w_\zeta-w^*) \quad (66)$$

If we set $w_o=w_\zeta$, i.e., $\hat{w}_o=0$, then we find t_{II} from Eq. (65)

$$t_{II}=(\delta^2/4\bar{D}_2)[1+2\hat{D}\ln(1+\hat{w}/\hat{D})] \quad (67)$$

The mean moisture content $\bar{w}_{II}$ at time t_{II} can be expressed as:

$$\bar{w}_{II}=\frac{1}{2}(w_\zeta+w^*) \quad (68)$$

It follows from Eq.(62) that $\zeta(t_{II})=0$, since $\hat{w}_o(t_{II})=0$. Then Eqs.(57)-(58) reduce to a single equation

$$\rho\bar{D}_2\frac{\partial^2 w_2}{\partial x^2}=0 \qquad (0\le x\le\delta) \quad (69)$$

and the system of Eqs.(60)-(61) becomes the single equation

$$(w_2)_{t=t_{II}}=w^*+(w_\zeta-w^*)\frac{\delta-x}{\delta} \quad (70)$$

Eq.(69) with condition (70) describes the transport of moisture in the plate for $t\ge t_{II}$.

Solving problem (69)-(70), we obtain for $t\ge t_{II}$ of the second period

$$w_2=w^*+(w_o-w^*)\frac{\delta-x}{\delta} \qquad (0\le x\le\delta) \quad (71)$$

$$\dot{m}=\frac{\rho\bar{D}_2}{\delta}(w_o-w^*) \quad (72)$$

$$m=\rho\delta[w_{cr}-w^*-\frac{1}{2}(w_o-w^*)] \quad (73)$$

Then, using Eq.(51) and substituting into it expressions for $\dot{m}$ and m obtained from Eqs.(72) and (73), we arrive at the differential equation for w_o and t

$$(2\bar{D}_2/\delta^2)dt=-\frac{dw_o}{w_o-w^*} \quad (74)$$

Since $w_o(t_{II})=w_2(0,t_{II})=w_\zeta$, by integrating Eq.(74) we now obtain the following relationship between w_o and t,

$$(4\bar{D}_2/\delta^2)t=1+2[\hat{D}\ln(1+\hat{w}/\hat{D})-\ln\frac{w_o-w^*}{w_\zeta-w^*}] \quad (75)$$

The solution obtained above for the transport of moisture in a plate undergoing convective drying is thus seen to divide the second drying period of duration t into three intervals:

1. the interval $0\le t\le t_I$ (with the corresponding variation of the mean moisture content over the interval $w_{cr}\ge\bar{w}\ge\bar{w}_I$), during which the moisture content w_o at the center of the plate has still not fallen below w_{cr};

2. the interval $t_I\le t\le t_{II}$ ($\bar{w}_I\ge\bar{w}\ge\bar{w}_{II}$) during the course of which w_o decreases from w_{cr} to w_ζ;

3. the interval $t\ge t_{II}$ ($\bar{w}_{II}\ge\bar{w}\ge w^*$) during which w_o decreases from w_ζ to w^*.

4. DRYING RATE CURVES

The relationship $\dot{m}=f(\bar{w})$ is known as the drying rate curve. During the second drying period, we obtain the drying rate curves, using the results presented here in Sec. 3 for each of the three time intervals of this period.

For the interval $0\le t\le t_I$ ($w_{cr}\ge\bar{w}\ge\bar{w}_I$), we apply Eq.(50) to express $\delta-x°$ in terms of m and substitute the resulting expression into Eq.(52). We thus obtain:

$$\dot{m}=\bar{D}_2\rho^2(w_\zeta-w^*)^2[1+\hat{w}(2+\hat{w}/\hat{D})]\frac{1}{2m} \quad (76)$$

And since

$$m=\rho\delta(w_{cr}-\bar{w}) \quad (77)$$

we find the drying rate curve

$$\dot{m}=\frac{B}{w_{cr}-\bar{w}} \qquad (w_{cr}>\bar{w}>\bar{w}_I) \quad (78)$$

where

$$B=\rho\bar{D}_2(w_\zeta-w^*)^2\,[1+\hat{w}(2+\hat{w}/\hat{D})]/2\delta \quad (79)$$

w_{cr} is defined by Eqs.(41)-(42) and $\bar{w}_I$, from (55).

For the interval $t_I\le t\le t_{II}$ ($\bar{w}_I\ge\bar{w}\ge\bar{w}_{II}$), we proceed as follows. We find $\hat{w}_o(m)$ by solving the quadratic equation (64) for $\hat{w}_o$, substituting $\hat{w}_o(m)$ into Eq.(63), we find $\dot{m}=f(m)$ and using Eq.(77) we obtain in the final analysis

$$\dot{m}=\rho\bar{D}_1(w_\zeta-w^*)[i+\sqrt{i^2+\hat{D}(1-\hat{D})}\,]/\delta \qquad (\bar{w}_I\ge\bar{w}\ge\bar{w}_{II}) \quad (80)$$

where

$$i=i(\bar{w})=\hat{D}-(w_\zeta-\bar{w})/(w_\zeta-w^*) \quad (81)$$

$\bar{w}_{II}$ is determined according to Eq.(68).

The interval $t\ge t_{II}$ ($\bar{w}_{II}\ge\bar{w}\ge w^*$) is treated as follows. From Eqs.(72) and (73), we find $\dot{m}=f(m)$ and using Eq.77 we obtain:

$$\dot{m}=2\bar{D}_2\rho(\bar{w}-w^*)/\delta \qquad (\bar{w}_{II}\ge\bar{w}\ge w^*) \quad (82)$$

We consider the question of the signs of derivatives $d\dot{m}/d\bar{w}$ and $d^2\dot{m}/d\bar{w}^2$, on the basis of Eqs.(78)-(82).

We first find

$$\frac{d\dot{m}}{d\bar{w}}=\begin{cases} B/(w_{cr}-\bar{w})^2 & (w_{cr}\ge\bar{w}\ge\bar{w}_I) \quad (83)\\ \rho\bar{D}_1[1+i/\sqrt{i^2+\hat{D}(1-\hat{D})}\,]/\delta & (\bar{w}_I\ge\bar{w}\ge\bar{w}_{II}) \quad (84)\\ 2\rho\bar{D}_2/\delta & (\bar{w}_{II}\ge\bar{w}\ge w^*) \quad (85)\end{cases}$$

$$\frac{d^2\dot{m}}{d\bar{w}^2}=\begin{cases} 2B/(w_{cr}-\bar{w})^3 & (w_{cr}\ge\bar{w}\ge\bar{w}_I) \quad (86)\\ \dfrac{\rho\bar{D}_2(1-\hat{D})}{2\delta\,(w_\zeta-w^*)[i^2+\hat{D}(1-\hat{D})]^{1.5}} & (\bar{w}_I\ge\bar{w}\ge\bar{w}_{II}) \quad (87)\\ 0 & (\bar{w}_I\ge\bar{w}\ge w^*) \quad (88)\end{cases}$$

Since B>0, $i^2+\hat{D}(1-\hat{D})>0.25>0$, and the case i<0 is possible only at $\hat{D}<1$ [see Eq.81)], then it follows from analysis of Eqs.(83)-(85) that

$$d\dot{m}/d\bar{w}>0 \qquad (w_{cr}\ge\bar{w}\ge w^*) \quad (89)$$

and from analysis of Eqs.(86)-(88) that

$$\frac{d^2\dot{m}}{d\bar{w}^2}=\begin{cases} >0 & (w_{cr}\ge\bar{w}\ge\bar{w}_I) \quad (90)\\ >0 \text{ when } \hat{D}\le1;\ <0 \text{ when } \hat{D}\ge1 & (\bar{w}_I\ge\bar{w}\ge\bar{w}_{II}) \quad (91)\\ =0 & (\bar{w}_{II}\ge\bar{w}\ge w^*) \quad (92)\end{cases}$$

The sign of the first derivative $d\dot{m}/d\bar{w}$ indicates whether $\dot{m}$ decreases, increases or remains constant with the variation of $\bar{w}$. It follows from inequality (89) that $\dot{m}$ decreases with decreasing $\bar{w}$ during the entire second drying period.

The sign of the second derivative $d^2\dot{m}/d\bar{w}^2$ indicates whether $d\dot{m}/d\bar{w}$ decreases, increases or remains constant, with decreasing $\bar{w}$, which is uniquely related to the direction of the curvature of the curve $\dot{m}(\bar{w})$ relative to the $\dot{m}$-axis.

It follows from inequalities (90)-(92) that as $\bar{w}$ decreases from w_{cr} to w^*, the curve $\dot{m}(\bar{w})$ will, in general, be composed of the following segments:

1. the interval $w_{cr} \geq \bar{w} \geq \bar{w}_I$ - a concave segment ($d\dot{m}/d\bar{w}$ decreases).

2. the interval $\bar{w}_I \geq \bar{w} \geq \bar{w}_{II}$ - either (at $\hat{D}<1$) a concave segment or (at $\hat{D}>1$) a convex segment ($d\dot{m}/d\bar{w}$ increases), or (at $\hat{D}=1$) a linear segment ($d\dot{m}/d\bar{w}$=const).

3. the interval $\bar{w}_{II} \geq \bar{w} \geq w^*$ - a linear segment.

Thus, following this procedure, we can obtain all the typical drying rate curves found experimentally.

5. COMPARISON BETWEEN THE THEORETICAL SOLUTION AND EXPERIMENTAL DATA

Experimental results $\dot{m}(\bar{w})$ for a burned clay slab for different temperatures and humidities of the drying gas are given by Haertling in [2]. This publication also contains desorption isotherms of burned clay at 25°C and some data necessary for our theoretical calculations. The drying rate curve predicted analytically was compared with experimental drying rate curves at a temperature of the drying gas of 51.2°C and a humidity of 5.2 g/kg. In this case the average temperature of the material during the second drying period was 37°C, a temperature sufficiently close to that of the used desorption isotherm.

For this case Eqs.(78), (80) and (82) yield the following relationships:

$$\dot{m} = \begin{cases} 0.063/(0.13-\bar{w}) & (0.13>\bar{w}>0.065) \quad (93) \\ 0.022(i^2+\sqrt{i^2-0.31}\,) & (0.065>\bar{w}>0.0025) \quad (94) \\ 18.8(\bar{w}-0.001) & (0.0025>\bar{w}>0.001) \quad (95) \end{cases}$$

where

$$i=1.25+\frac{\bar{w}-0.004}{0.003}, \quad [\dot{m}]=kg/m^2h \qquad (96)$$

A comparison of the analytical predicted drying rate curve with the experimental curve [2] is shown in Fig. 5. Note that at $\bar{w}$=0.13 $\dot{m}=\infty$, that is connected with the assumption of the instantaneous decreasing of the moisture content of the plate surface from w_m to w_x. But $\dot{m}$ decreases very fast and beginning with $\bar{w}$=0.1 virtual agreement with the experimental data was obtained. As can be seen, our solution is in very good agreement with the experimental data, more precise values being obtained than those from the numerical solution presented in [2].

NOMENCLATURE

ρ	- density of dry plate	kg/m^3
ρ_L	- density of liquid	kg/m^3
ρ_v	- density of vapor	kg/m^3
ρ_s	- density of saturated vapor	kg/m^3
ψ	- porosity of plate	
ψ_v	- volume of vapor per unit volume of plate	
ϕ	- relative density of vapor	
$\phi(w)$	- desorption isotherm	
w	- liquid moisture content	kg/kg
w_v	- vapor content	kg/kg
$w°$	- initial moisture content	kg/kg
w_o	- moisture content in the center of the plate	kg/kg
w_δ	- moisture content on the surface of the plate	kg/kg
w^*	- equilibrium moisture content of the plate	kg/kg
w_m	- maximum hygroscopic moisture content	kg/kg
w_{cr}	- critical moisture content	kg/kg
w_ζ	- moisture content in the point of abrupt change in the slope of isotherm $\phi(w)$	kg/kg
$\bar{w}$	- mean moisture content of plate	kg/kg
D	- coefficient of diffusion of liquid moisture	m^2/s
D_v	- diffusion coefficient of vapor within the plate	m^2/s
I	- rate of vaporization within the plate	$kg/(m^3s)$
m	- removed moisture	kg
$\dot{m}$	- drying rate	$kg/(m^2s)$
x	- coordinate	m
x°	- coordinate of the moving boundary of the domain of influence	m
ζ	- coordinate of a point on the moisture-content profile with value w_ζ	m
t	- time	sec
t_m	- duration of first period	sec
δ	- half thickness of plate	m

Subscripts

v	- vapor
s	- saturated vapor
ζ	- point of abrupt change in the slope of isotherm $\phi(w)$
0	- moving boundary of the domain of influence, initial location, or location in the center of the plate
*	- equilibrium
1,2	- segments of the desorption isotherm and also of the moisture content profile
I,II	- specific drying times
^	- relative quantities
–	- mean quantities

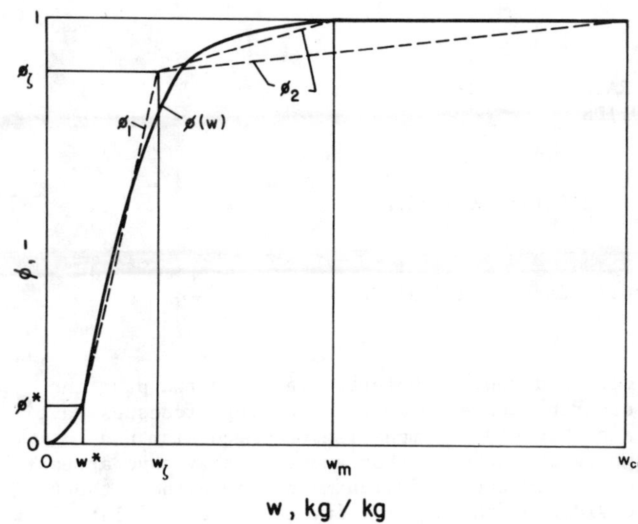

Fig. 1. Sorption isotherm

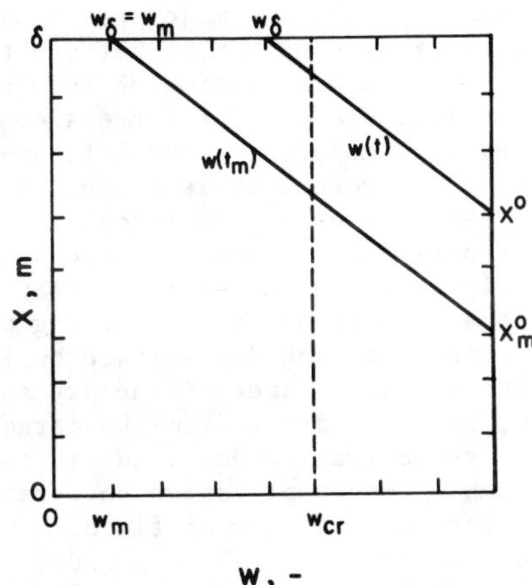

Fig. 2. Quasi-steady moisture content in a thick plate (first period)

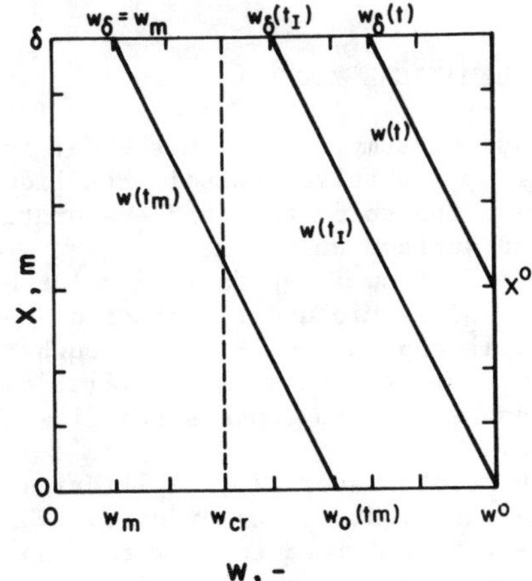

Fig. 3. Quasi-steady moisture content profiles in a thin plate (first period)

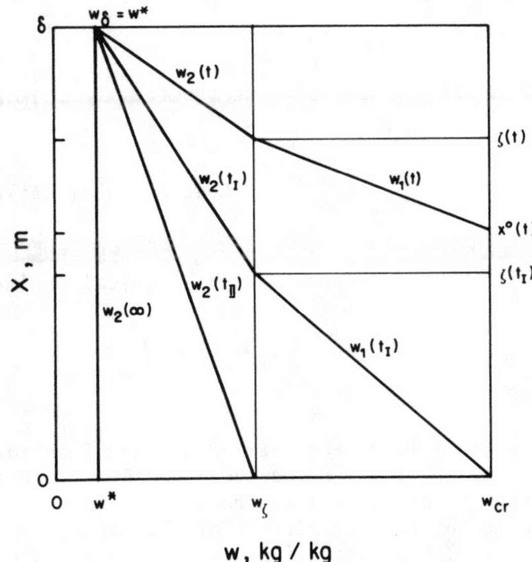

Fig. 4. Quasi-steady moisture content profiles in a plate (second period)

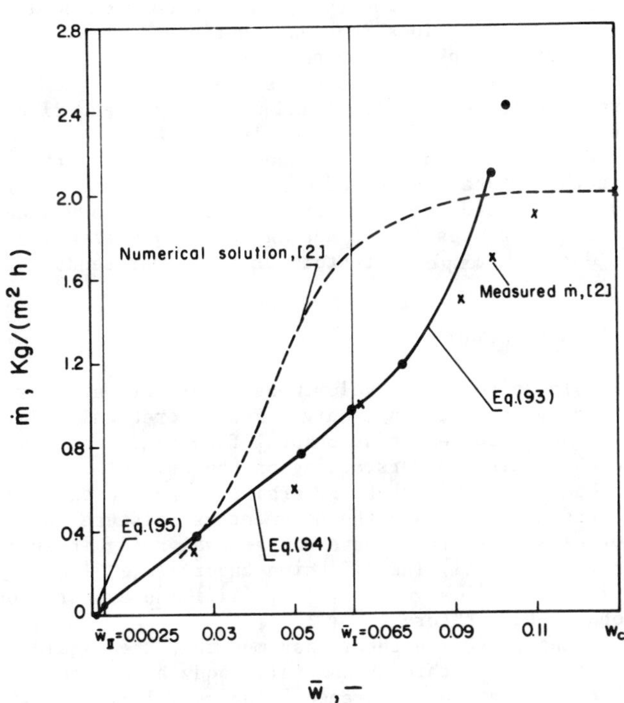

Fig. 5. Comparison of calculated and experimental drying rate curves of a burned clay slab

REFERENCES

1. Luikov, A.V., Theory of Drying, Energiya, Moscow (1968).
2. Haertling, M. Proc. 2nd Int. Symposium, Drying '80, vol. 1, 88 (1980).

AN ANALYSIS OF DRYING RATE MECHANISM IN COMPLEX SOLIDS

Gülsen Ergün, Tarık G. Somer and Bilgin Kısakürek

Department of Chemical Engineering, Middle East Technical University
Ankara, Turkey

ABSTRACT

In the falling rate period drying of solids, a new drying model which results in a linear partial differential equation is introduced. The main assumption of this model is that the capillary flow mechanism controls the liquid motion inside the solid body and evaporation takes place at the surface only. The analytical solution is possible by using Laplace Transformation technique for the simplified boundary conditions. The resultant drying equation reveals that the contribution of moisture concentration change due to convection is considerable besides the contribution of moisture concentration change due to diffusion. The applicability of this model is checked with the experimental data of coal and sunflower seed dried in a tunnel dryer. It is concluded that experimental results are in good aggreement with the theoretical values of coal. But in the case of sunflower seed, it is observed that the theory and experimental results do not aggree very well which is due to the complex structure of sunflower seeds.

1. INTRODUCTION

In spite of the voluminous literature studies in the area of drying, only a few theoretical models have yet resulted in successful fundamental approach to the general understanding of the drying mechanism. Among these several theories so far suggested to explain the movement of moisture in porous solids, three have won general recognition in the past: (i) The diffusion theory |1,2,3|, (ii) The capillary theory |4|, (iii) The evaporation condensation theory |5,6|.

The diffusion theory assumes that the liquid moisture moves through the solid body as a result of concentration difference. The capillary theory assumes that only the liquid is present in the capillaries of the solid and the flow of this liquid moisture through interstices and over the surface is caused by liquid-solid molecular attraction. The latest version of this theory is presented by the basic assumption that the moisture flux is proportional to the gradient of chemical potential of the moisture. The evaporation condensation theory assumes that the flow of moisture within the solid body takes place entirely in gaseous phase.

It can be said that the flow of moisture in the solid may follow one or combination of these models mentioned above. Unfortunately, in the past the research people on internal drying processes have mainly tended to emphasize one mechanism and ignore the others. For example, they have attemped to explain the diffusion mechanism alone without considering the possibility that the capillary flow theory may also be present at the same time or vice versa. Drying model proposed in this study considers both diffusional and capillarity effects.

It is said that porous solids can be considered as containing a very large number of capillaries of all sizes extending in all directions. In most of the models proposed, the major weakness lies in the absence of cross connections between the capillary tubes. An examination on thin section of sand stone indicates that the cross connections between pores is a major structural future of porous medium. Therefore it is correct to state that of all capillaries are interconnected. During the constant period of drying, these capillaries are completely full and the entire surface is covered with a film of water. As drying proceeds, water film evaporates and dry surface begins to appear. With the appearance of the dry speck on the surface, water is drawn from the large capillaries by the smaller ones, and it is carried to the surface. Long capillaries show relatively large resistance to the flow of fluid. This is most probably the case for thick solids. But for thin solids, the lengths of these capillaries are small and consequently, there is negligible resistance to the flow of liquid inside the body. When drying proceeds, the number of capillaries that carry water to the surface decreases |7|.

2. THE MATHEMATICAL MODEL

The basic assumption of this model is that capillary flow mechanism controls the liquid motion inside the solid body and evaporation takes place at the surface only.

For steady flow in uniform circular pipes running full of liquid under isothermal conditions, Darcy's law is applicable |8|. Although unsaturated flow is present in capillaries, Darcy's equation can be used to express the flow through capillaries.

In wet porous materials, capillaries retain water either as liquid or as hydrate. When drying proceeds, water is forced to flow through capillaries of smaller radii. This is simply due to the higher potential of the smaller ones.

Considering the single capillary as shown in

Fig. 1, by neglegting the density of the vapor as a result of force balance at the liquid gas interface, pressure drop from 0 to x is,

$$\Delta P_{xo} = \rho_1 x \frac{g}{g_c} \cos\alpha_n + \Delta P_f = 2\sigma \left(\frac{1}{r_{c_s}} - \frac{1}{r_c}\right)\cos\alpha \quad (1)$$

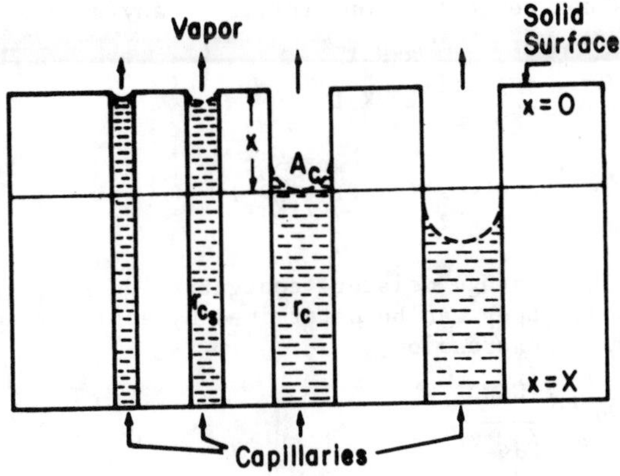

Fig. 1 Capillaries of the Solid

On the left hand side of this equation, first term corresponds to gravitational effect, second term corresponds to frictional effect of capillary walls and summation of these values is equal to pressure drop coming from surface tension at the liquid vapor interface.

The gradient of frictional pressure is evaluated by taking derivative of Eq.(1) with respect to x,

$$\frac{\partial P_f}{\partial x} = \frac{2\bar{\sigma}\cos\alpha}{r_c^{\,2}} \left(\frac{\partial r_c}{\partial x}\right) - \rho_1 g \cos\alpha_n \quad (2)$$

For purely viscous flow of incompressible fluids through porous media, Darcy's law is given as

$$\frac{P_1 - P_2}{X} = \frac{\alpha \mu u}{g_c} \quad (3)$$

where P_1 is absolute upstream pressure, P_2 is the absolute downstream pressure, X is thickness of the medium and u is the superficial velocity and $1/\alpha$ is called as permeability coefficient |8|.

Equation (3) can be written in differential form,

$$\frac{\partial P_f}{\partial x} = \frac{\alpha \mu u}{g_c} \quad (4)$$

Multiplying both sides of Eq.(4) by ρ, an expression for mass flux is obtained,

$$G = g_c\, B \frac{\partial P_f}{\partial x} \quad (5)$$

where B is the modified permeability coefficient and G is the mass flux through pores.

Substitution of Eq.(2) into Eq.(5) gives

$$G = 2\, B g_c \sigma \left(\frac{1}{r_c^{\,2}}\right) \frac{\partial r_c}{\partial x} \cos\alpha - B\rho_1 g \cos\alpha_n \quad (6)$$

On the other hand, a mass balance over a section of drying solid can be written as follows

$$\frac{\partial G}{\partial x} = \frac{\rho_s}{\varepsilon} \frac{\partial W}{\partial t} \quad (7)$$

where ρ_s is the solid density, ε is the total porosity and W is the moisture content of solid in dry basis.

At the same time total mass flux is equal to the summation of fluxes through each pore,

$$G = \sum_{i=0}^{r_c} G_i \quad (8)$$

or

$$G\, \varepsilon\, A = \sum_{i=0}^{r_c} u_{c_i} A_{c_i} \rho_1 \quad (9)$$

where u_{c_i} is the liquid velocity in the capillary.

The investigations on the fluid flow through porous media have revealed that in almost all natural seapage systems the flow regime is laminar. Therefore the liquid velocity through a given capillary radius can be written by the use of Poiseulle's law,

$$u_{c_i} = \frac{\partial P_f}{\partial x} \left(\frac{r_{c_i}^{\,2}}{8\mu}\right) \quad (10)$$

Combining Eq.(10) and Eq.(9), the following equation is obtained.

$$G = \frac{\pi \rho_1}{\varepsilon A} \left(\frac{1}{8\mu}\right) \frac{\partial P_f}{\partial x} \sum_{i=0}^{r_c} r_{c_i}^{\,4} \quad (11)$$

where $\frac{\partial P_f}{\partial x}$ is considered as constant and $A_{c_i} = \pi r_{c_i}^{\,2}$

An expression for modified permeability coefficient can be obtained by comparing Eq.(11) with Eq.(5) as follows

$$B = \frac{\bar{K}_1}{g_c} \sum_{i=0}^{r_c} r_{c_i}^{\,4} = \frac{\bar{K}_1}{g_c} \lim_{\Delta r_c \to 0} \sum_{i=0}^{r_c} r_{c_i} \frac{\Delta r_c}{\Delta r_c}$$

$$= \frac{\bar{K}_1}{g_c \Delta r_c} \int_0^{r_c} r_{c_i}^{\,4}\, dr_c \quad (12)$$

where $\bar{K}_1 = \frac{\pi \rho_1}{\varepsilon A} \left(\frac{1}{8\mu}\right)$ and Δr_c is the average difference in capillary radius between any two pores arranged in descending order according to their size.

If Eq.(12) is integrated from 0 to r_c, the modified permeability coefficient can be determined.

$$B = \frac{\bar{K}_1}{g_c\, r_c} \left(\frac{1}{5}\right) r_c^{\,4} \quad (13)$$

Substituting the expressions obtained for the permeability coefficient from Eq.(13) into Eq.(6), the general form of the drying equation can be obtained as given below

$$\frac{\partial G}{\partial x} = \bar{K}_2 \frac{\partial^2 r_c^4}{\partial_x^2} - \bar{K}_3 \frac{\partial r_c^5}{\partial x} \tag{14}$$

with

$$\bar{K}_2 = \frac{2}{5} \frac{\bar{K}_1 \sigma\cos\alpha}{\Delta r_c} \qquad \bar{K}_3 = \frac{\bar{K}_1}{5\Delta r_c} \rho_1 \left(\frac{g}{g_c}\right) \cos\alpha_n$$

as the constants of the differential equation.

Mass flux can be related to the moisture content of solid at any time as in Eq.(7). Combining Eq.(7) and Eq.(14) results

$$\frac{\partial W}{\partial t} = \bar{K}_4 \frac{\partial^2 r_c^4}{\partial_x^2} - \bar{K}_5 \frac{\partial r_c^5}{\partial x} \tag{15}$$

where

$$\bar{K}_4 = \bar{K}_2 \frac{\varepsilon}{\rho_s} \quad \text{and} \quad \bar{K}_5 = \bar{K}_3 \frac{\varepsilon}{\rho_s}$$

Let dimensionless moisture concentration, C be W/W_c where W is the moisture content at any time and W_c is the moisture content of solid at critical point.

The capillary radius r_c may be any function of the moisture concentration, C. As an approximation, r_c is taken as a power function of dimensionless moisture concentration,

$$r_c = K_4 C^m \tag{16}$$

where K_4 and m are constants which depend on the solid structure.

If these expressions are inserted into Eq.(15) the dimensionless form of the general equation would result in,

$$\frac{\partial C}{\partial t} = \bar{K}_4 K_4^4 \frac{\partial^2 C^{4m}}{\partial x^2} - \bar{K}_5 K_4^5 \frac{\partial C^{5m}}{\partial x} \tag{17}$$

By linearizing the above equation[(1)], a linear partial differential equation is obtained as such

$$\frac{\partial C}{\partial t} = K_1 \frac{\partial^2 C}{\partial x^2} - K_2 \frac{\partial C}{\partial x} \tag{18}$$

with

$$K_1 = \bar{K}_4 K_4^4 4m\, C_i^{4m-1}, \quad K_2 = \bar{K}_5 K_4^5 5m\, C_i^{5m-1}$$

Now, Eq.(18) can be solved analytically by using Laplace Transformation technique by introducing simplified boundary conditions.

(1) $C^P = C_i^P + P\, C_i^{P-1} (C - C_i)$ where P is any arbitrary constant.

In this way, the following boundary conditions may be described for the system,

$t=0$	$C=C_i=1$	(for all x)
$x=0$	$C=0$	(for all t)
$x=\infty$	C is finite.	

Analytical solution of the drying equation gives the concentration profile at any x as such,

$$C = \frac{1}{2} + \frac{1}{2}\operatorname{erf}\left(\frac{x-K_2 t}{2\sqrt{K_1 t}}\right) - \frac{1}{2}\exp\left(\frac{K_2 x}{K_1}\right)\left(1-\operatorname{erf}\left(\frac{x+K_2 t}{2\sqrt{K_1 t}}\right)\right) \tag{19}$$

The average moisture concentration through the solid body can be determined by applying the following expression,

$$\bar{C} = \frac{\int_0^X C\,dx}{\int_0^X dx} \tag{20}$$

where X corresponds to the total thickness of slab.

Finally, average moisture concentration profile is obtained as,

$$\bar{C} = \frac{x}{2} + \sqrt{K_1 t}\left(\frac{1}{\sqrt{\pi}}\left(\exp\left(-\left(\frac{x-K_2 t}{2\sqrt{K_1 t}}\right)^2\right) - \exp\left(-\frac{K_2^2 t}{4K_1}\right)\right) + \left(\frac{x-K_2 t}{2\sqrt{K_1 t}}\right)\operatorname{erf}\left(\frac{x-K_2 t}{2\sqrt{K_1 t}}\right) - \frac{K_2 t}{2\sqrt{K_1 t}}\operatorname{erf}\left(\frac{K_2 t}{2\sqrt{K_1 t}}\right)\right) + \frac{K_1}{2K_2}\left(\exp\left(\frac{K_2 x}{K_1}\right)\left(\operatorname{erf}\left(\frac{x+K_2 t}{2\sqrt{K_1 t}}\right)-1\right) - \operatorname{erf}\left(\frac{x-K_2 t}{2\sqrt{K_1 t}}\right) - 2\operatorname{erf}\left(\frac{K_2 t}{2\sqrt{K_1 t}}\right) + 1\right) \tag{21}$$

with $K_1 = \dfrac{\pi\rho_1\, \sigma\cos\alpha\, K_{4m}^4}{5\rho_s A\, \mu\Delta r_c\, W_i}$, $K_2 = \dfrac{5\, K_1 K_4 \cos\alpha_n}{8\,\sigma\cos\alpha}$

as the constants of the drying equation.

K_1 and K_2, the constants of the drying equation are mainly the functions of initial moisture content, solid structure and the temperature of the drying medium. $\sigma\cos\alpha$ can be considered as a constant in evaluating K_1 and K_2 values. When $\sigma\cos\alpha$ is calculated at the average capillary radius for total thickness, Eq.(22) results,

$$\sigma\cos\alpha = \frac{x}{2}\rho_1 r_{c\,avg.}\left(\frac{g}{g_c}\right)\cos\alpha_n \tag{22}$$

Inserting Eq.(22) in K_1 and K_2 expressions, the following final equations for the constants K_1 and K_2 are obtained.

$$K_1 = \frac{\pi\rho_1^2 K_4^4\, m\, X\, r_{c\,avg.} \cos\alpha_n}{10\, A\mu\rho_s \Delta r_c W_i} \quad |=| m^2/s$$

and

$$K_2 = 1.25\left(\frac{K_1 K_4}{X r_{c_{avg}}}\right) \quad |=| \ \text{m/s} \tag{23}$$

3. RESULTS AND DISCUSSIONS

The drying model developed for falling rate period drying of solids results in a linear partial differential equation which is given in Eq.(18). Inspection of this equation reveals some significant points. The first term of this equation describes the moisture concentration change due to diffusion and the last term describes the moisture transfer due to convection. When drying proceeds, the medium under consideration is not at rest. Then besides concentration changes due to diffusion, concentration changes caused by convection should be considered too. In the proposed model unidirectional flow against gravity is assumed and, the constants K_1 and K_2 represent effective diffusivity and convection velocity respectively. If K_1 is very large when compared to K_2 only diffusion is expected. When K_1/K_2 is close to zero, it can be said that the liquid motion through the capillaries is close to plug flow. As it is seen in Eq.(23), K_1 is inversely proportional with the viscosity. Since physical properties of the liquid is calculated at the wet-bulb temperature of the drying medium, it is expected that as wet-bulb temperature of the drying medium increases, K_1 will increase also. On the other hand there is a linear relationship between K_1 and total thickness, but K_2/K_1 value decreases as total thickness increases. From this point, it can be concluded that diffusion through the capillaries of solid is dominant when solid is thick. K_1 and K_2 values are also the functions of initial moisture content, W_i, apparent solid density, ρ_s and the properties related with the structure of the solid body which are K_4, m and Δr_c. As it is mentioned before, capillary radius, r_c is assumed to be a power function of moisture concentration, C as given in Eq.(16) where K_4 and m are constants. If natural logarithm of Eq.(16) is taken, Eq.(24) results,

$$\ln r_c = \ln K_4 + m \ln C \tag{24}$$

From Eq.(24), it is observed that there is a linearship between $\ln r_c$ and $\ln C$. Since capillary radius r_c changes from minumum capillary radius, $r_{c_{min}}$ to maximum capillary radius $r_{c_{max}}$, and the dimensionless moisture concentration changes from the equilibrium moisture concentration C_{eq} to the initial moisture concentration which is unity, the constants K_4 and m are obtained as given in Eq.(25).

$$K_4 = r_{c_{max}}, \quad m = \frac{1}{\ln C_{eq}}\left(\frac{\ln r_{c_{min}}}{\ln r_{c_{max}}}\right) \tag{25}$$

$r_{c_{max}}$ and $r_{c_{min}}$ values can be evaluated from mercury penetrometer data. Another constant Δr_c which is the average difference between two pore radii in descent order can be estimated approximately from the slope of the pore radius versus penetration volume curve.

It is seen that when K_2 is zero, Eq.(19) reduces to the solution of diffusion equation which is known as Fick's Law. Effect of K_2 values on drying equation is shown in Fig. 2 for different slab thicknesses.

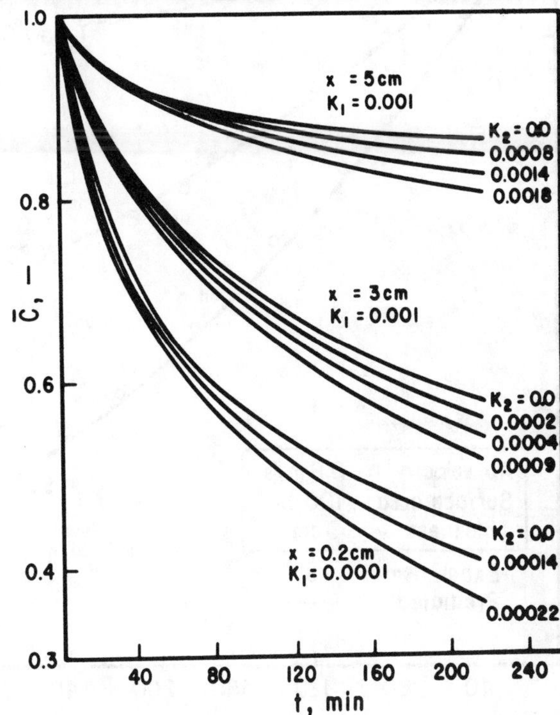

Fig. 2 Theoretical Moisture Concentration versus Time Profiles, Slab Thickness being a Parameter

Solids can be broadly classified according to whether they are homogeneous or heteregeneous (complex). Drying model proposed in this study is accomplished for homogeneous solids having uniform moisture distribution and physical properties. However in the industry, most of the solid products, especially foodstuffs which have to be dried are not homogeneous. They consist of layers conducting in series. This type of solids can be considered as complex solids. To examine the applicability of the model to the complex solids, sunflowerseeds are dried and coal is dried as an homogeneous solid.

In the case of coal drying, fine granular coal particles are put in slab shape with a surface area of $100 cm^2$ and a thickness of 3 cm. The drying experiments reported here are carried out in air temperatures of 50, 70, 90^oC flowing parallel to the evaporating surface at a rate of $0.8 ms^{-1}$. Experimental and predicted values are illustrated in Fig. 3. On the other hand, 250 seeds of sunflower having surface area of $70 cm^2$ and a thickness of 3 mm are dried in the same tunnel dryer where the air flow rate is $0.3 ms^{-1}$ Experimental runs are performed 50, 70, 85 and 110^oC air temperatures. Although theoretical results of coal are in good aggreement with its experimental data, it is observed that predicted values deviate from experimental points for sunflower seeds which could easily be seen in Fig. 4.

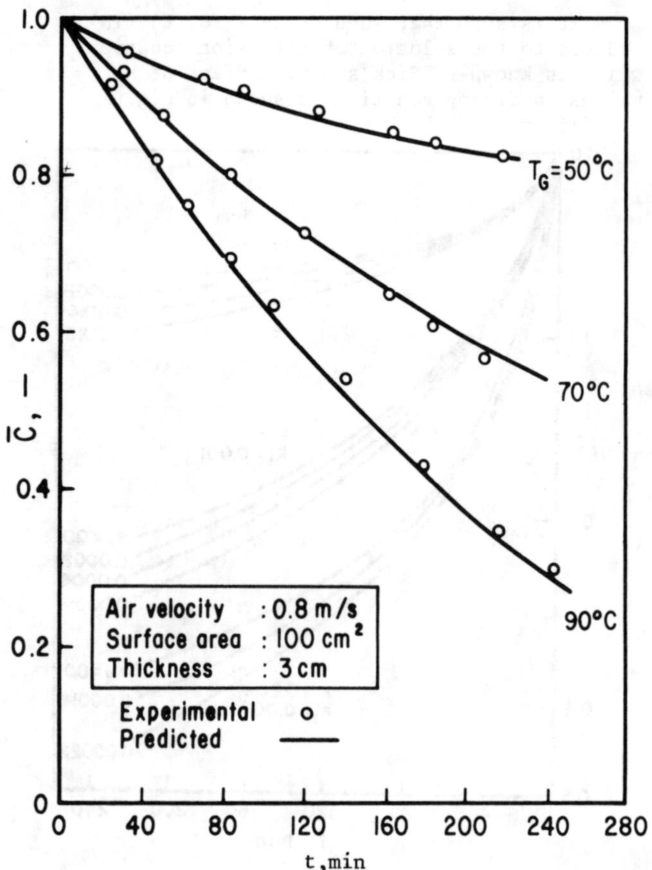

Fig. 3 Moisture Concentration Profiles for Coal

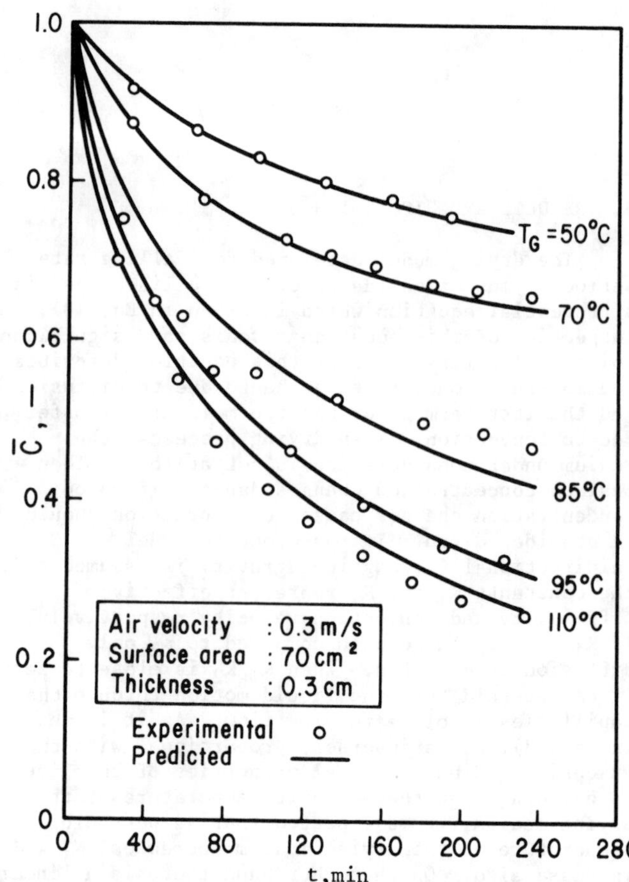

Fig. 4 Moisture Concentration Profiles for Sunflower Seeds

This event may be explained with the heteregeneous structure of the sunflower seed. Three layers exist in the sunflower seed; inner part of the sunflower seed, outer part of the sunflower seed (protective shell) and the air film between them. Initial moisture content (dry basis) and total porosity are found as 0.04 and 0.15 for inner part and 0.15 and 0.56 for the protective shell where the overall initial moisture content and total porosity of the sunflower seed itself are 0.09 and 0.39 respectively. From this point of view, it can be noticed that at different rates moisture transfer mechanism occur throughout the solid body during drying and as a result of series resistances within the sunflower seed, theoretical data may not fit in experimental data. K_1 and K_2 values for both products dried are tabulated in Table 1.

Table 1. K_1 and K_2 Values for Coal and Sunflower Seed.

Sample	T_G (°C)	T_{WB} (°C)	K_1 (m^2/s)	K_2 (m/s)
Coal	50	28	0.0008	0.0004
"	70	35	0.01	0.0048
"	90	42	0.011	0.0081
Sunflower seed	50	25	0.000018	0.00004
"	70	33	0.000030	0.00010
"	85	39	0.000120	0.00040
"	95	41	0.000140	0.00050
"	100	53	0.000180	0.00060

4. CONCLUSIONS

The mass transfer equation describing the mathematics of the capillary flow mechanism in drying is solved by Laplace Transformation technique. The coexistence of the diffusive and convective flow should be considered together. It is concluded that for thin solids, convective flow is dominant when compared to diffusional flow. In the proposed model K_1 and K_2 are constants and both are functions of initial moisture content, air temperature and solid structure.

In the case of coal drying, it is found that there is a single resistance to moisture transfer within the sample. But in the case where the seed is dried following resistances cause deviation from experimental data (i) resistance within the inner part, (ii) resistance within the outer part, and (iii) resistance through the air film between the inner and outer parts.

NOMENCLATURE

A	total area of the solid available for mass transfer	m^2
A_c	cross-sectional area of the capillary	m^2
B	modified permeability coefficient	s^{-1}
C	dimensionless moisture concentration of the solid	kg/kg
$\bar{C}$	average moisture concentration within the solid body	kg/kg
C_{eq}	equilibrium moisture concentration	kg/kg
C_i	initial moisture concentration of the solid body	kg/kg

G	mass flux of moisture	$kg/(m^2 s)$
m	constant	-
P_f	pressure drop due to friction in the capillaries	N/m^2
r_c	capillary radius	m
$r_{c_{max}}$	maximum capillary radius	m
$r_{c_{min}}$	minimum capillary radius	m
T_G	dry-bulb temperature of the surrounding fluid	^oC
T_{WB}	wet-bulb temperature of the surrounding fluid	^oC
W	moisture content of the solid (dry basis)	kg/kg
W_C	critical moisture content of the solid (dry basis)	kg/kg
X	total thickness of the solid	kg/kg

Greek Symbols

α	angle of curvature of liquid in the capillaries	-
α_n	angle between the drying surface and the normal to the force of gravity.	-
ε	total porosity	-
ρ_s	solid density	kg/m^3
μ	viscosity	kg/(s m)

REFERENCES

1. Sherwood, T.K., Ind.Eng.Chem, vol.22, 132 (1939).
2. Gilliland, E.R., Ind.Eng.Chem, vol.30, 506 (1938).
3. Newman, A.B., Trans.Inst.Chem.Eng, vol.27, 203 (1931).
4. Kirkwood, K.C. and Mitchell, T.J., J.App.Chem., vol.15, 251 (1965).
5. Lebedev, P.D., Int.J.Heat and Mass Tr., vol.13, 302 (1960).
6. Morgan, R.P. and Yerazunis, S., A.I.Ch.E.J., vol.13, 132 (1967).
7. Kısakürek, B., The Preparation of Generalized Drying Curves for Porous Solids, Ph.D. Thesis, Illinois Institute of Technology, Chicago (1972).
8. Perry, R.H. and Chilton, C.H. Chemical Engineers' Handbook, pp.5-21, McGraw-Hill, New York (1973).

SECTION III: DRYING OF GRANULAR SOLIDS

PARTICLE VELOCITIES IN PNEUMATIC CONVEYING DRYERS

David Reay and Richard E. Bahu

Engineering Sciences Division, A.E.R.E. Harwell,
Oxfordshire OX11 ORA, ENGLAND.

ABSTRACT

Steady state particle velocities were measured in pneumatic conveying tubes of diameters 50, 80 and 150 mm, using particles of mean diameters 0.35, 0.65 and 1.5 mm. Gas velocities and solids-gas loadings were typical of those used in pneumatic conveying dryers. The results confirmed that in tubes of this size the upward motion of the particles is slowed considerably by frictional impact on the walls. This will lead to more drying per unit length of tube than in a full-scale dryer operated at the same conditions. The measurements were compared with particle velocities predicted using Yang's parallel flow and random flow correlations for the solids-wall friction factor. A modified version of the parallel flow correlation was developed which gave reasonable agreement between measured and predicted velocities for the two smaller particle sizes.

1. INTRODUCTION

In recent years several authors have proposed mathematical models of pneumatic conveying dryers (1-5). Invariably, they have neglected particle-wall interactions when calculating particle velocities up the tube. These interactions have been studied by pneumatic conveying researchers and several correlations for the solids-wall friction factor have been proposed, among the most recent being those of Yang (6). Preliminary calculations using Yang's correlations suggested that particle-wall interactions can slow down the particles considerably in tubes of laboratory and pilot plant size, leading to errors in the interpretation of data and in scale-up if these interactions are neglected. There was some doubt about these calculations, however, because the data on which Yang's correlations are based were mostly obtained at much higher solids-gas loadings than are used in pneumatic conveying dryers. Therefore, measurements have been made of steady state particle velocities in tubes of various diameters under conditions typical of pneumatic conveying dryers and the results compared with, firstly, predictions neglecting particle-wall interaction, and secondly, predictions using Yang's correlations for solids-wall friction factor.

2. EXPERIMENTS

2.1 Equipment

The work was performed in three vertical tubes of diameters 50, 80 and 150 mm. The two narrower tubes were 7 m long and were built from 1 m long QVF glass sections. The air flow was metered through rotameters and entered the tubes through short flow-straightening sections fitted with honeycomb inserts. Above the flow-straightening sections solids from a sealed hopper entered the tubes via a rotary screw constant-volume feeder which could be swung to supply either tube. To encourage uniform flow the hopper was vibrated by a rotating eccentrically-mounted weight. At the top of the tubes the solids were separated from the air in a cyclone. The air was exhausted to atmosphere while the solids descended to a collection bin. Between the glass sections of the tubes there were stainless steel spacers containing pressure tappings connected to a bank of water-filled inclined manometers.

The wider tube was 6 m long. It was built from mild steel except for a short glass section mounted with its mid-point 5 m above the feed point. It was fed by a Vibrascrew constant-volume feeder. The conveying air flow rate was measured by a hot wire anemometer.

2.2 Test Materials

Three test materials were used:

(a) Small beaded silica gel having an arithmetic mean particle diameter of 0.35 mm and a particle density of 1200 kg/m^3.

(b) Large beaded silica gel having an arithmetic mean particle diameter of 1.5 mm and a particle density of 1200 kg/m^3.

(c) Glass ballotini having an arithmetic mean particle diameter of 0.65 mm and a particle density of 2350 kg/m^3.

The silica gel beads were supplied by W.R. Grace Limited and the glass ballotini by English Glass Company Limited.

2.3 Particle Velocity Measurement

Particle velocities were measured by a double flash technique. Two flash guns were connected to a remote control unit which fired them sequentially. The time interval between the flashes could be varied from 80 to 2000 µs to suit the particle velocity being measured. The flashes were coloured by green and red filters to give a photograph with the particles coloured green in the first position and red in the second. The background was yellow. A transparent scale was fitted to the tube and appeared in all the photographs. The velocity of each particle was obtained by dividing the distance between its two images by the interval between the flashes. The camera was a Nikon 35 mm SLR with a 135 mm lens on an extension bellows. The depth of field was about 10 mm at the centre of the tube.

For each experimental condition and position up the tube the average particle velocity was estimated from a sample of 100 particles. Increasing the sample size to 200 particles did not reduce the standard deviations significantly. There was no significant variation in particle velocity across the width of the tube at a given vertical position.

2.4 Operating Conditions

All the measurements reported here except one were made at ambient temperature. The air velocity up the tube was varied from 7.4 to 34 m/s and the solids-gas loading from 0.10 to 0.62 kg solid/kg gas.

3. RESULTS

The final steady state particle velocities are listed in Table 1. Acceleration lengths were less than 0.5 m in the 50 mm tube. In the 80 mm tube most of the acceleration was completed in the first 0.5 m, but the particles continued to accelerate slowly for some further distance up the tube. In the 150 mm tube it was assumed that the particles had reached their final steady velocity by the time they reached the measurement point 5 m above the feed point.

Table 1 Steady state particle velocities

Material	Tube Diameter (m)	Gas Velocity (m/s)	Solid-Gas Loading (kg/kg)	Steady Particle Velocity (m/s)	Standard Deviation (m/s)
Small silica gel	50	7.4	0.29	4.98	0.57
	50	14.0	0.15	8.82	1.28
	50	22.3	0.10	15.21	2.21
	80	7.4	0.29	5.85	0.43
	80	14.0	0.15	11.71	1.09
	150	14.0	0.15	11.80	1.66
	150	23.0	0.10	17.44	1.93
Large silica gel	50	10.5	0.62	1.89	0.40
	50	14.0	0.46	3.36	0.54
	50	22.3	0.29	5.76	1.03
	50	34.0	0.41	9.64	3.49
	80	10.5	0.62	3.17	0.58
	80	14.0	0.46	4.68	0.92
	80	14.0[a]	0.66	4.63	1.01
	150	14.0	0.46	4.45	1.50
	150	23.0	0.28	7.05	2.79
Glass ballotini	80	10.5	0.32	4.20	0.59
	80	14.0	0.24	6.73	1.06
	80	19.5	0.17	10.44	1.28
	150	14.0	0.24	7.88	1.01
	150	23.0	0.14	12.96	1.49

a = Gas velocity at 150°C

4. COMPARISON WITH PREDICTIONS

Inspection of Table 1 shows that:

(a) in all tube sizes the slip velocity between the particles and the gas increased with increasing gas velocity;

(b) at a given gas velocity, particle velocities were higher in the 80 mm tube than in the 50 mm tube, but particle velocities in the 80 and 150 mm tubes were fairly similar.

These observations suggest that particle-wall interactions must be having a significant retarding effect on the particle motion, particularly in the 50 mm tube.

In the absence of particle-wall interaction the difference between the gas velocity U_G and the particle velocity U_p at steady state equals the terminal free fall velocity U_t. It is given by a force balance as:

$$U_G - U_p = U_t = \sqrt{\frac{4\rho_p d_p g}{3\rho_G C_D}} \qquad (1)$$

The correlation for the drag coefficient C_D proposed by Clift, Grace and Weber (7) was used in equation (1) to calculate U_p for each of our experimental conditions. These values are compared with the measured values of U_p in Figure 1. The observed values of U_p are consistently lower than those predicted by neglecting particle-wall interactions. The discrepancy is particularly large for the large silica gel particles, which a high-speed cine-film showed to have a more zig-zag motion than the smaller particles and to strike the wall more often.

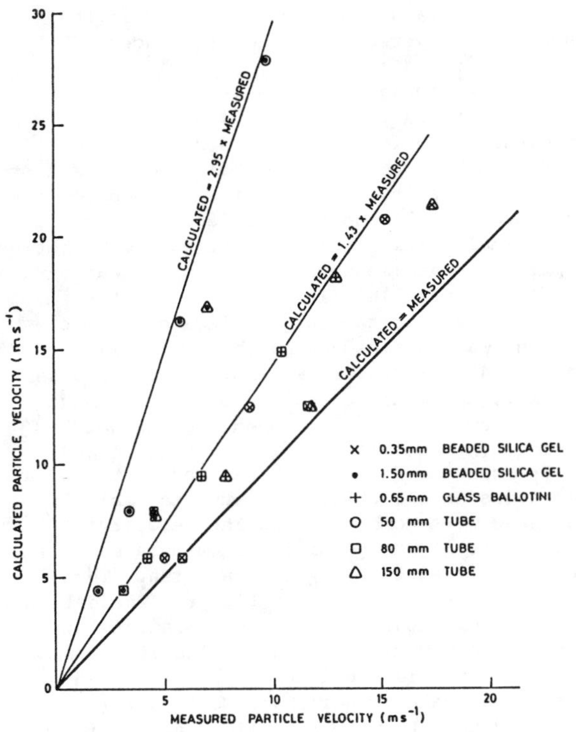

Fig. 1 Comparison of measured particle velocities with those calculated neglecting particle-wall interactions

Particle-wall interactions slow down the particles and increase the slip velocity $(U_G - U_p)$ beyond U_t. The slip velocity in the presence of particle-wall interactions may be formulated as:

$$U_G - U_p = U_t \sqrt{1 + \frac{f_p U_p^2}{2g D}} \quad (2)$$

where f_p is a solids-wall friction factor and D is the tube diameter. Using this formulation, Yang (6) proposed two correlations for f_p which he speculated might correspond to two different particle flow regimes called parallel flow and random flow. Parallel flow was said to occur when U_G is much higher than U_t and it had comparatively few particle-wall collisions; f_p is given by:

$$f_p = 0.0126 \frac{(1-\varepsilon)}{\varepsilon^3} \left(\frac{(1-\varepsilon)U_t}{U_G-U_p}\right)^{-0.979} \quad (3)$$

Random flow was said to occur when U_G was comparatively close to U_t and it had many more particle-wall collisions; f_p is given by:

$$f_p = 0.0410 \frac{(1-\varepsilon)}{\varepsilon^3} \left(\frac{(1-\varepsilon)U_t}{U_G-U_p}\right)^{-1.021} \quad (4)$$

The voidage ε is given by:

$$\varepsilon = 1 - \frac{4 W}{\rho_p \pi D^2 U_p} \quad (5)$$

where W is the solids mass flow rate. Solution of equation (2) for U_p with f_p given by either (3) or (4) is clearly an iterative process.

The materials and conditions used in the present work covered the following ranges of U_G/U_t:

small silica gel	6.5 - 16.1
large silica gel	1.6 - 4.0
glass ballotini	2.3 - 4.2

For each of the experimental conditions listed in Table 1, predicted values of U_p were calculated using equation (2) and both (3) and (4) for f_p. For the small silica gel the best agreement between the predicted and measured values of U_p was obtained by using the parallel flow equation (3) for f_p (see Fig. 2). For the large silica gel, best agreement was obtained using the random flow equation (4) for f_p (see Fig. 3). For the glass ballotini, the measured values of U_p lay about mid-way between the values predicted using equations (3) and (4) for f_p (see Figs. 4 and 5).

Pressure profiles were measured up the 50 and 80 mm tubes in all the runs listed in Table 1. When these were combined with the velocity measurements it was possible to calculate experimental values of f_p for each run. The values obtained are shown in Fig. 6. The data set on which Yang's parallel flow correlations are based was then modified by adding to it our small silica gel and glass ballotini f_p values and deleting the data of Hariu and Molstad (8) on the ground that they were obtained at conditions furthest removed from practical pneumatic conveying drying situations (very narrow tubes and very high solids-gas loadings). A least squares regression analysis then yielded the following modified parallel flow correlation:

$$f_p = 0.04 \frac{(1-\varepsilon)}{\varepsilon^3} \left(\frac{(1-\varepsilon)U_t}{U_G-U_p}\right)^{-0.9} \quad (6)$$

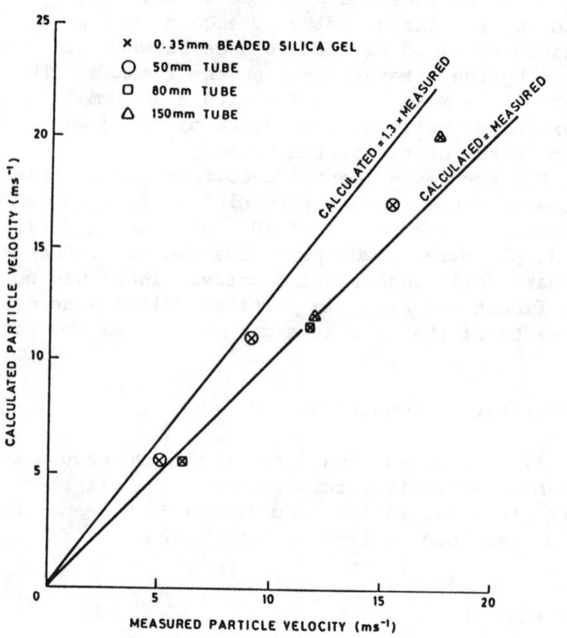

Fig. 2 Comparison of measured velocity of small silica gel beads with predictions of the parallel flow model

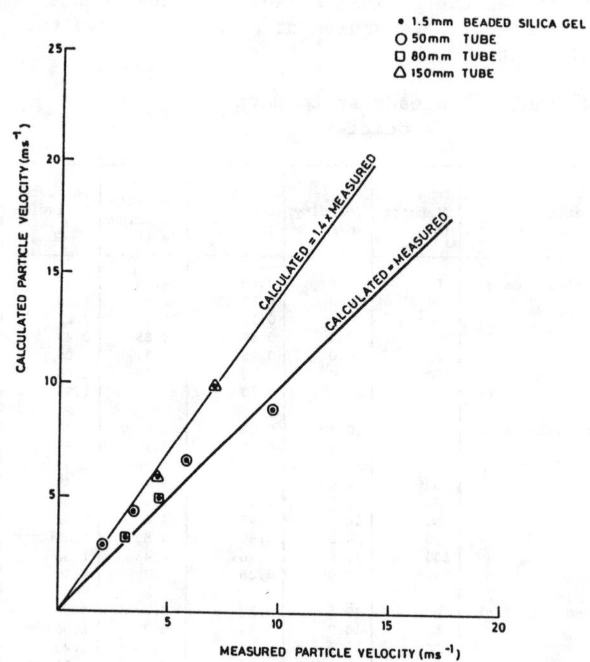

Fig. 3 Comparison of measured velocity of large silica gel beads with predictions of the random flow model.

When this was used with equation (2) to calculate values of U_p, improved agreement was obtained with the measured values of both the small silica gel and the glass ballotini (see Fig. 7). When our large silica gel f_p values were added to the data used by Yang in obtaining his random flow correlation, that correlation was not altered significantly.

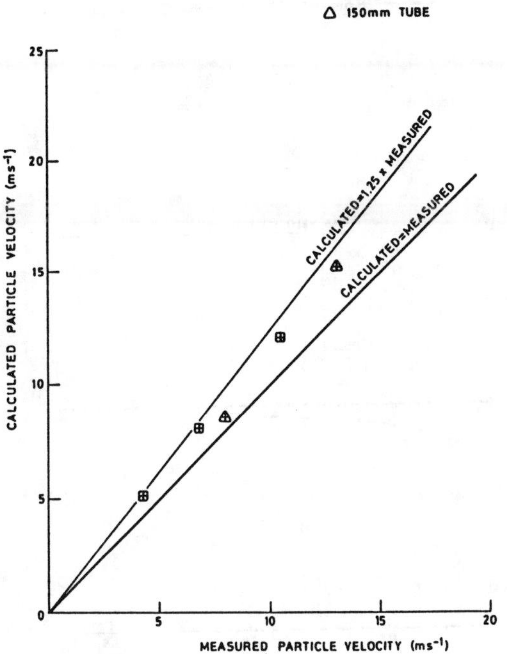

Fig. 4 Comparison of measured velocity of glass ballotini with predictions of the parallel flow model

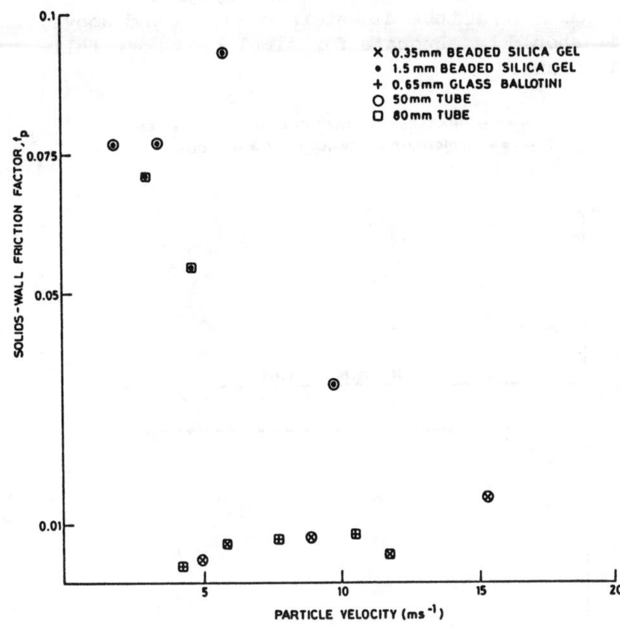

Fig. 6 Measured solids-wall friction factors

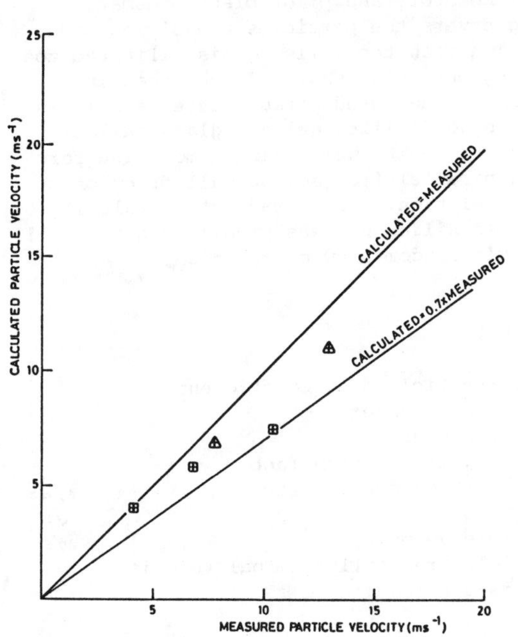

Fig. 5 Comparison of measured velocity of glass ballotini with predictions of the random flow model

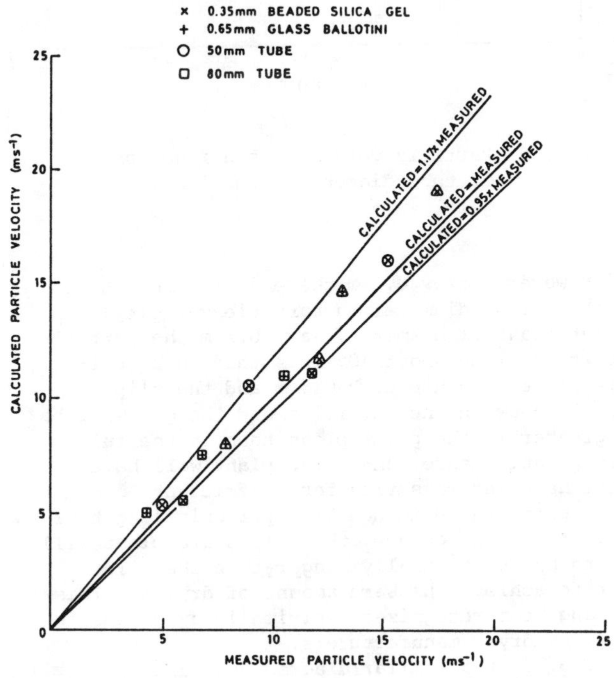

Fig. 7 Comparison of measured velocity of small silica gel beads and glass ballotini with predictions of the modified parallel flow model

5. IMPLICATIONS FOR SCALE-UP

In Fig. 8 the steady state velocity of small silica gel, calculated using equations (2) and (6), is plotted as a function of tube diameter for a gas velocity of 14 m/s. For comparison the particle velocity which would be calculated if particle-wall interactions are neglected, i.e. using equation (1), is also shown. The latter is a very good approximation at tube diameters of 0.5 m and above, so it should be adequate for simulating most full-scale plant.

---- NEGLECTING PARTICLE-WALL COLLISIONS
—•— MODIFIED PARALLEL FLOW MODEL

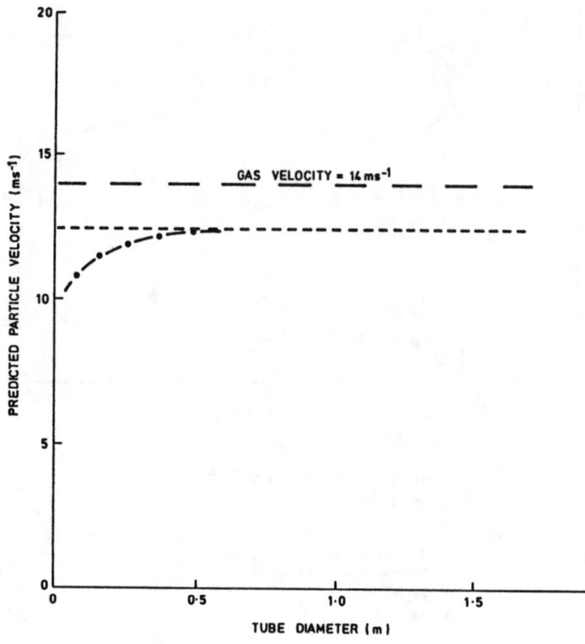

Fig. 8 Particle velocity as a function of tube diameter - small silica gel

However, below 0.5 m there is a substantial effect of tube diameter on particle velocity. In a pilot plant of diameter, say, 0.2 m the particle velocity will be about 10% less than in a full-scale plant of diameter $\geq$ 0.5 m and the slip velocity between the particles and the gas will be 60% greater in the pilot plant than in the full-scale plant. Hence, the pilot plant will have higher heat and mass transfer coefficients and a longer particle residence time per unit length of tube. As a consequence, the full-scale plant will have to be substantially longer than the pilot plant to achieve the same amount of drying. This has long been recognized empirically by pneumatic conveying dryer manufacturers.

Fig. 9 is a similar diagram for large silica gel with the values of U_p calculated using equations (2) and (4).

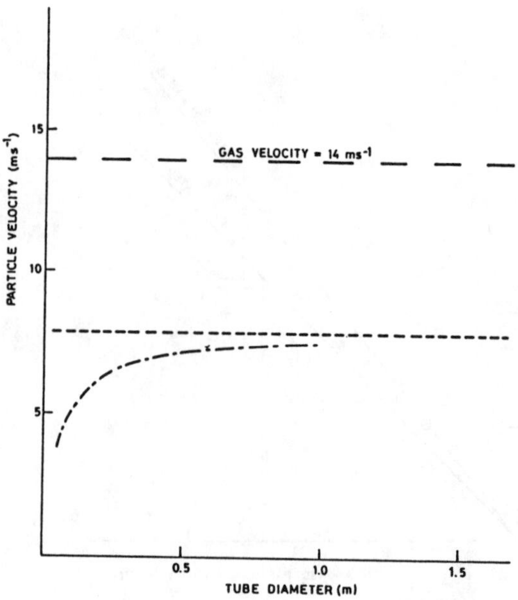

Fig. 9 Particle velocity as a function of tube diameter - large silica gel

6. CONCLUSIONS

In laboratory and pilot plant pneumatic conveying dryers the particles are slowed down by collision with the walls. This will lead to more drying per unit length of tube than in larger plant. Measured steady state velocities for small beaded silica gel and glass ballotini were predicted quite well using a modified form of Yang's parallel flow solids-wall friction factor correlation. The steady state velocity of large beaded silica gel was predicted quite well using Yang's random flow correlation.

NOMENCLATURE

C_D	fluid-particle drag coefficient	-
d_p	particle diameter	mm
D	tube diameter	mm
f_p	solids-wall friction factor	-
g	acceleration due to gravity	m/s^2
U_G	gas velocity	m/s
U_p	particle velocity	m/s
U_t	particle free fall terminal velocity	m/s
W	solids-gas loading	-

Greek

ε	voidage	-
ρ_G	gas density	kg/m^3
ρ_p	particle density	kg/m^3

REFERENCES

1. Andrieu, J. and Bressat, R., Proc. 3rd Int. Drying Symp., Birmingham, vol. 2, 10 (1982).
2. Martin, H. and Saleh, A.H., Verfahrenstechnik, vol. 16, 162 (1982).
3. Stein, W.A., Chemie. - Ing. - Technik, vol. 45, 1302 (1973).
4. Thorpe, G.R., Wint, A. and Coggan, G.C., Trans. Inst. Chem. Engrs., vol. 51, 339 (1973).
5. Babukha, G.L. and Shrayber, A.A., Heat Transfer Sov. Res., vol. 17, 119 (1975).
6. Yang, W.-C., A.I.Ch.E.J., vol. 20, 605 (1978).
7. Clift, R., Grace, J.R. and Weber, M.E., Bubbles, Drops and Particles, p.112, Academic Press, New York (1978).
8. Hariu, O.H. and Molstad, M.C., Ind. Eng. Chem., vol. 41, 1148 (1949).

DRYING OF GRANULAR PARTICLES IN A MULTISTAGE INCLINED FLUIDIZED BED WITH MECHANICAL VIBRATION

Masanobu Hasatani,
Norio Arai and Kiyoshi Hori

Department of Chemical Engineering, Nagoya University
Nagoya, 464 JAPAN

ABSTRACT

This paper describes the effect of mechanical vibration on the drying rate of granular particles in a multistage inclined fluidized bed(IFB).

A mathematical model for the drying rate of the particles in the multistage IFB is presented from the standpoint of simultaneous heat and mass transfer, with taking the effect of mechanical vibration added vertically into consideration.

Steady-state distributions for the temperatures and concentrations of the particles and the heating gas, and for the moisture content of the particles are numerically calculated based on the present model. The calculated results show fairly good agreement with the experimental data, which were obtained from the drying experiments of brick particles in a three-stage IFB using comparatively low temperature air (40 - 60°C) as a heating gas.

Within the range of the experimental conditions employed, it has been found that, the mechanical vibration added vertically enhances the over-all drying rate of the particles and its effect can be considered equivalent to an increase in the air velocity.

1. INTRODUCTION

In recent years the application of mechanical vibration has been increasingly remarked as one of potent methods for the improvement of heat and mass transfer processes, such as fluidized-bed dryers for wet granular particles. Transport phenomena, including heat and mass transfer in a vibro-fluidized bed and its practical use have been reviewed in detail by Mujumdar, Pakowski et al.[1-3]. They introduced some of typical advantages in applying the vibration to the commonly used fluidized-bed dryers: i) The behavior of fluidization of the particles, even being wetted, can be easily improved by manipulating amplitude and frequency of vibration, ii) The entrainment of undried fine particles is reduced because of decreasing air requirement for fluidization, and iii) The over-all drying rate increases due to breakup lumps of agglomerating wet particles.

As one of the most promising dryers for some kind of special solid materials, the authors have studied on the fundamental characteristics of an inclined fluidized bed(IFB) having a slightly inclined porous plate. So far, in order to obtain necessary data for practical application, mixing and flow of solid particles[4,5], heat transfer between particles and gas[6,7], and thermal decomposition of solids[8] have been examined in a single and a multistage IFB.

One reason of our consistent researches since 1968 is that, a global flow of the particles in the IFB is much closer to the piston flow than that of conventional fluidized beds. This might be one of the most desirable features, particularly in drying of foodstuffs, pharmaceutical materials, plastics and so on. In general, these solid materials are considerably delicate to heat and their final products are limited severely within a narrow range of the moisture content as well. Therefore, a precise drying-operation has been strongly required at a comparatively low temperature of heating gas, together with making residence time of each particle as uniform as possible.

In the present work, as a nearly final goal of our developmental project (IFB typed dryer), the drying rate of granular particles in a multistage IFB with mechanical vibration is deduced theoretically and demonstrated experimentally. Firstly, a mathematical model to predict the drying rate of the particles is presented from the standpoint of simultaneous heat and mass transfer. The steady-state distributions for the temperatures of the particles and the heating gas, and for the moisture content of the particles are numerically calculated on the basis of the proposed model. The calculated results are compared with the experimental data, which were obtained from the drying experiments of brick particles in a three-stage IFB with vertical mechanical vibration. Furthermore, the dimensionless correlation between the drying efficiency and the operation variables including the vibration intensity are obtained, and the effectiveness of multiplying the number of stages is also discussed.

2. THEORETICAL ANALYSIS

2.1. Mathematical Model and Basic Equations

Consider the general case in which the drying of moist granular particles takes place in a multistage IFB with mechanical vibration. In the present analysis, after simultaneous heat and mass transfer between the particles and the heating gas has reached steady state over the entire section of the IFB dryer, the temperature distributions of the particles and the gas, and the moisture content distribution of the particles are to be theoretically obtained by the following method.

The assumptions for the present theoretical approach are as follows:

(1) The particles are perfectly mixing in the direction of height of the bed, and the gas flowing through the bed and the space between the stages is of piston flow type.
(2) The temperature and the moisture content within each particle are uniform, because the particles used are comparatively fine[9].
(3) The effect of longitudinal particle mixing on the heat and mass transfer rates can be neglected.
(4) The heat losses from the side walls are negligibly small.

Based on the above assumptions, the drying model for a multistage IFB can be simplified as shown in Fig.1(a).

From an infinitesimal heat and mass balance in an arbitrary (2i+1)th stage, the following basic equations are derived:

Heat transfer equations

$$u_{s,2i+1}(C_p+C_w w_{2i+1})\frac{dt_{p,2i+1}}{dx} = h_{p,2i+1}a_p(\Delta t)_{av.,2i+1} + \lambda_{2i+1}u_{s,2i+1}\frac{dw_{2i+1}}{dx} \quad (1)$$

$$t_{g,2i+1} = t_{p,2i+1}+(t_{g,2i}-t_{p,2i+1})\exp(-\frac{h_{p,2i+1}a_p W}{GC_g \ell u_{s,2i+1}}) \quad (2)$$

Mass transfer equations

$$-u_{s,2i+1}\frac{dw_{2i+1}}{dx} = k_{p,2i+1}a_p\delta(\Delta H)_{av.,2i+1} \quad (3)$$

$$H_{g,2i+1} = H_{p,2i+1}+(H_{g,2i}-H_{p,2i+1})\exp(-\frac{k_{p,2i+1}a_p W\delta}{G\ell u_{s,2i+1}}) \quad \cdots\cdots\cdots (4)$$

The boundary conditions are:

$$\begin{aligned}
&i = n,\ x = 0\ ;\ t_{p,2n+1} = t_{p,0},\ w_{2n+1} = w_0 \ \text{(known)}\\
&i = 0,\ 0 \leqq x \leqq L\ ;\ t_g = t_{g,0},\ H_g = H_{g,0} \ \text{(known)}\\
&\left.\begin{array}{l} i \neq 0\ (i= 1,2,---,n-1,n) \\ x = L\end{array}\right\};\ t_{p,2i}=t_{p,2i+1},\ w_{2i}=w_{2i+1}\\
&\left.\begin{array}{l} i \neq n\ (i=0,1,---,n-2,n-1) \\ x = 0\end{array}\right\};\ \begin{array}{l} t_{p,2i+1}=t_{p,2i+2} \\ w_{2i+1} = w_{2i+2}\end{array}
\end{aligned} \quad (5)$$

$(\Delta t)_{av,2i+1}$ and $(\Delta H)_{av,2i+1}$ in Eqs.(1) and (3) are the mean temperature- and concentration-difference between the particles and the gas over a Δx- section, which are given as follows:

$$(\Delta t)_{av.,2i+1} = \frac{1}{\Delta x}\int_0^{\Delta x} (\overline{\Delta t})_{2i+1}\,dx \quad (6)$$

$$(\Delta H)_{av.,2i+1} = \frac{1}{\Delta x}\int_0^{\Delta x} (\overline{\Delta H})_{2i+1}\,dx \quad (7)$$

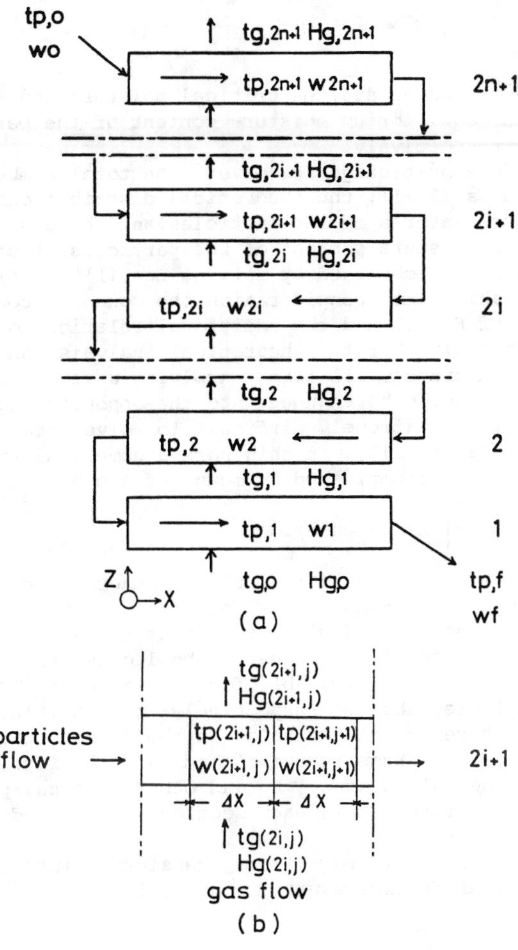

Fig.1 Drying model in a (2n+1)stages IFB

Here $(\overline{\Delta t})_{2i+1}$ and $(\overline{\Delta H})_{2i+1}$ in the above equations are expressed by the following forms, respectively.

$$(\overline{\Delta t})_{2i+1} = \frac{GC_g u_{s,2i+1}\ell}{h_{p,2i+1}a_p W}\{1-\exp(-\frac{h_{p,2i+1}a_p W}{GC_g u_{s,2i+1}\ell})\} \times (t_{g,2i}-t_{p,2i+1}) \quad \cdots\cdots (8)$$

$$(\overline{\Delta H})_{2i+1} = \frac{Gu_{s,2i+1}\ell}{k_{p,2i+1}a_p\delta W}\{1-\exp(-\frac{k_{p,2i+1}a_p\delta W}{Gu_{s,2i+1}\ell})\} \times (H_{p,2i+1}-H_{g,2i}) \quad \cdots\cdots (9)$$

The Eqs.(8) and (9) can be easily derived based on the assumption(1): the gas flowing through the bed being of piston flow type. H_p is the saturation absolute humidity at the particle-temperature t_p. δ in Eqs.(3),(4) and (9) is a characteristic parameter which should be determined from the drying rate curve of the particles employed. In fact, as described in detail in the previous paper[10], δ (=0 to 1.0) is assumed to be given as a function of only the moisture content of the particles as follows:

$$w \leqq w_c \ ; \ \delta = 1$$

$$w_e \leqq w \leqq w_c \ ; \ \delta = \frac{w - w_e}{w_c - w_e} \tag{9}$$

where w_c and w_e are the critical moisture content and the equilibrium moisture content of the particles, respectively.

In a multistage IFB dryer, the total number of stages is (2n+1); the theoretical distributions for the temperatures of the particles and the gas, and for the moisture content of the particles at steady state can be obtained by solving Eqs.(1) to (4), with taking into consideration the boundary conditions of Eq.(5) and the additional relations of Eqs.(6) to (10). For the theoretical analysis, on even numbered stages, all terms involving u_s in Eqs.(1) and (3) should be changed into the opposite sign. Since it is extremely difficult to solve these equations analytically, in this work a numerical analysis can be accomplished by means of a computer.

2.2. Numerical Calculation

The procedure of the present numerical calculation is approximately as follows:

(i) Each stage is divided into m sections(the length of each section is equal, labelled as Δx).

(ii) As shown in Fig.1(b), four variables(t_p, t_g, w and H_g) are labelled respectively: for example, the temperature of the particles in the j-th section of the (2i+1)th stage is symbolized as $t_p(2i+1,j)$.

(iii) The values of h_p, k_p and other physical properties of interest in each section are assumed to be constant.

(iv) Equations(1) to (4) are transformed into simultaneous difference equations as follows:

$$\begin{aligned} t_p(2i+1,j+1) = & \left\{ \frac{2-A\alpha(2i+1,j)\gamma(2i+1,j)}{2+A\alpha(2i+1,j)\gamma(2i+1,j)} \right\} t_p(2i+1,j) \\ & + \left\{ \frac{2A\alpha(2i+1,j)\gamma(2i+1,j)}{2+A\alpha(2i+1,j)\gamma(2i+1,j)} \right\} t_g(2i,j) \\ & + \left\{ \frac{2}{2+A\alpha(2i+1,j)\gamma(2i+1,j)} \right\} \\ & \times \frac{\lambda(2i+1,j)\gamma(2i+1,j)}{C_p} \\ & \times \{w(2i+1,j+1)-w(2i+1,j)\} \end{aligned} \tag{11}$$

$$\begin{aligned} t_g(2i+1,j) = & \, t_p(2i+1,j)+\{t_g(2i,j)-t_p(2i+1,j)\} \\ & \times \{1-\alpha(2i+1,j)\} \end{aligned} \tag{12}$$

$$\begin{aligned} w(2i+1,j+1) = & \, w(2i+1,j)-\frac{B\beta(2i+1,j)}{2}\{H_p(2i+1,j+1) \\ & + H_p(2i+1,j)-2H_g(2i,j)\} \end{aligned} \tag{13}$$

$$\begin{aligned} H_g(2i+1,j) = & \, H_p(2i+1,j)+\{H_g(2i,j)-H_p(2i+1,j)\} \\ & \times \{1-\beta(2i+1,j)\} \end{aligned} \tag{14}$$

where

$$\alpha(2i+1,j) = 1-\exp\{-\frac{h_p(2i+1,j)a_p W}{GC_g u_s(2i+1,j)\ell}\},$$

$$\beta(2i+1,j) = 1-\exp\{-\frac{k_p(2i+1,j)a_p \delta W}{Gu_s(2i+1,j)\ell}\},$$

$$A = \frac{GC_g \ell \Delta x}{WC_p}, \quad B = \frac{G\ell\Delta x}{W}, \tag{15}$$

$$\gamma(2i+1,j) = \frac{C_p}{C_p + C_w w(2i+1,j)}$$

$\lambda(2i+1,j)$ in Eq.(11) is the latent heat of vaporization at the temperature of $\{t_p(2i+1,j+1)+t_p(2i+1,j)\}/2$.

Figure 2 is a schematic flow chart of an iterative calculation adopted in this work, in order to obtain the theoretical distributions of t_p, t_g, w and H_g at steady state as their convergent solutions of Eqs.(11) to (14) with satisfying the boundary conditions of Eq.(5).

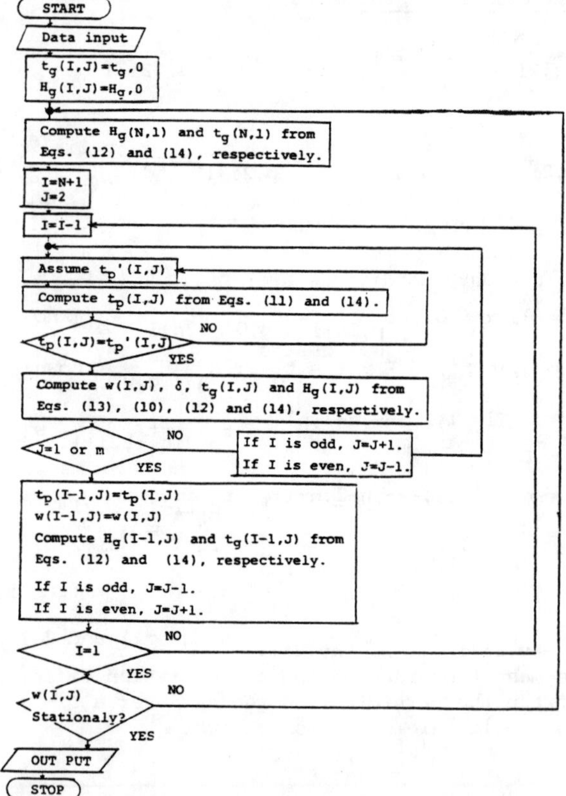

Fig.2 Flow chart of numerical calculation

3. EXPERIMENTAL APPARATUS AND PROCEDURE

Figure 3 shows a schematic set-up of the three stage IFB dryer employed in this work. A detailed configuration of each stage is sketched in Fig.4. As seen from both figures, the channel of the particles flow in each stage was sized with 500 mm in length and 150 mm in width. The porous plate, made of a perforated stainless steel plate(2 mmϕ perforations, 4% of free area ratio), was installed in each stage with an inclination angle of 3° against the horizontal surface. A variable amplitude rotary eccentric vibrator(Yasukawa Electric Co.Ltd., FEC-2.5-2 type) was used as a mechanical vibration source, by which a comparatively wide range of vertical vibration can be generated. In the present experiments, however, the frequency of vibration was fixed at 58.7 c.p.s.. The side walls of the dryer were insulated with glass fibre about 30 mm thick to minimize the heat loss to the surrounding air.

At the beginning of each experimental run, the heating air at a constant temperature and humidity was supplied continuously to the dryer with a constant flow rate. After the temperatures and the humidities of the air had reached at the designed constant value over the entire part of the dryer, the vibrator was driven. Instantaneously, the moist granular particles, whose moisture content had been adjusted at a constant, initial value(w_0), were fed to the entrance of the highest stage by a vibrofeeder. The moist particles thus fed were fluidized with the aid of mechanical vibration as well as aeration, and were gradually dried as flowing down to a lower stage. An auto-recording type thermometer was used to measure the time-changes of the temperatures of the heating air at the positions of x-marks on Fig.3. In fact, simultaneous heat and mass transfer between the particles and the air in the dryer was considered to reach steady state when all values of the air temperature thus measured became constant. After that time, a small portion of the flowing particles was sampled from the ●-marked positions in Fig.3 to determine the steady state distributions for the moisture content and the temperature of the particles.

The temperatures of the particles and the heating air were measured by chromel-alumel thermocouples of 300 μm in diameter. The humidity of the air at the entrance of the lowest stage was watched continuously by a hygrometer(made by Yokogawa Electric Works Ltd.;Dewcel 4031 type). The linear velocity of the particles(u_s) was measured by the same method as described in the previous paper[6].

Two kinds of brick particles (D_p=620 and 1190 μm, ρ_s=630 kg/m^3) were used as a sample of granular particles.

The drying experiments were conducted within the following ranges:

$t_{g,0}$=40-60°C, $H_{g,o}$=0.005-0.018 kg-H_2O/kg-dry.air, w_0=0.01-0.45 kg-H_2O/kg-dry·solid, $a_0\omega^2/g$=1.6-4.4, G=200-1300 kg/m^2·h and W=0.24-3.2 kg/h.

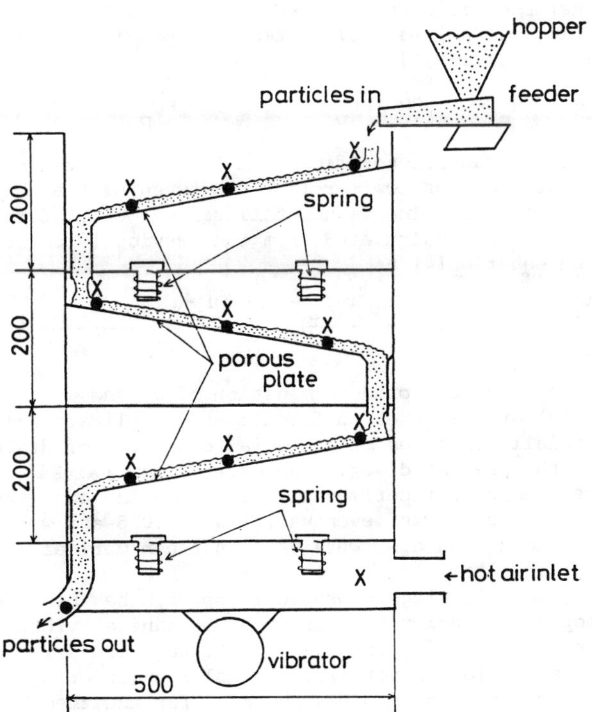

Fig.3 Experimental apparatus(three stage IFB dryer with mechanical vibration)

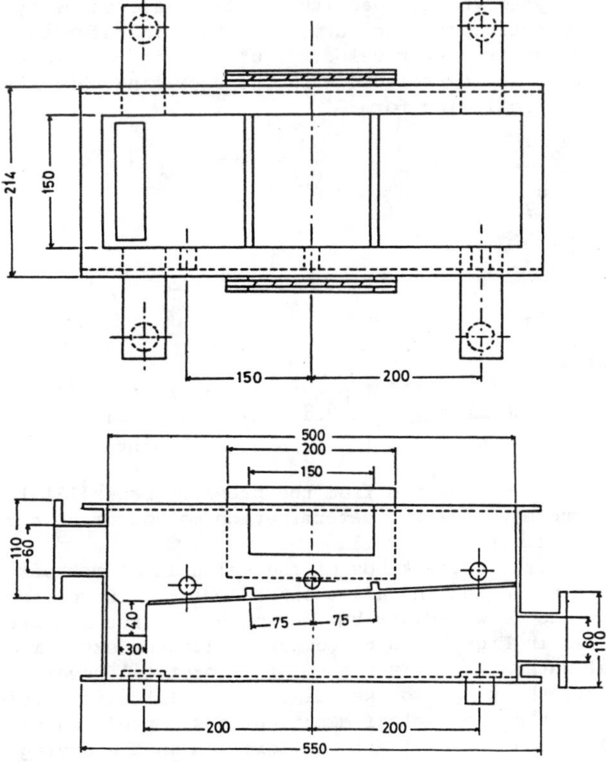

Fig.4 Detailed configuration of each stage

4. RESULTS AND DISCUSSION

4.1. Comparison of Experimental Data with Calculated Results

Representative examples of the comparisons of the measured data with the numerically calculated results are shown in Figs. 5 and 6, where the temperatures and the moisture content of the particles at steady state are plotted against distance from the entrance of the particles at the highest stage, x.

In the present numerical calculation; the heat

transfer coefficient between the particles and the heating air(h_p) was calculated from previous empirical correlation[6],

$$Nu_p = 0.008\ (Re_p')^{1.53} \qquad 1 < Re_p' < 100 \qquad (16)$$

The mass transfer coefficient(k_p) was estimated on the basis of the Lewis-relation between heat and mass transfer. The value of linear velocity of the particles was calculated from the previously correlated equation[4] as

$$\left(\frac{u_s}{u_g}\right) = 2.75\times10^{-3}\left(\frac{D_p^2 u_g \rho_s}{W}\right)^{-0.8}\left\{\frac{(u_g-u_{g,mf})D_p\rho_g}{\mu_g}\right\}^{0.575} \qquad (17)$$

Theoretical predetermination of w_c and w_e in Eq.(10) are extremely difficult and complicate[11], especially in cases of a special condition of drying like the present dryer. Therefore, both values were predicted experimentally based on the observed drying rate curves: eventually, $w_c = 0.025$ and $w_e = 0.014$ kg-H_2O/kg-d.s. were used in the numerical calculation.

The following theoretical approach has been developed to predict the effect of mechanical vibration on the drying rate in a multistage IFB dryer. In the previous paper[12], a simpler engineering, but successful method was proposed for the theoretical prediction of an augmentative effect of vibration on the particle-to-gas heat transfer as well as the linear velocity of the particles in a single stage IFB: its effect can be compared to an increase in a hydrodynamic quantity, that is, air velosity. A new characteristic parameter which is referred to as an apparent air velocity, u_g^* was introduced and it has been correlated with the operating variables by the following form:

$$\frac{u_g+u_g^*}{u_g} = 1.0 \qquad \Gamma < 3.5$$

$$\frac{u_g+u_g^*}{u_g} = 0.285\ \Gamma \qquad 3.5 < \Gamma < 15 \qquad (18)$$

where

$$\Gamma = \left(\frac{D_p^2 u_g \rho_s}{W}\right)^{0.3}\left\{\frac{\rho_g u_g^2}{(\rho_s-\rho_g)D_p g}\right\}^{-0.4}\left(\frac{a_0\omega^2}{g\sin\theta}\right)^{0.42}$$

Figure 7 was quoted from the previous paper[12] to concretely show the determination method of u_g^* and its definition as well.

Here, three kinds of the calculation methods were presented and all their calculated curves for t_p- and (w/w_0)-distribution in the x-direction are shown in Figs.5 and 6, compared with the experimental data. As a first step of theoretical development, the single broken lines(calc.III) were obtained as the theoretical distributions, premising that no augmentative effect of vibration on the drying rate appears. It was evident from both figures that, a considerably great difference exists between the calc.III and the experimental data and its difference increases with an increase in the vibration intensity($a_0\omega^2/g$). On the other hand, both the solid and the dotted lines(calc.I and calc.II, respectively) on Figs.5 and 6 were calculated by taking account of the effect of vibration as a function of

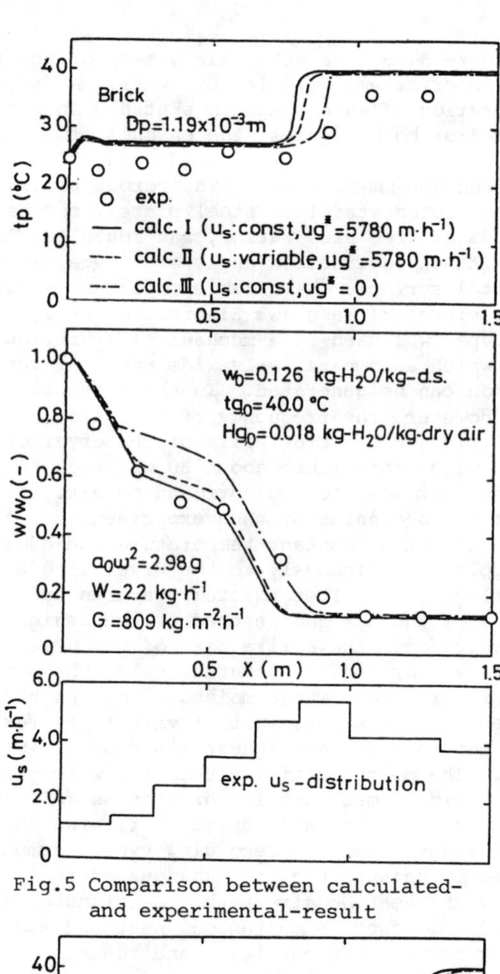

Fig.5 Comparison between calculated- and experimental-result

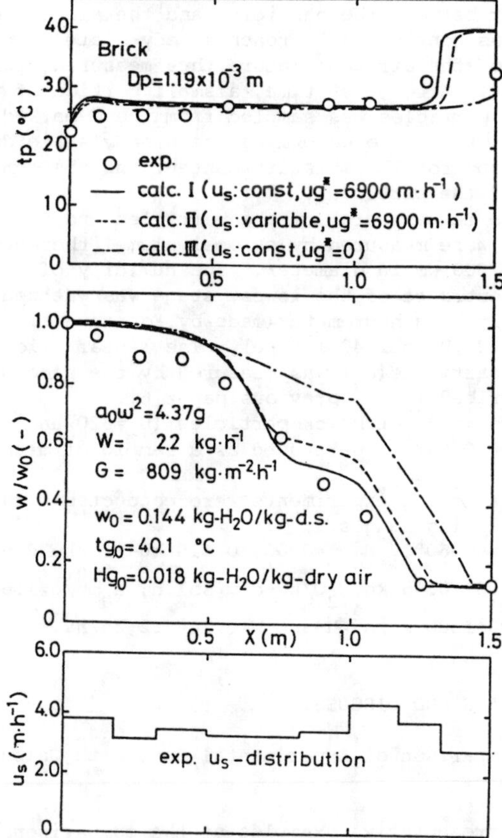

Fig.6 Comparison between calculated- and experimental-result

u_g^*. In fact, the values of h_p, k_p and u_s were calculated by substituting $(u_g+u_g^*)$ into u_g in the previous correlations in a non-vibrating system such as Eqs.(16) and (17). By using these values, the calc.I and calc.II were obtained, but only a difference between two calculations is that the latter was taken into account the measured u_s-distribution as well. As can be seen from these figures, both calculated results(calc.I and calc.II) agree well with the experimental data, particularly for the t_p- and (w/w_0)-distribution. Within the range of the present experimental conditions employed, the maximum deviation between calc.I and calc.II is about 5 %. In addition of this fact, the calc.I is in better

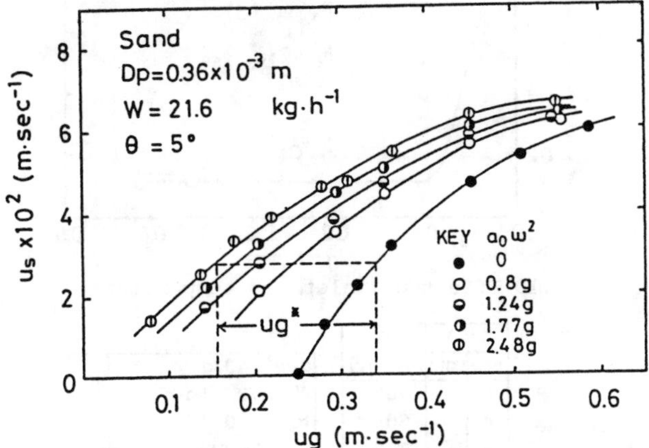

Fig.7 An illustrative definition of u_g^*

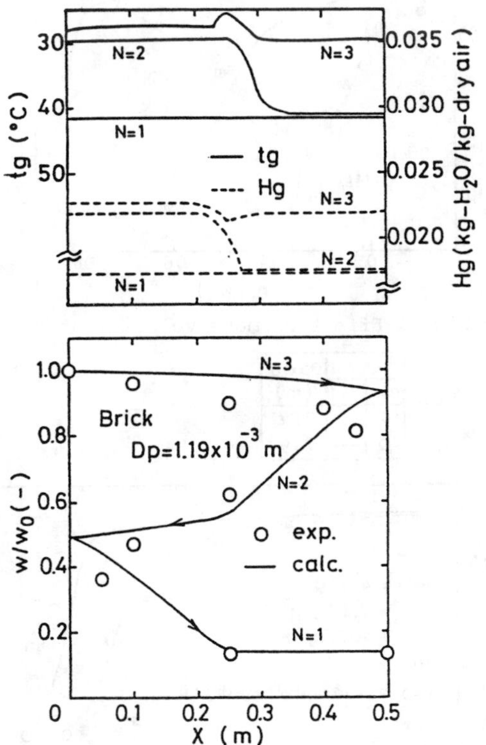

Fig.8 Calculated distributions of t_g, H_g and (w/w_0): calculated conditions are the same as in Fig.6

method because the measured u_s-distribution is unnecessary in the actual calculation. Therefore, the calc.I will be adopted in further-discussing the relationship between the drying rate and the operating variables, the drying efficiency and so on.

Apart from that aspect, when comparing between Fig.5 and Fig.6, in terms of the measured u_s-distributions, it was found that the application of mechanical vibration is advantageous, not only to accelerate the particle velocity, but also to uniformalize the u_s-distribution even the particles being wetted.

Incidentally, an inflection appears on the curves of (w/w_0) vs. x in Figs.5 and 6, though the temperature of the particles is held on a constant value at a nearly entire section of the dryer. In order to make clear this point, the caluculated distributions for the temperature and the humidity of the heating air as well as for the dimensionless moisture content of the particles are drawn in Fig.8, where the calculated conditions are the same as shown on Fig.6. In this figure, however, the distributions of t_g, H_g and (w/w_0) are plotted in a different manner than that in Fig.6. Such an illustration may give us much clearer information of simultaneous heat and mass transfer phenomena taking places in the dryer. It was seen from Fig.6 as well as this figure that, i) the flowing particles have being contacted with a locally different temperature and humidity air in each stage. This results in the appearance of an inflection, namely, in this type of dryer, the drying rate is fully localized, ii) the drying of the particles advances remarkably at a section of the dryer, in which both the t_g- and H_g-difference between the adjacent stages are extended: in the case of Fig.8, such an advancement occurs in the two ranges of x= 0 - 0.25 m at the first stage and x= 0.25 - 0.5 m at the second.

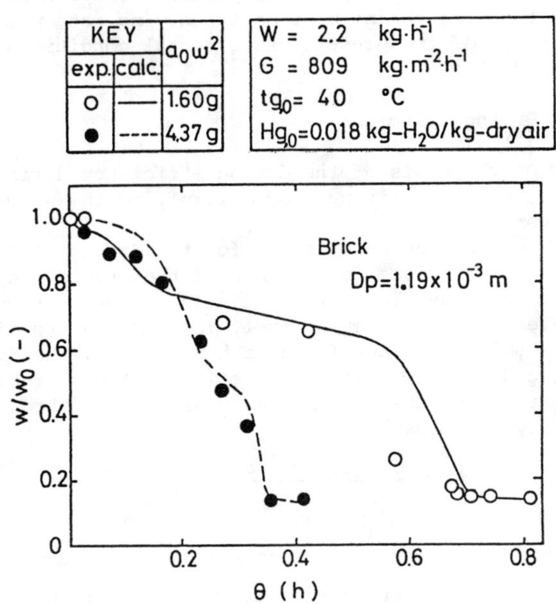

Fig.9 Effect of vibration intensity on drying rate

4.2. Effect of Operating Variables on Drying Rate

Figures 9 to 12 show the effects of the operating variables on the drying rate, where both the calculated and experimental dimensionless moisture contents (w/w_0) are plotted against a modified drying time θ (= x/u_s). All the calculated results hereafter are from the calc.I based on the above-mentioned discussion.

Effect of vibration intensity ($a_0\omega^2/g$)

A typical example for the effect of the vibration intensity is shown in Fig.9, where both experiments were done under the same drying conditions except $a_0\omega^2$. As seen from this figure, the total dead time of drying is to shorten by about 50% when the vibration intensity increases from 1.6 to 4.37. Such an enhancement of the drying rate by vibration has been already recognized in Figs.5 and 6. Besides, a fairly good agreement between the calculated and the experimental results in Fig.9 is very encouraging in discussing below.

Effect of the other variables

Representative examples for the effects on the drying rate processes of the inlet air temperature and the mass velocity of the air and the diameter of the particles are depicted in Figs.10 to 12, respectively. In these figures, the comparisons of the calculated results with the experimental data for the (w/w_0) vs.θ relations were made. The calculated results agree well with the experimental data in all cases. Furthermore, it was found from these comparisons that, in addition of the vibration intensity, the mass velocity of the air(G) and the inlet air temperature($t_{g,0}$) strongly affect to enhance the drying rate : for example, as obviously seen from Fig.10, the dead time of drying at $t_{g,0}$=60 °C is about a quarter of that at $t_{g,0}$=40 °C. However, the diameter of the particles seems to have a little effect on the enhancement of the drying rate within the range of the present experimental conditions covered.

4.3. Drying Efficiency

On the basis of the drying efficiency defined below, the effectiveness of multiplying the stages was confirmed.

So far many definitions for the drying efficiency, by which the performance of the dryers can be evaluated, have been proposed. In the present work, however, since a comparatively low temperature air (40 - 60°C) has been fed as a heating gas, the following drying efficiency focussing in the over-all change of the absolute humidity of the air through the dryer are used:

$$\eta = \frac{(H_{g,f})_{av.} - H_{g,0}}{(H_{p,f})_{av.} - H_{g,0}} \qquad (19)$$

where $(H_{p,f})_{av.}$ is the saturation absolute humidity at the mean temperature of the particles integrated over the entire part in the highest stage, and $(H_{g,f})_{av.}$ is the mean absolute humidity of the air leaving finally from the dryer. Figure 13 shows an example of the theoretically calculated results for the relationship between the drying efficiency(η) and the number of stages(N). In this calculation, a heat-flow ratio, A* (= $GC_g\ell L/WC_p$) was selected as a parameter, and α is assumed to be constant(α=0.5). Other calculation conditions are given on the figure.

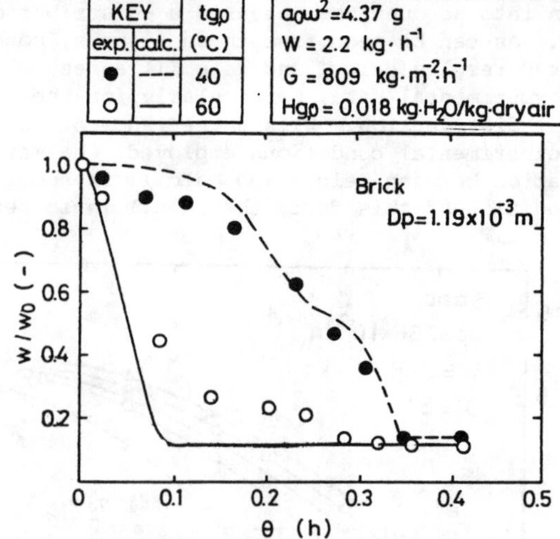

Fig.10 Effect of inlet air temperature

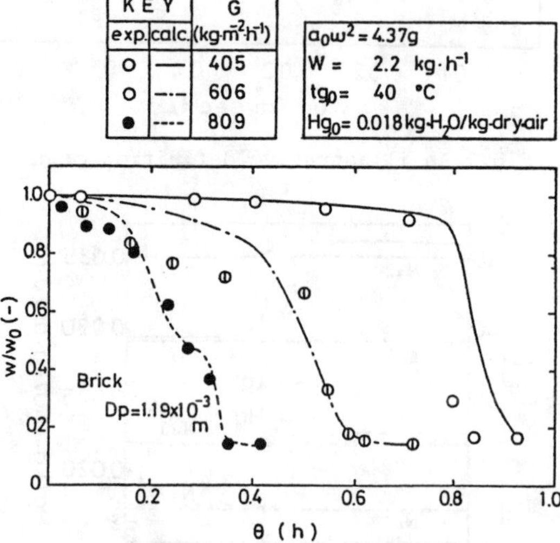

Fig.11 Effect of mass velocity of air

KEY exp.	KEY calc.	Dpx10³ (m)
●	---	0.62
○	—	1.19

Brick
$a_0\omega^2$= 4.37 g
W = 2.2 kg·h⁻¹
G = 405 kg·m⁻²·h⁻¹
t_{g0} = 40 °C
H_{g0} = 0.018 kg-H_2O/kg-dry·air

Fig.12 Effect of particle-diameter

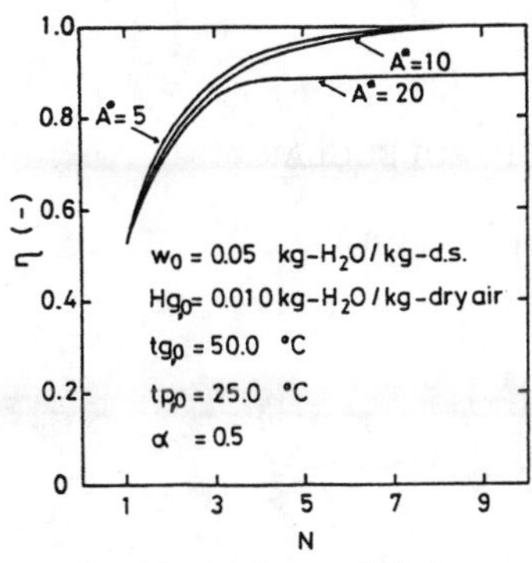

Fig.13 Drying efficiency (parameter calculation)

Effectiveness of multiplying the stages is confirmed by the calculated results shown in Fig.13. This is expected from the fact that heat of the heating air is used more effectively as N increases, even the low temperature air(40-60°C) being used. But the rate of increase of the drying efficiency decreases gradually with increasing the number of stages. Thus, it seems that an increase in the number of stages is not so effective under certain circumstanses. Besides η is virtually independent of A* when N is smaller than three. As long as the calculated results in Fig.13 are considered, a three-stage IFB would be recmmended as a promising dryer.

CONCLUSION

A mathematical model has been developed for the drying of granular particles in a multistage inclined fluidized bed (IFB) with mechanical vibration. The overall drying rate as well as, the steady state distributions for the temperatures of the particles and the heating gas and for the moisture content of the particles were calculated from numerical solutions of the basic equations based on the proposed model.

The experiments were conducted for the drying of brick particles in a three-stage IFB dryer with mechanical vibration.

The calculated results show satisfactory agreement with the measured data. In addition, the effectiveness of the application of mechanical vibration and multiplying the stages were evident from both experimentally and theoretically.

NOMENCLATURE

A = dimensionless parameter $(GC_g\ell\Delta x/WC_p)$ [-]
A^* = dimensionless parameter $(GC_g\ell L/WC_p)$ [-]
B = dimensionless parameter $(G\ell\Delta x/W)$ [-]
a_0 = half amplitude of vibration [m]
a_p = surface area per unit mass of particle $[m^2.kg^{-1}]$
C_g = specific heat of gas $[J.kg^{-1}.°C^{-1}]$
C_p = specific heat of particle $[J.kg^{-1}.°C^{-1}]$
C_w = specific heat of water $[J.kg^{-1}.°C^{-1}]$
D_p = diameter of particle [m]
G = mass flow rate of gas $[kg.m^{-2}.h^{-1}]$
H_g = absolute humidity of gas $[kg\text{-}H_2O/kg\text{-dry air}]$
H_p = saturation absolute humidity $[kg\text{-}H_2O/kg\text{-dry air}]$
h_p = heat transfer coefficient $[W.m^{-2}.°C^{-1}]$
k_p = mass transfer coefficient $[kg\text{-dry air}.m^{-2}.h^{-1}]$
L = total length of porous plate [m]
ℓ = width of porous plate [m]
N = total number of stages [-]
Nu_p = Nusselt number $[h_pD_p/\lambda_g]$ [-]
R_{ep}' = Reynolds number $[(u_g+u_s)D_p\rho_g/\mu_g]$ [-]
t_g = temperature of gas [°C]
t_p = temperature of particle [°C]
u_g = superficial velocity of gas $[m.h^{-1}]$
u_g^* = apparent superficial velocity of gas $[m.h^{-1}]$
$u_{g,mf}$ = minimum fluidizing velocity $[m.h^{-1}]$
u_s = linear velocity of particle $[m.h^{-1}]$
W = mass velocity of particle $[kg.h^{-1}]$
w = moisture content of particle $[kg\text{-}H_2O/kg\text{-dry solid}]$
x = distance from particle entrance [m]
α = dimensionless parameter $[1-\exp(-h_pa_pW/GC_g\ell u_s)]$ [-]
β = dimensionless parameter $[1-\exp(-k_pa_p\delta W/G\ell u_s)]$ [-]
Γ = dimensionless parameter [-]
γ = dimensionless parameter [-]
δ = dimensionless parameter [-]
η = drying efficiency [-]
θ = time or inclined angle [h or °]
λ = latent heat of vaporization $[J/kg\text{-}H_2O]$
λ_g = thermal conductivity of gas $[W.m^{-1}.°C^{-1}]$
μ_g = viscosity of gas $[kg.m^{-1}.h^{-1}]$
ρ_b = bulk density of bed $[kg.m^{-3}]$
ρ_g = density of gas $[kg.m^{-3}]$
ρ_s = density of particle $[kg.m^{-3}]$
ω = angular velocity $[rad.h^{-1}]$

Subscripts

0 = initial
av. = average
c = critical
e = equilibrium
f = final

REFERENCES

1. Gupta,R. and Mujumdar,A.S.,DRYING'80,Ed.A.S. Mujumdar,Hemishere-McGill-Hill,N.Y.,Vol.1,141 (1980)
2. Strumillo,C. and Pakowski,Z.,ibid.,211(1980)
3. Pakowski,Z.,Mujumdar,A.S. and Strumillo,C., Advances in Drying, Vol.3(1983)
4. Arai,N., Hasatani,M. and Sugiyama,S.,Kagaku Kogaku,Vol.35,565(1971)
5. Arai,N.,Fujishiro,T.and Sugiyama,S.,ibid.,Vol. 38,255(1974)
6. Sugiyama,S.,Hasatani,M.and Arai,N.,ibid.,Vol. 33,435(1969)
7. Arai,N.,Hasatani,M. and Sugiyama,S.,ibid.,Vol. 37(1973) also Heat Transfer Japanese Research, Vol.2,18(1973)
8. Arai,N.,Hasatani,M. and Sugiyama,S.,Kagaku Kogaku,Vol.34,649(1970)
9. Hasatani,M. and Arai,N.,Chem.Eng.Commun.,Vol. 10,223(1981)
10. Sugiyama,S.,Arai,N.,Shiga,A. and Kume,T., Kagaku Kogaku Ronbunshu,Vol.1,594(1975)
11. Shishido,I.,Ogino,M.,Suzuki,M. and Ohtani,S., ibid.,Vol.10,173(1984)
12. Arai,N.,Hasatani,M. and Sugiyama,S.,Kagaku Kogaku,Vol.36,181(1972)

COMBINED HEAT AND MASS TRANSFER DURING DRYING OF MOLECULAR SIEVE PELLETS

J. Schadl and A.B. Mersmann

Institute B for Chemical Engineering, Technical University of Munich, Federal Republik of Germany

ABSTRACT

For fundamental studies on sorption kinetics, gravimetrical experiments were carried out with spherical single pellets in a flow system to determine the rate limiting step (external mass transfer, macro- or micropore diffusion or heat transfer). By this way it is possible to investigate separately the influence of different parameters as pellet size, temperature, gas velocity and partial pressure jump. Good agreement with experimental results was obtained with a numerical calculated macropore diffusion-model taking into account the isotherm shape, which is favourable for adsorption but unfavourable for desorption. Mass transfer resistances in the boundary layer influenced only slightly the sorption kinetics. Here, heat effects are only of secondary influence as shown with numerical calculated and measured (with very thin thermocouples) temperature profiles versus time.

With regard to technical plants experiments were carried out in adiabatic fixed beds. At different heights of the column the temperature profiles (axial and radial) and concentration profiles were measured. Results for adsorption of H_2O and CO_2 from air and thermal desorption as well as purge gas desorption are presented. With high feed-concentrations a first mass transfer and second heat controlled zone were obtained. Here, for numerical calculation mass and energy balance as well as transfer equations must be solved simultaneously. With thermal desorption, cycle time, heat and air requirements for drying of ms-fixed beds can be calculated approximately with a heat balance.

1. INTRODUCTION

Molecular sieves (ms) are widely used in chemical plants. By drying of air or purification of buffer gas, for instance, CO_2 and H_2O are removed by ms-adsorbers. Often the expenditures for desorption are decisive for the economy of the whole process. For design of an adsorption process, both equilibrium and kinetics data are required. Contrary to adsorption, only a few studies on desorption, respectively drying, were carried out. In principle three different procedures exist: temperature swing, pressure swing and displacement desorption. In practice these processes are mostly overlapped and often it is difficult to distinguish between them. The objective of our experients was to obtain fundamental results for adiabatic adsorption as well as the purging step of pressure swing desorption and thermal regeneration.

For adsorption, diffusion in the macropores of the porous adsorbents as the rate controlling step is commonly confirmed [1,2]. The adsorption step resistance itself is negligible. For desorption, however, the findings are contradictory. Also the desorption step itself as the rate controlling factor is found [3,4]. Thermoswing processes are calculated with a heat balance only [5]. With the "equilibrium Theory", no mass and heat transfer resistances are considered [6,7].

2. DESCRIPTION OF MATERIAL

Molecular sieves belong to a class of compounds known as zeolites. In contrast to other adsorbents the micropores of zeolites are of uniform size. Fig. 1 shows the crystalline structure of zeolites type A and X [8].

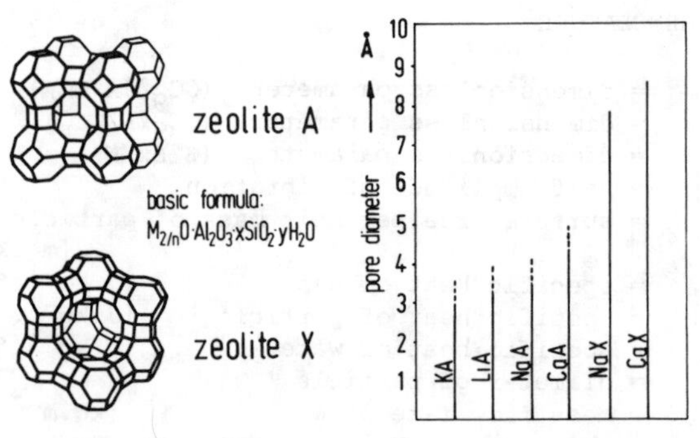

Fig. 1: Crystalline structure and micropore diameters of zeolites

Depending on the pore size, molecules may be adsorbed or completely excluded. Here, only ms 5 Å (CaA) is investigated.

For most technical uses of molecular sieves, the very small zeolite crystals (1-10 µm) are pelletized by means of a clay binder and are supplied in a bead form (dmr. 1-5 mm). By this production process, molecular sieve pellets obtain a very characteristic bimodal pore structure [9]. The adsorbate molecules diffuse through this macropore system. In figure 2 the macropore size distribution of two fractions is shown, measured by mercury penetration. The average pore diameter and free lenght of path are in the same order of magnitude. In this case, the pore diffusion coefficient is in the transition region between Knudsen (D_K) and free gas diffusion (D_G) [10].

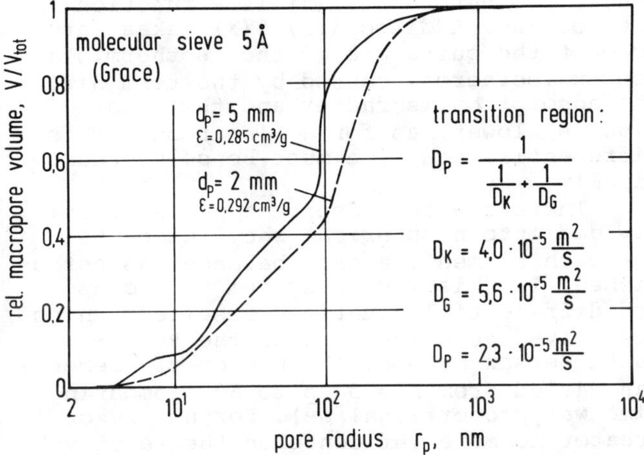

Fig. 2: Macropore radius distribution and diffusivity (values: H_2O in air, 200 °C, 1 bar, $\bar{r}_p$ = 80 nm)

3. EQUILIBRIUM ISOTHERMS

Adsorption systems are characterized at equilibrium by the field of isotherms. Besides the primary information about the adsorptive capacity X = f(p, ϑ) the curvature of the isotherm reveals the forces of attraction.

On account of the polarity of water, the H_2O-isotherms are highly convex. Because at ambient temperature, molecular sieves have an almost constant adsorptive capacity irrespective of the vapor pressure, with this system the purge gas desorption experiments must be carried out at higher temperatures. The very different driving force for desorption is revealed by the isosteres: for example, at ϑ = 50 °C and X = 0.02 kg ad/kg ms, the equilibrium partial pressure is about 10^{-3} mbar for H_2O whereas it is 3.5 mbar for CO_2. In the special case of H_2O-preadsorption the CO_2-isotherm (reduced CO_2-capacity) and therefore the sorption rates are influenced [11] In practice, this is important for thermal regeneration with humid air.

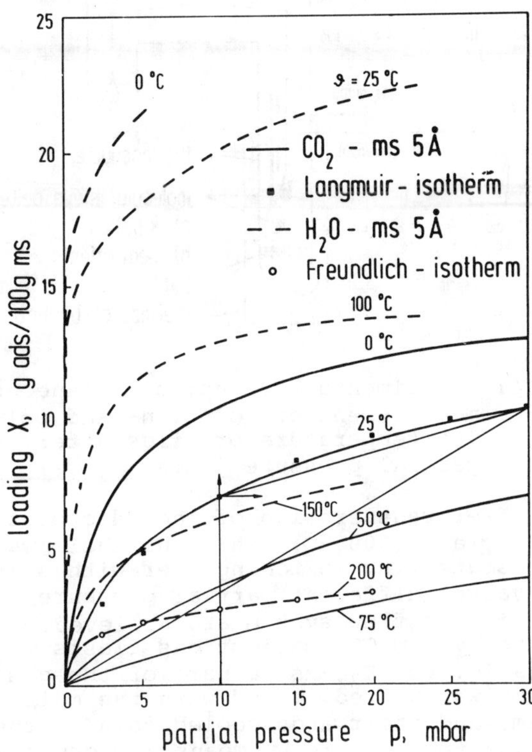

Fig. 3:
Adsorption isotherms of investigated systems
continuous and broken lines: measured points: correlated with Langmuir-equation (CO_2-ms 5 Å), respectively Freundlich-equation (H_2O-ms 5 Å)
thin lines: influence of pressure jump on the deviation of the isotherm from a linear one.

4. SINGLE PELLET STUDIES

4.1 Apparatus and Procedure

By the experiments with single pellets in a flow system at ambient pressure the sorption and temperature profiles were measured simultaneously. Commercial spherical ms-pellets (Grace) were used. The change of loading was measured by means of an electronic microbalance; the temperature change was measured with very thin thermocouples, imbedded within the area of fracture of split pellets, recemented after cautious wetting with the original mixture of molecular sieve powder and alumina binder. Very thin NiCr-Ni-wires (30 µm in diameter) with small mass compared to pellet mass and with low thermal conductivity were used. The preparation influenced the sorption kinetics of the pellet only very slightly [11].

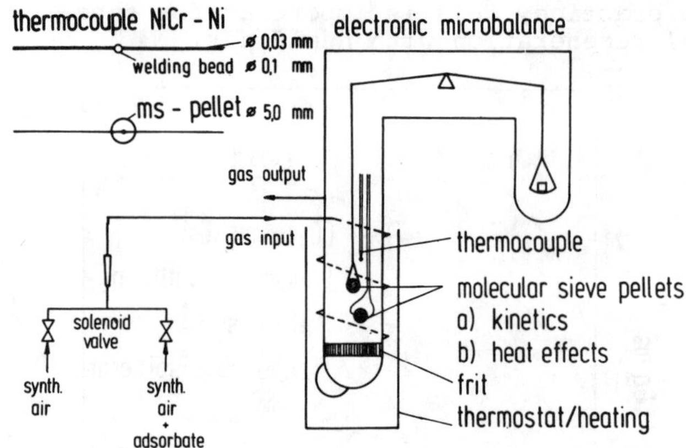

Fig. 4: Experimental set-up: simultaneous measurement of sorption- and pellet temperature profiles after partial pressure jumps

After regeneration of the adsorbent with dry air (300 °C, 3 h), the shell was thermostated. By switching over with solenoid valves different partial pressure jumps in the flow system are achieved. For the system CO_2-ms test gas flasks were available. For H_2O-ms, a part of the gas stream was wetted, mixed with the main stream, and afterwards cooled down to the dewpoint temperature by means of a cryostat. To achieve a step change (partial pressure jump) two parallel systems are necessary. Additionally the dewpoint was controlled by a dew-point meter (MBW). Changes in weight and temperature were recorded with a continuous-line-recorder.

The variables studied were: pellet diameter d_p (2.5 to 5 mm); superficial gas velocity v_G (0.05 to 0.25 m/s) temperature (25 to 200 °C) and partial pressure jump (CO_2: 0, 10, 30 mbar, H_2O: dry gas up to 20 °C dew point, reciprocal for desorption).

4.2 Nonsteady Transport Processes

Adsorption and desorption processes are caused by the nonequilibrium of the system. If a partial pressure jump takes place in the ambient atmosphere of a pellet, the adsorbate diffuses through the boundary layer in the pores, respectively to the surface of the pellet for desorption. The following steps can limit the sorption rate:

a) external mass transfer through the boundary layer,
b) diffusion in the macropore system,
c) diffusion in the micropores,
d) the sorption step itself,
e) internal or external heat transfer.

The effective molecular diameter of the gas components are nearly the same [12] $d_{m,CO_2} = 2.8$ Å, $d_{m,H_2O} = 2.6$ Å compared with a zeolite pore diameter of about 5 Å. Therefore the activation energy for zeolitic diffusion is very low.

Assuming macropore diffusion after a partial pressure jump, the equation for nonsteady diffusion in a sphere is valid. An approximate solution exists only for the spherical case of a linear isotherm [13].

However, for technical systems, the isotherm curvature is mostly very favourable for adsorption, but unfavourable for desorption. For these nonlinear systems Equation (2) is valid [1].

$$\frac{\partial X}{\partial t} = \frac{D_p/\mu_p}{1+\alpha(X)}\left[\frac{\partial^2 X}{\partial r^2} + \frac{2}{r}\left(\frac{\partial X}{\partial r}\right) + \beta(X)\left(\frac{\partial X}{\partial r}\right)^2\right] \quad (1)$$

$$\alpha(X) = \frac{\rho_{ad}}{\varepsilon_p}\frac{\mathcal{R}T}{M_{ad}}\cdot\frac{dX}{dp}; \quad \beta(X) = -\frac{d^2X}{dp^2}\Big/\left(\frac{dX}{dp}\right)^2$$

Instead of the constant slope for a linear isotherm here $\alpha(X)$ is a function of the loading. Additionally $\beta(X)$ takes into account the curvature of the isotherm. For convex isotherms, caused by the term $\beta(X)$ the adsorption is faster and the desorption is slower, as for a linear one. This differential equation must be solved numerically.

Therefore the adsorbent pellet was divided into n spherical shell elements. For each element, a mass balance was established. In this way, a system of n coupled differential equations resulted, which was solved numerically with the Runge-Kutta-Merson-method. The number of elements was varied from n = 5 up to 80 (computing time was proportional n^3). For n = 20 or greater no more influence on the relative loading X/X_{max} up to a three digit accuracy was determined.

The measured isotherms were approximated by a Freundlich - equation for the system H_2O-ms 5 Å, respectively by a Langmuir-equation for CO_2-ms 5 Å (see figure 3) to obtain a differentiable function for reason of computation.

In general, transport by gas phase diffusion and in the adsorbed phase is possible. Here only gas phase diffusion was considered. Caused by the production process, the walls of the macropores are mainly composed of the clay binder, where only small quantities of H_2O and CO_2 are adsorbed. Transport in the adsorbed phase, however, occurs only with higher degrees of loading. Also it was assumed no capillary condensation to occur.

The mass transfer in the boundary layer was calculated in analogy to heat transfer equations [14].

$$Sh = 2 + \sqrt{Sh_{lam}^2 + Sh_{turb}^2} \quad (2)$$

$$Sh_{lam} = 0.664\sqrt{Re}\;\sqrt[3]{Sc} \quad (3)$$

$$Sh_{turb} = \frac{0.037\,Re^{0.8}\,Sc}{1 + 2.44\,Re^{-0.1}(Sh^{2/3} - 1)} \quad (4)$$

4.3 Results and Discussion

As shown in figure 5, good agreement between experimental and theoretical results was obtained.
The tortuosity factor $\mu_p = 5$ was also found by other authors [1] to be in the same order of magnitude.

In case of pore diffusion controlled sorption, a proportionality

$$\Delta X/\Delta X_{max} \sim \sqrt{t} / R$$

for the change in saturation for short times is valid. Sometimes in the beginning there exist deviations.
Firstly, at very high mass transfer rates an influence of external mass transfer in the boundary layer is observed, which slows down the sorption process. Secondly, especially for the adsorption process at low temperatures, a part of the preadsorbed nitrogen from the carrier gas air is displaced by H_2O or CO_2. Therefore, no change in weight can be measured.

The remarkable slower desorption rate after identical partial pressure jumps in the ambient atmosphere, in comparison to adsorption, is explicable, if the numerically calculated local profiles of partial pressure and loading within the pellet are compared (figure 6). Assuming diffusion in the macropores and film resistance to be the limiting factors, very different profiles result, caused by the unfavourable isotherm curvature for desorption. ($d^2X/dp^2 < 0$)

At the same arbitrary chosen degree of loading change, the partial pressure profiles within the pellet illustrate that the pressure gradient is remarkably greater for adsorption in comparison to desorption. For desorption, the convex curvature of the Freundlich equation causes very flat loading profiles inside the pellet and a fast decrease of the driving partial pressure difference.

For further illustration the calculated partial pressure profiles within a pellet are shown at 20 time steps (Figure 7).

In contrast, for linear isotherms at the same degree of change in loading homologous profiles for local partial pressure and loading within the pellet are obtained [11].

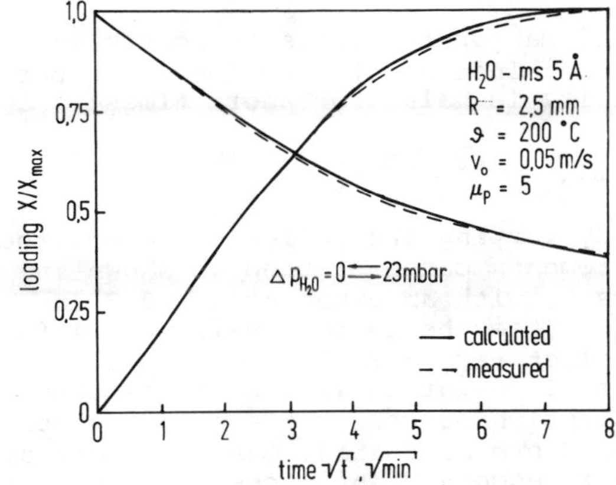

Fig. 5: Sorption profiles with spherical single pellets (calculated with combined macropore diffusion and boundary layer model)

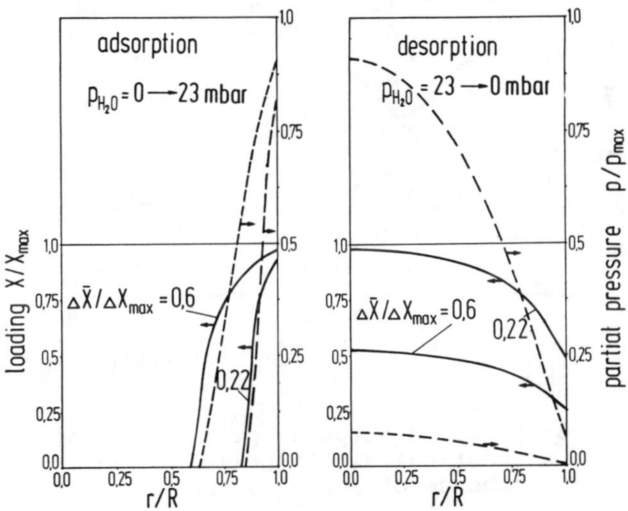

Fig. 6: Calculated local partial pressure and adsorbent loading profiles inside a single pellet, showing the influence of shape of adsorption isotherm (datas see Fig. 5)

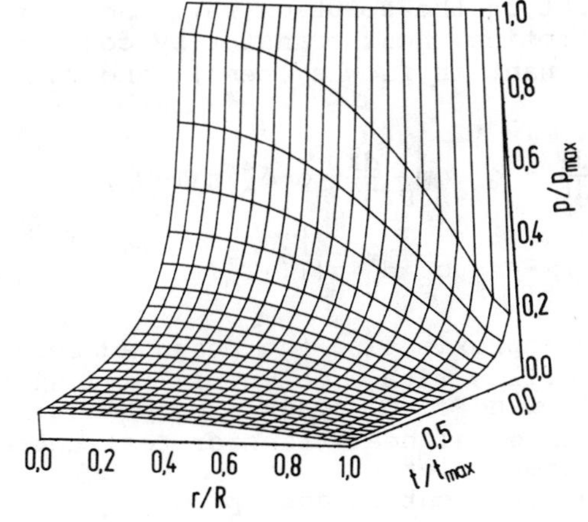

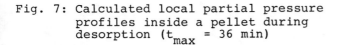

Fig. 7: Calculated local partial pressure profiles inside a pellet during desorption (t_{max} = 36 min)

4.4 Experimental Verification

If macropore diffusion controls the sorption kinetics, the following proportionality is valid for short times:

$$\Delta X/\Delta X_{max} \sim \sqrt{t} / R^n$$

By varying the pellet diameter, the measurements can be fitted as shown in figure 8, with an exponent n = 0.9 instead of the theoretical value n = 1 for the pellet radius R.
The small deviation also confirms that the sorption processes are mainly governed by macropore diffusion. Other effects are only of secondary influence. Analogous results are obtained with measurements carried out with the system CO_2-ms 5 Å.

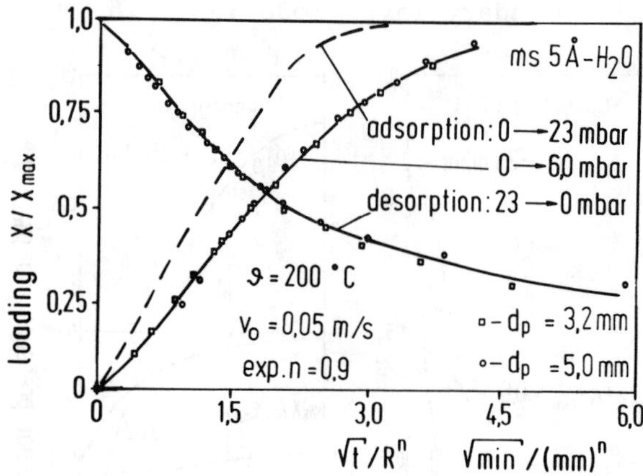

Fig. 8: Measured sorption profiles with spherical single pellets; influence of pellet diameter

4.5 Temperature Effects

As shown in figure 4 sorption and temperature profiles of single pellets are measured simultaneously. The change of temperature as a function of time is dependent on the sorption rate and heat of adsorption, heat transfer by convection (predominant in flow systems), and radiation.

$$m_p c_p \frac{dT}{dt} = q_{ad} m_p \frac{d\bar{X}}{dt} - \alpha d_p^2 \pi (T-T_o) - \varepsilon_p c_s d_p^2 \pi (T^4 - T_o^4) \qquad (5)$$

By approximating the measured sorption profiles with a polynomial and under the follqwing simplyfying assumptions:

a) no internal heat transfer resistence in the pellet
b) constant heat of adsorption
c) ideal flow of gas around the single pellet
d) constant temperature of the carrier gas

the temperature profile of the pellet was calculated numerically by the Runge-Kutta-method.

In contrast to experimental results in a pure gas atmosphere (vacuum systems), in flow systems the temperature effects, especially for desorption, are negligible because of the better heat transfer. An essential hindrance [11] or even limitation [15] of the sorption by the heat transfer as in a pure gas atmosphere, didn't occur with single pellets in the flow system. Figure 3 shows that for the adsorption of CO_2 (0 to 3% step change), a 5 °C change in the adsorbent temperature changes ΔX_{max} from 10 to 9.7 g/100 g only. To a first approximation, an isothermal calculation is permissible.

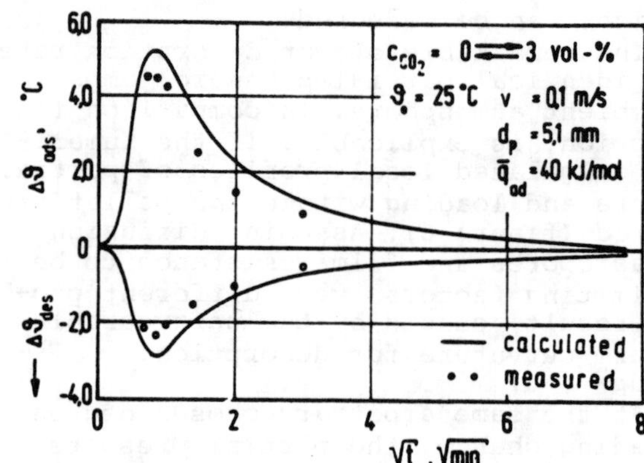

Fig. 9: Temperature profiles of single pellets during sorption

During sorption of H_2O ($\bar{q}_{ad}$ = 60 kJ/mol H_2O) (data see Figure 5) a maximum temperature rise of 11 °C, respectively -decrease of 6 °C was calculated.

5. ADIABATIC FIXED-BED STUDIES

5.1 Apparatus and Procedure

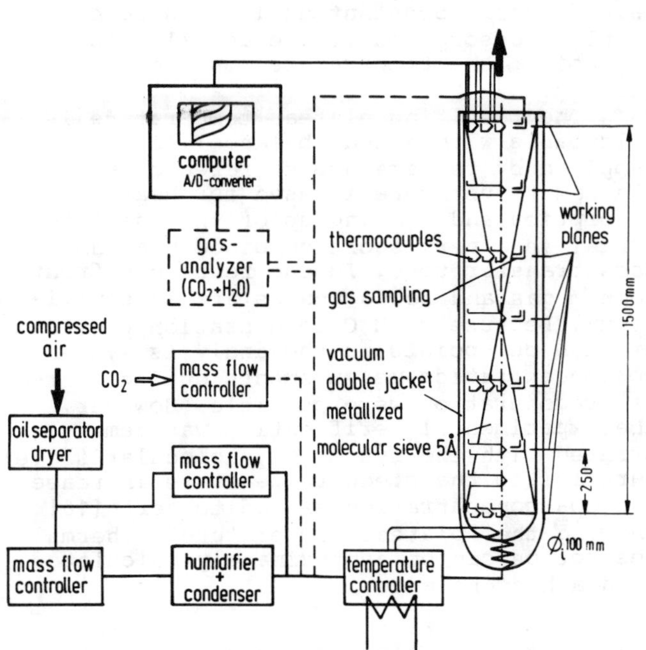

Fig. 10: Experimental set-up to measure concentration and temperature profiles (axial and radial) at different working planes

With regard to technical adsorbers an adiabatic fixed-bed was chosen for the experiments. Two different constructions were tested. On one hand a very thin stainless steel tube with small heat capacity and heat conductivity (thickness of wall 0.25 mm, diameter 100 mm, length 1.5 m) was insulated with different insulation materials (thickness of layer 100 mm). Secondly a modified Dewar-glass double jacket was used as shown in figure 10.

First of all, the adiabaticity of the column was tested. Therefore heat transfer experiments by heating up with a hot, dry air stream were carried out. In Fig. 11 the temperature profiles (axial and radial) versus time at different working planes are plotted.

The comparison of the two different constructions showed that the modified Dewar-flask was advantageous. Firstly, the measured heat losses, especially during thermal regeneration, were smaller. Secondly, within the layer of insulation material a temperature profile occured. Concerning the storage term in the heat balance it is necessary to integrate over the thickness of the layer. With the Dewar-flask it can be assumed that the inner jacket is at the fixed-bed temperature, whereas the outer jacket is nearly cold.

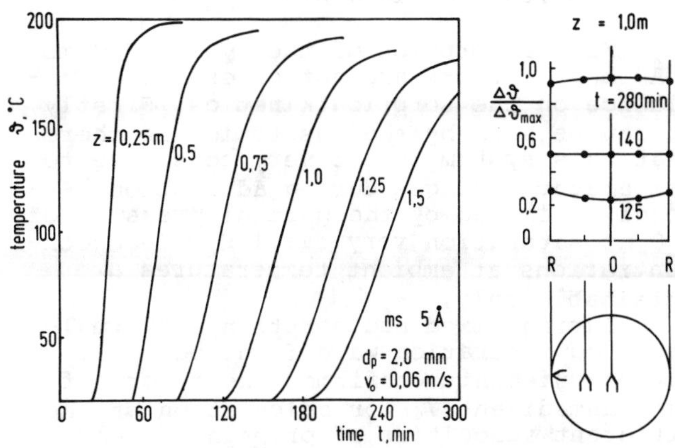

Fig. 11: Adiabacity of the column: axial and radial temperature profiles

With increasing fixed-bed length the slope of the temperature profiles becomes flatter, caused by the axial heat dispersion in the bed. The radial profiles show, that at first a faster temperature increase at the column wall takes place. The reasons are higher thermal conductivity of glas wall compared with the ms-pellets as well as marginal effects: higher gas velocity in the wall region ($\varepsilon \rightarrow 1$). Here, the superficial gas velocity is very low. With our sorption experiments smaller temperature differences occured.

The purified and dried air stream was adjusted with a mass flow controller. For adsorption, a part of the stream was wetted, respectively CO_2 fed in, and before entering the column the gas stream was thermostated. Alternately the feed concentration and the concentration profiles at the different working planes were measured. Axial and radial temperature profiles were recorded continuously.

At first, the temperature data were recorded with a 12-point-recorder, the concentration profiles with a line-recorder. Now an IBM-Computer is installed. With help of a multiplexed A/D-converter the analog signals of the thermocouples and the gas analyser are digitized and stored on disks. Also the solenoid valves of the switch selector are operated by the processor. In this way the plotting of experimental results can be done by the computer.

The variables studied were: pellet diameter $\bar{d}_p$ (2.0 and 3.8 mm), superficial gas velocity (0.05 to 0.4 m/s) and feed concentration (CO_2: 0.3 to 10 vol-%, H_2O: up to 20 °C dew-point).

5.2 Results and Discussion

With adsorption of the system H_2O-ms 5 Å, heat effects are not of essential influence on the sorption kinetics. Firstly, it can be seen by the adsorption isotherms that this system is not very sensitive to temperature changes during adsorption. Secondly, limited by the partial pressure of H_2O at saturation very small H_2O-feed concentrations at ambient temperatures are attainable only.

During fixed-bed sorption a thermal- and a concentration wave front occured. Our measurements confirmed the theory of Pan/Basmadijan [7] for calculation of the different velocities of propagation and maximum temperature changes.

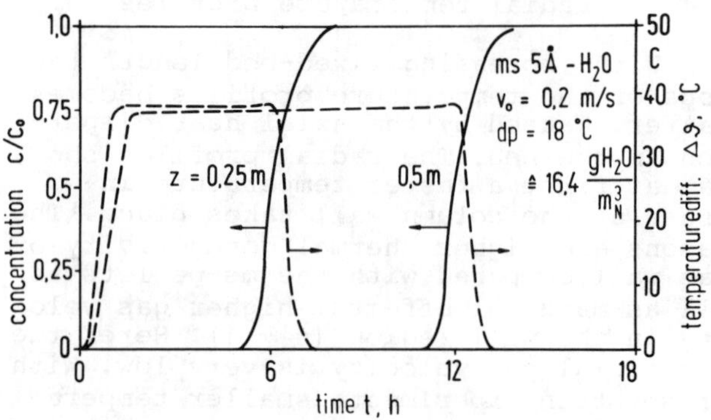

Fig. 12: Concentration and temperature profiles during adiabatic fixed-bed adsorption

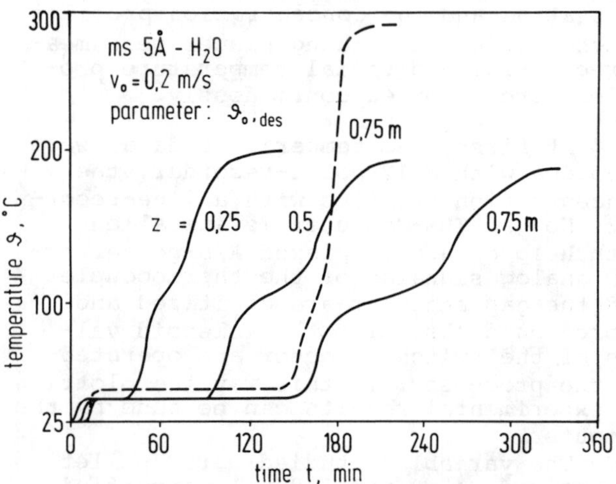

Fig. 13: Thermal desorption of an adiabatic fixed-bed (X = 22 g H_2O/ 100 g ms)

Caused by the polarity of the H_2O molecules, desorption, respectively drying, is only possible by thermal regeneration. After complete saturation, the cold, loaded fixed-bed was heated up with a hot, dry gas stream.

Resulting temperature profiles versus time at different drying temperature are plotted in Fig. 13.

Within a short time the bed temperature rose slightly. This temperature remained nearly constant until, caused by complete desorption at the actual measuring points, the temperature rose very steeply.

The resulting plateau temperature is calculable with a heat balance: All heat supplied by temperature difference feed - plateau temperature is used for heat of desorption and heating up of the inert material in a very short combined heat and mass transfer zone. In the rear and front zone, gas and solid phases are in equilibrium. Because of H_2O condensation problems at high dew-points in the analysis system no concentration measurements during thermal regeneration were possible. However, the experimental verification was demonstrated with the system CO_2-molecular sieve: parallel to the steep temperature increase the CO_2-concentration sagged to zero [16]. Knowing the "plateau temperature", thermal desorption can be approximately calculated with a heat balance.

5.3 Modelling of Fixed-Bed Sorption

For numerical calculation of the adiabatic sorption both mass and heat balance and the transfer equations must be solved simultaneously:

mass balance:

$$\varepsilon \, \rho_G \frac{\partial c}{\partial t} = - \dot{m}_G \frac{\partial c}{\partial z} + \varepsilon \, \rho_G \, D_{ax} \frac{\partial^2 c}{\partial t^2} - (1-\varepsilon)\,\rho_S \frac{\partial \bar{X}}{\partial t} \qquad (6)$$

The term for the axial dispersion is not necessary in physical view, but serves for the numerical calculation of the whole system of differential equations. A stable solution without changing the profile is attainable within shorter computing times

mass transfer:

$$\frac{\partial \bar{X}}{\partial t} = \frac{15 \, D_{eff}}{R^2} \left[X_B(c, \vartheta_S) - \bar{X} \right] \qquad (7)$$

In first approximation, with the simplified "driving force" model film resistances in boundary layer are neglected.

For calculation of the bed temperature the so-called single-phase-model is used; there heat transfer is included in the axial dispersion term

heat balance and transfer:

$$(1-\varepsilon)\rho_S\, c_S \frac{\partial\vartheta}{\partial t} = - \dot{m}_G\, c_G \frac{\partial\vartheta}{\partial z} +$$

$$\left(\lambda_o^{eff} + \frac{\dot{m}_G^2\, c_G^2}{\alpha\, a_p}\right) \frac{\partial\vartheta}{\partial z} - \rho_G(1-\varepsilon)\, \bar{q}_{ad} \frac{\partial\bar{X}}{\partial t} \qquad (8)$$

This coupled system of differential equations is solved with a library-programme on a Cyber 170/175.

Up till now a severe problem is to find a simple mathematical description for the field of isotherms for Eq. (7)

$$X_B = f(c, \vartheta)$$

With a very rough approximation for the system CO_2-ms

$$X_B = k\, c/\vartheta \qquad (9)$$

only qualitative agreement between theory and experiment is attainable. For more complicated equations, i.e. extended Langmuir-equation for CO_2-ms 5 A and Freundlich-equation for H_2O-ms 5 A, computingtimes are still too long. A more detailed report on the numerical solution will be published later.

Acknowledgement

Financial support of this project by the Deutsche Forschungsgesellschaft, Sonderforschungsbereich 153, is gratefully acknowledged.

NOTATION

a_p	specific surface area	1/m
c	concentration	kg/m^3
c_p	specific heat capacity	J/(kg K)
c_S	radiation constant	$W/(m^2\ K^4)$
d	diameter	m
D	diffusivity	m^2/s
D_{eff}	effective diffusivity	m^2/s
m	mass flux density	$kg/(m^2 s)$
M	molecular mass	kg/kmol
p	pressure	N/m^2
q	heat of adsorption	J/kg
r,R	radius	m
$\mathcal{R}$	universal gas constant	J/(mol K)
t	time	s
T, ϑ	temperature	K, C
v	superficial gas velocity	m/s
X	adsorbate loading	g/g
z,Z	height of fixed-bed	m
α	heat transfer coefficient	W/(m s)
β	mass transfer coefficient	m/s
ε	void volume fraction	-
μ_p	tortuosity factor	-
ρ	density	kg/m^3
λ_{ax}	axial heat conductivity	W/(m K)
Δ	difference	-

Indices

ad	adsorbate
ads	adsorption
des	desorption
p	pellet
o	ambient
B	bulk
G	gas
S	solid
max	maximum, equilibrium

$Re = v\, d/\nu$	Reynolds number
$Sc = \nu/D$	Schmidt number
$Sh = \beta\, d/D$	Sherwood number

REFERENCES

1. Jokisch, F.: Dissertation, TU Darmstadt, FRG (1975)
2. Mersmann, A., Münstermann, U., Schadl, J.: Ger. Chem. Eng., in print
3. Fukunaka, P., Hwang K.C., et.al.: I.E.C. Proc.Des.Dev. 7(2), 269-275, (1968)
4. Seewald, H., Jüntgen, H.: Ber. Bunsenges. phys. Chem., 81(7), 638-645, (1977)
5. Chi, C.W.: AIChE Symp. Ser., 74(179), 42-46, (1978)
6. Pan, C.Y., Basmadjian, D.: Chem. Eng. Sci. 25, 1653-1664, (1970)
7. Basmadjian, D., Ha, K.D., Proulx, D.: Ind. Eng. Chem. Proc. Des. Dev., 14(13), 340-347, (1975)
8. Breck, D.W.: Chem. Eng. Prog. 73(10), 44-48, (1977)
9. Ullmanns Enzyklopädie techn. Chemie, Vol. 2, Verlag Chemie, Weinheim (1972)
10. Satterfield, O.M.: Mass Transfer in Heterogenous Catalysis, M.I.T. Press, Cambridge, Mass. (1970)
11. Münstermann, U., Mersmann, A.; Schadl, J.: Ger. Chem. Eng. 6, 1-8, (1983)
12. Grubner, O., Jiru, P., Ralek, M.: Molekularsiebe, VEB Verlag, Berlin (1968)
13. Crank, J.: The Mathematics of Diffusion, Claredon Press, Oxford (1975)
14. Martin, H.: Chem. Ing. Techn. 52(3), 199-209, (1980)
15. Doelle, H.J.: Dissertation, TU Karlsruhe, FRG (1978)
16. Schadl, J., Mersmann, A.: Proc. 1st Int. Conf. Fundamentals of Adsorption, Ellmau, (1983) in print.

AIR DRYING KINETICS OF BIOLOGICAL PARTICLES

J.J. Bimbenet, J.D. Daudin and E. Wolff

ENSIA (Ecole Nationale Supérieure des Industries
Agricoles et Alimentaires)
INRA (Institut National de la Recherche Agronomique)
1 avenue des Olympiades - F 91305 Massy France

ABSTRACT

Drying kinetics of several biological products (carrots, potatoes, corn, plums, coconut, parsley, cassava, ...) have been measured in warm air of various characteristics (temperature was also determined in certain cases).

The obtained curves (drying rate vs product water content), as well as the evolution of the product temperature, showed no constant-rate period in most cases. Certain products exhibited a very marked initial temperature warming-up period, associated with an increasing drying rate. The classical interpretation of the falling-rate period in two distinct parts was not confirmed for most products.

Discussion of the factors of these kinetics :
- air temperature always has a strong influence which may often be interpreted by an Arrhenius law
- air humidity has very different influence according to the type of product : negligible in certain cases, important in others, but always more marked during the beginning of the drying
- air velocity influence may be described in comparable terms
- when possible, the influence of product size was measured and appeared to be strong
- but most of all, the nature of the product was the main factor : one can speak of a "personality" of each product as to its drying behavior.

In certain cases only the shape of the drying curve could be interpreted by a diffusion law with a constant diffusivity.

It is therefore concluded that no theory can presently predict drying kinetics of biological products. Experimentation remains essential in this field.

1. INTRODUCTION

Reasons for measuring drying kinetics may be :
- fundamental : give experimental support to modelization of transfer phenomena during drying
- practical : provide a starting point to calculation of industrial drying processes, specially by simulation (e.g. Daudin and Bimbenet -4-).

This paper considers the case of biological particles dryed in warm air.

2. METHODS

The basic idea is to perform a drying test in "pure conditions", i.e. all the product being submitted to the same conditions. Many types of equipment and procedure are used. Our own methods are described in the above-mentioned publication (4). They consist in drying a thin layer of product lying on a tray, weighing it at regular intervals during interruptions of the air flow. Automatic treatment of the data furnishes curves of the type shown further. The temperature of the product may be measured by means of an insulated thermocouple inserted in the center of a piece of product located in the warm air flow.

3. SHAPE OF DRYING RATE CURVES

3.1. Do we find a warming-up period ?

Certain curves given further (e.g. fig. 8, 10, 12) exhibit a very marked warming-up period. It must however be observed that this period generally concerns a short part of the drying process. In other cases, no such period is clearly found.

The existence and the importance of this period is related to :
- product size (the bigger the product, the longer that period). Comparison of fig. 1 and 2 is clear at that respect

- initial product temperature and air characteristics : the farther is the product temperature from the initial "equilibrium" product temperature, the more pronounced will be that period. Since the "equilibria" product temperatures are related to the position of iso-enthalpic lines on the wet-air diagram (Roth and Loncin -13-), for a given initial temperature of a cold product, the warmer and wetter the air, the longer the warming-up period.

Fig. 1 and 2 clearly show the influence of air humidity on temperature rise. It must also be noticed that this "warming-up" period may as well be a "cooling-down" period, especially in low-temperature and dry air (cf fig. 8)

- weighing frequencies : coconut albumen ($\theta_o \simeq 20°C$) in shells of about 1 cm thickness in air at 100°C should show a warming-up period. It does not ... because weighing period is one hour. Fig. 2 shows a warming-up period of about 5 min. This period does not appear on fig. 6, since weighing period was about 10 min. Same, the mode of calculation of derivatives may accentuate or "damp" that period on the drying rate curve.

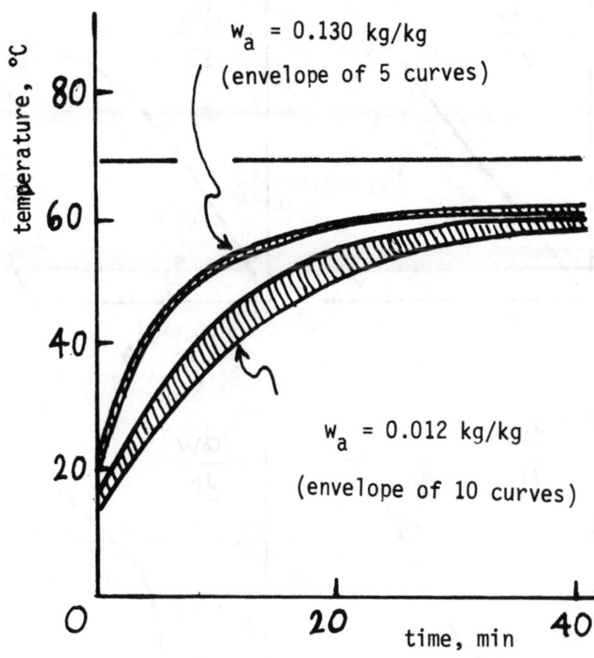

Fig. 1 Evolution of temperature at the center of product : prunes, 31-35 mm diameter, θ = 70°C, u = 0.7 m/s, two values of w_a (Daudin and Bimbenet -7-)

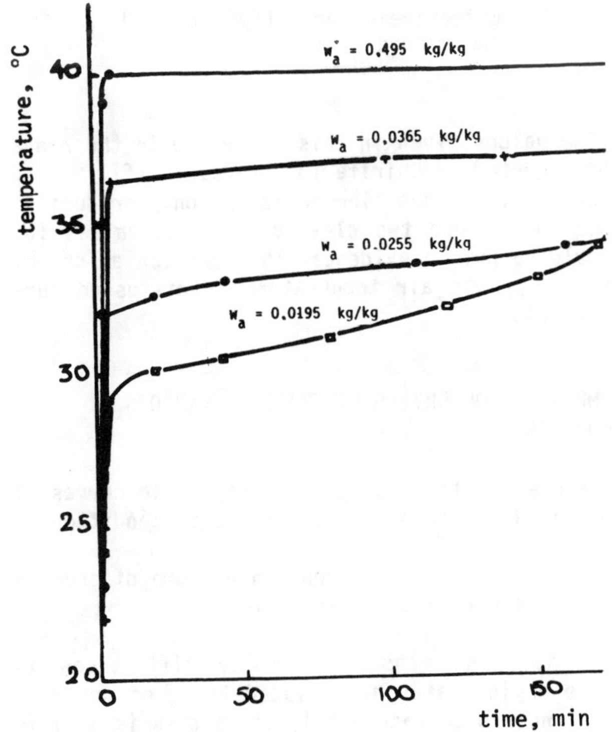

Fig. 2 Evolution of temperature at the center of product ; carrot cubes, initially 1x1x1 cm^3, θ = 40°C, u = 1 m/s, four values of w_a (Wolff -17-)

3.2. Do we find a constant-rate period ?

Such a period never appeared on the products we tried, even at low air temperature (40°C). However, the horizontal tangent to the curve we observe for most products may be interpreted as :

- the embryo of a constant-rate period, as showed by Fornell et al. (10) for sufficiently wet products (fruit and vegetables)
- the intersection between a warming-up and a falling-rate periods, in the case of corn, where the initial drying rate is far from a constant rate value.

We therefore completely agree with Chirife (3) for whom the existence of a constant rate period in food drying is rather exceptional.

3.3. The falling-rate period

Van Brackel (16) presents the various shapes of drying-rate curves. Our own experiments lead us to consider, so far, four types of curves (fig. 3) as illustrated by the further given curves. Shapes "a" and "c" seem to be the most common. Shape "b" has only been found in certain drying curves of prune and "d" for coconut albumen. The presence of fats, in this latter case, may explain that steep fall.

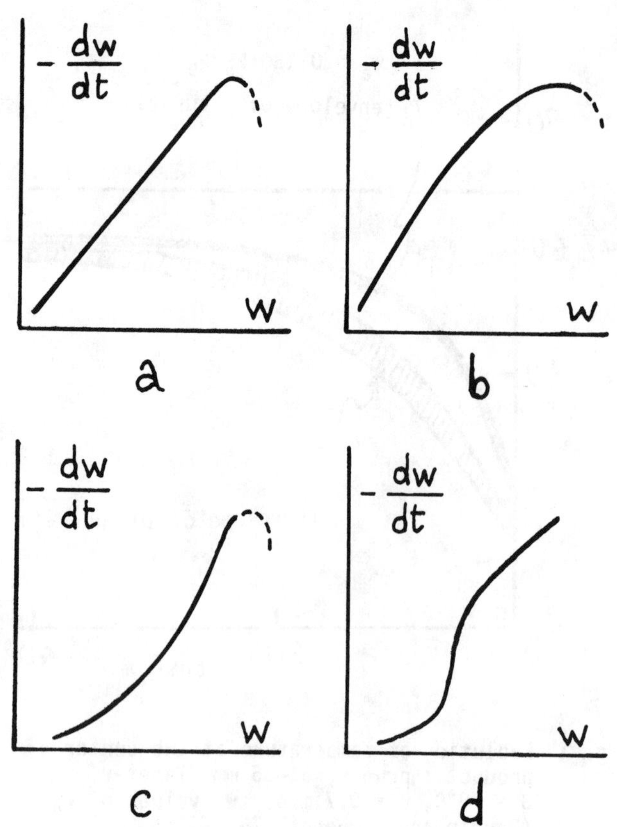

Fig. 3 Types of drying curves observed in the laboratory

	θ°C	D_{eff} m^2/s
Carrot cubes, initially 1x1x1cm^3	40	6.75 x 10^{-10}
	60	12.1 x 10^{-10}
	80	17.9 x 10^{-10}
	100	24.1 x 10^{-10}
Coconut albumen (calculation for 1 cm thickness) w = 0.6 kg/kg	45	4.6 x 10^{-10}
	60	7.7 x 10^{-10}
	70	8.2 x 10^{-10}
	80	10.4 x 10^{-10}
	90	10.8 x 10^{-10}
	100	11.8 x 10^{-10}
	110	12.8 x 10^{-10}
w = 0.2	45	1.0 x 10^{-10}
	80	3.3 x 10^{-10}
	110	6.6 x 10^{-10}

Table 1. Effective water diffusivities D_{eff} for two products.

Curves of shape "a" only can be interpreted by a diffusion law. Daudin (6) gives in details the hypotheses which allow the calculation of an effective water diffusivity. In addition to that, if the curves tend towards the origin of coordinates, which is practically the case for several products we tested, we have to consider that the equilibrium water content w_e is null. In fact, the use of an "exact" value for w_e is very difficult, since we have very few sorption curves at temperatures of the end of drying.

In those conditions, we calculated the effective water diffusivity for carrot cubes, from results of fig. 6 (table 1).

For coconut albumen, the product shape was not that simple. The calculation was done considering an infinite flat plate of 1 cm thickness, which is inexact. Worse, the shape of the curves does not correspond to the straight line of a diffusion law. However, we were interested in comparing effective diffusivity before and after the drying rate fall. So, considering the slope of straight lines from the origin, we obtained the values of table 1, in which appears a strong difference of diffusivity between the beginning and the end of drying.

The values given in this table are in the range of those quoted by Chirife (3) or Daudin (6).

However, coconut albumen is the only product we found so far, were two clearly distinct values for D_{eff} could be put in evidence. The question arisen by relating D_{eff} to air temperature is discussed further in 5.1.

4. COMPARISON OF DRYING KINETICS OF VARIOUS PRODUCTS

We drew on the same graph drying-rate curves of various products dried in same or close conditions.

The result (fig. 4) shows an extreme discrepancy in the magnitude of drying rates.

- The first reason for these differences is that the drying rates are related to kg of dry matter (and not to surface unit), since this is a basis very easy to measure and convenient to use in simulations. Therefore, large differences of specific surfaces (m^2/kg dry matter) result in large differences in drying rate, expressed per kg dry matter. When Fornell (9) expresses drying rates of apple cubes of various sizes as related to surface (fig. 5), the curves appear to be close to one another.

Vaccarezza et al. (15) take this same effect in account by plotting t/l^2 instead of t on their drying curves, l being a characteristic length of the product. This effet of specific surface explains for example the position of the parsley curve compared to the others.

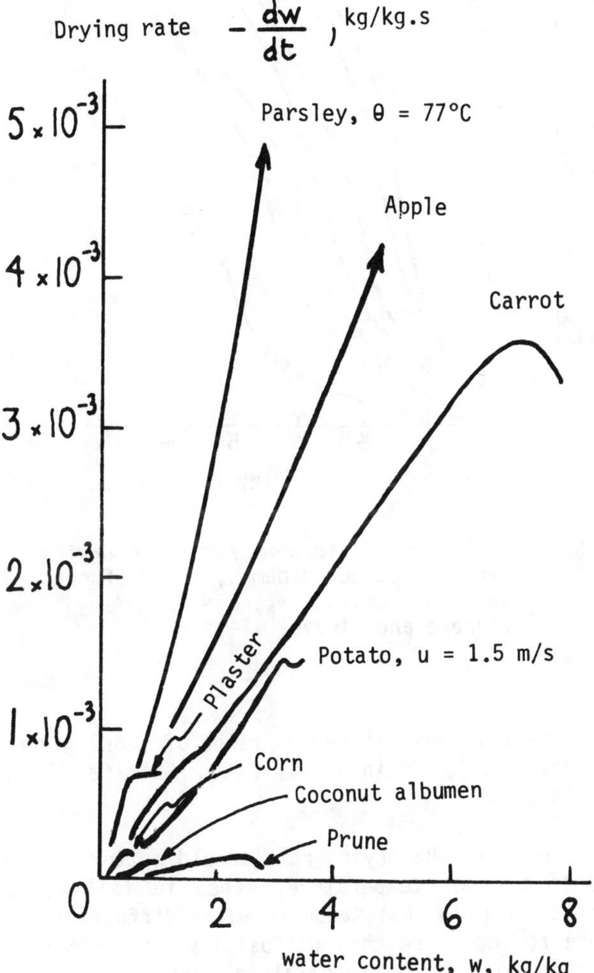

Fig. 4 Drying rate curves of various products dried in similar conditions : θ = 80°C, u = 1 m/s, ambient w_a, unless otherwise mentioned. Parsley : Daudin and Richard (5) ; Golden apple cubes (initially 1x1x1 cm^3) Fornell et al. (10) ; carrot cubes (initially 1x1x1 cm^3) : Wolff (17) ; potato parallelepipeds (initially 1x1x4 cm^3) : Pin (11) ; corn : Daudin et al. (4) ; coconut albumen (third nuts, dehulled) : Aurore and Clouvel (1) ; prune (caliber 31-35 mm) : Daudin (8) ; plaster cubes (1,5x1,5x1,5 cm^3) : Fornell (9)

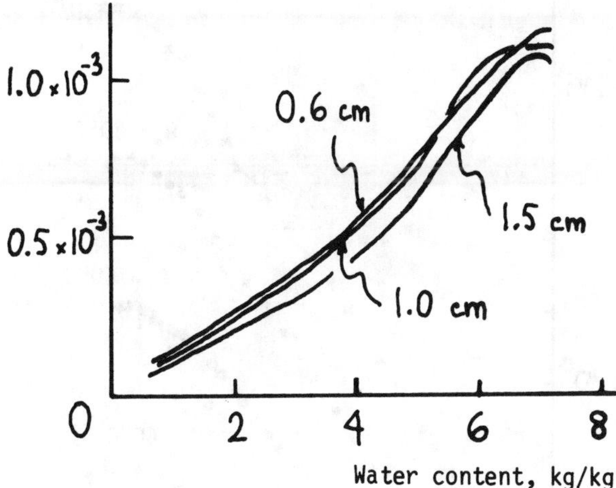

Fig. 5 Drying rate curves of apple cubes of various side lengthes, related to surface unit, θ = 80°C, ambient w_a, u = 3 m/s (Fornell -9-)

- Low water activity at given water content, i.e. mainly high sugar concentration, might explain low drying rates of products like prune. Recent results Riva and Peri (12) on grapes confirm this idea.

- Large concentration in fats, mainly at the end of drying, possibly accounts for the low drying rate of coconut albumen.

More generally, biological products exhibit at same water content and comparable dimensions, a much lower drying rate than a product like plaster. How to explain that low "dryability" and explain in the same time the general absence of constant-rate period ? We think that sugar content, sometimes fat content, but more generally the strong deformation of the product and peripherical cell squeezing account for that low drying. Can we say that life has given the organisms an "anti-drying" structure ?

5. INFLUENCE OF VARIOUS PARAMETERS ON DRYING RATES

5.1. Air temperature

This is by far the most important factor influencing the drying rates. One can find many results in literature. Daudin et al. (4) already published results for corn. Fig. 6 and 7 show similar sets of curves for carrot cubes and fragments of coconut albumen.

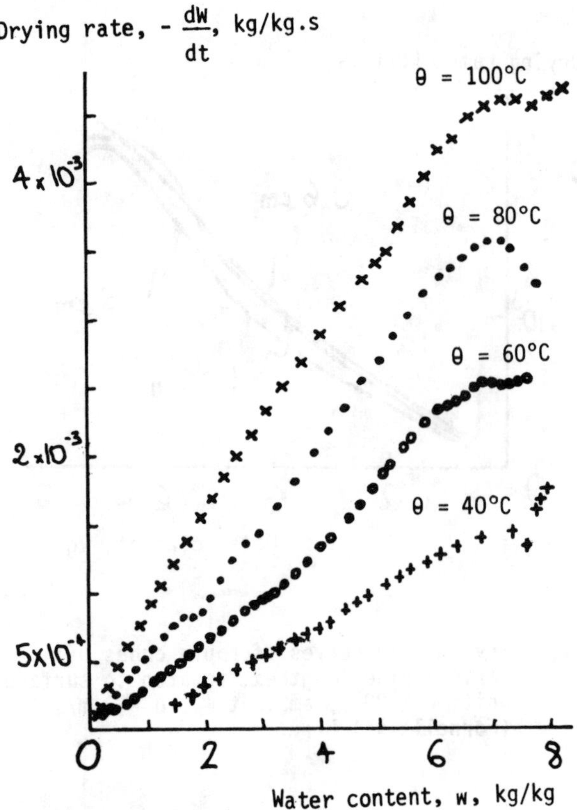

Fig. 6 Influence of air temperature on drying rate of carrot cubes (initially $1x1x1cm^3$), ambient w_a, u = 1 m/s (Wolff -17-)

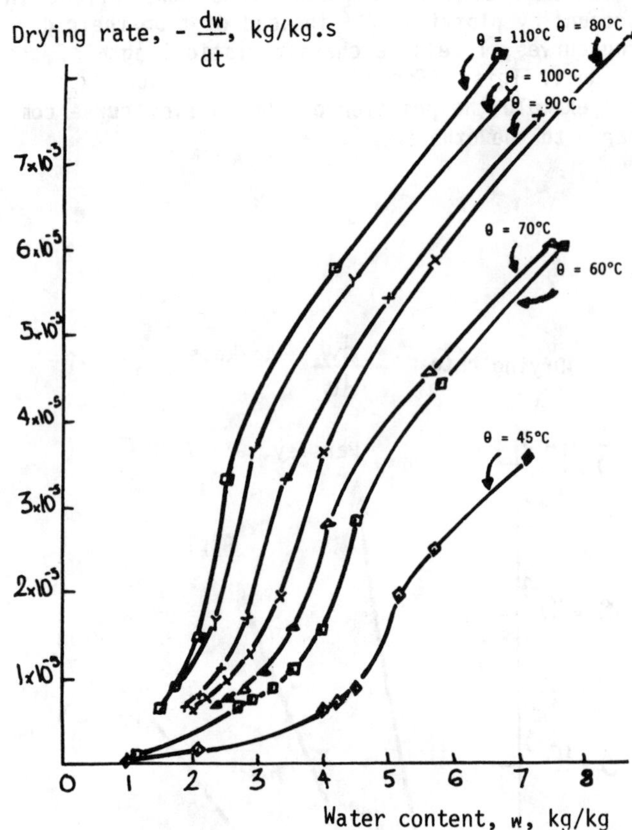

Fig. 7 Influence of air temperature on drying rate of coconut albumen, third of nuts, dehulled, ambient w_a, u = 1.6 m/s (Aurore and Clouvel -1-)

One generally interprets this influence in terms of an activation energy in an Arrhenius law. Despite the fact that there is very little theoretical ground to do so, it gives an easy basis for comparison.

This calculation of E may be performed on drying rates at a given water content (it may come to be the same for the whole range of water content) or from the effective water diffusivity (cf 3.3.), the result being the same. Our results are (table 2) :

	E kJ/mole
Corn (Daudin et al. -4-)	46
Carrot (Wolff -17-)	21
Coconut albumen (Aurore and Clouvel -1-):	
w > 0,4	13
w < 0,4	34

Table 2 : Activation energy in Arrhenius law for several products.

These values are in the range of those given by Schoeber (14), Bruin et al. (2), Chirife (3) and Daudin (6).

In fact, the drying rate should better be related to product temperature, since in falling rate period the main resistance is water diffusion in the product, and since this diffusivity is strongly dependent on product temperature. The relation with air temperature, however, is much more convenient for practical purpose.

5.2. Air humidity

Certain products show during drying a very low sensitivity to air humidity. This is the case for corn or coconut, and we can observe that these products have very low drying rates.

The other products are influenced at some extent by air humidity mainly in the beginning of drying, for potatoes (fig. 8) or carrots. It is clear from fig. 9 that this influence diminishes when air temperature is higher.

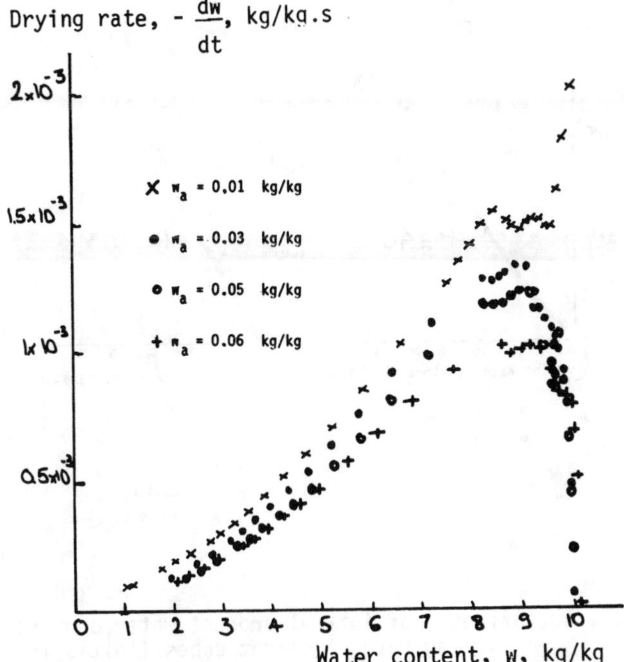

Fig. 8 Influence of air humidity on drying rate of potato parallelepipeds (initially 1x1x4 cm^3), θ = 80°C, u = 1.5 m/s (Pin -11-)

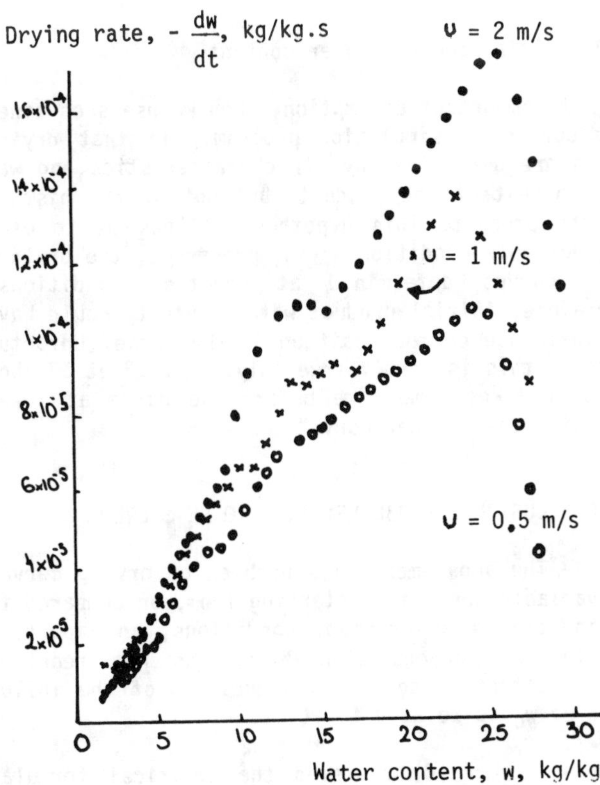

Fig. 10 Influence of air velocity on drying rate of prunes, caliber 31-35 mm diameter, θ = 80°C, ambient w_a (Daudin -8-)

It generally appears that air humidity has the stronger influence as drying is the more controlled by external diffusion.

5.3. Air velocity

We have similarly shown that slow-drying products (corn and coconut) were very little influenced by air velocity. This is however not true for prunes (fig. 10). On carrots, the interaction with temperature (fig. 11) is not as clear as for air humidity.

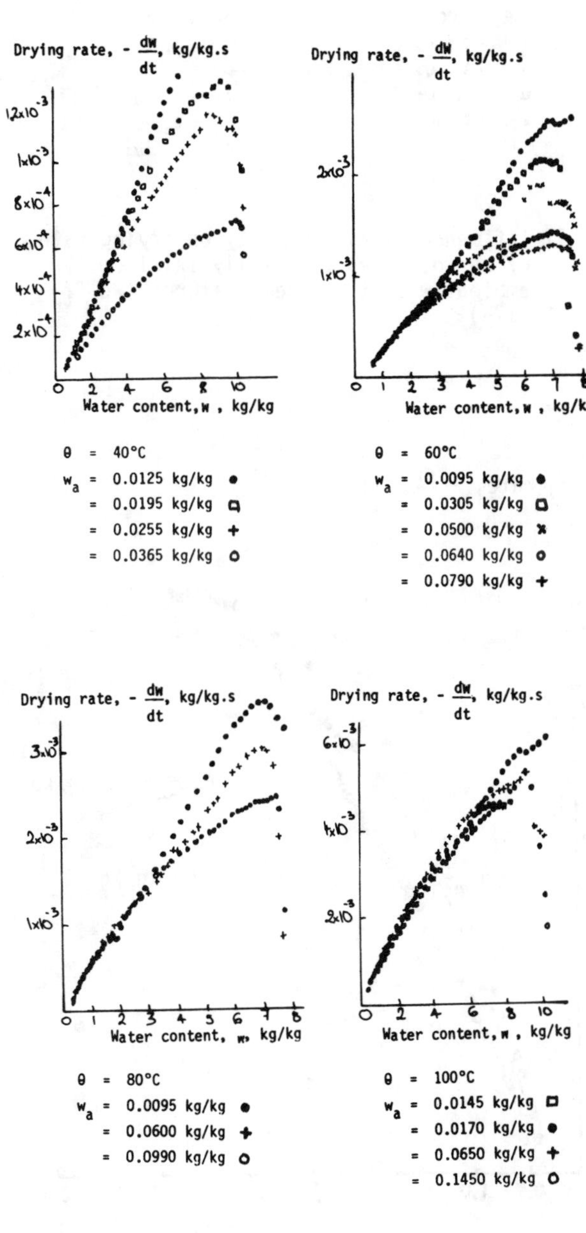

Fig. 9 Influence of air humidity on drying rate of carrot cubes (initially 1x1x1 cm^3) at four air temperatures, u = 1 m/s (Wolff -17-)

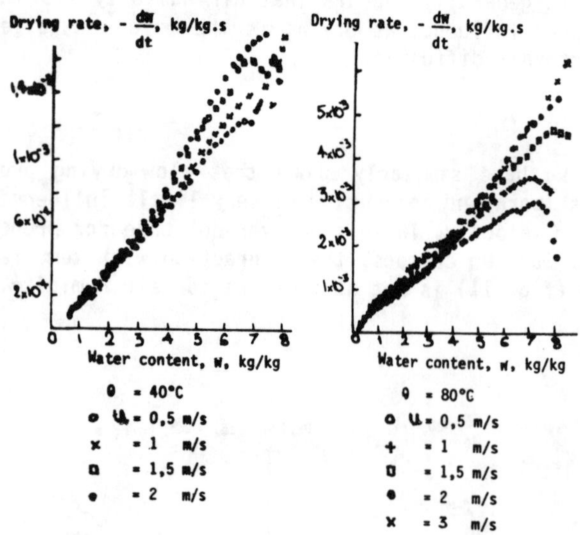

Fig. 11 Influence of air velocity on drying rate of carrot cubes (initially 1x1x1 cm^3), ambient w_a, at two temperatures (Wolff -17-)

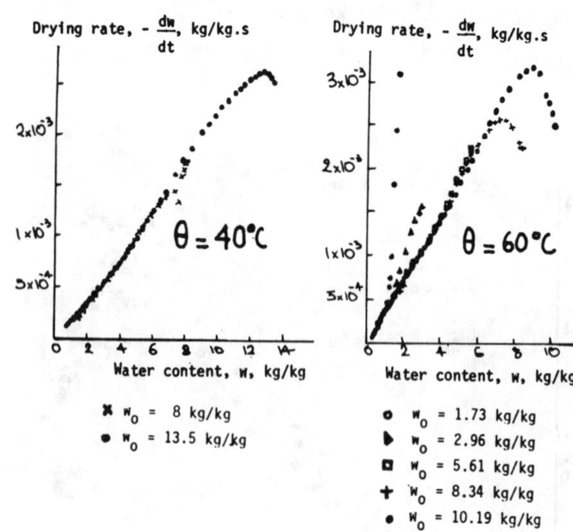

Fig. 13 Influence of initial product water content on drying rate of carrot cubes (initially 1x1x1 cm^3) at two air temperatures, u = 1 m/s, ambient w_a (Wolff -17-)

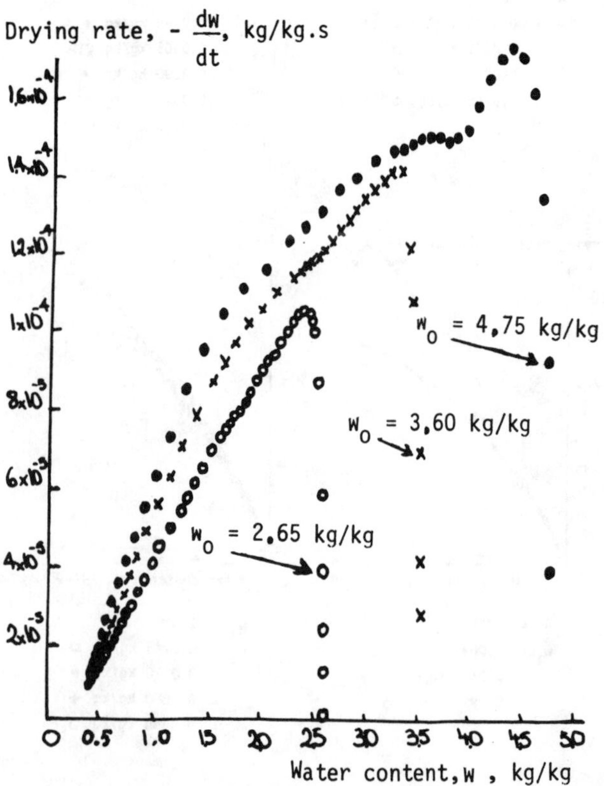

Fig. 12 Influence of initial product water content on drying rate of prunes, θ = 80°C, u = 0,5 m/s, ambient w_a (Daudin and Bimbenet -7-)

5.4. Initial product water content

An important assumption, when we use such kinetic curves in simulation programs, is that drying rates are determined by air characteristics and water content of the product, but not by the history of the product. This hypothesis allows us to use, for variable condition drying processes, the drying-rate curves determined at constant conditions. Therefore, initial product water content should have no influence on the position of the curve. Unfortunately, this is partly true only. Fig. 12 et 13 show that it takes some time before the curve at lower w_0 joins the "mother curve".

6. CONCLUSION : PROBLEMS STILL TO BE SOLVED

- The above-mentioned problem of drying curves at variable conditions starting from, or compared to drying curves at constant conditions, an important problem for dryers simulation, has not received enough attention so far. The question of the influence of w_0 is related to it.

- Daudin (6) reviewed the empirical formulas for drying curves of biological products : w = f(t). In some cases the coefficients of these formulas may be calculated from air temperature, sometimes air humidity and scarcely air velocity.

However the formulas useful for dryer simulation have no predictive power. There is much more to do before being able to predict a drying curve (or effective diffusivity) of a product knowing its biochemical and biophysical characteristics.

We must therefore consider as a "temporary truth" that each biological product has an unpredictable "personality" with respect to drying.

Since no theory is so far usable we must refer to empirical ways. We completely agree with Chirife (3) when he writes that experimental determinations remain essential. More drying curves have to be obtained, published, compared and reviewed. The presentation of these results should be homogenized : we propose (-dw/dt) vs w, and conditions precisely given (product, its shape, air characteristics, etc ...). Can we dream of a data bank on drying kinetics ?

NOMENCLATURE

D_{eff}	effective water diffusivity (m^2/s)
E	activation energy in Arrhenius law (J or kJ/mole)
t	time (s, unless otherwise mentioned)
u	air velocity (m/s)
w	product water content (kg water/kg dry material)
w_o	initial product water content (same unit)
w_e	equilibrium product water content (same unit)
w_a	air water content (kg water/kg dry air)
θ	air temperature (°C)
θ_o	initial product temperature (°C)

REFERENCES

1. Aurore, G. and Clouvel, P., Etude du séchage de l'amande de coco, Inst. Rech. pour les Huiles et Oléagineux, Rept., Paris, January (1984).
2. Bruin, S. and Luyben, K.Ch.A.M., Drying of Food Materials : A review of Recent Developments, in : Adv. Drying, ed. Arun S. Mujumdar, vol. 1, Hemisphere Publ. Corp., Washington, 155-215 (1980).
3. Chirife, J., Fundamentals of the Drying Mechanism During Air Dehydration of Foods, in : Adv. Drying, ed. Arun S. Mujumdar, vol. 2, 73-102, Hemisphere Publ. Corp., Washington (1983).
4. Daudin, J.D. and Bimbenet, J.J., Characteristic Drying curve of Shelled Corn and Simulation of a Vertical Corn Drier, in : Proc. Third Intern. Drying Symp., ed. J. Ashworth, Drying Research Ltd, Wolverhampton (1982).
5. Daudin, J.D. and Richard, H.M.J., Essais de séchage du persil, Sciences des Aliments, vol. 2, 405-410 (1982).
6. Daudin J.D., Calcul des cinétiques de séchage par l'air chaud des produits biologiques solides, Sciences des Aliments, vol. 3, 1-36 (1983).
7. Daudin, J.D. and Bimbenet, J.J., Facteurs physiques du séchage, in : Les mécanismes biologiques et technologiques impliqués dans le séchage de la prune d'Ente, C.T.I.F.L. Rept., Paris (1983).
8. Daudin, J.D., Compte-rendu des travaux effectués pendant la campagne 1982 à l'ENSIA sur le séchage de la prune d'Ente, C.T.I.F.L. Rept., Paris (1983).
9. Fornell, A., Séchage de produits biologiques par l'air chaud, Doct.-Ing. thesis, E.N.S.I.A. Massy (1979).
10. Fornell, A., Bimbenet, J.J. and Almin, Y., Experimental Study and Modelization for Air Drying of Vegetable products, Lebensm.-Wiss. u.-Technol., vol. 14, 96-100 (1980).
11. Pin, B., Séchage de produits solides en couche mince et en couche épaisse, Dipl. Etudes Approf. thesis, E.N.S.I.A. Massy (1982).
12. Riva, M. and Peri, C., Etude du séchage des raisins, Sciences des Aliments, vol. 3, 527-550 (1983).
13. Roth, T. and Loncin M., Fundamentals of Diffusion of Water and Rate of Approach of Equilibrium A_w, presented at ISOPOW III, Beaune (1983).
14. Schoeber, W.J.A.H., Regular Regimes in Sorption Processes, Doct.-Ing. thesis, Technische Hogeschool Eindhoven (1976).
15. Vaccareza,L.M., Lombardi, J.L. and Chirife,J., Kinetics of Moisture Movement During Air Drying of Sugar Beet Root, J. Fd. Technol., vol. 9, 317-327 (1974).
16. Van Brackel, J., Mass Transfer in Convective Drying in : Advances in Drying, ed. Arun S. Mujumdar, vol. 1, 217-267, Hemisphere Publ. Corp., Washington (1980)
17. Wolff, E., Sur l'étude de la cinétique de séchage de la carotte, Student Rept., E.N.S.I.A. Massy (1983).

SINGLE TUBE HEAT TRANSFER IN AERATED VIBRATED BEDS

Karun Malhotra and Arun S. Mujumdar

McGill University, Montreal, Canada

ABSTRACT

Experimental results are presented and discussed for heat transfer for single tubes mounted horizontally and rigidly in a vibrated bed of glass ballotini and molecular sieve particles. A dramatic increase in the heat transfer rate was found to occur during removal of surface moisture for molecular sieve particles. Below the critical moisture content the transfer rate approaches rapidly that for completely dry beds under identical operating conditions. Effects of vibrational amplitude / acceleration, aeration rate and particle size are also presented.

INTRODUCTION

Heat transfer between submerged surfaces and vibrated fluid beds (VFB's) is an area of significant industrial interest especially in the design of energy-efficient thermal processing of hard-to-fluidize, polydisperse and/or sticky granular solids which cannot be fluidized without mechanical assist. The heat exchange surfaces may be plane, vertical panels or multiple tubes arranged horizontally or vertically. Since the bed depth is limited due to rapid attenuation of vibrational energy within the bed vertical tube arrangements appear to have little industrial potential in VFB applications. Rigidly mounted horizontal tubes allow more heat exchange surface to be packed in a given VFB while permitting use of deeper beds as the tubes impart additional vibrational energy within the bed itself. Furthermore, in drying applications the particulate solids are in general sticky and evaporation takes place simultaneously. An attempt is made in this study to isolate the effects of stickiness from that of evaporation. Since the heat transfer process is governed by the bed aerodynamics, a flow visualization study was conducted in a two dimensional bed to examine, in particular, the flow of particulates around horizontal tubes (single as well as multiple) immersed in the VFB.

The present work constitutes an extension to our earlier studies reported by Pakowski and Mujumdar (1) which showed for the first time the dramatic diminution in cylinder heat transfer due to glycerine-induced stickiness in vibrated beds of glass ballotini. It was shown that this reduction is caused by inhibition of particle mixing and formation of larger stable air gaps around the cylinder as compared to beds of dry ballotini. Mujumdar (2) has summarized both the aerodynamic and heat transfer aspects of VFB's with emphasis on work carried out at McGill University. More recently, Malhotra and Mujumdar (3) have discussed the flow behaviour of dry and sticky particles in VFB's, including the effect of various pertinent parameters on the flow past single and multiple cylinders immersed in VFB's. While all of the prior work in conventional as well as VFB's is confined to circular cylinders, Mujumdar and Pakowski (4) have also examined the flow around various noncircular cylinders submerged in VFB's. This work is now being extended to heat transfer aspects. For further details or references to prior work the interested reader is referred to Mujumdar (3).

EXPERIMENTAL APPARATUS AND PROCEDURE

All experiments were carried out in a 20 cm x 20 cm plexiglas bed fitted with a perforated plate distributor of 16% open area. A fine Nylon cloth covered the grid to avoid fines from accumulating in the plenum. The bed height was held constant at 105-110 mm, depending on the particle type. Two sizes each of commercially available glass ballotini and molecular sieve were used as model particles. The particles were sieved and found to be very nearly spherical.

The bed was vibrated vertically by an eccentric drive arrangement; both vibration amplitude and frequency were varied, the former in discrete steps (1.50 mm, 2.75 mm, 4.25 mm) and the latter continuously from zero to about 16 Hz. Minimum fluidization velocities for particles examined were determined experimentally which compared favorably with those predicted empirically. Pressure drop data were taken over ranges of Γ ($= A\omega^2/g$); dimensionless vibrational acceleration) and aeration rates that are of interest industrially i.e. $0 \le \Gamma \le 4$ and $0 \le U/U_{mf} \le 1.20$. Effect of vibration has been proven to become increasingly insignificant with increase in U/U_{mf}. The heat transfer cylinder was made of copper fitted with Teflon ends of same diameter which connected it rigidly to the bed walls. The cylinder was always located in the center of the bed. It should be noted that the cylinder vibrated rigidly with the vessel i.e. it was not suspended from stationary support which is another possible configuration to be studied later. A steady state technique was employed for measurement of h. An electrical heater was embedded in the circular cylinder (37.5 mm diameter, 60 mm long) to provide a source of thermal energy (monitored with a wattmeter) which is transferred to the VFB. Strategically located thermocouples (5) measured the cylinder surface temperatures at various

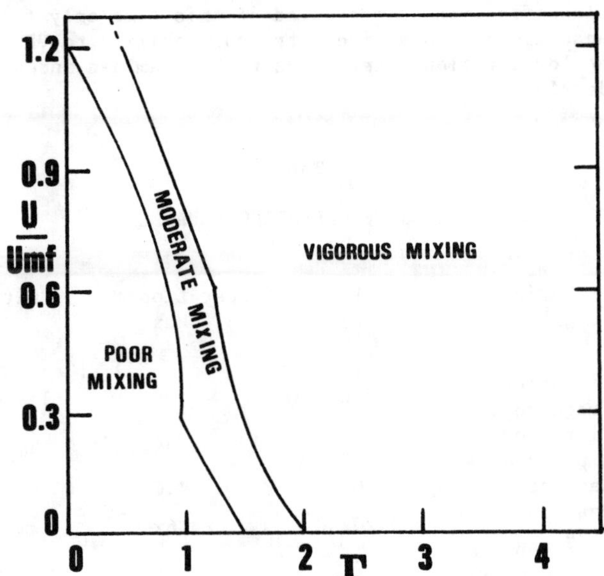

Figure 1. Map showing mixing characteristics in a vibrated fluidized bed.

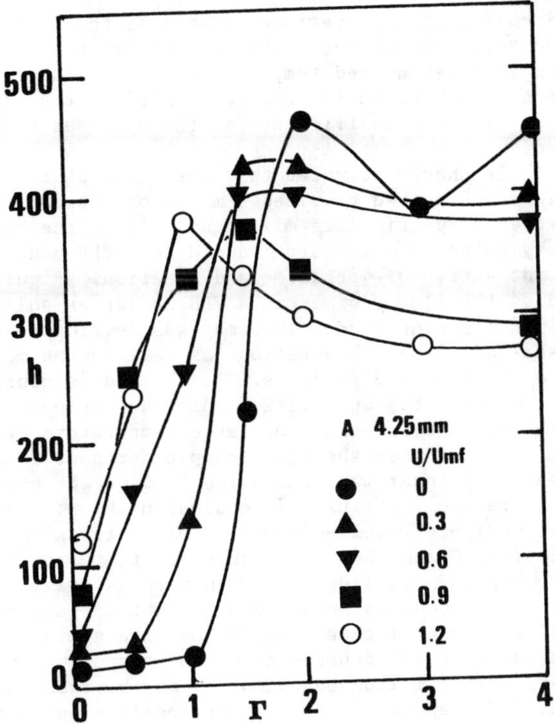

Figure 2. a. Heat Transfer coefficient for bed of glass ballotini, d_p = 0.325 mm

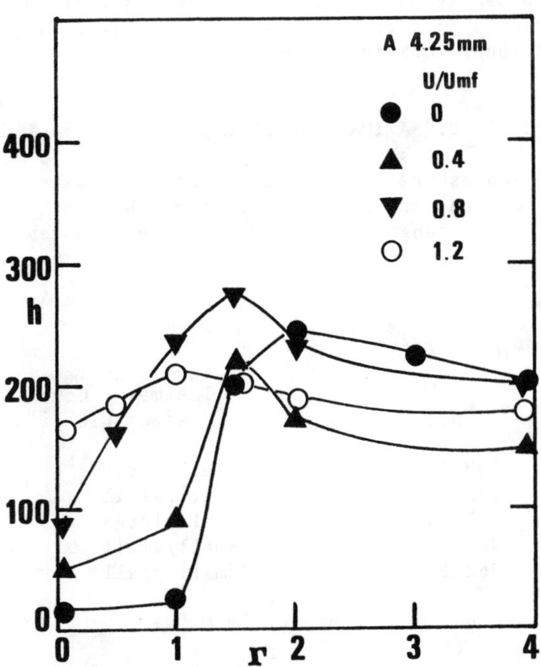

Figure 2. b. Heat Transfer coefficient for bed of glass ballotini, d_p = 0.670 mm

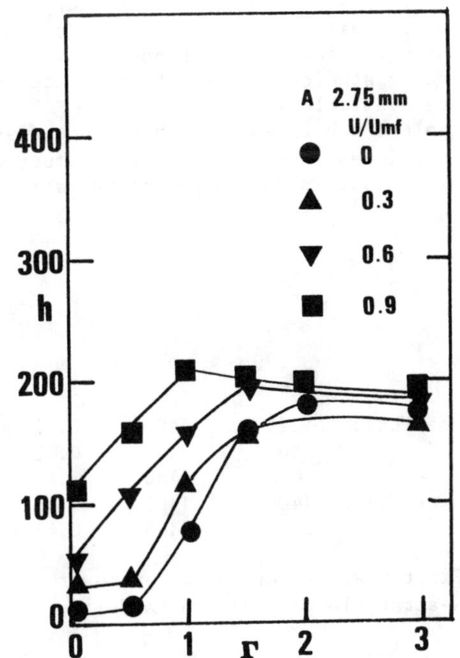

Figure 3. a. Heat Transfer coefficient for bed of Molecular Sieves, d_p = 1.4 mm

circumferential locations as well as those at various locations in the bed. The average cylinder surface temperature and bed temperature far from the cylinder were were used in the computation of h, the effective heat transfer coefficient for the cylinder to the bed.

It should be noted that the temperature variations in the bed provided a means of estimating the degree of particulate mixing as well as the circulatory patterns which are induced in a VFB under the joint action of vibration and aeration. Figure 1 shows the flow regimes in a VFB as far as solids mixing is concerned. This map was developed on the basis of visual observations as well as the measured bed thermocouple readings. If the bed is poorly mixed under the operating conditions employed (indicated by presence of large temperature gradients in the bed) then the thermocouple farthest from the heater cylinder was considered to give the bed temperature to be used in evaluating h. If the bed was moderately well-mixed all temperatures were within 2^0C and then an average bed temperature could be used with confidence. With vigorous mixing, the bed thermocouples measured uniform temperatures allowing use of an average value once again. Because of the inherent dependence of the computed h on the particulate mixing behaviour it may be expected that the data reported here are perhaps less applicable to continuous VFB's; indeed an effect on the vessel size may also be present under certain ranges of operation.

Details of the set-up can be found in Ref. (1). The reproducibility of typical runs was found to be better than $\pm$ 3% while the measurement uncertainty in the heat transfer runs is estimated to be about $\pm$ 8%. Air flow was measured with a dry gas flowmeter at low air velocities (< 20 cm/s) and with calibrated orificemeter at higher flow rates. Aerodynamic studies will not be presented here as they are available in Ref. (3).

Table 1 tabulates the key characteristics of model particles used in the present study while Table 2 summarizes the ranges of operating parameters. Note that Table 1 also includes particle classification according to the well-known Geldart (5) scheme; the particles belong to the B and D classes. Use of criteria developed by Jovanovic and Catipovic (6) for "thermal size" of the particles, places the particles in the thermally small to large range depending on the air flow rate employed. This classification is presented at this time only for academic purposes since its applicability to VFB's at low air flow rates needs to be examined independently.

TABLE 2

RANGES OF OPERATING PARAMETERS

Variable	Range of Operation Min.	Max.	Units
Vibration frequency, ω	0	100	1/sec
Vibration amplitude, A	0	4.25×10^{-3}	m
Bed height	100	110	m
Mean particle size, d_p	3.20×10^{-4}	23.6×10^{-4}	m
Bulk density	641	1440	kg/m^3
Particle density	1165	2480	kg/m^3
U/U_{mf}	0	1.2	-
Particle shape	spherical	spherical	-

RESULTS AND DISCUSSION

Effects of U/U_{mf}, amplitude of vibration, particle size as well as presence of moisture are discussed in the light of experiments carried out in ranges of parameters presented in Table 2. Additional work is currently in progress.

EFFECT OF VIBRATIONAL ACCELERATION

Representative h versus Γ data are given in Figures 2(a) and 2(b). Vibration enhances h over the entire range of Γ and U/U_{mf}; the enhancement is maximal at U/U_{mf}=0. Increase of U/U_{mf} results in a decrease in the effectiveness of vibration. This is expected since the vibrational energy imparted by the vibrating grid is not transmitted effectively to the bed as the bed voidage increases and the time of contact between the grid and the bed is progressively reduced with increase in U/U_{mf}.

TABLE 1

PARTICLE CHARACTERISTICS

Type	d_p, μm	ρ_p	ψ	ϕ_s	U_{mf}	C_p	λ_p	Thermal* size	Geldart's** Classification
Glass ballotini	325	2480	0.42	1	0.085	753.12	0.837	small/large	B
"	500	2480	0.42	1	0.201	753.12	0.837	small/small	B
"	670	2480	0.42	1	0.363	753.12	0.837	small/large	D
Molecular Sieve	1400	1165	0.45	0.9	0.361	962.0	-	small/small	D
"	2360	1165	0.45	0.9	0.480	962.0	-	small/small	D

* First classification is for lower limit of air flow and second for the upper limit (6).

** A-aeratable powders, B-sand-like, C-cohesive, D-spoutable (5)

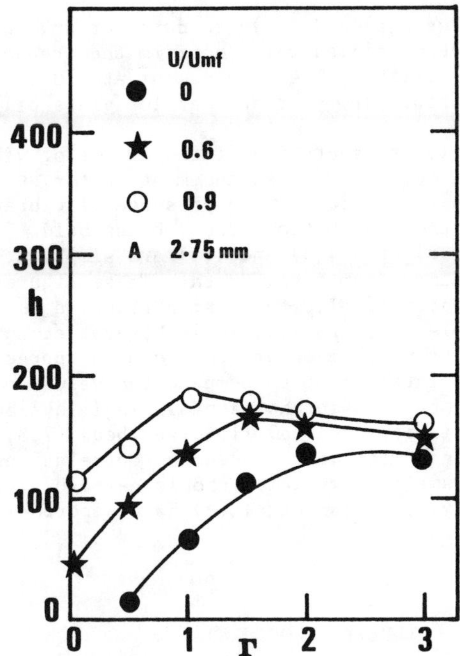

Figure 3. b. Heat Transfer coefficient for bed of Molecular Sieves, d_p = 2.36 mm

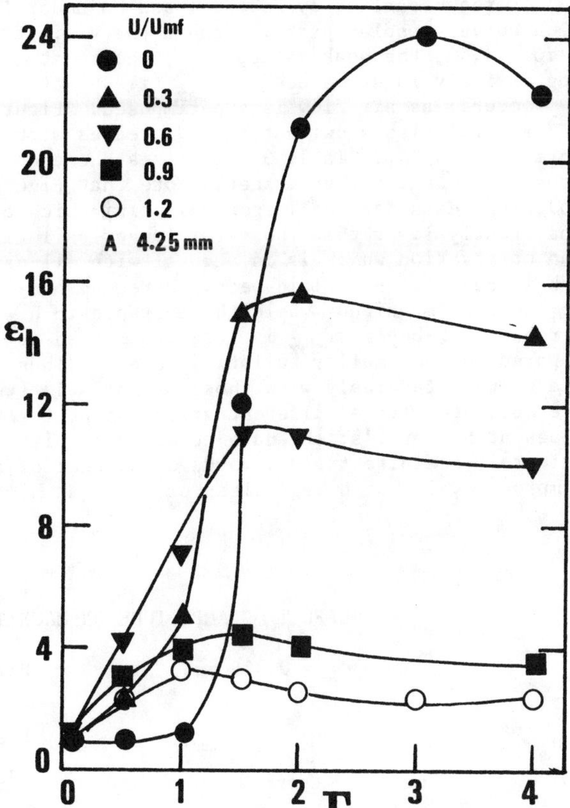

Figure 4. a. Enhancement factor for glass ballotini, d_p = 0.325 mm

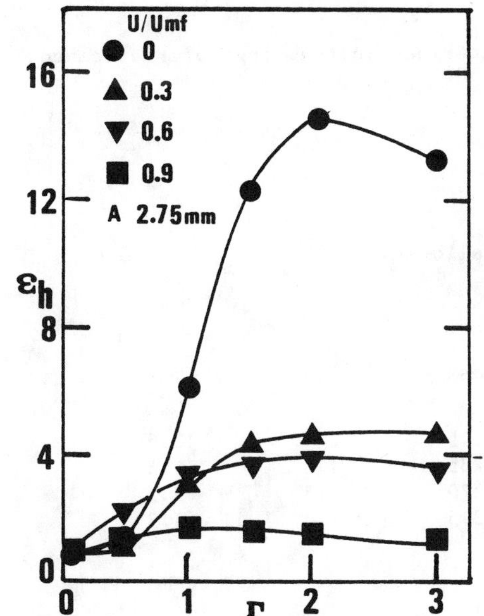

Figure 4. b. Enhancement factor for Molecular Sieves, d_p = 1.4 mm

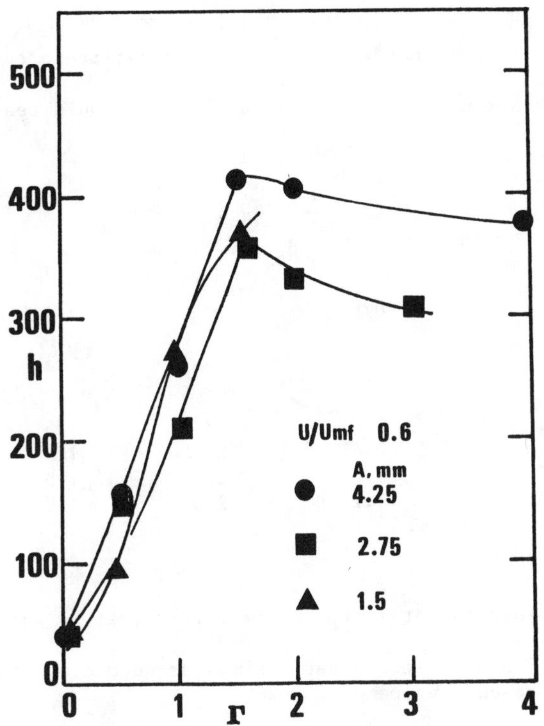

Figure 5. a. Effect of amplitude of vibration on heat transfer for glass ballotini, d_p = 0.325 mm

As shown earlier by Pakowski and Mujumdar (1) all h curves display peak values in the vicinity of $\Gamma=1.0 \sim 1.50$; the peak value, h_{max}, occurs at progressively lower values of Γ_{max} (i.e. Γ at which h_{max} occurs) as air flow is superimposed. Figures 2(a) and (b) also show that h_{max} decreases with increase in U/U_{mf}. Table 3 summarizes these observations in a quantitative manner. Note that Figure 2(b) gives data for the larger ballotini which are consistently lower than those for longer ballotini - an observation which is in accural with all earlier work in conventional fluid beds. Here the h_{max} peaks are lower in magnitude while the decrease in h with increase of Γ beyond Γ_{max} is less pronounced as compared to the smaller ballotini data. These data compare favorably with those of Ref. (1) (which were obtained with a different grid) except for the values at higher Γ's; the data reported earlier indicate a definite trend towards attainment of an asymptotic value of h regardless of U/U_{mf} at high Γ's.

Figures 3(a) and 3(b) are data for dry molecular sieve particles which confirm the trends discussed earlier. The enhancement at $U/U_{mf} = 0$ is considerably lower than that for glass ballotini.

A useful presentation of the effect of vibration is in terms of the enhancement factor, ε_h, defined as the ratio of h in presence of vibration to that without vibration, U/U_{mf} being held constant. Figures 4(a) and 4(b) are simply cross-plots obtained from h vs Γ data. These figures show more dramatically the augmentation in h attributable to vibration; it is highest at no air flow but its value even at U/U_{mf} can be appreciable.

It is interesting to compare the values of average h obtained for horizontal single cylinders submerged in conventional fluidized beds (7,8,9). Unfortunately most data (and hence correlations) apply for smaller particles and larger U/U_{mf} ratios. Fortunately, Chandran et al (8) have reported data

TABLE 3a

EFFECT OF AMPLITUDE ON MAGNITUDE OF h_{max} AND Γ_{max} (WHERE IT OCCURS)

$d_p = 0.325$ mm, Particle - Glass ballotini U_{mf} - 8.5cm/s

U/U_{mf}	A = 1.5 mm*			A = 2.75 mm			A = 4.25 mm		
	h_{max}	Γ_{max}	ε_{max}	h_{max}	Γ_{max}	ε_{max}	h_{max}	Γ_{max}	ε_{max}
0	265++	1.5	17.67	NA	NA	NA	455	2	25
0.3	NA	NA	NA	437	1.5-2	17.0	437	2	15.5
0.6	375	1.5	10	350	1.5	9.2	412	1.5	11
0.9	375	1.5	4.5	365	1.0-1.5	4.20	375	1.5	4.5
1.2	345	1.0	2.90	NA	NA	NA	380	1.0	3.25

* This is not a true maximum value; it is the highest value attained in the range of Γ attained.

++ Not maximum yet. Trend shows it would reach about 425

TABLE 3b

$d_p = 0.670$ mm Particle - Glass ballotini $U_{mf} = 36.3$ cm/s

U/U_{mf}	A = 1.5 mm*			A = 4.25 mm		
	h_{max}	Γ_{max}	ε_{max}	h_{max}	Γ_{max}	ε_{max}
0	237+	1.5	12	245	2	12.5
0.4	215	1.5	4.3	235	1.5	4.6
0.8	225	1.5	2.6	270	1.5	3.1
1.2	215	1.0	1.3	215	1.0	1.3

* This is not a true maximum value; it is the highest value attained in the range of Γ attained.

\+ Has not yet reached maximum as seen from the graphs, but is just near the peak. Probably it would reach a value of 250-260 for glass ballotini at U/U_{mf} = 1.2 and 1.6. These correspond very well to values obtained in this study. The h_{max} value obtained around $\Gamma \approx$ 1-2 at U/U_{mf} = 1.2 in the VFB in the range of h values attained at much higher air flow velocities ($U/U_{mf} \sim$ 5-7) in conventional beds. When higher air velocities are not desirable the advantage of VFB's over conventional FB's is clear as far as heat transfer rates are concerned.

EFFECT OF AMPLITUDE OF VIBRATION

Figures 5(a) through 5(c) portray the effect of A. It should be noted that this study was limited to lower Γ's at smaller A's since the upper frequency limit of the vibratory mechanism was limited. Further studies will extend this work. At higher U/U_{mf}, Γ values seem to couple effectively the combined influence of A and ω (Figure 5(c)). This is not the case with larger ballotini at U/U_{mf} = 0.8. Figure 5(b) shows that for $\Gamma < 1.5$ h is lower for smaller amplitudes; but with an increase in Γ a common asymptote is reached.

It is clear that the effect of amplitude of vibration needs to be examined in some detail. In the meantime, following convention, we present all data in terms of Γ.

EFFECT OF PARTICLE SIZE

Figures 6(a) through 6(d) display for VFB's the well-known trend verified repeatedly for conventional fluid beds viz h is higher for smaller particles. Examination of Figure 6(a), 6(b) and 6(d) leads to the conclusion that the decrease of h with increase in particle size is more significant at lower d_p values (d_p = 0.325 mm to d_p = 0.500 mm) as compared to that at higher values (d_p = 0.500 mm to d_p = 0.670 mm). This observation is in accord with data reported for conventional FB's.

Another way of examining the effect of d_p is in terms of ε_h. Figure (7) shows that ε_h decreases with increase of d_p, regardless of U/U_{mf}. The actual decrease depends on U/U_{mf} too. Thus larger particles are less sensitive to vibration-induced enhancement of heat transfer.

EFFECT OF MOISTURE

In drying applications it is important to examine the effect of moisture evaporation on h. No prior work appears to document this effect even in an exploratory manner. Pakowski and Mujumdar (1) and Malhotra and Mujumdar (3) examined the effect of interparticle adhesion (or stickiness) but without concurrent moisture evaporation. They reported rather severe adverse effect of stickiness as far as immersed surface heat transfer is concerned. In this work wetted molecular sieve particles were employed to examine the effect of moisture content of porous particles on h. Figure (8) is a typical h versus average particle moisture content, X, curve. Ambient air was employed to minimize the rate of drying so that quasi-steady state could be attained permitting measurement of h without resorting to a more complex transient technique. Further, since only the average X could be measured by sampling the bed during operation, the choice of Γ and U/U_{mf} was dictated by the requirement of good mixing in the bed. No data could be obtained at low Γ's and low U/U_{mf} values.

Figure (8) displays the appreciable increase in h due to presence of moisture. When surface evaporation takes place (X_c for molecular sieve is about 24% dry basis) evaporation of unbound water causes a significant rise in h. In the range $0.30 \leq X \leq 0.35$ vibration causes interparticle moisture to free itself. As it is removed physically the bed mobility rises which is probably responsible for the increase of h from about 475 $W/m^2.s.K$ at X = 0.35 to about 600 $kW/m^2.s.K$ at X = 0.30. With decrease of X, h also decreased markedly. Presence of X is expected to increase the effective thermal conductivity of the porous molecular sieve material. Note that both the moisture and thermal diffusivity values are expected to decrease with X which contribute to the fall in h with decrease in X. It is noteworthy that the decrease occurs in the vicinity of the critical moisture content of the particles. Below X = 0.20 the heat transfer values approach those for dry particles. Indeed in these experiments the measured values were slightly lower than those obtained earlier for dry particles; this is due to the formation of thin film of molecular sieve material on the test cylinder during the drying runs.

Further work in this area is continuing.

CLOSURE

Results are presented for immersed-surface heat transfer in VFB's which corroborate and extend prior work in this area. Data obtained with no vibration are in accord with the very limited data in the open literature for large particles around the minimum fluidization air flow rates in conventional fluidized beds. Application of vibration is shown to enhance heat transfer; this effect is more pronounced at lower U/U_{mf} values. Although not discussed in this paper (for reasons of space) it may be noted that the trends observed were consistent with visual observations on flow around immersed cylinders in a two dimensional VFB. Formation and extent of standing air gaps around the cylinder were found to correlate well with measured h variation when particle mixing in the bed was also considered. One of the most interesting results reported here for the first time is the variation of h with moisture content which follows well-known behaviour in drying of porous media.

ACKNOWLEDGEMENTS

The authors are grateful to Purnima Mujumdar for her prompt typing of this manuscript. We are also indebted to K.C.L. Law, R. Figueredo and R. Lozada for their help with the experimental work.

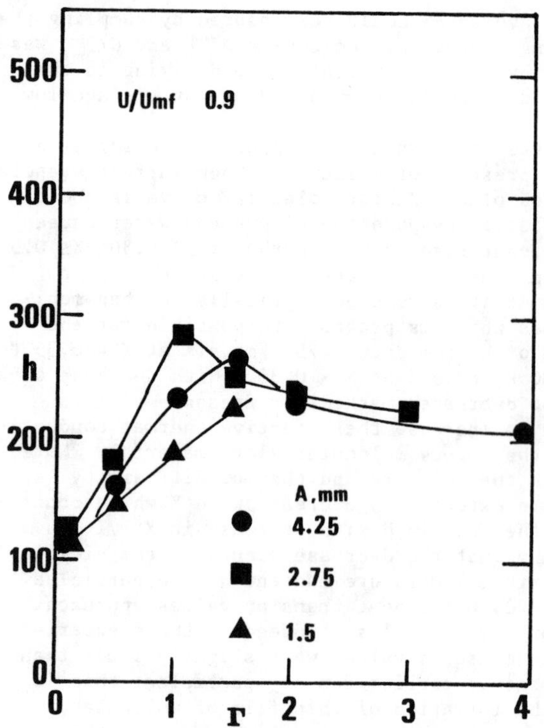

Figure 5. b. Effect of amplitude of vibration on heat transfer for glass ballotini, d_p = 0.670 mm

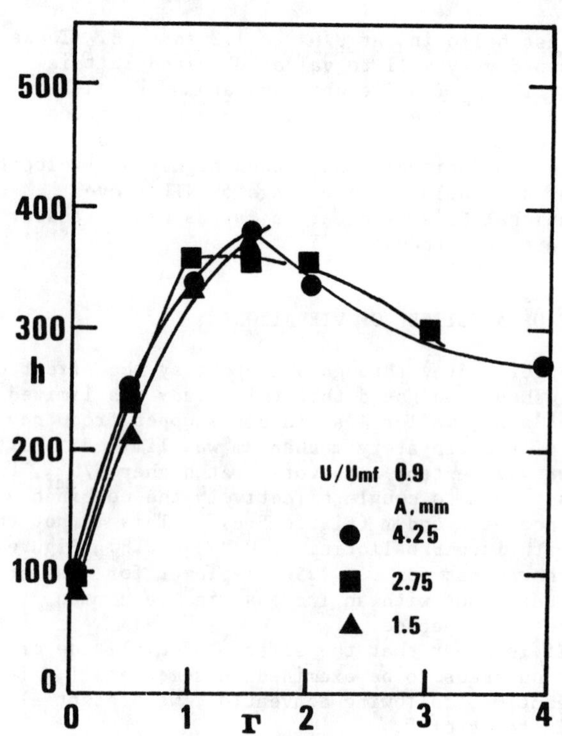

Figure 5. c. Effect of amplitude of vibration on heat transfer for glass ballotini, d_p = 0.325 mm

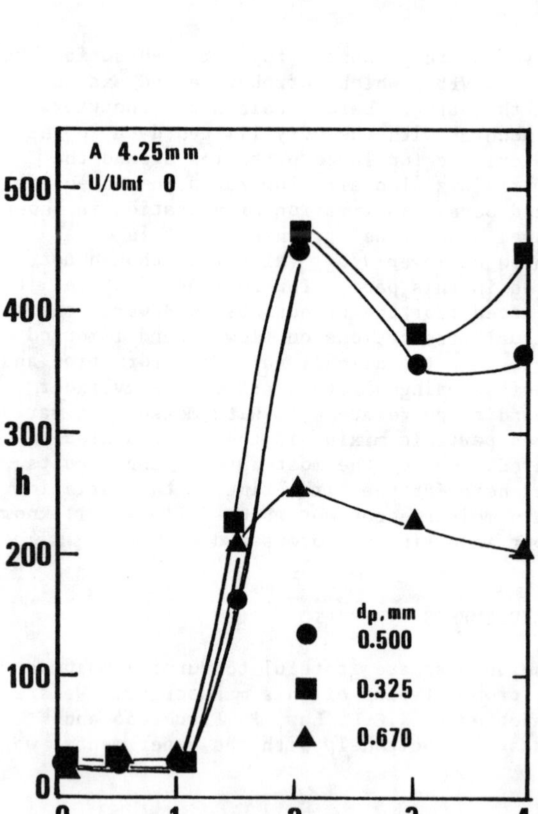

Figure 6. a. Effect of particle size on heat transfer coefficient for glass ballotini

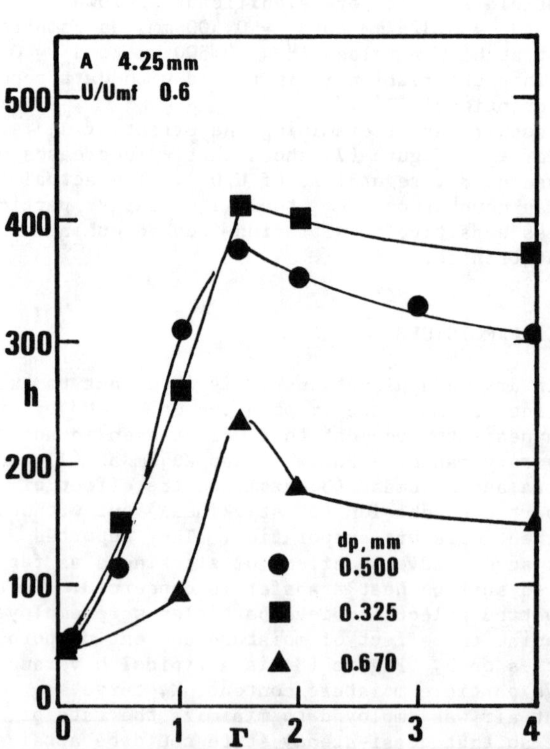

Figure 6. b. Effect of particle size on heat transfer coefficient for glass ballotini

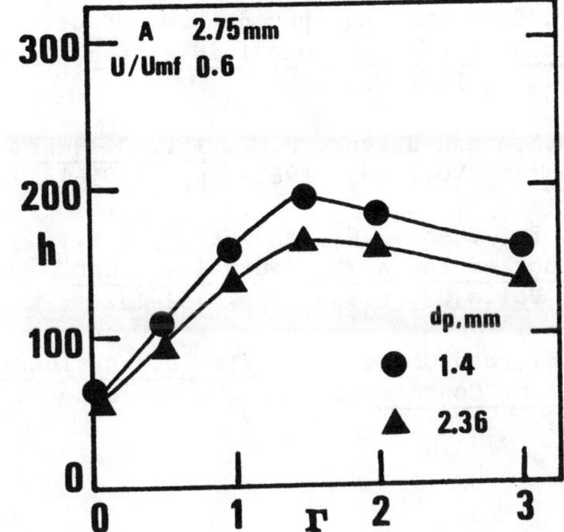

Figure 6. c. Effect of particle size on heat transfer coefficient for molecular sieves

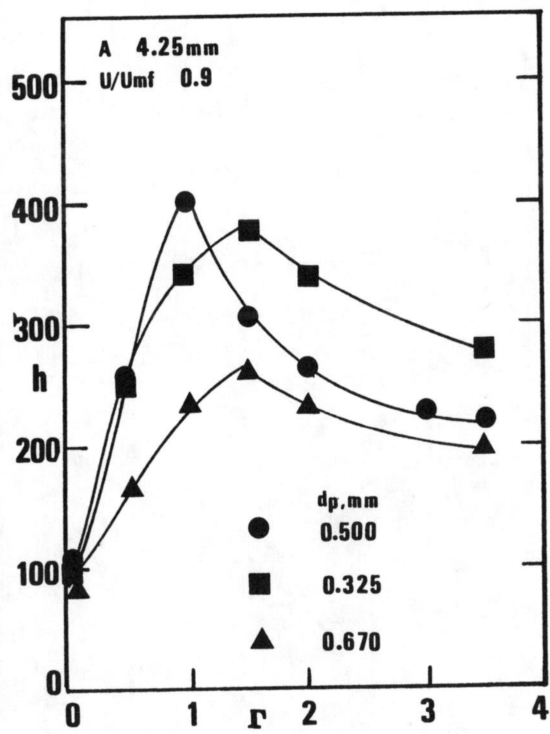

Figure 6. d. Effect of particle size on heat transfer coefficient for glass ballotini

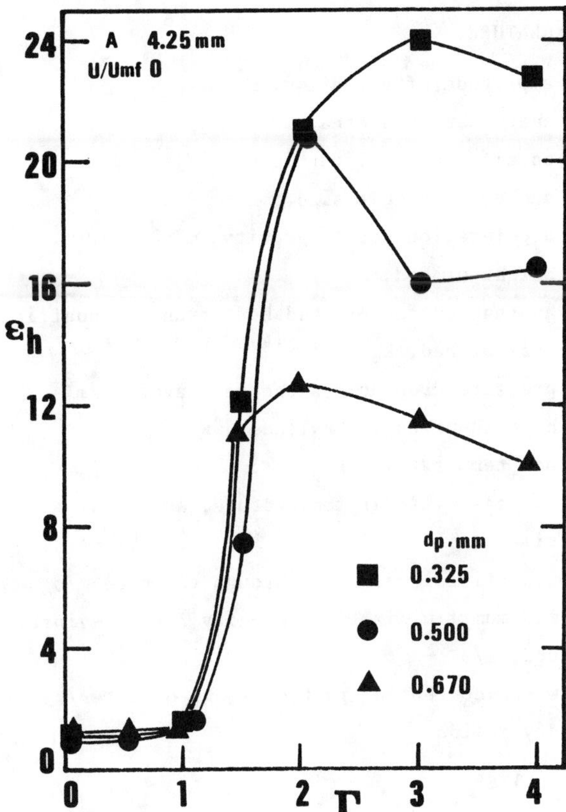

Figure 7. Effect of particle size on enhancement factor for glass ballotini

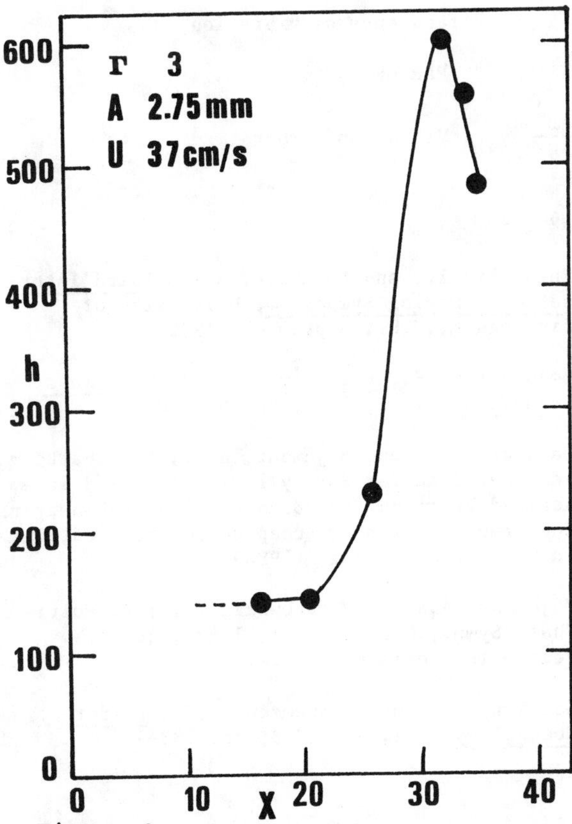

Figure 8. Heat transfer coefficient for moist molecular sieves, d_p=1.4 mm

NOMENCLATURE

A	amplitude of vibration, m
A_s	heater surface area, m^2
C_p	specific heat of particles, J/(kg K)
d_p	average particle size, m
g	acceleration due to gravity, m/s^2
H	height of bed, m
h	average surface-to-bed heat transfer coefficient (W/m^2 s K)
M	mass of bed, kg
ΔP	pressure drop across the bed layer, N/m^2
Q	heat input to the cylinder, W
T_b	bed temperature, K
T_s	average cylinder temperature, K
t	time, s
U	superficial velocity through the bed, m/s
U_{mf}	minimum fluidization velocity for non-vibrated bed, m/s
X	water content of particles, kg of water/kg of dry solids

GREEK LETTERS

ε_h	enhancement factor for heat transfer coefficient
λ_p	thermal conductivity of particles, W/m K
ρ_p	particle density, kg/m^3
ϕ_s	shape factor
ψ	void fraction
ω	angular frequency of vibration, s^{-1}

DIMENSIONLESS PARAMETER

$\Gamma = \frac{A\omega^2}{g}$ Vibration number

REFERENCES

1. Pakowski, Z., and Mujumdar, A.S., Third International Drying Symposium, University of Birmingham, U.K., September, 1982.

2. Mujumdar, A.S., Latin American J. Heat Mass Transfer, Vol. 7, 1983, pp. 99-110.

3. Malhotra, K., and Mujumdar, A.S., Flow Patterns and Heat Transfer for Cylinders Immersed in an Aerated Vibrated BEd, 33rd Chemical Engineering Conference, Toronto, accepted for publication in Can. J. Chem. Eng. (1984).

4. Mujumdar, A.S., and Pakowski, Z., Int. Multi-Phase Symp., Miami, April, 1983., Ed. T.N. Veziroglu, Hemisphere, N.Y.

5. Geldart, D., and Abrahamsen, A.R., Powder Technology, Vol. 19, 1978, pp. 133-136.

6. Jovanoviç, G.N., and Catipovic, N.M., Fourth International Conference on Fluidization, Tokyo, Japan, Vol. 1, 9, (1983).

7. Grewal, N.S., and Saxena, S.C., Int. J. Heat Mass Transfer, Vol. 23, (1980), pp. 1505-1519.

8. Chandran, R., Chen, J.C., and Staub, F.W., Transactions of the ASME, Journal of Heat Transfer, Vol. 102, (1980), pp. 152-157.

9. Biyikli, S., and Chen, J.C., 7th International Heat Transfer Conference, Munich, Hemisphere, N.Y., 1982.

INDIRECT HEATING OF SOLID PARTICLES

Satoru Abe and Takeshi Akao

Kikkoman Corporation
339 Noda, Noda-shi, Chiba, 278 JAPAN

ABSTRACT

Conductive heat transfer between particles and a wall was studied in continuous operation by horizontal well-stirred cylindrical vessels. An annular particle layer of 0.7 to 12 mm thickness on the wall was formed under the action of centrifugal force from blade rotation. Three sizes of vessels (82-150mm I.D., 450-1000mm L.) and three kinds of materials were used in this study.

An extended surface renewal theory, considering an unsteady distribution of the surface age at the initial stage, was introduced. This theory is very useful to evaluate the heat transfer performance between particles and a wall.

1. INTRODUCTION

Recently indirect dryer with agitation has been re-evaluated because of its high heat efficiency and less air pollution. However this type of dryer has usually so much hold-up of material that its residence time becomes long. Short residence time is necessary to avoid the thermal degradation of product, especially in food process. Therefore, we studied the short residence time apparatus of indirect dryer. Particles in the cylinder become an annular layer by the agitation at high speed. The layer thickness can be minimized by proper design of the apparatus. Three apparatuses (82mm I.D. x 450mm L., 82mm I.D.x 900mm L., 150mm I.D.x 1000mm L.) and three materials (wheat bran, glass beads, sands)were used in this study, and three types of blades were tested. The layer thickness ranged 0.7 to 1.2mm.

An extended surface renewal theory, considering an unsteady distribution of surface age at the initial stage, was introduced to this phenomena, because there was not proper theory applicable to such a short residence time apparatus. This theory had been originally introduced to gas absorption in wetted wall column by T.Tadaki [5]. This is very useful to predict the heat transfer performance between particles and a wall. This theory was also testified by Schlünder's data [1] and agreed well.

2. THEORY

The mean heat transfer coefficients during contact time t between a wall and semi-infinite stagnant solid is expressed by Eq.(2), using the temperature profiles of the solution of Eq.(1) with the initial and boundary conditions (1-1) [2].

$$\frac{\partial T}{\partial t} = a\frac{\partial^2 T}{\partial x^2} \qquad \text{-----(1)}$$

$$\left.\begin{array}{lll} \text{I.C.} & \text{at } t=0, T=T_0 & \text{for all } x \\ \text{B.C.1} & \text{at } x=0, T=T_w & \text{for all } t>0 \\ \text{B.C.2} & \text{at } x=\infty, T=T_0 & \text{for all } t>0 \end{array}\right\} \qquad \text{----(1-1)}$$

$$\frac{\alpha}{\frac{\lambda}{\sqrt{at}}} = \frac{2}{\sqrt{\pi}} \qquad \text{-----(2)}$$

The existance of heat transfer resistance($1/\alpha_w$) between a wall and the surface of the adjacent layer of solid particles was reported [1]. In this case the mean heat transfer coefficient is expressed by Eq.(3), using the conditions(1-2).

$$\left.\begin{array}{lll} \text{I.C.} & \text{at } t=0, T=T_0 & \text{for all } x \\ \text{B.C.1} & \text{at } x=0, -\lambda\frac{\partial T}{\partial x}=\alpha_w(T_w-T) & \\ & & \text{for all } t>0 \\ \text{B.C.2} & \text{at } x=\infty, T=T_0 & \text{for all } t>0 \end{array}\right\} \qquad \text{----(1-2)}$$

$$\frac{\alpha}{\frac{\lambda}{\sqrt{at}}} = \frac{\frac{\lambda}{\sqrt{at}}}{\alpha_w}\left[\mathrm{EXP}\left(\frac{\alpha_w}{\frac{\lambda}{\sqrt{at}}}\right)^2 \times \mathrm{erfc}\left(\frac{\alpha_w}{\frac{\lambda}{\sqrt{at}}}\right) - 1\right] + \frac{2}{\sqrt{\pi}} \qquad \text{-----(3)}$$

Eq.(2) and Eq.(3) are also available for heat transfer to moving solids, if the residence time is very short or the mixing is not so remarkable that adjacent materials are hardly swept away into bulk part.

On the other hand, at long residence time the supposition that the adjacent materials are replaced with the bulk parts at some renewal rate, would be necessary. Schlünder proposed Eq.(4) without the resistance and Eq.(5) with the resistance [1].

$$\frac{\alpha}{\frac{\lambda}{\sqrt{at_s}}} = \frac{2}{\sqrt{\pi}} \qquad \text{-----(4)}$$

$$\alpha = \frac{2\alpha_w}{\sqrt{\pi\tau_s}}\left[1 + \frac{1}{\sqrt{\pi\tau_s}}\ln\frac{1}{(1+\sqrt{\pi\tau_s})}\right] \qquad \text{----(5)}$$

$$\text{where } \tau_s = \frac{\alpha_w^2 \cdot t_s}{\lambda\, C_p\, \rho}$$

The assumption of Eqs.(4),(5) that the apparent resting times of whole material are uniform, would be not valid for the case at short residence time. And also the validity of Eqs.(2),(3) for the apparatus planned to study, whose residence time is relatively short, was not obvious because the valid duration of these equations were unknown. Then, to explain the state between Eqs.(2),(3) and (4),(5) the extended surface renewal theory [5] was applied to this heat transfer phenomena. It was originally proposed for gas-liquid mass transfer system, which explaining unsteady state in that the surface age distribution is not fully developed yet, while the original surface renewal theory [4] was based on the state of fully developed that.

According to this theory, taking the concept of the resting time distribution ($\bar{t}$: mean resting time), the heat transfer coefficient α is expressed by Eq.(6) with the conditions (1-1). (See Appendix)

$$\frac{\alpha}{\frac{\lambda}{\sqrt{a\bar{t}}}} = \left(1+\frac{1}{2\tau}\right)\cdot \mathrm{erf}\sqrt{\tau} + \frac{1}{\sqrt{\pi\tau}}\cdot e^{-\tau} \qquad \text{-----(6)}$$

where $\tau = t/\bar{t}$

$$\text{and when } \tau=\infty, \quad \frac{\alpha}{\frac{\lambda}{\sqrt{a\bar{t}}}} = 1 \qquad \text{-----(7)}$$

At long residence time t, Eq.(6) approaches Dankwerts' original Eq.(7).

Considering the resistance ($1/\alpha_w$), Eq.(8) is obtained with conditions (1-2) instead of (1-1).

$$\frac{\alpha}{\frac{\lambda}{\sqrt{a\bar{t}}}} = -\frac{\beta}{\beta^2-1} + \frac{\beta^2}{\beta^2-1}\left[\left(1-\frac{1}{2\tau}\right)\cdot \mathrm{erf}\sqrt{\tau} + \frac{1}{\sqrt{\pi\tau}}\cdot e^{-\tau}\right] + \frac{\beta^3}{(\beta^2-1)^2\cdot\tau}\left[e^{(\beta^2-1)\tau}\cdot \mathrm{erfc}(\beta\sqrt{\tau}) - 1 + \beta\cdot \mathrm{erf}\sqrt{\tau}\right] \qquad \text{---(8)}$$

$$\text{where } \beta = \frac{\alpha_w}{\frac{\lambda}{\sqrt{a\bar{t}}}}$$

when $\beta=1$ Eq.(8) becomes

$$\frac{\alpha}{\frac{\lambda}{\sqrt{a\bar{t}}}} = 1 + \frac{\tau}{2}\cdot \mathrm{erfc}\sqrt{\tau} + \left(\frac{1}{8\tau}-\frac{1}{2}\right)\cdot \mathrm{erf}\sqrt{\tau} - \left(\frac{\sqrt{\tau}}{2\sqrt{\pi}} + \frac{1}{4\sqrt{\pi\tau}}\right)\cdot e^{-\tau} \qquad \text{---(8')}$$

$$\text{when } \tau\to 0 \quad \frac{\alpha}{\frac{\lambda}{\sqrt{a\bar{t}}}} \to \beta \quad \text{i.e. } \alpha \to \alpha_w$$

$$\text{when } \tau\to\infty \quad \frac{\alpha}{\frac{\lambda}{\sqrt{a\bar{t}}}} \to \frac{\beta}{1+\beta} \quad \text{i.e. } \alpha \to \frac{1}{\frac{1}{\alpha_w}+\frac{1}{\frac{\lambda}{\sqrt{a\bar{t}}}}}$$

To make comparison between various heat transfer models, Eq.(2) through Eq.(8) are presented in Fig.1, where Eq.(3') comes from multiplying Eq.(3) by $\sqrt{\bar{t}/t}$ on both sides of it.

$$\frac{\alpha}{\frac{\lambda}{\sqrt{a\bar{t}}}} = \frac{e^{\beta^2\tau}\cdot \mathrm{erfc}(\beta\sqrt{\tau})-1}{\beta\tau} + \frac{2}{\sqrt{\pi\tau}} \qquad \text{---(3')}$$

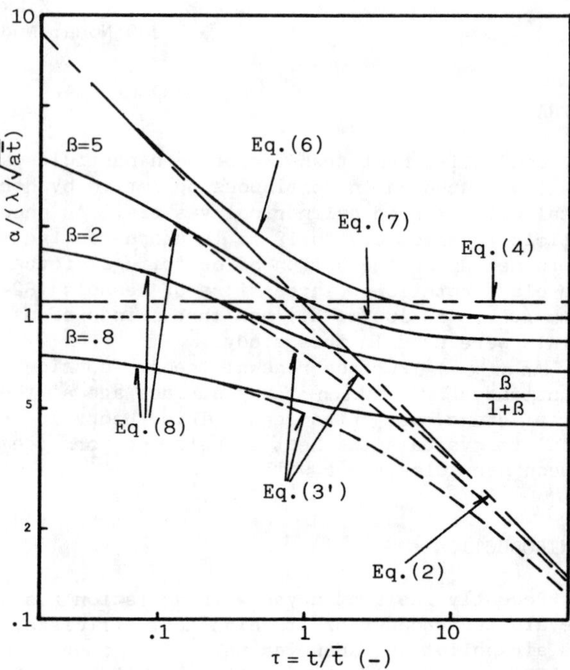

Fig.1 Comparison between various heat transfer models

This figure makes it clear that the valid duration of Eqs.(2),(3) are far less than the dimensionless residence time τ of 1 and the heat transfer coefficient α decreases toward the time-independent value, which could be expressed by Eqs.(4), (5), with residence time going on. And also this suggests that the heat transfer coefficients α are strongly affected by short contact time or residence time when β value is large. In other words heat transfer coefficients α would be almost constant at any contact/residence time t, when β value is small.

Fig.2 shows the effect of mean resting time $\bar{t}$ on heat transfer coefficient, using the results which come from multiplying Eq.(8) by $\sqrt{t/\bar{t}}$ on both sides of it.

By this figure the extent of variation of heat transfer coefficients, by varying $\bar{t}$ at fixed contact /residence time, can be predicted. For example, when a change of apparatus blades is planned to look forward to the higher heat transfer coefficient value, you can predict that only a extreme improvement on $\bar{t}$ could result in slight increase of α, if $\bar{\beta}$ value is small or $1/\tau$ is very large. In other words great improvement of α could not be looked forward to in that case.

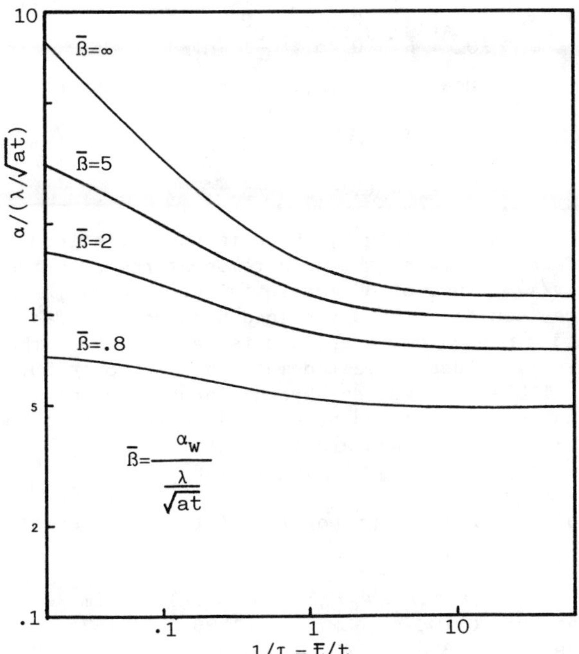

Fig.2 Effect of mean resting time $\bar{t}$ on heat transfer coefficient.

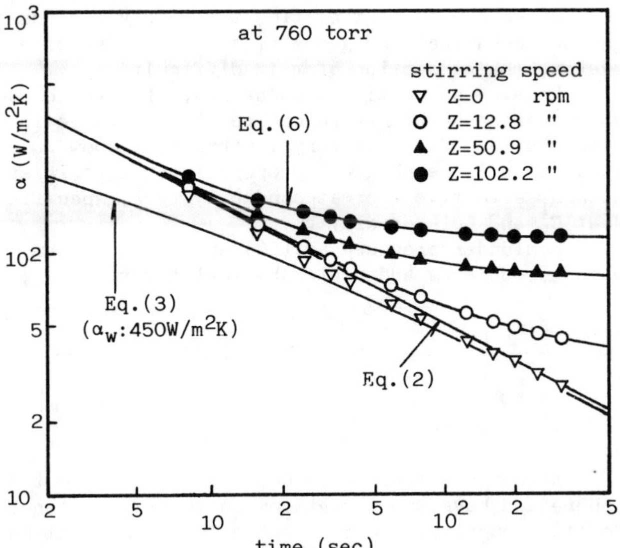

Fig.3 Comparison of results by Eq.(6) with the experimental data of glass beads at 760 torr, by Schlünder [1]

Table 1 Mean resting times $\bar{t}$ obtained by Eq.(6) using data from Fig.1

stirring speed	$\bar{t}$
Z=102.2 rpm	15.4 sec
Z= 50.9 "	33.2 "
Z= 12.8 "	160.0 "

(α_w:very large,indefinite)

Using Schlünder's experimental results [1] of glass beads at 760 torr and 1 torr in the stirred heating apparatus, the suitability of the above theory was examined. The measured heat transfer coefficients at 760 torr are shown in Fig.3. Higher the stirring speed is, larger the heat transfer coefficients are. They decrease with increasing contact time and then tend to move toward constant final values.

By the experimental results at Z=0 and Eq.(3), the determination of the resistance value ($1/\alpha_w$) was performed. But these results were very close to Eq.(2) (i.e. Eq.(3) with α_w:∞) , which left α_w value indefinite. For reference the values calculated by Eq.(3) with α_w:450 W/m^{2}K are lined in Fig.3, which suggested α_w is far larger than 450 W/m^{2}K.

The mean resting times obtained by Eq.(6) are listed in Table 1, provided α_w is infinite.

Fig.4 shows the measured heat transfer coefficients in exactly the same conditions as those of Fig.3 except its pressure of 1 torr. By the results at Z=0 and Eq.(3), the resistance ($1/\alpha_w$) could be determined at 0.0067 m^{2}K/W (i.e. α_w:150 W/m^{2}K).

The results calculated by Eq.(8) with the above value of α_w and the mean resting time $\bar{t}$ in Table 1, were added in Fig.4. The calculated results have good agreement with the experimental results at all contact times.

Regarding the point discussed as above, the surface renewal theory is applicable for the indirect heat transfer performance in stirred batch vessels and also in steady state plug flow vessels.

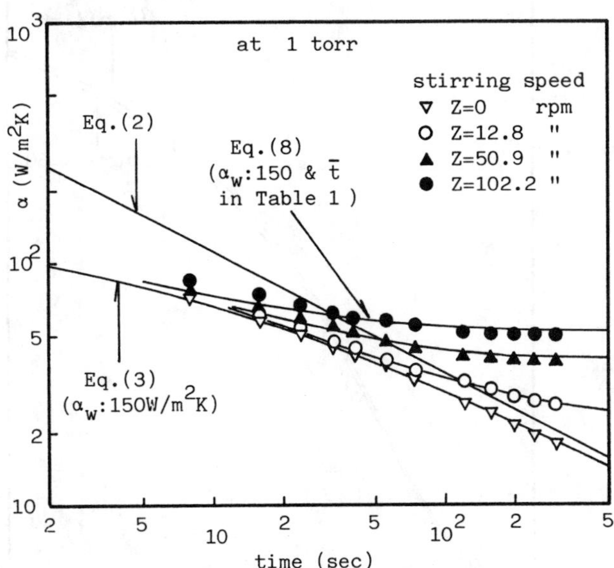

Fig.4 Comparison of results by Eq.(8) with the experimental data of glass beads at 1 torr , by Schlünder [1]

3. PHYSICAL PROPERTIES OF MATERIALS AND α_w

Since a mean resting time $\bar{t}$ strongly depends on the mechanical and geometrical properties of a system, the prediction of $\bar{t}$ is difficult.

However as mentioned as before, efforts to reduce $\bar{t}$ value sometimes do not results in large increase of α. So it is very important to know α_w value as well as physical properties of materials such as effective thermal conductivity λ, specific heat C_p and bulk density ρ.

Schlünder proposed Eq.(10) as to the system with homogeneous and real spherical materials [1] .

$$\alpha_w = \frac{2\lambda_g}{r_p}\left[(1+\frac{\delta}{r_p})\cdot \ln(\frac{r_p}{\delta}+1) - 1\right] + 4\varepsilon\sigma T_m^3 \qquad \text{---(10)}$$

$$\text{where } \delta = 2\Lambda(\frac{2}{\gamma} - 1)$$

Although actual processed materials are not sphere and have the variations in those diameters, Eq.(10) roughly indicates that a large material has a small α_w and a small one has a large α_w value.

We looked forward to obtaining α_w also by measuring performance of the effective thermal conductivities of tested materials such as wheat bran, sands (the average particle diameter: 0.7mm) and glass beads (the diameter: 0.29 to 0.42mm).

The effective conductivities λ and the resistances of tested materials can be obtained by axis temperatures for unsteady heat conduction in the cylinder (r=0.0545 m) at the constant wall surface temperature and Eq.(11) [6] .

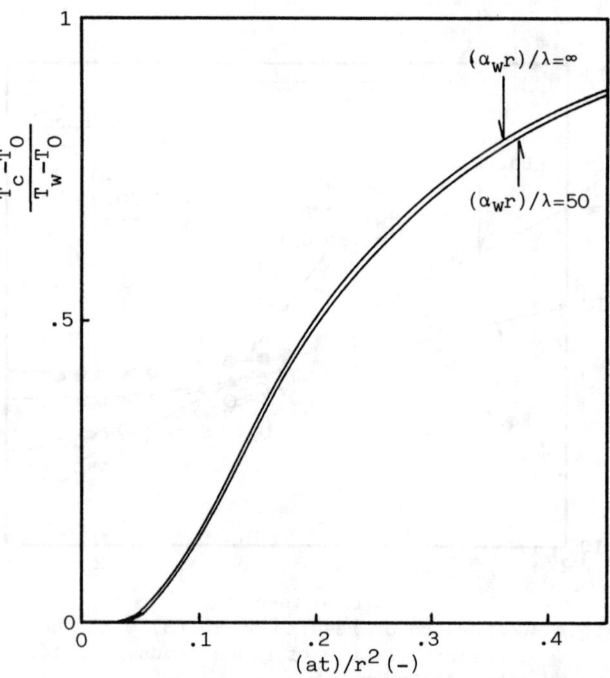

Fig.5 Axis temperature profiles for unsteady heat transfer conduction in a cylinder [6]

$$\frac{T_w - T_c}{T_w - T_0} = \sum_{n=1}^{\infty} \frac{2}{\xi_n} \cdot \frac{J_1(\xi_n)}{J_0^2(\xi_n) + J_1^2(\xi_n)} \cdot e^{-\xi_n \cdot \frac{at}{r^2}} \qquad (11)$$

where now ξ_n, n=1,2,..., are the roots of

$$\frac{\alpha_w r}{\lambda} = \frac{\xi J_1(\xi)}{J_0(\xi)}$$

As shown in Fig.5, there is not a distinct difference between the axis temperatures with the $(\alpha_w \cdot r/\lambda)$ values of 50 and infinite. Then, α_w values could not be determined because λ in Table 2 and α presented later in this paper suggest that $(\alpha_w \cdot r/\lambda)$ values of tested materials are over 50.

Table 2 shows the physical properties of tested materials obtained by Eq.(11) on the assumption that $(\alpha_w \cdot r/\lambda)$ values are infinite.

Table 2 Physical properties of tested materials

	λ (W/mK)	ρ (kg/m^3)	C_p (J/kg K)	a (m^2/s)
wheat bran	0.108	327	1285	2.57×10^{-7}
sands	0.300	1450	837	2.47 "
glassbeads	0.247	1500	879	1.87 "

4. EXPERIMENT I

The heating experiment to measure the heat transfer coefficient α at two different residence time but at the same mean resting time $\bar{t}$ was performed by two apparatuses having stirring blades of the identical pattern and the different length, which are shown in Fig.6.

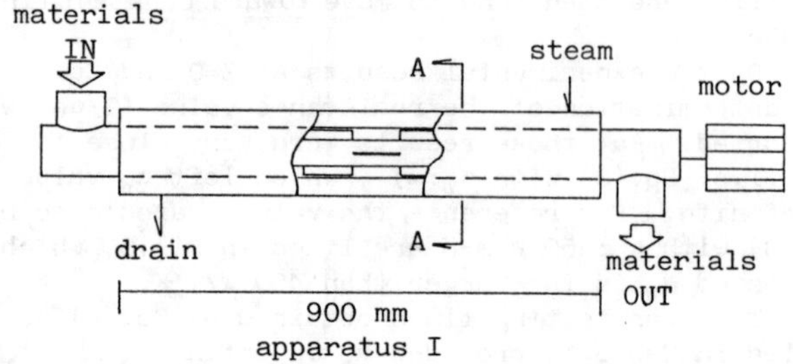

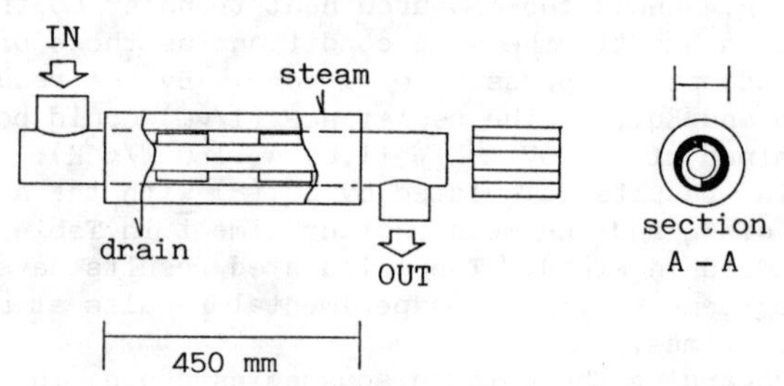

Fig.6 The apparatuses to measure heat transfer coefficients at two different t but at the same $\bar{t}$.

Material inlet and outlet temperatures were measured and heat transfer coefficients were obtained by Eq.(12).

$$\alpha = \frac{WC_p}{A} \ln \frac{(T_w - T_{in})}{(T_w - T_{out})} \quad \text{-----(12)}$$

Fig.7 shows its experimental results, and Table 3 presents the conditions of this experiment and the mean material thickness on the wall estimated from hold-up. The measured α of wheat bran are close to the calculated values by Eq.(2) and slope is nearly 1/2. This suggests that ß value is large and the residence times in this experiment are relatively short to the mean resting times by making reference to Fig.1.

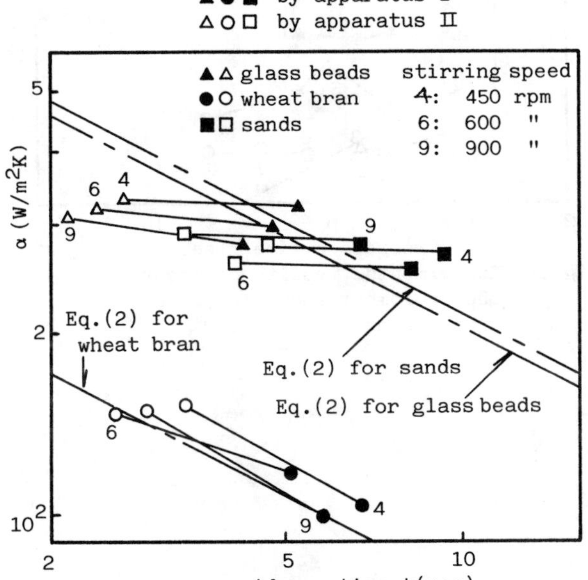

Fig.7 Experimental results by the apparatuses shown in Fig.6.

Table 3 Mean material thickness, estimated from hold-up

sample (feedrate)	stirring speed	thickness
wheat bran (100kg/hr)	450 rpm 600 " 900 "	2.5 mm 2.1 " 1.9
glassbeads (600kg/hr)	450 rpm 600 " 900 "	2.6 mm 2.3 " 1.9 "
sands (400kg/hr)	450 rpm 600 " 900 "	3.1 mm 2.2 " 1.8 "

Regarding to sands and glass beads, the slopes are nearly flat while the measured α values are close to Eq.(2), which indicates that those ß values are smaller than wheat bran's. In other words, the resistance ($1/\alpha_w$) of wheat bran may be neglected in comparison with $\lambda/\sqrt{a\bar{t}}$ value, but those of glass beads and sands may not be.

Here we note, the above conclusion does not mean that α_w value itself of wheat bran is larger than others.

5.EXPERIMENT II

Three kinds of blades were tested in the cylindrical vessel shown in Fig.8 (150mm I.D. x 1000 mm L., heat transfer area A: 0.5m^2). The pipe used for the cylinder has about ±0.8mm tolerance in diameter, and the mean clearances between the blade edges and wall are set at 1.5mm for selected three types, shown in Fig.9.

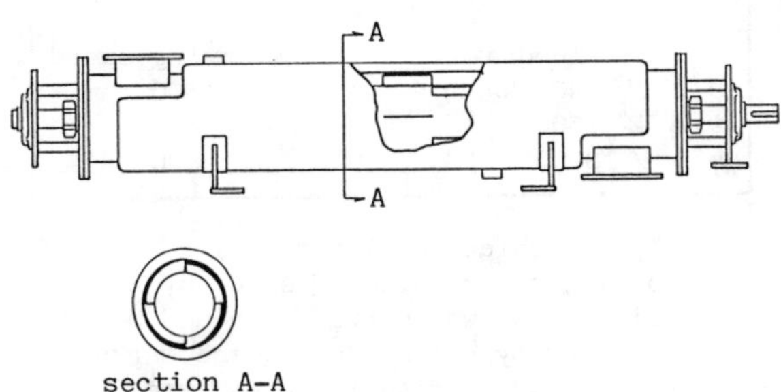

Fig.8 Cylindrical indirect heating/cooling equipment (150mm I.D. x 1000mm L., heat transfer area:0.5m^2)

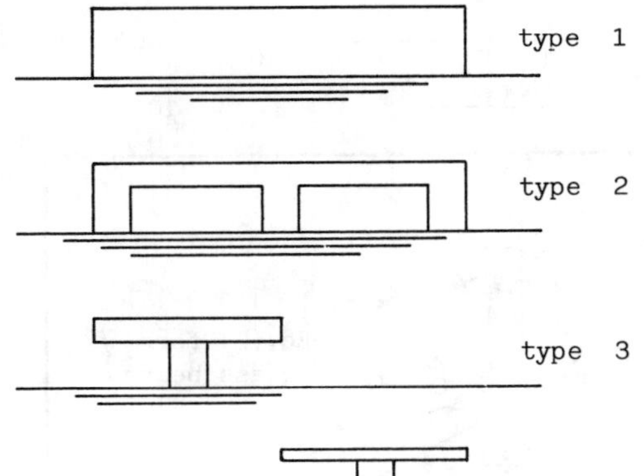

Fig.9 Shapes of the tested blades. (Average clearances between blade edges and wall are 1.5 mm)

The runs are performed with varying the stirring speed and feedrate, and changing the type of blade. The heat transfer coefficients were obtained by the same methods as those of experiment I.

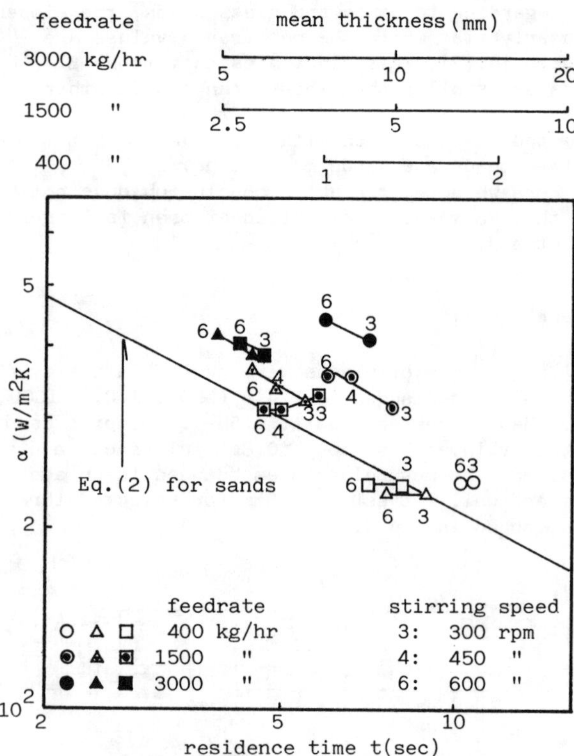

Fig.10 Experimental results of sands by the equipment shown in Fig.8. (round marks represent type 1 blade, square marks: type 2 blade, triangle marks: type 3 blade)

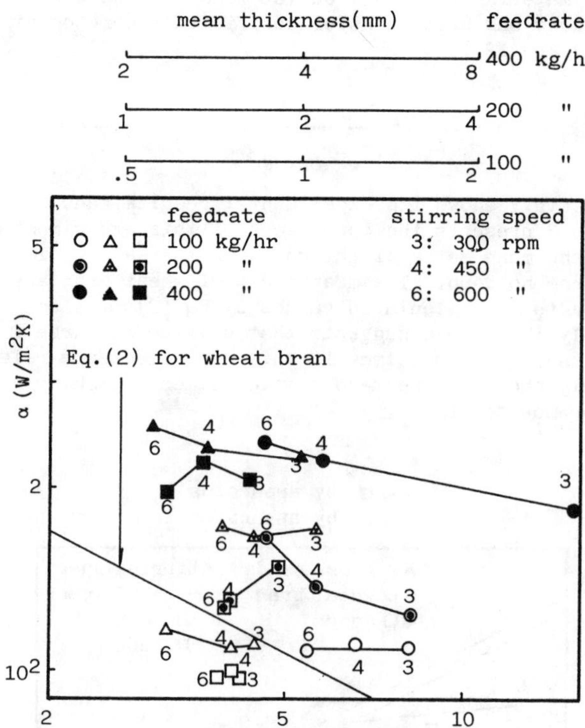

Fig.12 Experimental results of wheat bran by the equipment shown in Fig.8. (marks: same as in Fig.10)

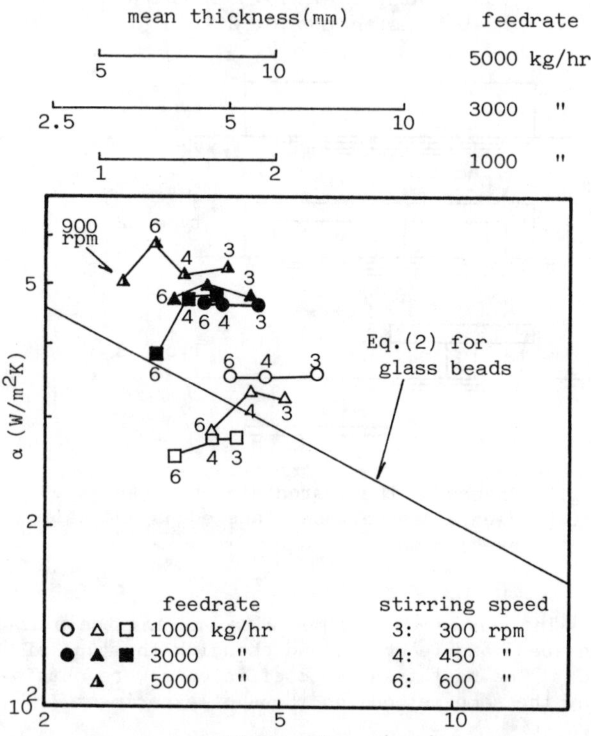

Fig.11 Experimental results of glass beads by the equipment shown in Fig.8. (marks: same as in Fig. 10)

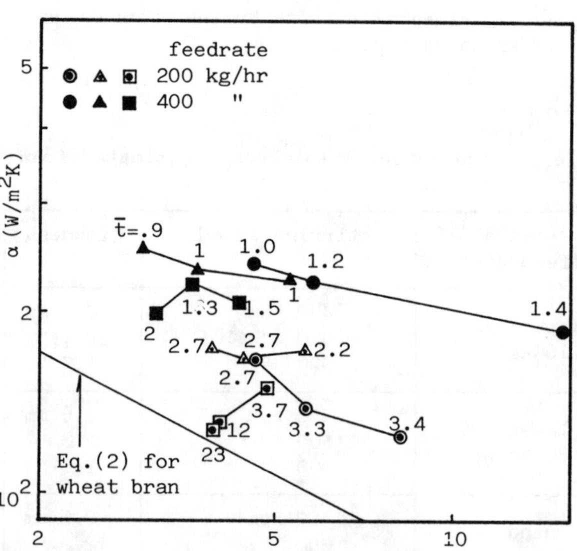

Fig.13 Mean resting times $\bar{t}$(sec) of wheat bran, obtained by Eq.(6).

Figs.10 and 11 present the experimental results of the heat transfer coefficients α of sands and glass beads respectively, and the scales above the figures indicate the mean thickness of materials calculated from hold-up. These hold-up and residence times depended on experimental conditions, such as feedrate and stirring speed. These figures indicate that the heat transfer coefficient α of sands and glass beads are strongly affected by varying feedrate, while they are not so affected by varying stirring speed and changing the type of blade.

Fig.12 presents the experimental results of the heat transfer coefficients α of wheat bran, and Fig. 13 presents the mean resting time $\bar{t}$ calculated by the above experimental results of wheat bran and Eq.(6), provided no resistance ($1/\alpha_w$). Also the feedrate affects the heat transfer coefficients of wheat bran. There are differences in the results by change of the stirring speed and blade type in comparison with sands and glass beads, although these differences are small. The mean resting time $\bar{t}$ range from 0.9 to 23 sec., while the heat transfer coefficients α range from 130 to 260 W/m^2K, concerning the experimental results at the feedrates of 200 and 400 kg/hr.

Regarding to the mean resting times of glass beads and sands, the calculation by Eq.(6) with no resistance gives the $\bar{t}$ values from 2 to 30 sec., although the experiment I suggested some existance of the resistance ($1/\alpha_w$). Therefore the above calculation would not be so accurate.

Among three tested materials, wheat bran has the biggest difference in α values, from the results calculated by Eq.(2) having individual physical properties, probably because of its largest β value.

On each material the highest value of α is within the region of twofold value of the smallest α, in spite of big difference in $\bar{t}$. So rough values of our apparatus could be obtained by Eq.(8) with the resting time $\bar{t}$ about 3 to 4 sec., and the physical property of various materials. Of course it is the matter to know α_w value in the case of large particle size.

CONCLUSION

The parameter of the mean resting time $\bar{t}$ based on the extended surface renewal theory was introduced to conductive heat transfer phenomena. The applicability of this hypothesis was testified by the good agreement of Schlünder's experimental results.

The conductive heat transfer experiment by horizontal well-stirred cylindrical vessels was performed with three kinds of materials and three types of blades. The heat transfer coefficient α value of 100 - 550 W/m^2K and the mean resting time $\bar{t}$ value of 0.9 - 30 sec. calculated by Eq.(8), were obtained.

NOMENCLATURE

d_t	mean material thickness on the wall	mm
F	resting time distribution function see appendix	
J_0	the first kind Bessel Function order 0	
J_1	the first kind Bessel Function order 1	
r	radius of a cylinder	m
r_p	radius of a particle	m
t	contact time in a batch vessel or residence time in continuous plug flow vessel	sec
t_s	apparent resting time in Eq.(4),(5)	sec
$\bar{t}$	apparent mean resting time in Eq.(6),(7),(8) see appendix	sec
T_0	initial temperature of material	K
T_c	material temperature on axis of cylinder	K
T_m	mean temperature in Eq.(10)	K
T_w	wall surface temperature	K
W	feedrate of materials	kg/hr
x	distance from wall surface at x=0	m
α_i	instantaneous heat transfer coefficient	W/m^2K
α_w	heat transfer coefficient between the wall and the surface of the adjacent layer of particles	W/m^2K
ß	$\alpha_w/(\lambda/\sqrt{a\bar{t}})$, a parameter in Eq.(8)	-
$\bar{ß}$	$\alpha_w/(\lambda/\sqrt{a\bar{t}})$, a parameter in Fig.2	-
γ	acomodation coefficient	
δ	modified mean free path	
θ	time for which a solid has been contacted with the wall, resting time	sec
Λ	mean free path of gas molecules	
λ	effective thermal conductivity of solids	W/mK
λ_g	thermal conductivity of gas	W/mK
ρ	bulk density of material	kg/m^3
τ	$t/\bar{t}$, the dimensionless contact/residence time in Eq.(6),(7),(8)	-
τ_s	the dimensionless resting time in Eq.(5)	-

LITERATURE CITED

1) Schlünder,E.U.and Wunschmann,J.:Int.Chem.Eng.,20, 555 (1980)
2) Carslaw,H.S.and Jaeger, J.C.:"Conduction of Heat in Solids,"chapter 2.4,Oxford University Press (1959)
3) Carslaw and Jaeger,op.cit.,chapter 2.7
4) Dankwerts,P.V.:Ind.Eng.Chem.,43,1460 (1951)
5) Tadaki,T.:Kagaku Koguku,27,864 (1963)
6) Bachmann,H.:"Tafeln über Abkühlungsvorgänge einfacher Körper" (Springer, 1938)

APPENDIX

The distribution function of the resting time is shown by Eq.(A1) in the steady state [4] .

$$F(\theta) = \frac{e^{-\frac{\theta}{\bar{t}}}}{\bar{t}} \qquad \text{----(A1)}$$

$$\text{where} \int_0^\infty F(\theta)\, d\theta = 1 \text{ and } \bar{t} = \int_0^\infty F(\theta)\, d\theta$$

In the unsteady state where the distribution of the resting time has not been fully developed, the resting time θ over the contact time t does not exist, then the distribution function is shown by Eq.(A2) [5] .

$$\left.\begin{aligned} F(\theta,t) &= \frac{e^{-\frac{\theta}{\bar{t}}}}{\bar{t}} + e^{-\frac{\theta}{\bar{t}}} \cdot \delta(\theta-t) \\ & \qquad \text{for } 0 \leqq \theta \leqq t \\ F(\theta,t) &= 0 \qquad \text{for } t<\theta \end{aligned}\right\} \qquad \text{----(A2)}$$

where δ: Dirac's delta function of area = 1

Using the above distribution function, instantaneous heat transfer coefficient α_i is defined as Eq.(A3)

$$\alpha_i = \int_0^t F(\theta,t)\cdot\left(-\lambda \left.\frac{\partial T}{\partial x}\right|_{x=0}\right) d\theta \times \frac{1}{T_w-T_0} \qquad \text{----(A3)}$$

So α_i without the resistance $(1/\alpha_w)$ and with the resistance are expressed by Eq.(A4) and Eq.(A5) respectively.

$$\frac{\alpha_i}{\frac{\lambda}{\sqrt{a\bar{t}}}} = \text{erf}\sqrt{\tau} + \frac{1}{\sqrt{\pi\tau}}\, e^{-\tau} \qquad \text{----(A4)}$$

$$\frac{\alpha_i}{\frac{\lambda}{\sqrt{a\bar{t}}}} = \frac{ß}{ß^2-1}\left[ß^2 \cdot e^{(ß^2-1)\tau} \cdot \text{erfc}(ß\sqrt{\tau}) -1 + ß\,\text{erf}\sqrt{\tau}\right] \qquad \text{----(A5)}$$

The mean heat transfer coefficient α during contact time is defined as Eq.(A6).

$$\alpha = \frac{1}{t}\int_0^t \alpha_i dt \qquad \text{----(A6)}$$

HEAT TRANSFER COEFFICIENT BETWEEN HEATING WALL AND AGITATED GRANULAR BED

Ryozo Toei, Takao Ohmori, Takeshi Furuta* and Morio Okazaki

Department of Chemical Engineering, Kyoto University
Kyoto, 606 JAPAN

ABSTRACT

The heat transfer coefficient between the heating wall and the mechanically agitated granular bed in small indirect-heat agitated dryer was measured. There is a clearance between the wall and the agitator blades in this type of the dryer. The effect of this clearance on the heat transfer is very significant. Based on "the particle heat transfer model" suggested by Schlünder and partly revised by Mollekopf and Martin, the heat transfer model in which the effect of the clearance was considered was proposed to correlate the observed heat transfer coefficients. The comparison between the experimental and the calculated heat transfer coefficients showed no serious deviation.

1. INTRODUCTION

The heat transfer between the heating wall and the mechanically agitated granular bed is an important process in many industrial operations: for example, in the drying, the desorption, the catalyst reaction and so on. The present work is concerned with the heat transfer in the indirect-heat agitated dryer. This type of dryer has been widely used recently. One of the advantages of this type of dryer is that the thermal efficiency is high. This is mainly due to the conductive-heating. However, the heat transfer mechanism which is basic information about the design of this dryer is not well known.

There are generally two ways to supply heat to materials to be dried in this type of dryer: "stationary heating-plane type" and "moving heating-plane type"[1]. This work deals with the former. In this type, the steam-jacketed wall oe the dryer works as a stationary heating-plane and the granular material moves on it with the motion of the agitator. And there is a clearance between the wall and the agitator blades in this dryer. It is noted that the effect of this clearance on the heat transfer is very significant. It becomes furthermore important to estimate quantitatively its effect on the heat transfer coefficient when the scale-up of the dryer is considered, because the width of the clearance can not be kept quite small as the shaft of the agitator may bend in the large scale dryer.

2. THEORY AND MODEL

"The particle heat transfer model" was proposed by Schlünder[2,3] and partly revised by Mollekopf and Martin[4]. They mention that this model can describe the heat transfer between packed, agitated and fluidized beds and the heating plane. The heat transfer is considered to consist of three basic mechanisms as follows[3].

1) wall-to-particle heat transfer
2) heat conduction in packed beds
3) heat convection by particle motion

The wall-to-particle heat transfer coefficient is given by the following equations providing that the shape of particle is spherical.

$$\alpha_p = 4(\lambda_g/d_p)\{(1 + 2\sigma/d_p)\ln(1 + d_p/2\sigma) - 1\} \quad (1)$$

$$\sigma = 2\,\frac{2-\gamma}{\gamma}\sqrt{2\pi RT/M}\,\frac{\lambda_g}{p(2c_{pg} - R/M)} \quad (2)$$

$$\alpha_s = \psi\alpha_p + (1 - \psi)\alpha_{2p} + \alpha_R \quad (3)$$

The details of these equations are described in the another paper presented in IDS'84 by the authors[5]. The surface coverage factor $\psi = 0.91$ if we assume that particles contacting to the heating wall directly are arranged in a hexagonal closest packing on it.

The heat transfer coefficient of the heat conduction in packed beds is expressed by "the penetration model" as follows under the condition that the temperature of the heating wall is constant.

$$\alpha_c = \sqrt{\lambda_e c_{pm}\rho_b/\pi t} \quad (4)$$

The effect of the heat convection by the particle motion on the heat transfer can be neglected if the bulk material particles are mixed perfectly when the agitator blade scrape them. It is noted that the models described below are only applicable under this condition.

When the resistance in the wall-to-particle heat transfer is connected in series to that of the heat conduction in packed beds, we can get the instantaneous heat transfer coefficient as follows.

$$\alpha_i = 1 / (1/\alpha_s + 1/\alpha_c) \quad (5)$$

As there is a clearance between the heating wall and the agitator blade, the heat resistance in this part must be taken into account in addition to the other heat resistances mentioned above(Fig.1). As shown in Fig.2(a), there may be some velocity distribution of particles in the clearance. It is difficult, however, to estimate this. Then, we introduce the hypothesis to this

*Now at Dept. of Food Science and Technology, Toa University, Shimonoseki 751, Japan

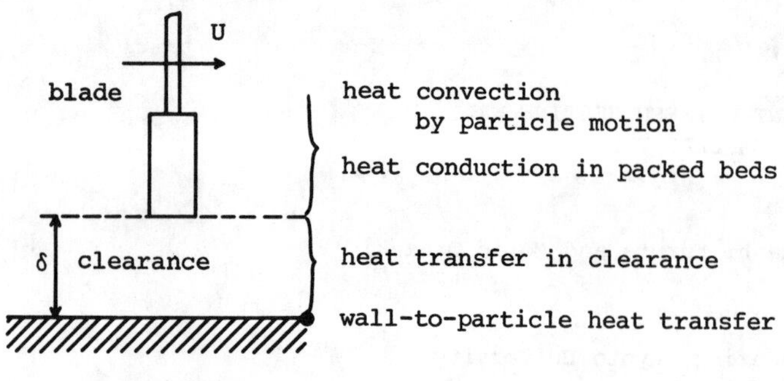

Fig.1 Heat transfer mechanism

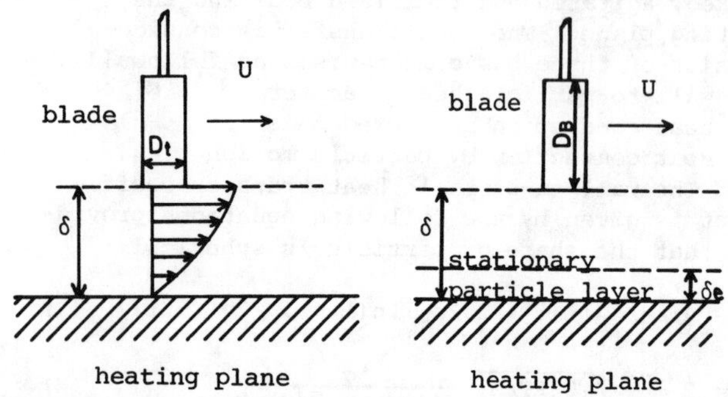

Fig.2(a) Flow between agitator blade and heating plane

Fig.2(b) Stationary particle layer

problem that there exists the stationary particle layer, which has some effective thickness δ_e, on the heating wall as shown in Fig.2(b).

According to this hypothesis, the heat transfer resistance in the clearance is equal to δ_e/λ_e and this additional resistance is connected in series to other heat transfer resistances mentioned above. So, we can get the following.

$$\alpha_i = 1 / (1/\alpha_s + \delta_e/\lambda_e + 1/\alpha_c) \tag{6}$$

Furthermore, it is assumed that the contact time in this case is equal to the interval of the scraping by the agitator blade. Therefore, the time-averaged heat transfer coefficient can be obtained by integrating during this cantact time.

$$\alpha = \frac{1}{\tau} \int_0^\tau \alpha_i \, dt$$

$$= \frac{2\alpha_s\lambda_e}{(\lambda_e+\delta_e\alpha_s)} \{ \sqrt{\pi\tau^\circ} - \ln(1+\sqrt{\pi\tau^\circ}) \}/(\pi\tau^\circ) \tag{7}$$

$$\tau^\circ = \alpha_s{}^2\lambda_e\tau/\{ (\lambda_e+\delta_e\alpha_s)^2 c_{pm}\rho_b\} \tag{8}$$

$$\tau = \pi(D - 2\delta) / U \tag{9}$$

Finally, we can calculate the heat transfer coefficient by Eqs.(1)∿(3) and (7)∿(9) in this case if δ_e is known.

In our previous model[1], there were two problems to be improved. One was the following assumption: the velocity profile of particles in the clearance(Fig.2(a)) did not change even if U changed. This hypothesis turned out to be unsuitable as the result of further experiments. The other was that our previous model included the parameter depending on the type of the agitator. To solve these problems, the following new model is proposed, which is not based on that hypothesis and which does not include such a parameter.

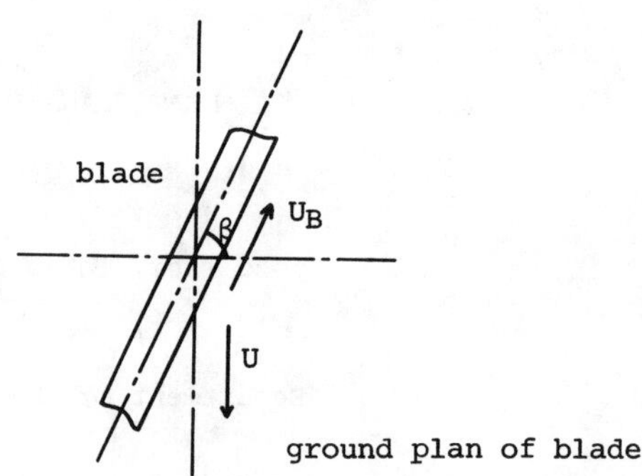

Fig.3 Relative velocity between blade and particles along side of blade

It can be thought that the effective thickness depends on the width of the clearance, the particle diameter, the fluidity of the granular bed, the thickness and the width of the agitator blade, the velocity of the agitator blade and the relative velocity between particles and the agitator blade along the side of the blade in this model(Fig.3).

$$\delta_e = f(\delta, d_p, \mu, D_B, D_t, U, U_B) \tag{10}$$

where μ is an angle of repose and it would be allowed to use as an index of the fluidity of the granular bed. However, μ,as well as D_B and D_t, did not vary so widely in the present experimental condition. And besides, the effects of D_B and D_t on δ_e might be insignificant unless they are so small. Therefore, we postulate the following form from Eq.(10) considering the extreme cases.

$$\delta/d_p \geqq 1 \quad \delta_e/d_p = 1 / (1/\xi + d_p/\delta) \tag{11}$$

$$\xi = a(\delta/d_p-1)^b/(U^c+dU_B{}^e) \tag{12}$$

$$U_B = U \sin\beta \tag{13}$$

$$0 \leqq \delta/d_p \leqq 1 \quad \delta_e = 0 \tag{14}$$

where a, b, c, d and e are constant which should be determined by fitting experimental data. The dimensionless variable ξ is an index indicating the variation of the effective thickness. If this value is equal to zero, $\delta_e = 0$ and $\delta_e = \delta$ if it becomes infinite as shown in Eq.(11). ξ is naturally depends on the same factors as δ_e in Eq.(10). So, we postulate the form of Eq.(12), where the dimensions of U and U_B are meter by second. U_B defined in Eq.(13) is the relative velocity between the granular material and the agitator blade along the side of the blade(Fig.3). This is also an index indicating the effect of the angle of the blade to the moving direction of it on δ_e. In Eqs.(11)∿(14), δ_e reaches zero if U becomes infinite and δ_e reaches δ if U becomes zero.

Through the experiments, a = 0.6, b = 0.5, c = 0.8, d = 3.5 and e = 0.45 were determined as described later in §4.

3. EXPERIMENTALS

3.1. Apparatus

The experiments were carried out with a small horizontal semi-cylindrical indirect-heat agitated dryer. Its length is 80 cm and the diameter is 20 cm. A schematic diagram of the experimental apparatus is shown in Fig.4.

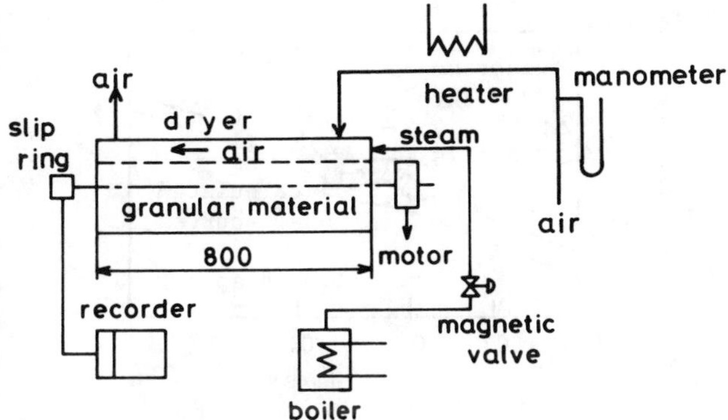

Fig.4 Schematic diagram of experimental apparatus

The temperature of the wall was kept at constant by adjusting the steam flow rate. To avoid the heat transfer from the granular material to air over the bed surface, hot air was blown over it. The heater for air was controlled to keep the temperature of air equal to that of the granular material.

The material temperature was measured at two points in a radial direction to make sure that no temperature distribution existed in that direction in the bulk material. (It had been confirmed beforehand that there was no distribution in an axial direction.) The temperatures of both wall and air were measured at four respective points and the averaged values were used in the calculation of the heat transfer coefficient. All temperatures were measured by thermocouples.

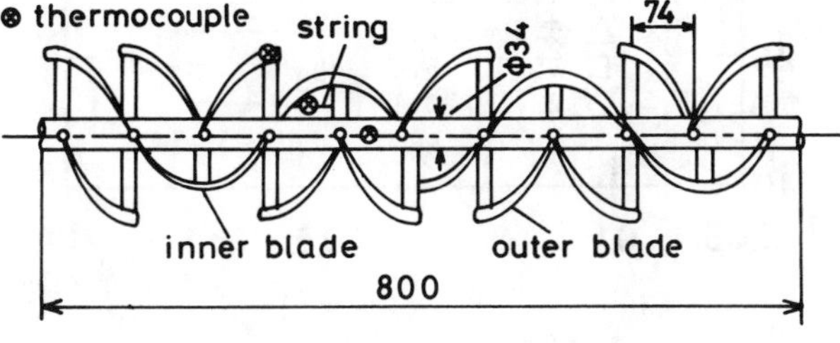

Fig.5 Double spiral agitator

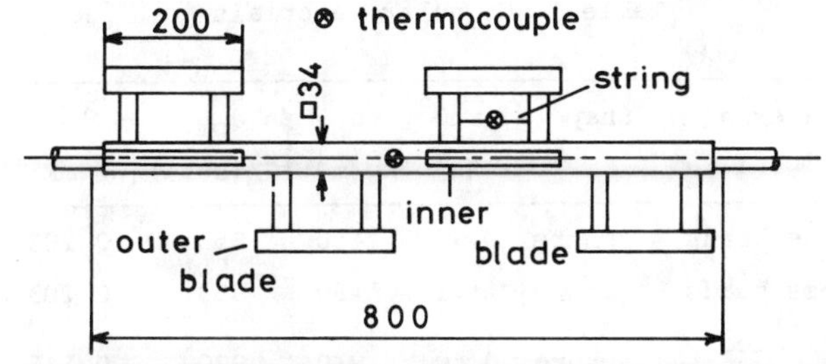

Fig.6 Flat bar agitator

Two types of agitators were used in this experiment. These are shown in Figs.5 and 6. The former shows the double spiral agitator, which consists of discontinuous outer blades and continuous inner blades. The width of the clearance between the wall and the outer blade was changed by cutting down the blades. The latter shows the flat bar agitator, which has four outer blades and four inner blades. With this unit, the width of the clearance can be changed at will by adjusting the four outer blades.

3.2. Procedure

The following equation can be obtained by the heat-balance in the system of both the granular material and the agitator.

$$W_m c_{pm} \frac{dT_m}{dt} + W_{sh} c_{psh} \frac{dT_{sh}}{dt} = \alpha A_w (T_w - T_m) - \alpha A_a (T_m - T_a) - Q_{loss} \quad (15)$$

where the first term on the left side is the heat accumulation rate of the granular material, the second is that of the agitator, the first term on the right side is the heat transfer rate from the wall to materials, the second is that from materials to air over the bed surface and the third is other overall heat loss.

As mentioned above, T_a was equal to T_m in the present experimental condition. So, the second term on the right side of Eq.(15) could be deleted. Furthermore, the third term Q_{loss} turned out to be negligible compared with the first as the result of the steady-state heat transfer experiment. Finally, the time-averaged heat transfer coefficient can be calculated by the following.

$$\alpha = (W_m c_{pm} \frac{dT_m}{dt} + W_{sh} c_{psh} \frac{dT_{sh}}{dt}) \,/\, \{A_w (T_w - T_m)\} \quad (16)$$

The hold-up of the material was about 55%. Because, the bulk material was not mixed perfectly if the hold-up was 100%. The heat transfer area A_w was determined by the observations.

After the dryer was filled with materials, the wall was heated to a given temperature by blowing steam into the jacket and the rotational speed of the agitator was set for a given one. Simultaneously, the flow of air was started. Then, the change of the temperatures of the material and the shaft of the agitator followed by the lapse of time were recorded. And then, the heat transfer coefficient was calculated by Eq.(16) using data obtained.

Table 1 Granular materials

Material	Shape	$d_p x10^3$ [m]	ρ_b [kg/m³]	c_{pm} [J/(kgK)]	λ_e [W/(mK)]
glass beads A	sphere	0.36	1450	853	0.203
glass beads B	sphere	1.1	1450	853	0.203
activated alumina	sphere	0.40	910	920	0.145
acrylic resin	sphere	0.57	690	1560	0.116

The possible overall error of the heat transfer coefficient introduced by the present experimental procedure is estimated at about 20%.

3.3. Granular materials

Four kinds of granular materials were used for the present experiments. Two of them were glass beads A and B, which have the same thermal properties and only differ in the particle diameter. The others were the spherical activated alumina and the spherical acrylic resin. The properties of these granular materials are tabulated in Table 1. In this table, the thermal conductivities were measured by the Q.T.M. rapid thermal conductivity meter(manufactured by Showa Denko Co., Ltd.;QTM-D1) and the apparent densities of the bed were obtained by our measurements.

4. RESULTS AND DISCUSSION

In Figs.7∿9, the observed heat transfer coefficients are plotted as a function of the circumferential velocity of the double spiral agitator. And in Figs.10 and 11, the observed heat transfer coefficients are plotted as the same one of the flat bar agitator. As the result of determining the parameters in Eq.(12) to provide the best agreement between data points and the curves calculated by the present model, a = 0.6, b = 0.5, c = 0.8, d = 3.5 and e = 0.45 were obtained. The calculated curves obtained in this way are also shown in Figs.7∿11.

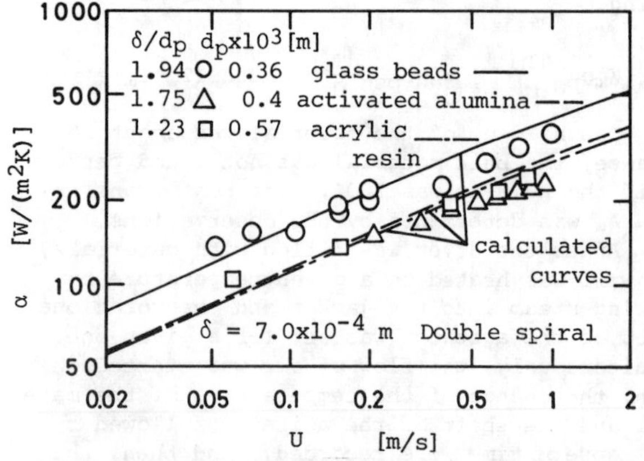

Fig.7 Heat transfer coefficient vs. circumferential velocity of double spiral agitator

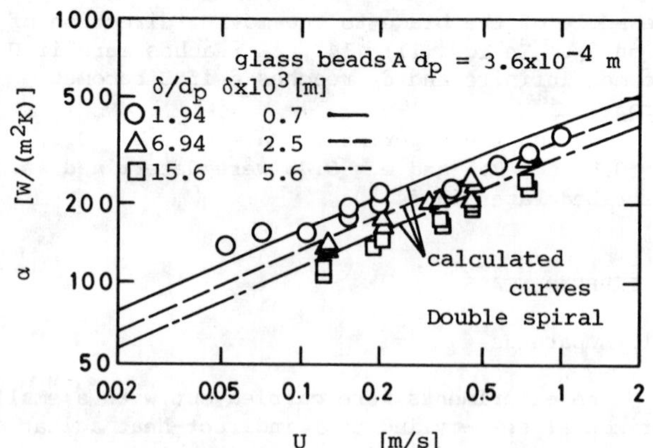

Fig.8 Heat transfer coefficient vs. circumferential velocity of double spiral agitator

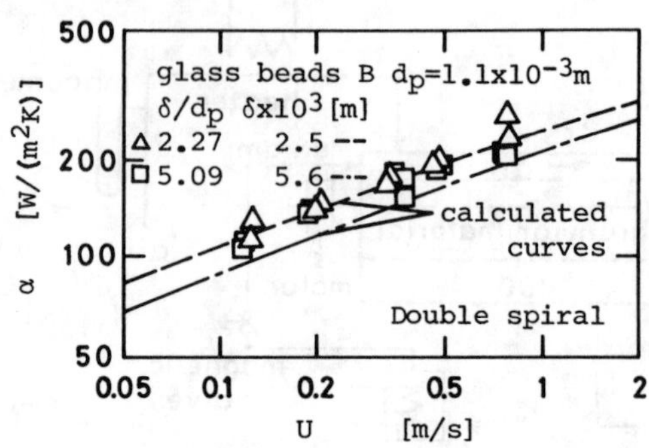

Fig.9 Heat transfer coefficient vs. circumferential velocity of double spiral agitator

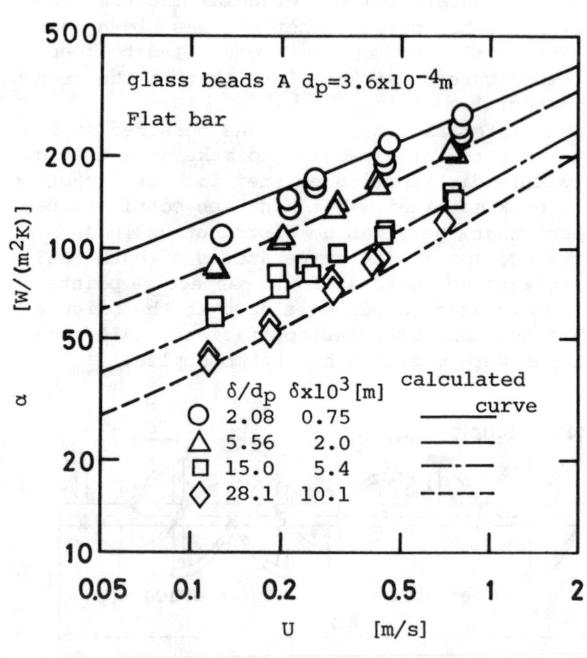

Fig.10 Heat transfer coefficient vs. circumferential velocity of flat bar agitator

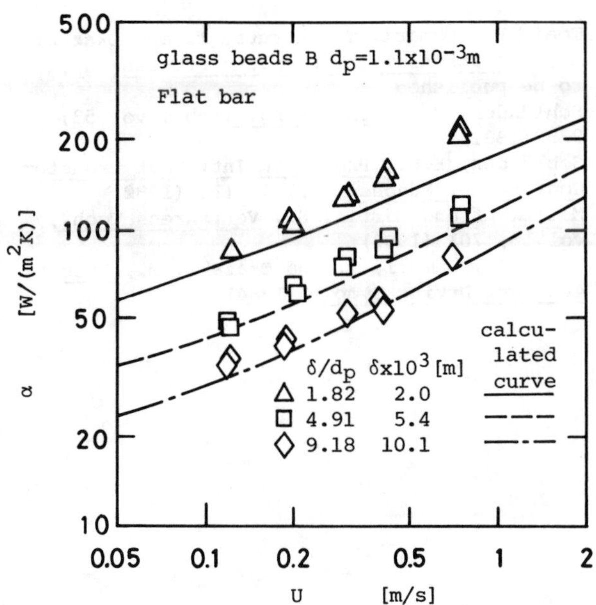

Fig.11 Heat transfer coefficient vs. circumferential velocity of flat bar agitator

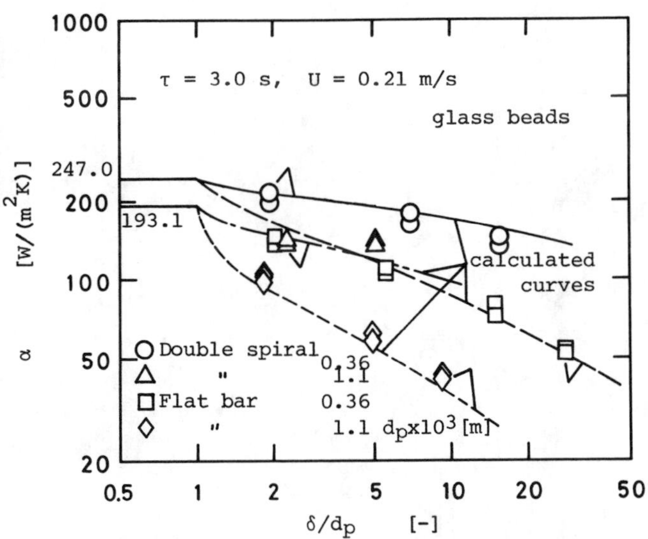

Fig.12 Heat transfer coefficient vs. dimensionless clearance

In order to illustrate the effect of the clearance on the heat transfer coefficient, the observed heat transfer coefficients for U = 0.21 m/s and U= 0.78 m/s are plotted as a function of the dimensionless clearance δ/d_p in Figs. 12 and 13.

In Fig.14, the calculated α vs. the experimental α are shown for all data. As shown, all data are almost correlated within ±25%. Therefore, it is concluded that the agreement between the calculated and the experimental heat transfer coefficients is good under the condition that the shape of particle is spherical.

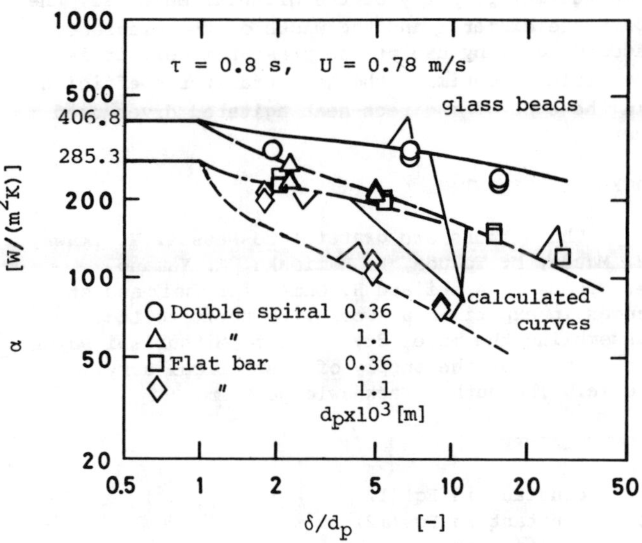

Fig.13 Heat transfer coefficient vs. dimensionless clearance

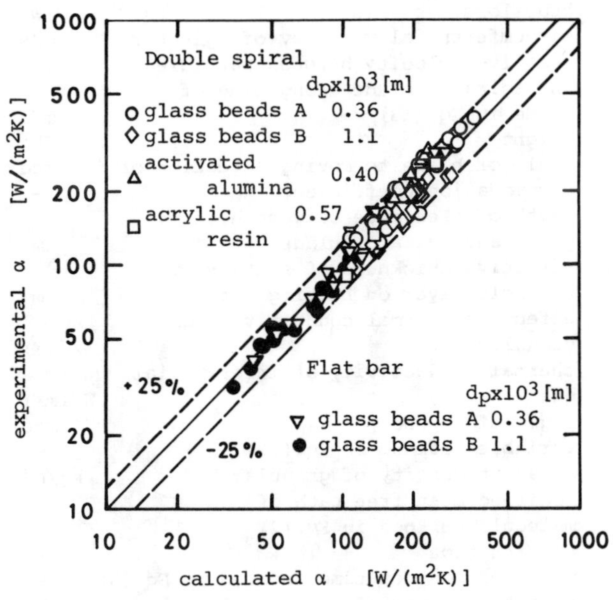

Fig.14 Experimental α vs. calculated α

5. CONCLUSION

The heat transfer coefficient between the heating wall and the mechanically agitated granular bed in the small indirect-heat agitated dryer was measured. Based on "the particle heat transfer model" suggested by Schlünder and partly revised by Mollekopf and Martin, the heat transfer model considering the effect of the clearance between the heating wall and the agitator blades was proposed. This model can describe the heat transfer coefficient dependence of the circumferential velocity of the agitator, the thermal property of the granular material, the particle diameter and the width of the clearance. Accordingly, by use of the present model, it is possible to estimate the heat transfer coefficient in the type of indirect-heat agitated dryer used here.

ACKNOWLEDGEMENT

The authors are grateful to Messrs. H. Ikawa, H. Maeda, M. Tokuda, K. Kamisaku, M. Yamano, K. Mutou, S. Araki and H. Ohmae for their assistances in experimental works, to Kurimoto Ltd. for assembling the model dryer and to Mitsubishi Rayon Co., Ltd. for the supply of a spherical acrylic resin. The authors acknowledge them.

NOMENCLATURE

a	constant in Eq.(12)	-
b	constant in Eq.(12)	-
c	constant in Eq.(12)	-
D	diameter of agitator	m
D_t	thickness of agitator blade	m
D_w	width of agitator blade	m
d	constant in Eq.(12)	-
d_p	particle diameter	m
e	constant in Eq.(12)	-
Q_{loss}	heat loss	W
U	circumferential velocity of agitator	m/s
U_B	relative velocity between particles and agitator blade along side of blade in Eq.(13)	m/s
W	weight	kg
β	angle of blade to moving direction of it	rad
γ	accomodation coefficient	-
δ	width of clearance between heating plane and agitator blade	m
δ_e	effective thickness of stationary particle layer on heating plane	m
λ_e	effective thermal conductivity of granular bed	W/(mK)
λ_g	thermal conductivity of interstitial gas	W/(mK)
μ	angle of repose	rad
ξ	variable defined in Eq.(12)	-
ρ_b	apparent density of granular bed	kg/m^3
σ	modified mean free path of gas molecule defined in Eq.(2)	-
τ	contact time	s
τ°	modified contact time defined in Eq.(8)	-
ψ	surface coverage factor	-

Subscripts

a	air over granular bed
c	conduction
g	interstitial gas
i	instantaneous value
m	granular material
p	particle
R	radiation
s	particle layer
sh	shaft of agitator
w	wall

REFERENCES

1. Toei, R., Ohmori, T., Furuta, T. and Okazaki, M., Verfahrens Tech.(Chem.Eng.Processing), to be published
2. Schlünder, E.U., Chem.-Ing.-Tech., vol 53, 925 (1981)
3. Schlünder, E.U., Proc. 7th Int. Heat Transfer Conference, München, vol 1, 195 (1982)
4. Mollekopf and Martin, H., Verfahrens Tech., vol 16, 701 (1982)
5. Toei, R., Ohmori, T. and Okazaki, M., Proc. 4th Int. Drying Symp., (1984)

HEAT TRANSFER FROM SUBMERGED BODY MOVING IN GRANULAR BED

Ryozo Toei, Takao Ohmori and Morio Okazaki

Department of Chemical Engineering, Kyoto University
Kyoto, 606 Japan

ABSTRACT

The heat transfer coefficient between the granular bed and the heating cylinder submerged in it was measured under the conditions that there existed the relative velocity between both ones and that air flowed through the bed. The velocity of through-flow gas was limited to less than the minimum fluidization velocity. The angle between the path-line of particles and the axis of the cylinder was changed. In order to measure the heat transfer coefficient, the heat flux generator was used, which was especially prepared for this measurement. Using this device, the heat transfer coefficient could be easily and quickly obtained.

The heat transfer model was proposed which was based on "the particle heat transfer model" suggested by Schlünder and partly revised by Mollekopf and Martin. Experimental heat transfer coefficients almost coincided with calculated ones. As the results, the effect of through-flow gas on the heat transfer coefficient turned out to be insignificant so long as its velocity was less than the minimum fluidization velocity but the relative velocity between the cylinder and the granular bed was not so small.

1. INTRODUCTION

The serious problem of the large indirect-heat agitated dryer, which has been widely used because of its high thermal efficiency, is that the power consumption required for rotating the heating and agitating devices is very large. Blowing gas through the granular bed is suggested as one method solving this and Ohashi reported that a little through-flow gas decreased the power consumption considerably[1]. However, the velocity of this through-flow gas must be less than the minimum fluidization one because the advantage that exhaust gas is little in this type of dryer is lost.

The present work is concerned with the heat transfer in "the moving heating-plane type"[2] of indirect-heat agitated dryer in which the device for the through-flow gas is installed. There is little information on the design of this type.

About the heat transfer between the granular bed and the body submerged in it, many studies have been made so far. They are classified as follows according to the state of the granular bed: the heat transfer in flowing or agitated beds [2∿7, for example] and in fluidized beds[8∿12, for example]. However, few studies on the heat transfer in the state mentioned above have been reported. So, the present work is concerned with the heat transfer in the flowing granular bed with the through-flow gas whose velocity is less than the minimum fluidization one in other words.

In order to measure the heat transfer coefficient in this case, the heat flux generator(H. F.G.) was used, which was especially prepared for this measurement. Using this device, the heat transfer coefficient can be easily and quickly obtained.

2. THEORY AND MODEL

"The particle heat transfer model" was proposed by Schlünder[13,14] and partly revised by Mollekopf and Martin[15]. They mention that this model can describe the heat transfer between packed, agitated and fluidized beds and the heating plane. The heat transfer is considered to consist of three basic mechanisms as follows[14].

1) wall-to-particle heat transfer
2) heat conduction in packed beds
3) heat convection by particle motion

The wall-to-particle heat transfer coefficient is given by the following equations providing that the shape of particle is spherical.

$$\alpha_p = 4(\lambda_g/d_p)\{(1 + 2\sigma/d_p)\ln(1 + d_p/2\sigma) - 1\} \quad (1)$$

$$\sigma = 2\,\frac{2-\gamma}{\gamma}\sqrt{2\pi RT/M}\;\frac{\lambda_g}{p(2c_{pg} - R/M)} \quad (2)$$

$$\alpha_s = \psi\alpha_p + (1 - \psi)\alpha_{2p} + \alpha_R \quad (3)$$

In these equations, α_p is the maximum achievable wall-to-particle heat transfer coefficient, α_s is the same wall-to-particle-layer one, α_{2p} is one between the heating plane and the second layer particle from the plane and α_R is one by radiation. The surface coverage factor $\psi = 0.91$ if we assume that particles contacting to the heating plane directly are arranged in a hexagonal closest packing on it. The second and third terms in Eq.(3) are negligible in comparison with the first term so long as it is under atmospheric pressure and temperature.

γ is an accomodation coefficient. In order to calculate its value for air, Martin[16] used the empirical equation based on the experimental data measured by Reiter et. al.[17], which is presented here in Eq.(4).

$$\lg(1/\gamma - 1) = 0.6 - (1000/T + 1)/2.8 \quad (4)$$

Using this equation, $\gamma = 0.87$ is given at 65°C for example.

The heat transfer coefficient of the heat conduction in packed beds is expressed by "the penetration model" as follows under the condition that the temperature of the heating plane is constant.

$$\alpha_c = \sqrt{\lambda_e c_{pm} \rho_b / \pi t} \tag{5}$$

As Schlünder mentioned, the heat convection by the particle motion has not been quantitatively analyzed because of the fact that the particle motion is not well known. However, its effect on the heat transfer can be neglected if the bulk material particles are mixed perfectly. It is noted that the model described below are only applicable under this condition.

When the resistance in the wall-to-particle heat transfer is connected in series to that of the heat conduction in packed beds, we can get the instantaneous heat transfer coefficient as follows.

$$\alpha_i = 1 / (1/\alpha_s + 1/\alpha_c) \tag{6}$$

Substituting Eqs.(3) and(5) into Eq.(6) and integrating it during the contact time τ, the time-averaged heat transfer coefficient between the heating plane and the granular material is finally obtained by the following equations.

$$\alpha = \frac{1}{\tau} \int_0^\tau \alpha_i \, dt$$

$$= 2\alpha_s\{\sqrt{\pi\tau^*} - \ln(1+\sqrt{\pi\tau^*})\}/(\pi\tau^*) \tag{7}$$

$$\tau^* = \alpha_s^2 \tau / (\lambda_e c_{pm} \rho_b) \tag{8}$$

We make the following assumptions furthermore to obtain this contact time.

(1) The particle flow near the cylinder wall is parallel to this wall and its velocity is constant and equal to that of bulk particles.
(2) There is no velocity distribution and no mixing of particles in the radial direction of the cylinder.
(3) Particles completely contact the semi-circle of the ellipse, which is the shape of the cross section of the cylinder when it is cut by the plane including the path-line of particles(Fig.1).

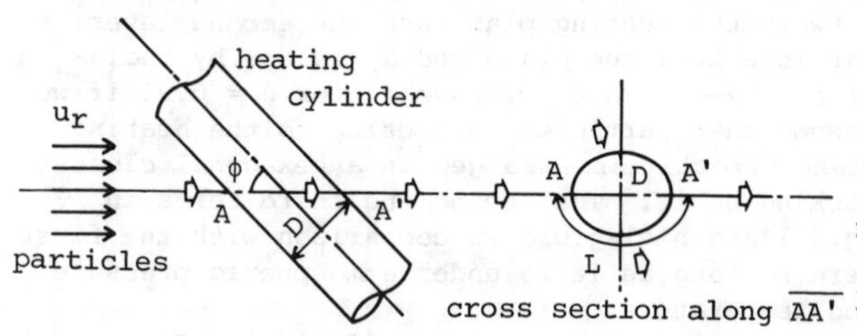

Fig.1 Motion of moving particles around inclined cylinder

Under these assumptions, the contact time can be given by the following equations.

$$\tau = L / u_r \tag{9}$$

$$L = \frac{D}{\sin\phi} \int_0^{\pi/2} \sqrt{1 - \cos^2\phi \sin^2\xi} \, d\xi \tag{10}$$

where D is the diameter of the cylinder and L is the contact length, which is equal to the semi-circle of the ellipse as mentioned above.

It is thought that the through-flow gas has the effect only on the effective thermal conductivity of the granular bed in the heat transfer. This value can be predicted by the following equation proposed by Yagi and Kunii[18].

$$\frac{\lambda_e}{\lambda_g} = \frac{\lambda_{e\ u_g=0}}{\lambda_g} + (ab)\left(\frac{c_{pg}\eta_g}{\lambda_g}\right)\left(\frac{d_p G}{\eta_g}\right) \tag{11}$$

where a and b are dimensionless parameters depending on the state of packing of particles. When particles are arranged in a closest packing, a = 0.179 and b = 1.

Finally, the heat transfer coefficient in this case can be calculated by Eq.(1)∿(3) and (7) ∿(11).

3. EXPERIMENTALS

3.1. Apparatus

A schematic diagram of the experimental apparatus is shown in Fig.2. The experiments were carried out with a vertical double-cylindrical type of trough. The outer diameter is 70 cm and the inner one is 30 cm. The height is 35 cm except for the wind box. The granular material was filled with the annular space of the trough. The H.F.G. and the mixing rod were inserted into the bed from the upper side as shown in Fig.2. Air was blown to the wind box installed in the lower part of the trough through the rotary joint and the center hollow shaft. The perforated plate and the metal mesh were held between the main body of the trough and the wind box. The material temperature was measured by the thermocouple at about 10 cm ahead of the H.F.G..

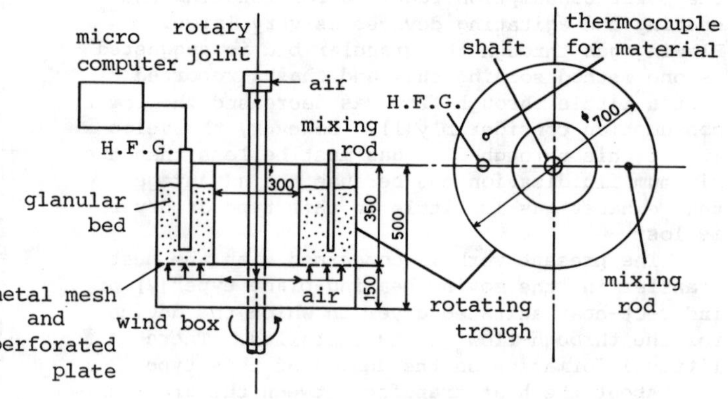

Fig.2 Schematic diagram of experimental apparatus

In this apparatus, the granular bed is moving, while the submerged heater is stationary. However, it is thought that this heat transfer coefficient is equal to that in the case which the heating plane is moving in the granular bed mixed perfectly.

The principle and dimension of the H.F.G. is shown in Fig.3. The surface temperature of the main heater was kept at constant by adjusting the electric power. In order that heat generated in the main heater may be transferred completely from its surface to the granular bed, the guard heaters and the devices for measuring the temperature difference(D.M.T.D.) are installed at both ends of the main heater. Each guard heater was controlled to keep the temperature difference indicated by the corresponding D.M.T.D. equal to zero. The D.M.T.D. consists of dozens of thermocouples, by which the electric power generated by the temperature difference of both sides of the D.M.T.D. is multiplied.

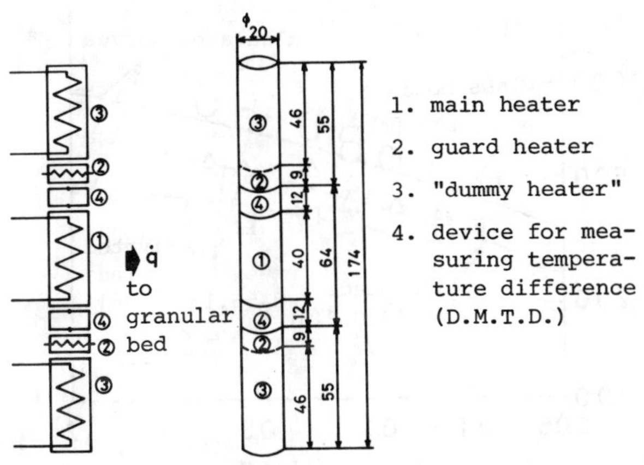

Fig.3 Principle and dimension of heat flux generator (H.F.G.)

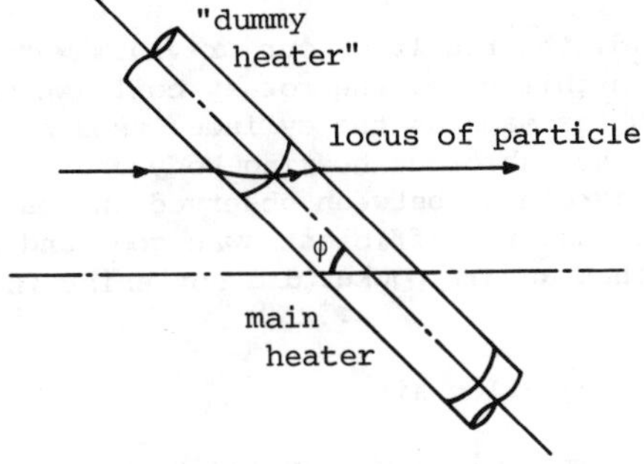

Fig.4 Entrance effect

More two heaters called "the dummy heater" are installed in the H.F.G.. These heaters prevent the error caused by so-called "the entrance effect" from getting into the measurement. When the angle between the path-line of particles and the axis of the cylinder is small to some extent, the particle flowing along the locus shown in Fig.4 reaches the main heater with keeping the temperature of the bulk unless "the dummy heater" is installed. This causes the over-estimation of the heat transfer coefficient. The surface temperatures of these heaters were controlled to keep them equal to that of the main hater. Each temperature in the H.F.G. was measured by the thermocouples. All heaters consisted of the nichrom wire were heated electrically and controlled by the micro computer.

The H.F.G. was calibrated by the experiment of the steady state heat transfer to flowing water. As the result, the possible error by using the H.F.G. was estimated within 10%.

3.2. Procedure

Using the H.F.G., the heat transfer coefficient can be calculated by the following equation as the electric power put into the main heater is equal to the heat flux from the heater surface to the granular bed.

$$\alpha = q / A_h(T_h - T_m) \qquad (12)$$

After the granular material was filled with the annular space in the trough, air was blown into the bed. Then, a given rotational speed was set up and the electric power was put into the main heater. After the surface temperature of the main heater reached a given one and other heaters were controlled adequately, the change of the electric power and the temperature of the material followed by the lapse of time were recorded. Then, the heat transfer coefficient was calculated by Eq.(12) using data obtained.

The possible overall error of the heat transfer coefficient introduced by the present experimental procedure is estimated at about 15%.

3.3. Granular materials

The glass beads was used for the present experiments. The properties are tabulated in Table 1. In this table, the thermal conductivity was measured by the Q.T.M. rapid thermal conductivity meter(manufactured by Showa Denko Co., Ltd.;QTM-D1) and the apparent density of bed was also obtained by our measurement.

Table 1 Granular material

Material	Shape	$d_p \times 10^3$ [m]	ρ_b [kg/m^3]	c_{pm} [J/(kgK)]	λ_e [W/(mK)]
glass beads	sphere	0.36	1450	853	0.203

4. RESULTS AND DISCUSSION

4.1. No through-flow air

In Figs.5∿7, the heat transfer coefficients are plotted as a function of the velocity between the heating cylinder and the granular bed under the condition of no through-flow air. The curves in the same figures are calculated by the present model explained in §2. In the case of ϕ =90° and ϕ =-45°(Figs.5 and 7), the comparison between experimental and calculated heat transfer coefficients shows no serious deviation so long as u_r is less than about 0.8 m/s. However, experimental data are located below the calculated curves in the range that u_r is more than about 0.8 m/s. This can be explained by the fact that an air pocket was observed behind the cylinder in this range. It is thought that the air pocket caused the heat transfer area to decrease and therefore the heat transferred from the heating surface to the granular bed was also decreased. If it is possible to estimate the real heat transfer area, experimental data in this range also coincide with calculated ones.

In the case of ϕ =+45° (Fig.6), as the size of the air pocket might be small even if it existed, the deviation in the range mentioned above is small. While in the range that u_r is low, observed data are located a little below the calculated curve. It is thought that this is caused by the fact that the contact length was extended by that the path line of particles turned to the bottom along the cylinder. Though it is thought that the path-line of particles might also turn to the surface of the bed in the case of ϕ =-45°, such a deviation did not appear in Fig.7. This may be explained by the effect of the gravity force.

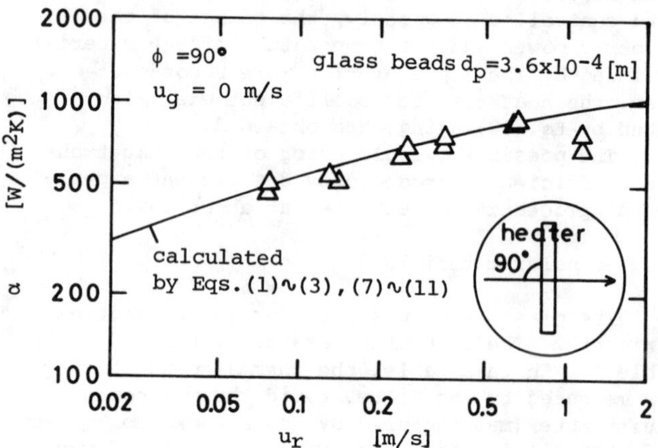

Fig.5 α vs. u_r with no through-flow air (ϕ = 90°)

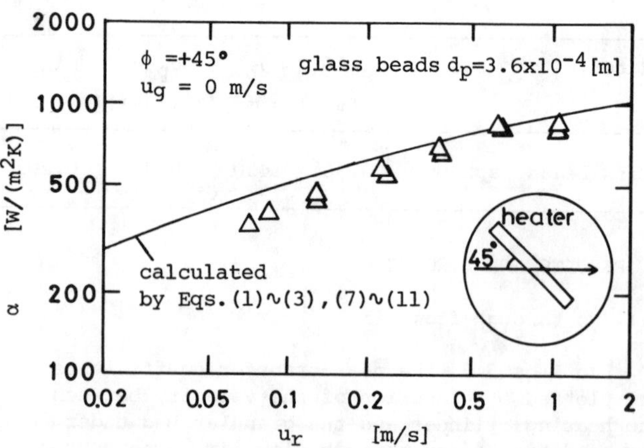

Fig.6 α vs. u_r with no through-flow air (ϕ =+45°)

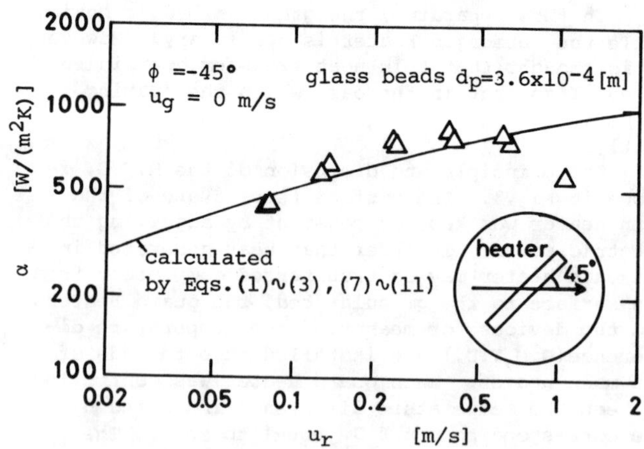

Fig.7 α vs. u_r with no through-flow air (ϕ =-45°)

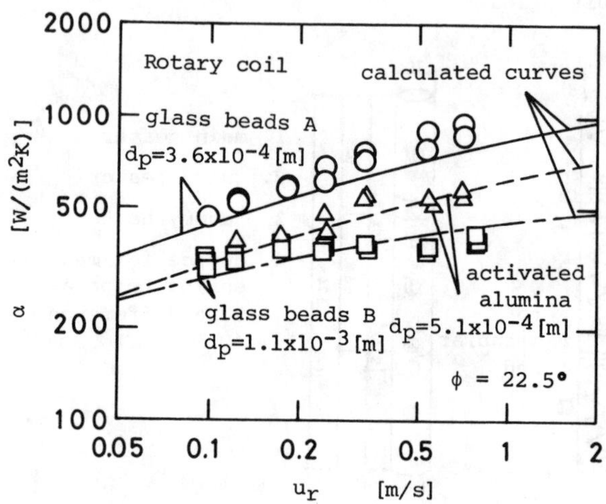

Fig.8 α vs. u_r (rotary coil) with no through-flow air [2]

In Fig.8, the result of our previous work [2] is shown. In this case, the rotary coil, whose diameter was the same as the cylinder used here, was used as the submerged heating body and ϕ = 22.5°. The agreement between observed and calculated heat transfer coefficients was good and it is thought that an air pocket did not exist in this case.

4.2. With through-flow air

The minimum fluidization velocity u_{mf} was determined to be equal to 0.148 m/s experimentally in this case. The heat transfer coefficients are plotted as a function of the relative velocity between the heating cylinder and the granular bed under the condition that ϕ = 90° and ug = 0.14 m/s in Fig.9. This velocity is a little less than the minimum fluidization one. The tendency of the agreement between the observed and calculated α is the same as that in Fig.5. But, the air pocket appeared when u_r was less large. In Fig.10, α vs. u_r is shown in the case of ϕ =+45°. As shown, the observed data are scattered when

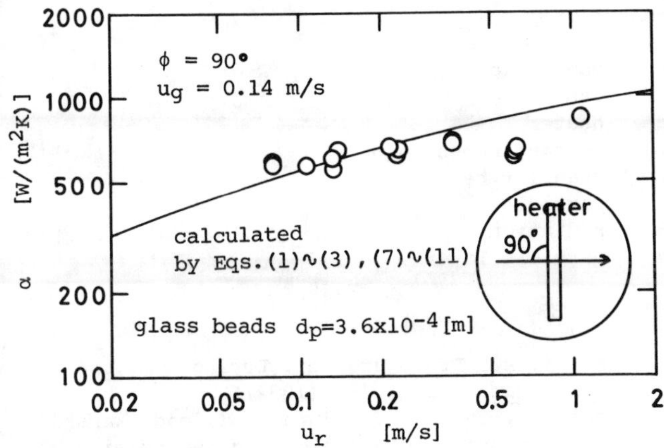

Fig.9 α vs. u_r with through-flow air (ϕ = 90°)

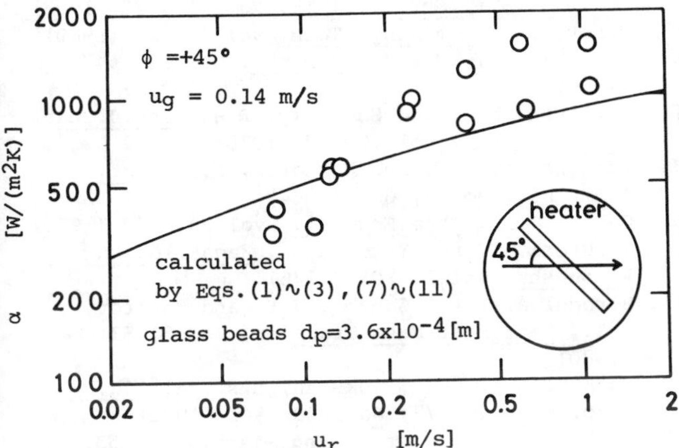

Fig.10 α vs. u_r with through-flow air (ϕ =+45°)

u_r is larger than 0.2 m/s. This was caused by the fact that the state of the granular bed near the cylinder was unstable. So, the discussion is limited to less than u_r = 0.2 m/s in the following.

In Figs.11∿13, the heat transfer coefficients are plotted as a function of the velocity of the through-flow air. On these same figures are given the results calculated by the present model. Those lines are nearly flat and experimental data are also not affected by the through-flow air in the range that its velocity is less than u_{mf}. When u_g is a little larger than u_{mf}(u_g =0.155 m/s), data do not suggest any definite tendency. However, nothing can be discussed about this because the fluidized state near the cylinder might not be uniform.

As known in Eq.(11), the effective thermal conductivity of the granular bed with the through-flow gas mainly depends on the velocity of it and the particle diameter. And the minimum fluidization velocity also depends on the particle diameter. Therefore, it might be thought that the effect of the through-flow gas is little because the particle diameter used here is small and that the larger the particle diameter is, the larger its effect on the heat transfer is. However, the wall-to-particle heat transfer coefficient α_p, which is independent of the effective thermal conductivity

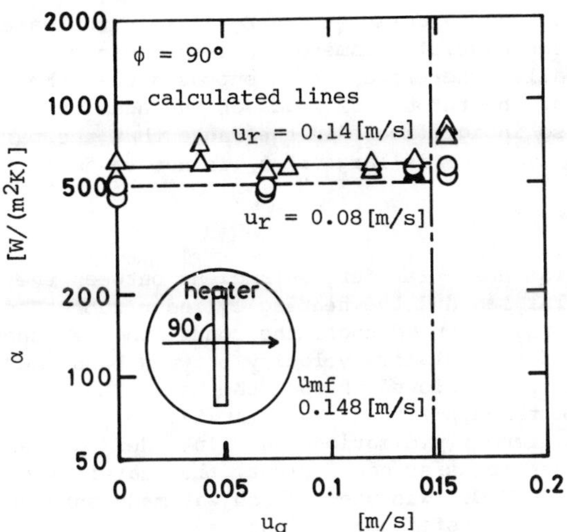

Fig.11 α vs. u_g (ϕ = 90°)

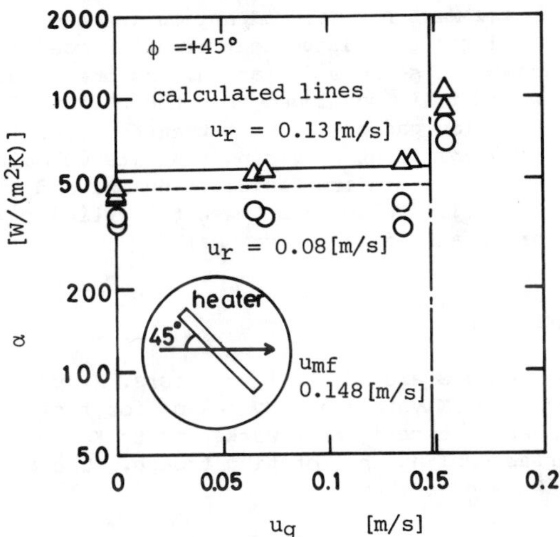

Fig.12 α vs. u_g (ϕ = +45°)

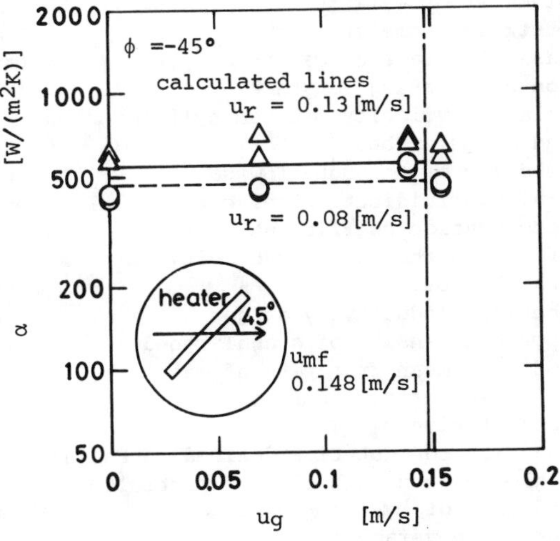

Fig.13 α vs. u_g (ϕ = -45°)

of the bed, is dominant (in Eq.(6)) in the case that the particle diameter is large unless u_r is so small. Therefore, it is supposed that the effect of the through-flow gas on the heat transfer is also insignificant in the range that the particle diameter is large if u_r is not so small.

5. CONCLUSION

The heat transfer coefficient between the granular bed and the heating cylinder submerged in it was measured under the condition that there existed the relative velocity between both ones and that air flowed through the bed. The velocity of the through-flow air was limited to less than the minimum fluidization one. In order to measure the heat transfer coefficient, the heat flux generator(H.F.G.) was used, which was made sure that the heat transfer coefficient was obtained with good accuracy.

The heat transfer model was proposed which was based on "the particle heat transfer model" suggested by Schlünder and partly revised by Mollekopf and Martin. The comparison between the observed and the calculated heat transfer coefficients showed no serious deviation. As the result of this work, it is concluded that the effect of the through-flow gas on the heat transfer coefficient is insignificant providing that its velocity is less than the minimum fluidization one and that the relative velocity between the cylinder and the granular bed is not so small.

ACKNOWLEDGEMENT

The authors are grateful to Messrs. H. von Saint Paul, M. Miyahara and A. Takami for their assistances in experimental works and to Kurimoto Ltd. for assembling the rotating trough. The authors acknowledge them.

NOMENCLATURE

a	constant in Eq.(11)	-
b	constant in Eq.(11)	-
D	diameter of cylinder	m
d_p	particle diameter	m
G	mass flow rate of gas	kg/(m^{2}s)
L	contact-heating length	m
u_r	relative velocity between cylinder and granular bed	m/s
u_g	velocity of through-flow gas	m/s
u_{mf}	minimum fluidization velocity	m/s
γ	accomodation coefficient	-
λ_e	effective thermal conductivity of granular bed	W/(mK)
λ_g	thermal conductivity of gas	W/(mK)
ρ_b	apparent density of granular bed	kg/m^3
σ	modified mean free path of gas molecule defined in Eq.(2)	-
τ	contact time	s
τ^*	modified contact time defined in Eq.(8)	-
ϕ	angle between path line of particles and axis of cylinder	rad
ψ	surface coverage factor	-

Subscripts

c	conduction
g	gas
h	heater
i	instantaneous value
m	granular bed
p	particle
R	radiation
s	particle layer

REFERENCES

1. Ohashi, k., Proc. 3rd Int. Drying Symp., Birmingham, vol 1, 467 (1982)
2. Toei, R., Ohmori, T., Furuta, T. and Okazaki, M., Verfahrens Tech.(Chem. Eng. Processing), to be published
3. Kurochkin, Yu.P., I.F.Zh., vol 2, 3 (1960)
4. Harakas, N.K., and Beatty Jr., K.O., Chem. Eng.Prog.Symp.Ser., vol 59, No. 41, 122 (1963)
5. Ernst, R., Chem.-Ing.-Tech., vol 32, 17 (1960)
6. Kurochkin, Yu.P., J.Eng.Phys., vol 10, 447 (1966)
7. Sullivan, W.N. and Sabersky, R.H., Int.J.Heat Mass Transfer, vol 18, 97 (1975)
8. Kharchenko, N.V. and Makhorin, K.E., Int. Chem.Eng., vol 4, 650 (1964)
9. Gabor, J.D., Chem.Eng.Sci., vol 25, 979 (1970)
10. Yasutomi, T. and Yokota, S., Kagakukogaku Ronbunshu, vol 2, 205 (1976)
11. Borodulya, V.A., Ganzha, V.L. and Zheltov, A.I., Lett.Heat Mass Transfer, vol 7, 83 (1980)
12. Decker, N.A. and Glicksman, L.R., A.I.Ch.E. Symp.Ser., vol 77, No. 208, 341 (1981)
13. Schlünder, E.U., Chem.-Ing.-Tech., vol 53, 925 (1981)
14. Schlünder, E.U., Proc. 7th Int. Heat Transfer Conference, München, vol 1, 195 (1982)
15. Mollekopf and Martin, H., Verfahrens Tech., vol 16, 701 (1982)
16. Martin, H., Chem.-Ing.-Tech., vol 52, 199 (1980)
17. Reiter, T.W., Camposilvan, J. and Nehren, R., Wärme-Stoffübertrag., vol 5, 116 (1972)
18. Yagi, S. and Kunii, D., A.I.Ch.E.J., vol 3, 373 (1957)

SECTION IV: SPRAY DRYING

EFFECT OF DROPLET SIZE ON THE PROPERTIES OF SPRAY-DRIED WHOLE MILK

T.H.M. Snoeren*, A.J. Damman, H.J. Klok and P.J.J.M. van Mil

Netherlands Institute for Dairy Research (NIZO) P.O.box 20, 6710 BA EDE, The Netherlands

*Present address: Nutricia, Zoetermeer, The Netherlands

ABSTRACT

The droplet size of the whole-milk concentrate spray formed by wheel atomization strongly affects the drying characteristics of the whole-milk powder particles. The droplet size depends on the viscosity of the concentrate. The concentrate viscosity can be changed by varying the pre-heat treatment of the milk (denaturation of whey protein), the dry matter content and by the homogenization pressure. By homogenization the fat globules are divided into smaller globules with new surfaces which are occupied by the protein present in the concentrate. As a consequence of homogenization the apparent volume fraction of the fat-protein complexes increases.

In this study it was found that the particle size of the powder is a function of the basic viscosity of the concentrate (i.e. the viscosity obtained by extrapolation to infinite rate of shear). We expect that the basic viscosity is relevant, since at the wheel there exists a very high rate of shear. Increasing the basic viscosity of the concentrate results in an increased moisture content of the powder, a drop in the free fat content and a rise in the insolubility of the powder. This last may be explained by the fact that larger particles spend a longer time in the critical moisture range in which heat damage to the protein may occur.

1. INTRODUCTION

The physical properties of milk powder, such as moisture content, bulk density and solubility, are important in storage, transport and reconstitution of the powder. Drying characteristics and therefore the properties of each particle depend, under fixed drying conditions, on the size of the droplet which is leaving the atomizing wheel. The viscosity of the concentrate affects that droplet size. The viscosity in turn is governed by the homogenization of the whole milk and the heat treatment (1, 2, 3).

The viscosity of a suspension, η, depends on two parameters only, the volume fraction Φ of the dispersed particles and the viscosity of the medium. An empirical but useful relationship is given by Eilers (4).

$$\eta = \eta_{ref} \left(1 + \frac{1.25\Phi}{1 - \Phi/\Phi_{max}}\right)^2 \qquad (1)$$

in which η_{ref} = viscosity of the serum (i.e. the solution containing the milk salts and the lactose) and Φ_{max} = the maximum volume fraction. For whole milk concentrate Φ_{max} was taken to be 0.79 (cf. Φ_{max} for skim milk (3)).

The volume fraction Φ of the concentrate is affected by the amount of casein, denatured whey protein, native whey protein (3), and the fat content. The heat treatment of the milk, as well as age-thickening, influences the viscosity of the concentrate since the voluminosity of the denatured whey proteins is much higher than that of the native whey proteins (3) and the voluminosity is increased by loosening of the casein micelles (5) (see Fig. 1). If whole milk is homogenized, the protein/fat ratio is not changed. However, the finely dispersed fat globules are covered by protein. The protein (casein micelles) on the globule surface is probably not a dense layer and will enhance the apparent volume fraction (Fig. 2). A higher homogenization pressure results

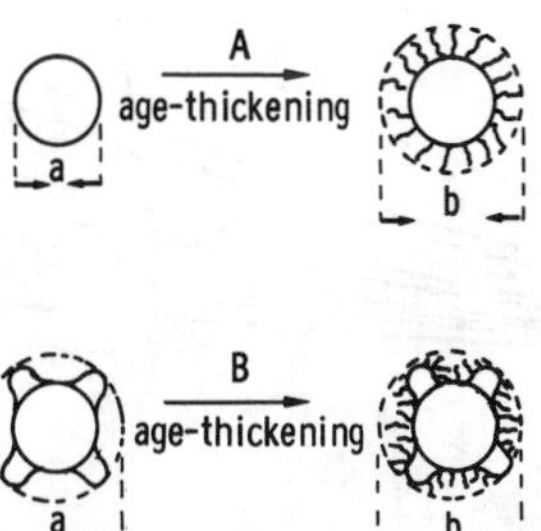

Fig. 1 Schematic representation of age-thickening process for whole-milk, non-homogenized (A) and homogenized concentrate (B).

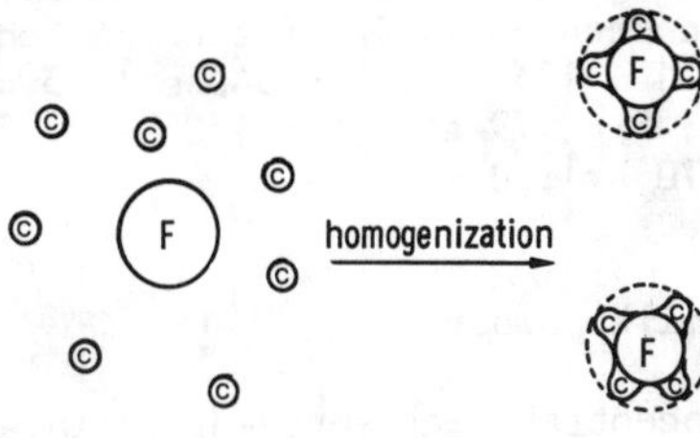

Fig. 2 The influence of homogenization on the volume fraction. The volume of fat and protein remains the same. The apparent volume, however, increases. F = fat; c = casein.

in smaller fat globules, and, in consequence, a larger surface area and thus an increase of Φ_{app}.

As a result of the increase of the volume fraction by the apparent volume fraction the viscosity of the concentrate will rise.

$$\eta = \eta_{ref}\left(1 + \frac{1.25\ (\Phi_{app}/\ \Phi)\Phi}{1 - \dfrac{(\Phi_{app}/\ \Phi)\Phi}{\Phi_{max}}}\right)^2 \qquad (2)$$

2. EXPERIMENTAL

2.1. Whole-milk concentrate

Standardized whole milk was pre-heated (2 min, 85 °C) and evaporated in a pilot plant falling-film evaporator (a combination of a two-stage Holvrieka and a two-stage Stork) to a dry matter content of about 54 %. If the concentrate was homogenized it was done by a Ranny-250 homogenizer at 50 °C at various pressures. The viscosity of the concentrate was measured, after various storage times, by a Haake Rotovisco RV2 MK500 at different shear rates. Fig. 3 shows the viscosity as a function of storage time and shear rate (a VTS diagram, viscosity, time and shear).

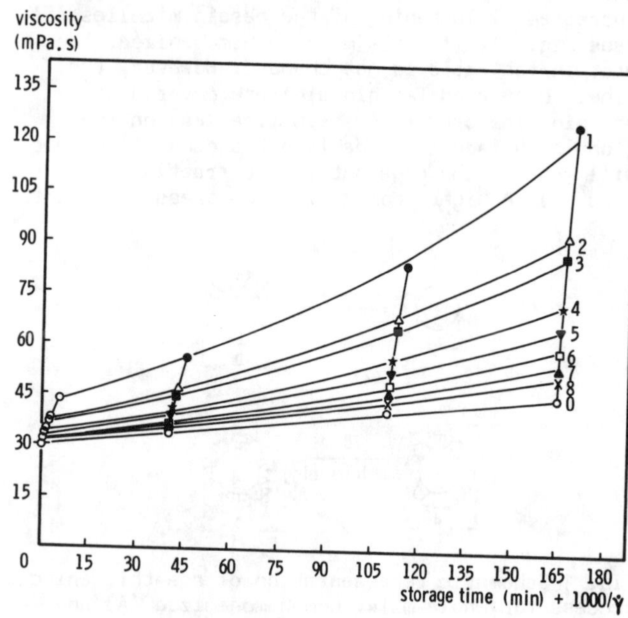

Fig. 3 VTS diagram (viscosity-time-shear) of whole-milk concentrate. Dry matter content 52.7 %. Shear rates 1 = 173 s^{-1}; 2 = 346 s^{-1}; 3 = 392 s^{-1}; 4 = 692 s^{-1}; 5 = 979 s^{-1}; 6 = 1385 s^{-1}; 7 = 1958 s^{-1}; 8 = 2770 s^{-1}; 0 = $\dot{\gamma} \to \infty$.

2.2. Whole-milk powder

The concentrate was spray-dried in a NIRO one-stage drier (25 kg water/h). The temperature of the concentrate was 50 °C. The inlet temperature of the drying air was 180 °C, the outlet temperature was varied or kept constant at 90 °C. The peripheral velocity of the wheel was 81 m/s.

2.3. Powder characteristics

The dry matter content of the concentrate and the moisture content of the powder were determined at 102 °C (6). The bulk density of the powder was obtained by an Engelsmann apparatus, whereas the particle density was measured with a Beckmann air pycnometer, model 930. The solubility index of the powders was determined by the method described by the American Dry Milk Institute (7), and the dispersibility by a method according to an IDF standard (8). Determination of the free-fat content was carried out by extracting with carbon tetrachloride (15 min) (9). Particle size analysis by a microscopic method is described elsewhere (1). The degree of homogenization was determined by a method developed by Mol (10).

3. RESULTS AND DISCUSSION

3.1. Non-homogenized whole milk

During storage of whole-milk concentrate the viscosity of the concentrate increases with time. This phenomenon, called age-thickening, is described elsewhere in full detail for skim-milk concentrate (5). The age-thickening rate can be calculated from the VTS diagram (Fig. 3).

The slope of curve (o), i.e. basic viscosity ($\dot{\gamma} \to \infty$) versus storage time, is the age-thickening rate ($\Delta\eta/\Delta t$). Equation (1) offers the opportunity to convert $\Delta\eta$ into $\Delta\Phi$, the change in voluminosity (($\Delta\Phi/\Phi$)/min, Table 1).

Table 1. The viscosity characteristics of concentrates of non-homogenized whole milk.

dry matter content (%, m/m)	viscosity at maximum shear $\eta_{\dot{\gamma}\to\infty}$ (mPa. s)	volume fraction dispersed particles Φ	shear dependence $\Delta\eta/\dot{\gamma}^{-1}.10^3$	age-thickening rate $\Delta\eta/\Delta t$	age-thickening $(\frac{\Delta\Phi}{\Phi}/\Delta t).10^4$
46,7	13,5	0,563	1,28	0,017	2,60
48,6	17,1	0,588	1,97	0,034	2,70
52,7	29,9	0,644	2,80	0,095	2,50
53,7	41,8	0,659	4,18	0,183	2,67
56,3	59,5	0,694	6,76	0,290	2,60
57,4	92	0,712	9,42	0,633	2,54
58,4	148	0,725			
58,9	143	0,732			

This ($\Delta\Phi/\Phi$)/min, the change in voluminosity, is a constant for certain types of milk and is for whole milk somewhat smaller than for skim-milk concentrate since the milk fat cannot be expected to participate in the age-thickening process (see Fig. 1B).

If the relation between viscosity of the whole-milk concentrate and droplet size during atomization is of the same order as it was for skim-milk concentrate we also may expect a relationship between viscosity of the concentrate and powder characteristics (1, 3, 11). Figure 4 shows the temperature of the outlet air as a

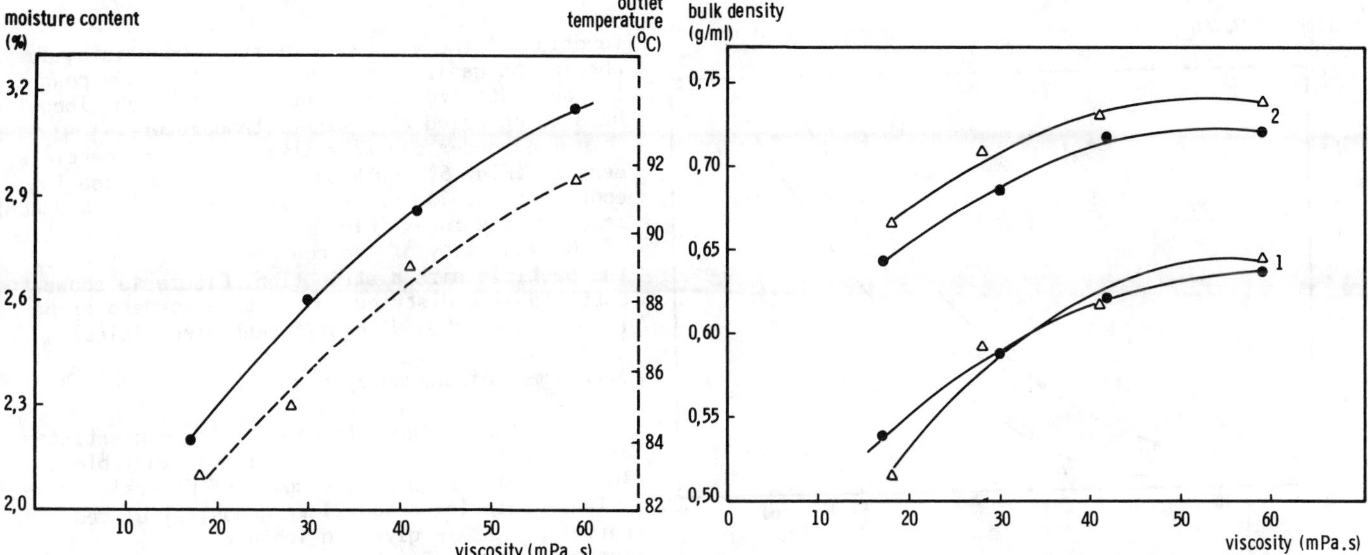

Fig. 4 Temperature of outlet air (Δ, constant moisture content) and moisture content at constant temperature of outlet air (●) as a function of the viscosity of the concentrate.

Fig. 6 Bulk density of whole-milk powder as a function of the viscosity of the concentrate 1 = no tapping of the cylinder; 2 = after tapping 1250 times (Δ and ●, see Fig. 5).

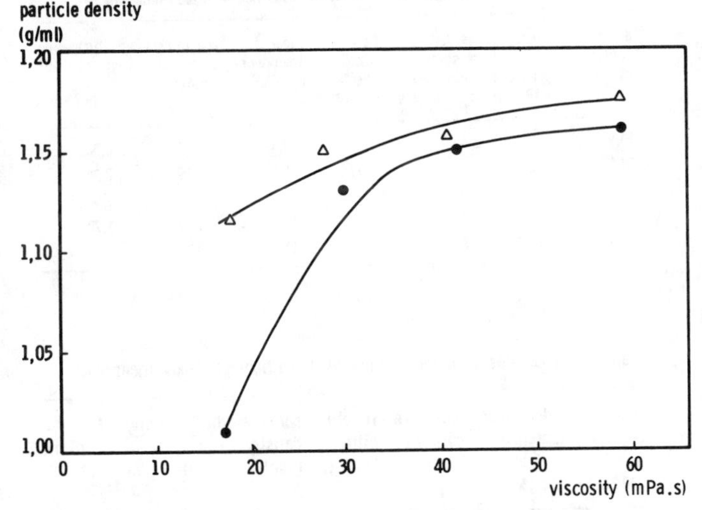

Fig. 5 Particle density of whole-milk powder as a function of the viscosity of the concentrate. Δ at constant moisture content of the powder; ● at constant outlet temperature.

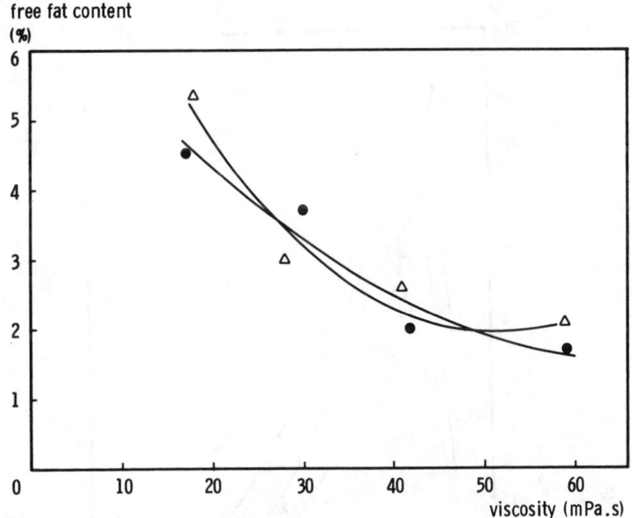

Fig. 7 Free-fat content of whole-milk powder as a function of the concentrate (Δ and ●, see Fig. 5).

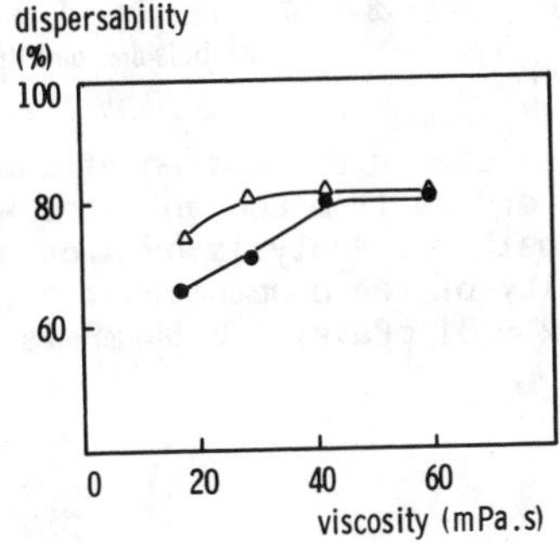

Fig. 8 Dispersibility of whole-milk powder as a function of the viscosity of the concentrate (Δ and ●, see Fig. 5).

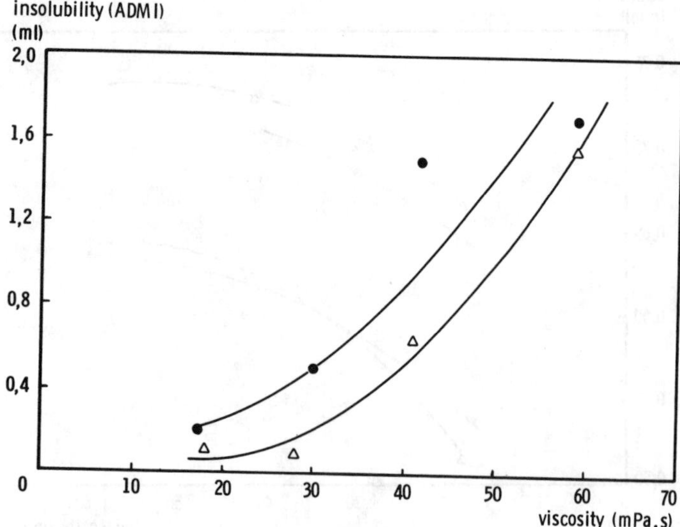

Fig. 9 ADMI insolubility index of whole-milk powder as a function of the viscosity of the concentrate (Δ and •, see Fig. 5).

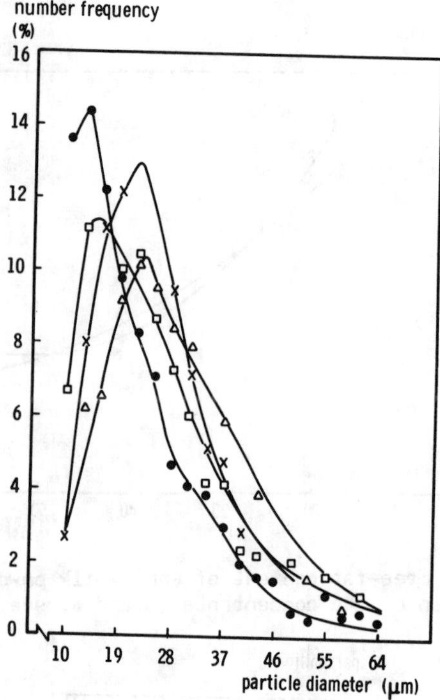

Fig. 10 Particle size distribution of some samples of whole-milk powders from concentrates with different viscosities. Analysis by microscopic method. Viscosity of the concentrate: • = 18 mPa.s; □ = 28 mPa.s; x = 31 mPa.s; Δ = 59 mPa.s.

function of the viscosity of the concentrate. To obtain the desired moisture content of the powder, the droplets from a concentrate with high viscosity have to be dried at a higher temperature.

The powder characteristics, such as particle density (Fig. 5), bulk density (Fig. 6), free fat content (Fig. 7), dispersibility (Fig. 8), and ADMI insolubility index (Fig. 9), seem to be a function of the viscosity of the concentrate or rather of the particle size distribution. Figure 10 shows the particle size distribution of some powders prepared from concentrates with different viscosities.

3.2. Homogenized whole milk

The basic viscosity ($\dot{\gamma} \rightarrow \infty$) for concentrates of homogenized whole milk are listed in Table 2. Together with a graphic display of Eilers' relationship (equation 2) Φ_{app} was calculated. The ratio Φ_{app}/Φ is given in Table 2. The homogenization effectiveness increases with homogenization pressure. Table 2 also shows the age-thickening rate and $(\Delta\Phi/\Phi)$/min, calculated from the VTS diagrams.

Table 2. The viscosity characteristics of concentrates of homogenized whole milk.

dry matter content (%, m/m)	homogenization pressure (MPa)	viscosity at maximum shear $\eta_{\dot{\gamma}\rightarrow\infty}$ (mPa. s)	homogenization effect Φ_{app}/Φ	shear dependence $\Delta\eta/\dot{\gamma}^{-1}.10^3$	age-thickening rate $\Delta\eta/\Delta t$	age-thickening $(\frac{\Delta\Phi}{\Phi}/\Delta t).10^4$
54,9	0	45,9	1	4,57	0,17	2,36
54,9	5	50,0	1,01	5,90	0,19	2,34
55,0	7,5	106	1,06	8,22	0,37	1,94
55,3	10	149	1,07	17,10	0,50	1,79
54,0	12,5	186	1,09	17,50	0,51	1,19

Table 3. Properties of whole milk powder from homogenized concentrate.

homogenization pressure (MPa)	degree of homogenization (%)	moisture content (%)	insolubility (ADMI) (ml)	particle density (g/ml)	bulk density 0x (g/ml)	bulk density 1250x (g/ml)	free fat content (%)
0	51	2,48	3,70	1,17	0,637	0,735	1,55
5	59	2,72	3,70	1,16	0,627	0,725	0,89
7,5	75	3,52	3,90	1,17	0,636	0,716	0,52
10	81	3,57	4,20	1,17	0,639	0,718	0,55
12,5	88	3,62	4,75	1,17	0,635	0,709	0,38

The characteristics of the powders obtained from concentrates of homogenized whole milk are listed in Table 3. The moisture content of the powder increases with homogenizing pressure, and the droplets leaving the atomizer wheel are larger at high viscosity, induced by the increased homogenizing pressure. The same effect of increased moisture content is observed too in the case of an increased dry matter content (Fig. 11). The free-fat content as a function of viscosity (Fig. 12) decreases with viscosity. The total surface area (and thus the free-fat content) is higher for the small particles resulting from low viscosity (Figs. 13, 14).

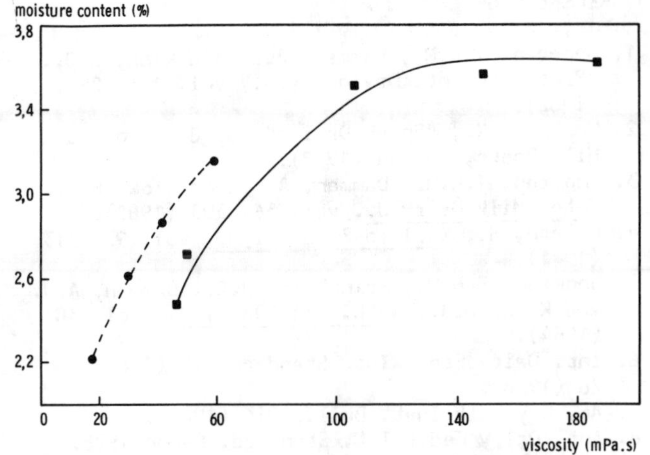

Fig. 11 Moisture content of homogenized whole milk powder as a function of the viscosity of the concentrate because of: ■ variation of homogenizing pressure; ● increase of dry matter content.

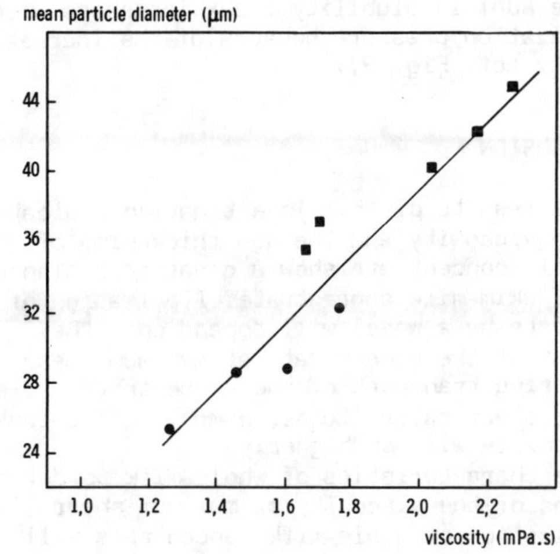

Fig. 13 Particle size of homogenized whole-milk powder as a function of the logarithm of the viscosity of the concentrate. For symbols see Fig. 11.

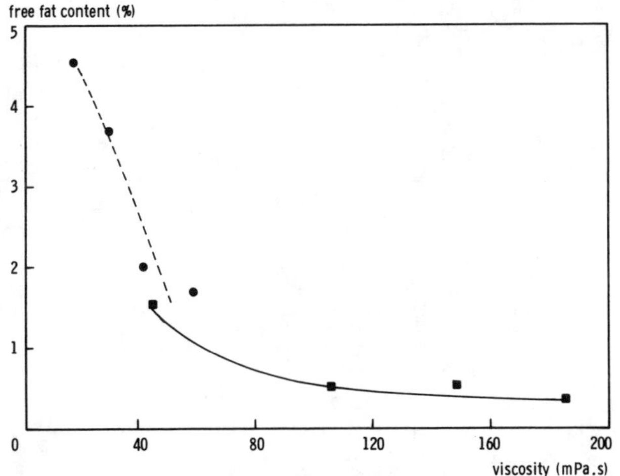

Fig. 12 Free-fat content of homogenized whole-milk powder as a function of the viscosity of the concentrate. For symbols see Fig. 11.

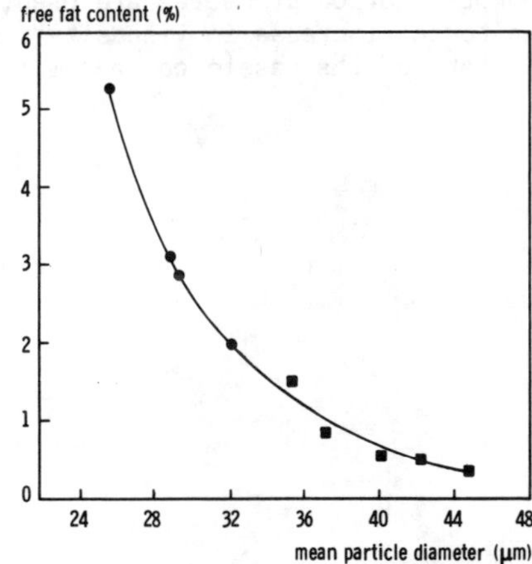

Fig. 14 Free-fat content of homogenized whole-milk powder as a function of particle size. For symbols see Fig. 11.

The ADMI insolubility index increases with homogenization pressure because of the increase in viscosity (cf. Fig. 9).

4. CONCLUSION

The results of this investigation indicate that the viscosity and the age-thickening of whole-milk concentrate show a great resemblance to those of skim-milk concentrate. Atomization of the concentrate by a wheel will depend upon the viscosity of the concentrate at maximum shear rate. During transport of the concentrate, however, at lower shear rates (tubes, pumps, buffer tanks) the viscosity will be higher.

The characteristics of whole-milk powder are functions of the viscosity at maximum shear (11). Homogenization of whole-milk concentrate will increase the viscosity of the concentrate because the apparent volume fraction of the dispersed particles has become larger. The age-thickening of these concentrates is faster than that of non-homogenized whole-milk concentrates, when expressed as increase of viscosity with time.

The powder characteristics change when different homogenization pressures are used, probably due to the increase in viscosity rather than to the state of the casein-covered milk fat globules.

5. REFERENCES

1. Snoeren, T.H.M., Damman, A.J. and Klok, H.J., "Proc. 3rd Int. Drying Symp." vol. 1 p. 290, (1982).
2. Masters, K., "Spray Drying", p. 87, Leonard Hill Books, London (1972).
3. Snoeren, T.H.M., Damman, A.J. and Klok, H.J., Neth. Milk Dairy J., vol. 36, 305 (1982).
4. Eilers, H., Kolloid-Z. & Polym., vol. 97, 313 (1941).
5. Snoeren, T.H.M., Brinkhuis, J.A., Damman, A.J. and Klok, H.J., Neth. Milk Dairy J., vol. 38 (1984).
6. Int. Dairy Fed., Int. Standard, 15 (1961) and 26 (1964).
7. Am. Dry Milk Inst. Bull., 916 (1964).
8. Int. Dairy Fed., Int. Standard, E-Doc 1978.
9. Lampitt, J.H. and Bushill, J.H., J. Soc. Chem.Ind., vol. 50, 45 T (1931).
10. Mol, J.J., 1963, Off. Org. Kon. Ned. Zuivelbond FNZ, vol. 55, 529 (1963).
11. Snoeren, T.H.M., Damman, A.J. and Klok, H.J., Zuivelzicht, vol. 73, 1004 (1981);Voedingsmiddelentechnologie, vol. 14, (26) 27 (1981).

AN ASSESSMENT OF STEAM OPERATED SPRAY DRYERS

C. T. Crowe, L. C. Chow and J. N. Chung

Mechanical Engineering Department, Washington State University
Pullman, WA 99164-2920

ABSTRACT

The performance of a spray dryer using steam in lieu of air is modeled numerically to assess the feasibility of steam as a drying medium. The numerical predictions suggest that steam is the more effective drying medium because of the higher heat transfer coefficient and specific heat. The higher heat transfer coefficient augments the gas-droplet heat transfer rate while the higher specific heat results in a smaller depression in the medium temperature due to droplet-gas thermal coupling.

1. INTRODUCTION

Spray drying is common to a variety of industries, ranging from the ceramic industry to the manufacturing of detergents and the production of dye-stuffs and pigments. There are also many applications of spray drying in the food industry, such as the production of milk powders. Spray drying is a process whereby liquid material is sprayed, dried, and collected, the residue representing the final product. It is especially attractive to drying materials that are heat sensitive because the drying process is so rapid that the materials are not subjected to high temperatures for a long period of time.

In almost all commercial spray dryers, air is the drying medium. Spray dryers are available in several different configurations. Co-current dryers are configurations in which the spray and drying air flow in the same direction and the product is collected in a cyclone separator. In counter-current dryers, the spray is injected at the top and the drying gas flows upward through the spray and out the top. The dried product falls into a bin at the bottom. Mixed flow dryers have the spray injected upward and the drying air flowing in from the top while both the air and the product exit from the bottom.

The main focus of the present work is to assess the performance of a spray dryer with steam as the drying medium. The performance evaluation will be based on comparing the drying characteristics for spray drying in air and spray drying in steam.

The evaporation rate of water from a horizontal water layer into a laminar stream of air or steam has been investigated by Chow and Chung [1]. They showed that, if the medium temperature is below 250°C, water will evaporate faster in air than in steam. However, the reverse is true if the medium temperature is above 250°C. The drying of a single drop containing suspended or dissolved solids in air or steam was studied by Trommelen and Crosby [2]. They also showed that, if the medium temperature is above 250°C, the drying rate could be higher if steam were used instead of air.

Drying with steam has numerous advantages over drying with air. Both Chu et al. [3] and Chow and Chung [4] have demonstrated that a considerable improvement in thermal efficiency is possible. Air drying of certain powdery materials may be harmful to the environment and the problem can be eliminated if steam is used because steam can be condensed to recover the powders. It was also noted that steam provides an inert environment in which to dry volatile materials, thus reducing the danger of an explosion.

A major disadvantage of using steam is that the medium temperature must be maintained at a temperature above the saturation temperature of water at the dryer pressure. This requirement may cause problems for materials that are temperature sensitive.

The time and distance required to complete drying of the droplet spray, is dependent upon the rate of heat and mass transfer between the droplets and the drying medium. Extensive analytic and numerical studies have been done on heat and mass transfer to pure droplets in uniform temperature, concentration and velocity fields. However, such "ideal" conditions do not exist in spray dryers. The local medium temperature and velocity continuously vary as the droplet is conveyed by the drying medium. Also, the physical features of the droplet change as the vapors are driven off and a particulate residue forms. This paper presents an axisymmetric numerical model which adequately accounts for the change in the medium and droplet properties as drying proceeds.

2. REVIEW OF PREVIOUS MODELS

Over the past thirty years, considerable effort has been devoted to the development of models to interpret spray dryer performance and to complement spray dryer design. The earlier models, by necessity, were based on simplified geometries and the results were usually expressed in the form of graphs or tables. Calculations had to be done by hand. With the development of the computer, more complex models have appeared which, hopefully, are more descriptive of the phenomena.

These models have been reviewed by Crowe [5]. The various numerical and analytical models were subdivided into one-dimensional, quasi one-dimensional, and axisymmetric models, and into two categories, those which neglect the effect of the droplets on the drying medium and those in which the coupling between the droplets and the drying medium need to be considered. The first category is known as one-way coupling while the second is identified as two-way coupling. Most of the one-dimensional models for spray dryers were based on one-way coupling. However, Dickinson and Marshall [6], as well as Parti and Palancz [7], devised one-dimensional models which accounted for the thermal coupling between phases.

It is not expected that the one-dimensional models will work well in the vicinity of the atomizer. Recognizing this shortcoming, Gluckert [8] developed a quasi one-dimensional model with one-way coupling in 1962. However, more recently, Katta and Gauvin [9] have proposed a quasi one-dimensional numerical scheme with thermal coupling. This particular scheme has been applied to several applications and appears to work well. The scheme, however, is based on empirical factors such as the expansion rate of a free jet. This scheme will not be applicable in situations where the spray-drying process cannot be considered as a free jet, such as in the counterflow dryer.

In 1982, O'Rourke and Wadt [10] presented a two-dimensional numerical model for spray dryers using the extension of the SOLA code (developed at Los Alamos) by Dukowicz [11] for gas-particle flows. This scheme is based on the unsteady formulation of the flow equations and the steady state solution is achieved by integrating in time until no further changes take place. This approach requires large storage capacity and run times.

3. THE DRYING PROCESS

The drying process in an air medium has been described by Crowe [12]. For spray drying in air, the evaporation rate is controlled by both the heat transfer and binary diffusion. The latent heat of evaporation is supplied by the air medium. For the case of steam drying, the evaporation rate is controlled by heat transfer only and is described below.

The process can be divided into two periods, the constant-rate period and the falling-rate period. During the constant-rate period, the slurry droplet behaves as if it were a liquid drop and the temperature of the droplet remains at the saturation temperature. The evaporation rate is quantified by the equation

$$h_{fg} \frac{dm}{dt} = -Nu\, k_g\, \pi D(T_g - T_p) \qquad (1)$$

where m is the droplet mass, Nu is the Nusselt number, k_g is the thermal conductivity of steam, D is the droplet diameter, h_{fg} is the latent heat of evaporation and (T_g-T_p) is the temperature difference between the steam medium and the droplet. The Nusselt number depends on the relative velocity between the droplet and the steam medium. Again, during this period, the temperature of the droplet is constant and is equal to the saturation temperature.

After a certain amount of liquid is removed, The evaporation rate decreases and the slurry droplet becomes a wet, porous solid or shell. Due to the small size of the droplet, it may be assumed that the temperature within the droplet is uniform and local thermodynamic equilibrium holds everywhere within the droplet. The mass loss rate depends on the relationship between the equilibrium moisture content, EMC, of the droplet and its temperature. There is very little information available on the relationship between the EMC and temperature of a droplet in a steam medium. In this paper, the relationship between the EMC and temperature for a wood particle is used [13]. During this period, the particle size is essentially unchanged and the temperature increases toward the steam temperature. The transformation from slurry droplet to porous solid may also lead to fracture of the solid into smaller fragments.

The primary concern in the design of a spray dryer is to dry the particle to a sufficient degree before removal. Particles which are not sufficiently dried may tend to cake on the walls of the dryer or cyclone separator.

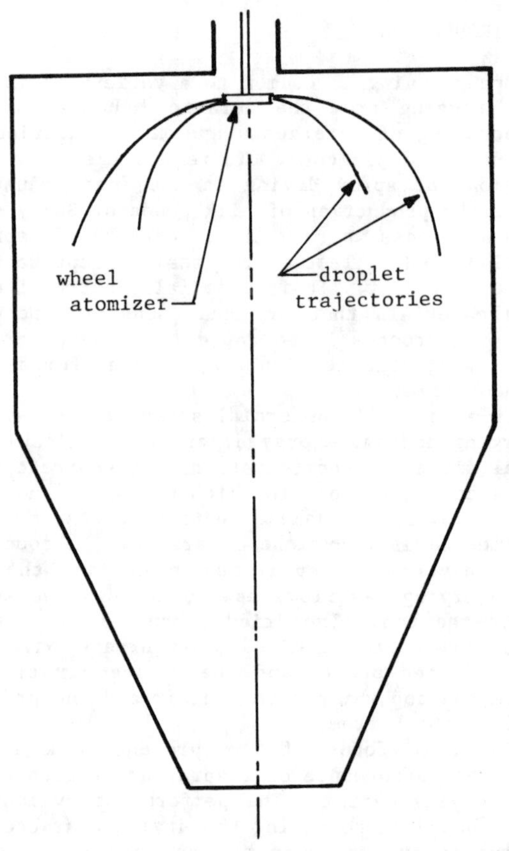

Figure 1. Co-current spray dryer configuration

The numerical model described in this paper is applied to the co-current dryer shown in Fig. 1. These types of dryers generally operate with wheel atomizers. The conical bottom suggests the possibility of a recirculation flow pattern.

4. NUMERICAL MODEL

The PSI-Cell (Particle-Source-in Cell) model [14] was developed at Washington State University in 1976. Since that time, it has been used to model a variety of applications including fire suppression, electrostatic precipitators, pneumatic transport, and coal particle and droplet combustion.

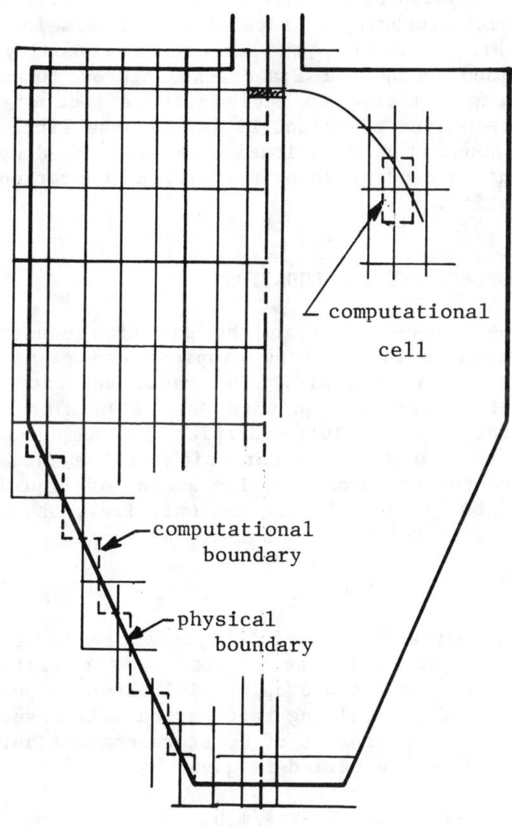

Figure 2. Computational grid system

The first task in the application of the PSI-Cell model is to divide the flow field into a series of grids defining computational cells as shown in Fig. 2. The conical section of the dryer is represented as a series of finite steps as shown in the figure. Each computational cell is treated as a control volume for the continuous phase and, as the droplets traverse the control volume, they provide a source of mass, momentum and energy to the gaseous phase.

The gas flow field is analyzed using the Eulerian approach in which finite difference equations for mass, momentum and energy conservation are written for each cell, incorporating the contribution to the droplet phase. These finite difference equations constitute a system of simultaneous algebraic equations that can be solved to yield the fluid properties in each cell. The numerical scheme used to solve the gas flow field is an extension of the TEACH program [15] developed at Imperial College of Science and Technology, London.

The droplet properties are obtained by using the Lagrangian approach in which the equations of motion for individual droplets are integrated to yield droplet trajectories in the flow field. Droplet temperature and mass are calculated along the trajectory by utilizing expressions for mass and heat transfer rates. The change in mass, momentum and energy of the droplets on traversing each cell provide droplet source terms for the gas-flow equations. In this fashion a solution is obtained which accounts for the interaction between the droplet and gas flow fields.

The droplet field is initiated by choosing starting locations for the droplet trajectories. Along each trajectory it is assumed that the number flow rate of the droplets is constant. If shattering or agglomeration occur, the number flow rate will change. This effect can easily be incorporated into the model if information is available on breakup or agglomeration rates.

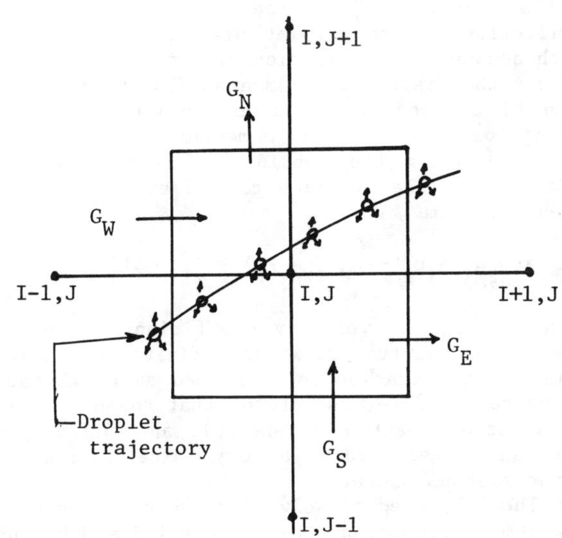

Figure 3. Evaporating droplet transversing continuity cell

4.1. Mass Conservation Equations

As illustrated in Fig. 2, the flow field is divided into a series of computational cells. A single computational cell enclosing node I,J is shown in Fig. 3. The four faces of the computational cell are identified as points on a compass: N, E, S and W. The continuity equation for steady flow of a gas droplet mixture through this cell control volume is

$$G_E + G_N - G_W - G_S + \Delta m_D = 0 \qquad (2)$$

where G_i is the mass flow rate through face i and Δm_D is the net mass efflux rate of gas due to droplets in the cell. If the droplets are evaporating, they represent an influx of mass to the cell and Δm_D is negative. One notes that the droplet mass term can be regarded as a source of mass to the gaseous flows which is the basis of the PSI-Cell model.

4.2. Momentum Equation

As noted in Fig. 3, the velocity needed to evaluate the mass flux across the west face of the cell lies midway between nodes I,J and I-1,J. Thus, the computational cell used for x-component of velocity is bounded by I-grid lines and lines midway between J-grid lines. The x-momentum control volume is displaced to the left of the continuity control volume but still identified by the node, I,J.

The momentum equation in the x-direction for the gas is written as

$$M_E^x + M_N^x - M_S^x - M_W^x + \Delta M_D^x = (P_W - P_P)A_W + F_D^x + S \qquad (3)$$

where M_i^x is the momentum flux of the gas in the x-direction across face i, ΔM_D^x is the momentum efflux due to mass transfer from the droplets, P is the pressure, F_D^x is the force on the gas in x-direction due to droplet drag and S is a term which accounts for variation of effective viscosity over the field. The momentum flux term is composed of the contributions due both to convection and diffusion [16]. The momentum flux terms are evaluated using the hybrid differencing scheme. By so doing, the equation can be expressed in an algebraic form

$$a_P u_P = \sum_{NESW} a_i u_i + A_W(P_W - P_P) + F_D^x - \Delta M_D^x + S \qquad (4)$$

where a_i are the velocity coefficients resulting from the momentum flux and finite difference forms. The summation is performed over all faces of the cell. Note, as before, that the net efflux of particle momentum in the cell and the droplet drag can be regarded as a body force term acting on the gaseous phase.

The cell used to solve for the y-component of momentum is located below the node I,J and bounded north and south by J-grid lines. The finite difference equations are of the same form as Eq. (5). These equations also have a source term representing efflux of momentum from the droplets and droplet drag on the gas.

4.3. Energy Equation

The conservation of thermal energy for steady flow through a computational cell can be written as

$$E_N + E_E - E_S - E_W + \Delta E_D = Q_D \qquad (5)$$

where E_i is the energy flux across face i (including both convection and conduction), ΔE_D is the energy efflux to gas due to droplet mass transfer, and Q_D is the heat transferred to gas from the droplets. The dissipation term is small in the application addressed in this paper and is neglected. The droplet heat transfer and energy flux represent a source of energy to the gaseous phase. The energy equation can be written in finite difference form in the same manner as the momentum equation.

Two more equations are needed to model the turbulence and dissipation rate to determine the effective viscosity in the flow field. The finite difference form of the equations is incorporated into the program to yield the kinetic energy of turbulence and dissipation rate according to the scheme proposed by Launder and Spalding [17]. Once the local turbulence intensity and dissipation rates are evaluated, the effective viscosity is determined using the Prandtl-Kolmogorov formula. No attempt is made to include the effect of the droplets on the turbulent field, because little is known about the quantitative effect of droplet size and concentration on turbulence generation or dissipation.

5. DROPLET/PARTICLE EQUATIONS

The source terms in the gas flow equations are evaluated by tracking droplets and particles through the flow field. The mass, velocity and temperature are updated with each step along the trajectory. The slurry droplet is composed of liquid and insoluble matter which will eventually comprise the dry product. The amount of liquid is quantified by the moisture content, i.e., the mass of liquid to the mass of solid,

$$W = m_\ell / m_s. \qquad (6)$$

This parameter is a useful dependent variable to define the state of the droplet. At a critical moisture content, the droplet will become a porous particle and the falling rate period will ensue.

The rate of change of moisture content during the constant-rate period is given by

$$\frac{dW}{dt} = -Nu\ k_g \pi D(T_g - T_p)/m_s h_{fg} \qquad (7)$$

The droplet diameter also decreases as moisture is removed according to the relation

$$dD^3/dt = D_s^3 (dW/dt)(\rho_s/\rho_\ell) \qquad (8)$$

where D_s is the diameter of dry solid (compacted) and ρ_s/ρ_ℓ is the solid/liquid material density ratio.

After the critical moisture content is reached or during the falling-rate period, the drying is assumed to proceed according to the equation

$$W = W_c \exp[-2.37\tanh((T_p - T_s)/24.6)] \qquad (9)$$

where T_p is the temperature of the droplet, T_s is

the saturation temperature and W_c is the critical moisture content. Equation (9) is a curve fit of the EMC-temperature relationship for a wood particle [13].

During this second stage, the particle size is assumed to remain constant.

The equation of motion for the droplet or particle is

$$d\vec{V}/dt = 3\pi\mu Df(\vec{U}-\vec{V})/m_s(1+W) + \vec{g} \tag{10}$$

where $\vec{U}$ is the local gas velocity and $\vec{V}$ is the gas viscosity. The factor f is the ratio of the drag coefficient to Stokes drag defined as

$$f = C_D Re/24 \tag{11}$$

where Re is the Reynolds number based on the relative velocity between phases. The drag coefficient is primarily a function of Reynolds number. A good representation for the factor f valid for Reynolds numbers up to 1,000 is [18]

$$f = 1 + 0.15\ Re^{0.67} \tag{12}$$

Other factors such as mass transfer, acceleration, shape, rotation, and so on can also influence the drag coefficient. These other effects are not significant in the application to spray dryers. If the slurry droplet dries to an irregularly shaped particle or fragments into nonspherical shapes, a correction would have to be made at the end of the constant rate period to account for the change in shape of the particle. This effect is easily included into the model. Integration of Eq. (10) yields the droplet velocity. A further integration yields the droplet trajectory.

The energy equation for the droplet or particle is expressed as

$$(m_\ell c_p + m_s c_{ps})dT_p/dt = Nu\pi k_g D(T_g - T_p) + \dot{m}h_{fg} \tag{13}$$

where c_p and c_{ps} are the specific heats of the liquid and solid phase, respectively, Nu is the Nusselt number, k_g is the thermal conductivity and h_{fg} is the heat of the vaporization of the liquid. The Nusselt number depends on the relative Reynolds number between phases. The Ranz-Marshall correlation [19] is generally used to account for Reynolds number effects. The energy equation can be rewritten as

$$(1 + Wc_p/c_{ps})dT_p/dt =$$

$$Nu\ \pi k_g D(T_g - T_p)/m_s c_{ps} + (dW/dt)h_{fg}/c_{ps} \tag{14}$$

During the constant rate drying period, the droplet temperature assumes the local saturation temperature. During the second stage, the temperature increases as the heat transfer term begins to dominate. Integrating this equation yields the droplet/particle temperature history along the trajectory.

During the second stage, the moisture content is a function of temperature as given by Eq. (9). Thus, the last term in Eq. (14) can be written as

$$\frac{dW}{dt}\frac{h_{fg}}{C_{ps}} = \frac{dW}{dT_p}\frac{dT_p}{dt}\frac{h_{fg}}{c_{ps}} \tag{15}$$

and the equation of particle temperature becomes

$$(1+W_{C_p}/c_{ps} - \frac{dW}{dT_p}\frac{h_{fg}}{c_{ps}})\frac{dT_p}{dt} = Nu\pi k_g D(T_g - T_p)/m_s c_{ps} \tag{16}$$

The rate of change of W with particle temperature is obtained by differentiation of Eq. (9) to yield

$$\frac{dW}{dT_p} = -0.096W_c \exp[-2.37\tanh\theta]\text{sech}^2\theta \tag{17}$$

where $\theta = (T_p - T_s)/24.6$ and T_s is the saturation temperature.

Integration of the energy equation yields the droplet/particle temperature history along the trajectory.

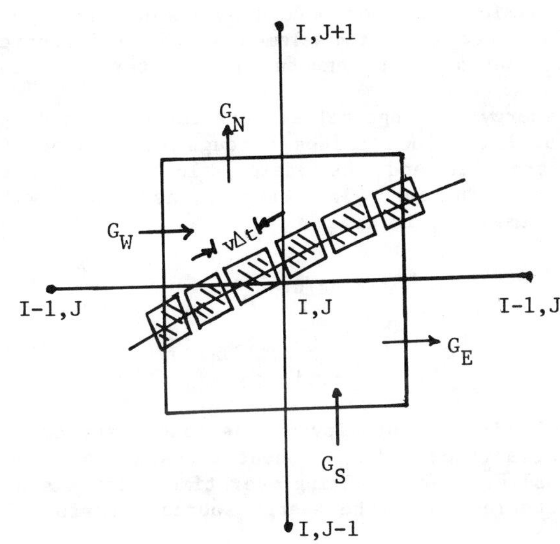

Figure 4. Decomposition of droplet path into series of contiguous time steps

6. SOURCE TERMS

The droplet source terms for the gas flow field are evaluated as the particles pass through the flow field. As a particle traverses a computational cell, it continually supplies mass, momentum and energy to the gas within the cell. Consider the trajectory shown in Fig. 4. The number of particles associated with a time interval, Δt, is $\dot{n}\Delta t$ where n is the number flow rate along a given trajectory. The mass efflux due to these particles is

$$S_M = \dot{n}\Delta t(dm/dt) \tag{18}$$

where dm/dt is the rate of change of mass for a single droplet or particle. If the droplet is evaporating, dm/dt is negative and the term represents an influx of mass to the cell. Summing over all time intervals associated with the particles' traverse of the cell yields the net mass source term for that trajectory. Summing over all trajectories which traverse the cell yields the source term for that cell.

$$\Delta m_D = \sum_{traj} \sum_{\Delta t} S_M \quad (19)$$

The momentum source terms are evaluated in the same way. The momentum source associated with time element Δt is given by

$$\vec{S}_{MOM} = \dot{n}\Delta t(\vec{V}dm/dt - \vec{f}_D) \quad (20)$$

$$\Delta\vec{M}_D - \vec{F}_D = \sum_{traj} \sum_{\Delta t} \vec{S}_{MOM} \quad (21)$$

where f_D is the aerodynamic force acting on a single droplet. The first term represents the momentum efflux associated with mass transfer from the droplet and the second term represents the aerodynamic force of the droplets on the gas. Summing over all time elements and trajectories yeilds the overall momentum source term for the gas.

Energy exchange between the gas and the droplet/particle cloud arises through the convective heat transfer and the energy flux due to mass transfer. The energy source term, associated with time element Δt is given by

$$S_E = \dot{n}\Delta t(h_g dm/dt - q_D) \quad (22)$$

$$\Delta E_D - Q_D = \sum_{traj} \sum_{\Delta t} S_E \quad (23)$$

where h_g is the enthalpy of the vapor leaving the droplet surface and Q_D is heat transfer to an individual droplet. Summing over time intervals and trajectories gives the energy source term for the cell.

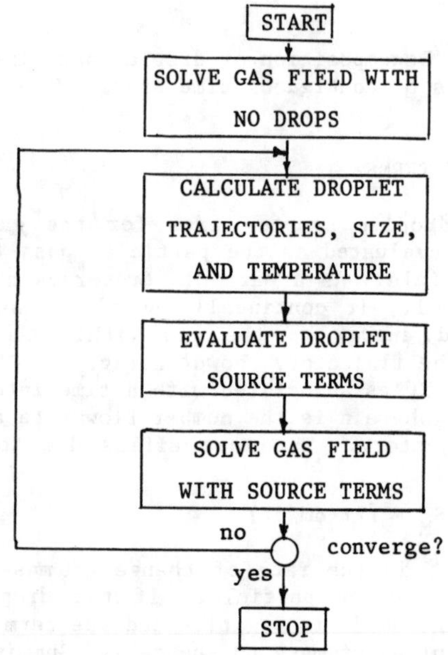

Figure 5. Flow chart for sequence of operations in numerical model

7. SOLUTION SCHEME

The flow chart for the numerical program is shown in Fig. 5. The procedure begins by obtaining a solution for the flow field with no droplets to give the first estimate of the gas velocity and temperature distribution. Droplet trajectories and size and temperature along each trajectory are then calculated. At the same time, the mass, momentum and energy source terms are accumulated for each cell. The gas flow field is recalculated, incorporating these source terms. If convergence is not achieved (which will not be the case for the first iteration), then the droplet trajectories and source terms are re-evaluated. The gas flow field is re-calculated using the updated source terms. When convergence is achieved, the calculation terminates with printout of the final values.

8. APPLICATION

The numerical model was applied to the co-current configuration shown in Fig. 1. The dryer was four meters in diameter and six meters high. The length of the cylindrical section was 2.8 meters. Both steam and air were considered as the drying media. It was assumed that the air entered at 200°C with an absolute humidity of 1 percent. The superheated steam was assumed also to enter at 200°C. The pressure and axial velocity at the entry port was 120 kPa and 20 m/s for both drying media. The nominal residence time of the dryer was 10 seconds. The inlet flow rate was assumed to be non swirling although the model is capable of including a swirl component of velocity, if needed.

The atomizer was a wheel atomizer which produced a mass median droplet size of 200 microns. The inlet droplet size distribution was discretized into five sizes which are provided in the following table.

Mass fraction	Initial droplet diameter (microns)
.2	80
.2	140
.2	200
.2	260
.2	400

Table. Initial droplet diameter distribution

The initial velocity leaving the wheel atomizer was 60m/s. The inlet mass flow rate of the slurry was 5 percent of the inlet mass flow of the drying medium.

The slurry was taken as an aqueous solution of an insoluable material with an initial moisture content of 1.2. The material density of solid components was assumed to be 550 kg/m³. The critical moisture content (transition point between drying periods) was taken as 0.2.

The comparative drying rate using steam or air as the drying medium for a slurry droplet

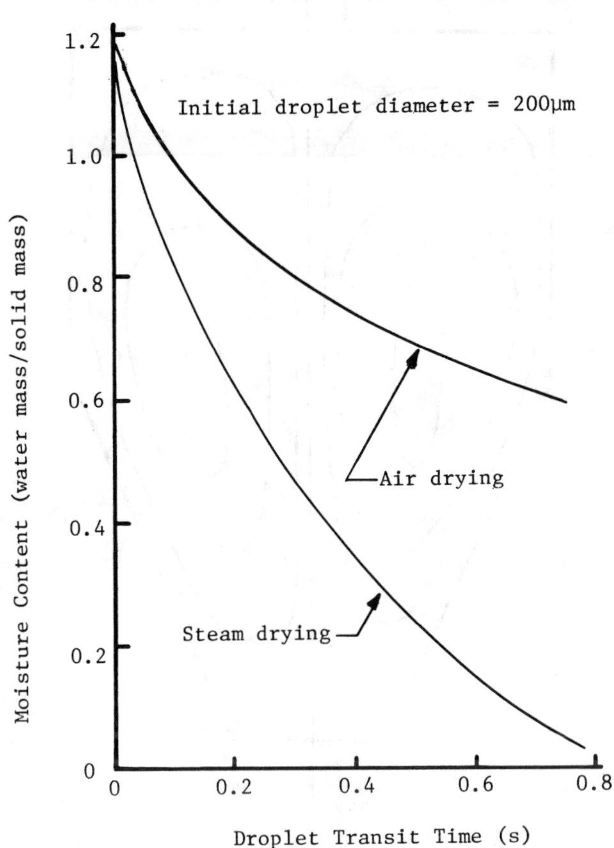

Figure 6. Moisture content vs. droplet transit time in spray dryer for air and steam media

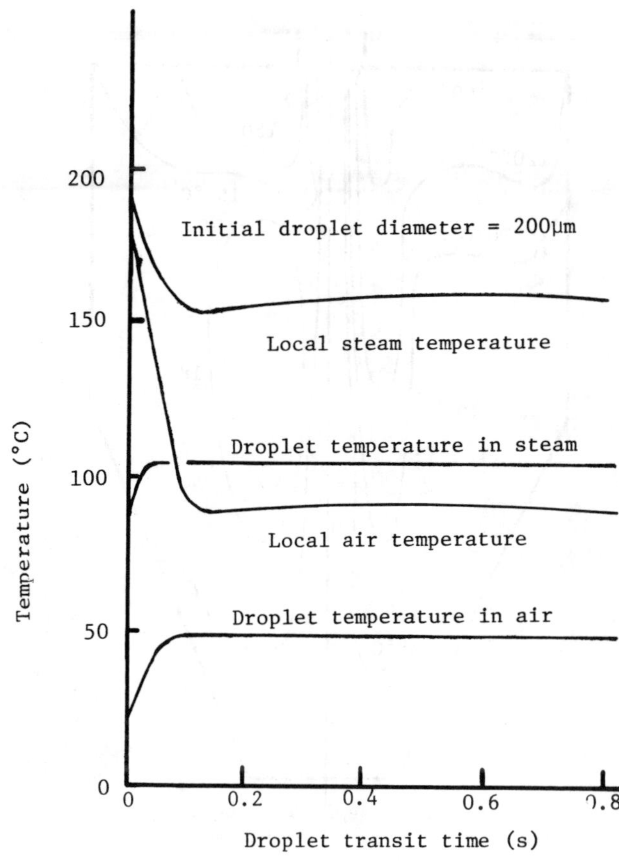

Figure 7. Droplet and local medium temperatures for air- and steam-drying

initially 200 microns in diameter is shown in Fig. 6. The ordinate is the moisture content and the abscissa is the transit time along the droplet trajectory. One notes a very definite faster drying rate with steam than with air. This same trend was observed for all the other drop sizes as well. This result was unexpected because previous work suggested that air-drying would be more effective at 200°C.

The reason underlying this result is shown in Fig. 7 where the droplet temperature and drying-medium temperatures are plotted as a function of droplet transit time. In air drying, the droplet temperature rapidly rises to the wet bulb temperature of 50°C and remains unchanged (until the critical moisture content is reached which is beyond residence time shown on the figure). The air temperature, however, drops to about 90°C due to the gas-droplet thermal coupling; i.e., heat transfer from the gas to the droplet to accomplish moisture evaporation. The droplet temperature in the steam immediately rises to the saturation temperature and remains at that temperature during the constant-rate period. The steam temperature drops to about 150°C due to thermal coupling. The fact that the temperature drop of steam is less than that for air drying is due to the higher thermal capacity of steam. Removal of a given quantity of energy from steam will result in a smaller temperature drop than removing the same quantity of energy from air. Even though the droplet-medium temperature difference for steam and air drying are comparable, steam drying is more effective because of the higher heat transfer coefficient [1].

The isothermal lines for steam drying and air drying are shown in Fig. 8. One notes that the steam temperature is uniformly higher than the air temperature. One also notes a significant temperature depression in the vicinity of the atomizer where the majority of the evaporation takes place.

The streamlines for steam and air drying are shown in Fig. 9. One notes similar recirculation patterns for each medium.

Droplet trajectories for steam and air drying are shown in Fig. 10. For air drying, the largest droplet penetrates almost to the dryer wall before it is redirected inward by the recirculating flow pattern. The outward radial velocity created by the recirculation zone near the exit forces the particle toward the wall, culminating in impact with the wall. The 400 micron droplet, released in the steam dryer, quickly impacts the wall.

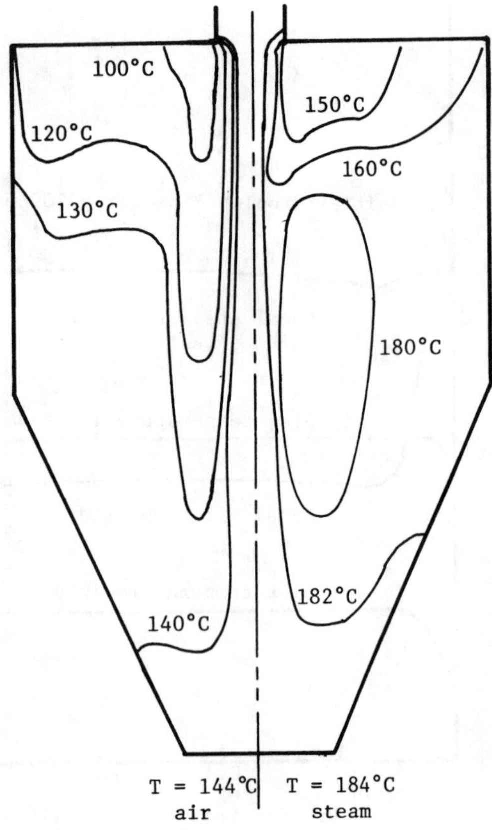

Figure 8. Predicted isotherms in spray dryer with air and steam flows

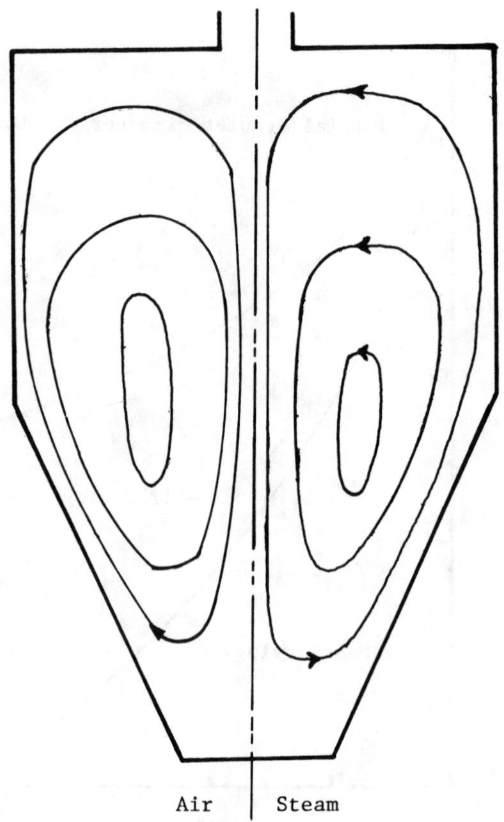

Figure 9. Predicted streamlines in spray dryer with air and steam flows

This is the result of the reduced viscosity in steam and smaller drag force on the particle. A dryer designed for steam operation would have to be somewhat larger in diameter to account for this effect.

9. CONCLUSION

The use of steam as the drying medium in a spray dryer looks attractive. The higher thermal capacity of steam compared to air leads to a higher operating temperature but results in an augmented drying rate of the droplet. The predicted improvement in performance, together with the other advantages in using steam as a drying medium, encourage further studies of its implementation in spray drying systems.

10. ACKNOWLEDGEMENTS

The support of the computer services and the WSU Mechanical Engineering Department is appreciated. The authors would like to thank Pat Martin for her excellent typing and Lisa Loney for her quality drawings.

REFERENCES

1. Chow, L.C. and Chung, J.N., Int J Heat Mass Transfer, vol. 26, 373-380, (1983).
2. Trommelen, A.M. and Crosby, E.J., AIChE Jnl, vol. 16, 857-867, (1970).
3. Chu, J.C., Lane, A.M. and Conklin, D., Ind Engr Chem, vol. 45, 1586-1591, (1953).
4. Chow, L.C. and Chung, J.N., ASME Paper 83-HT-2 presented at the 21st ASME/AIChE National Heat Transfer Conference, Seattle, (1983).
5. Crowe, C.T., Advances in Drying, Chapter 3, ed. A. Mujumdar, Hemisphere, New York, 63-99, (1980).
6. Dickinson, D.R. and Marshall, W.R., Jr., AIChE Jnl, vol. 144, 541-552, (1968).
7. Parti, M. and Palancz, B., Chem Engr Science, vol. 29, 355-362, (1974).
8. Gluckert, F.A., AIChE Jnl, vol. 8, 1460-1466, (1962).
9. Katta, S. and Gauvin, W.H., AIChE Jnl, vol. 21, 143-152, (1975).
10. O'Rourke, P.J. and Wadt, W.R., Los Alamos Rpt LA-9423-MS (1982).

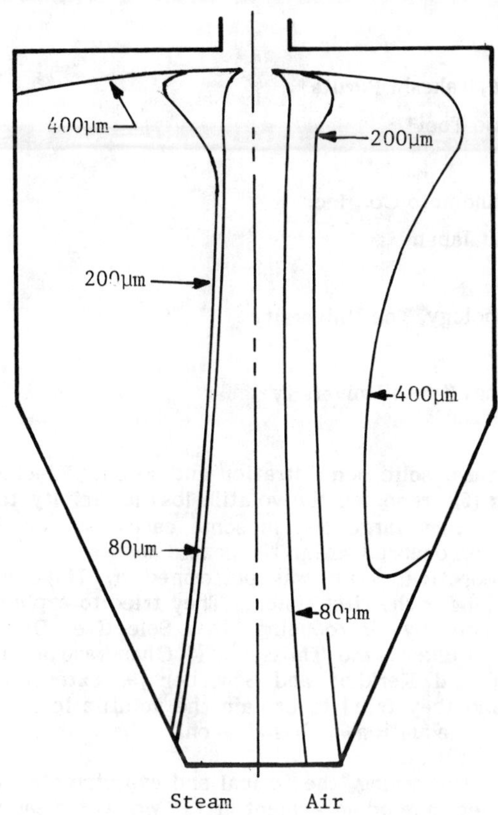

Figure 10. Trajectories of three droplet sizes in spray dryer for air and steam flows

11. Dukowicz, J.K., J. of Computational Physics, vol. 35, 229-253, (1980).
12. Crowe, C.T., Paper No. 41f, 1983 AIChE National Summer Meeting, Denver, (1983).
13. Kauman, W.G., Forest Products Jnl, vol. 6, 328-332, (1956).
14. Crowe, C.T., Sharma, M.P. and Stock, D.E., Jnl Fluids Engr, vol. 99, 325-332, (1977).
15. Gosman, A.D., and Pun, N.M., Lecture Notes, Imperial College of Science and Technology, London, (1973).
16. Patankar, S.V., Numerical Heat Transfer and Fluid Flow, Hemisphere, New York (1980).
17. Launder, B.E. and Spalding D.B., Mathematical Models of Turbulence, Academic, New York, (1972).
18. Clift, R. et al., Bubbles, Drops and Particles, Academic, New York, (1978).
19. Masters, K., Spray Drying Handbook, Halsted, New York, (1979).

VOLATILE LOSS IN SPRAY DRYING

Susumu Tsujimoto, Makoto Nishikawa, Takeshi Furuta*, Morio Okazaki**, and Ryozo Toei**

Central Research Laboratories, Ajinomoto Co., Inc.
Suzuki-cho, Kawasaki, 210 Japan

*Department of Food Science and Technology, Toa University

**Department of Chemical Engineering, Kyoto University

ABSTRACT

The volatile loss in spray drying was measured at several points in a spray dryer. The mechanism of the loss was discussed.

The measurment showed that within a certain range of solute concentration the volatile loss in vicinity to the spray nozzle was very large and larger than the estimated value by the Selective Diffusion Theory.

On the other hand, when the solute concentration was increased, the volatile loss in the nozzle zone became smaller and close to the value which was estimated by the Selective Diffusion Theory.

These results mean that the dominant mechanism of volatile loss exists in different stage according to operating condition (i.e.solute concentration) of drying.

In the practical operation of spray drying, prevention of the volatile loss is quite important, since the solute concentration generally falls in the region where the large volatile loss in the nozzle zone takes place.

The higher retention of volatile component was achieved not by increasing viscosity but by adding a small amount of food additive polymer without decreasing dry rate. It was considered that food additive polymer reduced the volatile loss in nozzle zone.

1. INTRODUCTION

Drying plays very impotant role in food industry to prevent the foodstuff from microbial spoilage in storage,to reduce transportation cost and so on.

From viewpoint of keeping good quality of dried food product,it is quite important to hold dilute volatile components in the product as much as possible.

Spray drying economical, but it is said to that retention of volatile components is lower in spray drying than in freeze drying. It needs some technical improvements.

Studies on the dissipation of volatile component in spray drying are as follows. Menting and Hoogstad [1] and Ban [2,3] observed the volatile loss from a single droplet fixed on a wire in a hot air stream. Rulkens and Thijssen [4] used the practical spray dryer and observed the relationship between volatile loss of dried powder and operating conditions such as liquid temperature, inlet air temperature, solid concentration and so on. Kieckbusch and King [5] reported the volatile loss in vicinity to the nozzle is very large and in some cases, 40 -60 % of volatile loss occured within the nozzle zone.

Theoretical work was performed by Thijssen and Rulkens [6] for the first time. They tried to explain the phenomenon by introducing the Selective Diffusion Theory. Rulkens and Thijssen [7], Chandrasekaran and King [8] and Kerkhof and Schoeber [9] extended this theory and they tried to explain the volatile loss by the diffusion equations based on the irreversible thermodynamics.

In spray drying, theoretical and experimental works have lacked in good agreement and it was not clear when dominant volatile loss happen in practical spray drying.

Authors [10, 11, 12] measured the gas-liquid equilibrium of dilute volatile component and diffusion coefficents of water and volatile component in food liquid, and could quantitatively explain the volatile loss from a single droplet of about 2 mm in diameter in the hot air stream. Authors [13] also performed the experimental and theoretical work on the volatile loss from a single droplet which was accompanied by expansions of droplet, and observed that gentle expansion did not affect the volatile loss very much.

The present study was performed by focusing on the behavior of the volatile loss in practical spray dryer by measurement of AR (Retention of volatile component) and WR (Retention of water) at several points from the spray nozzle, and the Selective Diffusion Theory was examined. At the same time, a new method to improve the retention of volatile component was proposed.

2. EXPERIMENT

2.1. Experimental apparatus and method

Figure 1 shows a schematic drawing of the experimental apparatus, which is a co-current spray dryer with 1.6 m in diameter about 15 m in heigt.

Pressure nozzle (Type SX,Spraying Systems Company) is used.

Standard operating conditions were settled on 27 m^3/min as drying air, 20 l/hr as feed rate of solution and 398K -458K as inlet hot air temperature. (①-④) in Fig. 1 are measuring points. Air flow rate, liquid flow rate and

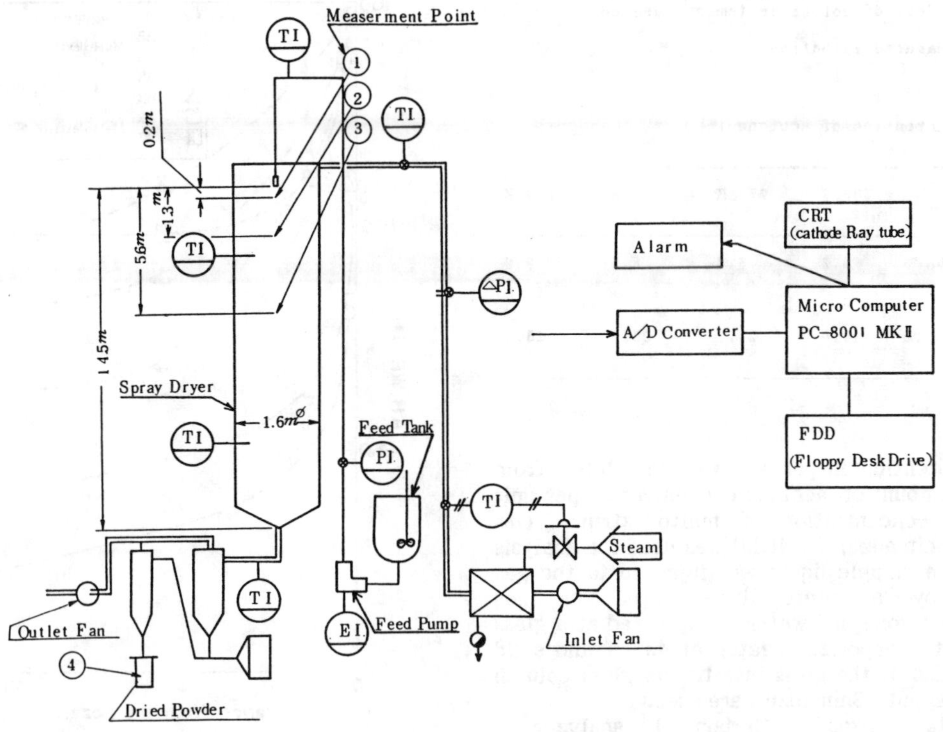

Fig. 1 Schmatic diagram of the spray dryer and measurement system

temperature at measuring points are watched by microcomputer. The operating conditions are controlled by manual when the operating conditions are fluctuated and the computer alarms.

Maltodextrin (Amicole No.1, Nichiden Kagaku Co.,) solution of 20-50 wt% with 1 wt% of acetone were used. The dextrose equivalent of the maltodextrin is about 17.

The sampling equipment mainly used for droplet sampling is shown in Fig. 2. The sample collector made from glass is cooled by dry ice and sampling is carried out using two air cylinder (reciprocal and rotary motion), which connected to silicone rubber stopper. Motion of the stopper is controled by a timer.

In case of sampling the droplets for measurment of volatile retention, about 20 ml of water is added in the sample collector through silicon rubber stopper by an injector just after droplet sampling. The cooling temperature of the sample collector was determined by the spraying test at room tempeature. Spraying test was performed by collecting droplets about 30 seconds at 1 m just under the spray nozzle. The relation between AR (Acetone Retention) and the cooling temperature of collector is listed in Table 1.

The figures in Table 1 displays the ratio of acetone remaining in droplets after traveling from the nozzle to the collector. Variation of the values implies the re-vaporization of acetone from the sample collector during sampling. Acetone retention increased with decreasing temperature and became constant when the cooling temperature was less than a certain temperature. This temperature was different at each solute concentration of feed liquid, but cooling below 220 K was enough to get reliable retention.

Kieckbusch and King [5] examined the methanol temperature for cooling the sampler and they showed that cooling less than 235 K is low enough to get the high acetone retention at 21 wt% sucrose solution.

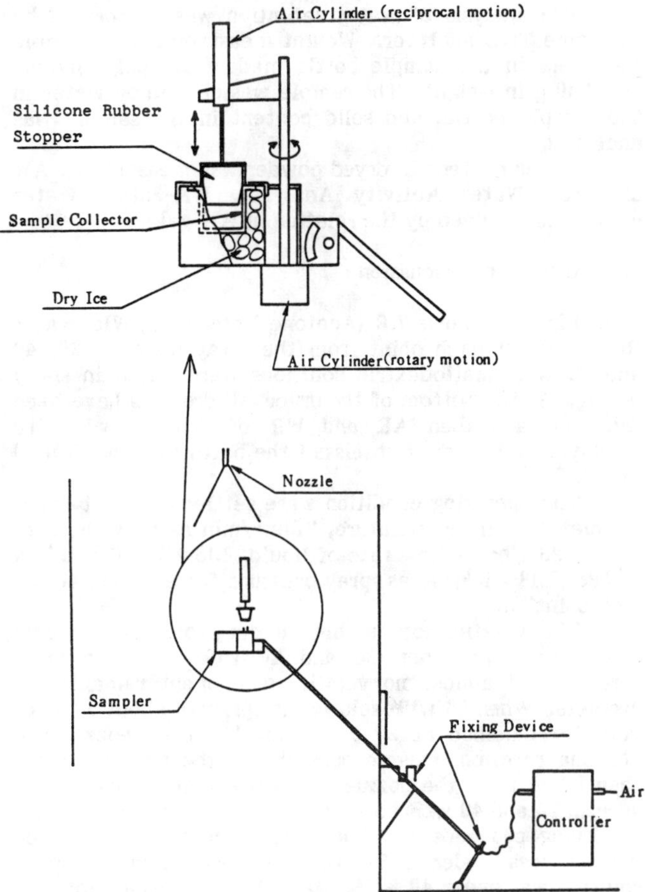

Fig. 2 Sampling equipment for measurement of acetone and water retention

Table 1 Effect of collector temperature on measured retention

Solution	Retention of acetone [%]			
	303 K	273 K	256 K	196 K
Maltodextrin 20wt%	4.2	6.9	8.5	8.9
Maltodextrin 40wt%	15.0	21.1	25.5	26.8

2.2 Analysis

Acetone retention in percent was calculated from the ratio of the amount of acetone concentration per unit mass of solids concentration of maltodextrin. Gas chromatograph (Shimadzu, GC-4CM) was used for analysis of acetone, and a sample liquid was injected to the gas chromatograph by a microcylinge. In the gas chromatograph, acetone and water is vaporized at a glass insert setted in the vaporizer heated at 433 K and solid component remains at the glass insert. 2m glass column and TSG packed agent (Shimadzu) are used.

TOC(Total Organic Carbon) analyzer (Shimadzu,TOC-10B) was used for analysis of carbon as weight of maltodextrin.

Sample liquid was diluted to measurable concentration and injected about 40μl to the analyser.

The analysis of water retention was performed by the same TOC analyzer. Weight measurement of sample was done in the sample bottle made from polyethylene about 10 g in weight. The sample was diluted by water in the sample bottle, and solid content in the sample was analyzed.

Water content in dryed powder was measured by AW analyzer (Water Activity Analyzer, WA-351). Water content is obtained by the relation of AW value.

2.3. Results and discussion

Figure 3 shows AR (Acetone Retention), WR (Water Retention) of each point from the spray nozzle. 20, 40 and 50 wt% maltodextrin solutions were used in spray drying. At the bottom of the dryer, all droplets have been dried up and then AR and WR of dried powder are displayed at 14.5 m of abscissa (the bottom of the dryer) in Fig. 3.

The operating condition were settled at about 457 K as inlet hot air temperature, 27 m^3/min as flow the rate of air, 20 l/hr as flow rate of liquid, 2.16 x 10^6[Pa], 1.77 x 10^6[Pa], 1.18 x 10^6[Pa] as spray pressure for each 20, 40, 50 wt% solution.

The volatile loss within 20 cm from spray nozzle was very large when 20 and 40 wt% solutions were sprayed and almost no volatile component remained in droplets. When 50 wt% solution is sprayed, about 50 % of volatile remained at 20 cm from the spray nozzle and acetone retetion became constant at the region farther than 1.3 m from the nozzle. This phenomenon was never seen in 20 and 40 wt% solutions, acetone continued to lose as drying proceeds. The acetone retentions of dried powder was under 1 % when concentration of spray solution was under 40 wt%. On the other hand, about 10 % of acetone retention could be obserbed at 50 wt% of solution.

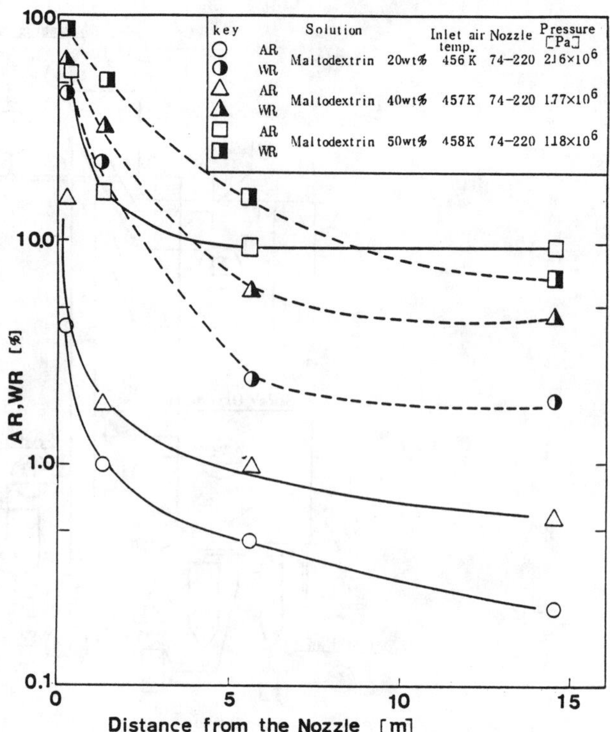

Fig. 3 Acetone retention (AR) and water retention (WR) of 20,40,and 50 wt% maltodextrin solutions

Figure 4 illustrates acetone retention at 458 K -398 K for inlet hot air temperature, when 40 and 50 wt% maltodextrin solution were sprayed.

For 40 wt% solution, the larger volatile loss could be seen in vicinity to the nozzle when the inlet air temperature is high. But, at the distance far from the nozzle, volatile loss became less, and AR at 457 K was larger than AR at 398 K. For 50 wt% solution, the volatile loss in the nozzle zone was almost independent of the inlet air temperature. The larger volatile retention could be obtained at the farther point from the nozzle when drying is proceeded at higher inlet temperature.

Figure 5 shows the relation between AR and WR of 40 and 50 wt% solution. For both solute concentrations, higher volatile retentions were obtained when dried by the higher inlet air temperature. Kieckbush and King [5] showed the similar result in sucrose solution, and concluded that the Selective Diffusion Theory is applicable except in vicinity to the nozzle.

2.4. Drying with additives

Figure 6 shows the AR and WR of 40 wt% solution with and without 1 wt% of gelatine (Nitta Gelatine Co., G-1114) dried at 458 K as inlet air temperature. WR was not affected by adding dilute gelatine, but AR appreciably increased and AR remained constant from 20 cm to the bottom of the dryer.

2.5. Comparing with Selective Diffusion Theory

It is difficult to make clear the effect of gelatine theoretically but the discussion on the effect will be

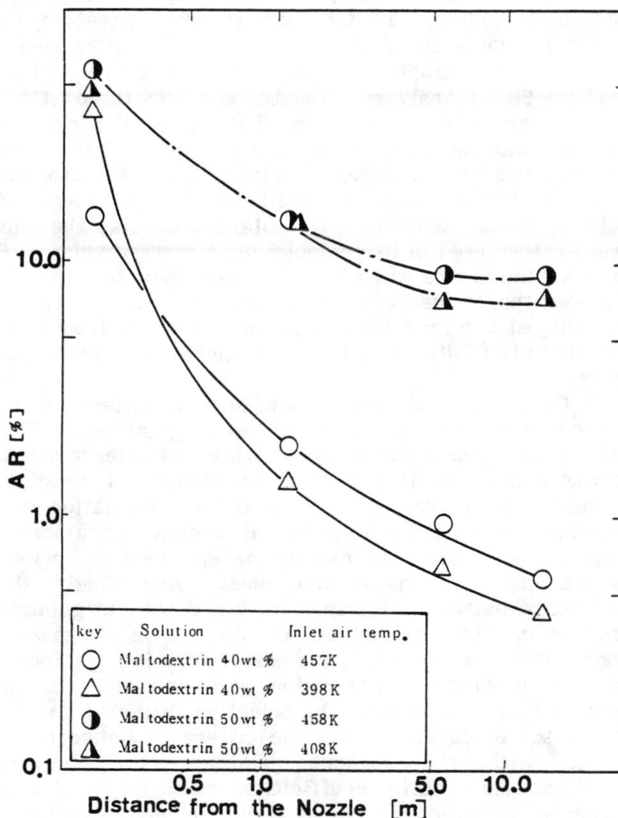

Fig. 4 Acetone retention of each distance from the nozzle when drying was performed at different temperature (40 and 50 wt% maltodextrin solutions)

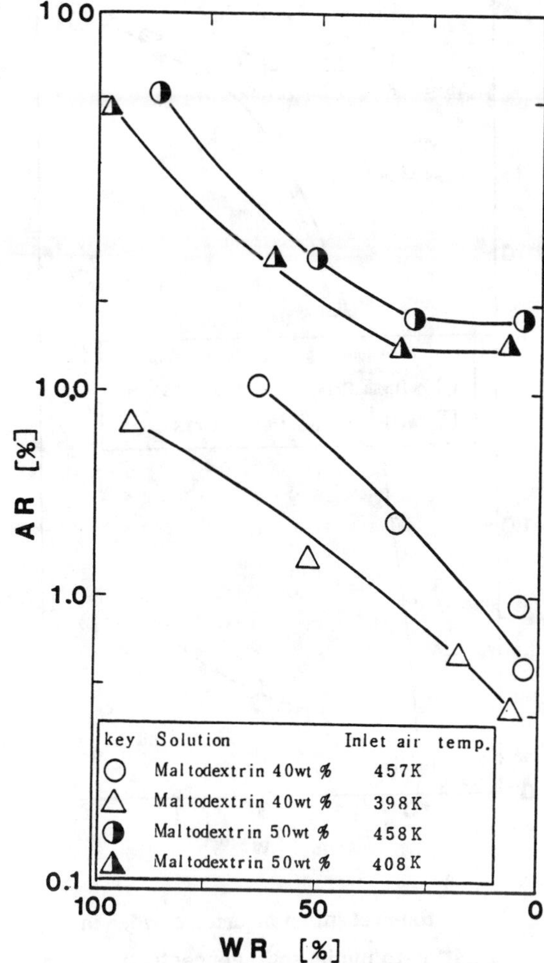

Fig. 5 Acetone retention in relation with water retention when drying was performed at different inlet temperature (40 and 50 wt% maltodextrin solutions)

useful to consider the mechanism of volatile loss and availability of the Selective Diffusion Theory.

Figure 7 shows the AR of dried powder vs. m_o (initial concentration of solution) when dried at about 403 K. AR was measured in two cases (1. with gelatine, 2. without gelatine). AR shows quite different pattern in each case of with or without gelatine, i.e. AR increases much when sprayed at 40-50 wt% of solution without gelatine. On the other hand AR is very high and increases little over 20 wt% solution when gelatine was added.

The experimental AR was compared with calculated results using the model [12].

The equations are as follows:

$$\frac{\partial C_w}{\partial t} = \frac{1}{r^2}\frac{\partial}{\partial r}\left(D_w r^2 \frac{\partial C_w}{\partial r}\right) \quad (1)$$

$$\frac{\partial C_a}{\partial t} = \frac{1}{r^2}\frac{\partial}{\partial r}\left\{r^2\left(D_a\frac{\partial C_a}{\partial r} + D_{wa}\frac{\partial C_w}{\partial r}\right)\right\} \quad (2)$$

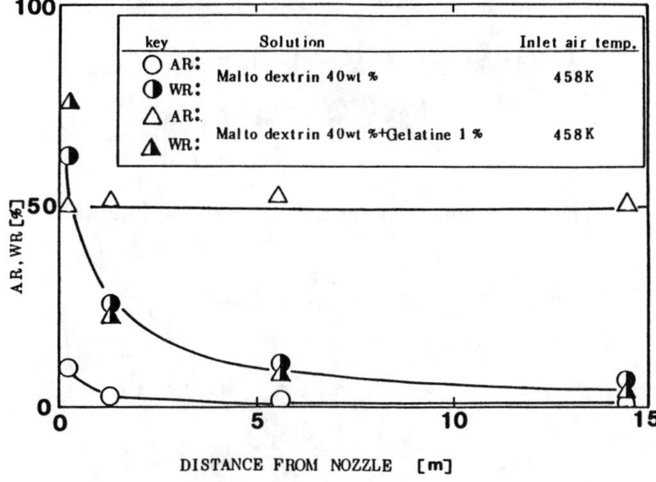

Fig. 6 Comparison of AR and WR with or without gelatine in 40 wt% maltodextrin solution

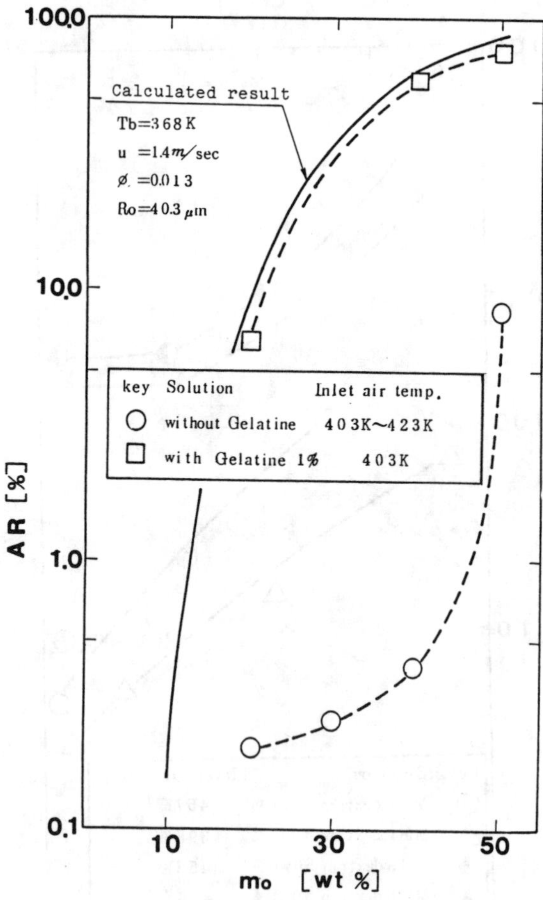

Fig. 7 Acetone retention of dried powder in relation to inicial solid concentration of maltodextrin (with gelatine or without gelatine) and calculated acetone retention using eq.(1)-eq.(4)

The heat balance on the droplet gives the equation for the droplet temperature as:

$$(RC_{pm}\rho_m/3)\,dT_m/dt = h_g(T_b - T_m) - k_{gw}(P_{wi} - P_{wb})\,r_w \qquad (3)$$

The initial and boundary condition are as follows:

$$\left.\begin{aligned}
&t=0,\ 0 \leqq r \leqq R_0\ ;\ C_w = C_{wo},\ C_a = C_{ao},\ T_m = T_{mo}\\
&t>0,\ r=0\ ;\ \partial C_w/\partial r = 0,\ \partial C_a/\partial r = 0\\
&r=R\ ;\ \frac{-D_w}{1-C_w\bar{V}_w}\frac{\partial C_w}{\partial r} = k_{gw}(P_{wi}-P_{wb})\\
&-D_a\left(\frac{\partial C_a}{\partial r} + C_a\frac{\partial \ln H_a}{\partial C_w}\frac{\partial C_w}{\partial r}\right) = k_{ga}(P_{ai}-P_{ab})
\end{aligned}\right\} \qquad (4)$$

The calculation was performed using the droplet of 80.5μm in diameter, 368K for drying air temperature, 1.4 m/sec for velocity of drying air and 0.013 kg/kg as air humidity. The droplet diameter was measured by Micro Particle-Size Analyzer (Leeds and Northrup Co.). Droplets are catched in silicon oil (0.1 Pa s) at 1 m under the spray nozzle and spraying were performed at the room temperature (293K). After catching droplets, the droplets size distribution was measured by the analyzer and volume mean diameter is obtained. As the air temperature used in this calculation, measured value at 6 m under the nozzle was adopted. The velocity of drying air was the terinal velocity of the droplet. The air humidity at 6 m from the nozzle was estimated from inlet air humidity (0.01 kg/kg) and additinally vaporized water by drying.

Calculated AR had a quite similar pattern to the experimental results of dried with gelatine. This calculation ignores the transient state just after sprayed. It was considered that gelatine stopped the convection currents in the droplets during droplet formation and gelatine prevented the large loss of acetone in the nozzle zone and the results of calculation appeared to be well agreed with the experimental ones. Accordingly, the difference between AR values with and without gelatine may be mainly regarded as volatile loss in the nozzle zone. The volatile loss in the nozzle zone is large for 40 wt% initial solute concentration by comparing it to the volatile loss calculated by the Selective Diffusion Theory. In low solute concentration, the difference between AR values with and without gelatine was relatively small,because diffusion coefficients are so large [11] that estimated retention became small and should be similar to the retention accompanied by the volatile loss in the nozzle zone. When 50 wt% solution was sprayed, this difference was also smaller. It was considered that volatile loss in the nozzle zone become less than that of lower solute concentration,because the dried film might be formed faster at the surface of the droplets.

In the practical operation of spray drying, it stands almost on the limit to spray at 50 wt% of solution by reason of high viscosity. Because the volatile retention estimated by the Selective Diffusion Theory is large enough at 30-40 wt% solution, it can be expected that the high volatile retention will be accomplished by reducing the volatile loss in the nozzle zone. Table 2 shows the comparison of viscosity of the each solution. 40 wt% solution with 1 wt% of gelatine had lower viscosity than that of 50 wt%, but much higher volatile retention could be achieved. Accordingly, it was more useful to add small amount (under 1 wt%) of the food additive polymer than to spray the solution of higher viscosity in order to get higher volatile retention

By developing the above idea, spray drying of 40 wt% solution with 0.35 wt% of sodium polyacrylate was performed and the high volatile retention of about 50 % could be achieved.

CONCLUSION

It was concluded that the Selective Diffusion Theory is basically applicable in practical spray drying. But the volatile loss in the nozzle zone also plays important role. In the low solute concentration, the volatile loss in nozzle zone consisted dominant loss in the total volatile loss in drying. Specially, experimental loss differed remarkablly when about 40 wt% solute concentration was spray dried.

Table 2 Viscosity of solution

Solution	Viscosity [Pa s]	Retention of acetone [%]
Maltodextrin 40wt%	$22.\times10^{-3}$	0.6
Maltodextrin 40wt% +Gelatine 1wt%	$28.\times10^{-3}$	50.2
Maltodextrin 50wt%	$58.\times10^{-3}$	9.1

-measured by rotational viscometer at 298 K

-retention of acetone is the the data from Fig.3,Fig.6 (dried at 458 K as inlet air temp.)

In practical spray drying, the operation condition should be sellected so that the high volatile retention can be obtained from the Selective Diffusion Theory and the volatile loss in the nozzle zone can be controled.

NOMENCLATURE

C	concentration	kg/m^3-solution
C_p	heat capacity of droplet	J/kg K
D	diffusion coefficient	m^2/s
D_{wa}	cross diffision coefficient	m^2/s
Ha	modified activity coefficient	
k	mass transfer coefficient	Kg/m^2sP_a
m	weight fraction of maltodextrin	
p	partial vapor pressure	P_a
R	radius of droplet	m
r	radial coordinate in dropet	m
T	temperature	K
t	drying time	s
n	velocity of drying air	m/s
V	partial specific volume	m^3/kg
ϕ	humidity of air	kg-steam/kg-dry air
γ	latent heat of evaporation of water	J/kg

SUBSCRIPT

o	inital value
a	acetone
b	bulk of air stream
g	gas phase
i	liquid-gas interface
m	droplet
w	water

REFERENCES

1. Menting, L.C., and Hoogstad, B., J. of Food Sci., Vol. 32, 87 (1967).
2. Ban, T., Kagaku Kogaku Ronbunshu (in Japanese), vol.4, 515 (1978).
3. Ban, T., Kagaku Kogaku Ronbunshu (in Japanese), vol.5, 213 (1979).
4. Rulkens, W.H., and Thijssen, A.C., J. of Food Technol., Vol. 7, 95 (1972).
5. Kieckbusch T.G., and King, C.J., AIChE Journal, Vol. 26, No. 5, 718 (1980).
6. Thijssen, H.A.C., and Rulkens, W.H, De Ingenieur, Vol. 80, Ch45 (1968).
7. Rulkens, W.H., and Thijssen, H.A.C., TRANS. INSTNCHEM. ENGRS., Vol. 47, T292 (1969).
8. Chandrasekaran, S.K., and King, C.J., AIChE Journal, Vol. 18, No. 3, 520 (1972).
9. Kerkhof, P.J.A.M. and Schoeber, W.J.A.H., Advances in preconcentration and dehydration of foods, p.349 (1974).
10. Tsujimoto, S., Matsuno, R., and Toei, R., Kagaku Kogaku Ronbunshu (in Japanese), vol.8,103 (1982).
11. Furuta, T., Tsujimoto, S., Makino, H., Okazaki, M., and Toei R., J. of Food Eng., to be published.
12. Furuta, T., Tsujimoto S., Okazaki, M., and Toei, R., J. on Drying Technology, to be published.
13. Furuta, T., Tsujimoto, S., Okazaki, M., and Toei, R., Proceeding of ICEF 3 at Doublin.

THE MATHEMATICAL MODELLING OF COCURRENT SPRAY DRYING

Wu Yuan*, Fu Jufu, and Zheng Chong

Department of Chemical Engineering, Beijing Institute of Chemical Technology, Beijing, CHINA

ABSTRACT

A mathematical model has been developed with the droplet size distribution being taken into account with a volume frequency function. The model is adequate in describing the overall behaviour of spray dryers. Three examples are given, and the calculated results are in agreement with the plant operating data.

1. INTRODUCTION

This paper reports a study on modelling cocurrent spray drying. Although the process is but a physical one, the problem is rather complex [1,2], and, up to now, the design of spray dryers is essentially based on experience.

The crux in designing lies in determining the time required to dry the whole spray. Some previous authors evaluated it on the basis of a Sauter mean or mass mean diameter of spray droplets [3-5] or of an average volume heat transfer coefficient [6], and natually the calculated results usually quite deviated from the actual operating data. Dlouhy and Gauvin [7] calculated the drying time by employing a step-by-step method advocated by Marshall [8], and reported good agreement between the experimental data and the calculated results. However, the droplet size data, on which their calculation being based, might be unreliable because no representative sample of droplets could be collected with the immersion cell used in their study owing to the fact that the small droplets tend to follow the streamlines round the cell, as pointed out by Ranz and wong [9] and Kim and Marshall [10]. The "good agreement" merely implies that the calculation method would be questionable. Charlesworth and Marshall [11] and Miura and Ohtani [12] also used the step-by-step method in their studies on this subject.

A spray dryer must be capable not only to deal with the amounts of heat and mass to be transferred but also to dry the largest particles to avoid caking on the wall [13]. Since the small droplets have much larger heat transfer coefficients and specific surface areas and will be quickly dried as soon as they come into contact with drying air, the intensities of mass and heat transfer are extremely nonuniform along the height of dryer. For making a good design, a drying rate based on any kind of mean diameter will not be satisfactory, even for a droplet group with a small range of sizes in a step-by-step procedure. The size distribution must be taken into account in developing a model of spray drying. Dickinson and Marshall [14] studied the evaporation rates of pure liquid sprays, using volume frequency functions. Of course, the model developed will not be applicable for spray drying of solutions or suspensions.

2. THEORETICAL CONSIDERATION

Since such informations as the radial distribution of droplets and the radial temperature profile of drying air in a dryer are usually unavailable, a one-dimensional model may be more practical and more preferable.

It is very difficult to deal with the size distribution and some other variables simultaneously even for developing a one-dimensional model. In order to deal in detail with the former, which plays the most important role in spray drying, other variables should be treated in a reasonably simplified way as possible.

The first simplification is made about the movement of droplets. When a pneumatic nozzle is used, the spray droplets are carried forward by the atomizing gas. The jet is initially at a very high velocity (over 300 m/s) and then moves deceleratively. The center velocity of the jet may be described by the Albertson's equation [15]

$$\frac{u}{u_o} = 6.2\,\frac{d_{eff}}{Z} \qquad (1)$$

Where u_o is the velocity of jet at the nozzle outlet. The whole spray is generally being dried before the jet velocity has been significantly slowed down. This suggests that the relative velocity between the droplets and the atomizing gas, and that between different diameter droplets, can be neglected.

The movement of droplets from pressure nozzle is somewhat similar to that from pneumatic nozzle. The momentum transfer from the spray to the surrounding air produces a jet of air [15,16], and the center line velocity of the jet was given by [15]

$$\frac{u}{u_{o1}} = 3.2\,\frac{d_{eff}}{Z} \qquad (2)$$

where

---

* Present adress: Department of Chemical Engineering, Wuhan Institute of Chemical Engineering, Wuhan, Hubei, China.

$$d_{eff}=d_{or}(\rho_l/\rho_g)^{0.5} \tag{3}$$

The droplets may also be considered to be carried forward by the jet of air, and so the velocity difference may be neglected too.

This assumption may also be reasonable for the case of using rotating atomizers, if the simplified scheme shown in Fig. 1 is considered. In the procedure of calculation it is not necessary to take the dried particles into account, as the heat required to raise their temperature further is negligible compared with that to evaporate water. From the scheme, it follows that the largest height difference between droplets to be dried at any time is rather small. This is shown by the following calculation.

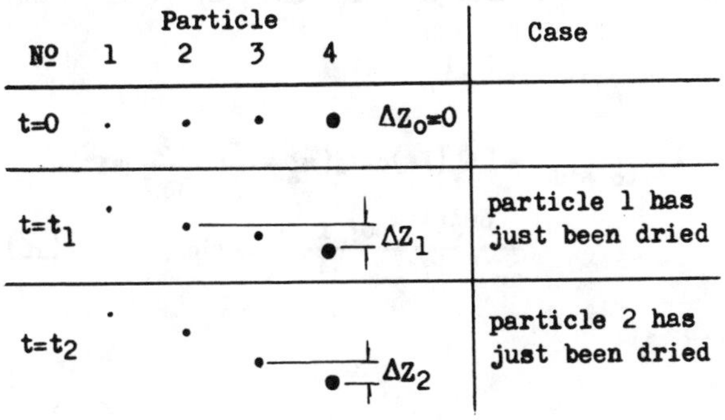

Particle № 1 2 3 4	Case
t=0 · • • ● $\Delta Z_0=0$	
$t=t_1$ (particles 2, 3, 4 remaining; height difference ΔZ_1)	particle 1 has just been dried
$t=t_2$ (particles 3, 4 remaining; height difference ΔZ_2)	particle 2 has just been dried

Fig. 1 Simplified scheme for falling movement of particles

Distance-time relationship for a droplet/particle with diameter X in a stationary air was given by Coulson and Richardson [17] as

$$Z=\frac{b_2}{b_1}t+\frac{u_{po}}{b_1}-\frac{b_2}{b_1^2}+\left(\frac{b_2}{b_1^2}-\frac{u_{po}}{b_1}\right)\exp(-b_1t) \tag{4}$$

where

$$b_1=\frac{18\eta_g}{X^2\rho_p} \quad \text{and} \quad b_2=\left(1-\frac{\rho_g}{\rho_p}\right)g_n \tag{5}$$

And in an air stream flowing cocurrently at an average vertical velocity of u_g, the relation can be written as

$$Z=\frac{b_2}{b_1}t+\frac{u_{po}}{b_1}-\frac{b_2}{b_1^2}+\left(\frac{b_2}{b_1^2}-\frac{u_{po}}{b_1}\right)\exp(-b_1t)+u_gt \tag{6}$$

In the case of using rotating atomizer, $u_{po}=0$, and b_2 can be approximated by g_n as $\rho_g \ll \rho_p$, equation (6) then becomes

$$Z=\left(u_g+\frac{g_n}{b_1}\right)t-\frac{g_n}{b_1^2}\left[1-\exp(-b_1t)\right] \tag{7}$$

For illustration, a set of results calculated for the case of a spray with the following characteristics: mass mean diameter of 40 µm, largest droplet diameter of 120 µm, ρ_p=1 206 kg/m^3; while u_g=0.3 m/s, and η_g= 16.63 10^{-6} kg/(m s), are given in Table 1, assuming no significant changes in diameters or densities of droplets. The largest diameter of the dried particles in the second column of Table 1 are calculated by our model (to be discussed later). From Table 1, it can be seen that the largest height difference between droplets to be dried at any time is relatively small. The smaller the mean diameter of spray droplets, the smaller the height difference.

Table 1 Results calculated for the movement of droplets *

t s	X x10^6 m	Z_x m	Z_{max} m	ΔZ_t m
0.10	53.2	0.040	0.060	0.020
0.20	66.2	0.092	0.145	0.050
0.30	76.5	0.154	0.228	0.074
0.40	84.0	0.224	0.315	0.091
0.50	91.2	0.303	0.402	0.099
0.60	97.2	0.390	0.488	0.098
0.70	102.3	0.482	0.575	0.093
0.80	108.1	0.588	0.662	0.074
0.90	113.6	0.703	0.749	0.046
1.015	120.0	0.849	0.849	0

* In the table, X is the largest diameter of dried particles; Z_x is the distance travelled by the droplets with a diameter of X, Z_{max} is that by the largest droplets of the spray; and $\Delta Z_t=Z_{max}-Z_x$, the largest height difference between the droplets to be dried.

Thus, the assumption of the velocity difference among the droplets to be dried in vertical direction being negligible is justified, at least, for the sprays of which the mean diameters are not too large. This assumption enables us to take the time, t, as a unique independent variable.

In a drying process, the change in temperature of a single droplet may be described as a stage-wise process, but it is unfeasible for the whole spray. For simplification, in principle, it may be assumed that each droplet is at a pseudo-mean temperature, T_{pm}, through the process. This temperature may be determined for certain specific conditions. This should be a better assumption than all the droplets being considered to have the same diameter.

In our previous paper [18], a simplified physical model has been proposed for drying of highly concentrated solutions of certain inorganic salts, where it was pictured that water is essentially evaporated at the boiling point temperature of the saturated solution. For those materials, instead of T_{pm}, the boiling point temperature, T_b, would be used in calculation.

To make the problem further simplified, another assumption introduced is that the diameters of all the droplets are considered to be constant during drying. For the case where the diameters change appreciably, a correction factor, ξ, which is to be discussed later, can be introduced.

3. MATHEMATICAL MODEL

The other assumptions made in the development, of which most are the conventional, are as follows: (1) as the droplets are small, the temperature difference between their surface and interior can be neglected, (2) the radial distribution of droplets and the radial conditions of drying air over any cross section of dryer are uniform, and a plug flow exists, (3) the ratio of heat lost through the wall to that consumed is constant along the height of dryer, (4) the effect of relative velocity between droplets and fluid and that of mass transfer on heat transfer are both neglected in evaluating heat transfer coefficient (these effects are in opposide directions), and hence the simplified Ranz and Marshall's equation [19]

$$Nu = \frac{\alpha X}{\lambda} = 2.0 \qquad (8)$$

can be used, and (5) the final moisture content of all the dried particles, w_{pf}, are the same.

3.1. Relationship Between Diameter of the Largest Dried Particles and Drying Time

considering a single droplet having diameter of X, the rate equation for heat transfer can be written as

$$\alpha_x A_x (T_g - T_{pm}) dt = -\frac{\pi}{6} X^3 \frac{\rho_1 \Delta h_v}{1+w_{po}} dw_p \qquad (9)$$

where the temperature of drying air, T_g, is a function of time, t. Combining Eq.(9) with Eq.(8) and the relation of $A_x = \pi X^2$, we have

$$(T_g - T_{pm}) dt = -\frac{X^2 \rho_1 \Delta h_v}{12\lambda(1+w_{po})} dw_p \qquad (10)$$

If the droplet begins to evaporate at $t_o=0$, and has just been dried at time t, integrating Eq.(10) gives

$$\int_0^t (T - T_{pm}) dt = \frac{X^2 \rho_1 \Delta h_v}{12\lambda(1+w_{po})} (w_{po} - w_{pf}) \qquad (11)$$

or

$$X^2 = \frac{12\lambda(1+w_{po})}{\rho_1 \Delta h_v (w_{po} - w_{pf})} \int_0^t (T_g - T_{pm}) dt \qquad (12)$$

Eq.(12) defines X as a function of the upper limit of the integral, t. It means that the droplets with diameter X have just been dried at time t and, consequently, those smaller than X in diameter have been dried before time t.

Differentiating Eq.(12) with respect to t leads to

$$\frac{dX}{dt} = \frac{C_1}{X} (T_g - T_{pm}) \qquad (13)$$

where

$$C_1 = \frac{6\lambda(1+w_{po})}{\rho_1 \Delta h_v (w_{po} - w_{pf})} \qquad (14)$$

and is a constant for a given process.

Eq.(13) represents the rate of change of diameter of the largest dried particles with time, and therein exists an unknown function, $T_g = T_g(t)$.

3.2. Heat Balance For Drying of the Spray

Examine the heat balance at the interval from t to t+dt. The drying air gives a differential heat dQ as its temperature drops by dT_g. This amount of heat is expended in three items: (1) drying the droplets larger than X in diameter, (2) raising the temperature of the vapor produced from droplets to that of drying air, and (3) loss through the wall.

In the differential time dt, the heat transferred to the droplets with diameters ranged from X to X+dX is

$$(dQ_1)_{X \text{ to } X+dX} = \alpha_X A_{X \text{ to } X+dX} (T_g - T_{pm}) dt \qquad (15)$$

If the volume frequency function of the spray is $\phi_v(X^*)$ and the calculation is on the basis of the amounts of materials per unit time, $A_{X \text{ to } X+dX}$ can be found by

$$A_{X \text{ to } X+dX} = [\phi_v(X^*) dX^*][m_s' \frac{1+w_{po}}{\rho_1} \frac{6}{\pi X^3}] \pi X^2 = \frac{6 m_s' (1+w_{po})}{\rho_1} \frac{1}{X} \phi_v(X^*) dX^* \qquad (16)$$

where

$$X^* = X/X_m \qquad (17)$$

and X_m is mass mean diameter of droplets. Substituting Eqs. (8) and (16) into Eq. (15), we have

$$(dQ_1)_{X \text{ to } X+dX} = \frac{12\lambda m_s'(1+w_{po})}{\rho_1} \frac{1}{X^2} \phi_v(X^*) dX^* (T_g - T_{pm}) dt \qquad (18)$$

Thus, the total heat transferred to all the droplets larger than X in diameter can be found by integrating Eq.(18) with respect to X* as

$$dQ_1 = \frac{12\lambda m_s'(1+w_{po})}{\rho_1 X_m^2} (T_g - T_{pm}) [\int_{X^*}^{X^*_{max}} \frac{1}{(X^*)^2} \phi_v(X^*) dX^*] dt$$
$$= \frac{12\lambda m_s'(1+w_{po})}{\rho_1 X_m^2} (T_g - T_{pm}) [I_1 - I(t)] dt \qquad (19)$$

where

$$I_1 = \int_{X^*_{min}}^{X^*_{max}} \frac{1}{(X^*)^2} \phi_v(X^*) dX^* \qquad (20)$$

$$I(t) = \int_{X^*_{min}}^{X^*} \frac{1}{(X^*)^2} \phi_v(X^*) dX^* \qquad (21)$$

and, I_1 is a constant for a given spray and I(t) is a function of t.

In Eq.(19), the term $[I_1 - I(t)]$ can be named as an "Effective Heat Transfer Factor "(EHTF) for drying, and I(t) a " Spent Factor " (SF), characterizing the change in drying rate.

A spray must have droplets of a non-zero minimum diameter and those of a finite maximum diameter. For various size distributions [1], the minimum diameter may be determined by

$$X_{min} = 0.05X_m \tag{22}$$

and the maximum by [21]

$$X_{max} = 3X_m \tag{23}$$

when mass mean diameter, X_m , is not too large or too small.

Strictly speaking, a size distribution can not be rigorously described with $\phi_v(X^*)$ since this function is generally defined in the region from zero to infinite and $\phi_v(X^*)>0$ for any value of X between these limits. However, it is practically adequate in describing the overall characteristics of size distribution and is convenient for calculation. Now taking a non-zero minimum diameter and a finite maximum one as the limits here means that only the major fraction of the total is taken into account. This is

$$I_2 = \int_{X^*_{min}}^{X^*_{max}} \phi_v(X^*)dX^* \tag{24}$$

To eliminate this inconsistency and to ensure that the average moisture content of all the material will converge to the given value, w_{pf}, as X to X_{max}, both the the amount of the material to be dried and that of drying air must be corrected by the factor I_2.

If the heat transferred, dQ_1 , causes a change of the average moisture content of the spray by dw_p ,then

$$dQ_1 = -I_2 m_s' \Delta h_v dw_p \tag{25}$$

Combining Eqs.(25) and (19) gives

$$\frac{dw_p}{dt} = -\frac{12\lambda(1+w_{po})}{I_2 \rho_1 \Delta h_v X_m^2}(T_g-T_{pm}) [I_1-I(t)]$$

$$= C_2(T_g-T_{pm}) [I_1-I(t)] \tag{26}$$

where

$$C_2 = -\frac{12\lambda(1+w_{po})}{I_2 \rho_1 \Delta h_v X_m^2} \tag{27}$$

The heat expended in raising the temperature of the vapor produced from droplets, dQ_2 , can be written as

$$dQ_2 = - c_{pw} I_2 m_s' (T_g-T_{pm}) dw_p \tag{28}$$

Thus, if the fraction of heat lost is expressed by q, the heat balance within the interval dt can be written as

$$-I_2(1-q)m_g' c_{pg} dT_g = dQ_1+dQ_2 \tag{29}$$

Substituting Eqs. (25) and (28) into Eq. (29) and rearranging the expression, we obtain

$$dT_g = \frac{m_s'}{(1-q)m_g' c_{pg}} [c_{pw}(T_g-T_{pm})+\Delta h_v] dw_p \tag{30}$$

Combining it with Eq. (26), Eq.(30) becomes

$$\frac{dT_g}{dt} = \frac{C_2\, m_s'}{(1-q)m_g' c_{pg}} [c_{pw}(T_g-T_{pm})+\Delta h_v] (T_g-T_{pm})[I_1-I(t)]$$

$$= \frac{C_3}{c_{pg}} [c_{pw}(T_g-T_{pm})+\Delta h_v](T_g-T_{pm})[I_1-I(t)] \tag{31}$$

where

$$C_3 = \frac{C_2\, m_s'}{(1-q)m_g'} = -\frac{12\lambda m_s'(1+w_{po})}{I_2(1-q)m_g' \Delta h_v \rho_1 X_m^2} \tag{32}$$

In Eqs.(26) to (31), the specific heat of drying air, c_{pg} , is a function of its humidity:

$$c_{pg} = c_{pa}+c_{pw}w_g \tag{33}$$

Using the relation of mass balance

$$dw_g = -\frac{m_s'}{m_g'} dw_p \tag{34}$$

and Eq. (26), we have

$$\frac{dw_g}{dt} = -\frac{m_s'}{m_g'} C_2(T_g-T_{pm}) [I_1-I(t)]$$

$$= C_4 (T_g-T_{pm}) [I_1-I(t)] \tag{35}$$

where

$$C_4 = -\frac{m_s'}{m_g'} C_2 = \frac{12\lambda m_s'(1+w_{po})}{I_2 m_g' \Delta h_v \rho_1 X_m^2} \tag{36}$$

And, combining Eqs. (33) and (35) leads to

$$\frac{dc_{pg}}{dt} = C_4 c_{pw}(T_g-T_{pm}) [I_1-I(t)] \tag{37}$$

At last, the differential equation for function I(t) can be obtained readily by differentiating Eq. (21) with respect to t as

$$\frac{dI(t)}{dt} = \frac{dI(t)}{dX^*}\frac{dX^*}{dX}\frac{dX}{dt}$$

$$= \frac{1}{(X^*)^2}\phi_v(X^*)\ \frac{1}{X_m}\frac{dX}{dt} \tag{38}$$

Substituting Eqs. (13) and (17) into Eq. (38), we have

$$\frac{dI(t)}{dt} = \frac{C_1}{X^3} X_m \phi_v(X/X_m)\ (T_g-T_{pm}) \tag{39}$$

3.3. Mathematical Model

Finally, we have the following set of non-linear ordinary differential equations:

$$\frac{dX}{dt} = \frac{C_1}{X}(T_g-T_{pm}) \tag{13}$$

$$\frac{dw_p}{dt} = C_2(T_g-T_{pm}) [I_1-I(t)] \tag{26}$$

$$\frac{dT_g}{dt} = \frac{C_3}{c_{pg}}[c_{pw}(T_g-T_{pm})+\Delta h_v]\,(T_g-T_{pm})\,[I_1-I(t)] \quad (31)$$

$$\frac{dw_g}{dt} = C_4(T_g-T_{pm})\,[I_1-I(t)] \quad (35)$$

$$\frac{dc_{pg}}{dt} = C_4 c_{pw}(T_g-T_{pm})\,[I_1-I(t)] \quad (37)$$

$$\frac{dI(t)}{dt} = \frac{C_1}{X^3}\,X_m\phi_v(X/X_m)\,(T_g-T_{pm}) \quad (39)$$

with the initial conditions

$$X(0)= X_{min} \quad (40)$$

$$w_p(0)= w_{po} \quad (41)$$

$$T_g(0)= T_{go\ eff} \quad (42)$$

$$w_g(0)= w_{go} \quad (43)$$

$$c_{pg}(0)= c_{pa}+c_{pw}w_{go} \quad (44)$$

$$I(0)= 0 \quad (45)$$

where

$$C_1= \frac{6\lambda(1+w_{po})}{\rho_1\Delta h_v(w_{po}-w_{pf})} \quad (14)$$

$$C_2 = -\frac{12\lambda(1+w_{po})}{I_2\Delta h_v\,\rho_1 X_m^2} \quad (27)$$

$$C_3= -\frac{12\lambda m_s'(1+w_{po})}{(1-q)I_2 m_s'\Delta h_v\,\rho_1 X_m^2} \quad (32)$$

$$C_4= \frac{12\lambda m_s'(1+w_{po})}{I_2 m_g'\Delta h_v\,\rho_1 X_m^2} \quad (36)$$

$$I_1= \int_{X^*_{min}}^{X^*_{max}} \frac{1}{(X^*)^2}\,\phi_v(X^*)dX^* \quad (20)$$

$$I_2= \int_{X^*_{min}}^{X^*_{max}} \phi_v(X^*)dX^* \quad (21)$$

Of the equations above, only four (Eqs.(13),(31), (35), and (39)) are independent.

In order to satisfy the overall heat balance, a " effective " initial temperature of drying medium corresponding to T_{pm} , $T_{go\ eff}$, should be used in calculation, and it can be determined by

$$T_{go\ eff} = T_{go} - \frac{m_s'(1+w_{po})c_{pl}}{(1-q)m_g'c_{pg}}(T_{pm}-T_{po}) \quad (46)$$

The set of differential equations can be solved readily by a numerical method, say, Runge-Kutta method. The step size for calculation must be small enough, especially for the initial stage of drying, because of strong non-linearity of the equations.

The calculated results have shown that the average moisture content of total material would converge to the given value for w_{pf} as X approaches to X_{max} . If a relative deviation Δ_w is defined as

$$\Delta_w = \frac{(w_{po}-w_{pf})-(w_{po}-w_{pf\ cal})}{w_{po}-w_{pf}} \quad (47)$$

usually Δ_w= 0.015 ~ 0.016 for the Kim-Marshall's logistic distribution [10] when calculation is made with $X_{min}=0.05X_m$ and $X_{max}=3X_m$.

In order to obtain accurate results, correct information on droplet size distribution of spray is essential.

Using the model above, the drying time of the largest droplets or the whole spray can be found at $X=X_{max}$. Combining the model with the movement equations of droplets in vertical direction, both the dryer height and the profiles of drying air temperature and other variables through the dryer can be determined. A number of such equations for spray droplets from various type atomizers have been reported[1, 2,13, 20]. The dryer diameter required may be determined in a way similar to that of Gluckert [13] or other authors [1, 2,20].

4. CORRECTION CONCERNING THE CHANGE IN DROPLET DIAMETER

Droplet diameters of many materials change considerably during drying. For these cases, a correction factor, ξ , is introduced to correct the drying time. If it can be assumed that the droplets fall and evaporate independently and no breaking-up of droplets happens, then both the characteristics of size distribution and the number of droplets would not change during the process. The correction then can be based on the information of the change in diameter of a single droplet. Since the surface area of the droplet with a diameter of X is proportional to X^2 , and the heat transfer coefficient to X^{-1}, the rate of heat transfer to a single droplet is proportional to X. Thus, the correction factor for drying time can be readily found from the ratio of the initial diameter of a droplet to the average one of the same droplet throughout the process. For example, for the case where the droplets simply contract, it can be obtained as

$$\xi = \frac{X_{mo}}{(X_{mo}+X_{mf})/2} = \frac{2X_{mo}}{X_{mo}+X_{mf}} \quad (48)$$

and a corrected drying time

$$t_{eff}=\xi t \quad (49)$$

will be used instead of t in the equations of the mathematical model.

5. NUMERICAL EXAMPLES

The following three examples show how the model works.

5.1. A Process Analysis

Study the effects of some variables in the spray drying of a highly concentrated solution of an inorganic salt. The operating conditions and the properties of the solution are as follows:

$m_s'=0.042$ kg/s $\quad\quad \rho_1=1\ 550$ kg/m^3

$w_{po}=0.852$ kg/kg $\quad\quad T_b=380.5$ K

$w_{pf}=0.02$ kg/kg $\quad\quad \Delta h_v=2\ 240$ kJ/kg

Assume a pneumatic nozzle is used and the size distribution of spray droplets can be described by the Kim-Marshall's equation [10]

$$\phi_v(X^*)=\frac{16.7\exp(-2.18\ X^*)}{1+6.67\exp(-2.18X^*)} \tag{50}$$

with $X_{min}=0.05X_m$ and $X_{max}=3X_m$.

For this case, instead of T_{pm} , the boiling point temperature T_b could be used in calculation, and the correction factor, ξ , can be considered to be unit as the diameters of droplets are essentially constant during drying as reported in [18] .

The computation was carried out with the Runge-Kutta fourth order forward method for seven sets of conditions listed in Table 2. The results are given in form of curves of T_g vs t and X vs t.

Table 2 Process conditions given

Nº	m_g' kg/s	X_m μm	T_{go} eff K	w_{go} kg/kg	q -
1	0.207	25.4	787	0.195	0.20
2	0.207	40	787	0.195	0.20
3	0.207	50	787	0.195	0.20
4	0.200	40	787	0.195	0.20
5	0.215	40	787	0.195	0.20
6	0.399	40	623	0.140	0.16
7	0.156	40	923	0.210	0.24

All the results show that most of the heat and mass transfer is carried out within a very short early stage, while the drying of the largest droplets would last a considerably longer time. This is certainly the case with the conventional cocurrent spray dryers.

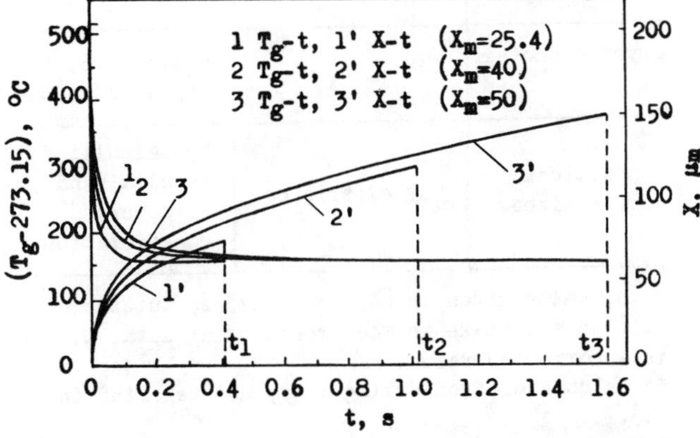

Fig. 2 Effect of the mean diameter of droplets

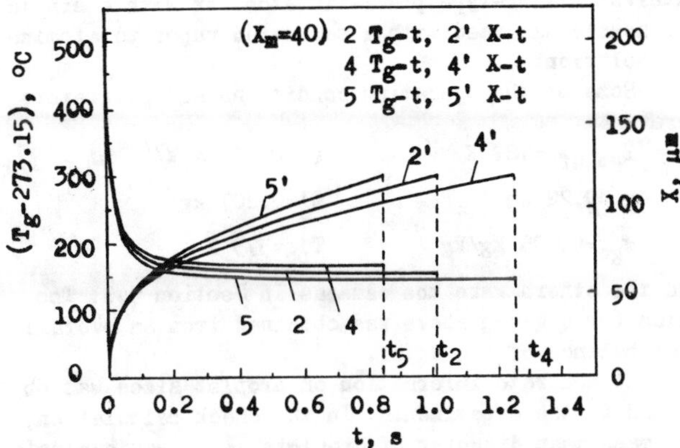

Fig. 3 Effect of the final temperature of drying air

Fig. 2 shows the effect of the mean diameter of droplets. The drying time increases approximately in proportion to X_m^2. The importance of the effect of the final temperature of drying air is shown in Fig. 3. The drying time required is almost inversely proportional to $(T_{gf}-T_b)$ when other conditions are essentially the same. The reason for this is obvious, since the drying of the largest droplets essentially proceeds in the drying medium at its final temperature. A proper final temperature therefore must be chosen. The effect of the initial temperature of drying air is much less important than that of final one, as shown in Fig. 4. Raising the initial temperature only increases the intensity of heat transfer in the initial stage of drying; but does not significantly shorten the time required to sufficiently dry the largest droplets.

The results obtained above show that the model simulates the real processes qualitatively well.

5.2. A Check Analysis

In this example, the operating data of a pilot plant are checked. The material processed was a mixed solution of Na_2HPO_4 and NaH_2PO_4 , and the dry product was used to produce tripolyphosphate of sodium. The dryer had a total height of 4.8 m, an effective height of 4.2 m, and a diameter of 1.6 m. A convergent

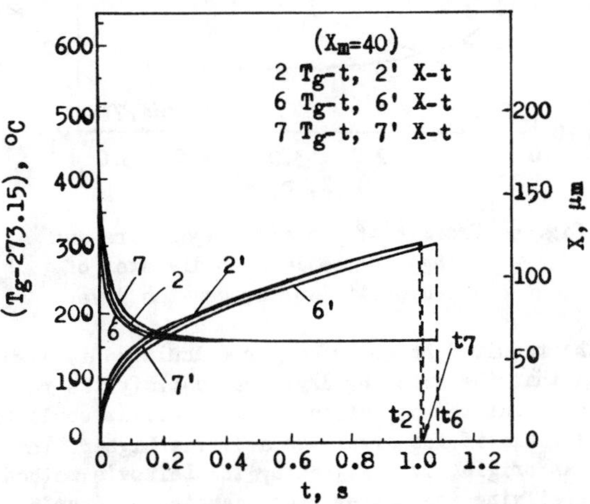

Fig. 4 Effect of the initial temperature of drying air

external mixing-type pneumatic atomizer with a single gas nozzle was used with a saturated vapor to atomize the solution.

Some of the operating conditions and parameters were

$T_{go\ eff}$=787 K　　　λ =0.031 4 W/(K m)

q =0.22　　　m_g'=0.207 kg/s

w_{go}=0.195 kg/kg　　　T_{po}=373 K

and the others were the same as in Section 5.1. The value for q given above was obtained from an overall heat balance of the dryer.

No accurate information on droplet sizes was obtained in the experiments. In the check calculation, the mass mean diameter of droplets, X_m , was evaluated by the Gretzinger and Marshall's equation [22] as 25.4 μm according to the atomizing conditions and the design of the atomizer. A value for X_m was also obtained by the equation of Kim and Marshall [10] and was only a little different from 25.4 μm. This value therefore was used in calculation. Also, it was assumed that the size distribution of the droplets could be described by Eq. (50) because the atomizer was of the same type as used in the study of Kim and Marshall [10].

In order to obtain the profile of air temperature through the dryer, the model developed above is combined with Eq. (1), which can be written as

$$\frac{dZ}{dt} = 18.95\ Z^{-1}$$

under the atomizing conditions used, with

$$Z = 0 \quad \text{at} \quad t = 0$$

The calculation was carried out as done in Section 5.1. with step sizes of 0.5μs for the initial stage and 0.001 s for the later stage. The results are shown in Fig. 5 in form of T_g vs Z and X vs Z curves. The temperature profile agrees well with that measured.

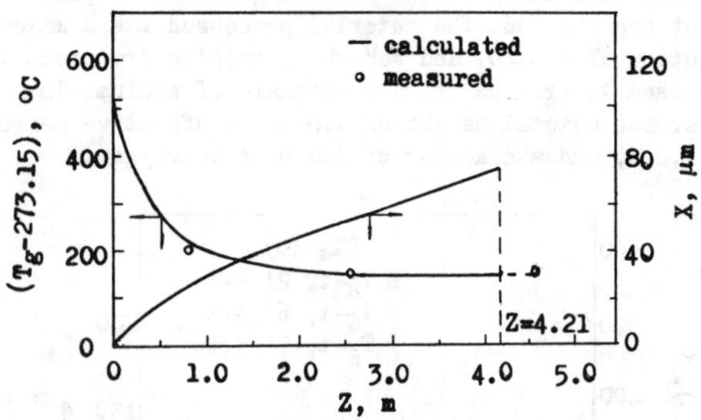

Fig. 5 Profile of the air temperature and change in the largest diameter of dried particles through the dryer

The required height of dryer calculated is 4.21m, showing that the existing dryer can satisfy the requirement, but the operation is critical. It would be more flexible if the dryer was a little higher. The dryer was originally designed by the Leikov's method [6], specifying the Sauter mean diameter of droplets to be 50 μm and the outlet air temperature 403 K. It could not operate under such conditions, and serious caking on the wall of the dryer occurred. Only after a number of improvements has been made then the operation became normal under the conditions given above.

This example shows the model is good for design of cocurrent spray dryers.

5.3. Examination of a Spray Dryer Design

Examine the design of a cocurrent spray dryer producing 300 tons per year of dry reduced blue dye RSN40.

A convergent external mixing-type atomizer with a single air nozzle was used, and so Eq. (50) may be applicable for describing the size distribution. The operating conditions were as the following:

m_s'=0.013 9 kg/s　　　m_g'=0.761 kg/s

w_{po}=4.0 kg/kg　　　w_{pf}=0.02 kg/kg

T_{go}=573 K　　　T_{gf}=373 383 K

X_{mf} required =15 μm

q is assumed to be 0.16

For this case, a pseudo-mean temperature of droplets, T_{pm}, must be used, and it is estimated to be 346 K from the drying curve [4]. And it is necessary to introduce the correction factor for drying time, ξ, concerning the change in diameters of droplets. If a volumetric summation rule is applicable to the aqueous dye suspension, a value of 27 μm for X_{mo} can be obtained from the density of dry dye and that of water, the value for X_{mf}, and the values of w_{po} and w_{pf} , and then, from Eq. (48), we have

$$\xi = \frac{2\,X_{mo}}{X_{mo}+X_{mf}} = \frac{(2)(27)}{15+27} = 1.286$$

Table 3 Comparison of the results obtained by two different design methods

terms	based on mean diameter of droplets [4]	calculated by the model of this paper
drying time s	0.11 *	1.18
dryer height m	1.81**	5.91
dryer diameter m	-	1.48∽1.69 =(1/4∽1/3.5)Z
practical dryer sizes	by experience	by calculated results and structure consideration

* The value given in [4] was 0.173 s, but there existed a mistake in the treatment of data. 0.11 is a corrected value.

** calculated from drying time, not reported in [4].

The computation results in t_{eff}=1.18 s and T_{gf} = 375.7 K. For the design conditions given, the dryer height required was found by integrating Eq. (1) to be

$$Z=(5.436\ 76)(t_{eff})^{0.5}=(5.436\ 76)(1.18)^{0.5}$$
$$=5.91\ m$$

The results calculated are compared with those given in [4] in Table 3.

The dimensions chosen for the actual dryer in the original design were: height: total=8 m, upper conical section=2.5 m, lower conical section=1.5 m, middle cylinder=4 m; and diameter=1.6 m. Comparing them with those listed in Table 3, it can be seen that the results calculated by the method of this paper are applicable, while those based on mean diameter of droplets are infeasible.

NOMENCLATURE

b_1	parameter defined by Eq. (5)	s^{-1}
b_2	parameter defined by Eq. (5)	m/s
C_1	constant defined by Eq. (14)	$m^2/(K\ s)$
C_2	constant defined by Eq. (27)	1/(K s)
C_3	constant defined by Eq. (32)	1/(K s)
C_4	constant defined by Eq. (36)	1/(K s)
I_1	integral value defined by Eq. (20)	-
I_2	integral value defined by Eq. (24)	-
I(t)	integral function defined by Eq. (21)	-
m'	mass flow rate	kg/s
q	heat lost fraction	-
X	droplet diameter, in Eqs. (2) to (5); also largest diameter of dried particles	m
X*	reduced diameter defined by Eq. (17)	-
X_m	mass mean diameter of droplets	m
X_{min}	diameter of minimum droplets in spray	m
X_{max}	diameter of maximum droplets in spray	m
Z	distance from outlet of the atomizer; also height of dryer	m

Greek letters

Δ_w	deviation defined by Eq. (47)	-
ξ	correction factor for drying time	-
ϕ_v	reduced volume frequency function for droplet size distribution	-

Subscripts

a	dry air
cal	calculated
eff	effective or corrected value
f	final condition
g	drying gas, its amount or humidity is based on dry air
l	solution or suspension to be dried
m	mean value
o	initial condition
or	orifice
p	droplet(s)/particle(s)
s	dry solid
w	water vapor
x	property or amount of a single droplet with diameter X

REFERENCES

1. Masters, K., Spray Drying, pp. 33-38, 241-310, Leonard Hill Books, London (1972).
2. Masters, K., Spray Drying Handbook, pp. 165-322, George Godwin Limited, London (1979).
3. Wong Xizhong, Huaxue Gongcheng, (Chem. Eng.) (Chinese), № 2, 77, (1976).
4. Wong Xizhong, Huaxue Gongcheng, (Chem. Eng.) (Chinese), № 3, 63, (1976).
5. Parti, M., and Palancz, B., Chem. Eng. Sci., Vol. 29, 355, (1974).
6. Leikov, M.V., Swushka Raspeileniem (Spray Drying) (Russian), Pishepromizdat (1955)
7. Dlouhy, J., and Gauvin, W.R., A.I.Ch.E.J., Vol. 6, 29 (1960).
8. Marshall, W.R., Jr., Trans.Am.Soc.Mech.Engrs., Vol. 77, 1377 (1955).
9. Ranz, W.E., and Wong, J.B., Ind.Eng.Chem., Vol. 44, 1371 (1952).
10. Kim, K.Y., and Marshall, W.R., A.I.Ch.E.J., Vol. 17, 575 (1971).
11. Charlesworth, D.H., and Marshall, W.R., A.I.Ch.E.J., Vol. 6, 9 (1960).
12. Miura, T., and Ohtani, S., Kagaku Kogaku Rombanshu (Japanese), Vol. 5, 130 (1979).
13. Gluckert, F.A., A.I.Ch.E.J., Vol. 4, 460 (1962).
14. Dickinson, D.R., and Marshall, W.R., A.I.Ch.E.J., Vol. 14, 541 (1968).
15. Albertson, M.L., Dai, Y.B., Jensen, R.A., and Hunter Rouse, Proc.Am.Soc.Civil.Engrs., Vol. 74, 1577 (1948).
16. Soo, S.L., Chem.Eng.Sci., Vol. 5, 59 (1956).
17. Coulson, J.M., and Richardson. J.F., Chemical Engineering, 3rd. ed., Vol. 2, pp. 104-108, Oxford Pergamon Press (1978).
18. Wu Yuan, Zheng Chong, and Fu Jufu, Hua Kung Shueh Pao (J.Chem.Indus.and Eng. (China)), № 4, 348 (1982).
19. Ranz, W.E., and Marshall, W.R., Chem.Eng.Prog., Vol. 48, 141,173 (1952).
20. Mochido, T., and Kukida, Y., Kagaku Sochi (Piant and Process) (Japanese), № 2, 57, № 3, 46, № 4, 42 (1972).
21. Litch, W., A.I.Ch.E.J., Vol. 20, 595 (1974).
22. Gretzinger, J., and Marshall, W.R., A.I.Ch.E.J., Vol. 7, 312, (1961).

CONTROL AND ENERGY ASPECTS OF MILK POWDER SPRAY-DRYING IN RELATION TO PRODUCT QUALITY

G.H. Niels, L.A. Jansen, A.J.B. van Boxtel

Netherlands Institute for Dairy Research (NIZO), Ede, Netherlands

ABSTRACT

Heat recovery in spray-drying of milk powders is possible without affecting product quality. The specific energy consumption of a spray-drier for whole milk powder can be decreased from 1.51 (without heat recovery) to 1.20 (with all possible heat recovery systems). The same economical effect can be achieved by increasing the TS content of the concentrate that leaves the evaporator. However, the latter is not possible nowadays without product degradation. A new method of controlling the moisture content in the production of milk powders is introduced. This control method compensates for the effect the concentrate viscosity has on the drying behaviour of the droplets.

1. INTRODUCTION

In 1982 almost 13 thousand million kilograms of milk were produced in the Netherlands. Twenty-eight percent of this quantity was used for the production of milk powders, and thirty-six percent for cheesemaking. In the latter process, for the Dutch situation, roughly the same amount of whey is produced, which served as a basis for making whey powder. The production methods for skim-milk powder, whole milk powder and whey powder are quite similar. Concentration of the thin liquid takes place in a multi-stage falling-film evaporator. The viscous concentrate leaving the evaporator is atomised and dried in a spray-drier with co-current air flow, whilst in many cases final drying and cooling of the powder are effected on vibrating or shaking fluid-bed driers. Nowadays, the capacity of a plant producing these powders is between 3 and 4 tonnes of powder per hour. This implies that approximately the same amount of water will be removed from the product in the spray-drying installation. As distinct from the energy-efficient water removal in a multi-stage evaporator, water removal in a spray-drier demands a lot of energy. This article gives the results of research work carried out at NIZO with the aim of reducing production costs without unnecessary risks to product quality. The first part of the article deals with what is possible and economical with heat recuperation in spray-drying. It is based on a study carried out in the Dutch dairy industry by Jansen et al. (1). The second part gives a new method for controlling the moisture content of milk powders. The work involved carried out by Alderlieste et al. (2) was started at a NIZO pilot plant and evaluated at two industrial driers; one for skim-milk and one for whole milk.

2. ENERGY SAVINGS

Fig. 1 is the diagram of a two-stage spray-drying installation for the production of whole milk powder. Ambient air is heated to 180 °C in an indirect, gas-fired air heater and enters the spray-tower at the top. The feed is atomised (wheel or nozzle) to between 44 and 52 % total solids. The outlet air temperature is 72 °C or higher. The milk powder has a moisture content of 5 to 6 % when leaving the spray-tower. On a fluid-bed drier and cooler the powder's moisture content is further reduced to less than the allowable maximum of 3 %.

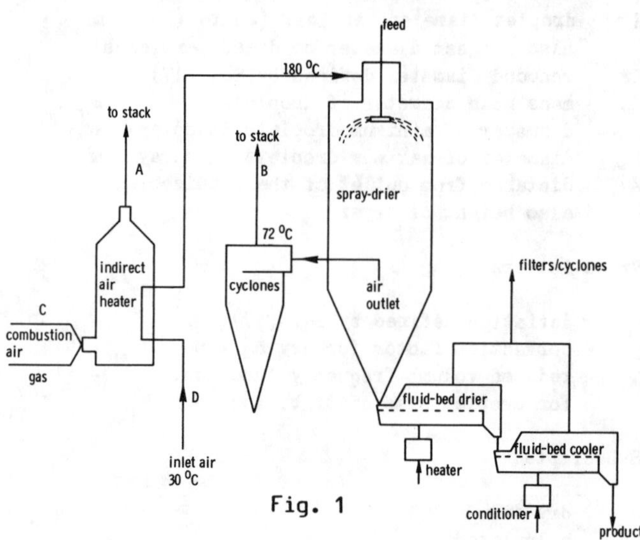

Fig. 1

Energy savings can be achieved in different ways; a rise in inlet air temperature of the drier, an increase of the TS content of the concentrate leaving the evaporator, and heat recuperation (the fluid beds are not taken into consideration).

2.1. Increase in inlet air temperature

This is the easiest way of saving energy, but it may cause thermal damage to milk products and thus impair product quality. In conventional spray-towers, the maximum inlet air temperature normally used is approx. 180 °C for whole milk powder and 240 °C for skim milk powder.

2.2. Increase in TS content of the concentrate

More water removal takes place in the multi-stage evaporator. A modern evaporator design will give a specific energy consumption of 0.10. This means that only 0.1 tonne of steam is needed to evaporate 1 tonne of water from the product. However, a high TS content gives the concentrate a high viscosity, which is a disadvantage in atomisation. Also, to reach the same moisture content in the powder the air outlet temperature has to be raised when the traditional control method is used, and this may affect product quality.

2.3. Heat recuperation

Heat recuperation is possible between the inlet air flows and the outlet air and/or flue gases of the air heater without any influence on the drying process itself. As an example we shall take a plant with a powder capacity of 3750 kg/h. At an inlet air temperature of 180 °C and an outlet air temperature of 72 °C, the air flow will be 82 000 kg/h.

Flue gas; combustion air (Fig. 2). A system using two heat exchangers and a circulating fluid as heat carrier is the most economical one. At the Dutch prices for gas (in 1982 Hfl. 13.6/GJ) and electricity (Hfl. 0.19/kWh) we found the lowest specific investment for a flue gas outlet temperature of 125 °C. This economical optimum, however, has no sharply defined limits. All temperatures between 110 °C and 140 °C will give about the same specific investment on condition that the recuperator has been designed for those temperatures. The specific energy consumption (SEC) decreases from 1.51 without heat recuperation to 1.47 with heat recuperation. The pay-out time (POT) depends on the number of production hours per year (N) and the gas price (Pg) (Fig. 3).

Flue gas; drying air (Fig. 4). It appears that in this case it is economical to cool down the flue gases to the dew point temperature and so to make use of the heat of condensation. Here again, the pay-out time depends on production hours and gas price (Fig. 5). The specific energy consumption decreases from 1.51 to 1.36.

Fig. 2

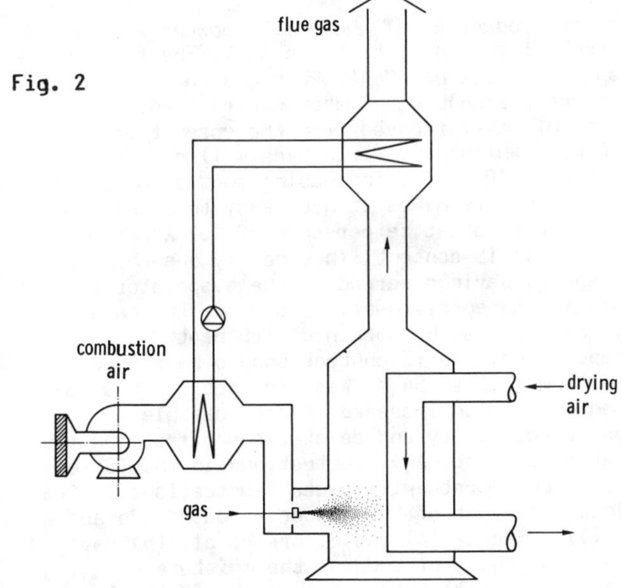

Fig. 3

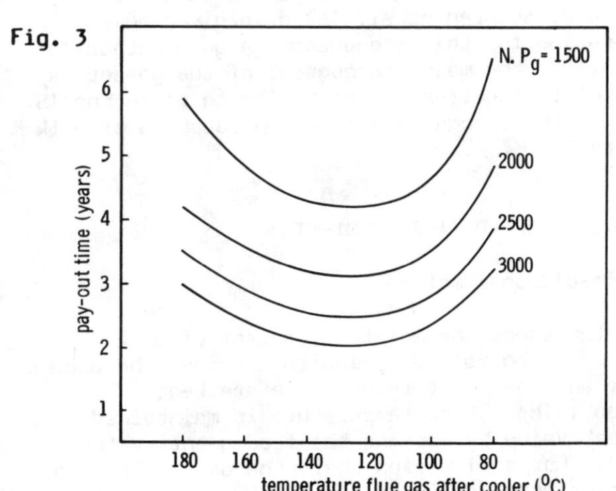

Fig. 4

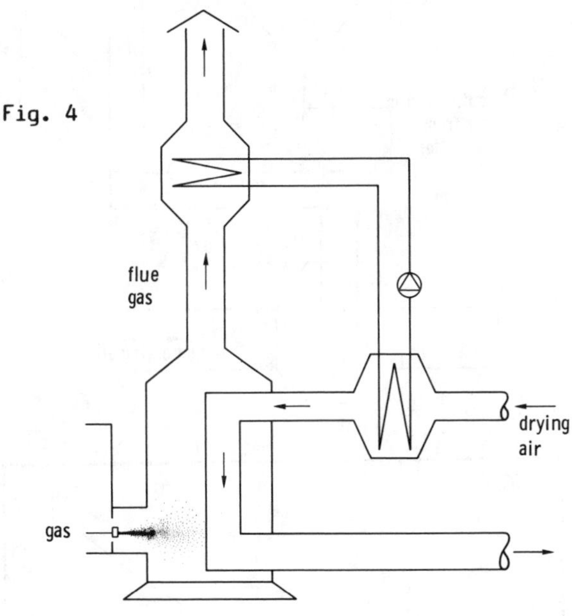

Fig. 5

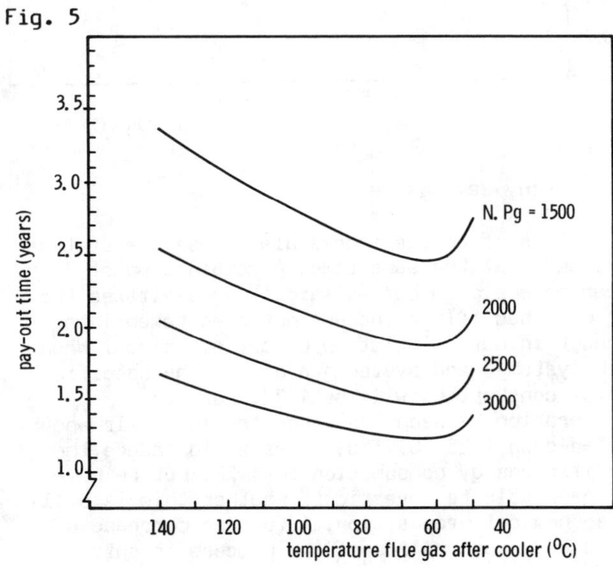

Inlet drying air; outlet drying air (Fig. 6). Heat recuperation from the outlet air is only possible if a bag filter is used for dust removal. Without such a filter, the heat transfer coefficient would decrease so much that no economical investment is feasible. The shortest pay-out time will be obtained at a degree of recuperation of 50 % (Fig. 7). With a dual-heat exchanger system as shown in Fig. 6, 70 % recuperation can be achieved. This degree of recuperation involves a longer pay out-time but gives much higher long-term savings. At 70 % recuperation the specific energy consumption of the drier decreases from 1.51 to 1.25.

Fig. 6

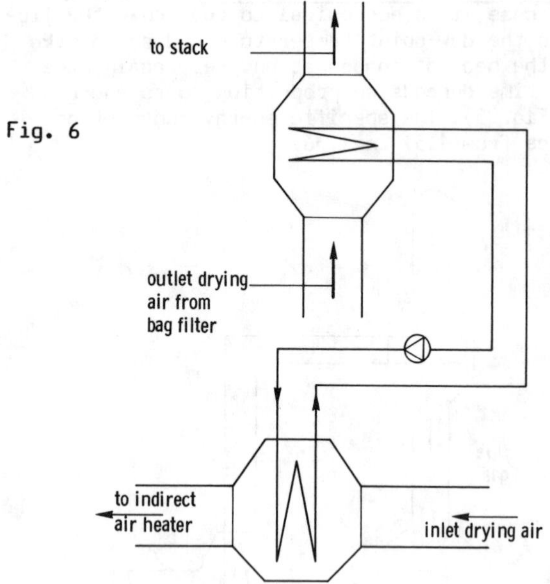

Fig. 7

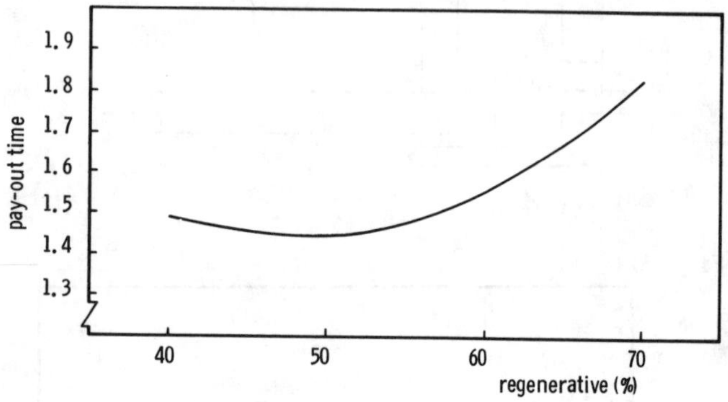

2.4. Energy savings

It is of course impossible to use the systems 1, 2 and 3 at the same time. A combination of 1 and 3 can be employed but as said it necessitates the use of a bag filter and has not been taken into account in our calculation of pay-out times. When both system 1 and system 3 are used the specific energy consumption will be 1.22. Further recuperation between the preheated inlet air and cooled-down (125 °C) flue gases would reduce the specific energy consumption to 1.20, but we have not been able to investigate whether this is still an economical proposition. A further decrease of the energy consumption of the process is only possible by the method mentioned in Sections 2.1. and 2.2.

Fig. 8 shows the effects of the decrease in specific energy consumption from 1.51 to 1.20, in relation to the TS content of the concentrate for the plant described. Calculation of the total energy costs has been based on a price of Hfl 40 per tonne of steam and on a specific energy consumption of 0.1 for the multi-stage evaporator.

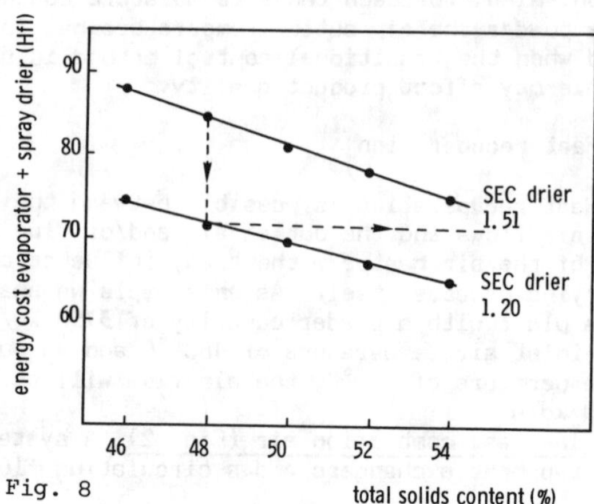

Fig. 8

For the production of whole milk powder a normal TS content of the concentrate is 48 %. The feed to the evaporator will be 30 440 kg/h and in the evaporator 22 830 kg/h water are removed. The amount of water removed from the spray-tower is 3760 kg/h whilst the fluid beds will dispose of an additional 100 kg/h. Increasing the TS content of the concentrate makes it necessary to raise the spray-tower outlet temperature of 1.2 K per percent increase in TS content. Thus nearly one-third of the energy savings earned in the evaporator is lost again in the spray-tower. To achieve the same savings that can be obtained with heat recuperation, the TS content should be raised from 48 % to more than 54 %! With the equipment used nowadays such an increase is not possible without loss of solubility and deleterious effects on the moisture content. Any fluctuations in the TS level of the concentrate will cause fluctuations in the moisture content of the powder produced. Jansen et al. (3), Pisecky (4) and Bloore et al. (5) have shown that the amplitude of the moisture fluctuations will be greater as the TS level increases. Snoeren et al. (6) have given an explanation for this phenomenon. A good method for controlling the moisture content of the powder is an absolute requirement for trying to raise the TS level of the concentrate in the production of milk powders.

3. CONTROL OF MOISTURE CONTENT

3.1. Traditional method

At present the moisture content of milk powders is controlled by indirect means. The amount of air and the inlet temperature are kept constant. The outlet temperature is maintained at a constant value by varying the feed quantity (see Fig. 9). For a situation where the powder leaving the drier is in equilibrium with the outlet air

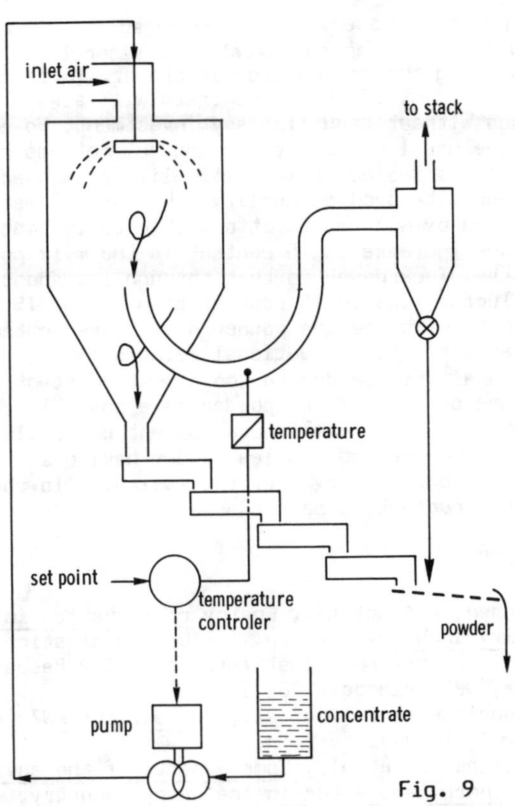

Fig. 9

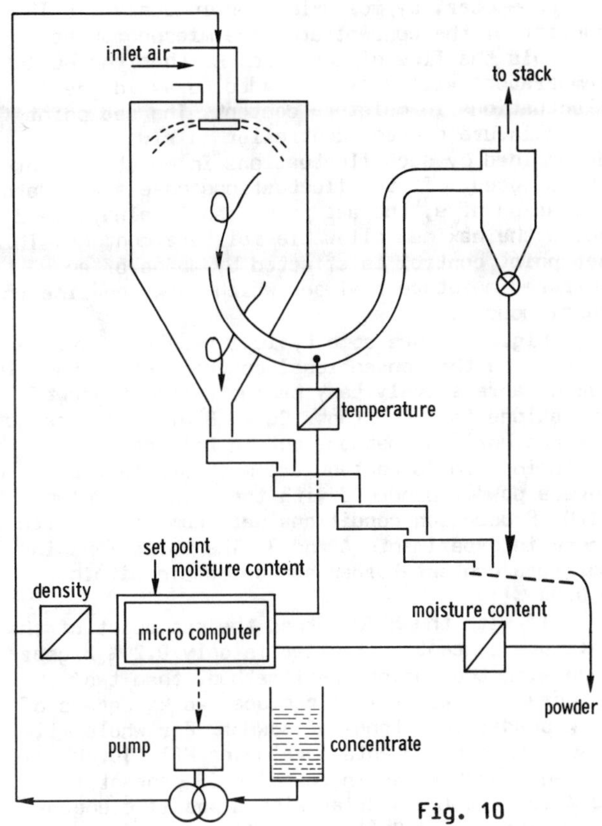

Fig. 10

this method would be satisfactory but in practice this never happens. Jansen (3) and Pisecky (4) have given a relation for the fluctuations in the air outlet temperature, ΔT_{ao}.

$$\Delta T_{ao} = f(1).\Delta T_{ai} + f(2).\Delta TS + f(3).\Delta X_p \ldots (a)$$

The functions f(1), f(2) and f(3) are product- and installation-dependent. Furthermore, they depend on the temperature and TS levels.

From this relation it is evident that at a constant inlet air temperature ($\Delta T_{ai} = 0$) and a controlled constant outlet air temperature ($\Delta T_{ao} = 0$), the moisture content X_p of the powder will fluctuate with variations in TS content of the concentrate.

3.2. NIZO control method (Fig. 10).

The control method developed has become possible by the introduction of a continuous on-line measurement of the moisture content of milk powders. Comparatively inexpensive process computers permit the use of equation (a) combined with a statistical analysis of the process conditions. This control method also uses a constant inlet-air temperature. Thus in relation (a) only the functions f(2) and f(3) are important and for a given installation and production method they merely depend on the temperature level and TS content of a given product concentrate. For two industrial spray drying installations, each producing approx. 4 tonnes of powder per hour and for three types of milk powder the functions f(2) and f(3) were determined around the normal working points. A programme describing the dynamic behaviour of the installation combined with the results of relation (a) was fed into the

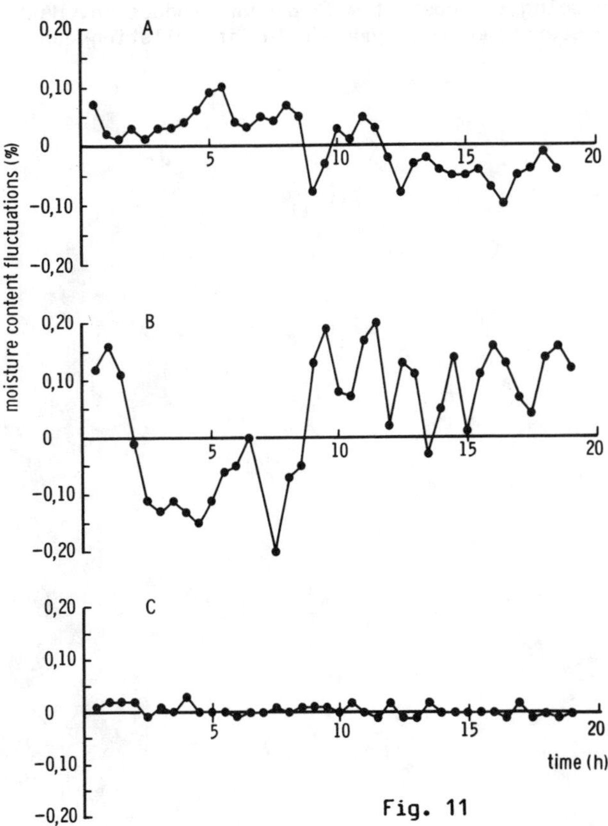

Fig. 11

microcomputer. By measuring continuously the TS content of the concentrate, the microcomputer controls the flow of the feed, so that the outlet temperature will vary as needed to avoid great fluctuations in moisture content. The set point for the moisture content controller is always determined by such fluctuations in moisture content as may occur. If the fluctuations have a standard deviation of s, the set point should always be 2.s below the maximum allowable moisture content. This set point control is effected by means of an infrared moisture analyser which works on-line and continuously.

Fig. 11 shows some typical results. Curve A relates to the conventional control method used in making a relatively easy product without great variations in TS content. Curve B also relates to the conventional method, but in this case the variations in TS content are much greater. Curve C covers powder produced with the method developed by NIZO. Production conditions were comparable with those in experiments A and B. The variations in moisture content turned out to remain within ± 0.04 %!

If with the NIZO method the set point of the moisture content controller is only 0.2 % higher than with the conventional method, then two kilograms of water will replace two kilograms of milk powder in 1 tonne of powder. For whole milk powder this represents a value of Hfl. 10.40 which is tantamount to an increase in TS content from 48 % to 54 % for a drier with a specific energy consumption of 1.2 (see Fig. 8).

4. CONCLUSIONS

This paper deals with three projects for reducing the cost of milk powder production. Heat recuperation in a spray-drying installation can bring down the specific energy consumption to a level as low as approximately 1.2 without influencing the drying process proper. Increasing the TS content of the concentrate will always give savings without requiring any investment. However, with the usual equipment and control methods such increase is subject to specific limits imposed by the requisite product quality. The control method developed by NIZO makes it possible to initiate a cautious increase in TS content in the milk powder industry. The reason is that through its adoption, any fluctuations in TS content at a higher TS level will not influence the powder's moisture content as much as with the conventional methods. An additional feature due to the almost constant moisture content of the powder is a possible raise in set point of the moisture content controller. Thus the new method enables powder having a moisture content very near the maximum allowable moisture content to be produced.

REFERENCES

1. Jansen, L.A. et al., Energy conservation in the dairy industry, 244 pp. (NIZO-communication M17), Netherlands Institute for Dairy Research, Ede, Netherlands (1983).
2. Alderlieste, P.J. et al., Zuivelzicht FNZ, vol. 76, 156(1984).
3. Baltjes, J. et al., Some aspects of the saving of energy and water in the dairy industry, 77 pp. (NIZO-communication M13), Netherlands Institute for Dairy Research, Ede, Netherlands (1979).
4. Pisecky, J., Dairy Inds. Intern., vol. 48, 21(1983).
5. Bloore, C.G. and Boag, I.F., N. Z. Jl. Dairy Sci. Technol., vol. 17, 103(1982).
6. Snoeren, T.H.M. et al., Zuivelzicht FNZ, vol. 75, 847(1983).

APPLICATION OF OPTIMAL CONTROL STRATEGY TO HYBRID MICROWAVE AND RADIANT HEAT FREEZE DRYING SYSTEM

T. N. Chang and Y. H. Ma

Department of Chemical Engineering
Worcester Polytechnic Institute
Worcester, Massachusetts, USA

ABSTRACT

The time-optimal control algorithm taking into consideration the bounds on control and state variables is applied to microwave freeze drying system. A general unsteady state mathematical model with time variant transport and dielectric properties is formulated as a stage wise operation problem. Microwave power, radiator temperature, water vapor pressure and air pressure are considered controllable. Constraints are set to avoid scorching, melting and gas breakdown. All control variables are kept constant during each stage.

With high non-linearity and moving boundary characteristics of this problem, an optimal sequence is computed by using direct search method.

The algorithm is illustrated in several cases. Results suggest that the hybrid microwave and radiant freeze drying can substantially reduce the drying time. It is further demonstrated that the algorithm can be used to devise an optimal policy for limited freeze drying.

1. INTRODUCTION

Freeze drying by microwave dielectric heating appears to be one of the most promising techniques to accelerate the dehydration process. Microwave dielectric heating supplies energy of sublimation volumetrically and thus overcoming the heat transfer problems generally associated with conventional freeze drying processes.

Early work in the microwave freeze drying process concentrates mostly on experimental investigations [1,2,3,4]. Early theoretical work was done by Copson [5] employing a simplified approach using a quasi steady state assumption. Recent more extensive mathematical analysis of microwave freeze drying was performed by Ang et al. [6,7], Ma and Peltre [8], and Ma and Arsem [9,10]. A comprehensive and critical review on microwave freeze drying was reported by Sunderland [11].

Previous researchers indicated that a significant improvement in drying rate could be achieved by the optimal combination of factors which affected the drying rate. The transport properties in dried product may be affected by the air and water vapor pressure in the dryer. Higher pressure can increase the heat conduction rate [12,13,14,15] but decrease the sublimed water permeation rate through the dried layer [12,16,17].

The objective of the present work is to apply the optimal control policy to microwave freeze drying to demonstrate the feasibility of using such a strategy to improve drying rate. It is further demonstrated that an optimal time policy of all controllable factors in a microwave freeze drying system can be used to provide policy for limited freeze drying.

2. MATHEMATICAL MODELLING

2.1 The Physical System

The material to be freeze dried is assumed to have the geometry of an infinite slab as shown in Figure 1. The left edge of the slab is insulated (or may be considered the centerline of a symmetric slab) while the right edge is exposed to a vacuum at temperature T_R and concentration C_R. As the sublimation proceeds, a dry outer layer forms while the frozen core retreats. In order to describe the physical system shown in Figure 1, it is necessary to make the following additional simplifying assumptions:

1. A sharp sublimation zone is formed during the drying process
2. The removal of water vapor from sublimation is by diffusion through the porous dried layer
3. Diffusion is described by Fick's equation characterized by an effective diffusivity
4. Heat transfer in the frozen region is by conduction while heat transfer in the dried layer is by conduction and convection
5. The dissipation coefficients for frozen and dried regions are functions of temperature.

For a detailed discussion of the assumptions, the reader is referred to the work by Ma and his co-workers [8,9,10,11].

2.2 The Mathematical Model

Based on the assumptions described above, a general unsteady state analysis has been used to derive the following mathematical model for the previously described physical system. A detailed derivation can be found in Chang [18].

Energy balance in the dried layer:

$$\rho_d C_{pd} \dot{T}_d = \nabla(k_d \nabla T_d) + K_d E^2 - C_{pw} \nabla(N_w T_d) \qquad (1)$$

Energy balance in the frozen region

$$\rho_f C_{pf} \dot{T}_f = \nabla(k_f \nabla T_f) + K_f E^2 \qquad (2)$$

Mass balance in the dried region

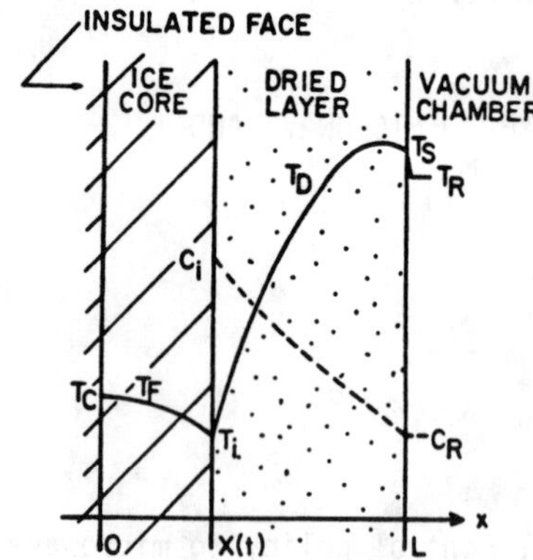

Fig. 1 Model with expected forms of the concentration and temperature profiles for the microwave freeze-drying process.

$$\sigma\dot{C} = (D\nabla C) \quad (3)$$

Air pressure variation in the freeze dryer

$$\dot{P}_A = -R_E P_A + R_L \quad (4)$$

Water vapor concentration variation in the freeze dryer

$$(V_R - A_s L)\dot{C}_R = N_{wR} A_s - k_c (C_R - C_R^*) A_c \quad (5)$$

where

$$E^2 = P_M / (\int_o^{sL} K_f dx + \int_{sL}^{L} K_d dx) \quad (6)$$

Solutions of Eqs. (1), (2) and (3) in conjunction with Eqs. (4) and (5) requires the following boundary conditions:

for $t > 0$

At the centerline of the slab, the symmetry gives

$$\nabla T_f = 0 \quad (7)$$

while at the interface, continuity of temperature is assumed

$$T_f = T_d = T_{int} \quad (8)$$

and equilibrium is assumed to exist at the interface

$$C = f_E(T_d) = C_{int} \quad (9)$$

and a mass balance at the interface gives

$$N_w = \rho_{ice} \sigma \dot{S} L = -D\nabla C \quad (10)$$

$$-N_w \Delta H_s = k_f \nabla T_f = k_d \nabla T_d \quad (11)$$

At the outer surface of the dried layer $x = L$, the radiant boundary condition is

$$-k_d \nabla T_d = \frac{B(T_p^4 - T_s^4)}{(1/\varepsilon_p) + (1/\varepsilon_s) - 1} \quad (12)$$

and

$$T_d = T_s \quad (13)$$

$$C = C_R \quad (14)$$

$$N_{wR} = -D\nabla C \quad (15)$$

$$C^*_R = f_E(T_c) \quad (16)$$

As it is necessary to minimize the drying time which is unbounded, it would be more convenient if the following change of variable is made

$$\frac{d}{dt} = \dot{S}\frac{d}{dS} \quad (17)$$

where $\dot{S} = \frac{dS}{dt}$ and S represents the dimensionless interface position. By utilizing the interface moving rate $\dot{S}$ as shown in Eq. (17), the system can be converted from a time varying problem to an S-varying one. The advantages of doing this are several fold:

1. The termination of the drying process can be exact since the value of S varies from unity to zero as the drying proceeds.

2. The system can be treated as a cascade operation which enables one to model the drying process as several stages in series. This makes the system more suitable for computer implimentation.

3. The objective function for time-optimal control can be represented as a function of the interface moving rate.

Eqs. (1), (2) and (3) are a set of partial differential equations which are to be solved with a moving boundary. After placing the equations and the associated boundary conditions into dimensionless forms and changing the variable indicated in Eq. (17), Landau transformation is used to convert the problem of a moving boundary to that of a fixed boundary to facilitate the solution. Furthermore, a spatial discretization shown in Fig. 2 reduces the problem to a lumped parameter system. This allows one to use the state space analysis to formulate a time optimal control policy for the freeze drying process. A detailed description of the algorithm used in the computation can be found in Chang [18].

3. APPLICATION OF THE OPTIMAL CONTROL POLICY

In order to test the developed algorithm, the algorithm was tested by using the policy reported by Arsem [10] and compared with his experimental drying curves. Some pertinent data employed in the simulation are tabulated in Table 1.

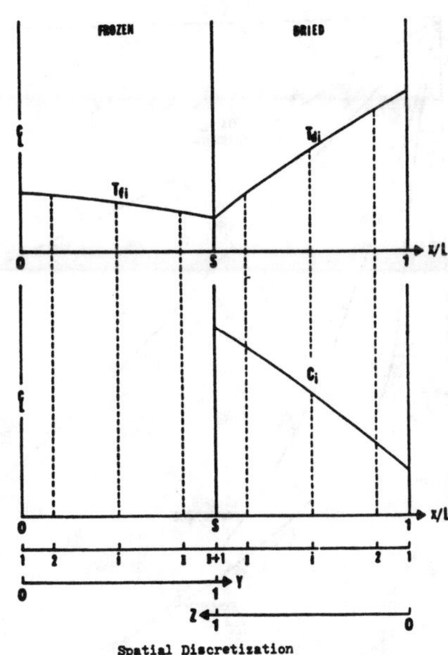

Fig. 2 Spatial Discretization

Table 1

Sample Parameters for Case Studies

Sample Thickness	1.3 cm
Dried Sample Density	0.32 g/cm^3
Frozen Sample Density	0.96 g/cm^3
Sample Surface Emissivity	0.7
Scorching Temperature	50°C
Melting Temperature	-2.5°C
Initial Sample Temperature	-40°C
Sample Water Content	73%
Sample Fat Content	0.9%

The comparison between the experimental drying curves and those obtained from the calculations using the operation policy specified by Arsem is shown in Fig. 3. The excellent agreement between the experimental data and the calculated curves is an indication of the validity of the model and the formulation of the control policy.

A direct search method was used to establish the time optimal control algorithm. The reason for choosing direct search method over the first and second derivative methods is, in part, due to the large number of decision parameters and the complexity of the problem involved which makes it difficult to provide analytical functions for the needed derivatives.

The time optimal control algorithm was applied to hybrid microwave and radiant heat freeze drying system. The process parameters and the control domain for the case studies are shown in Table 2. The temperature histories of external

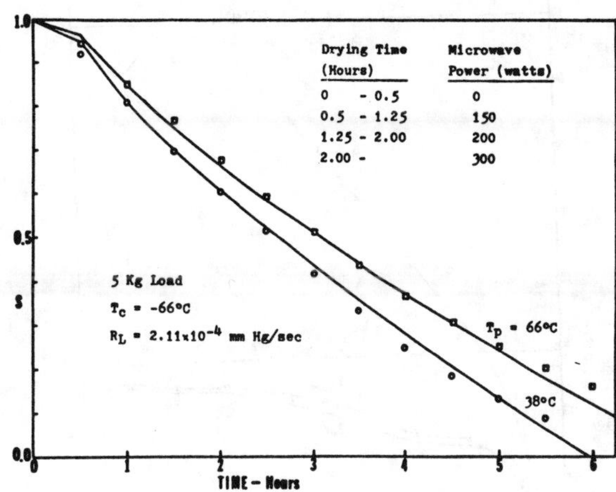

Fig. 3 Comparison Between Calculated and Experimental Drying Curves

Table 2

Process Parameters and Control Domain for Case Studies

Process Parameters for Case Studies	
Ambient Temperature	25°C
Chamber Volume	1350 liters
Microwave Frequency	2450 MHz
Condenser Surface Area	4000 cm^2
Air Evacuation Rate	1.62 x 10^{-2} sec^{-1}
Load Size	5 Kg
Radiator Surface Emissivity	0.83
Condensation Rate Constant	4 x 10^6 sec^{-1}
Control Domain for Case Studies	
Microwave Power	0 - 500 Watts
Radiator Temperature	25 - 130°C
Air Leaking Rate	2.11x10^{-4} mm Hg/sec
Condenser Temperature	-66 - -10°C

surface, ice core and interface are shown in Fig. 4. It is obvious that the control policy is to keep all three temperatures as high as possible without violating the constraints which are limited by scorching of the dried product and melting of the ice core. Fig. 5 shows the peak electric power and the computed breakdown power in the chamber. It is clear that the peak electric power is always less than the breakdown power as required by the specified constraint of no electric discharge. The large difference between the breakdown power and the peak power supplied to the sample is the direct result of introducing a safety factor in the formulation of the control policy due to the uncertainty and the complexity involved in the criteria for breakdown.

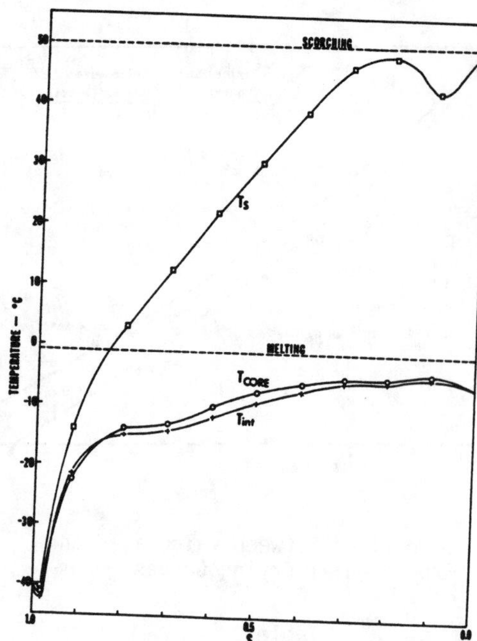

Fig. 4 Sample Temperature Histories

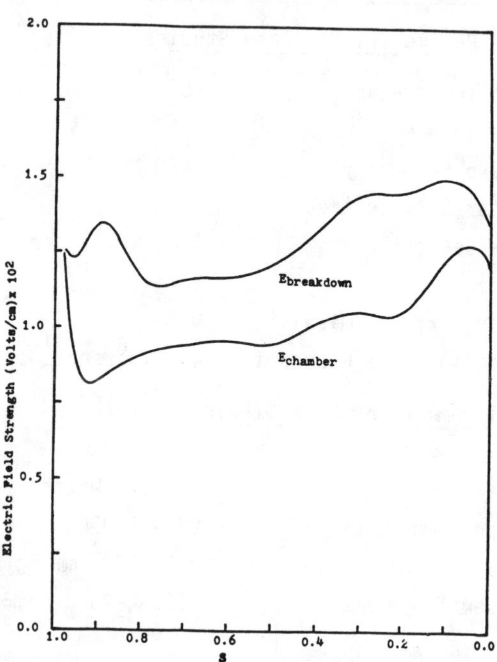

Fig. 5 Electrical Field Strength as a Function of Interface Position

An interesting application of the mathematical model to limited freeze drying is recently demonstrated by Chang [18]. The model was applied to provide optimal control policy for freeze dehydration of beef. A hybrid radiant and microwave system was considered. The distribution of moisture in the sample at the instant when the ice disappears is shown in Fig. 6. The numbers in the parentheses are the times required for the ice core to disappear completely. Curves 1 and 2 are the moisture distribution when an optimal policy is applied to microwave and radiant heating. As

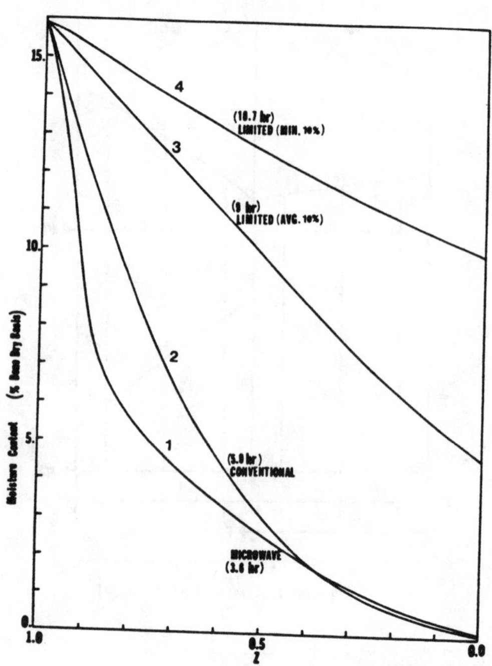

Fig. 6 Moisture Distribution in the Sample at the End of Drying Period

expected, the sample is quite dry at the exterior surface while the center still retains considerable amount of moisture. On the other hand, curves 3 and 4 represent cases where an optimal policy is applied to a hybrid microwave and radiant heating system. By judicially adjusting the microwave and radiant energy input, it is possible to obtain relatively uniform moisture content in the dried sample, Curve 3 indicates the case where the average moisture content in the dried region is kept at an average value of 10% during the entire drying process. Curve 4 represents the case where the minimum moisture content at any point in the dried region is kept at 10%. This evidently lengthens the drying time due to the imposed restriction which limits the power input. The results from the simulation of a limited freeze drying as shown in Fig. 6 clearly demonstrates the feasibility of application of hybrid system to obtain products at a specified moisture content. This is particularly useful for compression of freeze dried food as it eliminates the rehydration step generally required prior to compression.

4. CONCLUSION

Results from the computer simulation of the mathematical model developed for microwave freeze drying agree well with experimental data. Application of time optimal control policy to a hybrid microwave and radiant heating system enables one to obtain a better microwave energy efficiency and to reduce the cost of dehydration. Results on limited freeze drying using a hybrid system show that by judicially adjusting the microwave and radiant energy input it is possible to freeze dry samples to a prespecified moisture level.

5. ACKNOWLEDGEMENT

The work presented in this manuscript has been supported by the U.S. Army Natick Research and Development Laboratory.

NOMENCLATURE

B Stefan Boltzmann's constant ($W/(K^4m^2)$)

C water vapor concentration (Kg/m^3)

E cavity electric field strength (Volts/m)

f_E ice-vapor equilibrium function

ΔH_s sublimation latent heat (J/Kg)

k_c condensation rate constant (m/s)

k thermal conductivity ($W/(Km^2)$)

K microwave energy dissipation coefficient ($J/(sm^3(Volts/m)^2)$)

L half thickness of sample (m)

N water vapor flux ($Kg/(m^3s)$)

P_M microwave power input (Watts)

R_E gas evacuation rate (s^{-1})

R_L gas leaking rate ($N/(m^2s)$)

S interface position (dimensionless)

$\dot{S}$ interface moving rate (s^{-1})

x coordinate dimension

Greek Letters

ρ density

σ product porosity

Subscripts

A air

c condenser

d dried layer

f frozen layer

ice ice in frozen layer

int interface at sublimation front

p radiator

R chamber

s surface of dried layer

w water vapor

REFERENCES

1. Jackson, S., Rickter, S.L. and Chichester, C.O., Food Technol., vol. 11, 468 (1957).
2. Copson, D.A., Food Technol., vol. 12, 270 (1958).
3. Hoover, M.W., Markantonatos, A. and Parker, W.N., Food Technol., vol. 20, 103 (1966).
4. Gouigo, E.I., Malkov, L.S. and Kaukhcheshvili, I.E., Inst. Intern. du Froid, Commission X, Lausanne, Switzerland (1969).
5. Copson, D.A., Microwave Heating in Freeze-Drying, Electronic Ovens and Other Applications, The AVI Publishing Co. Inc., Westport, CT (1975).
6. Ang, T.K., Ford, J.D. and Pei, D.C.T., J. Food Sci., vol. 43, 648 (1978).
7. Ang, T.K., Ford, J.D. and Pei, D.C.T., Intern. J. Heat Mass Transfer, vol. 20(5), 517 (1977).
8. Ma, Y.H. and Peltre, P.R., AIChE J., vol. 21, 335 (1975).
9. Ma, Y.H. and Arsem, H.B., Drying '80, A.S. Mujumdar, ed., Hemisphere Publishing Corp., 196 (1980).
10. Arsem, H.B., Ph.D. Thesis, Chemical Engineering Department, Worcester Polytechnic Institute, Worcester, Massachusetts (1980).
11. Sunderland, J.E., J. Food Process Eng., vol. 4, 195 (1980).
12. Harper, J.C., AIChE J., vol. 8, 298 (1962).
13. Saravacos, G.D. and Pilsworth, Jr., M.N., Food Sci, vol. 30, 773 (1965).
14. Hill, J.E., Leitman, J.D. and Sunderland, J.E., Food Technol., vol. 21, 1143 (1967).
15. Massey, W.M. and Sunderland, J.E., Food Technol., vol. 21, 408 (1967).
16. Sunderland, J.E. and Dyer, D.F., Int. J. Heat Transfer, vol. 9, 519 (1966).
17. King, C.J., Sandall, D.C. and Wilke, C.R., AIChE J., vol. 13, 428 (1967).
18. Chang, T.N., Ph.D. Thesis, Chemical Engineering Department, Worcester Polytechnic Institute, Worcester, Massachusetts (1984).

CHARACTERISTICS OF VIBRO-FLUIDIZED BED FREEZE DRYING

Kanichi Suzuki, Masaaki Ikeda, Muneharu Esaka and Kiyoshi Kubota

Department of Applied Biological Science, Hiroshima University
Fukuyama, 720 JAPAN

ABSTRACT

The freeze drying of granular or chopped materials was carried out in a laboratory size vibro-fluidized bed. The heater, made of stainless steel pipe, was set in the bed for minimizing heat losses. General characteristics of the vibro-fluidized bed freeze drying process were discussed in this study.

Under the appropriate conditions of vibration, the drying processes proceeded with almost uniform bed temperature and moisture content due to thorough mixing by vibro-fluidization. The whole-bed temperatures were kept constantly at near sublimation points balanced to pressures in the drying chamber until the drying stages reached near the end points. The drying rates obtained were up to 6 kg-H_2O/m^2h, though these values were restricted by the abilities of equipment used such as the vacuum pump, and condenser. The thermal efficiencies of the heater were almost 100%. The overall heat transfer coefficient between heater and material depended on the bed moisture content. The values decreased from 300-500 W/m^2K in the initial period of drying to 60-80 W/m^2K in the latter half of the period in accordance with the progress of drying process, the heat generated in the bed by vibration contributed as significantly as the heat for drying. The values were measured when the intensities of vibration were higher than the acceleration due to gravity.

1. INTRODUCTION

The vibro-fluidized bed is one kind of fluidized bed which is vibrated mechanically to enhance the fluidity of powders or granular materials in the bed. The effect of the vibration on fluidizing the materials is remarkable when the air velocities are lower than the minimum fluidization velocity [1, 2]. The vibro-fluidized bed is also useful for fluidizing large particles or sticky materials that are difficult to be treated in the conventional fluidized bed [3, 4]. Further, when the appropriate conditions of vibration are applied to the bed, the granular materials fluidize smoothly even if there is no air flow into the bed [5-7].

In the vibro-fluidized bed, the heat transfer coefficient between the solid surface in the bed or inner wall of the chamber and the material is improved considerably in comparison with that in a bed which is not vibrated [8-10].

Because of the specific characteristics of the vibro-fluidized bed as mentioned above, it has been considered that the vibro-fluidized bed is suitable for use as a dryer for the granular materials. Because of this, many research studies on vibro-fluidized bed drying have been performed [10]. However, it seems that there is no report of the vibro-fluidized bed being used as a freeze dryer.

On the other hand, it is generally recognized that the freeze drying is an ideal method of food dehydration. But, the drying ability and the thermal efficiency of conventional freeze dryers, for instance the tray freeze dryer, are not good. Therefore, developing high ability and easily operable freeze dryers is an important subject of food preservation.

In this study, therefore, the freeze drying was carried out in the vibro-fluidized bed for the purposes of examining the possibility and usefulness of the vibro-fluidized bed as a freeze dryer. There are many materials which can be dried in the vibro-fluidized bed freeze dryer if it becomes useful. Although only granular or chopped materials can be treated in the vibro-fluidized bed.

The heater in the drying chamber was located in the bed designed so that the heat loss was expected to be small, because the heat from the heater conducts entirely to surrounding materials.

The general characteristics, drying rate, thermal efficiency of the heater, and some other experimental results of the vibro-fluidized bed freeze drying process are reported in this paper.

2. EXPERIMENTAL APPARATUS, SAMPLES AND METHOD

2.1. Experimental Apparatus

The experimental apparatus used is shown in Figure 1. The drying chamber was made of a stainless steel pipe and was installed inside the jacket for regulating the inner wall temperature of the drying chamber. The inner diameter of the drying chamber was 108 mm. The heater was made from a nichrome wire and a stainless steel pipe (outer diameter was 10 mm), and was placed in the bed to minimize the heat loss of the heater. The surface area of the heater was 59% of the cross area of the drying chamber. The distance between the lower side of the heater and the bottom plate of the drying chamber was 1.5 cm. The temperatures in the bed (1, 5, 10, 15, and 30 mm from the bottom plate), at the inner wall of the drying chamber, and at other positions were measured by using the copper-constantan thermocouples. The temperatures of the bottom plate and the inner wall of the drying chamber were controlled by adjusting the jacket temperature for measuring the thermal efficiency of the heater and the heat transfer coefficient between the heater and the material. The usefulness of the wall of the drying chamber to the heater was also studied by controlling the wall temperature. The vertical vibration was generated by using a pair of vibromoter. The intensity of vibration was regulated by adjusting both the amplitude and frequency of vibration. The frequency of vibration was measured by a tachometer, and a calibrated wedge-shaped rule having an error less than 0.025 mm was used to measure the amplitude of vibration.

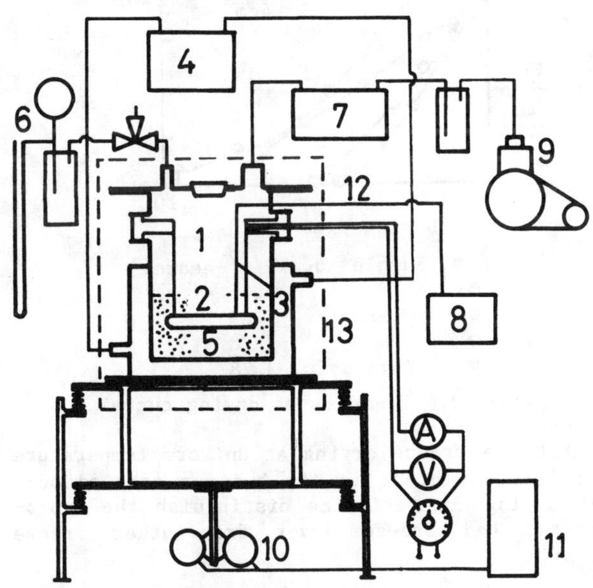

1: Drying chamber, 2: Sample, 3: Jacket, 4: Coolant circulator, 5: Heater, 6: Manometer, 7: Condenser, 8: Temperature recorder, 9: Vacuum pump, 10: Vibration motor, 11: Frequency converter, 12: Thermocouple, 13: Heat insulator.

Figure 1. Experimental apparatus of the vibro-fluidized bed freeze dryer.

2.2. Samples

Three samples, i.e., hydrated and chopped Koya-dofu, Okara, and ion exchange resin particles (Amberlite IR-120B, d_p = 590-710 µm, abbreviated as IER) were used in this study. Koya-dofu is a traditional Japanese freeze dried food made from Tofu (soy curd). It was chopped into granular state by an electric kitchen cutter after it was hydrated. Then chopped Koya-dofu was sieved to the diameter range from 1.4 to 2.4 mm. The moisture content was adjusted to be nearly 3.5 kg-H_2O/kg-d.m. Okara, which is an isolated cake in the expression process of the soybean milk or Tofu production, was sieved by the sieve (1.4 mm opening) and the under product of the sieve was stored after it was dried in a vibro-fluidized bed air dryer (air temperature was about 60°C) [1, 3]. The dried Okara particles were wetted so that the moisture content reached about 3 kg-H_2O/kg-d.m. before they were frozen. The moisture content of the dried IER at room temperature was adjusted nearly 0.6 kg-H_2O/kg-d.m. Below this moisture content, IER particles did not agglomerate. A desired amount of each sample was weighed and kept a polyethylene thin bag before freezing in a -20°C freezer. The bags were shaken several times to prevent the agglomeration of the granular materials during the freezing process.

2.3. Experimental Method

Before beginning the drying process, the temperature of the inner wall of the drying chamber, was adjusted at a value below 0°C by controlling the jacket temperature. The value was 1 or 2°C below the expected bed temperature which is nearly equal to the sublimation temperature balanced to the pressure in the drying chamber. The frozen sample was charged quickly in the drying chamber which was vibrating at the desired intensity of vibration. Then the drying chamber was evacuated. Soon after that, some desired amount of heat was supplied to the bed from the heater adjusting the jacket temperature so that the wall temperature was set to the value which was determined by the purpose of each drying experiment. The thermal efficiency of the heater and the heat transfer coefficient between the heater and the materials were studied under the condition where the heat loss from the material to the outside of the drying chamber seemed to be negligible. This was accomplished by controlling the wall temperature to be equal to the bed temperature. The usefulness of the bottom plate of the drying chamber to the heater was examined by measuring the increased drying rate when the bottom temperature was kept at some known degrees higher than the bed temperature.

For the purpose of measuring the change in the moisture content, a small amount of the material was collected from the upper portion of the bed at every time interval after the heater was switched off and the drying chamber was released to normal pressure. Then, the vacuum drying was continued under the same condition. It took about one minute for each of these procedures. But the time was very short compared with the total drying time. Thus, it was expected that there would be few effects of these procedures on the drying result. The uniformity or the distribution of the moisture content in the bed was estimated from the measured values of the moisture content at the upper portion of the bed by using the vibro-fluidized bed drying model [1].

3. EVALUATION OF THERMAL EFFICIENCY OF HEATER AND OVERALL HEAT TRANSFER COEFFICIENT BETWEEN HEATER AND MATERIAL

The heat balance in the drying chamber is expressed as

$$R\lambda_s + q_s = q_h + q_w + q_r - q_\ell \quad (1)$$

If the temperature of the material and the inner wall of the drying chamber are the same, then $q_w = 0$. And, $q_s = 0$ during the bed temperature is a constant value. Then, Equation 1 is reduced as

$$R\lambda_s = q_h + q_r - q_\ell \quad (2)$$

The value of q_h can be calculated from the electric power supplied to the heater. The q_r value is calculated by using next equation

$$q_r = \sigma A (T_1^4 - T_b^4) F_{ae} \quad (3)$$

Though, the value of F_{ae} in the drying chamber cannot be determined exactly, the value was estimated to be 0.2 in this study.

On the other hand, the overall heat transfer coefficient U between the heater and the material is calculated from next equation if all the heat supplied to the heater conducts to the material

$$q_h = U A_h (T_a - T_b) \quad (4)$$

When the thermal efficiency of the heater, η, is less than 100%, then the left hand term of Eq. (4) has to be replaced by $q_h\eta$. Experimentally, the decrease of η appears as the result of the increase of q_1 value, and only the apparent thermal efficiency of the heater defined by the following equation can be evaluated

$$\eta_{app} = (R\lambda_s/q_h) \times 100 \quad (5)$$

4. EXPERIMENTAL RESULTS AND DISCUSSION

4.1. Characteristics of Vibro-fluidized Bed Freeze Drying Process

Figure 2 shows an example of the drying results of the vibro-fluidized bed freeze drying process obtained in this study. In this case, the wall temperature T_w was kept higher than the bed temperature T_b so that the general situation expressed by Eq. (1) was realized.

There was no appreciable temperature distribution in the bed even though the temperatures were measured at several points in the bed. Furthermore, the temperature of the material in the bed remained almost constant at below 0°C until the drying process was nearly completed. The bed temperature was lower than the room temperature in most part of the drying process. Therefore, if the jacket was not installed, the wall temperature would be higher than T_b as shown in Figure 2.

On the other hand, the moisture content at the upper portion in the bed decreased in proportion to the drying time. This result indicated that the moisture content in the bed was almost uniform during the drying process [1]. These results indicate that the frozen material in the bed fluidized smoothly by applying vibration only,

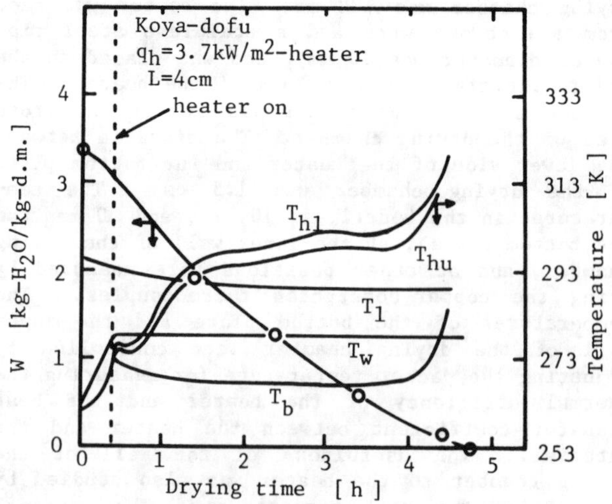

Figure 2. Graphical explanation of vibro-fluidized bed freeze drying.

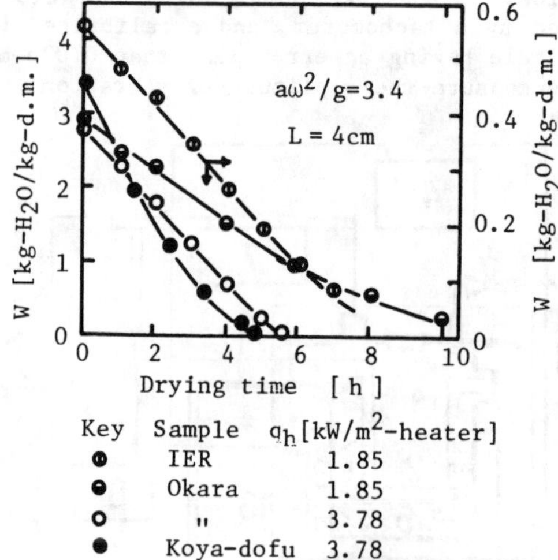

Figure 3. Examples of drying curve.

and that the freeze drying at uniform temperature and moisture content in the bed was carried out. These drying chacteristics distinguish the vibro-fluidized bed freeze dryer from other freeze dryers.

The surface temperature on the upper side of the heater was slightly lower than the temperature on the lower side. It was considered that the reason of this fact was because of mixing intensity of the material was different between upper and lower portion in the bed.

In this study, the heat supplied to the heater q_h was kept constantly during one drying process. Therefore, the bed temperature began to rise when the moisture content became low and the heat used for drying did not balance to q_h. Because of this, it should be noted that the value of q_h has to be determined while considering the drying properties of material and the dyring conditions.

Some drying curves are shown in Figure 3. These results show the moisture content in the bed decreased almost linearly or as the content drying rate period from the beginning to the end of the freeze drying process. Of course, measured valued of surface temperature of materials were all below 0°C until nearly end point of the drying process.

4.2. Heat Balance and Drying Rate

The experimental results of the apparent thermal efficiency of the heater, drying rate, and other thermal values are listed in Table 1. In these experiments, T_b was set to be equal to T_w so that q_w was negligible. But, the bed temeprature began to rise when the moisture content decreased to some value, for example about 0.4 kg-H_2O/kg-d.m. in the case of Koya-dofu as is shown in Figure 2. Some time after this, T_b became higher than T_w. Then, the heat loss from the material to the outside of the drying chamber has to be taken in consideration. Thus, two different apparent thermal efficiencies were calculated from Eq. (5). The first one η_{app} (W = 1.0) is the value for the drying period from the initial point to W = 1.0 kg-H_2O/kg-d.m. During this period, q_w = 0. The second one η_{app} (W = 0.1) is the value until W = 0.1 kg-H_2O/kg-d.m. This value may include the heat loss. In this calculation, q_s was assumed to be zero. The results show that both η_{app} (W = 1.0) and η_{app} (W = 0.1) were equal to or larger than 100%. Since the wall temperature of the drying chamber was kept at the bed temperature, it was considered that only two heat sources contributed to the drying when the thermal efficiency of the heater was examined. One is the heat conducted from the heater, q_h, and the other is the radiant heat from the lid of the drying chamber to the bed, q_r. But, the combined values of these two heats did not compensate for the drying heat calculated from the drying rates. Therefore, a new heat source has to be introduced to consider the heat balance in the drying chamber. The values were evaluated, as indicated as $-q_\ell$ in Table 1, when the drying process were carried out under the conditions of q_s = 0 and q_w = 0. These $-q_\ell$ values will be discussed later.

Table 1. Thermal efficiency, $-q_\ell$, drying rate and other information of drying results.

q_h	p	μ_{app}(W=1)	μ_{app}(W=0.1)	q_r	$-q_1$	R
2.77	240	138	130	0.024	0.61	2.90
3.70	253	124	112	0.018	0.49	3.46
4.62	267	116	104	0.018	0.51	4.04
5.55	307	117	100	0.018	0.55	4.91
6.47	320	120	105	0.014	0.74	5.84
7.39	333	109	100	0.016	0.38	6.09

Sample:Koya-dofu, $a\omega^2/g = 4.1$
q_h [kW/m^2-heater]
p [Pa]
μ_{app}(W=1) [%]
μ_{app}(W=0.1) [%]
q_r [kW/m^2-bed]
$-q_1$ [kW/m^2-bed]
R [kg-H_2O/h m^2-bed]

The drying rate was expressed as the whole drying time average value per unit cross area of the drying chamber for comparing with drying rates of other freeze dryers. Though the abilities of the vacuum pump and the water vapor condenser used in the experimental apparatus were not enough to obtain the maximum drying rate of the vibro-fluidized bed freeze dryer, the drying rates were obtained up to about 6 kg-H_2O/m^2h. The drying rate may increase slightly when the vacuum and condensing systems are improved. Nevertheless, these drying rates are very high in comparison with the 0.5-1.0 kg-H_2O/m^2h values of conventional freeze dryers, for example of the tray freeze dryer.

On the other hand, since the temperature of materials was kept constantly and at values below 0°C during the most part of the vibro-fluidized bed freeze drying process, as shown in Figure 2, it seems that the vibro-fluidized bed freeze drying is also very desirable for food dehydration from the stand point of the quality of the product.

4.3. Overall Heat Transfer Coefficient

The overall heat transfer coefficient between the heater and the material can be calculated from Eq. (4), when there is no need to consider the heat loss from the heater. The result is shown in Figure 4.

In this case, the average value of T_{hl} and T_{hu} was substituted for T_h. In the first stage of the drying process of Koya-dofu, whose initial moisture content is rather high, the value of U decreased in accordance with the decrease in moisture content of the bed. The U values then reached a constant value. But the U values of IER were nearly the same value of the constant value of the low moisture content region's Koya-dofu.

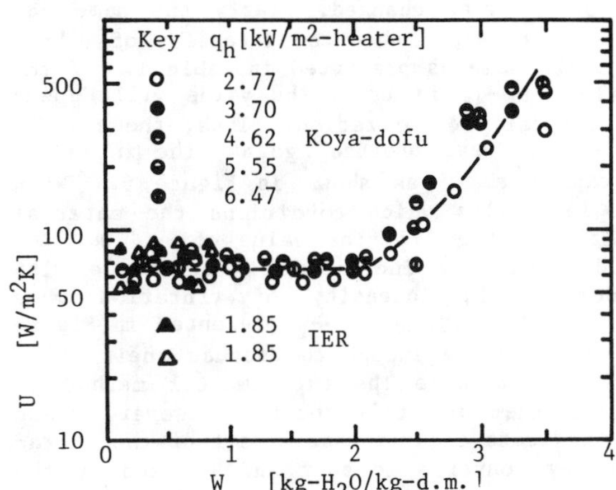

Figure 4. Heat transfer coefficient between heater and material.

On the other hand, the bottom plate and the wall of the drying chamber of the vibro-fluidized bed freeze dryer can also be used as a heater. Thus, the relationship between the increases of drying rate and the differences in temperature of the bed and the bottom plate were examined. The results are shown in Figure 5. Though the dependency of U of a horizontal heater on the condition

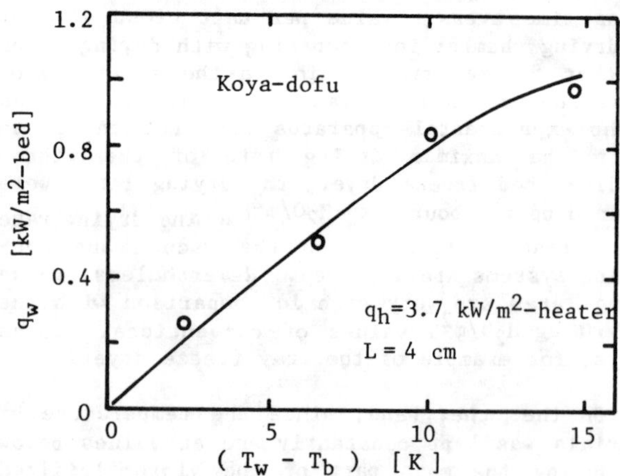

Figure 5. Relationship between heat conducted from wall to bed q_w and temperature difference $(T_w - T_b)$.

of vibration is different from that of a vertical heater in the vibro-fluidized bed [8], each heat transfer coefficient could not be studied separately because both temperatures of the wall and the bottom plate were the same. The average time values of U from the beginning to the end point of the drying process were in the range of 60 to 70 W/m^2K.

4.4. Heat Generated by Vibration

In this section, the discussion will focus on new heat source, $-q_\ell$, found in this study. It was considered that $-q_\ell$ is a heat generated by vibrating the bed. The physical meaning of $-q_\ell$ was assumed to be the heat converted from the kinetic energy of the bed when it collides with the bottom plate of the drying chamber after leaving the bottom plate by vibration. Though the values of q_h were changed, nearly the same $-q_\ell$ values were obtained when the intensity of vibration was the same as presented in Table 1. If the assumption of $-q_\ell$ is true, the value will depend on the intensity of vibration. Thus, the average values of $-q_\ell$ were plotted against the intensity of vibration $a\omega^2/g$ as shown in Figure 6. When $a\omega^2/g < 1$, under which conditions the material does not fluidize [9], the value of $-q_\ell$ was not measured. But, it increased in accordance with the increase in intensity of vibration when $a\omega^2/g > 1$. The values of $-q_\ell$ presented in Figure 6 are not accurate enough to discuss their absolute values, because the experimental method of evaluating them is still rough. However, these results indicate clearly that a part of the vibration energy converts to heat and is used by the bed for drying heat. Since the values of $-q_\ell$ were not small, it is an important subject to investigate this value more accurately in relation to the intensity of vibration. Therefore, we have tried to analyze the $-q_\ell$ values theoretically [see Appendix]. According to our theoretical and experimental examination using the material that has known physical properties, Figure 7, in which the experimental value of $-q_\ell$ was evaluated from the increase in temperature of the dry material charged in the insulated chamber during some given time interval of vibration, it seems that the results shown in Figure 6 are fairly reasonable. And, it is expected that over a half or most part of the vibration energy will be able to be recovered in the drying energy as a form of heat generated in the bed by vibration.

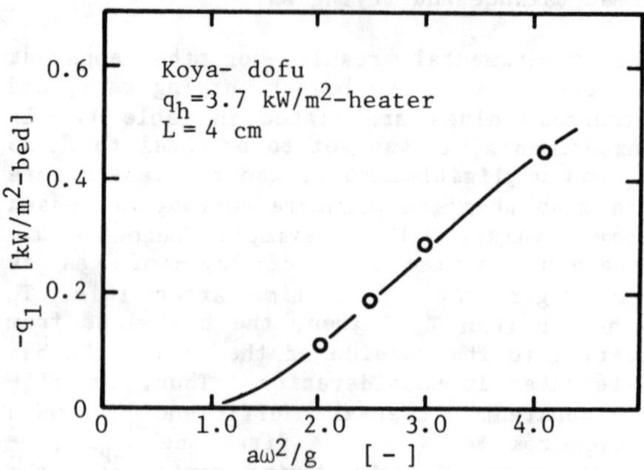

Figure 6. Dependency of heat generated by vibration $-q_\ell$ on intensity of vibration.

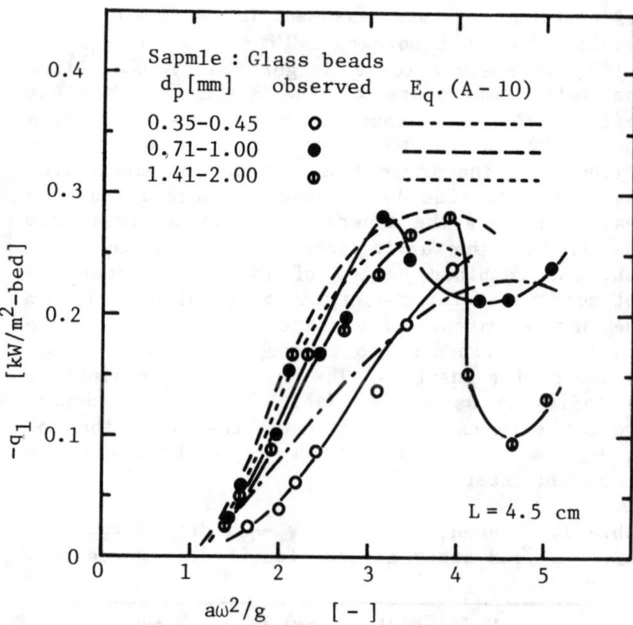

Figure 7. Comparison of theoretical values of heat generated by vibration $-q_\ell$ with experimental values.

APPENDIX (Mathematical Model of $-q_\ell$).

Figure A shows the motion of the bed and the bottom plate of the drying chamber [2].

Basic equations

$$\chi_d = a \sin \omega t \qquad \text{(A-1)}$$

$$\alpha_d = d^2\chi_d/dt^2 \qquad \text{(A-2)}$$

Condition at which the bed leaves the bottom plate

$$\alpha_d + g = 0 \qquad \text{at } t = t_s \qquad \text{(A-3)}$$

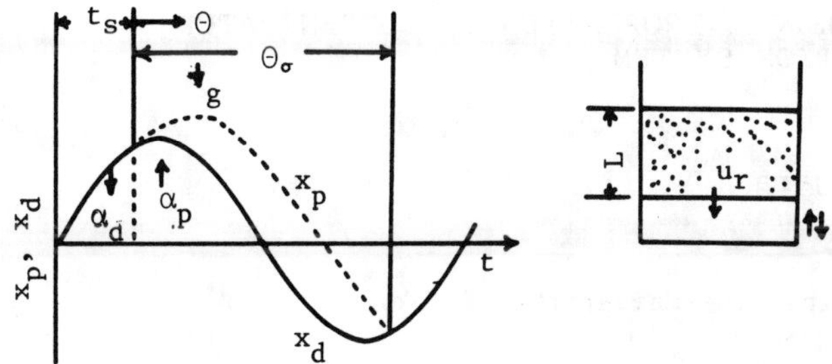

Figure A.

After that, the equation of motion of the bed

$$d^2\chi_p/dt^2 + \alpha_p + g = 0 \tag{A-4}$$

where,

$$\alpha_p = K_s u_r/\rho_p \equiv K\, u_r \tag{A-5}$$
$$K_s = 180\,\mu\,(1-\varepsilon)/\Phi_c{}^2 d_p{}^2 \varepsilon^3 \tag{A-6}$$
$$U_r = d\chi_p/dt - d\chi_d/dt \tag{A-7}$$

Solution

$$\chi_p = a \sin \omega t_s + \frac{1}{K} a\omega \cos \omega t_s \,(1 - e^{-K\Theta})$$
$$+ \frac{Ka\omega}{(K^2+\omega^2)} e^{-K\Theta} + \frac{Ka}{(K^2+\omega^2)} (K \sin w\Theta$$
$$- \omega \cos \omega\Theta) - \frac{g}{K^2} (K\Theta + e^{-K\Theta} - 1) \tag{A-8}$$

The time Θ_o at which the bed collides with the bottom plate again so that $\chi_p = \chi_d$ is calculated by the trial and error method. The relative velocity of the bed to the bottom plate V_r at $t = t_s + \Theta_o$ is expressed as

$$V_r = \frac{d\chi_d}{dt} - \frac{d\chi_p}{dt} \quad \text{at } t = t_s + \Theta_o \tag{A-9}$$

If the all collision energy converts to heat

$$-q_\ell = \frac{1}{2} f\, M V_r{}^2/A \quad [\mathrm{W/m^2}] \tag{A-10}$$

NOMENCLATURE

- A = cross area of drying chamber, m^2
- A_h = surface area of heater, m^2
- a = amplitude of vibration, m
- d.m. = bone dry material
- d_p = diameter of sample, mm
- Fae = overall emissivity factor
- f = frequency of vibration, s^{-1}
- g = acceleration due to gravity m/s^2
- K_s = defined by Eq. (A-6)
- K = K_s/ρ_p, s^{-1}
- L = bed height, m, cm
- M = charged weight of material, kg
- p = pressure in drying chamber, Pa
- q_h = heat supplied by heater, W, kW/m^2-heater
- q_ℓ = heat loss or heat generated by vibration, W, kW/m^2-bed
- q_s = sensitive heat, W
- q_w = heat from wall to bed, W, kW/m^2-bed
- R = drying rate, kg-H_2O/m^2h, kg-H_2O/s
- T_b = bed temperature, K
- T_h = temperature of heater, K
- T_l = temperature of lid, K
- T_w = temperature of wall, K
- t = time, s
- t_s = time defined in Figure A, s
- U = overall heat transfer coefficient, W/m^2K
- U_{mf} = minimum fluidization velocity, m/s
- U_r = defined by Eq. (A-7), m/s
- V_r = collision velocity, m/s
- W = moisture content, kg-H_2O/kg-d.m.
- χ_d = displacement of bottom plate, m
- χ_p = displacement of bottom of bed, m
- α_d = defined by Eq. (A-2)
- α_p = defined by Eq. (A-5)
- ε = bed voidage
- Θ = time defined in Figure A, s
- Θ_o = time defined in Figure A, s
- η = thermal efficiency of heater
- μ = viscosity of air, kg/ms
- λ_s = latent heat of sublimation, Ws/kg
- ρ_p = density of solid particle, kg/m^3
- σ = Stefan-Boltzmann constant, W/m^2K^4
- Φ_c = shape factor
- ω = angular frequency, s^{-1}

Subscript

- l = lower side of heater
- u = upper side of heater

REFERENCES

1. Suzuki, K., H. Hosaka, R. Yamazaki and G. Jimbo. J. Chem. Eng. Japan, Vol. 13, 117 (1980).
2. Suzuki, K., R. Yamazaki and G. Jimbo. Kagaku Kogaku Ronbunshu, Vol. 8, 307 (1982).
3. Suzuki, K., A. Fujigami, K. Kubota and H. Hosaka. Nippon Shokuhin Kogyo Gakkaishi, Vol. 27, 393 (1980).
4. Suzuki, K., A. Gujigami, R. Yamazaki and G. Jimbo. J. Chem. Eng. Japan, Vol. 13, 495 (1980).
5. Kroll, W., Chemie-Ing.-Techn., Vol. 27, 33 (1955).
6. Gutman, R. G., Trans. Instn. Chem. Engrs., Vol. 54, 174 (1976).
7. Fujigami, A., K. Suzuki, K. Kubota and H. Hosaka. J. Fac. Appl. Biol. Sci., Hiroshima Univ., Vol. 19, 147 (1980).
8. Bukareva, M. F., V. A. Chlenov and N. V. Mikhailov. Khim. Prom., Vol. 6, 432 (1968).
9. Gutman, R. G., Trans. Instn. Chem. Engrs., Vol. 54, 251 (1976).
10. Strumillo, C. and Z. Pakowski. Drying '80, Vol. 1, p. 211, Ed. by Mujumdar, A. S., Hemisphere Pub., Washington (1980).

FREEZE DRYING CHARACTERISTICS AND TRANSPORT PROPERTIES IN CONCENTRATED COFFEE SOLUTION SYSTEM

Yasuyuki Sagara

Department of Agricultural Engineering, The University of Tokyo
Tokyo, 113 JAPAN

ABSTRACT

Aqueous solutions of 29-45 % soluble coffee solid were freeze dried under drying conditions used in commercial operations. Data pertinent to the drying characteristics of the sample consisted of the changes in sample weight, drying rate, position of sublimation front and temperature distribution in the sample, and of the density as well as water content for both frozen and dried samples.

Scraping free surface of the sample after freezing was found to be successful for removing a vapor transfer barrier effect of this layer and thus the sample surface temperature was allowed to heat up to 70 °C or higher resulting in a markedly reduced drying time.

A mathematical model was developed to determine the thermal conductivity and permeability for the dried layer of the sample during drying process. Thermal conductivity decreased in proportion to the porosity of the dried layer, and its temperature and pressure dependances were not appeared. The permeability increased with increasing the porosity, pressure and temperature of the dried layer. The results indicate that in commercial operations the solute concentration is one of the critical processing factors since this factor decisively governs the structure of a solute matrix formed during freezing of coffee solutions and the transport properties mainly depend upon the nature of this structure during drying.

1. INTRODUCTION

Freeze drying has had a great impact upon the production of dehydrated food because of the superior quality of the product obtained and promises continued expansion of the number of applications. However, the process is only feasible if the cost of production can be lowered by optimum plant operations. This requires not only a good engineering design and control to minimize the drying time, but also detailed knowledge for the transport properties of material to be dried and its characteristic behavior during freezing and freeze drying. Since the rate of freeze drying is limited by heat and mass transfer rates across a dried material which surrounds the undried, frozen portion of product, the thermal conductivity and permeability of the dried layer and the effects of processing factors on these transport properties are fundamental informations to determine the drying rate.

Various method, both transient and steady state, have been used to determine the transport properties of freeze dried food. The former method generally permits data to obtain relatively easily, whereas the latter gives better accuracy. The transient method used in this study based on a quasi-steady-state analysis of actual drying data and is described by Lusk et al.(1), Massey and Sunderland(2), Hoge and Pilsworth(3), Stuart and Closset(4), Bralsford(5), Gaffney and Stephenson(6) and Sandall et al.(7). Available literature on the transport properties for the dried layer of coffee solution was limited. Quast and Karel(8) reported the permeabilities of 20 and 30 % coffee solutions. Sagara and Hosokawa(9) determined the thermal conductivity and permeability of 2-30 % solutions at constant surface temperatures ranging from 20 to 53 °C under the usual pressure range of commercial freeze dryers. However, economic consideration of freeze drying process make it desirable to prepare a concentrated solution of 30-50 % prior to freeze drying. Values of transport properties for this concentration range have not measured systematically, and the data for freeze drying characteristics were not appeared in the literature.

The objectives of this work were a) to measure both freeze drying characteristics of coffee solution in a concentrated form and the corresponding drying conditions, and b) to determine the thermal conductivity and permeability of the dried layer in connection with controllable factors such as the solute concentration, temperature and pressure of the dried layer.

Coffee solutions of 29-45 % were freeze dried at constant sample surface temperatures ranging from -7 - 70 °C under the chamber pressure range of 7-12 Pa using radiant heating upon the sample surface. The transport properties of the dried layer were determined by applying the drying data to a model based on heat and mass transport in the sample, and then the effects of processing parameters on transport properties as well as on the drying rate were discussed in connection with the freezing and freeze drying operations.

2. THEORETICAL MODEL

A model used to determine the thermal conductivity and permeability of the dried layer is described next and shown in Fig.1.

During a typical freeze drying process for liquid food, heat is applied to the surface of frozen material by radiation or conduction. As the sublimation proceeds, a dried layer forms on the heated side of material while the ice-front retreats. The heat necessary to sublime the ice must be conducted to the ice-front across an increasing thickness of dried material, in order for sublimation to continue, and the water-vapor from the sublimation at the ice-front must be removed by diffusing through a porous dried layer. To explain this drying behavior, a theoretical model based on the rates of heat and mass transfer within the sample was developed by employing some simplifying assumptions.

The sample is assumed to have the geometry of a semi-infinite slab and the dried layer is separated from the frozen layer by an infinitesimal sublimation interface retreating uniformly from the sample surface. These assumptions were provided for in the experimental procedure by applying heat from a radiant plate heater located directly above the exposed sample surface and by insulating the side and bottom of sample from heat conduction. The assumption of infinitesimal sublimation front was based on the fact that a sharp sublimation zone had been observed to separate the completely dried layer and the frozen layer in the previous experiments (9), and also this has been validated for some applications by Sandall et al.(7) and Ma and Peltre (10). In addition, it is assumed that drying proceeds under quasi-steady-state condition;namely, the movement of sublimation front and the changes in temperature as well as in pressure within the sample are negligible at any instant of drying time. Dyer and Sunderland (11) showed that this assumption results in an error of about 2 percent, and so this assumption is generally quite good for the radiant heating case; however,this is not applicable to microwave heating because of the rapid temperature transition occuring in the sample(10).

The bottom of the sample is insulated while the surface is exposed to a evacuated space at the temperature Θ_s and pressure p_s. The insulated sample bottom may be considered the center line of a symmetric slab heated from both surface and bottom. Since the thermal conductivity of the frozen layer is 20-50 times greater than that of the dried layer, the temperature of the frozen layer can be supposed to be uniform and same as that of sublimation front at any instant of time.

Equations representing the rates of heat and mass transfer to and from the frozen layer may be written with these assumptions. The expression for the rate of heat transfer across the dried layer is taken from Massey and Sunderland (2);

$$\dot{q} = \frac{\lambda}{X(t)}(\Theta_s - \Theta_f) - \dot{m}\int_{\theta_f}^{\theta_s} c_p \, d\Theta \qquad (1)$$

The specific heat of water-vapor c_p was assumed constant and equal to the value of pure water-vapor. The first term on the right side of this equation represents the heat conducted across the dried layer and the second term represents the energy absorbed by the water-vapor flowing through the dried layer. The temperature distribution in the dried layer is not linear as expressed by the second term on the right side of equation (1). Since sample calculations show that the influence of the nonlinear temperature distribution is very small, a linear distribution of temperature within the dried layer was assumed in the calculation of transport properties. Similarly, the mass flux may be expressed as

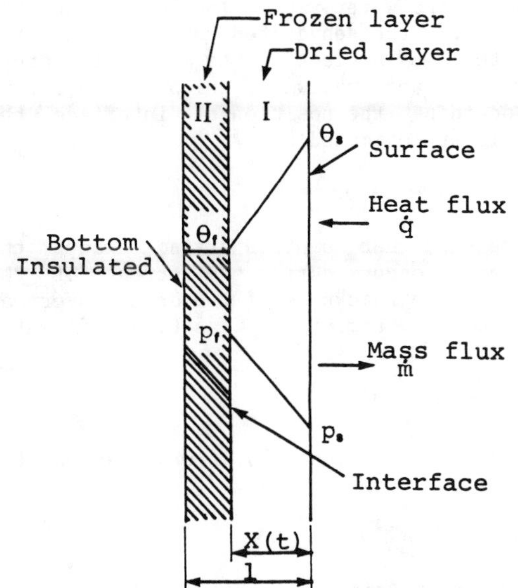

Fig.1 Freeze-drying model for transport properties analysis.

$$\dot{m} = \frac{KM_w}{R\,T\,X(t)}(p_f - p_s) \qquad (2)$$

where K is the permeability between the sample surface and the sublimation front with the partial pressure of water-vapor p_f.

An expression for the equilibrium vapor pressure (torr) of pure ice as a function of temperature is given by I.C.T. (12);

$$\log p_f = -2445.5646/T_f + 8.2312\log T_f - 0.01677006T_f + 1.20514 \times 10^{-5} T_f - 6.757169 \qquad (3)$$

As heat supplied across the dried layer may be considered to be dissipated as the latent heat of sublimation, the sublimation front acts as the heat sink which can be represented by

$$\dot{q} = \dot{m}\,\Delta H \qquad (4)$$

The mass transfer rate can be related to the drying rate as

$$\dot{m} = \rho_w l\left(-\frac{dX}{dt}\right) \qquad (5)$$

where ρ_w is the density of ice or frozen liquid and X is the fraction of internal water still remained in the sample. Sagara and Hosokawa (9) have presented the configuration of water within the sample during sublimation dehydration process. According to this configuration, the fraction of X be expressed as

$$X = (m_d + m_f + m_w)/m_0 \qquad (6)$$

where m_d is the residual water as the sample is fully dried, m_f is the water contained in the frozen layer, m_w is the water-vapor flowing through the dried layer, m_l is the dehydrated water and m_0 is the initial water. By neglecting the mass of water-vapor as well as residual water which does not take part in deciding the position of interface the value of X is approximated by

$$X = m_f / (m_0 - m_d) \tag{7}$$

If it is assumed that complete drying except the residual water occurs during the retreat of the frozen layer, the thickness of the dried layer may be expressed in term of the fraction of water remaining, X;

$$x(t) = (1 - X)l \tag{8}$$

Substitution of equation (5) into equation (4) gives

$$\dot{q} = \rho_w l \, \Delta H(-dX / dt) \tag{9}$$

Following a substitution of equation (8) into equations (1) and (2), equations (1) and (9) may be combined to give the fraction of dehydrated water;

$$(1 - X) = \frac{\lambda (\Theta_s - \Theta_f)}{\rho_w l^2 \left(\Delta H + \int_{\Theta_f}^{\Theta_s} c_p d\Theta\right)\left(-\frac{dX}{dt}\right)} \tag{10}$$

Similarly, from equations (2) and (5);

$$(1 - X) = \frac{K M_w (p_f - p_s)}{\rho_w l^2 R T_f \left(-\frac{dX}{dt}\right)} \tag{11}$$

Equations (10) and (11) were used to test drying rate data for conformity to the model and to deduce the transport properties from drying rate data. By rewriting equations (10) and (11), the following equations for the thermal conductivity and permeability are determined, respectively;

$$\lambda = \alpha \rho_w l^2 \left(\Delta H + \int_{\Theta_f}^{\Theta_s} c_p d\Theta\right) \tag{12}$$

$$K = \beta \rho_w l^2 R T_f / M_w \tag{13}$$

where,

$$\alpha = \frac{(1 - X)}{(\Theta_s - \Theta_f)/(-dX/dt)} \quad \beta = \frac{(1 - X)}{(p_s - p_f)(-dX/dt)} \tag{14}$$

In this study, the average transport properties in quasi-steady state periods were determined with the average values of α and β in these periods.

3. EXPERIMENTAL

3.1. Experimental Freeze Dryer

The vacuum chamber and adjacent parts of the apparatus are illustrated in Fig.2. The vacuum chamber was a cylindrical, thermally insulated iron enclosure of 92.2 mm in inside diameter by 470 mm long, with a Plexiglas door of 30 mm thick. This transparent door permitted the visual observation of the inside of the vacuum chamber during freeze

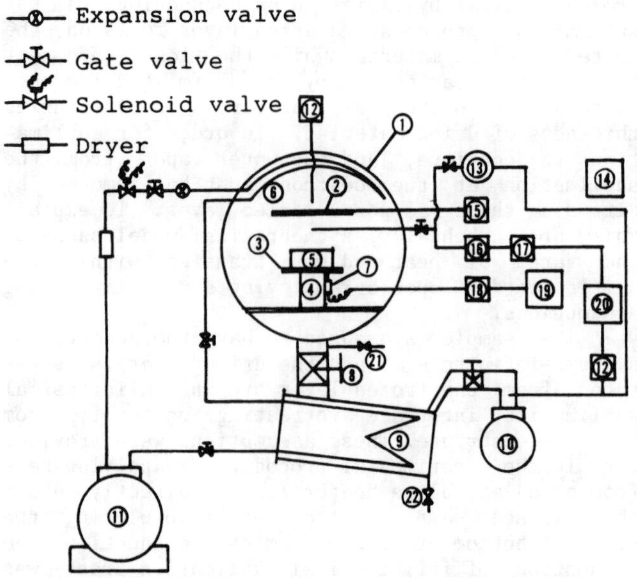

1.Vacuum chamber 2.Heater 3.Platen 4.Balance 5.Sample 6.Internal condenser 7.Displacement transducer 8.Main valve 9.External condenser 10.Vacuum pump 11.Refrigerator 12.Pirani-type vacuum gage 13.Diffusion pump 14.Thermo-recorder 15.Temperature controller 16.Diaphragm-type pressure sensor head 17.Pressure indicator 18.Displacement meter 19.Weight-recorder 20.Pressure recorder 21.Leak valve 22.Drain

Fig.2 Schematic diagram of experimental freeze-dryer.

drying process. A rotary vacuum pump was connected through a external condenser and a main valve to the lower part of the vacuum chamber. A diffusion pump was used to evacuate a diaphragm-type pressure sensor head and to adjust its zero-point. An internal condenser consists of 9 m of 12.7 mm diameter, refrigerating coil running the length of the vacuum chamber in zigzag manner, with the nearest surface of the coil 25 mm from the inside wall. These condensing systems were used to collect and freeze the water vapor sublimed from the sample and to prevent moisture from reaching the vacuum pump. A refrigerator for these condensers is a 1.5 kW (R-22 Refrigerant) air-cooled condensing unit with sufficient capacity to reduce the coil surface temperature to -45 °C. The surface temperature of the internal condenser was regulated by adjusting a hand-expansion valve in refrigerant entrance line. Radiant heat was supplied by electrically heated plate located about 40 mm above the sample surface. Then the sample surface temperature was controlled with the PID controller by regulating the electrical power to the radiant heater. The electrical leads as well as thermocouple wires put through to the back of the vacuum chamber using demountable lead-in seals.

The weight loss of the sample was followed by supporting a sample holder on a weight measuring device located in the center of the vacuum chamber and recorded as a displacement of a platen equipped with a balance against the drying time. This was accomplished by mounting an electrical differencial transformer (EDT) with the balance and by connecting the output of the EDT to a displacement indicater. Thus weight readings to within about +0.01 g could be made.

The chamber pressures were measured by using both Pirani- and diaphragm-type analyzers, and pressure indicaters mounted on the control panel gave the continuous recordings of changes in pressure during drying processes.

3.2. Sample holder

The sample holder and measuring locations of temperature in the sample were shown in Fig.3. The sample holder was a Plexiglas dish of 70.5 mm in inside diameter and 28 mm in height. To promote one-dimensional freezing and freeze-drying, a fiber-glass insulation was placed around the side of the sample holder. After the sample was frozen the sample holder was placed on a 50 mm by 150 mm square piece of Polyuretane foam fixed on the platen. All the exposed surfaces of insulating materials were covered with reflecting alminium foil to reduce radiant heat transfer to the side and bottom of the sample holder.

Six thermocouple probes were permanently placed in the center of the sample holder and equally spaced from the exposed surface of the sample, and made it possible to make temperature measurements at the same points in the sample from run to run. These thermocouples were made from 0.2 mm copper-constantan wire and calibrated with the standard mercury-in-glass thermometer. The thermocouple junctions for measuring and for controlling the sample surface temperature were placed just under the exposed surface of the sample to ensure a meaningful surface temperature reading, and the probes were run along the surface to the side of the sample holder to provide an isothermal path to minimize conduction errors along these probes. The leads were also shielded from the heater by passing them through the insulation in order to prevent direct radiant heat transfer to the wire.

3.3. Procedure

Instant coffee powder was experimentally convenient to control the solute concentration of the sample, and it was confirmed in our previous experiments that the freeze drying characteristics of instant coffee solution were almost the same as those of aqueous solution extracted from roasted coffee beans. Thus the samples of 29-45 % aqueous solutions of instant soluble coffee solid were used and each sample was prepared for freeze drying in the same manner. Instant coffee powder was thoroughly mixed with distilled water at temperature ranging 50-100 °C and allowed to cool to room temperature. Then it was poured into the sample holder and frozen with freezing-mixture (Dry ice-Ethanol) from the bottom of the sample holder using a freezing set-up as showed in the previous paper(9). Freezing continued until uniform temperature of the sample was obtained, resulting in a relatively fast freezing rate with crystallization period under 30 min. and a final sample temperature below -65°C within 60 min. This method of freezing produced a simple, capillary-type solute matrix with its grain orientation parallel to direction of heat and mass transfer.

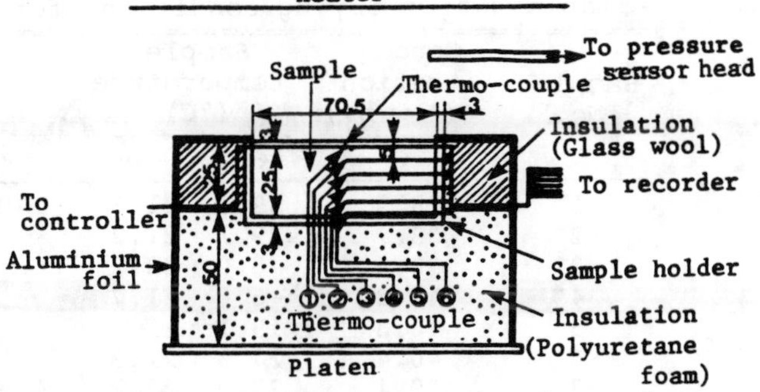

Fig.3 Sample holder and measuring points of temperatures.

As described later an expanded portion of frozen sample contained a concentrated film (See Fig.4) was scraped with a knife after freezing.

The frozen sample was then transfered into the vacuum chamber, and freeze drying began within about 6 min.. Freeze drying continued until constant weight of the sample was obtained, resulting in a final water content below 4 % w.b.. The final weight was measured as a check against the weight loss recorded during drying, and then the water content of the dried sample was determined by Karl Fisher titration method. The initial moisture content was calculated from this value using the data of sample weights.

4. RESULTS AND DISCUSSION

4.1. Drying characteristics

Table 1 presents the drying conditions for the coffee samples used in this study. The sample temperatures and vapor pressures indicate the average values obtained during quasi-steady-state periods and these values were used to calculate the transport properties. The vapor pressure at the sublimation front, p_f was calculated from the sublimation temperature, T_f which was measured with the thermocouple nearest to the sublimation front within the frozen layer. Heater temperature represents the maximum or minimum value appeared during drying process. The heater temperature was observed to be closely related to the sample surface temperature, indicating that increasing the controlling value of the surface temperature led to an increase in the heater temperature as indicated for the samples of about 40 % solutions.

The physical properties for the initial and dried samples were presented in Table 2. Porosity, volume, and density were assessed from cumulative composition using 0.625 specific volume for pure coffee solubles (15), since meaningful value of the volume scraped from frozen sample was appeared to be difficult to measure. The relationship between solute concentration and initial density obtained is consistent with that used in commercial plants operations. The data for the samples of 40 % solutions indicate that the scraped volume is about 10 per cent of the initial, and the dried sample or layer has a porosity of 0.70.

It was observed that during freezing the sample a concentrated film of solute developed at the free surface of the sample and this film forms an

Table 1 Drying conditions for the Coffee Samples

Sample No.	Concentration (% w.b.)	Sample temperature (°C)		Vapor pressure (Pa)		Heater temperature (°C)		Drying time (hr)
	C	θ_s	$-\theta_f$	p_s	p_f	θ_{hmax}	θ_{hmin}	t
1*	28.5	59.0	21.3	11.0	91.7	260.0	88.2	43.3
2*	33.7	29.1	21.2	6.9	92.6	138.7	35.5	44.7
3*	38.6	0.0	22.2	10.4	83.8	88.1	0.0	67.4
4*	40.2	-6.5	21.7	8.1	88.4	67.8	-6.8	69.1
5	40.3	19.1	21.9	9.2	86.5	66.2	19.3	38.0
6	40.4	24.3	21.3	9.3	91.7	74.5	24.8	36.8
7	40.4	28.9	22.2	12.4	84.0	77.4	31.6	29.2
8	40.6	35.2	21.3	9.1	91.7	77.4	35.5	27.7
9	40.5	39.5	21.3	8.7	91.7	83.6	39.8	32.0
10	40.2	49.7	18.9	9.9	115.5	90.0	48.3	45.3
11	40.3	54.5	20.3	9.1	101.0	119.7	54.8	22.3
12	40.0	60.5	18.2	10.1	123.5	128.7	58.5	25.1
13	40.1	70.7	18.6	10.0	118.9	130.7	68.0	22.4
14	45.0	19.6	20.3	9.5	101.0	76.9	21.3	32.0

Condenser temperature θ_c = -43.1 ~ -45.4 (°C)
*Without scraping

Table 2 Physical properties of initial and dried samples.

Sample No.	Volume (cm³)	Porosity (-)	Mass (g)				Water content (% w.b.)		Density (g/cm³)		
	V_0	Ψ	m_0	m_d	m_s	m_w	w_0	w_d	ρ_0	ρ_d	ρ_w
1	105.0	0.80	118.55	34.72	33.81	84.74	71.5	2.61	1.129	0.331	0.807
2	103.6	0.76	120.19	41.92	40.55	79.64	66.3	3.28	1.160	0.405	0.769
3	102.5	0.71	123.00	50.12	47.52	75.48	61.4	5.36	1.200	0.490	0.736
4	103.1	0.69	125.57	54.05	50.48	75.09	59.8	6.61	1.218	0.524	0.729
5	92.8	0.70	111.56	46.79	44.91	66.65	59.7	4.01	1.202	0.504	0.718
6	92.9	0.70	111.30	46.50	44.92	66.38	59.6	3.39	1.198	0.501	0.715
7	92.5	0.70	111.01	46.56	44.87	66.14	59.6	3.62	1.200	0.503	0.715
8	92.3	0.70	110.54	46.30	44.90	65.64	59.4	3.03	1.198	0.502	0.711
9	93.4	0.70	111.36	46.21	45.14	66.22	59.5	2.31	1.193	0.495	0.709
10	93.9	0.70	111.82	46.01	44.94	66.88	59.8	3.03	1.191	0.490	0.712
11	93.8	0.70	111.96	46.39	45.12	66.84	59.7	2.73	1.194	0.495	0.713
12	94.0	0.70	111.85	45.77	44.73	67.12	60.0	2.27	1.189	0.487	0.714
13	94.1	0.70	111.71	45.57	44.74	66.97	59.9	1.83	1.187	0.484	0.712
14	92.8	0.65	114.41	53.81	51.51	62.90	55.0	4.27	1.233	0.580	0.678

effective water-vapor barrier, which causes melting or puffing in the sample for the reason that the temperature of the frozen layer or the dried layer rise until the melting or collapse temperature is reached. The samples numbered from 1 to 4 were dried without scraping after freezing. As shown in these samples, the solute concentration exerts a great influence on the sample surface temperature. For example, the surface temperature of about 40 % solution was not be allowed to heat to above -6.5°C in order to avoid the puffing in the sample. Table 3 shows the concentration distribution in the frozen sample, which was determined by Karl Fisher titration method using three samples of about 40 % solution, and Fig.4 illustrates its measuring locations. The concentration of the surface layer which contains the film was determined to be about 1 % higher than other layers. Lambert et al.(13) mentioned the existence of a surface film or crust resistance to mass transfer, and a sample whose frozen surface was scraped prior to drying had a significantly higher permeability. Flink(14) interpreted that in most solutions this is due to the concentration of carbohydrates, and suggested several method for preventing or removing the surface layer effect. According to these suggestions, an

Table. 3 Solute concentration values for frozen samples

Specimen No.*	Concentration (% w.b.) Sample No.1	No.2	No.3
1	41.4	41.8	41.9
2	40.4	40.8	41.1
3	40.4	40.8	41.0
4	40.4	40.8	41.2
5	——	40.8	41.5
6	40.5	40.7	41.3

* Shown in Fig.

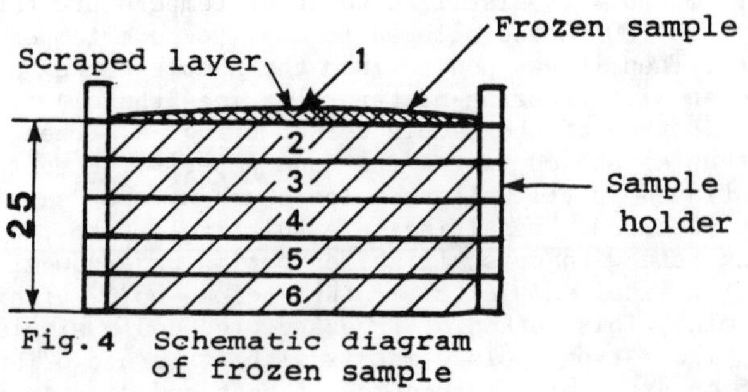

Fig.4 Schematic diagram of frozen sample

expanded portion was scraped with a knife after freezing as shown in Fig.4. This treatment was found to be successful for removing the surface film effect and thus the surface temperatures, which had been limmitted as low as -6.5°C for 40 % solution, were allowed to heat up to 70°C or higher, resulting in a reduced drying time by one-third as shown in sample No.13.

A typical freeze drying characteristics and corresponding drying conditions are shown in Fig.5. The former consists of the change in sample weight, drying rate, temperature distribution in the sample while the latter includes the surface temperatures of the heater and internal condenser coil as well as the chamber pressure. The sample surface temperature increased until it approached the control temperature and then remained constant. A uniform temperature in the frozen layer was observed during most of runs and the assumption employed in the model was confirmed to be satisfied. The temperature at any given location in the sample appeared to reach a minimum value just before it began to rise and then rose toward the surface temperature indicating the passage of retreating sublimation front. The temperature of sublimation front was found to decrease as the chamber pressure gradually decreased. Same behavior was observed for all samples used and this indicates the fact that the drying process was heat transfer controlled. For example, as shown in sample No.5-13, a variation in surface temperature of 19.1 to 70.7°C has little effect on the interface temperature;namely, the drying rate is much more sensitive to the thermal

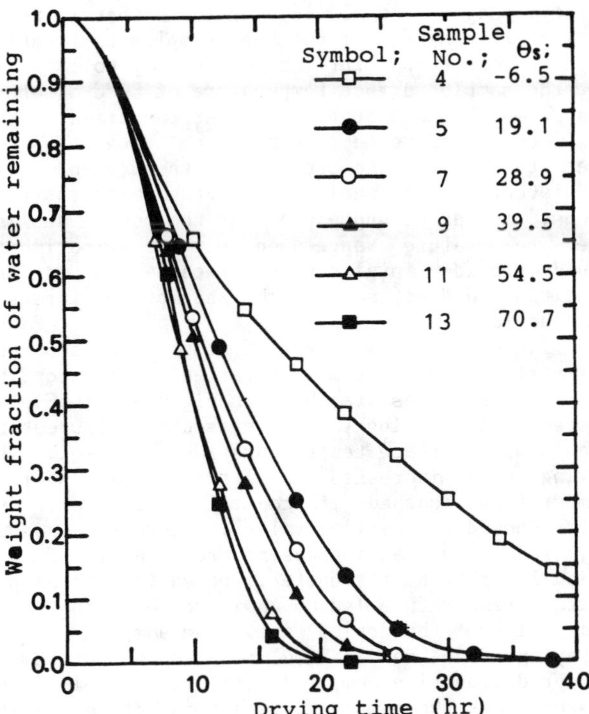

Fig.6 Experimental drying curves of 40% solutions at various surface temperatures.

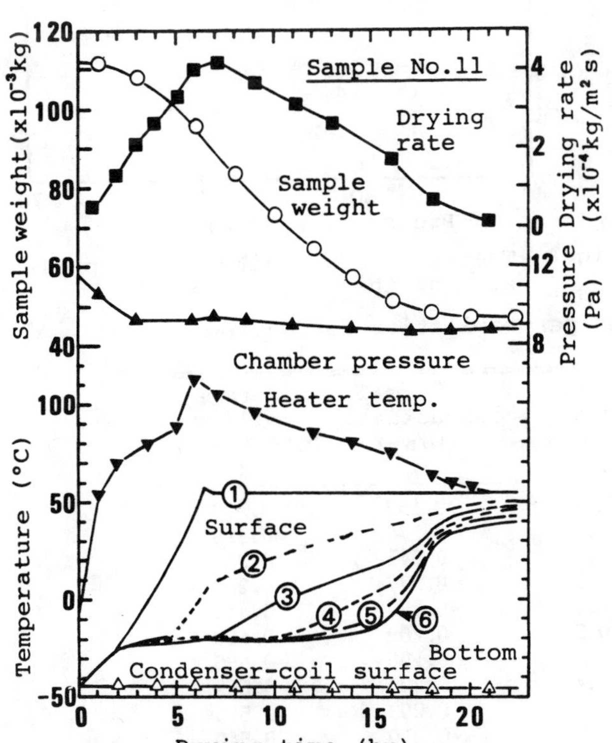

Fig.5 Experimental data obtained during freeze-drying of 25mm layer of a 40.3% aqueous solution of soluble coffee at surface temperature 54.5 °C.

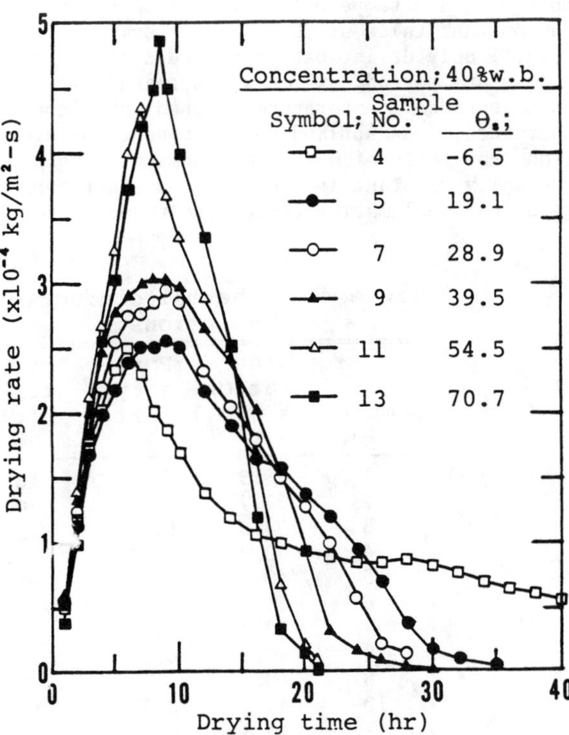

Fig.7 Drying rate curves of 40% solutions for various surface temperatures.

conductivity of the dried layer than to the permeability. Sagara et al.(16) mentioned that in their drying experiments for the beef samples the sublimation front temperature had a tendency to increase at the sample surface temperature of 80°C or above and then the heat supplied across the dried layer was considered to be dissipated as both sensible heat to rise the temperature of the frozen layer and latent heat of sublimation. From these observations, it may be suggested that the model applied over temperature range used in this experiments would provide invalid values because the drying condition did not satisfy the assumption expressed by equation (4).

The weight loss curves of about 40 % solutions at various surface temperatures and corresponding drying rate curves are shown in Fig.6 and in Fig.7, respectively. Drying time decreased with increasing the sample surface temperature as expected. The drying rate increased until the sample surface temperature reached its control temperature and after showed the maximum value it decreased gradually indicating an increasing resistance of the dried layer to heat transfer. Increasing the sample surface temperature led to an increase in the maximum value of the drying rate. The sample without scraping (See sample No.4) required an extremely longer drying time compared with other samples for reason that the sample surface temperature had to be kept lower as described above.

4.2. Transport properties

The thermal conductivity and permeability of the dried layer are given in Table 4. Because the sample surface temperature did not instantly jump to a constant value at the start of drying as shown in Fig.5, only drying data obtained during a quasi-steady-state period, which was appeared after the sample surface temperature reached the constant temperature, was applied to the model. During this period the ratio of heat flow density to mass was essentially constant indicating a constant temperature of the sublimation front. The temperatures and pressures listed in Table 4 indicate the average values for the dried layer obtained during the quasi-steady-state periods.

The thermal conductivity for the sample without scraping was found to increase with increasing the solute concentration, and its value for about 40 % solution was about 20 per cent greater than those of scraped samples. The relationship between thermal conductivity and pressure at various temperatures of the dried layer is shown in Fig.8. Effects of temperature and pressure of the dried layer on thermal conductivity were not appeared definitely under experimental condition and the average of 40 % solutions was assessed to be 0.203 W m^{-1} K^{-1}. The results obtained in the present and previous studies show clearly that thermal conductivity is markedly affected by solute concentration or the porosity of the dried layer as shown in Fig.9. A linear relationship between thermal conductivity and porosity was obtained and the equation for a regression line fitted to all of the data is also presented.

It was observed that in the samples dried without scraping a markedly low permeability was assessed as shown in sample No.4 and the effects of

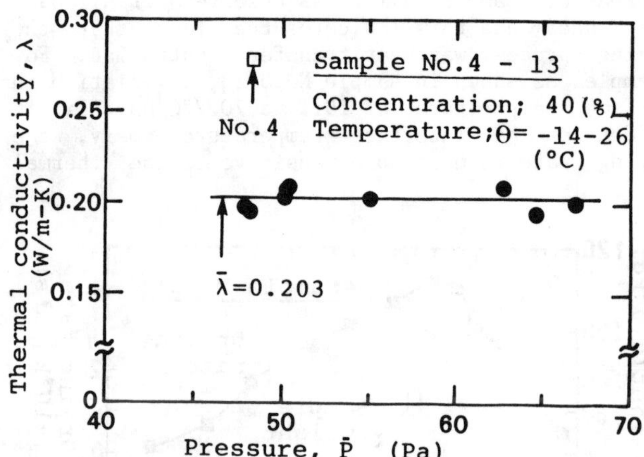

Fig.8 Thermal conductivity versus pressure of the dried layer.

Table 4 Thermal conductivities and permeabilities for coffee solutions.

Sample No.	Concentration (% w.b.) C	Porosity (-) Ψ	Temperature* (°C) $\bar{\theta}$	Pressure* (Pa) $\bar{P}$	Thermal conductivity (W/m-K) λ	Permeability ($x10^{-2}$ m^2/s) K
1	28.5	0.80	18.9	51.4	0.153	0.593
2	33.7	0.76	4.0	49.8	0.170	0.398
3	38.6	0.71	-11.1	47.1	0.241	0.293
4	40.2	0.69	-14.1	48.3	0.277	0.213
5	40.3	0.70	-1.4	47.9	0.197	0.419
6	40.4	0.70	1.5	50.5	0.209	0.460
7	40.4	0.70	3.4	48.2	0.196	0.555
8	40.6	0.70	7.0	50.4	0.208	0.562
9	40.5	0.70	9.1	50.2	0.203	0.585
10	40.2	0.70	15.4	62.7	0.210	0.540
11	40.3	0.70	17.1	55.1	0.203	0.649
12	40.0	0.70	21.2	66.8	0.202	0.554
13	40.1	0.70	26.1	64.5	0.195	0.625
14	45.0	0.65	-0.2	55.3	0.224	0.391

* Average value for the dried layer

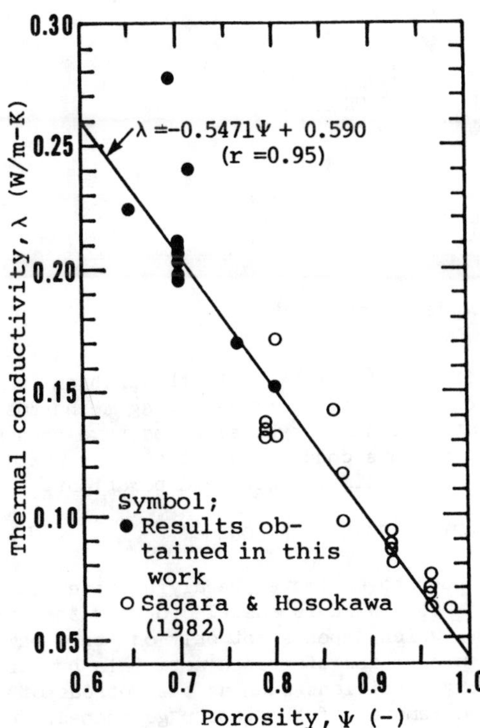

Fig. 9 Thermal conductivity versus porosity.

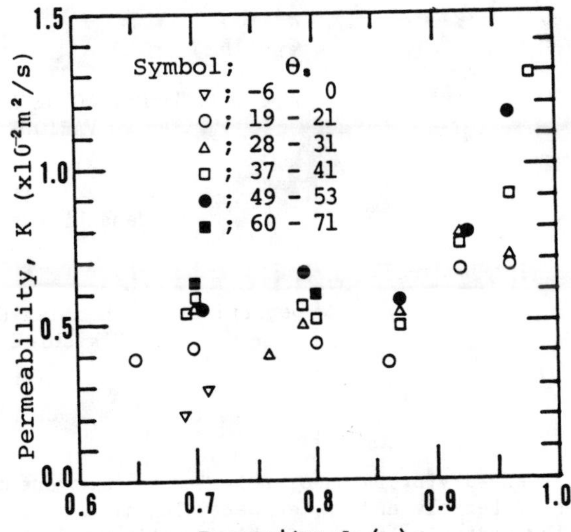

Fig.10 Permeability versus porosity for various surface temperatures.

concentrated surface film on permeability was decreased with decreasing solute concentration as indicated in sample No.1-4.

For all samples of about 40 % solutions whose surface layers scraped prior to drying, the permeability was found to increase with the pressure and temperature of the dried layer as shown in Table 4. This behavior is in good agreement with Mellor and Lovett's theoretical investigations (17) based on the collision theory, and also with their experimental results obtained for several kinds of solutions.

The relationship between permeability and porosity at various sample surface temperature is shown in Fig.10. The data obtained in the previous work(9) are also plotted on the figure. Permeability was found to depend mainly on the solute concentration or the porosity of the dried layer and then other factors such as temperature or pressure of the dried layer.

These results indicate that the transport properties mainly depend upon the structural nature of the dried layer and secondary on the operating factors such as pressure or temperature. In commercial plant operations the solute concentration is one of the critical processing factors since this factor decisively governs the structure of solute matrix formed during freezing of coffee solution and this structure is remained as that of the dried layer during sublimation dehydration process.

REFERENCES

1. Lusk,G.,Karel,M. and Goldblith,S.A., Food Tech., vol.18, 1625 (1964)
2. Massey,W.M. and Sunderland,J.E.,Food Tech.,vol. 21, 90A (1967)
3. Hoge,H.J. and Pilsworth,M.N., J. of Food Sci., vol.38, 841 (1973)
4. Stuart,E.B. and Closset,G., J.of Food Sci.,vol. 36, 388 (1971)
5. Bralsford,R., J. of Food Tech., 339 (1967)
6. Gaffney,J.J. and Stephenson,K.Q., Trans. of ASAE, vol.11, 874 (1968)
7. Sandall,O.C.,King,C.J. and Wilke,C.R., AIChE J. vol.13, 428 (1967)
8. Quast,D.G. and Karel,M., J. of Food Sci., vol. 33, 170 (1968)
9. Sagara,Y. and Hosokawa,A., Proc. 3rd Int.Drying Sym., Birmingham, vol.2, 487 (1982)
10. Ma,Y.H. and Peltre,P.R., AIChE J., vol.21, 335 (1975)
11. Dyer,D.F. and Sunderland,J.E., Trans. ASME, J. of Heat Transfer, 379 (1968)
12. Nat. Res. Council of USA,International Critical Table, vol.3, McGraw-Hill, New York (1928)
13. Lambert,J.B. and Marshall,W.R., Freeze-Drying of Foods, pp.105-135, Natl. Aca. Sci.- Nat. Res. Council (1962)
14. Flink,J.M., Application of freeze-drying for preparation of dehydrated powders from liquid food extract, in Freeze Drying and Advanced Food Technology, ed. Goldblith,S.A., Rey,L. and Rothmayr,W.W., pp.309-329, Academic Press, London (1975)
15. Sivetz,M. and Desrosier,N.W.,Coffee Technology, p.551, AVI, Westport (1979)
16. Sagara,Y., Kameoka,T. and Hosokawa,A., J. of Soci. Agr. Machinery, vol.44, 477 (1982)
17. Mellor,J.D. and Lovett,D.A., Vacuum, vol.18,625 (1964)

FREEZE-DRYING MECHANISMS AND DRYING RATES UNDER VARIOUS DRYING CONDITIONS

Masashi Asaeda and Masanobu Hara

Department of Chemical Engineering, Hiroshima University
Higashi-Hiroshima, 724 JAPAN

ABSTRACT

Freeze-drying rates of relatively homogeneous frozen materials and frozen particles in a shallow tray by radiative and conductive heating are studied both experimentally and theoretically. The drying mechanism of frozen materials of which initial ice content is less than 1.0 is quite different from the flat plane sublimation, which was found previously by one of the authors to become unstable under some conditions. The total drying times under various conditions are discussed for this kind of frozen materials and also for a packed bed of frozen particles using dimensionless heat and mass transfer equations to obtain some information for shorter drying time.

1. INTRODUCTION

Freeze-drying, which is probably the most sophisticated method of drying giving the highest quality of dried products, has been extended in application to various fields, especially to food processing. Many hours are usually required to dry materials by this method, however, because it has to be conducted below the freezing point of wet materials to be dried, which usually gives a quite small vapour pressure or a quite small driving force for mass transfer. In order to increase the mass transfer rate freeze-drying is usually conducted under vacuum, which gives, on the other hand, an inverse effect on drying rate due to a small thermal conductivity of dried-up zone of the frozen materials under drying.

For the cost reduction of freeze-dried materials it is necessary to establish economical design standards of freeze-dryers and to determine their optimum operating condition. For these purposes it is essential to descriminate the freeze-drying mechanisms under various conditions taking into account the physico-chemical properties of frozen materials. There are many physico-chemical factors which control the drying mechanism and the drying rate, for example, thermal conductivities of frozen and dried up layers, permeability of vapour in dried-up layer, conductance of evacuation system, condition of ice contained, initial ice content and so on. All of these must be taken into account to discriminate the drying mechanism and to find its optimum drying conditions for shorter drying time.

Because of the large variety of the properties it is difficult to just treat the problem in general. In this paper freeze-drying mechnisms and drying rates of three typical cases have been studied under various conditions to discuss the directions for the optimum design and perations.

2. THEORETICAL CONSIDERATION

In order to estimate the drying rate of frozen materials it is indispensable to know the drying mechanism which depends not only on the physico-chemical properties of frozen materials but also on the drying conditions such as the methods of heat supply and removal of the vapour generated. Freeze-drying by radiative heat transfer is a rather simple process, that is, uniform retreating of frozen front in most cases [1~4]. On the other hand the drying mechanism by conductive heat transfer is little complicated. In this paper freeze-drying mechanisms of three types of frozen materials are treated in a wide and shallow tray, heated simultaneously by radiation and conduction.

Classification of frozen materials to be dried is rather difficult because of large variety of their structures and physico-chemical properties. Three cases of frozen materials shown in Fig.1 are considered here: (a) frozen material of initial ice content φ_i =1.0 ,(b) packed bed of small frozen particles, and (c) frozen meterial with ice distributed homogeneously among the solid parts φ_i <1.0.

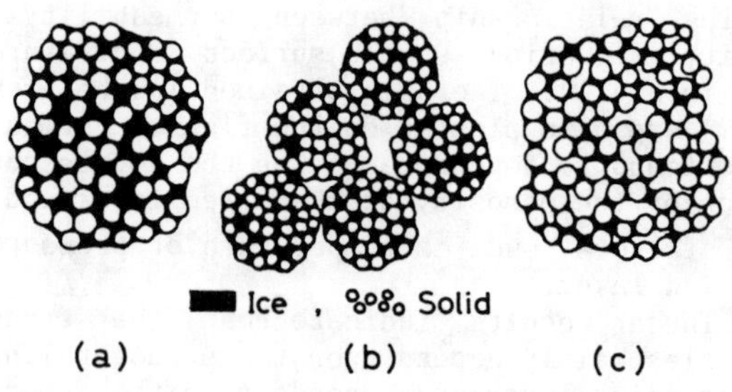

Fig.1 Three types of frozen materials

2.1 In the instance of φ_i =1.0

In this case the frozen front simply retreats to the heating plate for most cases and a relatively large drying rate can be expected because of a comparatively large thermal conductivity of frozen part through which most heat flows to the drying plane. The theoretical treatment for this simple process is given elsewhere [1∿5]. Under some conditions, however, the retreating frozen front becomes unstable, which will result in a large drying time. The criterion for the stability of the sublimation front is given in the literature [4,5] and it provides an important factor for the determination of optimum operating conditions.

2.2 In the instance of packed bed of frozen particles

One of the typical cases of frozen materials to be dried is a packed bed of frozen particles, case (b) in Fig.1. The drying mechanism of this case is quite different from the one in the previous section. Drying occurs from both sides of the frozen materials as shown in Fig.2. The vapour generated near the bottom can migrate through the interparticle space of the frozen zone up to the surface. As the drying proceeds, the frozen front in each particle retreats, leaving the dried-up shell as shown in the figure, which gives an increasing resistance to drying of the particle.

Assuming a pseudo-steady state, which is quite reasonable for such a slow process as freeze-drying under quite low pressure with a large latent heat of sublimation compared with the heat capacity of the frozen material, the rate equations for this case can be derived as follows:

from mass balance at z

$$\frac{d}{dz}(k_0\frac{dp}{dz}) = -K(p_s-p) \qquad (1)$$

from heat balance at z

$$\frac{d}{dz}(\lambda\frac{dT}{dz}) = \gamma K(p_s-p) \qquad (2)$$

where k_0 is the permeability of vapour in the interparticle space, λ the effective thermal conductivity, γ the latent heat of sublimation, p_s the saturated vapour pressure at temperature T and p is the pressure in the interparticle space. The coefficient K is defined here as the transfer coefficient of vapour in the dried up shell of frozen particles per unit volume of the packed bed. Assuming spherical particles of radius r, K can be given as

$$K = \frac{k_i}{r^2}\cdot\frac{3(1-\varepsilon)}{\varphi^{-1/3}-1} \qquad (3)$$

where k_i is the permeability of vapour in the dried-up shell of a particle, ε the porosity of the interparticle space and φ is the ice content of a particle. The permeability of vapour through the interparticle space, k_0, does not depend on the ice content in this case but depends on the total pressure if the transfer mechanism is in the transition or in the viscous region. The above equation is obtained assuming that the frozen surface of each particle retreats concentrically as the drying proceeds because of quite large thermal conductivity of frozen part compared with that of dried-up layer and because of relatively large permeability, k_0, in comparison with K. The total drying rate is given by

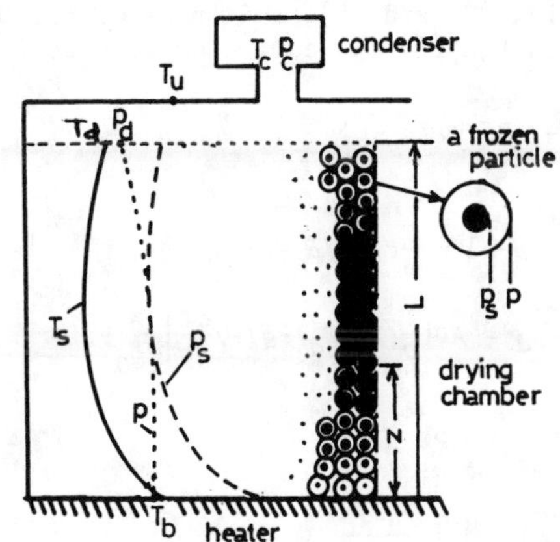

Fig.2 A drying model of frozen particles

$$R_d = \frac{C}{S}(p_d-p_c) = \frac{h_r}{\gamma}(T_u-T_d) - \frac{\lambda}{\gamma}\cdot\frac{dT}{dz}\Big|_{z=0} \qquad (4)$$

where C is the conductance of the evacuation system, p_d the pressure in the drying chamber, T_d the surface temperature of the frozen sample, T_u the temperature of the radiative heat source and h_r is the apparent radiative heat transfer coefficient given by

$$h_r = \sigma\phi(T_u^2+T_d^2)(T_u+T_d) \qquad (5)$$

The boundary conditions are given as follows:

at $z=L$

$$-k_0\frac{dp}{dz}\Big|_{z=L} = \frac{C}{S}(p_d-p_c) \qquad (6)$$

$$\lambda\frac{dT}{dz}\Big|_{z=L} = \frac{h_r}{\gamma}(T_u-T_d) \qquad (7)$$

at $z=0$

$$k_0\frac{dp}{dz}\Big|_{z=0} = 0 \qquad (8)$$

$$T = T_b \qquad (9)$$

where T_b is the temperature of the heating plate.

The above equations can be put in dimensionless forms using the following dimensionless variables:

$$\bar{z} = \frac{z}{L}, \quad \bar{T} = \frac{T-T_c}{T_b-T_c}, \quad \bar{p} = \frac{p-p_c}{p_b-p_c},$$
$$\bar{R}_d = \frac{R_d}{R^*_d}, \quad \bar{t} = \frac{t}{t^*}, \quad g(\varphi) = \frac{\lambda}{\lambda_0} \qquad (10)$$

where

$$R^*_d = C/S(p_b-p_c) \quad (11), \qquad t^* = m/(SR^*_d) \quad (12)$$

here m is the total weight of ice initially present and S is the sectional area. R^*_d is the maximum drying rate under given drying condition if the radiation is negligible and t^* is the corresponding shortest drying time. The dimensionless forms of Eqs.(1), (2) and (4), and those of the boundary conditions, Eqs.(6)-(9) can be written as follows:

$$\frac{d^2\bar{p}}{d\bar{z}^2} = -Ef(\varphi)(\bar{p}_s-\bar{p}) \tag{1'}$$

$$\frac{d}{d\bar{z}}\{g(\varphi)\frac{d\bar{T}}{d\bar{z}}\} = \frac{E}{B}f(\varphi)(\bar{p}_s-\bar{p}) \tag{2'}$$

$$\bar{R}_d = \bar{p}_d = AB\{\frac{1}{D}(\bar{T}_u-\bar{T}_d)-g(\varphi)\frac{d\bar{T}}{d\bar{z}}\Big|_{\bar{z}=0}\} \tag{4'}$$

$$-A\frac{d\bar{p}}{d\bar{z}}\Big|_{\bar{z}=1} = \bar{p}_d \tag{6'}$$

$$g(\varphi)\frac{d\bar{T}}{d\bar{z}}\Big|_{\bar{z}=1} = \frac{\bar{T}_u-\bar{T}_d}{D} \tag{7'}$$

$$\frac{d\bar{p}}{d\bar{z}}\Big|_{\bar{z}=0} = 0 \tag{8'}$$

$$\bar{T}_b = 1 \tag{9'}$$

The change of local ice content can be obtained by

$$d\Psi/d\bar{t} = -AEf(\varphi)(\bar{p}_s-\bar{p}) \tag{13}$$

where $f(\varphi)$ is a function of the ice content of a particle and is given as

$$f(\varphi) = \frac{3(1-\varepsilon)}{\varphi^{-1/3}-1} \tag{14}$$

The constants A,B,D and E in these equations are dimensionless contents given by

$$A = \frac{k_0/L}{C/S} \quad (15), \qquad B = \frac{\lambda_0(T_b-T_c)}{\gamma k_0(p_b-p_c)} \quad (16)$$

$$D = \frac{\lambda_0/L}{h_r} \quad (17), \qquad E = \frac{k_i}{k_0}\cdot(\frac{L}{r})^2 \quad (18)$$

The constant A, which is the ratio of the vapour conductance in the dried-up zone to the conductance of the evacuation system, can be an important factor in determining the dimension of the evacuation duct. The constant B, which is the ratio of the equivalent amount of heat transferred through the dried-up zone under the maximum temperature difference to the mass transfer rate through the dried-up zone under the possible maximum driving force corresponding to the maximum temperature difference, is referred to as a measure to determine the rate controlling step. The value of D is not strictly a constant because of the possible surface temperature change during drying but is assumed to be a constant so long as the effect of the surface temperature change to the apparent radiative heat transfer coefficient is relatively small. The constant E is the ratio of the permeability of vapour through interparticle space and that in the intraparticle of dried-up layer.

Using the dimensionless variables given above, the dimensionless vapour pressure of sublimating substance (ice) can be approximated as

$$\bar{p} = \bar{a}\bar{T} + \bar{b} \tag{19}$$

where $\bar{a}=a(T_b-T_c)/(p_b-p_c)$ and $\bar{b}=(aT_c+b-p_c)/(p_b-p_c)$. The values of a and b are dependent on temperature and given in the simplest expression of vapour pressure of ice, $p=aT+b$. Knowing the functions $f(\varphi)$ and $g(\varphi)$ theoretically or experimentally, the above equations can be solved numerically to obtain the drying rate, total drying time, local ice content and the temperature distribution under given drying conditions (C, S, T_b, T_u, T_c, L etc.).

2.3 Drying of macroscopically homogeneous frozen materials of φ_i<1.0

In this case, the ice distribution is considered relatively homogeneous in frozen materials of which solid parts do not have any pores in them, case (c) in Fig.1. Because of void space present in the frozen materials, dried-up layers are expected to appear on both sides of the frozen sample just as the case in the previous section. The boundary between dried and frozen parts was shown to be clear in the previous work [5]. A model of this case is shown in Fig.3. The rate equations for this case are also given by Eqs.(1) and (2), but the coefficients k_0 and K have somewhat different meanings from those in section 2.2. The vapour permeability k_0 is now dependent on the ice content and $K=\sqrt{M/2\pi RT}$ which is the coefficient in the Hertz-Knudsen's equation. The value of K being quite large in comparison with other transfer coefficients, the local equilibrium can usually be assumed. With this assumption the rate equations in section 2.2 can be simplified as follows: The equations (1) and (2) can be combined to give the following equation for the frozen part:

$$k(\varphi)\frac{dp}{dz} + \frac{\lambda(\varphi)dT}{\gamma\ dz} = \{k(\varphi)\frac{dp}{dT} + \frac{\lambda(\varphi)}{\gamma}\}\frac{dT}{dz} = -R_0 \tag{20}$$

where $k(\varphi)$ and $\lambda(\varphi)$ are the function of the ice content and p is now the saturated vapour pressure of frozen material at temperature T. The boundary conditions are given as

at $z=z_1$

$$-k_0\frac{dp}{dz}\Big|_{z=z_1} = \frac{C}{S}(p_d-p_c) = \frac{p_1 - p_c}{(L-z_1)/k_0+S/C} = R_d \tag{21}$$

$$\lambda_0\frac{dT}{dz}\Big|_{z=z_1} = \frac{T_u - T_1}{(L-z_1)/\lambda_0+1/h_r} = \gamma R_u \tag{22}$$

at $z=z_2$

$$\lambda_0\frac{dT}{dz}\Big|_{z=z_2} = \frac{\lambda_0(T_b-T_2)}{z_2} = R_b = R_d-R_u \tag{23}$$

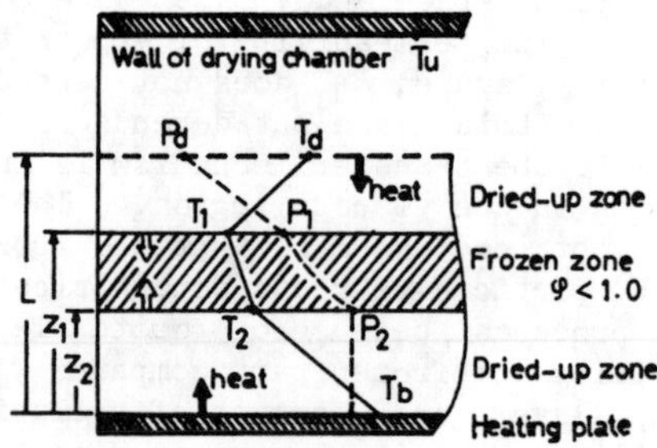

Fig.3
A drying model of homogeneous frozen materials

The rate of local ice content change is given by

$$\frac{d\varphi}{dt} = -\frac{1}{\gamma\varepsilon\rho}\frac{d}{dz}\{\lambda(\varphi)\frac{dT}{dz}\} \tag{24}$$

The above equations can be put in dimensionless forms using the same dimensionless variables as in section 2.2. Equations (20)-(23) give

$$-A\{h(\varphi)\frac{d\bar{p}}{d\bar{T}} + Bg(\varphi)\}\frac{d\bar{T}}{d\bar{z}} = \bar{R}_0 \tag{20'}$$

$$\bar{p}_d = A\bar{p}_1/(1-\bar{z}_1+A) = \bar{R}_d \tag{21'}$$

$$AB(\bar{T}_u-\bar{T}_1)/(1-\bar{z}_1+D) = \bar{R}_u \tag{22'}$$

$$AB(1-\bar{T}_2)/\bar{z}_2 = \bar{R}_b = \bar{R}_d-\bar{R}_u \tag{23'}$$

where $h(\varphi)=k(\varphi)/k_0$ and $g(\varphi)=\lambda(\varphi)/\lambda_0$. Equation (24) can be written as

$$\frac{d\varphi}{d\bar{t}} = -AB\frac{d}{d\bar{z}}\{g(\varphi)\frac{d\bar{T}}{d\bar{z}}\} \tag{24'}$$

Using Eq.(19) as the saturated vapour pressure relation, the above dimensionless equations can be solved numerically to simulate the drying process.

3. EXPERIMENTAL

Apparatus. A schematic diagram of the experimental apparatus is shown in Fig.4. The drying chamber comprises a glass bell jar(4) of diameter 22cm, inside of which are placed a sylindrical metal plate(3) cooled by a coolant from a refrigerator(6) and a heater(2) for radiative heating of the surface of sample box(1). In Fig.5 are shown the details of the sample box, to which is attached a heating plate at its bottom. Seven copper-constantan thermocouples are fixed to the box at about 5 mm intervals to measure the temperature distribution and to control the temperature of the heating plate. The pressure of the drying chamber is measured with an oil manometer(8). The weight loss of the sample was measured by switching the cold traps cooled by liquid nitrogen.

Samples. Sample particles of coffee solution (30 wt.% of coffee) were prepared by dripping or blowing the solution from a small nozzle onto a pool of liquid nitrogen. By this method frozen spherical particles of relatively homogeneous diameter were obtained.

Procedures. Before packing the frozen particles into the sample box they were kept for a while at temperature around -40°C in a container of the refrigerator. After placing the sample box in the right position in the dryer, evacuation was started to attain enough low pressure in the chamber and then the measurements started after the temperatures of the radiation plate and the heater at the bottom of the box were controlled at the specified values. For measurements of the ice content distribution the drying was interrupted after some hours and the sample box was taken out to a cooled chamber filled with nitrogen to avoid melting and condensation of vapour from the atmosphere to the frozen particles. The sampling of particles at various depths was done by sucking off the particles in the upper layer with a great care not to mix them up. The ice content was obtained by measuring the weight difference of the sampled particles before and after drying them at 60°C.

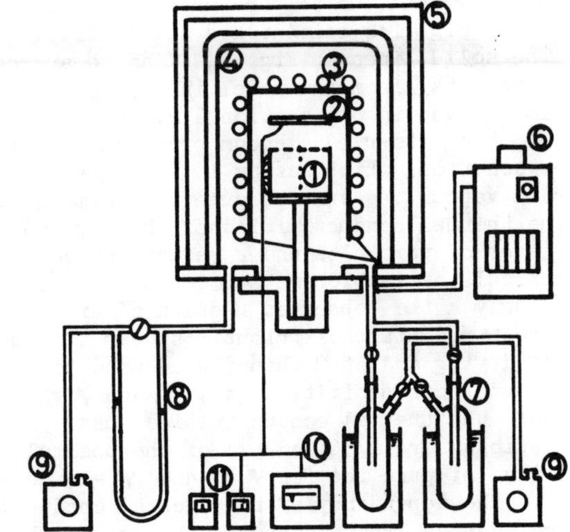

1:sample box 2:radiative heat source 3:cooling coil 4:glass bell jar 5:insulation box 6:refrigerator 7:cold traps 8:oil manometer 9:vacuum pumps 10:recorder 11:temperature controllers

Fig.4 Experimental apparatus

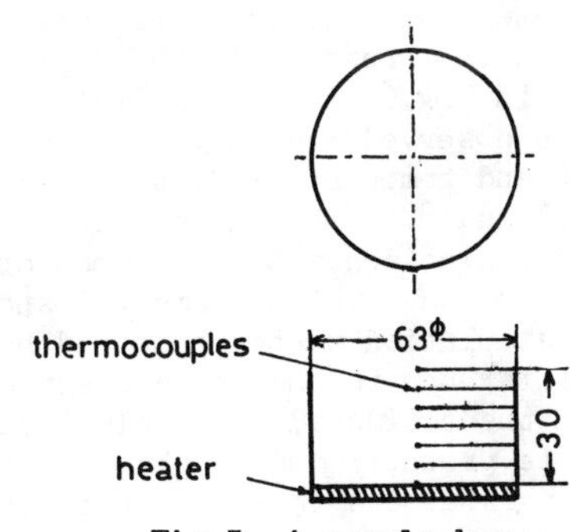

Fig.5 A sample box

4. EXPERIMENTAL RESULTS AND DISCUSSION

4.1 Experimental results and discussion for packed beds of frozen particles

An example of the experimental results for a packed bed of spherical frozen particles of coffee-water solution is shown in Fig.6, where the changes of the temperature, pressure and sublimated ice are shown against the drying time. In this case the average diameter of the frozen particles is 0.9 mm and the temperatures of the heater and the radiative heat source were kept constant at -20 C. The temperature distribution curves are quite different from those of a flat plane sublimation for a case of φ_i = 1.0 (4,for example). It is observed that a sudden temperature gradient exists just above the heating plate even at the beginning of drying. This is due to the quick decrease of ice content just above the heating plate, which leads to a low thermal conductivity. This can be more clearly seen from the observed results of ice content and temperature distributions shown in Figs. 7 and 8. From these figures two regions of low ice content can be reasonably seen near the surface and the bottom, and the temperature distributions at relatively high ice content are nearly flat.

The solid lines in Figs.6, 7 and 8 are calculated results from the model in Fig.2 with the equations given in the section 2.2. In the calculations the observed vapour pressure of frozen coffee solution, which was by 15 % smaller than the reported vapour pressure of pure ice, was used. As for the thermal conductivities of the dry and frozen layer of particles under reduced pressures no theoretical estimation method could be available. Accordingly, the observed results of drying rates and the temperature distributions near the beginning of drying and at 25 hours were used to obtain the thermal conductivities at $\varphi=1$ and $\varphi=0$. Values of the thermal conductivities used in the calculations are 1/50 and 1/5 of the thermal conductivity of pure ice for $\varphi=0$ and $\varphi=1$, respectively. The dependency of the thermal conductivity on ice content was assumed to be given by $g(\varphi)=g(0)(1-\varphi^3)$. The permeability of water vapour in the inter-particle space can be estimated by the method proposed in the literature [6]. The intra-particle permeability k_i is difficult to be determined either experimentally or theoretically. This value was determined to be 1/3000 of k_0 by curve fitting at 10 hours (see Fig. 8).

The calculated results for the weight of sublimated ice and the pressure in the drying chamber are shown in Fig.6 with the solid lines to compare with the observed results. And those of the ice content and temperature distributions are shown in Figs.7 and 8. It can be seen that a relatively good coincidence between the observed and calculated results, which seems to show the adequacy of the simulation method proposed here. A fairly large difference in the ice content distributions between the estimated and calculated results in Fig.7 is probably due to the experimental error, considering that the integration of the area above the observed curve gives the weight of sublimated ice smaller than the one obtained from the weight loss at the end of the run by about 7 %. On the other hand the calculated results of weight of sublimated ice show a good coincidence with the observed results, two small circles on the calculated solid line in Fig.6 which were obtained experimentally for the runs shown in Figs.7 and 8.

Figure 9 shows some calculated results of ice distributions for different intra-particle permeability k_i. For a small value of k_i or E the ice content distribution becomes rather flat. On the contrary, relatively sharp boundary between dried-up and frozen parts appers as the k_i or E increases and the drying mechanism is considered to approach the one in section 2.3 (see Fig.10 for experimental results in this case). The calculated results of the total drying time for these three cases are 67.5 hrs (case a), 98.6 hrs (case b) and 67.9 hrs (case c). These results were obtained simply by fixing other factors except k_0, of which dependency

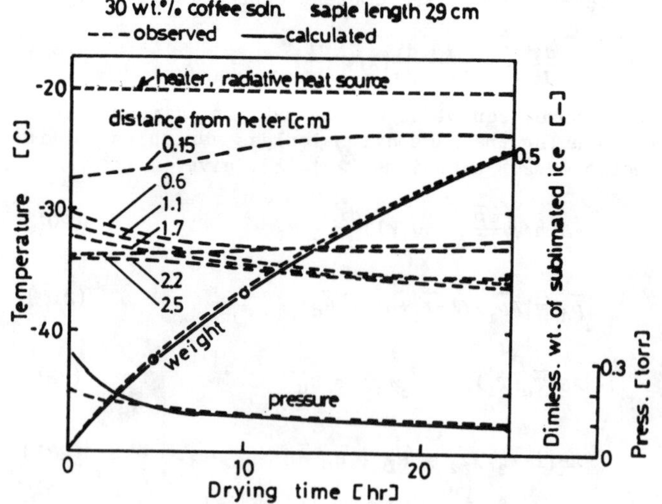

Fig.6 An example of observed results

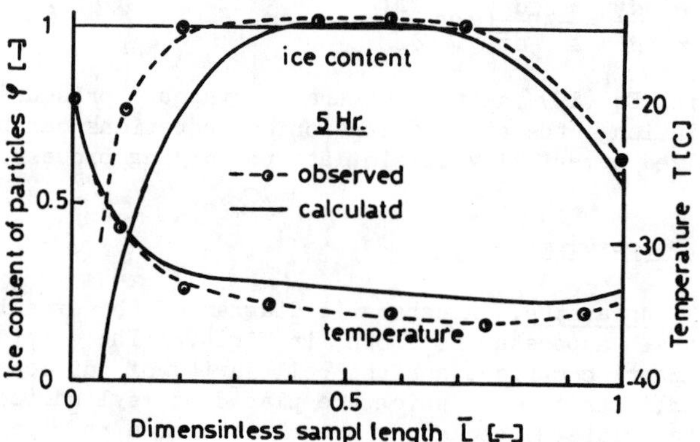

Fig.7 Ice content and temperature distributions after 5 Hrs

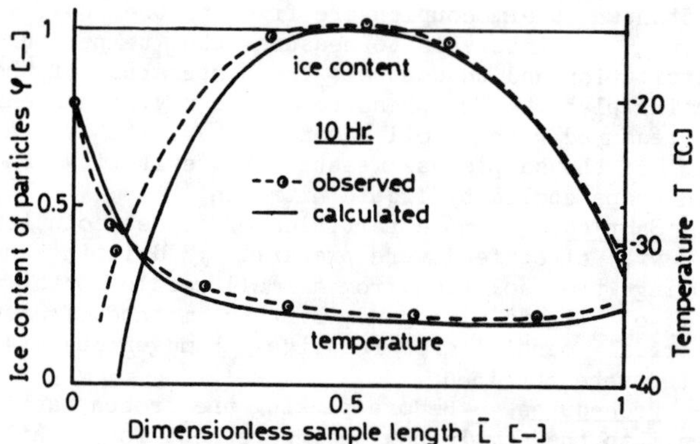

Fig.8 Ice content and temperature distributions after 10 Hrs

on temperature and pressure is fully known [6]. Little difference in drying time can be seen between cases (a) and (c). Freeze-drying, however, is a very complicated process, where a factor is combined dynamically with some other factors. A change in r_i, for example, gives changes in the inter-particle permeability and the thermal conductivity, which lead to a change in the temperature profile. Therefore it is quite difficult to give the detailed directions for shorter drying time without sufficient knowledge of transfer properties. In the following section some discussion is to be given for a simpler case.

4.2 Some further discussion on drying of macroscopically homogeneous frozen materials

An example of the calculated results for this case is compared with the experimental results in Fig.10. The initial ice content, which is defined as a fraction of ice to the total void if the dry material, is 0.25. The theoretical results are in good accord with the observed ones. In the calculations the transfer properties given in the literature [5] are used. Some numerical results of the effect of the dimensionless constant A on the total drying time are shown in Fig.11. The total drying time decreases as the value of A becomes smaller, and below a value of 0.01 there can be seen little effect of A on the drying time. This value can be the measure of determining the appropriate conductance of the size of the evacuation system for the economical design of freeze-dryers. One more important factor concerning the vapour removal is the temperature or the vapour pressure in the condenser. The vapour pressure of ice at -60° C is about 0.008 torr., which is low enough for usual freeze-drying.

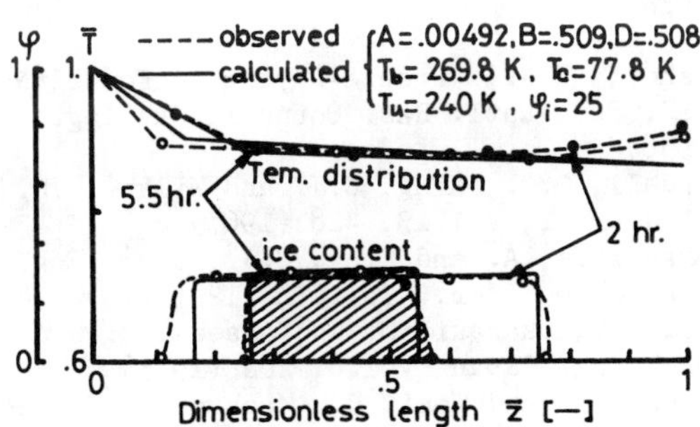

Fig.10 Comparison of calculated and observed results

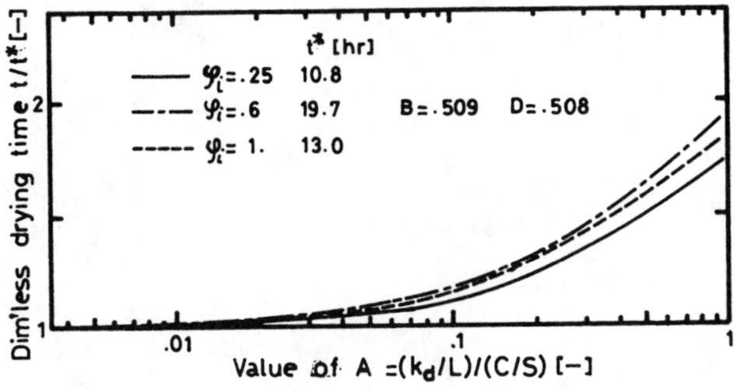

Fig.11 Dimensionless drying time vs. constant A

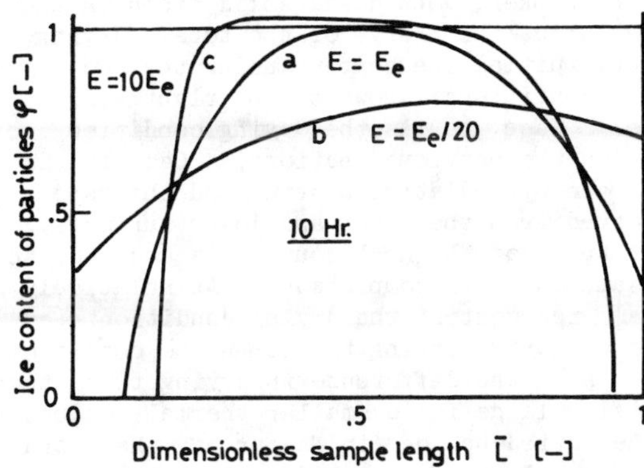

Fig.9 Calculated ice content distributions after 10 Hrs

In Fig.12 are some numerical results of the total drying time against the initial ice content and the value of B. In the calculations the value of A was taken as less than 0.05 for a negligible resistance of the evacuation system. A pair of decreasing curves for φ_i=0.6 and 0.25 were obtained for a given heat source temperature keeping the values of A and D constant. These curves show that the total drying time decreases with the increase of B. In the region where B is larger than 2 or 3 the rate controlling step is the mass transfer in the dried-up zone. Since the value of B depends largely on the temperature of the heat source and the condenser, no absolute value can be given here which generally defines the rate controlling step. It should be noticed, however, that the rate controlling step must be determined not by a simple ratio of λ_0/k_0 but by the constant B which includes temperature and pressure differences as their driving forces.

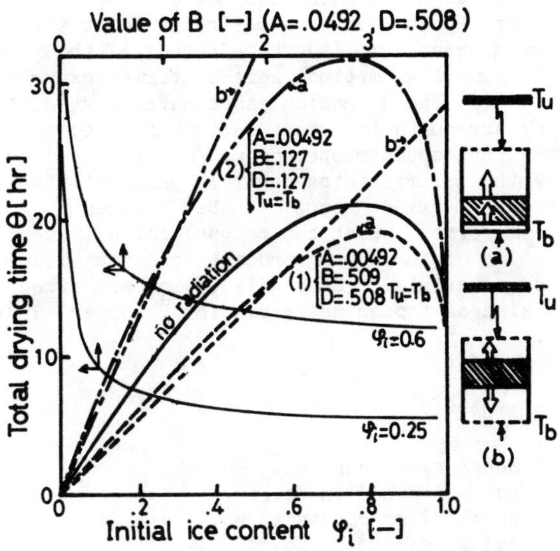

Fig.12 Some calculated results of drying time under various drying conditions

The broken, dotted and solid lines in the same figure show the change of the total drying time with the initial ice content under the two typical drying conditions shown on the right side of the figure. Case (a) is the drying condition considered in the previous sections. Case (b) is the usual one by radiative heating and the vapour can be removed from the upper and lower surfaces. The temperature of the heat sources is assumed to be the same for fair comparison. At relatively high initial ice content the drying condition (a) requires a shorter drying time than the condition (b) does. And the difference in drying time becomes especially large for a smaller thermal conductivity of the dried-up zone in favour of the condition (a). The values of A, B and D shown in Fig.10 were chosen as the basis of their calculations. All these curves show that the total drying time increases as the initial ice content becomes larger and the decrease after reaching the maximum at about φ_i =0.8, and that this tendency does not change with the magnitude of thermal conductivity of the dried-up zone. The maximum in the drying time goes higher and the value of φ_i giving the maximum decreases slightly as the thermal conductivity of dried-up zone decreases. It should be noticed that the maximum value of total drying time becomes larger than the one at φ_i =1, and the difference becomes quite large in case of small thermal conductivity of the dried-up zone. This obviously shows that instead of drying the material of φ_i =0.8 it is sometimes much more attractive from the view point of shortening the drying time. to add more water to make φ_i become 1.0 before drying.

5. CONCLUSION

A model of freeze-drying mechanism has been proposed for drying of frozen materials which have bi-disperse pore structures. Dimensionless heat and mass transfer equations based on this model were derived and solved numerically to obtain the following results: (1) The adequacy of the model and the numerical methods were confirmed experimentally. (2) The dimensionless constants A, B, D, E and φ_i are quite important factors for economical designs and optimum operations of freeze-dryers. (3) With a large intra-particle permeability the equations derived here can be reduced to the simpler forms, which can be used for a numerical simulation of the drying process for mono-disperse porous materials. Some discussion are given to this kind of frozen materials for shorter drying times.

NOMENCLATURE

A	dimensionless constant, Eq.(15)	
$\bar{a}$	a value defined in Eq.(19)	-
B	dimensionless constant, Eq.(16)	
$\bar{b}$	a value defined in Eq.(19)	kg/hrPa
C	conductance of evacuation system	-
D	dimensionless constant, Eq.(17)	
$f(\varphi)$	function of φ, Eq.(14)	-
$g(\varphi)$	thermal conductivity dependency on φ	-
h_r	defined by Eq.(5)	J/m^2hrK
E	dimensionless constant, Eq.(18)	-
K	value defined by Eq.(3)	kg/m^2hrPa
k	permeability in inter-particle space or dried-up zone	kg/mhrPa
L	sample length	m
p	pressure	Pa
$\bar{p}$	dimensionless pressure	
R_d	drying rate	kg/m^2hr
$\bar{R}_d$	dimensionless drying rate	
R^*_d	muximum drying rate, Eq.(11)	kg/m^2hr
R_0	value defined in Eq.(20)	kg/m^2hr
$\bar{R}_0$	$=R_0/R_d$	-
R_u	value defined in Eq.(22)	kg/m^2hr
$\bar{R}_u$	$=R_u/R_d$	-
S	sectional area of frozen sample	m^2
T	temperature	K
$\bar{t}$	t/t^*	
t^*	value defined in Eq.(12)	hr
z	distance from the heating plate	m
$\bar{z}$	$=z/L$	-
γ	latent heat of sublimation	J/kg
ε	porosity	-
$\lambda(\varphi)$	thermal conductivity at ice content φ	J/mhrK
λ_0	thermal conductivity of dried-up layer	J/mhrK
ρ	density of ice	kg/m^3
φ	ice content	-
φ_i	initial ice content	-
Subscripts		
1,2	values at 1,2 in Fig.3	
b	value at the heating plate	
c	value at the condenser	
u	value at the radiative heat source	
r	particle radius	m

REFERENCES

1. Fisher, F.R., Freeze-Drying of Foods, Natl. Acad. Sci.-Natl. Res. Council, Washington D.C. (1958)
2. Sandall, O.C., King, C.J. and Wilke, C.R., AI ChE Journal, Vol.13, 428 (1967)
3. Margaritis, A. and King, C.J., Chem. Eng. Prog. Sympo. Ser., Vol.67, 112 (1970)
4. Toei, R., Okazaki, M. and Asaeda, M., J. Chem. Eng. Japan, Vol.8, 282 (1975)
5. Asaeda, M. and Toei, R., Mechanism of Freeze-Drying of Porous Bodies by Conductive Heat Transfer, 18th ASME/AIChE National Heat Transfer Conf., San Diego, ASME Paper 79HT86 (1979)
6. Asaeda, M., Yoneda, S. and Toei, R., J. Chem. Eng. Japan, Vol.7, 93 (1974)

FREEZE-DRYING OF CONCENTRATED LIQUID FOODS BY BACK-FACE HEATING

Hitoshi Kumagai, Kozo Nakamura and Toshimasa Yano

Department of Agricultural Chemistry, University of Tokyo
Bunkyo-ku, Tokyo, 113 Japan

ABSTRACT

The sample solutions, milk and other liquid foods, with different water content, were freeze-dried by back-face heating. The drying rate gradually decreased at the early stage of drying when the heater temperature was kept constant, and its decrease became sharp after a certain time of drying elapsed. Thus the time for freeze-drying was divided into constant- and falling-rate periods.

The data of a contant-rate period could be analyzed with the uniform retreating ice front model where the higher resistance to vapor permeation was assumed at the surface of a frozen sample. The constants related to drying-rate were estimated, and the effect of initial water content was elucidated. In freeze-drying of the milk with higher initial water content, the transition of drying period started ealier before the water content of a drying sample reached that of the concentrated amorphous solution. This transition was accelerated as the back-face temperature became higher. The reason for this transition was supposed to be due to the structural change of the frozen part.

The time needed for sublimation of frozen water could be shortened by the stepwise decrease of the heater temperature.

1. INTRODUCTION

There have been many research works reported on freeze-drying of foods, but the physical properties related to heat and mass transfer are still limitedly available. The temperature of a freeze-drying food should be kept below a certain level, but the safe temperature has not well been informed.

In this study the sample solutions, milk and other liquid foods, with different water content, were freeze-dried by back-face heating. The solutions were frozen in the unidirection, and the surface of a frozen sample was supposed to have a tight structure due to freeze-concentration [1]. The surface resistance to vapor permeation was considered in the uniform retreating ice front model, and the data of a constant-rate period were analyzed with the modified U.R.I.F. model. The apparent thermal conductivity of a frozen part, the surface resistance, and the permeability of vapor in a dried porous part were estimated

The disturbance of a sublimating plane [2] and the structural change in a dried part or collapse[3] were the reasons suggested for a sharp decrease of drying rate. The reason may be different between the methods of heat delivery, heating through a dried part and heating through a frozen part. In the latter case the reason can be related to the state of water in a frozen part. Thus the weight ratio of frozen water was measured by differential scanning colorimetry, and the water content of the non-ice part named CAS was calculated.

The temperature of a sublimating plane increases as the ice front approaches the heater of which temperature is kept constant. The heater temperature should be controlled for effective freeze-drying. The stepwise decrease of the heater temperature was tried to complete sublimation of ice before a sharp decrease of drying rate.

2. DSC OF FROZEN LIQUID FOODS

2.1. Materials and Method

The liquid foods used in DSC as well as in freeze-drying were milk, concentrated coffee extract and orange juice. The milk and the orange juice were the comercial ones and the coffee extract was obtained from the company. The concentrated milk was prepared by dissolution of the freeze-dried milk into warm water. The aqueous solution of KCl was also used as a solution having a clear eutectic temperature.

The heat-flux differential scanning calorimetry was performed at a heating rate of 0.5°C/min after the preceding cooling down to -55°C using the low temperature type SSC 544 available from Seiko- Denshi- Kogyo Co. Ltd. The area bounded by the DSC curve, the corrected base line and the line derived in consideration of the time lag of heat response was numerically integrated, and converted to the amount of melted water with a constant value of heat of melting, 79.7 kcal/kg. The total amount of water was calculated to be the weight loss of the sample dried at 105°C.

2.2. Results and Discussions

Fig. 1 shows the amount of unfreezable water in the frozen milk. It is almost constant irrespective of the initial water content. This result is diffent from what Nagashima and Suzuki recently reported on the amount of unfreezable water in the frozen bean paste "miso", although they used a different technique, NMR, to measure it[4]. The amount of unfreezable water was measured on the other liquid foods, and the results are summarized in Table 1.

Fig. 2 shows the effect of temperature on the amount of frozen water divided by the total amount of freezable water and on the water content of concentrated amorphous solution, CAS. The water contained in CAS was assumed to consist of melted water and unfreezable water. The solution of KCl melts sharply at its eutectic temperature, -11°C, and each of the liquid foods starts melting at each specific temperature. The orange juice melts at the lower temperature than the other liquid foods and contains a larger amount of melted water at each temperature. This is a reason why the orange juice was not easily freeze-dried. The water content of CAS was measured using concentrated milk, and there was no effect of initial water content on it as expected by the phase equilibrium.

Table 1 Unfreezable water of liquid foods measured by DSC

Sample solution	Water content [kg-H_2O/kg-dry matter] Total	unfreezable
Coffee extract	1.76	0.38-0.40
Orange juice	2.21	0.53-0.60
Lactose	6.28	0.48
Milk	1.0-8.0	0.40

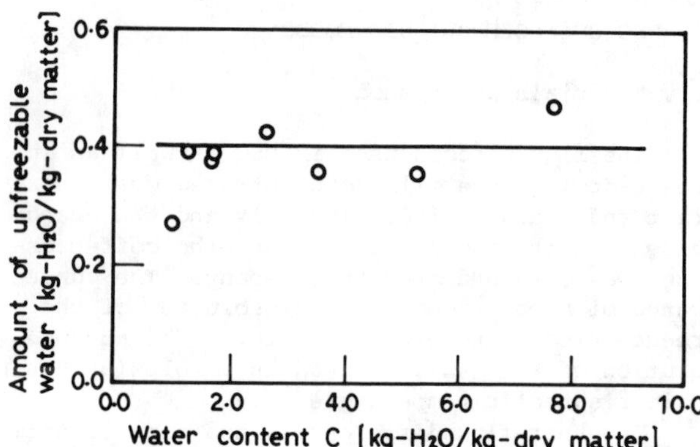

Fig. 1 Amount of unfreezable water in frozen milk

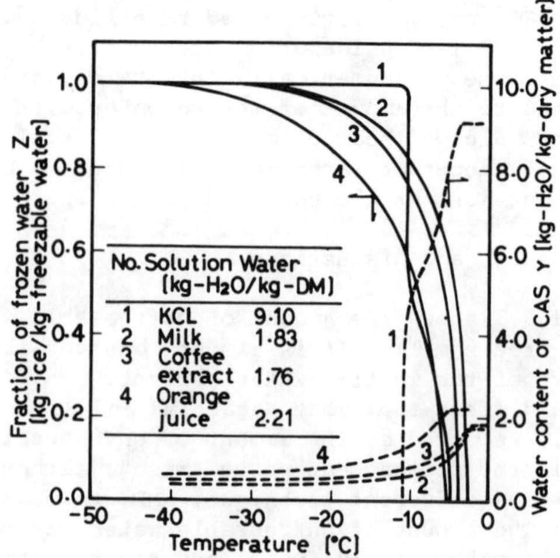

Fig. 2 State of water in frozen liquid foods

The volume fraction of ice, ε, was calculated with the water content of CAS, y, using Eq.(1) and used in the analysis of drying rate to be mentioned in the following section.

$$\varepsilon = \frac{C - y}{1 + C} \cdot \frac{\rho_o}{\rho_i} \tag{1}$$

3. FREEZE DRYING OF LIQUID FOODS

3.1. U.R.I.F. Model

When the liquid foods are frozen in the unidirection, freeze-concentration proceeds and the surface of a frozen sample located at a distance above a cold plate has a tight structure with high resistance to vapor permeation. The surface resistance was considered in U.R.I.F. model applied to analysis of the rate of freeze-drying by back-face heating.

Equation of heat transfer:

$$Q = \lambda \frac{T_b - T_s}{L - X} \tag{2}$$

Equation of mass transfer:

$$W = \frac{P_s - P_c}{(X/k) + R_a + R_s} \tag{3}$$

Relationship between Q and W:

$$Q = HW \tag{4}$$

Vapor pressure at sublimating plane:

$$P_s = f(T_s) \tag{5}$$

Equation used in conversion of sample weight to thickness of dried part:

$$X = (M_o - M) / (\varepsilon \rho_i A) \tag{6}$$

The weight of a sample, M, was continuously measured in the experiments of freeze-drying, so that the drying rate, W, and the thickness of dried layer could be calculated at each drying time. The temperature of a sample was measured at the four vertical positions, and the temperature of sublimation, T_s, was estimated by extrapolation. Then the apparent heat conductivity, λ, could be calculated using Eqs.(2) and (3). The relationship between X and $(P_s - P_c)/W$ was derived from Eq.(3) and plotting the data of this linear relationship made it possible to estimate the vapor permeability, k, and the resistance, $R_a + R_s$.

3.2. Experimental Apparatus and Method

Drying rate depends on the condition of freezing, and a sample was frozen in the vessel of Fig. 3 while the temperature of the bottom plate was controlled with the thermo module. The wall of the vessel was made of porous plastics coated with the adhesive of epoxy resin to prevent liquid from penetrating into the pores.

The vessel was transfered in the chamber of

freeze-drying soon after the freezing of a sample was completed. The chamber pressure was controlled to be 13.0 Pa using a Pirani gauge and a solenoid valve. The temperature of the heater, which was the bottom plate of the sample vessel, was also controlled during freeze-drying. The vacuum chamber was cooled by immersion of its bottom part in the circulating cooling agent and by circulation of cooling agent in the metal tube wound around the chamber. Thus the temperature in the chamber could be kept at about 0°C, and the heat conducting through the dried layer was supposed negligibly to contribute to heat of sublimation. The cold trap was cooled with acetone and dry ice.

The weight of a sample was continuously measured with an electric element detecting displacement of angle. The temperature of a sample and that of the chamber room were measured with C-C thermo couple.

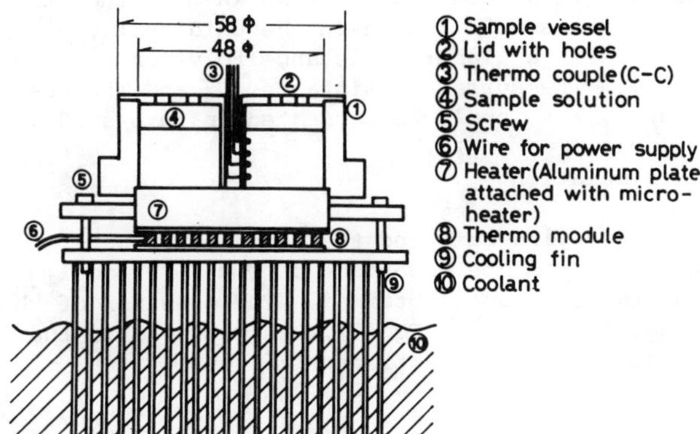

Fig. 3 Sample vessel and method of controlled freezing

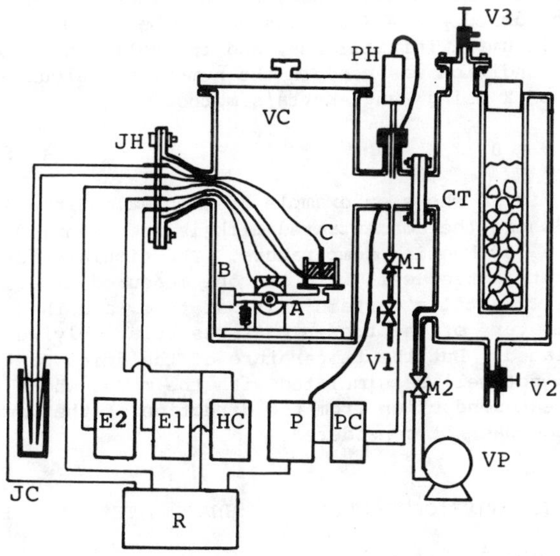

A: Element detecting displacement of angle
B: Counter-balance CT: Cold-trap
E1 and E2: Electric power for element A and heater
HC: Controller of heater temperature
JC: Cold junction JH: Hermetic connector
M1 and M2: Solenoid valve
PC: Controller of pressure P: Pirani gauge
PH: Head of Pirani gauge R: Recorder
VC: Vacuum chamber (0.25 mI.D.x0.3 m)
VP: Vacuum pump V1,V2, and V3: Valve

Fig. 4 Experimental apparatus of feeze-drying

3.3 Results and Discussions

The drying rate measured using milk and coffee extract maintained an almost constant value until a certain critical time, and thereafter decreased remarkably. The drying rate and the temperature of the frozen part could be analyzed in the early stage of freeze-drying with U.R.I.F. model, and the constants related to drying rate were obtained by the method mentioned in 3.1.

Fig. 5 shows the thermal conductivity of frozen milk. The conductivity estimated with the data of sublimation is approximately twice as large as the one measured at about -5°C with the method of steady state using Shotherm RTM-G5 available from showa- Denko Co. Ltd. The apparent thermal conductivity was also obtained when the

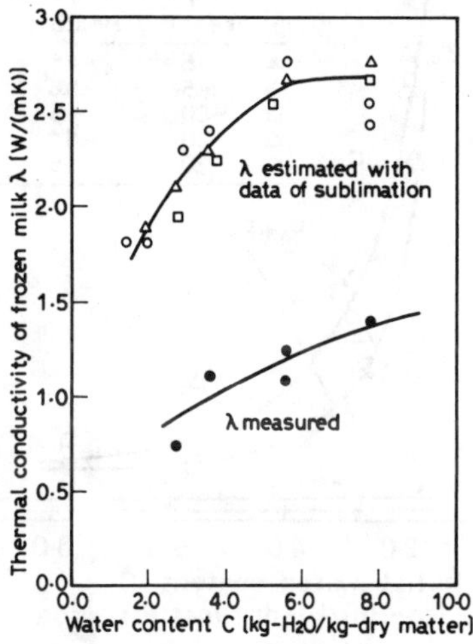

Fig. 5 Thermal conductivity of frozen milk (Temperature of heater: -5°C, -11°C and -16°C)

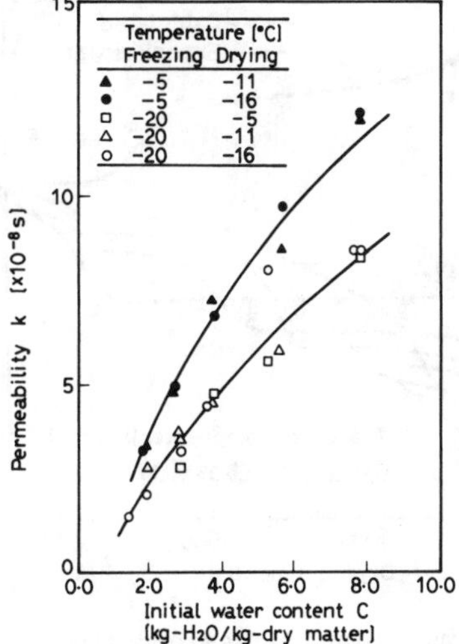

Fig. 6 Effect of initial water content on vapor permeability (milk)

water frozen in the free space of packed glass beads was freeze-dried, and agreed well with the one estimated using the model of heat conductivity of packed bed[5]. The contribution of heat conduction through the dried layer to the apparent heat conductivity, if any, is supposed to be less in milk than in water-glass beads system. The reason is not known at this moment for the deviation of the apparent heat conductivity of frozen milk from the one measured directly.

Fig. 6 shows the effect of initial water content on the vapor permeability, k, which was obtained by application of U.R.I.F. model using the apparent heat conductivity to the data of freeze-drying of milk. The vapor permeation, k, decreased as the initial water content became smaller, and it was larger in the higher freezing temperature. These effects of initial water content and freezing temperature are supposedly ascribable to the different structure of capillary pores left after sublimation of ice.

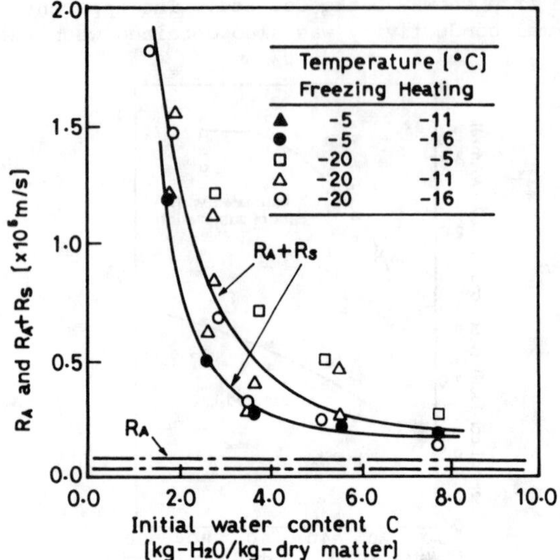

Fig. 7 Resistances of surface and evacuation system to vapor transfer

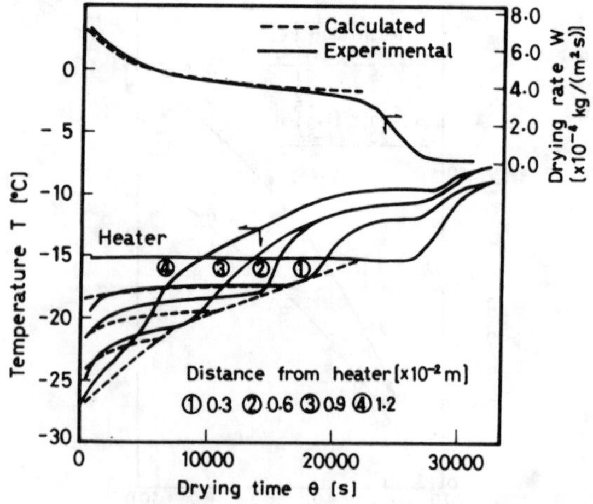

Fig. 8 Simulation of freeze-drying of milk (Water content C: 3.57 kg-water/kg-dry matter, Freezing temperature: -20°C, and Chamber pressure: 13 Pa)

Fig. 7 shows the effect of initial water content on the resistances to vapor transfer given by the surface of dried milk and the evacuation line, R_s and R_a. R_a was estimated by freeze-drying of ice where the thickness of dried part, X, and the surface resistance, R_s, were both zero. R_s became larger in freeze-drying of milk with decrease of the initial water content, while R_s was zero in freeze-drying of the frozen bed of water-glass beads due to no appearance of freeze-concentration. The surface resistance, R_s, was not always negligible, since the sum of R_s and R_a became equivalent to the resistance, X/k, at the thickness of dried part, X, corresponding to 10% of the total thickness, L. It is qualitatively known that the surface of the sample frozen in the unidirectional way exhibits a higher resistance to vapor permeation[1]. The surface resistance, R_s and the resistance, X/k, could be separated in this study.

3.4. Simulation

The drying rate and the temperature were simulated with the equations of U.R.I.F. model using the constants obtained. The equation of the retreating velocity, Eq.(6), was replaced in the simulation with the differential form.

$$\frac{dX}{d\theta} = \frac{W}{\varepsilon\rho_i} \tag{7}$$

It took some time for the experimental condition to arrive at the pseudo steady state. The thickness of the dried part, X was assumed to be δ (=0.0001 m) in this small initial time, and the following initial condition was used in the numerical calculation of X using Runge-Kutta's method.

$$\theta = 0 \ ; \quad X = \delta \tag{8}$$

Fig. 8 shows an example of simulation in a case where the concentrated milk is freeze-dried with a low heating temperature. The simulated drying rate agreed well with the one measured experimentally until a certain critical time, and the temperature of the frozen part was relatively well simulated. But the temperature of the dried part could not well be simulated with the model where the heat conduction from the direction of the sample surface was not considered.

4. TENTATIVE OPTIMIZATION OF HEATING TEMPERATURE

4.1. Decrease of Drying Rate

As shown in Fig. 8 the drying rate decreased remarkably at a certain drying time. The water content measured at this critical time was equal to that of CAS in case of Fig. 8. The critical water content was measured in the other experiments and shown in Fig. 9 indicating that the decrease of drying rate could occur before the sublimation of ice seemed to be completed. The drying rate of milk decreased earlier as the initial water content increased and as the heater temperature became higher.

There are several reasons suggested for the sharp decrease of drying rate. Toei, Okazaki and Asaeda presented the parameter to evaluate the condition under which the sublimating plane became unstable[2]. The critical time was caluladed using this parameter to be smaller than the time corresponding to a sharp decrease of drying rate. It means that the instability of the sublimating plane, if any, does not instantly lead to a fall of drying rate. Bellows and King proposed the amorphous viscosity theory of collapse and derived a characterislic time of the collapse which was assumed to occur by flow of concentrated amorphous solutin, CAS, into open capillaries left after sublimation of ice[3]. They considered a case where a sample was heated with radiant heat being applied from above, so that the collapse was assumed to occur in the matrix above the ice front. The collapse is, however, supposed to occur in the frozen part due to partial melting when a sample is heated from the back-face.

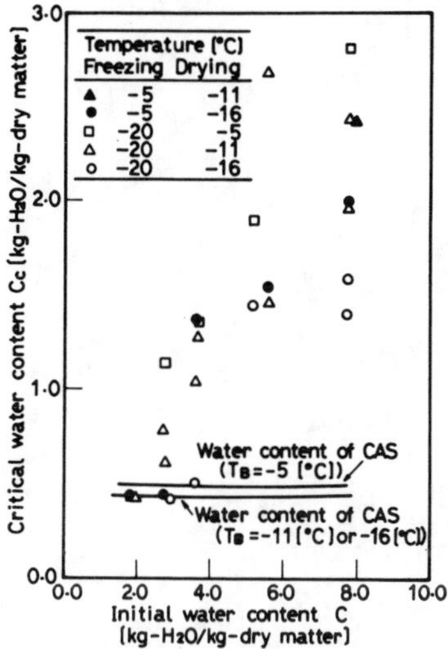

Fig. 9 Critical water content in freeze-drying of milk

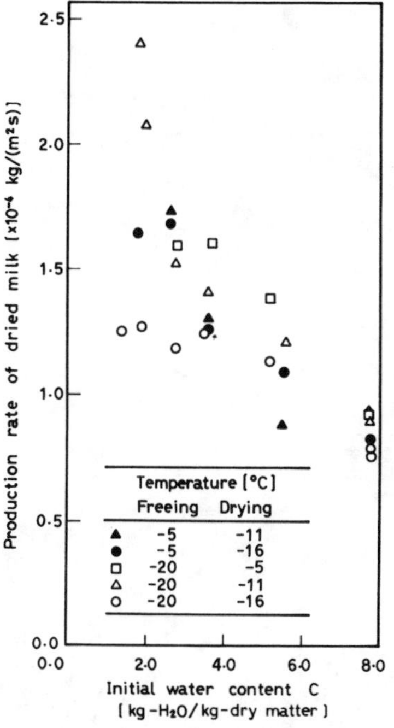

Fig. 10 Effect of initial water content on production rate of freeze-dried milk

The production rate of dried milk was calculated when the water content of product was set to be 0.2 kg-water/kg-dry matter. The results are shown in Fig. 11 indicating that the production rate increases with concentration of milk, although the extent of its increase is not so large as that expected simply by an inversely proportional relationship between the production rate and the amount of water to be removed. The drying rate of a constant-rate period decreased with concentration of milk, but the influence of this decrease to the overall drying rate was partly compensated by the extension of a constant-rate period. The longer constant-rate period is supposed to reflect the avoidance of collapse which is also favorable for such product quality as even dryness, aroma retention and rehydration.

4.2. Stepwise Change of Heater Temperature

It is important in freeze-drying by backface heating to keep a drying rate high without occurence of collapse. Thus the stepwise decrease of heater temperature was tried. The safe heater temperature in freeze-drying of milk is -16°C as read from the result of Fig. 9. The heater temperature was initially set at -11°C and lately decreased to -16°C as an example. The results of drying rate and sample temperature are shown in Fig. 11.

The drying rate observed before and after the step change was equal to the one of the drying conducted at each temperature. The same sharp decrease of the drying rate appeared after the heater temperature was lowered when the critical water content arrived at the almost same as that of CAS. It means that the sample of milk can be dried to the state of CAS without collapse in the stepwise decrease of heater temperature ealier

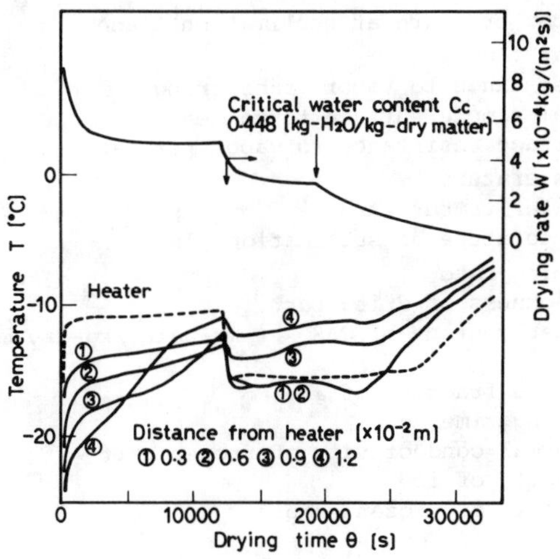

Fig. 11 Freeze-drying of milk by stepwise control of heater temperature (C=2.74 kg-water/kg-dry matter)

than in use of the constanly lower heater temperature.

CONCLUSIONS

The constant-rate period observed in freeze-drying of milk by back-face heating was analyzed with U.R.I.F. model where the surface resistance was considered. The surface resistance, the apparent heat conductivity of the frozen part and the vapor permeability could be obtained, although the apparent heat conductivity was deviated from the one measured directly. The drying rate and the temperature of the frozen part could be simulated with the equations of the model using the constants estimated.

The drying rate decreased with some sharpness at a certain drying time, and the reason was supposed to be due to partial melting of the frozen part elucidated by DSC-measurement of the state of water. Concentration of milk and use of lower heater temperature both contributed to extension of a constant-rate period. It was found that the stepwise decrease of the heat temperature was an effective method to decrease the water content of a sample to that of CAS in a shorter time than in use of the constantly lower heater temperature.

Acknowledgement

The authors thank to Prof. T. Fujita and Dr. Y. Maeda for their help in DSC measurement.

NOMENCLATURE

A	sectional area of sample	m^2
C	initial water content kg-water/kg-dry matter	
C_c	critical water content kg-water/ky-dry matter	
H	latent heat of ice sublimation	J/kg
k	vapor permeability	s
L	thickness of sample	m
M	sample weight	kg
M_o	initial sample weight	kg
P_c	vapor pressure in condenser	Pa
P_s	vapor pressure at sublimation plane	Pa
Q	heat flux	J/m^2
R_a	resistance to vapor transfering from sample surface to condenser	m/s
R_s	surface resistance to vapor permeation	m/s
T	temperature	K
T_b	heater temperature	K
T_s	temperature of sublimation plane	K
W	drying rate	$kg/(m^2 s)$
X	thickness of dried part	m
y	water content of CAS	kg-water/kg-dry matter
ε	volume fraction of ice	-
θ	drying time	s
λ	thermal conductivity of frozen part	W/(mK)
ρ_i	density of ice	kg/m^3
ρ_o	density of frozen sample	kg/m^3

REFERENCES

1. Quast, D.G., and Karrel,M., J.Food Sci., vol. 33, 170 (1968)
2. Toei,R., Okazaki,M., and Asaeda,M., J.Chem. Eng.Japan, vol.8, 282 (1975)
3. Bellows,R.J., and King,C.J., AIChE Symp. Series, 69(132), 33 (1973)
4. Nagashima,N., and Suzuki,E., Preprints of Symposium--Rheology and Physical Properties for Food Processing-, p.11 (Tokyo, 1982)
5. Kunii,D. and Smith,J.M., AIChE J. vol.6, 71 (1960)

DEVELOPMENT OF NEW REFRIGERATION SYSTEM AND OPTIMUM GEOMETRY OF THE VAPOR CONDENSER FOR PHARMACEUTICAL FREEZE DRYERS

Masakazu Kobayashi

Kyowa Vacuum Engineering, Ltd.
No. 18-8 Nishi-Shimbashi 2-chome, Minato-ku, Tokyo, 105 JAPAN

ABSTRACT

Features of a new refrigeration system called "Triple Heat Exchange Trap" system are that an array of the vapor condenser plates serves as a heat exchanger with three effects: heat transfer between water vapor and refrigerant(1), between water vapor and brine(2), and between refrigerant and brine(3). The condenser plate array also acts as brine cooler for shelf cooling. Since its development in 1980, this new system has been employed for more than 60 sets of phermaceutical freeze dryers, ensuring accurate temperature control, stabilized operation and high efficiency. In observing ice build-up patterns on the condenser and deterioration of the condenser capability through ice building, it was confirmed that comparing the observations with simplified mathematical model, both experimental and theoretical values are nearly equal.

1. INTRODUCTION

All freeze-dryers for pharmaceutical and biological-like products today use a refrigerated vapor trap or vapor condenser as a means to keep a desired vacuum in the drying chamber by trapping water vapor from the drying process. And prefreezing that precedes the drying process is performed by the cooling of the drying shelves. Prefreezing and water vapor trapping are two most important factors in a successful freeze-drying process, for which selection of an optimum refrigeration system has been of critical importance. Today most production dryers employ a refrigerating system based on compound refrigeration cycles (two stage) using a halocarbon refrigerant (R-502). Yet, it is the general opinion of both the users and the manufacturers that the development of refrigeration system required for freeze dryers has not kept pace with advances in other component area, and the reliability expectation of which is slightly below those in other fields.

Another important technical problem for the reliability and efficiency of the vapor-condensing system is the geometry of the refrigerated plates or coils. Though there is a certain consensus as to the area necessary for the refrigerated plates or coils, the design is still empirical. Consequently, there are various designs from various standpoints and not all necessary technical information concerning their performance is being supplied to the users.

This paper deals with two major problems of the refrigerated vapor-condensing system used in fine-chemical freeze-dryers for pharmaceutical and biological-like products, namely, 1) the design of the refrigeration system and 2) the geometry of the vapor-condensing surfaces (coils or plates). As for the refrigeration system, Powell [1] proposed a) a redundant-refrigeration system and b) a brine storage system, as possible solutions in his discussion on freeze-drying equipment. Rowe [2], on the other hand, proposed eight guidelines concerning the geometry and vapor-trapping capacity of the condensing system, while Mellor [3] summarized in his book the related studies so far reported. The author's study reported in this paper is different from the above-mentioned works in the following respects:

i) Up to now the vapor condenser was constituted as a heat exchanger between refrigerant and water vapor or between brine and water vapor. In the system developed in this study, however, an array of plates, which constitutes a vapor condenser, acts as the heat exchanger between refrigerant and brine, each of which also undergoes a direct heat exchange with water vapor independently of each other, and these heat-exchange plates are used at the same time as the brine cooler to cool the shelves.

ii) Guidelines for practical design are obtained through comparison of test results with a simplified mathematical model in which the ice build-up pattern on the cooling surfaces is determined by the relationship between the mass transfer resistance of water vapor flowing deeper along the cooling surfaces and the mass and heat transfer resistance of water vapor getting condensed on the cooling surfaces and reaching the refrigerant through ice layer.

2. NEW REFRIGERATION SYSTEM

2.1. Outline of New Refrigeration System

Advantages of new refrigeration system over the conventional systems. The standard performance specifications of today's biological and pharmaceutical freeze-dryers and the refrigeration systems to meet them have been outlined by Powell [1] as follows:

Shelf cooling (from +25°C down to -40°C within 1.0 - 1.5 hrs., final -50°C), vapor condenser load (from evaporating temperature -60°C for peak load to no-load temperature -70°C).

And a (R-502) 2-stage compression cycle is

prefered as a refrigeration equipment to meet the above specifications. In the refrigeration cycle capable of maintaining -60°C at peak load, however, if the load drops below half the peak level, operation at temperatures below the allowable limit of -70°C is required, while the vapor condensing load changes to no-load which is less than 1/10 of the peak load. Hence, the refrigeration compressor and the capacity control devices of a refrigeration system can hardly cope with such extensive variations. Also, the automatic control of refrigerant flow by a thermostatic expansion valve can not comply with such wide variations without very frequent readjustments made by the operator.

If a brine system is employed for the vapor condenser as the case with the shelves, then the condenser load and temperature can be controlled easily and accurately by mixing the brine with that of the shelves. And the operational conditions of the refrigeration system can be adjusted to meet the refrigeration compressor, thermostatic expansion valve and other specifications from the manufacturer, thus improving the operation reliability and the easy and accurate process control. However, to keep the condenser temperature at -60°C or -70°C, the evaporating temperature of the refrigerant must be lowered by 6 or 7°C. On top of that, the refrigeration capacity is reduced by more than 20 percent on account of the heat introduced by the pump circulating the brine to the condenser even when a brine temperature difference of 2°C between its inlet and outlet is permitted. Because of these two losses, the refrigeration equipment and energy consumption need to be both doubled. Furthermore, the evaporating temperature for condenser load drops below the lower limit of the R-502 2-stage compression cycle.

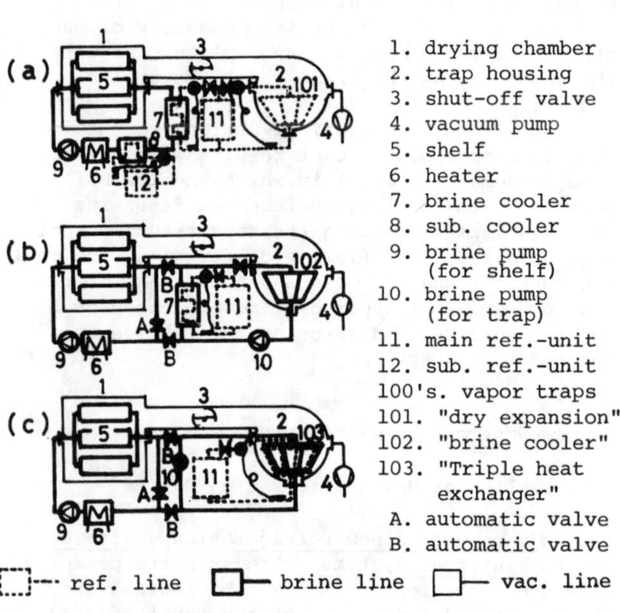

(a) conventional (dry expansion type)
(b) brine cooler type
(c) new "Triple heat-exchange" type

Fig. 1 Schematic diagrams of tipical freeze-dryers for pharmaceuticals and biologicals

The new refrigeration system. Fig. 1 shows the outline of the new refrigeration system [4], [5] & [6] together with two typical conventional systems. As is clear in Fig. 1, the new system is characteristic in that the "plate-type" refrigerant-brine heat exchanger is provided in the vapor-trap housing and its outer surfaces are used directly as the condensing surface. The array of plates acts not only as a brine cooler for shelf cooling, but also is itself a vapor condenser. For shelf cooling the brine circulates through a loop passing the shelves and plates, but during the drying process it is separated into two loops, one for the shelves and the other for the vapor condenser, and the adjustment of shelf temperature by cooling and the load-temperature control of the vapor condenser, as the case with the brine system, are performed by mixing the brines of the two loops through the operation of a thermostat.

As described above, the new refrigeration system has the same refrigerant-brine relationship as the conventional brine system, thus sharing the same advantages. However, the new system incorporates some new advantages in addition to the absence of need to double the refrigeration equipment size and energy consumption, a drawback of the conventional brine-system vapor condenser as shown in Fig. 1 (b).

2.2. Heat Exchange at the Plates

When the outer surface of a plate-type or coil-type refrigerant-brine heat exchanger provided in the vacuum chamber is used as the condensing surface, there are three choiced of heat exchange

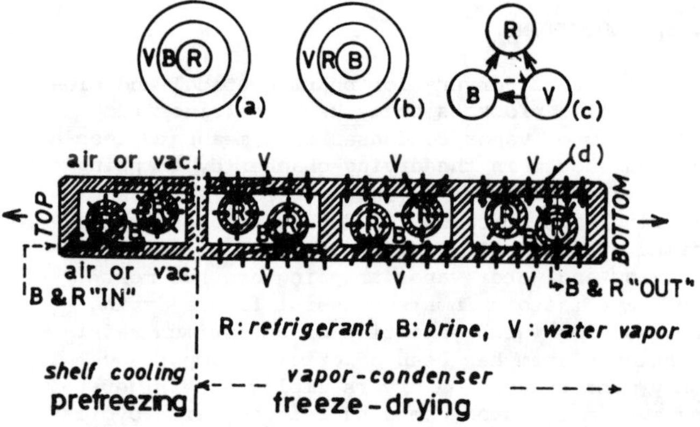

Fig. 2 Cross-section of "Triple heat-exchanger" and heat transfer relations among three mediums

concepts just as shown in Fig. 2. And the condenser load and temperature can be controlled by any of them. In Fig. 2 (a) Type, however, the heat exchange between refrigerant and water vapor, becomes indirect, thus causing a temperature loss, and if much brine is mixed for shelf-tempering, the condensing temperature becomes higher near the brine inlet and lower near the outlet with the result that the effective condensing surface is shifted to the outlet side if the flow rate is low. In Fig. 2 (b) Type, there is no temperature losses between refrigerant and vapor, but the wet refrigerant at the inlet must be vaporized into a suitable superheated gas at the outlet. Therefore, unless the thermostatic expansion valve is adjusted most accurately, or rather very slightly on the "open" side, the effective surface will be reduced to the neighborhood of the inlet. In the case of Fig. 2 (c), there exists direct heat exchange between any two of the three media of refrigerant, brine and vapor without mediation of another. This relationship represented by Fig. 2 (c) was realized by the structure shown in Fig. 2 (d). The figure shows schematically the heat exchange relations under the shelf-cooling load at pre-freezing in the top channel, those near the inlet under vapor condenser load in the next channel and those near the exit in the bottom channel. Since there is direct conduction between the refrigerant tube and the outer plate, under shelf cooling load the outer plate helps cool the brine as the fin and under vapor condenser load the refrigerant tube cools the whole plate uniformly, with the refrigerant and brine having the opposite tendencies between inlet and outlet supplementing each other.

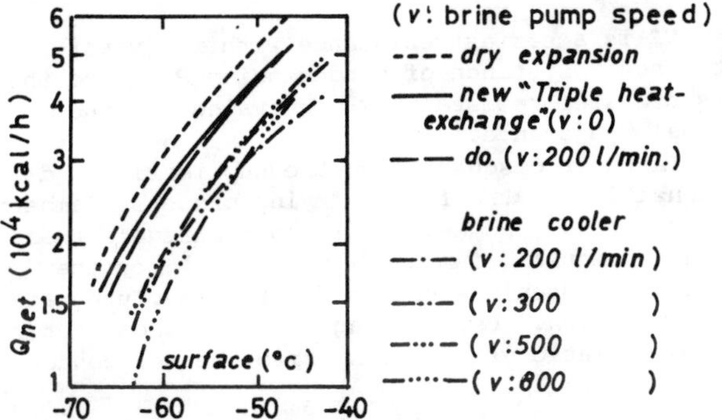

Fig. 3 Performances of typical condensers (calculated value)

The calculated values in Fig. 3 were derived on the assumption of steady-state heat conduction, through comparison of the refrigerant tube temperature with the performances of conventional brine cooler condenser and new system's condenser.

Then an economical plate structure and area capable of properly performing the two functions of shelf-cooling and vapor condensing was determined using the structure and heat flow model schematically shown in Fig. 2 (d). It was found as a result that the total condensing area of the plates is nearly equal to the total shelf area, the value of which satisfies the typical relationship between the shelf area and the condensing surface area pointed out by Rowe [2] and Powell [1]. The geometry of the plate arrangement was designed on the basis of the calculations discussed in the following section.

3. GEOMETRY DESIGN OF VAPOR CONDENSING SYSTEM

3.1. Mathematical Model

An actual freeze-drying process, as was pointed out by King [7], can be treated as a quasi-steady-state process with errors of within several percent.

The efficiency of vapor condenser is given by a small total pressure of different Δp between the water vapor pressure p_0 at the vapor inlet ($l = 0$) and the equilibrium water vapor pressure p_r^* at the temperature T_r of the circulating fluid to cool the condensing surface, or by a small total mass transfer resistance of the condensing system R_c which is defined by the following equation. (Q_{m0} the water vapor inflow rate)

$$\Delta p \equiv (p_0 - p_r^*) \equiv R_c\, Q_{m0} \tag{1}$$

From the viewpoint of water vapor flux to be caught after a progress of distance l, Δp is the sum of the pressure drop Δp_l of the water vapor flowing from $l = 0$ to l through the gap between condensing surfaces (Fig. 4), the pressure drop Δp_{con} at the vapor-ice interface at l and the water vapor pressure drop Δp_h equivalent to the temperature drop of the heat flow from the ice surface to the circulating fluid. If Δp_l^* is the sum of Δp_{con} and Δp_h at l, then $\Delta p = \Delta p_l + \Delta p_l^*$ is not dependent on l as long as T_r is constant in the direction of l. And if the mass transfer resistance per unit length of water vapor flow in the l direction is r, the condensing resistance (combination of mass and heat transfer) of the water vapor caught at l by the ice surface contained in the unit length of l, turned into heat flow and reaching the circulating fluid is r^* and the water vapor flow rate at l is Q_m, then

$$\int_0^l r\, Q_m\, dl - r^* \left(\frac{dQ_m}{dl}\right)_l = \text{independent of } l \tag{2}$$

Generally, r and r^* are the function of l if it is assumed that the water vapor flow Q_m flows in the l direction only and the heat flow resulting from condensation does not flow in the l direction.

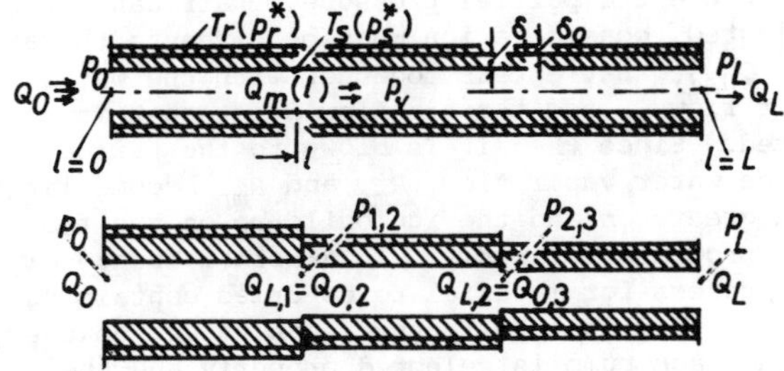

Fig. 4 Simplified Mathematical model (one dimentional) for geometry design of vapor condenser

However, if the total length L of the condensing system is divided and attention is given to the part where r and r^* are approximately constant, Eq. (2) can be written in a known form of differential equation.

$$\frac{d^2Q_m}{dl^2} = \frac{r}{r^*}\, Q_m \tag{3}$$

And the solution Q_m can be obtained. Now if the parts of the total length L are named in the vapor flow direction $L_1, L_2, \ldots L_n, \ldots L_N$ and the resistance when the water vapor flows from entrance to exit of each L_n without condensation is written $R_{m,n} = r_n L_n$, the resistance when the water vapor condenses uniformly over the cooling surface within L_n is written $R^*_{m,n} = r^*_n/L_n$, the water vapor flows at entrance and exit of each L_n are respectively $Q_{0,n}$ and $Q_{L,n}$, the condensation rates per unit length at entrance and exit of each L_n are respectively $q^*_{0,n}$ and $q^*_{L,n}$ and $\Delta P_n = (P_{0,n} - P^*_r)$ holds for the pressure $P_{0,n}$ at the entrance of each L_n, then the following equations hold for each L_n (Subscripts n are omitted in Eqs. (4) to (6).):

$$\Delta p = (R_m R^*_m)^{1/2}\{Q_0 \coth(R_m/R^*_m)^{1/2} - Q_L \operatorname{cosech}(R_m/R^*_m)^{1/2}\} \quad (4)$$

$$q^*_0 = -(\frac{dQ_m}{dl})_0 = (R_m/R^*_m)^{1/2}\{Q_0 \coth(R_m/R^*_m)^{1/2} - Q_L \operatorname{cosech}(R_m/R^*_m)^{1/2}\}/L \quad (5)$$

$$q^*_L = -(\frac{dQ_m}{dl})_L = (R_m/R^*_m)^{1/2}\{Q_0 \operatorname{cosech}(R_m/R^*_m)^{1/2} - Q_L \coth(R_m/R^*_m)^{1/2}\}/L \quad (6)$$

Also, the conditions for continuation at the joint between L_n and L_{n+1} are

$$Q_{L,n} = Q_{0,n+1} \ \& \ R^*_{m,n}\, q^*_{L,n} = R^*_{m,n+1}\, q^*_{0,n+1} \quad (7)$$

$Q_{L,N}$ is determined by the evacuation rate of the back pump (usually negligible). If $R_{m,n}$ and $R^*_{m,n}$ can be determined in a certain state, the condensation in each L_n: $w_n = (Q_{0,n} - Q_{L,n})$ will be determined. The ice build-up pattern on the condensing system and the change of R_c with time can be calculated approximately by repeating corrections to R_m and R^*_m according to the obtained results for each time interval Δt during which the increase of R_m and R^*_m caused by the ice thickness $\Delta\delta$ added by w_n is not too large.

Where the partial pressure of air can not be neglected, consideration must be given to the air flow Q_{0air} that enters together with the water vapor flow Q_0 and the air flow Q_{Lair} that is evacuated. Since the air is blown to the tail end by the water vapor flow, R_{mn} and R^*_{mn} become larger with greater n and the ice build-up on the front side progresses rapidly. In a normal freeze-drying process for material to be dried containing no solvent (e.g. alcohol) other than water, however, if the back pump is selected properly and the relative position of the condensing system and back pump inlet is proper, the partial pressure of air in the sublimation period can be neglected as long as air or gas bleeding is not performed for vacuum control. Therefore, finding a solution in neglecting permanent gases is also meaningful in practical applications.

In that case, calculations to determine the geometry of the condensing system and its housing to contain the increase of R_c within the allowable limits in the ice build-up process that realizes a desired condensation capacity W_{max} (kg per batch) must be performed in consideration of the condenser load characterized by the drying cycle: $Q_0[t]$; the performance of the back pump: $Q_L[P_L]$; and the performance of the incorporated refrigeration system: $T_r[Q_0 - Q_L]$, or the controlled $T_r[t]$. For this purpose, it is necessary to find the equation by which to determine or estimate R_m and R^*_m using the above geometry as well as such boundary conditions as Q_0, Q_L and T_r.

3.2. Equations for R_m and R^*_m

Based on Knudsen's experimental equation (known as an equation well applicable to the whole transition region from viscous flow to molecular flow in long cylindrical tubes), the following equation for resistance R_m of the vapor flow between parallel plates was developed by making necessary cylinder-to-parallel plate corrections and inserting the values of the viscosity and mean free path of water vapor as the functions of vapor temperature T_V. The plate width is B, the distance between the parallel plates is D, and the ratio of space occupied by the ice layer δ is $\xi = 2\delta/D$.

$$R_m = \frac{0.123\ (1 - 1.81\times10^{-3}T_V)\ T_V\ L}{\{P_V + 1.63\times10^{-2}/D(1-\xi)\}\{D(1-\xi)\}^3\ B} \quad [(\mathrm{N/m^2})\ \mathrm{sec/kg})] \quad (8)$$

R^*_m is a series resistance with the vapor-ice interface resistance of condensation R_{con} and the mass transfer resistance R^*_h equivalent to the thermal resistance.

In their discussion on the sublimation and condensation in the freeze-drying process, Lambert & Marshall [8] and Mellor [1] used Shrage's theory for a non-equilibrium condition (1953), on the basis of Knudsen's equation for the absolute rate of evaporation. Using Shrage's correction factor for condensation Γ_c, we can write the net condensation rate $\dot{m}_c$:

$$\dot{m}_c = \dot{m}_{s,abs}\left[\frac{P_V}{P_S{}^*}(\frac{T_S}{T_V})^{1/2}\ \Gamma_c - 1\right],$$

$$\dot{m}_{s,abs} = \sigma\ P^*_S(\frac{M}{2\pi R T_S})^{1/2} \quad (9)$$

where P_V: vapor pressure; T_S: ice surface temperature; $P_S{}^*$: equilibrium water vapor pressure at T_S; $m_{s,abs}$: absolute vacuum rate of sublimation from ice surface T_S; σ: coefficient of evaporation by Knudsen ($\lesssim 1$).

It is widely accepted that σ_{ice} is nearly 1.0 in the temperature range of below -60°C or -40°C. The correction factor Γ_c needed for Eq. (9), the author's own calculation according to Shrage's theory was adopted on condition that $\sigma \approx 1$ and $(T_S/T_V)^{1/2} \approx 1$).

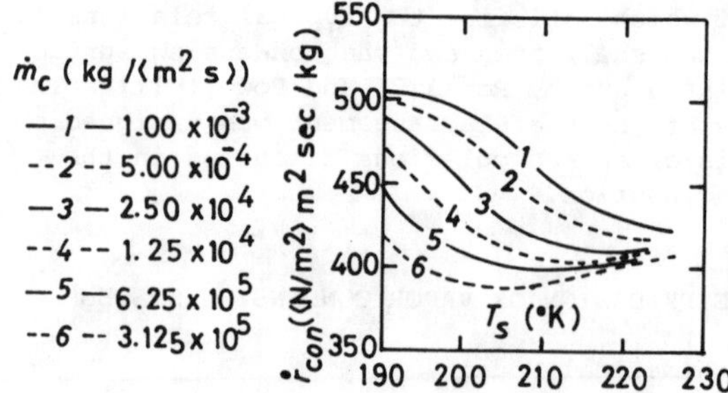

Fig. 5 Vapor-ice interface resistance per unit condensing area [calculated by eq. (9)]

Using Eq. (9), $\Delta p_{con} \equiv (p_v - p_s^*)$ at $\dot{m}_c$ can be calculated within the temperature and condensation rate ranges necessary for design computation (Fig. 5).

Under a quasi-steady state where the mass flow is equivalent to the heat flow and the pressure drop is equivalent to the temperature drop, then, the mass transfer resistance (R_h^*), equivalent to the thermal resistance (R_h), is written approximately

$$R_h^* = \frac{C_2 \; \Delta h_s' \; P_i^*}{T_i^2} R_h \tag{10}$$

where C_2 = 6 152.9(°K) for ice/vapor,

$\ell n \; P_i^* = C_1 - \frac{C_2}{T_i}$, C_1 = 28.939 for ice/vapor,

$\Delta h_s' = 2.9 \times 10^6$ [J/kg] for superheated water vapor,

$T_i = (T_s + T_r)/2$,

P_i^*[N/m^2]: equilibrium water vapor pressure at T_i.

Then we replace the heat transfer resistance between the outer surface of condensing plates and the circulating fluid by the equivalent ice layer thickness δ_0 and make a simplified expression of $\xi_0 \equiv 2\delta_0/D$ after the fashion of $\xi = 2\delta/D$. And using $\lambda_{ice} = 5.38\,(1-2.16\times10^{-3}T_i)$ [w/(°K m)] determining R_h^* from Eq. (10) and then adding R_{con}, we obtain

$$R_m^* = \{\overset{\circ}{r}_{con} + \frac{1.66\times10^9 \; P_i^* \; D(\xi_0+\xi)}{T_i^2(1-2.16\times10^{-3}T_i)}\}/2LB \tag{11}$$

$$[(\text{N/m}^2)\,\text{sec/kg}]$$

From Eqs. (4), (5), (6), (7), (8) and (11) and Fig. 5, we can obtain the changes of R_c with respect to T_r, Q_0, and Q_L.

3.3. Design Computation

Design computation was conducted for the first plant of the new "triple heat exchange" system. As a prerequisite, total condensing area of the plates was set at 1.8 to 2.0 m^2 to satisfy the need for a heat exchanger with shelf-cooling. Also in consideration of the internal structure of the plates, L_n=0.12 was adopted as the computation unit for the condensing system. Other conditions included 7.5 kW two-stage cycle (R-502) refrigeration equipment, $\bar{P} \approx$ (1.6N/m^2). $T_i \approx$ 213°K, and Q_L = negligible.

Figs. 6 and 7 show respectively the geometry employed, the computation results of parallel plates with equal inter-plate space for comparison, and the outline of the vapor condenser.

4. EXPERIMENTAL

4.1. Equipment

Principal experiments were conducted using equipment A and equipment B.

Equipment A. This is the first equipment made on an experimental basis using the new "triple heat-exchange" system. It employs the plates (6 channels) of the structure as shown in Fig. 2 (d) and has the geometry as shown in Figs. 6 and 7. As the refrigeration equipment a used (R-22) 2-stage compression type (11 kW) was initially utilized and later a new (R-502) 2-stage compression type (7.5 kW) was employed. (The refrigeration capacity was nearly the same for both types.). The vacuum pump system consisted of an oil rotary pump (300 ℓ/min) and oil diffusion booster. The brine was TCE.

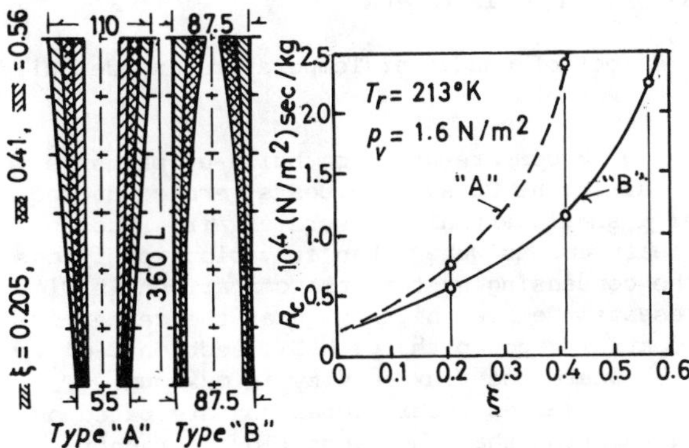

Fig. 6 Ice build-up patterns on the trap and deterioration of the trap capacity (calculated value)

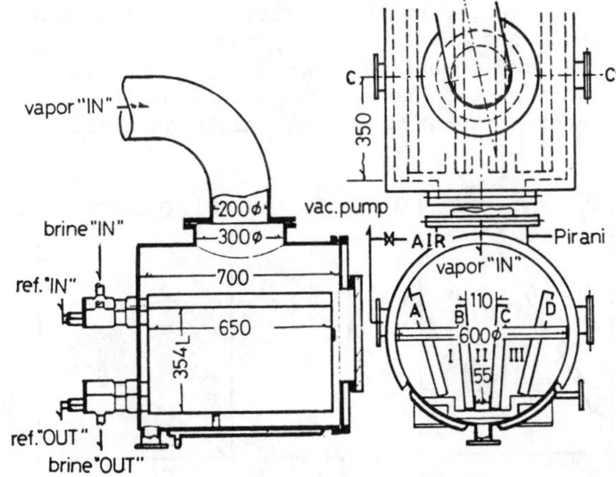

Fig. 7 Essential features of the vapor trap designed for equipment A

Equipment B. Though the plates and their arrangement were the same as equipment A, the water vapor inlet to the condenser was 0.2 m in diameter and the relative positions of the plate arrangement and the inlet were slightly different. The refrigeration equipment was the cascade type (R-13 & R-502) and the vacuum pump system consisted of an oil rotary pump (300 ℓ/min) and roots-type mechanical booster. The brine was TCE.

Others. Freeze-dryers of various specifications manufactured by Kyowa Vacuum Engineering, Ltd. between 1980 and 1983.

4.2. Methods

Detailed tests on the changes in ice build-up pattern and condenser efficiency were made using equipment A and equipment B, with an ice thickness measuring scale installed in the position of C-C' section in Fig. 7. And these tests were conducted under various condenser temperatures and partial air pressures.

The overall performance and functions of the triple heat-exchange system were tested not only with the equipments A and B but also during the manufacture, trial run and production run of all other equipments.

5. RESULTS AND DISCUSSION

5.1. Effect of Condenser Temperature on Ice Build-Up Pattern

Fig. 8 compares the ice build-up patterns under different T_r's. The tests were conducted under a small partial pressure of air. Though the measurements were taken at various positions of the condensing system, the comparison in Fig. 8 represents the ice thickness δ at the center of the central vapor path (path II, section C-C' in Fig. 7) where the flux density is maximum. The calculated values under comparison are based on the assumption that the vapor flow (per unit width) Q_0' to the observation point is constant. During the experiment, the total vapor flow to the condenser Q_{total} is held constant. However, $Q_0'/\bar{Q}_0'$ which shows the degree of concentration of the flux density to the center, is not constant but reduces generally. $Q_0'/\bar{Q}_0'$ can not be estimated from the one dementional model. Q_0' for this calculation needs to be estimated from the final ice layer patterns observed. Although there exists some difference between δ_{exp} and δ_{cal} on account of the simplified assumption (Q_0' = const), a good agreement is observed between the test values and calculated values except at the plate end at T_r=197°K where an air influence is not negligible.

Fig. 9 shows the rise of pressure P_0 at the condenser head with the increase in the amount of condensation $W = Q_{total}\ t$ (estimated from δ_{exp} in Fig. 8). If T_r is too low, the vapor flow will be blocked by the ice layer that concentrates on the condenser head, and as a result P_0 rises rapidly for small W. To achieve a maximum W under $P_0 \leq P_{permissible}$ (e.g. 2.0 N/m^2), the vacuum control by controlling the condenser temperature is effective that keeps T_r high at the limit where $P_0 = P_{permissible}$.

However, where the product amount is small and there is no fear of blocking, the drying chamber pressure can be controlled by the vapor flow throttle device; T_r and vapor-trap housing pressure can be kept at the minimum levels; and at an emergency like power failure the drying chamber pressure can be maintained at a safe level for long time. At the end point (at t=19hr) of [Exp. II-1], after turning off all the power to the equipment, the shelf temperature drop was small and sublimation progressed at an almost constant rate, but the chamber pressure remained 10.0 N/m^2 for 27 min. This is due to the condensing plates of the new system which has a heat capacity several times as large as that of other systems. The equipment B used for the tests has the condensing plates with heat capacity of 73x10^3 J/°K.

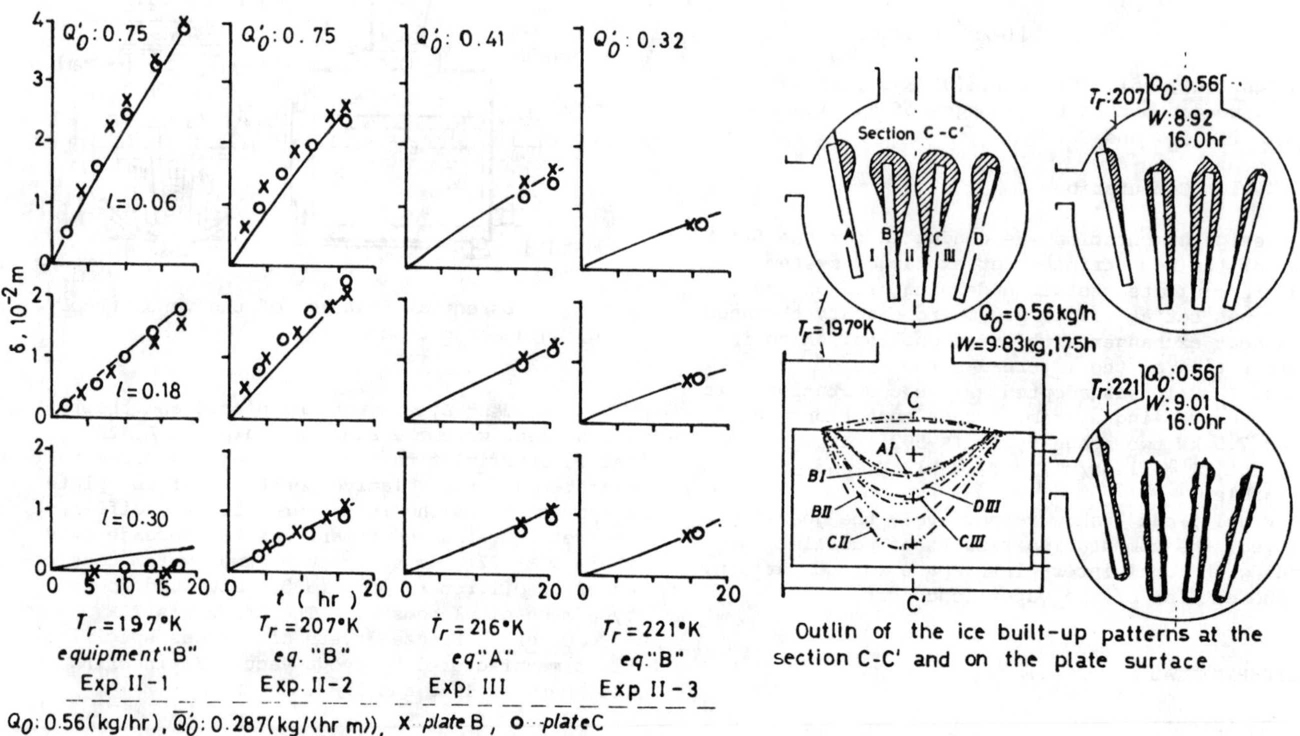

Fig. 8 Effects of condenser temperature on ice growth at the center of the vapor path II and various ice build-up patterns

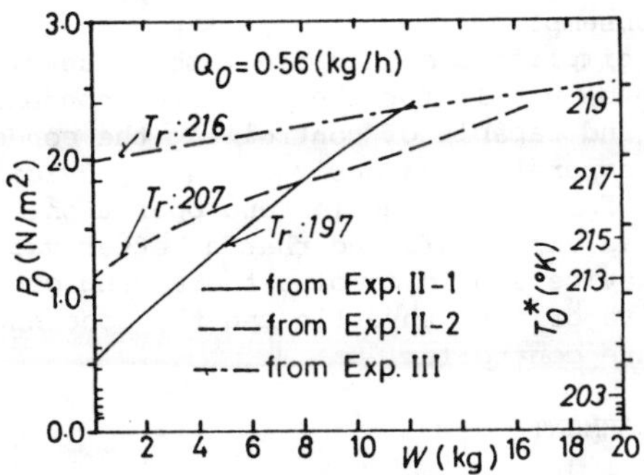

Fig. 9 Effect of condenser temperatures on deterioration of the condenser capacity with ice layer growth

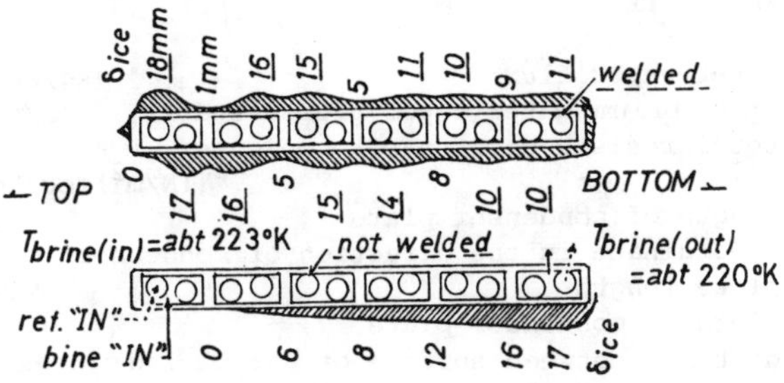

Fig. 10 Effect of direct heat conduction for "Triple heat exchanger" (between refrigerant pipe and outer plate)

5.2. Effect of Air on Ice Build-Up Pattern

Fig. 11 compares the ice build-up patterns under different effect of air. Air or non-condensable gas, as it accumulates in the vapor-trap housing, impedes the vapor flow with the result that the vapor transfer resistance and condensing interface resistance increase. In consequence, the ice build-up concentrates on the condensing surface near the inlet. When the oil-sealed rotary pump only is used with the booster pump stopped [Exp. I-2], there appears to be some effect of air. The effect becomes marked if the chamber pressure is controlled by a vacuum-bleed device. In [Exp. I-4], where there is the blocking by ice layer, the vacuum in the drying chamber is not recovered satisfactorily even if the vacuum-bleed device is closed.

5.3. Performances of New Refrigeration System

The new refrigeration system was incorporated into more than 60 production-scale freeze-dryers manufactured between '80 and '83. Their sizes

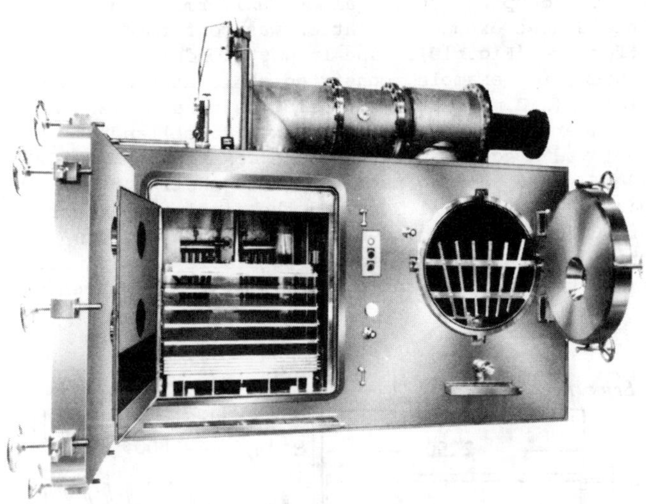

Fig. 12 (Photo) "Kyowa" Freeze Dryer, model: RL-2412BS (Shelf area: 12 m², vapor trap capacity 240ℓ)

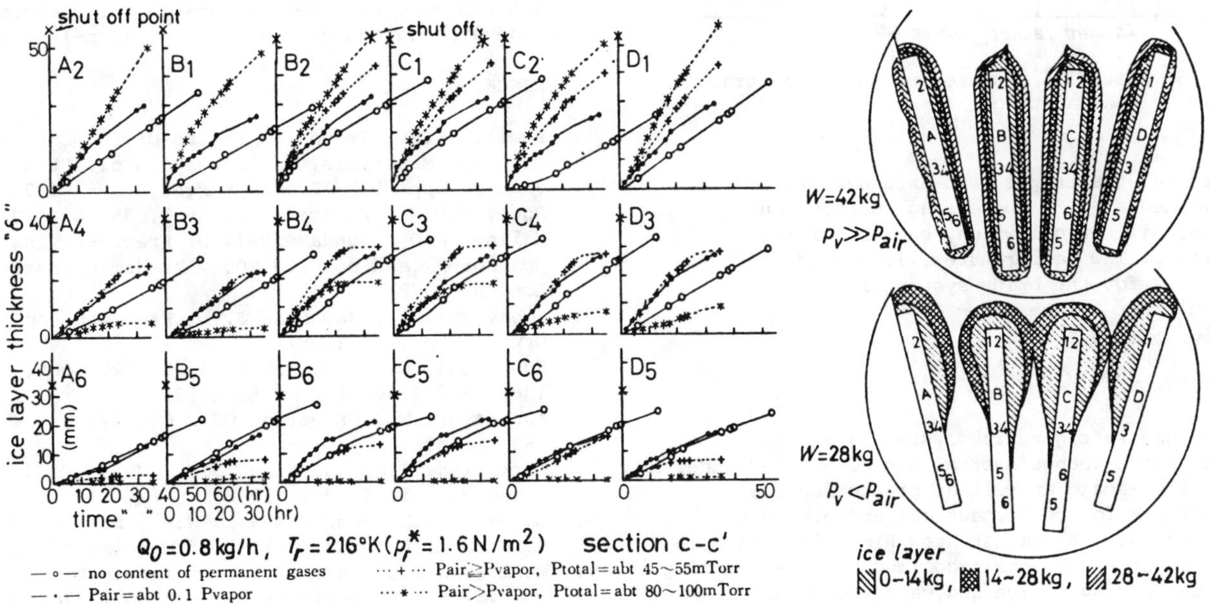

Fig. 11 Effects of partial air pressure on ice layer growth on the array of plates

varies from 1 m^2 to 43 m^2 in shelf area. Some of them are so designed that the vapor trap housing can also be administrated from the bio-clean room side as shown in Fig. 12. The refrigeration unit on the same scale as the conventional dry expansion system not only provides equivalent shelf-cooling rate, shelf-tempering performance and condensing efficiency, but also displays greatly improved reliability, ease of operation and controllability.

In [Exp. II-3], tests were conducted to keep T_r=221°K by brine mixing without performing capacity control of the refrigeration compressor. Accordingly, the temperature difference between the brine inlet and outlet turned out very large. Where the refrigerant tube was not welded to the outer plate, the ice build-up was marked and concentrated on the end area of the plate. In the structure as shown in Fig. 2 (d), however, the ice build-up was spread almost uniformly and the triple heat exchange relation was confirmed to be effective [Fig. 10]. The largest condensing system, for example, consisted of 10 heat-exchange plates (2.2 mB x 1.06 mL) held inside a horizontal-cylinder type housing (1.8 mϕ x 2.25 mL) and employed a screw-type two-stage compression (R-502) 90 kW. Its cooling rate of loaded shelf (43 m^2) was 25°C to -40°C/1.3 hr and the final temperature of -60°C, and the condenser temperature was controlled within ±0.5°C in the desirable temperature range between -55° and -65°C. $W_{max} \geqq$ 950ℓ/batch.

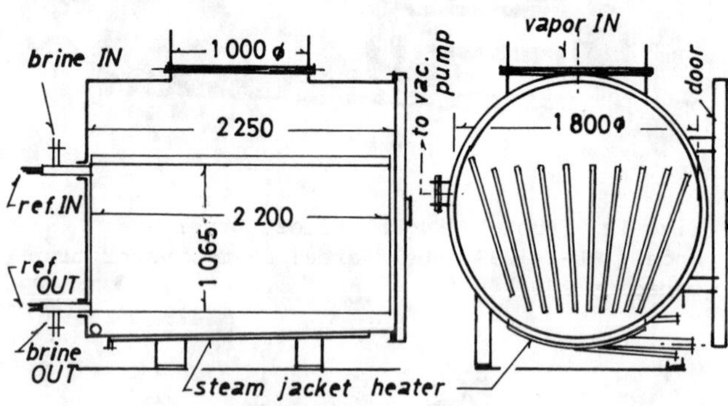

Fig. 13 Essential features of the vapor trap for large pharmaceutical freeze-dryers

Furthermore, the triple heat-exchange system accomplished a -65°C condensing surface control using the silicon oil which, because of the high viscosity at low temperatures, is considered inappropriate for the brine system of -60°C or below in other systems.

6. CONCLUSION

The new refrigeration system called the "triple heat-exchange system" can be applied widely to the pilot-scale as well as production-scale freeze-dryers for pharmaceutical and biological-like products. It demonstrates high efficiency in all the functions of shelf-cooling, shelf-tempering during the drying period and vapor condensing. And it realizes high reliability, easy operation and accurate control without the sacrifice of a large refrigeration compressor and excessive energy consumption.

The simplified mathematical model developed proved to be usable for the design of condenser geometry and capable of controlling the condenser surface temperature as desired, thus providing a yardstick for optimum design and operation.

It was also confirmed that a better vacuum control device should combine the advantages of both the vapor flow throttle and the condenser temperature controller.

ACKNOWLEDGEMENT

The author is grateful to his colleagues and users of "Kyowa" freeze dryers for their valued assistances in preparing this presentation.

NOMENCLATURE

- Q_m water vapor flux [kg/sec]
- P_x^* equilibrium water vapor pressure at T_x [N/m^2]
- R_c total resistance of vapor condenser [(N/m^2) sec/kg]
- L length of condenser plate [m]
- l co-ordinate in the direction of condenser plate length [m]
- B width of condenser plate [m]
- D distance between surface of paralleled plates [m]
- δ ice layer thickness [m]
- r vapor transfer resistance per unit length "l" [(N/m^2) sec/kg m]
- r_{con} vapor-ice interface resistance per unit length "l" for condensation [(N/m^2) m sec/kg]
- $\overset{o}{r}_{con}$ vapor-ice interface resistance per unit condensing area [(N/m^2) m^2sec/kg]
- r_h^* mass transfer resistance per unit length "l" equivalent to heat transfer resistance for condensation [(N/m^2) m sec/kg]
- r^* (=$r_{con}+r_h^*$) condensing resistance per unit "l"
- R_m (=r L) vapor transfer resistance from $\ell=0$ to L without condensation [(N/m^2) sec/kg]
- R_m^* (=r^*/L) condensing resistance equivalent to mass and heat transfer resistance for uniform condensing on the condenser [(N/m^2) sec/kg]

REFERENCES

1. Powell, H.R., Trends in Freeze-drying Equipment and Materials, in Develop. biol. Standard. vol. 36, pp. 117-129 (S. Karger, Basel 1977).
2. Rowe, T.W.G., Fd Trade Rev. vol. 35, 42 (1964)
3. Mellow, J.D., Fundamentals of Freeze-Drying pp. 66-67, and pp. 203-206. Academic Press, London (1978)
4. Kobayashi, M., Japanese J. of Freezing and Drying. vol. 27 (1981).
5. Kobayashi, M., U.S. Patent 4,353,222 (Oct. 1982) and 4,407,140 (Oct. 1983).
6. Kobayashi, M., UK Patent GB 2 061 474 B (Sep. 1983).
7. Zarkarian, J.A. and King, J.C., J. of Food Science. vol. 43,998 (1978).
8. Lambert, J.B. and Marshall, W.R., Nat. Acad. Sci., Nat. Res. Council, p105 Washington (1962).
9. Mellow, J.D., Fundamentals of Freeze-Drying pp. 48-51. Academic Press, London (1978)

DEFROSTING OF INTERNAL VAPOR CONDENSER DURING FINAL DRYING AND ITS APPLICATION TO FOOD PRODUCTION FREEZE DRYERS

Masakazu Kobayashi

Kyowa Vacuum Engineering, Ltd.
No. 8-18, Nishi-Shimbashi 2-chome, Minato-ku, Tokyo 105, JAPAN

ABSTRACT

A vapor condenser located internally in the drying chamber without shut-off valve having many advantages, it is believed that the condenser must be regenerated while the dryer is not in service.

It is, however, interesting to note that ice temperature for defrosting must be 0°c and its equilibrium vapor pressure corresponds to the relative humidity of 6.5 to 2.5% which is at the final drying temperature of foods, 45°c to 65°c, and is sufficiently low in humidity as compared with the water activity of 0.2 suitable for storage of dried foods. If, therefore, ice build-up on the condenser is melt off by use of low temperatured (near 0°c) vacuum steam, without affecting dried product with spray or mist incidental to defrosting, the condenser will be completely regenerated in the final drying stage. Based on this concept, we have developed a new defrosting system for the internal condenser.

In the final drying stage, small amount of defrost water is poured in the bottom of the drying chamber which is steam jacketed, and is heated to generate vacuum steam which defrosts the condenser coils set on both sides of the shelf stack. The moisture content of the dried food taken out from the chamber shows very satisfactory value and has no difference from that of dried products processed with the dryers with external condenser.

Food production freeze dryers of batch type with this new defrosting system are already in operation in large numbers.

1. INTRODUCTION

Freeze-dryers capable of processing far larger amount of product at low cost than those for pharmaceutical and biological products have been realized and are finding more and more application in the food industry.

Typical of such freeze-dryers are the plants for freeze-drying coffee, and various continuous plants were developed one after another in the latter half of 1960s. However, agitated freeze-drying for bulk materials, as was pointed out by Lorentzen [1][2] was not suitable for coffee and other food products because of such problems as fine powder caused by abrasion and its entrained scattering into the vapor flow. On the other hand, the freeze-drying of product in trays found popular use first as the tunnel type and then the compact moving tray type continuous plants [1], and is today being used for mass processing of coffee, meat and other single products.

Batch plants, on the other hand, process a variety of uncooked and cooked foods in various shapes and compositions, such as vegetables, meat, fish and other marine products, noodles and seasonings. They are processed simultaneously or undergoing frequent changes. According to Liebman [3], the broad application of freeze-drying in the food industry, with one massive exception of coffee, has not materialized despite the commercial expectation in the early 60s. In Japan, however, freeze-drying of varieties of foods other than coffee has come through a unique development since the early 70s. The total scale of the plants, excepting those of coffee, jumped from 1 500 m^2 in total tray area at the end of 1972 to about 9 500 m^2 at the end of 1982 and is still expanding. Such freeze-drying covers an increasing variety of food items, various shapes of same materials and those in the overlapped areas of food, biological and pharmaceutical products.

Continuous plants are considered suitable when the product to be dried is processed continuously at a constant rate for a period at least 20 times longer than the drying time of the product. Similarly, batch plants, which dry multiple or single variety of food in batches, are preferred when products to be dried are supplied intermittently in certain amounts. Therefore, both the continuous and batch types of freeze-drying must be developed with equal emphasis on volume freeze drying.

This paper discusses batch-type volume freeze-dryers. Specifically, it deals with the development of a new vapor condensing system. Ginnette, Kaufman [4] cited that 40 to 50% of the freeze-drying equipment costs is associated with refrigeration system for vapor trapping, while Lorentzen [2] attributed 65% of the energy consumption of freeze drying to the refrigeration system. With batch plants, especially, a large-capacity refrigeration is required because of the peak load in the initial stage of drying process, and in the latter half of the process a marked drop in energy efficiency will occur in operation of the large-capacity equipment under low load. Hence, the equipment cost for vapor condensing and the reduced energy consumption are two major factors in a successful batch plant.

Lorentzen [1] also discussed the vapor condensing system for industrial plants for foods as follows. The efficiency of the vapor trap is shown by a small total temperature difference ΔT between the saturation temperature for water vapor at the pressure in the drying chamber and the evaporation temperature of the refrigerant. In the conventional type equipment, the vapor condenser is built in a separate vacuum chamber connected to the drying chamber through a large pipe. The system efficiency is often low. Because there exist a high vapor resistance Δp of the connecting tube and an excessive ice thickness ΔT_{ice} due to the frontal ice build-up on the condensing surface near the connecting tube, more than one vapor condensing chamber can be provided through the shut-off valve, then frequent defrosting can be done alternately. This will reduce ΔT_{ice}, but raise the equipment cost. In another type, a vapor condensing system can be installed inside the drying chamber, for instance, in parallel on both sides of the shelf stack or in series behind it. In this type, Δp becomes negligibly small, and the ice layer thickness can be uniform, particularly in the former. Conventionally, however, the defrosting of this internal vapor condenser had to take place between the cycles. And it was believed that this gave a considerable reduction in drying capacity because of the time lost for defrosting. Especially with the type of vapor condensing system arranged on both sides of the shelf unit, it was used together with a mobile tray heater assembly, which was pulled out of the chamber at defrosting. This is because water flow or spray from top was conventionally used for defrosting.

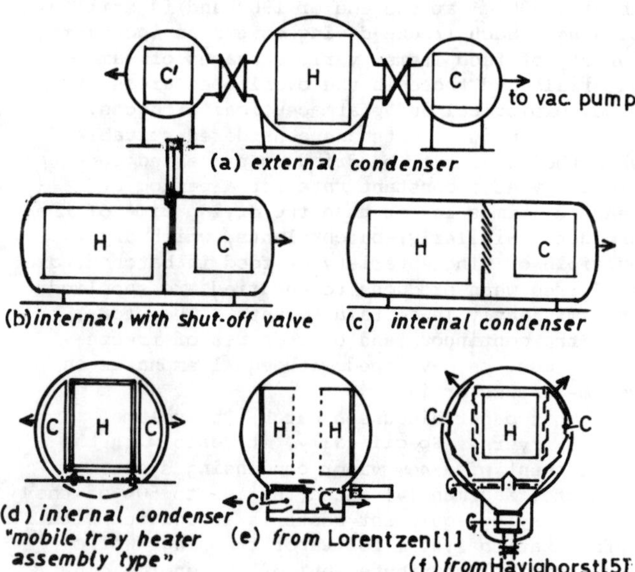

Fig. 1 Various types of freeze-dryers.
H: heating shelf stack, C: vapor condenser

Lorentzen [1], Havighorst [5], and King [6] have dealt with solutions to the disadvantages involved with the defrosting of the internal vapor condenser. In one of the solutions [1], (Fig. 1) is built into the bottom of the cabinet and divided into two sections. It is shut off from the drying chamber proper alternately by an overhead travelling gate, thus repeating the operation and defrosting alternately. Below each condenser section, there is a common bottom water tank via each shut-off valve. The water temperature is kept at about 20°c. Defrosting is performed by the vapor from the bottom tank. Therefore, this type is internal, but can be called a separate type. It is advantageous in that a thin wall structure can be employed because each section and bottom water tank are inside a common vacuum chamber and there occur an equilibrium vapor pressure of 20°C and a pressure difference of about 1/40 atm at the partition and gate valve between them. In another solution [5], a vertical plate type condenser of the brine-cooling system is provided on both sides of the shelf stack unit. And in the condensing plates, one after another warm brine is circulated momentarily and ice layer is peeled down onto the conveyor installed on the chamber bottom and discharged from a vacuum lock with a crasher. Still another solution [6] is a continuous type in which brine (calcium chloride and lithium chloride) is circulated in a condensing chamber, condensing the water vapor directly onto the surface of the falling brine film. The above alternatives can all be used in continuous processes.

Unlike the above-mentioned solutions, the method proposed in this study is characteristic in that in the final stage of the drying cycle the chamber pressure is held very close to the equilibrium vapor pressure of 0°C water and defrosting is carried out without impeding the final drying. Its use, therefore, is limited to batch plants only. However, since its structure is so simple as to have no thin walls, gates, or conveyor, etc., an available condensing area 5 to 7 times as large as that of the internal alternating closure defrosting system can be layed out in the chamber of an equal volume, thus solving the disadvantage of ΔT_{ice}. On the internal condenser (without shut-off valve) Lorentzen [1] made a comment "From an energy point of view it gives good conditions for the vapor flow (low Δ_p), but with only one deicing per cyele ΔT_{ice} may be relatively high". However, ΔT_{ice} is in proportion to the product of ice layer thickness δ_{ice} and heat flux density $\dot{q}$, while both δ_{ice} and $\dot{q}$ are in inverse proportion to condensing area A. Further, $d\delta_{ice}/dt$ is proportional to $\dot{q}$ which is very large in the initial stage of drying cycle. Thus, when the area A is 5 to 7 times larger, only one defrosting per cycle ΔT_{ice} is smaller than very frequent (25 to 49 times per cycle) defrosting ΔT_{ice} and that the needs an additional energy for recooling the condenser.

2. NEW DESIGN OF INTERNAL CONDENSER DEFROSTING SYSTEM

2.1. Basic Design Principle

The basic design principle of the internal condenser defrosting system in this study is based on the following conditions:

1. The water activity of dried food with optimal moisture content is about 0.2 at room temperature and more than 0.3 at permissible final temperature of the freeze-drying T_d 45°c to

65°c for a couple of hours. Accordingly, the saturated vapor pressure p_s of dried product at the final stage of the secondary drying (desorption) period is about $2.9x10^3 N/m^2$ (22 Torr) to $7.5x10^3 N/m^2$ (56 Torr).

2. Ice temperature for defrosting must be 0°c and its saturated vapor pressure is $6.1x10^2 N/m^2$ (4.6 Torr). Therefore it is theoretically possible to defrost the internal condenser without interruption while keeping the drying chamber pressure p_0 at very nearly $6.1x10^2$ (or 4.6 Torr).
3. Since p_0/p_s = 0.2 to 0.08, the vapor condenser can be regenerated in the drying cycle without much extending the freeze-drying process time if defrosting is carried out in 0.5 to 1.0 hr under p_0 of very nearly $6.1x10^2$ (4.6 Torr). For, roughly speaking, actual desorption rate Q_m at p_0 divided by maximum desorption rate Q_{abs} under absolute vacuum: $Q_m/Q_{abs} \approx (1-p_0/p_s)$.
4. An optimum method for p_0 to be very nearly $6.1x10^2 N/m^2$ must be to generate a uniform vacuum steam from a sufficiently wide water surface in the bottom of a horizontal cylinder type drying chamber. 1 kg of vapor can melt 6 to 7 kg of ice.

2.2. Application to Food Freeze-Dryers

After some experiments with a pilot plant, this new defrosting system was applied to freeze-dryers for foodstuff of production scale.

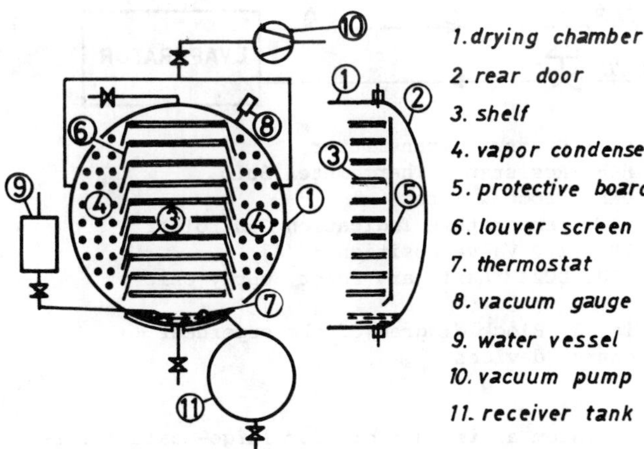

Fig. 2 Schematic diagram of the new defrosting system

There are already 16 freeze-dryers using this system. Half of them feature a tray area of more than 100 m^2 and vapor trap capacity of more than 2 ton/batch per unit drying chamber, whereas the largest one features a tray area of 226 m^2 and vapor trap of 4 ton/batch per unit chamber.

Design of vapor condenser with the defrosting system. Fig. 2 shows the schematic diagram. The operation, which is automatic and starts at a time signal about 0.5 hr before the end point of the drying cycle, is as follows: 1 The drying chamber pressure is raised to about $6.6x10^2$ to $1x10^3 N/m^2$ by the bleeding air. Evacuation is continued by the vacuum pump system. 2 Simultaneously with the stop of bleeding air, water is introduced in the bottom of the chamber.
Then the bottom jacket is heated by steam and the vacuum steam generated condenses on the ice layer of the condensing system. 3 The water from the melted ice overflows into the receiver tank with the water surface held at a constant level. As the water surface temperature reaches 2°C, heating is stopped and the condenser cooled again by the refrigeration system. 4 As the vacuum is broken by bleeding air or N_2 gas and the chamber pressure reaches 1 atm, the drain valve opens to drain all the water. After the automatic operation finished the drying chamber door is opened manually.

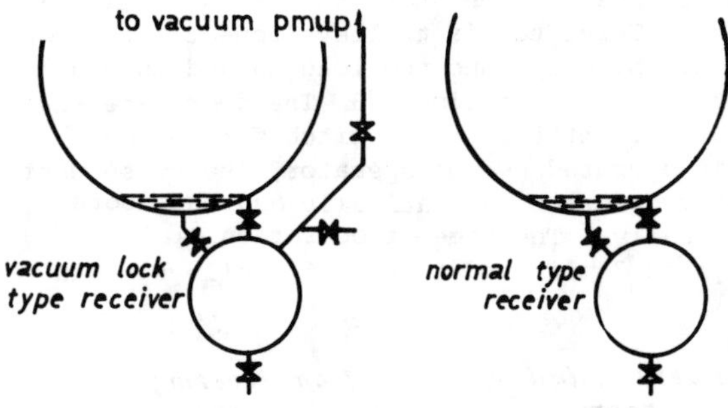

Fig. 3 Additional devices to the defrosting system shown in Fig. 2

There may be very rare cases where product of very low final water content is required, which is meaningless unless the product taken out needs to be kept in an environment with very low relative humidity (far below 20-30%). But if this is what is really desired, a shut-off valve can be set on the overflow piping and the chamber drain nozzle linked to the water receiver as shown in Fig. 3, whereby whole defrosted water is to be removed under vacuum from the drying chamber to the water receiver and the final drying can be continued.

The shelf stack unit is partitioned by a louver type screen from the condensing system on both sides for the purpose of thermal insulation and protection of product from the mist during defrosting, and protective boards are also provided at the front and back of the shelves.

Chamber feature and tray transport design. Optimum tray handling system and chamber features should be selected in consideration of main products to be handled and type and scale of the production to be planned. Kyowa Vacuum Engineering, Ltd. supplies batch plants for food industry on three to four standard designs, each of which have proponents for the design configurations. To satisfy these configurations, the condenser with the new defrosting system, like the case of Fig. 1(c), can be located in series with and in the back of the shelf unit in the horizontal cylinder type chamber. However, the tray slide system configurations developed to suit the new system dealt with in this study are as follows. Fig. 4 shows it schematically, the cross section of which being in agreement with that in Fig. 2 or Fig. 3.
1 The front door (loading side) opens to the

front on the floor rails and a tray pusher is collapsed into the back of the door. The rear door (unloading side) is suspended type and opens in rotation. The trolley with trays filled with prefrozen product carried manually in suspension from the overhead monorail is set between the front door and chamber, and a vacant trolley is set at the opposite end of the chamber. 2 By the operator's switch operation, the trays of frozen product are transferred from the trolley to the Teflon rail on the shelf by means of the pusher and from the opposite end the trays of dried product are pushed out onto a vacant trolley. The trolley vacant after the loading operation is sent to the unloading side manually to be set for receiving the next dried product. At the same time, the trolley of next frozen trays is set on the loading side. Three to six trolleys repeat this operation and by one round the loading and unloading complete at the same time. 3 The doors are shut immediately. If the start switch for the next cycle is operated by the operator, the subsequent process progresses automatically and the doors open manually. The time of operation of 3 trolleys is 3 min. and that of 6 trolleys 6 min.

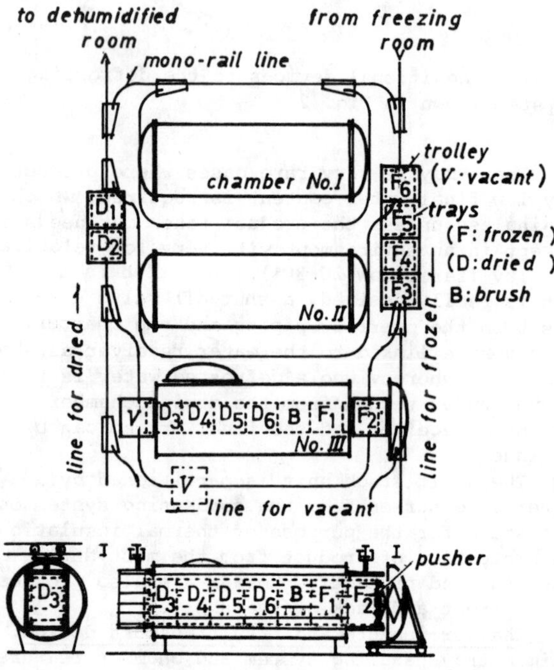

Fig. 4 Schematic diagram of freeze-dryer with tray slide system

In another layout, the unloading side is separated by a wall and protrudes into the dehumidified room. And the trolleys after the loading operation are recycled to the preprocessing room while the unloading trolleys are always kept in the dehumidified room.

<u>Cleaning</u>. In the next cycle, when one product is changed to another, the panels with brushes on top and bottom faces and on both sides (same width as tray, 2 to 3 times longer) are loaded on the next No. 1 trolley, and are to be pushed in while the frozen product trays are to be sent from No. 2 trolley. Operation of one trolley and 1 min addition of work will, therefore, sweep all the shelves with the chips and dust having fallen to the shelf surface being discharged together with the panel with brushes gently from the unloading side. When the operation is suspended and the chamber is not in service, washing with high-pressure water jet can be made as this type has no obstacles to the washing water flow.

<u>Heating system</u>. The radiant heating shelves are made from extruded aluminium and have been anodized. A Teflon tray rail is located on both sides of the shelves and the trays are transported on the rails. Heat carrier is circulated in the shelves, the inlet and outlet alternated between the odd-numbered and even-numbered shelves.

<u>Refrigeration system</u>. In Japan, which has frequent earthquakes, adsorption refrigeration plants using ammonia are not permitted. Thus, one of the following two types of refrigeration system is selected:

a) Refrigerant flooded, forced circulation system
b) Dry expansion system with super-heat control devices

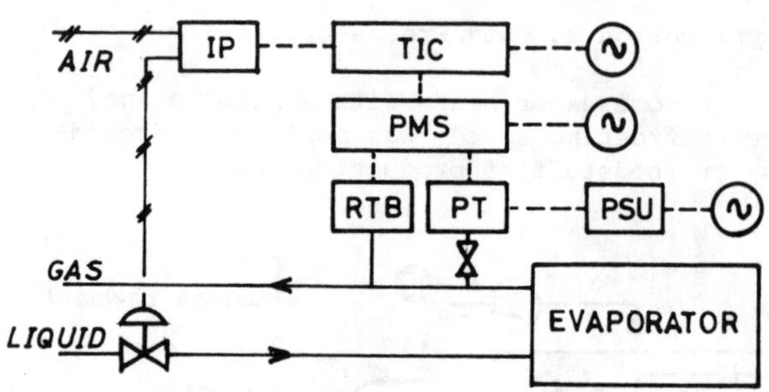

PT : Pressure Transmitter
RTB: Resistance Thermometer Bulb
PMS: Process Meter
TIC: Temperature Indication Controller
IP : I/P Valve Positioner
PSU: DC24V Auxiliary Power Supply Unit

Fig. 5 Block diagram of the superheat control devices

System a) is suitable for large-scale multi-chamber, staggered operation plants (e.g. total tray area 1 360 m^2, 6 chambers), In most cases liquid refrigerant is sent by the circulation pump from a common accumulator to condenser coils of each chamber.

On the other hand, System b) is recommendable for a middle scale plant with single or two/three chambers (e.g. total tray area 100 m^2 to several hundred square meters). This system developed by MAYEKAWA MFG. Co., Ltd. has been employed for freeze dryers through joint study with them.

The principles of operation are shown in Fig. 5. The system is applied to control the Remote Actuated Valve (R.A.V.) to keep refrigerant gas superheat constant, and performed as followed.

Firstly, refrigerant gas pressure and temperature at the outlet of evaporator, which are detected by the Pressure Transmitter (P.T.) and the Resistance Bulb (R.B.), are transmitted to the Process Meter (P.M.). Actual refrigerant gas superheat calculated automatically by P.M. is converted to the signal of 1-5VDC and transmitted to the Temperature Indication Controller (T.I.C.).

Secondly, manipulated variable for R.A.V., which is calculated from deviation of superheat signal as process variable and a set point variable, is transmitted to the I/P Valve Positioner (I/P.V.P.) attached to R.A.V. The stroke of R.A.V. is correctly and immediately controlled by means of air pressure which is controlled in the I/P.V.P. corresponding to the output current of 4-20mA from T.I.C.

The super heat control system, when applied to freeze dryers, acts accurately to meet big variations in the freeze drying load or in the condensing refrigerant temperature *cT* (10° to 20°C) due to the change of cooling water temperature between day and night or among seasons, without necessity of additional adjustment by the operator. And this makes possible utilization of about 10°C lower *cT* and about 20% energy saving on an annual average, as compared with the conventional thermostatic expansion system in which *cT* must be constant.

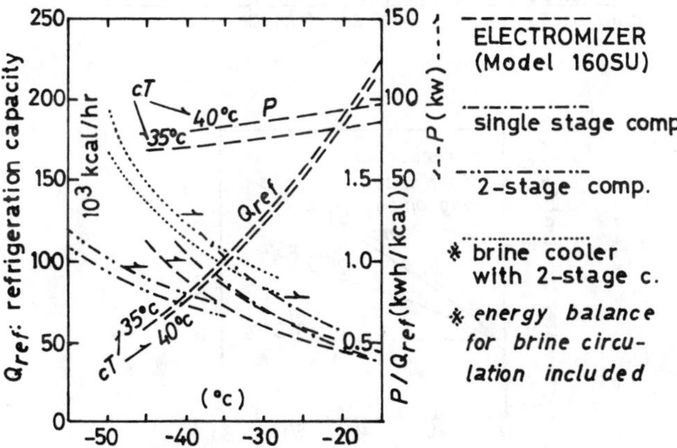

Fig. 6 Performances of typical refrigeration screw compressors

As the refrigeration compressor, the screw compressor with either of the following two systems is normally used.

a) 2-stage compression system ($eT \leq -33$°c)
b) ELECTROMIZER system (-50°c $\leq eT \leq$ -10°c)

The drying chamber pressure for System a) is normally 67 to 13 N/m^2 (0.5 to 0.1 Torr) while that for System b) 270 to 67 N/m^2 (2.0 to 0.5 Torr).

Since ΔT ($T_0^* - eT$) is very small, ELECTROMIZER system is recommended for wide applications of food freeze dryers. Its energy consumption is smaller as compared with any of single or two-stage system at -35°c $\leq eT \leq$ -10°c as shown in Fig. 6.

3. EXPERIMENTAL

3.1. Equipments

The experiment to determine whether the new system of defrosting the internal condenser in the final stage of freeze-drying affects the moisture content of dried product was conducted with a production plant.

The equipment, as shown in Fig. 7 and Table 1, featured a tray area of 118 m^2 and trap capacity of 2 tons. As shown in Fig. 7, the effective width of shelf was 1.0 m, the chamber diameter 2.2 m and the total condensing surface area (coils surface) 90 m^2.

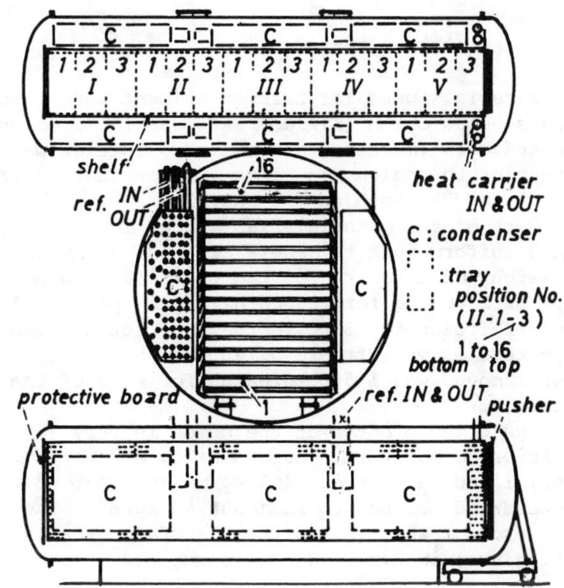

Fig. 7 Essential features of the freeze-dryer used for this experiment

Its refrigeration system comprises the "Dry expansion system with superheat control devices" and screw ELECTROMIZER compressor (100 kW), Model 160SU (Fig. 6). For the evacuation system, a combination of roots-type mechanical booster and water-ring pump with air ejector was used.

The main data on geometry of the condensing system is shown in Table 1.

Table 1 Main data on geometry of the condensing system used for this experiment

Item	Dimensions	Value
1 Chamber	(2.2ø x 8.2*L*)	V_1=31
2 Shelf stack	(1.1*W* x 7.5*L* x 1.65*H*)	V_2=13.6
total tray area		A_2=118
3 Louver board	(7.5*L* x 1.6*H*) x 2	A_3= 24
4 louver aperture for vapor flow	(A_3x1/3)	A_4= 8
5 Condensing system	(1.83*L* x 1.22*H* x 0.35*D*) x 6 sets	V_5=4.7
frontal area of condenser	(V_5/0.35*D*)	A_5=13.4
6 Condenser coil	(0.034ø x 880*L*)	V_6=0.8
surface area		A_6=94
7 space for ice build-up	(V_5-V_6)	V_7=3.9

ø:diameter; *L*:length *W*:width; *H*:height; *D*:depth, [m] *V*:volume[m^3]; *A*:area[m^2]

In the recording instrumentation of this production plant, the power input (kW) during the drying cycle of the refrigeration system was always recorded on a time base together with the product temperature, shelf temperature, condenser temperature and chamber pressure while the total energy consumption (kWh) was displayed digitally for each refrigeration system and others.

Hence, the operator can try various operational approaches with the purpose of not only quality control and shortening of the drying cycle time but also energy saving.

For reference purpose, 40 m^2 plants having the external condenser with a shut-off valve were used.

3.2. Materials and Methods

Materials used for this experiment were mainly Japanese long onion (*Allium fistulosum*) sliced to 4 mm and corn (maize) grain boiled. Initial water content of the onion was 91 to 92% and that of the corn 71 to 73% (both on wet basis).

Each of the materials was placed in 240 trays with a uniform bulk thickness of about 30 mm in the weight of 1 632 kg for the onion (6.8 kg per tray) or 2 064 kg for the corn (8.6 kg per tray). After prefrozen to -30°C in the prefreezing room, the materials were freeze-dried item by item. Water removed was 1 500 kg/batch for each of the materials.

The dried product of the onion is very sensitive to moisture. Bulk density of the frozen material bed ρ_f is about 460 kg/m^3 and that of the freeze-dried product ρ_d is about 40 kg/m^3. The dried product which has a large surface area contacting air, is extremely hygroscopic.

4. RESULTS AND DISCUSSION

4.1. Performance of Defrosting System and Moisture Content

Fig. 8 shows the typical recording chart of the defrosting process. Table 2 shows the moisture contents of dried product. Table 3 shows the moisture contents when the unloading door does not have a protective board on its back.

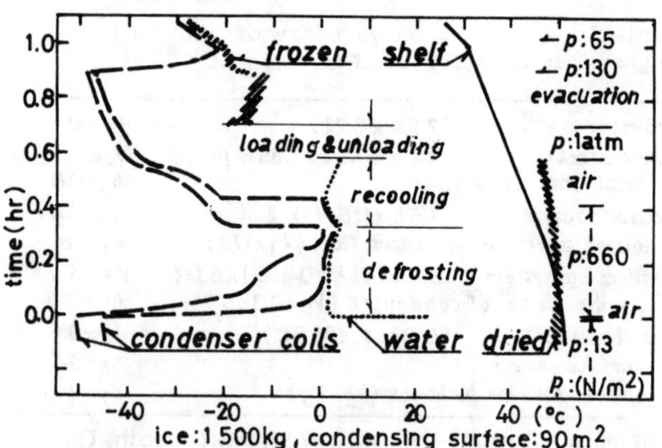

Fig. 8 Typical recording chart and other data on the defrosting process

Table 2 Average moisture content and standard deviation of the dried products

position No. / *sampling time	I-1-1,9&16 (tray end) (5 min.)	III-1-1,9&16 (8 min.)	V-3-1,9&16 (12 min.)	V-3-1,9&16 (2 min.)
Long onion				
run 1	4.29(0.58)	4.44(0.62)	4.90(0.79)	4.52(0.24)
run 2	4.83(0.28)	4.47(0.49)	5.14(0.55)	4.25(0.47)
run 3	4.68(0.37)	5.11(0.86)	5.08(0.56)	4.04(0.54)
run 4	4.94(0.93)	4.32(0.44)	4.25(0.44)	—
Corn				
※run 5	2.13(0.35)	1.63(0.23)	2.40(0.54)	—
※run 6	1.98(0.64)	1.72(0.20)	1.97(0.83)	—

Control Long onion: 5.54(0.80) ; Corn: 1.55(0.53) -- 8min
*after the chamber door opened for unloading
※protective board removed

Table 3 Effects of the protective board for defrosting

protective board removed (Long onion)

sampling position	from the end of tray No. I-1-1,9&16 0.0—0.1m	0.1—0.4m	0.4—0.5m
run 7	4.85(0.57)	4.50(0.62)	4.31(0.54)
run 8	5.73(0.93)	4.68(0.14)	4.38(0.72)
run 9	5.34(0.70)	4.41(0.66)	4.33(0.53)

Control: Table 2··(I-1-1,9&16) for (run 1—4)

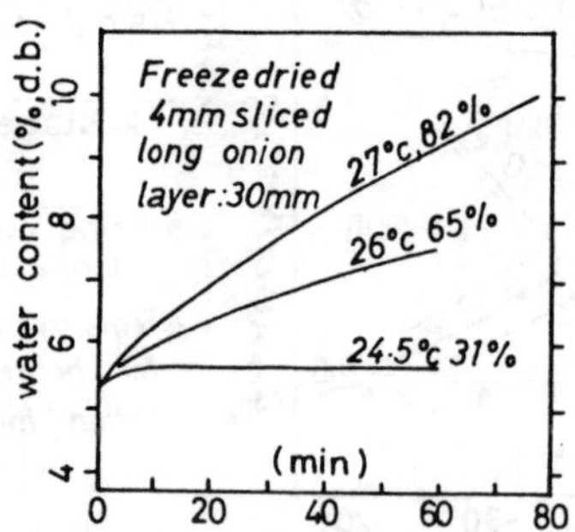

Fig. 9 Water adsorption kinetics of the freeze-dried long onion

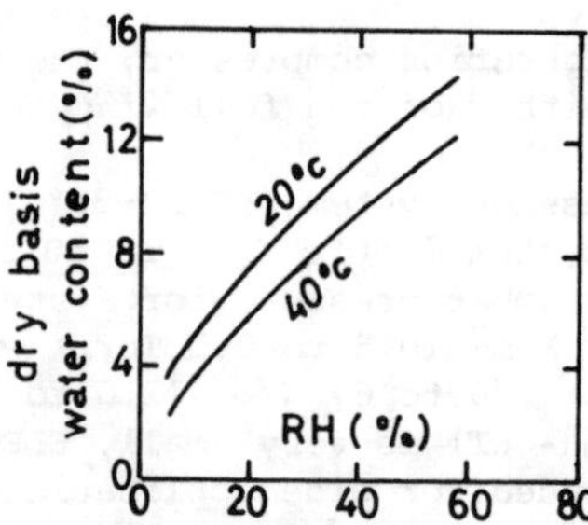

Fig. 10 Estimated sorption isotherm of the freeze-dried long onion

Fig. 9 and Fig. 10 show, respectively, the water adsorption kinetics when the unloaded dried product is left standing on the tray and the estimated sorption isotherm.

(1) Fig. 8 indicates that the water temperature was kept below 1°c during defrosting and the

defrosting was stopped at a +2°C signal. 1 500 kg of ice was melted in 20 min. A small amount of peeled ice pieces, a total of several kilograms, was caught by the condenser coils, but it did not cause any obstacle to the next cycle.

(2) The water contents of materials sampled from various parts of the shelves were all satisfactory, showing no significant difference when compared with those from different positions, those taken out without the defrosting process and the products of other freeze-dryers with external condenser.

(3) When the protective board on the unloading door side was removed, a slight difference was noticed only through repeated measurements as to the moisture content of the product placed within about 10 cm from the tray end. The products, however, was acceptable.

(4) The following troubles were also observed:
1 Without the process of raising the chamber pressure to above 5 Torr (600N/m^2) by air-bleeding, the water introduced gets frozen suddenly. This shock causes minute pieces of ice to scatter inside the chamber and sometimes to fall on the trays. 2 when the defrosting temperature was raised above 15°C, the material showed some evidence of caking.

4.2. Vapor Trap Efficiency

Main data of the typical drying cycle are shown in Table 4, Fig. 11 and Fig. 12. The sliced long onion (unfrozen) is one of the inhomogenius materials which are hard to have uniform bulk density and uniform bulk thickness when placed in a tray. It was found that there were many still-frozen spots left in the already-dried bed for a long time, even after a greater part of frozen layer was vanished from the material bed. The latter half of the drying cycle is, therefore, needed as shown in Fig. 11. Since the equipment used for this experiment was a single chamber plant with one unit of refrigerated compressor, marked drop in energy efficiency was noticed.

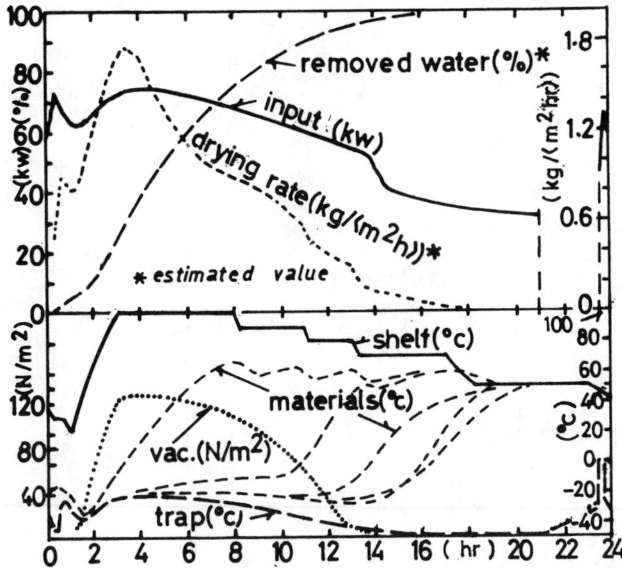

Fig. 11 Typical drying curves for sliced long onion's bed

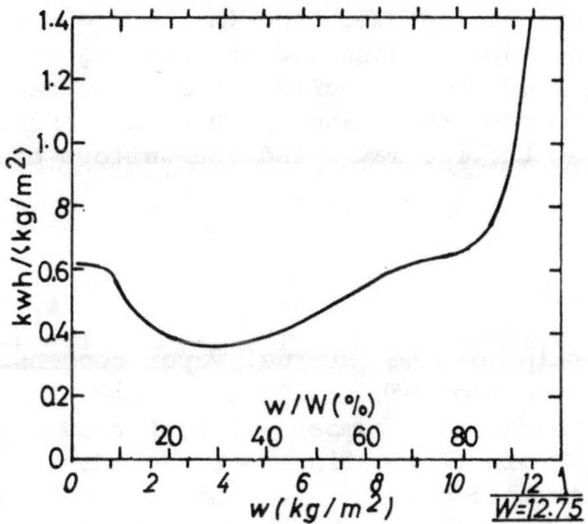

Fig. 12 Energy consumption per unit water removed in this experiment

However, thanks to the condenser geometry as shown in Fig. 7, Table 1, vacuum pressure loss Δp ($\equiv p_0 - p^*_{ref}$), p^*_{ref}: equilibrium water vapor pressure at eT. or temperature difference ΔT ($\equiv T^*_0 - eT$) being very small, it is assumed that the energy consumption will be reduced by further 20%, in case this plant is increased to double capacity by adding another drying chamber or is remodeled, instead, to have two refrigeration units of half the capacity each.

Table 4 Main data on the freeze-drying cycle of long onion

Total energy consumption for vapor condensing	1 178 kwh	(1 500 kg)	0.785 kwh/kg
the first half (0—12hr)	807	(1 400)	0.576
the latter half (12-24hr)	338	(100)	3.38
condenser recooling	33		
* *Tow chambers, staggered cycles*			0.58 kwh/kg
* *One chamber, tow ref. units*			0.65
Final ice layer thickness (average)	12.6×10^{-3} m		
area	129 m^2		
* *Max. ice surface temp.*: −18°c (δ_{ice}: 4.8×10^{-3}; ΔT_{ice}: 3.1°c; t: 4hr)			
* *Max.* ΔT_{ice}: 3.6°c (δ_{ice}: 6.8×10^{-3}; t: 5hr)			

* estimated value(from Fig. 11)

As seen from Fig. 11, Table 4, δ_{ice} is 6.8 mm for max. ΔT_{ice} and 4.8 mm for max. ice surface temperature (actual condensing temperature). In order to keep ΔT_{ice} and the ice surface temperature at the same level in the condensing area reduced to 1/5 - 1/7, δ_{ice} must be reduced to about 1 - 1.5 mm. In other words, defrosting must be carried out at 4 to 6 min. interval in the initial stage of drying cycle. Further, ΔT_{ref} increases about 1.5 times when the coil surface is reduced to 1/5 - 1/7 on the assumption of $\alpha_{ref} \propto \dot{q}^{0.8}$. Thus, on the batch plant, we can say the large condensing area is advantageous as compared with frequent defrosting.

Using the mathematical model of Eqs. (1), (2), (3) and (4) in the author's paper "Development of New Refrigeration System" also for IDS '84, the efficiency of the condensing system can be estimated from Table 1. In the region of food freeze-drying at temperature Tr = -45°c or above, corrections to the vapor-ice interface resistance and slip flow are negligible. And in the geometry of Table 1, the vapor resistance at existing condensation $W \leq 2\ 000$ kg is negligible as compared with the heat transfer resistance. This was confirmed by observing the ice layer and the uniform progress of defrosting.

5. CONCLUSION

Defrosting of the internal vapor condenser by the vacuum steam of 0°C to 1°C generated at the bottom of the drying chamber was performed during the 20 to 30 min of the final stage of drying, while the chamber pressure (vacuum) was held below 670 N/m^2 (5 Torr). And the records and deteiled measurements at 10 industrial plants showed that this system can produce dried food products of sufficiently low moisture content.

By use of this defrosting system, a maximum scale, for example, of 226 m^2 tray area and 4 000 kg ice capacity can be replaced by a compact design of a drying chamber of only 50 m^3 (the smaller scales reducible in the same proportion). Besides, the initial evacuation vacuum pump system can remain small enough. Thus, a compact, hygienic, high-efficiency batch-type volume freeze-dryer was developed without any other complicated attachments than the louver screens and the hydraulic tray pusher on the back of the front door.

ACKNOWLEDGEMENT

The author is grateful to his colleagues and users of "Kyowa" freeze dryers, especially ASAHI Food Industry Co., Ltd., for their valued assistances in preparing this presentation.

NOMENCLATURE

- p_0 drying chamber pressure
- T_0^* saturation temperature for water vapor pressure at p_0
- eT evaporating refrigerant temperature
- p_{ref}^* equilibrium water vapor pressure at eT
- T_d permissible final drying temperature
- p_s saturated vapor pressure of dried product at T_d
- cT condensing refrigerant temperature

REFERENCES

1. Lorentzen J., Industrial freeze drying plants for foods, in Freeze Drying & Advanced Food Technology, pp 429-443. Academic Press, London (1975)
2. Lorentzen J., XIV International Congress of Ref. (Moscow 1975) pp 140-147 & p. 269
3. Liebman H.L., Recent departures in volume freeze-drying, in Develop. biol. Standard. vol. 36, pp 33-40 (S. Karger, Basel 1977)
4. Ginnette L.F. and Kaufman V.F., Freeze-Drying of Foods, in The Freeze Preservation of Foods. 4th ed. vol 3, pp 377-403 (Avi Publishing Co., Westport, 1968)
5. Havighorst C.R.,: Fd. Engineering, pp 64-69 (Apr. 1974)
6. King C.J., Freeze drying of food products, in Freeze Drying & Advanced Food Technology, pp 333-349, Academic Press, London (1975)

FUNDAMENTALS OF DRYING OF FOODSTUFFS

G.H. Crapiste*, S. Whitaker** and E. Rotstein*

* PLAPIQUI (UNS - CONICET), 8000 Bahía Blanca, Argentina
** Dept. of Chem. Eng., U. of California, Davis, CA 95616, USA

ABSTRACT

The method of volume averaging is used to formulate a complete theory of mass transfer during the drying of cellular material in the stage in which the cellular structure prevails. The theory recognises the complex structure of the system and considers the different mechanisms of water transport. The volume-averaged transport equations for all the phases are added to obtain a total moisture transport equation in terms of water content and explicitly related with the shrinkage of the system. Then the assumption of local equilibrium is used to obtain the one-equation representation of water transport. A closure scheme is presented that allows for theoretical prediction of the effective mass transport coefficient. The general theory presented in this work would provide the basis to model water transport in different cellular materials.

1. INTRODUCTION

Drying of cellular materials is a problem of coupled mass and heat transport in a multiphase system which undergoes structural changes and shrinks during the process. As drying goes on, the mechanisms of water migration and all the transport properties change {1}. Nevertheless, most of the models describing the drying of foodstuffs were built on the basis of theories of common use in dealing with conventional porous media {2,3}, without acknowledging the features which make this problem inconventional.

The purpose of this work is to formulate a complete theory of mass transfer during the drying of cellular material. The analysis is confined to that part of the drying process during which most of the cell membranes are still intact and the cellular structure prevails. In this stage, a significant amount of the total water is released. The theory is built upon previous works on modeling food drying starting from a realistic picture of the cellular tissue {1,4,5,6} and applies the method of volume averaging to develop suitable macroscopic equations {7,8,9}.

A typical structure of a higher plant cellular tissue is illustrated in Fig. 1. This system has been described in considerable detail earlier {1} and, in a somewhat simplified picture four different phases are considered: vacuole, cytoplasm,

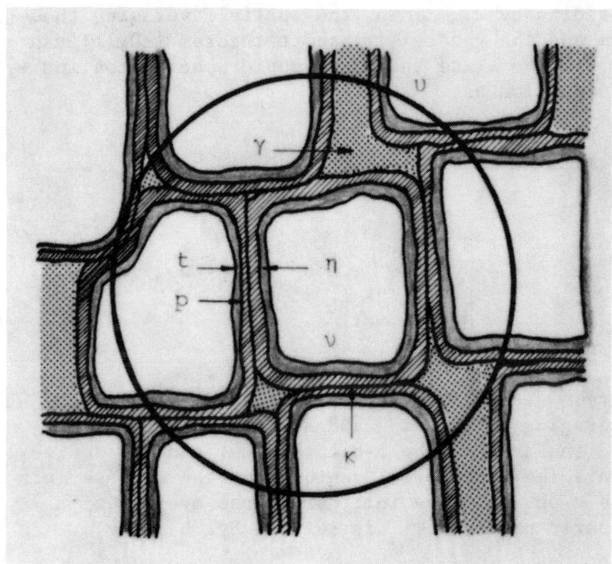

Fig. 1 Model of a cellular system

ν-phase: vacuole	η-phase: cytoplasm
κ-phase: cell wall	γ-phase: intercellular space
t: tonoplast	
υ: averaging volume	p: plasmalemma

cell wall and intercellular spaces. The vacuole (ν-phase) is an aqueous solution of different material such as sugars, organic acids and salts. The cytoplasm (η-phase) is a complex fluid or gel matrix containing cell organelles and reserve materials such as starch and proteins. It is surrounded by two thin semi-permeable membranes: tonoplast and plasmalemma. The cell wall (κ-phase) is composed mainly of cellulose and water. Among cells there are interconnected intercellular air spaces (γ-phase).

We begin the analysis in a general manner by considering the i-species continuity equation in a generic α-phase. In the absence of chemical reaction it can be expressed as:

$$\frac{\partial \rho_{i\alpha}}{\partial t} + \underline{\nabla} \cdot (\rho_{i\alpha} \, \underline{u}_{i\alpha}) = 0 \tag{1}$$

where ρ and $\underline{u}$ are the mass concentration and velocity of the species.

The method of volume averaging is based on the idea that we can associate an averaging volume υ to every point in space. Thus, macroscopic transport equations in terms of volume-averaged variables can be obtained by integration of the point equations over the averaging volume. To use the method of volume averaging we have a constraint on the length scales given by {7,8,9}:

$$l_\alpha << r_o << L \tag{2}$$

where l_α represents a characteristic length for the α-phase, r_o the radius of the averaging volume and L is a characteristic macroscopic length for the process.

The phase-averaged form of Eq. (1) can be obtained by integrating over $V_\alpha(t)$, the volume of α-phase contained in the averaging volume, and dividing by υ. Then, the spatial averaging theorem and the general transport theorem {10,11} can be used to write the macroscopic species continuity equation as:

$$\frac{\partial(\varepsilon_\alpha < \rho_{i\alpha} >^\alpha)}{\partial t} + \underline{\nabla}.(\varepsilon_\alpha < \rho_{i\alpha}\, \underline{u}_{i\alpha} >^\alpha) =$$

$$= - \sum_{\substack{\beta = 1 \\ \beta \neq \alpha}}^{\beta = M} \left(\frac{1}{\upsilon}\int_{A_{\alpha\beta}} \rho_{i\alpha}\, (\underline{u}_{i\alpha} - \underline{w}_{\alpha\beta}).\, \underline{n}_{\alpha\beta}\, dA\right) \tag{3}$$

Here ε_α is the volume fraction of α-phase in the averaging volume, $\underline{w}_{\alpha\beta}$ and $A_{\alpha\beta}$ are the velocity and the area of the α-β interface, and $\underline{n}_{\alpha\beta}$ represents the unit normal vector from the α-phase into the β-phase. The intrinsic phase average of a generic property $\psi_{i\alpha}$ is defined by:

$$< \psi_{i\alpha} >^\alpha = \frac{1}{V_\alpha(t)} \int_{V_\alpha(t)} \psi_{i\alpha}\, dV \tag{4}$$

2. NON-AQUEOUS MATERIAL AND WATER TRANSPORT EQUATIONS

Of the different components in the α-phase, one is water and we designate this species by $i = w$. The treatment of the non-aqueous components is based on the idea that they have nearly the same velocity in a given phase. We designate all the non-aqueous material by $i = s$. Now we can write Eq. (3) for the non-aqueous components and then we can sum the result over all phases in order to obtain the overall non-aqueous material continuity equation:

$$\frac{\partial < \rho_s >}{\partial t} + \underline{\nabla}.(< \rho_s > \underline{w}^*) = 0 \tag{5}$$

where the spatial average density of the non-aqueous material is given by:

$$< \rho_s > = \sum_{\alpha = 1}^{\alpha = M} \varepsilon_\alpha < \rho_{s\alpha} >^\alpha \tag{6}$$

and we have defined a mass average velocity of non-aqueous material by:

$$< \rho_s > \underline{w}^* = \sum_{\alpha = 1}^{\alpha = M} \varepsilon_\alpha < \rho_{s\alpha}\, \underline{u}_{s\alpha} >^\alpha \tag{7}$$

Notice that all the interfacial flux terms in Eq. (3) cancel when we sum over all phases.

In the usual analysis of diffusion problems one decomposes the species velocity in terms of a reference velocity and a diffusion velocity. In this analysis we decompose the water velocity $\underline{u}_{w\alpha}$ in terms of the non-aqueous material velocity $\underline{u}_{s\alpha}$ and a diffusion velocity relative to $\underline{u}_{s\alpha}$. Thus the water flux can be expressed as:

$$\rho_{w\alpha}\, \underline{u}_{w\alpha} = \rho_{w\alpha}\, \underline{w}^* + \rho_{w\alpha}\, \underline{\tilde{u}}_{s\alpha} + \underline{j}_{w\alpha} \tag{8}$$

where $\underline{j}_{w\alpha}$ represents the diffusive flux, and $\underline{\tilde{u}}_{s\alpha}$ is a spatial deviation of the velocity for the non-aqueous components which is defined as the difference between the point quantity and the mass average velocity given in Eq. (7).

Now we can write Eq. (3) for water and we can obtain the intrinsic phase average form of Eq.(8). Substitution of this result into the first equation allows us to write the water continuity equation for the α-phase as:

$$\frac{\partial(\varepsilon_\alpha < \rho_{w\alpha} >^\alpha)}{\partial t} + \underbrace{\underline{\nabla}.\,(\varepsilon_\alpha < \rho_{w\alpha} >^\alpha \underline{w}^*)}_{\text{Convective flux}} =$$

$$= \underbrace{- \underline{\nabla}.\,(\varepsilon_\alpha < \underline{j}_{w\alpha} >^\alpha)}_{\text{Diffusive flux}} \underbrace{- \underline{\nabla}.\,(\varepsilon_\alpha < \rho_{w\alpha}\, \underline{\tilde{u}}_{s\alpha} >^\alpha)}_{\text{Dispersive flux}} \underbrace{- \sum_{\substack{\beta = 1 \\ \beta \neq \alpha}}^{\beta = M} \left(\frac{1}{\upsilon}\int_{A_{\alpha\beta}} \rho_{w\alpha}\, (\underline{u}_{w\alpha} - \underline{w}_{\alpha\beta}).\, \underline{n}_{\alpha\beta}\, dA\right)}_{\text{Interfacial flux}} \tag{9}$$

Since our interest is in the total moisture transport during drying, we can sum Eq. (9) over all phases to obtain:

$$\frac{\partial < \rho_w >}{\partial t} + \underbrace{\underline{\nabla}.\,(< \rho_w > \underline{w}^*)}_{\text{Total convective flux}} =$$

$$= \underbrace{- \underline{\nabla}.\left[\sum_{\alpha = 1}^{\alpha = M} (\varepsilon_\alpha < \underline{j}_{w\alpha} >^\alpha)\right]}_{\text{Total diffusive flux}} -$$

$$- \underline{\nabla} \cdot \left[\sum_{\alpha=1}^{\alpha=M} (\varepsilon_\alpha < \rho_{w\alpha} \, \underline{\tilde{u}}_{s\alpha} >^\alpha) \right] \quad (10)$$

Total dispersive flux

where the spatial average density of water is defined as:

$$< \rho_w > = \sum_{\alpha=1}^{\alpha=M} \varepsilon_\alpha < \rho_{w\alpha} >^\alpha \quad (11)$$

When we add over all phases, interfacial fluxes in Eq. (9) cancel because we assume that the interfaces do not have a significant capacity to store water.

Drying of cellular material is a transient problem of water diffusion. Both the convective and dispersive fluxes are related with the shrinkage of the solid matrix during drying and they are the less significant terms in Eq. (10). It is an intuitive hypothesis to suppose that, in the drying stage which we are analyzing, all the solids shrink nearly in the same way following the cellular membrane in each cell. Thus, the non-aqueous velocity deviation should be small compared to both the point and average values and the dispersive transport of water in Eq. (10) can be safely neglected.

4. DIFFUSIVE TRANSPORT

The central issue at this point is to introduce appropriate constitutive equations for the diffusive fluxes to write the diffusive transport in terms of a single driving force. Neglecting the effect of body forces and thermal diffusion, the diffusive flux of water in each phase can be expressed as:

$$\underline{j}_{w\alpha} = - K_\alpha \, \underline{\nabla} \mu_{w\alpha} \quad (12)$$

where K is a mass conductivity and μ_w is water chemical potential.

Now we can follow the traditional line of analysis {7,8,9} and decompose the point values in terms of the intrinsic phase average and spatial deviations

$$\mu_{w\alpha} = < \mu_{w\alpha} >^\alpha + \tilde{\mu}_{w\alpha} \quad (13)$$

We can also decompose the intrinsic phase average as follows:

$$< \mu_{w\alpha} >^\alpha = \mu_w^* + \hat{\mu}_{w\alpha} \quad (14)$$

Here μ_w^* represents a characteristic equilibrium value for μ_w in the averaging volume, which may be the spatial average defined as:

$$< \mu_w > = \sum_{\alpha=1}^{\alpha=M} \varepsilon_\alpha < \mu_{w\alpha} >^\alpha \quad (15)$$

Use of decompositions given by Eqs. (13) and (14) and the spatial averaging theorem allows us to express the average form of Eq. (12) as:

$$\varepsilon_\alpha < j_{w\alpha} >^\alpha = -K_\alpha \left[\varepsilon_\alpha \underline{\nabla} \mu_w^* + \sum_{\substack{\beta=1 \\ \beta \neq \alpha}}^{\beta=M} \left(\frac{1}{\upsilon} \int_{A_{\alpha\beta}} \underline{n}_{\alpha\beta} \, \tilde{\mu}_{w\alpha} \, dA\right) + \varepsilon_\alpha \underline{\nabla} \hat{\mu}_{w\alpha} \right] \quad (16)$$

In obtaining Eq. (16) we have assumed that the variations of K_α are negligible within the averaging volume. We also have considered that averages can be treated as constants with regard to the integration over the averaging volume, which is satisfactory with the constraints on the length scales given by Eq. (2) {9}. Now we can return to Eq. (10), neglect the dispersive flux and use the earlier result for the diffusive flux in order to write the total moisture transport equation as

$$\frac{\partial < \rho_w >}{\partial t} + \underline{\nabla} \cdot (< \rho_w > \underline{w}^*) =$$

$$= \underline{\nabla} \cdot \left[\sum_{\alpha=1}^{\alpha=M} K_\alpha \left(\varepsilon_\alpha \underline{\nabla} \mu_w^* + \sum_{\substack{\beta=1 \\ \beta \neq \alpha}}^{\beta=M} \frac{1}{\upsilon} \int_{A_{\alpha\beta}} \underline{n}_{\alpha\beta} \, \tilde{\mu}_{w\alpha} dA \right) \right]$$

$$+ \underline{\nabla} \cdot \left[\sum_{\alpha=1}^{\alpha=M} K_\alpha \, \varepsilon_\alpha \, \underline{\nabla} \hat{\mu}_{w\alpha} \right] \quad (17)$$

The last term in Eq. (17) represents the non-equilibrium contribution to the diffusive flux. An important reduction in the magnitude of the problem is obtained using the simplifying assumption of local equilibrium. This kind of procedure has been successfully used to develop the one-equation model in the problem of heat transport in porous media {7,8,9}. The constraints that must be satisfied to make the assumption of local thermal equilibrium valid, were obtained elsewhere using order of magnitude analysis {8}. In addition, for a simplified model of mass transport in cellular tissue it has been shown {1} that it is plausible to assume local equilibrium, the validity of this assumption being related to the intervening length scales. In this case, we will consider that reasonable constraints exist, making it possible to apply the local equilibrium assumption and to neglect all contributions from the deviations $\hat{\mu}_{w\alpha}$.

5. TOTAL MOISTURE TRANSPORT EQUATION

The main problem now is to develop a closure scheme for the spatial deviations $\tilde{\mu}_{w\alpha}$ to complete the analysis of the total moisture transport equation. To do that we need to obtain the differential equations and boundary conditions to be satisfied by the spatial deviations and we must introduce constitutive equations for these functions. By using a general procedure {12} and standard order of magnitude analysis it can be shown that the governing differential equation for the spatial deviation is given by:

$$\nabla \cdot (K_\alpha \nabla \tilde{\mu}_{w\alpha}) = 0 \quad , \text{ in } V_\alpha \tag{18}$$

Assuming local equilibrium, the boundary condition at the α-β interface can be written as:

$$\underline{n}_{\alpha\beta} \cdot \left[K_\alpha \nabla \tilde{\mu}_{w\alpha} - K_\beta \nabla \tilde{\mu}_{w\beta} \right] =$$

$$= \underline{n}_{\alpha\beta} \cdot (K_\beta - K_\alpha) \nabla \mu_w^* \quad , \quad \text{at } A_{\alpha\beta} \tag{19}$$

For a simple interface, in addition to Eq.(19) we have the equilibrium condition:

$$\tilde{\mu}_{w\alpha} = \tilde{\mu}_{w\beta} \quad , \quad \text{at } A_{\alpha\beta} \tag{20}$$

and when there is a membrane between the phases the flux can be written as:

$$\underline{n}_{\alpha\beta} \cdot K_\alpha (\nabla \mu_w^* + \nabla \tilde{\mu}_{w\alpha}) =$$

$$= k (\tilde{\mu}_{w\beta} - \tilde{\mu}_{w\alpha}), \quad \text{at } A_{\alpha\beta} \tag{21}$$

where k is the membrane permeability.

At this point we follow the usual procedure for spatially periodic porous media {9} and represent $\tilde{\mu}_{w\alpha}$ as:

$$\tilde{\mu}_{w\alpha} = \underline{f}_\alpha \cdot \nabla \mu_w^* \tag{22}$$

where the function $\underline{f}_\alpha$ depends only on the structure of the system.

If we introduce the representation of the spatial deviations given by Eq. (22) into Eqs.(16) through (21) for all phases shown in Fig. 1, we obtain a closure scheme given by:

Differential equations

$$\nabla^2 \underline{f}_\alpha = 0 \quad , \quad \alpha = \nu, \eta, \kappa, \gamma \tag{23}$$

Boundary conditions

$$k_t (\underline{f}_\nu - \underline{f}_\eta) + K_\nu \underline{n}_{\nu\eta} \cdot (\nabla \underline{f}_\nu + \underline{\underline{I}}) =$$

$$= 0 \text{ , at } A_{\nu\eta} \tag{24}$$

$$k_p (\underline{f}_\eta - \underline{f}_\kappa) + K_\eta \underline{n}_{\lambda\kappa} \cdot (\nabla \underline{f}_\eta + \underline{\underline{I}}) =$$

$$= 0 \text{ , at } A_{\eta\kappa} \tag{25}$$

$$\underline{f}_\kappa = \underline{f}_\lambda \quad , \text{ at } A_{\kappa\lambda} \tag{26}$$

$$\underline{n}_{\nu\eta} \cdot \left[K_\nu (\nabla \underline{f}_\nu + \underline{\underline{I}}) - K_\eta (\nabla \underline{f}_\eta + \underline{\underline{I}}) \right] =$$

$$= 0 \text{ , at } A_{\nu\eta} \tag{27}$$

$$\underline{n}_{\eta\kappa} \cdot \left[K_\eta (\nabla \underline{f}_\eta + \underline{\underline{I}}) - K_\kappa (\nabla \underline{f}_\kappa + \underline{\underline{I}}) \right] =$$

$$= 0 \text{ , at } A_{\eta\kappa} \tag{28}$$

$$\underline{n}_{\kappa\lambda} \cdot \left[K_\kappa (\nabla \underline{f}_\kappa + \underline{\underline{I}}) - K_\lambda (\nabla \underline{f}_\lambda + \underline{\underline{I}}) \right] =$$

$$= 0 \text{ , at } A_{\kappa\lambda} \tag{29}$$

Now note that the water content in the averaging volume can be expressed as:

$$X^* = \langle \rho_w \rangle / \langle \rho_s \rangle \tag{30}$$

Using this definition, Eq. (5) and representations given by Eq. (22) in Eq. (17) we obtain the following total moisture transport equation:

$$\langle \rho_s \rangle \left[\frac{\partial X^*}{\partial t} + \underline{w}^* \cdot \nabla X^* \right] = \nabla \cdot (\underline{\underline{K}}^{eff} \cdot \nabla \mu_w^*) \tag{31}$$

where the effective mass conductivity tensor is given by:

$$\underline{\underline{K}}^{eff} = \sum_{\alpha=1}^{\alpha=M} K_\alpha \left[\varepsilon_\alpha \underline{\underline{I}} + \right.$$

$$\left. + \sum_{\substack{\beta=1 \\ \beta \neq \alpha}}^{\beta=M} \left(\frac{1}{\upsilon} \int_{A_{\alpha\beta}} \underline{n}_{\alpha\beta} \, \underline{f}_\alpha \, dA \right) \right] \tag{32}$$

In addition to Eq. (31) we also have that X* and μ_w^* are related by a theoretical or experimental equilibrium relationship of the type:

$$X^* = X^*(\mu_w^* , T^*) \tag{33}$$

where T* represents the equilibrium temperature.

Now we can assume that the mass of nonaqueous material m_s remains constant during drying so that the total water content is given by:

$$X(t) = \frac{\int_{V(t)} X^* \langle \rho_s \rangle \, dV}{m_s} \tag{34}$$

where V(t) represents the total sample volume.

Finally, to complete the formulation of the drying problem, there is a boundary condition at the sample surface, which can be expressed as:

$$\underline{n} \cdot \left[\langle \rho_s \rangle X^* \underline{w}^* - \underline{\underline{K}}^{eff} \cdot \nabla \mu_w^* \right] =$$

$$= \beta_e (\mu_w^* - \mu_{w\infty}) \quad , \quad \text{at } A(t) \tag{35}$$

where $\underline{n}$ is the outwardly directed unit normal and β_e the external mass transfer coefficient.

6. SHRINKAGE

At this point we must solve the problem of how to determine the average density of the nonaqueous material $\langle \rho_s \rangle$ and the mass average velocity $\underline{w}^*$. We can define a shrinkage bulk coefficient which relates the total volume sample $V(t)$ to the initial volume V_o by:

$$s_b = V(t)/V_o = \frac{< \rho_s >_o}{\frac{1}{V(t)} \int_{V(t)} < \rho_s > dV} \tag{36}$$

If we assume that the shrinkage is a unique function of the water content, there is a functional relationship which can be theoretically or experimentally determined {13}:

$$s_b = s_b(X(t)) \tag{37}$$

We will reduce the analysis to those cases of unidimensional shrinkage in which there is only one component of $\underline{w}^*$ given by:

$$w^* = \underline{w}^* . \underline{n} \tag{38}$$

and $< \rho_s >$ is independent of position. In these cases we can evaluate w^* by using the average continuity equation for the non-aqueous material, because Eq. (5) reduces to:

$$\frac{1}{\xi^{n-1}} \frac{d}{d\xi} (\xi^{n-1} w^*) = - \frac{1}{< \rho_s >} \frac{d < \rho_s >}{dt} , \quad 0 \leqslant \xi \leqslant L \tag{39}$$

where ξ represents a generalized spatial coordinate and n is a coefficient which depends on the geometry ($n = 1,2,3$ for plane, cylindrical and spherical shrinkage respectively). Since Eq. (36) simplifies to:

$$s_b = (L/L_o) = < \rho_s >_o / < \rho_s > \tag{40}$$

integration of Eq. (39) leads to:

$$w^* = \frac{\xi}{n} \frac{1}{s_b} \frac{ds_b}{dX(t)} \frac{dX(t)}{dt} , \quad 0 \leqslant \xi \leqslant L \tag{41}$$

Thus, given an expression for s_b we can evaluate $< \rho_s >$ and w^* by using Eqs. (40) and (41).

7. CONCLUSIONS

The method of volume averaging has been used to formulate a complete theory of mass transfer during the drying of cellular material in the stage in which the cellular structure prevails. It is important to note that while the model is applicable to any geometry, the shrinkage analysis is restricted to the special cases of planar, cylindrical or spherical shrinkage. To use this theory in the analysis of water transport in cellular material, the effective mass conductivity tensor must be measured experimentally or determined by theoretical means. This effective mass conductivity tensor is subject to direct theoretical determination by solution of the closure scheme obtained in this work, using some reasonable geometrical model for the system.

The general theory developed in this work should provide the starting point to study the drying of different cellular materials. Computational work is required in order to solve the model obtained and compare it with experimental results. Additional effort will lead to theoretical determinations of the effective mass conductivity tensor and to a fully predictive theory. Further work is in progress to complete the analysis along these lines.

NOMENCLATURE

Symbol	Description	Units
$\underline{f}$	vector field relating $\tilde{\mu}_w$ to $\nabla \underline{\mu}^*_w$	m
$\underline{\underline{I}}$	unit tensor	-
$\underline{j}_w$	diffusive flux of water	$kg/(m^2 s)$
k_p, k_t	permeabilities of plasmalemma and tonoplasto	$kg/(m^2 s\ J/mol)$
K	mass conductivity	kg/(m s J/mol)
$\underline{\underline{K}}_{eff}$	effective mass conductivity tensor	kg/(m s J/mol)
L	characteristic length of the sample	m
M	number of phases	-
$\underline{n}$	unit normal vector	-
n	coefficient dependent on the geometry	-
s_b	shrinkage bulk coefficient	-
υ	averaging volume	m^3
$\underline{w}$	non-aqueous material velocity	m/s
X	water content, dry basis	kg/kg
β_e	external mass transfer coefficient	$kg/(m^2 s\ J/mol)$
ε	volume fraction in the averaging volume	m^3/m^3
ξ	generalized spatial coordinate	m
ψ	generic variable	-

Subscripts and superscripts

i	generic component
o	initial state
s	non-aqueous material
w	water
α	generic phase
β	generic phase
$\alpha\beta$	α-β interphase
γ	intercellular space
η	cytoplasm
κ	cell wall
ν	vacuole

Special Symbols

$< \psi >$	spatial average
$< \psi_\alpha >^\alpha$	intrinsic phase average
ψ^*	local equilibrium value
$\tilde{\psi}$	deviation, $\psi_\alpha - < \psi_\alpha >^\alpha$
$\hat{\psi}_\alpha$	deviation, $< \psi_\alpha >^\alpha - \psi^*$

REFERENCES

1. Crapiste, G.H., Rotstein, E., and Urbicain, M.J., L.A.J.Ch.E.A.C., in press.
2. Fortes, M., and Okos, M.R., Drying Theories: Their Basis and Limitations as Applied to Foods and Grains, in Advances in Drying, ed. A.S. Mujumdar, vol. 1, pp. 119 - 154, Hemisphere Pub. Corp., Washington/New York/London (1980).
3. Bruin, S., and Luyben, K.Ch.A.M., Drying of Food Materials: A Review of Recent Developments, in Advances in Drying, ed. A.S. Mujumdar, vol. 1, pp. 155 - 215, Hemisphere Pub. Corp., Washington/New York/London (1980).
4. Rotstein, E., and Cornish, A.R.H., Proc. First Int. Symp. on Drying, Science, Princeton (1978).
5. Crapiste, G.H., Rotstein, E., and Urbicain, M.J., Proc. Third Int. Drying Symp., (1982).
6. Román, G.N., Urbicain, M.J., and Rotstein, E., A.I.Ch.E.J., vol. 29, 800 (1983).
7. Whitaker, S., Simultaneous Heat, Mass and Momentum Transfer in Porous Media: A Theory of Drying, in Advances in Heat Transfer, vol. 13, pp. 119 - 203, Academic Press, New York (1977).
8. Whitaker, S., Heat and Mass Transfer in Granular Porous Media, in Advances in Drying, ed. A.S. Mujumdar, vol. 1, pp. 23 - 61, Hemisphere Pub. Corp., Washington/New York/London. (1980).
9. Carbonell, R.G., and Whitaker, S., Heat and Mass Transport in Porous Media, in Mechanics of Fluids in Porous Media, ed. J. Bear and Y. Corapcioglu, Martinus Nijhoff, Brussels (1983).
10. Whitaker, S., A.I.Ch.E.J., vol. 13, 420 (1967).
11. Slattery, J.C., Momentum, Energy and Mass Transfer in Continua, Mc Graw-Hill, New York (1972).
12. Crapiste, G.H., Rotstein, E., and Whitaker, S., to be submitted to Chem. Eng. Sci.
13. Lozano, J.E., Rotstein, E., and Urbicain, M.J., J. Food Sci., vol. 48, 1497 (1983).

ADVANCES IN OPTIMIZATION OF FOOD DEHYDRATION WITH RESPECT TO QUALITY RETENTION

Marcus Karel, Israel Saguy and Martin A. Mishkin

Department of Nutrition and Food Science, Massachusetts Institute of Technology, Cambridge, Massachusetts U.S.A.

ABSTRACT

Feasibility of optimization of dehydration processes with respect to quality factors such as nutrient content, and organoleptic quality has been demonstrated for several cases typical of food dehydration. For the case of air-dehydration of potatoes, a simulation-optimization procedure based on the complex method was used to find the optimal dryer-temperature control for minimizing vitamin C loss for a given drying time. Optimal dryer-temperature control was also determined for minimizing drying time given a specified extent of vitamin loss. The procedures used kinetic models of ascorbic acid degradation and of heat and mass transfer during drying of the potatoes.

Optimization of ascorbic acid retention in potato drying was also undertaken for the case in which the extent of non-enzymatic browning is specified, and serves as a constraint.

In another case, optimal stage-duration and temperature control in a multi-stage dryer were found for minimizing ascorbic acid loss using a literature-derived kinetic model. In addition, enzyme (catalase) inactivation was considered in the staged process, and the process was controlled so that catalase activity was reduced below a specified level while minimizing ascorbic acid loss.

1. INTRODUCTION

Applications of optimization theory to food processes have lagged behind similar applications in chemical and related industries. This lag is due, in part, to the complex nature of food products, and in particular to the difficulty in defining food quality in quantitative terms. However, recently there has been a substantial increase in the interest in the field of quantitative description, modeling, and optimization of food processes. Recently, we have completed studies on application of modeling and optimization theories to control of quality changes in food dehydration [1]. In these studies we combined drying theory, and reaction rate kinetics to models of the course of a given deteriorative reaction during drying, and we used such models to optimize the process to achieve minimum deterioration given certain additional constaints imposed by economics or equipment characteristics (e.g. allowable drying times and temperatures).

We report here optimizations of air dehydration with the following objectives:

a) Minimizing vitamin C loss in white potatoes;
b) Minimizing browning in white potatoes;
c) Above two objectives when one of them is used as a constraint;
d) Minimizing vitamin loss while assuring a specific level of enzyme inactivation in a model system.

2. MODELS

2.1. Mass and Heat Transfer in Dehydration of White Potatoes

The drying behavior of white potato disks (4.0 mm thick and 4.1 cm in diameter) was studied and the drying behavior described using a model based on standard engineering practices [1]. The moisture transfer model was Fick's Law for unsteady unidirectional diffusion,

$$\frac{\partial M}{\partial t} = \frac{\partial}{\partial x}\left(D\,\frac{\partial M}{\partial x}\right) \qquad (1)$$

where: M is moisture content (g H_2O/g solids) and D is the effective diffusion coefficient (cm^2/sec). Edge effects were ignored due to the high ratio of diameter to thickness of the potatoe disks. Temperature dependence of the diffusion coefficient (D) was expressed by an Arrhenius expression, the parameters of which were determined experimentally.

The temperature of the potato disks (T_s) was modelled using an algebraic approximation to the heat balance where the temperature was assumed uniform,

$$T_s = \frac{\lambda_w m_s}{hA}\,\frac{(d\overline{M})}{dt} + T_{db} \qquad (2)$$

m_s is the mass of solids within the disk, (dM/dt) is the rate of vaporization for water. T_{db} is the air temperature and λ_w the latent heat of vaporization of water. Temperature uniformity was verified experimentally by inserting thermocouple probes at various depths within potato disks during drying. The surface area (A) and heat transfer coefficient (h) vary during drying due to shrinkage and other surface changes and have been estimated by a relation which assumed that the quantity (hA) varies linearly with M.

2.2. Mass and Heat Transfer in the Dehydration of a Cellulose Model System

The cellulose model system was based on work of Villota et al. [2,3,4]. The model system was slab 0.6 cm thick, 3.4 cm in width and 3.5 cm in length. The mass and heat transfer assumptions were similar to those used for the white potatoes. Effects of temperature and of shrinkage on heat and mass transfer were determined experimentally and incorporated in the appropriate equations.

2.3. Kinetic Model of Degradation of Ascorbic Acid (Vitamin C) in White Potatoes

The kinetic model representative of ascorbic acid degradation during air-drying of white potato disks was obtained using a dynamic test approach [1,5,6]. An empirical first-order kinetic model was used.

$$- \frac{dC}{dt} = kC \qquad (3)$$

where C is the concentration of ascorbic acid (normalized with respect to initial concentration). The first-order rate constant (k) has Arrhenius temperature dependence,

$$k = k_o \exp[-E_a/RT] \qquad (4)$$

where k_o and E_a have moisture functionality,

$$\ln k_o = (16.38) + (1.782)M + (1.890)M^2 \qquad (5)$$

$$E_2 = (14,831) + (241.1)M + (656.1)M^2 + (236.8)M^3 \qquad (6)$$

2.4. Model for Non-Enzymic Browning in White Potatoes

This model was based on literature data of Hendel et al. [7], which we also used in simulation of browning in drying in a previous study [8].

The model was pseudo-zero-order:

$$\frac{dB}{dt} = k_b \qquad (7)$$

where B is extent of browning

$$\ln k_b = \ln k_b - (Ea/R)(1/T - 1/T_o) \qquad (8)$$

where T_o is a reference temperature of 338°K

$$E_a = b_3M + b_4 \qquad (9)$$

$$\ln K_{bo} = b_1 \ln M + b_2 \qquad (10)$$

The parameters have the folllowing values:

for M ≤ 0.15% g/g solids : $b_1 = 1.122$; $b_2 = -7.072$; $b_3 = -97540$; $b_4 = 41780$

for M > 0.15 g/g solids : $b_1 = -0.5495$; $b_2 = -10.24$; $b_3 = -370.6$; $b_4 = 27200$.

2.5. Model for Ascorbic Acid Loss in a Cellulose Model System

Based on work of Villota et al. [2,3,4], the loss of ascorbic acid was described as a first order equation (eq. 3) with k dependent on moisture M (here as g/g sample) and T (°K) as follows:

$$\ln k = (17.94)M - (2.245 \times 10^8)T^{-3} - 33.33M^3 + 5921M^2T^{-1} - 1.58 \times 10^6\ MT^{-2} + 4.71 \times 10^8M^3T^{-3} - 2.339 \qquad (11)$$

2.6. Kinetic Model for Inactivation of Catalase in the Cellulose Model System

The kinetic model for inactivation of catalase in the model was assumed to be identical to Luyben et al. [9]. The first order equation for enzyme inactivation has the functionality in moisture M (g/g solds) and temperaure (°K):

$$k_c = k_1 \exp[E_a/RT] \qquad (12)$$

$$\ln k_1 = 90.36 - 83.07 \exp(-3.699\ M) \qquad (13)$$

$$E_a = 6.176\times10^4 - 4.916\times10^4 \exp(-3.699\ M) \qquad (14)$$

where: k_c is in min^{-1}
E_a in cal/mole

3.1. Dehydration Process Considered in Optimization

A drying process readily adaptable to optimization is air dehydration. The air flow rate and wet- and dry- bulb temperatures may be varied during drying as well as the shape, size, orientation and loading of the product. The optimization scheme may be implemented in a multistage process, in which the control of conditions within each stage is based on the optimal drying program. This problem may be attacked by assigning weights to each factor, and optimizing some derived objective function (e.g. dollar value of weighted quality factors), or it may be solved by optimizing one quality factor, while imposing constraints on other factors (e.g. defining minimum quality levels).

For the optimization, drying was treated as a batch process with dry bulb temperature (T_{db}) control. The complex method [6] was used to find the optimal temperature control path.

4. RESULTS

4.1. Maximizing Ascorbic Acid Retention During Drying of White Potatoes

The above models were used to simulate ascorbic acid degradation for specified drying conditions and provide a basis for selection of control schemes which improve ascorbic acid retention. The kinetic model developed for ascorbic acid kinetics was used in conjunction with the drying model to find optimal temperature control paths for maximizing ascorbic acid retention in potato disks. Numerical integrating of the system of simultaneous differential equations was accomplished using Gear's method [10]. Optimization was undertaken using either average moisture and local kinetics of ascorbic acid loss or average moisture content for the disk and kinetics based on this average condition. Both gave equivalent results. Optimizations were undertaken for several different constraints on drying time. The optimal control program for the air temperature for a 180 minute process is shown in Figure 1. This figure gives results using two different optimizaion procedures.

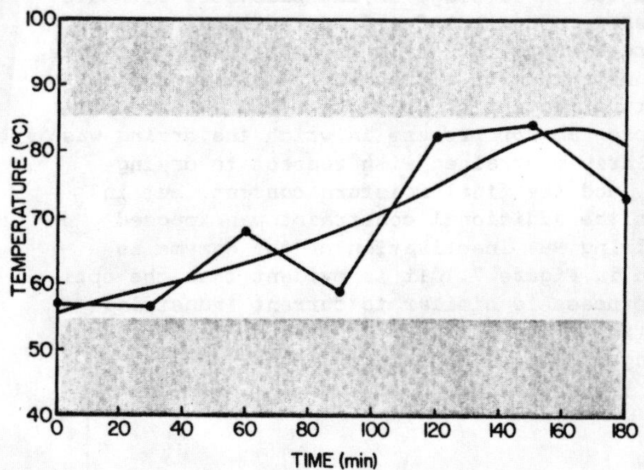

Figure 1. Optimal temperature control profile for minimizing ascorbic acid loss in an 180 minute drying process, using a polynomial (——) and a discrete-linear decision (o—o) optimization procedures.

4.2. Minimizing Non-Enzymic Browning in Dehydration of White Potato Disks

Figure 2 shows the optimal temperature control profile for minimizing browning in a drying process of 180 minutes, with browning kinetics based on the average and local moisture content. Similar results were obtained for processes of different drying durations. A constant-temperature process was also simulated at 73°C with an average final moisture cointent of 0.05 g/g-solids. The optimal process resulted in an 11% reduction in browning over the nominal process. However, the nominal process may already represent an improved process because it terminates when the average moisture content reaches the desired level, i.e. without overdrying the product. There is excellent agreement between the optimal control profiles based on the local- and average- moisture kinetics.

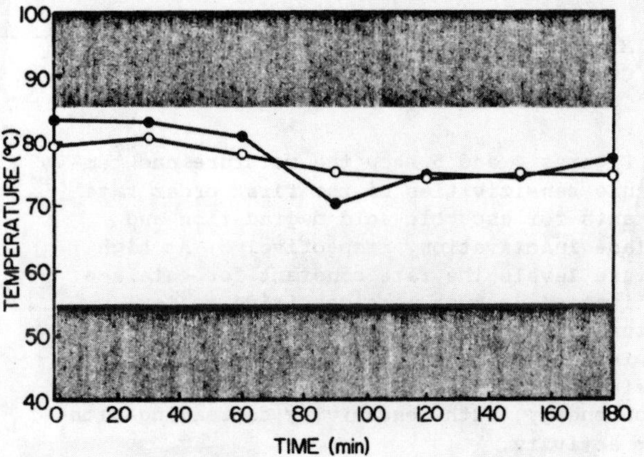

Figure 2. Optimal temperature control profile for minimizing browning in an 180 minute drying process. o—o based on average moisture content. o—o based on local moisture content.

4.3. Optimizing with Respect to Browning or Ascorbic Acid Retention with the Other Variable used as Constraint

Figure 3 illustrates the effect of simultaneous consideration of several objectives. Optimal profile A is for minimizing browning in an 180 minutes process without regard to vitamin loss. The resulting final quality was 0.021 Browning Units (arbitrary units) and 24.4% retention of vitamin C. Profile B is for maximizing vitamin C retention, while restricting browning to 0.025 units. Under these conditions vitamin C retention is 29.9%. Profile C is for maximizing vitamin retention with no restriction on browning. The resulting vitamin retention is 84.5% and the browning is 0.032 units.

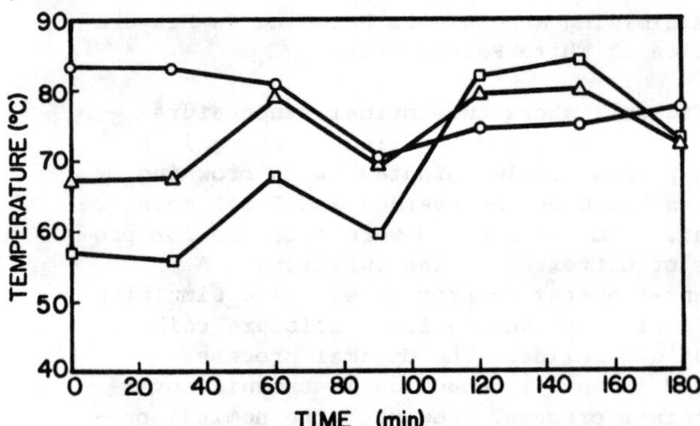

Figure 3. Optimal temperature control profile for an 180 minute drying process. A (o——o) minimizing browning. B (Δ——Δ) minimzing ascorbic acid loss with a constraint on browning. C (□——□) minimizing ascorbic acid acid loss without a constraint on browning.

4.4. Minimizing Ascorbic Acid Loss with a Constraint on Enzyme Inactivation in a Cellular Model System

Figures 4 and 5 show the moisture and temperature sensitivities of the first order rate constants for ascorbic acid degradation and catalase inactivation, respectively. At high moisture levels the rate constant for catalase inactivation is much more sensitive to temperature than that for ascorbic acid. This behavior is not unusual for enzymes. In fact, protein denaturation in general has a high activation energy, with sensitivity decreasing with water activity.

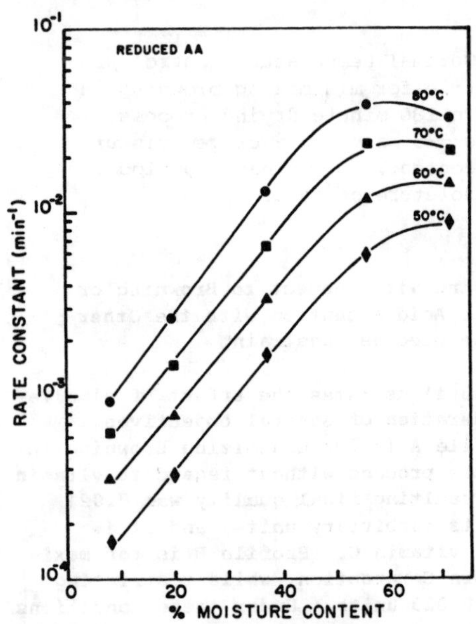

Figure 4. Moisture and temperature dependence of the first order rate constant for ascorbic acid degradation (reproduced with permission of J. Food Process. Preserv.).

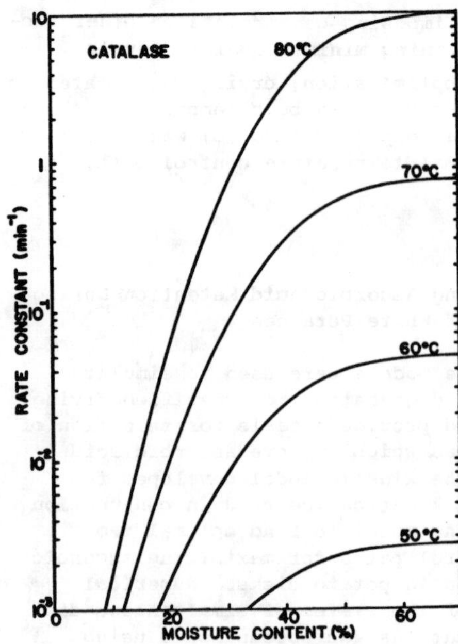

Figure 5. Moisture and temperature dependence of the first order rate constant for catalase inactivation. (reproduced with permission of J. Food Process. Preserv.).

As a consequence of this different temperature sensitivity the optimal process for retaining ascorbic acid without a constraint on the amount of enzyme to be inactivated is quite different from the optimal program in the presence of a constraint. We performed optimizations for three stage drying processes with air temperature control for different drying times. Figure 6 shows the optimal process when there is no constraint for a process specifying a final moisture content of 0.05 g/g solids. The optimal program for process in which the drying was similarly contrained with respect to drying time, and the final moisture content, but in which the additional constraint was imposed requiring 99% inactivation of the enzyme is shown in Figure 7. It is evident that the optimal process is similar to current industrial practice of separating the blanching and the dyring of enzyme-containing foods.

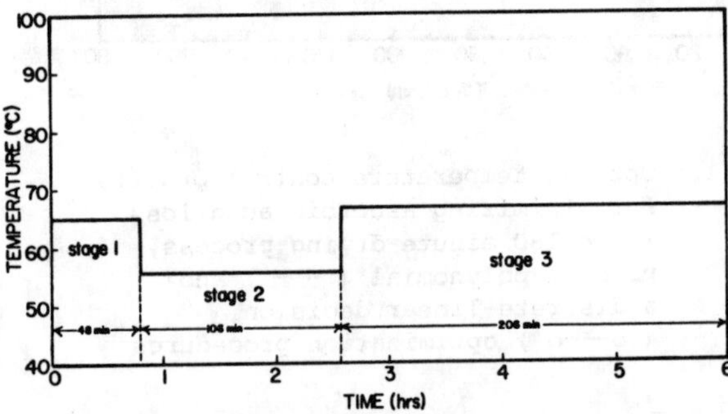

Figure 6. Optimal 3-stage drying process for minimizing ascorbic acid destruction for a 360 minute process. (final moisture = 0.05% g/g solids, no constraint on catalase.).

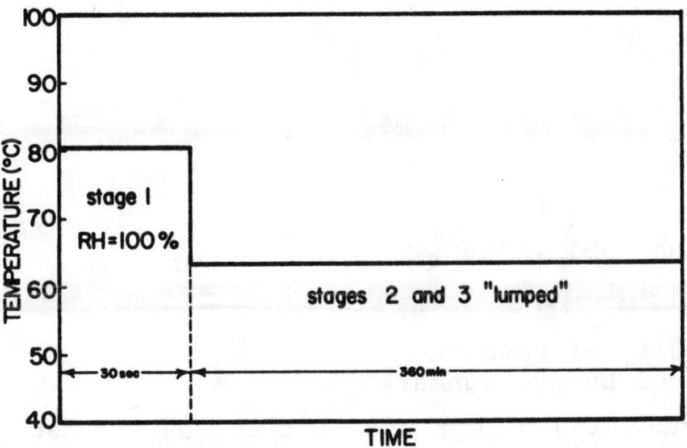

Figure 7. Optimal 3-stage drying process for minimizing ascorbic acid destruction with constrained final catalase activity (1%) for a 360 minute process (final moisture 0.05 g/g solids.).

5. CONCLUSION

It has been shown that experimentally determined kinetic models of quality changes during drying of foods may be combined with drying models describing temperature and moisture distribution in the food to allow optimization of drying processes. Further work will be required to make these optimizations applicable to a variety of food processes.

6. NOMENCLATURE

A	area $(cm)^2$
b	parameter
B	browning index
C	concentration
D	diffusion coefficient (cm^2/sec)
E_a	activation energy (cal/mol)
h	convective heat transfer coefficient (cal/°C-cm^2-min)
k	kinetic rate constant
M	moisture content (g/g-solids)
m_s	mass of solids in sample (g)
P	parameter
R	gas constant
t	time
T_{db}	dry-bulb temperature (°C)
T_s	sample temperature (°C)
x	normalized state variable or distance coordinate
λ_w	latent heat of vaporization for water (cal/g)

<u>Subscripts</u>

∘	initial
f	final
A	ascorbic acid
b	browning
c	catalase

7. REFERENCES

1. Mishkin, M.A., Dynamic modeling, simulation and optimization of quality changes in air-drying of foodstuffs, Ph.D. thesis, Massachusetts Institute of Technology, Cambridge, Mass. (1983).

2. Villota, R., Ascorbic acid degradation upon air-drying in model system, Ph.D. thesis, Massachusetts Institute of Technology, Cambridge, Mass. (1978).

3. Villota, R. and Karel, M., <u>J. Food Process. Preserv.</u>, vol. 4, 111 (1980).

4. Villota, R. and Karel, M., <u>J. Food Process. Preserv.</u>, vol. 4, 141 (1980).

5. Saguy, I., Mizrahi, S., Villota, R. and Karel, M., <u>J. Food Sci.</u>, vol. 43, 1861 (1978).

6. Mishkin, M., Karel, M. and Saguy, I., <u>Food Technol.</u>, vol. 36(7), 101 (1982).

7. Hendel, C.E., Silveira, V.G. and Harrington, W.O., <u>Food Technol.</u>, vol. 9, 433 (1955).

8. Aguilera, J.M., Chirife, J., Flink, M.M. and Karel, M., <u>Lebensm. Wiss. u. Technol.</u>, vol. 8, 12 (1975).

9. Luyben, K.Ch.A.M., Olieman, J.J. and Bruin, S., Enzyme degradation during drying processes, in eds. P. Linko et al., <u>Food Process Engineering: Enzyme Engineering in Food Processing</u>, vol. 2, pp. 192-209, Aopplied Science Publishers, Englewood, NJ.

10. Gear, C.W., <u>Commun. of the ACM</u>, vol. 14(3), 176 (1971).

8. ACKNOWLEDGMENT

This work was supported in part by grant CPE-8104582 from the Division of Engineering, National Science Foundation.

DRYING OF FOODS WITH FLUE GAS FROM NATURAL GAS RISK/BENEFIT

Poulsen, Kjeld Porsdal and Rubin, Jan

The Technical University of Denmark
Building 221, DTH, DK 2800 Lyngy, Denmark

ABSTRACT

In direct drying the process air is heated by mixing in hot flue gas, whereas by indirect heating the air is heated in a heatexchanger by flue gas or a heating medium. Energy consumption for direct drying is lower than for indirect and investment costs are also less. Further, direct drying equipment normally reacts more quickly than indirect, and shorter start-up and close-down phases may be achieved.

Since ancient times man has dried food by means of open fires, for some products even appreciating the smoked flavour. In the 1970's content of nitrosamines in beer was traced back to direct drying of malt and due to the known cancerogenic effect of these components direct drying was banned in many industries.

Natural gas is in general regarded as a very clean source of energy. Even so questions have been raised that foods produced by direct drying using natural gas may be contaminated by the contact with the nitric oxides of the flue gas.

Our examinations show that for many food products direct drying does not add to eventual contaminations. By direct drying of products like milk powder and malt low levels of contaminations with nitrite/nitrate and nitrosamines have been found. By use of "low-NO_x" burners instead of conventional burners the concentration of nitric oxides in the flue gas can be reduced considerably. The contamination caused by direct dyring will thereby be at such a level that it can not significantly be distinguished from that of other sources.

We suggest that regulations for direct drying of foods should include upper limits for nitrite/nitrate and nitrosamines in the final products.

Besides toxicological the present study involves economical aspects of direct and indirect drying methods.

1. INTRODUCTION

During the present period, use of natural gas as a fuel is being introduced in parts of Scandinavia. In Denmark considerations concerning direct use of flue gas for the drying of foods and grain, for baking bread and cakes, and roasting coffee and breakfast cereals are immediate. Despite the fact that direct contact has been common practice in a number of countries new installations are examined carefully before (and if) permission can be given by health authorities.

A requirement for minimizing the amounts of combustion generated oxides of nitrogen (NO_x) discharged from burners became apparent in 1978/79 following recognition that NO_x contained in air used in contact with foodstuffs can react with amines associated with proteins to produce cancerogenic nitrosamines. NO_x are the result of either thermal fixation of atmospheric nitrogen in the combustion air or of the conversion of chemically bound nitrogen in the fuel. For natural gas firering practically all NO_x derive from thermal fixation. When (crude) oil and coal is burned fuel-bound nitrogen can be the predominant NO_x generator. Only fuels with low content of nitrogen are acceptable for direct heating and natural gas fulfils this claim.

2. FORMATION OF NO_x AND NITROSAMINES

The formation of NO from air is a result of combustion of nitrogen and oxygen, but through a set of reactions in which atomic oxygen, atomic nitrogen and hydroxyl radicals are involved /1/:

$$N_2 + O \Leftrightarrow NO + N \quad (1)$$

$$O_2 + N \Leftrightarrow NO + O \quad (2)$$

$$OH + N \Leftrightarrow NO + H \quad (3)$$

The reactions leading to NO are endothermic and only at temperatures above 1600°C concentration of NO is becomming important. J. Kolar /2/ has illustrated the influence from temperature and residence time as in figures 1 and 2.

At temperatures below 600°C NO can react with O_2:

$$2\ NO + O_2 \Leftrightarrow 2\ NO_2 + 52\ kJ/mol\ NO \quad (4)$$

At room temperature the reaction is nearly

complete to the right side but preceding slowly so that the labile NO still normally will be found in more than trace concentration.

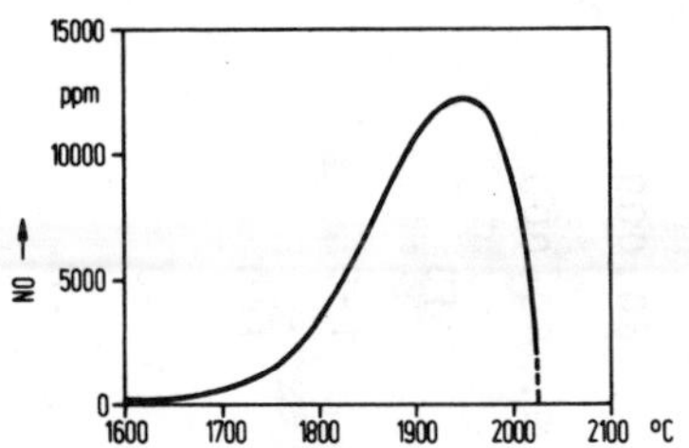

Fig. 1 Production of NO as a function of temperature.

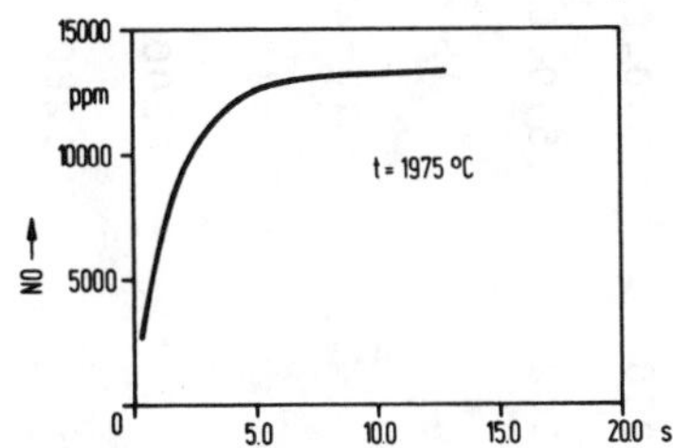

Fig. 2 Production of NO as a function of residence time.

The various gas fields produce natural gas (N-gas) with different compositions but all with methane as the major constituent, see table 1.

Formation of NO_x during combustion of N-gas is being studied at the Technical University of Denmark /3/. The kinetic models are very complex and uncertain to a large extent. Combustion of pure CH_4 is studied by use of a model with 83 different reactions including 29 components. The preliminary conclusions for reduction of NO_x are to lower

O_2 concentration
N_2 concentration
Temperature of combustion.
Residence time in combustion zone.

Craig and Pritchard /4/ determined equilibrium mole fractions of NO when H_2 was burned with air at various ratios:

25% excess air	6×10^{-3}
Stoichiometric air	3×10^{-4}
83% of theoretical air	5×10^{-5}

Table 2 Mole fractions of NO for combustion mixtures of H_2 and air.

	Tyra	Dan	Roar	Gorm	D.O.N.G. estimate	Ekofisk	Groningen	Kinsale
	Danish Fields					Norway	Netherlands	Ireland
Methane	90.0	92.9	90.3	82.5	91.1	84.6	81.2	99.15
Ethane	4.9	4.1	4.2	9.1	4.7	8.4	2.8	.26
Propane	1.9	1.2	1.3	5.7	1.7	2.9	0.4	87 ppm
n-Butane	0.7	0.5	0.4	0.8				17 ppm
i-Butane	0.4	0.5	0.4	0.4				25 ppm
n-Pentane	0.2	0.1	0.1	0.1		1.3	0.2	<10 ppm
i-Pentane	0.2	0.3	0.2	0.1				<10 ppm
C_6^+	0.3	0.1	0.2	0				<10 ppm
Nitrogen	0.3	0.3	2.1	0.6	0.6	0.9	14.4	.25%
Carbon dioxide	1.1	0	0.8	0.7	0.5	1.9	1.0	.34%

Table 1 Composition of N-gas from 4 Danish gas fields, the mixture which will be the Danish supply (D.O.N.G.) the Norwegian gas fields Ekofisk, the Dutch Groningen field and the Irish Kinsale field.

Nitrosamines can be formed when amines are brought into contact with NO_x in the air or nitrites in the product, nitrite being either added or reduced from nitrate /5/. The following reactions can take place /6/.

$$NO + NO_2 \Leftrightarrow N_2O_3 \quad (5)$$

$$2NO_2 \Leftrightarrow N_2O_4 \quad (6)$$

$$N_2O_3 + H_2O \Leftrightarrow 2HNO_2 \quad (7)$$

$$N_2O_4 + H_2O \Leftrightarrow HNO_2 + HNO_3 \quad (8)$$

$$\begin{matrix} R_1 \\ R_2 \end{matrix} > NH + HNO_2 \Leftrightarrow \begin{matrix} R_1 \\ R_2 \end{matrix} > N\text{-}NO + H_2O \quad (9)$$

R_1 and R_2 can be very different in molecular weight and composition. More than 130 different nitrosamines have been examined and more than 100 have been found to be cancerogenic in a number of different animals including rats, monkeys, snakes, fishes and birds. Cancer has been caused in most organs, e.g. liver, kidney, lung, brain, etc. Nitrosamines can also pass through placenta. One injection into pregnant rats has caused cancer not only in first generation but also in the following two generations /7/. Analysis for nitrosamines are difficult to make especially for non-volatile nitrosamines, and parts of the older literature can be influenced by incorrect results.

3. TECHNOLOGICAL AND ECONOMIC ASPECTS OF DIRECT PROCESSING

Formation of nitrosamines depends on a number of factors such as temperature, nature of the amines, and length of period for the process. If the foodstuff being processed has a content of protein, and nearly all foods have at least a few per cent, a NO_x concentration in the order of 0.05 to .15 ppm in the gases surrounding the product has been suggested as a maximum. This means that combustion gases at the burner mouth may average no more than 1 ppm of NO_x /1/. Methods available to achieve low levels of NO_x involve low temperature flames sometimes combined with a reduction of the level of oxygen available for reaction.

One method has a two stage combustion where the fuel gas is burnt with 50% air in the first step. After cooling in a heat exchanger final combustion takes place with a minimum of excess air. Another method is based on burning with a large amount of excess air at a low temperature and a short passage. The limiting temperature is set by the lower limit of inflammability of the gas with air. This system mixes gas and air in venturi tubes, flame stabilisers and an advanced automatic control system. The combustion products leaving

HEATING SYSTEM		Ind. gas	Dir. gas Maxon	Dir. gas Low Nox Maxon	Dir. gas Low Nox Urquhart *
Price supply heater and fan appr.	Dkr.	800.000	220.000	370.000	760.000
Plant total price appr.	Dkr.	10.000.000	9.420.000	9.570.000	9.960.000
Supply fan appr.	kW	22	15	15	15
PPM Nox appr.		-	50	10-20	1-5
Gas consumption appr.	kg/h	160	140	140	140
Evaporation appr.	kg/h	1250	1200	1200	1200
Powder rate appr.	kg/h	1320	1280	1280	1280

Table 3 Comparison of a direct and three indirect heated spraydrying plants designed for skimmed milk. 20000 kg supply air per hour. (Information from Niro Atomizer).

* Urquhart Engineering Company Limited supplies burners with a guarantee that NO_x concentration is lower than 1 ppm.

the burner are diluted by the air to be preheated and thus the NO_x level of the drying gases will be lowered to an extent controlled by NO_x in ambient air. NO_x levels up to 1 ppm can be found in ambient air in city areas /8/, typical values are about 50-100 ppb /9/. In order to evaluate direct versus indirect heating the above table 3 has been worked out by the kind assistance of Niro Atomizer, Copenhagen.

As can be seen from table 3 evaporation capacity is reduced 4% by changing from indirect to direct heating due to water formation in the combustion of gas. Consumption of gas is reduced 12.5% and fan power is also reduced. Investment can be reduced 5.8% by choosing the cheapest direct heated system. Urquhart low NO_x burners give the lowest NO_x level but at a total investment equal to indirect heating. For increased size of drying units savings by use of direct heating will increase more than proportionally /10/.

4. DISCUSSION AND CONCLUSION

Direct drying of foods is among the oldest ways of preservation. Hanging of meat and fish above open fires has not only dried foods but also given rise to a smoked flavour, which still is popular in many products. Identification of nitrosamines in beer could be traced back to direct drying of malt and instead of demanding better controlled burners, direct heating was banned in many contracts between maltsters and brewers.

Direct heating can give rise to increase in nitrite and nitrate content of foods. For these components upper limits exist in many countries, but with very few exceptions content of nitrosamines is only regulated voluntarily e.g. levels of 2.5 ppm in malt in Belgium, The Netherlands and Germany.

As stated earlier in the paper low NO_x burners can produce NO_x levels in the diluted air, which is surrounding the food to be processed, at the same order as in air used for indirect processing.

Direct heating gives lower variable costs than indirect and maintenance costs are also lower resulting in increased plant availability and productivity. The flexibility of direct fired systems is also higher meaning faster starting up and closing down.

Direct drying has been used to a large extent for many years in Western Europe, U.S.A., Canada and several other parts of the world for drying of milk, whey, malt, instant coffee, and potato starch, to roasting of coffee and breakfast cereals. Controlled experiments with the best low NO_x burners versus indirect drying gave no significant differences in volatile nitrosamines or benzo-a-pyrene.

The experience gained with direct drying indicates that the problem about legalization of this way of drying should be solved by establishment of maximum levels of contaminants rather than regulation of the processing.

P.S. When direct heating is being discussed it should be remembered that households and small bakers all over the world are traditional users of this practice /11/. Regulation in these areas has not been considered even if burners used are without control.

REFERENCES

1. Wheeler, W.H., The Chem. Eng., vol. 362, 693 (1980).
2. Kolar, J., gwf-gas/erdgas, Vol.124(2), 80 (1983).
3. Hadvig, S. et.al., NO_x-dannelse ved forbrænding af naturgas. Rept. Lab. for Varme- og Klimateknik, november (1983).
4. Craig, R.A. and Pritchard, H.O., Presentation at The Spring Technical Meeting of the Central States Institute, Oklahoma (1972).
5. Aizelmüller, K. and Thiele, E., Fette, Seife-Anstrichmittel, 83, 222 (1981).
6. Mangino, M.M., Scanlon, R.A. and O'Brien, T.J., N-nitrosamines in beer, ACS Symposium Series, Vol. 174, 229 (1981).
7. Gry, J., Meyer, O., Olsen, P. and Poulsen, E., Toxicological Investigation on Nitrite Treated Meat, Rept. Statens Levnedsmiddelinstitut, Copenhagen (1982).
8. Ann., Energi, hälse, miljö. Afsnit 2.5.5.1. Kväveoxider. Halter ovan tak. SOU 1977:68. Jordbruksdepartementet. Stockholm (1977).
9. Ritter, J.A., Stedman, D.H., and Kelly, T.J., Ground-level measurements of nitric oxide, nitrogen dioxide and ozone in rural air. In Grosjean, D., Nitrogeneous air pollutants. Chemical and biological implications. Ann Arbor Science Publishers (1979).
10. Edmondson, T. G., Urquhart Engineering Company Ltd., Perivale, England, Personal information.
11. Rubin, J. and Poulsen, K.P., Naturgas til direkte tørring af levnedsmidler, Rept. Technical University of Denmark (1983).

ACOUSTIC DEWATERING AND DRYING: STATE OF THE ART REVIEW

H. S. Muralidhara and D. Ensminger

Battelle's Columbus Laboratories,
Columbus, Ohio USA

1. INTRODUCTION

Suggestions to use ultrasonic energy for drying materials were made many years ago. Ultrasonics was especially appealing as a means of drying heat sensitive materials, such as pharmaceuticals, where rapid drying could prevent spoilage, if damaging heat levels could be avoided. Ultrasonic drying proved to meet both criteria--drying was rapid and temperatures seldom exceeded 1^{0}C above ambient. Although the ultrasonic method was effective it was also expensive and its usefulness was limited to these more exotic applications.

Recent developments in energy related areas of technology have caused a new search for more effective and more economical methods of drying. This search has produced a renewed interest not only in the possible use of ultrasound in drying but also in the related area of dewatering. Ultrasonics shows considerable promise as a potentially economical means of removing water from certain types of products to low levels, especially when it is used in combination with other techniques.

This paper reviews both past and recent literature on acoustic drying and dewatering and discusses the current status of the technology. The fundamental aspects and applications of the technique to a variety of materials are presented.

2. DRYING AND DEWATERING--DEFINITIONS

Drying in the present context refers to removing moisture from a product by vaporization, i.e., the water changes from liquid phase to vapor (or steam). Therefore, a significant amount of the energy for drying is consumed in supplying the heat of vaporization. Theoretically, the amount of energy required to evaporate a pound of water is approximately $1.055 \times 10^{+6}$ Joules.

Dewatering refers to removal of water from a product without producing a phase change in the water. It can be considered a "post filtration" [1] process. During the dewatering phenomenon it is important to identify different types of water that can be associated with a solid particle such as:

- o Bulk or free water
- o Micropore water
- o Colloidally bound water
- o Chemisorbed water.

The success of the dewatering process depends upon the content of the above types of water associated with solid particles. For example, water may be present in a solid in the following ratio:

Bulk water - 20%
Micropore water - 40%
Colloidally bound water - 20%
Chemisorbed water - 10%.

Conventional solid/liquid separation equipment such as filters and centifuges can separate out the bulk water to a limited extent. Pressure filter presses remove some of the water present in the micropores. However the colloidally bound and chemisorbed water cannot be released by any of the above techniques.

Dewatering may be used as an intermediate process prior to some form of thermal drying. Some of the water bound inside the capillaries or between the particles will be shaken loose or diffused out of the pores during the dewatering process. In most cases, the energy required for dewatering would be nominal, the only limitation being the level of solids concentration that can be achieved during the process. But this intermediate step may significantly reduce the heat load on the dryer and a substantial savings in the capital and operating costs is plausible. Obviously, it is desirable to remove as much water from a product as is economically feasible before the final energy-intensive drying stage.

2.1 Advantages and Disadvantages of Acoustic Drying

The advantages of the acoustic drying method have been exemplified by a number of workers. They are:

- Faster rates of drying
- Lower temperature operation
- Maintenance of product integrity.

The disadvantages of the acoustic drying techniques are:

- Development is in the experimental stage
- Applications to large throughputs have yet to be developed.

3. PAST WORK AND PRESENT STATUS

Burger and Sollner [2] were the first to suggest that sonic waves can be used to produce a drying effect. Brun and Boucher [3,4] credit Gregus and Mirkovics in Hungary as pioneering in ultrasonic drying techniques. In 1955, Gregus investigated the use of a 25,000 cycles/sec ultrasonic siren for drying textile fibers. In 1957 Brun and Boucher described ultrasonic drying as an economically promising commercial application. However, very little work has been done in acoustic dewatering. Many workers have alluded to pumping phenomena during application of acoustic fields, but it is not clearly stated or proven that the liquid occluded inside the pores diffuses out as a liquid or rarifies as vapor. At the present time the authors are conducting experiments to delineate the acoustic drying/dewatering mechanisms.

Brun and Boucher's description of mechanisms of both their own and of Gregus' work were limited to coupling energy through the atmosphere into the product. Also, according to their definition, drying included "all processes aimed at expelling liquids from solid materials, and since the liquid is usually water, it is frequently synonymous with dehydration". The definition makes no distinction between dewatering and drying.

Boucher's [5] description of the mechanism of drying is best understood as an air coupled phenomenon. It is as follows: "The rate of evaporation at a liquid/gas interface is largely controlled by the difference $\Delta P = P - p$, where P is the saturation vapor pressure of the liquid, and p is the vapor pressure of the surrounding atmosphere. Since acoustic waves are composed of compression and dilation regions, ΔP varies to a considerable extent at different points in the medium." They claim that the effect of dilation always predominates over that of compression, and as a result, water is continuously released either as a liquid or as a vapor. Water is mechanically eliminated in much the same way as if a sponge was subjected to a series of alternating pressures, related to the acoustic wave in intensity and compression frequency.

For Boucher's particular methods and for equal powers, effects over a frequency range of 10 to 33,000 cycles per second were essentially the same. Heat generated during the drying process is almost negligible--raising the temperature by approximately 1^{0}C above ambient. Materials dried were resins, pharmaceuticals, foodstuffs, etc. Boucher performed acoustic drying experiments with a number of materials. Some of his experimental data are shown in Figures 1, 2, 3, and Table 1. He summarized his results as follows:

- o Intensity level is the main factor governing the evaporation rate. A minimum of 145 db for industrial processing was suggested.
- o The thinner the bed, the faster the drying rate. He suggested a bed thickness of 25.4 to 50.8 mm. In Boucher's experiments, effective depth depends upon energy transfer from air into the material under treatment. Transfer from air into liquids or solids exceeding a wave length and depth is very low. The effective depth is also a function of attenuation of sound and of moisture conductivity or means of escape of vapors formed during the process.
- o Based on his experience, he indicated 6-10,000 cycles/sec as the optimum frequency.

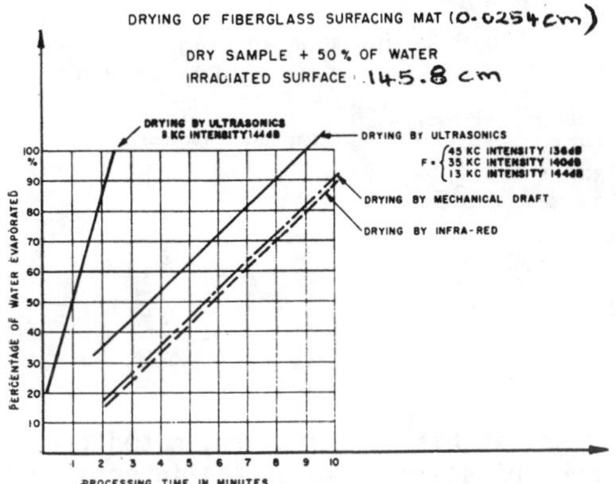

Figure 1 Drying of Fiberglas Mat

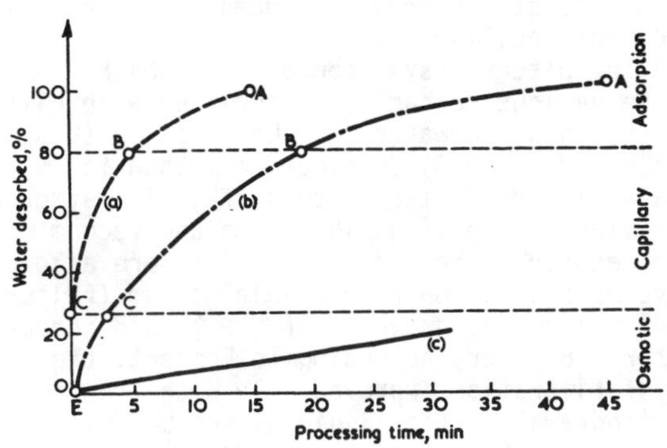

Figure 2 Comparison of three methods of drying silica gel (a) Direct irradiation by ultrasound (f = 8 kc/s, I = 152 dB) (b) Irradiation through a membrane by ultrasound (c) Vacuum drying

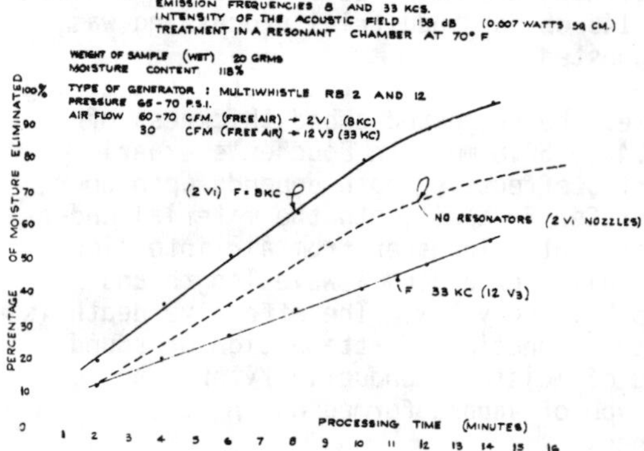

Figure 3 Acoustic Drying of Fermentation Sediment

Table 1 Hormone Drying

Weight of the Wet Sample	Processing Time	% of water Eliminated with Ultrasonics	Percentage of water Eliminated Through Vacuum Drying
30 grs.	0		
29.1	2 min.	3%	1%
28.6	4	4.7	1.7
28.1	6	6.3	2.7
27.6	8	8	3
27.1	10	9.7	3.3
26.5	12	11.6	3.7
26.1	14	13	4 3
25.6	16	14.6	4.7
25.2	18	16	5.3
24.7	20	17.6	5.7

Greguss, in his earlier work, noted that in the presence of a sonic field, the internal diffusivity of water increased. Sonic vibrations also reduced the viscosity of the diffusing liquid. Cavitation occurred due to sonic vibrations and caused the formation of bubbles which absorbed acoustic energy, expanded, and forced the liquid from capillaries.

A few patents have been granted which describe various mechanisms associated with ultrasonic drying and dewatering. Bongert [7] (U.S. Patent No. 3,970,552) developed a method to separate liquids from a mixture of solid matter and liquids by applying vibrations and vacuum simultaneously. The solid particles were allowed to move constantly on a permeable medium (filter), thus allowing good drainage of the liquid through a filter. However, according to Bongert, the rates of filtration improved only with the continuous withdrawal of the liquid by vacuum.

Sawyer (U.S. Patent No. 4,137,159) [8] developed a piece of equipment which used resonant rolls for applying vibrations through screens to the product to be dried. He claimed the energy level was sufficiently high to cause an intense state of vaporous cavitation in the liquid of the product to be dried, thus breaking the surface tension of the absorbed liquid. The exciting frequency was within the range 200-2000 cycles/sec (Hz).

Furedi [9] (U.S. Patent No. 4,055,491) developed an apparatus for removing fine particles from a liquid medium in the presence of sonic waves. This technique has been applied to materials such as algae from a solar or refuse pond, blood cells from blood, etc. The frequency of the generator is about 1.0 MHz. This is an interferometer technique of low intensity which utilizes the radiation pressure of a standing acoustic wave to cause small particles to migrate toward velocity nodal regions. The ultrasound is applied periodically, during which time the particles flocculate at the nodes. During the off periods, the particles are allowed to settle through baffle plates.

Gregus [6] claims drying in an ultrasonic field might be attributable to boiling inside a particle due to a "perfect vacuum" occurring on the "downstream side" of the particle. The explanation more nearly describes what happens when a high velocity jet of air or other gas is forced past a stationary particle. In an oscillating field, the "downstream side" of a stationary particle might alternate between the side toward and the side away from the source, or it might occur in equal intensity on both sides of the particle depending upon the particle size and shape, its physical characteristics and density, the nature of the incident wave and position of the particle in the field.

Boucher [10] claimed that sugar has been dried economically by ultrasonics to 1.2%. However, sugar will still agglomerate with 1.2% moisture making transportation of such materials difficult. An additional exposure to ultrasound for 16 minutes removed all of the moisture from sugar. For this reason, Gregus concluded that ultrasonics definitely has an effect on the moisture present in the pores or interstices of the particles.

Altenburg [11] postulated that intense ultrasonic irradiation can decrease the viscosity of water. His reasoning is based upon the 'hole' theory of water. When the water is subjected to large amplitudes of vibration, the number of holes increases thus causing the viscosity to decrease.

Tamas [12] demonstrated that the diffusivity may be increased many times by ultrasonic irradiation. The water filament may not be continuous; it is often interrupted by air bubbles. The bubble grows, or expands, due to abiabatic pressure changes which cause the water filament to migrate.

Huxsoll and Hall [13] studied the effect of sonic irradiation on drying rates of wheat and shelled corn. They found that drying of wheat and corn after 2, 8, and 60 minutes resulted in lower moisture content with the greater effect occurring in wheat. They hypothesized that if the movement of moisture occurred by surface diffusion, the diffusing water molecules move in groups instead of single molecules. Hence, they suggested that the diffusing species in the case of sonic drying might require fewer bonds to be broken per mole of diffusing water than in the case of conventional drying. The results are shown in Figure 4.

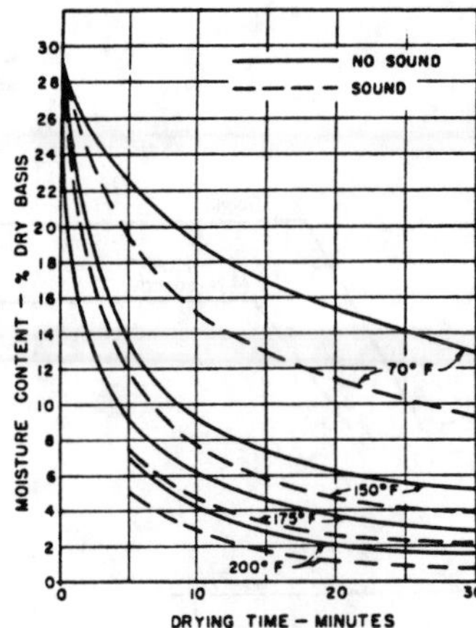

Figure 4 Moisture content versus time for crushed wheat at various ambient temperatures, sound at 165 db, 11.7 kc.

Kowalski, et al., [14] performed experiments using ultrasonic energy and flocculants to treat municipal sludges and mineral sludges. The ultrasonic frequency was 20,000 cycles/sec and the power into the transducer was 800 watts. The initial solids concentration of the sludges was 2 to 3%. Under the influence of the ultrasound with an exposure of 2 minutes and suitable use of flocculants, the moisture content of the sludges was reduced by as much as 75%. The positive effects of the ultrasound were obvious, of course, in improving the filtering properties of mineral sludges. The beneficial effects were decreased hydration forces, surface tension, adhesion and cohesive forces. They concluded that the combined effects of ultrasound and flocculants should result in multidirectional changes of properties of the particles in a multicomponent material and the flocculants are activated to promote hydration in the presence of the ultrasound.

Stephanoff [15] suggested that a material containing high moisture when placed in a shock region of a jet will dry due to the nearly perfect vacuum which is created on the downstream side of the material. According to him, an air jet produces an intense acoustic emission whose frequency is governed by:

$$F = \frac{C}{A} (K_1 + K_2/\sqrt{(R-R_c)}$$

where

- F = frequency of emission, kilocycles/sec
- A = nozzle orifice diameter, cm
- C = velocity of sound, cm/s
- R = ratio of reservoir pressure to atmospheric pressure
- R_c = critical ratio for which speed of sound is reached at orifice.

K_1 and K_2 are characteristic of the region under consideration. Figure 5 shows the ratio of acoustic pressure to atmospheric pressure for different intensities.

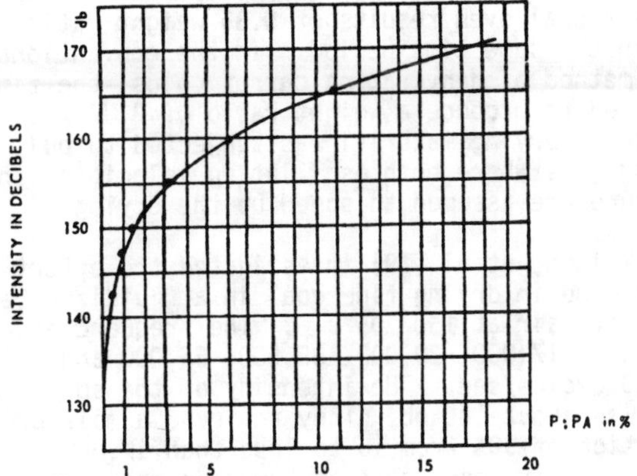

Figure 5 Ratio of the Acoustic Pressure "P" to Atmospheric Pressure "PΔ" For Different Intensities in a Standing Wave System. (Resonance)

Quite recently, Sonadyne Industries [16] has been working toward the development of a device which they call a valveless pulse jet engine system that has the capability of evaporating water from a large variety of materials. The materials include agricultural, municipal, industrial, and chemical products, for protecting process waste streams. The pulse jet drier can evaporate moisture from wet slimes with as little as 1% solids to as much as 99% solids. It can handle materials with 6.35 mm particle sizes. Since the residence time requirements are short, the solids never reach a temperature of more than 50^0C. The system can use any type of liquid or gaseous fuel including waste oils and digester gas. The average energy requirements to evaporate a pound of water is only 1.53×10^6 Joules input. The wet material is ground, centrifuged to remove any organic materials such as oils and excess water, and then pumped into the exhaust stream of the pulse jet operating at 250 cycles/sec and 1371^0C. Drying occurs in a short time as the sonic shock separates the feed from tiny particles.

According to Yang [17], the increase in drying rates he observed was due to the creation of an artificial turbulence. The jet stream blasts the dried product into a collector. Most of the particles settle by gravity while the other particles are removed in a cyclone separator.

Tamburello, et al [18] have developed a resonating pulse-combustion chamber to dehydrate fruits and vegetables. The combustion is an aerodynamically valved, U-shaped type with both inlet and outlet diverting the hot air through the food particles. Carrot chips 15 mm and 5 mm placed on racks were dehydrated after 10 minutes of drying time to a weight ratio of 0.37 compared with the conventional oven results of 0.85 weight ratio within the same time period. In the conventional oven method of dehydrating carrot chips, the time required to produce a weight ratio of 0.37 was 37 minutes. Drying material was subjected to pulsating jets where both oscillating velocities and pressure are assumed to speed up the drying process.

Wilson, et al [19] investigated the effect of ultrasound in drying fine coal in a fluidized bed mode with air at 150-205 ^{0}C. . The frequencies used were 17,000, 20,000, 35,500, 65,000 and 93,000 cycles/sec. THe intensity of the sound wave was about 135 db. They achieved a moisture reduction of 90% from 10 to less than 1% by weight. They observed that lower frequencies gave faster drying rates. The location of the ultrasonic horn with respect to the fluid bed appeared to play an important part in obtaining higher drying rates. They did not provide any information with regard to the amount of energy required to evaporate a pound of water.

Fairbanks [20,21,22] and his co-workers have done work related to ultrasonic and sonic drying of a variety of coals. They used a rotary drier as a means of drying coal fines. They observed that the presence of sound energy significantly increased the drying rates. The drying rates improved with a decrease in the particle size. This phenomenon is consistent with Boucher's experiments. They also suggested that a rotary driven design would be suitable for commercial applications. But during their experiments they encountered the balling of fine coals. There is no mention of the energy efficiency of this process. Since materials such as coals are used in large tonnages in industry, the drying process has to be energy efficient. Influence of ultrasound on the drying rate of powdered coal is shown in Figure 6. A sketch of the rotary drier is shown in Figure 7. The diameter of the vessel used was 250 mm. In some of Fairbank's work within an open and closed filtration circuit, the presence of the ultrasonic field increased the rates of filtration for oils passing through a steel filter. The improved filtration rates were attributed to vibration of the filter and the effects of the ultrasound on the liquid transport rate through the filter.

Takashi, et al [23] conducted experiments to dry coal tar by the application of ultrasonic waves at low temperatures. The frequencies used were in the range of 100-200,00 cycles/sec. They found that the breakage of emulsion was most effective higher ranges. The drying rate increased with increase in irradiation time. The drying rates were also higher at lower intensity and lower temperature.

Recently, Puskar [23] summarized the high intensity ultrasonics application in a variety of technologies including dewatering and drying. In his work, he refers to the work performed by Semmelinko. The position of the ultrasonic source

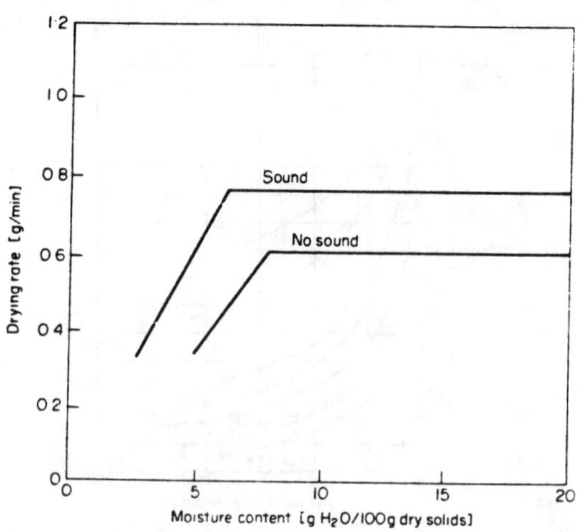

Figure 6 Typical curves showing the influence of ultrasound on the drying rate for powdered coal. Air temperature was 200 C.

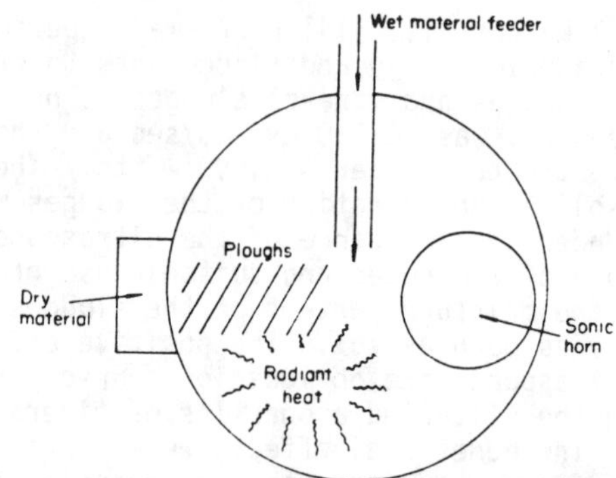

Figure 7 Rotary disc drier.

from the filter body was shown to be important. Semmilinko found that at a frequency of 20,000 cycles/sec, the filtration rate was increased and the particles were kept away from the filter surface. According to Puskar, the filtration efficiency depends on the size of the pore material and the acoustic parameters.

Muralidhara, et al [25], conducted experiments to dry green rice by the application of airborne sonic and ultrasonic fields. The objective of the research work was to perform low temperature acoustic drying to reduce the moisture content from 19 to 12 % moisture product with corresponding increase in the milling yields.

Figure 8 shows the relationship between percent moisture removed from green rice as a function of duration of exposure in the fluidized-bed. The circles o and ● represent the drying curves for 20.5 ^{0}C and 40 ^{0}C in the absence of an acoustic field. The squares □ and ■ represent the experimental data points using a whistle frequency of 12,000 cycles/sec at 20.5 ^{0}C and 40 ^{0}C. The triangles Δ and ▲ represent data points at 19,000 cycles/sec frequency at 20.5 ^{0}C and 40 ^{0}C.

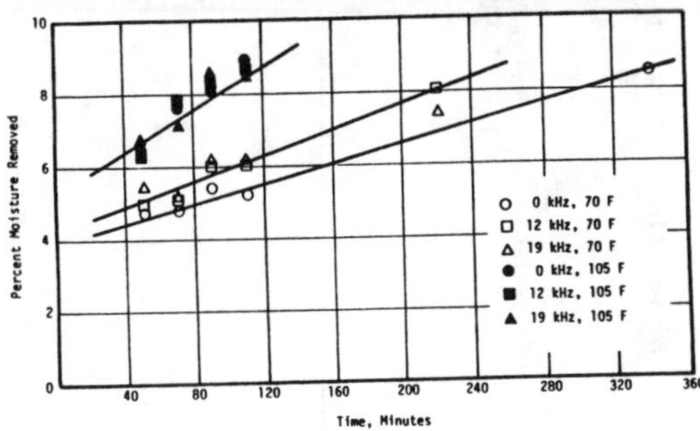

Figure 8 Variation of percent moisture removed in green rice as a function of time

The data indicate that at 40^0C the whistle frequency or degree of acoustic stimulation has little effect on the drying rate. However, at $20.5^0C$the drying rate is increased by use of a whistle.

The authors have also successfully tested other materials such as silica gel and pharmaceutical products using the acoustic drying technique. [26]

The initial moisture content of the feed material was 20 %. After subjecting the material to acoustic drying, the moisture content of the product was reduced to 5%.

White [27] conducted experiments to dewater the wet sheets of paper by employing a high frequency sound. The water in paper is held predominantly in pores. If this water has to be displaced, the interfacial forces must be overcome. White employed intense acoustic field to overcome the interfacial forces. He showed that time to reach equilibrium flow was greatly reduced in presence of the acoustic field and the final moisture content of the sheet of paper was also much lower.

Swamy, et al [28], studied the effect of soundwaves on the dewatering of granular materials prior to drying. The materials studied were calcium carbonate, magnesite, sand, and sawdust. The experiments were conducted in combination with a centrifuge. The variation in particle size of the materials seemed to have an effect on the dewatering characteristics.

Table 2 summarizes the applications of ultrasonic dewatering/drying techniques by a number of authors on a variety of materials.

4. PRINCIPLES OF ULTRASONIC DEWATERING AND DRYING

There are many mechanisms associated with ultrasonic irradiation. The mechanisms which are beneficial to dewatering in an acoustic field vary with the state of the material being dewatered or dried. Understanding the mechanisms of acoustic dewatering and drying is important in determining whether and how to apply such energy to these processes.

Previous research has shown that sonic or ultrasonic energy can enhance dewatering of various materials. Ultrasonic energy may be applied alone or it may be combined with other techniques in which case the total effect may be synergistic.

Systems requiring dewatering may be placed into one of two categories which relate to the nature of the water to be removed and to the state of the material to be dried. These major categories are (1) low moisture content systems in which the space between solids (particles, fibers, etc.) to be dried contains considerable air and (2) high moisture content systems or continuous phase systems in which all potentially void space is completely filled with moisture and the system has the characteristics of a liquid or a solid.

In the first category, the water may be in three forms--free, entrapped, or bound--but the free liquid appears in a thin film on the surface of the material to be dried. The material may have the appearance of a dry product.

In the second category, the liquid may also appear in the free, entrapped or bound forms but the amount of free water is appreciably greater than that of the first category in that it does fill the available volume of the specimen.

A very basic and critical factor in the effectiveness of any of the dewatering mechanisms is the intensity of the sound wave in the regions from which moisture is to be removed. The material attenuates the ultrasonic energy by various means such as interparticle friction, scattering, or viscous absorption. The energy thus removed from the ultrasonic wave is converted to heat which, of course, contributes to the drying process. However, heating by ultrasound alone is not justification for ultrasonic dewatering.

The attenuation of the ultrasonic energy in the medium is a function of frequency and of various geometrical (or structural) and acoustical characteristics of the material. Also the method of generating and coupling the energy into the material is decided by the acoustical impedance of the material and other factors previously mentioned. [29]

The effectiveness of ultrasonic energy in dewatering is related, of course, to the nature of ultrasonic (or sonic) wave propagation. Ultrasonic waves are stress waves. Two properties of a material are essential to the existence or transmission of these waves: elasticity and inertia. Stresses associated with high intensity ultrasound may be extremely high but the amplitude of displacements may be very low. Many effects may result from the high intensity of an ultrasonic wave. Intensely oscillating bubbles of vapor and

Table 2. Summary of Past Research in Acoustic Dewatering and Drying

Authors	Year	Type of Material Tested	Frequency, kcs/sec	Intensity, dB	Particle Size, μ-meters	Time of Drying, min.	% Initial Moisture	Amount of Material Tested, kg	% Moisture Reduction	Type of Sonic Device	Remarks
Brun and Boucher	1956	Metal Hydroxide	10		Colloidal	240	--	0.75	65	Multiwhistle	Control sample gave only 5% reduction.
		Carboxymethyl cellulose	34		Powder	240	--	0.75	40	Multiwhistle	Control sample gave 5% reduction.
Boucher	1959	Fermentation Sediment	8	138	Fine	14	118	0.02	100	Multiwhistle .95 x 10^5 - 4.83 x 10^5 Newtons/M^2	Air was drawn through the material kept in a thin layer; without resonator gave better results than 33 kcs.
		"	33	138	Fine	14		0.02	55		
		Hormone	8	152	NA	20		0.03	17	Monowhistle	With vacuum drying only 5.7% moisture was removed.
		Silica gel	6-8	145	NA	15		0.04	100	Monowhistle	Vacuum drying gave only 25% reduction.
		Fiberglas	8	144	250 (thick)	2-5			100	Monowhistle	The material was placed 254 mm from the source. Best results obtained at 8 kcs.
		Asbestos Paper	8	144	6,850 thick	5			100	Monowhistle	Drying by mechanical draft took 15 minutes.

Table 2.

Authors	Year	Type of Material Tested	Frequency kcs/sec	Intensity dB	Particle Size μ-meters	Time of Drying Min.	% Initial Moisture	Amount of Material Tested, Kg	% Moisture Reduction	Type of Sonic Device	Remarks
Boucher	1961	Gelatine	12	143-145	--	120	80	22.2/hr	52.5	Monowhistle	Control gave only 19% reduction.
		Yeast Cake	12	148	NA	20	70	22.2/hr	18	Monowhistle	Control gave 10% reduction wet at 37°C
		Granulated Sugar	12	152	NA	20	1.59	22.2 kg/hr	98.6	Monowhistle	
Greguss	1963	Silica gel	8	152	--	2.5	--	--	20	Ultrasonic Horn	Direct radiation gave better results than through the membrane
White	1964	Paper Sheet	10.5-12.2	148	Sheet-11.1 kg Filter Paper	8	--	--	--	Stem-Jet Whistle	--
Huxsoll Hall	1970	Wheat	11.5	165	NA	40	30	0.1 kg	27	Stem-Jet Whistle	Experiments performed in Drum Drier. Control gave 13% reduction.
Wilson et al	1971	Fine Coal	10	250 Watts Power	73	10	9	0.45 kg	89	Lead Zirconate transducer	Lower frequencies produced larger increase in drying rates
Fairbanks	1975	Fine Coal	20	150-170	147	5	30	0.5 kg	72	Ultrasonic Horn	Improper Mode of contact: Rotanj dryer was used for experiment
		Sand	20	150-170	833	55	--	--	5	"	
Kowlska, et. al	1978	Mineral Sludge	20	800 Watts Power	Fine	2	97	--	26	Piezo electric	The technique appears to be dewatering rather than drying. But it is not mentioned anywhere in the paper.
		Organic Sludge	20	800 Watts Power	Fine	4	83	--	5		

Table 2.

Authors	Year	Type of Material Tested	Frequency kcs/sec	Intensity dB	Particle Size μ-meters	Time of Drying Min.	% Initial Moisture	Amount of Material Tested, Kg	% Moisture Reduction	Type of Sonic Device	Remarks
Swamy, et al	1983	Calcium Carbonate Saw Dust	9.8	139	300-150 1680-420	-- --	22 --	0.75 0.9	85.5% 94.5	Stem-Jet Whistle	Dewatering in combination with centrifuge seems to be very effective.
Muralidhara, Ehsminger	1983	Green Rice	12 19	132 140	Coarse Coarse	100 180	20 20	0.4 kg 0.4 kg	40 40	Whistle Whistle	Lower frequency gave better results.
Muralidhara, Ensminger	1984	Pharmaceutical	12	130	Fine	10	20	0.100	80	Whistle	The experiments were conducted in a fluidized-bed mode.

gas (called cavitation) may be produced. Formation of cavitation is enhanced by the presence of solid particles. Cavitation has the effect of producing rapidly pulsating vacua which may draw fluid from a particle.

Cavitation also produces free ions of the liquid exposed and this phenomenon may be beneficial in a combination system such as ultrasonics and electrophoresis.

Cavitation is associated with high-intensity ultrasound. There are other mechanisms associated with both low intensity and high-intensity ultrasound of value to dewatering and drying. Some of these mechanisms are described as follows:

I. Effects associated with low-intensity ultrasound (noncavitating)--particle migration in a standing wave. Particles suspended in liquids subjected to ultrasound in standing waves migrate toward velocity nodes. The explanation is fairly simple. THe motion of the liquid is maximum at an antinode and zero at the node. Radiation pressure and viscous drag exert pressures on the particles causing them to be "thrown" by the acoustical motion of the liquid into a region of lower displacement. Since the motion at the node is zero, the particles remain at this position as the standing wave persists unchanged.

II. Effects associated with high-intensity ultrasound. Depending upon the nature of the materials and the fluid properties, the following effects may be produced:

(1) High-intensity cavitation produces free chemical ions.
(2) Cavitation erodes or breaks down susceptible particles (this can include metallic particles)
(3) High-intensity cavitation degases liquids
(4) High-intensity cavitation produces uniform dispersions of particulates, especially very small particles (very small particles may result from 2)
(5) High-intensity ultrasonics affects viscosity or the structural properties of fluids in different ways:
    a. Newtonian fluids - Newtonian fluids maintain Newtonian characteristics
    b. Dilatant materials - dilatant fluids tend to stiffen under dynamic (ultrasonic) agitation
    c. Thixotropic materials - thixotropic suspensions become more fluid under the influence of imposed ultrasonic shear stresses.

There is an effect produced by ultrasound on drag forces due to viscosity. The effect may be illustrated by using the simple geometrical configuration of a sphere suspended in a viscous fluid. If r is the radius of the sphere, η_i is the instantaneous viscosity coefficient which is a function of fluid particle velocity u (where u = $j\omega\xi$) and ξ is the fluid particle displacement, the viscous force (or Stokes force) on the particle is given by

$$F = 6\pi r \eta_i u.$$

The quantity $\eta_i u$ does not vanish during one cycle but has an average value

$$(¥_i u)_{ave} = \frac{\gamma - 3}{4} \frac{\eta_o}{c_o} u^2$$

where

γ is the ratio of specific heats
$¥_o$ is the ambient value of viscosity
c_o is the ambient value of the velocity of sound.

(The viscous effect on suspended particles is present in both low intensity and high intensity conditions. The considerations can become quite complex with changes in geometry, intensity, and fluid conditions.)

(6) High intensity ultrasound helps clear surfaces of debris allowing greater fluid contact. This mechanism is useful in enhancing filtration rate.
(7) Ultrasonics of high intensity tend to cause fibrous, clinging types of particles (such as wood paper pulp) to coalesce. Cavitation produces high accelerations to particles within the vicinity of the cavitation bubbles. Particles of materials such as wood pulps also show considerable drag resistance so that parts of the fibers remote from a cavitation bubble do not move at the same rate as the part exposed to the bubble and those particles that do accelerate due to cavitation slow down very rapidly. As a result, the fibers can become entangled and coalesce very rapidly. The principle can be used effectively in suspensions of low concentration.
(8) High-intensity ultrasound can increase the rate of evaporation at a gas/liquid interface. The rate of evaporation at a gas/liquid interface is controlled largely by the difference between saturation vapor pressures of the liquid, P_s, and the vapor pressure in the surrounding atmosphere, P_v (or $\Delta P = P_s - P_v$). At 160 dB in air at standard atmospheric pressure, the RMS value of pressure is 2 kPa. This means that the pressure in the wave swings above and below the ambient pressure by approximately 2.06 kPa at 160 dB. Intensities of 145 to 169 dB are recommended for dewatering when the energy is generated in air and the material to be dewatered is matched acoustically to air. Although the pressure excursion is very low, the high rate at which the pressure oscillates results in a rectified diffusion of moisture into the atmosphere.

(9) High-intensity ultrasound can cause rectified diffusion. Several mechanisms can be identified with this phenomenon depending upon the nature of the material. If ultrasonic energy is well coupled into the medium being dewatered, the material is subjected to a rapid series of contractions and expansions at the frequency of the impressed wave. Water is released much as it is when a sponge is squeezed. Each contraction releases a minute quantity of water until an equilibrium value has been reached. Again, rapid succession of compression and rarefaction produces a noticeable moisture migration.

This phenomenon has been observed in dewatering particulate, fibrous, and soft spongy materials in an air-coupled sound field when these materials were acoustically matched to the atmosphere.

The phenomenon also occurs even in a more densified mass which cannot be well coupled to air if an effective ultrasonic field can be coupled directly into the mass and the mass contains channels of escape for the liquid or vapor. Here, again, coupling and matching of impedances are very important and the manner in which the acoustic energy is applied determines its effectiveness. This generally includes introducing the energy into a cake through a surface under pressure. Stresses can be produced within a cake, such as condensed sewage sludge, that are much higher than those that can be generated in air. If the bias (static) pressure applied to the cake is not excessive the cake will "fracture" during rarefaction in a direction normal to wave propagation and during compression in a direction parallel to wave propagation forming many little channels through which fluid will escape. Thus, whereas an applied static pressure will seal pores and channels so that only a limited amount of water can be eliminated by such pressure, the alternating acoustic stress will make or maintain channels for escape. Thus the maximum solids content can be increased considerably by applying ultrasonic energy to the cake.

Amplitude of vibration, or intensity, is a very important factor in acoustic dewatering. Except for the effects of structural properties that affect attenuation of the wave, dewatering at frequencies within the range of 6 to 33 kHz at constant intensities is similar regardless of frequency.

Other factors influencing dewatering and drying include, under various conditions, such phenomena as high turbulence about particles, Oseen forces (a rectified force attributable to the nonlinearity of high intensity waves in air), and Bernoulli forces (forces of attraction due to reduced pressure in a narrowed passageway such as motion of gas between two stationary objects in close proximity).

Factors which influence the dewatering of particulate matter include:

(1) intensity of the ultrasonic energy at the surfaces of the particles
(2) ability of the vapors to be removed from the area of the surfaces from which they have been released
(3) time of exposure to intensities at effective dewatering levels
(4) wavelength of the sound relative to the particle dimensions, concentration, and thickness of layer of material to be dewatered
(5) particle dimensions or total area exposed to sonic energy
(6) acoustic impedance, density (or inertia) of the particles relative to the surrounding atmosphere
(7) ambient temperature and pressure and
(8) mechanical properties (shape, stiffness, etc.) of the particulate matter.

The inclusion of factors (1) through (3) is obviously logical. When sound energy propagates through a volume containing such particulate matter, the energy is attenuated by scattering and other causes mentioned previously. At a certain distance into the volume, the intensity may attenuate to a level at which it is no longer effective in dewatering. It is this distance that determines the effective dewatering thickness of the volume. The attenuation that determines the thickness is affected by the particle dimensions relative to a wavelength of the sound and by the concentration of particles (Item 4). It is therefore best to avoid using very high ultrasonic frequencies in dewatering such particulate matter for two reasons: (1) the medium in which the sound energy propagates has low acoustic impedance so that the energy is best radiated from a low-impedance, high-amplitude source in air and (2) attenuation both in air and by scattering increases rapidly with frequency in the ultrasonic range.

The rate at which evaporation occurs is directly proportional to the total surface area within a volume exposed to the sonic field.

With respect to Item 6, impedance and density of the particles, these relate to the reaction of the particles to the imposed sound field. In particular, if the inertia of the particles is high compared to the surrounding environment, when they are subjected to a sound field, relative motion occurs between the particles and the surrounding atmosphere. The result is cyclic pressure reduction (or partial vacuum) once on each side of the particles in the direction of wave motion during each cycle. This vacuum produces a rectified vaporization, i.e., the liquid vaporizes faster than it recondenses back onto the surface from which it was removed.

The rate at which evaporation occurs is dependent upon ambient temperature and pressure as discussed previously.

The attenuation as well as the reaction or motion of the materials to the sonic field are functions of the mechanical properties of the particles (Item 8) and, therefore, these factors influence the dewatering rate.

5. SUMMARY

Some very interesting results have been observed in applying ultrasonic energy to dewatering and drying materials. Although favorable drying results with suspended particles or fibrous materials in air have been demonstrated at intensities as low as 135 dB, intensities within the range 145 to 160 dB are preferable in materials that couple well to air acoustically. Air coupled systems may still be considered more applicable to exotic applications where the need for rapid drying is necessary to preserve a product, the product is heat sensitive, and its value justifies the cost.

The synergistic effects which have been observed when ultrasonics has been used in combination with other dewatering techniques, such as electrophoresis, offers very interesting possibilities to drying and dewatering. Ultrasonic, or acoustic, dewatering and drying in the past has been associated largely with air-coupled systems. Drying by these methods was expensive. Ultrasonics applied in combination with electrophoresis differs from the standard notion of acoustic drying, applies the mechanisms in new ways, and is potentially an economical process.

REFERENCES

1. Wakeman, R. J., Filtration Post-Treatment Processes, Elsevier Scientific Publishing Company, New York (1975).
2. Burger, F. J., and Sollner, K., Trans. Far. Soc., 32, 1958 (1936).
3. Brun, E., Boucher, R.M.G., JASA, 5, 163 (1957).
4. Gregus, P., Ultrasonics, 1, 83-86 (1963).
5. Boucher, R.M.G., Ultrasonic News, 4, 8 (1958).
6. Boucher, R.M.G., Ultrasonics News, 5(3), 7-11 (1961).
7. Bongert, W., U.S. Patent 3,970,552 (1976).
8. Sawyer, H. T., U.S. Patent 4,137,159 (1979).
9. Furedi, P., U.S. Patent 4,055,491 (1977).
10. Boucher, R.M.G., Chem. Eng., 20, 83 (1961).
11. Altenberg, K., Naturwissenshaften, 40, 289 (1953).
12. Tamas, Gy; Ronto, Gy; and Tarjan, T. M., Fizikai Folyoirat, 7, 407 (1959).
13. Huxsoll, C. C., and Hall, C. W., Trans ASAE 21-24 (1970).
14. Kowalska, E., Chmura, K., and Bien, J., Ultrasonics, 16, 183-185 (1978).
15. Stephonoff, N. N., U.S. Patent 2,297,726 (1938).
16. CE News Brief, Chem. Engr., 88, 25 (August 10, 1981).
17. Yang, A., U.S. Patent 2,344,754 (1942).
18. Tamburello, N. M., Hill, G. A., Goldmany Seigher A., Solomon, and Y. Timna, 13-1, 13-7, Pro. Pulse Combustion Applications, Atlanta, Georgia, (March 1982).
19. Wilson, J. S., Moore, A. S., and Bowie, W. S., Sonochemical Engineering, CEP Symposium Series (No. 109), (1971).
20. Fairbanks, H. V., Ultrasonics Int. Conf. Proc., IPC. Science Tech Press, London (1973).
21. Fairbanks, H. V., and Cline, R. E., IEEE Transonics and Ultrasonics SU-14 (4); 175-77 (1967).
22. Otsuka, T., Purdum, H., and Fairbanks, H. V., Ultrasonics Int. Conf. Proc. 1977, IPC Science Tech. Press, London, p. 91 (1977).
23. Takashi, F., Yoshinovi, N., Tanaka, Kato, Yasuo, Aromatikkusu, 20 (12), 677-683 (1968).
24. Puskav, A., Material Science Monographs 13, Chapter II, 30-34 (1983), Elsevier Publishing Company, New York, New York.
25. Muralidhara, H. S., Ensminger, D., Mack, G., Mink, W., Battelle Report, (November 1983).
26. Muralidhara, H. S., Ensminger, D., Battelle Report, (March 1984).
27. White, R. E., Tappi, 47 (8), 496 (1964).
28. Swamy, K. M., Rao, A.R.K., Narasimhan, K. S., Ultrasonics, 21 (6), 280-281, November (1983).
29. Ensminger, D., Ultrasonics, Low and High Intensity Applications, Marcel Dekker, New York (1973).

OVER-ALL DRYING-RATE EQUATIONS OF FOODSTUFFS

Kiyoshi Kubota, Kanichi Suzuki, Muneharu Esaka,
Hideki Araki and Mitsuo Nagai

Department of Applied Biological Science, Hiroshima University
Fukuyama, 720 JAPAN

ABSTRACT

Many studies on the drying-rate equations have been performed. They all divide the drying periods into a constant- and a falling-rate period. However, for several foodstuffs the drying phenomena are very complicated for reason of the surface case hardenings, the shrinkage with irregular shapes, the structure material changes and so on. Therefor the drying-rate curves do not always keep a constant shape, and we must take the approximating over-all drying-rate equations using some simple drying models.

In this study, we intend to report on the drying-rate equations based on the drying-shell and the uniform drying models which take in account the shrinkage phenomena, and to report too on the calculations methods of the rate parameters in these equations. From the results with shrinking root vegetables, the conclusion followed that these empirical over-all drying-rate equations are very useful.

1. INTRODUCTION

Drying equations of foods are among those researches that have attracted special attention recently. The reason is that they are so helpful for the improvement of food storage and transportation. In order to design and control various drying apparatuses, it is required first to determine the drying-rate equations and to obtain the rate parameters for these equations.

Basic studies of the drying-rate equations have been done dividing the drying periods into a constant- and a falling-rate period. However, for several foodstuffs the drying phenomena are very complicated for reason of the cell tissues, the complex components, the surface case hardening, the shrinkage with irregular shapes, the biological and chemical component changes and so on. Therefor, the drying-rate curves do not always keep a constant shape, and it is not easy to determine the exact critical moisture content and the exact theoretical parameters such as a diffusion coefficient and so on for the foodstuffs.

In order to design various drying apparatuses, we must take the approximating over-all drying-rate equations based on the some simple drying models, and must determine the rate parameters in these equations.

In this study, we report on the drying-rate equations based on the drying-shell models for the various shapes and the uniform drying models which took in account the shrinkage phenomena, and report too on the calculation method of the rate parameters in these drying-rate equations using a non-linear least square method.

2. DRYING-RATE EQUATIONS

2.1. Empirical rate equation

As types of drying-rate equations, we can consider two types. One type is a theoretical or semi-theoretical drying-rate equation based on various models. The other type is an empirical formula. In a previous papers, we have studied

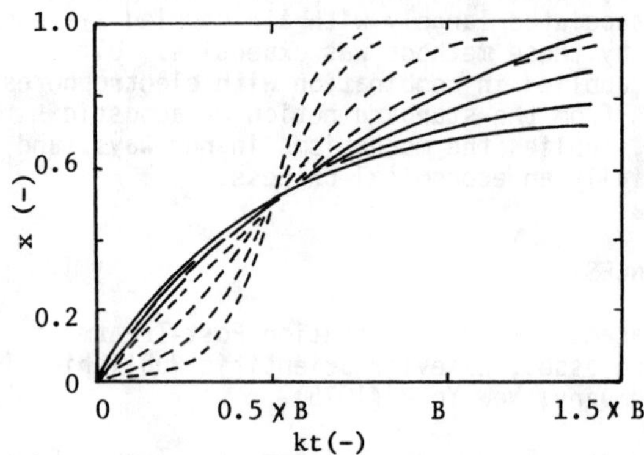

Fig. 1 Relations between the transforming ratio x and the time related rate parameters kt
Solid curves : Eq.(2)
from upper: n = 0.5,1.0,1.5,2.0
where, B = 1.17,1.39,1.66,2.00
Broken curves: Eq.(3)
from upper: α = 0.001,0.01,0.1,0.5
where, B = 13.81,9.25,4.25,1.85

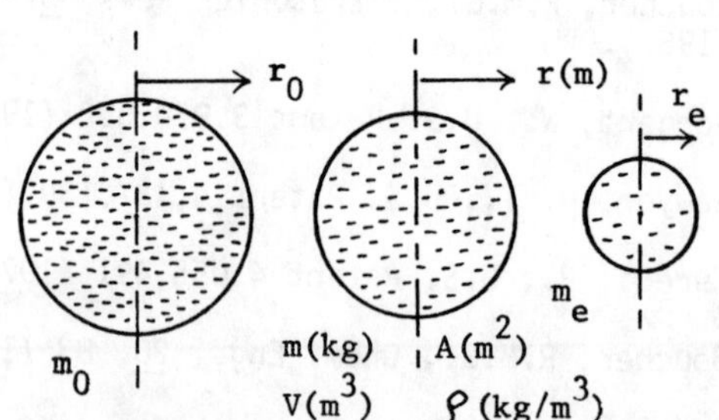

Fig. 2 Uniform drying model for sphere

the empirical rate equations [1] for the various chemical and physical transformations of foodstuffs such as for the chemical reactions, gelatinization [2], soaking [3,4], cooking [3-6], drying [7,8] and so on.

As most of the foodstuffs have complicated components and configurations, their real transforming mechanisms could not be elucidated throughly. However, the degree or extent y (example unit: kg) of chemical reaction, cooking, drying and so on of foodstuffs can be empiricaly expressed as the changing values of a simple property such as mass, reological property and so on.

Most of the transformations have two boundary values of y_0 and y_e correlated to the initial and equilibrium states, respectively. The transforming ratio x (-) as in the following equation is convenient for the various complicated transformations of foodstuffs.

$$x = |(y_0 - y) / (y_0 - y_e)| \qquad (1)$$

In general, the experimental data of the chemical and physical transformations of foodstuffs are obtained as the relationship of y or x vs. t, where, t (s) is the transforming time. Most of the relationships of x vs. t show a monotonous smooth or a S-shape curve. The empirical rate equations may be postulated as follows [1];

nth-order rate equation:

$$dx/dt = k\ (1-x)^n \qquad (2)$$

S-shape rate equation:

$$dx/dt = k\ (1-x)^n\ (x+\alpha) \qquad (3)$$

where, k, n and α are the rate parameters which can be obtained from the experimental data of x vs. t.

The values of n and α are interesting, because these values indicate the form of the curves. The curves in Fig. 1 are obtained from Eqs.(2) and (3) for the various values of n and α by using the time related k [1]. In the previous paper [9], we studied the relationships between the rate parameters n and α and obtained the results as follows: nth-order rate equation can be used for $\alpha \geq 0.75n$, but S-shape rate equation have to be used for $\alpha < 0.75n$.

The values of k can be indicated by using the Arrhenius equation for the chemical reaction, cooking and so on, but for the drying and so on

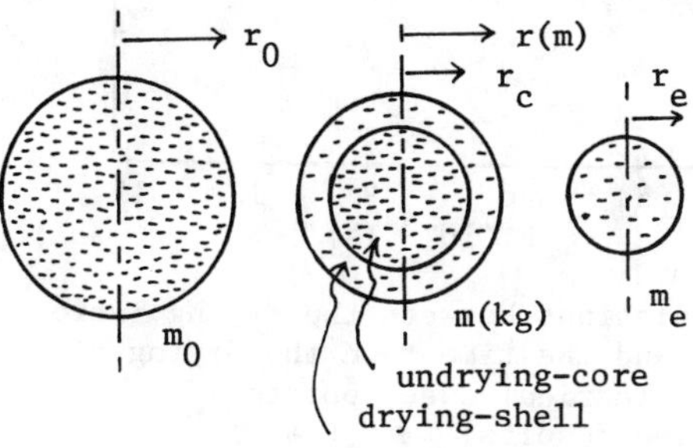

Fig. 3 Drying-shell model for sphere

the values have to be indicated by using the Reynolds number and so on.

2.2. Uniform drying model

As a simple model, we can postulate the uniform drying model [10,11] which has an uniform moisture content throughout the whole of the drying materials, and this moisture content decreases from the initial one to the equilibrium one. Figure 2 illustrates the case of a drying spherical material.

This model can be used for very slow drying of foodstuffs. We postulated the drying-rate equation as follows, assuming that: (1) the resistance of moisture internal diffusion is negligible compared to the resistance of moisture removal from the surface, and the rate of moisture removal from the surface seems to be proportional to the surface area A (m^2), (2) the internal moisture distribution is uniform, and the surface moisture content can be correlated to the exponent of (1-x) as shown in Eq.(2).

$$dx/dt = kA\ (1-x)^n \qquad (4)$$

$$\text{where,}\quad x = (m_0 - m) / (m_0 - m_e) = (w_0 - w) / (w_0 - w_e) \qquad (5)$$

$$w = (m - m_d) / m_d \qquad (6)$$

where, m (kg) and w (kg-H_2O/kg-D.M.) are the mass and the moisture content of the material, and the subscripts 0, e and d show the initial, equilibrium and bone drying states. k and n are the rate parameters which can be obtained from the data of x vs. t.

Next, we assumed that the shrinkage occurs by the removal of the moisture from the material.

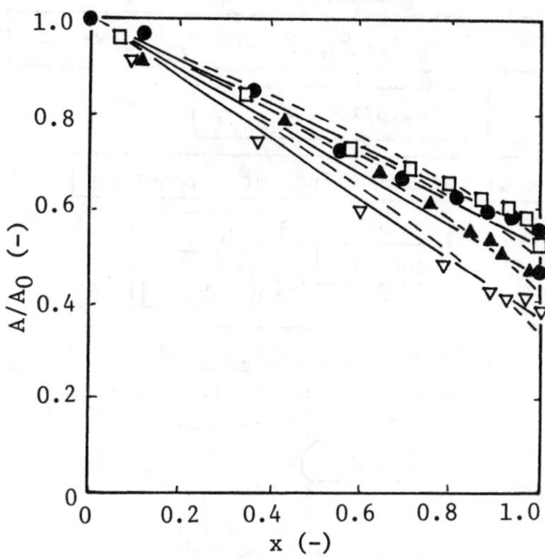

Fig. 4 Relations between the surface-area ratio A/A_0 and the drying ratio x on the drying of cylindrical sweet potato

Observed results: ● □ ▲ ▽

Calculated results:

Eq.(19): —— Eq.(7),(18): – – –

The drying-shrinkage equation for the spherical material can be expressed as follows [11].

$$A/A_0 = (-ax + 1)^{2/3} \tag{7}$$

where, $a = [(m_0-m_e)/\rho_H] / [(m_0-m_e)/\rho_H + m_e/\rho_e]$

where, ρ (kg/m^3) is the density of the material, and the subscript H shows the water' value. This equation can be written as follows.

$$A/A_0 = [(w+a) / (w_0+a)]^{2/3} \tag{8}$$

where, $a = m_e\rho_H/(m_d\rho_e) - m_e/m_d + 1$

When the value of a can be 0.8, this equation is similar to the equation postulated by Kilpatrick et al. [12].

2.3. Drying-shell model

As a another simple model, we may postulate the drying-shell model [11,13] which has a undrying-core and a drying-shell zone, and the former's and latter's moisture contents are postulated as the initial and equilibrium moisture contents, respectively. This model can be used for the very high-rate drying of foodstuffs. Figure 3 illustrates the case of a drying spherical material. This reverse model has been used for the cooking of rice and so on [3,6,14,15] as the soaking- or cooking-shell model.

We postulated the drying-rate equations as follows:

For sphere:

$$dx/dt = [4\pi r_c^2(C_c-C_g)/(m_0-m_e)] / [(r-r_c)(r_c/r)/k_m + (r_c/r)^2/h_m] \tag{9}$$

where, $r_c = (1-x)^{1/3}\, r_0 \quad (10)$

$$r = \{r_e^3 + [1-(r_e/r_0)^3]r_c^3\}^{1/3} \tag{11}$$

For long-cylinder:

$$dx/dt = [2\pi r_c \ell (C_c-C_g)/(m_0-m_e)] / [r_c \ln(r/r_c)/k_m + (r_c/r)/h_m] \tag{12}$$

where, $r_c = (1-x)^{1/2}\, r_0 \quad (13)$

$$r = \{r_e^2 + [1-(r_e/r_0)^2]r_c^2\}^{1/2} \tag{14}$$

For infinite-slab:

$$dx/dt = [2A(C_c-C_g)/(m_0-m_e)] / [(r-r_c)/k_m + 1/h_m] \tag{15}$$

where, $r_c = (1-x)\, r_0 \quad (16)$

$$r = r_e + [1-(r_e/r_0)]r_c \tag{17}$$

where, r (m), ℓ (m) and A (m^2) are the radius or half-thickness, length and one side surface area of materials; r_c (m) are the radius or half-thickness of undrying-cores; C_c and C_g (kg-H_2O/m^3-void) are the moisture concentrations at the undrying-core surface and the gas-film surface. h_m (m^3-void/m^2 s) and k_m (m^3-void/m s) are the rate parameters which can be obtained from the data. These parameters show the parameters in respect to the diffusions at the gas-film and the drying-shell parts of the materials, respectively. For the cases of a partly diffusion controlling only, we could integrate the drying-rate equations, and the parameters could be calculated in explicit function [13]. However, for the case of two

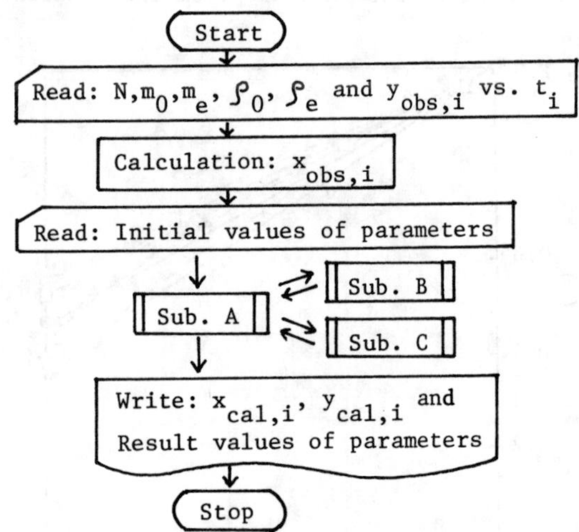

Fig. 5 Flow chart for calculation of parameters

Sub. A: non-linear least square method, Sub. B: Runge-Kutta-Gill method and rate equation, Sub. C: Gauss elimination method

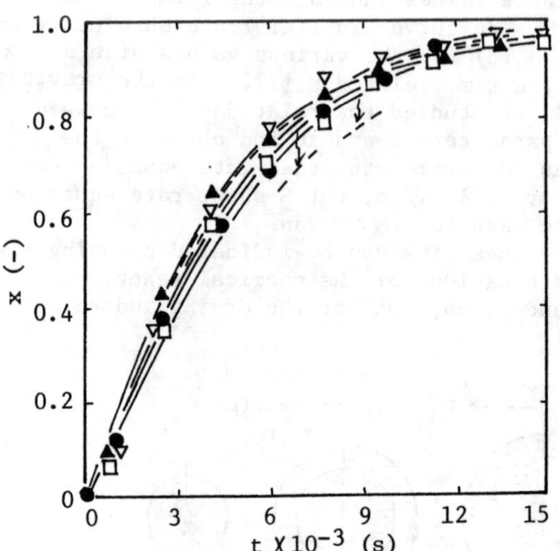

Fig. 6 Relations between the drying ratio x and the time t on the drying of cylindrical sweet potato

Observed results: ● □ ▲ ▽

Calculated results:

Eq.(12): —— Eq.(4),(7): – – –

part controllings, the parameters could not be calculated analytically so they should be calculated by using the method shown in a later section.

If we assume that the shrinkage occurs in this model, the drying-shrinkage equation for the spherical material can be expressed as follows [11].

$$A/A_0 = (-ax + 1)^{2/3} \tag{18}$$

where, $a = 1 - m_e \rho_0/(m_0 \rho_e)$
$= 1 - V_e/V_0$

where, V (m^3) is the volume of the material.

2.4. Drying-shrinkage equation

The drying-shrinkage equations can be obtained by using simple models as shown in Figs. (7) and (18). Another simple equation can be postulated as follows [11].

For $A = A_0(1-x) + A_e x$:

$$A/A_0 = (-ax + 1)^{2/3} \tag{19}$$

where, $a = 1 - A_e/A_0$

For $V = V_0(1-x) + V_e x$:

same to Eq.(18)

For $\rho = \rho_0(1-x) + \rho_e x$ [10]:

$$A/A_0 = [(-ax+b) / (-x+b)]^{2/3} \tag{20}$$

where, $a = [(m_0-m_e)\rho_0] / [m_0(\rho_0-\rho_e)]$
$b = a + m_e \rho_0/[m_0(\rho_0-\rho_e)]$

Other equations such as those based on the semi-core drying model and so on [16] have been postulated by authors too. The drying-shrinkage data are very scattered and show very complex curves as can be seen in Fig. 4 [8], therefore the more complex equations may be omitted. In Fig. 4, the differences between the data and the calculated values by Eqs. (17), (18) and (19) are very nearly, therefore, Eq.(19) is very useful as the approximating drying-shrinkage equation because this is a most simple equation.

In the shrinkages of the long-cylindrical and the infinite-slab food materials, the length and the one side surface area are not constant. Therefore, the following equations [8,11] were used as the approximating equations of ℓ in Eq. (12) and A in Eq.(15), respectively.

$$\ell = \ell_e + [1-(\ell_e/\ell_0)](1-x)\ell_0 \tag{21}$$

$$A = \left\{A_e^2 + [1-(A_e/A_0)^2](1-x)A_0^2\right\}^{1/2} \tag{22}$$

2.5. Calculation method of rate parameters

The experimental data are generally obtained as the integral data of x vs. t. The derivative values of dx/dt in the rate equations can not be obtained reliably from the data x vs. t. Therefore, the differential method is not better than the following integral method. The analytical integral method can be used for the simple rate equations such as Eq.(2), but not for the complex rate equations such as Eqs.(2) and so on.

The numerical integral method is most successful when using a digital electronic computer. Thus, the rate equations were integrated numerically using the Runge-Kutta-Gill method, and the rate parameters were calculated by a non-linear least square method [1,10,13]. (Correction at p.14 in [13]: B(1)=0.0 B(1)=1.0). The values of the following standard deviation σ (-) for the variable x were minimized.

$$\sigma = [\sum_{i=1}^{N} (x_{obs}-x_{cal})_i^2/N]^{1/2} \tag{23}$$

where, x_{obs} and x_{cal} are the observed and calculated values of x, and N is the total number of data. The flow chart for the calculation is shown in Fig. 5. We used the digital electric computer FACOM M-200 in the Computation Center of Nagoya University. The drying-rate data are

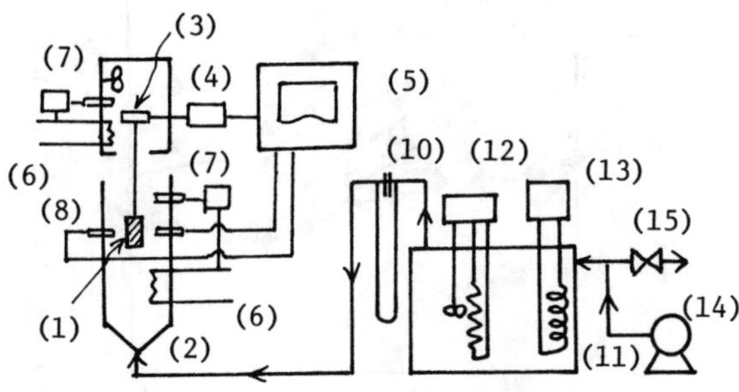

Fig. 7 Experimental apparatus on the heated air-flow dryer
(1) sample, (2) drying chamber, (3) strain gage, (4) strain amplifier, (5) recorder, (6) heater, (7) thermoregulator and transister relay, (8) dry-bulb thermocouple, (9) wet-bulb thermocouple, (10) orifice flow meter, (11) water bath, (12) heater unit, (13) cooler unit, (14) blower, (15) valve

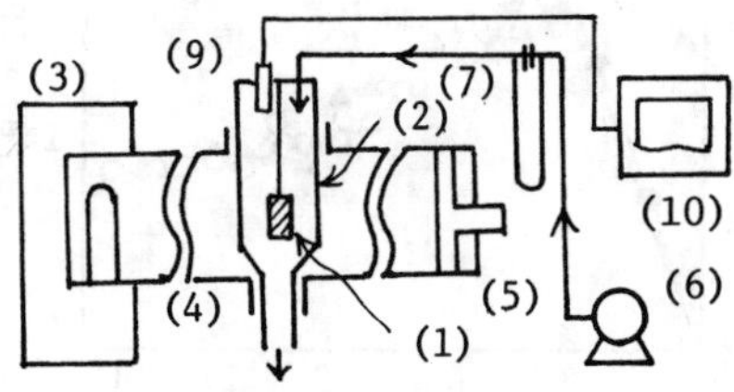

Fig. 8 Experimental apparatus on the microwave-heat dryer
(1) sample, (2) drying chamber, (3) microwave generator, (4) wave guide, (5) reflux plate, (6) air pump, (7) orifice flow meter, (8) air outlet, (9) thermocouple, (10) recorder

very scatterd and show very complex curves as appears in Fig. 6 [8].

3. EXPERIMENTAL

3.1. Heated air-flow drying

The apparatus used in the drying experiments by heated air-flow is shown in Fig. 7 [7,17]. The cylindrical dryer was made from acrylic resin, had a 11 cm inside diameter and 60 cm length. The upper part of the dryer was open, and had the strain gage chamber set upon. The sample was hung up in the dryer from the strain gage by using a fine thread. The mass change of the sample was determined by using a strain gage connected to a recorder. Velocity, temperature and humidity of the air were controled.

3.2. Microwave-heat drying

The apparatus used in the drying experiments by microwave energy is shown in Fig. 8 [17]. The wave guide was made from a copper box with a cross section of 5.9 X 10.9 cm and 77.0 cm length. The wave distribution in the wave guide can be changed by using a movable reflux plate. If the vapor of water is not removed, it accumulates on the guide wall and causes a reduction of microwave strength. The air was flown in by using a cylindrical drying chamber. The mass change of the sample was determined by using a chemical balance.

4. RESULTS AND DISCUSSIONS

In previous papers [7,8], we compared the calculated results between the nth-order rate equation and the rate equations based on the uniform drying model and so on. The differences between the results are very small as shown in Fig. 6, then the drying-rate equation which shown in Eqs. (4) and (19) is most useful as the approximating over-all drying-rate equation because this is the simpliest model equation.

In the over-all drying-rate equation shown in Eqs. (4) and (19), if the values of rate parameters k and n can be determined by the sample conditions such as the shape, diameter and so on, and the drying conditions such as the air flow rate, temperature, humidity and so on, we can use the drying-rate equations in order to optimum design various drying equipments of foodstuffs. However, we can not yet discuss these considerations because the data are still too little.

The results on the heated air-flow drying of the unbored and bored potatoes has been compared [17] by using the over-all drying-rate equation. The result was: no difference in both cases under our experimented conditions as shown in Fig. 9.

The microwave-heat drying phenomena are very complicated due to the temperature changes and the moisture content distributions can not be accurately observed. Consequently, the drying-rate equation based on the theoretical model can not be obtained. The results on the microwave-heat drying of potatoes can be expressed by using the over-all drying-rate equation as shown in Fig. 10 [17]. The values of n are found to be between 1.5 and 2.0 in our experimented and assumed conditions. The values are much larger than the values obtained to be between 0.60 and 0.65 on the heated air-flow drying. From this, we may infer that the drying rates of potatoes by microwave energy at the initial drying time are much larger than those ones by heated air-flow drying.

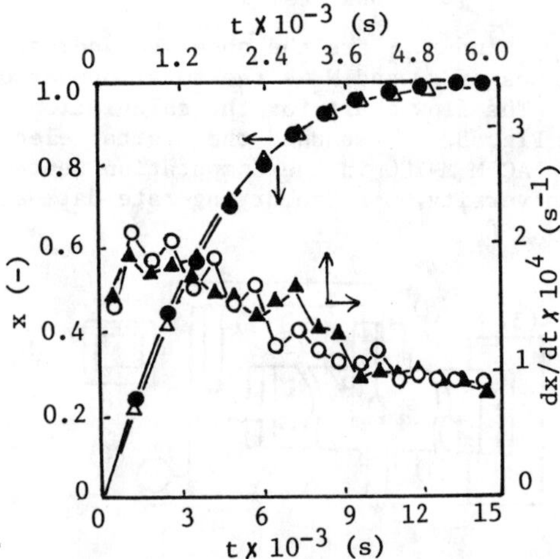

Fig. 9 Relations between the drying ratio x, the rate dx/dt and the time t on the drying potatoes by heated flowing air
Observed results (five average):
no boring: ● ○
boring : △ ▲
Calculated results: two curves

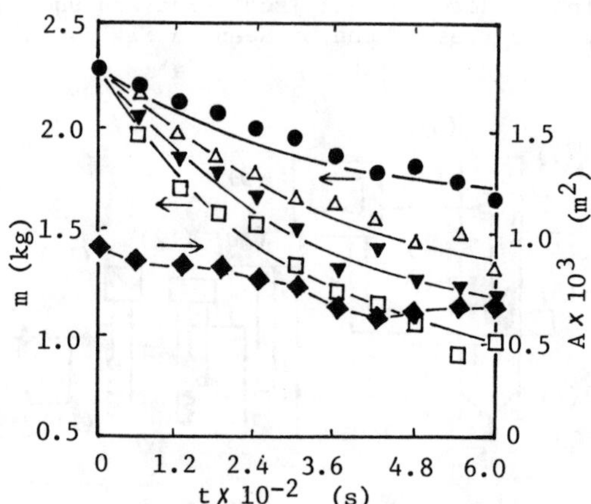

Fig. 10 Relations between the mass m, the surface area A and the time t on the drying potatoes by microwave energy
Observed results:
electric current on magnetron anode: 10 20 30 40 40 mA
● △ ▼ □ ◆
Calculated results: four curves

When the drying continued, the coloured section appeared as shown in Fig. 11. This phenomenon appeared too when using an electronic range on the

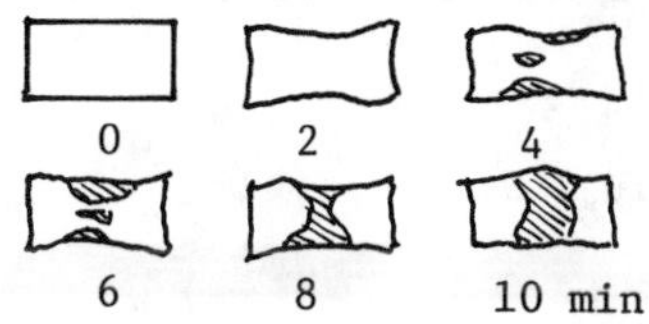

Fig. 11 Cross sections of dryed potatoes by microwave energy (40 mA)
lined area: coloured section

market (National Denki Co., Type NE-6310, 2450 MHz Input 1.15 kW, Output 240, 600 W) which microwave generated was stirred with a fun. In this paper, we did only drying in the initial drying region, since the coloured section appeared. Some considerations should be given to these countermeasures.

NOMENCLATURE

A	surface area or one side surface area	m^2
a, b	parameters in drying-shrinkage equations	
C	moisture concentration	kg-H_2O/m^3-void
h_m, k_m	parameters in drying-shell model	
k, n, α	parameters in drying-rate equations	
N	total number of data	-
r	radius or half-thickness	m
x	ratio or extent by transforming, such as drying	-
y	changing property by transforming, such as drying	kg etc.
σ	standard deviation	-

Subscripts:

0, e, d	initial, equilibrium and bone drying states
H, c, g	water's, undrying-core's, gas-film's values
obs, cal	observed, calculated values

REFERENCES

1. Kubota, K., J. Fac. Appl. Biol. Sci., Hiroshima Univ., vol. 18, 11 (1979).
2. Kubota, K., Hosokawa, Y., Suzuki, K. and Hosaka, H., J. Food Sci., voll. 44, 1394 (1979).
3. Kubota, K., J. Fac. Appl. Biol. Sci., Hiroshima Univ., vol. 18, 1 (1979).
4. Kubota, K., ibid, vol. 18, 161 (1979).
5. Kubota, K., Takasaki, K., Fujimotò, M., Suzuki, K. and Hosaka, H., Nippon Shokuhin Kogyo Gakkaishi, vol. 27, 157 (1980).
6. Kubota, K., Fujimoto, M., Suzuki, K. and Hosaka, H., ibid, vol. 27, 381 (1980).
7. Kubota, K., Matsumoto, T., Suzuki, K. and Hosaka, H., ibid, vol. 28, 491 (1981).
8. Matsumoto, T., Kubota, K., Suzuki, K. and Hosaka, H., ibid, vol. 29, 238 (1982).
9. Kubota, K., J. Fac. Appl. Biol. Sci., Hiroshima Univ., vol. 18, 171 (1979).
10. Kubota, K., Suzuki, K., Hosaka, H., Hosokawa, Y. and Hironaka, K., J. Fac. Fish. Anim. Husb., Hiroshima Univ., vol. 16, 131 (1977).
11. Kubota, K., Matsumoto, T., Suzuki, K., Hasegawa, T. and Hosaka, H., J. Fac. Appl. Biol. Sci., Hiroshima Univ., vol. 20, 99 (1981).
12. Kilpatrick, P.W., Lowe, E. and Van Arsdel, W. B., Advances in Food Reseach (Academic Press), vol. 6, 360 (1955).
13. Kubota, K., Suzuki, K., Hosaka, H., Hirota, R. and Ihara, K., J. Fac. Fish. Anim. Husb., Hiroshima Univ., vol. 15, 1 (1976).
14. Kubota, K., Suzuki, K., Hosaka, H., Hironaka, K. and Aki, M., ibid, vol. 15, 135 (1976).
15. Suzuki, K., Kubota, K., Omichi, M. and Hosaka, H., J. Food Sci., vol. 41, 1180 (1976).
16. Suzuki, K., Kubota, K., Hasegawa, T. and Hosaka, H., ibid, vol. 41, 1189 (1976).
17. Kubota, K., Araki, H., Nagai, M., Kintou, H., Suzuki, K. and Esaka, M., J. Fac. Appl. Biol. Sci., Hiroshima Univ., in press.

ACKNOWLEDGMENTS

This work has been supported in part by Grant-in-Aid for Energy Special-type Research No. 58045110 (in 1983) and Total-type Research (A) No.58360011 (in 1983) of Ministry of Education, Science and Culture of Japan.

ENZYME INACTIVATION DURING DRYING OF A SINGLE DROPLET

Shuichi Yamamoto, Masahiko Agawa, Hideo Nakano and Yuji Sano

Department of Chemical Engineering,
Yamaguchi University, Tokiwadai, Ube 755, Japan

ABSTRACT

The effects of various drying conditions on the enzyme inactivation during drying of a single droplet were investigated. Moisture, temperature and enzyme activity of the drop during drying were measured as a function of drying time. The enzyme inactivation rate constants and the diffusion coefficients were measured as functions of moisture content and temperature. By use of these values, the drying histories including the enzyme retention in the drop were computed and compared with the experimental results. Good agreements were observed between the experimental and calculated results. A simple correlation between the enzyme retention and the initial drop radius was derived theoretically and verified experimentally. Furthermore, the effects of oxidation and pH change during drying on the enzyme inactivation were examined experimentally by the addition of a reducing agent and also by using nitrogen as a drying medium. It was found that these additional effects as well as the thermal inactivation must be considered in order to predict the enzyme retention during drying.

1. INTRODUCTION

During drying of biological materials several desirable and undesirable biochemical reactions may occur. For example, in the production of proteases for washing detergents it is desirable to retain the initial enzyme activity at the end of the drying. On the other hand, some enzymes in food materials which catalyze undesired degradation reactions during storage must be completely inactivated. Although excellent studies have been presented by researchers in Agricultural University,Wageningen[1-3], quantitative studies on the enzyme inactivation during drying, in which calculations and experiments are compared, are rarely found except for the work of Liou[3]. Consequently, it is still difficult to predict the final residual enzyme activity of the dried products theoretically.

In our previous studies[4-5], the drying behaviour of a drop containing a dissolved solid such as skimmilk and polymer was investigated both experimentally and theoretically. The same procedure can be applied to the inactivation behaviour during drying of a drop containing enzyme if the inactivation rate equation is incorporated into the model and the inactivation rate constants are available as functions of moisture contents and temperature.

In this study, the effects of various drying conditions on the enzyme inactivation during drying of a single droplet were investigated. The drying experiments of a single droplet were carried out over a wide range of experimental conditions. The enzyme activity in the drop as well as the moisture content and temperature of the drop was measured as a function of drying time at a given set of drying conditions. The enzyme inactivation rate constants and the diffusion coefficients were measured as functions of moisture contents and temperature. By use of these data, a computer simulation of the drying experiment was done on the basis of a model which considered the combined heat and water transfer, and enzyme inactivation inside the drop.

Furthermore, the effects of oxidation and pH changes during drying on the enzyme inactivation were examined experimentally by the addition of a reducing agent and also by using nitrogen as a drying medium.

2. Materials and Methods

2.1. Materials

As a solute sucrose was employed due to its high solubility in water. The following enzymes were used: glucose oxidase (GOD) from Toyobo Co., alkaline phophatase (AP) from Sigma Co., and alcohol dehydrogenase(ADH) from Kojin Co. Three different buffer solutions(0.01M, pH 5.3 at 60°C) shown in Fig. 6 were used as a solvent. As a reducing agent, dithiothreitol(DTT) was employed. All the reagents were of analytical grade.

2.2. Measurement of enzyme activity

The GOD activity was measured by the indamine dye method which uses the peroxidase enzyme to utilize the hydrogen peroxide formed to convert the dye to an optically active state. The rate of colour production was measured at 510nm spectrophotometrically. The activity of AP was also determined spectrophotometrically at 420 nm with p-nitrophenyl phosphate as a substrate. Similarly, the activity of ADH was measured at 340 nm with nicotinamide adenine dinucleotide and ethanol as substrates.

2.3 Inactivation experiments

A sucrose solution of a desired concentration containing 3.5×10^{-5} wt% GOD was incubated in a constant temperature bath maintained constant within ±0.05°C. At different periods of time, aliquots were taken and the remaining activity of the enzyme was assayed as described above.

2.4. Drying experiments

A drying experiment of a single droplet was carried out by the same method as previously reported[4,5] except for the measurement of the enzyme activity. In order to measure the remaining activity of the enzyme in the drop during drying, the drop was suspended by a glass filament which was removable from the drying apparatus. At different lengths of drying time, it was removed and dissolved in an appropriate buffer solution and the remaining enzyme activity was measured as described above.

Fig.1 shows a schematic diagram of another drying apparatus employed for the investigation of the effect of oxidation on the enzyme inactivation during drying. A dessicated air or nitrogen gas is fed to the drying chamber having a volume of 500 cm^3 at a flow rate of 2 l/min. The weight of a drop during drying in this apparatus was measured by use of a micro-balance. Other techniques are the same as described above.

Dissolved oxygen in the solution employed for both the inactivation and drying experiments was purged with nitrogen prior to use. However, this procedure was found to have a minor effect on the inactivation of the enzyme employed in this study.

2.5. Method for the determination of diffusion coefficient

An isothermal drying of a slab was carried out by the apparatus shown schematically in Fig.2. A sucrose solution of desired concentration containing less than 1 wt% of agar-agar was poured into an aluminium dish (inner diameter 4.5cm, depth 5 mm). The weight of sucrose solution was measured as a function of time by using a micro-balance. In the case of high water concentrations, a capillary-cell method[6] was employed. In this method, the sucrose concentration was determined by an Abbe refractometer. From the experimental results obtained by these methods, the concentration dependent diffusion coefficient was determined by the method described elsewhere[7].

3. THEORETICAL CONSIDERATION

In our previous studies[4,5], the drying behaviour of a single droplet of skimmilk and of polymer was investigated on the basis of a model which considered the temperature and concentration dependencies of the diffusion coefficient and the water activity, and the shrinkage of the drop. This model can also handle the enzyme inactivation during drying of a droplet, if the inactivation rate equation is incorporated in the model. Under the assumptions that the drop shrinks due to the loss of water, keeping a spherical symmetry, the temperature of the drop is uniform, and the water movement inside the drop is governed by molecular diffusion only, the following set of partial and ordinary differential equations are obtained.

$$\frac{\partial u}{\partial t} = \frac{\partial}{\partial z}\left(D\rho_s^2 r^4 \frac{\partial u}{\partial z}\right) \quad (1)$$

$$\overline{\rho\hat{C}_p}\frac{dT}{dt} = \frac{3}{R}\left[\alpha(T_{air} - T) - j_{w,i}\,\Delta h_v\right] \quad (2)$$

where u is the water content (kg water/kg dry solid), T is the temperature of the drop and t is the drying time. D is the diffusion coefficient which depends on both temperature and water content. Although the equations in our previous studies[4,5] are given in stationary coordinates, in this study they are presented in reference component mass centered coordinates[1-3,8].

Initial and boundary conditions are given by

$$t = 0,\quad 0 \le z \le Z,\ u=u_0 \quad (3)$$

$$t > 0,\ z=0,\ D\rho_s^2 r^2(\partial u/\partial z)= 0 \quad (4\text{-}1)$$

$$z=Z,\ D\rho_s^2 R^2(\partial u/\partial z) = -j_{w,i} \quad (4\text{-}2)$$

$$j_{w,i} = k_g\,(a_w\rho_w^{sat} - \rho_{w,b}) \quad (5)$$

The heat and mass transfer coefficients are calculated by the correlations proposed by Toei et.al [9].

If the inactivation of enzyme can be represented by a first-order rate equation, then the residual enzyme activity in the drop, E/E_0, is given by

$$\ln(E/E_0) = -\int_0^t k(u,T)\,dt \quad (6)$$

where k is the first order inactivation rate constant which depends on water content and temperature. In the derivation of this equation, the velocities of enzyme and solute are assumed to be equal.

The above set of differential equations are solved numerically by the Crank-Nicolson finite difference method. In this calculation, the experimentally measured k and D are employed. For other physical properties, the values and equations presented in Liou's thesis[3] are used.

4. Results

4.1. Diffusion coefficient

The diffusion coefficient of sucrose-water system determined by the method proposed elsewhere[7] are shown in Fig.3. The experiments were carried out at 30°C and 50°C. The results were fitted by Eqs. (7)-(9), which several workers already employed for this type of concentration dependence[1,2,5]. The values extrapolated to 25°C by Eqs. (7)-(9) and the activation energy as a function of u are in good agreement with the values in the literature[10].

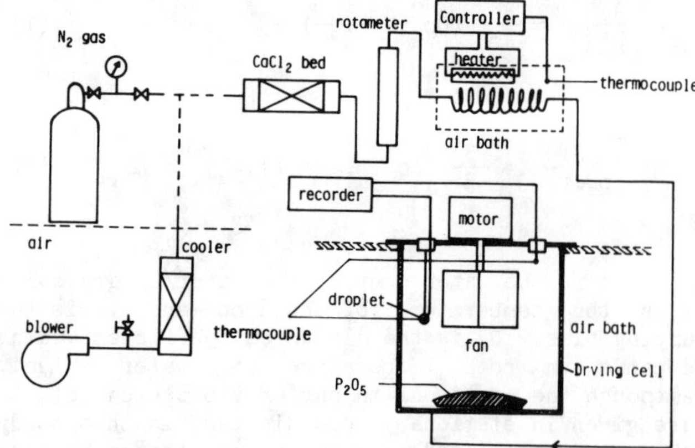

Fig.1 Drying apparatus with nitrogen as a drying gas

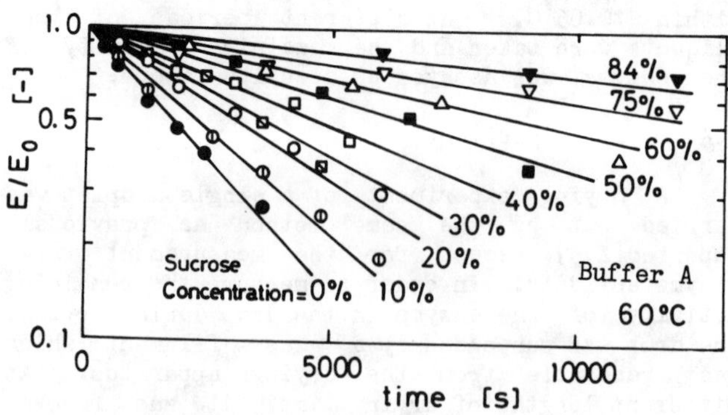

Fig.4 Residual activity of GOD as a function of time at various sucrose concentrations. Sucrose concentrations are expressed as weight%.

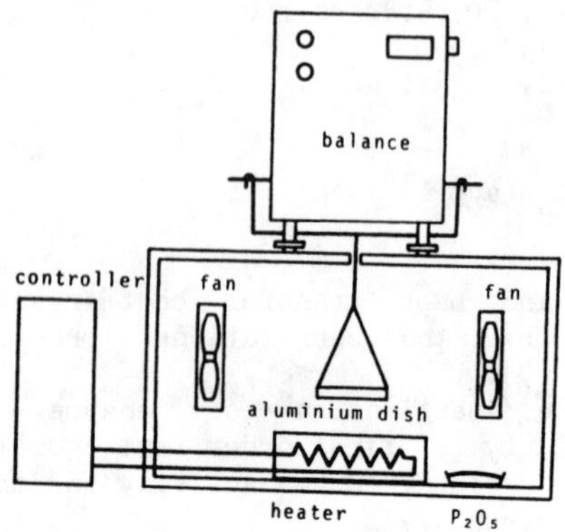

Fig. 2 Experimental apparatus for the measurement of the diffusion coefficient

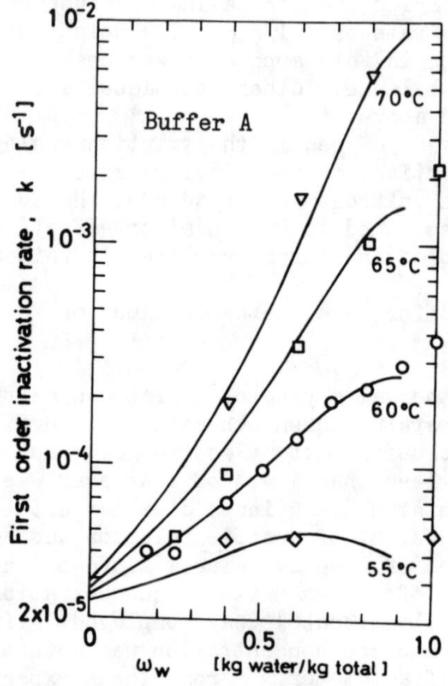

Fig. 5 First order inactivation rate constants as function of moisture content and temperature.

No appreciable difference was found between the experimental results with and witout DTT.

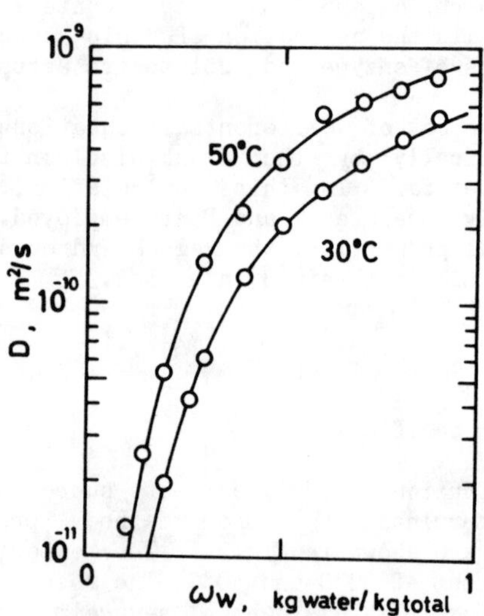

Fig. 3 Concentration dependence of diffusion coefficient of sucrose.

Table 1. Values for the parameters in Eqs.(10)-(12)

Buffer	T_R [°C]	k_1 [s^{-1}]	k_2 [s^{-1}]	p [-]	E_{a1} [kJ/mol]	E_{a2} [kJ/mol]	q [-]
A	60	2.5×10^{-5}	2.6×10^{-4}	0.85	10	340	0.50
B	60	1.5×10^{-5}	4.0×10^{-4}	0.46	46	390	0.32
C	65	3.0×10^{-5}	1.8×10^{-3}	0.50	40	300	0.80

4.2. Inactivation rate constants

Figure 4 shows the results of the inactivation experiment, in which the residual activity, E/E_0, was measured as a function of time at various sucrose concentrations. As is clear from the figure, the inactivation behavior of GOD is found to be expressed by a simple first order inactivation mechanism. As the sucrose concentration increased, the inactivation rate decreased.

In Fig.5, the first order inactivation rate constants obtained are plotted against a mass fraction of water, ω_w at various temperatures, T. k decreases strongly with decreasing ω_w, which means that GOD becomes stable at lower water concentrations. These experimentally measured values of k were then fitted by the following equations, which Luyben et al. [2] employed for the description of inactivation rate constants for several enzymes.

$$k = k_R \exp[-(E_a/R)(1/T - 1/T_R)] \quad (10)$$

$$\ln k_R = \ln k_2 + (\ln k_1 - \ln k_2) \exp(-pu) \quad (11)$$

$$E_a = E_{a2} + (E_{a1} - E_{a2}) \exp(-qu) \quad (12)$$

We also measured k for Buffer B and Buffer C. Since those values were similar to the results in Fig. 5, only the parameters in Eqs.(10)-(12) fitted to the experimental results are shown in Table I.

4.3. Drying experiments

Effect of the change of pH

Fig.6 shows the experimental results of drying of a single droplet at an air temperature T_{air}=70°C. Other conditions are shown in the figure. The calculated results for the average water contents, $\overline{u}$, and the temperature, T are in fairly good agreement with the experimental ones. The calculated average residual activity of GOD, $\overline{E}/E_0$, for three different buffers are indistinguishable and retains almost 95% of its initial value until the final stage of the drying. On the other hand, the observed value of $\overline{E}/E_0$ for Buffer A decreases with time. This might be ascribable to the effect of the change of pH since in Buffer A only acetic acid evaporate and pH might increase. The stability of GOD decreases with the increase of pH. On the other hand, both ethylenediamine and acetic acid in Buffer C evaporate and their boiling points are almost the same. The compositions of Buffer B, citric acid and sodium phosphate dibasic, do not evaporate. The experimental results for these buffers retained around 95% of its initial activity until the final stage of the drying. Oxidation is another possible factor which causes the inactivation of enzyme during drying. In order to examine the effect of oxidation, a reducing agent, dithiothreitol(DTT) at a concentration of 2mM was added to the sucrose solution of Buffer A. As shown in Fig.6, the enzyme with DTT retained higher residual activities than those without DTT. From the above results, we decided to employ Buffer C as solvent in the following experiments.

Effect of air temperature

Figs. 7 and 8 show the experimental results at T_{air}=80°C and 90°C, respectively. Other conditions are the same as in Fig. 6. From these figures, it is seen that the inactivation occurs during the period where the temperature of the drop changes from the wet-bulb temperature to the air temperature. Agreements between the calculated and experimental results are good. In Fig. 9, the residual enzyme activities at t=600 s for various air temperatures are shown. The residual activity decreased with the increase of the air temperature. The effect of the addition of DTT is remarkable and the results with DTT are in good agreement with calculated results.

Effect of the initial drop radius

In our previous studies[4,5], we have shown that the drying behavior can be represented similarly by use of the reduced time t/R_0^2, (R_0 is the initial drop radius) when the Reynolds numbers based on the initial drop radius, Re_0 are the same. This can be readily shown from Eqs.(1)-(5), if they are normalized by t/R_0^2. This relation was verified by the experimental results shown in Fig. 10, in which T and $\overline{u}$ for three different sizes of drops are plotted against t/R_0^2. However, $\overline{E}/E_0$ at the same t/R_0^2 decreases with the increase of R_0. In other words, under the above conditions small droplets can retain higher enzyme activities at the end of drying. This can be explained with the aid of Eq.(6) as follows.

$$\ln(E/E_0) = -R_0^2 \int_0^{t/R_0^2} k \, d(t/R_0^2) \quad (13)$$

Integrand in the right hand side of Eq.(13) is constant when Re, u_0 and T are fixed. Therefore, $\ln(E/E_0)$ is directly proportional to R_0^2. Fig.11 shows plots of $\overline{E}/E_0$ at $t/R_0^2 = 4 \times 10^8$ vs. R_0^2. It is clear from the figure that the relation predicted by Eq.(13) holds for these experimental results. Furthermore, the experimental results of GOD without DTT and of AP also obeyed this relation. Therefore, for an enzyme whose inactivation rates are unknown, the effect of R_0 can be readily predicted according to this plot when at least one experimental result for a certain R_0 is obtained.

Effect of oxidation

The results described in the preceding sections show that oxidation plays an important role in the inactivation of enzyme during drying. Therefore, we examined the effect of oxidation by using nitrogen as a drying gas and also by the addition of DTT to several enzymes.

Fig.12 shows the results of drying experiment carried out by the apparatus shown in Fig. 1. As is clear from the figure, the loss of $\overline{E}/E_0$ for nitrogen is smaller than that for air.

We examined the effect of the addition of DTT on the inactivation of several enzymes during drying. It was found that to enzymes containing sulphydryl groups such as GOD and ADH the addition of DTT was effective. On the other hand, the remaining activity of AP, which has no sulfhydryl group, was not changed by the addition of DTT.

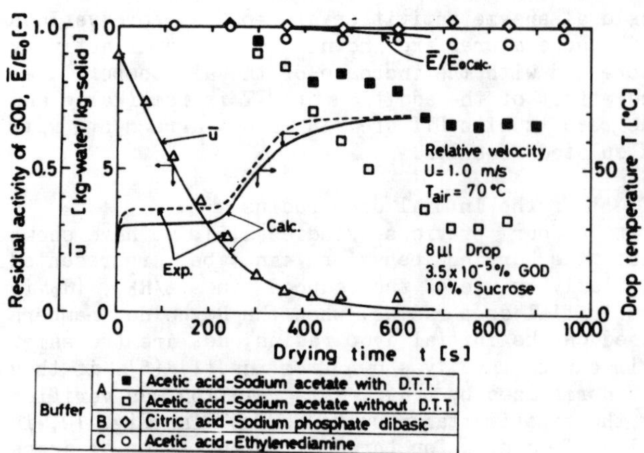

Fig. 6 Average moisture content, temperature and residual enzyme activity as a function of drying time.

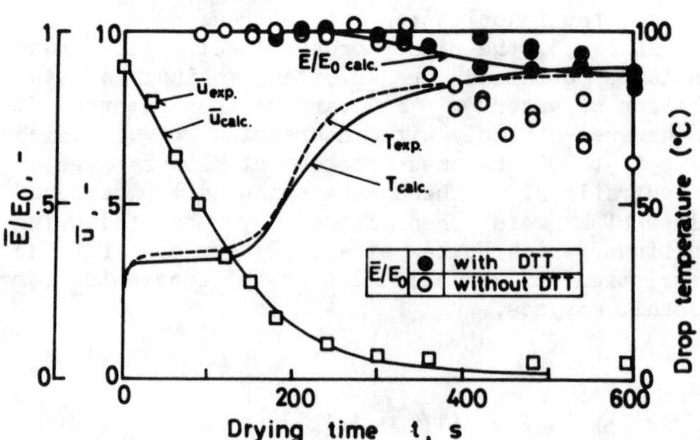

Fig. 8 Effect of air temperature on the enzyme retention during drying.

T_{air} = 90°C.

Other conditions are the same as in Fig. 6.

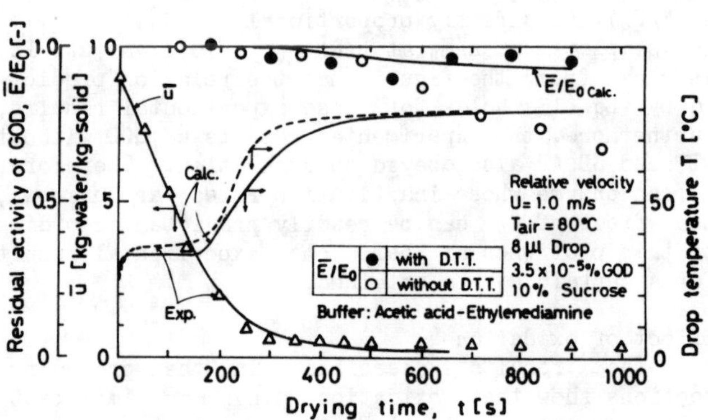

Fig. 7 Effect of air temperature on the enzyme retention during drying.

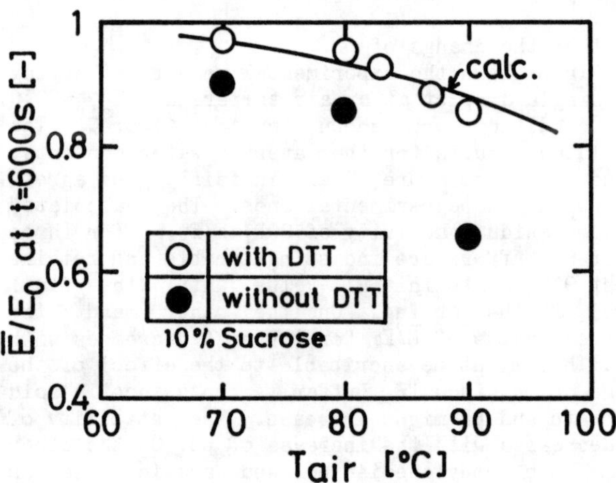

Fig. 9 Effect of air temperature on the enzyme retention at t=600 s.

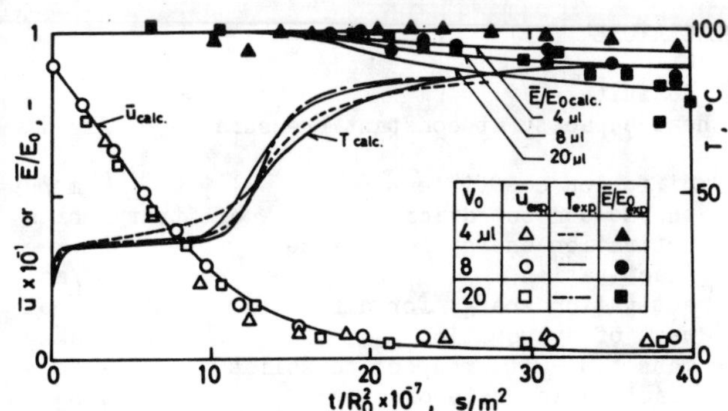

Fig. 10 Water content, temperature and enzyme retention as a function of reduced time. Vo is the initial drop volume in µl.

T_{air} = 90°C, R_0 x U = 1.241 x 10^{-3} m^2/s

Sample solution is the same as in Fig. 6.

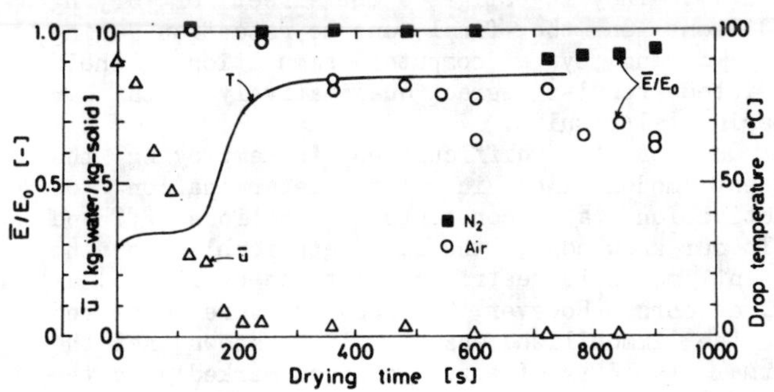

Fig. 12 Effect of drying medium on the enzyme retention

T_{air} = 83°C, 8 µl drop.
Sample solution is the same as that in Fig.6.

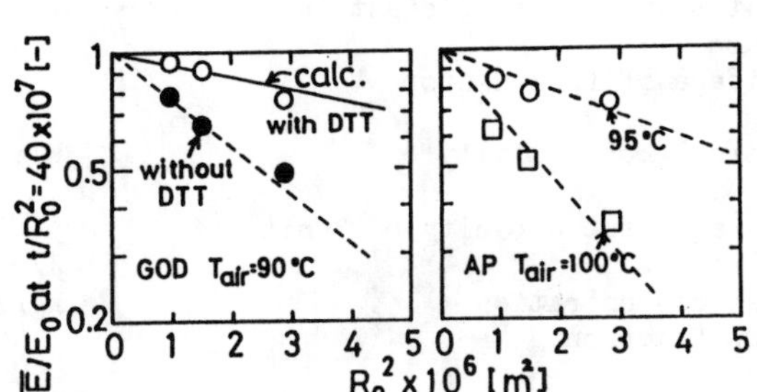

Fig. 11 Effect of intial drop radius on the residual enzyme activity.

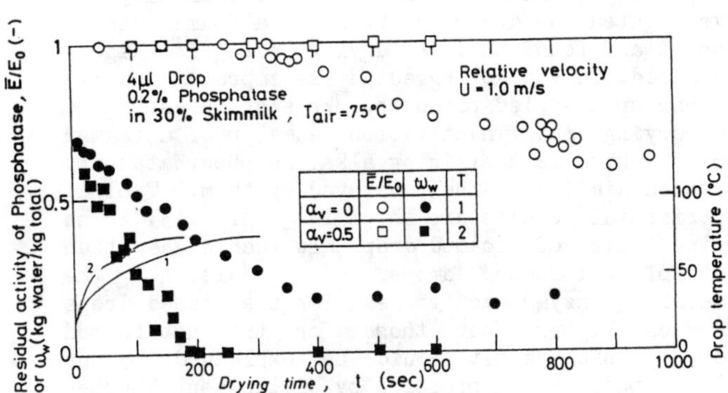

Fig. 13 Enzyme inactivation behaviour during drying of foamed drops and not-foamed drops.

4 µl drop of 30% skim milk containing alkaline phosphatase, T_{air} = 75°C, U = 1 m/s

---

Equations fitted to the diffusion coefficient of sucrose in water

$$D = D_R \exp[-\frac{E_D}{R}(1/T - 1/303.15)] \quad (7)$$

$$D_R = \exp[-(35+255.6u)/(1+12u)] \quad (8)$$

$$E_D = (100+190u)/(1+10u)\times 10^3 \quad (9)$$

5. DISCUSSION

During drying various physical and chemical changes may occur. It is known that aroma is lost during the constant rate period[11]. On the other hand, this study shows that the enzyme inactivation during drying takes place in the falling rate period characterized as the regular regime. This has already been stated by Wijlhuizen, Kerkhof and Bruin[1]. They investigated the effect of drying conditions on the final enzyme retention during spray drying by a computer simulation. Their predicted results agree qualitatively with our experimental results.

One of the difficulties in employing the present model lies in the determination of inactivation rate constants as functions of T and u. To our knowledge, the data applicable to the present model is restricted to those by Liou[3] as well as ours. However, his results were measured for the immobilized enzyme. It is known that the thermal stability of enzyme changes markedly by the immobilization technique.

Since the measurement of the inactivation rate constants k is time-consuming and laborious, simple rapid methods for the determination of k or some other functions which have predicting powers of inactivaion behavior of enzymes should be developed.

Another direction with which we concern ourselves is the development of a drying method by which the high retention of enzyme activity can be obtained. This is especially important from the point of view of the production of bulk industrial enzymes.

Foam spray drying first developed by USDA has been proved to be an effective way of drying by which quality loss of the drying matter is prevented[12]. Abdul-Rahman, Crosby and Bradley[13] investigated the drying behavior of a foamed single droplet and found that the drying rate is highly increased. They also suggested the improved quality of foam spray-dried products. We also investigated the drying and inactivation behavior of a foamed drop of skimmilk containing alkaline phosphatase by a method similar to that employed by them. Typical experimental results are shown in Fig. 13. The drying rate of foamed drops is much higher than that of not-foamed drops. In addition, the remaining enzyme activities for the foamed drops are much higher than those for the not-foamed drops. This result could be explained by the hollow-sphere model proposed by Wijlhuizen, Luyben and Bruin [1]. Theoretical analysis and full description on these results are now in preparation.

CONCLUSIONS

1) When the effects of pH change and oxidation are negligible, the enzyme inactivation behavior during drying of a drop can be predicted by the present model with the experimentally determined inactivation rate constants and diffusivity.

2) The effect of oxidation on the enzyme inactivation is highly dependent on the physical and biochemical properties of enzymes and often much more important than thermal inactivation.

ACKNOWLEDGMENTS

The authors are grateful to Professor R. Matsuno and Dr. K. Nakanishi, Dept. of Food Science and Technology, Kyoto Univ., Kyoto, for their helpful discussions and suggestions. They also owe Dr. Kawamura. Y., Kitazato Univ., many thanks for her valuable comments.

NOMENCLATURE

a	activity	-
$\hat{C}_p$	heat capacity at constant pressure	J/kg K
D	diffusion coefficient	m^2/s
E	enzyme concentration	arbitrary units
E_a	activation energy for enzyme inactivation	J/mol
E_D	activation energy for diffusion	J/mol
Δh_v	heat of evaporation	J/kg
j	mass flux with respect to solids	kg/m s
k	inactivation rate of enzyme	s^{-1}
k_g	mass transfer coefficient in the gas phase	m/s
p,q	parameter for enzyme inactivation	kg dry solid/kg
r	radial coordinate	m
R	radius of drop	m
R	gas constant in Eqs.(7) and (10)	J/mol K
t	time	s
T	temperature	K or C
u	water content	kg water/kg dry solid
U	relative velocity of air	m/s
z	$\int_0^r \rho_s r^2 dr$	kg
Z	$\int_0^R \rho_s r^2 dr$	kg

Greek

α	heat transfer coefficient in the gas phase	W/m^2K
α_v	voidage of foamed drop	-
λ	thermal conductivity	W/m K
$\overline{\rho\hat{C}_p}$	average heat capacity per unit of volume	J/m^3K
ρ	mass concentration	kg/m^3
ω	mass fraction	-

Subscripts

air	air
g	gas phase
i	interface
R	reference
s	solids
s,p	pure solid
w	water
wv	water vapor
0	value at t=0 or initial
1	value at u=0
2	value at u=∞

Superscript

sat	saturated value
-	average value

REFERENCES

1. Wijlhuizen,A.E., Kerkhof,P.J.A.M. and Bruin,S., Chem. Eng. Sci., vol.34, 651 (1979).
2. Luyben,K.Ch.A.M., Liou,J.K. and Bruin, S., Biotech. Bioeng., vol.24, 533 (1982).
3. Liou, J.K., Ph.D.Thesis, Agricultural University Wageningen, The Netherlands (1982).
4. Sano, Y. and Keey, R.B., Chem.Eng.Sci., vol.37, 881 (1982).
5. Sano, Y., Yamamoto.S.,and Keey, R.B., Proc. 3rd Pacific Chem. Eng. Cong.,(1983).
6. Robinson,R.A. and Stokes, R.H., R.B., Electrolyte Solutions, Chapt.10, Butterworths, London, (1965).
7. Yamamoto, S., Hoshika,M., and Sano, Y., Proc. 4th Int.Symp.Drying, Kyoto (1984).
8. Crank, J., The Mathematics of Diffusion, Oxford U.P., Oxford (1973).
9. Toei,R., Okazaki,M., Kubota,K.,Ohashi,K., Kataoka, K. and Mizuta K., Kagaku Kogaku, vol.30, 43 (1966).
10. Chandrasekaran, S.K. and King, C.J., AIChE J., vol.18, 513,(1972).
11. Kerkhof,P.J.A.M. and Schoeber, W.J.A.H., in Advances in Preconcentration and Dehydration of Foods, ed. by Spicer, A., Applied Science, p.349,(1974).
12. Hanrahan, F.P. and Webb, B.H., Food Eng., vol.31, 37 (1961).
13. Abdul-Rahman, Y.A.K., Crosby, E.J. and Bradley, Jr, R.L., J. Dairy Sci., vol.54, 1111 (1971).

FLAVOUR RETENTION ON DRYING OF A SINGLE DROPLET UNDER VARIOUS DRYING CONDITIONS

Takeshi Furuta, *Morio Okazaki and *Ryozo Toei

Department of Food Science and Technology of Toa University
15-2, Kusuno, Shimonoseki-shi 751 JAPAN
*Department of Chemical Engineering of Kyoto University
Sakyo-ku, Kyoto 606 JAPAN

ABSTRACT

The retentions of ethanol in a single droplet of aqueous solutions of some sugars and proteins were studied under various drying conditions, as a fundamental study on loss of trace flavour component during drying. In case of aqueous maltodextrin solution, the final retention of ethanol without expansion of droplet increased with increasing initial maltodextrin concentration, drying air temperature and velocity, and reducing air humidity. Experimental results were in good agreement with theoretical calculations which were based on Selective Diffusion Theory. Expansions of droplet affected more complicatedly the retentions, according to the behaviour of expansion. For aqueous protein solution, the ethanol in the droplet was lossed slowly even after the initial period of drying, which was major difference from that of sugar solutions. The addition of gelatin in aqueous maltose solution remarkably increased the final retention of ethanol.

1. INTRODUCTION

The spray drying is one of the most popular methods to make powder products from food liquid. On drying foodstuffs, it is necessary to take care to provide good quality to the products, as well as to remove water efficiently from the material. From energy saving point of view, it is desirable to make inlet air temperature as high as possible. However, many flavour components, presented at extremely dilute concentration in food liquid, show comparatively high relative volatility, so that they would be lost on drying, if the temperature of the material rises during drying. Therefore, it is important to find an optimum condition for spray drying of food liquid.

Thijssen and Rulkens [9] studied this subject on quantitative point of view. They proposed so-called "Selectve Diffusion Theory", in which they posturated that flavour components remained in liquid food because of formation of case hardening on its surface during drying. Recently, Thijssen and Rulkens [10], Chandrasekaran and King [3], and Kerkhoff and Schoeber [5] extended the theory on the basis of thermodynamics of irreversible process. On the other hand, Menting and Hoogstad [6] and Ban [1,2] studied experimentally the retention of flavour components on drying of a single droplet. They, however, have not compared their results with the theory.

The present study is focussed on the measurement of the behaviour of the flavour retention during drying of a single droplet under various drying conditions. Aqueous sugar and protein solution, in which a tarce amount of ethanol was dissolved as a model flavour component, were used as simulated food liquids. The loss of ethanol was traced during drying. The effect of drying condition - initial concentration of dissolved solids, air temperature, air velocity, and humidity - on retention of ethanol with and without expansion of droplet were investigated for aqueous maltodextrin solution. Retention of ethanol in aqueous protein solution and effect of addition of gelatin in some aqueous sugar solution were also studied under non-expanding condition of droplet.

2. THEORETICAL MODEL

The fundamental equations on loss of ethanol in a single droplet during drying were formulated on the basis of Selective Diffusion Theory. The droplet of initial radius R_0 is dried in a air of temperature T_b, as shown in Fig.1. If T_b is low, the droplet only shrinks during drying. However, if T_b is too high enough to make the inner pressure of the droplet above the atmosphere, the droplet expands on drying. At the instance of expansion, two improved models as to the concentration profiles of water and ethanol inside the spherical cell region of the droplet are considered [8]: perfect mixing model (PM-model) and non-mixing model (NM-model). For simplicity of calculations, following assumptions are made: i)the droplet is consisted three components and they are transferred by only molecular diffusion, ii)the additivity of volume yields among them, iii)expansion occurs abruptly and only once during drying, iv)the inner radius of the droplet R_{in} is unchanged after the expansion. Then the equation of transfer of water and ethanol are as follows:

$$\frac{\partial C_w}{\partial t} = \frac{1}{r^2}\frac{\partial}{\partial r}\left(D_w r^2 \frac{\partial C_w}{\partial r}\right) \quad (1)$$

$$\frac{\partial C_a}{\partial t} = \frac{1}{r^2}\frac{\partial}{\partial r}\left\{r^2\left(D_a\frac{\partial C_a}{\partial r} + D_{wa}\frac{\partial C_w}{\partial r}\right)\right\} \quad (2)$$

The equation of change of the droplet temperature is obtained as:

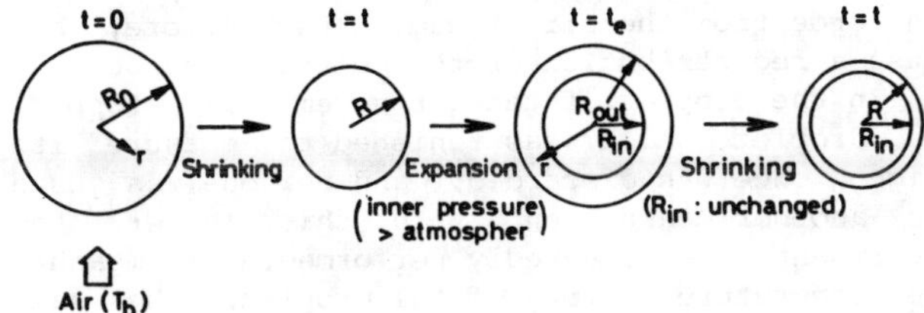

Fig.1 Drying Model of Droplet

$$\left(\frac{c_{pm}}{4\pi R^2}\right)\frac{dT_m}{dt} = \alpha_g(T_b - T_m) - \beta_{gw}(p_{wi} - p_{wb})\Delta H_v \quad (3)$$

where c_{pm} is the specific heat capacity of the droplet which is given by the following equation.

$$c_{pm} = \frac{4}{3}\pi R_0^2\{c_{ps}C_{so}+c_{pw}[C_{w0}-\rho_w(1-R^3/R_0^3)]\} \quad (4)$$

The initial and boundary conditions for non-expanding droplet are:

$$\left.\begin{aligned} &t=0,\ 0\le r\le R_0\ ; && C_w=C_{w0},\ C_a=C_{a0},\ T_m=T_{m0} \\ &t>0,\ r=0\ ; && \partial C_w/\partial r=0,\ \partial C_a/\partial r=0 \\ &r=R\ ; && \frac{D_w}{1-C_wv_w}\frac{\partial C_w}{\partial r} = \beta_{gw}(p_{wi} - p_{wb}) \\ & && -D_a\left(\frac{\partial C_a}{\partial r} + C_a\frac{\partial \ln\gamma_a}{\partial C_w}\frac{\partial C_w}{\partial r}\right) = \beta_{ga}(p_{ai} - p_{wb}) \end{aligned}\right\} \quad (5)$$

The diffusion coefficient of D_{wa} at the right hand side in Eq.(2) is so-called "cross diffusion coefficient", which results from the derivation of Eq.(2) by thermodynamics of irreversible process. Physically, the term $D_{wa}(\partial C_w/\partial r)$ indicates the amount of ethanol transferred by the moisture concentration gradient. Recently, Kerkhof and Schoeber [5] showed that the coefficient D_{wa} could be derived by diffusion coefficients of water D_w and ethanol D_a as follows:

$$D_{wa} = C_a\{D_a(\partial \ln\gamma_a/\partial C_w) - D_wv_w/(1 - C_wv_w)\} \quad (6)$$

where γ_a is the modified activity coefficient of ethanol which is related to the activity A_a as:

$$A_a = \gamma_aC_a \quad (7)$$

γ_a is a function of moisture content only.

In case of expansion of droplet, the same equations for transfer of water and ethanol are hold, and only boundary conditions at the instance of expansion should be changed. For PM-model, concentration profiles of water and ethanol inside the cell region become uniform at expansion, so that following conditions are valid:

$$\left.\begin{aligned} &t=t_e,\ R_{in}\le r\le R_{out}\ ; && C_w=C_{we},\ C_a=C_{ae},\ T_m=T_{me} \\ &t>t_e,\ r=R_{in}\ ; && \partial C_w/\partial r=0,\ \partial C_a/\partial r=0 \end{aligned}\right\} \quad (8)$$

where t_e is the time of expansion and C_{we} and C_{ae} are average concentration of water and ethanol at $t=t_e$, respectively. On the other hand, for NM-model, the concentration profiles of water and ethanol after expansion are similar to those before expansion, as described in the following equations:

$$\left.\begin{aligned} &t=t_e,\ R_{in}\le r\le R_{out}\ ; && C_w(r)=C'_w(r'),\ C_a(r)=C'_a(r') \\ &t>t_e,\ r=R_{in}\ ; && \partial C_w/\partial r=0,\ \partial C_a/\partial r=0 \\ & && (r'=(r^3-R_{in}^3)^{1/3}) \end{aligned}\right\} \quad (9)$$

where C'_w and C'_a are the concentration of water and ethanol just before expansion, respectively. R_{out} is determined by visual observations of drying experiments.

Crank-Nicolson's implicid method was applied to solve numerically Eqs.(1) to (9). At that time, to avoid the difficulties arisen from the shrinking of the droplet on drying, substantial coordinate σ was applied in place of r-coordinate, which is defined as:

$$\sigma = \int_0^r C_sv_sr^2dr \quad (10)$$

Using this coordinate, one can obtained the position of the surface of droplet as σ_0 (=constant) during drying. The amount of water and ethanol retained at any time in the droplet can be obtained as follows:

$$\Psi_w = \frac{3}{R_0^3C_{s0}}\int_0^{\sigma_0}\frac{C_w}{C_sv_s}d\sigma\ , \qquad \Psi_a = \frac{3}{R_0^3C_{s0}}\int_0^{\sigma_0}\frac{C_a}{C_sv_s}d\sigma \quad (11)$$

Dependences of diffusion coefficients D_w and D_a on water content and temperature, water activity of the solution, and activity of ethanol are reported in some papers [4,11]. Heat and mass transfer coefficients are given by well known equations proposed by Ranz and Marshall [7].

3. EXPERIMENTAL APPARATUS AND PROCEDURES

The experimental apparatus for drying droplet is schematically shown in Fig.2(a). The air was sent to a cooling tower, where the Raschig rings were packed and cooling water was sprayed from the top of the tower, to maintain the dew temperature of the air constant during the experiment. After that, to get a constant temperature air stream, the air was heated by electric heaters. The uniform velocity and temperature distribution were made by a contraction nozzle, and the turbulent velocity component was reduced through wire meshes

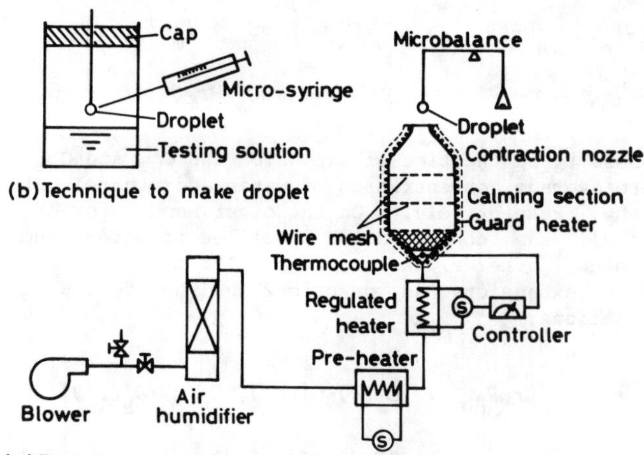

Fig.2 Experimental Apparatus for Drying of a Single Droplet

in the calming section. To prevent heat loss from the calming section and contraction nozzle, guard heaters were wound on the out side surface of them. The temperature and humidity of the air were measured by a thermocouple and a dew-point hygrometer, respectively.

The testing solution used were aqueous maltodextrin, protein (egg-albumin and nutrose) and skim milk solution in which trace amount of ethanol was added as a model flavour. Also, gelatin was added into some testing solutions to study the effect of addition of it. The preparation procedure is that maltodextrin, protein or skim milk powders weighted was dissolved with distilled water in a water bath, and degassed for several minutes. The solution was poured in a volumetric flask of 25 ml and mixed with carefully measured 5 µl ethanol. A droplet of 5 µl of the testing solution, formed carefully at a pointed end of a liquid micro-syringe, was suspended at the tip of a fine glass filament inside a small glass bottle, as shown in Fig.2(b). At the bottom of the bottle, the same testing solution was poured before. After one minute, the droplet was take off the bottle and placed at the center of the outlet of the nozzle, hanging from a small steel wire. This procedure for making a droplet was useful to prevent a initial loss of ethanol on creating the droplet, and also to calm the initial circulation of liquid inside the droplet. In fact, the reproduction of the data was improved greately by employing this technique. After some definite drying times, the droplet was removed and dissolved in distilled water of 100 µl. The ethanol concentration of this solution was analyzed by a gas chromatography with a fire ionized detector. The same procedure was also pursued with the non-drying testing droplet. The retention of ethanol was determined by the ratio of the concentration between them. The drying curve of the droplet, that is, the history of weight loss of the droplet, was measured by an electric microbalance (Cahn: type 2000). After the initial weight was measured, the droplet was exposed to the drying air stream for some minutes. At the time of weighing, the air was shut off to avoid the lifting force from the air stream. Furthermore, because a redistribution of the moisture was observed in the droplet if the air strem was stopped, the droplet which was finished to measure its weight once was discarded, and new one was used for another measurement. In this manner, the experiment was repeatedly performed. To measure the temperature history of the droplet, the same size of the droplet was attached at the junction of the thermocouple in the air stream. The change of configuration of droplet was observed by microscope and photographs. The outer diameter of the droplet after expansion was measured by them.

4. RESULTS AND DISCUSSION

4.1. Effect of Drying Conditions on Retention of Ethanol in Droplet Without Expansion

Fig.3 shows an example of the change of ethanol retention Ψ_a during drying of a single droplet of aqueous maltodextrin solution at $m_0=0.3$, together with the retention of water Ψ_w and the temperature history T_m of the droplet. A large part of the total amount of the ethanol loss appears in the initial period of drying for about one minute from the start of drying, where the temperature of the droplet is nearly constant during this period. After that, the rate of ethanol loss decreases rapidly, and then the retention remains almost unchanged till the end of drying. It was observed by microscope that a dry film (case hardening) began to cover the droplet surface rapidly at that time when Ψ_a becomes unchanged. Therefore, it is assumed that ethanol retention seems to have very intimate relation with the formation of this dry film on the surface

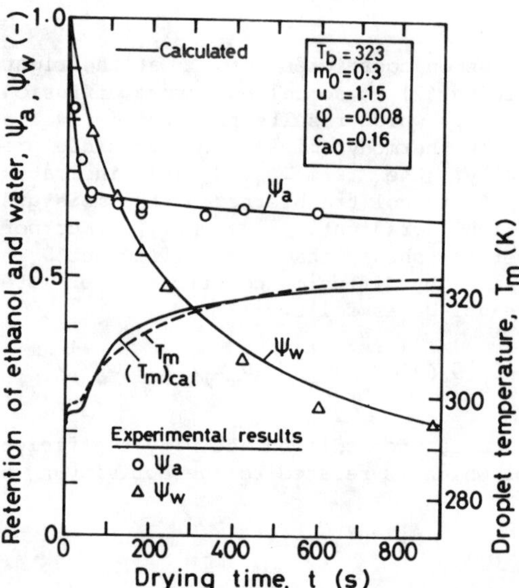

Fig.3 Experimental Results of Ψ_a Ψ_w and T_m During Drying of a Single Droplet of Aqueous Maltodextrin Solution

of the droplet. The solid lines in the figure are

the numerically calculated results with Eqs.(1) to (3). They are in good agreement with the experimental results. Fig.4 illustrates Ψ_a with various initial concentration of maltodextrin m_0, under the same drying condition in Fig.3. The curves of Ψ_a are the same for each initial concentration of maltodextrin. The calculations are illustrated by solid lines and they are in good agreement with experimental values, except for m_0=0.21 and 0.1. From visual observations, when m_0 was less than 0.2, the droplet was observed to deform into hemisphere in the course of drying, and this seems to be a possible reason that the experimental results of Ψ_a are different from that of calculations at

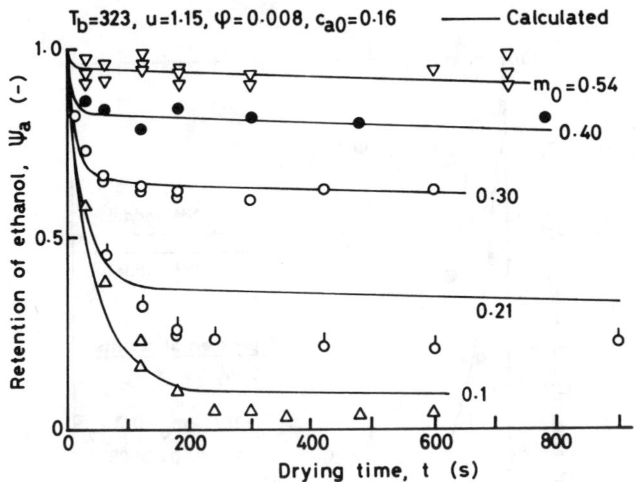

Fig.4 Effect of Initial Concentration of Maltodextrin on Ψ_a

m_0=0.21 and 0.1. Fig.5 shows the effect of air temperature on Ψ_a at m_0=0.3. The final values of Ψ_a increase as T_b increases. At high air temperature the case hardening begins to cover the droplet surface immediately after the droplet is ex-

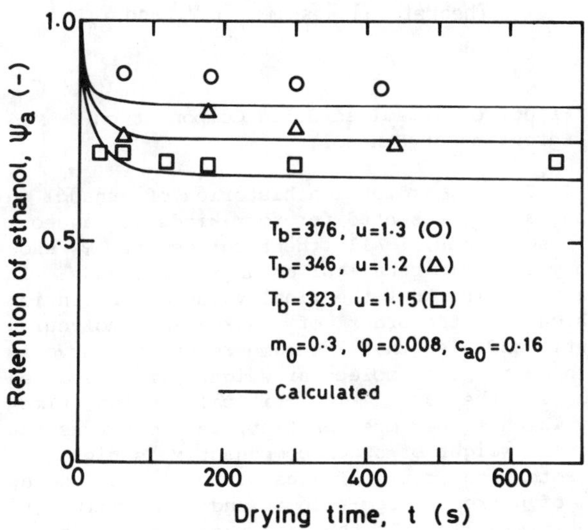

Fig.5 Effect of Air Temperature on Ψ_a

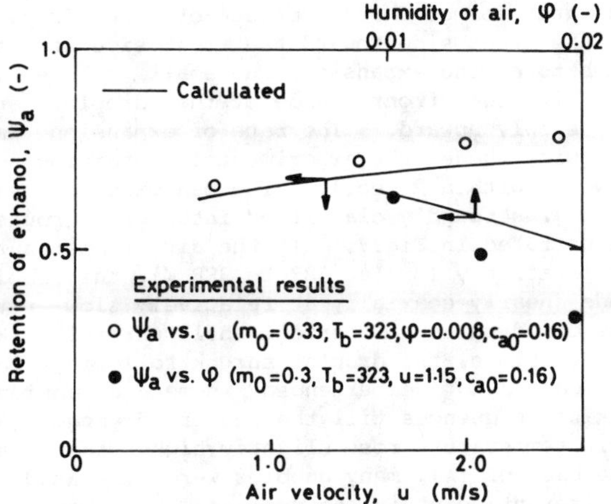

Fig.6 Effect of Air Velocity and Humidity on Ψ_a

posed to the drying air, which allows to permeate water selectively to result in high value of Ψ_a. The solid lines represent the numerical results, which are in good agreement with the experimental results. The effect of air velocity and humidity on the final value of Ψ_a are illustrated in Fig.6. The Ψ_a increases as the velocity of air increases and the air humidity decreases. It is found that the effect of the air humidity is significant on Ψ_a.

Above all, it can be concluded that the behaviour of the flavour retention in a single droplet dried in the air stream can be explained quantitatively by the selective diffusion theory. Furthermore, from visual observations, the flavour retention has very intimate connection with the formation of case hardening on the surface of droplet, and then inceases if the droplet is dried under the conditions which promote the generation of case hardening.

4.2. Ethanol Retention of Droplet With Expansion

If the droplet is dried in a high temperature air stream, it expands during drying because of rise of inner pressure of water above the atmosphere. Fig.7 shows a classification of change of configuration of aqueous maltodextrin droplet on drying accompanying expansion, which is arranged by visual observations with a microscope. In the

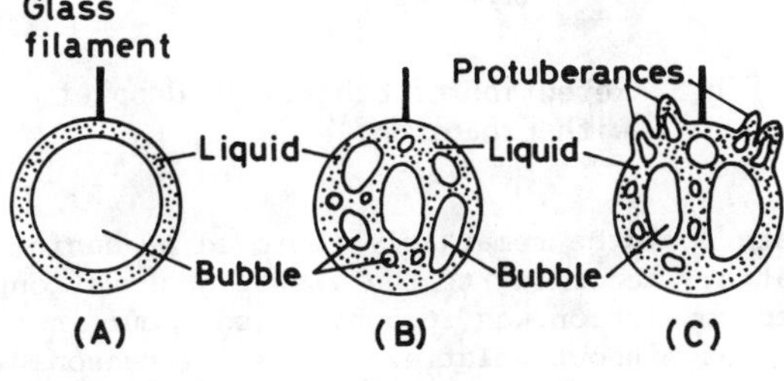

Fig.7 Behaviour of Expansion of Droplet

initial period of drying, the droplet shrinks nearly uniformly. At the end of this period, the droplet is covered with a case hardening and deep lines are formed on the surface of the droplet from the rear side toward the front side of it. Just before the expansion, a small bubble is formed in the front inside of the droplet and moves slowly upward. The rate of expansion was very quick under the experimental condition of this work (within 2 sec.). The behaviour of expansion was roughly classified into three groups, as illustrated in Fig.7. If the air temperature is comparatively low (at 383 to 388 K), the bubble expands nearly coaxially at relatively slow rate to be dried up as a spherical shell particle like (A). Sometimes, the droplet shrunk following the expansion, and experienced some expansion-shrinking sequences till the end of drying. As the air temperature rose slightly higher than that in the case of (A), many bubbles were generated at the different point in the droplet, and then the droplet became like a honeycome, as (B) in the figure. The expanding behaviour of (C) was observed if the air temperature was rather high (above 398 K). In this case, the droplet expanded violently and was suddenly burst to form some protuberances. The liquid inside the droplet was agitated and the inner liquid was exposed to the drying air when it was burst.

The experimental retention of ethanol of expanding droplet is shown in Fig.8. In this figure, Ψ_a for non-expanding droplet is also illustrated by a dotted line. The experimental results, except for those indicated by a symbol ●, are obtained for a testing solution added small amount of gelatin (about 0.07%). It is clearly found that the scattering and reproducibility of

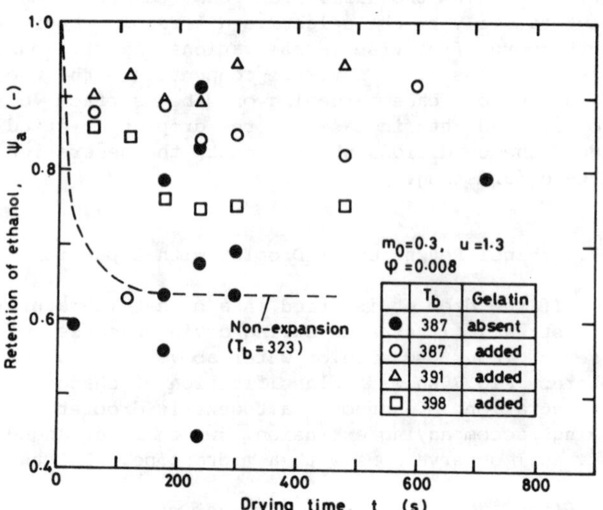

Fig.8 Retention of Ethanol in Droplet with Expansion

the data could be remarkably improved by addition of gelatin. According to the visual observations, a forced convection was recognized to occur inside the droplet without gelatin. It seems reasonable to assume that this convection resulted in the scattering of the data. From Fig.8, it is also found that Ψ_a increases as the air temperature rises if the expansion is not so violent. The decrease of Ψ_a at air temperature of 398 K is assumed to be caused by the exposure of inner liquid to the air at bursting. Fig.9 shows the comparison of the theoretical and experimental results of Ψ_a and Ψ_w under the drying condition of T_b=391 K and m_0=0.3. In this experiment, the expansion behaviour was belonging to type (A) in Fig.7. The calculated value of Ψ_a are lower than the measurements. It is noticed that the difference of calculations of Ψ_a between PM-model and NM-model is not so large. This seems to be the reason that the average water content at expansion become too low enough to protect the loss of ethanol. On the other hand the calculated result of Ψ_w is in good agreement with the experimental ones. In this case also, the difference between PM- and NM-model is very small.

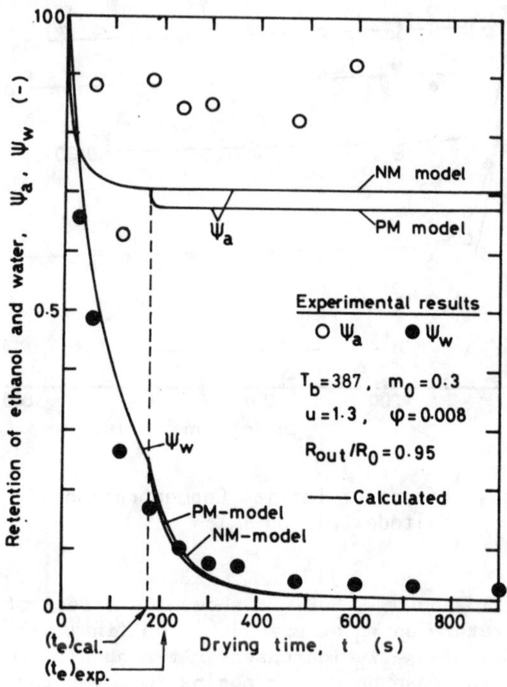

Fig.9 Comparison between Experimental and Theoretical Results of Ψ_a and Ψ_w

4.3. Effect of Dissolved Solid Component on Retention of Ethanol

In Fig.10, retention histories of ethanol on drying are represented for four kinds of aqueous sugar solutions. All these curves of Ψ_a show roughly the same behaviour as drying proceeds. It is interesting that the final value of retention is high in the order of increasing molecular weight, and also that the same retention curve is obtained if the molecular weight of dissolved solid is the same. The critical drying time, after which Ψ_a becomes constant, is shorter as the molecular weight of sugar component M is high.

Retention curve of ethanol for aqueous solution of nutrose, egg-albumin and skim milk are shown in Fig.11. In the same manner as for sugar solution, Ψ_a sharply decreases for a short time from onset of drying. However, Ψ_a does not remain constant, but decreases gradually, after the ini-

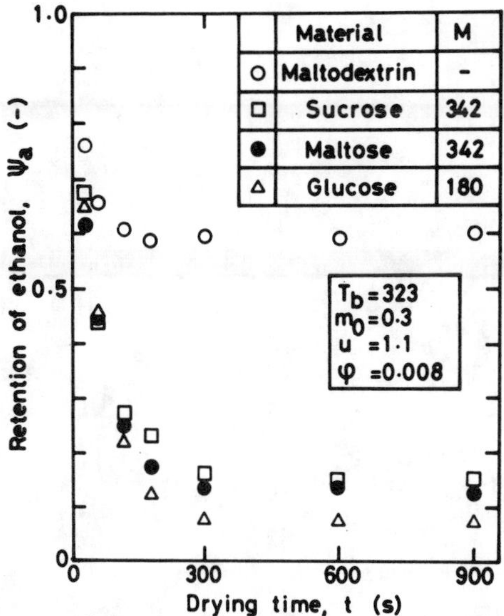

Fig.10 Retension of Ethanol on Drying Droplet of Various Sugar Solutions

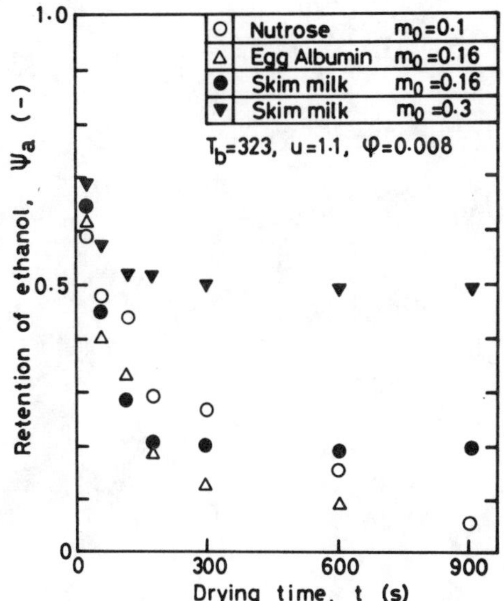

Fig.11 Retention of Ethanol on Drying Droplet of Various Protein Solutions

tial period of drying. This is sharply contrast behaviour to that of sugar solutions (see Fig.10). On the contrary, for aqueous skim milk solution, Ψ_a has nearly the same behaviour to maltodextrin solution. Skim milk powder used is consist mainly of sugars such as lactose (50%) and casein (30%). Then, it is regarded as a mixture of sugar and protein. Therefore, from Fig.10 and 11, it can be concluded that sugar components have an ability to protect loss of flavour components during drying.

The effect of addition of gelatin to aqueous maltodextrin and maltose solutions are shown in Fig.12. The total concentration of testing solution is initially 30 wt%. In case of maltodextin, the addition of small amount of gelatin (up to 2

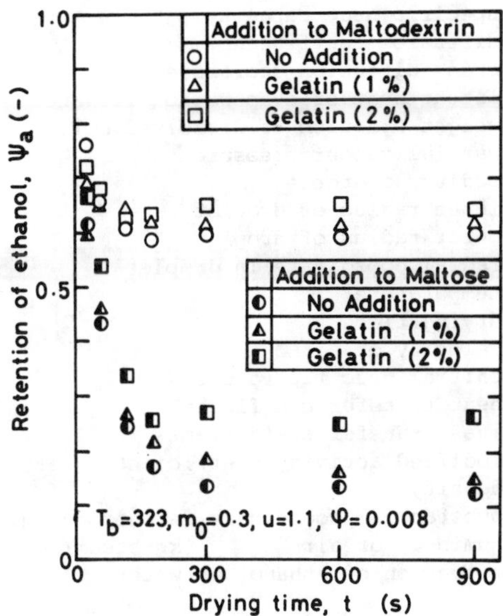

Fig.12 Effect of Addition of Gelatin on Retention of Ethanol for Sugar Solution

wt%) does not give no effect on Ψ_a curve. However, for maltose solution, 2 wt% addition of gelatin makes the final value of Ψ_a up to about 2 times as large as that of no addition. This is considered to result from the rapid formation of case hardening by addition of gelatin. From visual observation, the case hardening is recognized clearly in case of addition of 2 wt% gelatin, but not for no addition case.

CONCLUSION

Retention of ethanol in a single droplet of aqueous maltodextrin solution were measured under various drying conditions with and without expansion of droplet. It is concluded that the retention can be improved under the drying conditions which a case hardening is generated immediately, if the droplet did not expand. Under the drying condition accompanying expansion of droplet, the retention behaviour was much complex, and the formation of protuberances gave much effect on retention. In case of aqueous sugar solution, the final value of retention became large in the order of molecular weight of the sugar. The critical drying time, after which the retention remained unchange, was observed for the case of sugar solution. The retention curve on drying droplet of aqueous protein solution is a little different from that for sugar solution, in the point which no critical time mentioned above appeared and the retention decreased gradually after the initial period of drying. The addition of gelatin to aqueous sugar solution was greatly effective for maltose solution.

NOMENCLATURE

A_a : activity of ethanol

C : concentration (kg/m^3-solution)
c_p : specific heat capacity (J/kgK)
D : diffusion coefficient (m^2/s)
D_{wa} : cross diffusion coefficient (m^2/s)
Δh_v : latent heat of evaporation of water (J/kg)
m : weight fraction of dissolved solid
p : partial vapour pressure (Pa)
R : radius of droplet (m)
R_{in} : inner radius of droplet (m)
R_{out} : outer radius of droplet (m)
r : radial coordinate in droplet (m)
T : temperature (K)
t : drying time (s)
u : velocity of air (m/s)
v : partial specific volume (m^3/kg)
α_g : heat transfer coefficient (W/m^{2}K)
β_g : mass transfer coefficient (kg/m^{2}sPa)
γ : modified activity coefficient of ethanol
ρ : density (kg/m^3)
σ : substantial coordinate defined by Eq.(10)
ϕ : humidity of air (kg-steam/kg-dry air)
: retention of ethanol or water

<subscript>

0 : initial value
a : ethanol
b : bulk of air stream
e : values at expansion
g : gas phase
i : liquid-gas interface
m : droplet
s : disolved solid
w : water

REFERENCES

1. Ban, T., Kagaku Kogaku Ronbunshu (in Japanese), vol.4, 515 (1978)
2. Ban, T., Kagaku Kogaku Ronbunshu (in Japanese), vol.5, 213 (1979)
3. Chandrasekaran, S., and King, C. J., AIChE Journal, vol.18, 520 (1972)
4. Furuta, T., Tsujimoto, S., Makino, H., Okazaki, M. and Toei, R., in press in Journal of Food Engineering (1984)
5. Kerkhof, P. J. A. M. and Schoeber, W. J. A. H. Theoretical Modeling of the Drying Behaviour of Droplets in Spray Dryers, in Advances in Preconcentration and Dehydration of Foods, ed. A. Spicer, P.349-397, Applied Science Publishers, London (1974)
6. Menting, L. C. and Hoogstad, B., J. Food Sci., vol.32, 87 (1967)
7. Ranz, W. E. and Marshall, W. R., Chem. Eng. Prog., vol. 48, 141 (1952)
8. Sano, Y. and Keey, R. B., Chem. Eng. Sci., vol.37, 881 (1982)
9. Thijssen, H. A. C. and Rulkens, W. H., De Ingenieur, vol.80, Ch45 (1968)
10. Thijssen, H. A. C. and Rulkens, W. H., Trans. Instn. Chem. Engrs., vol.47, T292 (1969)
11. Tsujimoto, S., Matsuno, R. and Toei, R., Kagaku Kogaku Ronbunshu (in Japanese), vol.8, 103 (1982)

BELT TYPE FOAM MAT DRYERS

Junichi Uno and Nobuyuki Kobayashi

Kikko Foods Corporation

4-13 Koami-cho, Nihonbashi, Chuo-ku, Tokyo 103 JAPAN

ABSTRACT

Belt type foam mat dryers have been developed and operated successfully for fruit, vegetable and processed foods. In order to design commercial scale foam mat dryers, a series of foaming tests and drying rate studies were conducted. A pilot scale dryer was built to collect the data required for the commercial scale foam mat dryer design.

1. INTRODUCTION

The original idea of a foam mat dryer was first shown by Mink (1935) where egg white was foamed by a wire whipper and extruded onto a metal mesh as small tubular extrusions, similar to spaghetti. The spaghetti-like strings were then dried in a conventional batch type dryer oven, cooled, detrayed and crushed. Mink (1939) used 0.1 to 0.3 % of sodium laurylsulphate on a dry weight basis to make thermally stable foams of egg white and found better rehydration characteristics of the finished products.

Since then, some applications using this principle have been developed (Morgan et al. 1959, Annon. 1963 and Willis 1965). Quick drying is possible in foams due to a large surface area for water evaporation and easy movement of water by capillary action. Also, due to the honeycomb structure of the dried product, the rehydration occurs faster than products dried by other methods. These result in higher product quality.

In addition to these advantages of drying foams, USDA researchers (Ginnette et al. 1961 and Rockwell et al. 1962) modified the process whereby the raw material was foamed with foaming agents and spread onto metal perforated plates. Next, high velocity air was applied upward to make cones or crater in the foam. Then, the product was stack dried. This cratering of foam mats increases the surface area. The drying air flows through the crater providing better product contact. This cratering also shortens the water movement distance in the material. This particular process is now generally called 'Foam Mat Drying' but it really means 'Foam Mat Crater Drying'. This process was originally patented in The United States and later was applied for and granted in Japan.

For successful operations of foam mat dryers one of the critical requirements is to prepare a fine, homogeneous, stiff and thermally stable foam. Hart et al. (1963) and LaBelle (1966) studied the foam stability in connection with the various foaming agents and their amounts and foaming conditions.

Sjogren (1962) reported on a commercial tray type foam mat dryer which was similar to the USDA laboratory unit in principle. Sjogren reported the operating costs were low and the product quality was so excellent that taste panels seldom could distinguish between the quality of the starting material and that of the dried product.

2. PRELIMINARY STUDIES

Foams As previously stated, one of the key points of this particular process is to prepare a thermally stable foam. The raw material is required to have proper viscosity. Even though food concentrates do not show Newtonian behavior in general, the viscosity by a rotary viscometer at the range of 10,000 to 30,000 cp gives relatively stable foams.

Up to a few percent of pulp content in the raw material helps foaming performance of products with low soluble solid content. The addition of carbohydrates, such as dextrin, in pulp free foods may result in better foams.

Usually, prior to the drying tests, batch type foaming tests are conducted to choose a proper foaming agent and to examine the stability of the foam. The prepared foam is evaluated in three ways. First, its density is measured by weighing the foam in a small cup. Second, its viscosity is measured by a rotary viscometer. Finally, its thermal stability is tested by placing the foam in a heated oven.

In most of the foaming studies, GMS (glyceryl monostearate) stabilized foams show the most favorable foam characteristics.

Drying Rate The cratered foam mat of various fruit and vegetable concentrates were placed in a stack shown in Figure 1, where air flows at a constant temperature and velocity parallel or perpendicular to the mat. The temperature of the mat was monitored by the thermocouples located in the mat. The moisture change was measured by taking samples out of the mat at a proper time interval. Drying rate tests using apple, orange, peach, pineapple, Chinese cabbage, cabbage and tomato were conducted at the various mat thicknesses and air temperatures and velocities.

Figure 2 is an example of those test results which clearly shows the difference between the

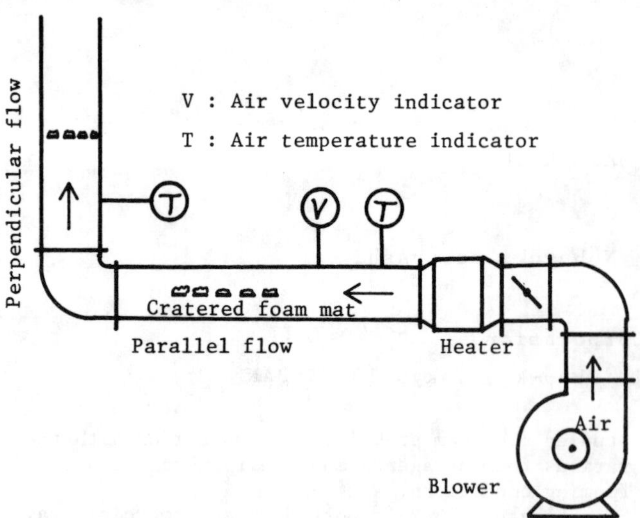

Figure 1. The scheme of drying rate tests

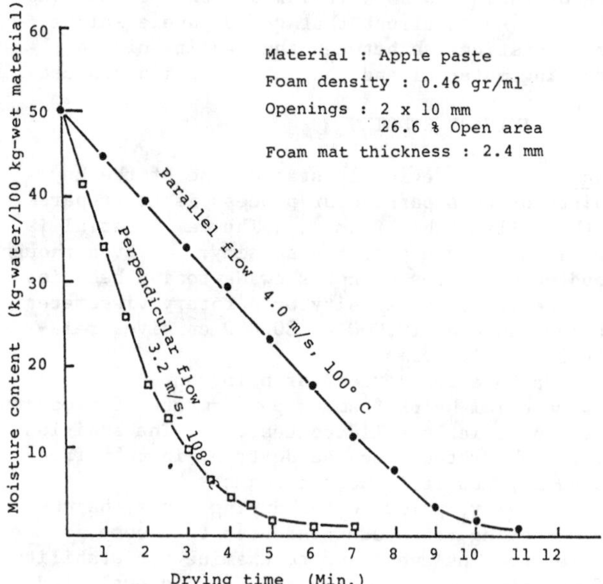

Figure 2. Moisture change in parallel and perpendicular flows

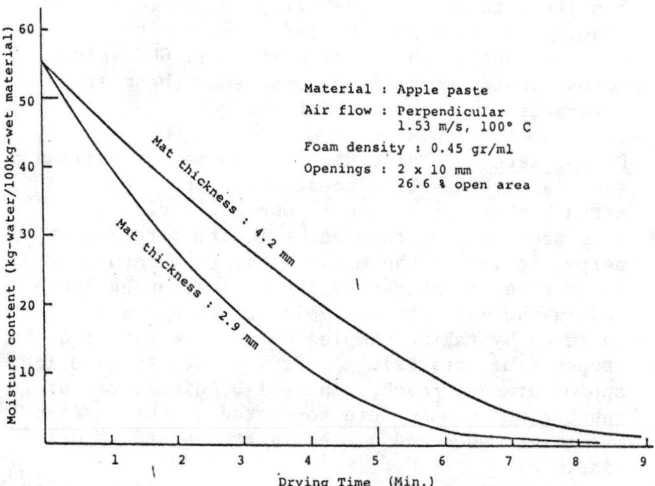

Figure 3. Effects of mat thicknesses on drying rates

parallel and the perpendicular air flows. There is a significant difference in the drying rates between the parallel and the perpendicular air flows, i.e., the perpendicular flow gives a much higher drying rate than the parallel flow. However, because of the huge volume of air required per unit area of the mat in case of the perpendicular air flow it is practical to apply the perpendicular air flow only for a limited period of the drying time, i.e., only for the initial stage of the drying where the moisture content is high.

Figure 3 is another example which shows two different thicknesses of the mats with the perpendicular air flow. A thinner mat drys at a faster rate.

Figure 4 shows the effect of the perpendicular air velocity on the drying rate. The higher air velocity gives the higher drying rate at relatively high moisture contents.

Figure 5 shows the effect of the shape and the opening area of the perforated plates on the drying rate. The circle opening gives a higher drying rate than the rounded-end rectangular opening and the greater percentage of the open area gives a higher drying rate.

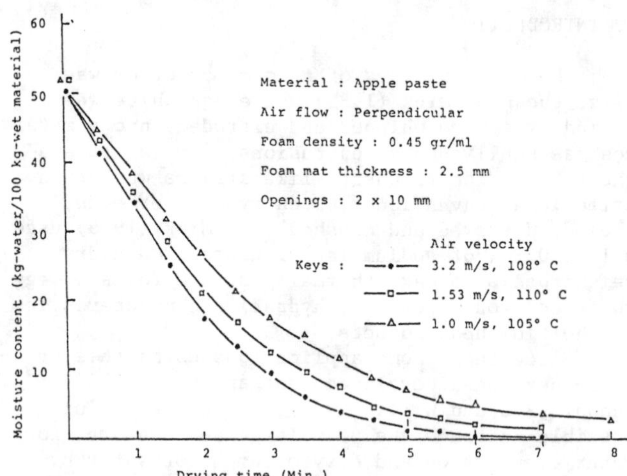

Figure 4. Effects of air velocity on drying rates

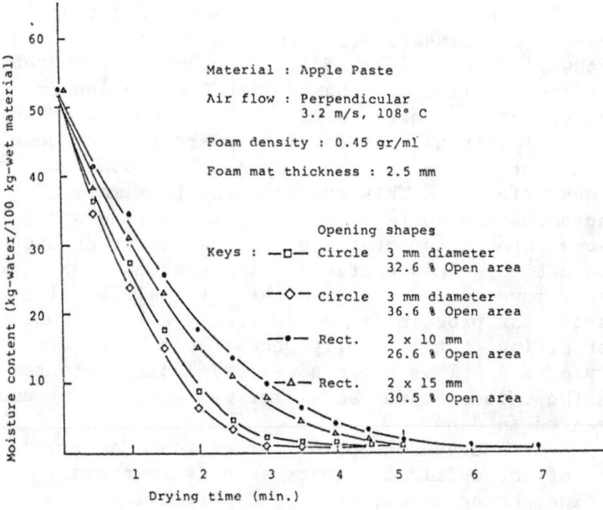

Figure 5. Effects of belt opening shapes on drying rates

3. COMMERCIAL SCALE FOAM MAT DRYERS

Continuous and perforated belt type foam mat dryers were developed at Kikko Foods Corporation, Japan, by utilizing the principle patented by Ginnette et al. (1961) and through a series of preliminary tests described above. Prior to the installation of the commercial scale foam mat dryer, a pilot scale foam mat dryer was installed by which the technical information of the dryer design specifications were collected. Even now, this small unit is frequently used to make preliminary tests with the particular food stuff prior to every commercial operation.

The small unit has a 0.6 mm thick perforated belt, 40 cm wide by 5.7 m long, with a 2 cm unperforated strip along each side.

The drying chamber is divided into two zones, a 2.9 m long first zone and a 3.1 m long second zone. The temperatures in each zone are independently controlled. The air flow is counter current to the belt in the first zone and co-current in the second. The drying time is adjustable from 15 to 30 minutes by the belt speed.

The two commercial scale foam mat dryers are almost identical in dryer features. Figure 6 shows the scheme of the dryer. The perforated belt is of stainless steel, o.8 mm thick, 1 m wide and 60 m long. The openings of the belt are 2 mm x 5 mm rectangular with rounded ends and has 27 % open area. The reasons why a rounded-ends rectangular opening is utilized in spite of a slower drying rate than a circle opening are, 1. the availability of punching tools. 2. the less curvature punching of the belt. 3. the reliable and smooth joint(s) of the belt.

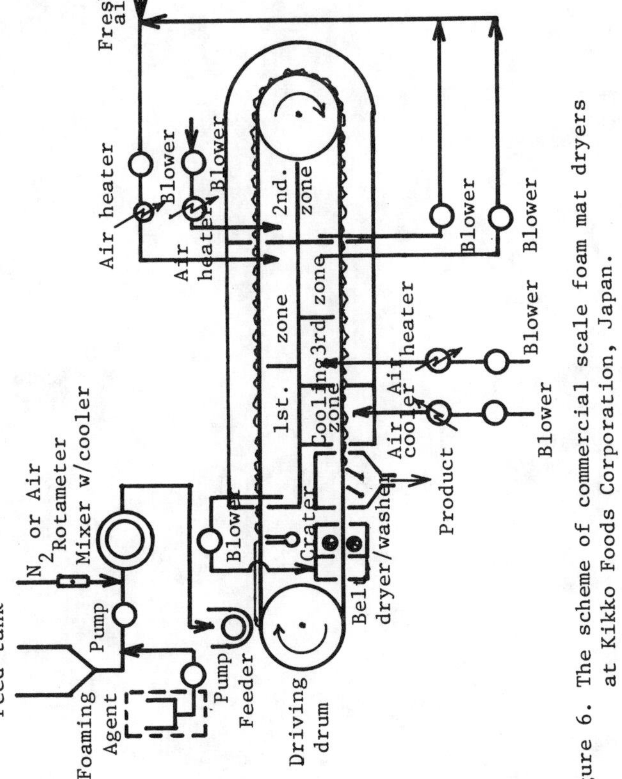

Figure 6. The scheme of commercial scale foam mat dryers at Kikko Foods Corporation, Japan.

The raw material is continuously mixed with the foaming agent slurry and nitrogen or air is injected into the product. Then the mixture is foamed in a specially designed continuous mixer, similar to an Oakes, Goodway or Mondo mixers. The mixer is equipped with a cooling jacket to produce a more stable foam, because the foam structure collapse more easily at elevated temperatures. The prepared foam then goes to the feeder which has a rotating roller to spread the foam uniformly onto the perforated belt. The thickness of the foam on the belt is usually 3 to 5 mm depending upon the stiffness and the density of the foams. Before the foam mat goes into the first drying zone the high velocity air is applied upward from the bottom of the perforated belt to make craters in the foam mat. The velocity of air from three slits in series is controlled independently to make the desired craters.

The dryer is divided into three compartments, first, second and third zones. The air flows are conter current in the first zone, co-current in the second and counter current in the third. The first zone is 18 m long, the second is 17 m and the third is 13 m. The belt is supported and driven by a set of drums with a 90 cm diameter. The driving drum is on the foam feeding side which is located on a movable frame. It is designed to have the constant tension of the belt regardless of the belt temperature. The temperatures of the drying air entering each zone are controlled independently. Each zone has two blowers for blow-in and exhaust. The total drying time is adjustable from 15 to 30 minutes by the belt speed. In the case of thermoplastic products, a cooling zone after the third drying zone helps to make the dried product crispy and promotes easy removal from the belt. The dried product of approximately 1 to 3 % moisture is scraped off by two sets of aluminum blades. The recovery of the product is 95 to 98 %. The belt is cleaned with a set of rotating brushes. The scraped product, which is usually flake shaped, goes through a shifter with a 10 mm opening and then may be disintegrated depending upon the product specification.

For energy saving, the exhaust air from the second and the third zones are recycled to the first zone. The exhaust air from the first zone is used to dry the belt after the belt washing zone.

4. EXAMPLES

Tomato Tomato paste of 35 % solid content was foamed in a continuous mixer with 1 % of Myveroll 18 00(glyceryl monostearate by Eastman Kodak) as a foaming agent. The foam was fed to the dryer at 125 kg/hr with a density of 0.6 gr/ml and was dried to 1.5 % moisture content in 15 minutes. The air temperatures at each zone were, 135 °C in and 69 °C out at the first zone, 112 °C in and 78 °C out at the second zone and 70 °C in and 72 °C out at the third zone. The exhaust temperature from the third zone was higher than the inlet temperature possibly because the divider between the second and the third zones was loose and the hot air came in from the second zone to the third zone. The moisture content was 28%

at the end of the first zone and 2.5 % at the end of the second zone.

Apple   Apple concentrate of 45 % solid content with pulp was foamed continuously with 1 % of Emulzee MS (glyceryl monostearate by Takeda). The foam had a density of 0.45 gr/ml and was fed to the dryer at 120 kg/hr. The product was dried in 25 minutes to 1.5 % moisture content. The temperatures were 135° C in and 86° C out at the first zone, 110° C in and 77° C out at the second zone and 40° C in and 45° C out at the third zone. The moisture contents were 18 % at the end of the first zone and 2 % at the end of the second zone. Samples were taken at several locations in the dryer to study the moisture change. Figure 7 shows the relation between the drying rate vs moisture content for tomato and apple.

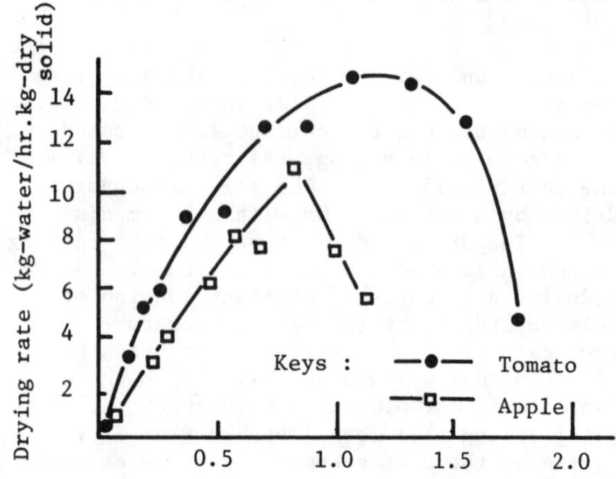

Figure 7. Drying rate curves for tomato and apple

5. CONCLUSION

Two belt type foam mat dryers of commercial scale have been successfully operated at Kikko Foods for fruit powder production, such as apple, orange, peach, pineapple, papaya, banana, lemon, grape and for vegetables, such as tomato, corn, cabbage, Chinese cabbage, lettuce, parsley, bean sprout, green onion, spinach, onion, bell pepper, pumpkin, taro, potato and other processed foods, such as Miso (fermented soy bean paste) and verious sauces.

The quality of the finished products is good. The drying cost is competitive to other conventional drying methods. The limiting factor of this dryer is that in order to prepare thermally stable foams of proper viscosity and stiffness, small amounts of foaming agents such as GMS are required. One of the future research on this drying concept should be aimed at developing the process of using product foams free of foaming agents.

REFERENCES

1. Annon., Florida tests foam-mat drying for citrus., Canner/Packer, Feb. (1963)
2. Ginnette, L. F., Graham, R. P., Morgan, A. I. Jr., Process of dehydrating foams. US Patent 2,981,629 (1961)
3. Hart, M. R., Graham, R. P., Ginnette, L. F. and Morgan, A. I. Jr., Foams for foam-mat drying. Food Techn. 17(10):90-92 (1963)
4. LaBell, R. L., Characterization of foams for foam mat drying. Food Techn. 20(8):89-94 (1966)
5. Mink, L. D., Treatment of egg whites. US Patent 2,200,963 (1935)
6. Mink, L. D., Egg material treatment. US Patent 2,183,516 (1939)
7. Morgan, A. I. Jr., Ginnette, L. F., Randall, J. M. and Graham, R. F., Technique for improving instants. Food Eng. 31(Sept):86-87 (1959)
8. Rockwell, W. C., Lowe, E., Morgan, A. I., Jr. Graham, R. F. and Ginnette, L. F., How foam-mat dryer is made. Food Eng. (Aug) (1962)
9. Sjogren, C. N., Practical facts of foam mat. Food Eng. (Nov) (1962)
10. Willis, R. Personal communication. (1965)

COMPUTER SIMULATION OF A MINTON DRYER

Jeffrey A. Hinds, Nancy R. Jackson, Amar N. Neogi

Weyerhaeuser Technology Center
Tacoma, Washington 98477 U.S.A.

ABSTRACT

Minton dryers are used commercially to dry heavy grade pulp sheets. An actual dryer on a machine producing bleached pulp has been simulated with a dynamic computer model which solves the basic transport equations using solution techniques reported earlier. The predicted drying rate and energy efficiency agree well with observed performance of the machine. The numerical solution also gives internal temperature, moisture and pressure profiles within the sheet. The phenomenon of blistering observed in a Minton dryer is a controlling factor in machine productivity. The simulation results shed considerable light on the mechanism of blister formation, heretofore unexplained. The mechanism of blister formation and conditions under which delamination occurs were further verified using a laboratory blister test unit. The computer model is thus demonstrated as an effective tool in optimizing drying rate, energy efficiency and productivity of paper machine dryers.

NOMENCLATURE

- b Basis weight of dry web, g fiber/cm^2
- C Water vapor concentration, g water/cm^3
- D Vapor diffusivity, cm^2/s
- F_D Transport coefficient for vapor diffusion, g fiber/cm/s
- F_L Liquid transport coefficient, (g fiber/cm^2)^{2}s
- F_P Transport coefficient for bulk vapor convection, (g water/cm^2/s/atm)(g fiber/cm^2)
- F_Q Transport coefficient for conducted heat, (cal/cm^2/s/C)(g fiber/cm^2)
- H_L Enthalpy of liquid water, cal/g
- H_s Enthalpy of fiber, cal/g
- H_v Enthalpy of water vapor, cal/g
- k Permeability, g water/cm/s/atm
- L Liquid flux, g water/cm^2/s
- M Moisture content, g water/g fiber
- P Total pressure, atm
- Q Heat transfer by conduction, cal/cm^2/s
- T Temperature, C
- t Time, s
- V Vapor flux, g water/cm^2/s
- x Distance from web surface, cm

INTRODUCTION

Conventional paper drying is carried out by passing the continuous web over a series of steam heated cylindrical cans. The rate of drying varies for different grades of paper and for different furnishes. Lightweight grades, such as writing papers, are thin enough such that heat and mass transfer occurs quite readily within the paper itself, and the drying rate is controlled by the heat and mass transfer at the surfaces of the paper. Heavier grades, such as pulp, are thick enough such that gradients in temperature and moisture content within the web are significant and the limit to drying rate becomes the ability to transfer heat into the center of the web and transport vapor out to the paper surfaces.

Vapor transport within the web occurs by two mechanisms: diffusion and convection. For a given web, the fluxes are governed by gradients in water vapor concentration and total pressure respectively. Drying under vacuum can promote the drying rate when internal resistances to heat and mass transfer are controlling by increasing the gradient in total pressure, shifting to the regime where the predominant mechanism for vapor transport is convective bulk flow.

Minton vacuum dryers (1) are used to dry heavy grades. They are identical to standard dryers except that the entire dryer is enclosed and maintained under vacuum. The increased drying rate achieved in a Minton dryer reduces the size of dryer compared with that which would be required if the web were dried under atmospheric pressure.

A web heated under vacuum reaches the boiling point of water at a lower temperature than it would at atmospheric pressure. Vaporization of water causes local pressure inside the web to passing over it. Liquid fluxes and local instantaneous temperatures and moisture contents were measured during the experiments. The correlations for transport coefficients and the methods used to develop them have been reported earlier (4) and will not be repeated here.

The unique feature of this model pertaining to drying under vacuum is inclusion of the bulk vapor flow term. In conventional dryers, large internal pressure gradients seldom develop, hence, diffusion plays the important role in the movement of vapor. In a Minton dryer, the significance of the pressure increase due to vaporization of water is much greater than in atmospheric pressure dryers. The flux depends upon the magnitude of the pressure gradient and the permeability of the web.

Permeability of a porous medium is a measure of its resistance to fluid flow. In the case of a moist paper web, the macroscopic porous structure is a result of the inter-fiber voids. The permeability of paper webs is dependent upon the fiber type, the fiber length and distribution of fiber lengths, the density of the mat and the local moisture content. For a given furnish, the density of the sheet and its moisture content are the significant variables in determining permeability.

For this investigation of drying on a specific machine, the permeability of the web to vapor flow was measured for several sheets of different densities and moisture contents. The resulting relationships between permeability and density and moisture content for a bleached softwood Kraft pulp are shown in Figures 2 and 3, respectively.

The above set of equations, together with the correlations for transport coefficients and appropriate boundary conditions, completely describe the drying of a fiber web (5). The numerical solution was obtained using a finite difference solution method employing a VAX 11/780 computer.

RESULTS

The computer model was used to simulate a Minton dryer, the operation of which had been thoroughly characterized. This machine produces bleached Kraft pulp of basis weight 0.05-0.07 g/cm^2. Production rate, web inlet solids content, steam conditions and other operating data were specified to match measured values.

Comparison of energy and drying rate results from the simulation with measured values is shown in Table 1.

Table 1
Comparison Between Observed and Predicted Dryer Performance

	Measured	Predicted
Outlet Solids Content, %	95	96
Drying Rate, g water/cm^2/s	0.0272	0.0275
Steam Usage, g steam/g water	1.33	1.29

Pressure and temperature profiles within the web were calculated with each numerical solution. It was observed that under certain operating conditions, high internal pressure gradients developed within the web. It is known that one of the factors limiting production in a Minton dryer is delamination of the web. This phenomenon, known as blistering, is postulated to occur when internal pressure gradients form rapidly within the web, and the resulting forces exceed the internal bond strength between fibers. This proposed mechanism is supported by the fact that in actual operation, blistering is prevented in the Minton dryer by reducing the vacuum (increasing the pressure). The overall pressure gradient and, consequently, the drying rate, are reduced.

In order to further investigate and confirm the mechanism of blistering, a laboratory blister test unit was constructed. The device consisted of a chamber enclosed on the top and bottom with metal platens. A screen suspended horizontally in the center of the chamber supported the web during a test. The platen temperature was controlled to a level selected to create the desired pressure within the chamber.

For each test, a thermocouple was inserted into the web before it was placed in the blister test unit. Another thermocouple monitored the temperature in the chamber. A reservoir of water was placed in the chamber, the device was sealed, and the unit heated slowly to the target temperature. When the desired temperature was reached, the force holding the chamber closed was quickly released, rapidly exhausting the steam to the surrounding atmosphere. The sheet was then examined for blistering. Tests were repeated at successively higher temperatures until blister formation was observed.

The critical parameter controlling blistering is the pressure difference between the inside and outside of the web. At the instant the pressure on the chamber is released, the pressure inside the web is equal to the vapor pressure of water at the web temperature, and the pressure outside the web is equal to that of the surrounding atmosphere.

Under the controlled conditions in the blister test unit there exists a critical pressure difference above which blistering will occur. Figure 4 shows the measured critical pressure for sheets formed under various conditions. This critical pressure is the difference between the internal web pressure and the pressure at the surface of the web for which blistering occurs. The higher the critical pressure for any sheet, the larger the pressure gradient it can withstand and hence, the higher that web's resistance to delamination.

For all sheets, the propensity for blister formation decreases as the web dries, due to development of stronger fiber-fiber bonds. Thus, blistering is most likely to occur in the early section of a Minton dryer.

increase. When pressure gradients develop between internal points or between an internal point and the paper surface, bulk vapor flow commences.

Heat transfer to the web is also promoted under the reduced pressure environment. Since the web temperature is lowered to the boiling point of water at reduced pressure, a larger driving force for heat transfer from the dryer can to the web exists.

Heat and mass transport external to the sheet as it passes over and between dryer cylinders can be quantitatively described using the well established correlations of Chilton and Colburn (2). The processes occurring within a sheet of paper during drying are less well defined. They are complex, interactive, and occur within a structure which is itself responding to local changes in moisture content.

Utilizing a laboratory technique for measuring the transport phenomena occurring within moist webs during drying (3), constitutive relationships describing the internal movement of liquid, vapor and heat have been deduced. This analytical description of the transport processes within the moist sheet, together with elemental heat and mass balances, has prompted development of a model capable of predicting the movement of moisture and heat throughout the sheet exposed to a variety of boundary conditions, including those encountered on a Minton vacuum dryer.

THEORETICAL BASIS

The mathematical description of the internal transport phenomena is obtained by balancing the input, output and rate of accumulation of moisture and heat for an elemental ply in the plane of the sheet with basis weight, db. This elemental ply is shown in Figure 1. Describing the position of the ply in terms of basis weight rather than a spatial variable simplifies the derivation of mass and energy balances as it avoids the necessity of accounting for the caliper change of the web as it responds to changes in temperature and moisture content.

Water can move into and out of the ply as either vapor or liquid. If the vapor flux into the ply is $\underline{V}$, the vapor flux leaving the ply is $(\underline{V} + (\partial\underline{V}/\partial b) \cdot db)$. Similarly, if the liquid flux into the ply is $\underline{L}$, the liquid flux leaving the ply is $(\underline{L} + (\partial\underline{L}/\partial b) \cdot db)$.

By applying the law of conservation of mass to the elemental ply and assuming that the quantity of water present as liquid far exceeds that present as vapor, the rate of change of moisture content in the ply, $\partial M/\partial t$, can be expressed as:

$$\frac{\partial M}{\partial t} = -\frac{\partial V}{\partial b} - \frac{\partial L}{\partial b} \tag{1}$$

The vapor flux, $\underline{V}$, is the the sum of contributions due to diffusion and convection, corresponding to the two terms in equation (2):

$$\underline{V} = -D\frac{\partial C}{\partial x} - k\frac{\partial P}{\partial x} = -D\frac{\partial b}{\partial x}\frac{\partial C}{\partial b} - k\frac{\partial b}{\partial x}\frac{\partial P}{\partial b} = -F_D\frac{\partial C}{\partial b} - F_P\frac{\partial P}{\partial b} \tag{2}$$

For a given web, the fluxes are governed by gradients in water concentration and total pressure repectively, and the transport coefficients D, local vapor diffusivity, and k, local permeability. The gradient in moisture content can be expressed in terms of basis weight by multiplying by the web density, $\partial b/\partial x$. The diffusivity and permeability are lumped together with the density and designated F_D and F_P. These coefficients are measured experimentally.

Liquid transport in a porous medium such as paper or board occurs due to differences in local moisture content. The liquid flux, $\underline{L}$, can be expressed as:

$$\underline{L} = -F_L\frac{\partial M}{\partial b} \tag{3}$$

by assuming that there is a direct relationship between local liquid flux and the gradient in moisture content. The gradient is again expressed in terms of basis weight, and F_L is the measured transport coefficient.

Heat can enter and leave the ply by convection with vapor and liquid and by conduction. Radiative heat transfer is assumed to be negligible at the temperatures typical of conventional dryers. The convective heat transfer is the sum of the vapor and liquid fluxes multiplied by their respective heat contents. If the heat flux into the ply by conduction is $\underline{Q}$, the heat flux leaving the ply in the same direction is $(\underline{Q} + \partial\underline{Q}/\partial b \cdot db)$. Applying the law of conservation of energy yields the following expression for the rate of change of heat content of the ply, $(\partial H_S/\partial t + \partial M \cdot H_L/\partial t)$:

$$\frac{\partial H_S}{\partial t} + \frac{\partial(M \cdot H_L)}{\partial t} = -\frac{\partial(V \cdot H_V)}{\partial b} - \frac{\partial(L \cdot H_L)}{\partial b} - \frac{\partial Q}{\partial b} \tag{4}$$

Heat transfer due to conduction is proportional to the local temperature gradient, again expressed in terms of basis weight, by the thermal conductivity of the fiber-water matrix, F_Q.

$$\underline{Q} = -F_Q\frac{\partial T}{\partial b} \tag{5}$$

The local and instantaneous movement of liquid, vapor and conducted heat depend upon local conditions such as moisture, temperature and the structure and composition of the paper or board. The complete description of the internal heat and mass transfer must include constitutive relationships describing how heat and mass fluxes depend upon local conditions. These relationships were characterized using a laboratory dryer which continuously measures the evaporation rate of water from a sheet which is held against a hot platen by a dryer fabric into a stream of hot air

The resistance to blistering is markedly higher for lighter basis weight sheets. Sheets labeled 1 and 2 in Figure 4 were of equal density but sheet 2 has a 14% higher basis weight than sheet 1. Although the two sheets are of the same permeability, the path length for vapor flow is longer for the thicker sheet, hence the impulse imposed by the work done by the vapor upon the web is higher and blistering is more likely to occur. Curve 3 in Figure 4 is for a sheet of equal density as sheet 4 but with lighter basis weight. As before, this sheet is observed to be less likely to delaminate compared with its thicker counterpart, due to the enhanced ability of the vapor to flow through the thinner sheet.

The effect of density upon critical pressure is determined by the balance between internal bond strength and permeability for each case. Pulp sheets 2 and 4 are of equal basis weight but different density. The denser web conducts heat more readily into the sheet, promoting rapid development of pressure gradients, and is less permeable to the water vapor which forms. However, this effect is offset by increased internal strength of the denser web. The resistance to blistering for a specific case depends upon the balance between the two effects.

To investigate the feasibility of increasing machine productivity, simulations were run for a heavier basis weight web of higher density and higher entering solids content which was to be dried under identical machine conditions as the previously described base case. The same machine speed was achieved with only a 1.4% drop in outlet solids content. The resulting energy consumption was reduced by 11% compared to the above results, due to the increased mechanical dewatering. However, the internal pressure gradients predicted for the heavier, denser web were significantly higher, suggesting that operation of a machine under these conditions would be subject to serious blistering problems.

SUMMARY

The phenomena of heat and mass transfer within and outside a web of paper as it dries under vacuum has been successfully simulated. Pressure gradients within the web, suspected to be responsible for blistering, have been predicted. Experiments using a blister test unit have confirmed that blistering does occur under certain conditions of pressure difference and moisture content for various webs.

The mechanism of blister formation has been shown to depend upon the strength of a web and the pressure impulse imposed upon it by drying conditions. The pressure impulse is a function of both the magnitude of the pressure gradient and the time period over which it is imposed. The strength of a web determines its resistance to blistering, and depends upon its properties and local conditions.

The computer model was used to predict pressure gradients for a specific application. Together with knowledge about the web's response to pressure gradients determined experimentally, the simulation provides considerable insight into the expected performance of a dryer after a proposed operational change. This approach is thus demonstrated to be useful in optimizing performance of a given machine.

ACKNOWLEDGEMENT

The authors wish to acknowledge the valuable assistance of David W. Kitts during the acquisition and analysis of the information presented in this report.

REFERENCES

1. Stephenson, J. N., Manufacture and Testing of Paper and Board, volume 3, pp. 319-321, McGraw-Hill, New York (1953).

2. Chilton, T. H. and Colburn, A. P., Ind. Eng. Chem., 26(11):1183 (1934).

3. Lee, P. F. and Hinds, J. A., "Optimizing Dryer Performance: A Technique for Measuring the Drying Characteristics of Fiber Webs," Tappi, 62(4):45 (1979).

4. Lee, P. F. and Hinds, J. A., "Analysis of Heat and Mass Transfer within a Sheet of Papermaking Fibers During Drying," Drying '82, pp. 74-79, Hemisphere Publishing Corporation (1982).

5. Hinds, J. A. and Neogi, A. N., "The Dynamic Computer Simulation of a Paper Machine Dryer," Tappi, 66(6):80 (1983).

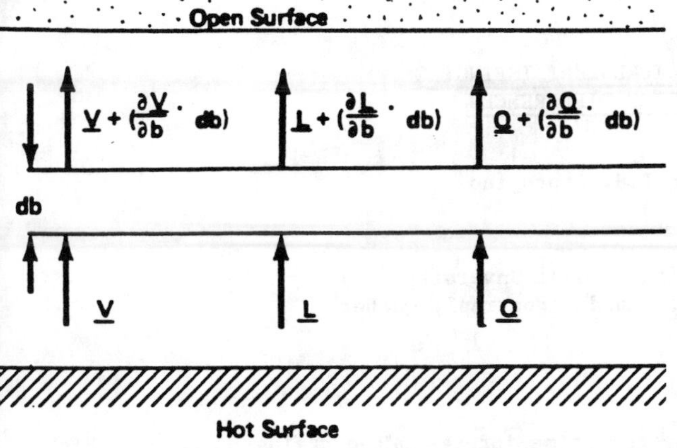

Figure 1. Mathematical model of paper drying.

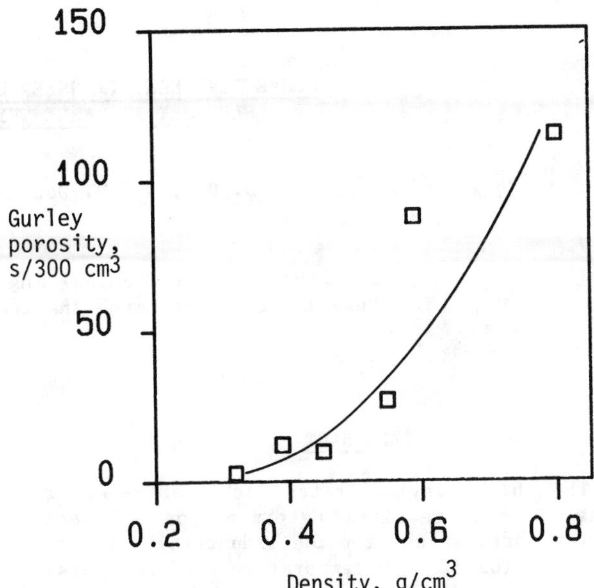

Figure 3. Permeability as a function of density for webs at a moisture content of 0.075 g water/g fiber at 100 C.

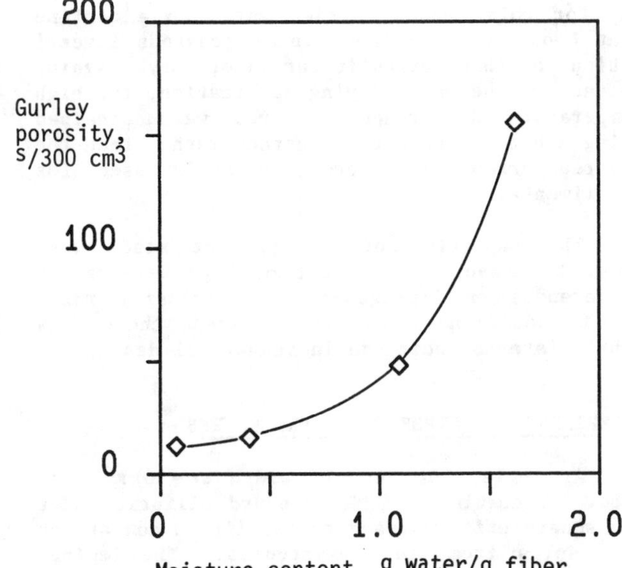

Figure 2. Permeability to bulk convection as a function of moisture content for one bleached kraft web with density = 0.39 g/cm^3 at 100 C.

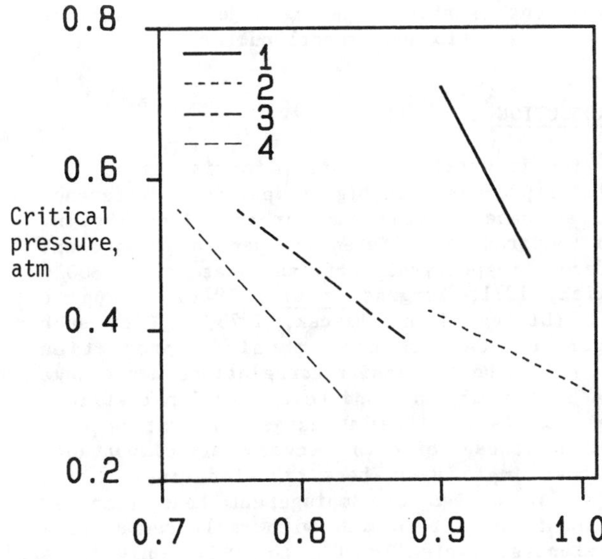

Figure 4. Critical pressure as a function of moisture content for various webs. Line 1 represents low density, low basis weight; 2 is low density, higer basis weight; 3 is higher density, intermediate basis weight; 4 is higher density, higher basis weight.

CONVECTIVE HEAT TRANSFER UNDER TURBULENT IMPINGING SLOT JET AT LARGE TEMPERATURE DIFFERENCES

D. Das, W.J.M. Douglas and R.H. Crotogino

Department of Chemical Engineering, McGill Unversity
Pulp and Paper Research Institute of Canada, Montreal, Quebec

ABSTRACT

The high drying rates achievable under impinging jets requires determination of heat transfer correlations for the industrially important condition of high temperature confined jets. The effect of high temperature differences on local and average heat transfer rates under a confined single slot jet was studied experimentally for a range of temperature differences (nozzle exit to impingement surface) from 50 to 300°C. While stagnation heat transfer coefficients varied 62% with this variation in T, an appropriately defined stagnation Nusselt number varied only by 6.5%. A general correlation was obtained for average Nusselt numbers as a function of jet Reynolds number, nozzle exit to impingement surface spacing, jet to surface temperature ratio and Prandtl number.

INTRODUCTION

The industrial potential for impinging jets includes processes at high temperature difference such as paper drying and turbine blade cooling with temperature differences, jet to impingement surface, respectively of somewhat over 300°C (Holick, 1971; Burgess et al., 1972) and up to 500°C (Livingood and Hrycak, 1973). Over such ranges of temperature, physical properties involved in heat transfer correlations may change by a factor of two, and heat transfer coefficients likewise. Thus at large temperature differences these effects become very important. Yet most impingement heat transfer studies, the basis for design of impingement heat transfer equipment, have been made with small temperature differences, typically 20°C to 70°C. Only three early studies, Perry (1954), Thurlow (1954) and Huang (1963), used temperature differences, ΔT, between jet, T_j, and impingement surface, T_s, of greater than 100°C. An analysis of these studies by Das (1982) indicates that, because of shortcomings concerning experimental techniques and large discrepancies between their data, these studies do not provide the reliable basis required for design and optimization of industrial processes.

The choice of flow configuration and major parameters for the present study was determined by our prime interest, the drying of paper. High velocity impingement dryers are used extensively on machines where production rates are limited by drying rates. The Yankee dryer for the drying of tissue products and the high velocity dryer on coating machines reflect current paper industry use of impingement drying. In the projected "Papridryer" process using impinging slot jets, impingement and throughflow drying may be combined to give drying rates for newsprint an order of magnitude larger than with conventional steam-heated cylinder drying (Burgess et al., 1972). Such process involve impinging jets which are confined by a hood, required for partial recirculation of the impingement air and for energy recovery. Thus, while most impinging jet studies have been of unconfined jets, the present study was for this reason carried out with a confinement hood, as were the several previous investigations of this series in our laboratory. Again, because of the paper drying application, the high temperature differences desired were obtained using a hot jet on a cool surface rather than the inverse arrangement more frequently used for experimental convenience.

The objective of the present study was, then, to measure the effect of high temperature differences on impingement heat transfer rates and to determine a way to represent these data appropriate to their use in industrial design.

EXPERIMENTAL EQUIPMENT AND PROCEDURES

A single slot jet, of width w = 6 mm, was used. A nozzle of ASME standard elliptic inlet and square exit was used to facilitate comparison of results from previous studies. The impingement flow was confined for a lateral distance of 136 mm (or 23 w) or both sides of the nozzle by a hood parallel to the impingement surface and flush with the nozzle exit. The spent flow exited through two 61 mm wide nozzle ports (i.e. 10 w) at the end of this confinement surface. These dimensions provided the same ratio of jet width to centerline spacing between inlet nozzle and exhaust ports as existed in the pilot plant Papridryer, i.e. a ratio of 0.035, conventionally termed as 3.5% open area for the jets in the confinement hood. The impingement surface could be spaced, H, from 30 to 72 mm (i.e. 5 w to 12 w) from the nozzle exit. Local heat flux at the impingement surface was measured with two extremely thin (0.18 mm), flush mounted heat flux sensors, 4.78 mm x 1.9 mm, i.e. 0.32 w in the nozzle width direction. The profiles of local heat transfer were quite precisely defined by lateral traversing, at intervals of 1 w, for 19 positions from the stagnation point out to x = 18 w.

Air from a 30 H.P. blower and a 100 KW duct heater entered the 190 mm long slot nozzle from a plenum chamber with honeycomb and screens giving a controlled, low turbulence level flow to the nozzle, thus providing jet flow characteristics comparable to those of previous studies. Except for the water cooled central section of the impingement surface (copper) the apparatus, including the nozzles, was fabricated of a calcium silicate based structural material, stable to high temperatures and with a low thermal expansion. The impingement surface was maintained isothermal, in the range of 30 to 11°C, with the largest variation of surface temperature of less than 2°C for the maximum jet temperature of 300°C. Complete experimental details are given by Das (1982).

The range of experimental parameters was:

Variable	Nominal Values Used
Re_j	1,000, 5,000, 10,000, 15,000, 20,000
H/w	5, 8, 10, 12
$T_j - T_s$, °C	50, 100, 150, 200, 250, 300
T_j/T_s	1.18, 1.35, 1.53, 1.71, 1.88, 2.06
x/w	0, 1, 2, 3, 4, ... 18 (19 values)

The range of Re_j was selected to include that of the mill trial Papridryer (1000 to 3000) as well as higher values expected to be of interest in the future and to provide comparision with previous studies at low ΔT. The range of impingement surface spacing, H/w, 5 to 12, was chosen to include the value H/w =8, the spacing generally found to given maximum Nusselt number at the stagnation line, and H/w = 5, the values determined by another study in this laboratory (Saad, 1981) to be the spacing for maximum average heat transfer rate for multiple confined impinging shot jets. A total of 104 runs were made, each with 19 sets of data to define a complete profile of local Nusselt number.

RESULTS

1. Effect of Temperature Dependent Physical Properties at High ΔT

Studies of this problem for confined and for unconfined flows include those by Douglas and Churchill (1956), Kays (1966), Petukhov (1970), Shah and London (1978). Two methods are used, principally, the temperature ratio and the reference temperature method. In the former, the ratio of Nusselt number at bulk temperature (for a confined flow) to that at the constant property condition is related to the ratio of the corresponding temperatures as:

$$\frac{Nu_b}{Nu_{cp}} = \left(\frac{T_b}{T_o}\right)^n$$

Usually surface temperature is used as the reference temperature T_o. By contrast, in the other method, all properties are evaluated at some reference temperature so that no temperature ratio is required. The "film" temperature, the mean between that of the surface and either the bulk temperature (for a confined flow) or a boundary condition temperature (for an external flow) is a frequent choice for this reference temperature.

For impinging jets, considered as an external flow, the temperature ratio relation is

$$\frac{Nu_j}{Nu_s} = \left(\frac{T_j}{T_s}\right)^n$$

Kays (1966) solved numerically for the case of a laminar two dimensional flow and obtained n = -0.1 for heating and -0.07 for cooking at the stagnation line.

The results at stagnation for $Re_j > 5000$ and H/w > 8 from the present study may be correlated equally well, to within ±5%, by either the temperature ratio method, as:

$$Nu_{oj} = 0.79\, Re_j^{0.485}\, (H/w)^{-0.134}\, (T_j/T_s)^{-0.115}\, Pr_j^{1/3} \qquad (1)$$

and by the reference method as:

$$Nu_{oT_{ref}} = 0.648\, Re_j^{0.485}\, (H/w)^{-0.138}\, Pr_j^{1/3} \qquad (2)$$

where the subscript, T_{ref}, indicates evaluation at the reference temperature which was determined to be:

$$T_{ref} = T_j - 0.2\,(T_j - T_s)$$

The subscript j on dimensionless numbers indicates evaluation of physical properties at nozzle suit jet temperature, T_j.

Of these two alternate correlations, the temperature ratio method is preferable in that less complex iterations are required in a typical application in design. Thus although either T_s or T_{ref} must in general be solved by iteration, the physical properties based on T_j remain unchanged throughout such iterations with the temperature ratio method which is therefore the method recommended.

2. Stagnation Heat Transfer

Stagnation Nusselt number passes through a maximum at a spacing around 8 w except for the laminar jet, Re_j = 1000, in which case Nu_o is independent of H/w. This maximum for turbulent jets from similar nozzles is well known from numerous studies, most recently that of Saad (1981). The maximum reflects the fact that Nu_o is dominated for lower spacings by the increase of turbulence with distance from the nozzle exit, and for larger spacings by the decay of mean axial velocity. Saad (1981) presented a quantitative analysis of the combined effect. The fact that in Equation (1) the exponent on H/w is -0.134 is a consequence of the experimental range of the data, i.e., spacings from 8 w to 12 w, and the occurence of a maximum at H/w = 8. thus, the higher the upper experimental limit of H/w, the larger the absolute value of this exponent. As for the laminar jet results, the data of Sparrow and Wong (1975) indicate that, for $Re_j = 950$, a jet will still be laminar for an impingement surface spacing $H/w < 10$. The numerical predictions of van Heiningen et al. (1976) show that Nu_o under a laminar jet, $Re_j = 950$, is insensitive to spacing, a prediction confirmed by the present experiments.

The exponent, 0.485, on Re_j in Equations (1) and (2) indicates that N_uo varies effectively with $Re_j^{0.5}$, as applies for laminar flow. The most recent verification that the boundary layer in the near vicinity of stagnation is laminar even for the case of a highly turbulent jet was given by van Heningen (1982).

The exponent of T_j/T_s in Equation (1), i.e. -0.115, closely approximates the value of -0.1 derived theoretically by Kays (1966) for a heating jet.

As noted, correlations (1) and (2) do not apply at the lowest value of impingement surface spacing, H/w = 5, or the lowest value of Reynolds number, $Re_j = 1000$, because Nu_o passes through a maximum at H/w = 8 while for the laminar jet at Re = 1000, Nu_o is independent of H/w over the tested range of spacings, 5 w to 12 w. In these ranges, i.e. $1000 < Re_j < 5000$ and $5 < H/w \quad 8$, Nu_o may be obtained from graphical representation of the data (Das, 1982).

3. Local Nusselt Number Profile

Two typical sets of profiles are reproduced in Figures 1 and 2. The combination of a sufficently low spacing and a sufficiently high Reynolds number produce an off-stagnation minimum and a secondary maximum, features displayed on Figure 1 for H/w = 5, $Re_j = 15000$. Under such combinations of spacing and Reynolds number the negative pressure gradient at the impingement surface is sufficiently strong to maintain the boundary layer laminar for some distance from stagnation even for a highly turbulent jet. The off-stagnation minimum marks the onset of transition to a turbulent boundary layer, with a corresponding large increase in transport rates, while the secondary maximum marks the completion of transition, after which Nu_j drops because of a thickening boundary layer and a decaying wall jet velocity. For a wide range of values of H/w, Saad (1981) demonstrated that the transition point minimum occurs at slightly below x/H = 1. This behaviour was confirmed by the results of van Heiningen (1982) and again in the present study where the transition minimums occur at x/H = 0.8. Beyond this minimum all three studies noted find that about 2.5 w is required from onset to completion of transition, i.e. the distance between the minimum and secondary maximum in Nu_j.

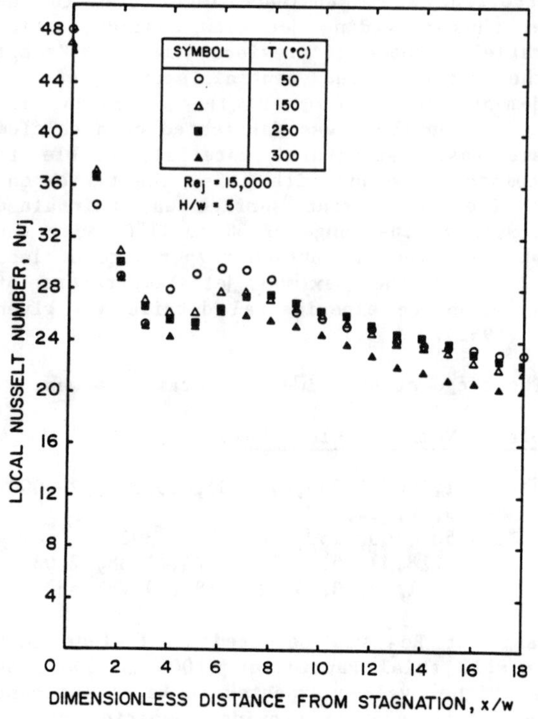

Fig. 1. Effect of ΔT on Local Nu Profile at H/w = 5, $Re_j = 15000$.

Figure 1 shows that over a wide range of ΔT, the profiles of heat transfer effectively superimpose when expressed as Nu_j. The small decrease of Nu_j with increasing ΔT at stagnation, consistent with the exponent -0.115 on T_j/T_s in Equation 1, is seen in Figure 1 to extend out to about the secondary maximum in Nu_j, i.e., for the distance over which the boundary layer retains some or all of the laminar nature it has at stagnation. Beyond the stagnation region, the profiles of Figure 1 are seen to cross randomly, and all results of the present study indicate that ΔT is no longer a variable.

Figure 2 displays a comparison of results at $Re_j = 10000$, H/w = 8, conditions for which the off-stagnation features described above do not occur. Gardon and Akfirat (1966) and Cadek (1974) used unconfined cooling jets, while Saad

(1981) used a confined cooling jet. The differences are attributed to the different effects of entrainment for confined and unconfined jets and for a cooling jet (their studies) and a heating jet (present study). Moreover, Obot (1980) demonstrated the sensitivity of impingement heat transfer to conditions upstream from the nozzle exit which affect the development of turbulence or the decay of mean velocity from nozzle exit to stagnation point. Thus the differences in results displayed on Figure 2 are not attributed to experimental error, as all these studies are considered reliable, but reflect the sensitivity of results to the variables noted.

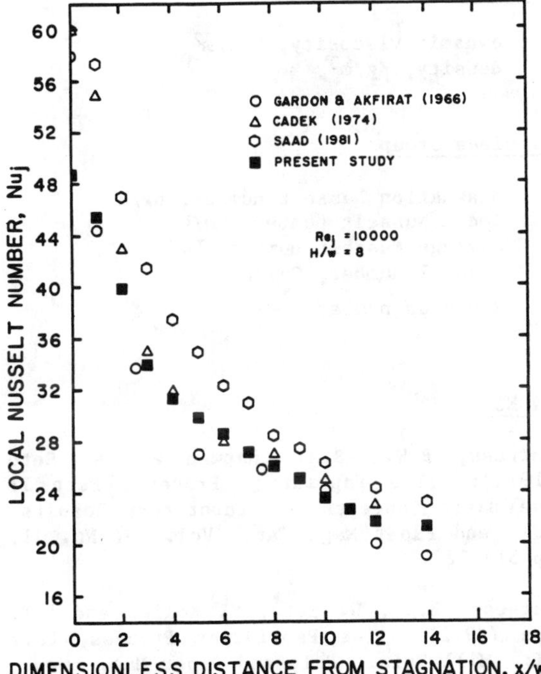

Fig. 2. Comparison of local Nu profile with previous results.

4. Average Nusselt Number

For use of experimental impinging jet results for industrial process design in applications such as the drying of paper and other wet webs, average rather than local values of heat transfer coefficient are required. Thus Nu_j was obtained by numerical integration for the 104 runs of the present study. A general correlation for Nu_j, incorporating the ratio method for effect of temperature difference, and valid for $5000 < Re_j$ 20,000, $8 < H/w < 12$, and $T_j/T_s < 2.1$ is:

$$\overline{Nu_j} = K\, Re_j^{\,a}\, (H/w)^b\, (T_j/T_s)^c\, Pr_j^{\,1/3} \quad (3)$$

The coefficient K and exponents a, b and c are functions of extent of heat transfer area x/w, as given on Figure 3 out to x/w = 18. The deviation of experimental data from this correlation is within ±15%, and appropriate tests indicated that this depression is uniform for all of the independent parameters thus further confirming the validity of the method to account for all variables.

The exponents display trends which may readily be related to known physical behaviour of impinging jets. Thus the exponent a for Re_j remains at about 0.5, the value characteristic for laminar heat transfer, out to the position where transition to a turbulent boundary layer begins, around x/w = 2. The limiting value of about 0.55 for this exponent at the maximum value of x/w = 18 reflects the averaging of the affects from the stagnation region, where Nu_j is high but where its dependence is with $Re^{0.5}$ because of the laminar boundary layer, with those from the wall jet region, where Nu is low but varies locally with $Re_j^{0.5}$ because the boundary layer is turbulent. Likewise, the decrease in importance of spacing as reflected by the approach to zero of the exponent b for H/w at the maximum value of x/w = 181 reflects the diminishing effect of H/w relative to the total distance travelled by the impingement flow, H/w + x/w. The dependence of Nu_j on T_j/T_s, represented by the exponent c, is small even for stagnation heat transfer, Equation (1), and Figure 3 indicates that c does not vary much with x/w. A value of c = -0.11 could be used for the entire range of x/w.

In considering operation at high levels of ΔT it is important to distinguish between the effects of ΔT on h and on Nu_j. Thus an increase in ΔT from 50°C to 300°C, corresponding to a change in T_j/T_s from 1.18 to 2.06, causes only a small change in Nu_j, a decrease by about 6.5%. However, allowing for the corresponding 60% increase in thermal conductivity, the heat transfer coefficient, h, increases by about 50%. Thus in refining to the effect of high ΔT or heat transfer it is necessary to be quite explicit as to the basis used.

The logarithmic linear correlation (3) does not apply for the laminar jet case, Re_j = 1000, nor for the lowest spacing, H/w = 5, for which the noted off-stagnation phenomena occur. Thus in the ranges of conditions $1000 < Re_j < 5000$, and $5 < H/w < 8$, Nu_j may be obtained from graphical representation of the Nu_j data (Das, 1982).

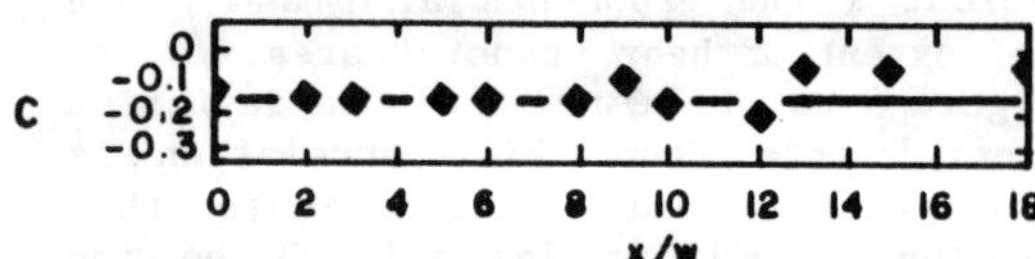

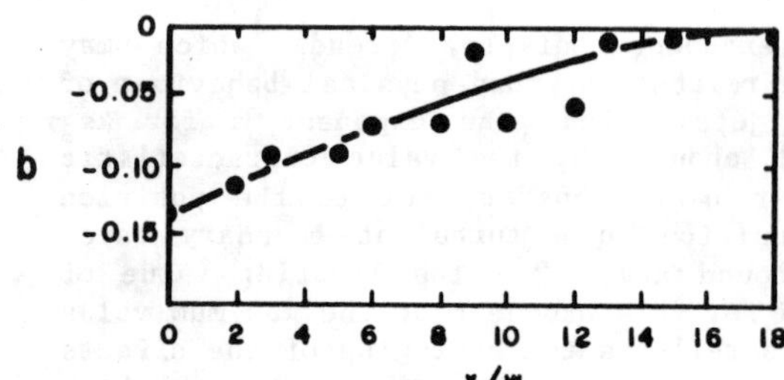

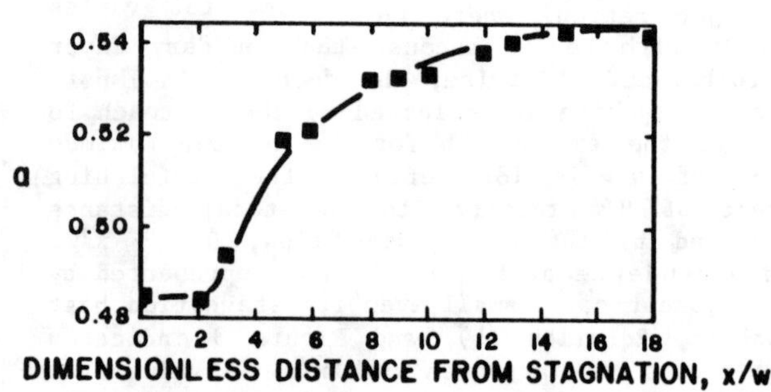

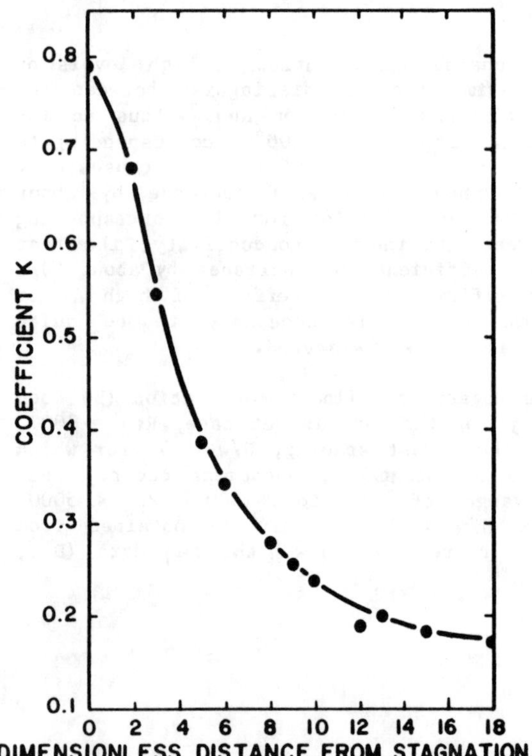

Fig. 3. Coefficient and exponents of Equation 3 as a function of x/w.

NOMENCLATURE

a	exponent of Reynolds number
b	exponent of dimensionless spacing
c	exponent of temperature ratio
K	coefficient of Equation (3)
k	thermal conductivity, W/m°C
T_j	jet exit temperature, K
T_{ref}	reference temperature, K
T_s	impingement surface temperature, K
ΔT	temperature difference between jet and impingment surface, K
v_j	velocity at jet exit, m/s
w	nozzle width, mm
x	lateral distance from stagnation, mm

Greek Letters

μ	dynamic viscosity, $N.s/m^2$
ρ	density, Kg/m^3

Dimensionless Groups

Nu_o	stagnation Nusselt number, hw/k
Nu_j	local Nusselt number, hw/k
$\overline{Nu_j}$	average Nusselt number, $\bar{h}w/k$
Pr	Prandtl number, $C_p\mu/k$
Re_j	Reynolds number, $\frac{w\, v_j C_j}{\mu_j}$

REFERENCES

1. Burgess, B.W., S.M. Chapman and W. Seto (1972): The Papridryer Process, Part I: The Basic Concepts and Laboratory Results, Pulp and Paper Mag. Can., Vol. 73, No. 11, pp 314-322.

2. Burgess, B.W., W. Seto, E. Koller and I.T. Pye (1972): The Papridryer Process, Part II: Mill Trial, Pulp and Paper Mag. Can., Vol. 73, No. 11, pp 323-331.

3. Cadek, F.F. and R.D. Zerkle (1974): Local Heat Transfer Characteristics of Two Dimensional Impinging Air Jets — Theory and Experiments, Proc. Fifth Int. Heat Transfer Conf., Tokyo, FC 1.4 pp 15-19.

4. Das, D. (1982): Convective Heat Transfer Under a Turbulent Impinging Slot Jet at Large Temperature Differences, M. Eng. Thesis, Chem. Engineering Dept., McGill University, Montreal.

5. Douglas, W.J.M. and S.W. Churchill (1956): Chem. Eng. Prog. Symposium Series, No. 18, Vol. 52, pp 23-28.

6. Gardon, R. and J.C. Akfirat (1966): Heat Transfer Characteristics of Impinging Two Dimensional Air Jets, Trans. ASME, J. of Heat Transfer, Vol. 88, pp 101-108.

7. Huang, G.C. (1963): Investigations of Heat Transfer Coefficients for Air Flow Through Round Jets Impinging Normal to Heat Transfer Surface, Trans. ASME, J. of Heat Transfer Vol. 85, pp 237-245.

8. Kays, W.M. (1966): Convective Heat and Mass Transfer, McGraw-Hill Book Company.

9. Livingwood, J.N.B. and P. Hrycak (1973): Impingement Heat Transfer from Turbulent Air Jets to Flat Plates — A Literature Survey, NASA TM X-2778.

10. Obot, N.T. (1980): Flow and Heat Transfer for Round Turbulent Jets Impinging on Permeable and Impermeable Surfaces, Ph.D. Thesis, Chem. Eng. Dept., McGill University, Montreal.

11. Perry, K.P. (1954): Heat Transfer by Convection from a Hot Gas Jet to a Plane Surface, Proc. Inst. of Mech. Engineers, Vol. 168, pp. 775-780.

12. Petukhov, B.S. (1970): Heat Transfer and Friction in Turbulent Pipe Flow with Variable Physical Properties, Advances in Heat Transfer, Vol. 6, Academic Press, New York.

13. Saad, N.R. (1981): Flow and Heat Transfer for Multiple Turbulent Impinging Slot Jets, Ph.D. Thesis, Chem. Eng. Dept., McGill University, Montreal.

14. Shah, R.K. and A.L. London (1978): Laminar Flow Forced Convection in Ducts; A Source Book for Compact Heat Exchanger Analytical Data, Academic press, New York.

15. Sparrow, E.M. and T.C. Wong (1975): Impingement Transfer Coefficients Due to Initially Laminar Slot Jets, Int. J. of Heat and Mass Transfer, Vol. 18, pp 597-605.

16. Thurlow (1954): Communication to Perry, Proc. Inst. of Mech. Engineers, Vol. 168, pp 781-784.

17. Van Heiningen, A.R.P., A.S. Mujumdar and W.J.M. Douglas (1976): Numerical Prediction of the Flow Field and Impingement Heat Transfer Caused by a Laminar Slot Jet, Trans. ASME, J. of Heat Transfer, Vol. 98, No. 4, pp 654-658.

18. Van Heiningen (1982): Heat Transfer Under An Impinging Slot Jet, Ph.D. Thesis, Chem. Eng. Dept., McGill University, Montreal.

MATHEMATICAL MODELING OF WOOD DRYING FROM HEAT AND MASS TRANSFER FUNDAMENTALS

Mark A. Stanish*, Gary S. Schajer, Ferhan Kayihan

Weyerhaeuser Technology Center
Tacoma, Washington 98477

ABSTRACT

A mathematical model is developed to simulate the drying of wood. In this approach, mathematical formulations for the transport rates are incorporated into one-dimensional partial differential material and energy balance equations which are coupled with algebraic relations for local phase equilibria. Spatial derivatives are approximated in finite difference form with variable spacing between mesh points in order to maintain accuracy. The resulting ordinary differential equations are solved numerically using a special solver for systems of stiff equations, with the solution constrained locally to satisfy the phase equilibria relations. The model simulation provides spatial distributions of each moisture phase (vapor, liquid, and bound) and of temperature and pressure as functions of time. Because this model is based on fundamental transport theory, it provides a tool to investigate the impacts on drying behavior both of process variables and of the material properties of wood.

Drying rate experiments were performed using a porous ceramic solid and were simulated using the non-hygroscopic version of the drying model. Calculated model predictions are in very satisfactory agreement with experimental results. Parametric sensitivity analyses for variables such as permeability reveal the role and relative importance of each in determining overall drying rate and behavior.

1. INTRODUCTION

Drying of moist solids is a complex process of simultaneous heat and mass transport, and as such may be influenced by many independent factors. Characterization of drying phenomena using purely experimental means is therefore difficult; our objective is to develop a mathematical model, or tool, to simulate drying behavior. In particular, we want to use this tool to extend experimental investigations of the drying of wood in lumber kilns in order to evaluate the impacts of process variables such as temperature, air velocity, and relative humidity, and of wood variables such as density, moisture content, and permeability.

*Current affiliation: Energy International, Inc., Bellevue, WA 98004

Mathematical modeling of drying phenomena has been and currently is the object of much research interest. A survey of such work is beyond the scope of this paper but information is readily available in the literature. Numerous drying topics of recent interest are covered in publications edited by Ashworth[1] and Mujumdar[2,3]. Rosen[4] reviewed works specific to lumber drying.

Our approach distinguishes itself in two respects. First, we retain a general description of the drying process in terms of a comprehensive set of fundamental heat-and mass-transfer mechanisms together with thermodynamic phase equilibrium. We include heat transfer via both conduction and convection, and mass transfer by binary gaseous diffusion, pressure-driven bulk flow of gas and liquid, and diffusion of bound water through the cell-wall matrix. Second, we describe bound water migration as diffusion of sorbed water driven by the gradient in its chemical potential. Although the concept was previously suggested by Siau[5] and others[6], we have developed a unique, explicit expression for flux in terms of temperature and vapor pressure gradients.

This paper describes the conceptual picture of drying on which the model is based, summarizes the governing equations which comprise the model, and outlines the methods employed in the numerical solution. Finally, it draws comparisons between model predictions and experimentally observed drying behavior.

2. MODEL DEVELOPMENT

2.1 Approach

The drying phenomenon is viewed as a process of simultaneous mass and energy transfer occurring both inside and outside the wood. Three moisture phases are recognized: free (liquid) water, bound (absorbed) water, and water vapor. Heat and mass transfer are coupled by the requirement that all phases remain in thermodynamic equilibrium at the local temperature. Thus evaporation rates are determined locally by the balance between heat flows, temperature changes, and moisture flows, all subject to phase equilibrium constraints. The overall drying phenomenon results from the simultaneous interaction of all of these processes.

The mechanisms of energy and mass transport are critical factors in such a conceptual description of drying. In this model, we try to identify all significant modes of transport and retain a quantitative mathematical expression for each. For example, an independent migration mechanism is postulated for each of the three moisture phases. Free water flows in bulk form driven by a gradient in the pressure within the liquid phase. Bound water migrates by a diffusion process that is driven by a gradient in the chemical potential of the sorbed water molecules. Water vapor and air are transported both by bulk flow driven by a gradient in total gas pressure and by binary diffusion driven by a gradient in the mole fraction of each component. Similarly, energy flows resulting from both conductive and convective mechanisms are included.

2.2 Assumptions

1. Mass and energy transport occur in only one dimension.

2. Moisture can exist in three different phases: vapor, bound (absorbed), and free (liquid).

3. Local thermal and phase equilibria are always obeyed. If free water is present, the vapor remains saturated and the bound water content remains at the fiber saturation point at the local temperature. In the absence of free water, the bound and vapor phase compositions obey the sorption equilibrium.

4. Conditions at the two boundaries are independent of each other, may vary with time, and are characterized by convective transfer coefficients.

5. Wood is a hygroscopic porous solid. Water, air, and solid are recognized as independent phases having unique properties. Physical properties vary in both space and time; transport properties which are functions of physical properties therefore also vary in space and time.

6. Bulk flows of liquid and gases follow Darcy's Law. Effective permeabilities to liquid and to gases are dependent upon the relative liquid saturation of the void space within the solid.

7. Bound water migration is a molecular diffusion process whose flux is proportional to the gradient in the chemical potential of the sorbed molecules; no assumptions regarding temperature-driven (Soret potential) diffusion are necessary because a contribution from the temperature gradient arises automatically from the chemical potential expression.

8. No a priori assumptions are made regarding the evolution of the drying process. For example, no wet line drying front is necessarily postulated. The drying mechanism follows solely from the solution of the governing equations.

2.3 Derivation of Model Equations

The model contains five main dependent variables (including four densities - those of air, water vapor, bound water, and free water - and temperature) which are functions of space and time. There are five governing equations: two mass balances (for air and water), one energy balance, and two equilibrium equations (liquid saturation and a bound water sorption relation).

<u>Mass balances</u>. In order to include mass transfer contributions from pressure-driven bulk flow, a mass balance on air must be satisfied, even though air density in itself has a negligible impact on measured moisture content. The conservation equation for air takes the form:

$$\frac{\partial}{\partial t}(\rho_a) = -\frac{\partial}{\partial z}(n_a) \tag{1}$$

where the effect of porosity has been taken into account by defining density in terms of unit volume of space.

By writing a balance equation for total water, we avoid having to specify explicitly the unknown rate of evaporation. The conservation of water requires that:

$$\frac{\partial}{\partial t}(\rho_v + \rho_b + \rho_f) = -\frac{\partial}{\partial z}(n_v + n_b + n_f) \tag{2}$$

<u>Energy balance</u>. Since we include convective contributions from mass flow in the energy equation, the rate of evaporation need not appear explicitly. The energy balance is:

$$\frac{\partial}{\partial t}(\rho_a h_a + \rho_v h_v + \rho_b h_b + \rho_f h_f + \rho_d h_d) = -\frac{\partial}{\partial z}\left(n_a h_a + n_v h_v + n_b h_b^* + n_f h_f - k\frac{\partial T}{\partial z}\right) \tag{3}$$

<u>Equilibrium relations</u>. If the local free water density is non-zero, the gas phase at that point must be saturated at the local temperature. Saturated water vapor densities from steam tables were correlated over the range 300-500 K to within 1.0% accuracy. The saturated water vapor density in wood is given by:

$$\rho_v^{sat} = (\varepsilon)\exp[-46.490 + 0.26179(T) - 5.0104\times10^{-4}(T)^2 + 3.4712\times10^{-7}(T)^3] \tag{4}$$

where the effective local void fraction is $\varepsilon = \varepsilon_d - (\rho_f/\rho_w)$ and the void fraction of dry wood is $\varepsilon_d = 1 - (\rho_d/1500)$.

In the absence of free water, the gas phase is assumed to be saturated with respect to the local bound water content and temperature. Bound water sorption data were correlated by Simpson[7]. We use the inverted form of his equation to explicitly relate vapor density to the bound water content and the temperature:

$$\rho_v = \rho_v^{sat}\{a_4 + [(a_4)^2 + \frac{1}{a_1(a_2)^2}]^{1/2}\} \tag{5}$$

where:

$$a_1 = -45.70 + 0.3216(T) - 5.012\times10^{-4}(T)^2 \quad (6)$$

$$a_2 = -0.1722 + 4.732\times10^{-3}(T) - 5.553\times10^{-6}(T)^2 \quad (7)$$

$$a_3 = 1417 - 9.430(T) + 1.853\times10^{-2}(T)^2 \quad (8)$$

$$a_4 = \frac{[1-(18\rho_d/a_3\rho_b)]}{2a_2} - \frac{[1+(18\rho_d/a_3\rho_b)]}{2a_1a_2} \quad (9)$$

Boundary conditions. Each of the three differential balance equations requires for solution an initial condition and two boundary conditions. To satisfy the first requirement, initial profiles must be specified for temperature and for air and moisture densities:

$$\text{at } t=0: \quad T(z,0) = T^o(z) \quad (10)$$

$$\rho_i(z,0) = \rho_i^o(z);\ i=a,v,b,f \quad (11)$$

The initial moisture and temperature profiles must satisfy the local equilibrium constraints given by Eqns. (4) and (5).

The boundary condition requirements are satisfied by specifying three relations at each surface of the solid. First, the total gas pressure at each surface must always equal the respective ambient pressure:

at z=0 and z=L:

$$P_a(z,t) + P_v(z,t) = P_a^\infty(t) + P_v^\infty(t) \quad (12)$$

where pressure is related to bulk density by $P=RT\rho/M\varepsilon$ and separate values of P^∞ may be specified at each surface.

The second relation at each surface requires that the total flux of moisture from within the solid be equal to the flux of water vapor through the external boundary layer:

at z=0 and z=L:

$$n_v + n_b + n_f = [x_v(N_v + N_a) + k_x(t)\left(\frac{P_v^\infty(t)}{P_a^\infty(t) + P_v^\infty(t)} - x_v\right)]\ M_v \quad (13)$$

where separate values of k_x may be specified at each surface. In general, the convective mass transfer coefficients depend on the mass transfer rate itself; correlations are usually derived from data for very low rates of mass transfer and are corrected if mass fluxes become significant.

The third relation at each surface equates the total energy flux through the solid to the total energy flux through the external boundary layer. Both conductive and convective contributions are included:

at z=0 and z=L:

$$n_ah_a + n_vh_v + n_bh_b^* + n_fh_f - k\frac{\partial T}{\partial z} = M_vN_vh_v + M_aN_ah_a + h(t)[T^\infty(t) - T(z,t)] \quad (14)$$

where separate values of h may be specified at each surface. Heat transfer coefficients are also dependent upon the rate of mass transfer and are corrected if necessary.

Flux terms. Quantitative expressions for the fluxes are derived from fundamental transport mechanisms. The total flux of air consists of contributions from pressure-driven bulk (Darcy) flow and from binary molecular diffusion:

$$n_a = -\left(\frac{\rho_a}{\varepsilon}\right)\left(\frac{K_g}{\eta_g}\right)\frac{\partial}{\partial z}(P_a + P_v) - \frac{M_a}{\varepsilon}\left(\frac{\rho_a}{M_a} + \frac{\rho_v}{M_v}\right) D^{eff}\frac{\partial}{\partial z}\left(\frac{P_a}{P_a + P_v}\right) \quad (15)$$

where bulk flow is characterized by the relative permeability, K_g, and molecular diffusion by the effective diffusivity, D^{eff}. The total flux of water vapor is given by a completely analogous expression with subscripts "v" and "a" interchanged.

The flux of bound water is assumed to be proportional to the gradient in the chemical potential of the bound water molecules:

$$n_b = -(1-\varepsilon_d)\ D_b\frac{\partial \mu}{\partial z} \quad (16)$$

where the proportionality is governed by a diffusivity, D_b, and the volume fraction taken up by the solid matrix. Since thermodynamic equilibrium is assumed to exist at every location, the chemical potential of the bound water equals the chemical potential of the water vapor. Therefore, we may express the bound water flux in terms of the water vapor chemical potential using the thermodynamic relation for gases $M\,d\mu = -SdT + VdP$ together with Eqn. (16):

$$n_b = -(1-\varepsilon_d)\ D_b\left[-\left(\frac{S_v}{M_v}\right)\frac{\partial T}{\partial z} + \left(\frac{\varepsilon}{\rho_v}\right)\frac{\partial P_v}{\partial z}\right] \quad (17)$$

Free water is assumed to migrate by bulk flow following Darcy's Law. The flux is therefore proportional to the gradient in the pressure within the free water phase:

$$n_f = -\rho_w\left(\frac{K_f}{\eta_w}\right)\frac{\partial}{\partial z}[P_a + P_v - P_c] \quad (18)$$

where the relative permeability for the liquid phase is K_f and the pressure within the liquid is equal to the local gas pressure less the capillary pressure, P_c, associated with the gas-liquid interface.

Enthalpy terms. The zero enthalpy reference state for each component was chosen to be 273.15 K and one atmosphere; heat capacities were taken to be constant over the temperature range of

For wood, an additional attenuation factor α is included in the effective diffusivity to account for closed pores resulting from pit aspiration. The effective gas diffusivity in wood is:

$$D_{eff} = \alpha\, D_{rp} = \frac{1.22\times10^{-4}\ (T)^{0.75}\ \varepsilon^3\ \alpha}{R\ (\rho_a/M_a + \rho_v/M_v)} \tag{29}$$

The relative permeabilities for bulk gas flow and bulk free water flow are dependent upon the permeability of the dry solid and the fraction of void space that is occupied by free water (relative saturation). For gas flow, the relative permeability decreases with increasing relative saturation; we assume a linear dependence of effective permeability on saturation:

$$K_g = K_g^d\ [1 - (\rho_f/\varepsilon_d\rho_w)] \tag{30}$$

For free water flow in wood, Spolek and Plumb[8] reported that the relative permeability is zero below a certain critical relative saturation (irreducible saturation) due to a loss of continuity in the liquid phase. Above the irreducible saturation, the relative permeability increases with increasing relative saturation. Assuming again a linear dependence of permeability on saturation, the effective permeability to free water in wood is given by:

$$K_f = \begin{cases} 0 & ;\ \rho_f < \varepsilon_d\rho_w s_{ir} \\ K_f^d[(\rho_f/\varepsilon_d\rho_w) - s_{ir}]/(1 - s_{ir}) & ;\ \rho_f \geq \varepsilon_d\rho_w s_{ir} \end{cases} \tag{31}$$

where s_{ir} is the irreducible saturation.

We assume that the thermal conductivity is adequately described for all moisture contents by [9]:

$$k = (\rho_d/1000)\,[0.20 + 0.50(\frac{\rho_v+\rho_b+\rho_f}{\rho_d})] + 0.024 \tag{32}$$

Spolek and Plumb[8] measured capillary pressure in southern pine and suggested the following dependence on relative saturation:

$$P_c = 10000(\rho_f/\varepsilon_d\rho_w)^{-0.61} \tag{33}$$

The entropy coefficient, S, for bound water diffusion may be derived from the thermodynamic relation:

$$dS = (\frac{1}{T})C_p\ dT - (\frac{R}{P})dP \tag{34}$$

For water vapor as an ideal gas at one atmosphere pressure and 298.15 K, Moore[10] lists the statistical and Third Law entropies as 185.3 and 188.7 J/mole-K, respectively. Using an average value of 187 J/mole-K at those conditions, integration of Eqn. (34) yields:

$$S_v = 187 + 35.1\ \ell n(\frac{T}{298.15}) - 8.314\ \ell n(\frac{P_v}{101325}) \tag{35}$$

interest. For air, free water, and solid, enthalpy is assumed to be a function of temperature only:

$$h_a = 1000(T - 273.15) \tag{19}$$

$$h_f = 4180(T - 273.15) \tag{20}$$

$$h_d = 1360(T - 273.15) \tag{21}$$

The enthalpy of water vapor is given by:

$$h_v = 4180(T_{dp} - 273.15) + \lambda_{dp} + 1950(T - T_{dp}) \tag{22}$$

Heat of vaporization data from steam tables were correlated with temperature using the polynomial:

$$\lambda = 2.792\times10^6 - 160(T) - 3.43(T)^2 \tag{23}$$

Combining Eqns. (22) and (23) gives the water vapor enthalpy as:

$$h_v = 1950(T) + 1.65\times10^6 + 2070(T_{dp}) - 3.43(T_{dp})^2 \tag{24}$$

Dew point temperature was correlated with partial water vapor pressure by:

$$T_{dp} = 230.9 + 2.10\times10^{-4}(P_v) - 0.639(P_v)^{1/2} + 6.95(P_v)^{1/3} \tag{25}$$

The differential enthalpy of bound water at any concentration, or bound water content, is equal to the free water enthalpy less the differential heat of sorption. For wood, we assume that the differential heat of sorption varies quadratically with bound water content and at zero bound water content is equal to 40 percent of the heat of free water vaporization:

$$h_b^* = 4180(T - 273.15) - 0.4\lambda(1 - \frac{\rho_b}{\rho_b^{fsp}})^2 \tag{26}$$

where ρ_b^{fsp} is the bound water density at fiber saturation. The average bound water enthalpy for any given bound water content is obtained by integrating Eqn. (26) and is given by:

$$h_b = 4180(T - 273.15) - 0.4\lambda\ [1 - \frac{\rho_b}{\rho_b^{fsp}} + \frac{1}{3}(\frac{\rho_b}{\rho_b^{fsp}})^2] \tag{27}$$

Transport properties. The effective random-pore gas diffusivity in a porous solid is $D_{rp} \simeq \varepsilon^2 D_{AB}$, where ε^2 accounts for both the volume occupied by the solid and the tortuosity of the void space. The bulk binary diffusivity for air-water vapor mixtures is dependent on both temperature and pressure as follows:

$$D_{AB} = 2.20\times10^{-5}\ (\frac{101325}{P_a + P_v})\ (\frac{T}{273.15})^{1.75} \tag{28}$$

Physical properties. Data from steam tables for the density of liquid water were correlated with temperature using:

$$\rho_w = 1157.8 - 0.5361(T) \quad (36)$$

The viscosity of an air-water vapor mixture is obtained from a linear combination of the component viscosities, weighted by the mole fractions in the mixture:

$$\eta_g = [(4.06\times10^{-8}(T) + 6.36\times10^{-6})P_a + (3.80\times10^{-8}(T) - 1.57\times10^{-6})P_v]/(P_a + P_v) \quad (37)$$

The viscosity of liquid water is given in the Handbook of Chemistry and Physics [11] as the following function of temperature:

$$\log_{10}(\eta_w) = -13.73 + \frac{1828}{(T)} + 1.966\times10^{-2}(T) - 1.466\times10^{-5}(T)^2 \quad (38)$$

3. NUMERICAL PROCEDURE

The five governing equations were solved using a finite space - continuous time approach. We approximated the spatial derivatives in finite difference form following the procedure described by Patankar[12] and then numerically integrated the resulting ordinary differential equations using a specialized solver for stiff equations. This approach combines the conceptual simplicity of a finite difference formulation with the numerical power of a robust solver to handle the stiffness of the equations. Since convective heat flows could be comparable to or larger than conductive flows, we used the hybrid upwinding scheme described by Spalding[13] to keep the discretization stable.

For this one-dimensional model, the finite difference mesh is just a single line of points. For each mesh point, the five dependent variables describe the local state of the material. All other quantities, for example, total pressure, can be derived from the five variables. After differentiation, the equations for a single mesh point are:

$$\begin{bmatrix} 1 & 0 & 0 & 0 & 0 \\ 0 & 1 & 1 & 1 & 0 \\ h_a & h_v & h_b^* & h_f & C \\ 0 & 0 & 0 & 0 & 0 \\ 0 & 0 & 0 & 0 & 0 \end{bmatrix} \frac{d}{dt} \begin{bmatrix} \rho_a \\ \rho_v \\ \rho_b \\ \rho_f \\ T \end{bmatrix} = \begin{bmatrix} A \\ M \\ E \\ V \\ B \end{bmatrix} \quad (39)$$

where A, M, and E are numerical approximations to the right-hand sides of Eqns. (1), (2) and (3). These values derive from the finite difference approximations of the flux Eqns. (15), (17) and (18). The fourth and fifth rows represent the equilibrium conditions (4) and (5). The quantities V and B are the residuals computed by substituting the actual dependent variable values in the homogeneous forms of the equilibrium expressions. When equilibrium is satisfied, the fourth and fifth rows both reduce to the trivial equation, 0 = 0.

Equation (39) can be very easily adapted to handle a non-hygroscopic porous material. In this case, there is no bound water, so there is one less dependent variable and one less equilibrium condition.

Eqn. (39) could be solved directly using a solver which accepts a set of coupled differential-algebraic equations[14]. We developed a more computationally efficient approach by reducing the number of variables per mesh point to three. The two equilibrium equations (4) and (5) are incorporated into the three balance equations (1), (2) and (3), rather than expressing them explicitly as in Eqn. (39). A new total moisture density variable $\rho_m = \rho_v + \rho_b + \rho_f$ replaces the three separate moisture variables. With this change of variables, Eqn. (39) becomes

$$\frac{d}{dt} \begin{bmatrix} \rho_a \\ \rho_m \\ T \end{bmatrix} = \begin{bmatrix} A \\ M \\ C \end{bmatrix} \quad (40)$$

where the value of C is computed by substituting the first two rows of Eqn. (39) and the equilibrium expressions (4) and (5) into the third row of Eqn. (39). Since Eqn. (40) is in explicit form, we used a less specialized ordinary differential equation solver, the LSODES program developed by Hindmarsh[15], to effectively handle the stiff system of equations encountered here. This particular solver uses a knowledge of the sparsity structure of the Jacobian matrix to avoid redundant computation of zero-valued Jacobian entries.

The formulation uses the same numerical procedure throughout the modeled domain; it does not force the development of a drying front. If a drying front develops, it does so as part of the solution. At a drying front, there is a very sharp change in bound and free water densities. To maintain discretization accuracy, we developed

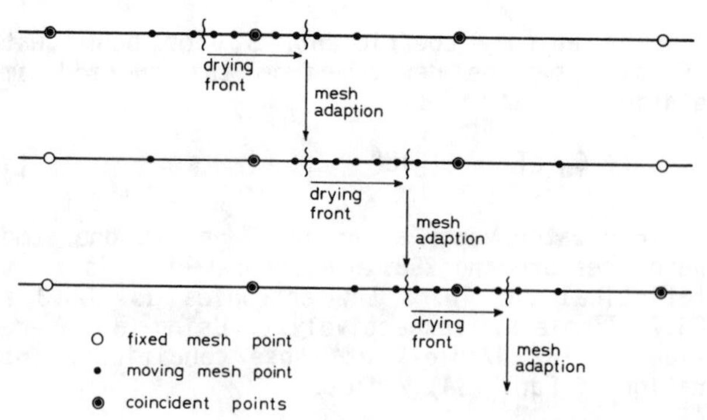

Fig. 1 Schematic diagram of the adaptive mesh scheme.

the adaptive finite difference mesh scheme shown in Fig. 1 to concentrate the mesh points and numerical effort around the drying front. Since the finite difference mesh is uniform around the drying front, the adaptive mesh scheme gives results essentially identical to those from a uniform non-adaptive mesh with the same fine mesh point spacing. However, the adaptive mesh scheme requires 1/2 to 1/5 of the computation time than the equivalent non-adaptive mesh.

4. MODEL RESULTS AND VERIFICATION

Testing of the model's computational performance and verification of its predictions are carried out in a progressive manner. We began by performing experimental drying tests using a porous ceramic solid and examining the nonhygroscopic version of the model. Model simulations were compared witn experimental data both to verify the model predictions and to derive appropriate transport parameter values. The results of these activities are summarized below. Further experimental testing using composite hygroscopic materials (e.g., particleboard), veneer, and lumber and model verification for the same are in progress and will be discussed in a future report.

Drying rate experiments were performed using a porous ceramic brick measuring 22.7x10.4x3.2 cm and weighing 1.55 kg dry. After repeated pressure-vacuum impregnation cycles, the brick absorbed a maximum of approximately 200 g of water. The saturated brick was inserted flush into a well that had been machined in a 40 cm-long section of dry 2x6 lumber, as shown schematically in Figure 2. The side and bottom surfaces of the brick were sealed with 3 mm-thick rubber sheets, ensuring a tight fit, preventing moisture loss from those five surfaces, and providing thermal insulation between the brick and wood. A thermocouple inserted to the center of the brick measured internal temperature. The brick assembly was suspended from a load cell in a specially designed drying chamber. As an air stream of controlled temperature, dew point, and velocity flowed over the exposed brick surface, the specimen weight and temperature were recorded with time.

Model simulation results are compared with experimental data for drying at 125°C and 75°C in Figures 3 and 4, respectively. Simulation parameters were identical in each case, as follows:

$$K_g^d = K_f^d = 5\times10^{-14}\ m^2$$

$$\alpha = 2.2$$

$$s_{ir} = 0.32$$

$$h = 60\ J/m^2/s/K$$

Simulation results and experimental data are in close agreement at both drying temperatures. The brick drying behavior exhibits two sharply distinct stages. Initially, the drying rate is very high and the brick temperature remains low.

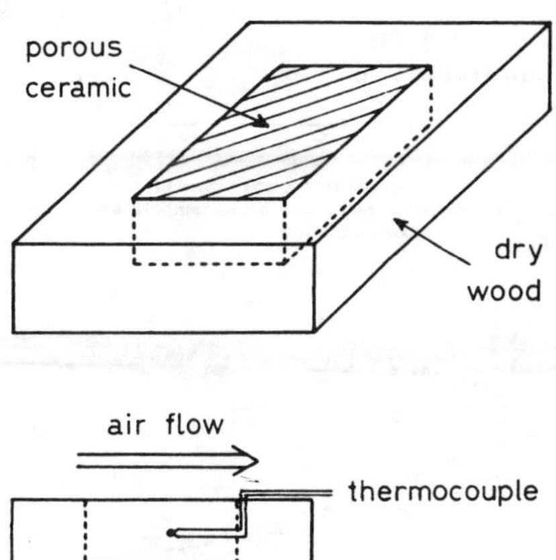

Fig. 2 Schematic diagram for drying experiments with porous ceramic brick.

In this first regime, liquid water migrates to the surface nearly as rapidly as it evaporates from the surface, and thus the drying rate is determined by the rates of external heat and mass transfer. However, when the surface moisture content reaches irreducible saturation, liquid can no longer be replenished at the surface. Shortly thereafter, the surface becomes dry and a drying front begins moving inward through the brick. Simultaneously, the brick temperature rises and the drying rate falls substantially. In this second regime, drying rate is determined primarily by the rates of heat and mass transfer between the external surface and the drying front.

The impacts of several transport parameters on model drying behavior predictions are explored in Figures 5, 6 and 7. Decreasing permeabilitiy (Fig. 5) causes the transition between the two drying stages described above to occur earlier in

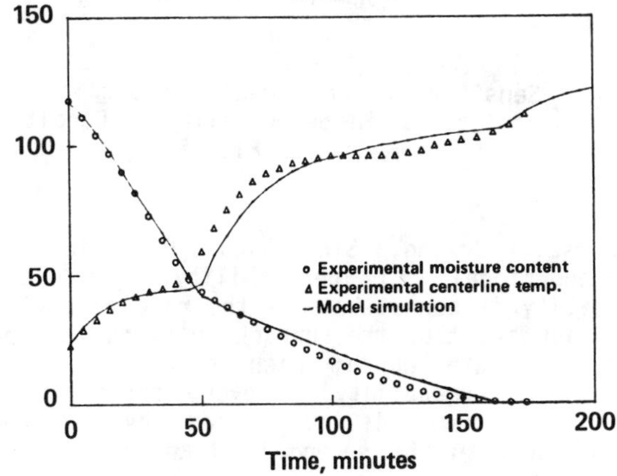

Fig. 3 Comparison of model simulation results wtih experimental data for brick drying at 125°C, 7 m/s air velocity, 10°C dew point.

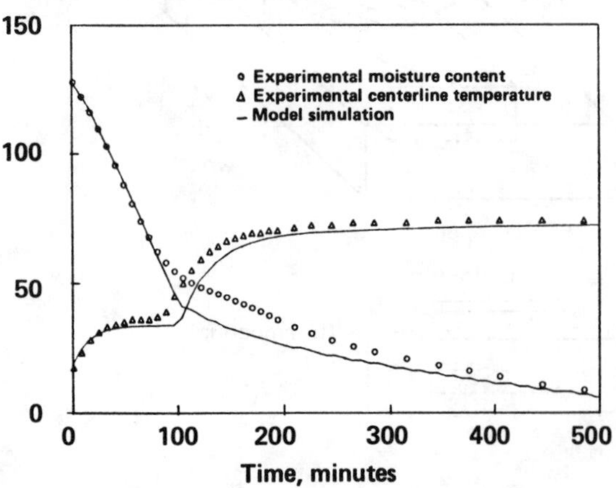

Fig. 4 Comparison of model simulation results with experimental data for brick drying at 75°C, 7 m/s air velocity, 10°C dew point.

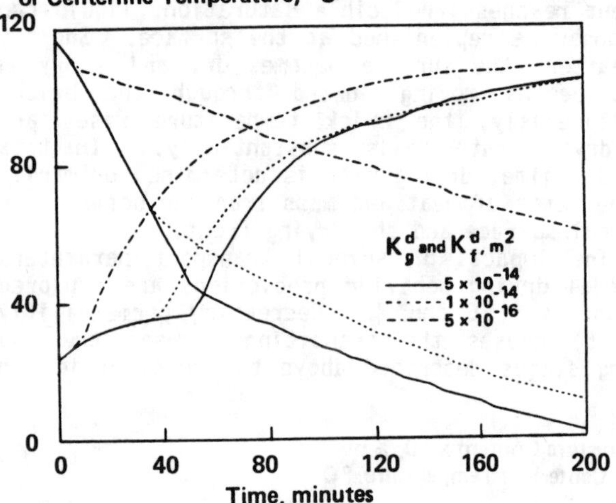

Fig. 5 Sensitivity of model predictions to variations in permeability. Conditions are the same as for Fig. 3.

the course of drying. Since liquid migration rate is proportional to permeability, at lower permeability it cannot balance the rate of surface evaporation and thus the surface dries out sooner. The drying rate during each stage is little affected by permeability, however, since in the first stage rate is controlled by external transport and in the second by transport through the dry layer.

Decreasing vapor attenuation factor (Fig. 6) likewise has little impact on first-stage drying rates. Second-stage drying rate decreases with decreasing attenuation factor because binary diffusion contributes to the rate of mass transfer through the dry solid layer. However, for very small attenuation factors ($\alpha<0.1$), pressure-driven bulk flow of gas dominates the internal mass transfer rate and further decreases in attenuation factor have little effect.

The irreducible saturation primarily affects the point of transition between the first and second stages of drying (Fig. 7). With the relatively high permeability ($5x10^{-14}m^2$) employed in these simulations, only a very small moisture content gradient arises during the first stage of drying. Therefore, the moisture content falls uniformly to the level of irreducible saturation, whereupon the surface dries out and the second stage of drying begins.

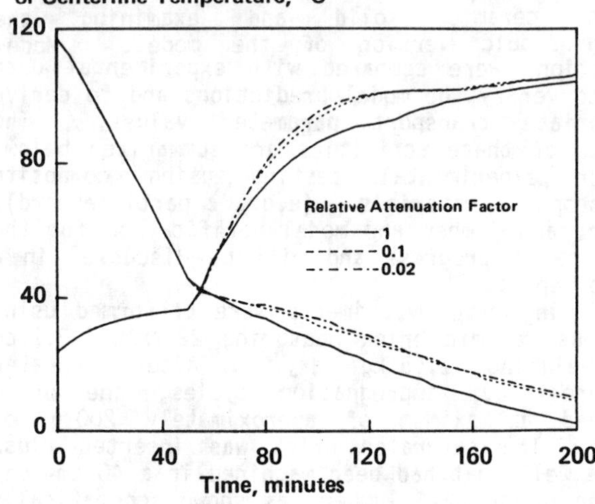

Fig. 6 Sensitivity of model predictions to variations in effective gas diffusivity. Conditions are the same as for Fig. 3.

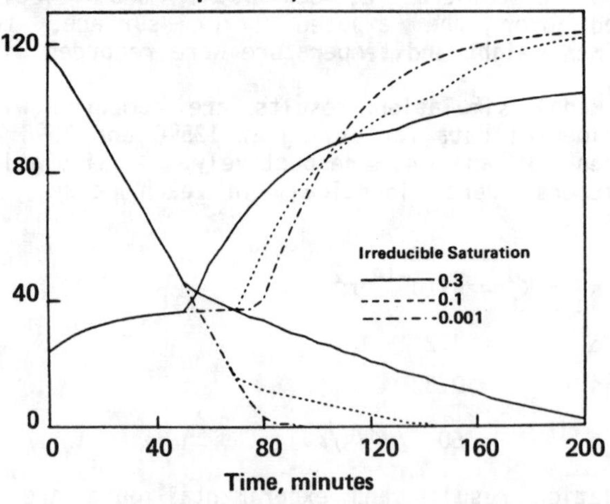

Fig. 7 Sensitivity of model predictions to variations in irreducible saturation. Conditions are the ame as for Fig. 3.

5. CONCLUSIONS

A unique modeling approach based on fundamental heat and mass transfer relationships and thermodynamic equilibrium provides a tool that is both effective in predicting drying behavior and also useful in exploring and understanding the impacts of important variables on the drying process. The inherently more complicated computational problem resulting from this approach was handled efficiently by developing a reduced but mathematically equivalent set of governing equations and numerically solving them using a specialized solver and an adaptive mesh scheme. Comparisons of model predictions with experimental drying data helped to verify the model for nonhygroscopic drying. Parametric sensitivity analyses provide insight into interpretation of drying characteristics in terms of important material and process variables.

6. NOMENCLATURE

- a_1, a_2, a_3, a_4 : parametric functions for bound water sorption correlation, given by Eqns. (6), (7), (8), and (9), dimensionless
- C : specific heat of wet wood, J/kg/K
- C_p : heat capacity, J/mole/K
- D_{eff} : effective gas diffusivity, m^2/s
- D_{AB} : bulk binary diffusivity for air-water vapor mixtures, m^2/s
- D_b : bound water diffusivity, $kg/s/m^3$
- D_{rp} : effective random-pore gas diffusivity, m^2/s
- H : enthalpy, J/mole
- h : enthalpy, J/kg
- h^* : differential enthalpy, J/kg
- h : external heat transfer coefficient at the solid surface, $J/m^2/s/K$
- K : effective permeability of moist solid, m^2
- K^d : permeability of dry solid, m^2
- k : thermal conductivity of moist solid, J/m/s/K
- k_x : external mass transfer coefficient at the solid surface, $moles/m^2/s$
- L : solid thickness, m
- M : molecular weight, kg/mole
- N : molar flux at the solid surface, $moles/m^2/s$
- n : flux, $kg/m^2/s$
- P : pressure, Pa
- P_c : capillary pressure, Pa
- P^∞ : ambient partial pressure at end of solid, Pa
- R : gas constant, J/mole/K
- S : entropy, J/mole/K
- s_{ir} : irreducible saturation, dimensionless
- T : temperature, K
- T^o : initial temperature profile within the solid, K
- T^∞ : ambient temperature at end of solid, K
- t : time, s
- V : molar volume, $m^3/mole$
- x_v : mole fraction of water vapor in the gas at the solid surface, dimensionless
- z : spatial variable, m
- α : attenuation factor for vapor diffusivity in wood, dimensionless
- ε : void fraction, dimensionless
- η : viscosity, kg/m/s
- λ : heat of vaporization, J/kg
- μ : chemical potential, J/kg
- ρ : density, kg/m^3
- ρ^o : initial density within the solid, kg/m^3

7. REFERENCES

1. Ashworth, J. C., Proceedings of the Third International Drying Symposium, v.1 & 2, Drying Research Limited, Wolverhampton, England (1982).
2. Mujumdar, A. S., Advances in Drying, v.1 (1980), v.2 (1983), Hemisphere Publishing, Washington.
3. Mujumdar, A. S., Drying '80, v.1 & 2 (1980) and Drying '82 (1982), Hemisphere Publishing, Washington.
4. Rosen, H. N., Recent Advances in the Theory of Drying Lumber, North Central Forest Experiment Station, Forest Service, U.S. Department of Agriculture (1983).
5. Siau, J. F., Wood Sci. Technol., 17, 101-105 (1983).
6. Kawai, S., Nakato, K., and Sadoh, T., Mokuzai Gakkaishi, 24(5), 273-280 (1978).
7. Simpson, W. T., For. Prod. J., 21(5), 48-49 (1971).
8. Spolek, G. A., and Plumb, O. A., Wood Sci. Technol., 15, 189-199 (1981).
9. Wood Handbook, U.S. Forest Products Laboratory, Forest Service, U.S. Department of Agriculture (1955).
10. Moore, Walter J., Physical Chemistry, Fourth Edition, Prentice-Hall, Englewood Cliffs, New Jersey (1972).
11. Handbook of Chemistry and Physics, 56th edition, Robert C. Weast, Ed., CRC Press, Cleveland (1974).
12. Patankar, S. V., Numerical Heat Transfer and Fluid Flow, McGraw Hill, New York (1980).
13. Spalding, D. B., International Journal of Numerical Methods in Engineering, 4(4), 551-559 (1972).
14. Hindmarsh, A. C., ACM-SIGNUM Newsletter, 15(4), 10-11 (1980).
15. Hindmarsh, A. C., ODEPACK, a systematized collection of ODE solvers,, Proc. 10th IMACS World Congress on Systems, Simulation and Scientific Computation, Montreal (1983).

STOCHASTIC MODELING OF LUMBER DRYING IN BATCH KILNS

Ferhan Kayıhan

Weyerhaeuser Technology Center
Tacoma, Washington USA

ABSTRACT

A Monte-Carlo simulation model is developed to predict the drying behavior of lumber in batch kilns. The green moisture contents and the drying rates of individual boards are represented by statistical functions reflecting theoretical analyses and experimental observations. The initial moisture contents are characterized by the Weibull distribution function. The drying rates are approximated by a novel combination of high and low moisture asymptotic rates which lend to a simple correction procedure to compensate for the temperature and the humidity variations.

The model predicts the moisture content distribution transient of the boards for any specified drying schedule. It serves as an efficient tool for understanding the effects of the process variables (such as the green moisture distribution, the board dimensions, the temperature and humidity schedule, the sticker thickness, the air velocity and reversal, and the equilibration or the conditioning periods) on the drying time and the final moisture distribution. The model will also be used for the kiln process control studies.

1. INTRODUCTION

Drying of lumber in batch kilns is a dynamic process which varies in its behavior and end product from load to load even for the same type and size boards. The inherent property differences among the boards coupled with the operating characteristics of a kiln (i.e., the degradation of the drying potential across the stack) are the main causes for the non-repetitive dryer performance. Present practice is to follow a set schedule for the complete drying cycle and to use the final results for any schedule modifications to be applied for the next cycle. A clear improvement on this technology is to monitor the kiln behavior as the drying progresses and to achieve the desired end product for each unique batch process through an appropriate process control scheme. This paper summarizes the efforts towards the understanding and the prediction of the dynamic behavior of batch lumber kilns which will provide the needed background for the process control studies.

The dynamic response of a kiln depends mainly on its size, its load, the heat exchange equipment, the air circulation rate and direction, the heat losses, and on the physical arrangement of the stacks and the boards. We are interested in predicting the moisture content transient of the boards during the drying cycle. What happens to the boards is directly related to the air properties flowing through the stacks. Where, in turn, the air properties depend on all of the factors listed above. This study is limited to the dynamic behavior of a course of boards in a stack as they respond to a specified schedule, i.e., pre-determined dry-bulb and humidity (or wet-bulb) variation with time. Our present efforts include the extension of the model to incorporate the dynamics of the air properties such that the modified model will predict the response of the kiln to the changes in the set points of its control loops.

In Fig. 1 a simplified schematic diagram of a kiln is shown. As depicted in Fig. 2 we define the course as the sequence of boards on each side of an air channel which is assumed to represent the general behavior of the boards in that stack. The model described in this study predicts the board-by-board moisture content transients of a course during the drying cycle. Monte-Carlo simulation approach is used to accommodate the stochastic nature of the board green moisture contents and the wood drying characteristics. The rest of the manuscript will develop the working equations and the principles of the model and show some simulation results.

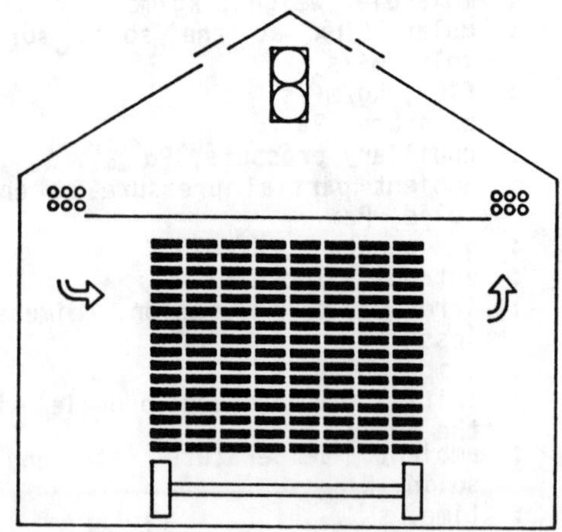

Fig. 1 A simplified schematic diagram of a lumber kiln

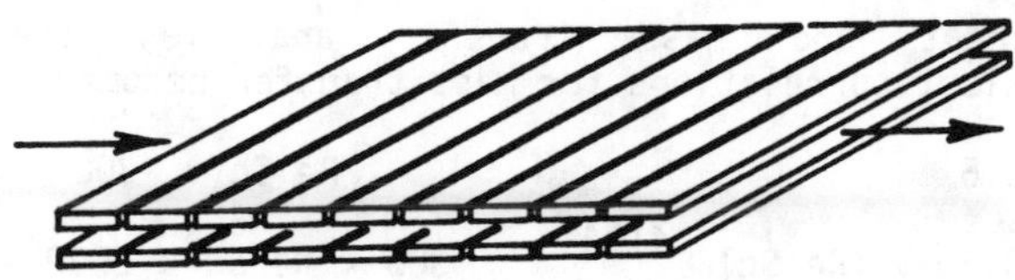

Fig. 2 A course in a stack consists of the two layers of boards on each side of an air channel

2. MATERIAL AND ENERGY BALANCES

2.1 Geometry

For the development of the model equations the COURSE is defined to include a single stream of the drying air with the contacting halves of the upper and lower boards. For practical purposes the boards are assumed to be exposed to symmetrical boundary conditions.

Material and energy balances are developed for the case when air flows from left to right. Opposite direction flow is handled in the computer program by the appropriate change of the reference indexes. Fig. 3 shows the major terms of the material and energy balances around each board segment.

2.2 Balances on Boards

Let L_b, H_b, and W_b represent the board length, height, and width respectively. Then the total mass of a board is

$$m_b = L_b H_b W_b \rho_b (1 + u)$$

The rate of change of the board mass due to drying is

$$\frac{dm_b}{dt} = -2 L_b W_b \text{ (drying flux)}$$

The empirical expression and the correlation for the drying flux will be discussed in Section 5. The factor 2 in the above expression is due to drying from both sides.

The convective heat transfer from the air stream to the board surface is described by

$$q = h L_b W_b (T_g - T_{bs})$$

where T_{bs} is the board surface temperature. In this model we are not solving the detailed internal transport equations for the boards and therefore have to be satisfied with an estimate of T_{bs}.

The convective and conductive resistances to heat transfer from the air to the wood are in series and are related through the evaporative heat loss as

$$h\, a (T_g - T_{bs}) - \frac{dm_b}{dt} \lambda = (\frac{k_b}{\delta})\, a (T_{bs} - T_b)$$

where k_b is the thermal conductivity of wood, δ is the penetration depth in wood representative of the bulk temperature level, and $a = 2 L_b W_b$ is the total surface area. Solving for T_{bs} gives

$$T_{bs} = \frac{hT_g + k_b T_b/\delta - (dm_b/dt)\,\lambda/a}{h + k_b/\delta}$$

which provides the needed estimation on T_{bs}. We used $\delta = \frac{H_b}{8}$ and

$$k_b = [\ \rho_b (0.20 + u/2) + 24\ ]\ X10^{-6}\ \text{kJ/m s K}$$

for computations.

The average board temperature is computed from

$$c_{pb}\, m_b \frac{dT_b}{dt} = \frac{k_b}{\delta}\, a (T_{bs} - T_b)$$

In this model we generate only an estimate of the board temperatures which help quantify the overall thermal response of the course. Since we do not use the board temperatures in computing the drying rates, which are correlated only in terms of the air properties, the proposed approximate computation for T_b is satisfactory.

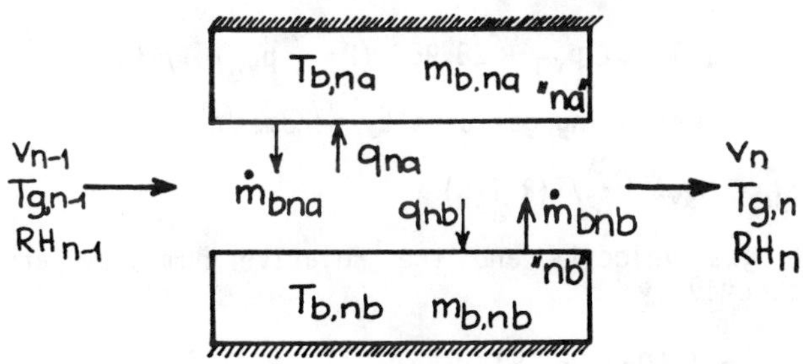

Fig. 3 Material and energy balance components for the n th board location

2.3 Gas Phase Balances

Gas phase balances are performed for sections which correspond to the board locations. Each section is assumed to have uniform property which changes with time. The following equations are derived for an arbitrary section n, and the air is assumed to be flowing from the n-1 st section as shown in Fig. 3.

Water vapor balance gives

$$Y_n = Y_{n-1} + [\ \frac{d}{dt}(m_{bna}) + \frac{d}{dt}(m_{bnb})\ ] / G_{air}$$

where Y is the absolute humidity, G_{air} is the constant air flowrate, and

$$\frac{d}{dt}(m_{bna}) \text{ and } \frac{d}{dt}(m_{bnb})$$

are the drying rates of the upper and the lower boards respectively.

Total gas balance gives the increase in flow rate as

$$G_n = G_{n-1} + [\frac{d}{dt} (m_{bna}) + \frac{d}{dt} (m_{bnb})]$$

Change in gas temperature is calculated through an enthalpy balance which gives

$$T_{gn} = \{ G_{n-1} C_{pgn-1} T_{gn-1} - (q_{na} + q_{nb}) +$$

$$T_{gn} [\frac{d}{dt} (m_{bna}) T_{bsa} + \frac{d}{dt} (m_{bnb}) T_{bsb}] c_{pv} \}$$

$$/(G_n c_{pgn})$$

where $(q_{na} + q_{nb})$ represents the total convective heat transfer to the boards and

$$[\frac{d}{dt}(m_{bna}) T_{bsa} + \frac{d}{dt}(m_{bnb}) T_{bsb}] c_{pv}$$

reflects the sensible heat of the evaporated moisture. The specific heat of the humid gas is

$$c_{pg} = (P_T + 0.20 \, p_v) \, / \, (P_T - 0.38 \, p_v) \qquad \text{kJ/kg K}$$

The molecular weight of the air stream changes as it becomes more humid. From the ideal gas law we get

$$M_{avn} = [18.01 \, p_{vn} + 28.967 \, (P_T - p_{vn})] / P_T$$

The corresponding gas density becomes

$$\rho_{gn} = M_{avn} P_T \, / \, (R \, T_{gn})$$

The gas velocity and the relative humidity are respectively

$$v_n = G_n \, / \, (\rho_{gn} L_s L_b)$$

and

$$RH_n = p_{vn} \, / \, p_{vs} (T_g)$$

3. HEAT AND MASS TRANSFER COEFFICIENT CORRELATION FOR TURBULENT FLOW BETWEEN PARALLEL PLATES

Handbook of Heat Transfer [1] presents, in a graphical form, the dependence of Nusselt number on the Peclet number for fully developed turbulent flow between parallel plates. For practical purposes we simplified the functional dependence in terms of three linear segments as

$$Nu = 5.5 \qquad Pe < 300$$

$$Nu = 0.5029 \, Pe^{.4194} \qquad 300 < Pe < 2000$$

$$Nu = 0.1244 \, Pe^{.6032} \qquad 2000 < Pe$$

where $Pe = Re \, Pr$

For air-water vapor system Pr (and Sc) is approximately equal to 1 and the above ranges for Pe also correspond to similar ranges in Re. Using the heat and mass transfer analogies the equivalent correlations for mass transfer become

$$Sh = 5.5 \qquad Re \, Sc < 300$$

$$Sh = 0.5029 \, (Re \, Sc)^{.4194} \qquad 300 < Re \, Sc < 2000$$

$$Sh = 0.1244 \, (Re \, Sc)^{.6032} \qquad 2000 < Re \, Sc$$

The following correlations represent air properties at sufficient accuracy for engineering calculations

$$Pr = 0.711 - 6.5 \; 10^{-5} \, T(C)$$

$$Sc = 0.608 - 3.0 \; 10^{-5} \, T(C)$$

$$\nu = (12.5 + 0.11 \, T(C)) \times 10^{-6}$$

$$k = (24.4 + 0.068 \, T(C)) \times 10^{-3}$$

$$D = 1.2 \times 10^{-9} \, T(K)^{1.75}$$

It is also convenient to relate the heat and the mass transfer coefficients to each other through a simple expression. Point-by-point computation of both coefficients and a linear correlation give the following simple dependence on temperature:

$$h \, / \, k_\rho = 1100 - 1.8 \, T(C)$$

This expression is valid for the 25 C to 200 C range.

4. SPECIFIC HEAT OF THE AIR-WATER VAPOR MIXTURE

Specific heats of air and steam in kJ/kg K are tabulated as

T (K)	T (C)	c_p air	c_p steam
300	27	1.005	1.87
400	127	1.013	1.93
500	227	1.029	1.99

For all practical purposes we can assume constant specific heats with respect to temperature and assign

$$c_{p \; air} = 1.0 \quad \text{and} \quad c_{p \; steam} = 1.9$$

At different humidities the air-water vapor mixture will have

$$c_s = 1.0 + 1.9 \, Y \qquad \text{kJ/kg air K}$$

where c_s is defined as the humid heat and Y is the humidity expressed as the vapor to air mass ratio, i.e., kg water vapor/kg dry air.

Another convenient specific heat expression is the one based on mixture rather than on air.

$$c_{p \; mix} = (1.0 + 1.9 \, Y) \, / \, (1 + Y) \qquad \text{kJ/kg mix K}$$

5. DRYING FLUX CORRELATION

Correlation of the drying rate behavior of wood is traditionally done in terms of the % moisture change per unit time. The reason for this is obvious. The data from drying experiments are almost always available in terms of weight or % moisture as a function of time and % moisture (dry base) is a customary variable in wood characterization. Therefore, drying rates are usually evaluated as the rate of % moisture change. Although this is a satisfactory way of quantifying a particular drying rate, it has potential limitations in generalizing the drying behavior of different samples for correlation purposes. For example, consider two samples of different thicknesses, but with the same heat and mass transfer surface areas, i.e., one-dimensional drying of two different thickness boards. During the initial high moisture drying period, when the external convective transport controls the drying rate, the two samples dry identically (if we ignore the temperature transient effects). That is, both lose the same amount of moisture per unit surface area per unit time. However, since the dry base moisture is defined as the ratio of water to dry wood, the two samples will have different rates of moisture change: larger sample moisture changing slower than the smaller sample. Thus, the drying flux (kg moisture loss/m^2 hr) becomes an absolute measure of the drying behavior compared to the drying rate (% moisture/hr) which is a relative measure. Thus, to maintain a consistent approach in characterizing the wood drying rates, in this model, the drying flux is analyzed and correlated rather than the drying rate.

Churchill and Usagi [2] suggest a generalized correlation form for diffusional processes where the extreme asymptotic behavior are known.

If $y(z)$ is the desired information and the low value ($z \to 0$) and the high value ($z \to \infty$) asymptotes are known as $y_0(z)$ and $y_\infty(z)$, then a power form combination is possible such that

$$y(z)^n = y_0(z)^n + y_\infty(z)^n$$

or

$$y(z)^{-n} = y_0(z)^{-n} + y_\infty(z)^{-n}$$

where n is a positive real number.

A more convenient form in terms of dimensionless variables is

$$Y = (1 + Z^n)^{1/n}$$

or

$$Y = (1 + Z^{-n})^{-1/n}$$

where $Y = y(z) / y_0(z)$

and $Z = y_\infty(z) / y_0(z)$

The correlation procedure is rather simple: plot Y and Z on a dimensionless graph similar to the one shown in Fig. 4 and identify the most appropriate parameter n.

For drying, we want to correlate the drying flux as a function of the dry-base moisture content u. We recognize that, for all practical purposes, at high moisture levels drying is controlled by external resistances and occurs at a constant flux. Let us call this flux A. At low moisture levels the drying flux is controlled by internal diffusion which can be approximated by $B(u - u_{emc})$ where u_{emc} is the equilibrium moisture content at the drying conditions.

Using these asymptotes the correlation form becomes

$$\left(\frac{dm_b}{dt}\right)/a = -\frac{B(u - u_{emc})}{\left\{1 + \left[\frac{B(u - u_{emc})}{A}\right]^n\right\}^{1/n}}$$

The choice between the positive and the negative power forms became apparent only after the data are plotted. All of the drying data analyzed indicate that n = 3 is an appropriate power. Drying rates predicted using this third power rule gave satisfactory results for lumber, sawdust, and flake drying behavior.

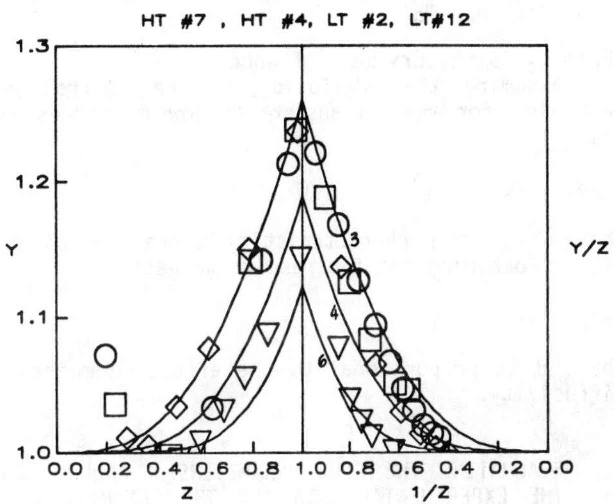

Fig. 4 Experimental results on the dimensionless plot show the preferance on the correlation parameter "n"

6. PHYSICAL INTERPRETATION OF THE DRYING FLUX PARAMETERS A AND B

At high moisture contents drying occurs at almost steady-state conditions when the rate is controlled by surface evaporation. This is equivalent to the wet-bulb conditions and the drying flux is given by

$$\left(\frac{dm_b}{dt}\right)/a = -(M_w/RT_{av})k_\rho\,[\,p_{vs}(T_{wb}) - p_v\,]$$

or

$$= -A$$

where the second equality reflects the high temperature asymptote of the correlation. Therefore, we have

$$A = (M_w/RT_{av})\,k_\rho\,[\,p_{vs}(T_{wb}) - p_v\,]$$

indicating that the parameter A is a function of temperature, velocity, humidity, and geometry.

T_{av} stands for the average film temperature defined as

$$T_{av} = [\, T_{db} \text{ (or } T_{gas}) + T_{wb} \,] / 2$$

where T_{wb} is the wet-bulb temperature [3] computed from

$$T_{wb} = T_{db} - 0.655\, \lambda\, (T_{wb})\, [\, p_{vs}\, (T_{wb}) - p_v \,]$$

$$\text{with } \lambda\, (T) = 267\, (374 - T \text{ C})^{0.38} \qquad \text{kJ/kg}$$

At low moisture contents the drying flux becomes proportional to the moisture content, i.e.

$$(\frac{dm_b}{dt})/a = -B\, (u - u_{emc})$$

or, in terms of moisture

$$du/dt = -(B\, a / m_o)\, (u - u_{emc})$$

where m_o is the dry mass of wood.

Assuming that diffusion is the controlling mechanism for mass transfer at low moistures we have

$$du/dt \sim D_{eff} / l^2$$

where l is the characteristic length for diffusion. Combining the two results we get

$$D_{eff} \sim (B\, a / m_o)\, l^2$$

Thus, B is proportional to the effective moisture diffusivity.

7. EVALUATING THE FLUX PARAMETERS A AND B FROM THE EXPERIMENTAL DATA AND THE DEFINITION OF THE STANDARD CONDITIONS

From any drying experiment, starting with a sufficiently high moisture content, we can get a pair of parameters: A and B. Different experiments will give different parameters. The differences will be due to the temperature, velocity, relative humidity, and geometry variations (as we showed through the physical interpretation arguments) as well as due to the wood structural and otherwise unaccountable variations in the wood or in the environment. In order to compile the experimental results in a unified manner we need to identify and remove the known variabilities on the parameters.

As shown before, A is approximately equal to (or proportional to)

$$(M_w/RT_{av})\, k_\rho\, [\, p_{vs}(T_{wb}) - p_v \,]$$

where p_{vs} is the vapor pressure evaluated at the wet-bulb conditions corresponding to T and p_v of the ambient conditions.

By choosing an arbitrary standard state we can evaluate from any given A value what it would have been at the standard conditions. Let us define the standard conditions as

100 C (T_{db})
20% RH
5 m/s velocity
20 mm sticker thickness

when the corresponding wet-bulb and average temperatures are

T_{wb} = 62.8 C
T_{av} = 81.9 C

At these conditions A would be approximately $(M_w/R)\, k_{\rho st}/14{,}794$ where the value of $k_{\rho st}$ (k_ρ at standard state) would depend on the correlation used for the particular operating conditions; i.e., suspension particle drying would require a different correlation than lumber drying in a kiln.

Therefore, to evaluate the corresponding standard state A, let us call this A_s, the following transformation becomes necessary.

$$A_s = A\, (k_{\rho st}/14{,}794)/\{k_\rho/T_{av}[\, p_{vs}(T_{wb}) - p_v \,]\}$$

A and the terms in the denominator are, of course, evaluated at the experimental conditions.

As expected, the A_s values obtained from individual experiments would still deviate from each other since the above analysis is not exact and the boards differ from each other. Results obtained from Western hemlock drying experiments are correlated as

$$A = a_o + \varepsilon(\sigma_A^2)$$

where $\varepsilon(\sigma_A^2)$ is a random error with zero mean and σ_A^2 variance. a_o and the variance were calculated from the data as $A = 1.5 + \varepsilon\ (0.03)$.

B is proportional to the effective moisture diffusivity (i.e., bound water) and is expected to be strongly dependent on the absolute (ambient) temperature through the Arrhenius expression as

$$B = B_o \exp\, (- E_b/RT)$$

A linear regression on semi-logarithmic coordinates provides the parameters B_o and E_b as well as the random variance as

$$\ln B = \ln B_o - E_b/RT + \varepsilon(\sigma_{\ln B}^2)$$

The random variance for this case is mostly due to the variability of the wood structure like the grain orientation and the growth time which are not accounted for in the correlation.

Our experiments with western hemlock gave B_0 = 1200, E_b/R = 2700 and $\sigma_{\ln B}^2$ = 0.006.

8. UPDATING THE CORRELATION PARAMETERS DURING SIMULATION

The drying flux correlation parameters A and B are not fixed constants. They change with the operating conditions. Therefore, throughout the simulation both A and B must be updated to represent the changing drying behavior.

At the beginning of the simulation each board is assigned a set of A and B values evaluated at the standard conditions and including the random errors according to the designated variances. These become the base values, A_s and B_s for each board, for the rest of the simulation. The actual parameters which represent current operating behavior are then computed from A_s and B_s.

The transformation for A is essentially the inverse of the operation which is necessary to compute the standard state values from experimental results. From a given A_s and the air stream properties the local A is computed by

$$A = A_s\ (14{,}794/k_{\rho st})\ (k_\rho/T_{av})\ [p_{vs}(T_{wb}) - p_v]$$

B varies with temperature only. Thus, it is updated by

$$B = B_s \exp\left[- E_b/R\ (1/T - 1/373) \right]$$

9. WEIBULL DISTRIBUTION OF GREEN MOISTURE CONTENTS

The statistical nature of the green moisture contents of wood are traditionally represented through the Weibull distribution for which the density and the cumulative distribution functions are respectively

$$f(x) = \left(\frac{x-c}{b}\right)^{a-1} \left(\frac{a}{b}\right) \exp\left[- \left(\frac{x-c}{b}\right)^{a} \right]$$

and

$$F(x) = 1 - \exp\left[- \left(\frac{x-c}{b}\right)^{a} \right]$$

There are software packages available to estimate these three parameters from the x vs f(x) or F(x) data. These give very reliable parameters. There is also an approximate way of estimating the parameters which is sufficient for simulation purposes. This will be briefly described here.

Assume that the experimental data on green moisture content are available in the x vs F(x) form. The location parameter c is then equal to or slightly less than the minimum x. To evaluate a and b approximately we need two points from the cumulative distribution function or table. Let these points be x_1,F_1 and x_2,F_2 such that $x_1 < x_2$ and $F_1 < F_2$. With these two points available a set of algebraic equations can be constructed to solve for a and b. The solution becomes

$$a = (Y_2 - Y_1)/(X_2 - X_1)$$

$$b = \exp\ (X_1 - Y_1/a)$$

where X and Y are defined as

$$X = \ln\ (x - c)$$

and

$$Y = \ln\left[\ \ln\ (1/(1-F))\ \right]$$

10. MONTE-CARLO SIMULATION AND THE GENERATION OF RANDOM NUMBERS

Board green moisture contents and the drying flux correlation parameters are known as distributions rather than as fixed (deterministic) values. In order to represent the random nature of the process during simulations we use the stochastic approach known as the Monte-Carlo simulation. Here, we assign random values (i.e., initial moistures, A and B parameters) board by board according to the known or assumed variations. Of course, the individual values are assigned such that a statistical analysis would essentially give the same variations (or distributions) used in assigning the random values. Proceeding with the rest of the simulation while keeping track of the individual (board) behavior then generates the statistical information which predicts the expected nature of the process response. This forms the basis of the Monte-Carlo simulation where instead of a single value for any process variable a distribution is generated giving a good measure for the expected deviations around the average values.

In generating a random value with any particular distribution property we essentially need to generate a uniformly distributed random number between 0 and 1, and know the cumulative distribution function for the desired random behavior. If we generate a large number of uniformly distributed random numbers between 0 and 1 and from the distribution curve find the corresponding independent variables, then we would create a collection of random numbers which would obey the distribution law represented by the cumulative probability curve. For standard distribution functions this inverse calculation procedure can be done analytically. For arbitrary functions, it can be done either graphically or numerically. The IMSL[3] package has standard routines for what we needed in this model: the Weibull distribution for the green moisture contents and the normal (gaussian) distribution for the random deviations (errors) of the flux parameters.

11. SIMULATION RESULTS

The model predictions were validated using a variety of experimental data. Among these are the data obtained from our laboratory kiln where detailed information was gathered for a series of drying runs. Here, we will compare the modeling results with two of the laboratory kiln trials.

During the experiments 2x4 (40 mm x 100 mm) Western hemlock boards were dried in courses of 24 boards using 3/4-inch (19 mm) stickers. The green moisture content distributions were obtained through board-by-board measurements. A load cell was used to monitor the weight change of the complete charge as a function of time. The end-point moisture content distribution was also determined by individual board measurements.

The two cases analyzed here represent two different drying schedules. A constant temperature and a CRT schedule as depicted in Figs. 5 and 6 respectively. The experimental drying curve for the first one is shown in Fig. 5a along with the model predictions. Simulation results are in terms of ranges because they represent multiple predictions.

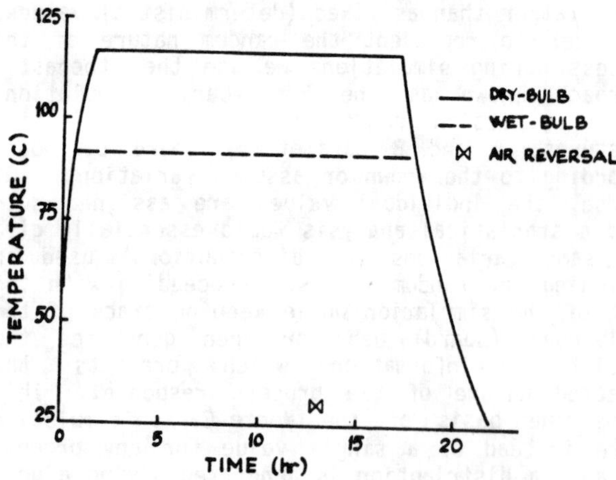

Fig. 5a Constant temperature schedule for kiln drying

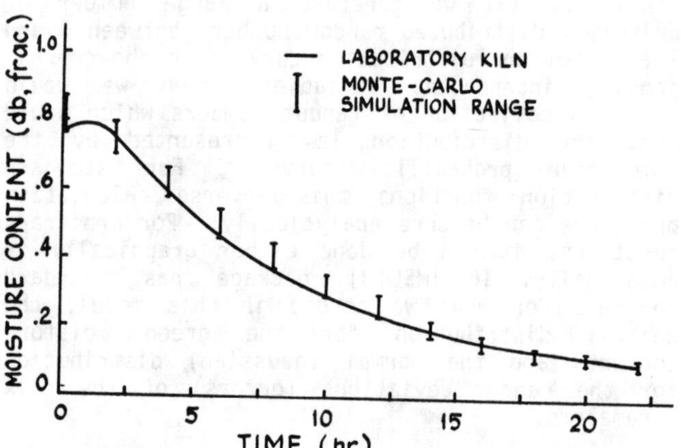

Fig. 5b Drying curve of the charge

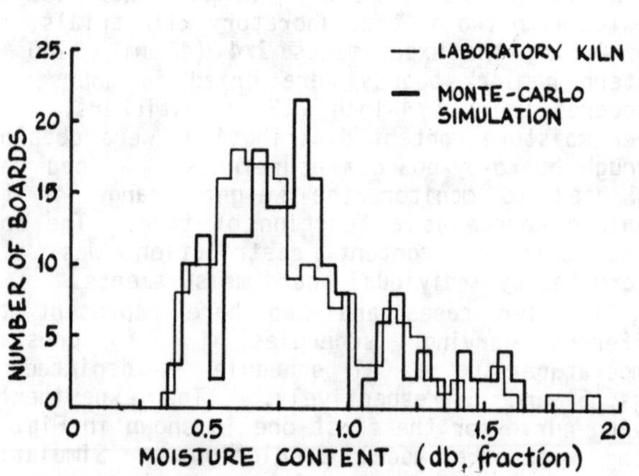

Fig. 5c Green moisture content distribution

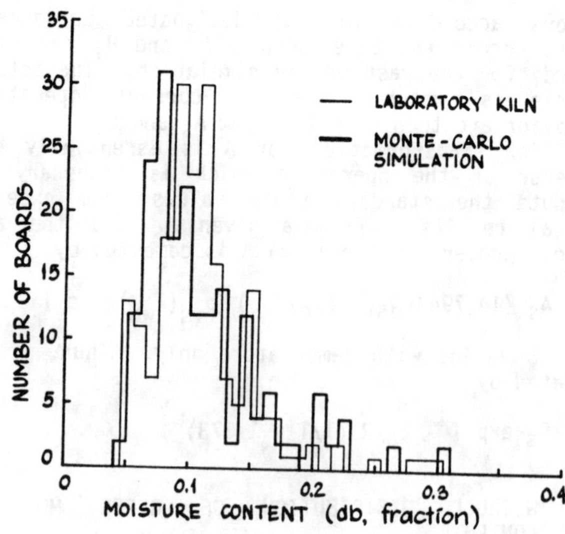

Fig. 5d Dry moisture content distribution

The stochastic nature of the model is such that every time the computer program is executed it generates a slightly different set of results. This is a desirable feature of the model which was planned in order to reflect the actual case, i.e., no two courses in a stack dry exactly alike. The program initiates the variations by generating and using different sets of random numbers, belonging to the desired statistical populations, every time it is run.

For both of the experimental cases the green and the dry moisture content distributions are given in Figs. 5c,d and 6b,c. The model predictions are also presented on the same diagrams. The theoretical histograms reflect multiple (four) simulation results.

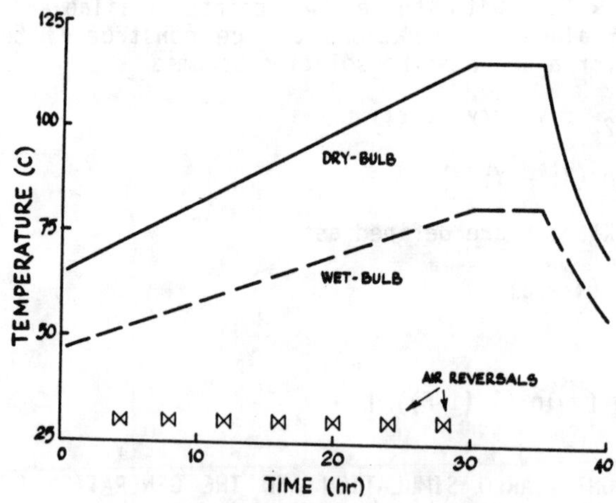

Fig. 6a CRT schedule for kiln drying

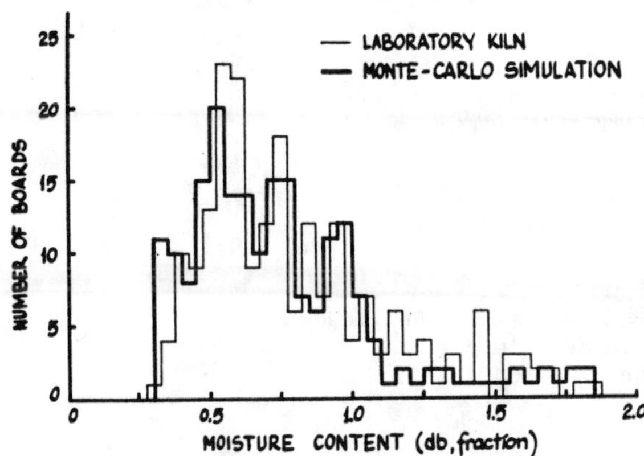

Fig. 6b Green moisture content distribution

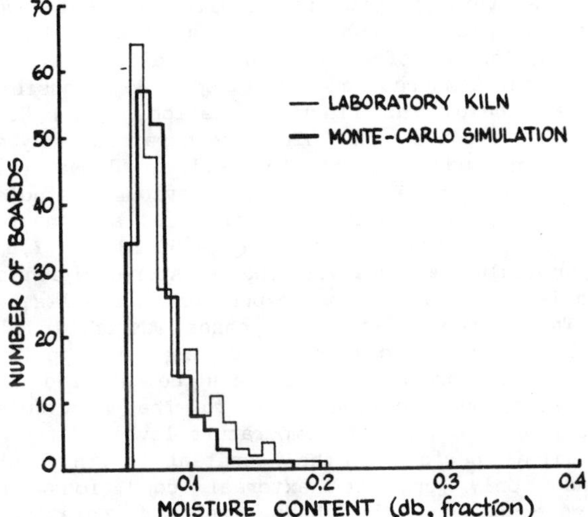

Fig. 6c Dry moisture content distribution

12. CONCLUSIONS

The stochastic drying model developed from the fundamental material and energy balance principles adequately represents the statistical drying behavior of lumber in batch kilns. Monte-Carlo simulation techniques were used to accommodate the board-by-board moisture content and the drying behavior variations. A novel drying flux correlation procedure was introduced and shown to be effective for process simulation purposes. The correlation parameters were easily calculated from single-board drying experiments. The model discussed here forms the basis of the more sophisticated kiln dynamics, process control and optimization models.

NOMENCLATURE

a	heat and mass transfer surface area (m^2)
c_{pb}	specific heat of wood (kJ/kg C)
c_{pg}	specific heat of gas (kJ/kg C)
c_s	humid heat (kJ/kg air K)
D	diffusivity (m^2/s)
E_b	activation energy (kJ/kmol)
G	total moist air (gas) flow rate (kg/s)
G_{air}	dry air flow rate (kg/s)
h	heat transfer coefficient (kJ/m^2 s K)
H_b	height of board (cross section) (m)
H_p	distance between two parallel plates (m)
k_b	wood thermal conductivity (kJ/m s K)
k_ρ	mass transfer coefficient (m/s)
L_b	length of board (m)
L_s	sticker thickness (m)
M_{av}	average molecular weight (kg/kmol)
m_b	board total mass (kg)
M_w	molecular weight of water (kg/kmol)
Nu	Nusselt number (h 2 Hp/k)
Pe	Peclet number (Re Pr)
Pr	Prandtl number (Cp μ/k)
P_T	total pressure (atm)
p_v	water vapor partial pressure (atm)
p_{vs}	vapor (saturation) pressure (atm)
q	heat transfer flux (kJ/m^2 s)
R	gas constant (0.08206 m^3 atm/K kmol)
Re	Reynolds number (2 H_p v /V)
RH	relative humidity (p_v/p_{vs})
Sc	Schmidt number (ν/D)
Sh	Sherwood number (k_p 2 Hp/D)
t	time (s)
T_b	board average temperature (C)
T_{bs}	board surface temperature (C)
T_{db}	dry-bulb temperature (C)
T_g	air stream (gas) temperature (C)
T_{wb}	wet-bulb temperature (C)
u	board moisture content (kg water/kg dry wood)
u_{emc}	equilibrium moisture content (kg water/kg dry wood)
v	gas (air) velocity (m/s)
W_b	width of board (cross section) (m)
Y	(absolute) humidity (kg vapor/kg air)
δ	characteristic length (m)
ε	error term
λ	heat of vaporization (kJ/kg)
μ	viscosity of gas (kg/m s)
ν	kinematic viscosity (m^2/s)
ρ_b	density of dry wood (kg/m^3)
σ	variance

REFERENCES

1. W. M. Rohsenow and J. P. Hartnett, Eds., "Handbook of Heat Transfer", McGraw-Hill, New York, 1973.

2. S. W. Churchill and R. Usagi, A Generalized Correlation Procedure, AIChE Journal, v18, 1121, 1972,

3. F. Kayihan, Low and High Temperature Psychrometric Relationships for the Air-Water Vapor System and the Equilibrium Moisture Contents for Wood, Weyerhaeuser Technology Center, Technical Note 045-4402-05, 1983.

4. IMSL Library, Edition 9, IMSL Inc., Houston, 1982.

HIGH TEMPERATURE CONVECTIVE DRYING OF SOFTWOOD AND HARDWOOD : DRYING KINETICS AND PRODUCT QUALITY INTERACTIONS

Christian MOYNE and Christian BASILICO

Laboratoire d'Energétique et de Mécanique Théorique et Appliquée
Ecole Nationale Supérieure de la Métallurgie
et de l'Industrie des Mines, Parc de Saurupt, 54042 NANCY CEDEX
FRANCE

ABSTRACT

High temperature convective drying of hardwood (beech) is experimentally investigated in the following conditions of incipient flow : steam or moist air, temperature range : 90-150°C, velocity 6-18 m/s. A comparison is made with the previously published results for softwood (fir). The influence of the different parameters is thoroughly examined. The conclusions agree very well with the moisture migration mechanism already proposed.

To evaluate the industrial interest of such a process, the mechanical properties of the dried wood are also investigated. For both softwood (fir) and hardwood (beech) the modulus of rupture and elasticity are not affected by high temperature drying. Furthermore the shock resistance is increased.

1. INTRODUCTION

Among the new industrial processes which are used in wood drying, high temperature convective drying (above the boiling point of water) seems to be one of the more attractive. Two reasons can be put forward :

- for a material which dries very slowly ensuring a better agreement between supply and demand is an essential factor from an economical point of view : the drying time is reduced to a few hours compared to days for softwood or even weeks for hardwood
- the energy consumption of the processes which are used at the present time is between two and ten times the latent heat of vaporization of water : at high temperatures the operation can be conducted with superheated steam. So the fluid can be always recirculated. More the latent heat of the evaporated moisture can be recovered by steam recompression. So this process is very propitious to energy savings.

High temperature drying is not a new subject. As early as 1867 an US Patent was granted for seasonning lumber by superheated steam. Nevertheless the first studies on physical fundamentals of high temperature drying can be attributed to Kollmann and Schneider in 1960 (1). Furthermore the latter has recently investigated temperatures as high as 180°C (2).

But this technique was first industrially used in Australia during the 70th years. The results for softwood with fluid velocity u=5m/s and the temperature T = 120°C are very satisfactory (3). The very propitious results for softwood lead to adapt such a process for hardwood (4). In spite of their greater difficulty to dry, high temperature drying seems to be possible at least below the fiber saturation point. More recently studies on high temperature drying have been made in USA (5) and in Japan (6).

In our laboratory we have previously studied the drying of softwood (fir) in superheated steam in the temperature range 150-190°C (7,8). In particular we analyse the moisture migration mechanism according to the experimental results.

The three classical phases described by Krischer about convective drying of various hygroscopic porous material can be applied but with some modifications due to the particular anatomy of wood and the temperature level :

First period : the constant drying rate depends only on the external conditions and is governed by the heat transfer relations. The temperature at the board surface is close to 100°C. The moisture migration is essentially in liquid phase with the capillary pressure as driving force.

Second period : it begins when the drying rate decreases. The moisture at the board surface falls down the fiber saturation point and the temperature increases above the boiling point of water. Simultaneously the total pressure in gaseous phase inside the wood increases quickly due to the saturated vapor pressure increase with temperature. Two regions can be distinguished in the board :

* a surface region where no more free water exists. Moisture migrates according to a diffusion-like movement described by Stamm (9).

* a inner region where due to the increase of the gaseous pressure (which is not negligible compared to the capillary pressure), the capillary water moves according to Darcy's law in the longitudinal direction, the transversal permeability value being considerably lesser than the longitudinal one. This transport is the most important during this period.

Third period : the entire board lies in the hygroscopic range and moisture moves according to Stamm's diffusion-like mechanism.

From a practical point of view the very much shorter drying times obtained for softwood (about three hours) lead us to study the feasibility of such a process for hardwood more especially as hardwood dries with conventional schedules much slower and as much more time can be earned. So we investigate here the drying kinetics of beech in comparison with the previous results obtained for fir.

Nevertheless the product quality is one of the major factor for evaluating the interest of a new process. Among the several criteria which can be considered, we have choosen to see how the mechanical properties for both fir and beech are affected by high temperature drying.

2. EXPERIMENTAL APPARATUS

Experimental investigations are made in an aerodynamic return flow wind-tunnel (Figure 1). Measurement and regulation devices permit us to adjust the fluid temperature in the range 20, 200°C and the dew-point temperature in the range 20, 100°C with an accuracy of ± 2°C. The fluid velocity lies in the range 2,20m/s. The external pressure is the atmospheric one. The test section dimensions are 0,25mx0,25mx1,20m.

Samples are boards of softwood (Fir : Abies Alba) or hardwood (Beech) with the following dimensions : 0,027 m x 0,180 m x 1,00 m. The board is placed in the horizontal plane of symmetry of the test section : the fluid direction corresponds to the longitudinal direction of wood, the thickness to the radial one. Wood is sapwood.

The board weight is recorded continuously by an electronic balance.

The drying rate curves are determined in the following manner : the mass versus time curve is considered by interval of an odd number of points. On each interval, the curve is smoothed by a parabolic approximation. The curve slope at the midpoint is assumed to be the drying rate at this point. The choosen number of points does not affect significantly the result.

3. EXPERIMENTAL RESULTS

3.1. Constant drying rate period

Previously we investigated the drying kinetics of softwood in superheated steam for the following conditions (7, 8) :

Fluid velocity 6 m/s $< u_\infty <$ 18 m/s

Temperature 150°C $< T_\infty <$ 180°C

The same range of experimental parameters is first studied with hardwood. The corresponding drying rate curves are drawn on Figure 2 for the same steam velocity u = 12 m/s.

For the fluid temperatures T_∞ = 180 and 160°C in accordance to the classical description of drying, a first period with a constant drying rate is observed. For T_∞ = 150°C, the curve looks like the other one but does not exhibit a true constant rate. For lower fluid temperature as it would be seen latter this period disappears.

To be sure that this first period corresponds to an unhindered evaporation from the board surfaces, we have demonstrated that the average mass flux density for softwood can be accurately theoretically calculated by use of classical heat transfer relations (8). We must consider simultaneously the development of a boundary layer over a flat plate and the radiation transfer which due to the high temperature level and the radiative properties of steam is one third of the total heat transfer.

An experimental evidence of the first period reality is shown on Table 1 which compares the measured mass flux density for the same conditions (u_∞ = 12 m/s, superheated steam) for beech and fir during this period. For beech at 150°C the value corresponds to the maximum value because no constant value is observed.

Steam temperature	fir	beech
	Mass flux density (10^{-3}kg/m^{2}s)	Mass flux density (10^{-3}kg/m^{2}s)
150°C	0,82	0,80
160°C	0,95	0,94
180°C	1,20	1,28

Table 1 : Constant drying rate in the first period of drying for fir and beech in superheated steam (fluid velocity u_∞ = 12 m/s).

It must be noted that if the expression of the drying rate in %/h seems adequate from a practical point of view, the physical analysis of the phenomena is easier if we prefer the mass flux density with kg/m^{2}s as unit.

The very good agreement between the results for the two wood species demonstrates clearly that the constant drying rate in this period which must correspond to unhindered evaporation does not depend on the internal physical properties of the material. As a matter of fact the saturation moisture contents of the two wood species are very different : slightly higher than 200 % for fir and than 100 % for spruce. The moisture migration mechanisms above the fiber saturation point (the limit of the hygroscopic range, about 22 % at 100°C) are different. For capillary water longitudinal circulation, hardwood as beech is equiped with a natural network of vessels while in softwood the flux flows from a fiber to another one by small openings which are called pits. So the longitudinal permeability of beech seems to be twenty time greater than the permeability of fir. The transversal permeability is much smaller (about twenty thousands smaller for both fir and beech) than the longitudinal one due to the fiber geometry and to the migration only through the pits (10).

From examination of Fig 2 we conclude that for beech when the fluid temperature decrases, the first period disappears. As when this temperature increases the mass flux density increases too, it indicates that at lower temperature, the water inside the wood migrates much more slowly since water does not come sufficiently quickly in the transverse direction to the board surface which temperature is equal to the boiling point of water. No physical evidence seems to explain clearly this result.

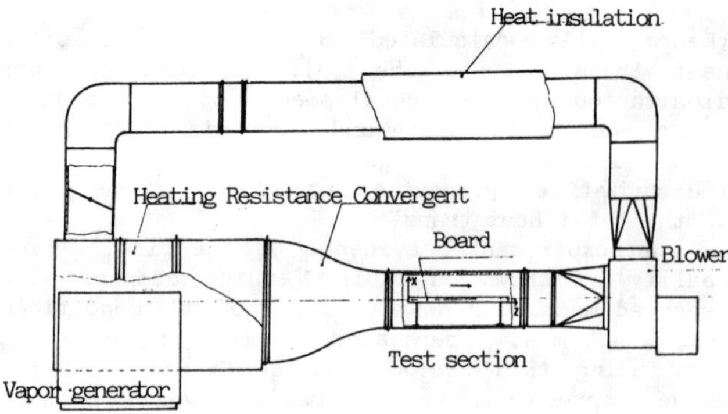

Figure 1 : Aerodynamic wind- tunnel

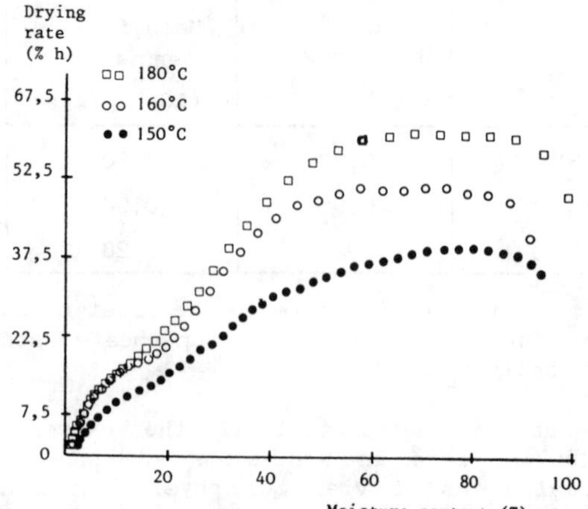

Figure 2 : Drying curves of beech for various temperatures T_∞ = 150, 160, 180°C in superheated steam v_∞ = 12 m/s

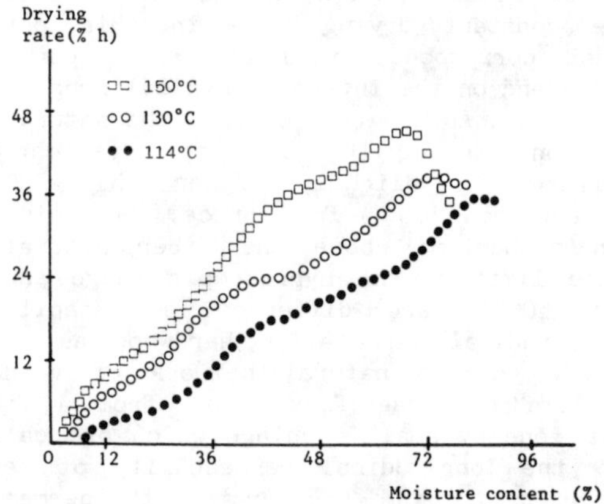

Figure 3 : Influence of fluid temperature on drying kinetics for beech in superheated steam (u_∞ = 12 m/s)

Nevertheless at temperature higher than 150°C the aspect of wood (color degradation) is too bad for practical use. So for hardwood we have investigated the temperature range 110-150°C and steam velocity range 6-18 m/s.

3.2. Influence of fluid temperature on drying kinetics

The Figure 3 shows the influence of the fluid temperature on the drying kinetics for a velocity u_∞ = 12 m/s. As stated before no constant rate period is observed at such temperatures. The fluid temperature affects the drying rate at any moisture content value.

During the second period, the drying rate decreases but capillary water is still present at the center of the board. The moisture migration is essentially longitudinal and is influenced by the development of a total pressure gradient in gaseous phase. If the temperature increases this pressure gradient becomes more important and so the drying rate increases.

In the third period, the wood being entirely in the hygroscopic range, as stated before the Stamm's diffusion-like mechanism predominates. The transerve diffusion coefficients are close together and 2.5 time lesser than the longitudinal one. But due to the board sizes, this mechanism is essentially transversal. The "diffusion" being thermally activated the drying rate increases with the fluid temperature. Lastly the temperature affects the equilibrium moisture content.

3.3. Influence of fluid velocity

On Figure 4 we investigate the influence of the fluid velocity for a fluid temperature T_∞ = 130°C.

At the beginning of drying when the external conditions govern for an important part the process an increase of the fluid velocity increases the heat transfer coefficient. More interesting is the velocity action at lower moisture content. Increasing the heat transfer coefficient leads to higher pressure gradient inside the board (7) and of course to higher drying rate. Nevertheless two resistances hinder the mass transfer : first the heat must be transferred from the bulk fluid to the inside of the sample and then the vapor must go out. If we accelerate the heat transfer, the mass transfer will increase until it reaches a value close of its maximum value and then increasing the fluid velocity will be of no effect.

Nevertheless up to a certain limit increasing the fluid velocity can increase the drying kinetics even in the hygroscopic range in a very significant manner. Those conclusions are in accordance with Schneider's investigations (11).

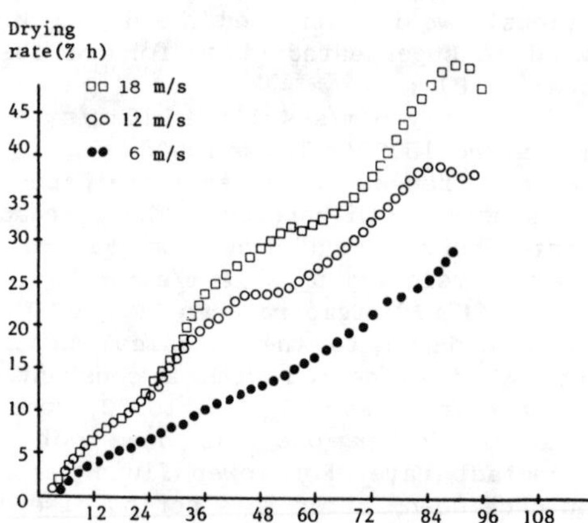

Figure 4 : Influence of fluid velocity on drying kinetics for beech in superheated steam (T_∞ = 130°C)

3.4. Some additional remarks

The temperature profiles for beech are of the same shape as those previously determinated for fir (7, 8). Because the first period in the investigated range does not exist, the surface temperature of the board increases quickly above 100°C. Inside the wood and successively in the depth after a stage at 100°C the temperature increases to be lastly close to the fluid temperature.

Due do the high longitudinal permeability of beech compared to fir, the pressure profile inside the wood cannot be measured. But as we point it out in (9), it does not mean that the filtration movement is unimportant. We demonstrate that if the heat transfer governs the process the product (overpressure in gaseous phase x Permeability) remains nearly constant inside the body although the mass transport under the influence of the pressure gradient predominates.

3.5. Influence of the fluid nature

In this part we want to compare the drying kinetics for superheated steam and moist air. To give a sense to this comparison the experiments are made in conditions which would give the same mass flux density in the period of constant drying rate (if such a period exists). So we keep the same difference ΔT = 50°C between the wet and dry bulb temperatures of the fluid and the same heat transfer coefficient. If we admit for sake of simplicity that the boundary layer over the flat plate remains laminar (in fact the laminar-turbulent transition can occur along the plate) and if we choose as reference state superheated steam at 150°C and 12 m/s, the Table 2 gives the experimental conditions which satisfy the two foregoing requirements.

Fluid	Dry temp. (°C)	Wet temp. (°C)	Fluid Velocity (m/s)
Steam	150	100	12
Moist air	130	80	11,6
Moist air	110	60	11,3
Moist air	90	40	11,2

Table 2 : Comparison of drying kinetics between moist air and superheated steam : experimental conditions

The Figure 5 shows the drying rate curves for all the tested conditions.Once more the curves do not present a constant rate period. Nevertheless for high moisture content due to the variations of the board dry weight,the drying rates are comparable. At lower moisture content the drying rate increases with the temperature. The slope of the drying curve (dw/dt versus w) indicates the ease with which water moves inside the material (for example if a diffusion model could be valid, the slope would be approximatively proportional to the diffusion coefficient). This slope increases with increasing dry bulb temperature due to the importance of the total

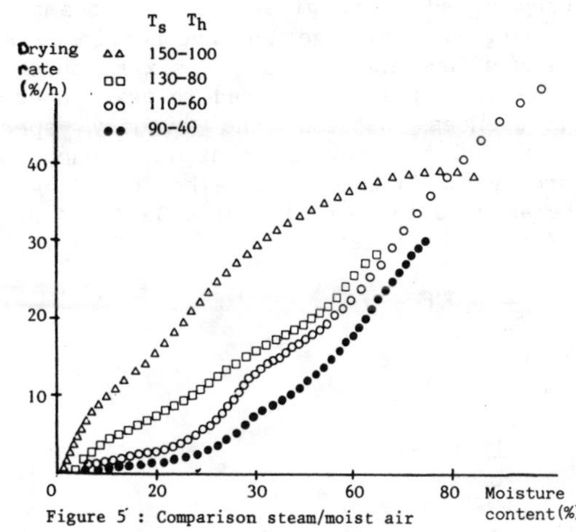

Figure 5 : Comparison steam/moist air

gaseous pressure gradient and perhaps to the activation of the bound water migration. In particular above the boiling point of water, the gaseous transport is a filtration while under this point the phenomena is essentially diffusive and so the migration is more difficult.

4. MECHANICAL PROPERTIES OF HIGH TEMPERATURE DRIED WOOD

To test the quality of high temperature dried wood its mechanical properties are investigated.

4.1. Fir (13)

Two samples group are been made. The first one is dried at 50°C, the second one at 170°C, 12 m/s in superheated steam. The experiments were conducted in Centre Technique du Bois by application of French Normalization. the differences betwen high temperature and traditional drying are listed below :

- Maximum crushing strength : - 1 %
- Static bending modulus of rupture : -0,5 % (with central loading)
- Modulus of elasticity : - 4 %
- Absorbed energy : + 5 % (single blow impact test)

We can conclude that high temperature drying does not damage in a significant manner the mechanical properties of fir.

4.2. Beech

4.2.1. Static bendic test

The modulus of elasticity and the modulus of rupture for both conventional and high temperature dried wood are investigated using a "four points" bending machine (Fig. 6) (14). The principal advantage of such a method is that in the central part of the test sample between the two internal points, the flexural moment is constant. The sizes of the sample are : tangential direction 20 mm, radial direction 20 mm, longitudinal direction 320 mm. The load is applied tangential to the annual rings.

Every dried board gives six test beams which after drying are alltogether reconditioned. The given values are the average of six measurements. More these values are corrected to take in account the differences between the ovendry specific gravity ρ_o and the moisture content of each sample w. According to Kollmann (15), the following relation are assumed to be valid for $0{,}08 < w < 0{,}20$:

- Modulus of elasticy :

$$E = \frac{k_E}{\rho_o} (0{,}8 - w)$$

where k_E is a constant.

- Modulus of rupture :

$$\sigma_r = \frac{k_r}{\rho_o} (0{,}45 - w)$$

where k_r is a constant.

All the values are expressed for a wood with a specific gravity $\rho_o = 650\ kg/m^3$ and a moisture content w = 12 %.

All the test conditions are figured in Table 3. By "conventional drying" we mean a drying with recommended drying schedules for industrial practise as cited in (16).

The Figures 7 and 8 show the modulus of elasticity and the modulus of rupture for the conditions listed in Table 3. The modulus of elasticity E is determined by the slope of the linear part of the curve strength versus deflection at the center of the beam. The modulus of rupture σ_r is a conventional one determined by use of linear relations theoretically valid only in the elastic range.

4.2.2. Single blow impact test

The blow of a pendulum hammer acts tangentially to the annual rings of a specimen (20 mm x 20 mm x 280 mm) with a span length equal to 240mm.

Each board gives two samples and the average absorbed energy for complete failure of the specimen is measured. Due to the lack of reliable data in the literature no correction is made. Nevertheless the measured value seems not to be related to moisture content. The results are drawn on Figure 9.

4.3. Conclusions

From the analysis of Figures 7, 8, 9 we can conclude that high temperature convective drying does not affect significantly the mechanical properties of wood. The dispersion between different specimens (although they are corrected for moisture or specific gravity differences) seems to be greater than the dispersion due to drying. Only the values of the absorbed energy for complete failure are larger (about 30 %) for beech for high temperature drying than for conventional drying.

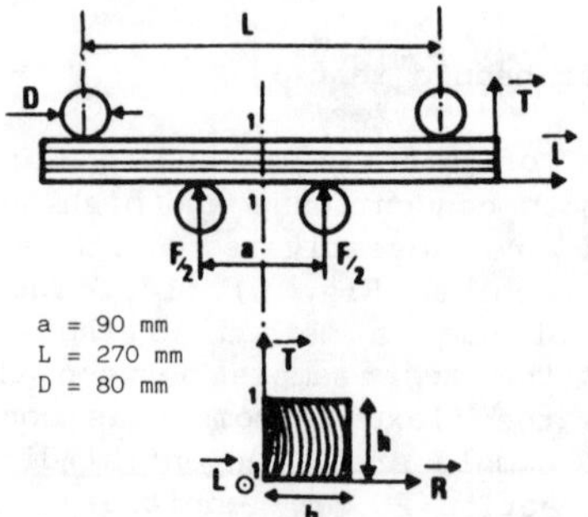

Figure 6 : Schematic static bending test apparatus

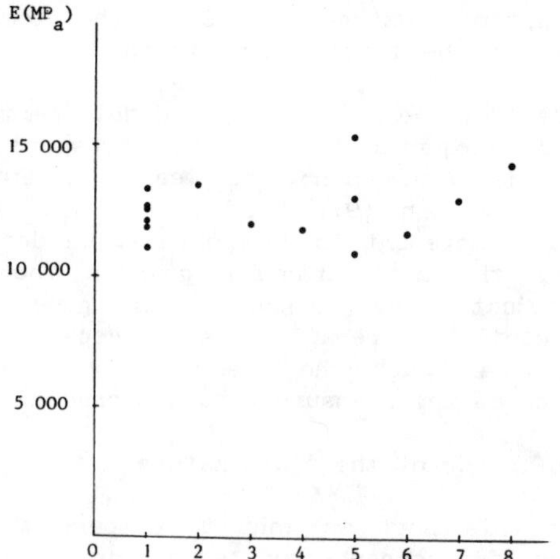

Figure 7 : Modulus of elasticity of dried beech

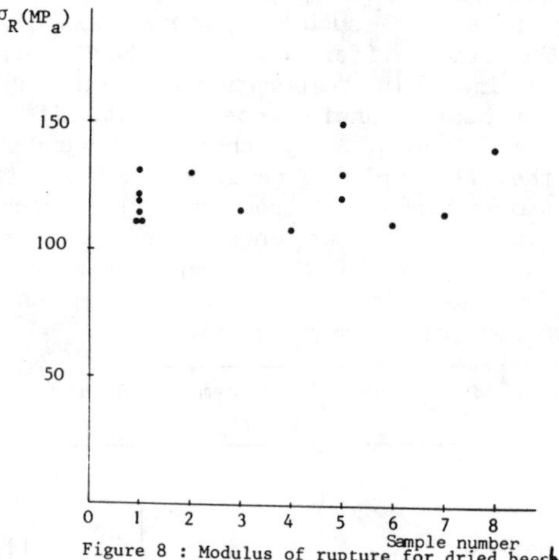

Figure 8 : Modulus of rupture for dried beech

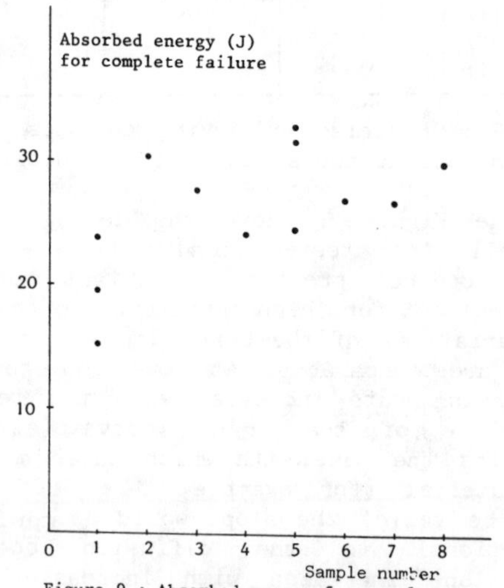

Figure 9 : Absorbed energy for complete failure for dried beech

Sample Number	Temperature	Fluid	Fluid velocity
		conventional drying	
1	40 - 60°C	Moist air	# 2 m/s
2	90°C	Moist air Dew Point T_{dp} =31,8°C	12 m/s
3	110°C	steam	6 m/s
4	130°C	steam	12 m/s
5	130°C	steam	12 m/s
6	130°C	Moist air Dew Point T_{dp} =80°C	12 m/s
7	130°C	steam	18 m/s
8	150°C	steam	12 m/s

Table 3 : Drying conditions for mechanical tests of beech

5. CONCLUSION

The high temperature convective drying of hardwood (beech) has been investigated and compared with previous published results for softwood (fir). Several points may be outlined :

- when the steam temperature decreases the first period with constant rate disappears,
- compared to softwood, drying must be conducted at lower temperatures (about 130°C). The drying time is then reduced to 4h while with the conventional schedules it lasts two or three weeks for a 27 mm thick board,
- the analysis of the drying kinetics for various temperatures, fluid velocity and the comparison between moist air and supherheated steam confirm the moisture migration mechanism(and especially the development of a total pressure gradient in the longitudinal direction during the second period which is the most important driving force) previously presented,
- from an industrial point of view the mechanical properties of fir and beech are comparable in conventional and high temperature drying. These conclusions are very propitious to an industrial development of such a process.

REFERENCES

1. Kollmann, F. and Schneider, A., Holz als Roh une Werkstoff, pp. 461-478 (1961)
2. Schneider, A., Holz als Roh und Werkstoff, Erste Mitteilung, pp. 382-394 (1972), Zweite Mitteilung, pp. 198-206 (1973)
3. Christensen, F.J. and Northway, R.L., Division of Building Research, Reprint 868, CSIRO, Australia (1979)
4. Fung, P.Y., Australian Forest Industries Journal pp. 46-50, April (1976)
5. Rosen, H.N., Bodkin, R.E. and Gaddis, K.D., Forest Products Journal, Vol. 33, PP. 17-24 (1983)
6. Sumi, H. and Mc Millen J., Forest Products Journal, pp. 25-33 (1979)
7. Basilico, C., Moyne, C. et Martin, M., Third International Drying Symposium, Vol. 1, pp. 46-55, Birmingham (1982)
8. Basilico, C. et Martin, M., Int. J. of Heat and Mass Transfer, (to be published)
9. Stamm A.J., Wood and Cellulose Science, Ronald New York, (1964)
10. Siau, S.I., Flow in Wood, Syracuse University Press, New-York, (1971)
11. Schneider, A., Wagner, L., Holz als Roh und Werkstoff 41, pp. 455-458 (1983)
12. Moyne, C. et Degiovanni, A., 4th International Drying Symposium, Kyoto (1984)
13. Basilico, C., Rapport final DGRST (1982)
14. Moussli, D., Thèse Docteur-Ingénieur, Nancy, (1983)
15. Kollmann, F. and Côte, W.A., Principle of wood Science andtechnology, Springer Verlag, (1968)
16. Anonymous, Conseils pratiques pour le séchage des bois, nº 56, Centre Technique du Bois, Paris, (1977)

COUPLED DRYING AND DEVOLATILIZATION OF LOW RANK COALS IN FLUIDIZED BEDS: AN EXPERIMENTAL AND THEORETICAL STUDY

Pradeep K. Agarwal, William E. Genetti and Yam Y. Lee

Department of Chemical Engineering, University of Mississippi
University, MS 38677 U.S.A.

ABSTRACT

A model is proposed for the coupled drying and devolatilization of large particles of low rank coals in fluidized beds. The model is based on the individual models developed and tested experimentally for drying and for devolatilization (using predried coal) and reported earlier. The temperature profile in the dry shell generated by the movement of a receding drying front (constituting a moving boundary problem) is assumed to characterize the rate of devolatilization during the first stage of the coupled phenomena when drying and devolatilization occur simultaneously. The temperature profile at the end of drying is used as an initial condition for solving analytically the transient heat conduction equation (with a convective boundary condition) for characterizing the remaining possible devolatilization after the drying is completed.

An empirical correlation is proposed for the estimation of the total time required for the coupled drying and devolatilization process.

The model predictions are compared with experimental data collected for Mississippi lignite in a three inch fluidized bed.

1. INTRODUCTION

Since fluidized bed combustion of coal is economical and environmentally acceptable, research efforts have been focused on this technology world wide. Fluidized bed combustors have been found to offer several advantages over conventional pulverized coal combustors (equipped with flue gas scrubbing units) including in-bed capture of SO_2, lower temperature operation reduces NO_x formation and the slagging/fouling of heat exchange surfaces, and higher heat transfer coefficients making more compact design possible[1,2].

Low rank coals such as lignite contain appreciable amounts of moisture in their 'as-mined' state. In view of the large amount of energy required, complete water removal from these coals is seldom done commercially. Thus, if 'as-mined' or wet lignite is to be used as combustor fuel, it would undergo several phenomena before and during the burning process. These would include, among others, drying (possibly with shrinkage) devolatilization and combustion of volatiles and residual char. All these processes are expected to occur in overlapping time periods with complex interactions which are not well understood. Major research has focused on the devolatilization and combustion of pre-dried pulverized coal. With the increasing emphasis on the use of low grade coals for fluidized bed combustion, it becomes increasingly important to consider the coupled behavior of the drying, devolatilization and combustion of such coals in fluidized beds.

In this paper a model is presented for the coupled drying and devolatilization of coal particles in inert atmosphere fluidized beds. The model predictions are compared with experimental data collected for Mississippi lignite. It is felt that this analysis would facilitate the coupling of the other phenomena, namely combustion of volatiles and residual char, and thus help in obtaining a more complete understanding of fluidized bed combustion of wet coal.

2. BACKGROUND

In a previous investigation[3], the drying of coal in an inert atmosphere fluidized bed, with reference to practical combustor operating conditions, was analyzed. A model was proposed in which heat transfer - to and through the particle - was assumed as the rate limiting step. A transient - in one dimensional spherical coordinates - heat conduction equation was set up. Immobilization of the receding drying front by change in space variable and the use of the heat balance integral approach led to the governing equation for the drying. The model predictions were found to be in good agreement with data for Mississippi lignite with an initial moisture content of 0.6 - 0.65 gms/gm dry coal.

In subsequent papers[4,5], the devolatilization of predried coal particles in inert atmosphere fluidized beds was examined. These models, safely applicable to low rank coals, assumed heat transfer - to and through the particle - and the chemical kinetics as the possible rate limiting steps for devolatilization. In the more general model[4] the kinetics of coal decomposition was assumed to be modeled by the non-isothermal expression proposed by Anthony _et al._[6,7]. The temperature distribution as a function of time was determined from the analytical solution[8,9] of the transient heat conduction equation (in spherical coordinates) with a convective boundary condition. Numerical integration led to the characterization of time dependent devolatilization. The model predictions were compared, for a wide range of particle sizes and for different temperatures, with the data reported by Morris and Keairns[10] with good agreement. Subsequently, a simplified model was proposed[5] for large particles. In this model, the rate determining steps for devolatilization were assumed to be the heat transfer to and through the coal particle, with only the pseudo-asymptotic yield of volatiles at lower temperatures being controlled by the chemical kinetics. The model parameters (which could be estimated from Proximate Analysis of the coal to determine devolatilization dependence on temperature) were two temperatures. The first T_{v1}, was the temperature at which devolatilization commenced and the other T_{v2}, at which it was substantially complete. Using these temperatures, and assuming a linear dependence of devolatilization on temperature in the intermediate region, volatile evolution rate was determined. This simplification led to the interpretation of the chemical kinetics controlled residual volatiles amount of lower temperatures in terms of the parameters T_{v_1} and T_{v_2}. Additionally, it was possible to derive analytical expressions to describe the devolatilization, in contrast to the general model which requires extensive numerical integration procedures.

The model predictions (from the simplified as well as the general model) were compared with fluidized bed devolatilization data of Mississippi lignite at lower temperatures with good agreement.

In the present paper, we use the simplified expressions for devolatilization of large coal particles[5] alongwith the equations for drying[3] to develop a model for the coupled drying and devolatilization of coal particles for operating conditions of interest in fluidized bed combustion (particle sizes >1 mm, superficial fluidizing velocity 1-2 m/s).

The advantage of this model lies in its ability to present analytical expressions for determination of devolatilization characteristics in the presence of drying.

3. THE MODEL

In this model for coupled drying and devolatilization of coal in fluidized beds, it is assumed that mass transfer is rapid. Heat transfer to and through the particle are rate controlling mechanisms for drying as well as devolatilization. The magnitude of the residual volatiles at lower temperatures of operation depends on the parameters T_{v_1} and T_{v_2} along with the bed temperature. The moisture and volatiles are assumed to be evenly distributed within the coal matrix. Devolatilization is assumed to be thermally neutral. The spherical coal particle is assumed to retain its shape during the drying and devolatilization process.

From the time the wet coal is introduced into the fluidized bed till the time all the volatiles evolution possible at a particular bed temperature T_a is completed, there would be two distinct stages. In the first stage, the particle would dry and devolatilize simultaneously. In the second stage, after the drying is completed, the particle could still devolatilize. The extent of these two stages would depend on the moisture content, type of coal as well as the operating conditions. We analyze the two stages separately in the following.

3.1 STAGE I

When the wet coal particle is introduced into the fluidized bed, drying would commence almost immediately with the drying front moving inwards. The heat conduction equation with a constant effective thermal diffusivity may be written, for the dry shell region, as

$$\frac{\partial T}{\partial t} = \frac{a}{r^2}\frac{\partial}{\partial r}\left(r^2 \frac{\partial T}{\partial r}\right) \qquad r_e \le r \le R_o \tag{1}$$

At the particle surface

$$\lambda_s \left.\frac{dT}{dr}\right|_{r=R_o} = \alpha(T_a - T_s) \qquad \text{for finite } \alpha \text{ and } T_s \neq T_a$$
$$= b(t) \qquad \text{in general} \tag{2}$$

At the receding drying front, a heat balance leads to

$$\lambda_s \left.\frac{dT}{dr}\right|_{r=r_e} = \lambda' C_o \rho_s \frac{dr_e}{dt} \tag{3}$$

where λ' is the sum of the sensible heat required to raise the temperature of the liquid and solid at the interface from the original temperature T_o to the constant evaporation temperature T_e, and the latent heat of vaporization. Thus

$$\lambda' = (T_e - T_o)(C_{pw} + C_p/C_o) + \Delta h_v \tag{4}$$

The moving boundary is immobilized by use of a transformed space variable $\emptyset = (r-r_e)/R_o-r_e)$. A quadratic temperature profile is assumed with constants evaluated from initial boundary conditions

$$T(r,t) = T_e + (T_s - T_e)(2\phi - \phi^2) + b(t)\,(R_o - r_e)/\lambda_s)(\phi^2 - \phi) \tag{5}$$

The use of the temperature profile in the heat balance integral approach requires the knowledge of the surface temperature. It is assumed that this temperature can be predicted, as an approximation, by a pseudo steady-state forumation of the same problem[11]. The pseudo steady-state formulation leads to the following equations to describe the drying process:

$$1 - \theta_s = \frac{3\,Bi}{Bi + 4}\phi_{ms}^{\;2} - \frac{2(Bi - 2)}{Bi + 4}\phi_{ms}^{\;3} \tag{6}$$

$$\frac{T - T_e}{T_a - T_e} = \frac{Bi\,(r/R_o - \phi_{ms})}{2\phi_{ms} + Bi(1 - \phi_{ms})\,r/R_o} \tag{7}$$

Use of equations (5), (6) and (7), integration and subsequent manipulation leads to the governing equation for drying.

$$A_1\phi_m^{\;3} + (A_4 - A_5 - A_2)\phi_m^{\;2} + (A_3 - A_2 - A_4)\phi_m + A_2 - A_3 + A_5 - A_1 \times A_6 = 0 \tag{8}$$

where $\phi_m = \frac{r_e}{R_o}$ and A_1, A_2, A_3, A_4, A_5 and A_6 are coefficients tabulated in Table 1. Complete details have been reported[3].

TABLE 1.

COEFFICIENTS FOR THE DRYING EQUATION (Equation 8)

$A_1 = (\frac{T_e + L}{2}) - \frac{T_e + 5\,T_s\vert_{\theta=1}}{12} + \frac{Bi\ \phi_{ms}\vert_{\theta=1}\ (T_a - T_e)}{12(2\phi_{ms}\vert_{\theta=1} + Bi\ (1 - \phi_{ms}\vert_{\theta=1}))}$	
$A_2 = \frac{\phi_{ms}\ Bi\ (T_a - T_e)}{12\ (2\phi_{ms} + Bi\ (1 - \phi_{ms}))}$	
$A_3 = \frac{T_e + 5\ T_s}{12}$	$A_4 = \frac{T_e + T_s}{4}$
$A_5 = (\frac{T_e + L}{2})$	$A_6 = (\frac{\phi^2_{ms} - 1}{\phi^2_{ms}\vert_{\theta=1} - 1})$
$L = \frac{\Delta h_v C_o}{C_p}$	$\lambda' = \Delta h_v + (T_e - T_o)(C_{pw} + C_p/C_o)$

With the drying front moving inwards, the radial position r_{v_1} and r_{v_2} corresponding to the temperatures T_{v_1} and T_{v_2} respectively, would also move inwards. Then the volumetric fractional average devolatilization may be written as

$$X_{avg} = \frac{3}{R_o^3}\int_0^{R_o} X\ r^2\ dr = \frac{3}{R_o^3}\left[\int_0^{r_{v_1}} X\ r^2\ dr + \int_{r_{v_1}}^{r_{v_2}} X\ r^2\ dr + \int_{r_{v_2}}^{R_o} X\ r^2\ dr\right] \tag{9}$$

The use of linear dependence of final yield of devolatilization on temperature between T_{v_1} and T_{v_2} leads to the following equation.

$$X = \begin{cases} 0.0 & T \leq T_{v_1} \\ (T-T_{v_1})/(T_{v_2}-T_{v_1}) & T_{v_1} < T < T_{v_2} \\ 1.0 & T \geq T_{v_2} \end{cases} \tag{10}$$

During the drying stage, the temperature profile in the dry shell is given by equation (5). Using equation (10), one may evaluate the integrals on the RHS of equation (9), which may then be rewritten as

$$X_{avg} = \frac{3}{(T_{v_2}-T_{v_1})(1-\frac{r_e}{R_o})^2}\left[(T_s - T_e)\ B_1\ (\frac{r}{R_o})^3 + B_2 B_3 (\frac{r}{R_o})^3\right]_{r_{v_1}}^{r_{v_2}} - \left[(\frac{T_{v_1}-T_e}{T_{v_2}-T_{v_1}})(\frac{r_{v_2}^3 - r_{v_1}^3}{R_o^3})\right] + \left[\frac{R_o^3 - r_{v_2}^3}{R_o^3}\right] \tag{11}$$

where coefficients B_1, B_2 and B_3 are tabulated in Table 2.

TABLE 2.

COEFFICIENTS FOR EQUATIONS 11 & 12

$B_1 = \frac{1}{2}\ (\frac{r_v}{R_o}) - \frac{1}{S}\ (\frac{r_v}{R_o})^2 + \frac{r_e}{3R_o}\ (\frac{r_e}{R_o} - 2)$	
$B_2 = \frac{1}{5}\ (\frac{r_v}{R_o})^2 - \frac{1}{4}\ (\frac{r_v}{R_o}) - \frac{1}{4}\ (\frac{r_v}{R_o})(\frac{r_e}{R_o}) + \frac{1}{3}\ (\frac{r_e}{R_o})$	
$B_3 = \frac{\phi_{ms}\ Bi\ (T_a - T_e)}{2\phi_{ms} + Bi\ (1 - \phi_{ms})}\ (1 - \frac{r_e}{R_o})$	
$C_1 = B_3 - (T_s - T_e)$	$C_2 = 2(T_s - T_e) - B_3$

The complete solution to the devolatilization characteristics for this stage is obtained by the specification of r_{v_1} and r_{v_2}, which may be calculated from the following equation - derived from equation (5) - by making appropriate substitutions, i.e.

$$\emptyset_v = \frac{r_v - r_e}{R_o - r_e} = \frac{C_2 \pm \sqrt{C_2^2 + 4C_1(T_v - T_e)}}{2C_1} \tag{12}$$

Equation (12) has only one root in $0 \leq \emptyset_v \leq 1$, hence the required values may be calcuated. Coefficients C_1 and C_2 are tabulated in Table 2.

3.2 STAGE II

Once the drying is completed, the particle would still devolatize. To analyze this stage, we can write the heat conduction equation, for finite Biot numbers, with time t' defined as $t' = t - \tau$, where τ is the total drying time

$$\frac{\partial T}{\partial t'} = \frac{a}{r^2}\frac{\partial}{\partial r}(r^2 \frac{\partial T}{\partial r}) \qquad 0 \leq r \leq R_o \tag{13}$$

The boundary conditions may be written for the now dry particle as

$$\lambda_s \left.\frac{dT}{dr}\right|_{r=R_o} = \alpha (T_a - T_s)$$
$$\left.\frac{dT}{dr}\right|_{r=0} = 0 \tag{14}$$

To evaluate the initial condition (i.e. the temperature profile for $t' = 0$) we make use of equation (5) with $r_e = 0$. Then

$$T(r,t'=0) = T_e + \left[(T_s|_{\theta=1} - T_e)(2(\frac{r}{R_o})^2 - \frac{r}{R_o})\right] + \left[b(t)|_{\theta=1} (\frac{R_o}{\lambda_s})((\frac{r}{R_o})^2 - \frac{r}{R_o})\right] \tag{15}$$

Following the analysis methods described by Jakob[8], equations (13-15) may be solved as

$$T(r,t') = T_a - \sum_{i=1}^{\infty} N_i \frac{\sin \beta_i \cdot r/R_o}{\beta_i r/R_o} e^{-\beta_i^2 a t'/R_o^2} \tag{16}$$

where β_i's are the roots of the transcedental equation

$$\beta \cos \beta = (1 - Bi) \sin \beta \tag{17}$$

and $N_i = 2\left[D_1E_1 - (2D_2-D_3)E_2 - (D_3-D_2)E_3\right]/E_4$.

The expressions for the coefficients D_1, E_1, D_2, E_2, D_3, E_3 and E_4 are tabulated in Table 3.

TABLE 3.

COEFFICIENTS FOR EQUATION 17

$D_1 = (T_a - T_e)$
$D_2 = (T_s\|_{t'=0} - T_e) = \frac{Bi\ (T_a - T_e)(1 - \phi_{ms})}{2\phi_{ms} + Bi\ (1 - \phi_{ms})}$
$D_3 = b(t)\|_{t'=0} \cdot R_o/\lambda_s = \frac{\phi_{ms}\ Bi\ (T_a - T_e)}{2\phi_{ms} + Bi\ (1 - \phi_{ms})}$
$E_1 = \sin\beta_i - \beta_i \cos \beta_i$
$E_2 = \cos \beta_i (\frac{2}{\beta_i} - \beta_i) + 2 \sin \beta_i - \frac{2}{\beta_i}$
$E_3 = \cos \beta_i (\frac{6}{\beta_i} - \beta_i) + 3 \sin \beta_i\ (1 - \frac{2}{\beta_i^2})$
$E_4 = \beta_i - \sin \beta_i \cos \beta_i$

Then using equation (9), (10) and (16),

$$X_{avg} = \frac{1}{R_o^3}\left[(R_o^3 - r_{v_2}^3) + \left[(T_a - T_{v_1})/(T_{v_2} - T_{v_1})\right](r_{v_2}^3 - r_{v_1}^3)\right] - \frac{3}{R_o^3}\frac{(T_a - T_o)}{(T_{v_2} - T_{v_1})}\left[\sum_{i=1}^{\infty} \frac{f_i \sin b_i r}{b_i^2} - \sum_{i=1}^{\infty} \frac{f_i r \cos b_i r}{b_i}\right]_{r_{v_1}}^{r_{v_2}} \tag{18}$$

where $f_i = N_i(R_o/\beta_i)e^{-\beta_i^2 a t'/R_o^2}$

and $b_i = \beta_i/R_o$.

Though the results presented here are for the case when $T_a > T_{v_2}$, the above results may be easily modified to take into account the residual amount of volatiles when $T_a < T_{v_2}$.

The model, thus, permits formulation of analytical expressions to describe the devolatilization of initially wet coal.

4. EXPERIMENTAL APPARATUS

The fluidized bed used was 7.6 cm in diameter. The bed was filled to a static height of 6-10 mesh alumina beads. Preheated air/nitrogen, introduced into the bed through a perforated steel plate distributor, was used to fluidize the bed. The air/nitrogen flow rate into the bed and the bed temperature were measured by an orifice meter and the thermocouple respectively.

Large chunks of Mississippi lignite were broken to 4-7 mesh size. Each test batch of lignite particles was put into a cylindrical cage shaped sampler constructed from woven steel mesh with aperture size of 10 mesh. The sampler was then inserted into the fluidized bed for the desired time period. At the end of the required time, the lignite particles were quenched, weighed and analyzed for residual moisture and volatiles in a Fisher Proximate Analyzer. Complete details of the experimental apparatus and procedure are presented elsewhere[12].

5. RESULTS AND DISCUSSION

In the following, first the results of the parametric studies made with the model are presented. An approximate expression to determine the total time required for coupled drying and devolatilization is presented. The model predictions are then compared with the data collected experimentally.

5.1 Parametric Studies

To demonstrate the effects of change in operating conditions, a set of parameters is chosen as standard, and then each of the parameters is varied in turn. The parameters chosen as standard are $T_{v_2} = 950°K$, $T_{v_1} = 525°K$, $a = 0.1\ mm^2/s$, $C_o = 0.6$ gm moisture/gm dry coal, Bi = 4, d = 2 mm, $T_a = 800°K$. The normalized drying curves are represented by solid lines (—) and the normalized devolatilization curves are represented by broken lines (— - —).

In Figure 1a, the effect of change in bed temperature is shown. The devolatilization becomes asymptotic to $V_{res} = (T_{v_2} - T_a)/(T_{v_2} - T_{v_1})$ when $T_a < T_{v_2}$. For $T_a > T_{v_2}$, complete devolatilization occurs. As expected, the drying time decreases with increase in temperature.

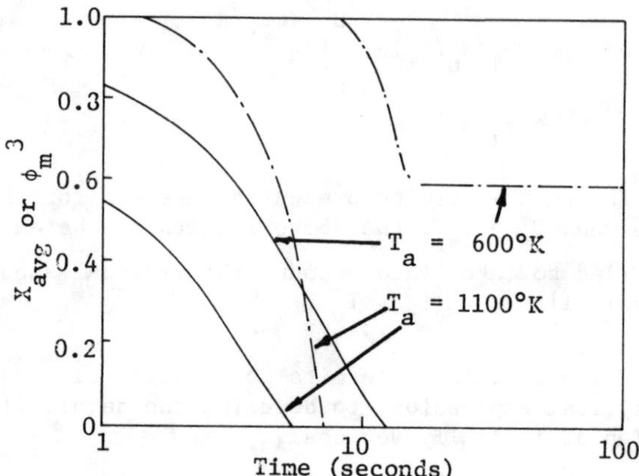

Fig. 1a. Effect of bed temperature

The effect of change in moisture content is shown in Figure 1b. The time lag for devolatilization introduced by the presence of moisture is apparent. Since $T_a < T_{v_2}$, regardless of the moisture content, the devolatilization curves becomes asymptotic to the same value of residual volatiles.

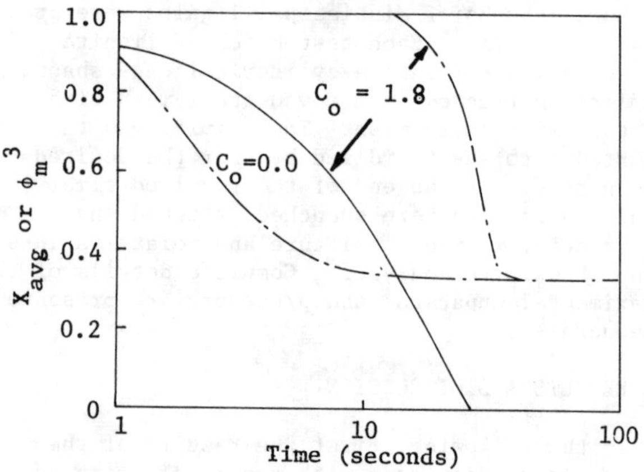

Fig. 1b. Effect of initial moisture content

In Figure 1c, the effect of change in T_{v_2} is shown. Since the drying is not effected by devolatilization parameters, the drying curve is the same. In addition, since $T_a < T_{v_2}$, the time required to reach the asymptotic volatile yield is the same; the magnitude of volatile the residual volatiles, of course, depends on T_{v_2}.

In Figure 1d, the effect of the variation in heat transfer Biot number is shown. As expected, for lower Biot numbers, the drying and devolatilization times are much larger. Also for lower Biot numbers, extent of devolatilization is comparatively small during the drying stage. In fact, for lower Biot numbers (~1.0), it is possible to decouple drying and devolatilization without significant error.

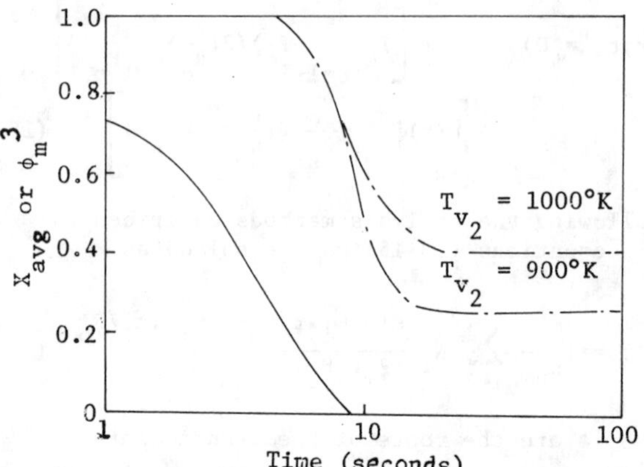

Fig. 1c. Effect of Coal Decomposition Parameters

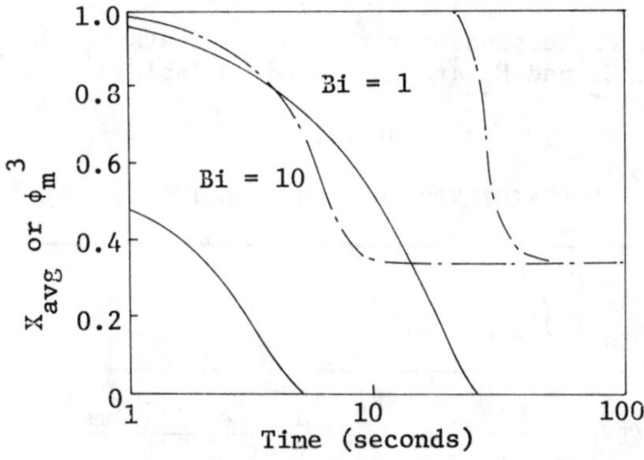

Fig. 1d. Effect of Biot Number

In Figure 1e, the effect of the particle size is shown. The normalized devolatilization and drying curves are plotted against $Fo = at/d^2$ (instead of time) and the curves for different diameters are seen collapse to the same curve.

Extensive numerical testing reveals that it is possible to estimate the time required for the completion of the coupled drying and devolatilization process (i.e. Stage I as well as Stage II as discussed in an earlier section) by the following approximate equation

$$t_{wv} = \frac{0.0175}{a}\left[\left(2.1\ C_o^{\ 1.25}\ \left(\frac{Bi + 4}{Bi}\right)\left(\frac{\Delta h_v}{C_p(T_a - T_e)}\right)\right) + \left(1 + \left(\ln\frac{1}{\psi}\right)\frac{15}{\beta_1^{\ 2}}\right)\right]d^2 \quad \text{for } T_a > T_{v_1} \tag{19}$$

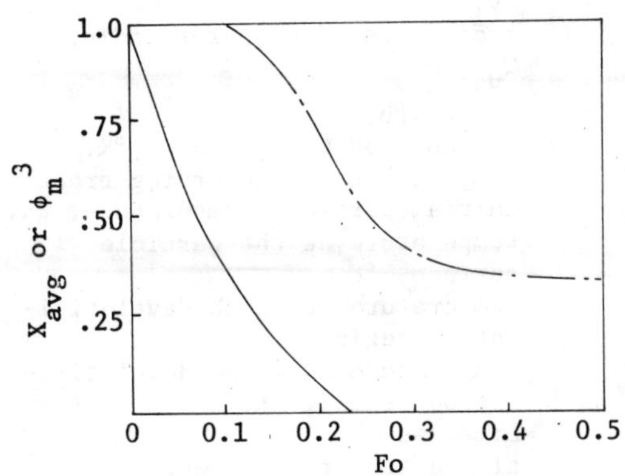

Fig. 1e. Effect of Particle Size

$$\text{where } \psi = \frac{T_a - T_{v_2}}{T_a - T_o} \qquad \text{for } T_a > T_{v_2}$$

$$\psi = 0.01 \qquad \text{for } T_a \leq T_{v_2}$$

and β_1 is the first root of equation (17). It may be pointed out that, for $C_o = 0.0$ i.e. predried particles, equation (19) reduces to the approximate expression proposed earlier[5].

The proposed empirical expression has been tested against model predictions for Bi = 1 to Bi = 10, $C_o = 0.0$ to $C_o = 1.8$ and has been found to predict detailed model calculations to within 10%. It must also be pointed out that this expression is valid only for $T_a > T_{v_1}$, i.e. the equation will not in general, predict particle drying times when there is not devolatilization.

The analysis presented indicates that for initially wet coal it is not possible, in general, to assume that devolatilization occurs instantaneously at the coal entry port in the overall modeling of fluidized bed combustion[13]. A similar result has been obtained by Stanmore[14].

5.2 Experimental Results

The data for coupled drying and devolatilization is compared with the model predictions in Figures 2, 3 and 4. An encouraging aspect of the model is that the thermophysical and coal decomposition parameters used have been obtained independently in the investigations for drying without devolatilization and devolatilization of predried coal; no further adjustment was required. The relevant operating conditions and the parameters used are summarized in Table 4.

As may be seen, the model is in good agreement with the data. The largest discrepancies are seen to occur in the earlier stages of drying. These discrepancies may be due to the simplifications made in the analysis. However, it must be pointed out that the moisture content drops very rapidly in the early stages of drying, and a small delay in the quenching of coal particles may result in an experimental error of the same order of magnitude. In making measurements related to the moisture content, the temperature has to be brought down rapidly to below 100°C. In comparison, devolatilization would be sufficiently quenched for particle temperatures below 200°C. Hence, it is felt that the devolatilization data may be more accurate. The results of the devolatilization predictions, which are based on the temperature profiles and movement of the drying front are seen to be in good agreement even during the early stages of drying.

TABLE 4.

I. Experimental Operating Conditions		
Fig. 2	U = 2.0 m/s T_a = 265°C Bi = 4.5	Particle Size = 4-7 mesh
Fig. 3	U = 1.5 m/s T_a = 340°C Bi = 4.0	Particle Size = 4-7 mesh
Fig. 4	U = 1.5 m/s T_a = 440°C Bi = 3.0	Particle Size = 4-7 mesh
II. Thermophysical and Coal Decomposition Parameters[3,5]		
a = 0.1 mm²/s	T_o = 283°K	Δh_v = 570 cal/gm
ρ_s = 1.25 gm/cm³	T_e = 373°K	C_p = 0.3 cal/gm°K
T_{v_1} = 475°K	T_{v_2} = 850°K	

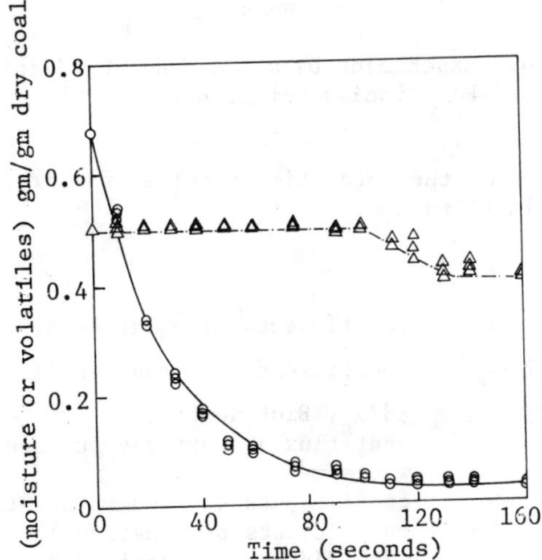

Fig. 2. Experiment Data and Model Predictions For Mississippi Lignite

6. CONCLUSIONS

A model has been proposed for the coupled drying and devolatilization of low rank coals. This model, valid for large particle sizes of interest in Fluidized Bed Combustion, enables analytical expressions for devolatilization in the presences of moisture. The model predictions are seen to be in good agreement with the data for Mississippi lignite. Based on extensive numerical computations an approximate expression is proposed for the

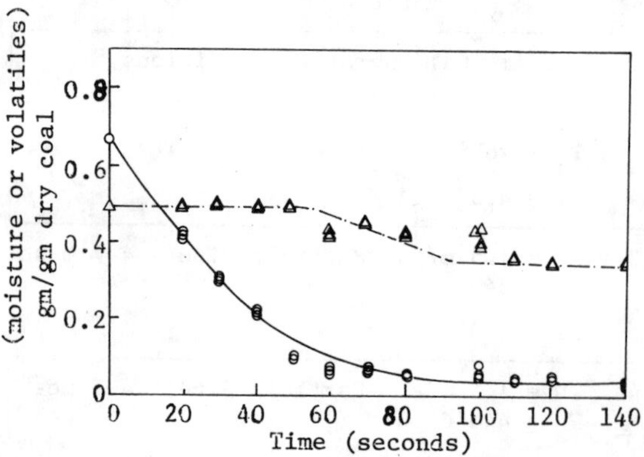

Fig. 3. Experiment Data And Model Predictions For Mississippi Lignite

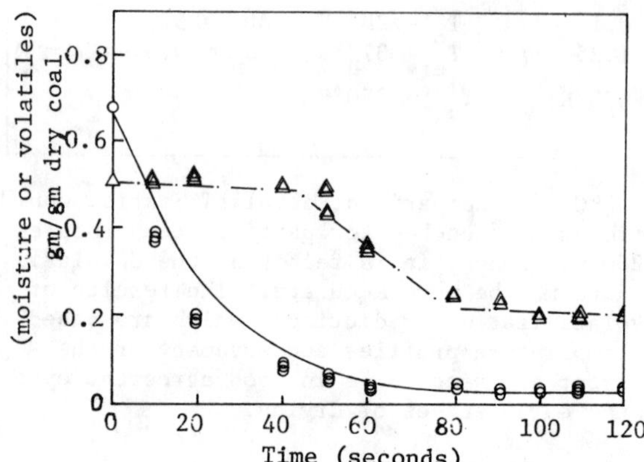

Fig. 4. Experiment Data And Model Predictions For Mississippi Lignite

estimation of the total time required for drying and devolatilization.

NOMENCLATURE

A_i (i = 1 to 6) coefficients of equation 8.
B_i (i = 1 to 3) coefficients of equation 11.
Bi $\alpha d/\lambda_s$, Biot number.
b(t) heat flux at particle surface, watts/m^2.
b_i coefficients defined in equation 18.
C_i (i = 1, 2) coefficients of equation 12.
C_p specific heat of coal, J/Kg°K.
C_{pw} specific heat of water, J/Kg°K.
C_o initial moisture content of coal, Kg/Kg dry coal.
D_i (i = 1 to 3) coefficients of equation 17.
E_i (i = 1 to 4) coefficients of equation 17.
f_i coefficients defined in equation 18.
L defined in Table 1.
N_i (i = 1 to ∞) coefficients in equation 16.
R_o initial particle radius, m.
r radial position, m
r_e position of the drying front, m.
r_{v_1} radial position corresponding to T_v, m.
r_{v_1} radial position corresponding to T_{v_1}, m.
r_{v_2} radial position corresponding to T_{v_2}, m.
T temperature, °K.
T_a ambient/bed temperature, °K.
T_e temperature at the drying front, °K.
T_o initial particle temperature, °K.
T_s temperature at the particle surface, °K.
T_{v_1} temperature at which devolatilization begins, °K.
T_{v_2} temperature at which devolatilization begins, °K.
t time, sec.
t′ time after drying, sec.
t_{wv} time required for the completion of drying and devolatilization, sec.
U superficial fluidized velocity, m/s.
V_{res} residual fractional amount of volatiles for $T_a < T_{v_2}$.
X extent of fractional devolatilization at any temperature T.
X_{avg} volumetric average fraction devolatilization for the particle.

Greek Symbols

β_i (i = 1 to ∞) roots of the transcedental equation 17.
ρ_s particle density of coal, kg/m^3
λ_s thermal conductivity of coal, w/m K.
λ′ coefficient defined in equation 4.
φ space transformation variable.
ϕ_m dimensionless position of the drying front.
ϕ_{ms} dimensionless position of the dry-front predicted by the pseudo steady state model.
ϕ_v dimensionless position corresponding to temperature, T_v.
θ t/τ, dimensionless time.
θ_s t/τ_s, dimensionless time in the pseudo steady state model.
ψ defined in equation 19.
τ total drying time, sec.
τ_s total drying time predicted by the pseudo steady state model, sec.

REFERENCES

1. Yaverbaum, L.H., Fluidized Bed Combustion of Coal and Waste Materials, Noyes Data Corpotion, 1977.
2. Selle, S.J., Honea, F.I. and Sondreal, E.A. in 'New Fuels and Advances in Combustion Technology', Institute of Gas Technology, Chicago, 1979.
3. Agarwal, P.K., Genetti, W.E., Lee, Y.Y. and S.N. Prasad to appear in Fuel.
4. Agarwal, P.K., Genetti, W.E. and Y.Y. Lee to appear in Fuel.
5. Agarwal, P.K., Genetti, W.E. and Y.Y. Lee to appear in Fuel.
6. Anthony, D.B. and Howard, J.B., AIChE J., 1976, vol. 22(4), 625.

7. Anthony, D.B., Howard, J.B., Hottel, H.C. and Meissner, H.P., Fifteenth Symp. (Int.) Combustion, Combustion Institute, Pittsburgh, 103 (1975).
8. Jakob, M., Heat Transfer, John Wiley and Sons, New York (1959).
9. Kutateladze, S.S., Fundamentals of Heat Transfer, Academic Press, New York (1963).
10. Morris, J.P. and Keairns, D.L. Fuel, Vol. 58, 465 (1979).
11. Agarwal, P.K., Genetti, W.E. and Y.Y. Lee, to appear in Chemical Engineering Communications.
12. Agarwal, P.K., Fluidized Bed Combustion of Wet Low Rank Coals, Ph.D. Thesis, University of Mississippi, University, Miss (1984).
13. Park, D., Levenspiel, O. and Fitzgerald, T.J., AIChE Symp. Series, Vol. 77, 116 (1981).
14. Stanmore, B., Personal Communication (1983).

A NEW FLUIDIZED BED DRYER FOR AMORPHOUS SLAG

Takashi MORIYAMA, Hiroshi ITAYA, and Akihiko NAKAMURA

Denka Consultant & Engineering Co., Ltd.
Yurakucho 1-4-1, Chiyoda-ku, Tokyo 100, Japan

ABSTRACT

The authors have developed a blow tank injection feeder with fluidized bed for solid gas two-phase flow. Based on experience gained on this technology within Denki Kagaku Kogyo and tests conducted in test plants, they have been able to develop several types of fluidized bed dryer. One type is a continuous feed fluidized bed dryer with special highly heat-efficient air distributers to the bed.

An application of this is in drying amorphous slag from blast furnaces in steel making companies which is supplied to cement companies as an ingredient of BF slag cement. Granular amorphous slag is very important in fuel conservation in the cement industry, but handling and pretreatment before drying are not easy. This is because coating, choking and blocking occur in wet conditions much faster when hot air free from CO_2 is used in place of waste combustion gas for the drying process. The authors have adopted a fluidized bed dryer which uses special air distributers and waste hot air from the cement air quenching cooler. A material was chosen for the air distributers which was not subject to abrasion from the dust contained in the hot air. In this dryer, wet slag is fed into the dryer on the bed by gravity flow and the inside surface of the freeboard is heated by hot air.

The result has been a successful and highly heat-efficient fluidized bed dryer with a capacity of 21 tons per hour for a bed area of 6 square meters. The design method was proven using data from test plants and data from actual operation which was stored in a desktop computer (HP 9845B).

1. INTRODUCTION

The authors have developed a blow tank injection feeder for solid gas two-phase flow.[1-3]

Based on experience gained within Denki Kagaku Kogyo on similar dryers and tests conducted on test plants, they have been able to develop several types of fluidized bed dryer. One is a thermal conducting horizontal mechanical agitating dryer with a fluidized bed for plastics and chemicals. Figure 1 shows the test plant.

Seven of these dryers have been constructed for batch processing and two for continuous processing; the largest constructed is a 15 cubic meter volume continuous dryer for BS resin.

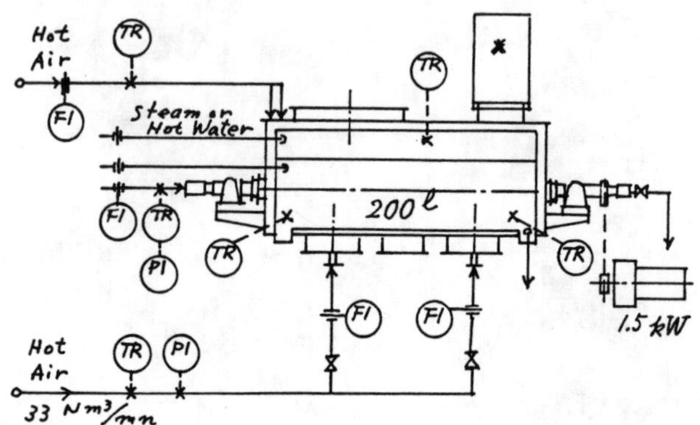

Fig. 1 Testing plant for thermal conducting horizontal mechanical agitating dryer with fluidized bed

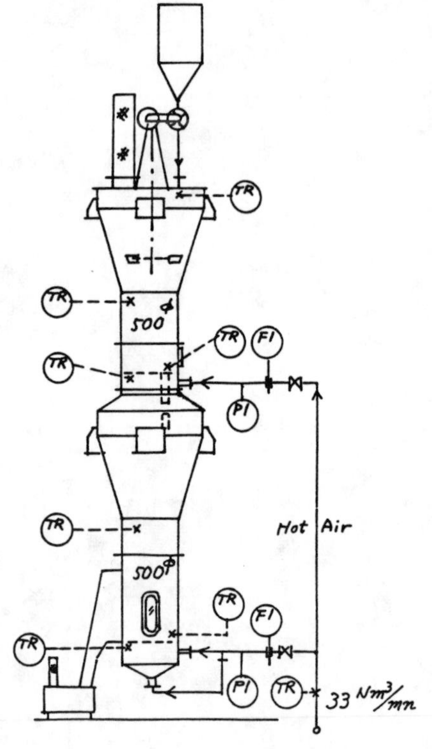

Fig. 2 Testing plant for multistage continuous fluidized bed dryer

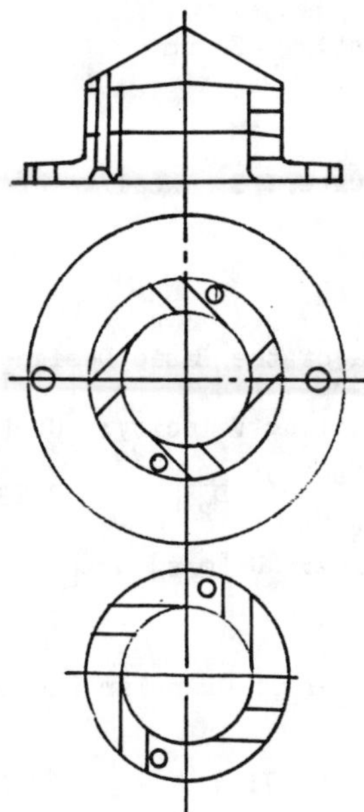

Fig. 3 A distributer of the fluidized bed

Photograph 1 The distributers

A second type is a continuous feed type fluidized bed dryer with special highly heat-efficient air distributers to the bed. Figure 2 shows the test plant and figure 3 and photograph 1 show the distributers.

One application is in drying granular amorphous slag from blast furnaces of steel companies which is supplied to cement companies as an ingredient of BF slag cement. This paper introduces the granular amorphous slag dryer.

2. FEATURES OF THE DRYER

Granular amorphous slag is very important for fuel conservation in the cement industry and makes up from 5 to 70 percent of the composition of BF cement. However, handling and pretreatment before drying are not easy. This is because coating, choking and blocking occur in wet conditions much faster by alkilation when hot air free from CO_2 is used in place of waste combustion gas for the drying process.

However, because of the constraints of the layout and the waste heat balance, dust-laden hot air from the cement air quenching cooler was selected as the hot gas used for slag drying.

Hot gas temperatures of 110 degrees C and 240 degrees C from the cement air quenching cooler were compared. The authors found that, as the cement plant has a waste heat power station supplied from the New Suspension Preheater (NSP) outlet gas and the air quenching cooler, if a temperature of 110 degrees were adopted, an increase in generating power would be accompanied by a corresponding increase in the electricity consumption of the hot gas blower.

3. DRYER TECHNICAL DATA AND TEST RESULTS

Capacity: 21 t/h wet base

Moisture: 15% wet base to below 1%

Bulk density: 1.12

Specific heat: 0.3 kcal/kg°C

Heat source: Hot air 240°C with 30 mg/Nm³x0.1mmø cement clinker dust

Granular size distribution:	+2.5 mm	11.1%	$\bar{D}p$ = 1.5mm
	+1.2 to -2.5	23 to 29%	
	+0.6 to -1.2	18 to 43%	
	-0.6	6 to 35%	

Photographs 2 and 3 show the shape of the granular amorphous slag.

Photograph 2 Wet granular slag

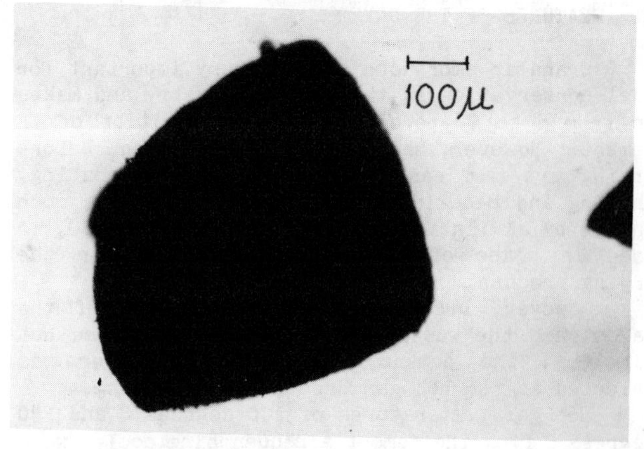

Photograph 3 Dry granular slag

4. CALCULATION AND SPECIFICATION OF THE BASIC DRYER DESIGN

4.1 Nomenclature

Material name:	B		BF slag
Wet base capacity:	E	kg/h	20,750
Wet base inlet water contents:	X	%	15
Wet base outlet water contents:	Y	%	1
Dry base capacity:	W	kg/h	17,640
Dry base inlet water contents:	F1	-	0.176
Dry base outlet water contents:	F2	-	0.010
Critical water contents:	Fc	-	0.017
Feed temperature:	Tf	°C	10
Granular size:	$\bar{D}p$	m	1.5×10^{-3}
Bulk density:	Ob	kg/m³	1.1×10^{3}
Net density:	Os	kg/m³	2.8×10^{3}
Hot gas temperature:	T1	°C	240
Hot gas density:	Og	kg/m³	0.665
Wet bulb temp.:	Tw	°C	53
Hot gas viscosity:	Ng	kg-m/s	2.7×10^{-5}
Hot gas thermal conductivity:	Dg	kcal/mh°C	0.037 (0.154kJ/mh°C)
Hot gas specific heat:	C_H	kcal/kg°C	0.245 (1.026 kJ/kg°C)
Static bed height:	Z	m	0.25
Gas velocity coefficient:	M	-	0.4
Latent heat of water:	S	kcal/kg	570 (2,386kJ/kg)

4.2 Output Data of the Basic Design

1) Terminal Settling Velocity: Ut [m/s]

$$Ut = \left[\frac{384}{225} \frac{(\rho_p - \rho_g)^2}{\mu_g \cdot \rho_g} D_p^3 \right]^{1/3} = 7.29$$

2) Space Velocity: U [m/s]

$$U = M\ Ut = 2.9$$

3) Reynolds Number of Granular Particle: Re

$$Re = \frac{D_p \cdot U \cdot \rho_g}{\mu_g} = 71.4$$

4) Volumetric Heat Transfer Coefficient : Hc [kcal/m³h°C]

$$Hc = \frac{24 \times 10^{-3} \rho_b \cdot D_g Re^{1.5}}{\rho_s D_p^2} = 93{,}500 (392{,}000 kJ/m^3h°C)$$

5) Actual Volumetric Heat Transfer Coefficient : He [kcal/m³/h°C]

He = 0.5 Hc for Dp = 1.5×10^{-3} of Reference 5 Fig. 7.10

6) Dry Material Temperature: Tm [°C]

$$Tm = t_1 - \left[\frac{\gamma_w F_2 - C_s(t_1 - t_w)(F_2/Fc)}{F_c \gamma_w - C_s(t_1 - t_w)} \right]^{\frac{F_c \gamma_w}{C_s(t_1 - t_w)}} (t_1 - t_w)$$

$= 59$

7) Required Heat Q_T [kcal/h]

Required heat in constant rate period Qc
falling rate period Qd

(i) Fc > F₂

$$Q_c = W/[(F_1 - F_c)\gamma_w + (C_s + F_c)(t_w + t_F)]$$
$$= 1{,}763{,}000\ (7{,}382{,}000\ kJ/h)$$

$$Q_D = W/[(F_c - F_2)\gamma_w + (C_s + F_2)(t_m - t_w)]$$
$$= 92{,}600\ (388{,}000\ kJ/h)$$

$$Q_T = Q_c + Q_D = 1{,}856{,}000\ (7{,}770{,}000\ kJ/h)$$

(ii) Fc < F₂

$$Q_c = W/[(F_1 - F_2)\gamma_w + (C_s + F_2)(t_w - t_F)]$$

$$Q_D = 0$$

$$Q_T = Q_c$$

8) Hot Gas Velocity: G [kg/m²h]

$$G = (3600)U.\rho_g = 6{,}943$$

9) Temperature Difference between Hot Gas and Material: T [°C]

$$\Delta T = \frac{t_m - t_w}{\ln\left(\dfrac{t_1 - t_w}{t_1 - t_m}\right)} = 184$$

10) Fluidized Bed Area: A [m²]

Fluidized Bed Area in constant rate period : Ac [m²]

Fluidized Bed Area in falling rate period : A_D [m²]

$$A_C = \frac{Q_C(G.C_H + He.Z)}{He.Z.C_H(t_1 - t_w)} = 5.94$$

$$A_D = \frac{Q_D(G.C_H + He.Z)}{He.Z.G.C_H.\Delta T} = 0.32$$

$$A = A_c + A_D = 6.26$$

11) Hot Gas Flow Rate: V [Nm³/min]

Inlet $V_1 = \rho_G.G.A. \div 60 = 578$

Outlet $V_2 = V_1 + \rho_G.W.(F_1 - F_2) \div 60 = 619$

12) Outlet Gas Temperature: t_2 [°C]

$$t_2 = t_1 - \frac{Q_T}{GA_T C_H} = 66$$

13) Pressure Loss into Gas Distributing Nozzle : P [mmH₂O]

$$\Delta P_T = \Delta P_N + \Delta P_C + \Delta P_D = 85 + 100 + 50$$

$$P = \Delta P_T + Z.\rho_b = 510 \ (5{,}100 \ N/m^2)$$

14) Minimum Fluidization Velocity: Umf [m/s]

$$Umf = \frac{\mu}{\rho_g . D_p}\left[(637.65 + 0.0651A_v)^{\frac{1}{2}} - 25.25\right] = 1.43$$

$$\text{as } A_v = D_p^3 \rho_g(\rho_s - \rho_g)\ g/\mu^2 = 84{,}480$$

15) Thermal Efficiency: [%]

$$\eta = \frac{t_1 - t_2}{t_1 - t_3} \times 100 = \frac{240 - 66}{240 - 10} \times 100 = 75.6$$

These calculation methods and actual figures were verified in the test plant and in the actual plant.

4.3 Comparison between Actual Data and Basic Design Data

	Data	Basic design	Data	Basic design
E	18,880	←	21,140	←
X	7.9	←	8.4	←
Y	0.01	←	0.03	←
Fc		1.0		1.0
Tf	20	←	←	←
$\bar{D}p$	1.5×10^{-3}	←	←	←
Ob	1.1×10^{3}	←	←	←
Os	2.8×10^{3}	←	←	←
S	0.2	←	←	←
t_1	240	←	←	←
Tw	53	←	←	←
Ng	2.7×10^{-5}	←	←	←
Dg	3.7×10^{-2}	←	←	←
Z	0.27	←	←	←
M		0.3		0.275
A	6.0	5.88	6.0	6.07
V_1	300	410	311	389
V_2		441		419
t_2	78	75	65	72
Tm	123	132	90	120
P	490	532	510	532
Ut		7.3		7.2
U		2.9		2.0
Umf		1.4		1.4

5. DETAILED DRYER DESIGN PARAMETERS

1) Nozzle velocity: 40 m/s

2) Sliding velocity of solid particles on bed : 3.6 m/s

3) Fluidized bed area: 6 m²

4) Number of distributers: 50, 48, 64 for sections 1 to 3

5) Dryer hot gas pipe size: 150, 600, 800, 800ø for sections 0 to 3

6) Dimensions of cyclone: 2700ø x (2600+4500) x 1

7) Inlet velocity: 15 m/s

8) Dust content of outlet: 3 mg/Nm³

9) Abrasion resistant material: Lining of fused tungsten carbide for distributers and lining of fused ceramics for the bed; lining of fused ceramic cast bricks for inside of dryer and cyclone. These were selected based on their successful use in hammer mills and pneumatic conveyors for CDQ coke under the same conditions by the authors.

6. RESULTS

The basic engineering design calculations were successful and they were stored in a desktop computer (HP 9845B).[10] The detailed hardware design was slightly modified later to improve wet slug handling. At the first stage, the authors adopted a mechanical agitator for the first section but problems were found with coating, choking and blocking by alkilation in the hot air. The problem was resolved by first removing the agitator, followed by the construction of a jacket and air curtain for the shute and wall in the dryer. The hot air and temperature distributions were measured using a hot-wire anemometer and temporary thermometers, and a fixed restricter for the gas distribution was installed. Figure 4 and photograph 4 show the dryer. Photograph 5 in the last page shows a typical fluidization by the distributers.

Photograph 4 The dryer

The outlet solid temperature was controlled by a single loop digital controller with sequencer. The measured variable is the outlet gas temperature; the actuating variable is the speed of rotation of the wet slug table feeder. The sequence conditions are; hot gas temperature and opening angle of the induced draft fan damper.

7. FUTURE WORKS

For multi-stage fluidized bed dryers, the authors recommend the double seal method shown in figure 2 for small dryers, and the cascading mono-stage type using serial gas counter flow shown in figure 5 for big dryers. For the spouted bed type bed coal gasifier as shown in figure 6, the authors recommend using distributers and gravity flow.[9]

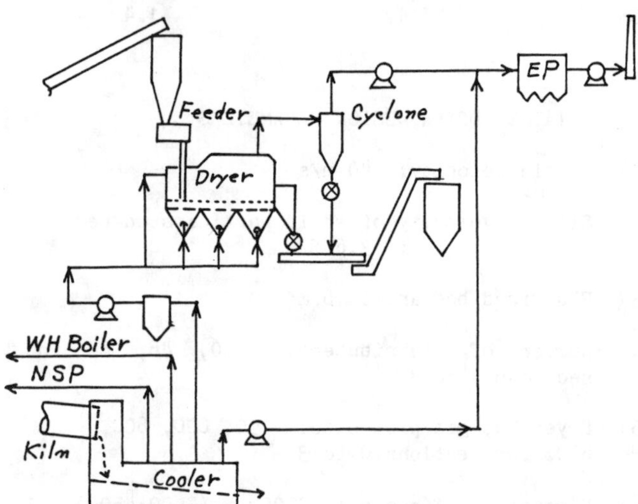

Fig. 4 Flow sheet of the amorphous slag dryer

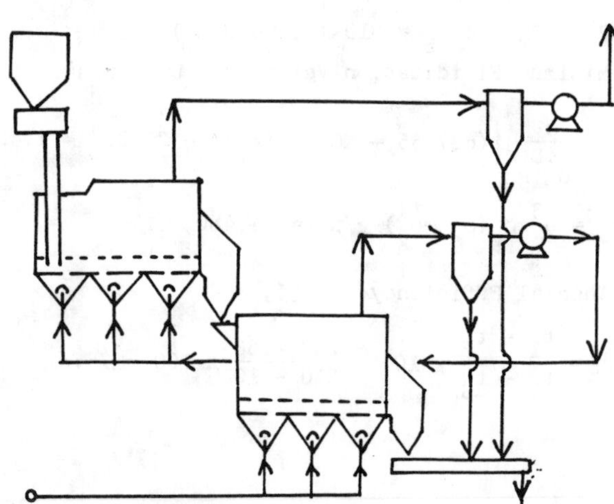

Fig. 5 Flow sheet of the two stage continuous fluidized bed dryer

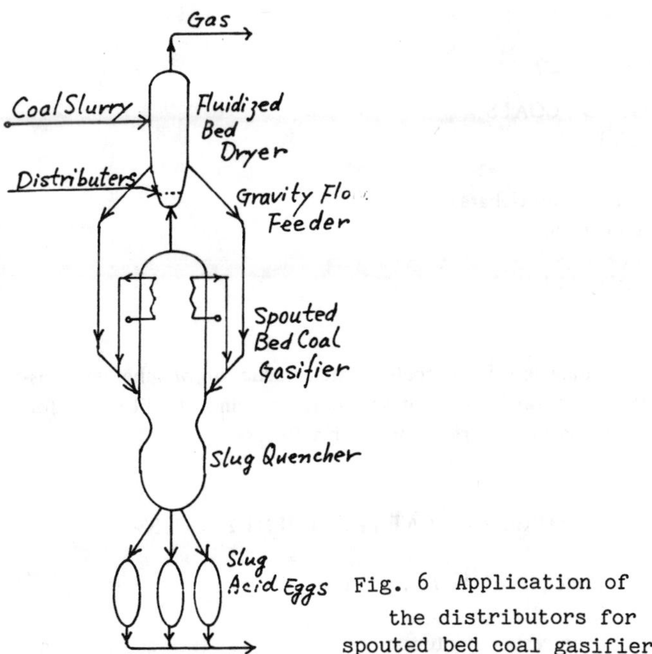

Fig. 6 Application of the distributors for spouted bed coal gasifier

8. CONCLUSIONS

8.1 The distributers were successful in improving the heat efficiency of the dryer.

8.2 It was found possible to successfully protect the distributers against abrasion by the dust-laden hot gas.

8.3 The distributers, bed and wall lining proved successful for dust and solid fluidization.

8.4 It is assumed that the solids in the fluidized bed caught the dust.

8.5 The design calculations for the dryer were proven in practice.

8.6 The authors' method for the hot air drying of amorphous slag was put into practice for the first time in Japan. They were able to successfully improve the dryer using a jacket air curtain to eliminate coating, choking and blocking problems in the hot air due to alkilation.

ACKNOWLEDGEMENT

The authors wish to express their grateful thanks to Professor Toei for his indispensable textbook[5], and to the staff of Denki Kagaku Kogyo's Aomi Plant for making it possible to realise this dryer, and also for the many stimulating and helpful discussions that took place on improvements to the dryer.

The authors also wish to thank Professor Okazaki for his patient forebearance in the preparation of this paper.

REFERENCES

1. Moriyama, T. et al: May 1982, "A New Pulverized Coal Injection System for Blast Furnaces KDP-I", Pneumatech-1/Stratford upon Avon, pp 1-22, (in English)

2. Moriyama, T.: April 1983, "A New Pulverized Coal Injection System with Applications in Several Industries", MICONEX '83, (in English)

3. Moriyama, T.: Aug. 1983, "Mass Flow Measurement and Control of Solid Gas Two-phase Flow Injection for Energy Conservation"; ROC-Japan Joint Symposium of Powder Technology (in English)

4. Moriyama, T.: Dec. 1983, "Development of a Control and Measuring Method for Mass Flow Rates in a Solid Gas Two-phase Distributing Injection System", Multiphase Flow Symposium, Science Council of Japan (in Japanese)

5. Toei, R.: Oct 1970, Dryer, Nikkan Kogyo Press

6. The Society of Chemical Engineers Japan: Oct. 1978, Chemical Engineer's Handbook, Maruzen

7. Perry, R.H. and Chilton C.H.: 1973 Chemical Engineer's Handbook, McGraw Hill

8. Herron, D. et al: Jan. 1981, "How to Select Polymer Drying Equipment", CEP

9. Moriyama, T.: April 1984, "Measuring and Control for Hot Solid Gas Two-phase Gravity Flow", JSME Symposium (in Japanese)

10. Moriyama, T.: March 1983, "DENKA's Power Handling Technology", TS-CAD Powder & Industry (in Japanese)

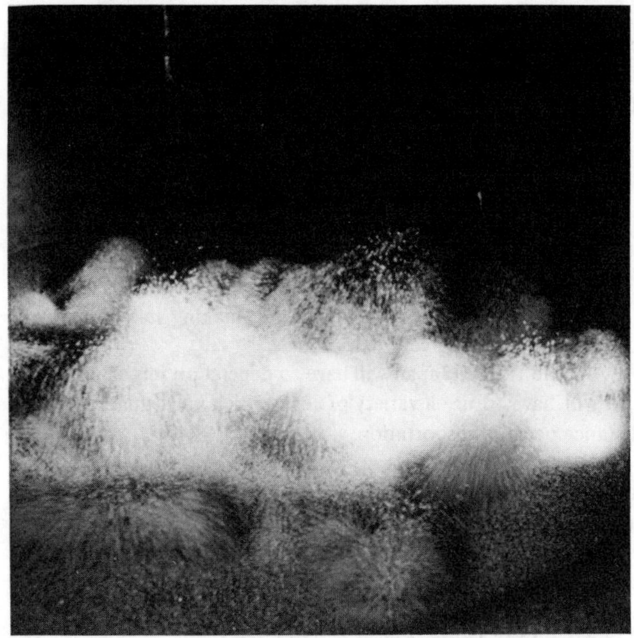

Photograph 5 Fluidization on the distributers

UPGRADING OF LOW-GRADE COALS
BY HEAT TREATMENT

Kiyomichi Taoda, Yoshifumi Ito, Seibi Uehara
Fumiaki Sato and Takeo Kumagaya

Mitsubishi Heavy Industries, Ltd.

ABSTRACT

Low-grade coals, such as subbituminous coal, are generally high in inherent moisture and even if they have been dehumidified they absorb moisture again and dried coals have property of high spontaneous ignition. They, therefore, are not suitable for long distance transportation. Because of this, they have been mostly used for power generation in their native districts. To cope with the recent energy situation there is a growing demand to use these coals as new energy resources by improving their quality.

In order to improve their quality, we have been developing a new drying process.

Test results have shown that in contract to raw coal, the coal dried by the process at temperatures of 300 – 350°C has a decreased inherent moisture, an increased calorific value and lower spontaneous combustibility which contribute to marked improvement in its capacity for long distance transport and storage.

1. INTRODUCTION

Low-grade coals (low-rank coals) such as subbituminous coal and lignite exist in large volumes at numerous locations throughout the world and, even though they can be mined open air easily, they are used mainly as fuel for power generation at mining sites because of the following reasons.

(1) The calorific value of these coals, on moisture-free basis, reaches 5000–7000 kcal/kg but the raw coals contain moisture of 15% or more, it is uneconomical to transport them as a fuel.
(2) Dehumidified low-grade coal is highly reactive and very liable to spontaneous combustion, making its transportation and storage very dangerous.

In the future, Japan will have to depend on imported coal, and will have to use a variety of coals which must undergo long distance ocean transportation.

As a consequence, greater attention will be paid to the development of technology which will upgrade coals through some processes such as advanced drying or liquefaction processes. Among these processes, there should be a simple, practical method of improving the utility of coal through appropriate, large-volume drying.

Mitsubishi Heavy Industries, Ltd. has been exerting efforts to develop a new drying process (heat treatment process) for upgrading low-grade coals through reduction of inherent moisture, and suppression of spontaneous combustibility. The following is a summary report of this process.

2. FEATURES OF HEAT TREATMENT PROCESS

2.1 Principles of the new process

The carbon content of low-grade coals such as subbituminous coal is about 70–80% C, and its inherent moisture reaches 10–25%. If this coal is dried by a conventional drying method at around 100°C, nearly all of its moisture can be removed, but by exposure to high humidity it reabsorbs moisture to the previous inherent moisture level.

Reducing the inner pores of coal is effective in resisting against moisture reabsorption and in lowering its spontaneous combustibility.

Generally, when coal is heated the following changes occur.

About 200°C or less .. Dehydration and removal of free gases
About 200°C–450°C .. Formation of tar and decomposition of carboxylic groups (Decarboxilation)
About 450°C or above .. Formation of char

When low grade coals were heated, they do not have softening and fusing tendencies as observed on bituminous coal. It is, however, believed that when they are heated rapidly, part of them (such as the resin component) exists in fused phase. And the temperature at which volatile matter is lost shifts to higher level.

The basic principle behind our new process is to rapidly heat coal at temperatures below that at which char is formed, and after the formed tar filled up the fine pores in the coal and coated the coal surface, the coal is quickly cooled down to be fixed in that condition.

Fig. 1 illustrates a model of the coal surface after the coal has undergone the heat treatment.

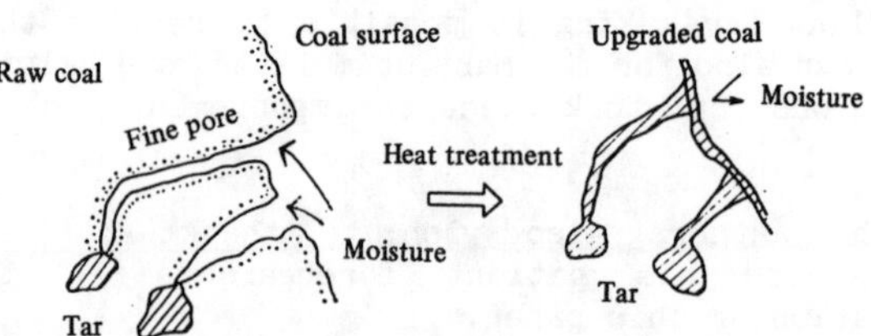

Fig. 1 Model of coal surface

Utilizing these changes in coal properties, we have devised an upgrading process for low-grade coals as shown in Fig. 2.

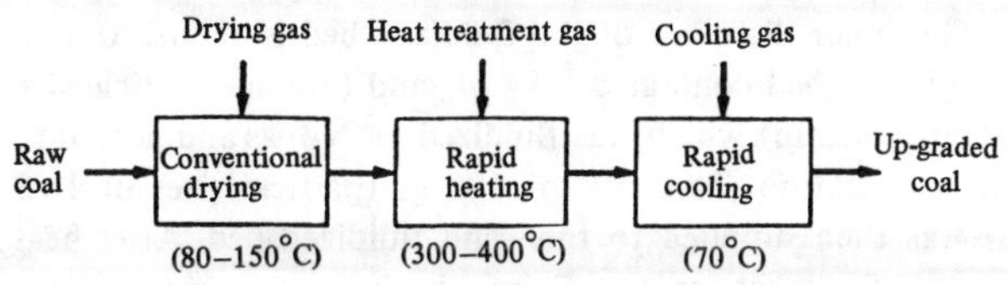

Fig. 2 Simplified diagram of the new drying process

2.2 Comparison with conventional drying process

In the fluidized bed, as the heat transfer rate between gas and particles is great, it is suitable for rapid heating or cooling of particles at a even temperature. Moreover, fluidized bed dryer is suited for coal treatment in a large volume.

A system incorporating a fluidized bed was thus employed in this process. Table 1 presents a comparison of the main differences between conventional drying and our new drying process.

Table 1 Comparison of coal drying processes

	New Process	Conventional process
Process	Conventional drying plus heat treatment at 300–400°C	Surface moisture removal through drying at a gas temperature of 80–150°C
Special features expected of dried coal	(1) Significantly lower inherent moisture (low moisture reabsorption and high calorific value than in raw coal.) (2) Suppressed spontaneous combustibility	(1) Similar inherent moisture to that in raw coal. (Moisture is reabsorbed) (2) Higher spontaneous combustibility than of raw coal

3. HEAT TREATMENT TESTS

At the first stage, fundamental tests were carried out to observe the effectiveness of the new drying process and to determine heat treatment conditions. Based on the results of the fundamental tests, mini-pilot plant tests were run to examine the practicality of the process.

3.1 Properties of tested coals

The coals used in the tests had high inherent moisture contents (10–20%) and they were selected from among the coals which can be imported to Japan from Northwest America and Australia. Table 2 gives an outline of their properties.

3.2 Methods of coal analysis

The methods of analysis used in the tests are summarized in Table 3.

Table 2 Properties of tested coals

Coal	Coal A	Coal B	Coal C	Coal D
Source.Coal rank	Northwest America. Subbituminous			Australia. Subbituminous
Inherent moisture (wt%)	12.9	15.0	14.8	16.0
Ash (wt%)	12.1	5.6	5.3	11.1
Volatile matter (wt%)	32.9	38.3	35.9	24.9
Fixed carbon (wt%)	42.1	41.1	44.0	48.0
Calorific value (kcal/kg)	5440	5680	5900	5000
Ultimate analysis				
C (wt% d.a.f)	77.9	74.2	76.4	77.7
H (wt% d.a.f)	5.9	6.6	6.6	4.1
N (wt% d.a.f)	1.6	1.0	1.2	1.6
Total S (wt% d.a.f)	0.5	0.4	0.3	0.5
O (wt% d.a.f)	14.0	17.8	15.5	16.1
Hardgrove index	32	–	–	–

Table 3 Methods of analysis

Item	Method or test facility
Proximate analysis	JIS M 8812
Ultimate analysis	JIS M 8813
Inherent moisture	JIS M 8811
Calorific value	JIS M 8814
Specific surface area	N_2-BET method
Spontaneous ignition time	Spontaneous ignition test apparatus
Pulverization level	Percentage of produced particles below the minimum particle size of test coal

3.3 Fundamental tests

Basic tests by heating tube. A heating tube type apparatus was used to study the features of coal heat treatment.

Several grams of 2~4mm grains of coal were placed in the qualtz tube. The coal was heated by a radiant heater up to a set test temperature feeding N_2 gas in the tube. After the heat treatment, the coal was cooled and analized.

The heating rates were from 10°C/min to 150°C/min and the cooling rate was approximately 60°C/min.

The test results are shown in Table 4 and Fig. 3.

Table 4 The test results by heating tube

	Test coal	Coal A	Coal D
Treatment conditions	Temperature (°C)	350–400	←
	Heating rate (°C)	150	←
Properties	Inherent moisture (wt%)	5.2/(11.0)	6.9/(16.0)
	Calorific value (kcal/kg)	6080/(5700)	5430/(5000)

Note: Figures in parenthesis are for raw coal.

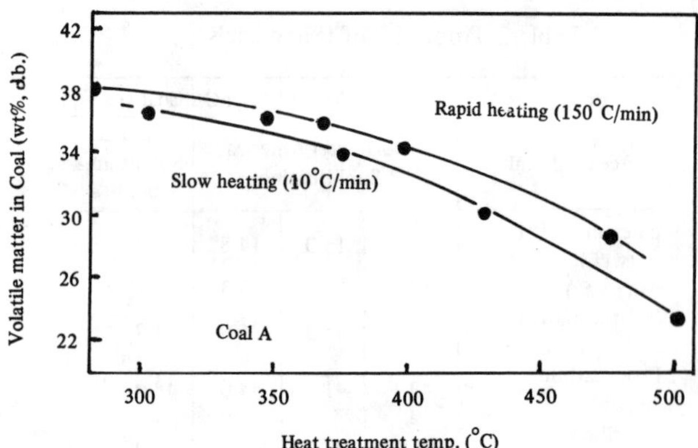

Fig. 3 Correlation between volatile mater content and treating temperature

From the Table 4, it is clear that at a heating rate of 150°C/ min and a heating temperature of 350–400°C, the inherent moisture content in coals was significantly reduced and the calorific value increased by several hundred kcal/kg, confirming the effectiveness of the heat treatment process.

And in the rapid heating, a loss of the volatile matter is less than that in the slow heating. It suggests that in case of rapid heating, temperatures range of volatalization shifts towards high temperature level.

When the heating temperature exceeded approximately 400°C, the tendency of pyrolysis was clearly recognized. This tendency was confirmed by TGA and DTA shown in Fig. 4.

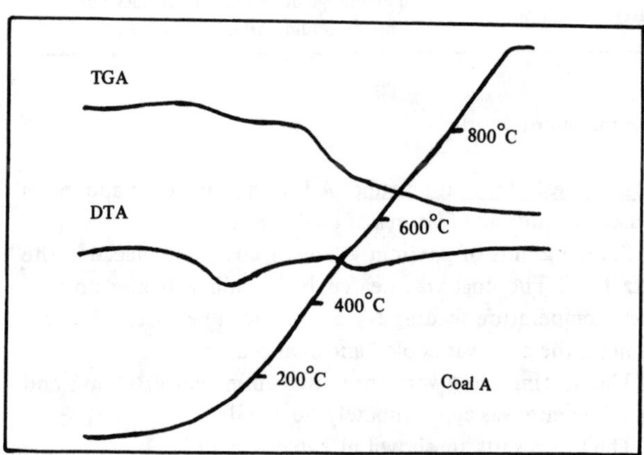

Fig. 4 TGA and DTA result

Small fluidized bed tests. This test was conducted to study the applicability of the fluidized bed and determine the treatment conditions in mini-pilot plant tests.

Fig. 5 is a schematic diagram of the test apparatus.

The inner diameter of the fluidized bed is about 70 mm. The fluidized bed contained 1 kg of sand (average particle size of about 0.7 mm) which was fluidized by N_2 gas and heated to a set temperature. Test coal of 100 gr (particle sizes of 1–3 mm) was then supplied to this sand fluidized bed. After heat treatment for a set length of time, both the test coal and the sand were transferred to the cooling fluidized bed and cooled by N_2 gas. The heat treated test coal was separated from the sand through a 1 mm sieve, and the effectiveness of the heat treatment was determined. In the tests, the coal heating temperature and the heating time were used as parameters.

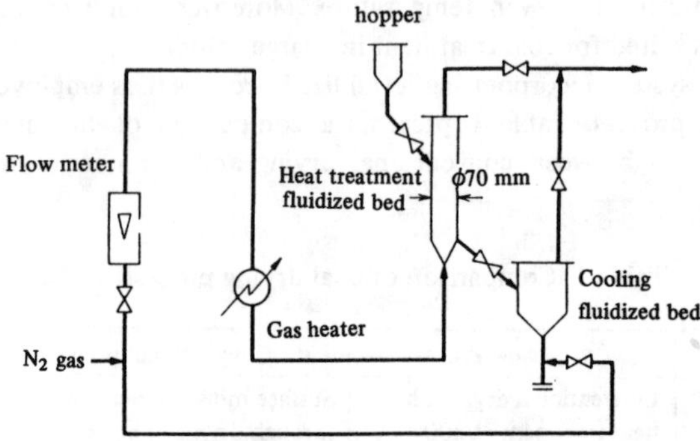

Fig. 5 Small fluidized bed test apparatus

Typical test results are shown in Table 5. Fig. 6 shows the effect of heat treatment temperature on the upgrading of coal A, B and C. Fig. 7 shows the effect of heat treatment on spontaneous combustibility.

By this fluidized bed test, the same effects of the heat treatment observed in the heating tube tests were confirmed.

Optimum conditions of the heat treatment vary little with the type of coal. The heat treatment temperature should be 300–380°C and the heating time should be 1–3 min. Upgrading effects of coal include a reduction of 6–8% inherent moisture content, and improvement in the calorific value by 300–500 kcal/kg. And Spontaneous ignition time of treated coal was delayed 2–3 times compared with raw coal.

Table 5 Example of small fluidized bed test results

	Test coal	Coal A	Coal B	Coal C	Coal D
Treatment conditions	Particle size (mm)	1–3	2–3	2–3	1–3
	Temperature (°C)	330–410	310–390	300–380	350–380
	Treatment time (min)	1–2	2–3	2–3	1–2
Properties	Inherent moisture (%)	5.5(11.2)	8.0/(15.0)	7.5/(14.8)	8.0/(16.0)
	Calorific value (kcal/kg)	6200(5880)	6200/(5680)	650.0/(5900)	5500/(5000)
	Pulverization level (wt%)	5–9	25–40	10–35	15–17
	Spontaneous combustion time (h)	215/(76)	510/(195)	460/(170)	–

Note: Figures in parenthesis are for raw coal.

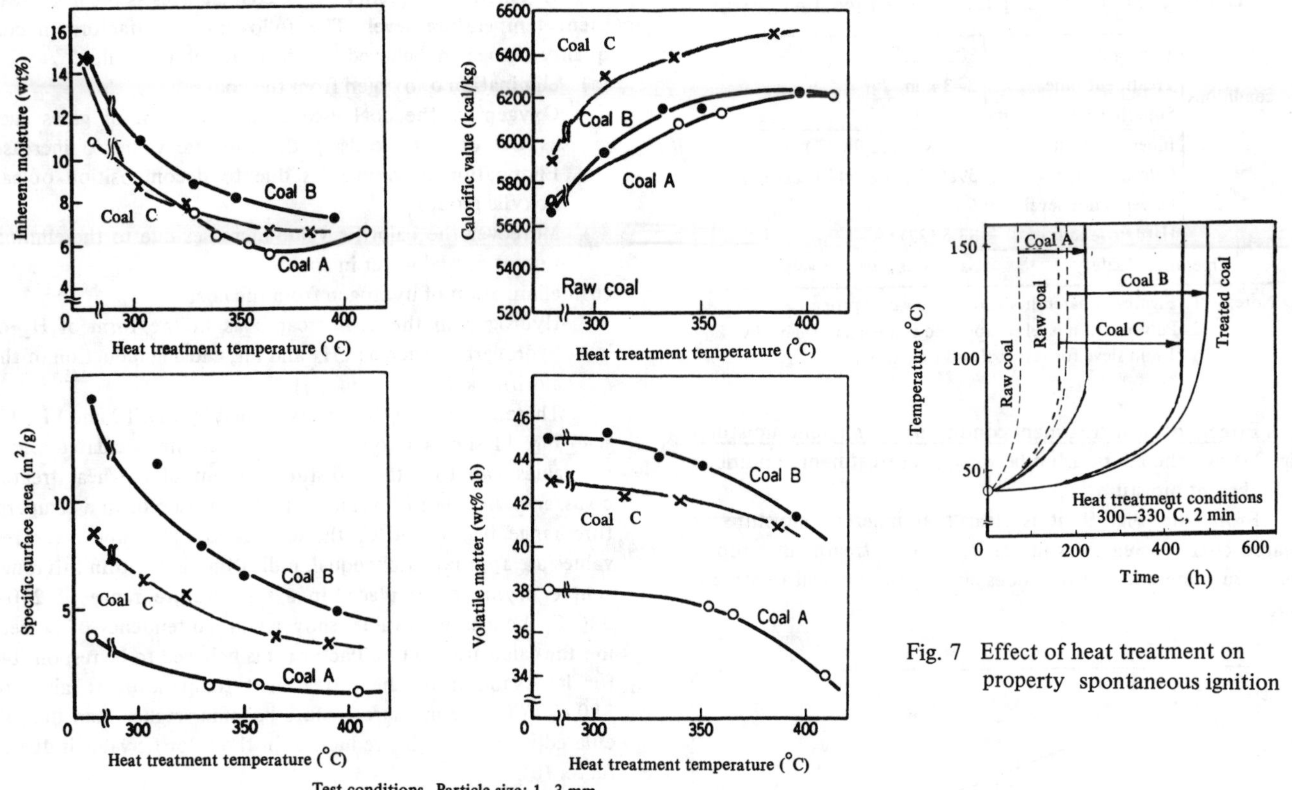

Fig. 7 Effect of heat treatment on property spontaneous ignition

Fig. 6 Influence of heat treating temperature

However, for coal B and C, which had a high inherent moisture content, cracking and pulverization of coal were observed.

Based on the results of the fluidized bed tests, mini-pilot plant tests for coal A were conducted.

3.4 Mini-pilot plant test

Mini-pilot plant tests were performed using coal A to determine heat treatment conditions, confirm the quality of heat treated coal, and check the operatabilities, safety and other features of the heat treatment plant.

Fig. 8 is a schematic diagram of the mini-pilot test plant. The plant consists of two fluidized beds, one for heat treatment and the other for cooling. The former has a bed area of 400 x 250 mm and height of 5000 mm and the latter has a bed area of 400 x 300 mm and a height of 4800 mm. Test coal was fed constantly to the heat treatment fluidized bed. After undergoing drying and heat treatment, the test coal was cooled in the cooling fluidized bed, then discharged outside the plant system. The plant is designed for a treatment capacity of 1 t/h of raw coal.

Kerosene combustion gas was used as the heating gas. The gas flowed out the heat treatment fluidized bed was cooled by a water scrubber after the coal dust was separated by a cyclone, and then fed into the cooling fluidized bed. The greater portion of flue gas from water scrubber returned to the hot gas generator, where it is mixed with high temperature combustion gas and was recycled as heating gas. The product coals were analysed on inherent moisture, calorific value, spontaneous combustibility, etc. Problems involved in the heat treatment process such as coal handling properties, safety measures against coal ignition and dust explosion, etc. were discussed and examination was made on the design features of facilities.

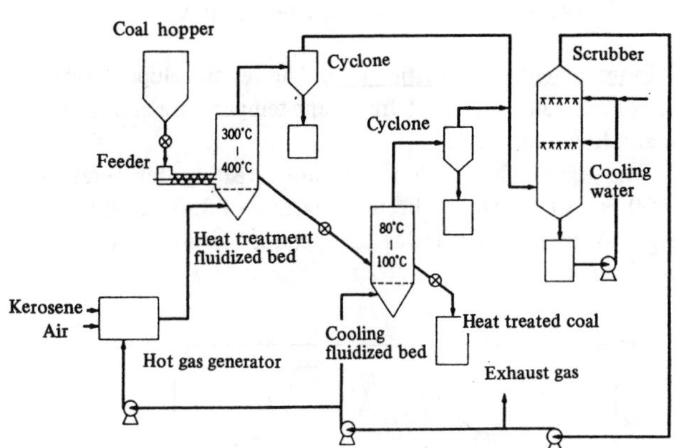

Fig. 8 Schematic diagram of mini-pilot test plant

Test coal was an air-dried coal in size of 10 mm or less. A summary of the major properties of the raw coal and of the treated coal as well as the heat treatment conditions is given in Table 6.

Table 6 Results of mini-pilot plant test (coal A)

Treatment conditions	Temperature	300–350°C
	Treatment time	2–3 min
	Superficial gas velocity	3–5 m/s
Properties	Inherent moisture	6–7 wt% (12.9 wt%)
	Calorific value	5950 kcal/kg (5440 kcal/kg)
	Pulverization level	6 wt% (3 wt%)
	Hardgrove index	33.5 (32)
Operational safety		Safe at below O_2 5 vol%

Note: 1. Figures in parenthesis are for raw coal properties.
2. Pulverization level is expressed as a percentage below 1 mm sieve for test coal and final product.

Effects of heat treatment conditions on inherent moisture. Fig. 9 shows the relationship between heat treatment conditions and inherent moisture.

From the Fig. 9, it is clear that inherent moisture of treated coals decreases as increasing of heat treatment temperature. But inherent moisture does not depend on heat treatment time.

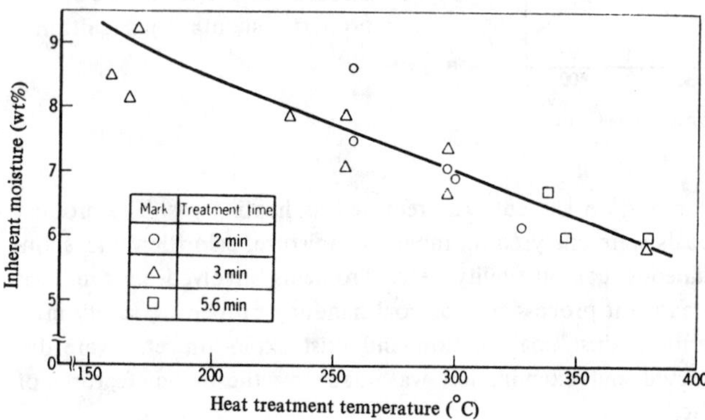

Fig. 9 Correlation between inherent moisture and treating temperature (mini-pilot plant test, coal A)

Improvement of calorific value. The relationships between the calorific, value and heat treatment temperature and heating time are shown in Fig. 10.

This figure shows that optimum treating temperatures are 300 ~ 350°C. Above the temperatures of 350°C, especially when heat treating time was long, reduction of the calorific value was observed.

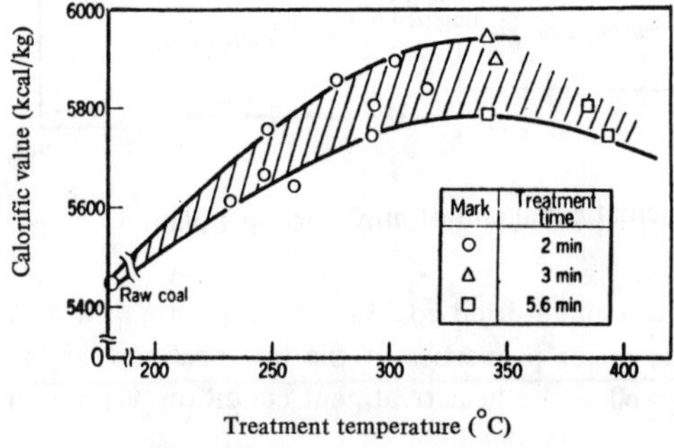

Fig. 10 Correlation between calorific value of coal and treating temperature (mini-pilot plant test, coal A)

In general, the quality of the coal changes as the heat treatment temperature level. The following two factors in coal quality change are believed to affect the calorific value.

(a) elimination of oxygen from the coal

Oxygen in the coal escapes in the form of gases such as CO or CO_2, causing the calorific value to increase. Elimination of oxygen is due to decomposition of carboxylic groups.

Moreover the calorific value increases due to the elimination of crystal water in coal.

(b) elimination of hydrogen from the coal

Hydrogen in the coal escapes as in the form of H_2 or hydrocarbon such as CH_4 and tar, causing reduction of the alorific value.

These factors (a), (b) are explained by Fig. 10 and Fig. 11.

Fig. 11 shows a comparison between the calculated calorific values based on the moisture content of the heat treated coals and the calorific values actually measured. In a temperature range below 260°C, the calculated values and measured values are approximately equal, indicating that within this range simple drying takes place. In a temperature range of 290–340°C, the measured values show a marked tendency of exceeding the calculated values. Factor (a) is believed to be responsible for it. Again, if the heat treatment temperature is raised to 380–390°C, the measured values become smaller than the calculated values and this reduction in the calorific value is due to factor (b).

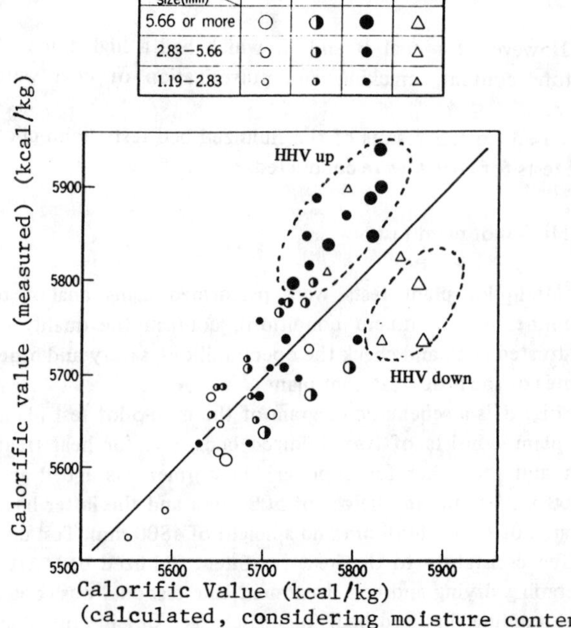

Fig. 11 Gross calorific value of heat treated coal (coal A)

Moisture reabsorption of treated coals. Fig. 12 shows the change in moisture content of the heat treated test coal when it was placed in a constant humidity chamber. Data on conventionally dried coal A is also included in the figure. The heat treated coal reached an inherent moisture of about 6% after one week and did not change in the moisture content after 20 days. Conventionally dried coal reached a moisture content of 9–

10%, close to the inherent moisture value, within about 2 days.

When treated coal was submerged in water for one day and then placed in a constant humidity chamber, assuming there was rainfall, a moisture content level reached to 7–8%, whereas conventionally dried coal under the same conditions reached a 12% moisture content. These data indicate that the effect of heat treatment on coal A is great, that as heat treatment conditions the temperature should be about 300–350°C and the heating time period should be 2–3 min. Moreover, for coal A, pulverization or cracking of coal by the heat treatment was little observed.

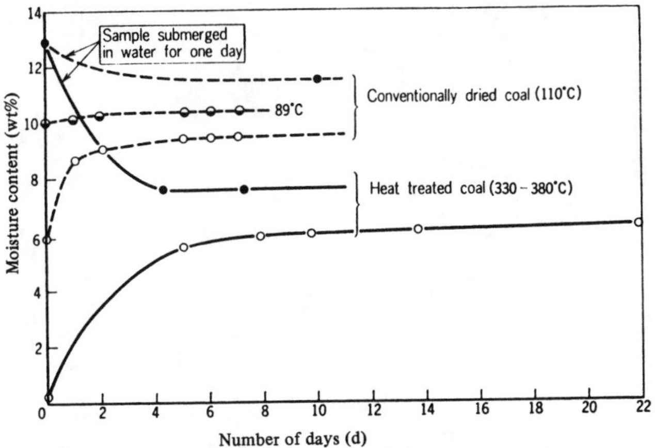

Fig. 12 Comparison of stability between heat treated coal and conventionally dried coal (coal A)

In planning the mini-pilot plant test, preliminary examinations, including elemental tests, were carefully executed to ensure safety of the test, especially prevention of coal dust explosion. In the operations of the mini-pilot plant, O_2 content of the process gas and recycled gas was kept below 5%. As a result, the mini-pilot plant tests proceeded without encountering any operational problem.

4. CONCEPT OF COMMERCIAL PLANT

Fig. 13 is a schematic flow diagram of the new coal drying process. The main components are four fluidized beds; one for drying, one for heat treatment and two for cooling.

Coal is fed continuously into the drying fluidized bed. After the surface moisture and a part of the inherent moisture of the coal is eliminated by drying gas from the heat treatment fluidized bed, the coal is transferred to the heat treatment fluidized bed. In this bed the coal is heated to 300–350°C by high-temperature gas(about 600°C) from the hot gas generator. At the same time that the inherent moisture is removed, upgrading occurs through a light degree of pyrolysis. Next, treated coal undergoes preliminary cooling by exhaust gas (about 150°C) from the drying bed, then it is further cooled by cold air to less than 70°C, and made ready for shipment. Waste gas from this process is treated by either a wet or dry type exhaust gas treatment unit. Fine particles of coal in the waste gas are collected and used as fuel for the hot gas generator. For this application, an electrostatic precipitator or bag filter should be used rather than a wet scrubber.

This process which handles coal at high temperatures must incorporate measures to prevent ignition of the coal and explosion of coal dust. Exhaust gas from the process is recycled through the hot gas generator after fine coal dust has been removed. The O_2 content of gas within the process should be controlled at less than about 5 vol%.

Conventional coal drying mainly aims at removing surface moisture from coal to increase the combustion efficiency and to reduce transport costs. The new heat treatment process not only accomplishes these purposes but also heightens the value of coal through, for example, augmenting its calorific values and suppressing its spontaneous combustibility.

We evaluate that this process is commercially feasible, and has economical merits in comparison with conventional drying process.

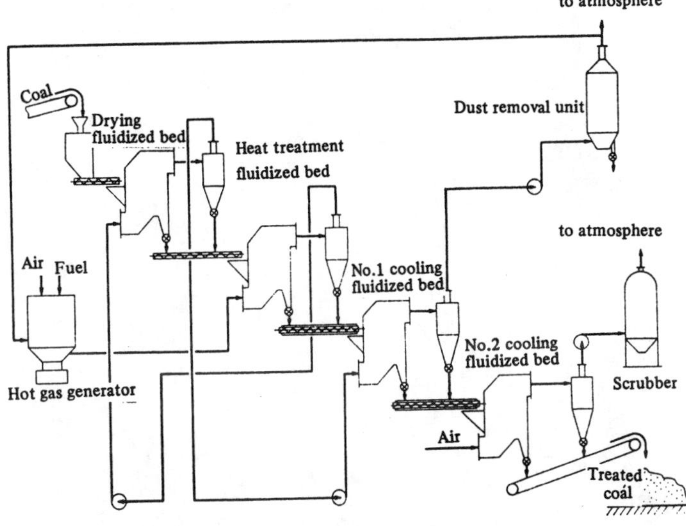

Fig. 13 Schematic flow diagram of new coal drying process

5. SUMMARY

Mitsubishi Heavy Industries, Ltd. has been very much concerned about the problems involved in utilizing low-grade coals, such as subbituminous coal, and as one method for improving their usability, we have devised an upgrading procedure whereby low-grade coal is rapidly heat treated at a high temperature (300–350°C).

The results have shown that in contrast to raw coal, the heat treated coal has a significantly low inherent moisture, high calorific value, and low spontaneous combustibility. It will be due to the reduction of the specific surface area, the dehydration and the decarboxilation of the coal.

There are some advantages of transportation savings and prevention of spontaneous combustion.

Mini pilot plant tests provided various data necessary for designing commercial heat treatment facilities. However, further examination should be done to confirm the effect of heat treatment on coals of larger diameters and its applicability to other types of coals than those tested.

In the development of the process, we intend to determine its applicability to various types of coal, especially lignite, and to make careful advances towards commercialization of the process through testing with a large-scale test plant.

This research was carried out with the cooperation of Mitsubishi Corporation, Ltd., and we wish to extend our appreciation to its personnel concerned.

DEWATERING AND UTILIZATION OF HIGH MOISTURE BROWN COAL

Takao Kamei, Fuminobu Ono, Keiichi Komai,
Takeshi Wakabayashi and Hayami Itoh

Kawasaki Heavy Industries, Ltd.
Kobe, 650-91 JAPAN

ABSTRACT

This paper introduces our research and development work for utilizing brown coal effectively by DK process which dewaters brown coal without evaporation.

1. INTRODUCTION

1.1. Object of Researches

Brown coal, in spite of its enormous reserves, is not enough utilized except in the vicinity of its mine, because it contains high moisture and is liable to ignite spontaneously. KAWASAKI HEAVY INDUSTRIES, LTD. has continued to develop technics to utilize brown coal as a new energy source for Japan. This technics comprises brown coal dewatering at mine-site which we introduce here, as well as its transportation from there and its final utilization in Japan.

1.2. Importance of Brown Coal

Brown coal occupies about a quarter of global coal reserves, 2.4 TT(tera-tonne), and a considerable percentage of it is buried thick and shallow and can be mined easily at low cost. Moreover, most of brown coal contains few such impurities as ash and sulfur.

In La Trobe Valley, Victoria State, Australia 300m thick coal layers extend 80km east and west and 40km north and south under only several ten meter overburden[1]. The mining cost is said to be A$2 - 3 per ton of raw coal, and its ash and sulfur contents are only 1% and 0.3% respectively.

1.3. Importance of Dewatering

Victorian brown coal has 26.2 MJ/kg dry basis heating value but contains water as high as 200%. For example, a 1 000 MW power station in Japan requires about 2.1 MT(mega-tonne) dry coal a year. But this coal is accompanied with 4.2 MT valueless water. Moreover, when this water evaporates in a boiler, it takes away latent heat equivalent to the heating value of 1.5 MT of raw coal. Consequently, if not dewatered, 7.8 MT raw coal must be carried yearly from mine to Japan.

Because of this high water content, Victorian brown coal is mined only 30 MT a year and 0.6 GT summed up in this century, but its reserves exceed 100 GT(giga-tonne)[1], and so an effective dewatering is important for expanding its utilization.

1.4. Characteristic Features of Brown Coal

Brown coal is young coal and little metamorphosed from its plant debris through the geochemical reaction, but its definition varies with countries[2-4]. We define it in a practical sense as all kinds of coal requiring dewatering at its mine-site, and it nearly corresponds to Japanese brown coal and lignite, American subbituminous and lignitic coal, and International Classes 7, 8 and 9 (Refer to Fig. 1).

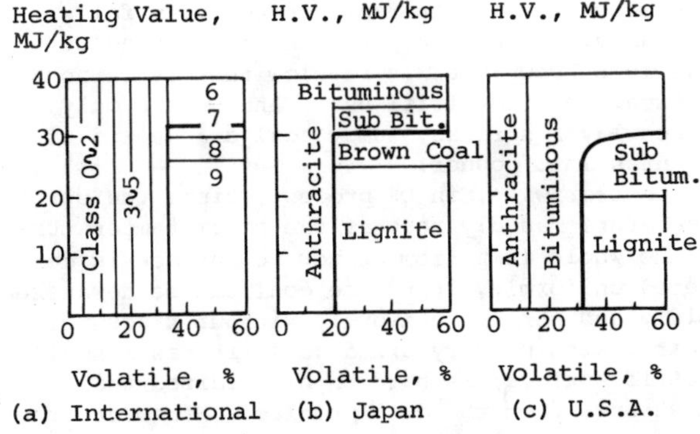

Fig. 1 Coal Classification

These kinds of coal (1) contain much volatile matter, ignite easily and become dangerous if dewatered, (2) contain much oxygen and, consequently, are low in heating value on dry-ash-free basis, (3) have loose structure both chemically and physically with large capillary volume filled with water, and, (4) contain much hydrophilic functional group especially carboxyl (-COOH) group.

1.5. Dewatering Technics

Dewatering technics we are developing applies the following phenomenon, i.e., if the temperature of brown coal is elevated in the ambience where water can not evaporate, water is removed from coal in liquid form by the phisical and chemical changes of coal.

We call our technics as DK process. Fundamentally, in 1927, Dr. Fleissner developed the so-called Fleissner Process, using saturated steam as heating medium[5]. The process of using hot water as heating medium which was studied at Melbourne University[6] also belongs to this kind.

We select this system because (1) its heat consumption is as small as half of that of evaporative drying, (2) it can process lumpy coal of 100 - 200mm top size, (3) there is no danger of igniting during dewatering, since coal is heated in steam, (4) not only water is removed, but also coal quality itself is improved.

To contact hot gas directly with brown coal is dangerous because of the high volatile content in coal. Danger is large especially in the bag filter for separating coal from gas. Therefore this type dryer is scarcely applied to brown coal except for feeding the boiler with dried coal and gas without separation.

The tube dryer which heats brown coal by steam through heat transferring wall is often used as pre-dryer for briquetting[7]. It is not so degerous because heating is mild and dust collection is not difficult. But it has poor thermal efficiency and requires 1.2 - 1.5 times as much steam as removed moisture. In DK process steam consumption is about 0.6kg/kg of removed water.

As is well known, in usual thermal dryers, moisture evaporates only from the surface of the particle, and the surface temperature does not rise above wet-bulb temperature, and soon low moisture layer is formed on the surface. When brown coal loses moisture, its thermal conductivity falls remarkably. Accordingly the thermal drying of a large lump brown coal is so slow that it is not practical for industrial use. And, since only surface layer shrinks, lumpy coal cracks and after all turns into powder.

Contrariwise, in DK process, since the surface temperature rapidly reaches the steam temperature, and the whole lump from center to surface is dewatered uniformly, lump size coal can be dewatered rapidly and shrinks uniformly without fracture. Compared with powdery dried coal, it has a small specific surface, is hard to be oxidized and is almost free from the dust problem. Moreover, the coal quality is improved in many points.

Concerning these matters we introduce hereunder a part of the results we have actually confirmed through our experiments.

2. TEST PROCEDURE

2.1. Schedule

Table 1 shows the schedule for the major test items. In the preliminary research we studied the system of handling fine powder brown coal dried evaporatively by a mill. Because of the danger of the system we paid our attention to the merit of the non-evaporative process. We have already completed the laboratory test and the pilot test, and at present we are preparing for the demonstration plant.

Besides the items mentioned in the table we are conducting many tests such as the basic laboratory test to improve the process, the laboratory test to investigate the dewatering, characteristics of new kinds coal, the pilot test to produce dewatered coal sample for the handling test, and the test to treat waste water from the process.

Table 1 Schedule

	'75	'76	'77	'78	'79	'80	'81	'82	'83	'84
Preliminary Research		═								
Loboratory Test			═							
Pilot Plant (Takehara)										
Engineering and Construction				═						
Yallourn Coal Test				═	═					
Beluga Coal Test					═	═				
Pilot Plant (Wakamatsu)										
Engineering and Construction								═		
Yallourn Coal Test									═	
Morwell Coal Test									═	

2.2. Tested Coal

The typical properties of the main tested coal are shown in Table 2. The largest number of tests is conducted with Victorian brown coal, especially with Yallourn coal. As to subbituminous coal of Alaska, samples from Capps Field and Chuitna Field (both in the Beluga area) were tested. Regarding the coal of the Western States, U.S.A., subbituminous from Powder River Basin, Wyoming was tested.

Table 2 Tested Coal

	Yallourn Aus.	Beluga U.S.A.	P.R.B. U.S.A.
Water	190∿240%	25∿40%	35∿55%
Ash	1.1%	18.7%	6.8%
Volatile	51.2%	43.5%	44.5%
Fixed Carbon	47.7%	37.8%	48.7%
C	67.1%	71.3%	74.5%
H	4.8%	5.3%	5.3%
O	27.2%	22.1%	18.6%
N	0.6%	1.0%	1.1%
S	0.3%	0.3%	0.5%
H.H.V. (Dry)	26.2MJ/kg	23.2MJ/kg	28.5MJ/kg

2.3. Laboratory Test Equipment

The laboratory test equipment was set up in 1977 at KHI's Akashi Technical Institute. It consists of an autoclave and an electric boiler as shown in Fig. 2. Brown coal is filled in a basket and then sealed up in the autoclave. Usually a basket of 3ℓ or 8ℓ is used. Saturated steam or hot water of the highest temperature 573 K (saturated steam pressure is 5.7 MPa) can be generated in the boiler and supplied to the autoclave. Steam temperature and pressure, the temperature in coal lump and the weight of the basket can be measured continuously. Hot water generated in the autoclave (the moisture removed from coal and the condensate of steam) can be discharged from the bottom and collected with the water-cooled condenser. Water can be stored in the autoclave and heated by an electric heater, if required.

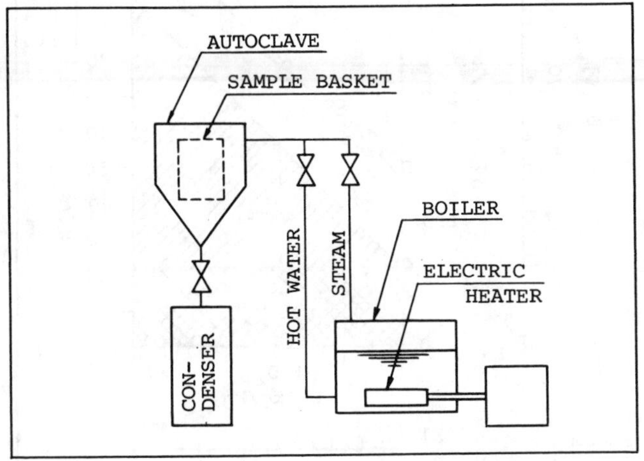

Fig. 2 Laboratory test equipment

2.4. Pilot Plant

This plant was set up by KHI in 1978 in Takehara Power Station of Electric Power Development Co., Ltd. Later, it was disorganized and re-constructed in Wakamatsu Power Station of EPDC.

As shown in Fig. 3, the plant consists of 4 autoclaves. The autoclave is 0.6m in inside diameter and 2m in height , dimensionally 1/5 of the practical plant (1/125 in capacity). Brown coal is processed batchwise in each autoclave by the steam supplied from the power station. Since raw coal is supplied to the 4 autoclaves through the rotary chute, the plant as a whole processes brown coal semicontinuously. If it operates continuously for 24 hours, it can process 10 T of raw coal per day.

Hot water generated in the autoclave is stored in the condensate tank attached to each autoclave. The autoclave which has completed dewatering reduces pressure and then discharges dewatered coal to the bunker below. The steam and the hot water discharged from each autoclave and condensate tank at depressurization stage can be used.for preheating other autoclaves.

A number of remodellings were made, when the plant was re-constructed in Wakamatsu, incorporating the experiences gained at Takehara and the results of subsequent researches, but its fundamental constitution remains unchanged from the illustration in Fig. 3.

3. RESULTS OF DEHYDRATION TEST

3.1. Decrease in Weight

Fig. 4 shows decrease in coal weight due to dewatering, specifying unit weight per kilogram of dry coal measured at the laboratory test of

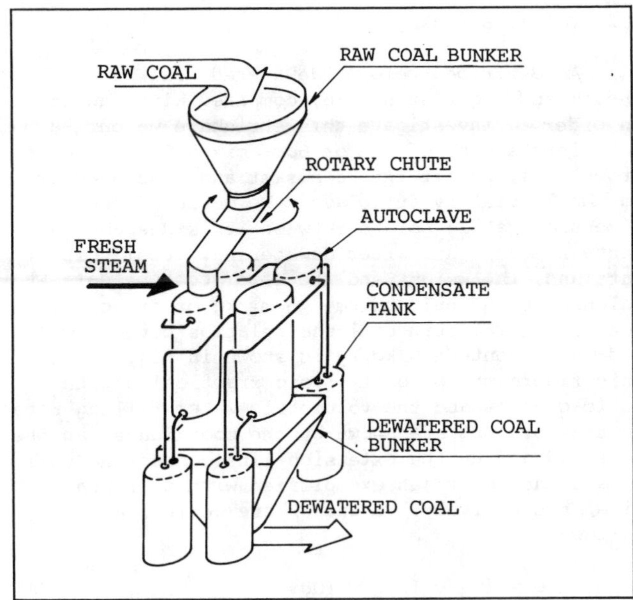

Fig. 3 Pilot Plant

Yallourn coal, Victoria. The unit weight of raw coal (water content 220%) was 3.2kg/kg, but was reduced to about half when heated with saturated steam. After autoclave was depressurized, brown coal was taken out and cooled down to room temperature, and then unit weight turned to 1.25 kg/kg, i.e., water content turned to 25%. The use of hot water as heating medium led to almost the same result, but it was necessary to expose brown coal from hot water into steam before depressurization. If brown coal was left immersed in water during depressurization (cooling) stage, the coal became heavier than it was when heating was completed. Since the moisture of coal can not be vaporized during raising temperature, it is removed in the form of liquid. During pressure relieving stage, the moisture of coal was vaporized by the sensible heat of brown coal. If coal is immersed in the water at this stage, not only this evaporation is repressed but also coal readsorbed water.

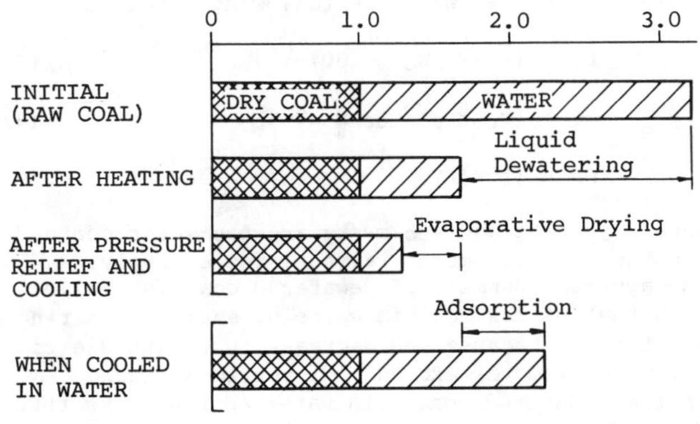

Fig. 4 Weight decrease during dewatering

3.2. Volume Shrinkage

As described before, dewatered coal is of shrunk and tight structure, compared with raw coal. In order to investigate this shrinkage we cut raw coal into a cube of 4cm or 6cm, mixed it with other brown coal, put it into a basket and dewatered it in the laboratory autoclave. We measured the dimensions at 15 points between its sides and face centers before and after dewatering. From this data and, the weight and the water content data we calculated the unit volume v_u m^3/kg of dry coal. Concerning Yallourn coal the relation between v_u and water content w kg/kg is shown in Fig. 5. In this figure the porosity (void fraction) of the cubic coal ψ% and the volumetric water filling rate in the pore WF% are shown by the coordinates at the left end and by the extension line connecting pole P_1 and each plot (as exemplified with a broken line) respectively. ψ and WF are defined as follows.

$$\psi = (v_u - v_t)/v_u \times 100\% \tag{1}$$

$$WF = v_w \cdot w/(v_u - v_t) \times 100\% \tag{2}$$

where $v_t = 0.65 \times 10^{-3}$m^3/kg means true specific volume of dry coal and $v_w = 10^{-3}$m^3/kg means specific volume of water.

Raw coal has large pore volume (ψ = 78%), which is filled with water almost completely (WF $\cong$ 100%), and this is the cause of its high water content. In dewatered coal, ψ and v_u has become small, but the empty pore volume has increased. This is considered to be chiefly the pore left by the moisture evaporated at depressurizing stage. Assuming that volume does not change at the pressure relieving stage, we calculated the unit volume immediately after the completion of heating and the results showed that even at that time WF does not reach 100%. But we think this is due to the thermal expansion of water at the high temperature, and, the pore is completely filled with water actually when heating is completed. For example, the volumetric thermal expansion of water at 500 K is 20%.

In Fig. 5 are indicated further the dewatering ratio DR% with the upper coordinates, the volumetric shrinking ratio Sv% with the right side coodinate and the apparent density of cubic coal ρ_a kg/m^3 on the extension line connecting pole P_2 with each plot (as exemplified with a broken line). These are defined with the following formulas.

$$DR = (w_o - w)/W_o \times 100\% \tag{3}$$

$$S_v = (v_{uo} - v_u)/v_{uo} \times 100\% \tag{4}$$

$$\rho_a = (1+w)/v_u \tag{5}$$

Here, w_o kg/kg and v_{uo} m^3/kg are the water content and the unit volume of raw coal respectively. The apparent density of dewatered coal is smaller than that of raw coal in spite of shrinkage during dewatering, because the decrease in weight due to the water removal exceeds volumetric shrinkage. In the dewatered coal with water content less than 0.6 kg/kg, ρ_a is about 0.8 x 10^3 kg/m^3 regardless of w.

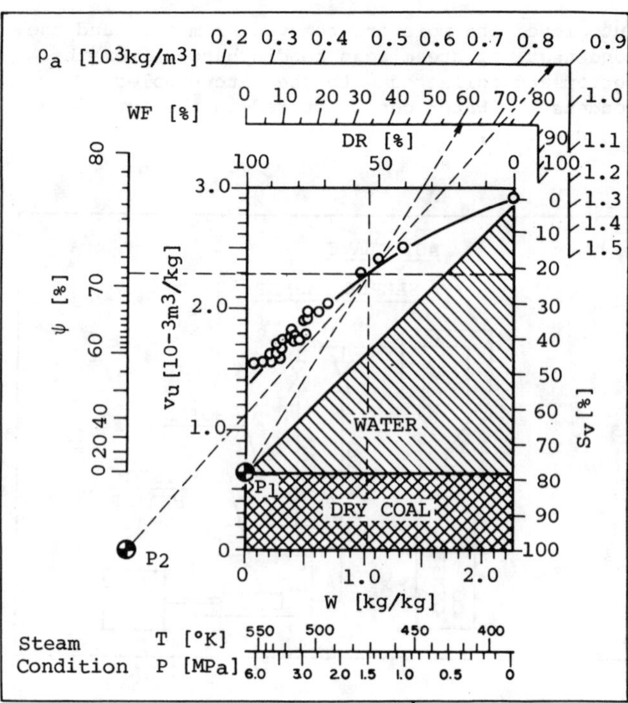

Fig. 5 Shrikage during dewatering

3.3. Dewatering Process

The lowermost scales in Fig. 5 indicate the rough steam condition (pressure , temperature) for obtaining the dewatered coal with respective water contents. But the steam condition changes not only with the kinds of raw coal but also with many parameters, such as coal size distribution, processing time, etc. We could get the fundamental relation between these parameters in the laboratory test and practical process parameters in pilot plant test, and develop an improved process.

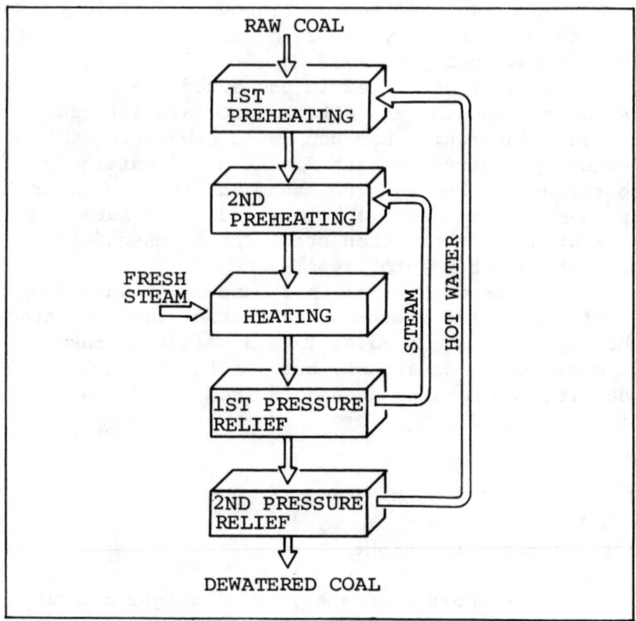

Fig. 6 Fundamental process sequence

The fundamental process sequence is shown in Fig. 6. Raw coal charged in an autoclave, at the 1st preheating stage, is warmed by hot water discharged from the condensate tank of another autoclave which is at the 2nd pressure relieving stage just then, and at the 2nd preheating stage, is warmed by steam discharged from another autoclave at the 1st pressure relieving stage, and finally is heated with fresh steam. Then the autoclave is depressurized in two stages and the dewatered coal is taken out.

This process is an excellent method as it is, but we tried to make it still more economical and efficient especially for the coal with high moisture such as Victorian, and succeeded to develop a new process.

The chief improved points are as follows. (1) We improved the pressure relieving stage by depressurizing the autoclave quickly to restrain water readsorption (cf. Fig. 4) and improve the dewatering rate largely. (2) Since useless time was minimized, securing enough preheating time autoclave of the same size can process much more coal per hour. (3) We arranged to feed autoclave with superheated steam after completion of saturated steam heating. As the result, we could achieve a high dewatering ratio without overchanging coal quality or applying excessive steam pressure. Since consumed superheated steam is used as saturated steam for another autoclave, we can maintain almost as low heat consumption as the process using saturated steam only.

4. EVALUATION OF DEHYDRATED COAL

In the foregoing sections we used water content w kg/kg of dry coal according to the custom of drying engineering, but in this section we use moisture M [%] on wet coal basis according to the custom of coal analysis.
w and M are related with each other, as follows:

$$M = W/(1+W) \times 100\% \qquad (6)$$

4.1. Heating Value

Even if coal quality does not change during dewatering, the heating value increase I_h% and moisture M of dewatered coal are related with each other, as follows:

$$I_h = (M_o-M)/(100-M_o) \times 100\% \qquad (7)$$

where M_o% is the raw coal moisture.
The transportation cost saving by dewatering CS $/T of raw coal is related with I_h, as follows:

$$CS = C \times I_h/(100+I_h) \qquad (8)$$

Here, C $/T is the total transportation cost from mine-site to Japan. These relations are illustrated in Fig. 7. The data for C $/T used here was investigated in 1981.

If Victorian coal is dewatered up to M = 20%, the heating value increases to over double of that of raw coal ($I_h \approx 130\%$). CS for Victorian coal is, of course, large (about 15 $/T). It is notable that the Western States coal , USA has approximately as large CS $/T as that of Victorian when dewatered up to M = 5%, although I_h% is not so large. This is because its transportation cost is high and occupies more than 60% of its CIF Japan cost.

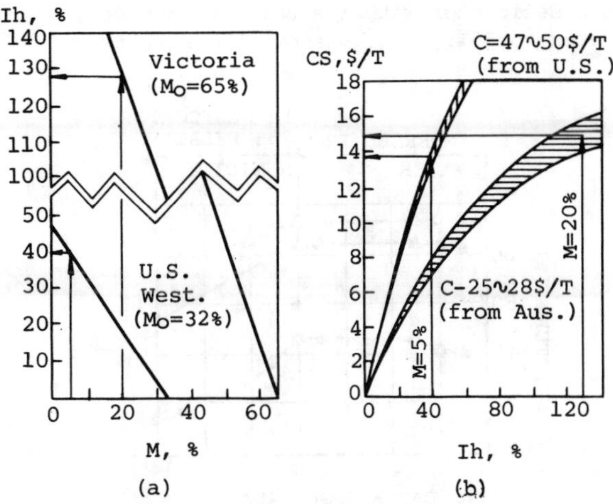

Fig. 7 Heating Value Increase and Cost Saving

Actually the heating value on dry basis also increases through DK process because of the change of coal quality. The results of experiments in Victorian coal are exemplified in Fig. 8(a). Another important thing as the effect of dewatering is the decrease of latent heat taken away by remaining moisture in the boiler. Taking these two effects into consideration, we revised Fig. 7(a) into Fig. 8(b). We find that I_h has increased further to over 100% with M = 20%.

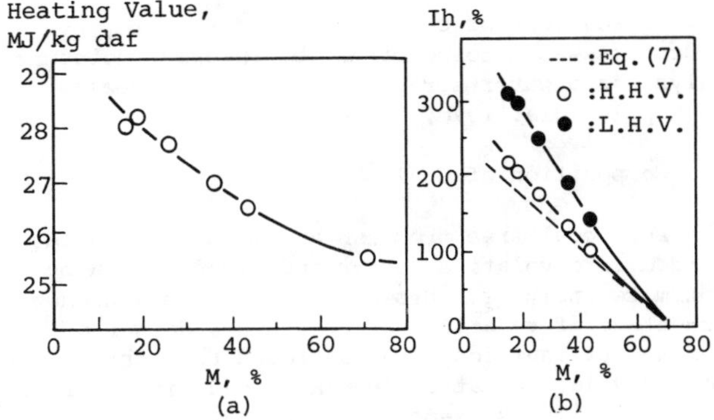

Fig. 8 Heating Value

4.2. Water Readsorption Resistance

In order to utilize dewatered brown coal after its transportation over a long distance to Japan, it is desirable that it does not readsorb water through humidity in the air and rain in the course of the transportation and during the period of its storage.

In order to evaluate this we conducted the experiment of immersing a lumpy brown coal sample in the water for 8 days and then taking it out into a desiccator with its relative humidity adjusted to 75%. Fig. 9 shows the results.

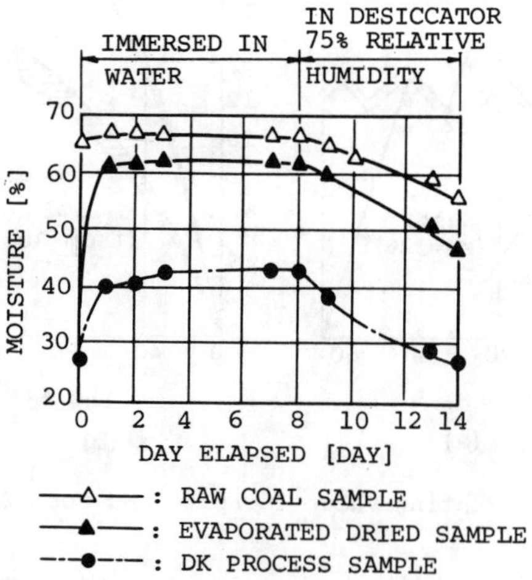

Fig. 9 Waterproof Test

Since the pore in raw coal is almost filled with water, its moisture M% scarcely changes even if it is immersed in the water. The sample dried evaporatively up to M = 20% in our electric furnace readsorbed water approximately to raw coal level in the water only one day later. Even 6 days after taken out into the desiccator its moisture does not return to the initial level. On the other hand, the sample dewatered by DK process reached far smaller M than that of raw coal even after its immersion in the water for 8 days, and when it has taken out into the desiccator, M dropped quickly and 6 days later it returned to the initial moisture.

Improvement of coal waterproof property by DK process was also confirmed by inherent moisture analysis and water spray test on a small-scale stock pile (dia. 1.7m, height 0.6m).

4.3. Composition of Coal

Pure coal free from ash and moisture can be divided into volatile matter and fixed carbon by proximate analysis. Brown coal contains a higher percentage of volatile matter and has danger of spontaneous ignition. But we found that through DK process volatile matter decreased and fixed carbon increased. It is exemplified in Fig. 10.

The main elements of coal are carbon(C), hydrogen(H) and oxygen(O). Through the ultimate analysis we found that the remarkable decrease of O and the accompanying relative increase of C were caused by DK process. H scarcely changed. In order to investigate this relation more clearly we plotted the data on a coordinate system consisting of its longitudinal axis: atomic ratio of H to C and its latitudinal axis: atomic ratio of O to C as is shown in Fig. 11(a). This coordinate system was first adopted by D.W. von Krevelen to investigate the reaction of natural coalification (Fig. 11(b)).

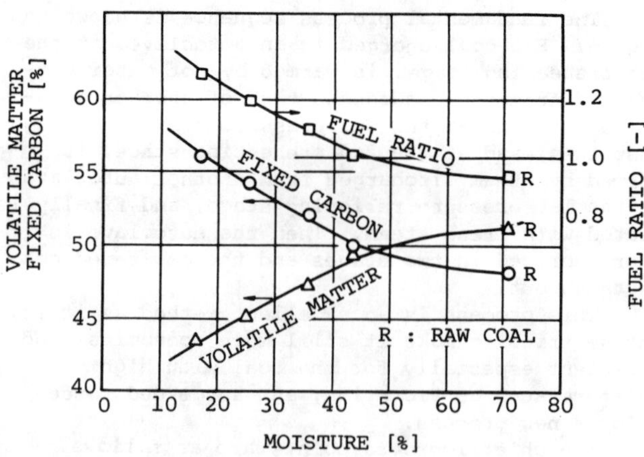

Fig. 10 Proximate Analysis

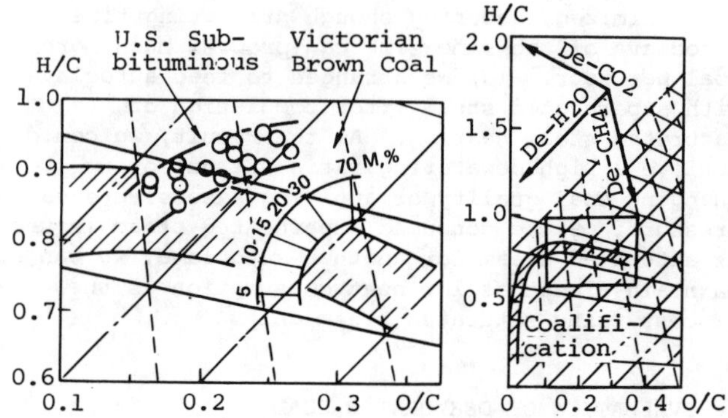

Fig. 11 Change in Coal Band

From this result it was found out that the change of coal quality due to DK process resembles the natural coalification, but with a far narrower range, and that the chief reaction is De-CO_2. This was confirmed also through the functional group analysis. The change was most remarkable in carboxyl group (-COOH) as shown in Fig. 12. This quantity of decrease almost agreed with the decrease of oxygen and also with the quantity of the gas (mostly CO_2) generated in the dewatering process.

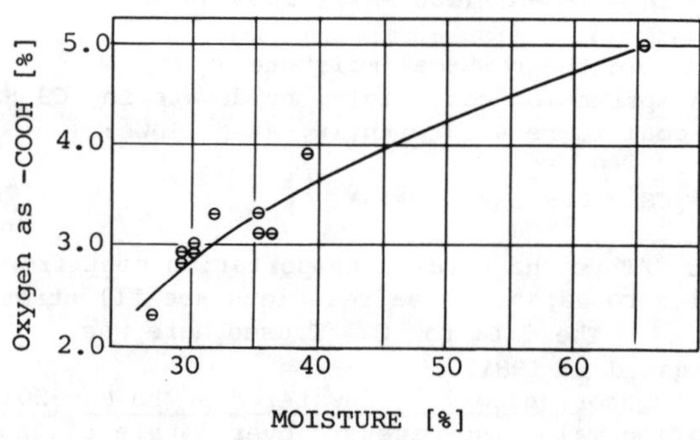

Fig. 12 Change of Carboxyl group

The decomposition of the hydrophilic carboxyl group is an important motive power of non-evaporative dewatering and, at the same time, one of the reasons for improving the waterproof property.

4.4. Coal Cleaning

Although most of brown coal contains few impurities, sometimes it contains an unusually large quantity of ash or sulfur. Sometimes it contains Na and Cl, at a high rate, and there are problems of fouling and corrosion in the boiler.

As the result of the analysis it was found out that many of these impurities, e.g., Na, Cl and S are reduced considerably by DK process. Most of other impurities, such as total ash, was found to decrease evidently, though not so largely , by DK process. Fig. 13 shows the results of the analysis made of certain brown coal containing especially much impurities.

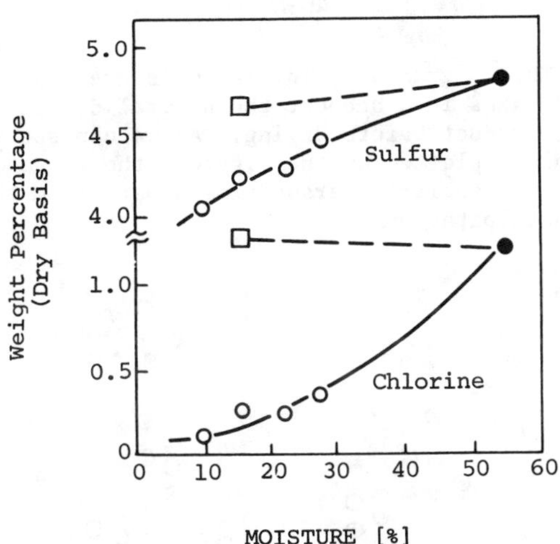

Fig. 13 Change of Impurities

5. CONCLUSION

From the researches we have made so far, we have obtained the following results.

(1) The fundamental mechanism of non-evaporative dewatering was made clear.

(2) Based upon Fleissner Process, a highly reliable dewatering process (DK process) was developed.

(3) The improved DK process is excellent in thermal efficiency and dewatering performance and increases capacity by shortening the processing time.

(4) The effectiveness and the economy of brown coal dewatering by DK process was confirmed.

(5) The coal quality improvement by DK process and its effect was confirmed.

ACKNOWLEDGEMENT

The present researches are conducted jointly with the Electric Power Development Co., Ltd., and we are granted a subsidy from the Japanese Government (MITI) via the Coal Technological Research Institute, a juridical foundation.

We express our heartfelt thanks to all those concerned for the kind cooperation and support they have extended to us.

NOMENCLATURE

Vu	Unit volume of wet coal per kg of dry coal	m^3/kg
WF	Volumetric water filling percentage in coal cappilarly	%
DR	Dewatering ratio	%
Sv	Volume shrinkage ratio	%
M	Percentage of moisture expressed on wet coal basis	%
I_h	Percentage increase of heating value due to dewatering	%
CS	Amount of transportation cost saving due to dewatering (per tonne of raw coal)	$/T
C	Coal transportation cost from mine to Japan	$/T

REFERENCES

1. Ministry of Fuel and Power, Victoria States, Australia, Brown Coal In Victoria (1977).
2. Kimura, H. and Fujii, S., Sekitan-Kagaku to Kogyo (Coal-Chemistry and Industry), Sankyo-Shuppan, Tokyo (1977).
3. Krevelen, D.W., Coal, Elsevier, London (1961).
4. Francis, W, Coal-Its Formation and Composition, Edward Arnold, London.
5. Fleissner, H., Method of Drying Coal and the Like, U.S. Patent No.1 632 829.
6. Evans, D.G., Separation of Water from Solid Organic Materials. Australia Patent No. 430 626.
7. Herman, H., Brown Coal, The State Electricity Commission of Victoria, Melbourne (1952).

DRYING OF CROP PRODUCTS WITH A SHELL : EXPERIMENTAL APPROACH AND MODELLING. ITS APPLICATION TO HAZELNUTS

J.C. Batsale and J.R. Puiggali

Laboratoire Energétique et Phénomènes de Transfert
ENSAM-ERA CNRS 1026
Avenue des Arts et Métiers 33405 Talence Cedex FRANCE

ABSTRACT

The aim of this study is to describe the convective drying of a crop product with a shell, especially hazelnuts.

We have looked at the resolution of the drying problem of concentric homogeneous stratums of various but simplified physical characteristics.

The drying kinetics of hazelnuts under various air conditions were recorded. A behaviour equation was set up taking into consideration the initial moisture content of the product and the relevant drying fluid parameters (ie. Temperature, Relative humidity, Flow rate).

Also considered is a theoritical modelling of the coupled heat and mass transfer. This is based on the profiles of temperature and moisture content inferred from the mean values in each stratum and the temperature and mass flows at each boundary. The specific coefficients allowing the description of the processes come from the litterature.These two approaches permit the classification and quantification of behaviour according to the initial configuration or to the previous changes undergone by this kind of product.

The results permit a practical use (abacus) and give justification of the interpretation of physical phenomena.

1 . INTRODUCTION

This paper is a synthetic set out of two complementary approaches to study convective drying as a function of the required understanding level.

We investigate a crop product with a shell, an almond and an intermediate air stratum, particularly hazelnuts.

On the one hand we carry out an appropriate thin layer equation with experimental results of drying kinetics [1], on the other hand we establish a heat and mass balance model to quantify the whole product drying according to the drying of each constituent [2].

2 . DRYING KINETICS

Kinetics study consist of making systematical tests with known air drying conditions (T_a, u_a, HR_a).

Experimental apparatus and set up are described in [3].

2.1 Appropriate thin layer equation

Moreover classical parameters we use in the kinetics equation we set out two caracteristic grandeurs describing initial state of the product ($\overline{w}_i$) and drying history of the product ($\overline{w}_f$) [4]

$$\frac{d\overline{w}}{dt} = f\left(\frac{\overline{w}-\overline{w}_f}{\overline{w}_i-\overline{w}_f}, \overline{w}_i\right)$$

The $\overline{w}_i$ grandeur, initial moisture content (d.b.) takes into account the natural development of the product before drying. As we can see with the points plotted on the figure 1 the value $\overline{w}_i$ decreases regularly versus time a part from the rain precipitations

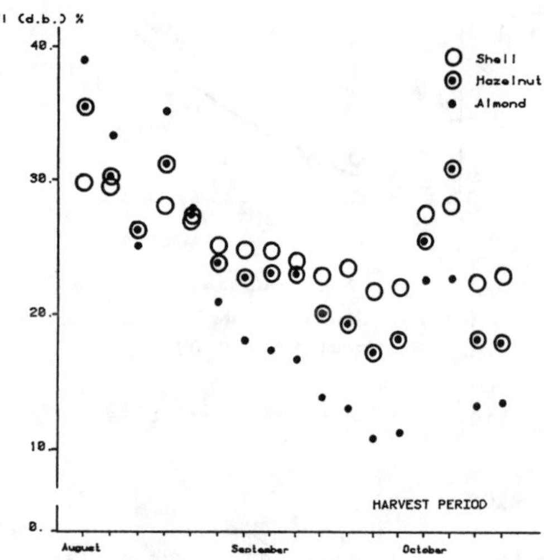

Fig. 1 : Natural moisture content evolution versus harvest period

The moisture content in the product is not spread out in the same proportion, for each constituent at the beginning and at the end of the harvest, figure 1. Therefore, we observe various drying curves. The thin layer equation carried out in this paper does not take into account the earliest harvested products.

Presuming that biological products are advancing, according to its environment and as we can assume that responses to a slow process as desorption are sharp different to a fast one as convective drying. Also, we introduce as an energetic criterion the grandeur, $\overline{w}_f$, final moisture content. $\overline{w}_f$ is estimated from experimental drying curves.

The table I summarizes the experimental moist air range and initial moisture content of the hazelnuts.

$\overline{w}_{n,i}$ % d.b.	13	30
T_a °C	30	50
w_a % d.b.	8	12
u_a $m\,s^{-1}$	.5	1.5

TABLE I

The convenient thin layer equation we obtain is expressed by :

$$\frac{d\,\overline{w}_n}{dt} = -\,u_a^{\alpha}\,e^{-A/T_a+B}\left(\frac{\overline{w}_n - \overline{w}_{n,i}}{\overline{w}_{n,i}-\overline{w}_{n,f}}\right)^{\beta}$$

where : α is linked with surfaces transfers around a sphere $\alpha = 0.5$

β describes the convexity of the drying curves $\beta = 1.32$.

Two linear functions of the initial moisture content are used to explain the diffusion coefficient e^{-A/T_a+B}

$$A = 19\ 177.\ \overline{w}_{n,i} - 7593$$

$$B = -\ 54\ \ \overline{w}_{n,i} + 19$$

$\overline{w}_{nf}$ the final moisture content is correlated with initial moisture content $(\overline{w}_{n,i})$ and with relative humidity (HR_a) and temperature (T_a) of the air.

(with our experimental moist air range it was impossible to obtain a significant evolution versus T_a)

$$\overline{w}_{n,f} = -\ 2.25\ 10^{-3} + 0.188\ \overline{w}_{n,i} + 0.254\ 10^{-2}\ HR_a - 0.501\ 10^{-4}\ HR_a^{\ 2} + 0.507\ 10^{-6}\ HR_a^{\ 3}$$

2.2 Uses of the thin layer equation

We used this experimental description in two ways.

First, this thin layer equation may be introduced in a deep bed model, taking a lot of precautions, thickness of thermal and evaporation front for instance.

Second, it may be operated roughly by means of abacus to determine dryer running.

We present now an example.

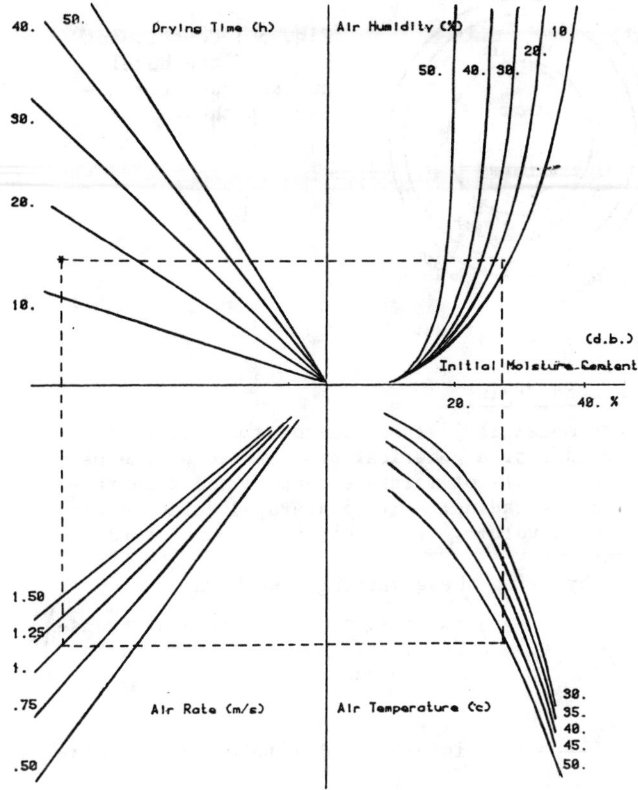

Fig. 2 : Abacus for dryer running. $\overline{w}_{n,f} = 10\%$

Let us suppose a layer of initial moisture content $\overline{w}_{n,i} = 27\%$. When will be the product at a final moisture content $\overline{w}_{a} = 10\%$ (conservation moisture) assuming that the inlet air characteristics are $T_a = 50°C$, $HR_a : 13\%$, $u_a = 1\ ms^{-1}$ We obtain $t_f = 16$ *hours* dashedline on the figure 2).

3 . BALANCE EQUATIONS

3.1 Model designing

To obtain the whole hazelnut drying model we survey three elementary sets of equations deduced from a physical analysis of the heat and mass balances at three scale levels :

- Almond drying
- Air stratum as a heat and mass transfer contact resistance
- Shell as a bulk volume for surface transfers

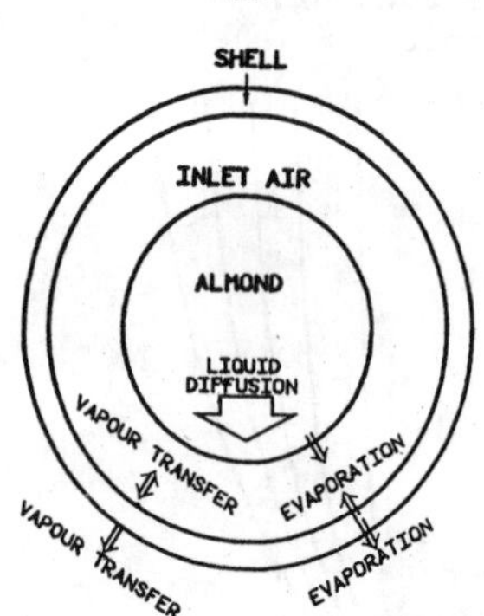

Fig. 3 : Constituent parts of the hazel - nut and mass trans - fer hypothesis

Almond drying

The model is given by Luikov's model, writen in the case of a spherical product. We assume parabolic profile of moisture content and tempera - ture versus radius. We take averaged value, over the almond volume, of temperature and moisture content.

So by using these variables we have :

$$\frac{d\overline{T}_{al}}{dt} = f\left(\begin{array}{l}\text{internal and boundary heat transfer , latent heat of evaporation}\\ \text{at the surface}\end{array}\right)$$

$$\frac{d\overline{w}_{al}}{dt} = f\left(\text{internal and boundary mass transfer}\right)$$

The two equations we obtain are considered as classical and appear in many papers with slight differences [5,6]. (*For the exact form see at the end of the paper Appendix*)

Coefficients of mass transfer are identified from kinetics curves of almond drying.

Thermophysics parameters and equilibrium isotherm of the product derived from the litterature [7,8] are fitted to our problem.

Intermediate air stratum

We assume air inertia be negligible as density and specific heat of a free air are slight compared with the characteristics of the two other constituents.

So one must consider air stratum as a skin resistance from the heat and mass transfer viewpoint.

We obtained two relations for averaged temperature and vapour pressure of the inlet air :

$$\overline{T}_a = f(\overline{T}_{al}, \overline{T}_s)$$
$$\overline{P}_{v,a} = f(\overline{P}_{v,al} , \overline{P}_{v,s})$$

These two functions roughly describe the transfer and allow us to pick out the main solid surface. (*For the exact form see at the end of the paper Appendix*)

Protective shell

The protective shell is the constituent where heat and mass exchanges with the outside medium (outlet drying air) take place, according to inlet supply of its internal side.

Inlet and outlet surface exchange coefficients are used to specify the important part of the shell.

However outlet heat and mass transfer coefficients are larger than the inlet one.

Hence the equations are :

$$\frac{d\overline{T}_s}{dt} = f(\text{boundary heat transfer and evaporation})$$

$$\frac{d\overline{w}_s}{dt} = f(\text{boundary mass transfer})$$

(*For the exact form see at the end of the paper Appendix*)

Whole nut drying

From these three sets of equations it is convenient to go further into the drying knowledge of each constituent.

Moreover it is convenient to appreciate the bulk behaviour of the hazelnut by using statistic datas inferred from laboratory trials :

$$\overline{w}_n = 0.65\ \overline{w}_a + 0.35\ \overline{w}_s$$

In the next part of this paper we present some uses of this model.

3.2 Results of simulation

We give here a general survey of results,we obtain with the previous model [2].

Experimental and numerical drying curves for the whole hazelnut are in good agreement as we can

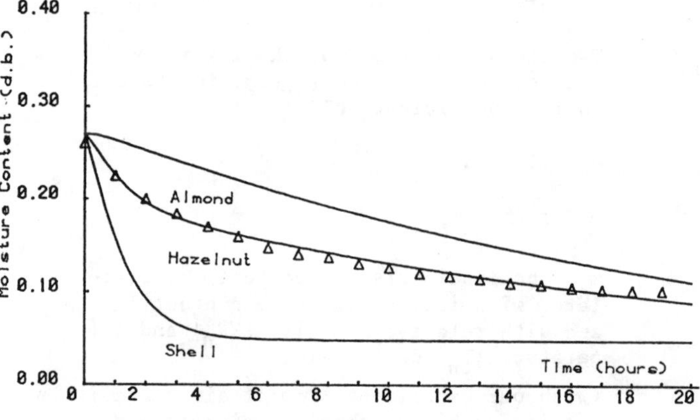

Fig.4 : Drying curves versus time (— model, Δ experiment)
(moist air conditions $T_a = 45°C$ $u_a = 1 m\,s^{-1}$ $HR_a = 18\%$)

see on the graphs of the figure 4. We note the strong difference between the drying kinetics of the two nut's constituents. Such experimental results are difficult to obtain when almond and shell are coupled up.

As we noted on the figure 1 relative moisture content of each constituent are not in the same proportion during the harvest period. The figure 5 visualize different responses versus the initial

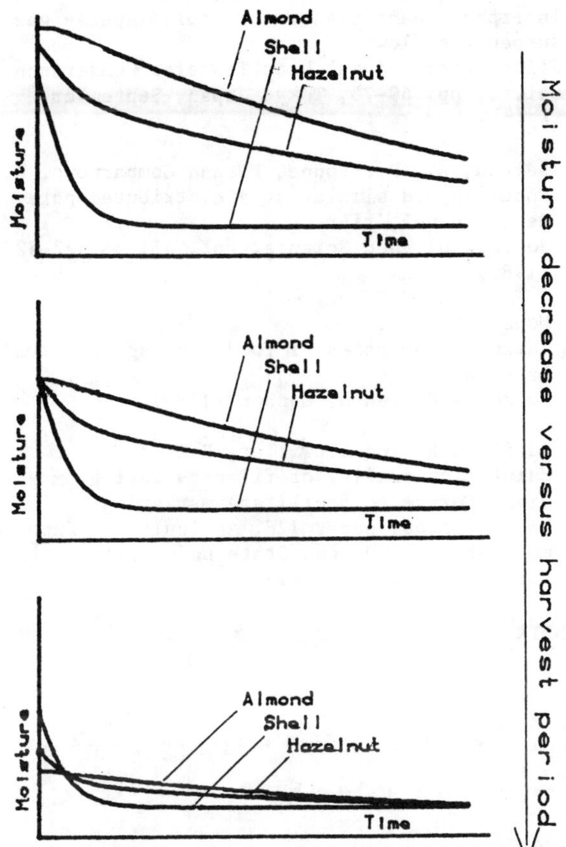

Fig. 5 : Drying curves versus time for various initial state during the harvest period. (moist air conditions $T_a=45°C$; $u_a=1ms^{-1}$ $HR_a = 18\%$)

state. The shell reaches quickly the equilibrium. Whereas the hight vapour pressure in the shell blocks the inlet transfer, the initial drying rate of the almond may be very low.

At the end of the drying the moisture is hardly linked to the initial one.

Therefore we note that the final moisture content used in the kinetic description does not appear in the physical model (Equilibrium isotherm). However we joined up two drying times: the kinetics drying time and the numerical drying time needed to reach a given moisture content on the figure 6 .

We illustrate with this figure the physical meaning of the kinetic grandeur , final moisture content.

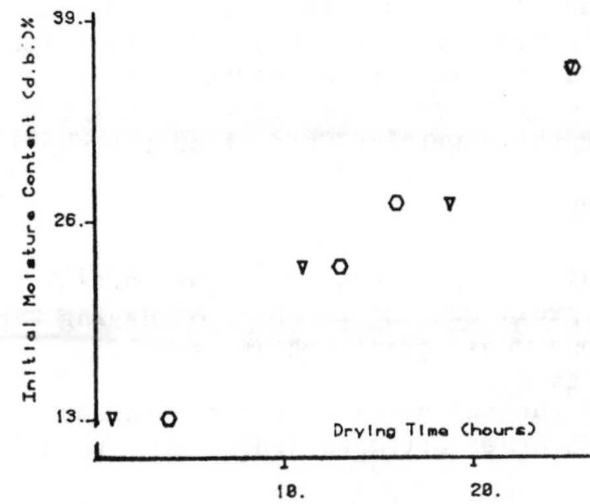

Fig. 6 : Initial moisture content and drying time to reach a final moisture content ($w_{n,f} = 10\%$).
(Δ numerical results, O kinetics results)

The illustration we give now is the response of the hazelnut to various moist air conditions(figure 7). It is a roughly approach of the deep bed

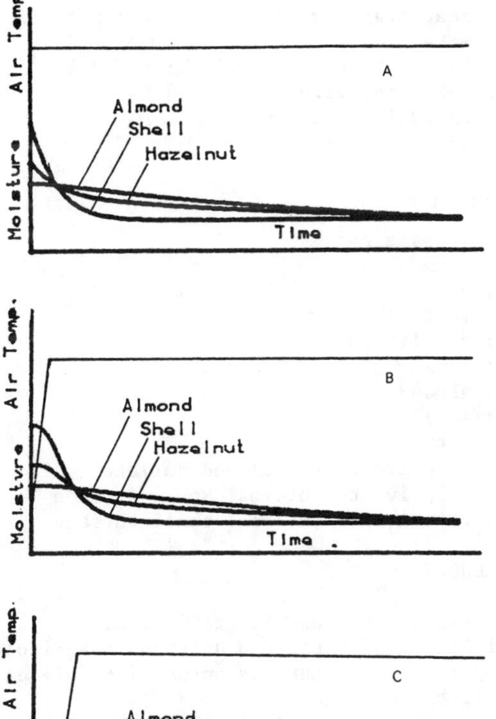

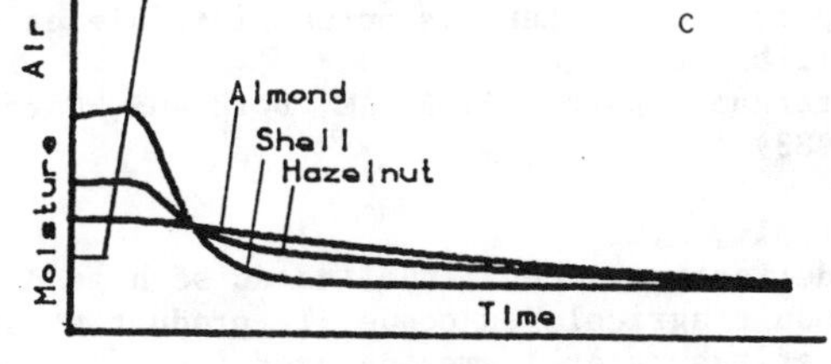

Fig. 7 : Drying cruves versus time for various moist air fronts.

A constant air moisture and temperature
B decreasing moisture and increasing temperature
C step air: decrease moisture and increase temperature

study. We can observe that these drying conditions induce a dephased response versus the step with temperature decreasing and moisture increasing so the deep of the bed. Moreover the model takes into account the rewetting effect on the shell.

4. CONCLUSION

This synthetic presentation of two different approaches spread out the interest of drying kinetics equations whose grandeurs are specified by physical balances.

A dryer running needs only a convenient knowledge but dryer optimisation needs more understanding.

NOMENCLATURE

w	: moisture cont.	kgw/kgd
t	: time	s , h
T	: temperature	°C
Hr	: relative humidity	%
P_v	: vapor pressure	P_a
P_{vs}	: saturated vapor pressure	P_a
u_a	: air rate	m/s
c_p	specific heat	$J\ kg^{-1}\ °C^{-1}$
R	: radius	m
h	: heat transfer coef.	$w°C^{-1}\ m^{-2}$
k	: mass transfer coef.	$kg\ s^{-1}\ m^{-2}$
DR	: latent heat at vaporisation	$J\ kg^{-1}$
W_A	: water activity	%
λ	: thermal conductivity	$w\ m^{-}\ °C$
S	solid density	$kg\ m^{-3}$

Subscript

— space average

Underscript

a	: air
i	: initial
e	: equilibrium
f	: final
al	: almond
s	: shell
n	: nut
1	: relative to the almond surface
2	: relative to internal shell surface
3	: relative to external shell surface

REFERENCES

1 Batsale, J.C., and Puiggali, J.R.
Cinétique de séchage d'un produit agricole : importance des teneurs en eau initiale et d'équilibre dynamique
Internal report LEPT-ENSAM, Bordeaux, March (1983)

2 Batsale, J.C.
Modélisation à deux échelles du séchage d'un produit agricole en coque :le produit et ses constituants et l'amas de produits
Internal report, LEPT-ENSAM, Bordeaux, February (1984)

3 Puiggali, J.R. and Batsale, J.C.
An experimental air drying apparatus and its use
3rd International Drying Symposium, vol. 1 pp. 30-38 , Birmingham, September (1980)

4 . Allen, J.R.
Application of grain drying of maize and rice
J. of Agri. Eng. Res., vol. 5, pp. 363-385 (1960)

5 . Babukha, G.L. and Schraiber, A.A.
Interphase heat transfer in polydisperse gaz suspension flows.
Fifth International Heat Transfer Conference vol. 5, pp. 69-73, Tokyo Japan, September (1974).

6 . Bertin, R., Pierronne, F. and Combarnous, M., Modeling and simulating a distributed parameter tunnel drier
Journal of Food Science, vol. 45, pp.122-125 (1980)

7 . Mohsenin, N.N.
Thermal properties of foods and agricultural materials
Gordon and Breach, London (1980)

8 . Cecil, S.R. and Litwiller, E.M.
Maintining quality of filberts during extensive storage to facilitate marketing
Miscellaneous paper 128, Agricultural Experiment station, Oregon State University, Corvallis, April (1962)

APPENDIX

Shell balance

$$\rho_s\ c_{p_s}\ \frac{d\overline{T}_s}{dt}\ \frac{(R_3^3 - R_2^3)}{3} = h_3\ R_3^2\ (T_{a_3} - \overline{T}_s) + h_2\ R_2^2\ (T_{a_2} - \overline{T}_s) + Dh\ \rho_s\ \frac{(R_3^3 - R_2^3)}{3}\ \frac{d\overline{w}_s}{dt}$$

$$\rho_s \left(\frac{R_3^3 - R_2^3}{3}\right) \frac{d\overline{w}_s}{dt} = k_3\ R_3^2\ (P_{va_3} - A_{w,s}\ \overline{P}_{vs,s}) + k_2\ R_2^2\ (P_{va_2} - A_{w,s}\ P_{vs,s})$$

Almond balance

$$\rho_{al}\ c_{p_{al}}\ \frac{d\overline{T}_{al}}{dt} = \frac{3}{R_1}\ \frac{1}{\left(1 + \frac{h_1 R_1}{5\lambda_{al}}\right)}\ (T_{a_1} - \overline{T}_{al}) + Dh\ \frac{R_1}{3}\ \rho_{al}\ \frac{d\overline{w}_{al}}{dt}$$

$$\rho_{al}\ \frac{d\overline{w}_{al}}{dt} = \frac{3}{R_1}\ e^{(-a/\overline{T}_{al} + b)}\ (w_e - \overline{w}_{al})$$

Air contact resistance

$$\overline{T}_{a_{1,2}} = (\tau)\ \overline{T}_{al} + (1-\tau)\ \overline{T}_s$$

$$\overline{P}_{va_{1,2}} = (\gamma)\ \overline{P}_{val} + (1-\gamma)\ \overline{P}_{vs}$$

NUMERICAL VALUES

$\rho_s = 880\ kg/m^3$; $c_{p_3} = 3000\ J°C^{-1}kg^{-1}$; $R_2 = .009\ m$; $R_3 = .01\ m$; $h_s = 100\ w°C^{-1}m^{-2}$
$h_2 = 10\ w°C^{-1}\ m^{-2}$; $Dh = 2{,}4\ 10^6\ J\ kg^{-1}$; $k_3 = .5\ 10^{-8}\ kg\ Pa^{-1}\ m^{-2}\ s^{-1}$;
$k_2 = .25\ 10^{-8}\ kg\ Pa^{-1}\ m^{-2}\ s^{-1}$; $\rho_{al} = 900\ kg/m^3$; $c_{p_{al}} = 3500\ J\ °C^{-1}\ kg^{-1}$;
$R_1 = .005\ m$; $\lambda_{al} = 4\ w\ m^{-1}\ °C^{-1}$; $R_1 = 10\ w\ °C^{-1}\ m^{-2}$; $a = 6157\ kg\ s^{-1}\ m^{-2}$;
$b = 8.7\ kg\ s^{-1}\ m^{-2}$; $\tau = 1/2$; $\gamma = 1/2$

SORPTION ISOTHERMS

$$Aw_{al} = -65.37 - 0.033\ \overline{T}_{al} + .01 + 3\ 771.1\ \overline{w}_{al} - 39\ 200\ \overline{w}_{al}^2 + .153\ 10^6\ \overline{w}_{al}^3$$

$$Aw_s = -13 - \overline{T}_s + (1045 - 8.9\ \overline{T}_s)\ \overline{w}_s + (-3380 + 68.8\ \overline{T}_s)\ \overline{w}_s^2 + (4250 - 128.8\ \overline{T}_s)\ \overline{w}_s^3$$

INDUSTRIAL APPLICATIONS FOR NEW STEAM DRYING PROCESS IN FOREST AND AGRICULTURAL INDUSTRY

Claes Svensson

Swedish Exergy Technology Inc., Gothenburg Sweden

A new drying technique has been developed by Chalmers University of Technology and Swedish Exergy Technology, in Sweden. This process, called "steam drying" utilizes the drying capacity of superheated steam to remove moisture from porous material such as paper pulp, different hog fuel and agricultural products. The first commercial pulp dryer based on this technique is installed at Rockhammar Bruk in Sweden. The first commercial drying unit operating on hog fuel is installed at Husum, Sweden, to dry 58% moisture content fuel to 35% moisture content for "on the grate" firing in the power boiler.

Sugar beet pulp is dried from 78% to 10% moisture in a dryer started up September/October 1983. Applications under review include the drying of peat, distillers grain residue, orange and pinapple pulp, grape and apple pomace and resudues for various end uses including the use of residues as combustible material in small boilers.

THE BASIC SYSTEM

The steam dryer is a closed, pressurized system (Figure 1 and 2) where the material to be dried is exposed to indirectly heated superheated steam. In the dryer, which consists of transport pipes, heat exchangers, cyclone and fans, superheated steam circulates at a pressure of 2 to 6 bar. The steam serves as a transport gas for material in the dryer. The heat for drying is transmitted by the drying medium from the heat exchanger tube surfaces to the suspended material. Primary heating steam is condensed, usually at pressure between 8 and 15 bar, on the shell side of the heat exchangers.

Dried material and steam are separated in a cyclon and the steam is recirculated. The excess steam generated during drying is continuously bled off from the system. Thus, if material is dried from 45-50% dry content to 85-90%, about 1 ton of steam per ton of material is generated in the dryer. This steam is available as process steam at a pressure of 2 to 6 bar. Alternatively, generated steam can be used as heating steam for the dryer after passing a vapour compressor.

Average drying rates in superheated steam are 2 to 3 times greater than by conventional air drying due to higher heat transfer coefficients for steam. Typical residence time for the material is 10-30 seconds in the steam dryer. Transport velocities within the dryer are fixed at values If 20 to 40 m/s to minimize fan operating costs and the capital cost if heat exchangers. Maximum drying throughout capacities are determined by the size of feeder available for specific materials and the physical characteristics of these materials. Material of construction is in most cases stainless steel.

FIGURE 1 STEAM DRYER

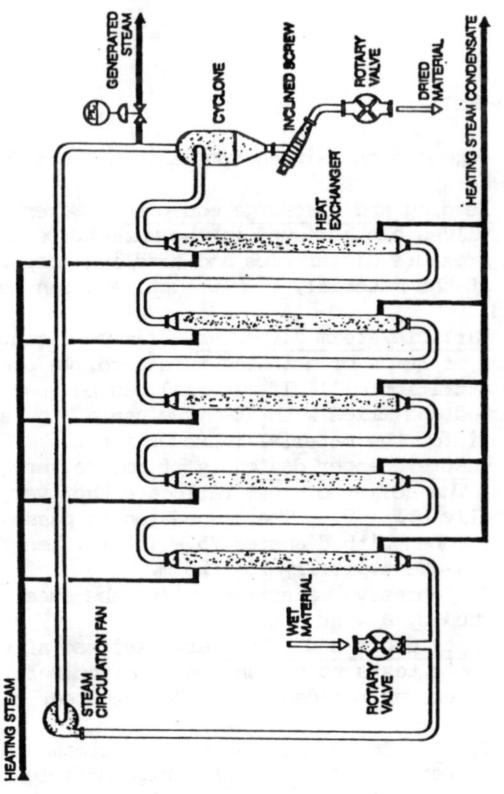

FIGURE 2 PRINCIPLE FOR DRYING

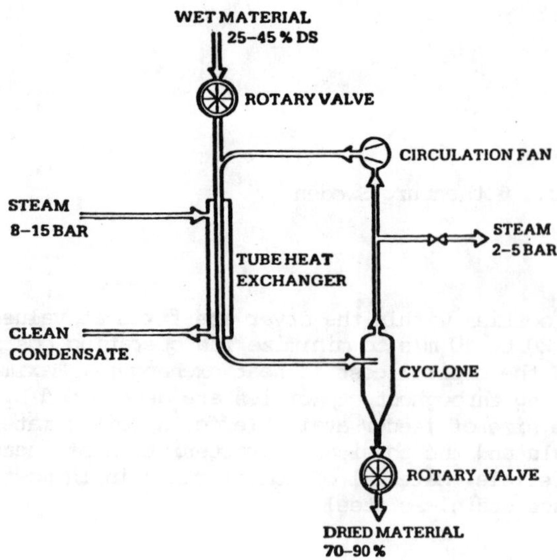

FIGURE 2

EQUIPMENT

Some comments on main equipment in a Steam Drying Process:

Feeding and discharge equipment. Several rotary valves can be found in the market. Depending upon pressure differences and handling properties of the material, 5 - 10 suppliers can be found.

Particle/steam disperser. If some disintegration of lumps or grinding is needed, we prefer a disc attrition mill. If material can be just distributed and mixed with steam we use a high speed venturi for the material introduction.

Heat exchanger design is of course very critical. The solid to steam ratio are kept very low, typically 0,1 - 0,8. The suspension is passed inside tubes with diameter 75 - 150 mm. Length of tubes are typically 10 - 20 m.

For abrasive materials inlet tube-sheet is protected by a wear plate.

Empty ducts and elbows are designed after full scale tests to prevent material to settle but to give enough residence time between heat exchangers.

Cyclone for separating carrier steam and dried material is designed for high efficiency performance. Although some recycle cannot be avoided. Where generated steam is bled-off we provide another inertia separator.

Recycle fans are pressurized centrifugal fans with mechanical seals or labyrinth seals. For high pressure drops fans with axial inlet can be provided. Efficiency typically is in the range of 70 - 85%.

Pre-superheater for recycle carrier steam is used when sticky material is dried. Enough superheat is needed for evaporation of surface moisture in some cases where protein and starch are present.

FIGURE 3 TUBE HEAT EXCHANGER, STEAM DRYER

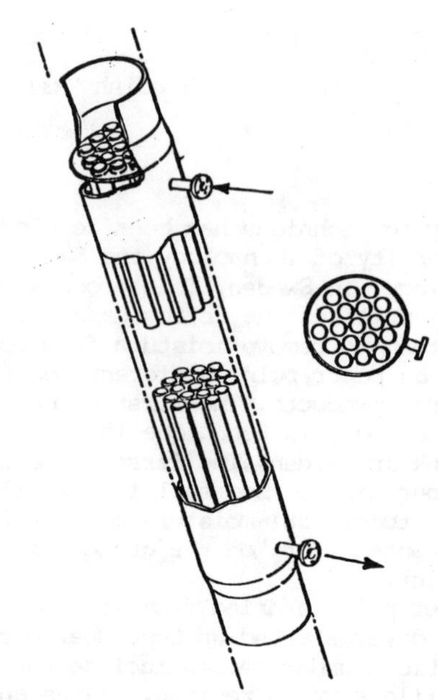

MAIN FEATURES OF THE STEAM DRYER

Compact Vertical Design

Pressurized operation

- enables the use of vaporized moisture as process steam
- or internal use as heating steam after vapor compression
- very high heat transfer rates

No air

- Very high mass transfer rates even in porous materials
- no oxidation
- short residence time, 5 - 60 sec
- no risk for fire or explosion reduced product losses in some cases
- controlled and reduced emission of pollutants

Easy to control and high flexibility since only pressure control valves are used for evaporative capacity and final moisture control.

The disadvantages with our present design are mainly:

- we need relatively small particles, 5 - 10 mm

although the distribution can be wide

pressurized equipment and in many cases stainless steel

depending upon operating conditions and available boiler/turbine capacity we would prefer using vapor compressor but power consumption for compressor is higher than loss of cogeneration. In some cases cost for electric power is detrimental for using vapor compression.

OPERATING INSTALLATIONS

Rockhammar. This is the first full scale dryer started up in 1978/79.

Mechanically dewatered pulp is dried from about 50-55% moisture to 10-15% moisture depending on end user. Dried pulp is baled and moisture is measured with IR-sensor to give set-point for carrier steam temperature. Steam temperature controls the heating steam pressure.

Generated 3 bar steam is used for bark drying (boiler fuel) and instead of fresh steam for heating process water.

Husum was first for drying hog fuel. Bark from birch, pine and spence is dried for firing on grate. Depending upon seasonal changes bark is dried from 60-70% moisture down to 30-40% moisture. Dryer is operating at maximum capacity all the time, i.e. no moisture control. Generated steam is condensed in a reboiler where heating steam condensate is vaporized to form 4,5 bar process steam. Contaminated bark condensate is used for heating.

Köpingebro, the dryer where sugar beet is dried after sugar recovery. After mechanical dewatering beet will be dried from about 78-80% moisture down to 10% moisture. Only a part of the beet flow is steam dried, screened and grinded to form a new market fibre.

Molasses and sugar juice can be added before or after drying depending upon end user.

Generated steam will be condensed in a reboiler and the produced process steam used for sugar juice evaporation.

Laxå Bruk, another Swedish company who will use the back pressure dryer. This dryer is now under construction and will be started-up in September 1984.

The steam dryer will be able to dry mineralwool from 25-30% moisture to 1%. By getting the material this dry, big improvements of quality will be achieved.

FIGURE 4 BARK DRYING, HUSUM

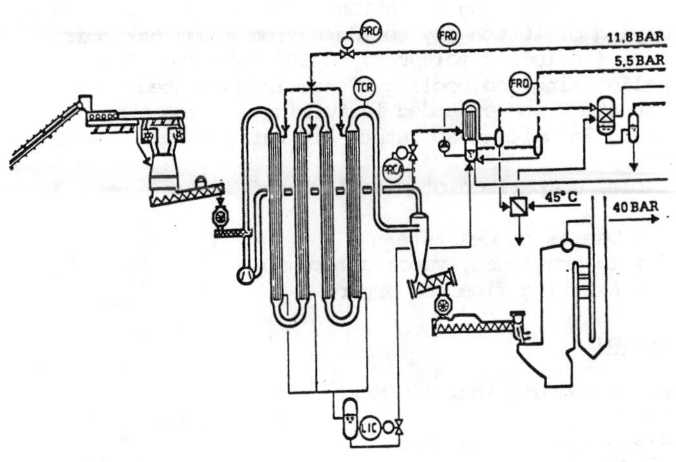

TEST FACILITIES AND EVALUATION

In co-operation with Chalmers University of Technology, Gothenburg, and Swedish Board of Technical Development we have extended test facitilities.

Our pilot dryer can evaporate 0,1 - 0,5 tons of water per hour depending upon material. In connection with our testing for dryer design we include tests for mixing, feeding and discharge equipments.

Usually 2 - 4 hours test at stationary conditions is enough. If clients wish to have more product for testing we can run 24 hours a day.

Evaluating the dryer we investigate: Deposits and sticking properties / Type of feeder and discharge equipment / Mixing and dispersion equipment / Need for superheater / Heat transfer rates for computerized dryer design

Depending upon dryer configuration we calculate heat transfer from tubes to steam and from superheated steam to particles.

A Comparison:

EXTERNAL STEAM USAGE - VAPOR COMPRESSION

For a sugar mill operation these two alternatives might be considered:

1. Generated steam is condensed in a reboiler where process steam for sugar juice evaporation is used.

2. Generated steam is condensed in a reboiler and clean steam is compressed to heating steam and condensed in the dryer.

3. Existing operation is based upon flue-gas drying of 10 ton dry solids/h from about 65% moisture (including molasses) to 10% moisture after pelletizing and cooling. Steam is generated at 24 bar 420°C and expanded in turbine for cogeneration. In juice evaporators 3,0 bar steam is used.

Table, comparison of case 1, 2 and 3

1 = Use as process steam
2 = Internal use, vapor compressor
3 = Existing flue gas dryer

	1	2	3
Drying			
Steam consumption, 1,0 MPa, ton/h	21,5	(21,5)	-
Steam consumption, 0,3 " , ton/h	-	-	1,0
Power consumption, (fans etc), kW	640	640	730
Power consumption, compressor, kW	-	1480	-
Produced process steam, ton/h	18,5	(18,5)	-
Fuel oil consumption, m^3/h	-	-	1,8
Evaporation			
Steam consumption 0,3 MPa, ton/h	55	55	55
Boiler			
Steam generation, ton/h	58	55	56
Power generation, kW	5015	5775	5880
Fuel oil consumption, m^3/h (36 GJ/m^3)	4,90	4,64	4,73
Total:			
Fuel oil consumption, m^3/h	4,90	4,64	6,53
Power, purchased, kW	775	1495	Base

Calculations above are based on 90% boiler efficiency, 84% turbine/ generator efficiency.

The compressor/boiler arrangement needs to have about 40% efficiency, based on compressed heating steam i.e. 270 kWs/kg or 75 kWh/ton of steam.

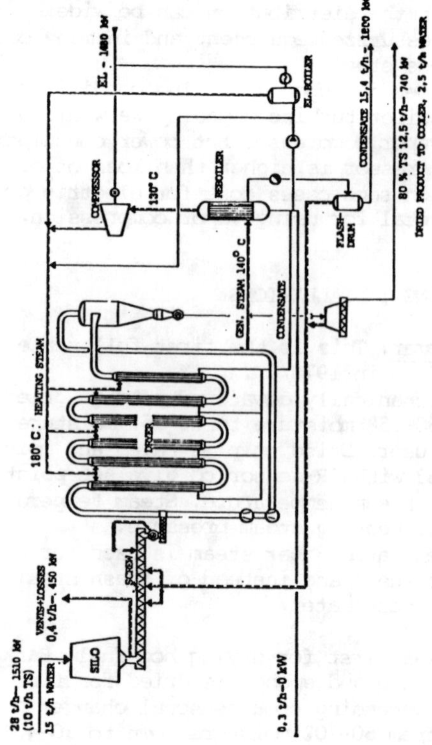

FIGURE 5 PRINCIPLE DRAWING - STEAM DRYER

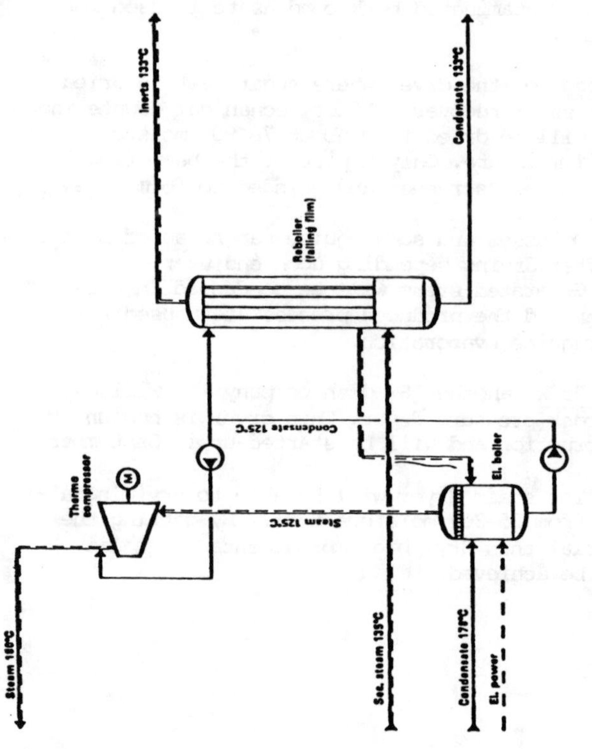

FIGURE 6 THERMOCOMPRESSION FOR THE STEAM DRYER

SUMMARY - INVESTMENT AND OPERATING COSTS

For economical justification, each case has to be considered separately. Available boiler capacity, price for purchased and produced power vary as well as cost for gas or oil.

Operating advantages with vapor compressor are mainly:

- undependent operation from boiler and turbine
- dryer available for drying grass, luzerne, hog fuel etc. between campaigns

Major disadvantages are:

- compressor are sensitive and expensive process equipment
- compressor gives less flexibility for the dryer.
- purchased power has to 50% less expensive than cogenerated power

Our example, for Swedish conditions, presents the following economic evaluation:

Price for dry product:	USD	186.-/ton
" " electric power	USD	20.-/MWh
" " fuel oil	USD	213.-/m^3
Operating hours		2.160 h/year

Savings in relations to existing operation

	1	2
Energy Savings USD/year	752 000	870 666
8% Increased product yield USD/year	322 666	322 666
Improved accuracy of Product	?	?
Moisture Control, 2%	80 000	80 000
Reduced Maintenance	?	?
Reduced Personnel	?	?
Improved Environment	?	?
Improved Quality	?	?
Costs for power, USD	33 333	64 000
Net savings (for beet campaign):	1 188 000	1 209 333

With the low price of hydroelectric power we have in Sweden, the two alternatives gives about the same savings. However, investment cost for alternative 2 is higher. Rough numbers for turn-key installation are USD 2 533 000 for alt. 1 and USD 2 733 333 for alt. 2.

If the dryers in this example can be used more than for just drying sugar beet, pay-back time can be reduced further.

EXERGY EVALUATION IN GRAIN DRYING

Koro Kato

Department of Agricultural Engineering, Kyoto University
Kyoto, 606 JAPAN

ABSTRACT

In Japan, about 800,000 kl of fuel oil are consumed for drying farm products every year. Consequently, saving energy in the grain drying process has become an important agricultural problem. To reduce the energy consumed in grain drying and to establish a method of evaluating energy requirement based on exergy, theoretical and experimental studies were carried out. The energy supplied for grain drying is finally converted into the energy of grain moisture concentration difference, and is stored in grain. This energy is based on the difference in moisture content of grain compared with surroundings, and is released as heat of wetting when the moisture absorption occurs. In this study, we propose a method of exergy evaluation for grain drying, taking into account this energy of moisture concentration difference.

First, we measured the thermal properties such as heat of wetting and equilibrium moisture content of rice. The exergy of moist grain can be obtained by integrating the exergy of the equilibrium vapor pressure of the grain moisture. Taking into account the moisture concentration exergy of moist grain and moist air, we investigated the flow and efficiency of energy and exergy of conventional heated air rice dryer. The efficiency of converting supplied energy to moisture concentration energy (excepting the latent heat of vapor) is 2 to 4% for the heated air rice dryers. The efficiency of converting supplied energy to moisture concentration exergy is extremery low: about 0.5%. The efficiency of converting heated air exergy to grain moisture concentration exergy is 10 to 15%. The irreversible loss of the heated air producer with oil burner is the dominant loss. In the case of rice drying, the suitable temperature of heated air is low: 35° to 50°C, but a large quantity of heated air is required, so that the exergy efficiency of the heated air producer of the rice dryer is only 3 to 5%. In this case, both from the thermodynamic and from the safety point of view, it is better to use a heat pump instead of an oil burner as a heated air producer. Then we made a rice drying system with a heat pump using atmospheric and exhaust heat as heat sources and experimented with the system. The flow of energy and exergy, and high efficiency of the system were determined.

1. INTRODUCTION

In Japan, 5,000,000 kl of fuel oil are consumed every year in agriculture. Japan is a major oil-consuming nation and most of the oil is imported. Therefore the problem of energy saving in agriculture becomes important. Since the humidity is high in Japan, heated air dryers are mostly used for drying farm products, and 800,000 kl of fuel oil are consumed annually for this purpose. Of this amount, some 450,000 kl are used for drying rice and wheat.

In general, the measure for energy savings consists of: (1) utilization and development of new energy or alternative energy sources,
(2) improvement and development of processes,
(3) utilization of exhaust heat, and
(4) reduction of total energy by systems analysis.

With regard to energy used for drying farm products, the following approaches have been studied and applied partially in practice:
(1) the utilization of rice-husks for fuel,
(2) the use of solar energy, and
(3) the use of atmospheric or exhaust heat, increasing the temperature with heat pump

Subsequently, the improvement and development of drying processes are necessary, and the establishment of a method of energy evaluation for the grain drying process becomes important. The purpose of this study is to establish a method of energy evaluation for grain drying, especially a method based on exergy.

For the grain drying process, the energy of grain moisture concentration difference has an important significance, because the energy supplied is finally converted into the energy of grain moisture concentration difference and stored in the dried products. Accordingly, it is necessary to investigate the flow of energy and exergy in the dryer, taking into account the moisture concentration energy.

2. METHOD OF EXERGY ANALYSIS IN CONSIDERATION OF MOISTURE CONCENTRATION

To investigate the exergy flow in a heated air dryer for grain, in addition to the exergy of fuel[1], it is also necessary to know the exergy of moist air which functions as a thermal medium, and the exergy of raw and dried grain as a function of moisture concentration. The exergy of moist grain and moist air are derived from the following equation[2] for the exergy of a general mixture.

$$e = \sum_k y_k \cdot \hat{e}_k \qquad (1)$$

where $\hat{e}_k$ is partial specific exergy of component k

and given by

$$\hat{e}_k = \hat{h}_k - \hat{h}_{ku} - T_u(\hat{s}_k - \hat{s}_{ku}) \qquad (2)$$

2.1 Exergy of Moist Air for Drying Process

When moist air is considered as a mixture of two components: dry air (a) and water vapor (v), from Eq.(1), the exergy per 1 kg of dry air or (1+x) kg of moist air is:[2]

$$\begin{aligned} e_x &= \hat{e}_a + x\,\hat{e}_v = (c_{pa} + xc_{pv})(T - T_u - T_u \ln T/T_u) \\ &\quad + T_u(R_a \ln p_a/p_{au} + x\cdot R_w \ln p_v/p_{vu}) \\ &= (c_{pa}+xc_{pv})(T-T_u-T_u \ln T/T_u) + T_u\{R_a \ln \frac{x_u+\varepsilon}{x+\varepsilon} \\ &\quad + xR_w \ln \frac{x_u+\varepsilon}{x+\varepsilon}\cdot\frac{x}{x_u}\} + T_u(R_a + xR_w)\ln P/P_u \end{aligned} \qquad (3)$$

The first, second and third terms in Eq.(3) describe the exergy of temperature, humidity and total pressure respectively. When the surrounding condition is standard air, the exergy of the moist air calculated from this equation is expressed in Fig. 1.

The exergy also varies with the surrounding condition (u). Fig. 2 shows the exergy of moist air when surrounding condition is saturated air at 20°C. In case of exergy evaluation of air for drying, it is proper to consider the surrounding condition to be state saturated with water vapor. In this case, the exergy of low moist air can represent the drying ability of the air.

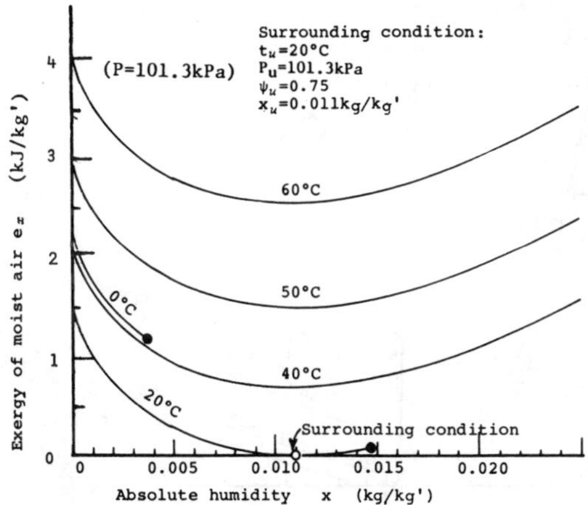

Fig. 1 Exergy of moist air at 1 atm (Surroundings: Standard air)

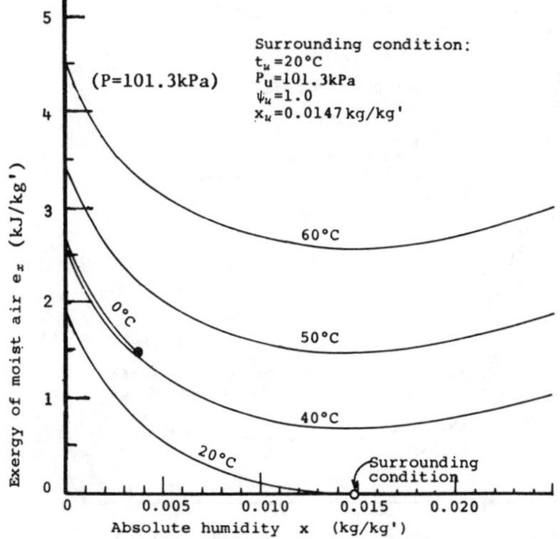

Fig. 2 Exergy of moist air at 1 atm (Surroundings: Saturated air at 20°C)

2.2 Grain Thermal Properties Related to Exergy

A hygroscopic material like grain produces the heat of wetting when water absorption occurs. The produced calorific value can be expressed as a function q(m,T) of the moisture content m and the temperature T of the grain. It is possible to consider this heat as the moisture concentration difference energy between the grain and surrounding water. The relationship between the differential heat of wetting per latent heat of vaporization of pure water r and the moisture content of rice m is shown in Fig. 3.

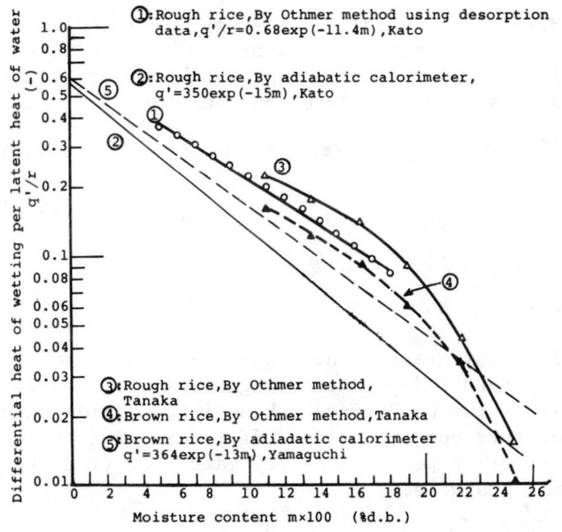

Fig. 3 Differential heat of wetting of rice

In this study the following experimental equation is used as the differential heat of wetting for rice.

$$q'(m,T) = 0.68\,r\,e^{-11.4\,m} \qquad (4)$$

On the other hand, there is a definite relationship between the moisture content m of a hygroscopic material and the relative humidity, and this can be expressed as the equilibrium relative humidity $\psi(m,T)$. Fig. 4 shows the equilibrium moisture content of rice. In this study the following experimental equation, which has the same form as that proposed by Strohman and Yoeger[3], is used for the equilibrium relative humidity of rice.

$$\ln\varphi(m,T) = 0.68\,e^{-11.4\,m}\cdot \ln(p_s/133) - 7.63\,e^{-14.4\,m} \qquad (5)$$

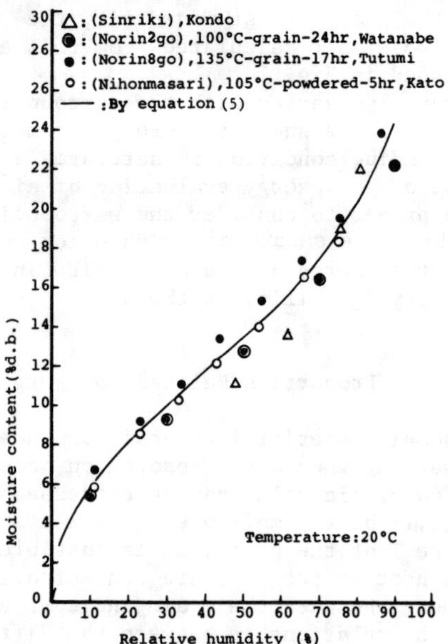

Fig. 4 Equilibrium moisture content of rough rice (Desorption data exept mark Δ)

Moreover, the following equations in relation to the heat adsorption of vapor L(m,T) and differencial heat of wetting of hygroscopic material are formed from Clausius-Clapeyron equation:

$$L(m,T) \doteqdot R_w \frac{T_1 \cdot T_2}{(T_2 - T_1)} \cdot \ell n(p_{v2} / p_{v1}) \tag{6-1}$$

$$q'(m,T) \doteqdot R_w \cdot \frac{T_1 \cdot T_2}{(T_2 - T_1)} \cdot \ell n\{\varphi(m,T_2)/\varphi(m,T_1)\} \tag{6-2}$$

2.3 Exergy of Moist Grain

If moist grain is considered as a mixture of two components: dry matter of grain and moisture, the exergy can be derived taking into consideration the moisture concentration difference. The exergy of moist grain per 1 kg of dry matter or (1+m) kg of moist grain is:

$$e_g = \hat{e}_d + m\hat{e}_w \tag{7}$$

Where $\hat{e}_d$ is partial specific exergy of dry matter and $\hat{e}_w$ is partial specific exergy of grain moisture. In a steady flow system, these are given by

$$\hat{e}_d = (\hat{h}_d - \hat{h}_{du}) - T_u(\hat{s}_d - \hat{s}_{du}) \tag{8}$$

$$\hat{e}_w = (\hat{h}_w - \hat{h}_{wu}) - T_u(\hat{s}_w - \hat{s}_{wu}) \tag{9}$$

The relation between the partial specific quantities of grain moisture, the above described differential heat of wetting $q'(m,T)$ and the equilibrium relative humidity $\psi(m,T)$ are:

$$\hat{h}_w = c_{pw}t_g - q'(m,T_g) + (h_{ow}) \tag{10}$$

$$\hat{s}_w = c_{pw}\ell n T_g/T_o - q'(m,T_g)/T_g - R_w \ell n\,\varphi(m,T_g) + (s_{ow}) \tag{11}$$

If the equilibrium moisture content of the surrounding atmosheric air is expressed by m_u, the partial specific quantities of the surrounding moisture is:

$$\hat{h}_{wu} = c_{pw}t_u + r(T_u) + (h_{ow}) \tag{12}$$

$$\hat{s}_{wu} = c_{pw}\ell n T_u/T_o + r(T_u)/T_u - R_w \ell n\varphi(m_u,T_u) + (s_{ow}) \tag{13}$$

From these equations and Eq.(9), the partial specific exergy of the moisture of grain is:[6]

$$\hat{e}_w = c_{pw}(T_g - T_u - T_u \ell n T_g/T_u) + R_w T_u \ell n\{\varphi(m,T_u)/\varphi(m_u,T_u)\}$$

$$= c_{pw}(T_g - T_u - T_u \ell n T_g/T_u) + R_w T_u \ell n\{p_{vg}(m,T_u)/p_{vu}\} \tag{14}$$

Using the same procedure, the partial exergy of dry matter is:[6]

$$\hat{e}_d = c_d(T_g - T_u - T_u \ell n T_g/T_u) + R_w T_u \int_{m_u}^{m} \ell n\varphi(m,T_u)dm - R_w T\{m\,\ell n\varphi(m,T_u) - m_u \ell n\varphi(m_u,T_u)\} \tag{15}$$

Finally, from Eqs.(7),(14) and (15), the exergy of moist grain becomes:[6]

$$e_g = (c_d + m c_{pw})(T_g - T_u - T_u \ell n T_g/T_u) + R_w T_u \int_{m_u}^{m} \ell n\{\varphi(m,T_u)/\varphi(m_u,T_u)\}dm \tag{16}$$

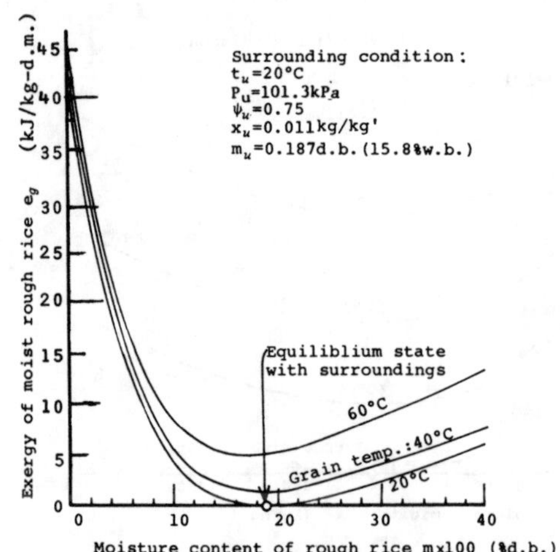

Fig. 5 Exergy of moist rough rice (Surroundings: Standard air)

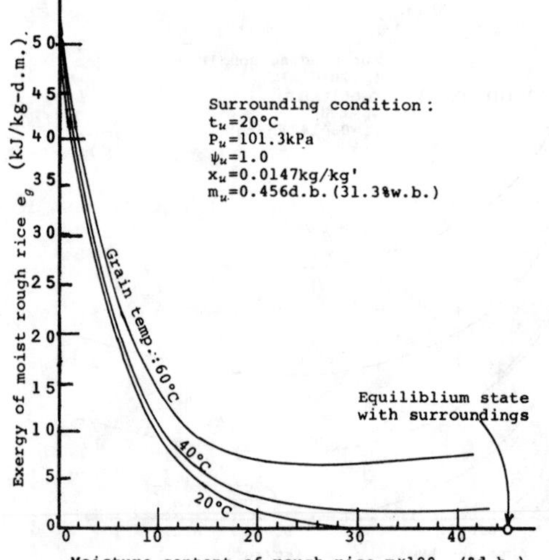

Fig. 6 Exergy of moist rough rice (Surroundings: Saturated air at 20°C)

The first term represents exergy based on the grain temperature. The second term represents the exergy based on the moisture concentration difference between the grain and the surroundings, and is expressed as follows:

$$e_{gc} = R_w T_u \int_{m_u}^{m} \ell n\{p_{vg}(m,T_u)/p_{vu}\}dm \qquad (17\text{-}1)$$

This is a definite integral of the exergy of the equilibrium vapor pressure of the grain moisture.

Figs.(5) and (6) show the exergy of moist grain when the surrounding condition is standard air and saturated air at 20°C respectively. In the case of exergy of moist grain, it is proper to consider the surrounding condition to be saturated state, too. In this case, the exergy of grain increases with the progress of drying.

From Eq.(6-1), the exergy of heat adsorption of vapor agrees with Eq.(17-1), when only the vapor pressure of grain $p_{vg}(m,T_g)$ in Eq.(17-2) exists in equilibrium with the surrounding vapor pressure p_{vu}.

$$\begin{aligned} e_L &= \int_m^{m_u} \frac{T_g - T_u}{T_g} L(m,T_g)dm \\ &= R_w T_u \int_m^{m_u} \ell n\{p_{vg}(m,T_g)/p_{vg}(m,T_u)\}dm = e_{gc} \end{aligned} \qquad (17\text{-}2)$$

3. ENERGY AND EXERGY FLOW OF ACTUAL RICE DRYER

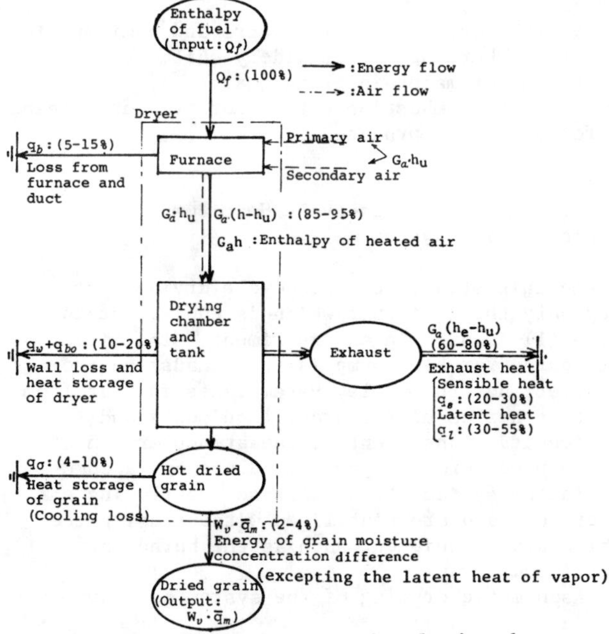

Fig. 7 Energy flow of conventional rice dryer

Fig. 7 shows the experimental energy flow data obtained in the most popular heated air dryers of rice in Japan. Since the humidity is high in Japan, most of the rice dryers are air heating type with oil burner. The heated air at 30 to 50°C is blown into the grain layer with rate of 0.3~1 m³/s per ton of rice for about 12 to 24 hours for drying. Grain fissure occurs if the temperature of the heated air is higher than this limit. It became clear that, with the enthalpy of fuel taken as 100%, 2 to 4% is stored finally in the grain as the energy of moisture concentration difference (exeping the latent heat of vapor).

The exhaust loss is great (60 to 80%), in which sensible heat loss is 20 to 30% and the latent heat loss is 30 to 55%. The rest are wall losses and heat storage of dryer and grain. Accordingly, the efficiency of converting the supplied energy to the energy of the moisture concentration difference is 2 to 4% for heated air rice dryers which are popular on Japanese farms.

Fig. 8 shows the exergy flow of the heated air dryer calculated from Eqs.(3) to (17) and the same experimental data. When the exergy of the fuel is taken as 100%, the exergy of heated air for drying is reduced to 3 to 5%. The irreversible loss in the furnance results from the loss due to the combustion of fuel (30 to 40%), and the loss due to the mixing of the cool secondary air (55 to 65%).

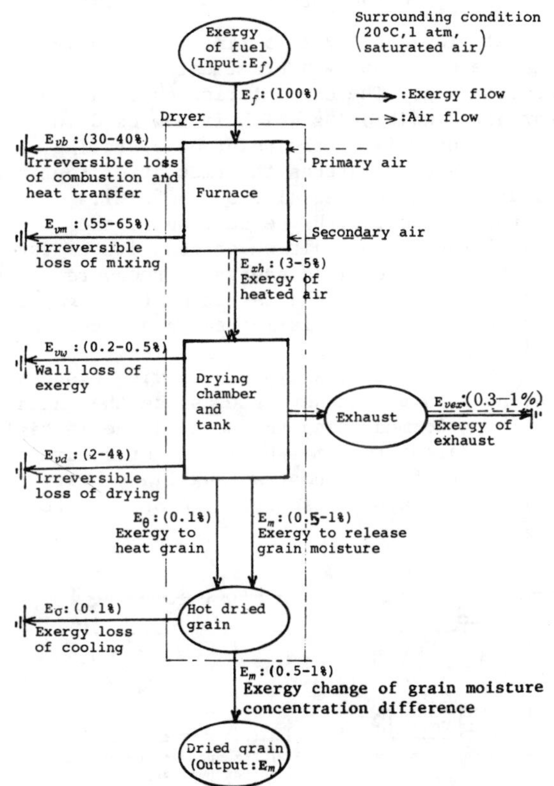

Fig. 8 Exergy flow of conventional rice dryer

Subsequently, the irreversible loss in the drying chamber is about 2 to 4%, and finally, only 0.5 to 1% is stored in grain as the exergy of moisture concentration differece. Accordingly, the exergy efficiency of the heated air rice dryer is extremely low (0.5 to 1%).

4. RICE DRYING BY HEAT PUMP AND EXERGY FLOW OF THE SYSYEM

As described in 3, the suitable temperature of heated air in rice drying process is low, but a large quantity of heated air is required, so that the exergy efficiency of the heated air producer with an oil burner is only 3 to 5%. In this case, both from the thermodynamic and from the safety point of view, it is better to use a heat pump instead of an oil burner as a heated air producer. Then we made a rice drying system with a heat pump using atmosperic and exhaust heat as heat sources and experimented with the system.

4.1 Method of Rice Drying Using Heat Recovery by Heat Pump

In general, when the temperature of the exhaust heat source is lower than the desired temperature, the operation of heat recovery and increase of temperature using the heat pump are planned. The coefficient of performance of the reverse Carnot cycle is:

$$\epsilon h_c = Q/L_c = T_h/(T_h - T_l) \quad (18)$$

A heat flow about 3 to 5 times the required power can be drawn up by a practical heat pump for air conditioning. The characteristics of a heat pump for air conditioning which is now used in practice, especially on the point of operating temperature, easily matches the rice dryer (optimum temperature of the heated air: 30 to 50°C).

In this method, with the heat pump, the enthalpy of the exhaust air which constitutes the major part of the supplied energy is recovered into the inlet air including the latent heat of evaporation and the temperature is increased and the exergy is regenerated.

Fig. 9 shows various types of drying systems usig the heat pump. In any case, since the initial heating is performed using the heat of the initial exhaust equivalent to atmospheric air or accumulating the power of heat pump, an auxiliary heat source is not always necessary. The open cycle heat pump dryer (I)[5] is suitable when the atmospheric temperature is high and humidity is low, because the drying capacity of the atmospheric air can be utilized.

On the other hand, when the temperature of the atmospheric air is low and humidity is high, the closed cycle heat pump dryer (II)[5] is suitable.

In system (V),[7] the flow rate of circulated exhaust air can be controlled to utilize the merits of both systems (I) and (II). System (III) is the method of cold air drying by the low humidity air.

The application of the heat pump in rice drying equipment has a number of merits: energy saving, security, operation and quality of dried grain. More specifically,

1) The required energy is reduced due to the utilization of the heat of atmospheric air and recovery of exhaust heat.

2) With respect to operating temperature, the heat pump easily matches the requirement of the rice dryer, therefore exergy loss during the production of heated air is small.

3) The rice dryer is used at night in Japan. A dryer operating without using fire improves the security and automatic operation at night is possible.

4) An ignition system including the supply and addition of fuel are not necessary and operation is simple.

5) The control of temperature and humidity is possible. Therefore it is widely suited to the requirement of maintain grain quality.

6) Since combustion gas is not present, clean air for drying is produced.

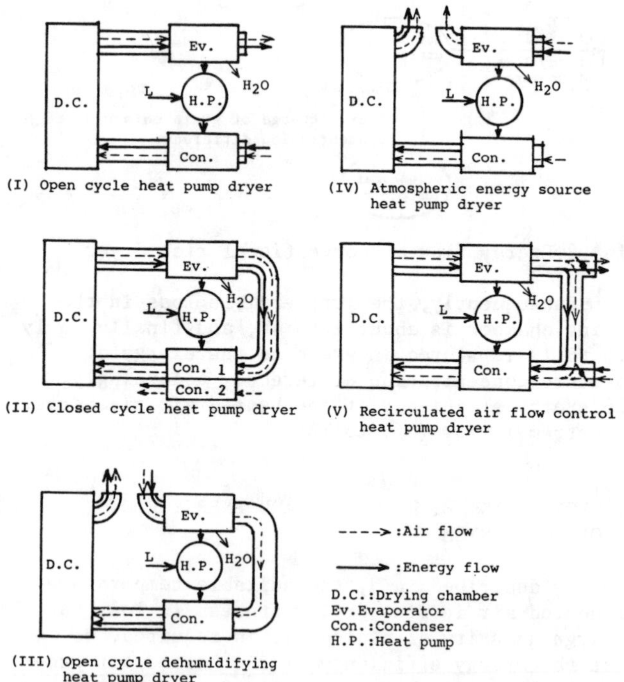

Fig.9 Variety of heat pump dryer

4.2 Performance of Open-cycle Heat Pump Rice Drying System

In this study, the open cycle drying system using only the heat pump which is equivalent to Fig. 9 (I) is used in an experiment. The direct heat source for heat pump is the exhaust heat of the dryer, but the whole system takes in heat from the atmospher as heat source, then cools and releases it. Consequently, the atmospheric heat and the power of heat pump circulate and accumulate in the system, then increase the flowing air temperature and are reutilized without using the usual heating equipment such as the burner or electric heater.

Aschematic drawing of the system is shown in

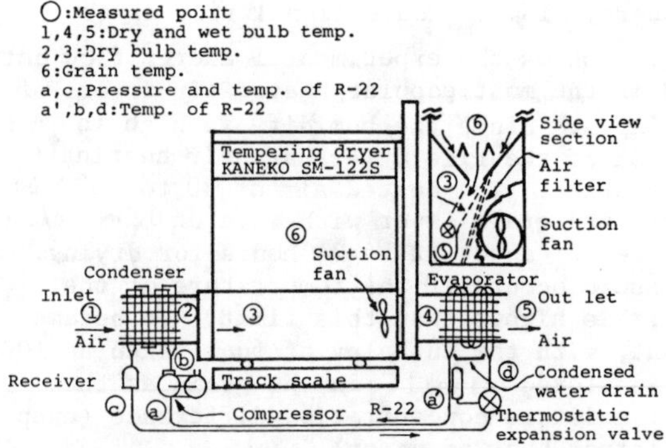

Fig. 10 Schematic drawing of the experimental open- cycle heat pump drying system

Table 1 Experimental condition and main results

			Experimental values
Grain	Kind of grain		Rough rice(Nihonmasari)
	Initial moisture cont.	m_1	24.9%d.b.
	Initial grain temp.	t_{g1}	20.6°C
	Initial grain weight		600kg
	Dry matter weight		480kg
Ambient air	Temperature	t_1	21.2°C
	Absolute humidity	x_1	0.007kg/kg'
	Relative humidity	ϕ_1	0.448
Drying condition	Air flow rate	Q_a	0.243 m^3/s
		G_1	1037kg'/h
	per 100kg-grain		0.0405m^3/s·100kg-grain
	Grain flow rate		2700∿3130kg/h
	Drying air temp.	t_2	48.8°C
Test results	Drying time		6.1h
	Final moisture cont.	m_2	16.8%d.b.
	Final grain temp.	t_{g2}	31.6°C
	Evaporated water from grain		38.8kg
	Evaporating speed	W_v	6.36kg/h
	Average drying speed		1.33%d.b./h
	Condensed water		12.7kg
	Exhaust air temp		
	Outlet of grain layer	t_4	27.8,(19.6∿32.0)°C
	Outlet of evaporator	t_5	16.3,(12.5∿18.0)°C
	Heat pump		
	Heat rejected at condenser	Q_c	8.22 kW
	Heat absorbed at evaporator	Q_e	6.16kW
	Output energy of compressor	Q_p	2.08 kW
	Power consumption of motor	L_m	2.2kW
	Refrigerant flow rate	G_{re}	132.5kg/h
	Coefficient of	$eh'=Q_c/Q_p$	3.95
	performance	$eh''=Q_c/L_m$	3.72
	Condensing pressure	P_H	18.9kg/cm^2abs.
	temperature	t_H	47.6°C
	Evaporating pressure	P_L	6.0kg/cm^2abs.
		t_L	4.8°C
	Degree of superheat		20.7°C
	Degree of subcooling		5.9°C
	Energy consumption per 1kg-water removed(except fan)	L_m/W_v	0.346kWh/kg-H_2O
	Energy efficiency	$\eta_1=W_v(r+\bar{q}_m)/L_m$	210%
		$\eta_2=W_v\cdot\bar{q}_m/L_m$	13%
	Exergy efficiency	$\eta_e=E_m/L_m$	2.3%

Fig. 10. The capacity of the dryer is 1.2 tons, a scal which is commom to small farms in Japan. The power of the heat pump is 2.2 kW and the refrigerant is R-22.

Table 1 describes the experimental condition and the main results. The experiment is carried under the design condition, the heat flow recovered by the evaporator is 6.14 kW, the heat flow rejected at the condenser is 8.22 kW, and the coefficient of performance compared with the required electric power is good:3.7, because the enthalpy and the temperature of exhaust heat source is high. The energy consumption per 1 kg of removed water is 0.346 kWh/kg.H_2O (1245 kJ/kg.H_2O),and this is about 1/4 of a common heated air rice dryer.

4.3 Change of State and Exergy of Flow Air

The change of state of the flowing air is illustrated on the psychrometric chart shown in Fig. 11. The circled numbers describe the state of the measured points which are shown in Fig. 10, the straight lined arrow express the change of state along the air flow at each period; and the dotted lined arrow shows the time dependent change at each measured point. The inlet air state ① is heated to state ② at constant absolute humidity by the heat pump only. The heated air ③ passes through the grain layer, the change is slightly inclined from the iso-enthalpy line. Subsequently the exhaust air ④ of the grain layer releases the latent and sensible heat into the evaporator, then the air discharged at state ⑤.

Fig. 12(lower) shows the mean values of the exergy of flowing air during the process. In this case, as described above, in order to analyze the drying process, the normarized surrounding condition

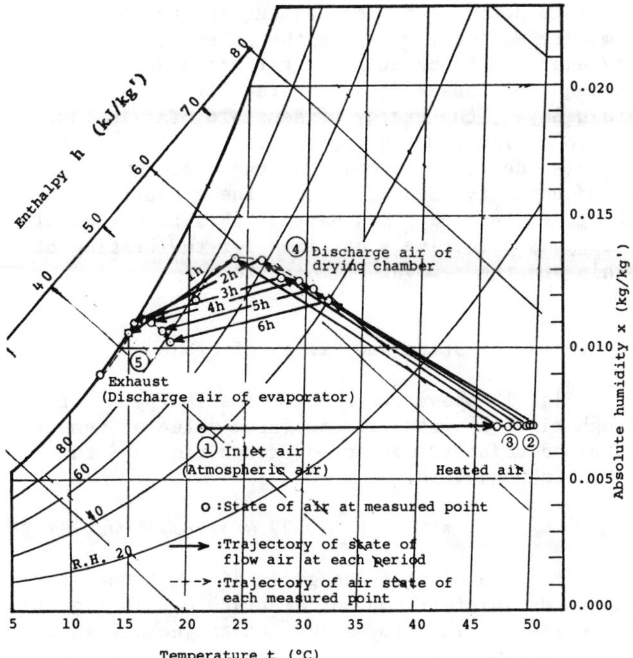

Fig. 11 Change of state of flow air

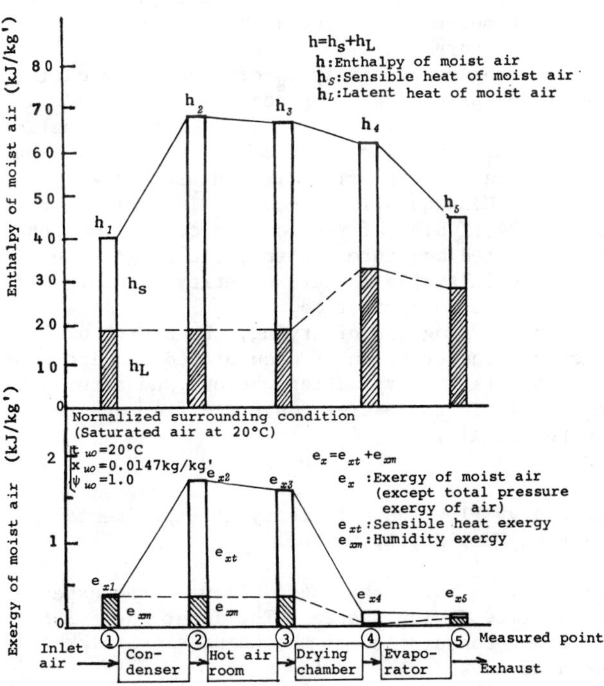

Fig. 12 Enthalpy (upper), sensible heat exergy and humidity exergy of flow air (lower)

is assumed saturated air at 20°C. The quantity e_{x1} is the exergy of the inlet atmosphere during the experiment; because the humidity is lower than the assumed saturated surrounding condition it possesses the exergy of low humidity e_{xm}. It can be used for evaluating the drying capacity of the atmosphere during the experiment, and it signifies the possibility of atmospheric ventiration drying.

The quantities e_{x2} and e_{x3} are exergies of heated air; and they possess the exergy of sensible heat e_{xt} due to heating and the exergy of humidity e_{xm} equivalent to the inlet air. These are exergies to be utilized for grain drying. During the passage of the heated air through the grain layer, the exergy of sensible heat is lost due to cooling; moreover, the exergy of humidity e_{xm} also decreases because of the approach to saturated humidity, and it becomes the exhaust air of the grain layer e_{x_4}. A part of this reduction of exergy is effectively utilized for the heating of grain and the removal of moisture.

4.4 Change of State and Exergy of Grain

Fig. 13 shows the change of enthalpy h_g of rough rice obtained from measured value of temperature and moisture content of rough rice and the following equation.

$$h_g = (c_d + c_{pw} \cdot m) t_g - \int_0^m q(m,T)\, dm + (h_{od} + m h_{ow}) \quad (19)$$

It is found that the grain of low temperature and high moisture content at state 0 h in Fig. 13 increases its enthalpy along the trajectory shown by the arrow, and approaches the grain state in equilibrium with the heated air marked by ○. The enthalpy of the grain increases by 28.9 kJ/(kg of dry matter) before and after drying. Of this, the energy of moisture concentration difference is 13kJ/(kg of dry matter).

The change of exergy e_g of rough rice during the drying process and its components e_{gt} and e_{gc} are shown in Fig. 14. In this case, the normalized surrounding condition is equal to the surrounding condition as for the flow air: the saturated state at 20°C. Consequently, since the initial moisture content 24.9% d.b. of the rough rice is lower than the saturated moisture content, the rough rice possesses initially a small quantity of exergy of low moisture concentration.

In the progress of drying, the exergy of low moisture concentration is accumulated and increases by 2.76kJ/(kg of dry matter) before and after drying. This is the exergy converted from one part of the heated air.

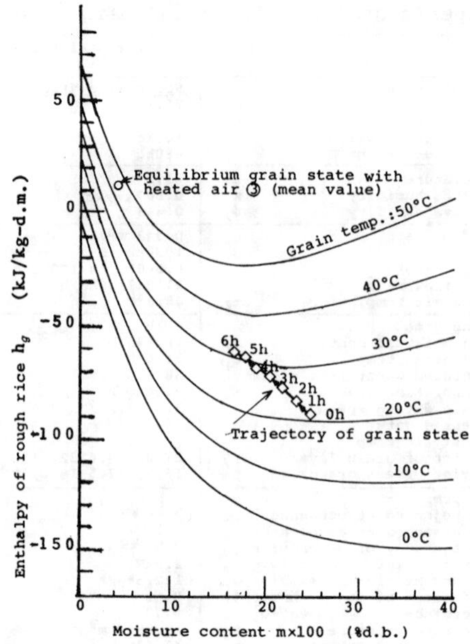

Fig. 13 Change of grain state during drying period (Enthalpy-moisture diagram of rough rice)

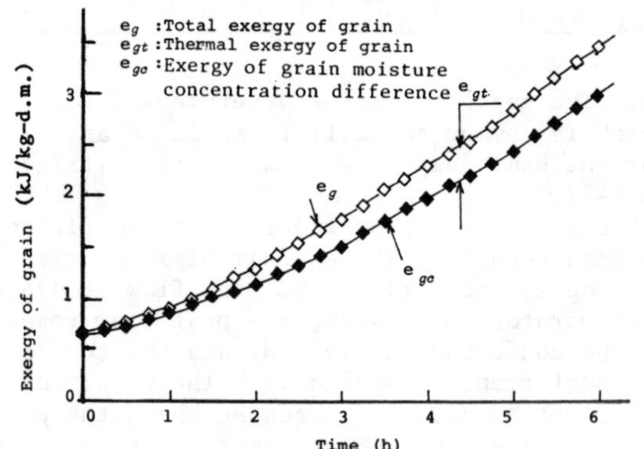

Fig. 14 Change of exergy of grain during drying period

4.5 Energy Flow and Efficiency of Experimental Heat Pump Drying System

Fig. 15 Shows the energy flow of the experimental rice drying system. The input energy: electric power consumption of the compressor (2.2kW) is taken as 100%.

Of the heat flow rejected at condenser (8.22 kW) to heat the fresh air, the heat recovered by evaporator from exhaust occupies about 3/4.

The heat loss of exhaust is a small quantity compared with conventional rice dryer.

The efficiency of converting the supplied energy to the energy of grain moisture concentration difference (excepting the latent heat of vapor) is 13 % in this heat pump drying system.

4.6 Exergy flow and Efficiency of Experimental Heat Pump Drying System

Fig. 16 shows the exergy flow of the experimental rice drying system. The elecrtic power consumption of the compressor which is the input exergy of the system is taken as 100%.

The exergy efficiency of the heated air producer of the heat pump using exhaust heat as heat source is 18%.

The efficiency of converting the exergy of heated air to the exergy of grain moisture concentration difference is 13%.

Finally, the efficiency of converting the supplied exergy to the exergy of grain moisture concentration difference for this system is 2.3%, a value 4 times that of the exergy efficiency of ordinary heated air rice dryer with oil burner.

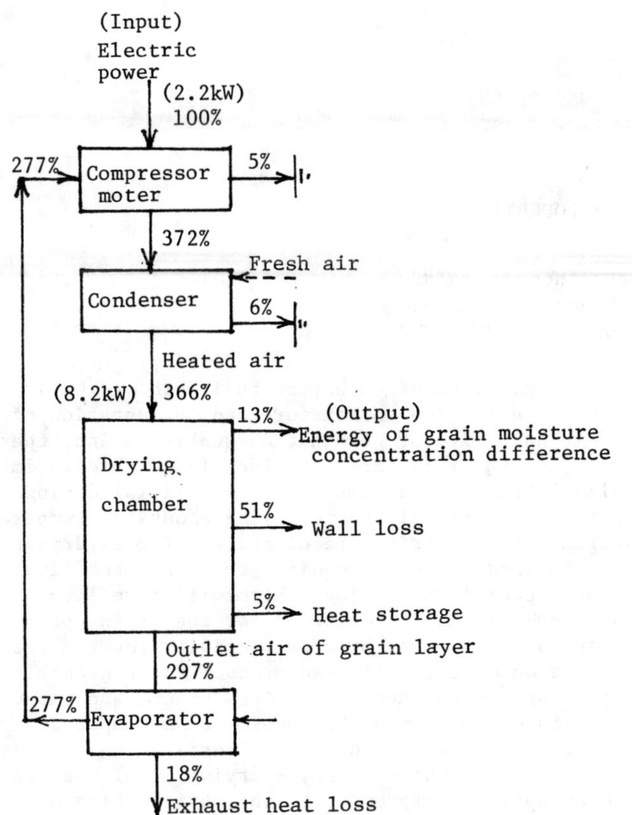

Fig. 15 Energy flow of the heat pump drying system

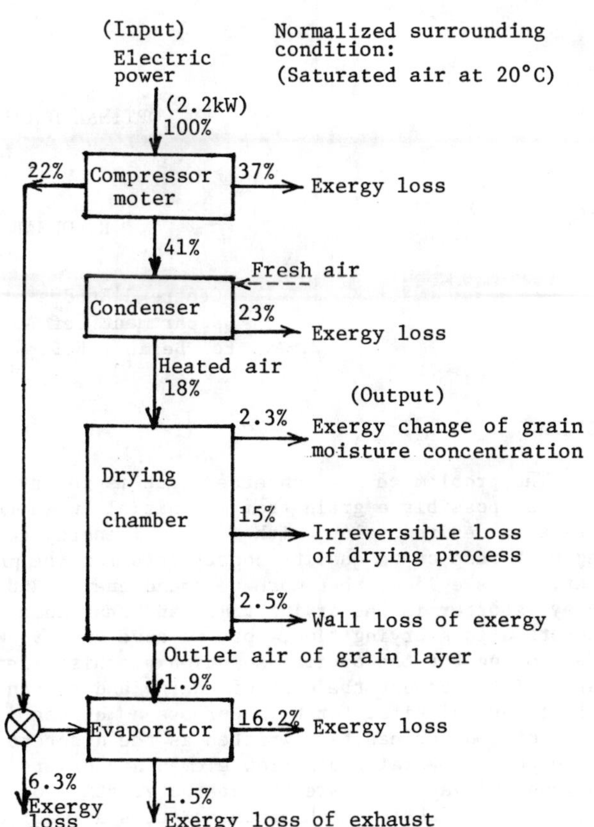

Fig.16 Exergy flow of the heat pump drying system

NOMENCLATURE

e	specific exergy	kJ/kg
e_x	specific exergy of moist air	kJ/kg'
e_g	specific exergy of moist grain	kJ/kg-d.m.
$\hat{e}_k$	partial specific exergy of component k	kJ/kg
y_k	mass fraction of component k	-
ε	$R_a/R_w=0.622$	-
c_d	specific heat of dry matter of rice:	1.256 kJ/kg K
P	total pressure	Pa
p_s	saturated vapor pressure of water	Pa
$p_{vg}(m,T)$	vapor pressure of grain moisture at state of (m,T)	Pa
P_{vu}	vapor pressure of surroundings	Pa
$L(m,T)$	heat adsorption of vapor to grain	kJ/kg-H_2O
r	latent heat of vapor	kJ/kg
$q'(m,T)$	differntial heat of wetting for grain at state of (m,T)	kJ/kg-H_2O
$\psi(m,T)$	equilibrium relative humidity of grain moisture at state of (m,T)	-
m	moisture content d.b.(decimal)	-
x	absolute humidity	kg/kg'
e_{gc}	exergy of moisture concentration difference	kJ/kg-d.m.

Sufix

^	partial specific quantity
k	k component
u	surrounding condition
0	standard condition
v	vapor
w	water or moisture
g	grain
d	dry matter
a	dry air

REFERENCES

1. Rant, Z., Brennstoff-Wärme-Kraft, Vol. 12, 1, (1960)
2. Moebus, W., Luft-und Kältetechnik, Vol. 8, 3, pp.125 (1972)
3. Strohman, R.D. and Yoerger, R.R., Trans. ASAE Vol. 10,657 (1967)
4. Othmer, D.F. and Huang, H., Ind. Eng. Chem. Vol. 57, 42 (1965)
5. Kato, K. and Matsuda, R., J. Soci. Agric. Machin. Jap. Vol. 37, 613 (1976) and Vol.38, 385 (1976)
6. Kato, K., J. Soci. Agric. Machin. Jap. Vol. 43, 443 (1982), Vol. 44, 69 (1982) and Vol. 45, 85 (1983)
7. Kato, K., and Yamashita, R., Proc 40th Anual Meeting of Soci. Agric. Machin. Jap. Vol. 40, 122 (1981)
8. Kato, K., and Yamashita, R., J. Soci. Agric. Machin. Jap. Vol. 43, 589 (1982)
9. Ishitani, K., et al., Text of the 68th Lecture Meeting of the J.S.M.E.(Kansai Branch) pp.1-66 (1979)

OPTIMAL DRYING CONDITIONS OF GRAINS

R. OUHAB and A. LE POURHIET

Centre d'Etudes et de Recherches de TOULOUSE
Department of Automatic Control (CERT-DERA)
2, Av. Ed. Belin - B.P. 4025 - 31055 TOULOUSE CEDEX - FRANCE -

ABSTRACT

The problem we are concerned with is to dry as quick as possible a grain from an initial known moisture to a desired value, with a minimal energy cost and with respect to quality constraints for the product. It is evident that more we spend energy and money, shorter is the drying time, and that more one accepts a long drying time, more we save energy. We are looking for the optimal solution of this dilemma, taking into account the cost of energy and an equivalent cost of time. For this purpose a mathematical model of a dryer has been written as the discrete sequence of several cross-flow elementary dryers. The control variables are the characteristics of drying airs entering each of these steps as well as the global drying time. Their optimal values are computed by using dynamic programming. This optimisation includes the renewing of some used airs if they keep some interesting drying capacities. Numerous abacus and simulation curves show the main results of this study, and applications to optimal design and automatic control of dryers are discussed at last.

I. INTRODUCTION

Everyone knows the important part of drying in the final cost of most industrial products such as textile, skin, paper, plaster, brick, ceramics, food products. This is mostly due to the high cost of energy and the necessity of making profitable the costly investment of a drying unit for example. Moreover, the natural slow process of evaporating compels to find means of accelerating this process, and therefore, increase the output of the unit in terms of quantity. The artificial drying will only be studied here. It is obtained when the humid product is subject to the action of a hot air current the characteristics of which (temperature, relative humidity, speed) determine its drying capacity. Many experiments have been carried out, -often authoritatively- in order to increase the productive capacity of industrial units for a same amount of energy. This care is always present in mind as everywhere else in the food industry and in the cereal drying in particular where, for some time, a willing to get a better output has been noticed, not only as far as the energy is spent but also as far as the existing productive units are concerned.

Many food products such as grains (cereals, oil seeds, protein seeds) are humid when harvested and must rapidly be dried before their storage to avoid deterioration, due to moisture and fermentation, of their nutritive or germination qualities and, therefore, of their commercial value. This rapidity implies straight away the use of artificial drying instead of natural static drying at outside temperature. The industrial use of the artificial drying in the food industry requires the implementation of a new specific technology which will take into account the quality criteria for the drying product, cost saving for the process, rapidity for the preservation of the collected product and reasonable cost for the purchase of the equipment. Whereas the quality criteria must be considered as imposed constraints, minimizing the criteria which include all factors (cost - energy - drying time) must lead to an optimal compromise to the satisfaction of all farmers, industrialists and customers.

This is why that for several years and especially since the brutal increase of the cost of fossile fuel, the drying professionals devoted themselves to make clever technological adjustments and tried to reach this optimum criterion. The important savings which have been obtained by means of intuitive designs of new dryers allow to foresee than even better results which could be obtained by using analytical optimising methods. It is therefore attracting for the Automatic scientist to intervene in a field where his methods can give a favourable answer to the requirements imposed by the economic conjoncture. The problem raised is to dry a grain as fast as possible until a final imposed humidity (w_f) is obtained and, at the same time, keep the energy cost of the process as low as possible and respect the quality constraints for the product. It is obvious that any increase of energy means a reduction of the drying time and we are faced with a dilemmma : which is the optimal compromise between the drying time and the spent energy which leads to the best satisfaction of the dryer user.

For this purpose a generalized cost has been defined as

$$J = a t_d + b W \qquad (1)$$

which should be minimized. On figure (5) it is clearly shown that an optimal policy exists.

In this study the complete dryer will be divided into N stage as shown on figure 1. The advance of the grains from a stage to the next one is made periodically when the grain of the lowest stage is extracted. Then the grain stays in each stage during a time $\Delta t = t_d/N$. Each stage is equivalent to an elementary crossflow dryer the length of which, L, depends on the internal structure of the dryer.

Its inlet airflow is the ambient outside air to which thermal and mecanical energies have been supplied ($T_{a_{in}}$ and u_a).

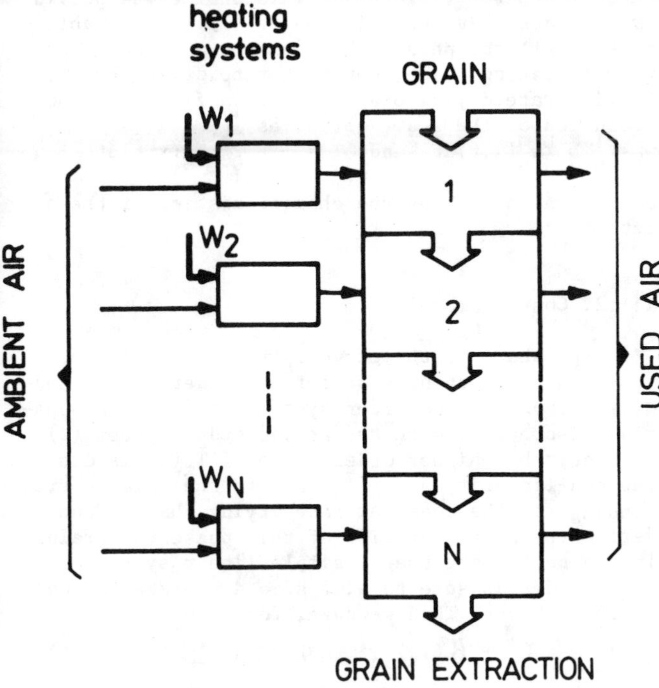

Fig. 1 - General scheme of the dryer

In the following the so-called control variables are the external parameters the user is able to tune : $\Delta t, u_a, T_{a_{in}}$. At the opposite the internal parameters are, for example, the nature of the grain, its initial moisture, the characteristics of outside air, the heating capacity of the energy source,... etc...; when these parameters have constant values, the optimisation is the computation of the control variables which minimize J. When this calculation is made at any time and takes into account any disturbances on the internal parameters one speaks on the automatic control of the dryer ; if the changes on internal variables cannot be directly mesured, they have to be estimated from all the available measurements, and this is generally the main difficulty of any automatic control problem [12]. The automatic control needs a small computer to process the measured data and determine at any time the best strategy to apply. Automatic control includes the optimisation phase and it is therefore necessary to achieve a proper optimisation scheme before going any further.

If the design of the dryer is free, some parameters describing its geometry and its structure can be added to the control variables. So, optimisation of the design can be performed for internal variables assigned to their most probable values [8].

II - OPTIMIZATION PRINCIPLES

The problem we are concerned with is to find the optimal control of each stage which minimizes J. The natural way to start this would be to do a sensitivity study to evaluate to what extent changes of control variables are acting on J. But the numerous experiments which would need to be realized, would surely be a serious and even unbearable disturbance in the normal use of the dryer.

So, a mathematical model is written which describes heat and mass transfers along the whole dryer; this model is so elaborated that the computed values must match with measurements which would be made on the dryer in a wide range of drying conditions (including shrinking). So it becomes a good representation of the real process, and it may be used instead of the dryer for the optimal control design. By considering model responses to different inputs, one decides then of the best policy to be applied to the real process. In fact, the sensitivity study can now be replaced by specific and more powerful mathematical methods to achieve computations of the optimal strategies.

Computing optimal control needs always very long and complicated calculations. So, the models which are used for this purpose must be as simple as possible, and therefore must be identified models (figure 2) where involved coefficients do not need to have any physical significance.

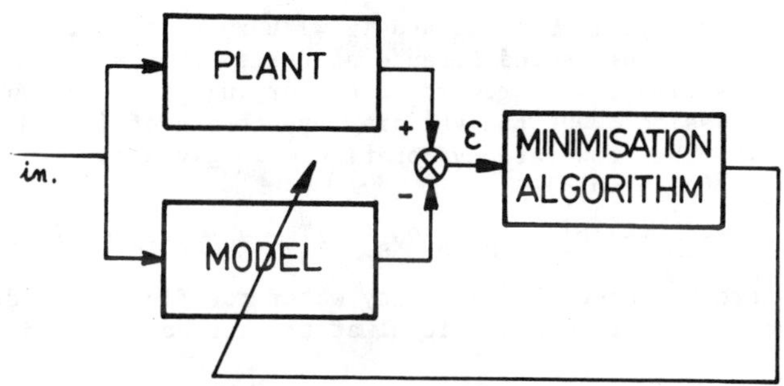

Fig. 2 - Identification principle

However, the more simple is this model, the less it is able to represent the process for different drying conditions ; then it is necessary to ensure its representativity by operating a new identification from time to time. When internal parameters are subject to unknown distrubances it is quite necessary to identify the process with a real-time algorithm which cannot be separated from the model itself : the control structure is then referred as "adaptative".This "self adaptation" is the price paid in exchange of the poor structure of the model. In other words the impossible computations made on the basis of a physical model is replaced by possible computations on the basis of a simple model, with increased computations due to self-adaptation. In such a control model, only control and state variables are writen explicitely.

The identification algorithms will be implemented on the computer controlling the dryer and their usefulness has just been clearly pointed out. To elaborate them off-line it is not necessary to dispose of real measurements, with the obvious over mentionned disturbances for the process. It is equivalent and more easy to write a descriptive model (i.e. simulation model) able to describe acurately any dryers ; the role of such a model is only to give realistic data instead of non yet available measurements. Moreover it can be used for many simulated drying conditions, and then helps for the best choice of the control model structure. A simulation model has not to be a quite physical model, its purpose being only to give evolution shapes close to the real ones.

III - DRYING MODELS

III.1. Simulation model

The simulation model describes the mean values of humidity and temperature inside each grain following its position on the path of the air flow ($0 \leq x \leq L$) :

$$\begin{cases} \frac{\partial}{\partial x}(G_a w_a) = -\sigma \dot{m} \\ \frac{\partial}{\partial x}(C_a^* G_a T_a) = -\alpha\sigma(T_a - T_g) \\ (1-\varepsilon)\frac{\partial}{\partial t}(C_g^* \rho_g T_g) = \alpha\sigma(T_a - T_g) + \sigma \dot{m}\left[C_g(T_a - T_g) + \Delta h_v\right] \\ (1-\varepsilon)\frac{\partial}{\partial t}(\rho_g w_g) = \sigma \dot{m} \end{cases} \quad (2)$$

Derivatives of w_a and T_a with respect to time have been neglected because of the shortness of the cross-time $u_a L$ in regard to the drying time. For more discussion about the validity hypothesis of (2), see [1]. The flow rate evaporation $\dot{m}$ is given by

$$\begin{aligned} \dot{m} &= \beta(p_{v_a} - p_{v_g}) \qquad (3) \\ &= 4.62\ 10^{-3}\beta\left[\phi_a(T_a, w_a).\rho_{v_s}(T_a).T_a - a_w(w_g, T_g).\rho_{v_s}(T_g).T_g\right] \end{aligned}$$

where the so-called activity water coefficient a_w depends on water energy level at the surface of grains

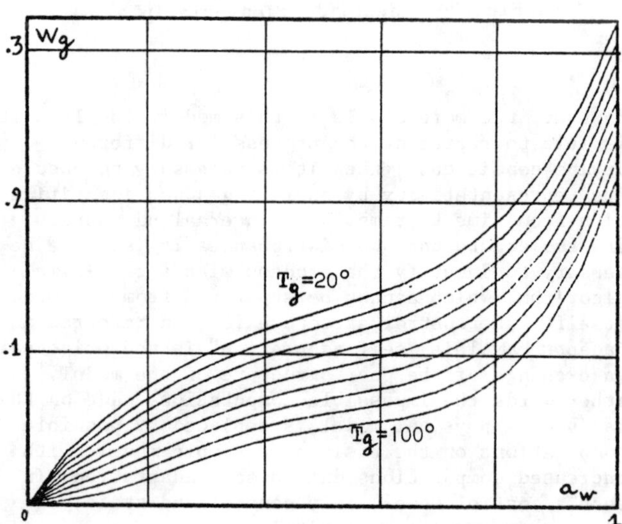

Fig. 3 - Water activity coefficient

[2]. At the equilibrium one has $\dot{m}=0$ and $T_g=T_a$ whence $a_w(w_g,T_a)=\phi_a(T_a,w_a)$ which is the equation of sorption isotherms ; in our simulation model we have choosen $a_w(w_g,T_g)$ to be given by abacus similar to these isotherms (figure 3); this choice was guided by the fact that the solution of (2) is then able to show all the phases which can be oserved in practice : heating phase, constant rate drying phase, falling rate drying phase [2-7]. On figure 4, the evolution of the state variables

$$\bar{w}_g = \frac{1}{L}\int_0^L w_g(x,t)dt \quad \text{and} \quad \bar{T}_g = \frac{1}{L}\int_0^L T_g(x,t)dt$$

are shown as well as the characteristics of the outlet air.

III.2. Control model

The shape of the curve $\bar{T}_g(t)$ is clearly the sum of steps responses of a first order system and of a delayed second order system ; this is an apparent discrepancy with the second order system (2). In fact, this higher order shape of $\bar{T}_g(t)$ is due to the non-linearity $a_w(\bar{w}_g,\bar{T}_g)$ and to the delay corresponding to the constant rate drying phase. Considering the short duration of this phase for grains it can be thought that a simple linear system of third order is able to give a good representation of the process. Then we take, for given u_a,

$$\dot{\mathcal{X}} = \mathcal{A}.\mathcal{X} + \mathcal{B}.(T_a)_{in}, \quad \mathcal{Y} = \mathcal{C}.\mathcal{X} + \mathcal{D}.(T_a)_{in} \quad (4)$$

with

$$\mathcal{X} = \left[\bar{w}_g, \dot{\bar{w}}_g, \bar{T}_g\right]^T, \quad \mathcal{Y} = \left[(\phi_{a_{out}} - \phi_{a_{in}}), T_{a_{out}}\right]^T$$

$$\mathcal{A} = \begin{bmatrix} 0 & 1 & 0 \\ a_{21} & a_{22} & a_{23} \\ 0 & a_{32} & a_{33} \end{bmatrix}, \mathcal{B} = \begin{bmatrix} 0 \\ b_2 \\ -a_{33} \end{bmatrix}, \mathcal{C} = \begin{bmatrix} 0 & c_{12} & 0 \\ 0 & c_{22} & 1-d_2 \end{bmatrix}, \mathcal{D} = \begin{bmatrix} 0 \\ d_2 \end{bmatrix} \quad (5)$$

The nine unknown coefficient of these matrix are identified (by using a gradient method [11-13]) to get the best matching of the solutions of (2) and (4). The result is shown on figure 4 for several values of α. A perfect matching can be achieved when all the elements of the $\mathcal{A}$ $\mathcal{B}$ $\mathcal{C}$ $\mathcal{D}$ matrix are unknown, the price of this beeing then an increased identification time [9]. In this study we have used the structure given in (5). In spite of its simple mathematical expression, we note that most of the existing drying or dryers exponential models are peculiar cases of (4), which is then a very powerful tool of investigation.

The dryer is divided into N stages, the outlet grains of any stage being the inlet of the next one. We are looking for the optimal temperature of the inlet air of each of them. Two different ways can be thought of :

A) Only one linear model (4) is used for the whole dryer and the optimal values of control variables for each stage are deduced from the whole found strategy. For this purpose the maximum principle may easily be used [10]; such a model is sufficient for the optimization purpose for specified species and initial moisture but in practice it is not possible to use it towards the automatic control of the dryer, since this needs a specific model for every stage.

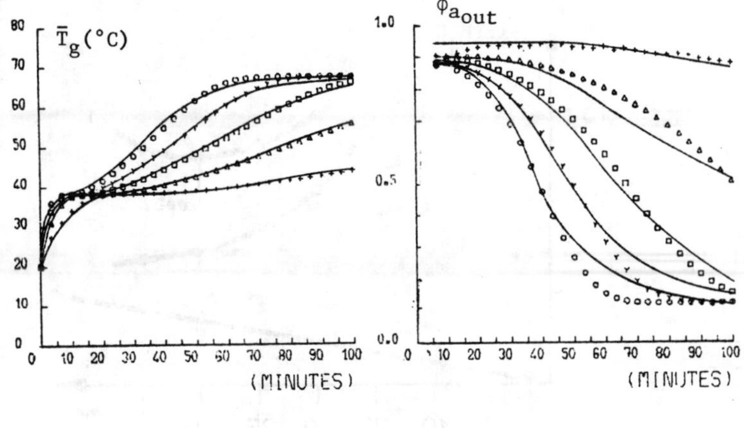

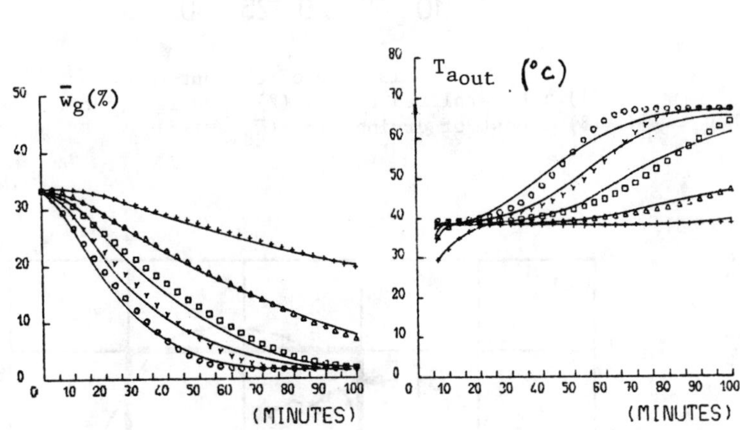

$\alpha/\beta = 0.43\ 10^{-5}$

$L = 15$cm

$(T_a)_{in} = 67°C$

$(\phi_a)_{in} = 15\%$

——— = simulation model

O, Y, □, Δ, + = identified model ; α = 20, 15, 10, 5, 2

Fig. 4 - Drying evolution curves

B) A specifically identified model of the type (4) is used for every stage ; so it is possible to consider disturbances on the grain species or on the initial moisture ; moreover this accurate approach is necessary to take into a better account a realistic and industrial evolution which may be globally different of figure 4 . Obviously this needs and identification procedure for each model but allows the automatic and optimal control of the whole dryer. The integration of (4) gives for each stage ($\mathcal{J}$ = unit matrix) :

$$\mathcal{X}_{out} = e^{\mathcal{A}\Delta t}.\ \mathcal{X}_{in} + \mathcal{A}^{-1}.\left[e^{\mathcal{A}\Delta t} - \mathcal{J}\right]\mathcal{B}.(T_a)_{in} \quad (6)$$

IV - OPTIMAL CONTROL OF THE DRYER

IV.1. The algorithm

On B-type model the best method to compute the optimal control in the general case is the discrete dynamic programming derived from Bellman's optimality principle [11-13]. This method is very convenient when constraints are present on state variables as well as on control variables, and when the models are different for each stage.

Instead of minimizing the generalized cost (1), we have choosen a=0 and $\hat{J}$ has been computed for several values of t_d. $\hat{J}$ is nothing else than a minimum energy with assigned values for t_d and $\bar{w}_f(t_d)$. The choice of the control policy is made after considering the curve $\hat{J}$ vs t_d.

The principle of the discrete dynamic programming [11] is to define a grid of admissible values for the outputs $\bar{w}_g(i\ \Delta t)$ (i=1,...,N) ; remembering that grains leaving the stage n-1 are entering the stage n, the cost reads

$$\hat{J}^n_{\ell} : \min_k J^{n-1}_k + j^{n-1,n}_{k,\ell}$$

where $j^{n-1,n}_{k,\ell}$ is the elementary energy cost to reach the ℓ point on the grid n, from the k point on the grid n-1. This cost is computed starting with the wet air formula

$$\Delta W = K\ u_a\ \Delta t(T_{a_{in}} - T_{a_o}) \quad (7)$$

and (6). $\hat{J}^n_{\ell}$ is the minimum cost to reach the ℓ point on the grid associated with the stage n. All the grids are peculiar for each stage and are limited by the values of constraints : there is only one point on the first and the last grids (the initial moisture is known and the final value is prescribed) and we have $\bar{T}_{g_{out}} \leq T_{g_{max}}$ according to quality criteria for the product. Constraints on rate drying can be introduced through assigned values of $\bar{w}_g$ at the output of some stages.

All these computations are made as soon as the N identifications have been performed ; the N found optimal policy are applied to each stage until new identifications are available ; then other optimal policies are computed again, starting from these new and more matched models which take into account the last effects of disturbances. We suggest to choose Δt as the time interval between two computations.

IV.2. Results and conclusions

For applications, we used a 5-stages dryer, with the same model for each stage and without disturbances.

The control model is very close to the simulation model ; so, one has verified that the optimal control computed from (4) lead; effectively to $\bar{w}_g(t_d)=\bar{w}_{gf}$ when it is applied to (2). On figure (5) the minimum energy cost versus t_d is drawn. It can be seen that when an equivalent cost a is associated to the drying time optimal value is found for $\hat{t}_d$.

For a given value of t_d, the figure (6) shows that the optimal inlet airflow temperatures are a little lower for the two highest stages. This result can be explained from the observation of the wet air diagram : p_{v_a} is independent on T_a then so does $\dot{m}$ at the beginning of the process, and then it is not useful to heat the inlet airflow during the first minutes. If one would look for a continuous optimal control (N=∞) one would find a curve identical to the (C_2) curve on figure (6) : (C_1) is obviously a N pieces discrete approximation of (C_2). A simulation has been done with a constant temperature 110° for the inlet airflows, equal to the mean value of (C_1), and it has been observed that the drying time to reach $\bar{w}_{gf}$ is increased by about 6%, with obvious credit for optimal policy.

Other simulations have been made with recovering of the heat of the used air ; it has proved that savings of non less than 15% can be made on the generalized cost for short time dryings with high temperatures (figure 7).

All this results are in the way of a good feasability for the automatic optimal control of the dryer. Efforts must be focused now on considering u_a as a second control variable, and on the self-adaptation of the control model. This is nothing else that automatic identification of the model parameters of every stage according to the distance between predicted values and the corresponding mesured values. This study will lead to the design of real time identification algorithms which will be implemented on the process computer controlling the dryer [12].

With this adequate auto-corrected simulation model, associated with a performing optimization algorithm, the chances of achieving the projected savings should be quite high.

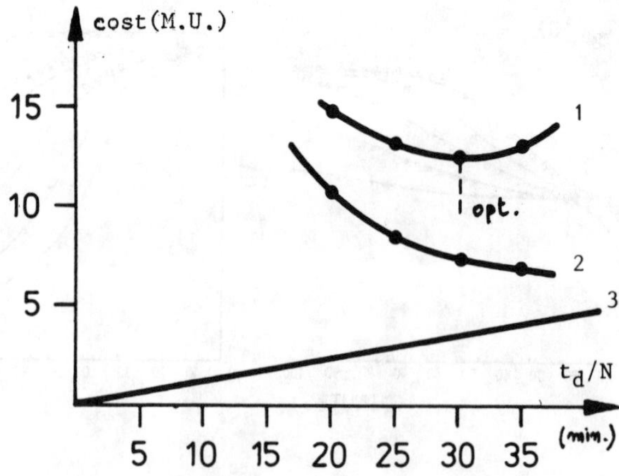

Fig. 5 - Existence of the optimum
(1) : Generalized cost ; (2) : energy cost ; (3) : cost of drying time ($\bar{w}_{g,f}$=15%).

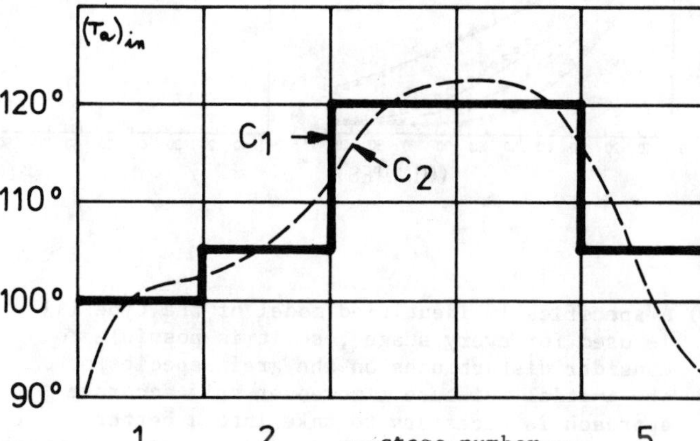

Fig. 6 - Optimal inlet temperatures (α=2 W/°K.m², T_{a_o}= 20°c, ϕ_{a_o}= 60%, a=0, t_d= 100 minutes, $\bar{w}_{g,f}$= 15%, $T_{g max}$ = 55°C).

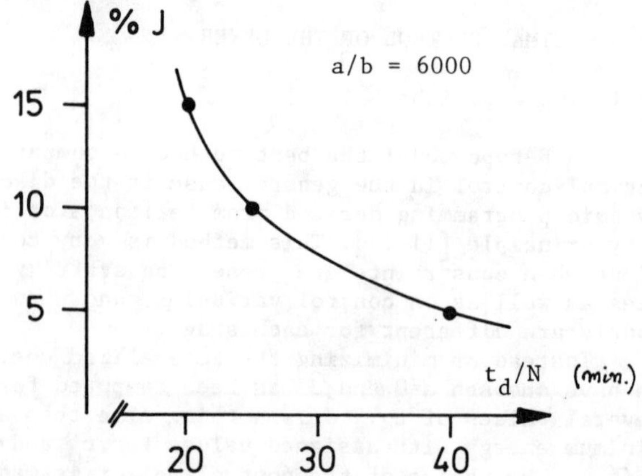

Fig. 7 - Realized savings with heat recovering

NOMENCLATURE

a equivalent cost of time (M.U./h)
a_w activity coefficient of surface water
b energy cost (M.U/w.h)
C specific heat capacity (J/°K.Kg)
G_a airflow rate (Kg/m².s)
J cost to be minimized
L equivalent dryer length in the airflow direction (m)
$\dot{m}$ mass flux density (Kg/m².s)
N number of stages
p partial pressure (N/m²)
T temperature (°K)
t_d drying time (s)
u_a airflow velocity ($=G_a/\rho_a$)(m/s)
w moisture (dry basis)(%)
W energy (J)

α heat transfer coefficient (W/°K.m²)
β mass transfer coefficient (m/s.°K)
ε void ratio
ϕ relative humidity
ρ mass density (Kg/m³)
Δh_v specific latent heat of vaporisation (J/Kg)=439.7 - 0.58T

SUPSCRIPT

a air
g grain
v_s saturated vapour
v vapour
in inlet, input
out outlet, output
o outside
f final

SUPERSCRIPT

* wet
\- average , mean
^ optimal

REFERENCES

1. F.W.BAKKER-ARKEMA et al.,"Grain Dryer Simulation", Research Report n°224,from the Michigan State University, Farm Science,Jan.1974.

2. J.J. BIMBENET, "Le séchage dans les industries agricoles et alimentaires", Cahier du Génie Industriel Alimentaire , SEPAIC,Paris (1978).

3. E. KRAUSE, "Le séchage en céramique,Principes et Techniques", éditions Septima,Paris (1977) (translation)

4. W. MUHLBAUER,"Recherches sur le séchage du maïs-grain et étude d'un dispositif de séchage à co-courants", Editions du CNEEMA, Paris, (1976). (translation)

5. O. KRISHER and K. KROLL,"Technique du séchage", Ed. CETIAT, Paris (translation)

6. J.D. DAUDIN and J.D. BIMBENET, "Characteristic drying curve of shelled corn and simulation of a vertical corn drier", Proc. of the third international Drying Symposium,Birmingham(G.B.)(Drying 82),vol.1, pp.337-346.

7. M. HAERTLING, "Prediction of Drying Rates", Drying 80, vol.1,pp.88-98,Hemisphere Publishing Corp., Washington.

8. R.C. BROOK, F.W. BAKKER-ARKEMA, "Design of multi-stage corn dryers using computer optimization", Michigan State University, Department of Argricultural Engineering,(1980).

9. R. OUHAB,"Simulation et optimisation du fonctionnement d'un séchoir à grain continu", Thèse ENSAE, Toulouse,(1984).

10. A.R.M. NOTON, "Introduction to variational Methods in Control Engineering", Pergamon Press,(1965).

11. R.L. ZAHRADNIK,"Theory and Techniques of optimization for practicing engineers", Barnes and Nosh ed.,(1971).

12. A.P. SAGE, "Optimum Systems Control", Prentice Hall,(1968).

13. R. BOUDAREL, J.DELMAS, P.GUICHET,"Commande optimale des Processus",Dunod ,Paris(1967).

THIN-LAYER DRYING AND REWETTING MODELS TO PREDICT MOISTURE DIFFUSION IN SPHERICAL AGRICULTURAL PRODUCTS

Odílio Alves-Filho and Thomas R. Rumsey

Higher School of Agriculture of Lavras – ESAL
Lavras, Minas Gerais 37200 – Brazil

ABSTRACT

Single layer drying tests were performed using whole English walnuts. The data were analyzed to estimate apparent diffusivities for both drying and rewetting. Diffusivities were different for drying and rewetting, and depended on air temperature and initial walnut moisture. A linear relationship between diffusivity and moisture content did not fit the data well.

1. INTRODUCTION

A key input to deep bed drying simulation models is the rate equation for moisture removal or uptake. Empirical "thin-layer" rate equations have been favored as they tend to be easy to implement [1]. A thin-layer equation for drying English walnuts has been developed [2]. Diffusion models are also popular in describing the moisture removal rate. Single component models have been developed for such products as peanuts [3], corn [4], and grains [5]. Multicomponent models for peanuts [6], rice [7] and ear corn [8] have also been proposed.

The purpose of this work was to see if the thin layer drying and rewetting of English walnuts could be modeled with the liquid diffusion equation in spherical coordinates. Rewetting is included because some of our deep bed modeling work involves ambient air drying with relatively high air humidity conditions.

2. DIFFUSION EQUATION

The diffusion equation in spherical coordinates assuming symmetry with respect to the origin and diffusivity a function of moisture content is

$$\frac{\partial D}{\partial M}\left(\frac{\partial M}{\partial r}\right)^2 + D\frac{\partial^2 M}{\partial r^2} + \frac{2D}{r}\frac{\partial M}{\partial r} = \frac{\partial M}{\partial t} \tag{1}$$

The boundary conditions used in this analysis are

$$M(a, t) = M_e \tag{2}$$

$$\frac{\partial M}{\partial r}(0, t) = 0 \tag{3}$$

and the initial condition is

$$M(r, 0) = M_o \tag{4}$$

Two models for diffisivity were tried: 1) diffusivity a constant and 2) a linear function of moisture (similar to that tried by Whitaker [9]).

$$D = D_O + D_1 M \tag{5}$$

For constant diffusivity, the analytical solution to equations (1)–(4) can be found in Crank [10]. This solution is integrated to find the average moisture content:

$$\frac{M(t) - M_e}{M_o - M_e} = \frac{6}{\pi^2}\sum_{n=1}^{\infty}\frac{1}{n^2}\exp\left(-\frac{n^2\pi^2 Dt}{a^2}\right) \tag{6}$$

In some cases, the product was not quite dried to equilibrium. Denoting the final drying time by t_f, equation (6) can be used to find the moisture ratio for large values of t_f

$$\frac{M(t) - M(t_f)}{M_o - M(t_f)} = \frac{6}{\pi^2}\,\frac{1}{1 - \frac{6}{\pi^2}\exp\left(-\frac{\pi^2 Dt_f}{a^2}\right)}$$

$$\times\left[\sum_{n=1}^{\infty}\frac{1}{n^2}\exp\left(-\frac{n^2\pi^2 Dt}{a^2}\right) - \exp\left(-\frac{\pi^2 Dt_f}{a^2}\right)\right] \tag{7}$$

If the product is dried for a sufficiently long time, the exponential term involving t_f will go to zero and equation (7) reduces to equation (6).

For the case of diffusivity a linear function of moisture, an explicit finite difference technique was used to solve equation (1). The resulting equation for interior points is

$$M(r, t + \Delta t) = M(r,t) + \frac{\Delta t D_1}{\Delta r^2}\left(M(r + \Delta r, t) - M(r, t)\right)^2$$

$$+\frac{\Delta t}{\Delta r^2}(D_O + D_1 M(r, t))\,(M(r + \Delta r, t) - 2M(r,t) + M(r - \Delta r,t))$$

$$+\frac{2\Delta t}{r\Delta r}(D_O + D_1 M(r, t))\,(M(r + \Delta r, t) - M(r, t)) \tag{8}$$

The boundary conditions are given by

$$M(0, t + \Delta t) = M(0, t) + \frac{6\Delta t}{\Delta r^2}$$

$$\times\ (D_O + D_1 M(0, t))\,(M(\Delta r, t) - M(0, t)) \tag{9}$$

and

$$M(a, t) = M_e \tag{10}$$

The initial condition is

$$M(r, 0) = M_O \tag{11}$$

The resulting solution was integrated numerically to calculate the average moisture content.

$$M(t) = \frac{3}{a^3} \int_0^a r^2 M(r, t)\, dr \qquad (12)$$

FORTRAN subroutines were written to solve equations(6) to (12). These were used in a nonlinear regression program, BMDPAR [11], to obtain values for the diffusivity.

3. EXPERIMENTAL METHODS

Samples of approximately 2.7 kg each were placed on 0.46 m x 0.46 m screen trays. The trays were placed in test chambers and supported by a Statham Model UL4–10 load cell. Conditioned air at an average velocity of 0.81 m/s was supplied to the test chambers by an Aminco Aire unit (Model J45590A).

Air dry bulb and wet bulb temperatures were continuously measured with thermocouples. Data were recorded using either a strip chart recorder or a Campbell Scientific (Model CRS) data logger.

Moisture contents were determined before and after each test. A sample of eight nuts were ground in a blender, and 50 gram subsamples were then dried in a Thelco (Model 29) vaccum oven for 24 hours at 90°C.

To obtain a larger sample of initial moisture contents, drying data given by Anigbankpu [2] were also analyzed. The experimental dryer used for these tests was slightly different, and is described in his paper. Moisture contents were determined in the same manner. Conditions for each of the tests used in this study are given in Table 1.

Table 1 — Test Conditions

Run No.	Dry Bulb °C	Rh (%)	M_0 (% d.b.)	M_f (% d.b.)
1[a]	32.0	47.0	56.4	15.5
2[a]	32.0	47.0	27.9	8.3
3	32.0	31.0	16.6	8.1
4	32.0	23.0	13.0	5.9
5[a]	32.0	48.0	11.4	6.8
6	43.0	26.0	50.3	5.4
7	43.0	28.0	29.5	5.2
8	43.0	30.0	18.6	4.1
9	43.0	20.0	16.1	4.2
10[a]	43.0	28.0	13.5	4.3
11	16.0	91.0	4.2	11.1
12	16.0	72.0	3.7	8.0
13	11.0	89.0	5.9	11.3
14	13.0	70.0	8.1	9.1

[a] Data from Anigbakpu [2].

4. DISCUSSION OF RESULTS

Plots of moisture ratio versus time for test runs at 32°C and 43°C are given in Figures 1 and 2, respectively. It is apparent from Figure 1 that the drying rate is depend on initial moisture content at the lower temperature.

At the higher temperature. Results for tests at 38°C showed the same characteristics: slower drying at higher initial moisture and a drying rate in between those for 32°C and 43°C.

Temperature and humidities used in the rewetting were limited by the operating range of the air conditioning unit. Plots of moisture ratio versus time for the rewetting tests are given in Figure 3. The curves are similar over the range of temperature used.

Apparent diffusivities based on a sphere radius of 1.9 cm were calculated for each of the test runs using the BMDPAR program along with equation (7). A total of five terms in the series were found to give acceptable results [12]. Using any more terms did not change the calculated diffusivities significantly. Plots showing measured and predicted moisture for runs 1 and 11 are shown in Figure 4. The calculated diffusivities are listed in Table 2. The apparent diffusivity decreases with initial moisture content. Similar results were found by Chittenden [4] for corn. Diffusivity values also increased with temperature. At a constant initial moisture content, the diffusivity followed on Arrhenius type equation [12].

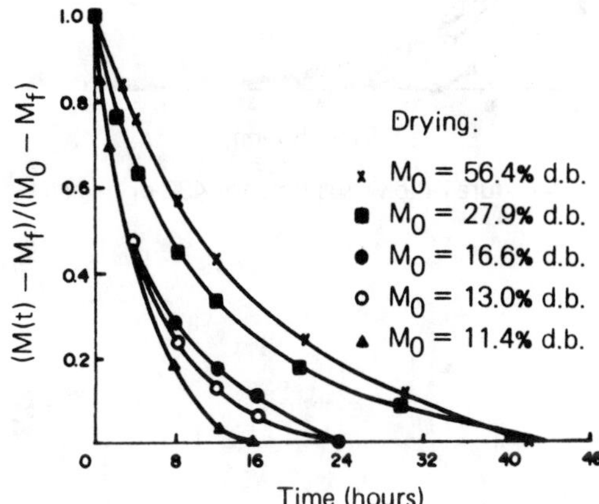

Fig. 1 Moisture ratio versus time for 32C drying air

The difusion type equation appears to describe the rewetting runs adequately.

The finite difference equations were checked out by comparing them to the solution for the constant diffusivity given by equation (6). A time step of 0.025 hr and distance step of 0.127 cm gave stable results and were in agreement with the analitycal solution for average moisture.

Table 2 — Apparent Diffusivities

Run No.	Diffusivity x 10^6 (m^2/hr)	Number of Observations	Mean Square Error x 10^3
1	0.687	8	12.80
2	1.330	8	4.78
3	2.220	7	5.88
4	3.000	7	6.67
5	2.210	7	15.50
6	1.430	9	10.30
7	3.360	10	2.38
8	4.140	9	2.95
9	4.560	8	3.26
10	5.270	7	7.46
11	2.210	39	16.40
12	2.450	38	10.50
13	2.050	41	8.51
14	2.290	38	11.90

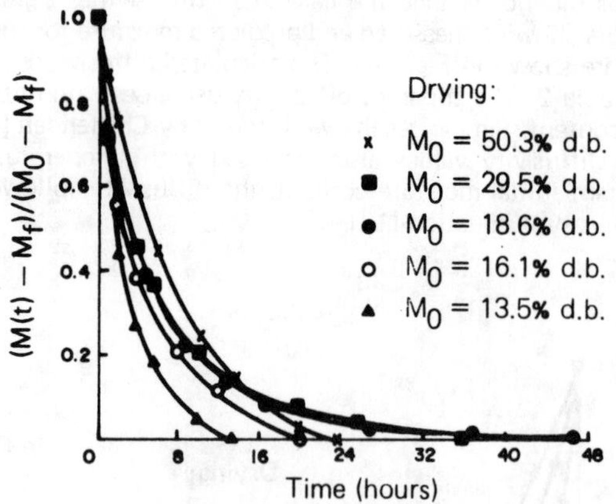

Fig. 2 Moisture ratio versus time for 43C drying air

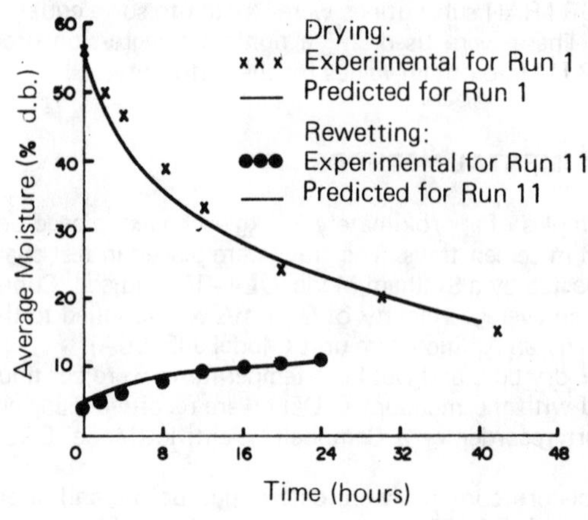

Fig. 4 Constant diffusivity predictions for druing and rewetting

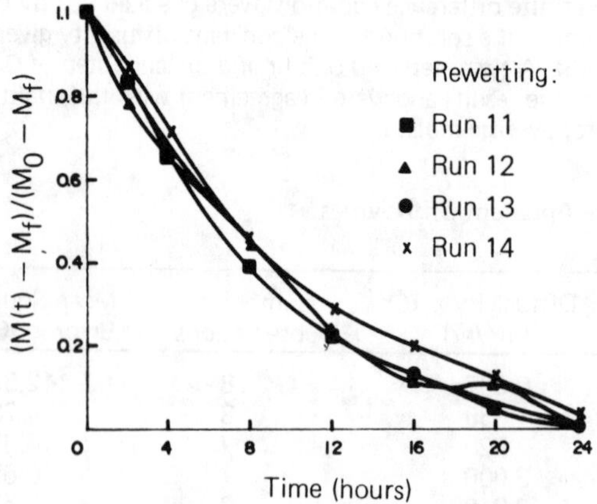

Fig. 3 Moisture ratio versus time for rewetting

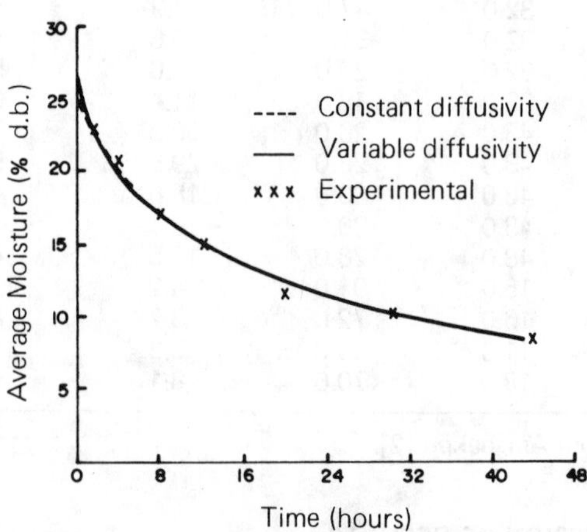

Fig. 5 Constant and variable diffusivity predictions for run 2

Runs 1, 2 and 4 were analyzed to see if the linear diffusivity-moisture relationship fit the data. Results for this analysis are given in Table 3. The linear relationship did not fit the data as well as the constant diffusivity, and did not give consistent values for the two parameters D_0 and D_1. Plots of predicated and observed results for run 2 are given in Figure 5.

Table 3 – Apparent Linear Diffusivities

Run No.	$D_0 \times 10^6$ (m^2/hr)	$D_1 \times 10^7$ $\left(\frac{m^2/hr}{100\ kg/kg}\right)$
1	0.00	0.454
2	0.00	1.170
4	4.15	0.000

5. CONCLUSÕES

The rate at which English walnuts dry depends on air temperature and moisture content. A diffusivity that is a function of temperature alone will not fit all the data. Assuming that diffusivity was a linear function of moisture content did not fit the data well. An exponential relationship between diffusivity and moisture content as used by Husain et al [13] for rice should be tried. The multicomponent models such as those obtained by Chhinnan and Young [6] should also be investigated.

6. REFERÊNCES

1. Brooker, D.B., F.W. Bakker-Arkema and C.W. Hall. 1974. Drying Cereal Grains. The AVI Publishing Company, Inc., Westport, Conn.

2. Anigbankpu, C.S., T.R. Rumsey and J.P. Thompson. 1980. Thin-layer drying and equilibrium moisture content equations for Ashley walnuts. ASAE Paper No. 80-6507. Presented at ASAE Winter Meeting, Chicago, IL.

3. Young, J.H. and T.B. Whitaker. 1971. Evaluation of the diffusion equation for describing thin-layer drying of peanuts in the hull. Transactions of the ASAE, 14(2):309-312.

4. Chittenden, D.H. and A. Hustrulid. 1966. Determining drying constants for shelled corn. Transactions for the ASAE, 9 (1):52-55.

5. Suarez, C., J. Chirife and P. Viollaz. 1982. Shape characterization for a simple diffusion analysis of air drying of grain. J. Food Science, 47:97-100.

6. Chhinnan, M.S. and J.H. Young. 1977. A study of diffusion equations describing moisture movement in peanut pods-I. Transactions of the ASAE, 20(3):539-546.

7. Steffee, J.F. and R.P. Singh. 1980. Liquid diffusivity of rough rice components. Transactions of the ASAE, 23(3): 767-774, 782.

8. Kumar, A., J.L. Blaisdell and F.L. Herum. 1981. Generalized analytical model for moisture diffusion in a composite cylindrical body. Transactions of the ASAE, 25(3):752-758.

9. Whitaker, T., H.J. Barre and M.Y. Handy. 1969. Theoretical and experimental studies of diffusion in spherical bodies with a variable diffusion coefficient. Transactions of the ASAE,12 (3):668-672.

10. Crank, J. 1975. The mathematics of diffusion. Second Edition. Oxford University Press, London.

11. Dixon, W.J. 1981. BMDP Statistical Software. University of California Press, Berkeley, CA.

12. Alves-Filho, O. 1982. Thin-layer drying and rewetting models to predict moisture diffusion in Ashley walnuts. Unpublished Master of Science Thesis, Agricultural Engineering Department, University of California at Davis, CA.

13. Husain, A., C.S. Chen and J.T. Clayton. 1973. Simultaneous heat and mass diffusion in biological materials. Jour. Agric.

PROPERTIES OF BROWN RICE KERNEL FOR CALCULATION OF DRYING STRESSES

Shinkichi Yamaguchi, Kaichiro Wakabayashi and Shingo Yamazawa

Department of Chemical Engineering, Toyama University
Takaoka, 933 JAPAN

ABSTRACT

Some properties required for the calculation of the internal stresses of a rice kernel during the drying process have been measured. The mechanism of moisture movement in a rice kernel has been discussed experimentally and the average equivalent-radius and the moisture diffusivity of a brown rice kernel have been determined. By examining the experimental results on volume change with moisture content of brown rice kernels, the coefficients of cubical and linear hygroscopic expansion have been estimated. The compression and stress relaxation tests were made by using test pieces of the rice endosperm formed into cylindrical shapes under various temperatures and moisture contents. Empirical equations for the compresive and shear stress relaxation moduli of rice endosperm have been obtained as a function of time, temperature and moisture content. Furthermore, equations for estimating the master curves on the moduli have been derived by assuming that the rice endosperm was a thermo- and hydro-rheologically simple material.

1. INTRODUCTION

There have not been enough properties published about rice kernels to calculate the internal stresses. In the present study, some properties for the kernel (a variety: Hōnen-wase) have been reported to supply the lock of the properties.

It is known that the thermal and moisture stresses are caused due to the inequality of temperature and moisture content, respectively, in the rice kernel. It is considered that the thermal stress in the rice kernel is negligibly small as compared with the moisture stress because the coefficient of thermal expansion is remarkably smaller than that of hygroscopic expansion for rice endosperm. An estimation method for the moisture distribution during the drying process has been discussed in order to calculate the internal stresses of a rice kernel. Then the average equivalent-radius and the moisture diffusivity have been determined.

Other mechanical properties for rice kernel, such as the coefficients of hygroscopic expansion and the stress relaxation moduli, have also been measured experimentally.

2. ESTIMATION OF MOISTURE DISTRIBUTION

2.1. Analytical Model

Heat transfer rate should be reflected on moisture movement in a rice kernel during the drying process. It is known, however, that the heat transfer rate is considerably higher than the moisture movement rate in the kernel. Thus it is considered that the inner temperature of the kernel is nearly equal to that of drying air and nearly uniform through the kernel during the drying process.

Based on the analyses of transient heat transfer problems [8], Chuma et al. [1] have presented an approximate solution for the moisture transfer in some cereal grains during the drying process. When air film mass-transfer resistance on grain surface is negligibly small, the change of moisture content w is given except for the earliest period of time as

$$\frac{w - w_e}{w_I - w_e} = C \exp\left(-\frac{G\pi^2 Dt}{\ell^2}\right) \tag{1}$$

Where the geometry index G is defined as a function of semi-thickness ℓ, semi-length ℓ_a and semi-width ℓ_b of a kernel (see Fig.1) by

$$G = \frac{1}{4} + \frac{3}{8(\ell_a/\ell)^2} + \frac{3}{8(\ell_b/\ell)^2} \tag{2}$$

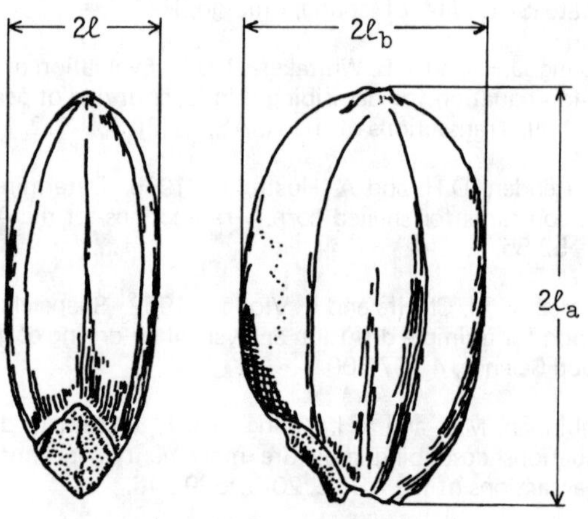

Fig. 1 Schematic figure of a brown rice kernel

If the slop -k on the drying curve (see Fig. 2) is obtained except for the earliest period of drying, the moisture diffusivity D of the kernel can be estimated from Eq.(1) as

$$D = \frac{k\ \ell^2}{G\ \pi^2} \tag{3}$$

Since the geometry index of sphere is unity, the equivalent radius of the kernel b is given by

$$b = \frac{\ell}{\sqrt{G}} \tag{4}$$

Therefore, the moisture distribution in the kernel can be calculated by assuming that the kernel is a homogeneous sphere. For example, the moisture content w is given by the following well known equation during the drying process [2].

$$\frac{w - w_e}{w_e - w_I} = \frac{2b}{\pi r} \sum_{N=1}^{\infty} \frac{(-1)^N}{N} \sin(N\pi \frac{r}{b}) \exp(-\frac{N^2 \pi^2 D t}{b^2}) \tag{5}$$

2.2. Equivalent Radius

The moisture content of a number of brown rice kernels were adjusted to three lebels of 0.18, 0.22 and 0.31 kg/kg(dry stock) by putting them into air conditioned chambers. Each of the 30 kernels was chosen arbitrarily among each moisture adjusted kernels and their sizes of ℓ, ℓ_a and ℓ_b were measured by using a digital meter (SONY-G 50).

The measured results were averaged and shown in Table 1. It is obtained that the average equivalent-radius of brown rice kernels is 1.44×10^{-3} m by substituting the total averaged values of ℓ and G in Table 1 into Eq.(4).

Table 1 Diameters and geometry data of brown rice kernel (Hōnen-wase)

moisture content [kg/kg(d.s.)] w	diameters [m] x 10^3 $2\ell_a$	$2\ell_b$	2ℓ	geometry index [-] G
0.31	5.18	2.95	2.05	0.490
0.22	5.08	2.88	2.01	0.492
0.18	5.03	2.80	2.01	0.498
total average	5.10	2.88	2.02	0.492

2.3. Diffusivity

Experiments of the thin-layer drying for brown rice kernels were carried out at constant air conditions. Here the thin-layer drying system means a through flow drying system in which all kernels in the layer are dried by the same air condition as the inlet. The initial moisture content w_I was adjusted about 0.32 kg/kg(dry stock) and the inlet air temperature T_f and relative humidity ϕ were kept nearly constant at 308 ∿ 309 K and 32 ∿ 35 %, respectively, through all the experiments. On the other hand, the air velocity u_f was turened into four steps of 0.5, 1.1, 1.9 and 2.7 m/s.

From preliminary experiments, it has become apparent that after 40 hours of drying time the mass of the kernels did not change. Then, it was thought that the moisture content reached equilibrium value after more than 40 hours of drying time.

The experimental results are shown in Fig.2. It is found that the drying curves are superimposed

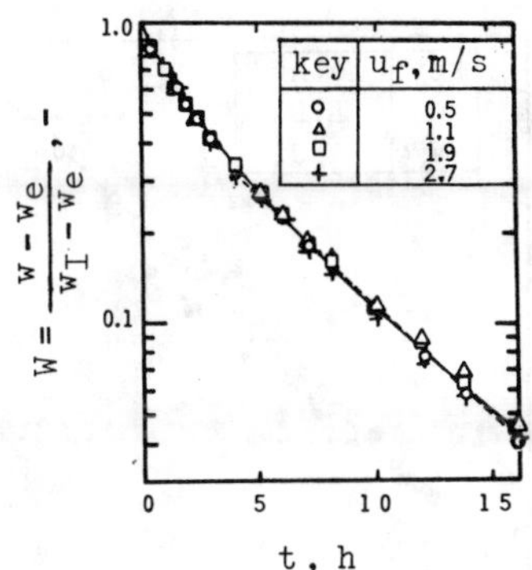

Fig. 2 Drying curves for brown rice kernels

upon each other despite the alteration of air velocity. These results show that the air film moisture-transfer resistance on the kernel surface is negligible and the surface moisture content always keeps the equilibrium value during the drying period.

It is possible to draw a straight line on the drying curve except for the earliest period of the drying time as shown by dotted line in Fig.2.

It is obtained that the slop -k of this straight line is 4.61×10^{-5} s^{-1} and then, by substituting those values of -k, G and ℓ into Eq.(3), the moisture diffusivity of the rice kernel D is 9.72×10^{-12} m^2/s.

3. COEFFICIENTS OF HYGROSCOPIC EXPANSION

3.1. Measurements of Density and Specific Volume

A large number of brown rice kernels were prepared as experimental samples by controlling the moisture content ranging from about zero to 0.43 kg/kg(dry stock). The density of these samples ρ was determined by means of the liquid-displacement method and the air comparisom pycnometer method at a constant temperature, 308 K.

In the present study, the specific volume v [ℓ/kg(dry stock)] based on dry stock of rice kernel has been defined by

$$v = \frac{1 + w}{\rho} \tag{6}$$

The results are plotted in Fig.3. All the data fall nearly on a straight line up to a moisture content of 0.08 kg/kg(dry stock). From the straight line, the following empirical equations for the specific volume and the density are obtained as

$$v = 0.672 + 0.945w \tag{7}$$

and

$$\rho = \frac{1 + w}{0.672 + 0.945w} \tag{8}$$

respectively.

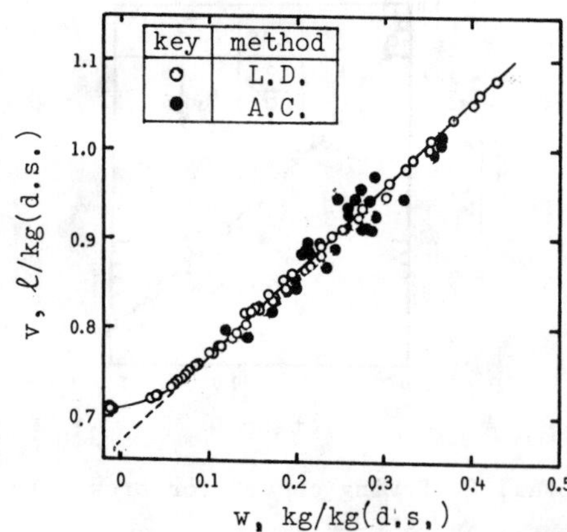

Fig. 3 Relationship between w and v for brown rice kernels (L.D. : liquid-displacement method, A.C.: air-comparison pycnometer method)

3.2. Coefficient of Cubical Hygroscopic Expansion

In the present study, a coefficient of cubical hygroscopic expansion γ_w has been defined in an analogous manner for the thermal expansion by

$$\gamma_w = \frac{d}{dw}\left(\frac{v}{v_0}\right) \tag{9}$$

Where v_0 is an unreal value of the specific volume, 0.672 ℓ/kg(dry stock), at $w = 0$ (see Eq.(7)).

From Eqs.(7) and (9), it has been determined that the coefficient of cubical hygroscopic expansion of the rice kernel γ_w was 1.42 $kg^{-1}\cdot$kg(dry stock). Then Eq.(7) is rewritten as

$$v = v_0(1 + \gamma_w w) \tag{7'}$$

3.3. Coefficient of Linear Hygroscopic Expansion

We suppose a cube which has a side of L [m] in a rice kernel. If the rice kernel is homogeneous, then the specific volume ratio v/v_0 is equal to $(L/L_0)^3$. Where L_0 is an unreal length of a side of the cube at $w = 0$. Then a coefficient of linear hygroscopic expansion α_w can be defined by

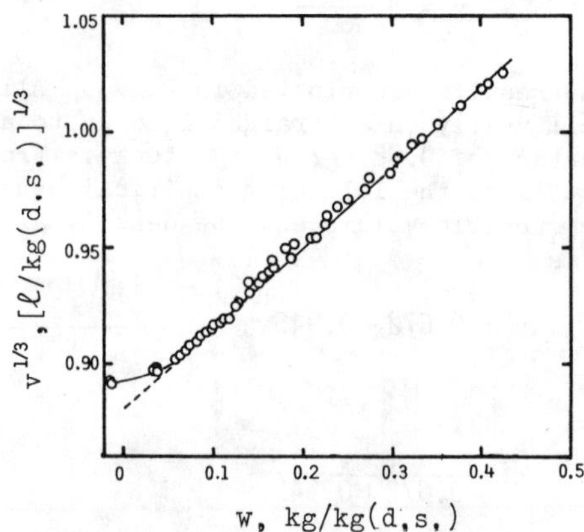

Fig. 4 Relationship between w and $v^{1/3}$ for rice kernels

$$\alpha_w = \frac{d}{dw}\left(\frac{L}{L_0}\right) = \frac{d}{dw}\left(\frac{v}{v_0}\right)^{\frac{1}{3}} \tag{10}$$

Fig.4 shows the relationship between w and $v^{1/3}$. Up to a moisture content of 0.08 kg/kg(dry stock), the following equation is obtained from Fig.4 as

$$v^{\frac{1}{3}} = 0.881 + 0.344w \tag{11}$$

From Eqs.(10) and (11), it has been determined that the coefficient of linear hygroscopic expansion α_w was 0.390 kg/kg(dry stock).

3.4. Discussion

Eq.(9) can be rewritten as

$$\gamma_w = \frac{d}{dw}\left(\frac{L}{L_0}\right)^3 = 3\left(\frac{L}{L_0}\right)^2 \frac{d}{dw}\left(\frac{L}{L_0}\right) \tag{12}$$

From Eqs.(7'), (10) and (12) the following equations are derived as

$$\gamma_w = 3\alpha_w(1 + \gamma_w w)^{\frac{2}{3}} \tag{13}$$

$$\alpha_w = \frac{\gamma_w}{3}(1 + \gamma_w w)^{-\frac{2}{3}} \tag{14}$$

These relations show that if one of the coefficient of γ_w or α_w is independent of w then another one should be dependent on w.

It is well known that both the coefficients of cubic thermal expansion γ_T and linear thermal expansion α_T are usually regarded as independent of temperature T and the relation $\gamma_T = 3\alpha_T$ is held. On the other hand, in the case of the hygroscopic expansion, it is noted that $\gamma_w(1.42)$ is not equal to $3\alpha_w(3 \times 0.390 = 1.17)$.

4. RELAXATION MODULI

4.1. Test Piece

A number of rice kernels were formed into a cylindrical shape and their moisture contents were controlled.

Fig.5 shows the working process for the cylindrical test piece of rice endosperm schematically. At first, both ends of a kernel are cut down (see Fig.5(a)). The rest of the piece is held between tow sticks of piano wire by using a reformed watchmaker's vice (see Fig.5(b) and Fig. 6). The contact planes of the piece with piano wire are pasted together with an adhesive (Aron Alpha). The shadowed portion shown in Fig.5(b) is carefully filed off by operating a very fine file and grindstone along the lateral surface of the piano wire. Then the piece is removed from the piano wire and put on the V-shaped groove of an iron block (see Fig.5(c)). The end surface planes of the piece are finished by operating a very fine grindston.

The diameters of the finished pieces were about 1.5 millimeter and the length of them was about 1.5 times of the diameter.

Prior to testing, the pieces were each equilibrated at several humidity atmospheres to give a range in moisture content of 0.07 to 0.35 kg/kg(dry stock).

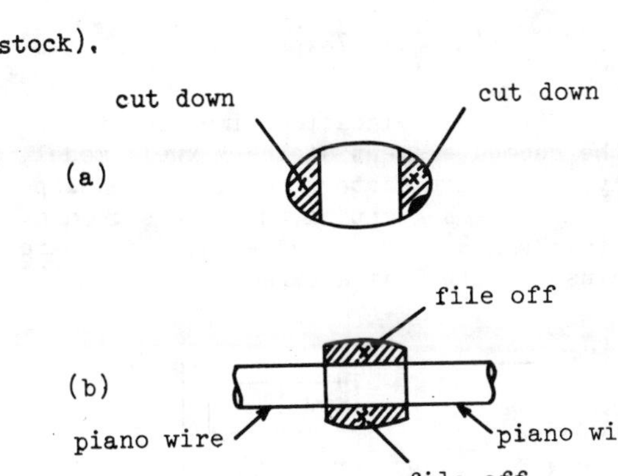

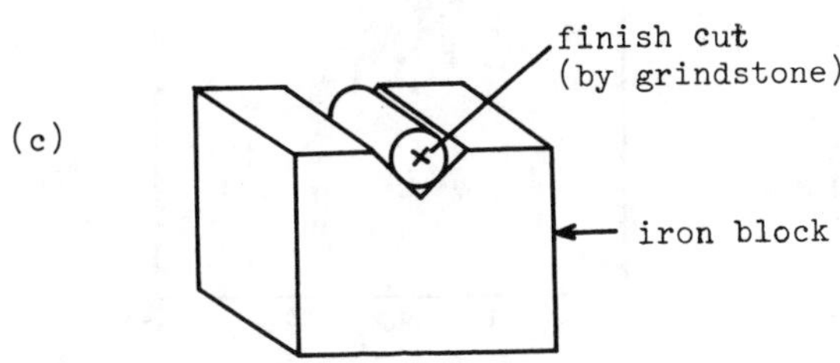

Fig. 5 Working process of cylindrical test piece of rice endosperm

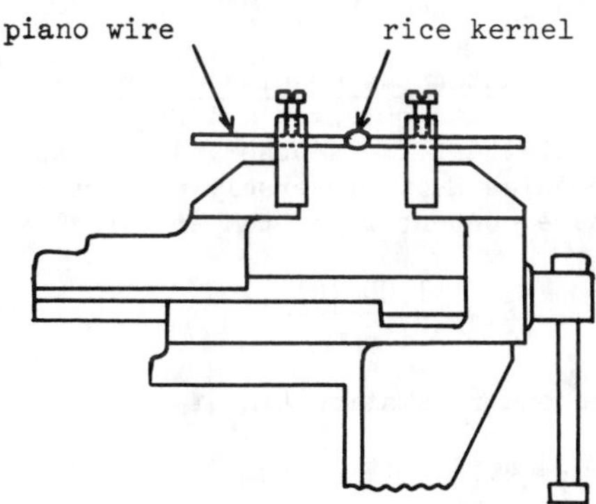

Fig. 6 Watchmaker's vice

4.2. Compression and Stress Relaxation Tests

The compression and stress relaxation tests for the test pieces of rice endosperm have been performed by using a material testing machine (INTESCO 2005).

The compression test. A test piece was tested in axially direct compression by the material testing machine. Each test was made under a constant crosshead speed and temperature.

Fig.7 shows some stress-strain curves under the conditions of $w = 0.11$ kg/kg(dry stock), T=293 K and six steps of strain rates $\dot{\varepsilon}$. It is found that the initial part of the stress-strain curves is convex towards the stress axis. This is probably due to imperfect end conditions of the piece since it is not possible to obtain perfectly flat and parallel ends for the test piece formation.

Except the initial part, it is found that the stress-strain curves first have a short linear-portion, then concave towards the stress axis and finally diverge according to the strain rates. Consequently, it is considered that the property of rice endosperm is expressed in a Maxwell model [6]. In the case of the Maxwell materials, the stress relaxation modulus can be estimated by differentiating the stress-strain curve [3]. However, the modulus for rice endosperm can not exactly be estimated from the above method because of the irregularity on the initial part of the stress-strain curves.

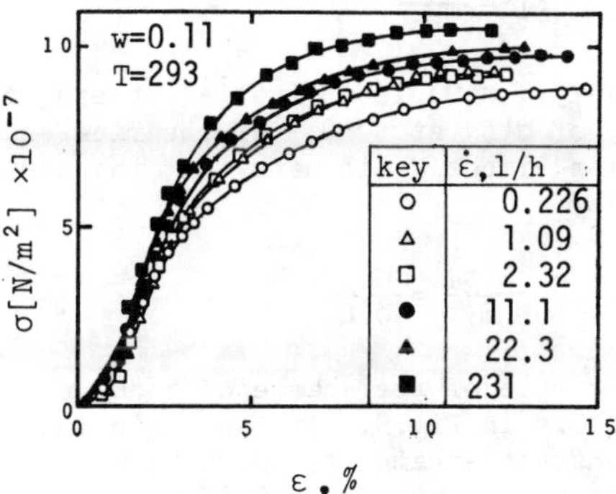

Fig. 7 Stress-strain curves on compression tests under various strain-rates

From the results of the compression test, the instantaneous modulus (the relaxation modulus at zero time) has only been determined by assuming that it is given by the maximum slope of the stress-strain curve. The time-dependent relaxation modulus has been obtained from the results on the stress relaxation tests described later.

Fig.8 shows the relationship between the instantaneous modulus E_I and moisture content w with temperature T as a parameter.

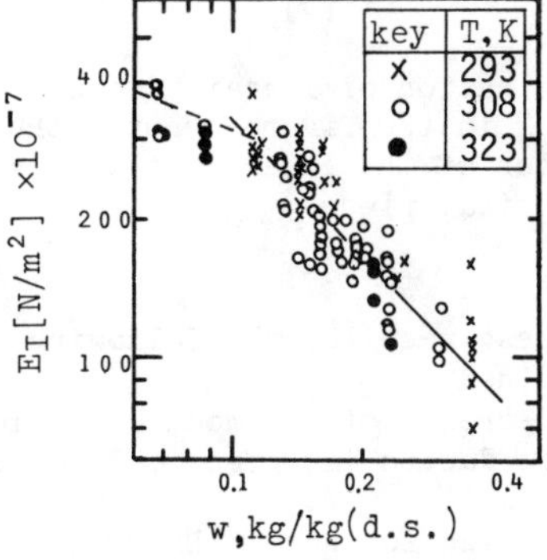

Fig. 8 Instantaneous elastic modulus for compression

The stress relaxation test. A test piece was suddenly brought to a given compression strain ε_0 in a stress relaxation test. Then the change of time-dependent stress $\sigma(t)$ which was required to hold the strain constant was recorded.

The compressive relaxation modulus $E(t)$ and the instantaneous modulus E_I are defined as

$$E(t) = \frac{\sigma(t)}{\varepsilon_0} \tag{15}$$

and

$$E_I = \frac{\sigma(0)}{\varepsilon_0} \tag{16}$$

respectively. $\sigma(0)$ is an initial stress, that is, the value of $\sigma(t)$ at $t = 0$. If the rice endosperm is a linear viscoelastic material, the modulus ratio is given by

$$\frac{E(t)}{E_I} = \frac{\sigma(t)}{\sigma(0)} \tag{17}$$

An example of the change of modulus ratio with time is shown in Fig.9. At the beginning of this test a constant strain ($\varepsilon_0 = 0.24$ %) was instantly applied to a test piece (w = 0.15 kg/kg(dry stock)) at T = 308 K. The modulus ratio for Repeat No.1 of the test decreased rapidly as time elapsed and approached zero. After the modulus ratio reached nearly zero, the test for Repeat No.2 was done by applying instantly the same strain 0.24 % to the test piece again. The test was repeated 5 times in a similar manner. The changes of the modulus ratio for Repeat No.2, 3 and 4 roughly agreed with each other but that for Repeat No.5 did not.

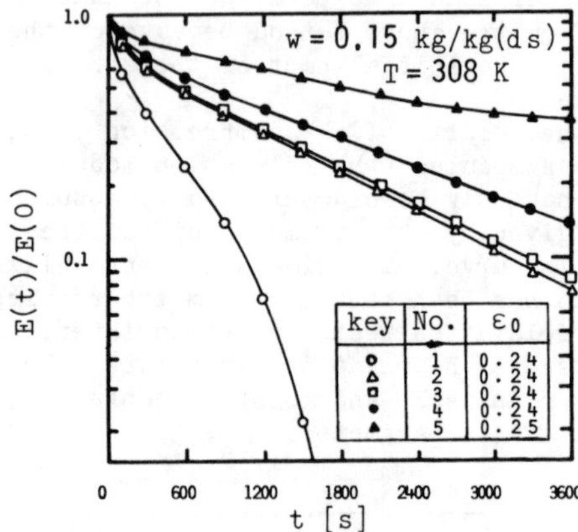

Fig.9 Examination of linear viscoelasticity ("No." in the figure denotes the repeat number of the experiment with an identical piece)

From the test results, the following considerations are obtained.

(1) The quick decrease of the modulus ratio for Repeat No.1 may be due to imperfect end conditions of the test piece.

(2) Since the curves for Repeat No.2, 3 and 4 are in agreement with each other and expressed in a straight line up to about ten minutes, it can be assumed that the rice endosperm has a linear viscoelastic property showing a generalized Maxwell model.

(3) If the total strain exceeds about 1 % then the linear viscoelastic property is no longer valid, so that the curve for the Repeat No.5 does not agree with others.

Based on the method of successive residuals for analysis of stress relaxation [6], the property of rice endosperm has further been studied using all the test results. Then it was found that the compressive stress relaxation modulus E(t) is expressed in a Maxwell model composed of two elements:

$$E(t) = E[0.3\exp(-\frac{t}{\tau_1})+0.7\exp(-\frac{t}{\tau_2})] \tag{18}$$

τ_1 and τ_2 are the relaxation times for the first and the second element of the Maxwell model, respectively. τ_2 can be determined from the slope of the straight line segment on the stress relaxation curve (see Fig.9). Obtained values of τ_2 are plotted against w with T as a parameter in Fig.10.

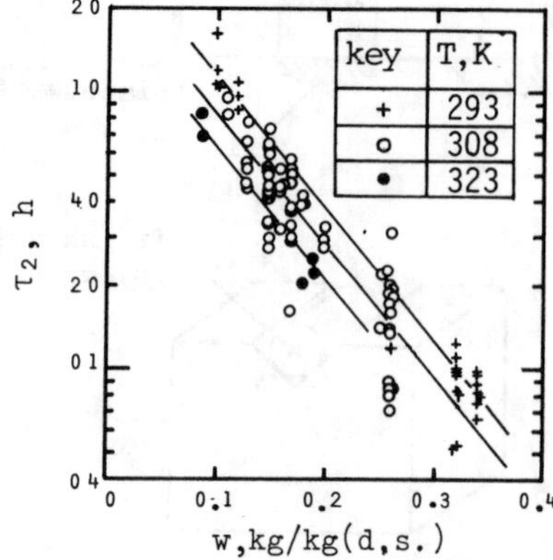

Fig.10 Correlation of τ_2 and w with T as a parameter

4.3. Relaxation Modulus

<u>The instantaneous modulus and relaxation time.</u> Although the plots on Fig.8 and Fig.10 show considerable scatter, it is observed that the instantaneous modulus E_I is inversely proportional to the moisture content w and temperature T as

$$E_I = \frac{1.00\times10^{11}}{T w} \tag{19}$$

and the second relaxation time τ_2 is given by

$$\tau_2 = 4.32\times10^6\exp[-(0.02T + 11w)] \tag{20}$$

Through the examinations of all the test results, it is found to be $\tau_1 = 0.042\tau_2$ then

$$\tau_1 = 1.81\times10^5\exp[-(0.02T + 11w)] \tag{21}$$

<u>The compressive relaxation modulus.</u> An empirical equation for the compressive relaxation modulus E(t) of rice endosperm is obtained by substituting Eqs.(19), (20) and (21) into Eq.(18) as

$$E(t) = \frac{1.00\times10^{11}}{T w}\{0.3\exp[-5.51\times10^{-6}(0.02T + 11w)t] + 0.7\exp[-2.31\times10^{-7}(0.02T + 11w)t]\} \tag{22}$$

It is obvious from Eq.(22) that the relaxation modulus E(t) is dependent not only on time t but also on temperature T and moisture content w.

It has been postulated that the rice endosperm was thermo- and hydro-rheologically simple material. Then it was defined in the present study that the reference temperature T_0 is 293 K and the reference moisture content w_0 is 0.17 kg/kg(dry stock).

The thermorheological nature of a material indicates that an increase in temperature corresponds to an increase in time. Similarly, in a hydrorheological nature, an increase in moisture content would correspond to an increase in time [3]. To relate time to temperature and moisture content, a dimensionless time-temperature shift factor a_T and a time-moisture shift factor a_w are defined as

$$a_T = \frac{t}{\xi_T} \tag{23}$$

and

$$a_w = \frac{t}{\xi_w} \tag{24}$$

respectively. Where ξ_T is a reduced time for temperature and ξ_w is a reduced time for moisture. In other words, ξ_T is the time required to observe the same phenomena at reference temperature T_0 and ξ_w is the time required to observe the same phenomena at the reference moisture content w_0. Furthermore, a time-temperature-moisture shift factor a is defined as

$$a = \frac{t}{\xi} = a_T \cdot a_w \tag{25}$$

Where ξ is the reduced time for both temperature and moisture content.

From Eqs.(22) ∿ (25), following equation are derived.

$$a_T = \frac{\exp(0.02T_0)}{\exp(0.02T)} = 351\exp(-0.02T) \tag{26}$$

$$a_w = \frac{\exp(11w_0)}{\exp(11w)} = 6.49\exp(-11w) \tag{27}$$

$$a = a_T \cdot a_w = 2280\exp[-(0.02T + 11w)] \tag{28}$$

$$\xi = \frac{t}{a} = 4.39\times10^{-4}\exp(0.02T + 11w)t \tag{29}$$

Then the modulus $\hat{E}(\xi)$ expressed as a function of reduced time ξ instead of t is given from Eqs.(22) and (29) as

$$\hat{E}(\xi) = \frac{1.00\times10^{-11}}{Tw}\cdot\{0.3\exp(-1.25\times10^{-2}\xi) + 0.7\exp(-5.28\times10^{-4}\xi)\} \tag{30}$$

The modulus $E_0(\xi)$ at T_0 and w_0 is given as

$$E_0(\xi) = \frac{Tw}{T_0w_0}\hat{E}(\xi) = 2.01\times10^{9}\{0.3\exp(-1.25\times10^{-2}\xi) + 0.7\exp(-5.28\times10^{-4}\xi)\} \tag{31}$$

A master curve for the compressive relaxation modulus of rice endosperm is given by Eq.(31).

Fig.11 shows the relaxation curves under various conditions of T and w. Each curve is representative of several test result with the same conditions.

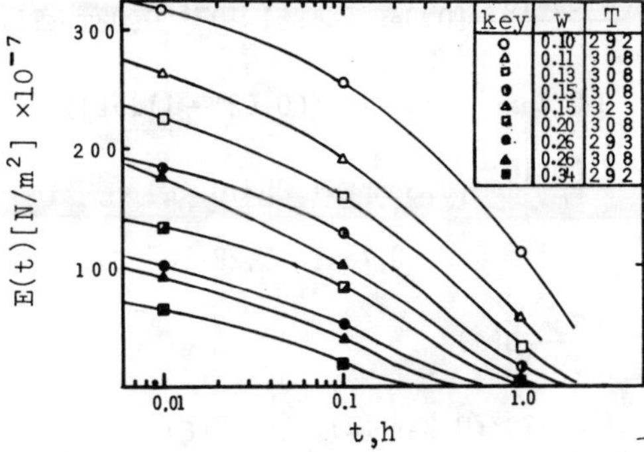

Fig.11 Relationships between compressive relaxation modulus E(t) and time t under various temperatures and moisture contents

The same data in Fig.11 is shown in Fig.12 by converting the ordinate into $E_0(\xi)$ and the abscissa into ξ. It is clear that in Fig.12 the curves converge considerably near the master curve (dotted line).

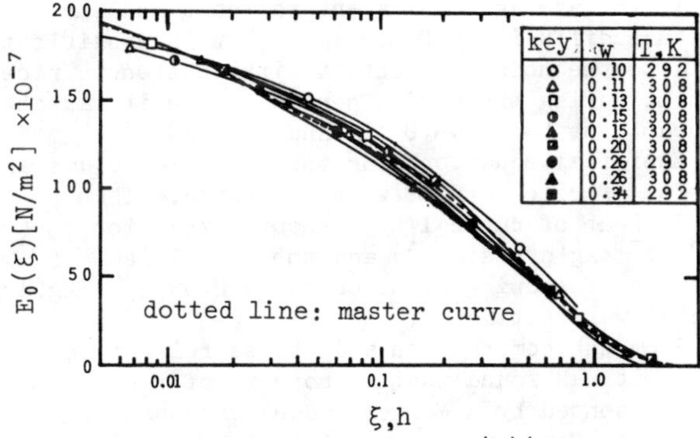

Fig.12 Relationships between $E_0(\xi)$(compressive relaxation modulus at T_0 and w_0) and reduced time ξ under the same conditions as shown in Fig.11

The shear relaxation modulus. It should be considered that the shear relaxation modulus G(t) and Poisson's ratio ν(t) for rice endosperm depend not only on time t but also on temperature T and moisture content w, analogous to the compressive relaxation modulus E(t). If $E^*(p)$, $G^*(p)$ and $\nu^*(p)$ are the Laplace transforms of E(t), G(t) and ν(t), respectively, p being the transform parameter, then the relation is held [7] as follows.

$$G(p) = \frac{E^*(p)}{1 + p\nu^*(p)} \tag{32}$$

Kawamura [4] and Kobayashi et al. [5] have shown that the measured values of Poisson's ratios for rice kernels were 0.21 and 0.258, respectively, and were nearly constant. In the present study, it is assumed for convenience that Poisson's ratio of rice endosperm is 0.25, independent of time, temperature and moisture content. Then the relation, $G(t) = E(t)/1.25$, is obtained from Eq.(32) and the following equations are derived.

$$G(t) = \frac{8.00\times10^{10}}{Tw}\{0.3\exp[-5.51\times10^{-6}(0.02T + 11w)t] + 0.7\exp[-2.31\times10^{-7}(0.02T + 11w)t]\} \quad (33)$$

$$\hat{G}(\xi) = \frac{8.00\times10^{10}}{Tw}\{0.3\exp(-1.25\times10^{-2}\xi) + 0.7\exp(-5.28\times10^{-4}\xi)\} \quad (34)$$

$$G_0(\xi) = \frac{Tw}{T_0w_0}\hat{G}(\xi) = 1.61\times10^{9}\{0.3\exp(-1.25\times10^{-2}\xi) + 0.7\exp(-5.28\times10^{-4}\xi)\} \quad (35)$$

These equations can be applied in the range of $293 < T < 323$ K, $0.10 < w < 0.35$ kg/kg(dry stock) and $\varepsilon < 1$ %.

5. CONCLUSION

Some properties of brown rice kernels have been measured and the following results have been obtained.

(1) if the average equivalent-radius b and the moisture diffusivity D are known, it is possible to estimate the moisture content distribution in rice kernels during the drying period. Then it was obtained to be $b = 1.44\times10^{-3}$ m and $D = 1.89\times10^{-11}$ m^2/s.

(2) Empirical equations for the specific volume and density of rice kernels were presented. Then the coefficient of cubical hygroscopic expansion γ_w (= 1.42 kg$^{-1}\cdot$kg(dry stock)) and the coefficient of linear hygroscopic expansion α_w (= 0.39 kg$^{-1}\cdot$kg(dry stock)) were estimated.

(4) From the compression and stress relaxation tests, it was found that a property of rice kernels is represented by a Maxwell model composed of two elements. Consequently, empirical equations for estimating the master curves on the compressive and shear relaxation moduli of rice kernels were derived by assuming that the rice kernel was a thermo- and hydro-rheologically simple material.

NOMENCLATURE

a	time-temperature-moisture shift factor	-
a_T	time-temperature shift factor	-
a_w	time-moisture shift factor	-
b	average equivalent-radius	m
$E(t)$	compressive relaxation modulus	N/m^2
G	geometry index defined by Eq.(2)	-
$G(t)$	shear relaxation modulus	N/m^2
ℓ	semi-thickness of a kernel	m
ℓ_a	semi-length of a kernel	m
ℓ_b	semi-width of a kernel	m
α_w	coefficient of linear hygroscopic expansion	kg$^{-1}\cdot$kg(d.s.)
γ_w	coefficient of cubical hygroscopic expansion	kg$^{-1}\cdot$kg(d.s.)
ε	nominal strain	-
$\nu(t)$	Poisson's ratio	-
ξ	reduced time for both temperature and moisture	h,s
ξ_T	reduced time for temperature	h,s
ξ_w	reduced time for moisture	h,s
$\sigma(t)$	nominal stress at time t	N/m^2
τ_1, τ_2	relaxation times	h,s

REFERENCE

1. Chuma, Y., Murata, S. and Iwamoto, M., J.Soc. Agric.Mech., Japan, vol.31, 250 (1969).
2. Crank, J., The Mathematics of Diffusion , Oxford at the Clarendon Press, 86 (1956).
3. Hammerle, J.R. and Mohsenin, N.N., Trans.ASAE, vol.13, 372 (1970).
4. Kawamura, N., J.Soc.Agric.Mach., Japan, vol.12 43 (1951).
5. Kobayashi, H., Miwa, Y. and Torii, T., Proceedings of 35th Soc.Agric.Mach.Meeting, Japan, 101 (1976).
6. Mohsenin, N.N., Physical Properties of Plant and Animal Material, Vol.1 , Gordon and Breach Science Publisher, 110, 136 (1970).
7. Muki, R. and Sternberg, E., Trans ASME, Series E, vol.28, 193 (1961).
8. Smith, R.E., Nelson, G.L. and Henrickson, R.L., Trans . ASAE, vol.10, 236 (1967).

CHANGE OF CRACKED RICE PERCENTAGE AND INTERNAL STRESS OF BROWN RICE KERNELS DURING DRYING OPERATION

Shinkichi Yamaguchi, Kaichiro Wakabayashi and Singo Yamazawa

Department of Chemical Engineering, Toyama University
Takaoka, 933 JAPAN

ABSTRACT

In order to investigate the cracking mechanism of a rice kernel, some experiments and analyses have been made.

The changes in cracked rice percentages with time have been observed during drying and preserving processes for brown rice kernels and the following findings have been obtained: (1) The crack was generated near the center of a kernel. (2) In the drying process, the cracked rice percentage increased after about 2 hours. (3) In the preserving (after drying) process, the cracked rice percentage increased remarkably at first and reached a final value within several hours. (4) The final percentage in the preserving process increased with the drying-period of the kernels and reached a maximum value at about a 3 hour drying-period.

A method of the numerical calculation for moisture stresses in a brown rice kernel has been proporsed by assuming that the kernel was a homogeneous and linear-viscoelastic sphere. Based on this method the changes in internal stresses of a brown rice kernel during the drying and preserving processes have been calculated under the same condition that the crack observations have been done. From the results of the crack observation and stress calculation, it has been recogniged that the changes in the cracked rice percentage were closely related to that in the calculated stresses.

1. INTRODUCTION

It is known that the cracked rice leads to reduction in yields of milled rice and the lowering of market value. To find the optimun condition without cracking for the rice drying, one has to clarify the cracking mechanism of the rice kernel. However, the mechanism has not yet thoroughly been investigated.

The changes in cracked rice percentages with time during the drying and preserving processes for brown rice kernels have been observed experimentally. In order to discuss the cracking mechanism, it was attempted to calculate the internal stresses of the rice kernels and compare the changes of the cracked rice percentage with the internal stresses calculated.

Arora et al. [1] and Mannapperuma [12] have estimated the internal stresses of a rice kernel by assuming that the rice kernel is an elastic material. However, it has already been reported [7,8,9, 15] that the nature of the kernel is not elastic but viscoelastic. In the present study, a numerical method for analysis of the moisture stresses in a rice kernel has been proposed by assuming that the kernel is a homogeneous and linear-viscoelastic sphere. This method is characterized by the fact that the rate of a deviatoric strain in a generalized relaxation integral law can be calculated by using the rate of a deviatoric strain in an elastic sphere.

The cracking mechanism of the rice kernel has been discussed by comparing the changes in cracked rice percentage with the calculated internal stresses of the rice kernels.

2. OBSERVATIONS OF CRACKED RICE

According to a measuring method for the cracked rice percentages previously proporsed by the present authors [16], some experiments were carried out for the observation of the cracked rice. Sound, uncracked brown rice kernels were selected for the experiments and its moisture content was adjusted to 0.27 kg/kg(dry stock).

2.1. Preliminary Experiments

To examine the effect of thermal and moisture stresses on the crack generation in a rice kernel, the observations of the rice cracking have been made under two conditions; (i) the moisture content of the kernel is variable at a constsnt temperature and (ii) the temperature of the kernel is variable at a constant moisture content. Then it has been observed that the crack is apparently generated under the former condition but not under the latter. Therefore, it can be considered that the internal cracks of a rice kernel are not brought by the thermal stress but by the moisture stress.

It has been noticed from the crack observations that the crack is generated near the center of a rice kernel.

2.2. Changes of Cracked Rice Percentage

Although it has been observed [2,14] that the cracked rice increases with time during not only the drying period but also the preserving period, the changes in cracked rice percentage with time have not been reported in detail. Thus those changes have been observed during both drying and preserving processes in the present study.

The brown rice samples each of about 200 kernels were placed in the drying chamber of which the air velocity, temperature and relative humidity

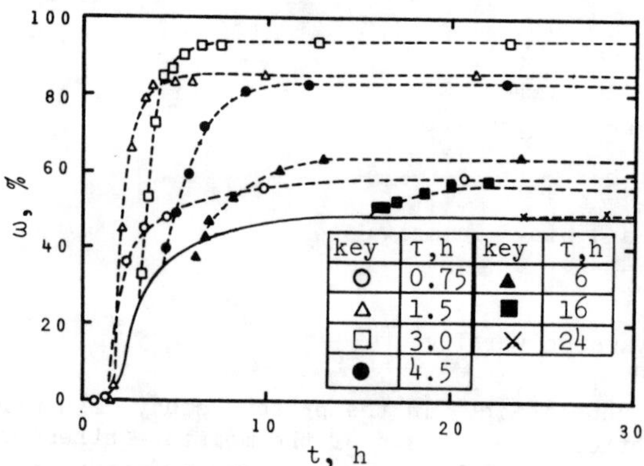

Fig.1 Experimental result for observation of changes of cracked rice percentage

were kept 0.8 m/s, 308 K and 35 %, respectively. After a given drying-period τ [h], each sample kernels was filled and sealed in the kernel holders which were made up placing a perforated rubber plate having 211 kernel-sized holes between two clear glass plates. The visual counting of cracked rice kernels was done immediately after the sealing of the sample and repeated at given time intervals under room temperature.

An example of the results is shown in Fig.1. The solid line in the figure represents the relationship between cracked rice percentage ω observed immediately after drying (drying-period τ) and drying time t. It can be considered that the solid line shows the change of the percentage during the drying process. The dotted lines in Fig.1 show the changes of the percentage ω with preserving time t_p after a given drying-period τ. It is obvious from Fig.1 that in the preserving processes the cracked rice percentages reach final values ω_f in several hours according to a drying-period τ.

2.3. Considerations

From the experiments the following results have been obtained.

(1) The cracks of rice kernels are not brought by the thermal stress but by the moisture stress.

(2) The crack is generated near the center of a rice kernel.

(3) In the drying process, the cracked rice percentage increases after about 2 hours and reaches an equilibrium percentage in about 15 hours.

(4) In the preserving process after a given drying-period τ, the cracked rice percentage increases remarkably at first and reaches a final percentage ω_f within several hours.

(5) The final percentage ω_f in the preserving process increases with the drying-period τ first and reaches a maximum value of about 3 hours of τ. Having reached the maximum percentage, the final percentage ω_f decreases with τ.

3. METHOD FOR STRESS ANALYSIS

3.1. Preliminary Considerations

The analytical model. To analyze the internal stress of a rice kernel, some assumptions have been made as follows:

(1) The brown rice kernel is a homogeneous and linear-viscoelastic sphere.

(2) The temperature in the kernel is always uniform and constsnt (equal to drying air temperature T_f) during the drying and preserving processes. Thus the time-temperature-moisture dependent modulus of stress relaxation can be expressed in the time-moisture dependent modulus.

(3) The moisture diffusivity D is invariable. Thus it is possible to calculate analytically the moisture content distribution in the kernel.

(4) The rate of strain in the spherical rice kernel is equal to that in the elastic sphere of which the elastic modulus is indentical with the instantaneous modulus of the rice kernel.

The time-moisture dependent moduli. The compressive and shear relaxation modulus of a rice kernel have previously been measured by the present authors [17] as functions of time t, temperature T and moisture content w. If the temperature is kept a constant T_f in the kernel, the moduli $E(t,T,w)_{T=T_f}$ and $G(t,T,w)_{T=T_f}$ are expressed in the time-moisture dependent moduli E(t,w) and G(t,w), respectively, as follows.

$$E(t,w)=E_I(w)\{0.3\exp[-5.51\times10^{-6}(0.02T_f+11w)t] +0.7\exp[-2.31\times10^{-7}(0.02T_f+11w)t]\} \quad (1)$$

$$G(t,w)=G_I(w)\{0.3\exp[-5.51\times10^{-6}(0.02T_f+11w)t] +0.7\exp[-2.31\times10^{-7}(0.02T_f+11w)t]\} \quad (2)$$

Where the instantaneous moduli $E_I(w)$ and $G_I(w)$ are given as

$$E_I(w)=\frac{1.00\times10^{11}}{T_f w} \quad (3)$$

$$G_I(w)=\frac{8.00\times10^{10}}{T_f w} \quad (4)$$

By assuming that the rice kernel is a thermo- and hydro-rheologically simple material, equations for the master curves with respect to the stress relaxation moduli can be obtained. Based on the reference temperature T_0=293 K and the reference moisture content w_0=0.17 kg/kg(dry stock), a time-temperature-moisture shift factor a(w) and a reduced time ξ(t,w) are given as follows.

$$a(w)=2280\exp[-(0.02T_f+11w)] \quad (5)$$

$$\xi(t,w)=4.39\times10^{-4}\exp(0.02T_f+11w)t \quad (6)$$

It is obvious from Eq.(6) that ξ(t,w), for fixed w, is a monotone increasing factor of t. Hence it can be considered that t is a function of ξ and w as

$$t=f(\xi,w) \quad (7)$$

Suppose F(t,w) is any function of time and moisture content. Then, to avoid ambiguity, we shall consistently adopt the notation

$$F(t,w)=\hat{F}(\xi,w)=F[f(\xi,w),w] \quad (8)$$

It should be noticed that F and $\hat{F}$ are distinct function unless ξ=t. Eq.(2) is rewritten by using the notation of Eq.(8) as

$$\hat{G}(\xi,w)=\frac{8.00\times10^{10}}{T_f w}\{0.3\exp(-1.25\times10^{-2}\xi)+0.7\exp(-5.28\times10^{-2}\xi)\} \quad (9)$$

The modulus $G_0(\xi)$ at the reference temperature T_0 and moisture content w_0 is given as

$$G_0(\xi)=\frac{T_f w}{T_0 w_0}\hat{G}(\xi,w)$$

$$=1.61\times10^9\{0.3\exp(-1.25\times10^{-2}\xi)+0.7\exp(-5.28\times10^{-2}\xi)\} \quad (10)$$

Eq.(10) represents the master curve with respect to the shear stress relaxation modulus.

The above equations can be applied in the range of $293 < T < 323$ K, $0.10 < w < 0.35$ kg/kg(dry stock) and $\varepsilon < 1$ %.

The moisture distribution during drying process. In the drying process, it is assumed that the mass transfer resistance of air film on a brown rice kernel is negligibly small and the surface moisture content of the kernel is kept an equilibrium value w_e. If the initial moisture content w_I is uniform through out the kernel, then the moisture content w(R,t) is given [3,4] as

$$w(R.t)=w_e+\frac{2(w_e-w_I)}{\pi R}\sum_{N=1}^{\infty}\sin(N\pi R)\exp(-\frac{N^2\pi^2 D}{b^2}t) \quad (11)$$

At the center of kernel, R=0, as

$$w(0,t)=w_e+2(w_e-w_I)\sum_{N=1}^{\infty}(-1)^N\exp(-\frac{N^2\pi^2 D}{b^2}t) \quad (12)$$

Where b is the equivalent radius of the kernel and R the dimensionless radial coordinate (R=r/b, $0 \leq R \leq 1$), respectively.

The moisture distribution during preserving process. In the preserving process after a given drying-period τ, it is considered that there is not any moisture transfer through the surface of a brown rice kernel. Then the moisture content w(R,t) is given [3] as

$$w(R,\tau,t)=\phi(\tau)+\frac{4(w_I-w_e)}{R}\sum_{N=1}^{\infty}\frac{\sin(\beta_N R)\Gamma(\tau,N)}{\sin\beta_N}\cdot\exp[-\frac{\beta_N{}^2 D(t-\tau)}{b^2}] \quad (13)$$

Where β_N denotes the Nth root of the transcendental equqtion

$$\tan\beta_N=\beta_N \quad (14)$$

$\phi(\tau)$ and $\Gamma(\tau,N)$ are given as follows.

$$\phi(\tau)=w_e+\frac{6(w_I-w_e)}{\pi^2}\sum_{N=1}^{\infty}\frac{1}{N^2}\exp(-\frac{N^2\pi^2 D\tau}{b^2}) \quad (15)$$

$$\Gamma(\tau,N)=\sum_{N=1}^{\infty}\frac{\exp(-L^2\pi^2 D\tau/b^2)}{L^2\pi^2-\beta_N} \quad (16)$$

3.2. Theoretical Considerations

Deviatoric strain and stress. Spherical coordinates (r,θ,ϕ: $0\leq r\leq b$, $0\leq\theta\leq\pi$, $0\leq\phi\leq 2\pi$) are introduced to analyze the moisture stresses in a spherical rice kernel having a radially symmetric moisture distribution. Then the dimensionless radial coordinate R=r/b ($0\leq R\leq 1$) is often used instead of r in the present study. In the expression of a strain or stress tensor, two subscript letters are used; for example $\varepsilon_{k\ell}$ (k and ℓ=r,θ or ϕ), the first indicating k the direction of the normal to the plane under consideration and the second indicating ℓ the direction of the component of the tensor.

The deviatoric strains and stresses are expressed as follows [13].

$$e_{rr}=-2e_{\theta\theta}=-2e_{\phi\phi}=\frac{2}{3}(\varepsilon_{rr}-\varepsilon_{\theta\theta})$$
$$e_{r\theta}=e_{\theta\phi}=e_{\phi r}=0 \quad (17)$$

$$S_{rr}=-2S_{\theta\theta}=-2S_{\phi\phi}=\frac{2}{3}(\sigma_{rr}-\sigma_{\theta\theta})$$
$$S_{r\theta}=S_{\theta\phi}=S_{\phi r}=0 \quad (18)$$

The equilibrium equation for stresses. In the absence of body forces, and under neglect of the inertia forces, the equations of motion reduce to the equilibrium equation as

$$\frac{\partial}{\partial R}\sigma_{rr}(R,t)+\frac{2}{R}[\sigma_{rr}(R,t)-\sigma_{\theta\theta}(R,t)]=0 \quad (19)$$

Eq.(19) is rewritten in view of Eq.(18) as

$$\frac{\partial}{\partial R}\sigma_{rr}(R,t)=-\frac{3}{R}S_{rr}(R,t) \quad (20)$$

Since $\sigma_{rr}(1,t)=0$ on the boundary, Eq(20) is integrated with respect to r, so that

$$\sigma_{rr}(R,t)=3\int_R^1\frac{S_{rr}(R',t)}{R'}dR' \quad (21)$$

The relaxation integral law. In the special case that moisture content (and temperature) is invariable in a kernel, that is, the relaxation modulus G(t) is a function of time only, the relaxation integral law is expressed [5,13] as

$$S_{k\ell}(R,t)=\int_0^t G(t-t')\frac{\partial}{\partial t'}e_{k\ell}(R,t)dt' \quad (22)$$

However, the moisture content in the rice kernel ought to be variable with position and time during the drying and preserving processes and, accordingly, Eq.(22) is not applicable to the analysis for the internal stresses of a rice kernel. In the present study, a generalized relaxation integral law has been derived in order to analyze the stresses in the rice kernel by applying the moisture-time equivalence hypothesis.

Now we suppose that the change of moisture content with time at an arbitrary radius position R is expressed in solid line of Fig.2. Then we assume that the moisture content decreases stepwise as shown in the chain-line of Fig.2, so that

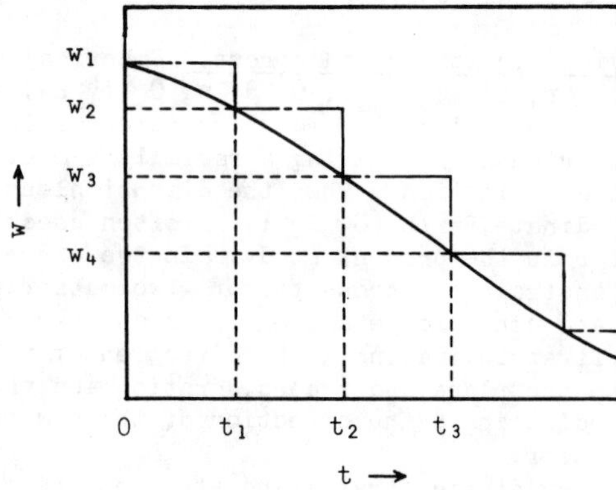

Fig.2 A schematic illustration for moisture distribution at an arbitrary position R

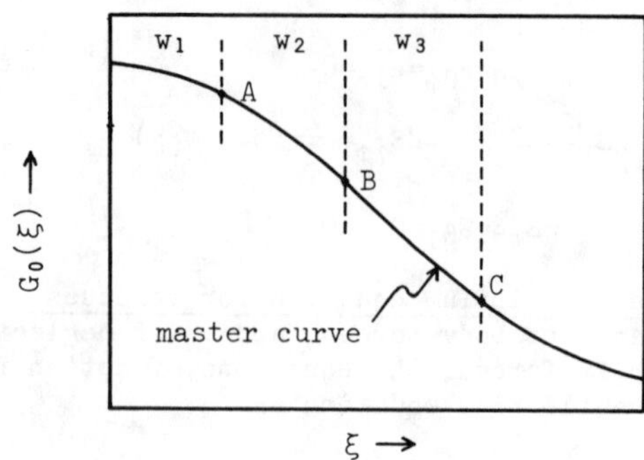

Fig.3 A schematic illustration for master curve

the moisture content is held at a constant w_1 within a short period of $0 \leq t \leq t_1$, and then a constant w_2 within $t_1 \leq t \leq t_2$, a constant w_3 within $t_2 \leq t \leq t_3$ and so on. The relaxation modulus corresponding to that moisture change can be expressed in the master curve of Fig.3.

In the period $0 \leq t \leq t_1$, the reduced time is given in view of Eq.(6) by

$$\xi(t,w_1)=\frac{t}{a(w_1)} \tag{23}$$

and the relaxation modulus is given in view of Eq.(10) and the master curve of Fig.3 by

$$\hat{G}(\xi,w_1)=\frac{T_0w_0}{T_fw_1}G_0(\xi) \tag{24}$$

Then $G_0(\xi)$ reaches point A on the master curve of Fig.3 at $t=t_1$.

In the period $t_1 < t \leq t_2$, $G_0(\xi)$ changes from point A at $t=t_1$ to point B at $t=t_2$ on the master curve. Thus reduced time and the relaxation modulus are given as follows.

$$\xi(t,w_2)=\frac{t_1-0}{a(w_1)}+\frac{t-t_1}{a(w_2)} \tag{25}$$

$$\hat{G}(\xi,w_2)=\frac{T_0w_0}{T_fw_2}G_0(\xi) \tag{26}$$

Similarly, the following equations are held in the period $t_2 < t \leq t_3$.

$$\xi(t,w_3)=\frac{t_1-0}{a(w_1)}+\frac{t_2-t_1}{a(w_2)}+\frac{t-t_2}{a(w_3)} \tag{27}$$

$$\hat{G}(\xi,w_3)=\frac{T_0w_0}{T_fw_3}G_0(\xi) \tag{28}$$

These concepts can be extended to the original change of moisture content as shown in the solid line of Fig.2 and the following relations are obtained.

$$d\xi=\frac{dt}{a[w(R,t)]} \tag{29}$$

$$\xi=\int_0^t \frac{dt'}{a[w(R,t')]} \tag{30}$$

$$\hat{G}[\xi,\hat{w}(R,\xi)]=\frac{T_0w_0}{T_f\hat{w}(R,\xi)}G_0(\xi) \tag{31}$$

According to Eq.(22), the relationship between the deviatoric stress $\hat{S}_{rr}(\xi,R)$ and strain $\hat{e}_{rr}(\xi,R)$ is given by using the notation of Eq.(8).

$$\hat{S}_{rr}(R,\xi)=\frac{T_0w_0}{T_f\hat{w}(R,\xi)}\int_0^{\xi}G_0(\xi-\xi')\frac{\partial}{\partial\xi'}\hat{e}_{rr}(R,\xi')d\xi' \tag{32}$$

The following relations are obtained from Eqs.(6) and (29).

$$\frac{d\xi}{dt}=\frac{1}{a(w)},\quad \frac{dt}{d\xi}=a(w),\quad \frac{dt}{d\xi}=\left(\frac{d\xi}{dt}\right)^{-1} \tag{33}$$

Thus Eq.(32) can be rewritten again by using the expressions of Eqs.(7), (8) and (33) as

$$S_{rr}(R,t)=\frac{T_0w_0}{T_fw(R,t)}\int_0^t G_0[\xi(R,t)-\xi(R,t')]\cdot\frac{\partial}{\partial t'}e_{rr}(R,t')dt' \tag{34}$$

3.3. Stress Analysis

The deviatoric stress and strain in an elastic sphere. We suppose an elastic sphere having a constant Poisson's ratio ν, a constant coefficient of linear hygroscopic expansion α_w and a moisture-dependent Young's modulus $E_I(w)$ given by Eq.(3). If this sphere is dried under an arbitrary radially symmetric-moisture-distribution, then the deviatoric strain $e^e_{rr}(R,t)$ in the sphere is given according to the study of Fujii [6] as follows.

$$e^e_{rr}(R,t)=\frac{2(1+\nu)\alpha_w}{3(1-\nu)}[w(R,t)-\overline{w}(R,t)] \tag{35}$$

Where $\overline{w}(R,t)$ is a mean moisture content defined as

$$\overline{w}(R,t)=\frac{3}{R^3}\int_0^R(R')^2w(R',t)dR' \tag{36}$$

Modification of the generalized relaxation integral law. In the light of previous discussions, the generalized relaxation integral law is modified to obtain the internal stresses of rice kernel (viscoelastic sphere).

It is assumed that the rate of the deviatoric

strain for a viscoelastic sphere $\partial e_{rr}(R,t)/\partial t$ is exacyly equal to that for an elastic sphere $\partial e^{e}_{rr}(R,t)/\partial t$. This rate which is obtained from Eq.(35) can be substituted into Eq.(34). Thus Eq.(34) is modified in view of Eqs.(6), (10) and (30) as follows.

$$S_{rr}(R,t)=\frac{\beta}{w(R,t)}\int_0^t[0.3\exp\{-\psi_1\int_{t'}^t\exp[11w(R,t")]dt"\} +0.7\exp\{-\psi_2\int_{t'}^t\exp[11w(R,t")]dt"\}] \cdot[\frac{\partial w(R,t')}{\partial t'}-\frac{\partial \overline{w}(R,t')}{\partial t'}]dt' \quad (37)$$

Where the constants ψ_1, ψ_2 and β are

$$\psi_1=5.49\times10^{-6}\exp(0.02T_f)$$
$$\psi_2=2.31\times10^{-7}\exp(0.02T_f) \quad (38)$$

$$\beta=\frac{2(1+\nu)\alpha_wT_0w_0G_{I0}}{3(1-\nu)T_f} \quad (39)$$

The instantaneous modulus G_{I0} which is $G_0(t)$ at t=0 is equal to 1.61×10^9 N/m^2.

It is noticed that the deviatoric stress $S_{rr}(R,t)$ is given by Eq.(37) as a function of the moisture content w(R,t) only. Therefore, now it is possible to estimate the deviatoric stress if the change of the moisture distribution is known. However, the integration of the type of Eq.(37) is analytically impossible. In order to estimate the deviatoric stress, thus, the integration of Eq.(37) should be practiced numerically.

A numerical solution for Eq.(37) during the drying process. From Eqs.(11) and (36) the following equation is derived for the drying process.

$$\frac{\partial w(R,t)}{\partial t}-\frac{\partial \overline{w}(R,t)}{\partial t} =\frac{2D(w_I-w_e)}{\pi b^3R^3}\sum_{N=1}^{\infty}\frac{(-1)^N}{N}\exp(-\frac{N^2\pi^2D}{b^2}t) \quad (40)$$

At the center of the kernel the following relation is held.

$$\frac{\partial w(R,t)}{\partial t}-\frac{\partial \overline{w}(R,t)}{\partial t}=0 \quad (41)$$

The deviatoric stress in the spherical kernel of a brown rice during the drying process is given by substituting Eq.(40) into Eq.(37) as

$$S_{rr}(R,t)=\frac{\alpha_1}{R^3w(R,t)} \cdot\int_0^t[0.3\exp\{-\psi_1\int_{t'}^t\exp[11w(R,t")]dt"\} +0.7\exp\{-\psi_2\int_{t'}^t\exp[11w(R,t")]dt"\}] \cdot[\sum_{N=1}^{\infty}\frac{(-1)^N}{N}\exp(-\frac{N^2\pi^2D}{b^2}t)\{(N^2\pi^2R^2-3) \cdot\sin(N\pi R)+3N\pi R\cos(N\pi R)\}]dt' \quad (42)$$

Where the constant α_1 is given by

$$\alpha_1=\frac{4(1+\nu)\alpha_wT_0w_0G_{I0}D(w_I-w_e)}{3\pi(1-\nu)b^2T_f} \quad (43)$$

To obtain a numerical solution for the internal stresses of a rice kernel, the dimensionless radial coordinate $0\leq R\leq 1$ is divided into m equal increments and the time required to obserye the stresses, $0\leq t'\leq t$, is divided into n equal increments. The ith dimensionless radial coordinate is named R_i (i=0,1,2,$\cdots$,m), so that $R_i=i\Delta R$, and the jth time is named t'_j (j=0,1,2,$\cdots$,n), so that $t'_j=j\Delta t$.

Then a numerical solution for Eq.(42) is expressed as

$$S_{rr}(R_i,t)=\frac{\Delta t\alpha_1}{2R_i^3w(R_i,t)}\sum_{j=1}^{n}[A(R_i,t'_j)+A(R_i,t'_{j-1})] \quad (44)$$

Where A is given by

$$A(R_i,t'_j)=\{0.3\exp[-\Omega_1(R_i,t'_j)]+0.7\exp[-\Omega_2(R_i,t'_j)]\} \cdot\{\sum_{N=1}^{\infty}\frac{(-1)^N}{N}\exp(-\frac{N^2\pi^2D}{b^2}t'_j) \cdot[(N^2\pi^2R_i^2-3)\sin(N\pi R_i)+3N\pi R_i\cos(N\pi R_i)]\} \quad (45)$$

Ω_1 and Ω_2 are given as follows.

$$\Omega_1(R_i,t'_j)=\frac{\psi_1\Delta t}{2}\sum_{\lambda=j}^{n-1}\{\exp[11w(R_i,t"_\lambda)] +\exp[11w(R_i,t"_{\lambda+1})]\} \quad (46)$$

$$\Omega_2(R_i,t'_j)=\frac{\psi_2}{\psi_1}\Omega_1(R_i,t'_j) \quad (47)$$

A numerical solution for Eq.(37) during the preserving process. From Eqs.(13) and (36) the following equation is obtained for the preserving process after a given drying-period τ.

$$\frac{\partial w(R,\tau,t)}{\partial t}-\frac{\partial \overline{w}(R,\tau,t)}{\partial t} =\frac{12D(w_e-w_I)}{b^2R^3}\sum_{N=1}^{\infty}\exp[-\frac{\beta_N^2D}{b^2}(t-\tau)] \cdot\frac{\Gamma(\tau,N)}{\cos\beta_N}\{\frac{R^2\beta_N^2-3}{3\beta_N}\sin(\beta_NR)+R\cos(\beta_NR)\} \quad (48)$$

The deviatoric stress is given by substituting Eq.(48) into Eq.(37) as

$$S_{rr}(R,\tau,t)=\frac{\alpha_2}{R^3w(R,\tau,t)} \cdot\int_0^t[0.3\exp\{-\Omega_1\int_{t'}^t\exp[11w(R,\tau,t")]dt"\} +0.7\exp\{-\Omega_2\int_{t'}^t\exp[11w(R,\tau,t")]dt"\}] \cdot\{\sum_{N=1}^{\infty}\frac{\Gamma(\tau,N)}{\cos\beta_N}\exp[-\frac{\beta_N^2D}{b^2}(t'-\tau)] \cdot[\frac{R^2\beta_N^2-3}{3\beta_N}\sin(\beta_NR)+R\cos(\beta_NR)]\}dt' \quad (49)$$

Where the constant α_2 is given by

$$\alpha_2=\frac{8(1+\nu)\alpha_wT_0w_0G_{I0}D(w_e-w_I)}{(1-\nu)b^2T_f} \quad (50)$$

A numerical solution for Eq.(49) is expressed as

$$S_{rr}(R_i,\tau,t)=\frac{\Delta t\alpha_2}{2R_i^3 w(R_i,\tau,t)}\sum_{j=1}^{\infty}[B(R_i,\tau,t_j')+B(R_i,\tau,t_{j-1}')] \quad (51)$$

Where B is given by

$$B(R_i,\tau,t_j')=\{0.3\exp[-\Omega_1(R_i,\tau,t_j')]+0.7\exp[-\Omega_2(R_i,\tau,t_j')]\}\cdot\{\sum_{N=1}^{\infty}\exp[-\frac{\beta_N^2 D}{b^2}(t_j'-\tau)]\frac{\Gamma(\tau,N)}{\cos\beta_N}\cdot[\frac{R_i\beta_N-3}{3\beta_N}\sin(\beta_N R_i)+R_i\cos(\beta_N R_i)]\} \quad (52)$$

<u>The stress components.</u> A numerical solution for the radial stress component $\sigma_{rr}(R_i,t)$ is expressed in view of Eq.(21) by

$$\sigma_{rr}(R_i,t)=\frac{3\Delta R}{2}\sum_{\kappa=1}^{m-1}[\frac{S_{rr}(R_\kappa,t)}{R_\kappa}+\frac{S_{rr}(R_{\kappa+1},t)}{R_{\kappa+1}}] \quad (53)$$

Hence the following relations are held at R=0.

$$[S_{rr}(R_i,t)]_{i=0}=0$$
$$[\frac{S_{rr}(R_i,t)}{R_i}]_{i=0}=0 \quad (54)$$

If the distribution of the deviatoric stress at t, $S_{rr}(R_i,t)$, is calculated from Eq.(44) or (51), then the radial stress component $\sigma_{rr}(R_i,t)$ can be calculated by Eq.(53). Moreover, the tangential stress component $\sigma_{\theta\theta}(R_i,t)$ is calculated according to Eq.(18) by

$$\sigma_{\theta\theta}(R_i,t)=\sigma_{rr}(R_i,t)-\frac{3}{2}S_{rr}(R_i,t) \quad (55)$$

4. STRESS CALCULATION

4.1. Conditions Set up for Calculation and Properties of Rice Kernels

We suppose that a lot of brown rice kernels are dried and preserved under a similar condition to the experimental condition for the observation of the cracked rice percentage mentioned previously (see Fig.1). Under this condition as shown in Table 1, the calculation of the stresses in the kernels have been performed by using a digital computer, FACOM 230-45S.

The viscoelastic property of the rice kernel is expressed in Eqs.(1) ∿ (10). The other properties are given from the report of the present authors [18] as shown also in Table 1.

Table 1 The conditions set up for the stress calculation and properties of rice kernels

air temperature (=kernel temperature): T_f=308 K
initial moisture content: w_I=0.27 kg/kg(d.s.)
equilibrium moisture content: w_e=0.13 kg/kg(d.s.)
drying-period: τ=0.75,1.5,3.0,4.5,6.0 and 10.0 hours
moisture diffusivity: D=9.72×10^{-12} m^2/s
equivalent radius of a rice kernel: b=1.44×10^{-3} m
coefficient of linear hygroscopic expansion: α_w=0.39 $kg^{-1}\cdot$kg(d.s.)
Poisson's ratio: ν=0.25
instantaneous modulus at T_0 and w_0: G_{I0}=1.61×10^9 N/m^2

4.2. Calculated Results of Moisture Content

At first the changes of the moisture in a rice kernel during the drying and preserving processes have been calculated analytically from Eqs.(11) and (13), respectively. Fig.4 shows the relationship between dimensionless moisture content W and dimensionless radius R with drying time t as a parameter during the drying process. Fig.5 shows the relationship between W and R with preserving time t_p (=t-τ) as a parameter during the preserving process after the drying-period τ=3.0 h.

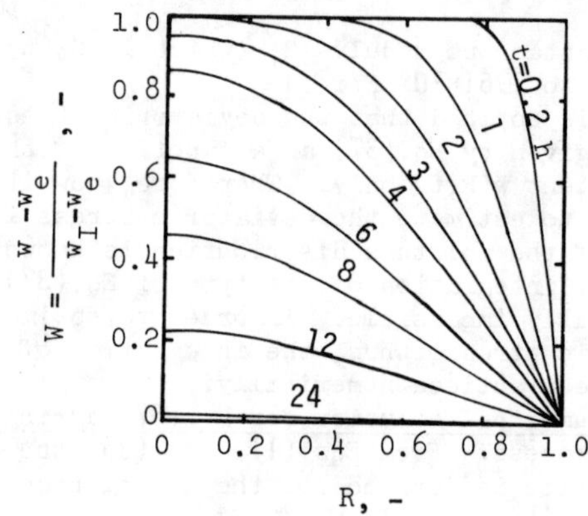

Fig. 4 Calculated result of moisture content distribution in a rice kernel (homogeneous sphere) during the drying process

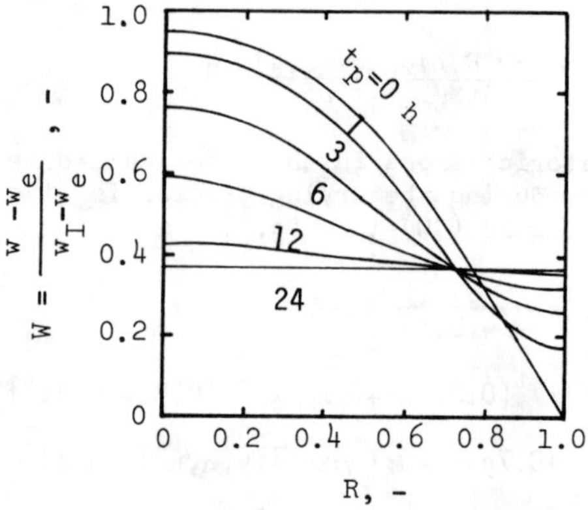

Fig. 5 Calculated result of moisture content distribution in a rice kernel (homogeneous sphere) during the preserving process after the drying-period τ=3 h

4.3. Calculated Results of Stresses

The calculated results of the radial and tangential stresses in a rice kernel during the drying process are shown in Figs.6 and 7, respectively. These figures show the relationships between the stresses and drying time t with R as a parameter. The positive and negative values of the stresses designate the tensile and compressive stresses, respectively.

At the center of the kernel where the crack is generated, the radial stress should be always equal to the tangential stress from Eqs.(18) and (54). The changes of the radial or tangential stress at the center of the rice kernel during the drying process (solid line) and the preserving processes (dotted lines) are shown in Fig.8.

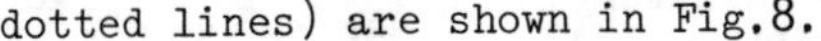

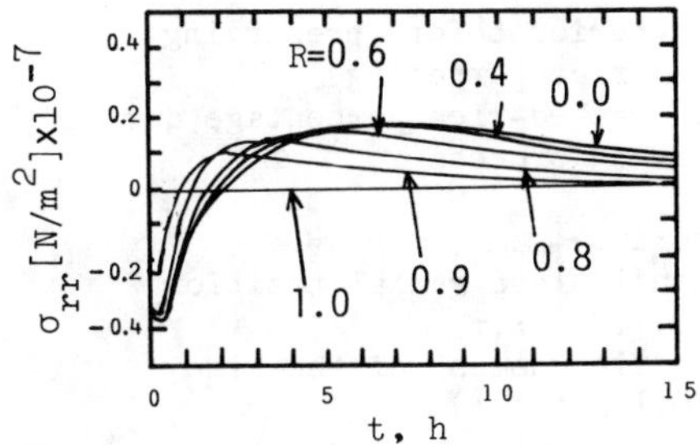

Fig.6 Calculated result of radial stress during drying process

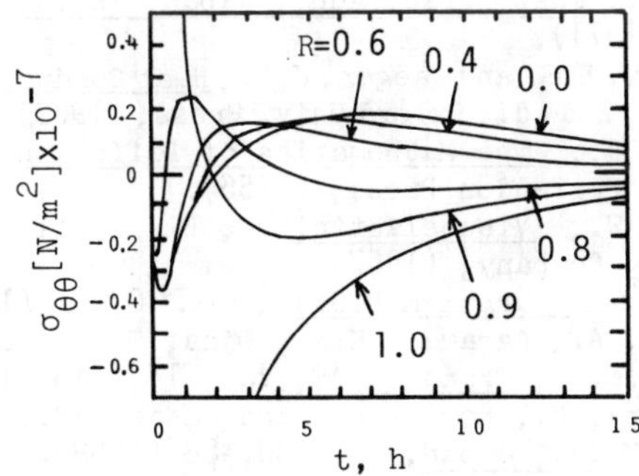

Fig.7 Calculated result of tangential stress during drying process

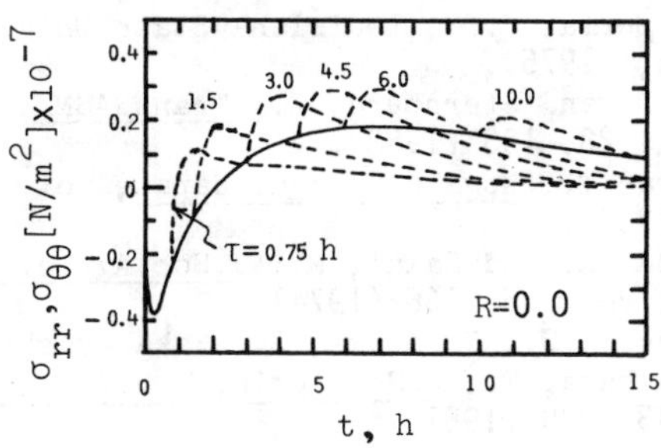

Fig.8 Calculated result of stresses in the center of the kernel during drying and preserving processes

5. DISCUSSION

5.1. Strength of Rice Kernel

The strength (or the limiting stress at which failure occurs) of the rice kernel has been researched several times.

Shimizu et al. [15] have reported, from the bending tests, that the tensile strength of the rice endosperm is remarkably small as compared with that of the surface layer (the protective tissues) of the kernel.

Kawamura et al. [8] have measured the tensile and compressive strengths of the rice endosperm formed into a hexahedral shape. They have reported that the tensile strength S_t is far less than the compressive strength and is decreasing with increasing moisture content w; S_t is about 2×10^6 N/m^2 at w=0.13 and about 8×10^5 N/m^2 at w=0.27.

The other researches[1,10,11] have also reported the tensile strength of sound rice kernels of which the surface layer is not eliminated. Since the cracks are generated in the endosperm near the center of the rice kernel, their values of strength have not been adopted to discuss the cracking mechanism for rice kernels in the present study.

5.2. Discussions on Drying Process

Although the stresses near the surface (for example R=0.9) of a rice kernel show relatively large positive value within about 2 hours of the drying process, the stresses at the center or near the center show a negative value (see Fig.6.7). Thus, in this period, it is considered that the cracks in the rice kernel are not generated because the surface layer is too tough to generate cracks and the center layer can withstand the compressive (negative) stress.

Within about 2 ∿ 5 hours of the drying process the stresses near the center of the rice kernel are converted into tensile (positive) stress. Then the tensile stress increase rapidly and exceed the strength of the endosperm. Thus it should be considered that the rice kernel begins to form cracks and the cracked rice percentage increases rapidly in this period.

The tensile stress in the rice kernel shows maximum value untill about 7 hours, then decreases with increasing drying time and reaches less value than the strength in about 15 hours of drying time. Thus it is considered that the cracked rice percentage is no longer increasing in about 15 hours.

Therefore, the cracking pattern observed during the drying process can be well explained by the results of the stress cslculation for a rice kernel. This may also be illustrated by the results of the moisture content shown in Fig.4.

5.3. Discussions on Preserving Process

The changes of the stresses at the center of a rice kernel during the preserving processes after given drying-periods are shown as dotted lines in Fig.8. In initial periods of the preserving processes, the increasing rates of these stresses are higher than that at τ hours of the drying process and these stresses reach maximum values in about 1 hour of the preserving time.

Then the stresses begin to decrease and reach a lesser value than the strength of the rice endosperm within several hours of the preserving time. Thus it should be considered that (i) the increasing rates of the cracked rice percentage for the initial short-period of the preserving process are greater than that of the drying process and (ii) the period in which the cracked rice percentage reaches an invariable during the preserving process is shorter than that during the drying process.

It is found out from Fig.8 that the maximum stress during the preserving process is dependent on the drying-period τ. The maximum stress increases with τ until it reaches the greatest value (when τ is about 6 hours) and then decreases with increasing τ. In a τ less than 6 hours, the maximum stress is not higher than the greatest value, however, the moisture content is higher (see Fig.4). Therefore, even if the maximum stress is less than the greatest value, the final cracked rice percentage may sometimes be greater than that when the maximum stress is the greatest value, because the tensile strength of the rice endosperm decreases with increasing moisture content. For example, the final cracked-rice percentage at τ=3 hours shows the greatest value as shown in Fig.1.

Consequently, all the patterns observed on the changes of the cracked rice percentage can be reasonably explained by the results of the stress calculations.

6. CONCLUSION

Some interesting behavior about the changes of the cracked rice percentage during the drying and preserving processes have been observed experimentally.

On the other hand, a method of the numerical calculation for the moisture stresses in a brown rice kernel has been proposed by assuming that the kernel is a homogeneous and linear-viscoelastic sphere. Based on this method the calculation of these stresses have been performed under the same conditions as the experiment for the cracked rice observation.

From the results of the cracking observations and the stress calculations, it has been shown that the changes in the cracked rice percentage are closely related to the internal stresses of the rice kernel during the drying and preserving processes.

NOMENCLATURE

a	time-temperature shift factor	–
b	equivalent radius of a rice kernel	m
D	moisture diffusivity	m^2/s
E	compressive relaxation modulus	N/m^2
E_I	instantaneous elastic modulus for compression	N/m^2
E_0	compressive relaxation modulus at $T=T_0$ and $w=w_0$	N/m^2
e	deviatoric strain	–
e^e	deviatoric elastic-strain	–
G	shear relaxation modulus	N/m^2
G_I	instantaneous elastic modulus for shear	N/m^2
G_0	shear relaxation modulus at $T=T_0$ and $w=w_0$	N/m^2
R,R'	dimensionless radial-coordinate, $=r/b$	–
r,r'	radial coordinate	m
S	deviatoric stress	N/m^2
T	temperature	K
T_f	air temperature ≒ kernel temperature	K
T_0	reference temperature, =293	K
t,t',t''	time	s,h
W	dimensionless moisture content, $=(w-w_e)/(w_I-w_e)$	–
w	moisture content	kg/kg(d.s.)
$\overline{w}$	a mean moisture content defined by Eq.(36)	kg/kg(d.s.)
w_e	equilibrium moisture content	kg/kg(d.s.)
w_I	initial moisture content	kg/kg(d.s.)
w_0	reference moisture content, =0.17	kg/kg(d.s.)
α_w	coefficient of linear hygroscopic expansion	$[kg/kg(d.s.)]^{-1}$
ε	strain	–
θ,ϕ	spherical coordinates	rad
ν	Poisson's ratio	–
ξ,ξ'	reduced time	s,h
σ	stress	N/m^2
τ	drying-period before preserving	s,h
ω	cracked rice percentage	%
ω_f	final cracked-rice percentage during preserving process	%

<Subscript>

i,κ	a dimensionless radial position ($i,\kappa=0,1,2,\cdots,m$)
j,λ	a specified moment of the time ($j,\lambda=0,1,2,\cdots,n$)

REFERENCES

1. Arora, V.K., Henderson, S.M. and Burkhardt, T. H., Trans.ASAE, vol. 16, 320 (1973).
2. Ban, T., Inst.Agric.Mach., Japan, Tech.Rep. No.8 (1971).
3. Carslaw, H.S. and Jaeger, J.C., Heat Conduction in Solid , 2nd ed., Oxford Univ. Press, (1959).
4. Crank, J., The Mathematics of Diffusion , Oxford Clarendon Press, (1956).
5. Fluge, W., Viscoelasticity , Blaisdell Publishing Company, (1967).
6. Fujii, I., Journal M.E.S.J., vol. 10, 321 (1975).
7. Husain, A., Agrawal, K.K., Ojha, T.P. and Bhole, N.G., Trans. ASAE, vol. 14, 313 (1971).
8. Kawamura, N., Horio, H., and Sasaki, Y., J.Soc. Agric.Mach., Japan, vol. 30, 88 (1968).
9. Kobayashi, H., Miwa, Y. and Matsuda, R., J.Soc. Agric.Mach., Japan, vol. 37, 551 (1976).
10. Kunze, O.R., and Choudhury, M.S.U., Cereal Chem., vol. 49, 684 (1972).
11. Lee, K., Texas Univ., Ph.D.Thesis, (1972).
12. Mannapperma, J.D., Louisiana State Univ., M.S. Thesis, (1975).
13. Muki, R. and Sternberg, E., Trans.ASME, Series E, vol. 28, 193 (1968).
14. Okamura, T., Nogaku-Kenkyu, Japan, vol. 27, 166 (1937).
15. Shimizu, H. and Sakai, M., J.Soc.Agric.Mech, Japan, vol. 36, 166 (1974).
16. Yamaguchi, S., Yamazawa, S., Wakabayashi, K. and Shibata, T., J.Soc.Agric.Mech., Japan, vol. 43, 239 (1981).
17. Yamaguchi, S., Yamazawa, S. and Wakabayashi, K., J.Soc.Agric.Mech., Japan, vol. 43, 239 (1981).
18. Yamaguchi, S., Wakabayashi, K. and Yamazawa, S., Proceedings for the 4th Int.Drying Symp., Kyoto, Japan, (1984).

DRYING OF SLURRY IN BED OF PARTICLES FLUIDIZED BY PURE WATER VAPOR

Noriyuki Takahashi*, Yasuyuki Satoh*, Toshio Itoh*,
Toshinori Kojima, Toshiyuki Koya, Takehiko Furusawa and Daizo Kunii

Department of Chemical Engineering, University of Tokyo
*Mizusawa Industrial Chemicals,Ltd.

ABSTRACT

The drying of solid materials with a high water content in a stream of pure water vapor allows for the re-utilization of the latent heat as heat source for another dryer in the multple-effect drying system. In this paper a thermally efficient fluidized bed dryer as a part of the multiple-effect drying system is proposed.

In order to minimize the amount of steam necessary for fluidization, and also to guarantee adequate surface area for heat transfer, a tall fluidized bed dryer with an internal draft tube was used. Dry silica particles were poured into the dryer at the beginning of the experiment. Superheated steam was sent to the bottom distributor and the peripheral one as fluidizing gas. A sheathed electric heater simulated the heat exchange bundles was used as a heat source. Silica slurry of 8 weight percent at feed rate 1.3 kg/h was dried continuously as long as 11 hours at a range of 476-498 K without any serious problems.

Smooth circulation of solid particles was observed and the necessary amount of thermal energy was conveyed to the part where the drying of slurry was taking place. Most of dried silica particles were caught by the attached cyclone collector. The size distribution of bed particles did not change appreciablly during the steady state of operation.

1. INTRODUCTION

In order to remove water from any type of slurry, it is usual to apply some of mechanical dehydration processes. There are a number of materials, however, which still remain considerable amount of water after the mechanical treatment. Lignite, sewage slurry, and wet municipal waste are examples of the above.

When such materials are dried by hot air/gas stream, enormous amount of thermal energy is consumed to vaporize the water contained. On the other hand, the drying of such materials in a stream of pure water vapor allows for the re-utilization of the latent heat as the heat source for another dryer, therefore composing so-called multiple-effect drying system. Kunii[1] and Potter[2] proposed the multiple-effect drying systems by application of fluidized bed dryers, in which pure water vapor was flowing.

It is the purpose of this paper to propose a suitable unit, which is thermally efficient for the drying of dilute slurry, and to bring attention to a fluidized bed dryer, which could be developed as part of the multiple-effect drying system.

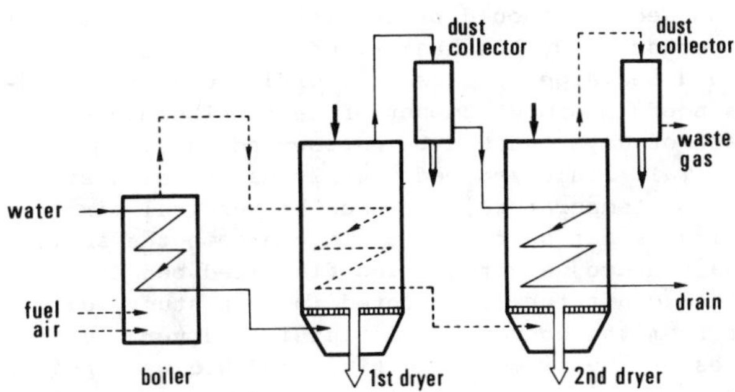

Fig. 1 Double effect drying system operated under atmospheric pressure

2. PROPOSED SYSTEM AND DRYER

2.1. Proposed System

As the simplest case of multiple-effect drying system, a double-effect drying system is shown in Fig. 1, which is composed of a boiler and two fluidized bed dryers. Solid particles in the first dryer is fluidized by pure water vapor under atmospheric pressure. Thermal energy necessary for drying is mainly supplied by the flue gas from the boiler, which flows through the heat exchange tube bandles positioned in the first dryer. The water vapor to fluidize the first one is supplied from the boiler. The fluidizing water vapor and that produced by the drying of wet materials in the first dryer as well, are introduced into the heat exchange tube bandles positioned in the second fluidized bed dryer. The water vapor is condensed in the tube of the second dryer and drained away. The flue gas from the heat exchange tube bandles in the first dryer is introduced to the bottom of the second dryer as the fluidizing gas, so that the second bed is fluidized by the hot air under atmosphere. The humid flue gas from the second dryer is exhausted to conventional abatement processes to prevent air pollution.

Some of the dry solid particles are discharged from the bottom of the dryer, and the others are separated from the flue gas or the water vapor stream by application of conventional collector, e.g., cyclone collector or bag filter.

2.2 Proposed Drying Unit

In the present system, fluidized bed is adopted to be the main dryer, because heat supply must be indirect across the heating surface in order to keep the purity of water vapor, and then high heat transfer capacity is required. The fluidized bed dryer has various advantages, for instance, uniformity of temperature in the bed, feasibility of continuous feeding and discharging, possibility of high drying capacity and so on.

To minimize the consumption of water vapor to fluidize the particles in the first dryer, its cross section should be as little as possible. On the other hand, the heat transfer surface area should be large, because the drying of wet materials needs enormous amount of heat. Therefore, the shape of dryer should be narrow and tall.

Tall fluidized bed usually has axial distribution of temperature, because the vertical mixing of solids is not sufficient. To overcome the above disadvantage, a circulating fluidized bed dryer with a draft tube is adopted in this study, different from the conventional fluidized dryer. Advantages of the circulating fluidized bed were pointed out by Yang [3], particularly in gasification of coal or oil. This type of circulation system is also very advantageous in drying of very wet solid materials.

1. In maintaining uniform bed temperature. Induced circulation of particles reduces temperature difference in a fluidized bed [3].

2. In feeding wet materials to the bed. Wet materials are distributed throughout the bed without serious problems of distribution and/or agglomeration.

3. In saving fluidizing water vapor. Water vapor from the wet materials, evaporated inside the draft tube, accelerates the circulation of solid particles.

3. EXPERIMENTAL PROCEDURE

3.1. Circulating Fluidized Bed Dryer with Draft Tube

A fluidized bed dryer of 80 mm inner diameter was used in the present experiments, which had an aspect ratio of considerable size, specifically (static height of bed)/(inner diameter of the dryer) = 6. The dryer is shematically shown in Fig. 2.

Most part of the dryer is made of stainless steel, SUS 304, while the wall of the bed is made of glass so that it is possible to observe visually the behaviour of the bed from two perspectives, one just above the fluidized bed, i.e., at the top of the draft tube, the other at the wall surface located at an intermediate level of the bed. Through the upper perspective behaviours of fed droplets and overflowing solids from the draft tube to the annular part of the bed were observed. Through the lower one in addition, behaviour of decending solids and ascending bubbles near the wall were visually observed.

The bed was equipped two types of distributors; a bottom distributor of 16.12 cm^2 with 18 orifices (diameter of 0.5 mm , opening area of 0.22 percent) and a peripheral distributor of 31.97 cm^2 with 6 orifices (diameter of 0.5 mm , opening area of 0.04 percent) positioned at a level 70 mm higher from the bottom distributor. The detailed configuration around the distributors is shown in Fig. 3.

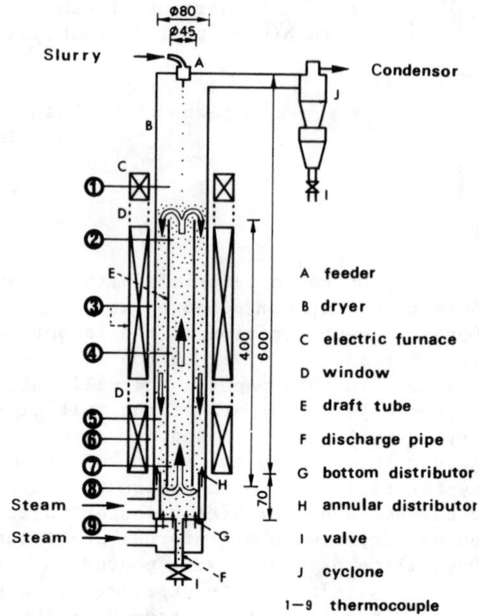

Fig. 2 Circulating fluidized bed dryer with draft tube

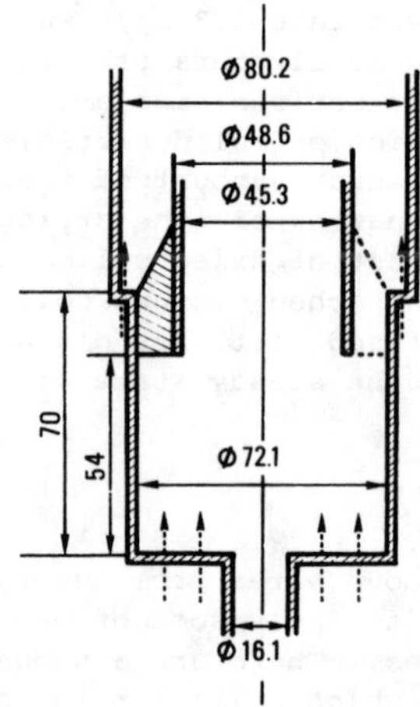

Fig. 3 Detailed configuration around distributor

An internal draft tube of 45 mm I.D./49 mm O.D. and 400 mm in length was positioned in the bed. The lowest end of the tube was located at 54 mm above the bottom distributor.

The top of the bed was closed by a plug of silicone rubber, through which a vertical stainless steel pipe (0.4 mm I.D./0.7 mm O.D. or 1.0 mm I.D./2.0 mm O.D.) is inserted. The bottom end of the pipe was located 220 mm above the top end of the draft tube.

An external cyclone collector (30 mm I.D. and angle of lower cone 19.3°) was connected to the top of the dryer through a pipe of 6 mm I.D. A double lock hopper system was connected to the bottom distributor of the fluidized bed through a pipe of 18 mm I.D.

An electric heater was placed around the dryer to compensate the heat loss, and was controlled by a conventional on-off system. A sheathed electric heater (total length being 2 m) was positioned around the draft tube, to heat the solid particles descending the annular space. A number of sheathed chromel-alumel thermocouples were placed at the various locations in and around the bed, which positions were shematically shown in Fig. 2.

3.2. Materials Dried

Dilute slurry consisting of very fine silica particles (R Mizukasil, 0.1 to 5 μm) was used as an example of highly hydrated material.

Slurry of the fine silica particles was dried by conventional air dryer to get flakes of the agglomerated fine silica particles. The flakes were disintegrated and sieved to get the suitable size for fluidization (at ordinary operation, mean size of the fluidized particles was 100 to 1700 μm). The coarse particles above prepared were used to be the fluidized solids in the dryer.

In some experimental runs, micro spherical alumina particles were used instead of the above particles, in order to study the influence of the physical properties of different fluidized particles. Major difference in the properties between the silica and the alumina particle is particle density as shown in Table 1. These particles are classified to group A defined by Geldart [4].

Table 1 Physical properties of bed material

material		silica	alumina
size distribution	[μm]	100-1700	50-350
mean size	[μm]	360	151
particle density	[kg/m^3]	216	1540
incipient fluidization			
velocity (steam, 473K)	[cm/s]	1.00	1.26
(air , 473K)	[cm/s]	0.66	0.83
heat capacity	[J/Kkg]	753	921

3.3 Materials Flow

The feed slurry containing 0 to 8 weight percent of silica was introduced to the thin feeder pipe at the rate of 0 to 1.3 kg/h by use of a roller pump. The pulsation of the slurry flow originated in the pump was smoothed by connecting a buffer device to the silicone tube through which slurry was flowing. The slurry was poured upon the bed vertically downward from the end of the thin feeder pipe. A chain of small droplets was dispersed around the top of the fluidized bed surface by vigorous motion of bed particles.

Water was introduced to a evaporator at the rate 4.45 to 7.65 g/min by use of another roller pump. The generated water vapor was heated up to around 470 K and introduced to the bottom distributor, establishing the superficial vapor velosity (based on the cross sectional area of draft tube) to be 10.2 cm/s to 17.5 cm/s, which corresponded to 10.2 to 17.5 u_{mf} at 473 K. Water vapor was also introduced to the peripheral distributor similarly to the previous one. The superficial vapor velocity was kept to be 1.0 cm/s to 1.3 cm/s based on the cross sectional area of annular part (corresponding to 1.0 to 1.3 u_{mf}).

In some experimental runs, the air was used instead of water vapor. The air was also heated up and introduced similarly to the previous procedure. The fluidizing gas stream from the bottom distributor resulted in good circulation of solid particles upwards within the draft tube. The other gas from the peripheral distributor made the solid particles in the annular space move downwards very smoothly.

The fluidizing gas, together with water vapor produced from the wet material flowed out of the dryer, and were introduced into the cyclone collector. Most of the dry fine particles elutriated from the bed were separated from the gas stream by the cyclone collector and piled up in a bin. Fine particles collected in the bin were taken out periodically. The gas stream from the cyclone collector was introduced to a condenser. Small amount of fines still exist in the gas stream because the experimental cyclone cllector was inefficient in this study. They were caught in the condenser accompanied with the drain.

The bed materials and agglomerated coarse particles of silica as well were discharged from the bottom through the double lock hopper system, when the increment of the bed height was observed.

3.4. Preliminary Experiment

Preliminary experimental runs were performed to search the effects of the following experimental conditions, i.e., bed materials, water content of the slurry, circulating behaviour of solid, and temperature distribution in the bed. Both water vapor and the air were used as fluidizing gas respectively in different runs, simulating the first and second dryers in the system proposed previously.

The experimental procedures were as follows;

1. About 2000cm^3 of bed material was weighed exactly and poured into the bed.
2. The bed was heated up and adjusted to keep the operating temperature (390 to 505 K) by use of the external electric heater.
3. The superheated water vapor (or air) of the planned flow rate was introduced to the bed.
4. Water was fed into the bed at the same rate with operating rate of slurry through the feeder pipe of 1.0 mm I.D.
5. Surplus bed material was withdrawn from the bottom through a double lock hopper system and weighed.
6. After getting the stable temperature distribution, the silica slurry was fed into the dryer at a planned feed rate in place of water.
7. Temperature distribution inside the dryer was measured after getting the steady state.

The procedures were iterated for various operating conditions and feed rates.

3.5. Experimental Long Run

Several experimental runs were performed to test the physical properties of products by the superheated water vapor drying in a fluidized bed . The procedures were almost the same as the preliminary ones except the following ones.

1. An internal heater was used in addition to external heater so that the capacity of the dryer was increased. The heat for drying was mainly supplied from the internal heater and the external heater was used to prevent the heat loss from wall. Namely the external heater was controlled so as to keep the temperature of the outside bed wall surface to be the same temperature as that in the annular part of the fluidized bed. The total electric power input in the internal heater was measured by means of a conventional technique.

2. A feeder pipe of 0.4 mm I.D./0.7 mm O.D. was used instead of that of 1.0 mm I.D. so that the smaller diameter of slurry droplet was formed. Diameter of droplets from feeder pipe was measured by a photographic technique.

3.6. Physical Properties of Dried Silica

Particle size distributions of bed materials before and after each run and the silica discharged from the bottom of the bed as well were measured by sieving. Further more that of fine silica powder caught in the cyclone collector was measured by use of a Coulter Counter.

Quality of fine silica powder from the cyclone collector were tested as follows; namely, specific surface area by B.E.T. method, bulk density by JIS K 6220-1977, dispersivity and oil absorption by JIS K 5101-1978 and matting effect by Mizusawa's standard.

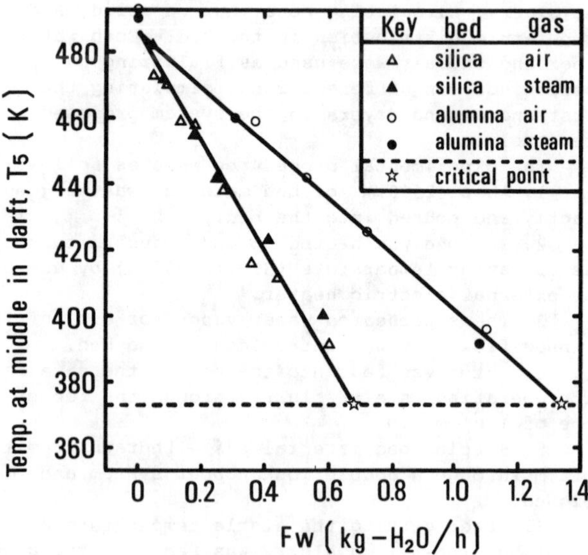

Fig. 4 Temp. at middle in draft part bed

4. EXPERIMENTAL RESULTS

4.1. General Observation

Smooth circulation of fluidized particles was observed for both alumina particles and silica particles. The fluidization in the draft tube was almost similar to a slugging bed, while in the downcomer it was dense bubbling bed. Some amount of particles in the draft tube were splashed and overflowed into the annular part every few seconds. When the water or slurry was introduced from the feeder, the bed expansion was observed and the circulation was accelerated. No apparent difference was observed at the circulating behavior for two fluidizing gases, namely water vapor and air, as far as the circulating behavior concerned. However, more intense pressure fluctuation was observed in the case of water vapor, which might be originated in some unstable boiling phenomena within the experimental water vaporizer. In some unusual runs, coarse agglomerates were formed at the level of peripheral distributor where the opening area of annular space was reduced, and disturbed the smooth circulation.

4.2. Effects of Water Feed Rate on Temperature in Draft Tube

Temperature at the middle level in the draft part (4 in Fig. 2) is plotted in Fig. 4 referring to the feed rate of water. Experimental conditions are summarized in Table 2. Temperature of the external surface of dryer (glass pipe) was maintained at 486 K. While the water was not fed into the bed, the temperature of the fluidized solids within the draft tube was nearly equal to that of external surface of the dryer. Temperature of the bed within the draft tube decreased with increase of the water feed rate. Physical property of bed materials seemed to be effective on the performance of such a drying system.

Table 2 Experimental conditions of Fig. 4

fluidizing gas	u_D/u_{mf} [-]	u_A/u_{mf} [-]	T_w [K]	T_b [K]	T_p [K]	bed material
Steam	11.2	1.93	486	473	453	Alumina
Air	14.3	1.76	486	473	473	Alumina
Steam	12.5	0.91	486	473	458	Silica
Air	11.9	1.29	486	510	473	Silica

As a criterion to evaluate the heat transfer capacity of this circulation system, a index showing the critical feed rate of water, F*, was defined here. This is a hypothetical value of feed rate, which corresponds to a critical operating condition, where the temperature of the fluidized solids within the draft tube would become to 373 K, namely the boiling point of water under atmospheric pressure. The value of F* for alumina particles was nearly twice as much as that for silica particles. This result may be originated in the difference of heat capacity per unit bulk volume. In other words heat transfer from the annular space to the inside of the draft tube by the solid circulation seems to play an important role in this type of drying system. Regardless of the above observation, silica particles were used as bed material in almost all experimental runs. It is common to use the same kind of solid particles with the product fines, in case when one manufactures fine particles without contamination.

4.3. Circulating Mass Flux Density of Fluidized Particles

Assuming that the vaporization occurs only within the draft tube and that the heat transfer across the wall surface of the draft tube is negligible, mass flux density of fluidized particles was calculated from the heat balance equation between two locations; namely at the lowest level within the annular tube (denoted by a superscript of *) and the highest level of the draft tube (2 in Fig. 2) as follows.

$$F_w(h_1-h_0) = \dot{m}A_A c_P(T^*-T_2)+(uA\rho)_D(h_9-h_2)$$
$$+(uA\rho)_A(h_8-H^*) \qquad (1)$$

Unfortunately the temperature at the lowest location, T*, was not measured, because of the configurational difficulties. However it may be reasonable to postulate that T* would be in the range between T_5 and $(4/3)(T_5-T_2)+T_2$. The calculated mass flux density based on the annular cross sectional area were shown in Fig. 5 for the two extreme cases of T*. It was apparent that the mass flux density was strongly dependent on the fluidizing gas velocity ratio between two locations, namely within the draft tube and the annular space.

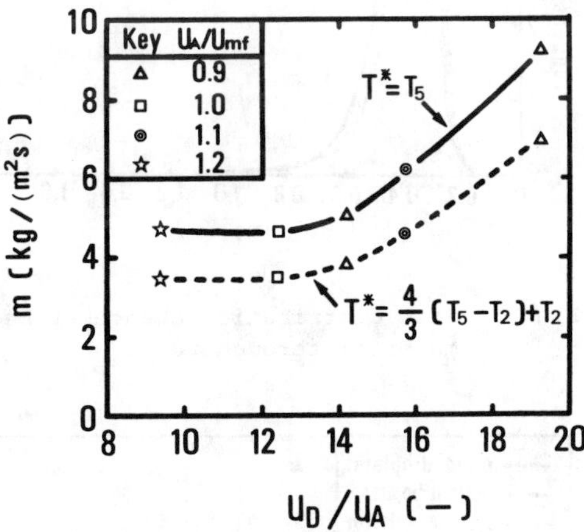

Fig. 5 Calculated mass flux density of circulating particles

Using the bulk density in annular space of 140 kg/m^3 (which is calculated from the total weight of particles divided by total volume of the bed under the fluidizing condition), the mass flux density of 6 kg/(m^2 s) corresponds to particle velocity of 4 cm/s. This value agrees fairly well with the observed value through the lower perspective.

4.4. Effects of Slurry Concentration

Effects of slurry concentration on the temperature distribution and maximum feed rate of slurry are shown in Fig. 6. Maximum feed rate is defined to be the maximum rate where the smooth circulation was observed. In case of the evapolation of water in the fluidized bed dryer, the resistance for mass and heat transfer inside the wet droplet is small, so that temperature of solid within the draft tube nearly reached to 373 K, namely to the boiling point of water at atmospheric pressure.

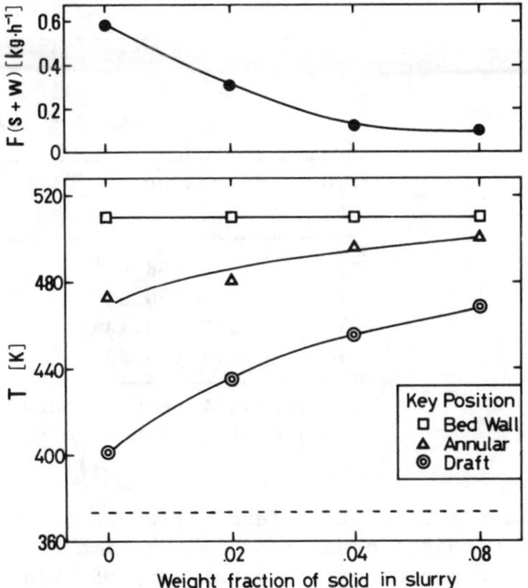

Fig. 6 Local temp. and max. feed rate
- - - boiling point of water at 10^5Pa

When slurry was fed into the dryer, however, the maximum feed rate above defined decreased tremendously with the increase of the weight fraction of solids in the slurry. When the 8 percent silica slurry was fed into the dryer at a rate larger than 0.1 kg/h, wet silica particles bigger than 2 mm were observed to lie on the bottom distributor, and disturbed the smooth circulation in the dryer. Furthermore compared with the vaporization of water, the temperature difference between solids in annular space and those in draft tube was observed to be smaller with the increase of solid content in the slurry. Referring a result from other fundamental study on drying by TAKAHASI et al. [5], it was confirmed that the internal temperature within a wet particle composed of fine silica, 20 mm O.D. was kept to be around 373 K (boiling temperature) throughout the drying.

Based on above results, it may be conceived that the resistances inside and around the wet silica particle was so big that the large temperature difference between fluidized bed and wet particles was needed. Therefore the initial diameter of slurry droplet should be the smaller the better. Hereafter, the feeder pipe of 0.4 mm I.D. was employed instead of that 1 mm I.D. in order to reduce the slurry droplet size.

4.5. Material Balance in Experimental Long Run

The 8 wieght percent slurry was continuously dried without any trouble by choosing suitable experimental conditions. The maximum capacity of the bed has increased to 1.28 kg - slurry/h at the bed temperature of 505 K, by introducing the inter-

Table 3 Experimental conditions of long term run

run	gas	u_D/u_{mf} [-]	u_A/u_{mf} [-]	T_w [K]	T_b [K]	T_p [K]	$F_{(S+W)}$ [kg/h]	t [h]
1	Steam	17.5	1.11	483	463	451	0.60	6.6
2	Steam	16.1	1.05	505	478	465	1.28	10.1
3	Steam	10.2	1.30	498	502	457	0.44	11.1
4	Air	23.9	4.52	390	409	380	0.42	4.5

Table 4 Product ratio of long term run

run	discharge [g]	cyclon [g]	condenser [g]	undetected [g]	feed silica [g]
1	-19.0 (-6.0)[a]	236.5 (74.8)	32.5 (10.3)	66.0 (20.9)	316.0
2	-2.9 (-0.3)[a]	822.1 (79.4)	121.7 (11.8)	93.9 (9.1)	1034.8
3	13.2 (3.4)	262.8 (67.0)	78.5 (20.0)	38.0 (9.6)	392.5
4	51.7 (33.9)	80.9 (53.1)	——[b]	19.8 (13.0)	152.4

( ) ; weight percent based on feed silica. a ; bed weight has decreased.
b ; most of fines released from condenser.

nal heater and the thin feeder pipe of 0.4 mm I.D. The experimental conditions and the amount of dried material from the experimental runs of relatively long duration are listed in Table 3 and 4.

Dried product was mainly taken out from the cyclone collector, whereas the amount of particles discharged from the bottom of the bed was very small. Some amount of the dried product was not caught by cyclone collector. It was collected in the condenser, and the small amount of the product could not be recovered. The dried silica particles had not a hard structure, and they shrinked easily by attrition during the circulation in the fluidized bed, especially for the case of rapid circulation. When the circulation rate of fluidized particles was relatively slow, e.g. in run 3 in Table 3, more solid particles were discharged from the bottom. In case of air drying, more amount of dried solid was discharged from the bottom.

4.6. Particle Size Distribution

Fig. 7 shows the distribution of fluidized particle size before and after the run 2 of long duration. Through the run 2, 0.94 kg of dried silica products were obtained and its amount corresponded to more than three times of the initial weight of silica particles loaded into the bed. Nonetheless, remarkable change in the size distribution of the fluidized particle is not seen in Fig. 7. Same results as above were also obtained from other experimental runs.

The size distributions of dried silica fine particles from the cyclone collector and coarse particles from the bottom were shown in Fig. 8. The distribution of coarse particles from the bottom hopper was almost same as that of fluidized particles in the bed, except the fraction of larger particles (d_p>0.5 mm).

In Fig. 8 the estimated size distribution of dried particles from droplets of slurry is also shown, assuming that the droplets had been dried without any breakage throughout the drying process and had the same density with that of fluidized particles in the bed. It is clearly shown in Fig.

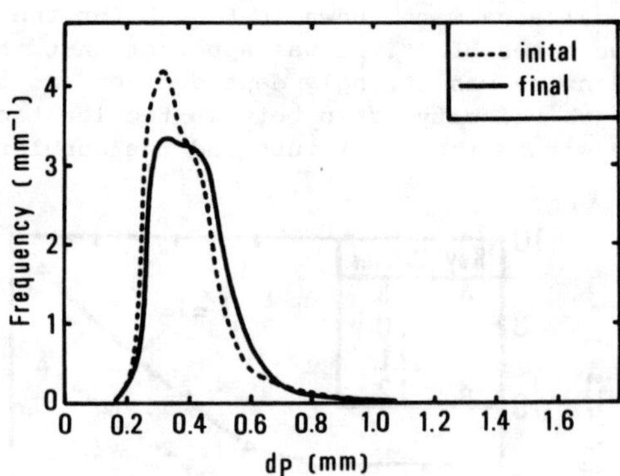

Fig. 7 Size distribution change of bed material through run 2

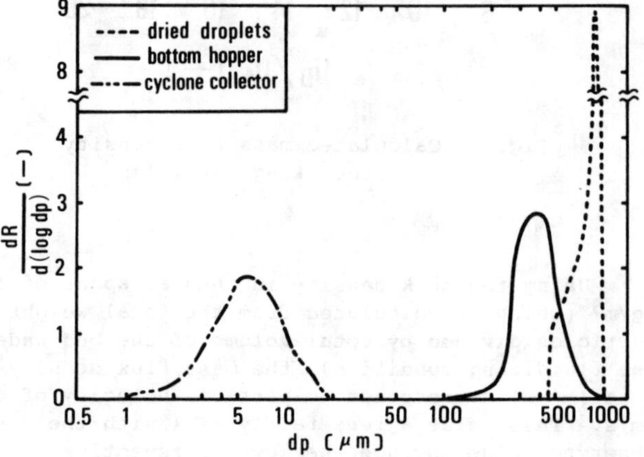

Fig. 8 Size distribution change of silica dried solid (run 3)

8 that the feed droplets were fairly well dispersed and the dried particles were broken fairly well in the bed to form fine particles.

To know the formation process of fines from slurry, the following experiment was performed. 300 cm^3 of violet colored silica slurry was fed into the bed, the whole material in the bed was discharged. A lot of particles were observed to be colored violet not only around the surface but also in the whole particle. This result indicates that some amount of silica in the slurry once formes coarse fluidized particles, which were successively broken into the fine powder by attrition and that the rest were broken into fine powder directly from the feed slurry.

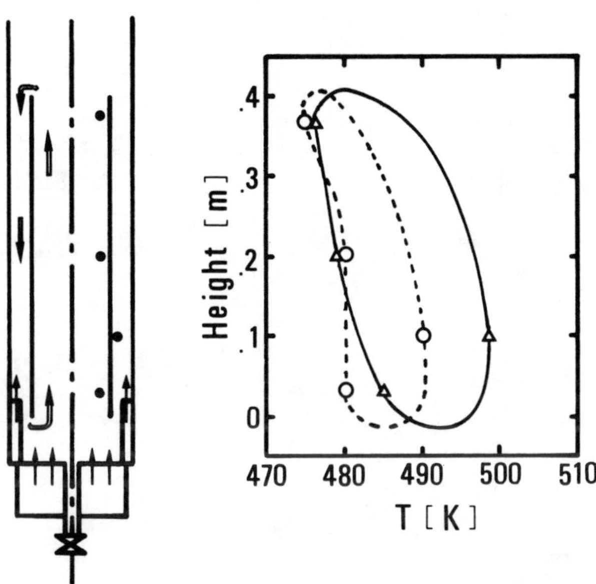

Fig. 9 Temp. distribution in bed
△; run 2 ; u_D/u_A=15.3 ; 8% slurry F_w=1.18kg/h
○; run 5 ; u_D/u_A=17.0 ; water F_w=1.03kg/h

4.7. Temperature Distribution in Draft Part of Bed

The steady state temperature distributions in the bed are schematically shown in Fig. 9 for run number 2. The measurement points of temperature in the draft tube were also shown. Fig. 9 indicates that the wet silica particles were descending downwards against the ascending fluidized particles by the difference in specific gravity, and were dried while descending within the draft tube.

The steady state temperature distribution under the same conditions as those in run 2 (but water was fed) is also shown in Fig. 9. Only the temperature at the upper part of the draft tube was lower than those at the other parts of draft tube. It suggests that water was evapolated only at the upper part of the fluidized bed. It might be attributed to the smaller intradroplet drying resistance and higher dispersiveness of water than those of slurry.

4.8. Heat Balance

All heat balance data are shown in Table 5. In these experimental runs, the external surface of the glass wall was maintained at same temperature as that in the annular part within the error of 7 K. It is apparent from Table 5 that most of the heat to dry the slurry was given from the inner heater to the solid particles in the annular space,

Table 5 Heat balance of long term run

run	input [W]		output [W]
	from inner heater	from gas	for drying
1	480	2	446
2	976	-1	921
3	303	-6	315

and then from the space to the inner part of the draft tube through the circulation of solid particles.

4.9. Quality of Product

The product of the fines from cyclone collector was tested on several quality items. The results were compared with that of air dried silica powder commercially produced and were shown in Table 6. No appreciable difference was found.

5. CONCLUSIVE REMARKS

A thermally efficient fluidized bed dryer with a draft tube is proposed in this paper for the superheated water vapor drying of solid materials with high water content. Silica slurry was successfully dried in the proposed dryer and silica fines of good qualities were produced. It is concluded that the present system is sufficiently capable of drying the dilute slurry and producing pure water vapor for further utilization in, for example, the multiple-effect drying system.

Through the experimental runs, several items

Table 6 Quality of products

run	specific surface area [m^2/g]	bulk density [g/cm^3]	dispersity [μm]	oil absorption [ml/100g]	matting [-]
2	99.3	0.130	24-29	235	38.1
3	96.8	0.129	24-29	235	36.8
std.	100-140	0.124	25-30	240	37.3-38.4

were found to need improvement as follows;

1. The reduction in the cross sectional area of annular part at the level of peripheral distributor should be avoided or minimized to prevent the obstruction by coarse silica particles at that level. Also the configuration of the draft tube legs should be improved.

2. Diameter of slurry droplet should be reduced by modifying the design of the feeder structure in order to minimize the intraparticle resistance in drying process.

3. The shape of the bottom distributor should be improved to avoid the accumulation of coarse particles and to facilitate the discharge of coarse particles.

NOMENCLATURE

A_A	cross sectional area of annular part	$[m^2]$
A_D	cross sectional area of draft part	$[m^2]$
c_p	specific heat capacity of fluidized particle	[J/(K kg)]
d_p	diameter of particle	[μm]
F^*	critical feed rate of water	[kg/h]
F_S	feed rate of solid in slurry	[kg/h]
F_W	feed rate of water in slurry	[kg/h]
R	cumulative residue on sieve	[-]
T_b	temperature of fluidizing gas from peripheral distributor	[K]
T_p	temperature of fluidizing gas from bottom distributor	[K]
T_w	temperature of bed wall	[K]
T^*	critical temperature	[K]
u_A	fluidizing gas velocity in annular part	[cm/s]
u_D	fluidizing gas velosity in draft part	[cm/s]
u_{mf}	incipient fluidizing gas velocity	[cm/s]

Subscripts

1-9 location in Fig. 2

REFERENCES

1. Kunii, D., Netuteki Tan-isosa Ge, p. 417, Maruzen, Tokyo (1978).
2. Potter, O.E., Beeby, C.J., Fernande, W.J.N., and Ho, P., Proc. 3rd Int. Drying Symp., Birmingham, vol.2, 115 (1982).
3. Yang, W.C., and Keairns, D.L., AIChE Symp. Ser., vol. 70, (141) 27 (1974).
4. Geldart, D., Powder Technol., vol. 14, 264 (1975).
5. Takahashi, N., Satoh, Y., Inoue, H., Kaneda, T., and Kunii, D., Preparing.

DIMENSIONING HEAT PUMP DRIERS FOR FISH PRODUCTS

O.M. Magnussen, I. Strømmen and S. Puntervold

The University of Trondheim, The Norwegian Institute of Technology
Division of Refrigeration Engineering
Trondheim, Norway

ABSTRACT

A laboratory heat pump drier has been used for drying of different fish products. Air temperatures and humidities can be varied within a wide range and kept constant during experiments. Quality and drying properties have been continuously evaluated during the experiments.

From the drying data a mathematical model of drying velocity is developed. The model has been combined with a simulation model of a tunnel drier and drying capacities calculated for actual conditions. For a given tunnel and heat pump the drying capacity will depend on type and size of fish and processing like splitting, filleting etc. and on inlet air temperature. Due to increased drying surface, split up fish or fillets will have a higher drying capacity than gutted whole fish. The heat pump should therefore be designed from the actual usage of the tunnel.

The inlet temperature should be as high as possible from the quality point of view. Simulations of a tunnel drier indicate that the production may be increased with 50% by changing inlet temperature from 10°C to 20°C and doubled when changing from 10°C to 30°C. The tunnel length is important for humidity at air outlet and thereby drying capacity and energy consumption per kilo water removed for a given heat pump. Simulation for the actual products and capacities needed, gives possibilities for optimal design of tunnel and heat pump.

1. INTRODUCTION

In a world with an increasing population and shortage of food, it is today, more than ever, important to take good care of the raw materials. In fish drying the products have traditionally been produced outdoors. The quality was thereby left to chance. It might be good if the weather was good, but flies, worms, birds and rainy weather could detoriorate and make sour fish and thereby completely destroy it.

Heat pumps for production of salted and dried fish and for unsalted fish, |1|, was introduced in Norway at about 1980. The heat pumps can make the desired air conditions for most kinds and size of fish and have a 100% utilization of a high quality product, independent of the outdoor air.

From a great number of drying experiments this paper presents a basis for dimensioning heat pump driers for unsalted fish of different kind, size and preparation.

2. DRYING EXPERIMENTS

For dimensioning heat pumps for drying purposes the temperature and the relative humidity of the drying air must be known. The humidity depends on the drying rate of the fish at different water content. The purpose of the drying experiments is to determine the drying rate at different temperatures, relative humidity and air velocity and to evaluate the quality aspects.
The first experiments was done in an already existing heat pump pilot plant, described in |1|. To increase the capacity and the possible size of the fish a new drier was built. This one is shown in Fig. 1a. The fish is placed in the drying section and the humid air having passed the fish, is lead to the evaporator, E1, where it is cooled down and dehumidified. Through the condensor, C1, the air is heated to the desired temperature.

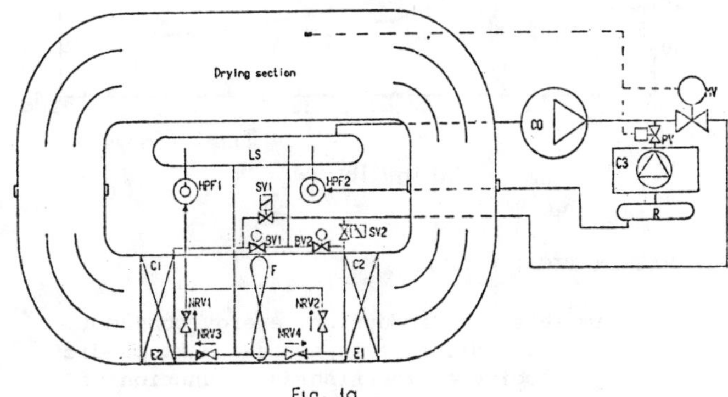

Fig. 1a
Heat pump pilot plant

The liquid refrigerant from the condensor is throttled into the liquid separator, LS, through the high pressure float, HPF1. To obtain uniform evaporation temperature gravity feed is used. The suction pressure, and thereby the dew point of the evaporator, is controlled by the compressor CO. The compressor capacity is regulated by changing the motor speed with frequency regulation. The excess heat in the system is taken out in the condensor, C3, placed outside the drier. The temperature is controlled with the motor valve, MV, and the pilot valve PV. The relative humidity of the drier is regulated by the capacity of the compressor. With temperatures in the drying section below about 10°C the water will freeze on the evaporator surface and defrosting is necessary. By operating the solenoid valves, SV, ball valves, BV and non return valves, NRV, the condensor and the evaporator is changing their operation and gives a very effective defrosting.

The temperature in the drying section can be regulated between 0°C and 30°C, relative humidity between 30% and 100%, and air speed between 0,5 m/s and 4,0 m/s.

The fish types used in the experiments were Cod, Ling and Tusk, whole gutted, split up or as fillets, and changing in size from about 300 g up to 6 kg. Fig. 1b shows some typical results from the drying experiments. Here is plotted the weight of whole gutted, 1 kg, cod versus time at different temperatures when the relative humidity is kept constant at 50%. For further information about the drying experiments and the results we are referring to |2|.

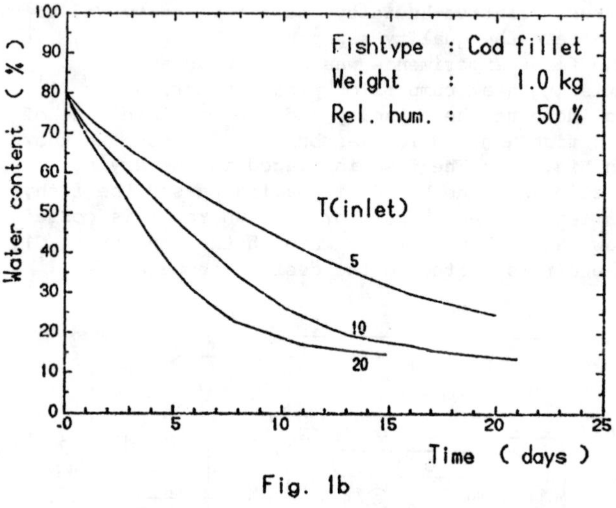

Fig. 1b

3.- MATHEMATICAL MODEL

To simulate the production, energy consumption etc. of a certain drier it is necessary to predict the drying velocity of the fish as a function of it's type, size, preparation, water content, air condition and air speed.

From the drying experiments a simplified mathematical model has been made. The model is described in detail in |3|. The basic equations in the model are presented in the following. The drying is divided into two phases. In the first we have drying from a wet surface and in the second a dry layer of zero water content is formed. The drying velocity in the first period can be found from:

$$G_d = \frac{\beta \cdot A}{R_d T} \cdot (p_{d,s} - p_{d,a}) \qquad (1)$$

In the second period the drying velocity is calculated from:

$$G_d = \frac{1}{R_d T} \cdot \frac{A}{\frac{1}{\beta} + \frac{\mu \cdot s}{D}} \cdot (p_{d,i} - p_{d,a}) \qquad (2)$$

The evaporator of water is in this period assumed to take place from a drying front inside the fish and water vapour are being transported by diffusion through the dry layer. From the drying curves recorded at constant drying conditions, the duration of the first period and other "fish dependent" parameters are determined. The water vapour diffusion resistance, for instance, is found as a function of the water content in the fish. The mass transfer coefficient is connected to the heat transfer coefficient and the air speed through the Lewis relation. On this basis a theoretical drying velocity can be calculated

SIMULATION OF INDUSTRIAL HEAT PUMP FISH DRIERS

From the drying model a simulation program of an industrial heat pump tunnel drier is developed. In Fig. 1c the drier is shown principally. The fish is placed on shelfs on wheeled racks and in counterflow with the air which is assumed to follow the isenthalp through the drier. The theoretical process is shown in Fig. 1c.

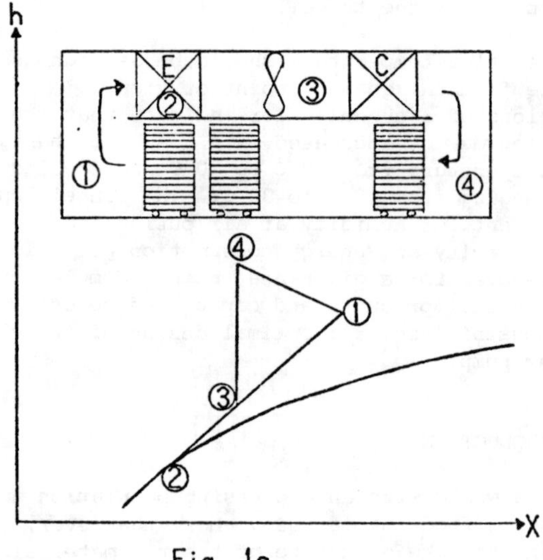

Fig. 1c

The air temperature at the inlet of tunnel is given from quality aspects and a humidity is assumed. The program calculates the temperature, humidity and water content in fish through the tunnel by stepwise iteration.

Table 1 is giving a complete list of the input and output parameters in the program.

Table 1. Input and output parameter list of the simulation program.

Input parameters:

- Tunnel dimensions
- Number of racks in length
- Number of racks in one row
- Drying procedure
- Weight of fish on one rack
- Weight of one fish
- Fish type and preparation
- Air temperature at the inlet of the drying section
- Air speed
- Cooling capacity of the heat pump
- Water content of the fish when finished

Output parameters:

- Production
- Air temperatures in the drier
- Relative humidities of the air in the drier
- Water content of the fish in each row of racks
- Thermal efficiency of the drier

With contiuous drying one row of racks with fish is taken out when finished. The other racks is pushed forward and new wet fish is put into the drier. In batch drying, the whole tunnel is emptied and filled in one time. The drying velocity used is based upon the drying experiments mentioned earlier. The condition of the air close to the evaporator surface is designed to be 17°C below the inlet temperature in the drying section.

The production is calculated in tons of dried product per unit time. The thermal efficiency of the drier is giving how many kWh cooling capacity that is needed to condense 1 kg of water on the evaporator surface.

In the simulations we have used Cod, whole gutted, split or as fillets. The size of the whole gutted Cod is varied between 0,5 kg and 5,0 kg. The Cod fillet size is varied between 0,25 kg and 2,0 kg and the split Cod between 0,5 kg and 3,0 kg. Temperatures in the tunnel are varied between 0°C and 40°C at the inlet of the drying section. Refrigeration capacities of the heat pump is varied between 100 kW and 500 kW. During the simulations we have used 5 wheeled racks in width and varied the number of rows in length between 10 and 50. Mainly continuous drying with water content of 20% of the dry fish have been used. The weight of wet fish on one rack is 500 kg. In most of the simulations we have used air velocity of 2 m/s, referred to the empty cross section of the drying tunnel. The simulation time interval used is 100 days assuming wet fish in the whole tunnel at start.

RESULTS

In Figs. 2, 3 and 4 we have plotted the production in tons of dried Cod fillets of 1,5 kg wet weight per 100 days as a function of the refrigeration capacity and the length of the tunnel. We have used 5 racks of fish in the row which are varied between 10 and 50. Inlet temperature in the drying section is 10°C, 20°C and 30°C. As expected the production increases with increasing length of the tunnel. At an inlet temperature of 20°C and refrigeration capacity of 300 kW the production is

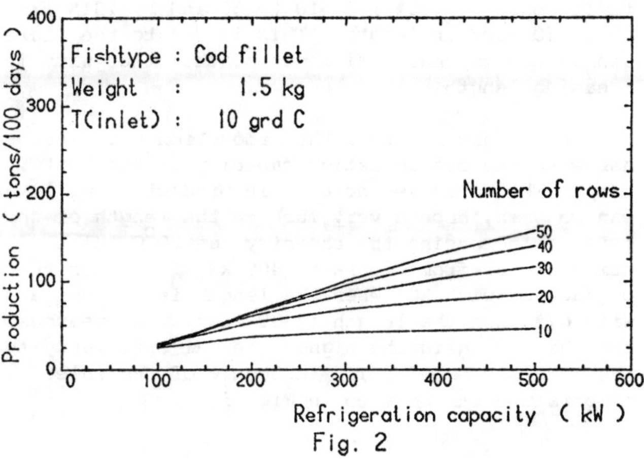

Fig. 2

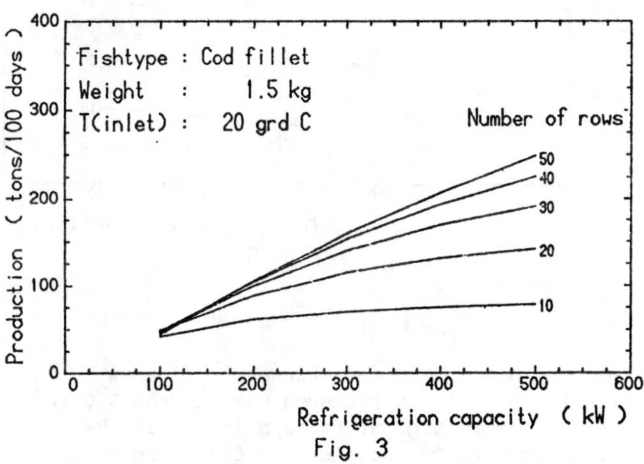

Fig. 3

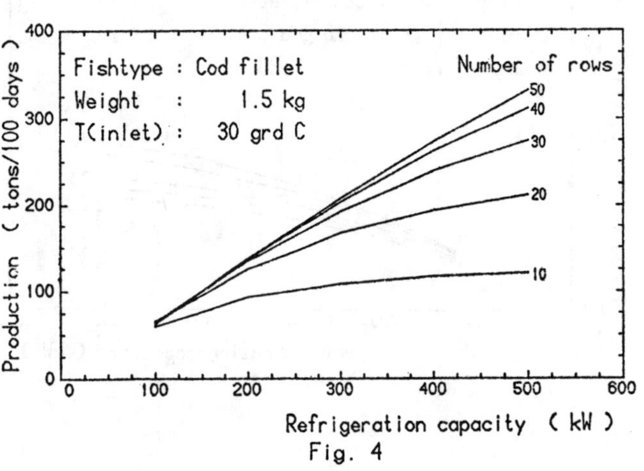

Fig. 4

increased by 64% by increasing the length from 10 to 20 rows, by 100% from 10 to 30 and by 131% from 10 to 50 rows in length. This is due to the increase in the thermal efficiency of the drier with increasing length.

For a given tunnel the production will increase with the refrigeration capacity of the heat pump. The relative increase in production will, as can be seen, depend very much on the length of the drier. Increasing the capacity, at 20°C inlet temperature, from 100 kW to 400 kW, increases the production with 20% when the length is 10 rows and with 67% when the length is 30 rows. The reason for this is again the higher thermal efficiency for the largest tunnel. The influence of the inlet temperature can be seen in Fig. 5.

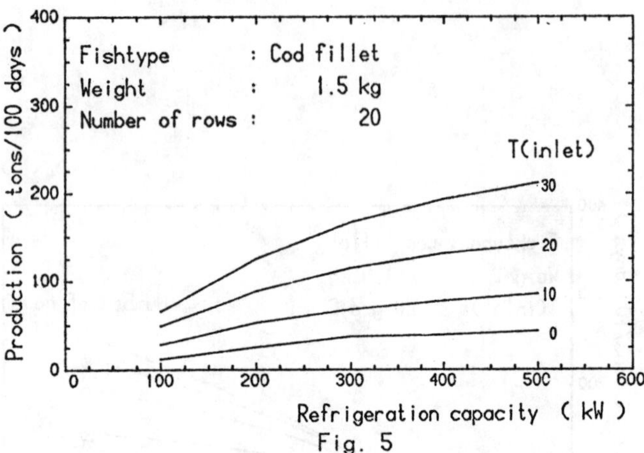

Fig. 5

Here we have looked to a tunnel of 20 rows in length at inlet remperatures varied from 0°C to 30°C. At a refrigeration capacity of 300 kW the production can be increased by 68% by increasing the temperature from 10°C to 20°C and with 145% by increasing the temperature from 10°C to 30°C.

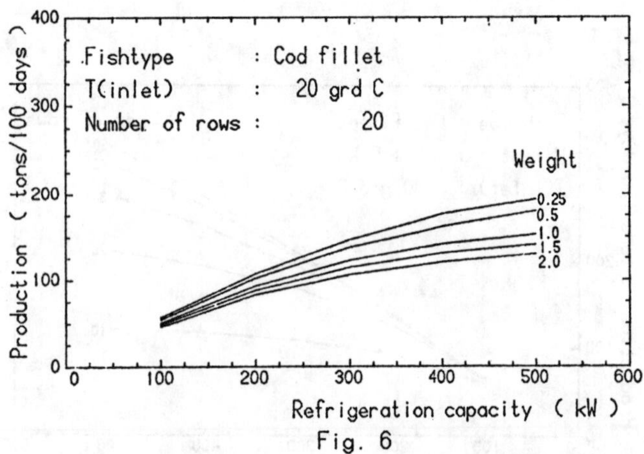

Fig. 6

In Fig. 6 the influence of the size of the Cod fillet is studied. At 20°C inlet temperature, 20 rows in length and 300 kW capacity, the production is increased by 20% when the size of the fillet is reduced from 1,5 kg to 0,5 kg. This is due to the increased surface per kg weight for the smallest fillets. We have taken into account the different weight of the racks when the fish size is changing. In Fig. 7 the production of whole gutted Cod is plotted versus the refrigeration capacity and the length of the tunnel. As for fillets the production is increased with increasing length of the drier. An increase in capacity from 100 kW to 300 kW gave an increase in production of 20% at 10 rows in length and with 200% at 30 rows in length.

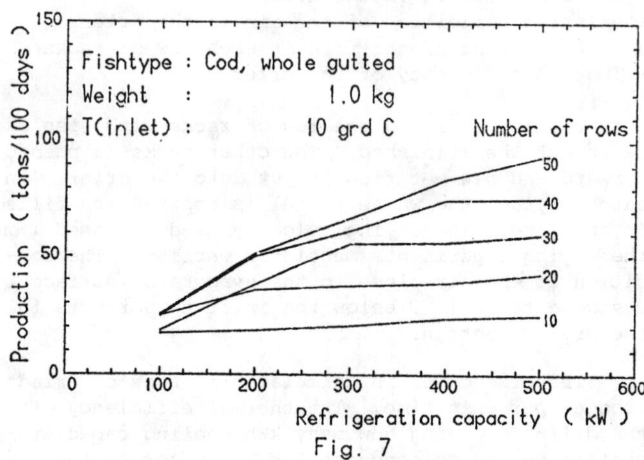

Fig. 7

Influence of preparation of the Cod is shown in Fig. 8. With split fish or fillets in the drier we can produce 2 to 3 times more than with whole gutted cod. The reason is mainly the increased surface per kg weight of split fish and fillet.

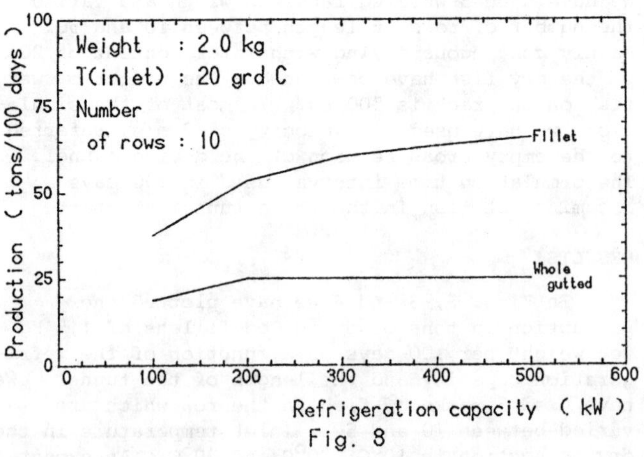

Fig. 8

Fig. 9 shows the influence of air velocity.

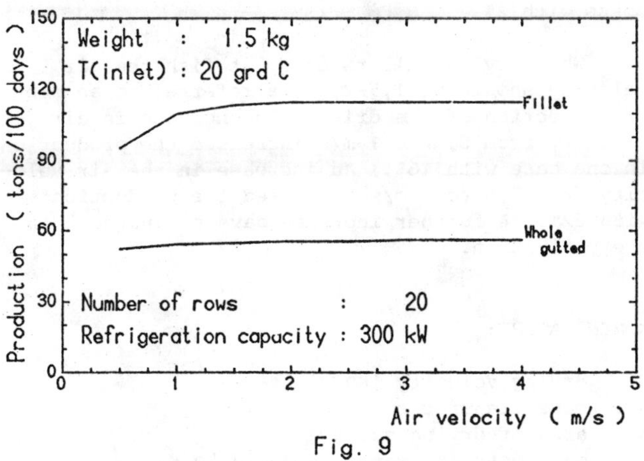

Fig. 9

An increase in air velocity from 0,5 m/s to 1,0 and 2,0 m/s increases production of fillets by 16% respective 22% and for whole gutted by 5% respective 7%. Higher velocity have little effect since the main water vapour resistance is in the dry fish. However, when choosing velocity the influence on heat transfer coefficient of the evaporator and the condensor, pressure drop etc. must be taken into consideration. Usually air velocities of 1,5-2,0 m/s is recommended.

ECONOMICAL CONSIDERATIONS

Energy consumption and production rate are calculated from simulations dependent on fish parameters and tunnel and heat pump size. In addition to energy cost also the investments in heat pump and tunnel must be considered. Here total cost for an installation on an actual site in Norway (Table 2) is used. |4|.

Table 2. Total costs of the heat pump drier equipment at different refrigeration capacities.

Capacity (kW)	100	200	300	400	500
Costs (1000 Nkr)	299	506	690	874	1.035

In addition tunnel costs for a common construction is about 340 Nkr/m^2 |5|. Energy consumption of fans is assumed proportional to tunnel length and heat pump capacity. The calculated total cost per kg dry fish dependent on the most important parameters is shown in Figs. 10, 11 and 12. The number of racks in a row is 5, the deprecation factor used is 0.20 and 250 days of operation is assumed. As can be seen the economical optimal heat pump size, for the very short tunnel is less than 100 kW. Increased tunnel length will decrease minimal total costs at a higher heat pump capacity and the cost will be less dependent on varations in capacity. Twice the normal Norwegian energy cost (0,25 Nkr/kWh) will increase total cost but have little effect on optimal heat pumps. However, drying of fish with lower drying rate will result in increased costs and lower optimum capacity.

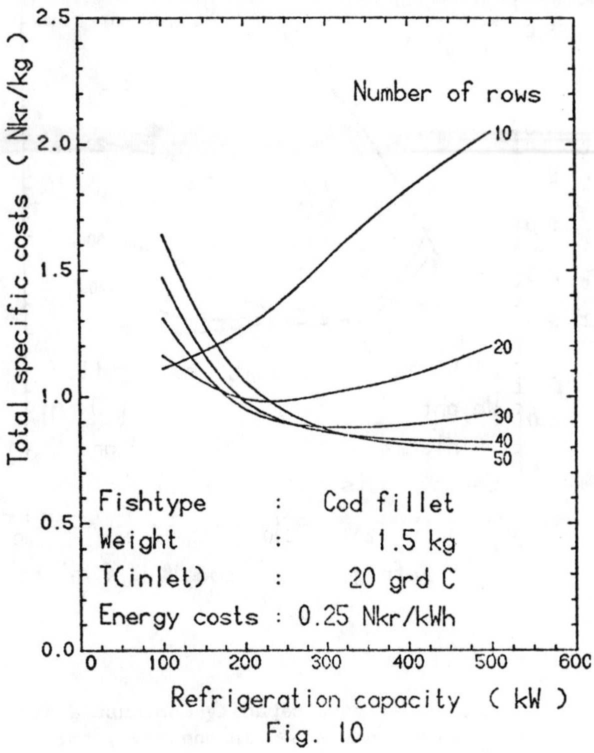

Fig. 10

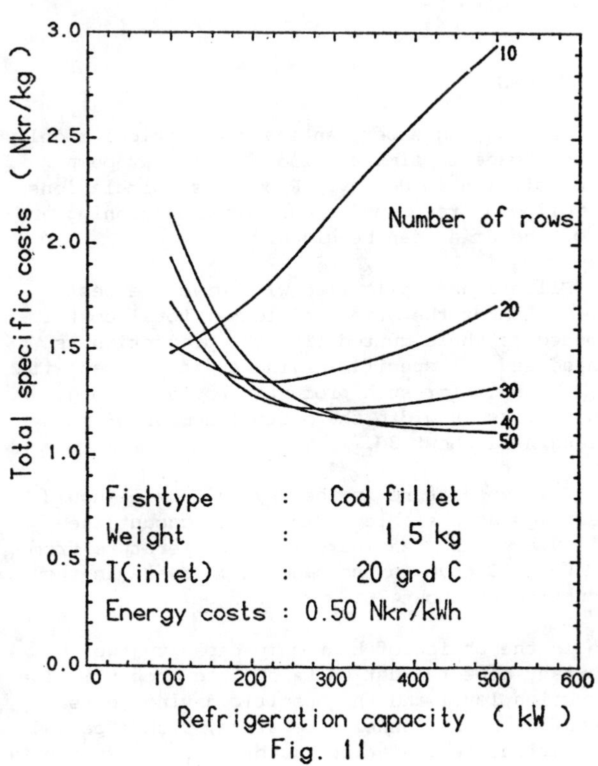

Fig. 11

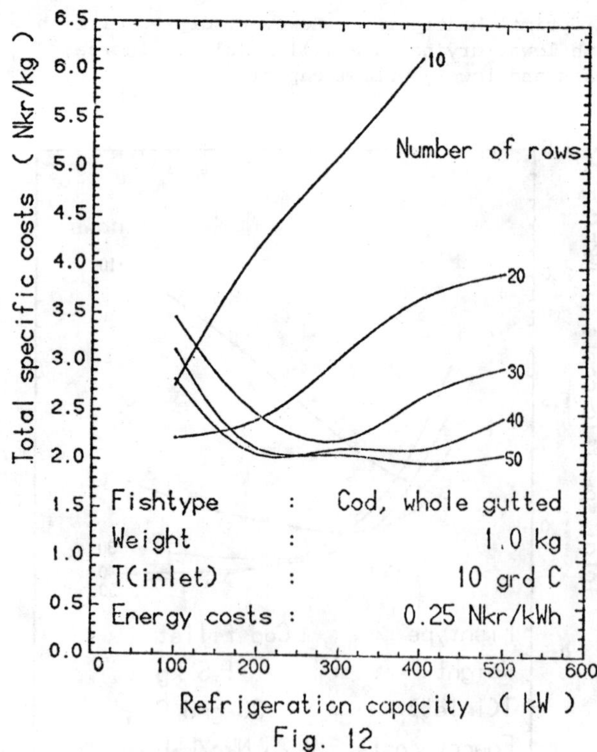

Fig. 12

Choosing a heat pump capacity one must also look to the production rate. In one case, for instance, an increase in the capacity from the optimum value of 250 kW to 300 kW increased to production with 30% while the specific costs only increased with 2%.

CONCLUSIONS

From drying experiments a mathematical model has been made to simulate industrial heat pump drying of fish in Norway. From these simulations the following recommendations for dimensioning and use of the drier can be given.

Fillets and split fish will give the best utilization of the drier and lowest total costs. Compared to whole gutted fish the production is doubled and the specific drying costs lowered with 50%. Tunnels for such products should be long, with 5 racks in width the optimal number of rows in length is about 30.

The temperature in the drying tunnel should be as high as possible taking into account the quality aspects. An increase in temperature from $10^{o}C$ to $20^{o}C$ gave in one case an increase in the production of about 70%.

In the choice of heat pump capacity two important aspects must be taken into account. The production level and the specific drying costs. Depending on the tunnel dimensions, fish type and preparation, depreciation factor of invested capital and energy costs, there will be an optimal capacity with respect to the specific costs. An increase in the capacity from this optimal value will, however, increase the production. In one case an increase in the capacity with 20% from the optimal value increased the production with 30% and the specific costs with 2%.

When drying fillets or split fish the air velocity should be 1,5-2,0 m/s referred to an empty cross section of the drier. An increase in air velocity from 0,5 to 1 m/s increased the production in one case with 16%. An increase in the air velocity from 0,5 to 2 m/s increased the production with 22%. A further increase gave no increase in the production.

NOMENCLATURE:

G_d: drying volocity |kg/s|
β^d: mass transfer coefficient |m/s|
A : area of drying surface |m^2|
R_d: gas constant, water vapour |J/kg K|
T : absolute temperature |K|
p_d: water vapour pressure |N/m^2|
μ : water vapour diffusion resistance |-|
s : thickness of dry layer |m|
D : diffusion constant, water vapour in air |m^2/s|

Subscripts:

s: on the outer surface
a: in the surrounding air
i: on the inside drying front

REFERENCES:

1. Strømmen, I., New equipment in fish drying. Proc. of the 3rd Int. Drying Symposium. Birmingham, England (1982).
2. Puntervold, S., The production of dried fish. The University of Trondheim, The Norwegian Institute of Technology. Division of Refrigeration Engineering. Internal Report. Trondheim, Norway (1984).
3. Strømmen, I., Drying of heavily salted codfish. Proc. of the 2nd Int. Drying Symposium, Montreal, Canada (1980).
4. Information from a Norwegian supplier of heat pump equipment.
5. Information from a Norwegian supplier of isolation and building materials.

THE APPLICATION OF MICROPROCESSOR TECHNOLOGY TO AUTOMATIC CONTROL OF TEXTILE DRYING PROCESSES

G. V. Barker* and J. H. Christie†

* Wool Research Organisation of New Zealand (Inc.)
Christchurch, NEW ZEALAND
† G. W. Streat Ltd
P.O. Box 8269, Christchurch, NEW ZEALAND

ABSTRACT

A worthwhile advance in the direct control of an industrial textile drying process has been achieved through the development of a multi-variable microcomputer-based controller known as *DRYCOM*. This paper describes the background to this development, with reference to earlier control systems and their shortcomings.

Successful industrial trials of the new controller have led to commercial exploitation of the system. Tangible cost savings have been demonstrated through reduced energy usage, increased productivity, and reduced reprocessing of incorrectly dried material.

1. INTRODUCTION

1.1. Background

The process of drying is encountered at various stages of textile processing. Material forms range from loose fibres to yarns and fabrics, of both natural-fibre and synthetic-fibre types. Drying generally follows processes such as scouring/washing [1], dyeing, and wet-finishing treatments.

The control system described in this paper was developed principally for controlling the process of drying loose scoured wool [2]. A wider range of applications for the system could be contemplated in the future, e.g., for slipe wool (wool removed from sheepskins in abattoirs and fellmongeries), loose dyed stock, and yarns.

1.2. The Wooldrying Process

Loose wool is dried with hot air (recirculated) in countercurrent air-flow systems, the wool fibres being transported either on slowly rotating perforated 'suction' drums or on a metallic conveyor [2]. The evaporation rates for these two types are generally in the range of 25-30 kg/(h m^2) and 10-12 kg/(h m^2), respectively. While drum dryers require much more electrical energy for the fans for air circulation (300-500 kJ/kg evaporation, compared with 100-200 kJ/kg), they have proved to be more economical overall.

Drum dryers with up to 10 drum sections (1.8 m wide) have been built, and recently dryers with a working width of 2.4 m have been developed to provide evaporation capacities of up to 1000 kg/h.

The combustion of natural gas has been applied for the direct heating of these dryers.

2. OVERALL DRYER CONTROL REQUIREMENTS

There are three inter-related requirements for the control of textile dryers, namely:

(i) product outlet moisture content on a dry mass basis (known as "regain" in the textile industry);

(ii) maximum evaporation capacity; and

(iii) optimum thermal efficiency.

The need to achieve closer control over the outlet moisture content of the product has recently increased and control to within ± 1-2 units of regain is now often called for. There is also a need to achieve a greater dryer capacity to satisfy demands for higher productivity (and profit), coupled with the objective of maximising the efficiency of energy usage. The latter has been influenced by the rapid increases in fuel costs over the last decade.

3. PROBLEMS WITH CONTROL OF MOISTURE CONTENT

A wide range of variables affects the wool-drying process and hence influences the outlet moisture content of the product [3]. The high variability of this parameter has been studied in detail [4 - 6], and has been observed over three identifiable time intervals, as shown in Figures 1 and 2.

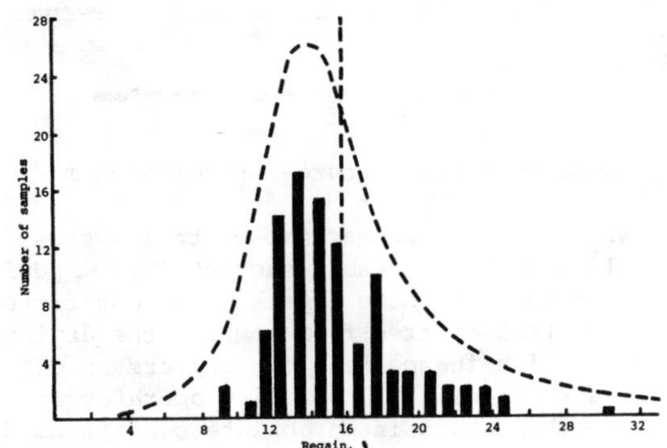

Fig. 1 'Spot' variations: distribution of outlet wool regain tests on 10-20-g samples taken over a total period of one minute

$$\text{Regain, \%} = \frac{\text{mass of water}}{\text{mass of dried wool}} \times 100 .$$

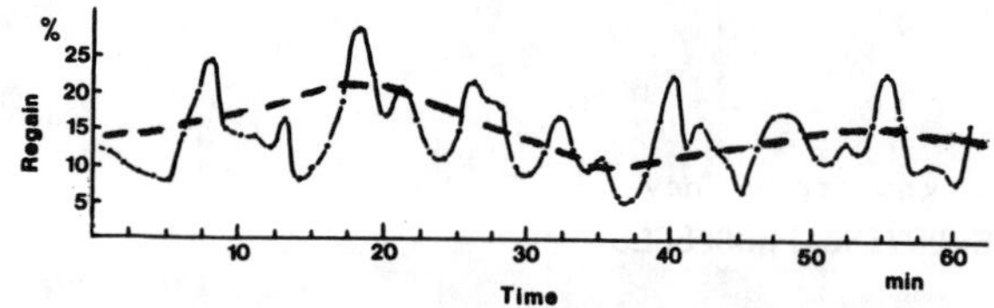

Fig. 2 Short and long-term variations in outlet wool regain (sampling rate - one per minute)

The variations of the main parameters affecting the outlet wool moisture content for a particular drum dryer produced the following linear correlation relating outlet regain (w, %) to wool flowrate (m, kg/min), air outlet temperature (T, ^{o}K), and absolute humidity (Y, kg/kg) [2].

$$w = 2.6\,m - 0.30\,(T - 273) + 53\,Y - 5.3$$

This illustrates, for example, that a variation of ± 10 % in wool flowrate would give rise to a variation of ± 2.6 units of regain (uncontrolled plants can experience flowrate variations of up to ± 40 %).

4. PREVIOUS WOOLDRYER CONTROL SYSTEMS

4.1. Temperature Control

This has been applied to textile dryers for many years with the aim of achieving a simple control of energy usage and some degree of control of moisture content (Fig. 3).

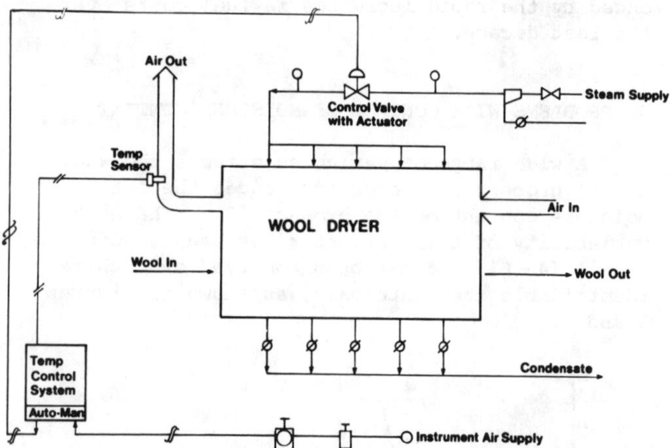

Fig. 3 Wooldryer temperature-control system

A wide range of commercial control systems has been applied with reasonable success [7, 8]. Unfortunately, in the past some dryers have been fitted with the control omitted from some of the drying sections or with independent controllers on different drying sections, all requiring operator attention. In other cases inappropriate control hardware has been used.

4.2. Humidity Control

Various attempts have been made to improve the control of product moisture content through regulation of the humidity of the circulating air (Fig. 4)

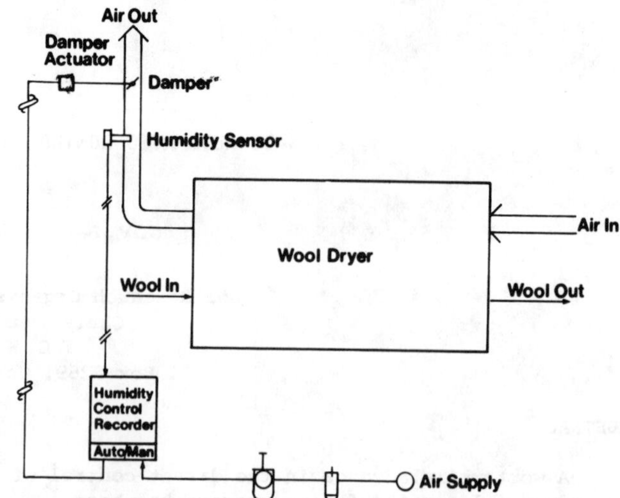

Fig. 4 Humidity control of a wooldryer

in combination with temperature control [9, 10].

While improvements in the control of the moisture content can be achieved with these systems, the operator is required to make personal judgements and trial-and-error selections of the set-point values of the controller.

The use of wet-bulb temperature sensing as an indirect measurement of humidity generally proved too difficult for operators to manage. Thus the 'high-temperature' direct-reading relative-humidity sensors (developed in the mid-1970s) were then applied to wooldryers. One of these, the Vaisala Humicap sensor [11] has been found to be fast in response and sufficiently accurate for humidity control in wooldryers, but problems with a drift in calibration (possibly due to contamination of the sensing element) and susceptibility to damage during brief 'over-shoots' in temperature have reduced the popularity of this device.

Other more sophisticated, direct-reading sensors have been developed for high-temperature environments. For example, the Mahlo absolute humidity sensor has been widely applied to some textile dryers [12-18]. Drawbacks would seem to be its high cost and rather complex design.

The *DRYCOM* Dryer Controller incorporates a much simpler humidity-measuring system which is described in section 7 [3].

4.3. Material Flowrate Control

The material throughput of a wooldryer can now be controlled with the application of a weighbelt controller, located at the inlet end of the wool-processing system [19]. This can be operated so as to improve the control of the product moisture content. However, the selection of the setpoint of this controller must also be made by the plant operator.

5. MICROPROCESSOR TECHNOLOGY FOR PROCESS CONTROL

Early process control systems were simple and basically mechanical in design. These were superseded by the use of pneumatically actuated systems which included P, I and D modes of control action. While pneumatic systems continue to be used today, particularly for actuation of the final control, solid-state electronic control systems now predominate. Also, for a decade or more, large processing plants have been placed under the control of electronic computers.

Advances in electronic devices over the last 8-10 years have brought a dramatic increase in the sophistication of computers and their peripherals, coupled with a large reduction in their cost.

The microprocessor-based computer technology has brought about a 'revolution' of new opportunities. These now include advances in process control and instrumentation; the provision of scope for a wide range of design features which could not be adopted in earlier control systems; and the opportunity to apply computer control to individual processing items rather than to the plant as a whole.

6. ON-LINE MOISTURE CONTENT MEASUREMENT

The most important parameter to be continuously monitored in a comprehensive dryer control system is the moisture content of the outlet material flow. The measurement of this parameter on loose wool and other textile materials has been investigated or reviewed by a number of workers, e.g., [20 - 22]. While a wide range of schemes has been developed, most have proved to be inadequate.

During the development of the *DRYCOM* Dryer Controller it was found desirable to design a new moisture meter [3], taking advantage of a number of new ideas which in part were made possible by the use of microprocessor technology. The design of the new system was based on the measurement of the electrical conductivity (in the range of 1.1×10^{-11} to 1×10^{-11} mhos) of the moving, and generally non-homogeneous, flow of wool from the dryer. The conversion of the measurements to percentage moisture content values was one of the functions for the microprocessor in this instrument, using experimentally determined calibration data programmed into the system.

7. FEATURES OF THE *DRYCOM* DRYER CONTROLLER

The new controller was purpose-designed for controlling a textile dryer, with special application to loose wool. Its primary function was for the direct control of the outlet moisture content. It was also designed to control and optimise the efficiency of energy usage of the dryer, and to automatically manage and maximise the material throughput within the capability of the plant.

The controller receives four input signals, functions under four alternative operating modes (plus a stand-by mode), and provides output-control signals for the adjustment of three parameters. The layout of the *DRYCOM* System is shown diagrammatically in Figure 5, an overview of its operation in Figure 6, and a block diagram of the design detail of the controller in Figure 7.

The system incorporates a carefully designed 'operation interface' with switches for mode selection, thumbwheel switches for data entry, and an 'information' section consisting of a 40-characters word display and a line printer. The last two items are both commercially available and function with the aid of their own microprocessor controls.

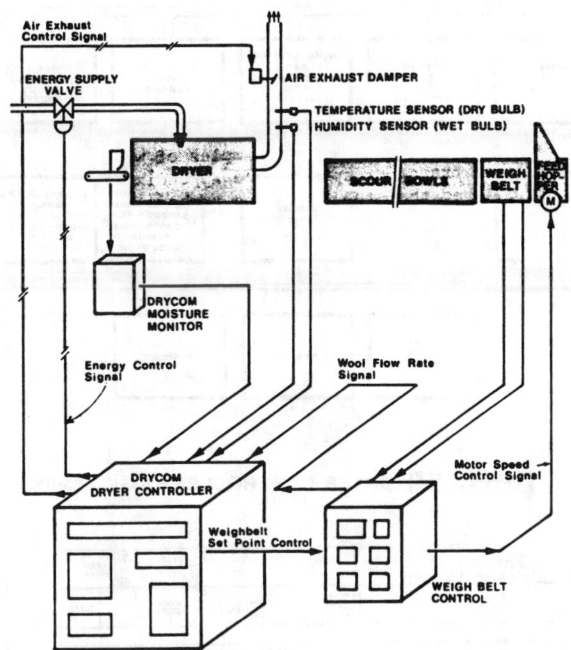

Fig. 5 Diagram of the layout of the *DRYCOM* System

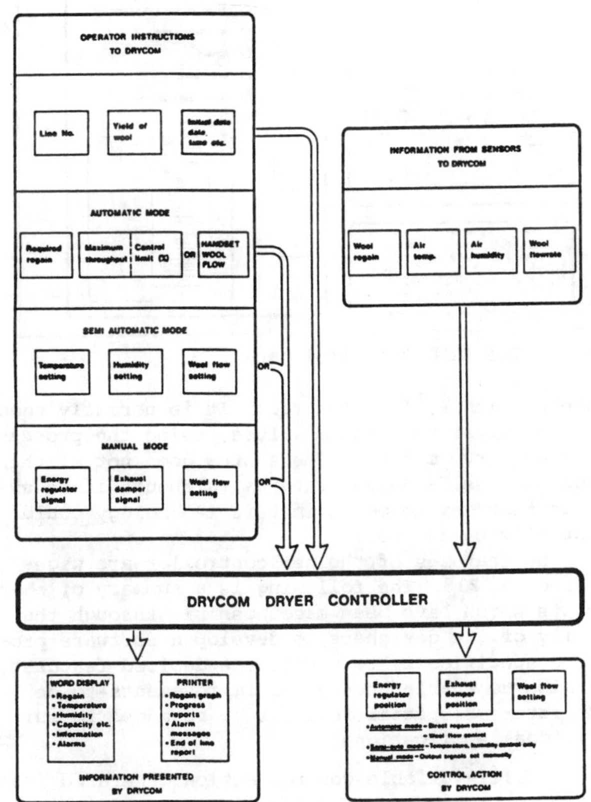

Fig. 6 Operation of the *DRYCOM* System

The layout of the control panel of the controller is shown in Figure 8. The blank square at the lower right is the line printer.

The control action for each output signal for both the automatic and the semi-automatic modes is

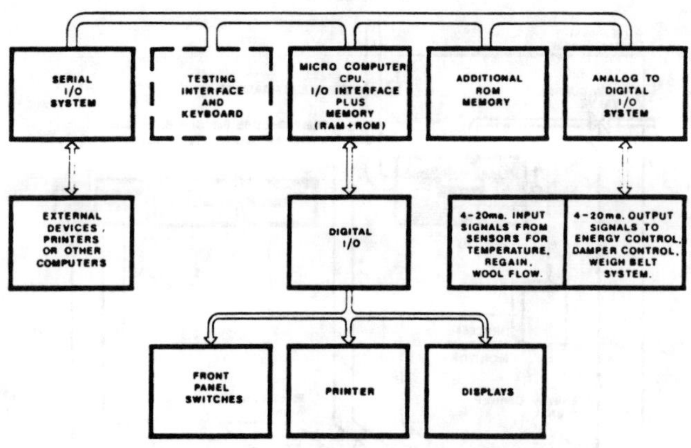

Fig.7 DRYCOM DRYER CONTROLLER – BASIC DESIGN LAYOUT

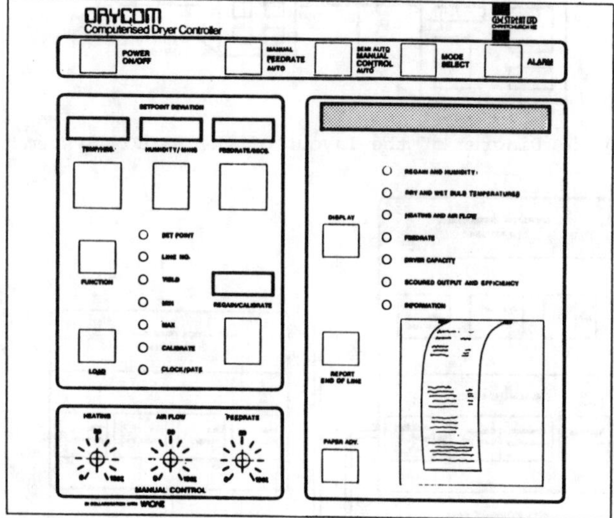

Fig. 8 The *DRYCOM* control panel

achieved with P, I, D control. It is normally tuned for each dryer to optimum values, using the process reaction curve method. The system does not at this stage have self-tuning features, although it is expected that developments in this technology could eventually be adopted.

The features of the new controller are wide-ranging [3, 23]. The following is a summary of those aspects which have been made possible through the ability of the designers to develop a software programme specifically to create the desired features, and also to enable changes or improvements to be made later or special features to be added to suit individual applications.

(i) Multi-variable control action generated from one controller.

(ii) Continuous calculation of information from input signals: for example, the determination of humidity is made from measurements made by simple wet- and dry-bulb sensors, overcoming the problems encountered with alternative humidity sensors, referred to in section 4.2. Other workers have reported on similar systems [24-26].

(iii) Mathematical treatment of input signals to achieve more appropriate data smoothing and noise suppression. This proved to be essential for the highly variable signals for the outlet moisture content.

(iv) Data storage to RAM-type memory to facilitate various control requirements: e.g., for the wool flowrate signal, the values of which must be 'stored' to allow for the 10-12 min time interval between the wool passing the weighbelt and entering the dryer.

(v) Opportunity to use word displays and computer printers, etc. (instead of simple digital indicators and recorders) to present information to the operator.

(vi) Facility to include a range of diagnostic features, e.g., checking of faults with input signals and of the system's own operation by self-checking routines.

(vii) Opportunity to communicate with other computer devices.

(viii) Clarity of design, e.g., multiple use of switches, and the ability to detect out-of-range operator-selected settings and prevent them from being used.

8. EVALUATION OF THE PERFORMANCE OF *DRYCOM*

The testing of the prototype system was initially carried out in the development laboratory and later on an industrial wooldryer under commercial operating conditions [23]. Further design features were implemented through minor alterations to the software. Finally, all aspects of the performance of the controller were evaluated, i.e.:

- Reliability of the controller and its ancillaries.
- Calibration and assessment of the new *DRYCOM* Moisture Meter.
- Evaluation of the capability of *DRYCOM* to perform its automatic control functions efficiently.
- Overall review of the advantages demonstrated by the controller - this was made to define its subsequent cost/benefit features.

An example of the output of the printer is given in Figure 9 and illustrates the detailed reports that are printed and the automatic control of outlet moisture content and productivity that result from the operation of the controller.

9. BENEFITS OF THE *DRYCOM* SYSTEM

The evaluation of the *DRYCOM* Controller established that a number of benefits could be achieved to the extent of providing worthwhile economic justification for its use [23, 27].

The achievement of direct control of the product moisture content was shown to lead to a substantial reduction in the need for costly reprocessing of over-dried or under-dried material. The frequency of the moisture content being within ± 1.5 regain units of the desired value was raised

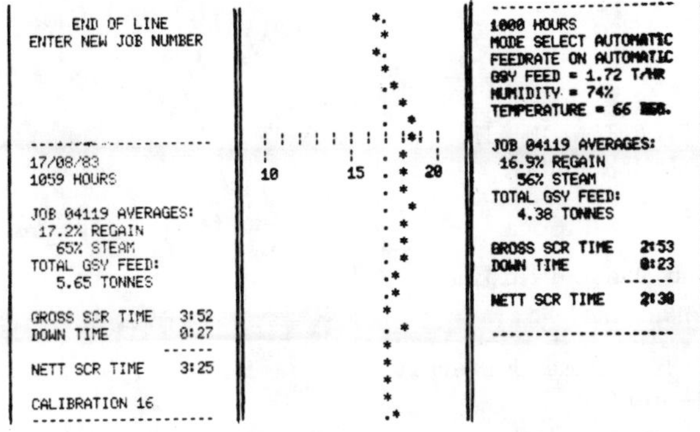

End-of-line report | Regain graph | 2-hourly report

Fig. 9 Example of the printout from *DRYCOM* (the regain graph is plotted at intervals of 3 min and is to be read upwards)

from 43 % to nearly 70 %, while the frequency of the results being outside the limits of ± 2.5 regain units was reduced from 30 % to less than 10 %.

The improved overall control led to an increase in the efficiency of energy usage and increased plant productivity. The value of the latter can be shown to cover the cost of the *DRYCOM* System in as little as a year for some installations.

The impact of the new controller on the overall plant operation and management was considerable. The operators indicated that they enjoyed using the system, while management confirmed that there were worthwhile benefits associated with the additional information printed out for each lot of material processed.

The *DRYCOM* Dryer Controller and its associated Moisture Meter have been refined for the manufacture of commercial models to meet a strong market need from the wool-processing industry.

10. EPILOGUE

The dryer control system described in this paper has been successfully developed through the application of background knowledge of earlier research and development studies. It has been made possible through the application of the latest computer technology to incorporate design features which would not have been feasible in conventional controller design. It should now be possible for this approach to be successfully applied to a wide range of drying and other process-control problems.

REFERENCES

1. Stewart, R. G., Woolscouring and Allied Technology, Wool Research Organisation of New Zealand, Christchurch (1983).
2. Barker, G. V., *WRONZ Commun.* Nos 18, 20, 25 (1973-74).
3. Barker, G. V., and Christie, J. H., *WRONZ Commun.* No. C85 (1983).
4. Peryman, R. V., and Barker, G. V., *WRONZ Commun.* No. 14 (1973).
5. Dixie, J. A., *Text. J. Aust.*, vol. 36, 890 (1961).
6. Mackay, B. H., Shanahan, A. G., and Hall, W. B., *J. Text. Inst.*, vol. 56, T409 (1965).
7. Walter, L., *Textile World*, August, 100 (1950); *Textile Mfr*, vol. 75, 228, 282, 328 (1949).
8. Barker, G. V., *WRONZ Reports* Nos 12, 12C (1972-73).
9. Barker, G. V., *WRONZ Commun.* No. 16 (1973) (Proc. Conf. Automatic Control and Instrumentation Society of New Zealand, 1973).
10. Higgins, J. J., and Keey, R. B., *J. Text. Inst.*, vol. 64, 574 (1973) (Proc. N.Z. Inst. Engineers Conf., 1972).
11. Lee, G. A., *Dyer*, vol. 156, no. 4, 186 (1976).
12. Beckstein, H., *Dyer*, vol. 155, no. 4, 155 (1976); *Textilveredlung*, vol. 4, 401 (1974).
13. Merritt, D. S., *Amer. Dyest. Reptr*, vol. 71, no. 9, 28 (1982).
14. Burgholz, R., *Textil-Praxis*, vol. 15, 1151, 1271 (1960).
15. Schellenberger, G., *ITB Dyeing/Printing/Finishing*, vol. 3, 281 (1974).
16. Pabst, M., *Melliand Textilber. (Eng. Ed.)*, vol. 3, 291 (1974).
17. Pleva, G., *Textilbetrieb*, no. 12 (special issue), 282 (1982).
18. Mohr, P. R., *Textilveredlung*, vol. 9, 409 (1974).
19. Barker, G. V., and Stewart, R. G., *WRONZ Commun.* No. 61 (1977); *Proc. Int. Wool Text. Res. Conf., Pretoria*, vol. III, 17 (1980).
20. Bennett, J. M. *et al.*, *Proc. Text. Inst. Ann. Conf. (Lucerne)*, pp. 96-111 (1972).
21. Roberts, M. H., and Worthington, A. P., *Shirley Inst. Bull.*, vol. 40, no. 5, 190 (1967).
22. Nicoll, S. R., *WRONZ Reports* Nos 31 (1975), 65 (1979).
23. Barker, G. V., *WRONZ Commun.* No. C86 (1983).
24. Nantou, Y., and Suzuki, S., *IEEE Trans. Instrumentation and Measurement*, vol. Im-30, no. 2, 98 (1981).
25. Fisher, P. D. *et al.* *IEEE Trans. Instrumentation and Measurement*, vol. Im-30, no. 1, 57 (1981).
26. Raudszus, G., *Feinwerktech and Messtech*, vol. 56, part 7, 300 (1978).
27. Streat, G. W., Ltd, Report on Economic Justification of *DRYCOM* Dryer Control System, Technical Brochure, December (1983).

FEASIBILITY STUDY ON DEHUMIDIFICATION OF AIR BY THIN POROUS ALUMINA GEL MEMBRANE

Masashi Asaeda, Luong Dinh Du and Masao Ushijima

Department of Chemical Engineering, Hiroshima University
Higashi-Hiroshima, 724 JAPAN

ABSTRACT

Some modules of thin alumina gel membrane supported on the outer surface of a coarse porous ceramic cylinder were made to apply for continuous dehumidification of air. The principle of water vapour removal from the air flowing around the membrane is that the water vapour condenses in quite small pores of the membrane due to the capillary condensation and migrates through the membrane to its inner surface, where it evaporates due to the evacuation of the inside of the cylinder. As the heat of condensation generated on the outer surface can be used as the heat of evaporation just inside the thin membrane, dehumidification is to be done under nearly isothermal condition, which is preferable from the view point of energy conservation. Because of blocking of pores with liquid water a quite large separation ratio was obtained. The separation mechanism, separation ratio, dehumidification rate and feasibility of this kind of membrane as a practical dehumidifier are discussed on the basis of some experimental results.

1. INTRODUCTION

Dehumidification of air is one of the important operations in various fields, from home airconditioning in the countries of high humidity in summer to the industrial air conditioning such as in food industries, chemical industries, precision industries and so on. Because of its importance various methods have been developed and are generally classified in two groups, condensation method by cooling or by compression and adsorption or absorption method with porous solids or chemical dessicants. The condensation method has been widely used for dehumidification of large amount of air in industries and also for home air conditioning in a small scale. This method has some merits when low humidity air at relatively low temperature is required, but has less advantages when dehumidification of air is required at relatively high temperatures, which occurs, for example, in a closed system dryer. It usually requires extra energy for undesirable cooling of the air and reheating of the air is sometimes necessary for further use. It is not so efficient either when air of quite low wet bulb temperature is required because of formation of frost on the cooled surface.

The adsorption method, on the other hand, has a merit on this context. It doesn't lead to any cooling of the air leaving adsorption columns, where heat of adsorption is generated because of its exothermic process. In order to raise adsorption the heat generated in adsorption columns must be removed by the non-adsorptive fluid flowing through the columns or usually by employing cooling coils inserted in the columns when a high concentration of water vapour in an air stream is present. High loads to an adsorption dehumidification system operating with a conventional thermal swing adsorption-desorption cycle require large amount of inventories and frequent regenerations, which require excessive energy. Because of high efficiency of this method at low water content a combination of these two methods is usually used for dehumidification of atmospheric air to obtain dry air of extremely low humidity.

Dehumidification of air at high temperatures without lowering the temperature is quite desirable especially in drying processes which usually have a very low thermal efficiency. Both the above methods are not very efficient for the dehumidification of air at high temperatures. In this papar a ceramic membrane modules are made for dehumidification of air at high temperatures. In the following sections are discussed the separation mechanism of water from humid air by a membrane dehumidifier, separation efficiency, dehumidification rates and the feasibility of this kind of membrane as a practical dehumidifier.

2. GAS SEPARATION BY POROUS MEMBRANE

From the view point of energy conservation separation by membrane is quite attractive and has recently studied vigorously. Most of them are concerned with liquid phase separations by various polymer membranes and a great success has been achieved. Separation of gaseous mixtures, however, has not yet studied very much. Organic polymer membranes for gas separation seem not to be so effective for their relatively low separation ratios at high fluxes and deformations at relatively high temperatures. On the other hand inorganic porous solid membrane seems preferable for this purpose.

Mechanisms of separation of gaseous mixtures by porous solid membranes can be classified into three as shown in Fig.1: (a) separation by the Knudsen diffusion, (b) separation by surface diffusion and (c) separation by capillary condensation. In the Knudsen diffusion gas molecules can only collide with the pore wall and the intermolecular collisions are negligibly small. In this case the separation ratio is limited to the value obtained theoretically from the kinetic theory of gases or the square root of molecular weight ratio of the gases to be separated. The application of this mechanism to industrial separation processes is clearly quite limited to gaseous systems of large molecular weight ratios, though it has been applied to the separation of ratio isotopes.

Surface diffusion can be hopeful for gaseous separation. Molecules adsorbed on the pore wall can diffuse on the surface due to the concentration gradient in the adsorbed phase, while molecules in the gas phase can also diffuse in the pore space, Fig.1-(b). The mechanism of surface diffusion has not been made clear thoroughly.

Separation by capillary condensation is not new at all but has been widely applied to separation processes using porous solid adsorbents such as silica gel and activated alumina. If such porous adsorbents of quite fine pores are made in a thin membrane, an efficient and truly continuous separation of condensable gases from non-condensable gases will be possible and its separation ratio can be expected high because the pores can be blocked with the condensable component. A schematic figure of this case is shown in Fig.1-(c). Suppose that humid air is led around a fine porous membrane supported with a coarse porous substrate cylinder, inside of which is evacuated. The water vapour in the air condenses in the fine pores of the membrane, flows or migrates through the pores to the inner surface of the membrane and evaporates there because of the low pressure in the inside of cylinder. The heat generated near the surface as heat of condensation can be used as the heat of evaporation just inside the thin membrane and the process is considered nearly isothermal.

3. EXPERIMENTAL

3.1 Preparation and characteristics of membrane

The preparation of a membrane module is quite important for this kind of separation. The membrane module must be strong enough for the pressure loaded across it and it is made in a cylinder of diameter 0.012 meter, 0.13 meter in length and 0.002 meter in thickness. A thin membrane of alumina was formed near the surface of a coarse porous ceramic cylinder, which was made of kaolin by firing at about 1230°C for about 24 hours. In order to make the porous substrate highly porous about 30 weight % of graphite powder was added to the kaolin powder and mixed well with small amount of alumina sol to form it in a cylinder. The cylindrical porous substrate was connected to a glass tube with alumina cement as shown in Fig.2-(a) for later use for the separation experiment. After washing the cylindrical substrate in clean boiling water for about 2 hours, it was dipped in alumina sol and dried at about 80°C. Then it was pyrolyzed to about 450°C. These procedures were repeated several times to obtain moderately fine pores in the membrane. The preparation method of alumina sol from alumina alkoxides is given elsewhere [1].

An example of the results of air permeation at 21°C is shown in Fig.3. Even at this stage the

a) Knudsen diffusion

b) Surface and Knudsen diffusion

c) liquid flow, surface and Knudsen diffusion

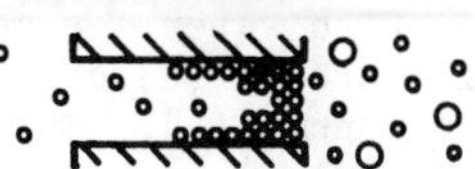

Fig.1 Three mechanisms of gaseous diffusion in fine pores

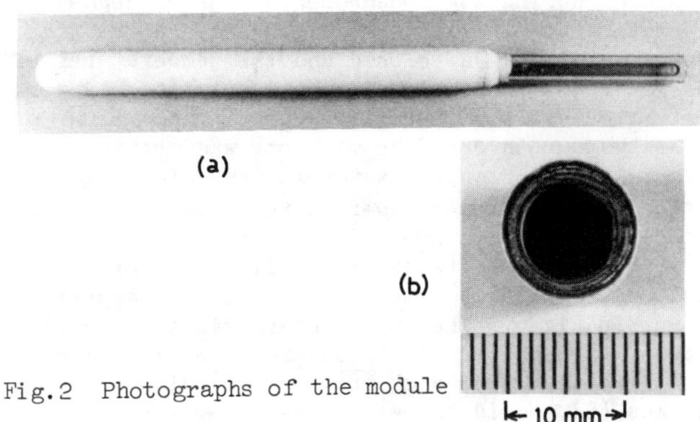

Fig.2 Photographs of the module

permeability of humid air through the membrane was quite sensitive with the humidity. But the leak of air was still too large to be served for gas separation and further treatment had to be added to the membrane to decrease the pore diameter. An alumina alkoxide (aluminum iso-propoxide, for example) dissolved in an organic solvent was used to fill the large pores in the membrane. The membrane was finally treated in steam for several hours in two ways: (I) after pyrolyzation to about 450°C (Module I) and (II) with dilute solution of sodium silicate after pyrolyzation (Module II). A photograph of the section of membrane module finally obtained is shown in Fig.2-(b). The module was dyed by pouring methylene blue in it and drying it at room temperature. The molecule of methylene blue could not penetrate the membrane and its concentrated layer was observed just inside of the thin membrane (white thin shell of about 0.0002 meter thickness).

An example of permeability of dry air in the finally obtained membrane is shown also in Fig.3. As can be seen from these results the flow is in the Knudsen region even at atmospheric pressure, which shows that the pores in the membrane are less than a few hundred Angstroms at the most. In Fig.4 is shown an adsorption isotherm of water in the membrane at 32° C. The measurements were done for a hard and thin shell of the membrane module which was obtained by scraping off the relatively soft substrate. Pore diameter is seen to be less than 95 Angstrom with two peaks at about 8 Angstroms and 40 Angstroms. From the permeability tests described below the leak of air form the atmosphere was quite small. These results show that the most of the pores have constriction of diameters less than about 10 Angstroms and the pores can be considered to have a structure shown in the same figure.

3.2 Apparatus and procedure

A schematic diagram of the apparatus for permeability measurements of pure water vapour is shown in Fig.5. The apparatus comprises a still (1) with water, a module of porous membrane (2), cold traps (3) cooled by liquid nitrogen and a vacuum pump (4). The upper part of this still is submerged in a thermal insulation box (5) of which temperature is controlled to be constant at specific values by a controller (6). The water in the still is stirred well but not to a extent to splash any sprays and its temperature is also controlled at the specific values to obtain various vapour pressures during the measurements. After degassing the water at a relatively high temperature the vacuum cock (7) was closed and the temperatures of the water and the inside of the insulation box were controlled at the specific values. The flux of water vapour through the membrane was obtained by weighing the trapped water in one of the cold traps. The measurements were done at various temperatures of the membrane and the water.

The measurements of dehumidification of air of various humidities were done with an apparatus shown in Fig.6. The apparatus consists of a flow meter (1), a heater (2) to control the air temperature, a boiler (3), a square duct (4) with membrane modules (5), cold traps (6), a blower (7) for leak measurements of air and a vacuum pump. The temperature and the humidity of flowing air were controlled at specified values and measured by dry and wet thermocouples inserted in the duct. In this case the rates of air leak were measured besides the fluxes of water through the membrane.

4. EXPERIMENTAL RESULTS AND DISCUSSION

4.1 Experimental results

An example of time dependency of water vapour flux during the treatment with steam is shown in

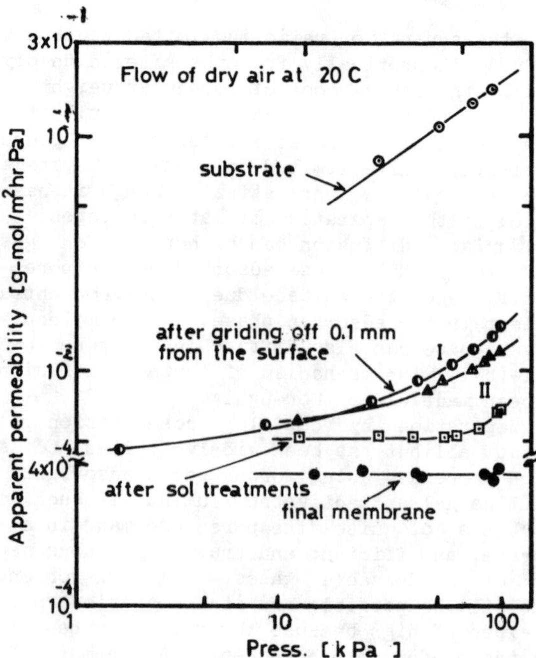

Fig.3 Flow permeabilities through membrane and substrate

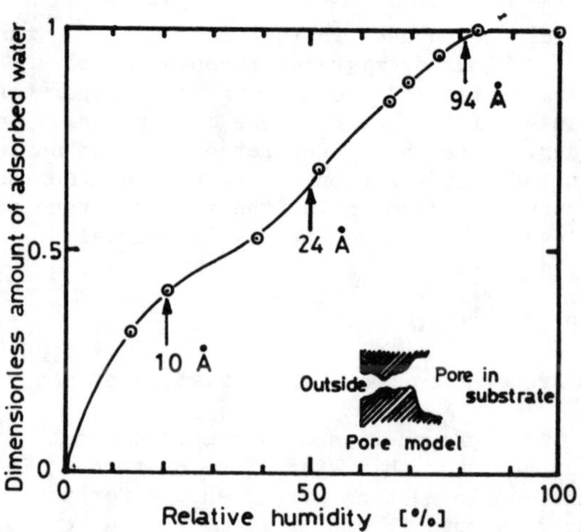

Fig.4 Adsorption isotherm of water on membrane at 32°C

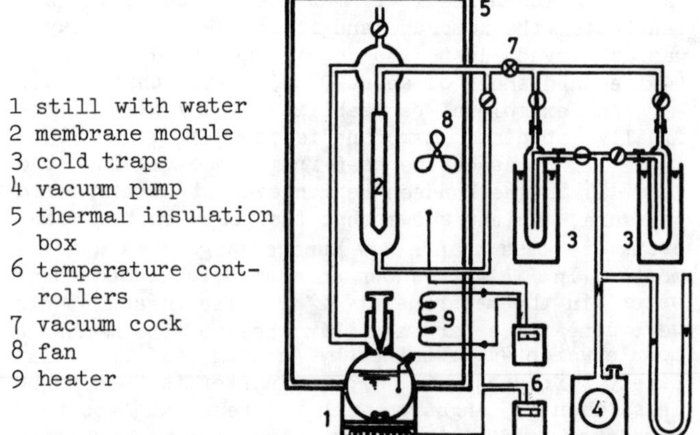

Fig.5 Experimental apparatus for water permeation

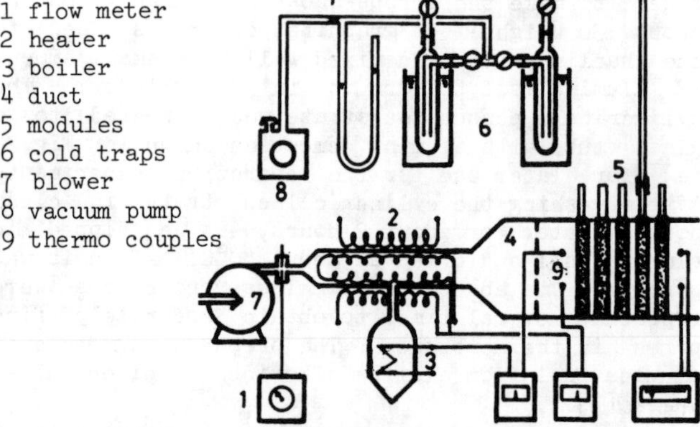

Fig.6 Apparatus for dehumidification of air

Fig.7 for Module I. The flux decreases with the time and approaches a constant after about 30 minutes. All the following data for Module I were obtained after the attainment of this constant flux. Some observed fluxes of pure water vapour under various conditions are shown in Fig.8, where fluxes are shown against the percentage saturation of vapour (defined as $100xp/p_0$) with the parameters of temperature and pressure of water vapour p outside of the membrane. The solid lines are those for isothermal conditions and the broken ones for isobaric conditions. The flux increses with the temperature increase largely due to the increase of driving force. The dotted line near the bottom of this figure is the flux of water vapour (at 88.4°C) estimated from the permeability data shown in Fig.3 assuming that the vapour permeates through the membrane in gaseous state or without condensing in the pores. This value is negligibly small in comparison with the observed results, which shows that the high flux is due to the condensation effect of the vapour in the pores of the membrane. Under isobaric conditions or the same driving force the flux decreases with the decrease of percentage saturation or with the increase of temperature. This can be seen more clearly from Fig.9, where some apparent permeabilities are plotted against the temperature. This tendency can be understood simply considering that, according to the Kelvin's capillary condensation theory, Eq.(1), the critical pore diameter of condensation decreases with the decrease of percentage saturation and with the temperature rise at constant percentage saturation.

$$ln\,\frac{p}{p_0} = \frac{2\gamma V}{rRT} \qquad (1)$$

Some results of dehumidification experiment with Module I are shown in Fig.10, where the fluxes of water vapour and air are shown against the relative humidity of the air streaming outside of the membrane. At relatively high humidities above 50 % the leak of air is quite small but it increases as the humidity decreases and approaches the value of dry air. The apparent permeabilities of water vapour and the air are also shown in the same figure. The separation ratio which, is defined here as the ratio of the apparent permeabilities, is seen quite large, larger than 7 at the smallest for the cases shown in this figure. Module I, however, was found not to be so stable. After many hours of measurements the leak of air increased about two times.

Module II was quite stable and the performance is almost equal to that of Module I as shown in Fig. 11, where fluxes and apparent permeabilities of water vapour are shown against the percentage saturation. The observed fluxes of dehumidification of air at various temperatures and humidities are shown in Fig.12. Some of the apparent permeabilities are calculated from the experimental results shown in Figs.11 and 12 and are shown in Fig.13 against the relative humidity or percentage saturation. All the permeabilities obtained for pure water vapour and for humid air are seen to decrease as the percentage humidity or the relative humidity decreases. And the permeabilities obtained for pure water are larger than those for

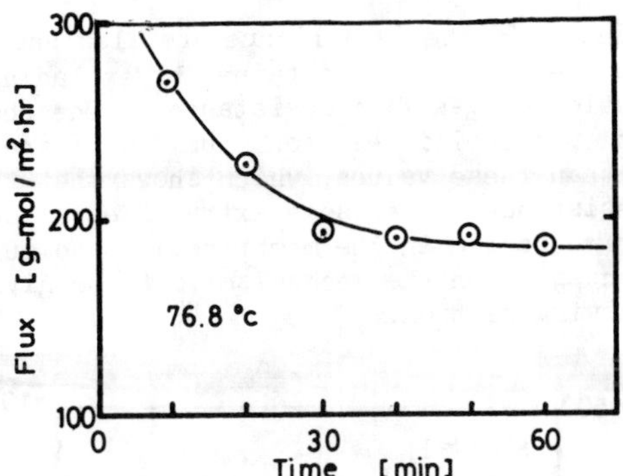

Fig.7 Time dependency of vapour flux for Module I

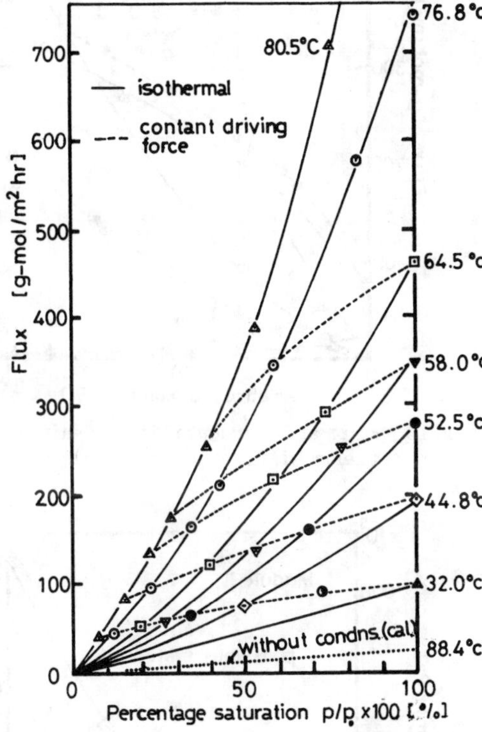

Fig.8 Flux of pure water vapour through Module I

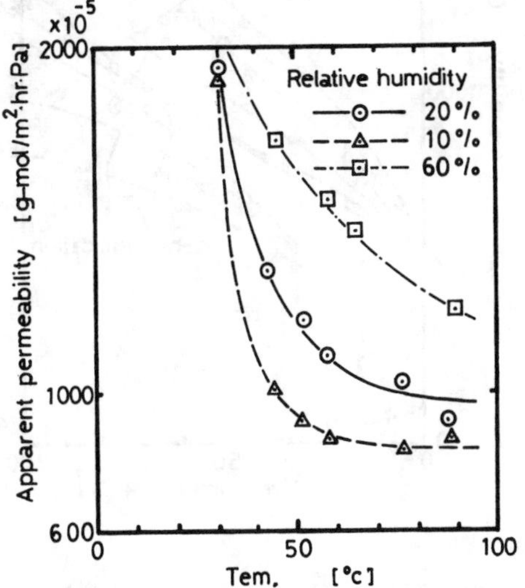

Fig.9 Apparent permeability vs. temperature

humid air. In the same figure are also shown the apparent permeabilities obtained by excluding the contribution of gas film resistance. The observed apparent permeabilities for humid air are even smaller than these values, which shows that the gas film resistance gives some extra effects on the transfer mechanism in the membrane. Some qualitative discussion on the mechanism is to be given in the following section.

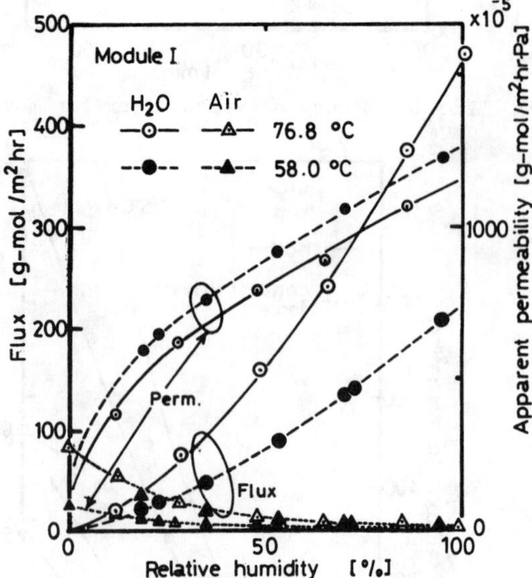

Fig.10 Flux and permeability of air and water (dehumidification of air)

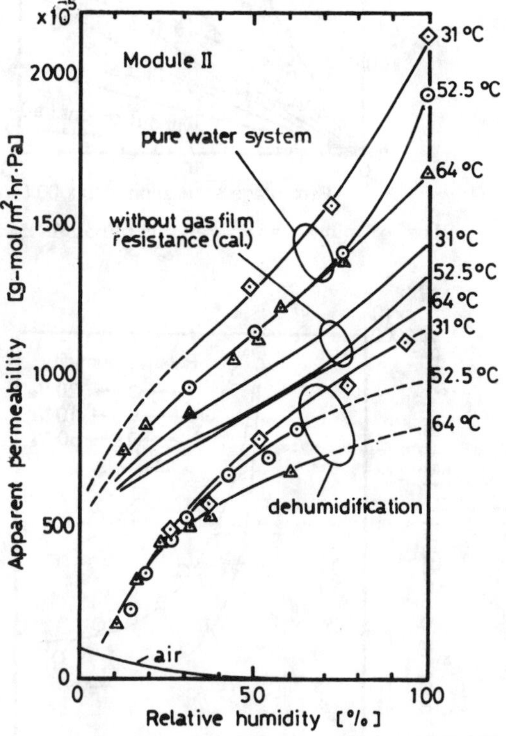

Fig.13 Apparent permeability vs. relative humidity (dehumidification and water permeation)

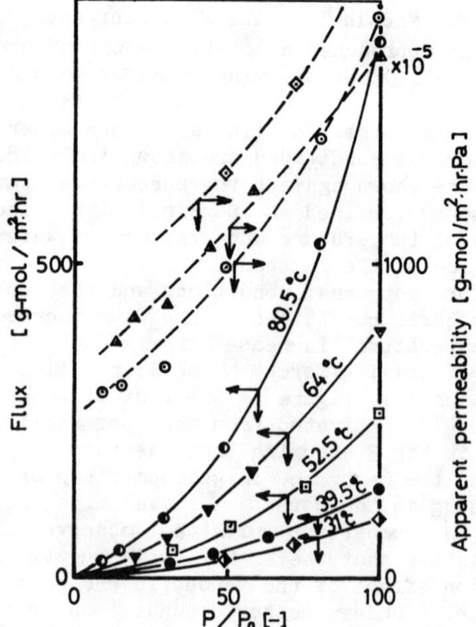

Fig.11 Flux of water through Module II (water permeation)

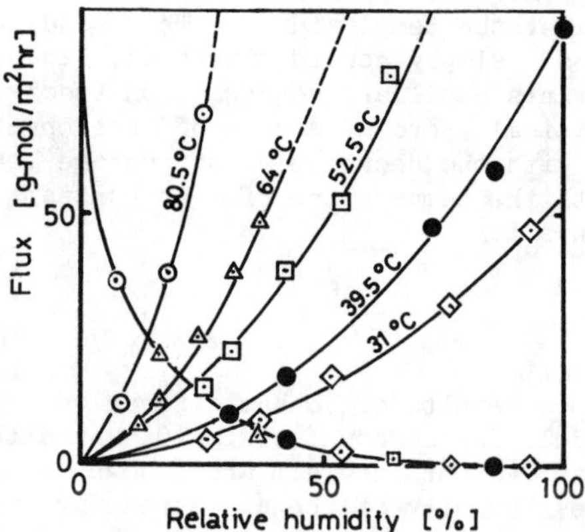

Fig.12 Flux of water and air through Module II (dehumidification of air)

4.2 Discussion on separation mechanism

A model of transfer mechanism in the membrane is shown in Fig.14. The water vapour condenses in the pores of the membrane and migrates through the fine pores to the inner surface of the membrane and evaporates there. The heat of condensation generated at the surface of the membrane can be used as the heat of evaporation at the inner surface. The vapour evaporated at the inner surface flows through the coarse porous substrate and is removed by condensation or evacuation.

Five probable resistances can be considered for mass and heat transfer, (1) diffusion resistance in the gas film at the surface, (2) flow resistance of condensate in the fine membrane pores, (3) flow resistance of vapour in the fine pores, (4) flow resistance of vapour in the coarse porous substrate and (5) heat transfer resistance in the thin membrane. Because of a quite small thickness of the membrane (less than about 0.0002 meter) the heat transfer resistance can be neglected in comparison with the others. The temperature drop across the membrane was estimated to be less than 0.5 °C at the most for the largest flux in the observation.

Two kinds of driving forces for the flow of the condensate can be considered, one of which is the total pressure difference across the membrane. A total pressure difference of about 10^5 Pascal was found experimentally to give a negligible contribution to the flux. The other driving force, which is probably decisive, is the capillary suction pressure difference due to the difference in the meniscus sizes of the condensate formed in a pore. Under equilibrium conditions the meniscuses formed in a pore have the same size on both sides of the condensate as shown in Fig.14-(b) but have different sizes when transfer occurs, Figs.14-(c) and (d). A quite large suction pressure difference can be expected because of quite small pore diameter provided that the usual theory of capillary suction pressure can be assumed in such a small pore as considered here.

Since the pemeability of gases in the membrane is quite small in comparison with those of the fresh substrate, the resistance (3) will give a large effect on the flux despite of its quite small length. In Fig.14-(c) is shown a case of permeation of pure water. Since the transfer is occuring, the meniscus of the condensate is somewhat different from that in the equilibrium condition (b). As the temperature rises, the length of the condensate decreases because the critical pore diameter for capillary condensation decreases with temperature rise. A larger length of the vacant part in the fine pores at a higher temperature will give a smaller apparent permeability than that at a lower temperature. This tendency also appears when the percentage saturation is decreased. The presence of gas film, case (d) in the figure, accelerates this tendency and at higher temperature or at a lower relative humidity the leak of air increases, because the number of pores which can not hold any condensate in themselves is considered to increase under these conditions. This model of the transfer mechanism can be a qualitative explanation of the results shown in the previous section. A further quantitative discussion will be given elsewhere.

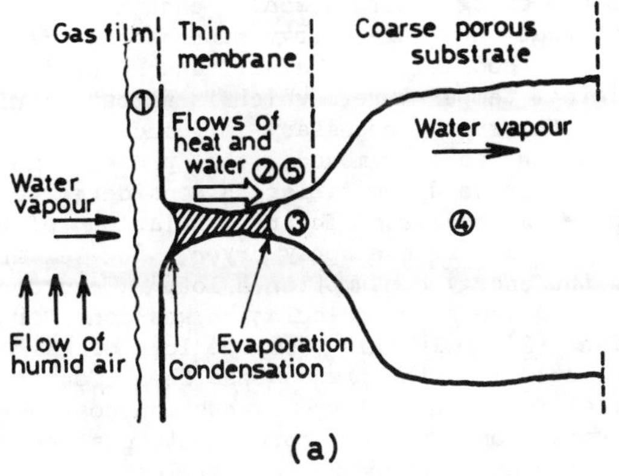

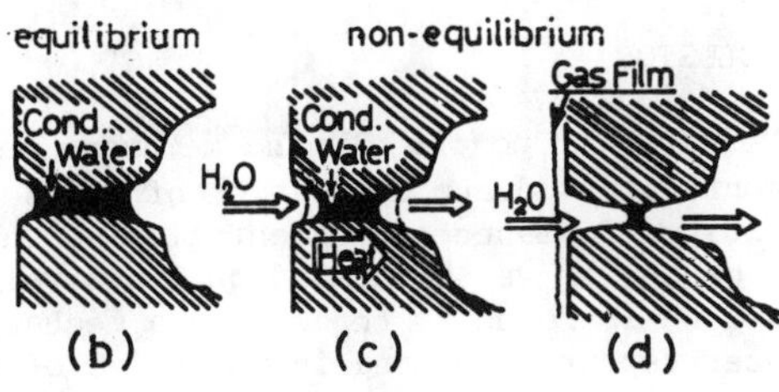

Fig.14 A model of transfer mechanism

4.3 Some calculated examples for membrane dehumidifier

In Fig.15 are shown some calculated examples of the performance of a membrane dehumidifier which is assumed to have membrane area of 3000 m^2. The dehumidifier is assumed to be used for dehumidification of large amount of air 48000 m^3/hr which is to be fed to a spray dryer, for example. The following two conditions of the air to be treated are assumed:

(1) temperature: 31°C, relative humidity: 70%

(2) temperature: 64°C, relative humidity: 13.5%

The calculations were done on the basis of the observed results shown in the previous section. In the figure are shown the relative humidity changes and the acumulated amounts of water removed and air leak. The air of condition (1) is to leave the apparatus at a relative humidity of 31% and temperature of 31°C (the absolute humidity: 0.0081 kg-water/kg-dry air). With this apparatus about 700 kgs of water can be removed in an hour. The leak rate of air is about 400 kgs/hr, which is less than 1 % of the inlet dry air. For air of condition (2) the outlet relative humidity is 8 % at temperature 64°C (the absolute humidity: 0.011 kg-water/kg-dry air). The rate of air leak for cindition (2) is about 2800 kgs/hr. The running coast for conditon

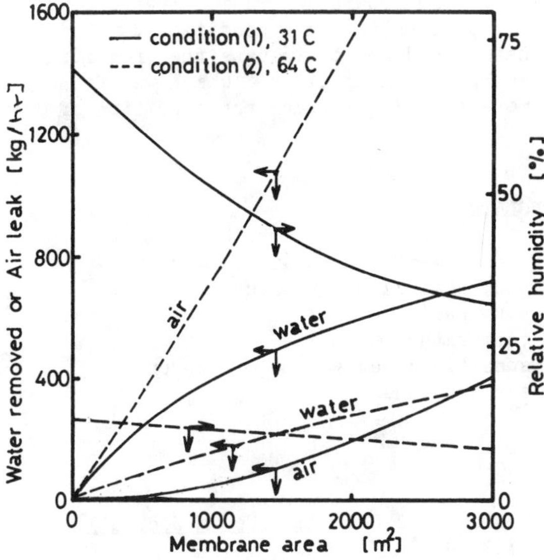

Fig.15 Some calculations on membrane dehumidifier

(1) was estimated to be about 80 % of that of a dehumidifier using a large scale condenser. Considering that the extra energy must be consumed for undesirable cooling of the air and reheating to the original temperature, which is almost equivalent to the energy necessary for condensation of water vapour to be removed, the present method using a membrane dehumidifier is donsidered to be much more preferable. The recirculation of the exaust hot air from a spray dryer, for example, reduces the energy consumption a lot. In order to assess the process the calculation was done for the condition (2). In this case the leak of air is quite large and the process has been found not economical for a quite large evacuation cost. For this purpose some further improvements must be done to the membrane to decrease the air leak.

5. CONCLUSION

Modules of porous alumina membrane were made for continuous dehumidification of humid air to study its performance experimentally and the separation mechanism was discussed qualitatively. The separation is found to be done by a mechanism that the water vapour condensed in the pores of about 10 Angstroms in a membrane and migrates through the membrane and evporates at its inner surface due to the evacuation of the inside of the module. The membrane prepared in this paper was found quite efficient for dehumidification of air of high humidity, but was not so efficient at low humidity and at high temperature because of relatively large air leak. The separation ratio, however, was large (above 7.0) even in these conditions. Some calculations were done to evaluate the feasibility of the membrane dehumidifier, which was found to be much more economical under the condition of low air leak.

NOMENCLATURE

p_0	pressure of condensate in gas phase	Pa
p	saturated vapour pressure	Pa
r	pore radius	m
T	temperature	K
V	specific molar volume of condensate	m^3/mol
γ	surface tension	N/m

REFERENCES

1. Yoldas, B.R., Amer.Ceram.Soc.Bull., vol.23, 803(1973)

EXPERIMENTAL STUDY ON APPLICATION OF PADDLE DRYERS FOR SLUDGE CAKE DRYING

Yusai Yamahata, Hidetaka Izawa
and Keizo Hasama

Chiba Laboratory, Mitsui Engineering and Shipbuilding Co., Ltd.
Ichihara, JAPAN

ABSTRACT

Drying tests were conducted on sludge cake using two types of paddle dryers. It was observed that polymer conditioned sludge cake was easier to dry than sludge cake conditioned with inorganic chemicals. Sludge in the dryer was separated into two zones, a paste zone upstream and a lump zone downstream. The heat transfer coefficient of each zone was strongly influenced by moisture content in the sludge and varied from 20 to 600 W/($m^2 \cdot K$).

1. INTRODUCTION

The widespread application of sewage sludge treatment in recent years has rapidly increased the generation of sludge cake (dewatered sludge). Most of this cake is finally disposed of by such measures as ocean dumping, incineration and landfill.

The advantages gained by effective drying of the sludge cake are a reduction in volume, lower fuel costs for incineration and recycling as fertilizer. The authors conducted experimental studies on the drying characteristics, heat transfer coefficient and other properties of sludge to develop an efficient sludge dryer.

2. EXPERIMENT

2.1. Apparatus

Sludge cake is usually highly viscous and gives off a strong odor during the drying process. In the author's prior experience, a steam tube rotary dryer and a single shaft dryer proved ineffective for proper drying of sludge cake because they were not equipped with a crushing device. Therefore, a configuration of indirect dryers agitated by twin shafts was chosen, and two paddle dryers of different capacities were fabricated by the authors for the tests.

A schematic and specifications of the paddle dryers are shown in Fig.1 and Table 1, respectively. The dryers were constructed with hollow shafts and paddles. The heat transfer media passes through each of these shafts and paddles as well as the jacket of the shell walls. The heating area of the pilot-scale dryer was increased four times over that of the bench-scale dryer, anticipating a proportional improvement in drying efficiency. Dried sludge is discharged from the overflow weir installed at the opposite side of the sludge cake inlet. The overflow weir is set at the same level as the center of the shaft to allow the sludge to remain in the lower part of the dryer.

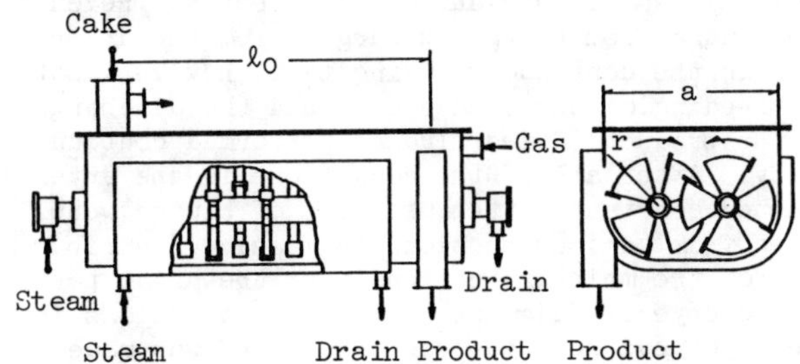

	ℓ_0 [m]	a [m]	r [m]
Bench-scale dryer :	1.22	0.65	0.13
Pilot-scale dryer :	2.10	1.15	0.32

Fig. 1 Schematic view of test paddle dryers

Table 1 Specification of test paddle dryers

		Bench-scale	Pilot-scale
Expected capacity F (from 80 to 40% wet base moisture)	kg/h	100	400
Heating surface location		Shell, Paddle, Shaft	
total area Ao	m^2	4.9	19.4
per unit volume	m^2/m^3	19.9	16.0
Number of shafts		Twin	
Paddle rotation	rpm	40	

2.2 Procedure

Figure 2 shows a flow diagram of the test dryer facility. Sludge cake is supplied to one end of the dryer by a snake pump and is dried while being transferred to the other end. Water evaporated during the drying process leaves the dryer in a carrier gas (hot air) and is introduced to a condenser to be discharged outside the system in the form of condensed water. After moisture is removed from the carrier gas it is passed through a gas heater and heated to approximate 360 K, and then returned to the dryer. Saturated steam was used as the heat transfer media for drying at a pressure of about 0.6 MPa.

In the batch drying test, a fixed amount of sludge cake was fed to the dryer, followed by rotating the paddle and counting the drying time.

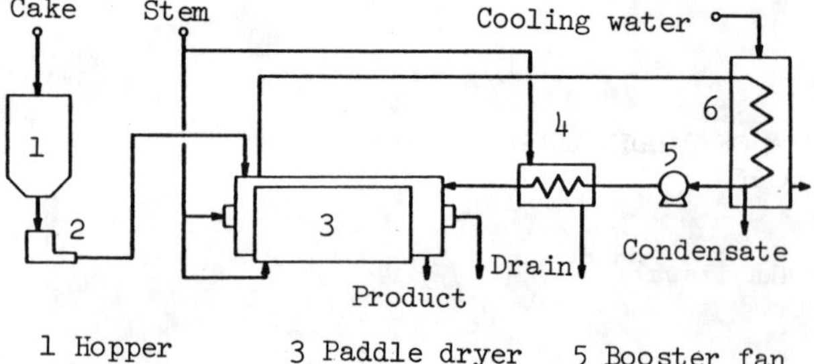

Fig. 2 Flow diagram of test dryer facility

Samples of the sludge were then taken periodically from the dryer to determine the relationship between drying time and moisture content. The overflow weir was closed during this test to prevent the sludge from being discharged from the dryer.

In the continuous drying test, however, both the feeding of the sludge cake and the discharge of the product (dried sludge) performed continuously. After all of the conditions in the dryer reached constant states, samples of sludge were taken from various sections to measure distribution of the moisture content over the axial path of the dryer. Judgment of whether conditions in the dryer had reached constant states was made when the moisture content of the product being discharged and the sum of the amounts of product and condensate reached constant levels. The moisture content in the sludge was measured using a Kett Moisture Tester.

Drying tests were performed on three types of sludge cake as shown in Table 2. Cakes A and B were dehydrated sludge conditioned with polymer agents, and Cake C was dehydrated sludge conditioned with inorganic chemicals. All Cakes A, B and C were generated in different sewage treatment plants.

Table 2 Types of sludge cake used in tests

	Cake A	Cake B	Cake C
Moisture content (dry base)	3.8-4.4	3.6-4.1	4.3-4.6
Conditioning agent	Polymer	Polymer	Ferric chloride Lime
Dewatering process	Belt-press	Belt-press	Vacuum filter

Table 3 Effect of moisture content on sludge conditions (dry base moisture content)

State	Cake B	Cake C
Viscous	3<	3.5<
Paste	1.5- 3	1.5-3.5
Large lump	1 -1.5	1.2-1.5
Small lump	0.4- 1	0.5-1.2
Powder	<0.4	<0.5

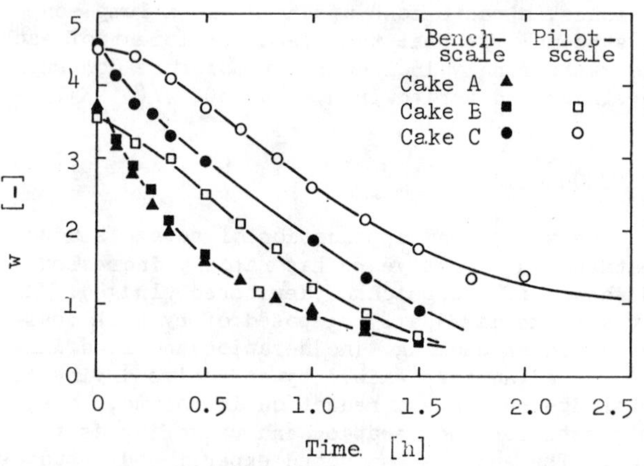

Fig. 3 Drying-time curves obtained from batch tests

3. RESULTS AND DISCUSSION

3.1. Batch Drying Test

Although the sludge cake was of the consistency as wet mud, its fluidity was improved temporarily by stirring due to its thixotropic characteristics. Along with the decrease in moisture content as the sludge was being dried, its properties remarkably changed. These changes in the properties of the sludge during batch tests are shown in Table 3.

For the short time after the drying operation is started, the sludge stays between the paddles in the form of a homogeneous body, and its movement is slow due to fluidity. As drying progresses, the sludge starts to rotate with the paddles, forming lumps of about 50 - 100 mm which are sheared by the crushing action of paddles of another shafts. As drying continues, these lumps are further crushed to about 10 - 20 mm, and are finally powdered at the end of the drying process. A difference was observed in the state changes between Cake B and C. Cake C tended to easily solidify.

Figure 3 shows the drying-time curves obtained by the batch drying test. Figs. 4 and 5 were drawn by plotting the drying rates, which were estimated by applying graphical differentiation to Fig. 3, against moisture content. Cake A and B, both of which are polymer conditioned sludge, indicated similar drying characteristics, although they were generated in different sewage sludge treatment plants. Cake C, an inorganic conditioned sludge, indicated somewhat different drying characteristics from the other two, and tended to dry at a slightly slower rate.

Remarkable differences in the drying characteristics by type of dryer rather than differences in the properties of sludge are observed in Figs. 3, 4 and 5. Although rate-of-drying curves describing a warming-up period and a falling-rate period were obtained with the pilot-scale dryer, those obtained with the bench-scale dryer were different and didn't describe the warming-up period. One of the rasons for such differences should be the difference in a start of the steam supply. In the bench-scale dryer test, the steam had already been supllied before feeding sludge in the dryer.

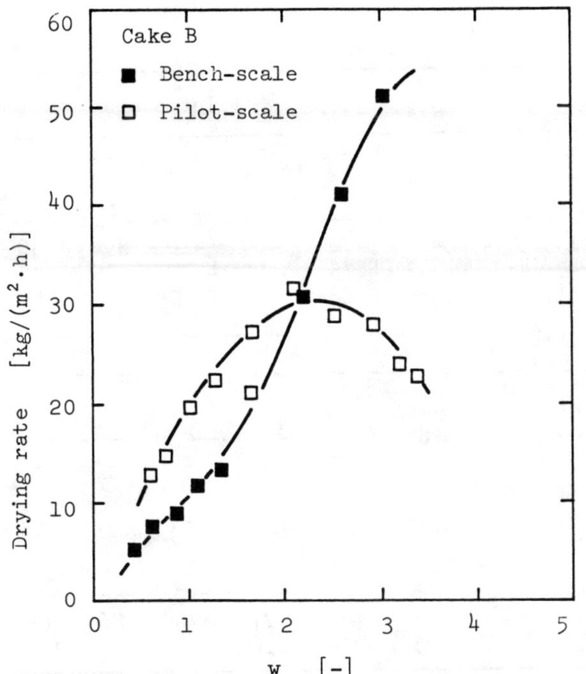

Fig. 4 Rate-of-drying curves obtained from batch tests

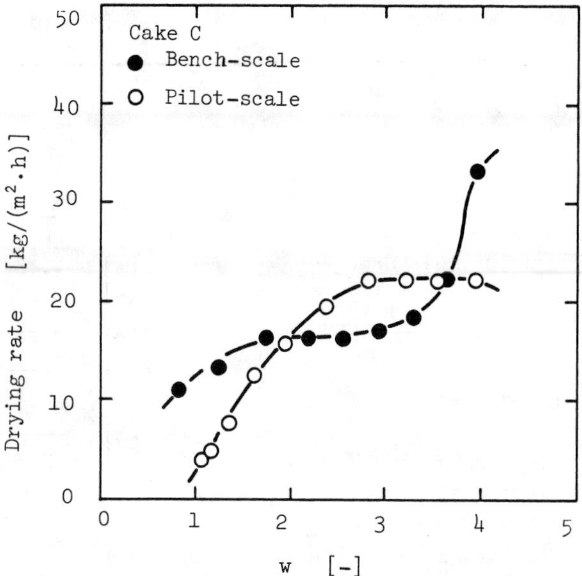

Fig. 5 Rate-of-drying curves obtained from batch tests

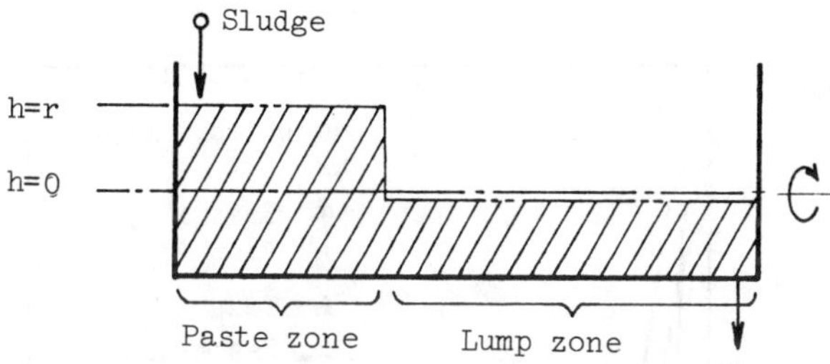

Fig. 6 Packing height of sludge in continuous dryer

3.2. Continuous Drying Test

Table 4 presents the experimental conditions and results of the test. In the continuous drying test, sludge is transferred continuously from the feeding end of the dryer to the product discharge at the other end as it is being dried. However, the state of the sludge in the dryer can be divided into two zones as shown in Fig. 6 according to observations of the packing density of the sludge. One of these zone is the "paste zone" which is located at the beginning half of the process, and the other is the "lump zone" located in the latter half. The border between the two zones was clearly identified by observation. A description of these zones is presented in the following:

Paste zone: As shown in Fig. 6, sludge is filled to the upper end of the paddles to form a homogeneous body. Sludge in the beginning half of the paste zone is in the form of a viscous fluid, and that in the latter half of the zone is in a plastic stage. Sludge is broken into pieces at the end of this paste zone which changes to the lump zone.

Lump zone: The height of filled sludge is decreased to the level of the overflow weir mounted at the discharge of the product. This is about a half of the packing height of the sludge in the paste zone. The form of the sludge is changed into lumps of a large size, 50 - 150 mm, at the beginning of the zone, which are subsequently reduced to 10 - 30 mm in the latter half of the zone.

The paste and lump zones appeared in both the bench-scale and pilot-scale dryers, regardless of the nature of the sludge cake. Table 4 shows the fractional distance along dryers occupied by paste zone.

Examples of moisture distributions during continuous drying tests are shown in Figs. 7 and 8. The solid lines and dotted lines shown in these figures indicate the paste zone and lump zone, respectively. Figure 7 shows an example of test results obtained from the bench-scale dryer. As can be seen in the figure, the larger the amount of sludge cake fed to the dryer, the longer the length of the paste zone and the larger the moisture content. Figure 8 shows example of test results obtained from the pilot-scale dryer. As can be seen in the figure, drying efficiency varies according to the nature of sludge cake. As in the case of the batch drying test, it seems that polymer conditioned cakes are more easily dried than inorganic conditioned cakes.

Assuming that temperature T_i and moisture w_i of the sludge cake fed to the continuous drying system were changed to T and w respectively after being heated in the system, the heat balance in this case is expressed as follows:

$$F \cdot C_p \cdot (T-T_i) + F \cdot \frac{w_i - w}{w_i + 1} \cdot \lambda = \alpha \cdot A \cdot (T_s - T) + \beta_g \cdot A_g \cdot (T_g - T) \quad (1)$$

Table 4 Data of continuous drying tests

Test dryer		Bench-scale						Pilot-scale			
Test no.		1	2	3	4	5	6	7	8	9	10
Sludge											
Type		Cake A						Cake B		Cake C	
Feedrate, F	[kg/h]	68.0	90.2	114	92.4	94.8	94.0	400	410	391	330
Inlet moisture, w_i	[-]	4.08	3.85	4.43	4.38	4.43	4.75	4.00	4.10	4.29	3.42
Discharge moisture, w_e	[-]	0.37	0.51	0.61	0.54	0.64	0.66	0.80	0.77	1.25	0.80
Heating steam pressure	[MPa]	0.6						0.6			
Paddle rotation	[rpm]	40						40			
Carrier air											
Flow rate	[Nm³/h]	92			90	83	83	300			
Inlet temperature	[K]	383	383	387	389	381	385	-	-	-	-
Fractional distance of paste zone along dryer	[-]	0.24	0.37	0.45	0.37	0.37	0.37	0.33	0.33	0.33	0.26
Heat transfer coefficient	[W/(m²·K)]										
Average in paste zone	α_p	576	482	545	500	566	507	581	617	558	528
Average in lump zone	α_ℓ	20	35	46	48	45	48	77	78	186	86
Average throughout dryer	α_o	157	199	272	210	235	213	228	241	258	177

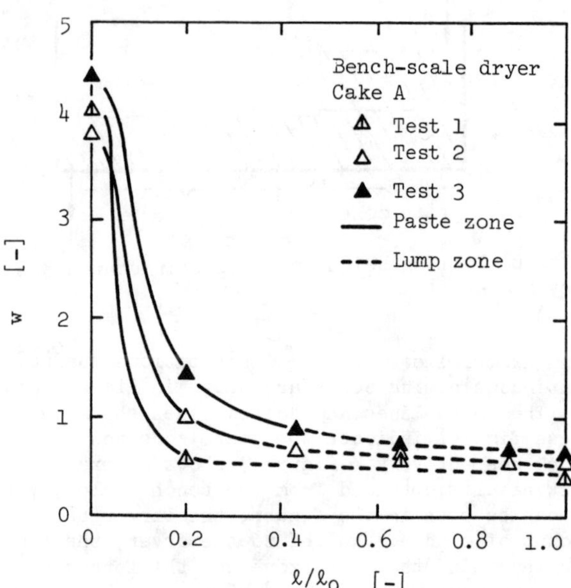

Fig. 7 Distribution of moisture content in sludge along dryer during continuous drying tests

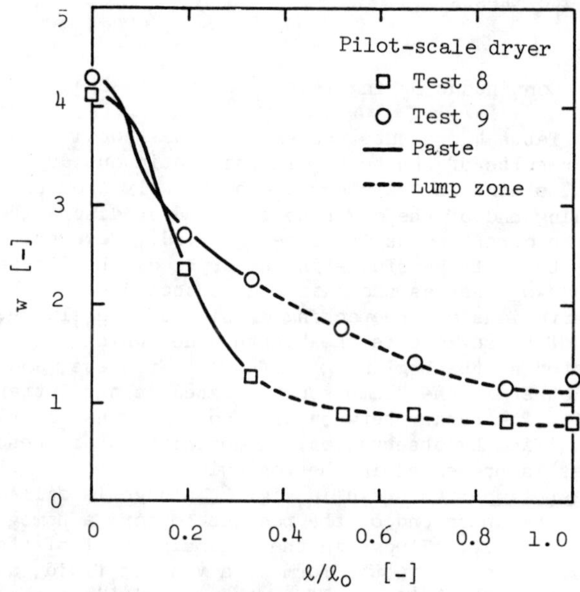

Fig. 8 Distribution of moisture content in sludge along dryer during continuous drying tests

Although the second term of the right side of the Equation (1) represents the quantity of transferred heat between the carrier gas and sludge, this is negligible and may be ignored compared with the first term. Therefore, the average overall heat transfer coefficients in the paste zone and lump zone are calculated as follows from Equation (1):

$$\alpha_p = F\cdot\left\{C_p\cdot(T-T_i) + \frac{w_i-w_b}{w_i+1}\cdot\lambda\right\} / \left\{A_p\cdot(T_s-T)\right\} \quad (2)$$

and in the lump zone:

$$\alpha_\ell = F\cdot\frac{w_b+w_e}{w_i+1}\cdot\lambda/\left\{A_\ell\cdot(T_s-T)\right\} \quad (3)$$

where:

A_p: Heating area in paste zone when h = r as shown in Fig. 6

A_ℓ: Heating area in lump zone when h = 0

Although a slight increase in the temperature of sludge, T, was observed when moisture content was reduced below 0.7, this was regarded constant for the sake of convenience in making the calculations.

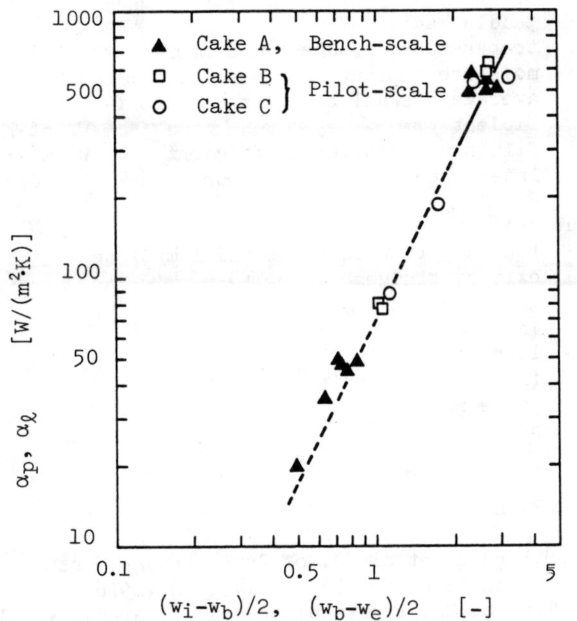

Fig. 9 Relationship between overall heat transfer coefficient and moisture content in sludge during continuous drying tests

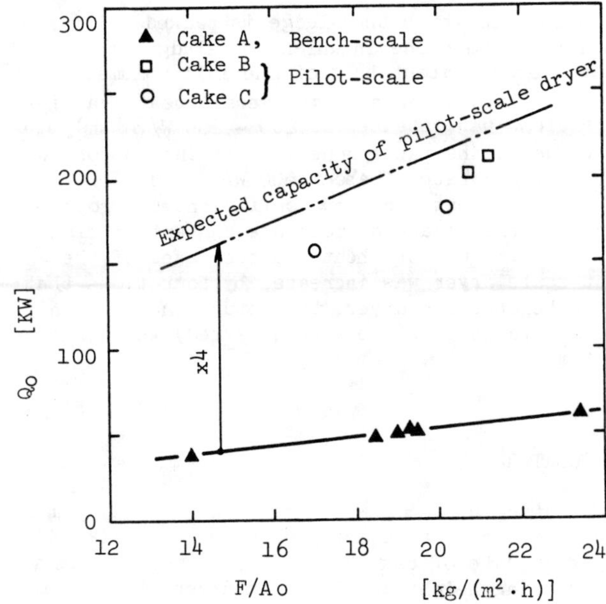

Fig. 10 Comparison of bench-scale and pilot-scale dryer capacities

Figure 9 shows the relationships between α_p, α_ℓ and averaged moisture content, $(w_i+w_b)/2$ or $(w_b+w_e)/2$. The solid line and dotted line represent the paste zone and lump zone respectively. These relationships were found to be linear, regardless of the types of sludge and scale of the dryers. The following equation could be obtained from the line in the figure,

$$\alpha \propto w^2 \qquad (4)$$

One of the reasons for the remarkable reduction in heat transfer coefficient along with the reduction of moisture content should be the reduction in the effective heat transfer surface due to formation of dried porous lumps of sludge.

Although the inside of the dryer is composed of two zones which can be clearly identified by observation as shown in Fig. 6, Figure 9 suggests that the overall heat transfer coefficients continuously decrease as drying progresses from the paste zone to lump zone.

When designing the paddle dryers, it is difficult to project the ratio to be occupied by the paste zone, ℓ_p / ℓ_o, and the moisture content at the border of the two zones. It was, therefore, convenient to design the dryer based on an average overall heat transfer coefficient covering the whole dryer rather than on α_p and α_ℓ. The authors attempted to calculate α_o based on the data obtained in this test. α_o can be obtained from Equation (2) by inserting $w_b = w_e$ and $A_p = A_o$. A_o is the value of the heating area in the dryer of which the length is ℓ_o at h = r. α_o = 200 - 230 and α_o = 230 - 260 W/(m²·K) were obtained in the bench-scale dryer and in the pilot-scale dryer, respectively, except in Test 1, 3 and 10, where the feeding rates were outside the normal range. It is thought that such slight differences between the two are due to differences in the moisture in products treated in both dryers, taking the tendency observed in Fig. 9 into consideration.

Several reports on α_o when drying the sludge cake in the paddle dryer have been made available. For example, M. Hiraoka[1] and T. Kato[2] reported values of 90 - 210 and 140 - 210 W/(m²·K), respectively. The value measured by authors were somewhat larger.

As shown in Table 1, the heat transfer area A_o of the pilot-scale dryer is four times larger than that of the bench-scale dryer. Consequently, the authors expected that the capacity of the pilot-scale dryer should also be four times greater than the bench-scale dryer. Figure 10 shows comparisons of their capacities by the total heat transferred Q_o. Q_o corresponds to the left side of Equation (1). As a result, it was found that the capacity of the pilot-scale dryer was about 3.6 times of that of the bench-scale dryer, lower than the expectation of the authors. It was considered that one of the reasons for this reduced value is that the heat transfer area per unit volume of the pilot-scale dryer is 80 percentage of that of bench-scale dryer.

4. CONCLUSION

Experiments were conducted to clarify drying characteristics of sludge cake and to obtain heat transfer coefficients in two types of paddle dryers with heat transfer areas in the ratio of 1 : 4. The following results were obtained:

(1) Polymer conditioned sludge is somewhat easier to dry than inorganic conditioned sludge.

(2) The state of the sludge changes from a viscous fluid state to a powder state along with the decrease of moisture content.

(3) During the continuous drying process, the inside of the paddle dryer is divided into a

paste zone in which the sludge is packed fully and into a lump zone in which the sludge is packed to only about half of the full volume.

(4) Although an average overall heat transfer coefficient was obtained at 200 - 260 W/($m^2 \cdot K$), the value in the paste zone was obtained a considerably larger rate of about 600 W/($m^2 \cdot K$).

(5) The heat transfer coefficient is proportional to the square of moisture content in sludge.

(6) Although the heat transfer area of the pilot-scale dryer was increased by four times that of the bench-scale dryer, the actual increase in drying capacity (the heat transferred) was about 3.6 times.

NOMENCLATURE

A	heating area	m^2
C_p	specific heat capacity of cake	J/(kg·K)
F	feed rate of cake	kg/h
h	packing height of sludge in dryer	m
ℓ	distance from inlet along dryer	m
Q	heat transferred	J/h
r	paddle radius	m
T	temperature, sludge temperature	K
w	moisture content in sludge (dry base)	-
α	average overall heat transfer coefficient	W/($m^2 \cdot K$)
β	film heat transfer coefficient	W/($m^2 \cdot K$)
λ	latent heat of water evaporation	J/kg

<Subscripts>

b	boundary between paste and lump zone
e	exit of sludge
g	gas side
i	inlet of cake
ℓ	lump zone
o	from inlet to exit
p	paste zone
s	steam

REFERENCES

1. M.Hiraoka et al, J. of Japan Sewage Works Association, vol.13, No.143, 22 (1976)
2. T.Kato, Environmental Creation (Japan), vol.13 No.3/4, 10 (1983)

A NEW MEASUREMENT METHOD OF LIQUID TRANSPORT PROPERTIES

Ikuro Shishido, Mutsumi Suzuki
and Shigemori Ohtani

Department of Chemical Engineering, Tohoku University
Sendai, 980 JAPAN

ABSTRACT

Liquid transport properties, such as suction pressure, permeability and apparent moisture diffusivity, were measured for wide range of degree of liquid saturation within wet material to be dried. Wet sample was pressurized by N_2 gas and drained water through membrane filter was measured continuously by a micro processor system. Suction pressure was determined from the equilibrium data. Permeability and difusivity were estimated from the transient change of the drained water volume, by using a linearized transport equation in each small increment of pressure.

Obtained data of both permeability and apparent moisture diffusivity show a strong dependence of saturation.

Validity of the present method is confirmed by comparing with the previous steady state drying method.

1. INTRODUCTION

Various mechanisms of moisture transport within wet material and various mathematical models of drying have been proposed by many researchers in order to develop theories of drying. Among them, the capillary theory is widely used in analysis of water movement within non-hygroscopic porous material. According to this theory, moisture movement is represented by the following Darcy´s equation:

$$q = -\frac{K}{\mu}\frac{dP_c}{dx} \qquad (1)$$

where Pc and K are the capillary suction pressure and permeability, respectively. When the suction pressure is represented by a unique function of saturation, equation (1) is rewritten as;

$$q = \left(-\frac{K}{\mu}\frac{dP_c}{dS}\right)\frac{dS}{dx} = D\frac{dS}{dx} \qquad (2)$$

where D is an apparent moisture diffusivity within wet material. We can, then, express the transient change of moisture within wet material by the following non-linear diffusion equation;

$$\frac{\partial S}{\partial t} = \frac{\partial}{\partial x}\left(D\frac{\partial S}{\partial x}\right) \qquad (3)$$

Solution of this equation under some kinds of initial and boundary conditions can be obtained, in principle, if the apparent difusivity or both the permeability and the suction pressure are known.

Hence, many researchers have tried to measure such transport properties by various techniques.

Kamei[1], Wakabayashi[2], Hayashi[3], and Okazaki[4] measured the transient change of moisture profile within drying specimen and estimated the apparent moisture diffusivity. Wakabayashi[5] Evans[6] and Endo[7] calculated the diffusivity from steady moisture profiles under steady state drying experiments. Krischer[8] and Ohtani[9] measured the suction pressure under the gravitational field or centrifugal force, and then estimated the diffusivity by using a capillary boundle model. Recently, Schoeber[10] and Liou[11] have reported a determining method for the moisture diffusivity by a regular regime drying curve. In the previous paper[12], we have proposed a determination method of the moisture diffusiity from transient change of surface moisture content or average critical moisture content.

Most of these methods require many laborious experimental measurement at various moisture levels and/or various drying conditions. Simple and quick measurement technique, thus, has been desired for a long time.

In this paper, a new experimental technique is presented for the measurement of the suction pressure, the permeability and the apparent diffusivity.

2. PRINCIPLE OF MEASUREMENT

2.1. Suction pressure

Suction pressure was measured, as a

function of saturation, by an ordinary pressurized method (Figs. 1 and 2). The wet specimen was set on a semi-permeable membrane filter in a cell and pressurized by an inactive gas such as Nitrogen. The water held in the sample, then, drained through the membrane until an equilibrium state was attained. The saturation at the equilibrium state was calculated from the drained water volume and initiated moisture content.

2.2. Permeability

Now let us consider a sample of wet material having a volume V, and a height L, which is placed on a membrane filter in a pressure cell. Suppose the initial pressure in the cell being P and saturation within wet material being in equilibrium with pressure in the cell. At time t=0, the pressure in the cell is increased by a small amount ΔP. Due to the increment of the pressure, liquid flows out of wet material until equilibrium is attained (cf. 1.1).

Though the permeability is, in general , a function of moisture content, we can assume it constant in the small increment of moisture change.

$$K(S) = \text{constant} \tag{4}$$

A linear relationship between capillary pressure and saturation can be also assumed in the small increment of moisture content as;

$$P_c = a'+b'S \tag{5}$$

The total pressure of gas phase in the porous material is assumed to be uniform;

$$\frac{\partial P}{\partial x} = 0 \tag{6}$$

and an equilibrium pressure balance is held between gas and liquid phase;

$$P = P_l + P_c \tag{7}$$

Then the saturation can be related with the liquid pressure by the following linear equation

$$S=a+bP_l \tag{8}$$

In such a situation, the transient change of liquid pressure relating to liquid flow may be described by the following partial differential equation with uniform initial condition (for one-dimensional vertical flow in pressure cell)

$$\frac{\partial P_l}{\partial t} = \frac{K}{\mu b}\frac{\partial^2 P_l}{\partial x^2} \tag{9}$$

$$P_l = \Delta P \quad at \quad t=0,\ 0\leq x\leq L \tag{10}$$

Since the drainage occurs only at the bottom of cell through the membrane filter, the boundary conditions are expressed by;

$$\frac{\partial P_l}{\partial x} = 0 \quad at \quad x=0 \tag{11}$$

$$K\frac{\partial P_l}{\partial x} + \frac{K_m}{L_m}P_l = 0 \quad at\ x=L \tag{12}$$

where K_m and L_m represent permeability and thickness of membrane filter respectively.

The solution of Eq.(9) with above initial and boundary conditions is easily obtained. The solution, then, is converted to the saturation distribution by use of Eq.(8). By integrating the saturation distribution, drained liquid volume is expressed as follows;

$$\begin{aligned} W(t) &= W_i - a\Psi_i V - 2b\Psi_i V\Delta P\sum_{n=1}^{\infty}\frac{\lambda^2}{\alpha_n^2\{\lambda(\lambda+1)+\alpha_n^2\}}\exp\left(-\frac{K}{\mu b}\alpha_n^2\frac{t}{L^2}\right) \\ &= A + B\sum_{n=1}^{\infty} f(\alpha_n)\exp(-C\alpha_n^2 t) \end{aligned} \tag{13}$$

where α_n are the positive roots of

$$\lambda = \alpha_n \tan\alpha_n = \frac{K_m L}{K\ L_m} \tag{14}$$

and A, B and C are the auxiliary parameters defined by

$$\left.\begin{aligned} A &\equiv W_i - a\Psi_i V \\ B &\equiv -2b\Psi_i V\Delta P \\ C &\equiv \frac{K}{\mu b L^2} \end{aligned}\right\} \tag{15}$$

The unknown parameters, A,B and C, in above Eq(13) can be determined in the following way.

Except for very short time range, infinite series (13) converges rapidly in a few terms.

The parameters, A, B and C in Eq.(13) can be determined by a nonlinear parameter estimation method, in which the following residual function is minimized

$$Res(A,B,C) \equiv \sum_{i=1}^{N}\{W(t)_{exp} - W(t;A,B,C)\}^2 \tag{16}$$

where $W(t)_{exp}$ is the experimental data of the drained liquid volume. Permeability K can be calculated by eliminating b from the determined parameters B and C.

2.3. Apparent moisture difusivity

Furthermore, the apparent moisture diffusivity, which is defined as

$$D = -\frac{K}{\mu}\frac{dP_c}{dS} \qquad (17)$$

can be calculated by

$$D \simeq \frac{K}{\mu}\frac{\Delta P}{\Delta S} = C\ L^2 \qquad (18)$$

3. EXPERIMENTAL APPARATUS AND PROCEDURE

Experimental set up is illustrated in Fig.1. Details of pressure cell is shown in Fig.2. Pressure was controlled within 50 Pa by two electro magnetic valves and semi-conductor pressure transducer. Temperature of circulating water was measured by thermocouples and controlled within ±0.1 K by an electric heater and a solid state relay. Weight of exhausted liquid water was measured by an electronic weight balance at every 5 seconds, accuracy of the balance was 0.1 mg. These control and data acquisition procedures were supervised by a BASIC program on a micro computer CBM 3016.

When the output of the weight balance shows no change, the computer interrupts the procedure and stores data into floppy disk, then increases the pressure again by a small amount. This routine was repeated until the pendular state was attained. Afterwards, all data were transmitted to another high speed computer, and then were processed to determine the suction pressure, the permeability and the diffusivity.

In order to check up the influence of thickness of wet material, height of wet bed was varied. From this experiment, we can find that the effect of thickness is negligible if the thickness of wet bed is less than 2 cm.

Used samples were crushed fire brick (mean diameter of which was about 200 μm) and spherical glass beads (with a mean diameter of about 150 μm).

4. EXPERIMENTAL RESULTS AND DISCUSSION

Figure 3 shows the experimental data of suction pressure of glass beds. It can be seen from this figure, that the initial saturation has great influence on the capillary suction pressure.

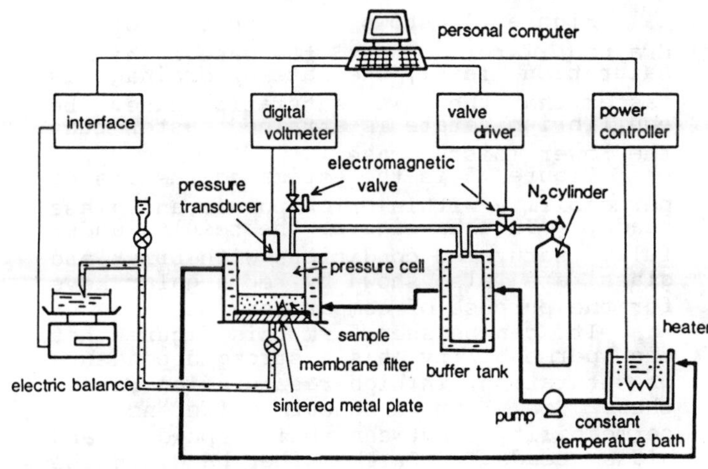

Fig.1 Schematic diagram of experimental apparatus

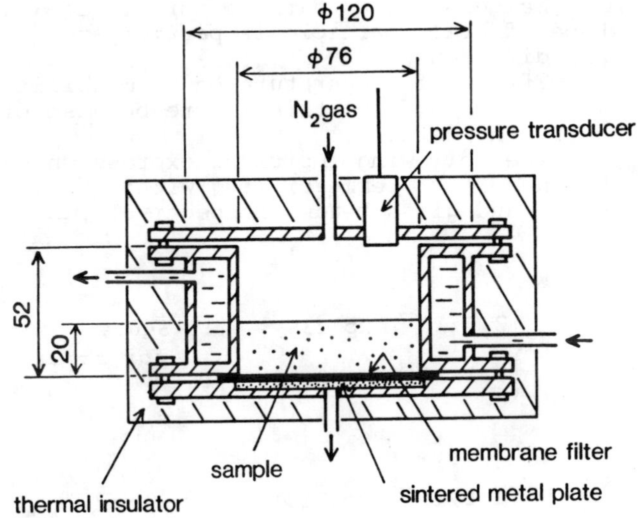

Fig.2 Details of the pressure cell

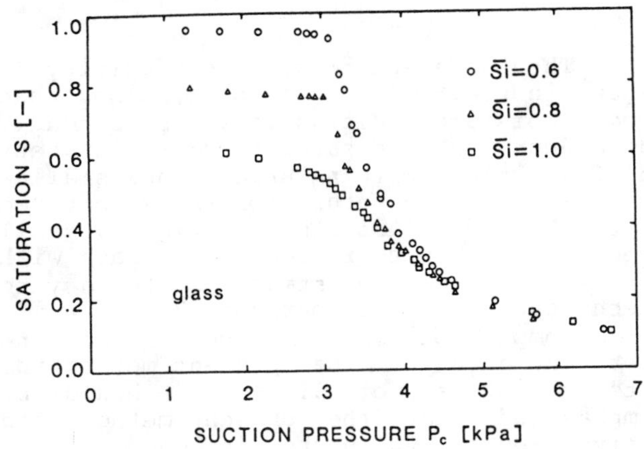

Fig.3 Experimental results of capillary suction pressure(glass beads).

Figure 4 shows the changes of the drained water. At the case of higher saturation level, the rate of drainage is faster than the lower saturation case. The equilibrium state is attained faster than the lower moistue case.

Figure 5 is the estimated results of permeability within brick powder and glass beads. In this figure, the results under fully saturated conditions with brick and glass are also shown by rectangular keys for the purpose of comparison.

It can be seen form this figure that the permeability has a strong dependence on saturation. In higher saturation range, there is not a clear difference in permeability between brick powder and glass beads. On the other hand, in the lower range, permeability within brick powder is considerably smaller than glass beads.

Although the reason is not clear, it may be considered that the more irregular shape of brick particle is participated in this difference.

Effect of temperature on permeability is not clear in this figure because of scattering data.

The following empirical expression is obtained for permeability within brick powder and glass beads, respectively,

$$K(S)=\begin{cases} 2.7\times10^{-17}\exp(23S) & 0.15<S\leq0.4 \\ 8.1\times10^{-15}\exp(8.2S) & S>0.4 \end{cases} \quad \textit{for brick} \tag{19}$$

$$\begin{cases} 1.0\times10^{-16}\exp(22S) & 0.04<S\leq0.3 \\ 1.1\times10^{-14}\exp(7.5S) & S>0.3 \end{cases} \quad \textit{for glass} \tag{20}$$

The apparent moisture diffusivity is shown in Fig.6. It can be seen, also, that the moisture diffusivity is strongly dependent on the saturation. Apparent diffusivities within brick show smaller values in whole saturation range compared with those of glass. In this figure, small rectangular keys represent the data with brick obtained by steady state drying technique at room temperature[7].

Comparison shows a good agreement between data of the present method and steady state method. It should be emphasized that the present method need only about 20 hours in one run of experiment through wide range of saturation, which is about 30 times faster than the steady state method.

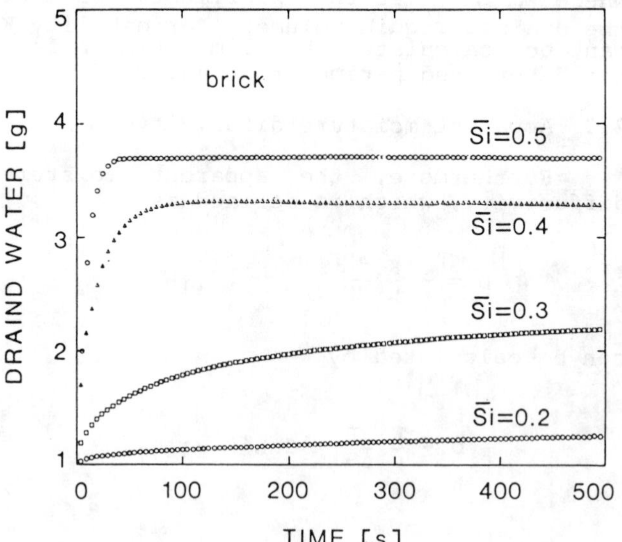

Fig.4 Measured the change of the drained water.

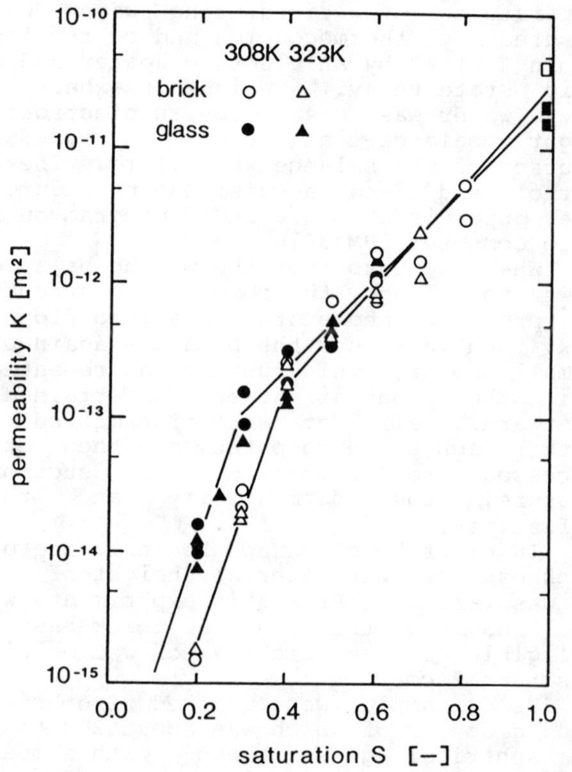

Fig.5 Estimated results of the permeability within brick and glass.

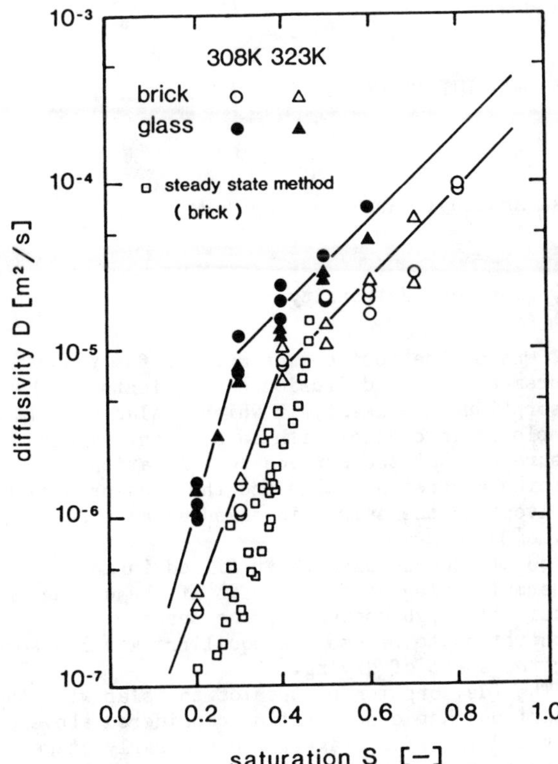

Fig.6 Estimated the apparent moisture diffusivity within brick and glass.

From this figure, the apparent moisture diffusivity can be correlated with saturation as follows;

$$D(S)=\begin{cases}1.3\times10^{-8}\exp(16S) & 0.15<S\leq0.4 \\ & \textit{for brick} \\ 7.2\times10^{-7}\exp(6.1S) & S>0.4\end{cases} \quad (21)$$

$$\begin{cases}1.2\times10^{-8}\exp(22S) & 0.04<S\leq0.3 \\ & \textit{for glass} \\ 1.5\times10^{-6}\exp(6S) & S>0.3\end{cases} \quad (22)$$

CONCLUSION

A new measurement technique is presented for liquid transport properties, such as suction pressure, permeability and apparent diffusivity within wet material to be dried.

Obtained data show that both permeability and apparent diffusivity have a strong dependency of saturation.

This new method requires only 20 hour experimental time. The validity of this unsteady state method is confirmed by comparing with the steady state method.

NOMENCLATURE

D :	apparent moisture diffusivity	[m²/s]
K :	permeability	[m²]
L :	length	[m]
P :	pressure	[Pa]
S :	saturation	[-]
t :	time	[s]
V :	volume of specimen	[m³]
W :	volume of drained water	[m³]

Greeks

μ :	viscosity	[Pa·s]
ψ :	initial moisture content	[kg/kg]

Subscripts

c : capillary
i : initial state
l : liquid phase
m : membrane
v : vapor

REFERENCES

1. Kamei,S. and T. Ushirohara, Kogyo Kwagaku Zasshi, 36, 1561 (1933)

2. Wakabayashi,K., Kagaku Kogaku, 28, 33 (1964)

3. Hayashi, S.Ph. D. thesis, Kyoto Univ., (1965)

4. Toei, R., M. Okazaki, K. Shioda and K. Masuda, Proc. of 1st Pacific Chem. Congr., part 3, 33 (1972)

5. Wakabayashi, K., Kagaku Kogaku, 33, 102 (1964)

6. Evans, A.A. and R.B.Keey, Chem. Eng. J., 10, 135 (1975)

7. Endo,A., I. Shishido, M. Suzuki and S. Ohtani, AIChE Symp. Ser. No.163, 73, 57 (1977)

8. Krischer, O., Die wissenschaftlichen Grundlargen der Trocknungstechnik, Springer-Verlag, 2/e (1963)

9. Ohtani, S. and S. Maeda, Kagaku Kogaku ,31, 463 (1967)

10. Schoeber, W.J. and H.Thijssen, AIChE Symp. Ser. No. 163, 73, 12 (1977)

11. Liou, J.K.,Ph. D. thesis, Univ. Wabeningen, (1982)

12. Shishido, I., M. Suzuki and S. Ohtani, Proc. of 1st Int. Symp. on Drying, 30 (1978)

DETERMINATION OF CONCENTRATION DEPENDENT DIFFUSION COEFFICIENT FROM DRYING RATES

Shuichi Yamamoto, Munehiro Hoshika and Yuji Sano

Department of Chemical Engineering, Yamaguchi University,
Ube 755, Japan

ABSTRACT

When the diffusion coefficient decreases with decreasing concentration, an isothermal drying process(desorption process), in which the surface concentration holds its equilibrium value, can be divided into two stages. In the first stage, where the process can be considered as diffusion in a semi-infinite medium, the initial desorption rate is usually approximated to an integral average diffusion coefficient. However, this is not a good approximation when the concentration dependence of the diffusion coefficient is strong. In the second stage, the effect of the initial concentration is negligible and the desorption rate is governed by the average concentration only as shown by Schoeber. A method for the determination of the concentration dependent diffusion coefficient from the drying rates of both stages is proposed. The desorption curves were numerically calculated with various types of the concentration dependence of the diffusion coefficient. Then, the desorption rates were correlated with a weighted mean diffusion coefficient. It was found that in the first stage the desorption rates were closely related to a single type of weighted mean diffusion coefficient proposed by Crank. In the second stage, another single type of weighted mean diffusion coefficient was found to represent the desorption rates. An approximate calculation procedure for the whole desorption curve is presented, in which the above two weighted mean diffusion coefficients are used. The calculated results by this method were in good agreement with those by the finite-difference method.

1. INTRODUCTION

It is known that the diffusion coefficients of solvent in polymers and of water in foodstuffs such as carbohydrates and proteins show a sharp decrease with decreasing solvent (or water) concentration[1-2]. In the analysis of the drying behaviour of foods and polymers, it is necessary to know such a concentration dependent diffusion coefficient over a wide range of concentration. However, the concentration dependence of the diffusion coefficient for the above-mentioned substances is so strong that there are several difficulties in determining it.

One of the most common methods for the measurement of the diffusion coefficients is the (de)sorption experiment, in which a slab or film of a sample is in contact with a constant vapour pressure so that the surface concentration maintains a constant value and the loss or uptake of solvent by the sample is measured as a function of time[3].
This condition can also be fulfilled in an isothermal drying of food stuffs if the drying intensity is high enough for the surface concentration to become the equilibrium value soon after the start of drying.

The (de)sorption process of the slab with the constant surface concentration considered above can be divided into two stages. In the early stage, the concentration of the center of the slab still holds its initial value as shown in Fig.1-a. In this period, the (de)sorption process can be considered as diffusion in a semi-infinite medium. We call this period the "penetration period" as Schoeber[4] does.

After this period, the center concentration gradually varies with time as shown by Fig.1-b. In this part of diffusion process, Schoeber[4] found that when the diffusion coefficient decreases with decreasing concentration,the desorption rate is not dependent on the initial concentration but governed by the average concentration only. This period is referred to as the "regular regime".

In the conventional (de)sorption method[3], only one information is obtained from the penetration period of (de)sorption experiments. That is, an apparent diffusion coefficient is obtained from the initial (de)sorption rate or half time of the (de)sorption experiment. Then, this apparent diffusion coefficient is approximated as an integral average diffusion coefficient. However, as Crank stated in his publication[3], this first approximation is only accurate when 1) The concentration dependence is not so strong, 2) The initial rates of both desorption and sorption are available, and 3) The effect of shrinkage (or swelling) is negligible. If these conditions are not fulfilled, the next step such as the iteration and the use of the correction curve is needed to obtain accurate values[3]. These methods seem to be rather laborious and tedious. As already stated, the concentration dependences of the diffusion coefficient for solvent-polymer, water-carbohydrate and water-protein systems are very strong. For some polymers, only the desorption process shows a Fickian behaviour and therefore, the sorption rate can not obtained[5,6].

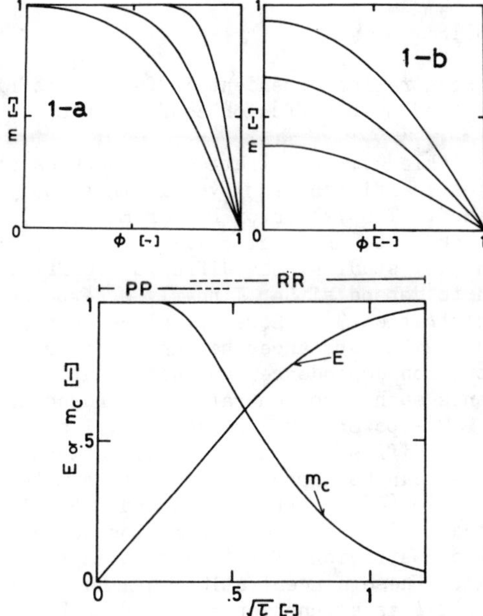

Fig. 1 Concentration profiles and desorption curve for a slab with constant surface concentration and constant diffusion coefficient. E is a fractional loss of moisture, m a reduced concentration and m_c a center concentration. PP and RR means the penetration period and the regular regime, respectively.

In addition, since the diffusion coefficient is necessary for a wide range of concentration, experiments have to be carried out at a relatively high solvent concentration, where the effect of shrinkage (or swelling) can not be ignored. These conditions require the re-examination of Crank's method.

On the other hand, several workers proposed a method which extract considerable amount of information from a single (de)sorption curve[4,7]. However, the method proposed by Duda and Vrentas[7] is difficult to apply to the strong concentration dependent diffusion coefficient as they stated in their later report[2]. At present, to our knowledge Schoeber's method[4], in which the regular regime curve is used, seems to be the best way to obtain a large amount of meaningful data from a single desorption experiment of food materials and polymers.

The purpose of this study is to present relationships between concentration averaged diffusion coefficients and the desorption rates (drying rates). This will be useful both for the determination of diffusion coefficient from the desorption rate and for the prediction of the desorption rate when the diffusion coefficient is known as a function of concentration.

We employed various types of the concentration dependence of the diffusion coefficient such as linear, exponential , power-law and fictitious dependence for food materials. Then, the (de)sorption curves for a slab were calculated numerically under the conditions of the initially uniform concentration, the constant surface concentration and the uniform constant temperature.

We employed a weighted mean diffusion coefficient proposed by Crank[3] as an mean value over the range of concentration involved, since it contains only one parameter and has been reported to be a good approximation of several concentration dependent diffusion coefficients.

The apparent diffusion coefficients obtained from the initial (de)sorption rates and those from the regular regime curve were compared with the weighted mean diffusion coefficient of different values of the weighting parameter.

Another purpose of this study is to present a simple approximate calculation procedure for the desorption process in which the diffusion coefficient decreases with decreasing concentration. On the basis of the assumption that the transition between the two periods is smooth which has been confirmed by Schoeber[4], desorption curves were calculated by use of the weighted mean diffusion coefficient for both stages and compared with numerical solutions.

2. THEORY

Consider a desorption process of a slab , initially having a uniform concentration profile. As soon as the desorption process is started, the surface concentration reaches its equilibrium value. (hereafter the equilibrium value is taken to be 0 for desorption.) Furthermore, the temperature inside the slab is uniform and constant during the (de)sorption process. Under these conditions, the transport of water (or solvent) in the slab is given by Eq.(1) in the reduced form.

$$\frac{\partial m}{\partial \tau} = \frac{\partial}{\partial \phi}[D_r \frac{\partial m}{\partial \phi}] \tag{1}$$

Initial and boundary conditions are

$$m=1,\ 0\leq\phi\leq 1,\ \tau=0 \tag{2}$$
$$m=0,\ \phi=1 \quad \tau>0 \tag{3-1}$$
$$\partial m/\partial\phi=0,\ \phi=0 \quad \tau>0 \tag{3-2}$$

The dimensionless variables in Eq.(1) are shown in Table 1. Eq.(1) can also be used in a shrinking (or swelling) slab provided that a reference component mass centered coordinate (shrinking coordinate) instead of the usual volume fixed coordinate is employed[3,4,7,8]. Basically, the symbols in this report are the same as those employed by Liou[8].

Table 1. Dimensionless variables

	Stationary coordinate	Shrinking coordinate
ϕ	r/R	z/Z
m	ω_w/ω_{w0}	u/u_0
D_r	D/D_0	$D\rho_s^2/D_0\rho_{s0}^2$
τ	$\frac{tD_0}{R^2}$	$\frac{tD_0\rho_{s0}^2}{d_s{}^2R_s^2}$

Integration of Eq.(1) between $\phi=0$ and $\phi=1$ yields

$$F = \partial E/\partial\tau = -D_r \frac{\partial m}{\partial \phi}\Big|_{\phi=1} \qquad (4)$$

where $E = 1 - \int_0^1 m \, d\phi = 1 - \overline{m}$

The early stage of diffusion process, i.e., the penetration period, can be considered as diffusion in a semi-infinite medium as already stated. Then, Eq.(1) can be transformed into Eq.(5) with the aid of the Boltzmann's similarity variable $\eta = (1-\phi)/\sqrt{4\tau}$

$$\frac{\partial}{\partial\eta}[D_r \frac{\partial m}{\partial\eta}] + 2\eta \frac{\partial m}{\partial\eta} = 0 \qquad (5)$$

Integration of Eq.(5) between $\eta=0$ and $\eta=\infty$ gives Eq.(6)

$$D_r \frac{dm}{d\eta}\Big|_{\eta=0} = \int_0^1 2\eta dm = \text{constant} = \beta \qquad (6)$$

Eq.(8) can be obtained together with Eqs.(4),(6)-(7).

$$D_r \frac{\partial m}{\partial\eta}\Big|_{\eta=0} = -2\sqrt{\tau} D_r \frac{\partial m}{\partial\phi}\Big|_{\phi=1} \qquad (7)$$

$$\frac{dE}{d\tau} = -D_r \frac{\partial m}{\partial\phi}\Big|_{\phi=1} = \frac{\beta}{2\sqrt{\tau}} \qquad (8)$$

$$dE/d\sqrt{\tau} = \beta$$

From Eq.(8), it is found that there is a period in which E is proportional to $\sqrt{\tau}$ regardless of the concentration dependence of D [2,4].
If D is constant, that is $D_r=1$,

$$dE/d\sqrt{\tau} = \beta = \sqrt{4/\pi} \qquad (9\text{-}1)$$

$$D_r = \frac{\pi}{4}\beta^2 = 1 \qquad (9\text{-}2)$$

Eq.(1) was solved numerically by the Crank-Nicolson finite difference method. For the calculation of β by Eq.(5) we need the upper and lower bounds of β. The method proposed by Shampine[9] was employed in most cases. In some cases, the assumed upper and lower bounds were read off from the corresponding figure (Figs.7-10). Together with these bounds, Eq.(5) was calculated by the 4-th order Runge-Kutta-Gill method. The condition for the termination of the integration of Eq.(5) and the method of the check of the accuracy were essentially the same as those presented by Shampine[9].

3. RESULTS

3.1. Concentration dependent diffusion coefficients and weighted mean diffusion coefficients

In Table 2, concentration dependences of the diffusion coefficient employed in this study are summarized. The values of Dr for Eqs.(D-1) to (D-3) are taken to be unity at m=1.

In many studies, the diffusion coefficient is assumed to depend either linearly or exponentially on concentration[3]. However, since we are concerned with substances having a strong concentration dependence, another equation which represents such a concentration dependence is needed. The power-law dependence will be useful since some diffusivities of the above-mentioned substances can be approximated by this relation as shown by Liou[8]. Luyben, Liou and Bruin[10] presented a fictitious concentration dependent diffusion coefficient for food materials given by Eq.(D-4). These concentration dependences are shown in Figs. 2-4 as a function of a and in Fig. 5 as a function of ω_w.

Crank[3] reported that the initial rate of (de)sorption is closely related by the weighted mean diffusion coefficients given by Eq(10) for sorption and by Eq.(11) for desorption.

$$\overline{D_{r_p}} = p m'^{-p} \int_0^{m'} m^{p-1} D_r \, dm \qquad (10)$$

$$\overline{D_{r_q}} = q m'^{-q} \int_0^{m'} (m'-m)^{q-1} D_r \, dm \qquad (11)$$

If p=1 or q=1, then these equations reduce to an ordinary integral average diffusion coefficient.

$$\overline{D_r} = \frac{1}{m'} \int_0^{m'} D_r dm \qquad (11')$$

However, the applicability of Eqs.(10) and (11) to the concentration dependence shown by Eqs.(D-3) and (D-4) has not yet been verified.

This weighted mean diffusion coefficient seems to be useful since it contains only one weighting parameter. Fig. 6 shows the relations between D, $\overline{D_p}$ and $\overline{D_q}$, schematically. If D decrease with decreasing concentration, $\overline{D_p}$ gives a somewhat higher value than $\overline{D}$ and $\overline{D_q}$ lower than $\overline{D}$. In this study, the desorption is considered to be a process where the average concentration, m changes from 1 to 0 and the sorption is the reverse process. In the sorption process, u_0 or ω_{w0} means surface concentration.

In the 2nd, 3rd and 4th columns of Table 2., Several analytical solutions for $\overline{D_r}$, $\overline{D_{r_p}}$ and $\overline{D_{r_q}}$ are also shown.

Table 2. Concentration dependences of diffusion coefficients

Eq.	D or D_r	$\overline{D}$ or $\overline{D_r}$ 2)	$\overline{D_p}$ or $\overline{D_{rp}}$ 2)	$\overline{D_q}$ or $\overline{D_{rq}}$ 2)
(D-1)	$D_r=am+(1-a)$	$am/2+(1-a)$	$p/(p+1)am+(1-a)$	$am/(q+1)+(1-a)$
(D-2)	$D_r=\exp[a(m-1)]$	$(\exp[a(m-1)]-\exp(-a))/am$	NI 1)	NI
(D-3)	$D_r=m^a$	$m^a/(a+1)$	$p/(p+a)m^a$	$a!m^a/(q+1)(q+2)\cdots(q+a)$
(D-4)	$D=\exp[\frac{-34.2+138u}{1+6.74u}]$	NI	NI	NI

1) NI means numerical integration.
2) In the calculation of $\overline{D_r}$, $\overline{D_{rp}}$ and $\overline{D_{rq}}$, in Figs. 7-9, m is taken to be 1.

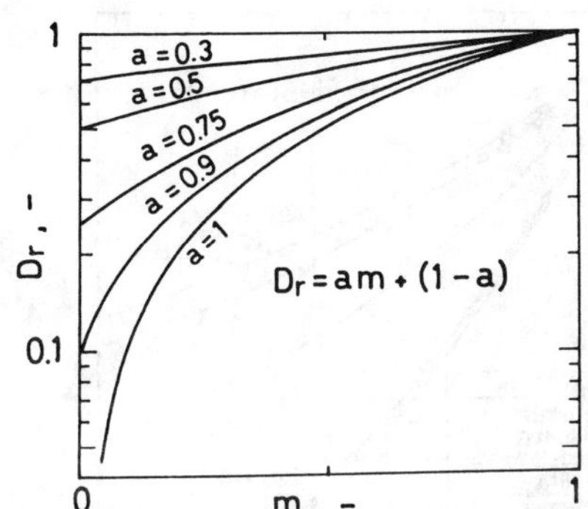

Fig.2 D_r vs. m for linear concentration dependence

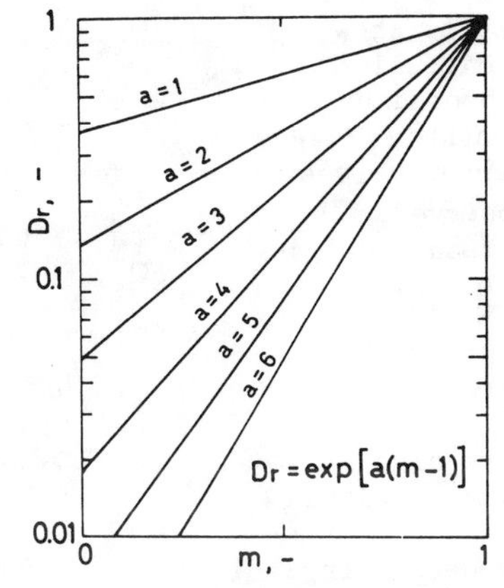

Fig. 3 D_r vs. m for exponential concentration dependence

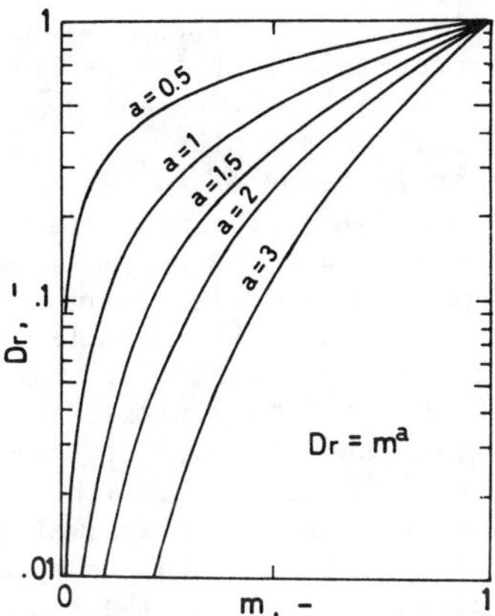

Fig.4 D_r vs. m for power-law concentration dependence

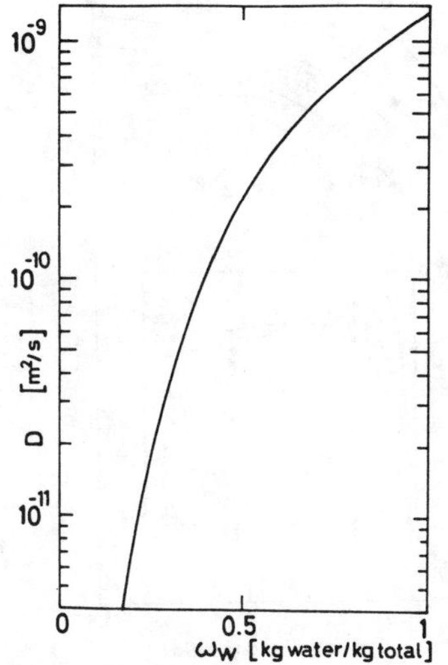

Fig.5 Concentration dependence of fictitious diffusion coefficient

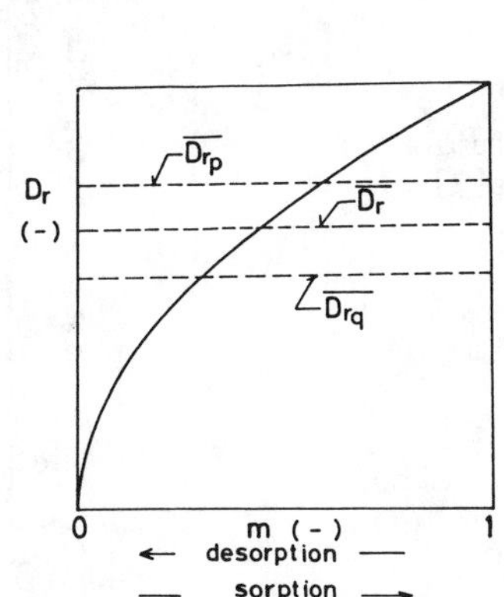

Fig.6 Schematic representation of relations between D, $\overline{D_p}$ and $\overline{D_q}$.

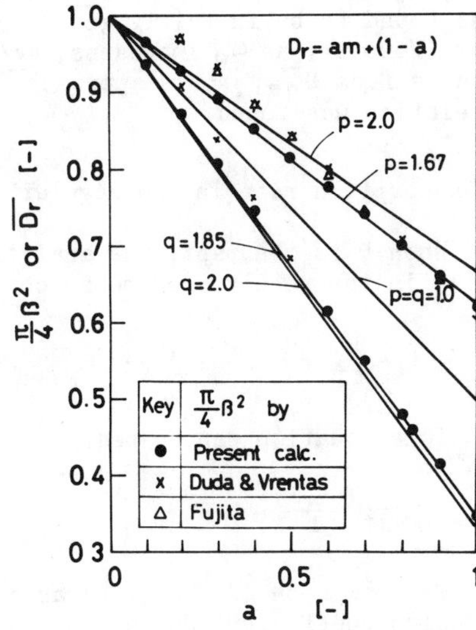

Fig. 7 Apparent diffusion coefficients vs. a

3.2. Initial (de)sorption rate ((de)sorption rate in the penetration period)

As already stated, in the penetration period E is proportional to $\sqrt{\tau}$. In the conventional (de)sorption method[3], an apparent diffusion coefficient, $\pi/4 \cdot \beta^2$ is obtained from the initial slope of a (de)sorption curve of this period and is approximated to the integral average diffusion coefficient as a first approximation. Therefore, We calculated $\pi/4 \cdot \beta^2$ for the diffusion coefficient expressed by Eqs.(D-1) - (D-3) for various values of a. The results are shown in Figs.7-9, together with the $\overline{D_r}$, $\overline{D_r}_p$ and $\overline{D_r}_q$ values as a function of a. For the values of p and q, the values proposed by Crank[3] (that is p=1.67 and q=1.85,) , p=1.0 and q=1.0 (the integral average diffusion coefficient) , and p=2.0 and q=2.0, are employed. In Figs.7-9, keys lie above the line for $\overline{D}_{p=q=1}$ are the results for sorption while those below that line are for desorption. The present calculated results show good agreements with the weighted mean diffusion coefficient, $\overline{D}_{p=1.67}$ and $\overline{D}_{q=1.85}$ proposed by Crank[3]. On the other hand, $\overline{D}$ deviates from $\pi/4 \cdot \beta^2$ with the increase of a. Although at larger values of a in Fig.8 a slight deviation is observed, the agreements between these weighted mean and $\pi/4 \cdot \beta^2$ in Figs. 7-9 are satisfactory.

Fig.10 shows the results for D by Eq.(D-4) in a shrinking coordinate, in which $\pi/4 \cdot \beta^2$ are plotted against the initial concentration for desorption and the surface concentration for sorption. Clearly, $\pi/4 \cdot \beta^2$ are also in good agreements with these weighted means . It is interesting to note that they are applicable not only to the monotonously decreasing function but also to the function which possesses an inflection point as shown by the inset of Fig.10.

We also examined the effect of shrinkage. $\beta = dE/(d\sqrt{t}/d_s R_s)$ calculated in the shrinking coordinate is transformed into $dE/(d\sqrt{t}/R_0)$ and compared with the weighted mean diffusion coefficient integrated against a mass fraction of water, ω_w. When ω_w is less than 0.1, $\pi/4 \cdot \beta^2$ was found to be in fairly good agreement with $\overline{D}_{q=1.85}$. However, as ω_w increases, $\pi/4 \cdot \beta^2$ deviates from $\overline{D}_{q=1.85}$ and approach $\overline{D}$. So no good correlation was found.

3.3 Desorption rate in the regular regime

When D is constant, the analytical solution of Eq.(1) in the regular regime is obtained as[3]

$$E = 1 - \frac{8}{\pi^2} \cdot \exp[- D_r \frac{\pi^2}{4} \tau] \qquad (12)$$

This equation can be rearranged to

$$D_r = \frac{4}{\pi^2} \cdot \frac{F}{(1-E)} \qquad (13)$$

We look upon $4/\pi^2 \cdot F/(1-E)$ as an apparent diffusion coefficient during the regular regime. We first calculated $4/\pi^2 F/(1-E)$ from numerically calculated desorption curves for the diffusion coefficient expressed by Eqs.(D-1)-(D-4). Then, the results were fitted with the weighted mean diffusion coefficient integrated between 0 and

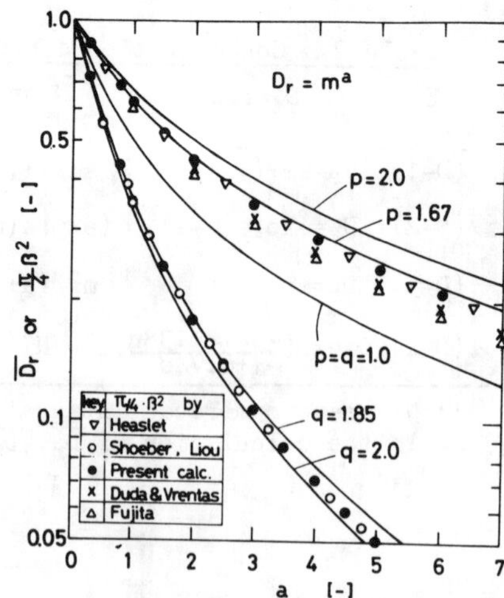

Fig. 8 Apparent diffusion coefficients vs. a

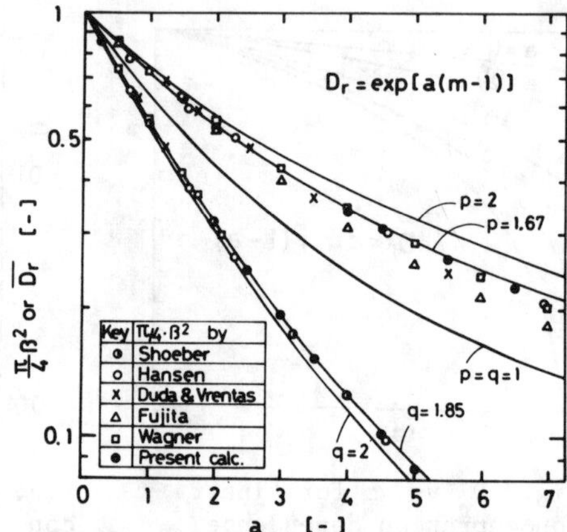

Fig. 9 Apparent diffusion coefficients vs. a

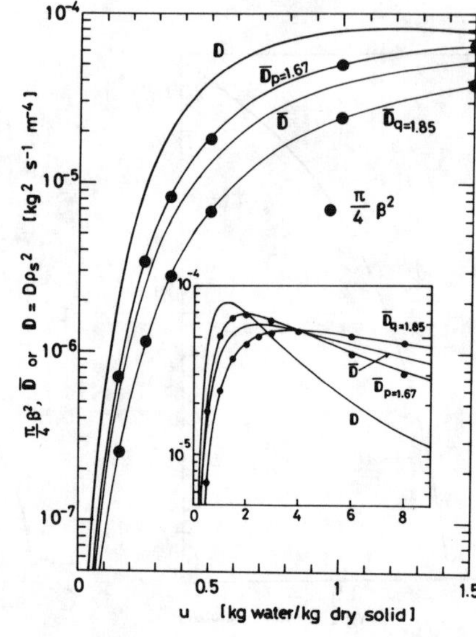

Fig.10 Apparent diffusion coefficients from the penetration period in shrinking coordinates vs. u.

the average concentration. It was found that the apparent diffusion coefficient, $4/\pi^2 \cdot F/(1-E)$ was closely approximated to the weighted mean diffusion coefficient given by Eq.(14) regardless of the type of the concentration dependence.

$$\overline{D_r}_p = p\overline{m}^{-p} \int_0^{\overline{m}} m^{p-1} D_r \, dm \qquad (14)$$

where p=1.4

Only the results for the shrinking coordinate system is shown in Fig. 11 and other results are not shown. In this figure, it is seen that $4/\pi^2 \cdot F/(1-E)$ for different initial concentrations merge into one master curve which is in good agreement with the weighted mean diffusion coefficient given by Eq.(14).

3.4. Determination of diffusion coefficient

The above findings indicate that the concentration dependent diffusion coefficient can be determined with accuracy from $\pi/4 \cdot \beta^2$ vs. u curve for the penetration period and $4/\pi^2 \cdot F/(1-E)$ vs. u curve for the regular regime by use of the shrinking coordinate. However, the use of the weighted mean diffusion coefficient for the determination of D from these curves is not easy especially for Eq.(11). A convenient way of using the weighted mean diffusion coefficient is the approximate of apparent diffusion coefficients by a polynomial equation as reported by Kishimoto and Enda[11]. We also applied this method to the desorption rates of polyacrylonitrile-dimethyl formamide system in the previous study[6].

When the apparent diffusion coefficient can be approximated to either linear or power-law functions, D can be readily determined since the analytical relation between D and $\overline{D_q}$ (or $\overline{D_p}$) are known as shown in the fourth and fifth columns of Table 2.

On the contrary, the weighted mean diffusion coefficient does not show an exponential dependence when the diffusion coefficient has that dependence. So when the apparent diffusion coefficient can be approximated to an exponential function, we first expand it into a polynomial equation. Then, it is inserted into Eq.(10) or Eq(11) and integrated by a procedure similar to that employed by Kishimoto and Enda[11]. The resulted equations are given by

$$D_r=[1 + (q+1)am + \frac{(q+1)(q+2)}{2!\,2!}(am)^2+\cdots+\frac{(q+1)\cdots(q+n)}{n!\,n!}(am)^n]e^{-a} \qquad (15)$$

$$D_r=[1 + \frac{p+1}{p}am + \frac{p+2}{2p}(am)^2+\cdots+\frac{1}{n!}\frac{p+n}{p}(am)^n]e^{-a} \qquad (16)$$

if apparent diffusion coefficients can be approximated by exp[a(m-1)].

Therefore, when the experimentally measured apparent diffusion coefficient can be approximated to one of the functions given above, the diffusion coefficient can be readily obtained.

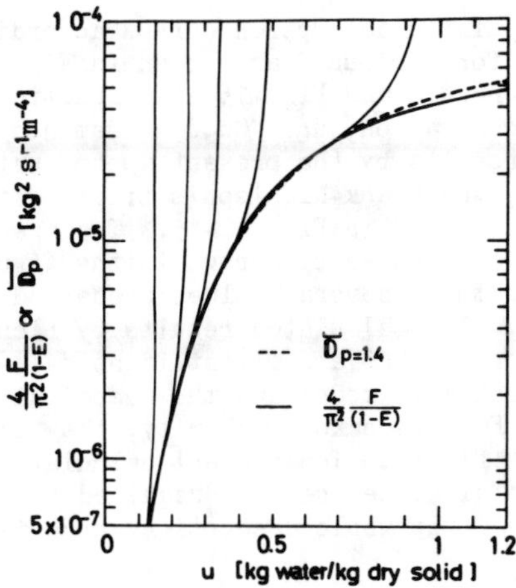

Fig.11 Apparent diffusion coefficients from the regular regime in shrinking coordinates for the diffusion coefficient given by Eq.(D-4)

3.5. A simple approximate calculation procedure of the desorption process

Schoeber[4] has shown that in the desorption process of a slab where the diffusion coefficient decreases with decreasing concentration the transition between the penetration period and the regular regime is smooth and therefore the desorption rates in the penetration period can be estimated from those in the regular regime. Later, based on this work Liou[8] proposed a simple calculation scheme for a desorption process in which D is expressed as $D_r=m^a$. In their method, first the transition point, from which the desorption rate in the penetration period is calculated, has to be determined.

On the other hand, we already know that the desorption rate in both periods can be represented by the weighted mean diffusion coefficient. So we attempted to calculate the desorption process based on the assumption that the transition between the two periods is smooth as described above. The calculation scheme is as follows. We first integrate Eq.(A-1) numerically (or in some cases analytically) by using Eq.(A-2) as F and at the same time F is also calculated according to Eq.(A-3). When the absolute difference between these two F values becomes minimum (usually less than 0.10), we determine this point as the transition point. After this point, Eq.(A-1) is calculated with Eq.(A-3).

$$dE/d\tau = F \qquad (A\text{-}1)$$

$$F = (2/\pi)\,\overline{D}_{q=1.85}/E \qquad (A\text{-}2)$$

$$F = (\pi^2/4)\overline{D}_{p=1.4}(1-E) \qquad (A\text{-}3)$$

The calculated desorption curves according to this scheme for various D are shown in Figs. 12-14 as a function of a and Fig. 15 as a function of the initial concentration, u_0. Good agreements between calculated results by the present approximate model and those by the Crank-Nicolson's finite difference method are observed in Figs.12-15. The transition points are represented by arrows in the figures and almost the same as several values presented by Schoeber[4]. The calculated results by Liou's scheme[8] for $D_r = m^a$ are almost identical with the present approximate model and the numerical solutions. For the sake of clarity, they are not shown in the figure. The present calculation scheme will be useful since the time required for the calculation of the whole desorption curve is highly shortened.

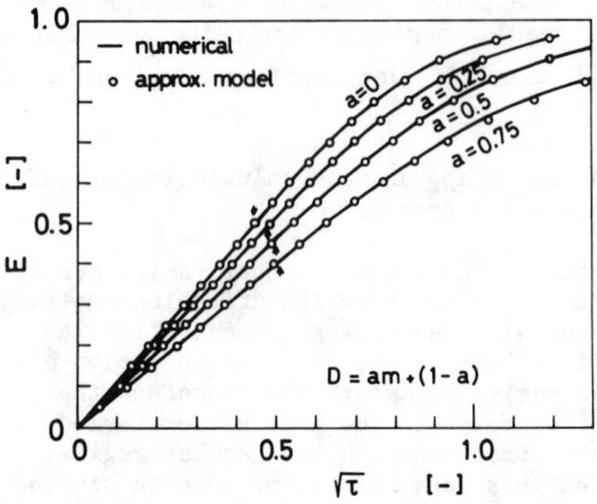

Fig. 12 Desorption curves for various values of a.

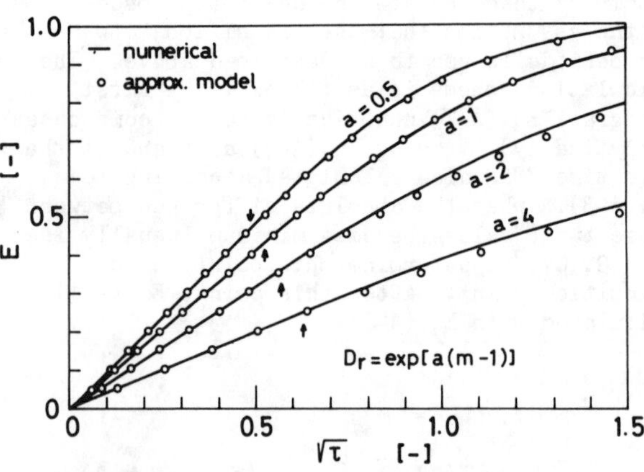

Fig. 13 Desorption curves for various values of a.

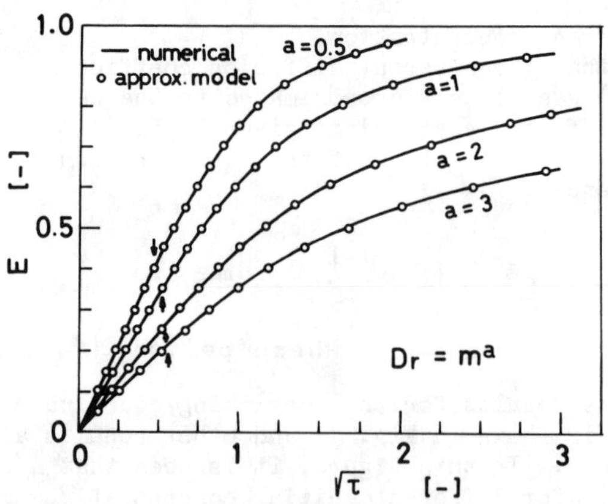

Fig. 14 Desorption curves for various values of a.

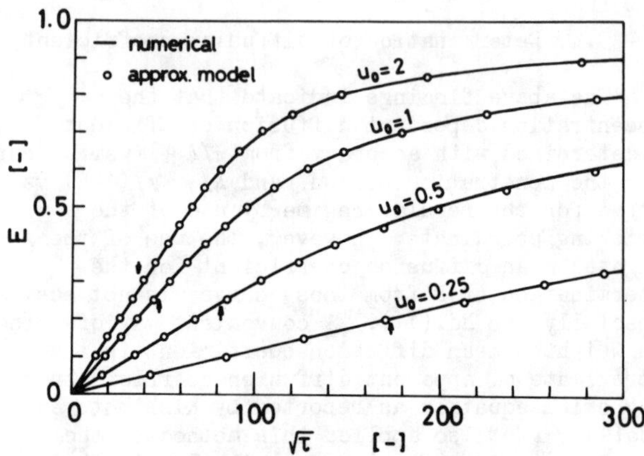

Fig. 15 Desorption curves for the fictitious diffusion coefficient.

4. DISCUSSION

Several $\pi/4\cdot\beta^2$ values derived from approximate solutions and numerically calculated results for the (de)sorption rate in the penetration period reported in the literature[12-15] are also shown in Figs.7-9. Their accuracy and range of applicability can be estimated from these figures. Fujita's method [12] is only applicable to diffusion coefficients which increase throughout the range of concentration involved and gives a considerable error. A similar tendency is observed for the method of Duda and Vrentas[3].

Luyben, Olieman and Bruin [16] determined diffusion coefficients of various materials such as glucose and skimmilk from the regular regime drying curve based on Schoeber's method[4]. Our method for the determination of diffusion coefficients from the regular regime curve is further simplified as compared with Schoeber's method[4].

The weighted mean diffusion coefficient proposed by Crank is found to be applicable to the power-law dependence and the fictitious diffusion coefficient for food materials given by Eq.(D-4) as well as the linear and exponential dependences.

It is interesting that a single type of weighted mean diffusion coefficient can be closely related to the desorption rate in the penetration period and another single one to that in the regular regime. Unfortunately, we could not find any mathematical proof for these relations.

The present method can also be applicable to the other experimental techniques such as a capillary-cell method as well as the drying experiments. A combination of the drying (desorption) experiment and the capillary-cell method permit the determination of concentration dependent diffusion coefficients over a wide range of concentration. We determined the diffusivity of sucrose-water system by the above experiment scheme according to the method described in this paper. The results are shown elsewhere [17].

5. CONCLUSIONS

In isothermal drying (desorption) of a slab with the constant surface concentration, 1) The desorption rates in the penetration period can be closely related to a single type of weighted mean diffusion coefficient and those in the regular regime to another single one.
2) A simple approximate calculation procedure for a desorption curve of slab is proposed in which the above two weighted mean diffusion coefficients are employed.

ACKNOWLEDGMENTS

The authors are grateful to Professor M. Okazaki, Dept. of Chemical Engineering, Kyoto Univ., Kyoto, for his helpful discussions and suggestions.

NOMENCLATURE

a	parameter in Eqs.(D-1)-(D-3)	-
D	diffusion coefficient	m^2/s
D	$= D\rho_s^2$ diffusion coefficient in shrinking coordinates	$kg^2\ s^{-1}m^{-4}$
d	density	kg/m^3
E	$=1-\overline{m}$	-
F	$=dE/d\tau$	-
m	dimensionless concentration	-
m'	upper bound of integral in weighted mean diffusion coefficient	-
p,q	weighting parameter in weighted mean diffusion coefficient	-
r	space coordinate	m
R	half-thickness of the slab	m
Rs	half-thickness of the slab in the absence of the solvent	m
u	concentration(kg solvent/kg dry solid)	
z	$\int_0^r \rho_s\, dr$	kg/m^2
Z	$\int_0^R \rho_s\, dr$	kg/m^2

Greek

β	$=dE/d\sqrt{\tau}$ initial desorption rate	-
η	Boltzmann's similarity variable	-
ϕ	dimensionless space coordinate	-
τ	dimensionless time	-
ω	mass fraction	-

Subscripts

r	reduced	
s	solid	
w	water	
0	value at t=0 or a reference value	-

Superscripts

—	average

REFERENCES

1. Bruin, S., and Luyben, K.Ch.A.M., in Advances in drying, ed. by Mujumdar, A.S., vol. 1, p.155, McGraw Hill (1979).
2. Vrentas, J.S., Duda, J.L. and Ni, Y.C., J. Polymer Sci., Polymer Phys. Ed., vol.15, 2039 (1977).
3. Crank, J., The Mathematics of Diffusion, Oxford U.P., Oxford (1973).
4. Schoeber, W.J.A.H., Regular regimes in sorption processes, Ph. D. thesis, Thechnical University Eindhoven, The Netherlands (1976).
5. Okazaki,M., Shioda,K., Masuda,K. and Toei,R., J. Chem. Eng. Japan, vol.7.99 (1974).
6. Sano,Y., Yamamoto,S., Watanabe,K. and Kimura,T., Kagaku Kogaku Ronbunshu, vol.9, 1 (1983).
7. Duda, J.L. and Vrentas, J.S., AIChE J., vol.17, 464 (1971).
8. Liou, J.K. and Bruin, S., Int. J. Heat Mass Transfer, vol. 25, 1209 (1982).
8' Liou,J.K., Ph.D.Thesis, Agricultural Univ. Wageningen, The Netherlands (1982).
9. Shampine, L.F., Q. Appl. Math., vol. 30, 441, vol.31,287 (1973).
10. Luyben, K.Ch.A.M., Liou, J.K. and Bruin, S., Biotech. Bioeng., vol. 24, 533(1982).
11. Kishimoto, A. and Enda, Y., J. Polym. Sci. A. vol. 1, 1799 (1963).
12. Fujita, H. and Kishimoto, A., Text. Res. J., vol. 22, 94 (1952).
13. Hansen,C.M., Ind.Eng.Chem.Fundamentals, vol.6,609(1967).
14. Heaslet, M.A. and Alksne, A., J. SIAM, vol.9, 584 (1961).
15. Wagner,C., J.Metals , vol.4, 91(1952).
16. Luyben,K.Ch.A.M., Olieman,J.J. and Bruin, S., Proc. of 2nd Int. Symp. Drying,(1980).
17. Yamamoto,S., Agawa,M., Nakano,H. and Sano,Y., Proc. of 4th Int. Symp. Drying, Kyoto (1984).

EFFECT OF DRYING CONDITION ON STRUCTURE
OF DRIED PRODUCT

Kazuyoshi Mochizuki, Hideo Shinagawa and Yuji Kawamura

Department of Chemical Engineering, Hiroshima University
Higashi-Hiroshima, 724 JAPAN

ABSTRACT

Radiation-convection type of drying experiments have been carried out for wetted layer of SiC fine powder. Variation of the appearance of surface in the cource of drying was observed with the naked eye, and the coagulation characteristics of the particles on the surface of dried product (including freeze-dried one) have been observed by scanning electron microscope. On the basis of observation, the effects of drying condition on drying characteristics and structure of dried product were examined.

1. INTRODUCTION

The knowledges relating to coagulation of particles during drying process for wetted layer of fine powder with sub-micron size are important to ceramic or semi-conductor industry. In spite of its practical importance, to our knowledge, there have been reported yet few works concerning to the problem.

As the first step to clarify the mechanism of the coagulation mentioned above, the present work was undertaken to examine experimentally the effect of drying condition(dryind rate and initial water content) on structure of drying material during and after radiation-convection type of drying. Taking it into consideration that little shrinkage occurs in the freeze-dried product, and freeze-drying is pointed out to be an useful process to the production of ceramic raw material [1 2], the structure of freeze-dried product was also examined for the purpose of comparison with that by radiation-convection type of drying (hereafter, we abbreviate the term to radiation-drying).

2. EXPERIMENTAL APPARATUS AND PROCEDURE

2.1. Radiation-Drying

The drying apparatus used is schematically shown in Fig. 1. It consists of wooden rectangular duct (79 mm high, 105 mm wide) covered with polystyrene insulating mat. Heat for drying was supplied by an infrared ray lamp. Drying rate was varied by controlling voltage to the lamp. Weight loss of the sample was measured by an automatic chemical balance (sensitivity 1 mg). The temperature in the sample was measured with three copper-

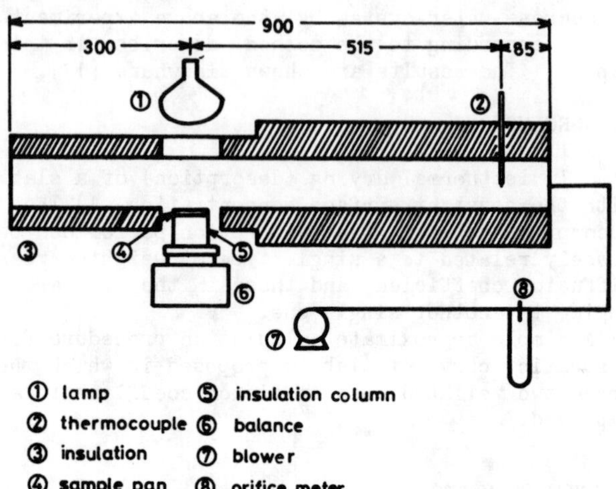

Fig. 1 Schematic diagram of radiation-dryer

constantan thermocouples of 100 μm diameter. Air at room temperature was flown to remove moisture. Sample pan is 58 mm in diameter, 10 mm high.

Prior to drying run, electric current at constant voltage was passed to the lamp and air was flown for over 2 hours, then the sample pan filled with sample was put on the insulation column. The weight of sample was measured at every prescribed time, simultaneously the appearance of sample surface was observed by the naked eye. Drying run was stopped when the weight loss rate decreased to less than 10 mg/hr.

Air flow velocity in the duct was 0.46 m/s and the air temperature was 4 °C∿11 °C. The difference between weight loss of water measured by the balance equipped in the apparatus and that obtained from the weight of sample before and after drying run was within 1 %.

2.2. Vacuum Freeze-Drying

Apparatus for vacuum freeze-drying consists of evaporation chamber of 200 mm in diameter and evacuation tube of 2 inch in diameter. Heat necessary for sublimation was supplied by passing hot water in the copper tube and jackets equipped with the evaporation chamber. The sample was frozen on the plane cooled by circulating coolant kept at low constant temperature. The sample pan was the same one as used in radiation-drying.

The sample pan was hung on the hook in the evaporation chamber after freezing the sample, then

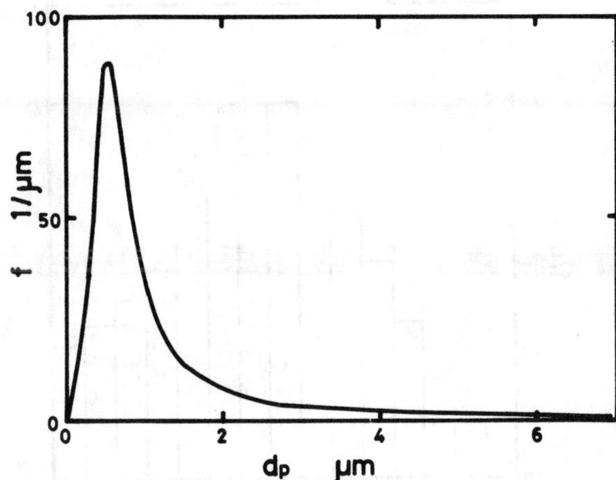

Fig. 2 Particle size distribution of sample powder

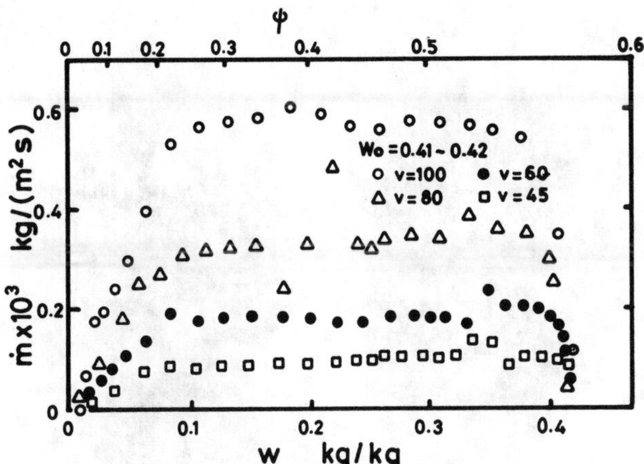

Fig. 3 Drying characteristic curve for radiation-drying

vacuum freeze-drying run was carried out till the weight loss of the sample became negligibly small.

The temperature of coolant was varied to -10 °C, -50 °C, temperature of evaporation chamber was kept at 58 °C.

The surfaces of dried product by radiation-drying and freeze-drying was observed by scanning electron microscopy (hereafter, we abbreviate the term to SEM). The dried product was ball-milled for 3 min., and was classified by screen to examine the degree of coagulation of particles.

SiC powder was used as sample because it is widely used in ceramic industry and has a larger size distribution frequency in sub-micron zone [3].

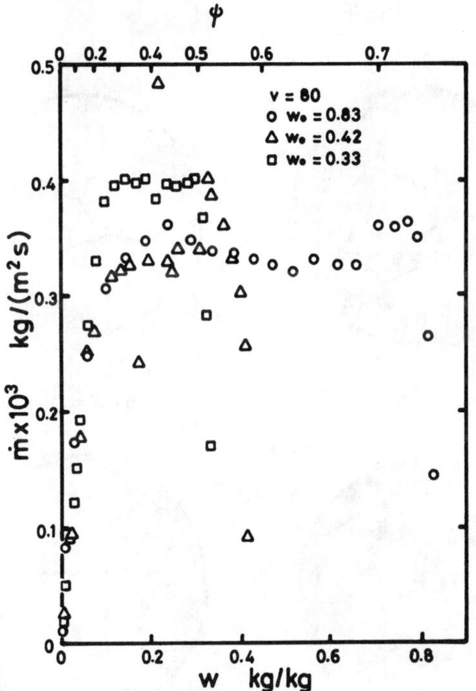

Fig. 4 Drying characteristic curve for radiation-drying

3. EXPERIMENTAL RESULT AND DISCUSSION

3.1. Drying Characteristics

Fig. 2 shows size frequency distribution of sample powder measured by the centrifuge-sedimentation method. The average particle size was a mode diameter of about 0.5 μm.

The drying characteristic curves are shown in Figs. 3 and 4. From these figures it seems that each curve can be separated roughly into a pre-heating period, a constant rate period and a falling rate period. The scatters of data points may be caused by variations of air temperature and convection condition resulting from unavoidable movement of balance and shrinkage of sample in the drying run. From the naked eye observation, it was observed that the cracks associated with shrinkage appear at first around the brim of sample pan in the pre-heating period, subsequently in the inner surface of the sample in the middle stage of constant rate period, and the shrinkage and cracks was almost completed by the later middle stage of drying run. It might be supposed that these shrinkage and cracks does not influence significantly on drying rate.

3.2. Coagulation Degree of Dried Product

<u>Macroscopic Observation</u>. Photographs of roughly crushed appearance of dried product are shown in Fig. 5, (a) in this figure shows the freeze-dried product frozen at θ_c=-10 °C, (b) shows the radiation-dried product in the case of V=60 V. Initial water content w_0 of both samples was 0.42 kg/kg. The coagulation degree of sample of these products were felt dissimilar intuitively, that is, the freeze-dried product was slipped out with 3 mm to 5 mm layered block, and then the blocks were broken into granules. It was suggested that the product was broken along the direction which ice crystals grew. Conclusively, it was evident that the freeze-dried product was very fragile and could be powdered easily.

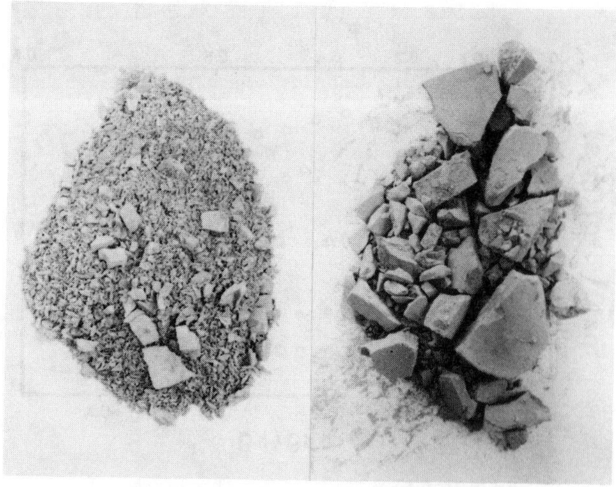

(a) freeze-drying (b) radiation-drying

Fig. 5 Roughly crushed appearance of dried product

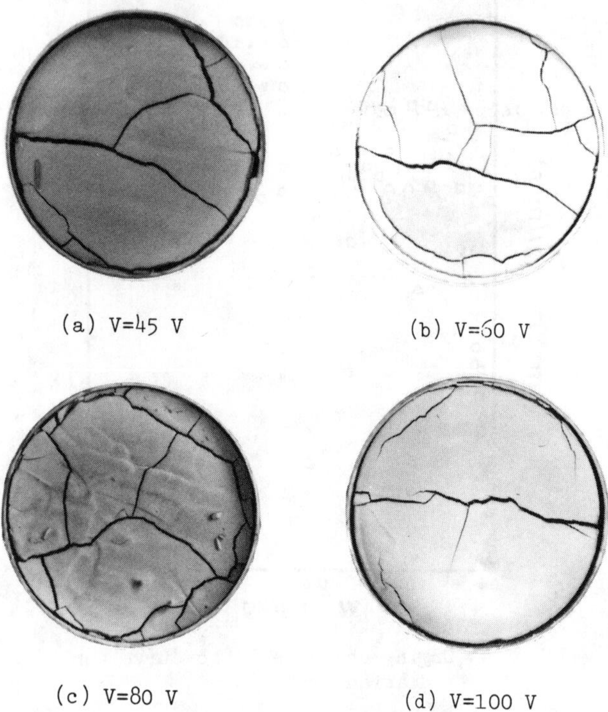

(a) V=45 V (b) V=60 V

(c) V=80 V (d) V=100 V

Fig. 7 Surface of dried product (w_0=0.42 kg/kg)

The radiation-dried product shown in Fig. 5-(b) was firmer than the freeze-dried one and could not be broken without a certain degree of force. It was observed that the radiation-dried product was broken isotropically. It can be pointed out that the difference of crushing characteristic is attributed to coagulation degree.

Both dried products was crushed under the same condition and particle size distribution was measured using a classifing screen test, the results

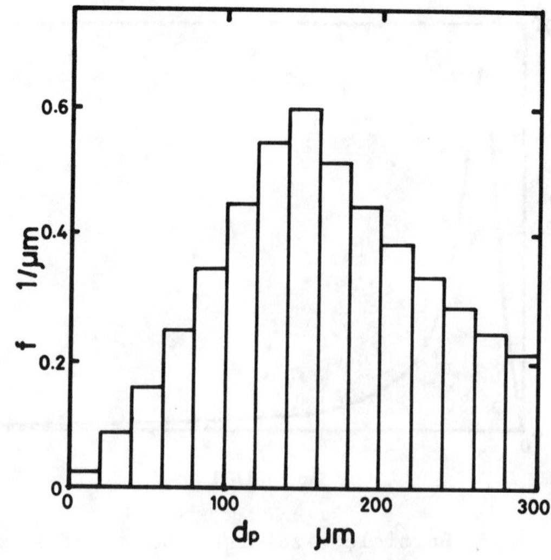

(a) freeze-drying

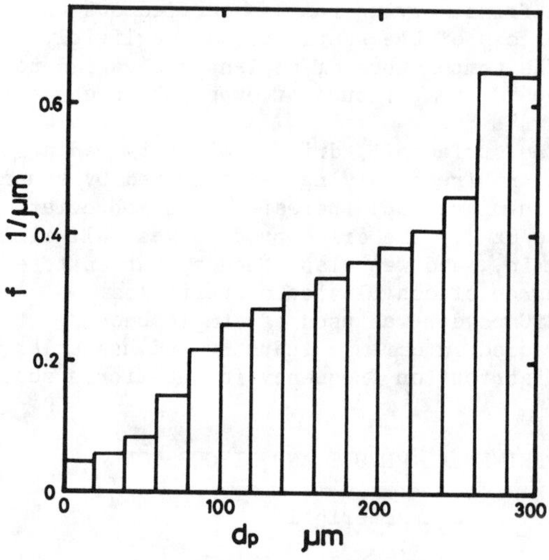

(b) radiation-drying

Fig. 6 Size distribution for ball-milled particles of dried product

are shown in Fig. 6. (a) in Fig. 6 shows the result for the freeze-dried sample of w_0=0.41 kg/kg and θ_c=-50 °C. The distribution profile has approximately a mode diameter of 150 µm, and was nearly symmetrical with regard to centering mode diameter. That in the case of the radiation-dried product under w_0=0.42 kg/kg, V=80 V was shown in Fig. 6-(b). The distribution profile has a mode diameter of about 270 µm and shifts to larger size than that of (a). It can be recognized that the result mentioned above attributed to **coagulation degree** of radiation-dried product than that of freeze-dried product.

The variation of surface appearance with radiation drying rate for the sample of w_0=0.42 kg/kg is shown in Fig. 7. The state of surface seems to be independent of drying rate. The variation with initial water content is shown in Fig. 8. The increase of initial water content caused marked

(a) freeze-drying
w_0=0.42 kg/kg

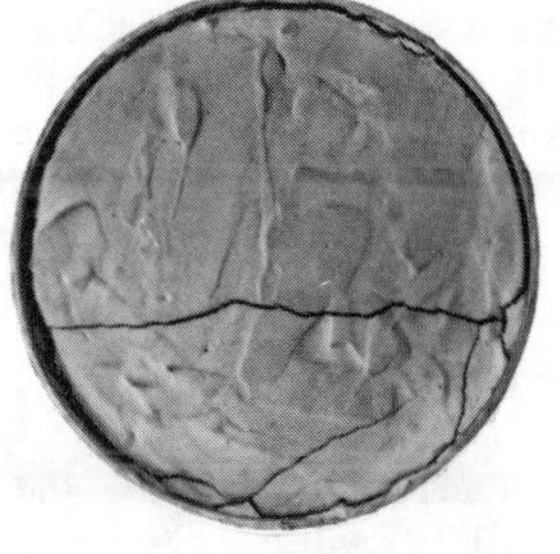

(b) radiation-drying
w_0=0.33 kg/kg

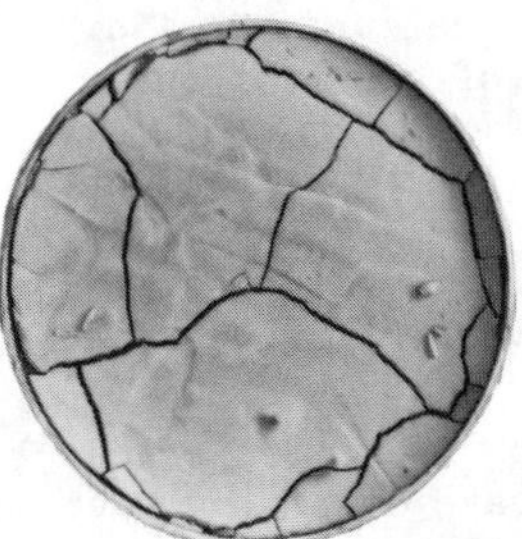

(c) radiation-drying
w_0=0.42 kg/kg

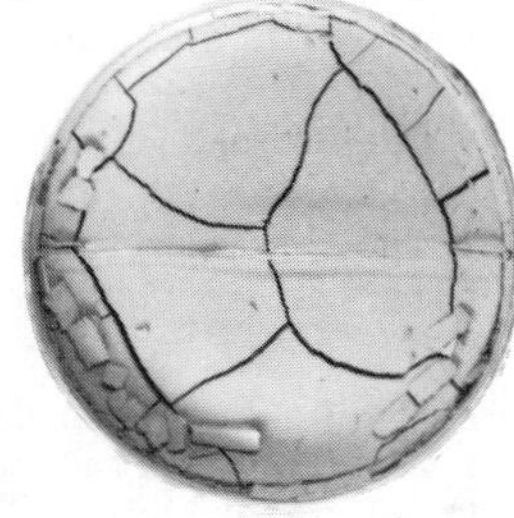

(d) radiation-drying
w_0=0.83 kg/kg

Fig. 8 Surface of dried product (V=80 V)

(a) freeze-drying

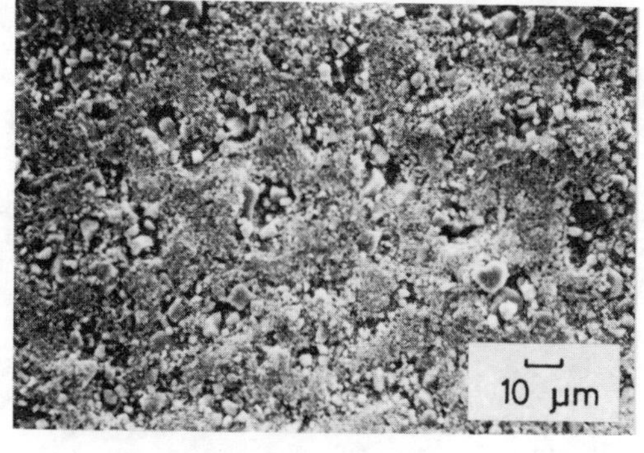

(b) radiation-drying

Fig. 9 Electron photomicrograph of surface for dried product

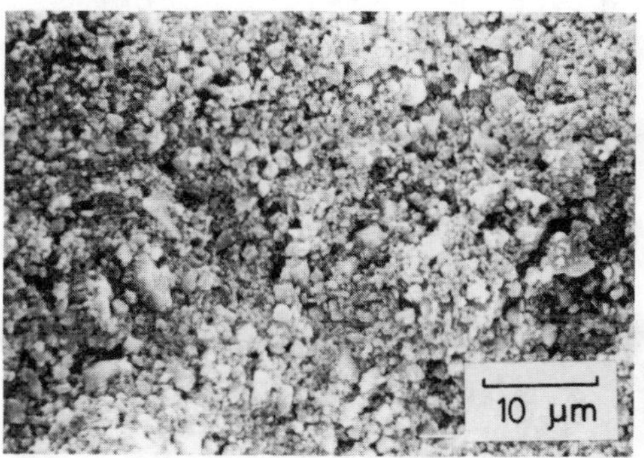

(a) V=45 V

(b) V=80 V

(c) V=100 V

Fig. 10 Electron photomicrograph of surface for radiation-dried product (w_0=0.42 kg/kg)

cracks on the surface and significant shrinkage. The volume of dried product in the case of w_0=0.83 kg/kg was observed to be 2/3 of initial one.

Photograph of freeze-dried product under the condition of w_0=0.42 kg/kg, θ_c=-50 °C was illustrated in Fig. 8-(a). The width of crack is narrow to a great extent. The cracks took place during freezing, therefore, the type of crack is different substantially from one in the case of radiation-drying.

Surface Structure. Electron photomicrograph of the surface by SEM is shown in Fig. 9. Cracks in the surface for the freeze-dried product (w_0=

(a) w_0=0.33 kg/kg

(b) w_0=0.42 kg/kg

(c) w_0=0.83 kg/kg

Fig. 11 Electron photomicrograph of surface for radiation-dried product (V=80 V)

0.42 kg/kg, θ_c=-50 °C) is narrow and net-worked.

The surface structure of radiation dried product (w_0=0.42, V=100 V) is shown in Fig. 9-(b), it can be observed that holes exist in the surface. These holes are generated around the larger particles, in the neighborhood of the holes there exist finer particles. It is obvious that the difference between surface structure is attributed to coagulation mechanism.

The variation of surface structure with drying rate in the case of w_0=0.42 kg/kg is shown in Fig. 10. The lower drying rate, the more closely the particles in the surface is compacted.

The variation of surface structure with the initial water content in the case of V=80 V is shown in Fig. 11. It can be observed that the increase of initial water content causes the coarser structure, especially in the case of w_0=0.83 kg/kg the existence of larger particles is significant.

4. CONCLUSION

Radiation-drying and freeze-drying for wetted fine SiC powder layer were carried out. From the examination of effects of drying condition on the drying characteristics and structure of dried product, the following conclusion was drawn.

1) The cracks associated with shrinkage of drying layer almost appeared on the surface of drying layer by later middle stage of constant rate period.

2) The cracks might not influence significantly on drying rate.

3) The coagulation degree of radiation-dried product is larger than that of vacuum freeze-dried product.

4) The coagulation degree of radiation-dried product increases with the decrease of drying rate and initial water content of sample.

NOMENCLATURE

d_p	particle diameter	μm
f	distribution frequency	1/μm
V	voltage	V
w_0	initial water content	kg/kg

Greek letter

θ_c	temperature of coolant	°C

REFERENCES

1. Roehig, F.K. and Wright, T.R., J. Vac. Sci. Technol., vol. 9, 1368 (1972).
2. Wheat, T.A., J. Can. Ceramic Soc., vol. 46, 11 (1977).
3. Enomoto, R., Ceramics (in Japanese), vol. 17, 828 (1982).

BULK THERMAL DIFFUSIVITY MEASUREMENT OF GRAIN RICE BY NEW FAST METHOD

Sei-ichi Oshita

Department of Agricultural Engineering, Mie University
Tsu, 514 JAPAN

ABSTRACT

The fast method of measuring thermal diffusivity was proposed. It is based on the newly developed technique to estimate temperature distribution of the material at desired time without the knowledge of initial temperature distribution.

This method can save much time for determining thermal diffusivity mainly because of unnecessity of considering the initial condition.

Bulk thermal diffusivities of rough rice and hulled rice were determined by use of this method. From the results, it was found that the effect of moisture on the bulk thermal diffusivity depends on the temperature.

1. INTRODUCTION

Bulk thermal diffusivity of grain rice is one of the essential properties when studying grain drying and storage. Although bulk thermal diffusivity of rough rice was previously determined indirectly by Wratten et al. [1] and Morita and Singh [2] as a function of moisture content, the range of temperatures used was not mentioned.

For the effect of temperature, Hosokawa and Masumoto [3] determined bulk thermal conductivity of rough rice as a function of temperature, but the range of temperatures was limited. Thermal properties of other grains were investigated by several researchers (Kazarian and Hall [4], for example), however, it looks quite difficult to find the study on the effect of temperature except the work done by Moysey et al. [5]

As described above, little published information is available on the bulk thermal diffusivity of grain rice, especially, on the relationship between temperature and bulk thermal diffusivity.

One of the major causes for this is that most of the measurement methods reported in the past were not suited for studying the change of thermal diffusivity with temperature. In addition, difficulty in arranging packed condition will possibly be another cause. Among the factors affecting on the magnitude of bulk thermal diffusivity, the posture of each kernel under packed condition can not be reproducible. For eliminating the effect of this packed condition on the measured values, the statistical data processing is quitely requested for a lot of values obtained from the experiments.

The objectives of the investigation reported in this paper were to:
1) Develop the method of measuring bulk thermal diffusivity suited for studying the change of thermal diffusivity with temperature and for obtaining many values within a given time.
2) Determine bulk thermal diffusivity over a wide range of temperature and moisture conditions.

2. ESTIMATION TECHNIQUE OF TEMPERATURE DISTRIBUTION

2.1. Theoretical Consideration

The outline of the estimation technique of temperature distribution of the material on which the proposed method of measurement in this paper is based can be described below.

Fig. 1 illustrates temperature variations at boundary surfaces of an infinite slab schematically, that is, the curve (A)-(C) for the variation at the surface at $x=0$ and the curve (B)-(D) at the other insulated surface at $x=\delta$ when the surface at $x=0$ is heated periodically with a period of t_0. In order to estimate the temperature distribution of Z at $t=t_0$, both boundary surface temperatures are supposed to be varied along the curves of (A)-(A') and (B)-(B'), while each of them is symmetric with respect to the line of $t=t_0$ shown in Fig. 1.

Since these temperatures are periodic functions with a period of $2t_0$, the temperature within the slab can be considered to be varied as quasi-steady-state. Therefore, when the analysis is performed under these conditions, those temperature distributions can be obtained, which are governed only by the boundary temperatures. In this case, temperature distribution of Z' at $t=t_0$ agrees well with that of Z if a period of t_0 is selected to be a suitable length.

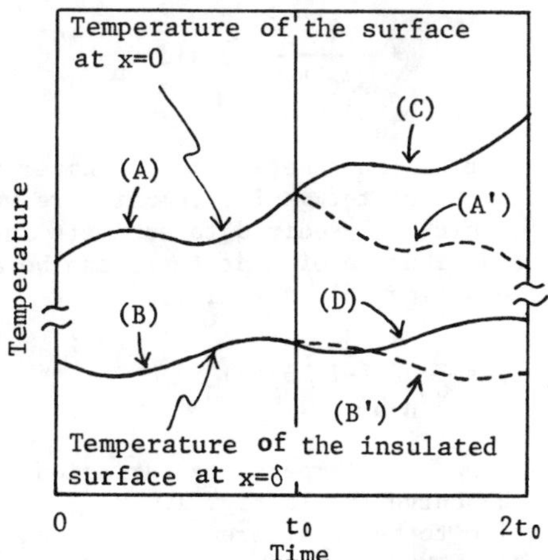

Fig. 1 Illustration of temperature variations

This idea can be validated theoretically by considering the following problem.

First, let us obtain the original temperature distribution of Z at $t=t_0$. The heat conduction equation can be given as:

$$\frac{\partial\theta(x,t)}{\partial t} = a\frac{\partial^2\theta(x,t)}{\partial x^2} \tag{1}$$

The initial and boundary conditions are

$$\theta(x,0) = \theta_i \tag{2}$$

$$\theta(0,t) = \cos\omega t, \quad \omega = 2\pi/t_0 \tag{3}$$

$$\left.\frac{\partial\theta(x,t)}{\partial x}\right|_{x=\delta} = 0 \tag{4}$$

where t_0 = period; δ = thickness of an infinite slab.

Introducing the transformation as:

$$\theta_n(t) = \int_0^{\delta} \sin C_n x\cdot\theta(x,t)\,dx \tag{5}$$

where

$$C_n = \frac{\pi}{\delta}\left(n+\frac{1}{2}\right), \quad n=0,1,2,---$$

the following equation can be obtained.

$$\theta(x,t) = \frac{2}{\delta}\sum_{n=0}^{\infty}\theta_n(t)\sin C_n x \tag{6}$$

where

$$\theta_n(t) = \left(\frac{\theta_i}{C_n} - \frac{a^2C_n^{\ 3}}{a^2C_n^{\ 4}+\omega^2}\right)\mathrm{EXP}(-aC_n^{\ 2}t) + \frac{a^2C_n^{\ 3}}{a^2C_n^{\ 4}+\omega^2}\left(\cos\omega t + \frac{\omega}{aC_n^{\ 2}}\sin\omega t\right) \tag{7}$$

The temperature distribution of Z at $t=t_0$ can be derived from equation (6) as follows:

$$\theta(x,t_0) = \frac{2}{\delta}\sum_{n=0}^{\infty}\left\{\left(\frac{\theta_i}{C_n} - \frac{a^2C_n^{\ 3}}{a^2C_n^{\ 4}+\omega^2}\right)\mathrm{EXP}(-aC_n^{\ 2}t_0) + \frac{a^2C_n^{\ 3}}{a^2C_n^{\ 4}+\omega^2}\right\}\sin C_n x \tag{8}$$

Second, let us find the estimated temperature distribution of Z' at $t=t_0$. The temperature variation at $x=\delta$, which is required to estimate the temperature distribution of Z at $t=t_0$, can be also derived from equation (6) as follows:

$$\theta(\delta,t) = \frac{2}{\delta}\sum_{n=0}^{\infty}(-1)^n\theta_n(t) \tag{9}$$

Equation (9) shows the temperature variation corresponding to the curve (B) in Fig. 1.

The problem to be considered for estimating temperature distribution can be expressed as:

$$\frac{\partial\theta(x,t)}{\partial t} = a\frac{\partial^2\theta(x,t)}{\partial x^2} \tag{10}$$

with boundary conditions

$$\theta(0,t) = \cos\omega t \tag{11}$$

$$\theta(\delta,t) = \frac{2}{\delta}\sum_{n=0}^{\infty}(-1)^n g_n(t) \tag{12}$$

where, from equation (7),

$$g_n(t) = \theta_n(t-2mt_0)$$

$$\text{for } 2mt_0 \leqq t \leqq (2m+1)t_0, \quad m=0,1,2,---$$

and

$$g_n(t) = \theta_n\{2(m+1)t_0-t\}$$

$$\text{for } (2m+1)t_0 \leqq t \leqq (2m+2)t_0, \quad m=0,1,2,---$$

and the initial condition diminishes as $t \to \infty$.

Here, introducing the transformation as:

$$\theta_k(t) = \int_0^{\delta}\sin\frac{k\pi}{\delta}x\cdot\theta(x,t)\,dx, \quad k=1,2,3,--- \tag{13}$$

the temperature distribution of Z' at $t=t_0$ is obtained as follows:

$$\theta(x,t_0) = \frac{2}{\delta}\sum_{k=1}^{\infty}\left\{(-1)^{k+1}\frac{2\pi k}{\delta^2}\sum_{n=0}^{\infty}(-1)^n\frac{1}{\left(\frac{k\pi}{\delta}\right)^2-C_n^{\ 2}}\cdot\left(\frac{\theta_i}{C_n}-\frac{a^2C_n^{\ 3}}{a^2C_n^{\ 4}+\omega^2}\right)\mathrm{EXP}(-aC_n^{\ 2}t_0) + \frac{a^2\left(\frac{k\pi}{\delta}\right)^3}{a^2\left(\frac{k\pi}{\delta}\right)^4+\omega^2}\right\}\sin\frac{k\pi}{\delta}x \tag{14}$$

Equation (14) is based upon the assumption of

$$\mathrm{EXP}(-k^2\pi^2F_0) \simeq 0, \quad F = \frac{at_0}{\delta^2}, \quad k=1,2,3,--- \tag{15}$$

Considering the equation (15),

$$\mathrm{EXP}(-k^2\pi^2F_0) < 1/1000 \tag{16}$$

can be derived when

$$F_0 \geqq 0.7 \tag{17}$$

When equation (17) is satisfied, equation (14) showing the estimated temperature distribution of Z' becomes nearly equal to equation (8) showing the original temperature distribution of Z.

2.2. Numerical Consideration

Although the analytical estimation of temperature distribution was done only on the case that the surface temperature at $x=0$ shows a cosine wave in the previous paragraph, this estimation technique is valid for the case that the surface temperature shows the other complicated wave. In that case, however, it is quite difficult to obtain the analytical solution for estimating temperature distribution.

Then the numerical estimation was tried to be conducted by using an implicit finite-difference scheme of the weighted average approximation type.

First, the simulation for obtaining the original temperatures was conducted. Analysis conditions for simulation is shown in Table 1.

Table 1 Analysis conditions for simulation

Heat conduction equation	$\dfrac{\partial\theta(x,t)}{\partial t} = a\dfrac{\partial^2\theta(x,t)}{\partial x^2}$
Initial condition	$\theta(x,0) = 0,\ 0\leq x\leq\delta$
Boundary conditions	$\theta(0,t) = 3\ ^\circ C,\ 2mt_0\leq t\leq(2m+1)t_0$ $\theta(0,t) = 0\ ^\circ C,\ (2m+1)t_0\leq t\leq(2m+2)t_0$ $\left.\dfrac{\partial\theta(x,t)}{\partial x}\right\|_{x=\delta} = 0$

(Note) m = 0,1,2,---

Finite difference expression for heat conduction equation can be as:

$$\theta_{i,j+1}-\theta_{i,j} = R\{\zeta(\theta_{i+1,j+1}-2\theta_{i,j+1}+\theta_{i-1,j+1}) +(1-\zeta)(\theta_{i+1,j}-2\theta_{i,j}+\theta_{i-1,j})\} \quad (18)$$

where

$$R = \frac{a\Delta t}{(\Delta x)^2}\ ;\ \zeta = \text{weight}$$

and for the boundary condition at $x=\delta$ can be

$$\frac{\theta_{n+1,j}-\theta_{n-1,j}}{2\Delta x} = 0,\quad \frac{\theta_{n+1,j+1}-\theta_{n-1,j+1}}{2\Delta x} = 0 \quad (19)$$

Here the subscript i designates the point with x-coordinate and the subscript j designates the point with time-coordinate, and the subscript n designates the point with x-coordinate at the insulated surface. The weight ζ was determined from the equation:

$$\zeta = 0.5178 - 0.08754/R \quad (20)$$

Table 2 Numerical analysis conditions

Thermal diffusivity : a	$1.20 \times 10^{-7}\ m^2/s$
Thickness : δ	6.0 mm
Period : t_0	144 s
Number of divisions of δ	12
Number of divisions of t_0	40

which was previously proposed by the auther [6].

Numerical analysis conditions are shown in Table 2. For the purpose of investigating the case that the surface of an slab is heated periodically and the temperature of which rises gradually, the temperature of the internal surface at x=1.5 mm obtained from the results of the simulation can be assumed to be the surface temperature of the slab to be considered.

Thus, Fig. 2 shows temperature variations of the slab from the surface at x=0 indicated with (A)-(C) to the other insulated surface at x=4.5 mm indicated with (B)-(D) and the original temperature distribution of Z at $t=t_0$.

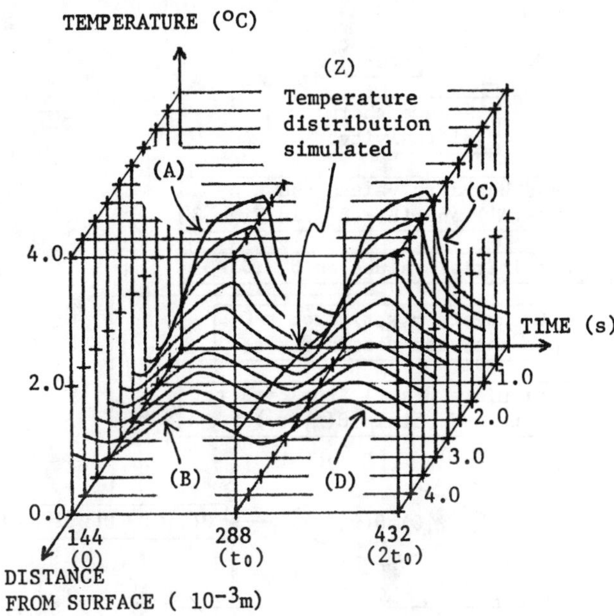

Fig. 2 Original temperature distribution

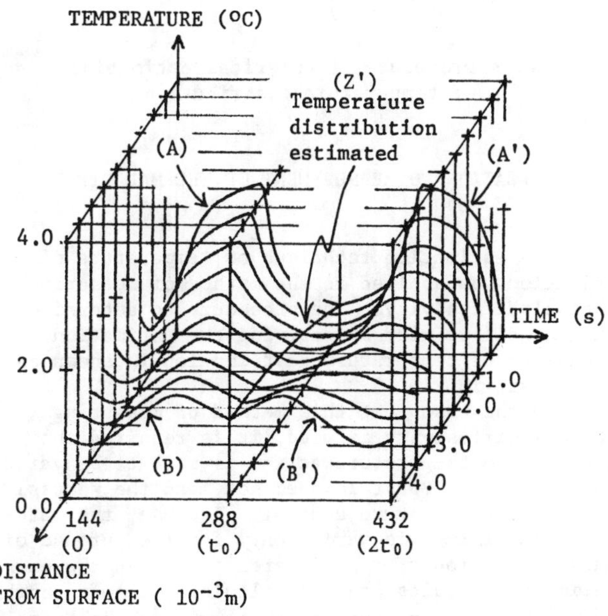

Fig. 3 Estimated temperature distribution

These are corresponding to the temperatures of the internal surface at x=1.5 mm to the ones of the insulated surface at x=6.0 mm which were obtained from the results of the simulation.

Fig. 3 shows the temperature distribution of Z' obtained from the numerical estimation of Z by using both boundary temperatures indicated as (A)-(A') and (B)-(B') in Fig. 3. From this results, it was observed that the estimated temperature distribution of Z' agrees well with the original temperature distribution of Z.

The procedure of this numerical estimation of temperature distribution of the material is shown as a flow chart in Fig. 4.

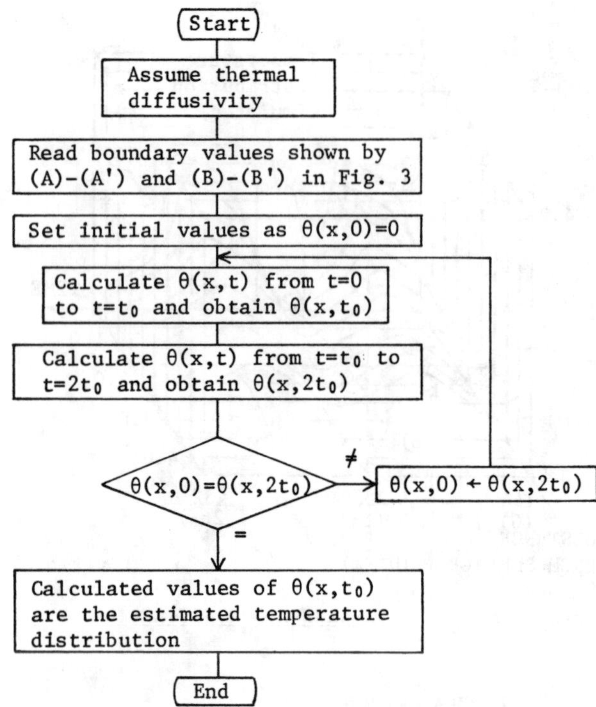

Fig. 4 Procedure of numerical estimation of temperature distribution

3. APPLICATION TO MEASUREMENT OF THERMAL DIFFUSIVITY

The estimation technique of temperature distribution was applied to the method of measuring thermal diffusivity. The procedure to determine thermal diffusivity by applying the estimation technique of temperature distribution is shown as a flow chart in Fig. 5.

In the case that this method of measuring thermal diffusivity is used, it is required that the surface temperature at x=0 is made to be varied periodically. But this does not mean the necessity of strict temperature control. Because, the material temperature is varied only for the purpose of easy evaluation of the difference between the experimental results and the calculated results both of which are expressed as $\theta(\delta,t)$ within the Decision mark in the flow chart shown in Fig. 5.

This measurement method was validated by following numerical examination. Supposing that the temperatures indicated by (A), (B), (C) and (D) in Fig. 2 are the measured values, this method was applied to determine thermal diffusivity of the slab, and as a result, the value of 1.20×10^{-7} m^2/s was obtained.. This obtained value of thermal diffusivity coincided with the value in the 3 significant digits given in Table 2.

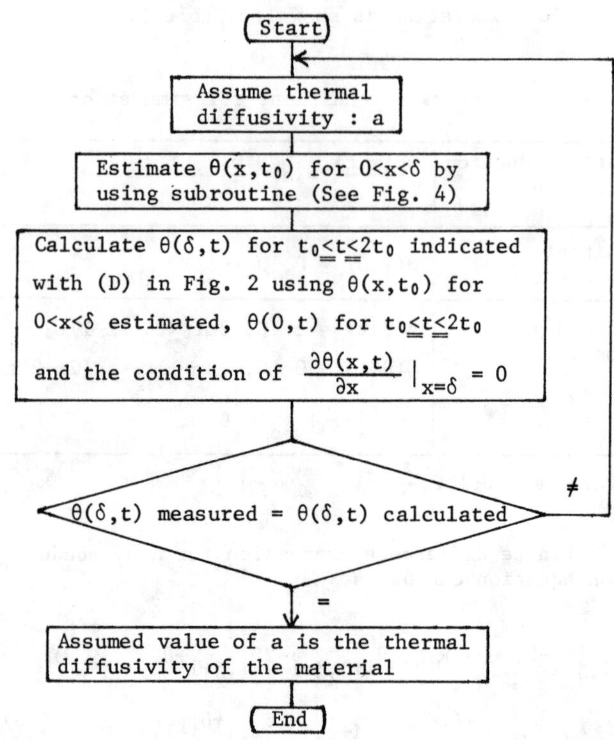

Fig. 5 Procedure for determining thermal diffusivity

4. EXPERIMENT

4.1. Experimental Apparatus

The apparatus used is shown in diagrammatic form in Fig. 6. Sample was contained to thicknesses ranging from 5.56 to 7.33 mm in the sample box, which was composed of a 0.3 mm thick and 100 mm square copper bottom surface and 15 mm height acrylic resin lateral surfaces. The bottom surface was large enough to neglect the end effects possibly caused by the heat input/output through lateral surfaces. The upper surface of sample was insulated with 50 mm thick styrofoam. The sample installation part shown in Fig. 6 was covered with a plastic box to prevent itself from the outside air flowing in.

The bottom of sample box was heated periodically by the heater of about 54 W set under the sample box. The average temperature of sample was controlled by adjusting the temperature of cooling water.

Temperatures at the bottom and the upper surfaces of sample were measured using copper constantan thermocouple sheets (Philips PR6452A) and those values were stored at 4-second intervals in the storage device.

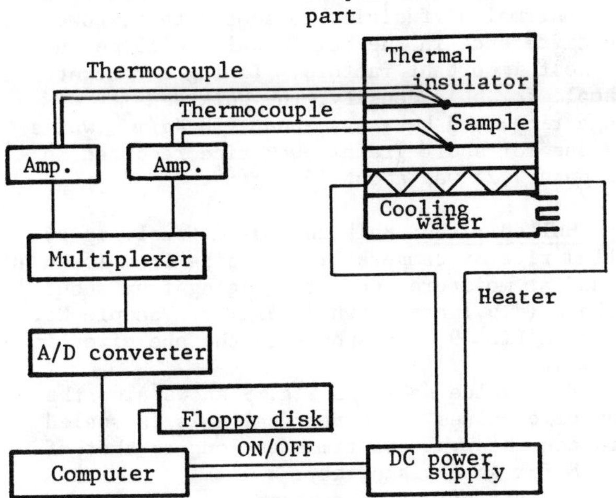

Fig. 6 Schematic diagram of experimental apparatus

4.2. Sample and Measuring Conditions

The short-grain rough rice and hulled rice used in this study were of the Ozora variety (harvested in Tsu, Japan). Initial moisture content of the samples was about 19 % (d.b.). Samples of about 150 g each were conditioned either by adding water or by drying in an oven at 30 °C to obtain different levels of moisture content ranging from about 9 to 25 % (d.b.). After conditioned, samples were stored at 5 °C for about two months prior to test to assure uniformity.

Conditions of samples were shown in Table 3 for rough rice and in Table 4 for hulled rice.

Moisture contents were determined by drying duplicate samples in a ventilated oven at 135 °C for 24 hours, and all results were calculated as a percentage of the dry weight.

Table 5 shows the measuring conditions of a set of experimental run.

Table 3 Conditions of rough rice

Sample No.	Moisture content % (d.b.)		Weight $\times 10^{-3}$ Kg	Thickness $\times 10^{-3}$ m	Bulk density Kg/m^3
	before used	after used			
1	9.08	9.35	24.89	5.67	550
2	15.31	14.91	—	6.34	—
3	21.05	20.15	28.44	6.64	536
4	23.07	22.90	29.28	6.64	552
5	25.06	23.61	31.34	6.98	562

Table 4 Conditions of hulled rice

Sample No.	Moisture content % (d.b.)		Weight $\times 10^{-3}$ Kg	Thickness $\times 10^{-3}$ m	Bulk density Kg/m^3
	before used	after used			
1	9.78	10.07	37.09	5.56	835
2	15.72	15.80	49.82	7.33	851
3	18.51	18.25	40.28	5.95	848
4	20.55	20.10	40.30	6.14	822
5	22.62	22.22	37.02	5.67	817
6	24.39	23.74	37.00	5.80	799

Table 5 Conditions of a set of experimental run

Period s	Sampling interval s	Number of temperature values to be measured	
		per one period	per one set
384	4	96	485

5. RESULTS AND DISCUSSION

5.1. Measured Temperatures and Data Processing

Fig. 7 shows an example of measured temperatures during one set of experimental run which are on the sample number 5 shown in Table 4.

To determine one value of bulk thermal diffusivity, the measured values of temperature during two periods were required to be processed.

Although sample temperature rose gradually as observed from Fig. 7, it can be assumed that thermal diffusivity does not vary with temperature within this range. Then the mean of 6 values

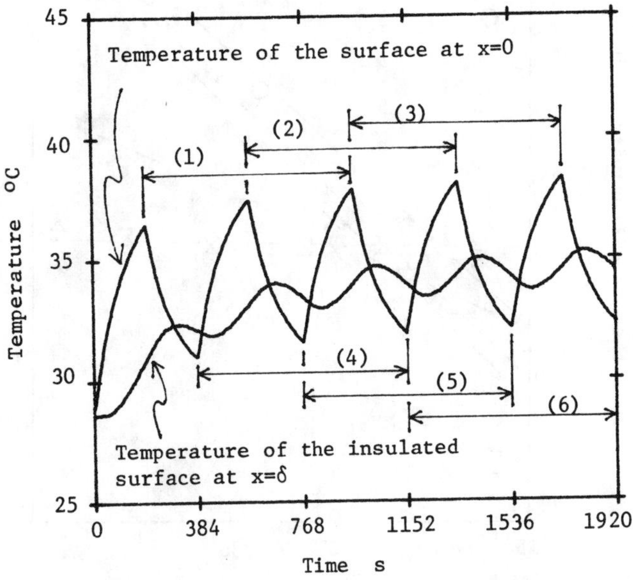

Fig. 7 An example of measured temperatures during one set of experimental run

calculated from 6 groups of data shown in Fig. 7 as (1) to (6) was determined to be the bulk thermal diffusivity at the average temperature over the interval.

5.2. Bulk Thermal Diffusivity of Rough Rice and Hulled Rice

Rough rice. Bulk thermal diffusivity of rough rice at temperatures ranging from 15 to 40 °C and at moisture contents ranging from about 9 to 25 % (d.b.) are shown in Fig. 8.

Each value shown in Fig. 8 is the mean of 6 values determined as described in the previous paragraph. Sample No. shown in Fig. 8 corresponds to the one given in Table 3.

A trend toward higher values at the higher temperatures is apparent for the moisture contents above 15 % (d.b.). The rate of increase with temperature of bulk thermal diffusivity was larger for higher moisture content than for lower moisture content. But at the moisture content of 9.08 % (d.b.), it was observed that bulk thermal diffusivity was independent from the temperature.

For the relationship between moisture content and bulk thermal diffusivity, Wratten et al. [1] and Morita and Singh [2] found that increase in the moisture content decreases the bulk thermal diffusivity. Their works do not agree with the results reported here for sample temperatures above 25 °C. But if it allows the author to extend each line jointing the obtained points toward the lower temperature range in Fig. 8, the same trend can be obserbed for the sample temperatures below 15 °C as they reported.

The values of bulk thermal diffusivity obtained here were lower than those reported by them. This is considered that the difference in variety of rough rice and in the measurement method and also the imperfect insulation at the insulated surface account for the difference in bulk thermal diffusivity to some extent, however, the difference in the bulk density will be one of the most important factors. Taking a concrete example for bulk density, the bulk density of rough rice used here was about 550 Kg/m^3, while the one for short-grain rough rice reported in the paper [2] was about 650 Kg/m^3.

Hulled rice. Bulk thermal diffusivity of hulled rice at temperatures ranging from 15 to 40 °C and at moisture contents ranging from about 9 to 25 % (d.b.) are shown in Fig. 9. Sample No. shown in Fig. 9 corresponds to the one given in Table 4.

Each value shown in Fig. 9 shows also the mean of 6 values. The vertical axis is scaled with the interval two times as long as that of Fig. 8 for good visibility.

For the effect of temperature on bulk thermal diffusivity, Fig. 9 shows a similar tendency to Fig. 8. But the rate of increase with temperature of bulk thermal diffusivity for each moisture content was less than one-half of the one for rough rice, and for the moisture content of 9.78 % (d.b.), it showed a negative value.

For the effect of moisture content, bulk thermal diffusivity decreased as the moisture con-

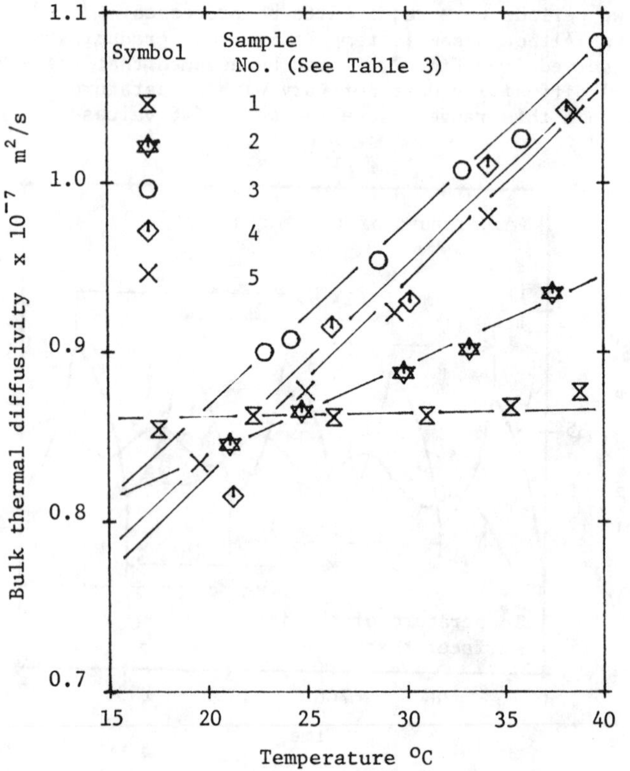

Fig. 8 Bulk thermal diffusivity of rough rice

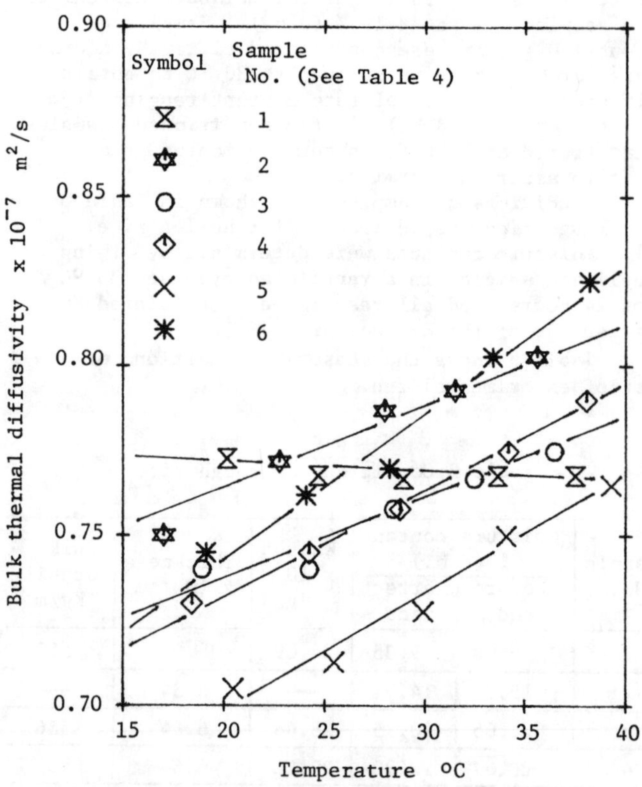

Fig. 9 Bulk thermal diffusivity of hulled rice

tent increased up to the temperature of about 22 °C except for at the moisture content of 24.39 % (d.b.). Beyond that temperature, inverse relationship between moisture content and bulk thermal diffusivity was observed for moisture contents of 9.78, 15.72, 18.51 and 20.55 % (d.b.).

It was found that the bulk thermal diffusivity of hulled rice was smaller than that of rough rice.

6. CONCLUSION

The new method of measuring thermal diffusivity which does not require the information on initial temperature distribution was proposed and it was found by theoretical and numerical considerations that this method was valid.

The experimental results as described in paragraph 5.1. showed that this method required no long time for obtaining values which were used for calculating bulk thermal diffusivity.

The effect of temperature on the bulk thermal diffusivity of rough rice was that the increase in temperature increased the bulk thermal diffusivity for moisture contents ranging from about 15 to 25 % (d.b.). This tendency was diminished with the decrease in moisture content and no effect of temperature was recognized at the moisture content of 9.08 % (d.b.).

These characteristics were approximately the same as for hulled rice. However, the bulk thermal diffusivity of hulled rice was lower than for rough rice.

For the effect of moisture content, the assumption that the increase in moisture content decreased the bulk thermal diffusivity of rough rice can be derived from Fig. 8 for the temperatures below 15 °C. But in the higher temperature range, this assumption was not valid. The same thing was more or less true of hulled rice except for the moisture content of 24.39 % (d.b.).

REFERENCES

1. Wratten, F.T., W.D.Poole, J.L.Chesness, S.Bal and V.Ramarao, TRANSACTIONS of the ASAE, vol. 12(6), 801-803 (1969).
2. Morita, T. and R.Paul Singh, TRANSACTIONS of the ASAE, vol. 22(3), 630-636 (1979).
3. Hosokawa, A and H.Masumoto, J. of the Agric. Machinery, Japan, vol. 32(4), 302-305 (1971).
4. Kazarian, E.A. and C.W.Hall, TRANSACTIONS of the ASAE, vol. 8(1), 33-37,48 (1965).
5. Moysey, E.B., J.T.Shaw and W.P.Lampman, TRANSACTIONS of the ASAE, vol. 20, 768-771 (1977).
6. Oshita, S., K.Nakagawa and K.Horibe, J. of the Agric. Machinery, Japan, vol. 44(3), 469-476 (1982)

ON THE VAPOUR TRANSFER RATE IN POROUS MEDIA UNDER A TEMPERATURE GRADIENT

Masashi Kuramae

Department of Sanitary Engineering, Hokkaido University
Sapporo, 001 JAPAN

ABSTRACT

This paper deals with the problem of the vapour transfer rate in porous media such as granular bed under a temperature gradient. The microscopic temperature and temperature gradient distribution in porous media were calculated by numerical relaxation method for two regular packed models. These results revealed that the value of ξ which defined as the ratio of the temperature gradient in air filled void to that of the overall porous media was 1.5-2.0 for usual packed bed. The mass transfer experiments under a constant temperature gradient were achieved In this experiments, granular beds of glass sphere particles that contained sublimate substance, for example, naphthalene were used as the sample and the quantities of this substance transfered as vapour state in a designated period were measured by quantitative analysis of gas chromatography. As the results, it was found that the experimentally obtained vapour transfer rate was generally larger than that of the calculated value by a past simple mass transfer model and these gave support to our theory.

1. INTRODUCTION

Drying is a heat and mass transfer phenomenon on or in porous media.[1-3] It is composed of several phenomena such as (1) heat transfer to material surface, (2) vapour transfer from a surface to an ambient air, (3) liquid vaporization on or in material by dissipating latent heat, (4) heat transfer into inner material for temperature rising or inner vaporization, (5) liquid transfer in material up to the vaporizing surface, (6) inner vaporized vapour transfer up to the material surface,etc.

In general, these transport mechanisms exert an influence on each other in a complex way depending on the material characteristics and drying process. So the transport properties describing these phenomena are essentially characterized not only by their material properties but also apparently by temperature and moisture content, and often show hysteretic behaviour.[4,5]

Especially for vapour(gas) transfer in porous media descrived above (6), a similar form of transfer appears in reaction engineering, soil science, heat transfer engineering, etc in addition to the drying (in particular, conduction drying and freeze drying). Its transport mechanism has been divided into (a) molecular diffusion, (b) Knudsen diffusion, (c) flow due to pressure gradient, (d) surface diffusion (e) thermal diffusion, etc.[6] However, for a partially saturated granular bed under a temperature gradient, partial vapour pressure in an air filled void of the bed is thought to be saturated and vapour flow rate in the bed has been estimated by the following equation.[7,8]

$$G_{DCAL} = -\beta \frac{D_v}{RT} \varepsilon (1-S) \frac{P}{P-P_s} \frac{dP_s}{dT} \frac{dT}{dx} \qquad (1)$$

This equation is derived by applying the Stefan's equation of unidirectional diffusion, by considering the rate of effective diffusional area, $\varepsilon(1-S)$ and by applying a correction factor, β, to account for the zigzag nature of the vapour flow path.

Yet, a question is raised concerning the temperature gradient dT/dx which is regarded as the driving force of the vapour transfer in Eq.(1). That is, as the vapour transfers in void space of the porous media, the dT/dx should be estimated by the temperature gradient of the air filled void of the bed. Nevertheless, dT/dx has been hitherto evaluated approximately by $\Delta T/L$ (in which ΔT is the temperature difference of both ends of the bed and L is the bed depth) or by interpolation of internal temperature measured by several thermocouples embedded in the bed.

In a previous paper, the author[9] has pointed out that the average temperature gradient of the overall bed was different from the average value of the temperature gradient of the air filled void of the bed. This problem was investigated by moisture transfer experiments under a temperature gradient in glass beads packing using NaCl as a tracer substance. This paper deals with a similar problem from another viewpoint by using such sublimate substances as naphthalene as a transfer substance to reinvestigate the vapour transfer mechanism.

2. GENERAL CONSIDERATION

As represented in Fig.1, a system composed of two materials is considered, of which the thermal conductivities are different. If a temperature difference is given in parallel with those layers as shown in Fig.1(a), the temperature distribution and the temperature gradient distribution in heat flow direction will be built up as shown in this Figure. These profiles will not change even if the intervals of these species become narrow.

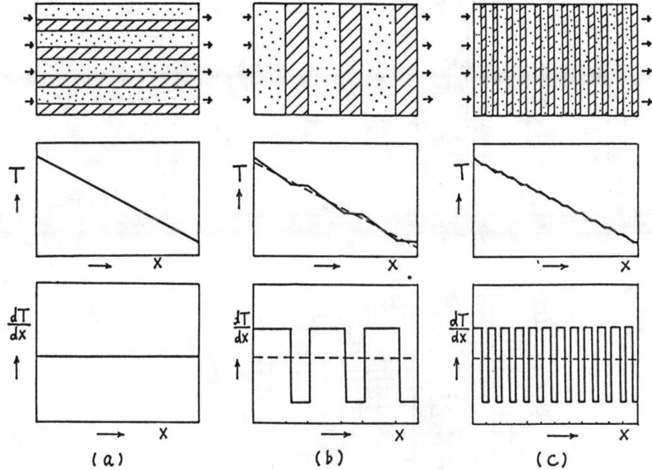

Fig. 1 Temperature and temperature gradient distribution profiles of a system composed of two materials

In contrast, if a temperature difference is given perpendicular to these layers as shown in Fig.1(b), the temperature distribution curve will not become smooth and the tempereture gradient distribution curve will not become continuous at the boundary of the two species. Moreover, if the intervals of these species become narrow, the temperature distribution curve will approach a smooth curve. This corresponds to the fact that the temperature distribution curve can be actually regarded as continuous if such the microscopic scale as particle diameter is considerably smaller than the scale of the bed. However, for the temperature gradient distribution, whatever intervals become narrow, its noncontinuity at the boundary of both phases can't be removed and they will be different from each other(in Fig.1(c)). Therefore, the average temperature gradient in an air filled void of a porous media will be generally different from that of a solid phase and also from that obtained by interapolation of the macroscopic temperature distribution of a porous bed.

Krischer[10] suggested a heat conduction model of a partially saturated granular bed as represented in Fig.2. Its thermal effectivity λ_e is given by

$$\lambda_e = \frac{1}{\frac{a}{\lambda_{ep}} + \frac{(1-a)}{\lambda_{es}}} \qquad (2)$$

in which

$$\lambda_{ep} = (1-\varepsilon)\lambda_g + S\varepsilon\lambda_\ell + (1-S)\varepsilon\lambda_f \qquad (3\text{-}1)$$

$$\lambda_{es} = \frac{1}{\frac{(1-\varepsilon)}{\lambda_g} + \frac{S\varepsilon}{\lambda_\ell} + \frac{(1-S)\varepsilon}{\lambda_f}} \qquad (3\text{-}2)$$

in which λ_g, λ_ℓ, λ_f are thermal conductivities of the solid, liquid and gas phase respectively.
In a previous paper[9], the author defined the temperature gradient ratio ξ as

$$\xi = \frac{(\text{ average temperature gradient of void })}{(\text{ temperature gradient of overall bed })} \qquad (4)$$

By the Krischer model ξ is expressed in the following Equation.

$$\xi = \frac{1}{1 + \frac{\lambda_{ep}(1-a)}{\lambda_{es}\,a}} + \frac{1}{\frac{a}{(1-a)}\frac{\lambda_f}{\lambda_{ep}} + \frac{\lambda_f}{\lambda_{es}}} \qquad (5)$$

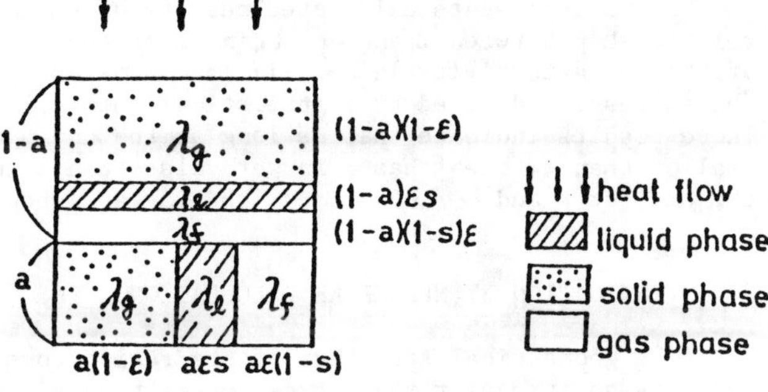

Fig. 2 Krischer model

Eq.(5) follows that the ξ of a partially saturated granular bed depends upon the moisture content and temperature of the bed.[9] In the case of a two phase system(S=0), Eq.(5) is

$$\xi = \frac{\frac{a}{(1-a)} + (1-\varepsilon)\frac{\lambda_g}{\lambda_f} + \varepsilon}{\frac{a}{(1-a)} + (1-\varepsilon)^2 + \varepsilon^2 + \varepsilon(1-\varepsilon)\left(\frac{\lambda_g}{\lambda_f} + \frac{\lambda_f}{\lambda_g}\right)} \qquad (6)$$

On the other hand, ξ can be derived by a statistical method. Firstly by definition the thermal effectivity λ_e is given by

$$-Q = \lambda_e \overline{\nabla T} \qquad (7)$$

in which Q is the average heat transfer rate and $\overline{\nabla T}$ is the average temperature gradient of the cross section. By a heat balance at that section, the following relationship is established.

$$-Q = \lambda_g \int_g \nabla T\, dS + \lambda_f \int_f \nabla T\, dS = \lambda_g \overline{\nabla T_g}(1-\varepsilon) + \lambda_f \overline{\nabla T_f}\,\varepsilon \qquad (8)$$

Average temperature gradient is defined by

$$\overline{\nabla T} = (1-\varepsilon)\overline{\nabla T_g} + \varepsilon\overline{\nabla T_f} \qquad (9)$$

$\overline{\nabla T_g}$ is eliminated by Eqs.(8)-(9) and obtained

$$\lambda_e = \lambda_g - \lambda_g \varepsilon \frac{\overline{\nabla T_f}}{\overline{\nabla T}} + \lambda_f \varepsilon \frac{\overline{\nabla T_f}}{\overline{\nabla T}} \qquad (10)$$

Accordingly, the next relationship is established.

$$\xi = \frac{\overline{\nabla T_f}}{\overline{\nabla T}} = \frac{\lambda_g - \lambda_e}{(\lambda_g - \lambda_f)\varepsilon} \qquad (11)$$

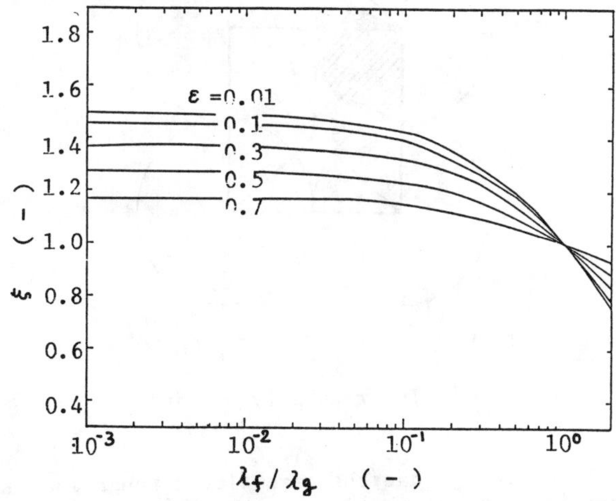

Fig. 3 Calculated results of ξ by statistical model

Fig.3 represents calculated results of the relationship between ε and λ_f/λ_g in which the value of thermal effectivity in Eq.(11) is estimated by the expression derived by a statistical theory[11]. These results indicate that as long as the λ_f/λ_g is smaller than 1, the ξ has a larger value than 1 and the smaller ε and λ_f/λ_g becomes, the larger ξ becomes.

3. INVESTIGATION BY NUMERICAL CALCULATION

The geometrical structure of the real porous media is so irregular that it is scarcely possible to analyze the microscopic temperature distribution mathematically. Accordingly in this case, the heat conduction problem of two regularly packed beds is investigated by a numerical relaxation method.[12]

These packings are

(1) Cubical packing (ε=0.474, in Fig.4(a))

(2) Hexagonal packing (ε=0.395, in Fig.4(b))

For each packing, a unit cell is chosen according to the pattern ABCA'B'C' (see Fig.4) and this is divided into several elements.

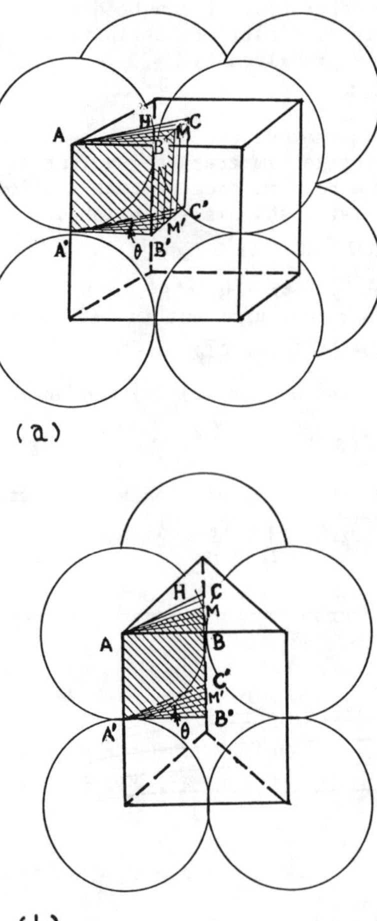

Fig. 4 Two regularly packing

By assuming that the heat flow through each side of neighbouring elements is negligible, the problem is approximately reduced to that of a two dimensional heat conduction for the element; that is, the plane AMM'A'. For the numerical procedure, each element was divided into several squares. The following relationship is given by the heat conduction equation.(see Fig.5)

$$T_{j,k}=\frac{\dfrac{T_{j,k-1}}{R1_{j,k-1}}+\dfrac{T_{j,k+1}}{R1_{j,k}}+\dfrac{T_{j-1,k}}{R2_{j-1,k}}+\dfrac{T_{j+1,k}}{R2_{j,k}}}{\dfrac{1}{R1_{j,k-1}}+\dfrac{1}{R1_{j,k}}+\dfrac{1}{R2_{j-1,k}}+\dfrac{1}{R2_{j,k}}} \qquad (12)$$

Boundary conditions are

$$T_{1,k}=1, \qquad T_{NB+1,k}=0 \qquad (13\text{-}1)$$

$$T_{j,1}=T_{j,2}, \qquad T_{j,NC+1}=T_{j,NC} \qquad (13\text{-}2)$$

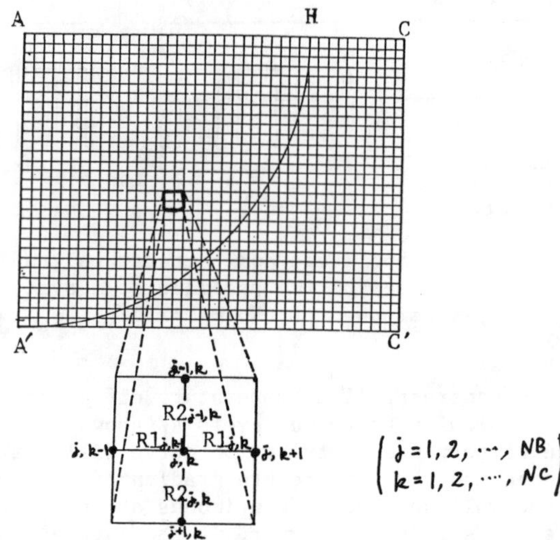

Fig. 5 Relaxation network

The normarized temperature of each nodal point is calculated one after another until the difference between corrected values and previous ones becomes within 10^{-5} for all nodal points.

The calculated temperature and temperature gradient distribution curves of some elements are represented in Figs.6-8. Figs.6 and 7 are of different elements in a unit cell, but distinctive features of the temperature and temperature gradient distributions are not observed. A comparison between Fig.6 and Fig.8 shows that the thermal conductivity ratio of the gas and solid phases has an influence upon the temperature and temperature gradient distributions of both phases. In particular, the constant temperature gradient curve shows a discontinuous pattern at the boundaries of both phases and indicates a great value near the contact point of two particles.

In the next step, by the above calculated results, the average value of the temperature gradient at each point of void space(that is equivalent to ξ) was calculated for each packing. Fig.9 shows these results with those of the Krischer and statistical models. In the Krischer model, ξ converges to a constant value irrelevant to the value of the parameter of a as λ_f/λ_g approaches 0. This fact signifies that when the structure of a real porous media is expressed by the Krischer model with a parameter a, the value of a is dependent upon λ_f/λ_g and must be estimated as approximately 1 as λ_f/λ_g decreases. Though the results obtained by a statistical model don't always coincide with those of the real packed structure model, this fact points out that the statistical model is based on a particular averaging manipulation and can't represent

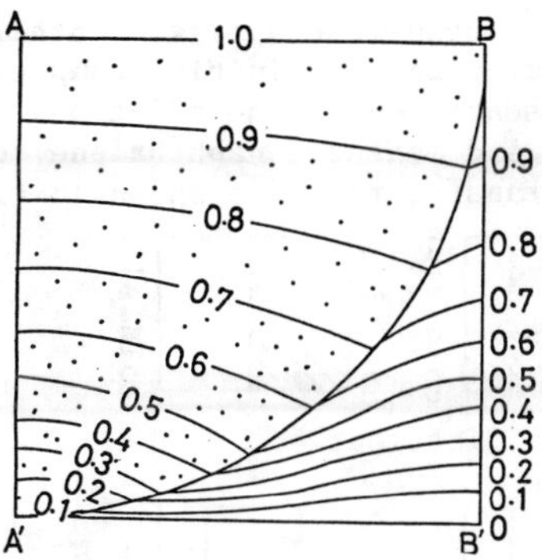

(a) temperature distribution

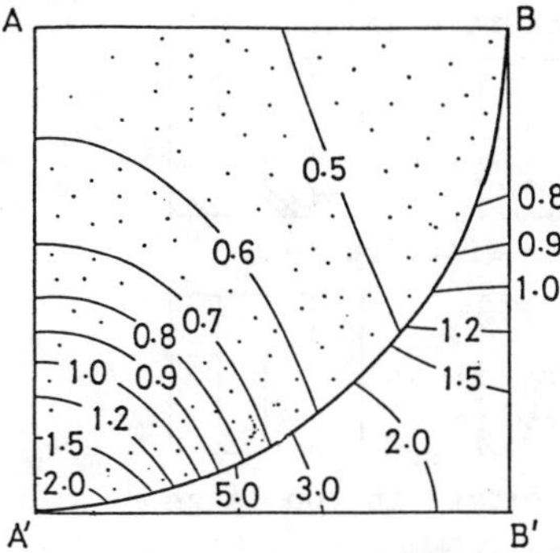

(b) temperature gradient distribution

Fig. 6 Calculated temperature and temperature gradient distribution of an element (θ =0 deg, λ_f / λ_g=0.1)

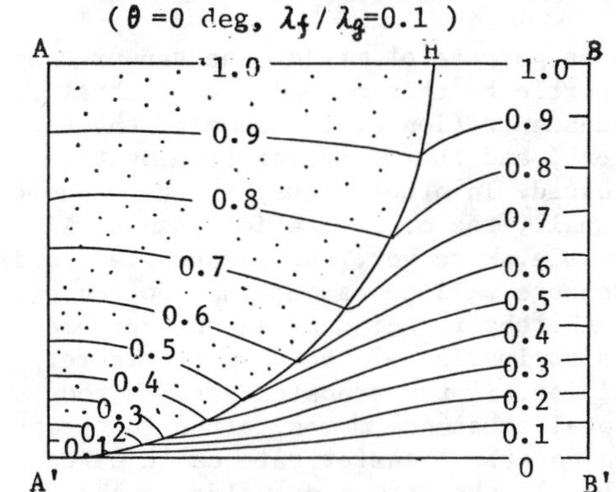

(a) temperature distribution

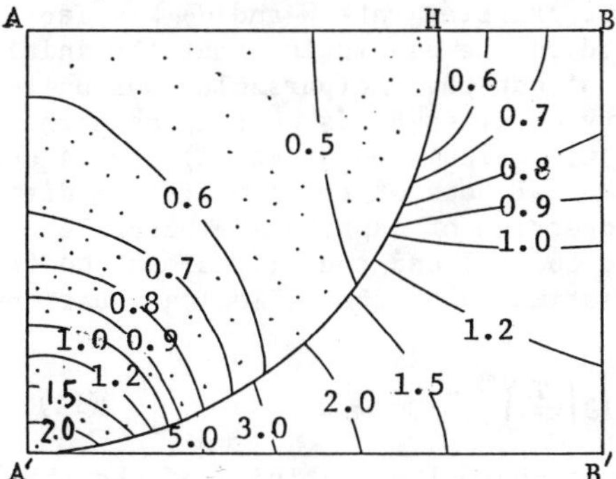

(b) temperature gradient distribution

Fig. 7 Calculated temperature and temperature gradient distribution of an element (θ=45 deg, λ_f / λ_g=0.1)

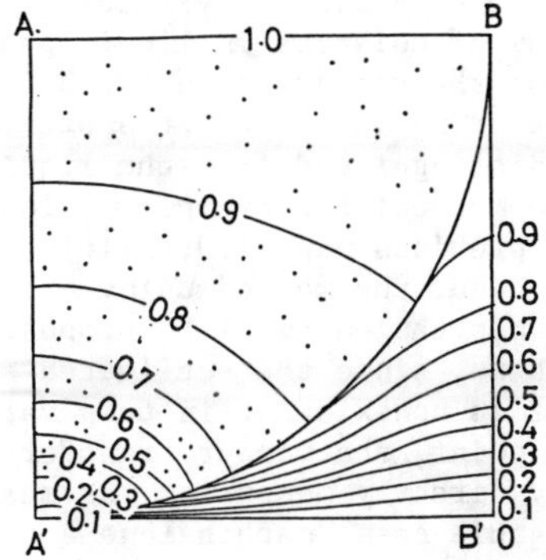

(a) temperature distribution

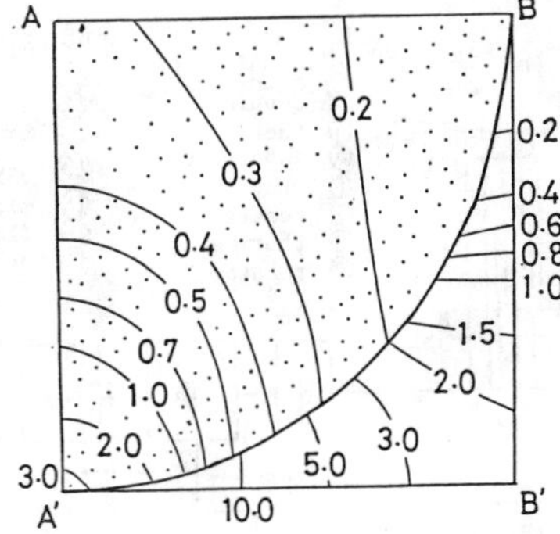

(b) temperature gradient distribution

Fig. 8 Calculated temperature and temperature gradient distribution of an element (θ =0 deg, λ_f / λ_g=0.001)

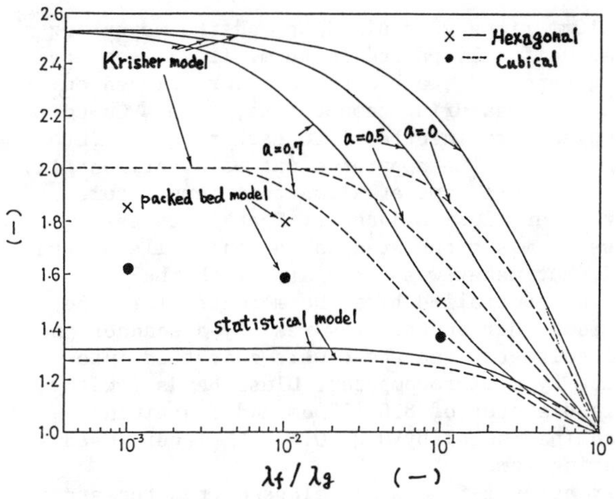

Fig. 9 Relationship between ξ and λ_f / λ_g

the geometrical features of a real porous media.

According to the above results, as in real porous media, the value of λ_f / λ_g is 10^{-2}-10^{-3}, and the corresponding value of ξ is thought to be about 1.5-2.0.

4. EXPERIMENTAL INVESTIGATION

In general, if a temperature difference is

applied to both ends of a partially saturated granular bed, the flow of moisture in liquid phase accompanies that of the vapour phase.[8,13,14] Therefore, for measuring the vapour transfer rate only, it must be distinguished from the liquid flow rate by a suitable method; for example, using NaCl as a tracer as in previous paper.[9,15,16]

On the other hand, the method using a sublimate substance such as naphthalene will be proposed for that purpose. That is, since the sublimated substance liquid phase is nonexistant in temperature under the melting point, the vapour transfer rate may be obtained by directly measuring the mass transfer rate. In this case, naphthalene and parazole, which have a sublimate nature in usual temperature, are chosen for the transfer substances.

Experimental apparatus is shown in Fig.10.

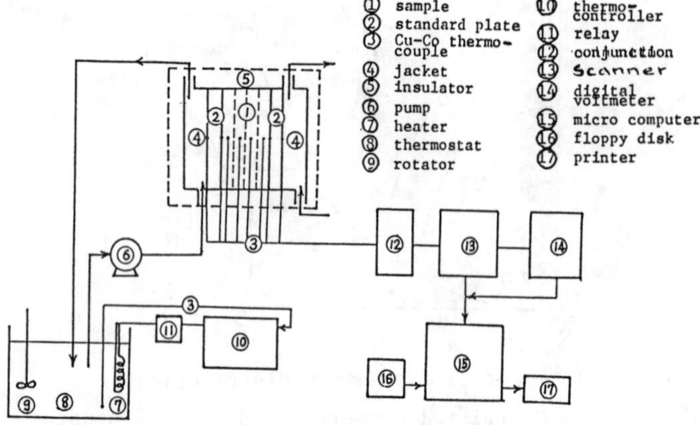

Fig.10 Experimental apparatus

780mmϕ x 10mm rings were piled up and the measuring sample was uniformly packed in them. It was sealed by winding tape and two blass plates were placed on both ends. For measuring temperature, 0.1mmϕ Cu-Co thermocouples were inserted into each ring and into the glass plates. The apparatus was held horizontally and both ends were kept at a constant temperature by use of water flowing jackets. That is to say, cooling water was circulated in one end while at the other end, hot water was circulated with the temperature being controlled by a thermocontroller. The e.m.f of each thermocouple was read by a scanner and a digital voltmeter and these were converted into temperature by a microcomputer. Glass beads (with a particle diameter of 830-1130μm and containing naphthalene(parazole) by 0.02-0.07 wt%) were used for measuring sample.

After about 1-6 days had elapsed from the start of the experiment, the granular bed was cut apart into each ring. Then a part of this sample was diluted by an amount of aceton and the naphthalene (parazole) was dissolved in the solvent. The concentration of naphthalene was measured by a quantitative analysis of gas chromatography and the vapour transfer rate was calculated from this result.

5. EXPERIMENTAL RESULTS AND DISCUSSION

Table.1 represents the experimental condition and the obtained results. A combination of pure glass beads and glass beads containing naphthalene (parazole) was prepared for the initial sample. The schematic representation of the relationship between the initial and measured naphthalene concentrations and their transfer rates is shown in Fig.11.

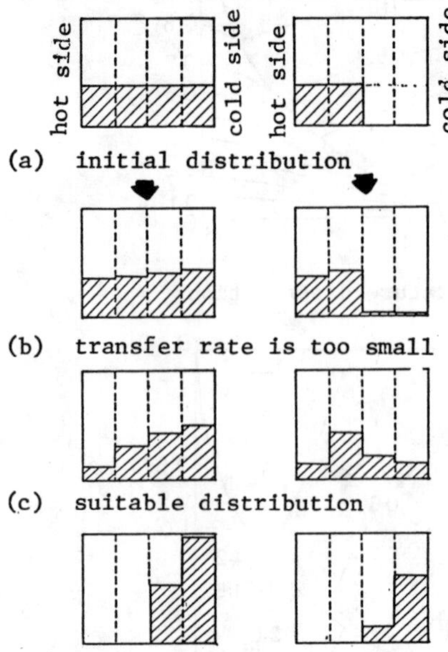

Fig.11 Schematic representation of the relationship between the initial and measured naphthalene(parazole) concentrations and their transfer rates

Namely, for the purpose of making the vapour transfer rate measurble by this method, the initial naphthalene concentration of the sample, the temperature gradient, and the measuring time must be suitably adjusted. In other words, if the transfer rate is too small, the concentration change will not be observable. Moreover, the unsaturated region of vapour pressure will appear at the low concentration side, thereby reducing to an unfavorable situation the estimation of the vapour transfer rate. In addition, if the transfer rate is too large, the total substance moves to the low temperature region and the transfer rate can't be calculated. In Table.1, the former examples appear in No.N-101, etc and later in No.P-102, etc.

The vapour transfer rate W and G_{Dexp} in Table.1 were calculated on the assumption that the initial concentration of naphthalene(parazole) was uniform in the category of glass beads containing naphthalene. And G_{DCAL} was calculated by Eq.(1) making use of the experimental data of the temperature distribution and properties of naphthalene(parazole).[17] β was assumed to be 1 and the diffusional coefficient D_v was estimated by the following equation.[18]

$$D_v = 0.0513\left(\frac{T}{273}\right)^{1.75} \text{ (cm}^2\text{/s)} \tag{14}$$

Furthermore, the thermal effectivity of the samples was measured by a relative method from the temperature drop of the standard plates. The temperature distribution in the measuring samples reached a steady state in about 1 hr which was such a short

Table. 1 Experimental condition and results

Sample No.	Sample	content (g/g)	transfered W (g/g)	T (°C)	dT/dx (°C/cm)	G_{DCAL} (mol/cm²s) ×10⁻¹⁰	G_{Dexp} (mol/cm² s) ×10⁻¹⁰	ξ(β= 1) (-)	ξ(β = (1.57)⁻¹) (-)
N-001	G + N	0.00701	0.00045	27.7	7.2	0.98	3.97 ※	4.05	6.36
τ =36hr	G + N	659	49	35.4	7.2	1.92	4.43 ※	2.31	3.62
λ_e=0.44	G + N	680	74	42.3	7.2	3.37	6.68 ※	1.98	3.11
	G + N	581							
N-101	G	1	1	23.1	6.3	0.77	0.05		
τ =48hr	G	1	-		6.3	1.38	0.09		
λ_e=0.33	G	57	2	35.6	8.7	3.52	2.26		
	G + N	541	59	44.3					
N-102	G	0	0	23.1	4.0	0.44	0		
τ =72hr	G	0	0	27.1	6.8	1.23	0		
λ_e=0.72	G	13	13	33.9	5.5	1.55	0.35		
	G + N	306	10	37.4	7.2	3.15	0.33		
	G + N	301		44.6					
N-103	G + N	325	66	26.8	4.1	0.65	0.95 ※	1.46	2.29
τ =130hr	G + N	362	170	30.9	7.3	1.90	2.40 ※	1.26	1.98
λ_e=0.54	G + N	321	233	38.2	7.3	3.42	3.30		
	G + N	26		45.5					
N-104	G	81	81	26.6	4.9	0.78	1.03 ※	1.32	2.07
τ =144hr	G	25	106	31.5	5.1	1.28	1.36 ※	1.06	1.67
λ_e=0.88	G + N	112	104	36.6	7.2	2.94	1.33		
	G + N	10		43.8					
P-101	G	121	121	22.4	4.1	4.37	4.64 ※	1.06	1.67
τ =48hr	G	136	257	26.5	6.9	11.34	9.83 ※	0.87	1.36
λ_e=0.57	G + P	361	618	33.4	6.4	17.38	23.72		
	G + P	0		39.8					
P-102	G	138	138	21.9	5.1	5.33	5.53 ※	1.04	1.63
τ =46hr	G	46	184	27.0	5.0	7.77	7.37		
λ_e=0.63	G	0	184	32.0	7.9	20.24	7.37		
	G + P	0		39.9					
P-103	G	62	62	24.1	5.6	7.16	4.7 ※	0.66	1.03
τ =24hr	G	81	143	29.7	6.0	12.26	10.96 ※	0.89	1.40
λ_e=0.54	G + P	449	295	35.7	5.7	17.53	22.6		
	G + P	3		41.4					

λ_e(kcal/mhr C)

G glass beads
N naphthalene
P parazole

time span in comparison with the measuring time that the transient effect can be considered negligible. Thus, the experimentally calculated value of ξ depends upon the estimated value in Eq.(1). Although there are several points of view regarding the value of β which is never larger than 1[19,20], ξ was calculated for β=1[8] and β=(1.57)⁻¹ [21]

Fig.12 shows the relationship between G_{DCAL} and G_{Dexp} for the data represented in Table.1 for which the vapour transfer rate is considered to be correctly measured.(designated by ※ in Table.1) Although this correlation shows some scattering, the value of ξ is for the most part 1.0-2.0. This result suggests that the vapour transfer in a porous media under a temperature gradient is carried out by the mechanism proposed in this paper and therefore supports the conclusion of the previous paper[9] from a different viewpoint.

It follows that the driving force of the vapour transfer in a partially saturated porous media under a temperature gradient will not be the temperature gradient of the porous media but the average temperature gradient of the void space.

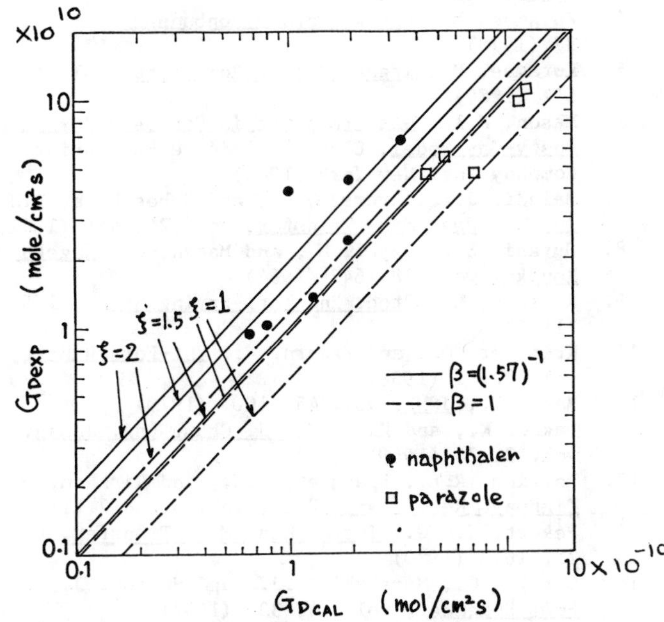

Fig.12 Relationship between G_{DCAL} and G_{Dexp}

NOMENCLATURE

a	parameter	-
D_v	diffusion coefficient	m^2/s
G	vapour transfer rate	kg/ms
L	depth of the bed	m
P	vapour pressure	N/m^2
P_s	saturated vapour pressure	N/m^2
Q	heat transfer rate	J/m^2s
R	gas constant	N/mol°C
R1,R2	resistance of heat conduction	-
S	saturation	-
T	temperature	°C
x	distance	m
β	coefficient	-
ε	porosity	-
λ	thermal conductivity	J/ms°C
λ_e	thermal effectivity	J/ms°C
ξ	temperature gradient coefficient	-
θ	angle	deg
τ	time	s

(Subscript)

p	parallel
s	series
g	solid
l	liquid
f	fluid
exp	experimental
cal	calculated

REFERENCES

1. Keey, R.B., Drying Principles and Practice, Pergamon Press, Oxford (1972).
2. Luikov, A.V., Heat and Mass Transfer in Capillary Porous Bodies, Pergamon Press (1966).
3. Chou, W., and Whitaker, S., Proc. 3d Int. Drying Symp., 135 (1982)
4. Kuramae, M., Kagaku Kogaku Ronbunshu, vol. 4, 87 (1978)
5. Kuramae, M., Kagaku Kogaku Ronbunshu, vol. 6, 591 (1980)
6. Mason, E.A., Gas Transport in Porous Media: The Dusty-Gas Model, Elsevier Science Publishing Company Inc, New York (1983)
7. Baladi, J.Y., Ayers, D.L., and Schoenhals, R.J., Int. J. Heat Mass Transfer, vol. 24, 449 (1981)
8. Ohtani, S., Suzyki, M., and Maeda, S., Kagaku Kogaku, vol. 28, 642 (1964)
9. Kuramae, M., Proc. 2nd Int. Drying Symp., 195 (1980)
10. Krischer, O., and Esdorn. N., VDI-Forsch Heft, vol. 22, 1 (1956)
11. Hase, T., JSME, vol. 45, 1003 (1979)
12. Wakao, N., and Kito, K., J. Chem. Eng. Japan., vol. 2, 24 (1969)
13. Rollins, R.L., Spangler, M.G., and Kirkham, D., Highway. Res. Board. Proc, vol. 33, 492 (1954)
14. Eckert, E.R.G., Int J. Heat Mass Transfer, vol. 23, 1613 (1980)
15. Gurr, C.G., Marshall, T.J., and Hutton, J.T., Soil Science., vol. 74, 335 (1952)
16. Philip, J.R., and de Vries, D.A., Trans. Amer. Geophs. Uni., vol. 38, 222 (1957)
17. Sogin, H.H., and Providence, R.I., Trans. ASME., vol. 25, 61 (1958)
18. Perry, R.H., and Chilton, C.H., Chemical Engineers' Handbook, 5th Edition, McGraw-Hill, Kogakusha (1973)
19. Brakel, J.V., and Heetjes., Int. J. Heat Mass Transfer, vol. 17, 1093 (1974)
20. Prager, S., Physica, vol. 29, 129 (1963)
21. Krischer, O., Chem. Ing. Tech, vol. 34, 154 (1962)

ACKNWLEDGEMENT

The author wish to express his thanks to prof. M.Suzuki (Tohoku Universury) for helpful advices and suggestions.

BEHAVIOR OF INVERSION POINT TEMPERATURE AND NEW APPLICATIONS OF SUPERHEATED VAPOR DRYING

Tomohiro Nomura and Tsutomu Hyodo

Department of Mechanical Engineering, Osaka City University
Osaka, 558 JAPAN

ABSTRACT

This report summarizes new fields of applications of both superheated vapor drying and highly humid and high temperature air drying. And then it describes fundamental research in developing a controlled superheated vapor dryer.

New applications of the vapor drying to domestic uses, in particular, drying of instant foods for home cooking and of washed clothing are illustrated. Through the fundamental research, by calculating from the well-known heat transmission formulae and by performing the experiment, the existence of the inversion point temperature has been confirmed, and the locus of its changing point with drying variables has been found. Moreover, for a vapor drying system controlled by humidity sensing, a direct measurement of the water content in a mixed gas of air and superheated vapor in the temperature range 373-573K has been also performed by using the ZrO_2-MgO porous ceramic sensor.

1. INTRODUCTION

Many studies of the evaporation of water in superheated vapor and of the use of its vapor as a drying agent have been done by various investigators[1-3, for example] for about thirty years. Successively, superheated vapor drying has been applied in many new drying fields in addition to industrial drying field, such as drying of instant foods for home cooking, and drying of washed clothing in our daily life, and so on.

The most significant reason for this wide range of applications of superheated vapor is that more water evaporates in this vapor or highly humid air than in dry air above the inversion point temperature. This is because the evaporation rate curve of these agents in high temperature and in low temperature regions cross at one point, as was established experimentally by our previous study[4].

Moreover, superheated vapor has many unique drying merits which cannot be found in dry air, such as its effects on the drying materials of a porous condition for easy grinding and dissolution into liquid.

On the other hand, as compared with these wide practical applications, theoretical research concerned with the inversion point temperature and fundamental research for determining controlled operating conditions or optimum design conditions for a superheated vapor drying system have not been sufficiently performed.

From this viewpoint, by calculating from the well-known heat transmission formulae (using a horizontal plate surface by Pohlhausen) and in experimenting, by changing drying variables, such as the mass velocity or heat transfer of the vapor, the authors confirmed the existence of the inversion point temperature and found the locus of the temperature. The behavior of the locus enables one to specify the drying variables for a controlled drying system. Furthermore, for this system, which is controlled by humidity sensing, a new direct measuring method of the vapor content in a mixed gas of air and superheated vapor over a high temperature range is also described using a ceramic sensor, which consists of a ZrO_2-MgO ceramic body with a surrounding heater.

2. RECENT SUPERHEATED VAPOR DRYING ADVANCES

Superheated vapor and highly humid and high temperature air are attractive drying agents, because of their merits which cannot be obtained with air drying , as stated above , but its applications have been limited to industrial purposes. Table 1 shows examples of recent applications of drying vapors.

Table 1 Examples of recent applications of drying vapors and their merits

Drying agent	Material	Merits of these drying
Super-heated vapor	Noodles[(a)] (for instant food)	*closed circuit dryer[(c)(d)] (less heat consumption)
	Washed clothing	*speedy drying
	Sterilized materials[(b)(d)]	*porous condition
	Soy sause lees[(b)]	*less surface hardness
	By-product of food industry (for feed)[(c)(d)]	*less oxidation
Highly humid & High temperature air	Waste molasses	*less explosion
	Fowl droppings	*effect of deodorization
	Sluges[(d)] (beer, pulp, etc.,)	*effect of sterilization[(b)(d)]
		*soft & fluffy finish (for clothing)
		*reuse of exhaust gas
		*less cost of exhaust equipment

(a)House food Ind. Co., Ltd. (b)Kikkoman Corp. (c)Gulf Mach. Co., INC.(USA) (d)Okamara Mfg. Co. Ltd.

As shown in Table 1, highly humid air has been used widely in many fields as a drying agent for the materials which can be processed at high temperature. On the other hand, at present, new fields of superheated vapor application have been extended to foodstuffs for home cooking, and to the drying apparatus which is used to simplify our everyday life and to make it more convenient, as illustrated in Table 1.

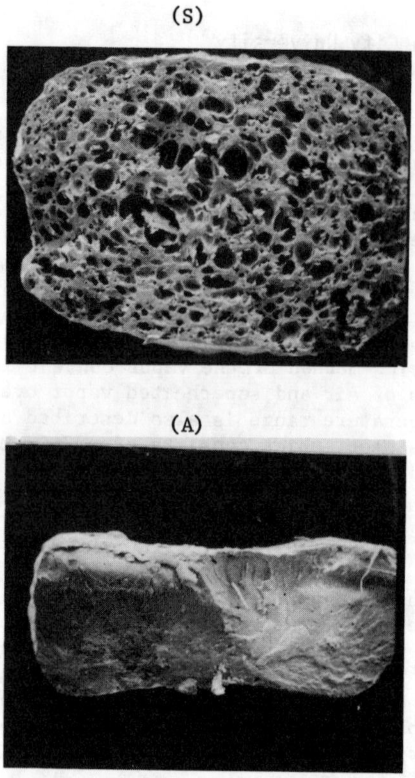

(S)

(A)

Fig. 1 Superheated vapor causes greater porosity as shown in instant food, top (S). Bottom (A) dried with hot air only.

(S) (A)

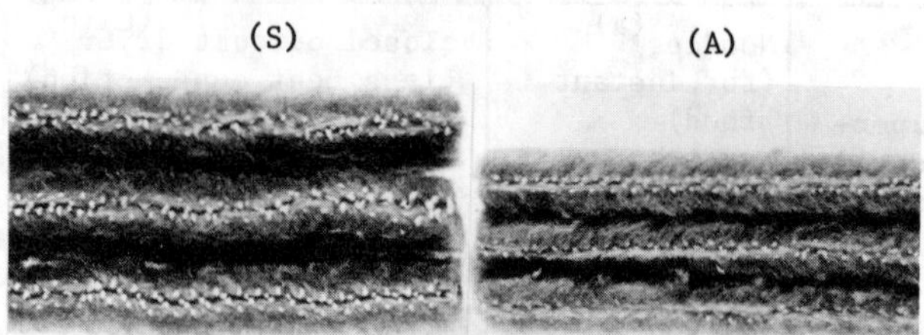

Fig.2 Soft and fluffy blanket (S) after superheated vapor drying, (A) dried with hot air only.

Figs. 1 and 2 photographically illustrate cross sectional features of the dried materials, which are in this case noodles, a so-called instant food for home cooking in Fig. 1, and everyday washed clothing in Fig. 2. These specimens are obtained by each of the methods of the new fields described above. The cross sections shown in Figs. 1(S) and 2(S) are for superheated vapor drying. As indicated in Fig. 1(S), when noodles are dried quickly in superheated vapor, they become porous, and then become quickly softened in hot water. But when noodles are dried in air as shown in Fig. 1(A), they become hard.

The drying of the washed blanket shown in Fig. 2(s) was performed by superheated vapor. After drying, the blanket becomes soft and fluffy, and in addition, cleaner, due to the sterilizing properties of superheated vapor.

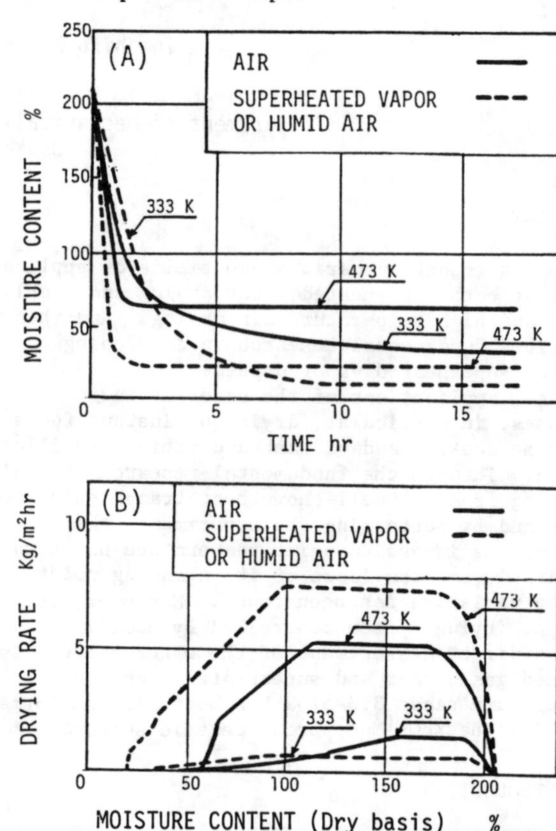

Fig. 3 Decreasing weight curves (A) and drying curves (B) of waste molasses

When a viscous liquid such as adhesive is dried by air, a thin skin grows on the surface and diffusion on the inside liquid becomes difficult. Fig. 3 shows the drying of a adhesive waste molasses in highly humid air and in superheated vapor. As shown in Fig. 3, the superheated vapor drying without the growth of thin skin is effctive for drying of the waste molasses.

3. BEHAVIOR OF INVERSION POINT TEMPERATURE

3.1. Inversion Point Temperature

The phenomena concerned with the inversion point temperature, which has been previously ascertained in study[4], is shown in Fig. 4. Above this temperature, more water evaporates in humid air than in dry air, and below this temperature, the water evaporation rate becomes the opposite. In order to confirm this phenomenon, the values of the water evaporation rate were calculated by using the well-known formulae and by using the fundamental model of the vapor dryer.

Fig. 5 shows schematically the fundamental

construction of the vapor dryer, which is a basis of the calculation of the rate. As shown in Fig. 5, the plate is taken as the representative drying material.

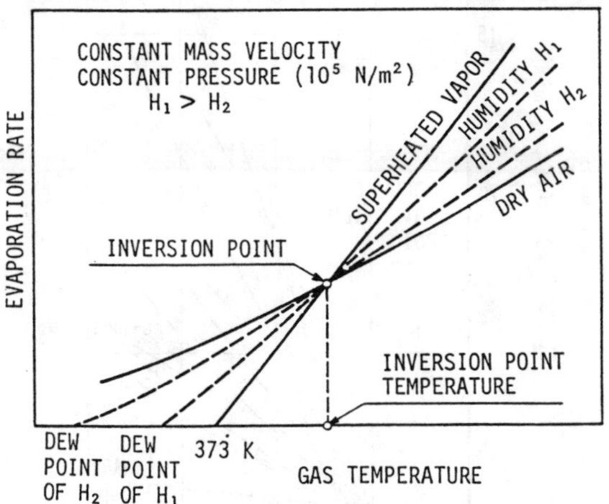

Fig. 4 Relation between evaporation rate and gas temperature

The calculations contain the following assumptions;
1. the drying material is water and the distribution of the temperature of the water is constant,
2. the drying agents except dry air emit radiant heat, and
3. the conductive heat from the water vessel is not considered.

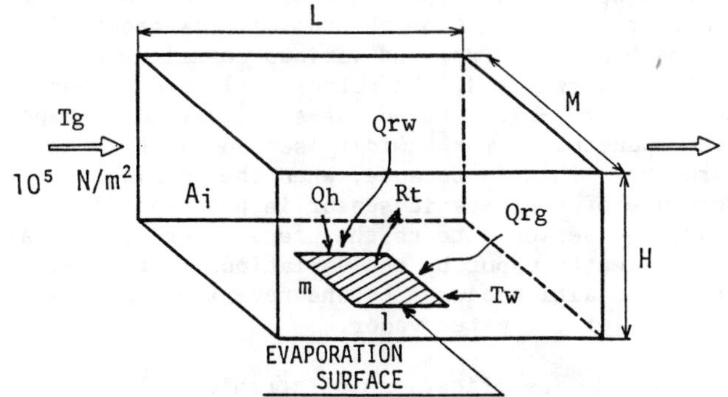

Fig. 5 Fundamental construction of vapor dryer

A total water evaporation rate Rt kg/m^{2}hr is calculated by the following equations:

$$Rt=(Qh+Qr)/r_w' \quad \cdots\cdots\cdots\cdot(1)$$

$$Qh=\lambda\cdot Nu(Tg-Tw) \quad \cdots\cdots\cdots\cdot(2)$$

$$Qr=Qrw+Qrg \quad \cdots\cdots\cdots\cdot(3)$$

here, r_w' is the latent heat of water surface, allowing the Ackermann effect[5],
Tg is the temperature of drying vapor,
Tw is the temperature of water surface,
Qh is the convection heat from drying vapor,
Qrw is the radiant heat from the dryer wall, and
Qrg is the radiant heat from the drying vapor.

Fig. 6 shows the calculated results of the water evaporation rate Rh (=Qh/r_w') from a horizontal plate surface by using Pohlhausen's transmission formula of Nu number, and the evaporation rate from the surface of a sphere by using the Ranz and Marshall formula.

Where, the transmission formula by Pohlhausen is

$$Nu=0.664Re^{1/2}\,Pr^{1/3} \quad \cdots\cdots\cdots\cdots(4)$$
(Pr>0.6, for laminar flow region),

and formula by Ranz and Marshall is

$$Nu=2+0.6Re^{1/2}\,Pr^{1/3} \quad \cdots\cdots\cdots\cdots(5)$$
($0.6<Pr<380$, $1<Re<10^5$).

These are recognized by almost all investigators. As noted in Fig. 6, it was observed that the curves of the water evaporation rate cross at the temperature of the same point, namely the inversion point temperature.

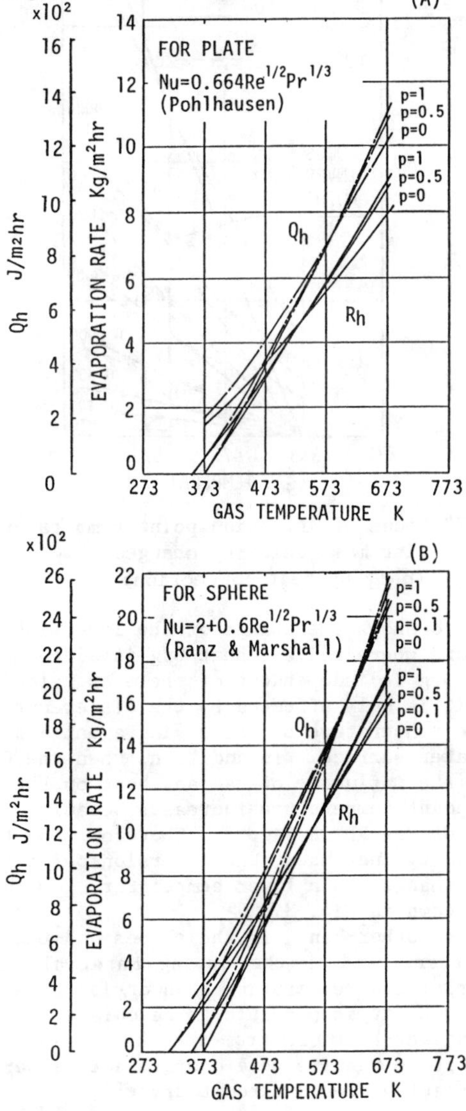

Fig. 6 Inversion point temperature of a horizontal plate surface and of a sphere

3.2. Locus of Inversion Point Temperature

The locus is determined by changing the drying variables, in which the mass velocity G kg/m^2hr and the radiant heat Qrg of the superheated vapor are dealt with, which are applied in order to decide the controlled operation or optimum design conditions of the vapor dryer.

The locus of the inversion point temperature for a plate, as regards the water evaporation rate Rh or the convection heat Qh from the drying vapor, changed when the mass velocity was changed as shown in Fig. 7. As seen from Fig. 7, the inversion point temperature for the plate remained at a fixed temperature (about 533K) regardless of the mass velocity G.

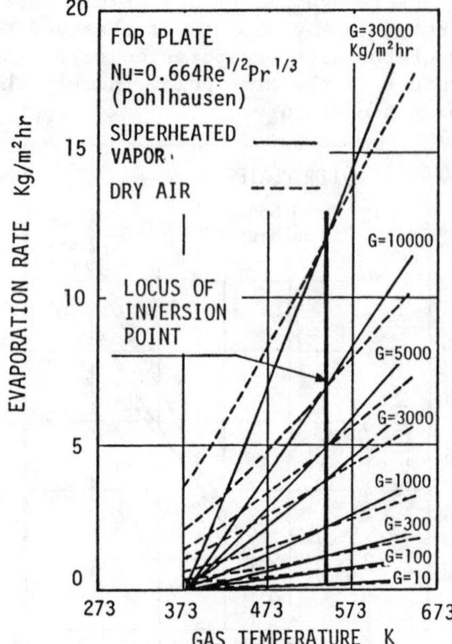

Fig. 7 Locus of inversion point temperature with mass velocity changed (only by heat convection)

Fig. 8 shows the change of the inversion point temperature where radiant heat Qrg is added to Qh.

The Qrg is independent of the mass velocity G, because it is only affected by the difference between the temperature of the drying agent and that of the water surface. In addition, when the G is smaller, the influence of radiant heat on the inversion point temperature increases because of the increase of the Rh in Fig. 7. Therefore, as the mass velocity increases, the inversion point temperature changes to a lower temperature towards 373K as shown in Fig. 8.

On the other hand, if the shapes and the sizes of the dryer, and of the drying material, and the material and temperature of a drying chamber's wall are set, it is possible to calculate the value of the radiant heat Qrw from the chamber's wall.

The above results were confirmed by experimental use of a closed circuit dryer[6].

4. HUMIDITY SENSING IN HIGH TEMPERATURE

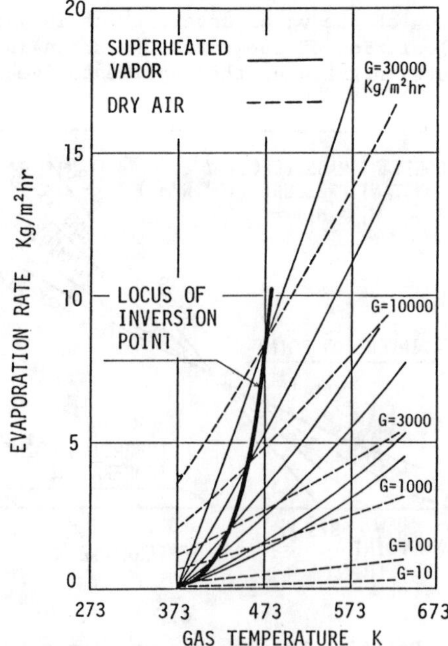

Fig. 8 Change of inversion point temperature where Qrg is added to Qh

In order to keep a good quality of the materials placed in the superheated vapor dryer, the control of humidity and temperature in the dryer is imperative.

At present, much about humidity sensors has been reported, and recently, ceramic sensors have been used with improved results. However, the effective operating temperature of these sensors for a direct measurement of the humidity is about 423K at the highest. From the various ceramic sensors so far investigated, the authors selected composite seramics ZrO_2-MgO with porous structure and n-type semiconductor[7] for humidity sensing in high temperature. This is because, when the working temperature of the ceramic sensor is between 673 and 973K, the sensor detects the presence or concentration of water vapor by the variations in the electronic condition caused by the reversible chemisorption of the water vapor.

Table 2 Specifications of ceramic sensor

Sensor characteristics	values
Operating span	
Temperature (K)	253 - 973
Water vapor content (ppmw)	$10^5 - 10^6$
Working temperature (K)	773 - 973
Response time	
Adsorption (s) *1	<20
Desorption (s) *2	<100

*1: $10^4 \rightarrow 10^5$ ppmw, *2: $10^5 \rightarrow 10^4$ ppmw,
Size of ceramic sensor: 1.5x1.5x0.2mm

4.1. Sensor Structure

The construction of the sensor with specifications given in Table 2 is shown in Fig. 9. The sensor consists of a ZrO_2-MgO ceramic, heater coil, ceramic base, terminals and mesh cover. A scanning electron micrograph of typical fracture section is shown in Fig. 10. The fractured sections exhibit intergranular pores in combination with raised openings. The porosity is from 6 to 12 percent; the pore size is under 1μm, and the grain size from 0.1 to 0.5μm.

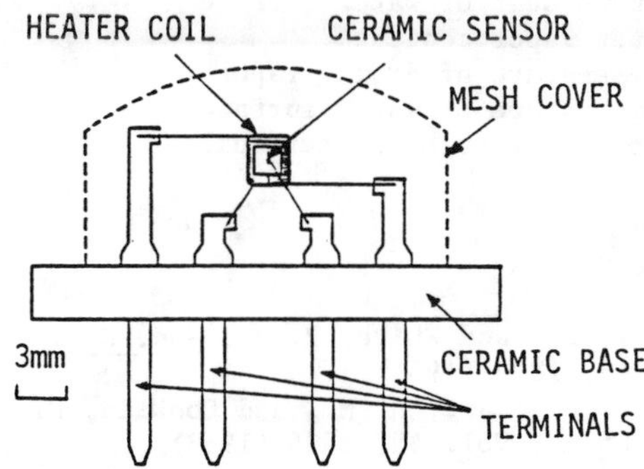

Fig. 9 Ceramic sensor

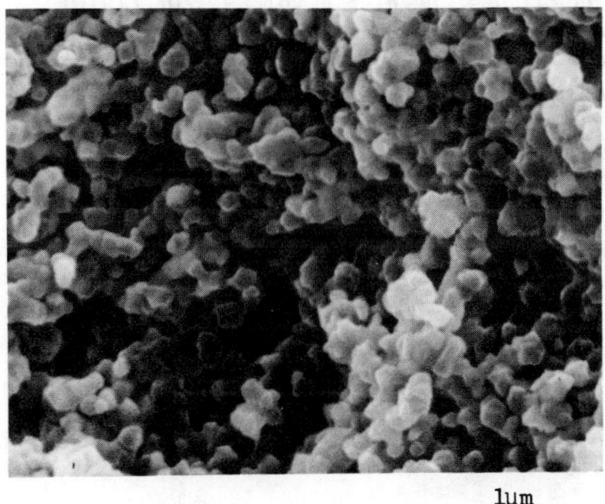

Fig. 10 Scanning electron micrograph of typical fracture section of ZrO_2-MgO ceramic

4.2. Sensor characteristics

As shown in Fig. 11, when the temperature of an ambient atmosphere is 523K constant, the electronic resistance is reduced by adsorption of water vapor in the atmosphere. In addition, as the working temperature becomes higher, the resistances decrease.

Fig. 12 shows the relationships between the water vapor content (ppmw) and the sensor resistance by changing the working temperature of the sensor. As the water vapor content is increased from 4×10^5ppm to 10^6ppm, the resistance decreases.

The relationships such as in Fig. 12 were experimentally confirmed within the following working temperatures (373 - 573K).

A characteristic line shown in Fig. 13 was obtained, independent of the ambient atmospheric temperature, by regulating the input voltage of the

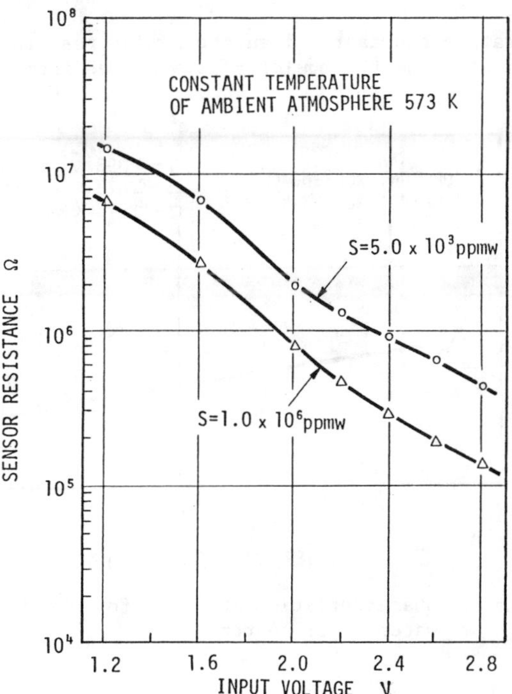

Fig. 11 Input voltage of heater coil depending on resistance in air containing water vapor (5×10^5 & 1×10^6 ppm)

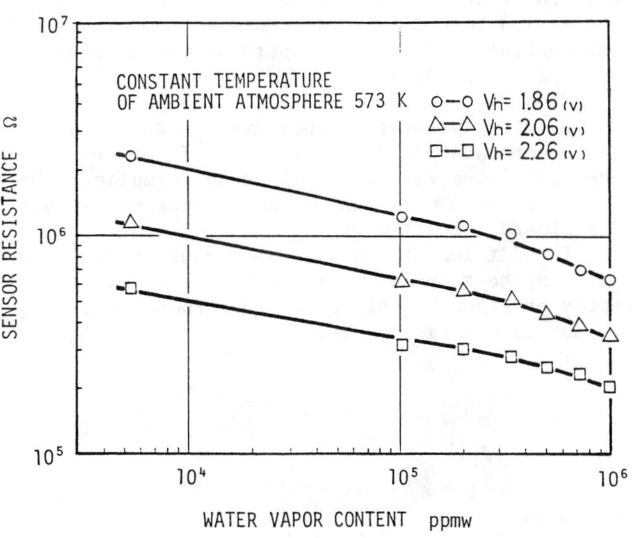

Fig. 12 Relations between water vapor content and sensor resistance by changing working temperature

heater coil so as to make a working temperature constant (823K in this test).

From these results, the selected sensor in this test has been to be adaptable to the detection of water vapor in humid air or of superheated vapor which has a water vapor content of 10^6 ppmw.

According to the characteristic line shown in Fig. 13, a method for obtaining the water vapor content is described. The ambient atmospheric temperature is first measured. The input voltage of the coil is regulated according to the atmospheric temperature in order to make the working temperature constant. Then the sensor resistance is measured and the humidity is obtained from Fig. 13.

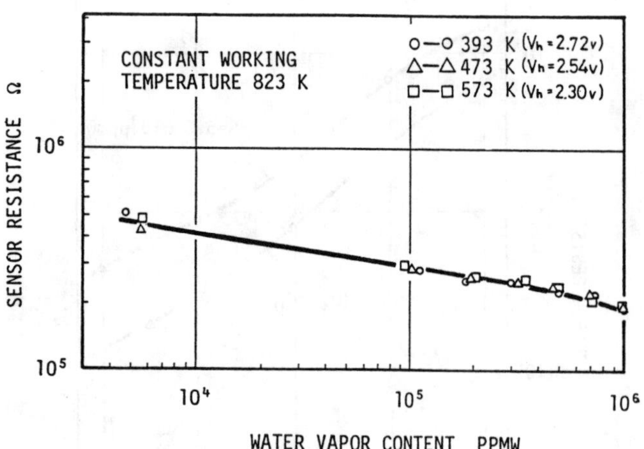

Fig. 13 Characteristic line used for obtaining water vapor content

5. CONCLUSIONS

Superheated vapor has been applied in many industrial drying fields because of its merits. In addition, recently, new fields of superheated vapor application have come to extended to foodstuffs for home cooking or to drying apparatus which is used to simplify our daily life.

From the standpoint of the fundamental research into superheated vapor drying, the inversion point temperature and the behavior of its locus were calculated with the well-known formulae. The desired results were confirmed by experimental use of a closed circuit dryer.

In addition, the water vapor content ($10^2 - 10^6$ ppmw) in the temperature range 373 - 573K was directly obtained by taking the resistance of a ZrO_2-MgO porous ceramic sensor.

NOMENCLATURE

Ai	Sectional area of dryer	m^2
G	Mass velocity	kg/m^2hr
Qh	Convection heat from drying vapor	J/m^2hr
Qr	Total radiant heat	J/m^2hr
Qrg	Radiant heat from drying vapor	J/m^2hr
Qrw	Radiant heat from dryer wall	J/m^2hr
Rt	Total water evaporation rate	kg/m^2hr
r_w'	Latent heat of water surface	J/kg
S	Water vapor content	ppmw
Tg	Temperature of drying vapor	K
Tw	Temperature of water surface	K
Vh	Input voltage of heater coil	V

REFERENCES

1. Wenzel, L., and White, R. R., Ind. Eng. Chem., Vol. 43, 1892 (1951).
2. Chu, J. C., Lane, A. M., and Conklin, D., Ind. Eng. Chem., Vol. 45, 1586 (1953).
3. Toei, R., Okazaki, M., Kubota, K., Ohashi, K., and Mizuta, K., Chem. Eng. Japan, Vol. 30, 43 (1966).
4. Yoshida, T., and Hyodo, T., Ind. Eng. Chem. Process Des. Develop., Vol. 9, 207 (1970).
5. Toei, R., Chem. Eng. Japan, Vol. 25, 65 (1961).
6. Nomura, T., Hiwatashi, H., and Hyodo T., Proc. Jpn. Soc. Mech. Eng., Vol. 834-7, 101 (1983).
7. Fukushima, F., Makimoto, R., Terada, J., and Nitta, T., "Humidity Sensor Neo-Humiceram", Nat. Tech. Rept., Vol. 29-3 Jun. (1983).

OPTIMIZATION OF HEAT PUMP DEHUMIDIFIER

Czesław Strumiłło
Romuald Żyłła

Institute of Chemical Engineering, Łódź Technical University
ul. Wólczańska 175, 90-924 Łódź, POLAND

ABSTRACT

The paper presents two mathematical models of the air drying with heat pump applied to an adiabatic dryer.

Elaboration of these models allows to optimize the system operation in order to decrease energy consumption of drying by adjusting refrigerant boiling temperature in the heat pump.

It was found that the difference between air dew point temperature and boiling temperature ($T_{DP}- T_{ev}$) correlates well the process data.

The conclusions drawn from the analysis of optimization results were checked out on the test rig. It was observed that the increase of system heat losses cause that the optimum value of the control parameter ($T_{DP}- T_{ev}$) moves towards its higher values in the region of 18-20 K and the specific energy consumption (SEC) values increase. Due to constant swept volume of the heat pump compressor tested it is necessary, however, to adjust the mass flow rate of air dried. From the experimental data it follows that good insulation and leak proofing of the system are of great importance for obtaining low (SEC) values.

1. INTRODUCTION

Convective drying is one of the most energy-consuming processes. Therefore, energy consumption is subject to thorough investigations and attempts at decreasing energy demands of the drying process are made.

One of numerous methods for heat recovery in a dryer is the application of a heat pump. However, this solution is profitable under specific operation conditions in a heat pump-dryer system.

The paper presents factors that influence the operation of a compression heat pump in a drying system. A method for optimization of this system operation is proposed, first by the solution of a mathematical model of this process, and then in the investigation of an experimental system.

2. THEORY

Heat pump due to power input W, can transfer heat Q_1 taken from the source of temperature T_{source} to heat sink of higher temperature T_{sink}. The amount of heat delivered to the sink is $Q_2 = Q_1 + W$, and the Q_2/W ratio is called the coefficient of performance (COP). The ratio of heat delivered Q_2 to power input for driving the heat pump is called the actual coefficient of performance $(COP)_A$. The efficiency $(COP)_A$ of the vapour compression heat pumps reaches about a half of the (COP) of Carnot cycle for the same temperature range [1].

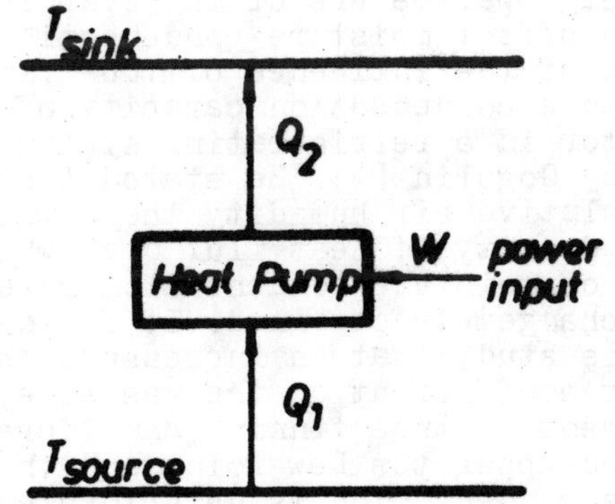

Fig. 1 Heat pump principle

It should be stressed that thermodynamic operation principles of heat pumps and refrigerators are the same, the only difference being the aim of their operation.

Among many known types of heat pumps the most widely used are vapour compression heat pumps because of the highest values of $(COP)_A$ obtained under comparable conditions. The subject of this paper is a vapour compression heat pump with a closed cycle of refrigerant.

Convective drying with the heat pump can be carried out either at an open drying gas cycle [2,3] or at a closed cycle, i.e. at a complete recirculation [4,5,6]. At the open cycle the heat pump takes heat from outlet gases discharged from the dryer and it heats up inlet gas to a much higher temperature. As shown in the previous paper [7] the drying system with the open gas cycle is characterized by worse power efficiency than the system with a closed gas cycle.

In the system with closed gas cycle which is discussed below, the basis of operation is dehumidification performed on the surface of heat pump evaporator and moisture removal in the liquid state out of the system. To make dehumidification possible, the evaporator surface temperature must be lower than the air dew point temperature. Heat of moisture condensation is transmitted by the heat pump to drying gas that flows to the dryer.

The mechanism of air dehumidification on a cooled surface was described by Goodman [8]. The process trajectory depends on heat transfer resistances on both sides of the heat exchanger tube and on the flow type of both media. From the point of view of dehumidification capacity the lowest possible temperature of the evaporator coil is advantageous. This is obtained at low temperatures of refrigerants and at high values of heat transfer coefficient on the cooling medium side. This condition is usually satisfied if the cooling medium is boiling refrigerant. On the other hand, when in the evaporator superheating of refrigerant vapour occurs then the evaporator tube temperature increases respectively and moisture condensation capacity decreases.

Design parameters of the evaporator may also affect moisture condensation. Description of the influence of tube finning on moisture condensation capacity of the evaporator in a refrigerating system is presented by Gogolin [9]. He stated that for given relative air humidity there was an optimum density of heat flux q at which a maximum of moisture condensation rate to heat exchanged is obtained. It follows also from this study that an increase of heat transfer coefficient on the gas side, or enhancement of tube finning can increase a so called Apparatus Dew Point and therefore moisture condensation is deteriorated. From the study carried out by Shaw [10] for several air conditioning heat exchangers it follows that the highest dehumidification capacity is characteristic for shallow heat exchangers of large face area and a short way of air. Calculations made for dehumidifiers with closed air circulation by Geeraert [6] prove that an increase of dried air temperature of the same relative humidities can decrease significantly the energy consumption for moisture condensation and thus for drying. Therefore, studies and applications of high-temperature heat pumps grow in number [11,12] etc.

3. ENERGY EFFICIENCY

In order to determine energy consumption in drying with a heat pump definitions of Specific Moisture Extraction Rate (SMER) [12,13] and of Specific Energy Consumption (SEC) [7,13] have been introduced.

$$\text{SMER} = \frac{\text{kg moisture evaporated}}{\text{kWh of power input}} \quad (1)$$

$$\text{SEC} = \frac{\text{heat pump power consumption}}{\text{dehumidification rate}}$$

$$\frac{\text{kW}}{\frac{\text{kg water}}{\text{s}}} \quad (2)$$

To compare the systems in which various types of dryer heating and heat pump compressor driving, i.e. electrical or gas engine are applied, the term Primary Energy Consumption (PEC) is also used [7,13].

Many authors [14,15,3,12] give as a typical specific energy consumption (SEC) 700-1500 kJ/kg moisture, lower power consumption being obtained in larger installations. Most authors state that energy saving reaches 40-60%. If, however, investment costs are taken into account, the payback period is usually 3 years and the main factor encouraging heat pump application is improvement of the dried product quality.

4. EXAMPLES OF HEAT PUMP APPLICATIONS

Heat pumps are applied in drying of wood, plaster blocks and ceramics, i.e. the materials which shrink, split and warp while being dried in severe drying conditions. Some case studies of applications of commercial heat pump system have been presented by Oliver [13].

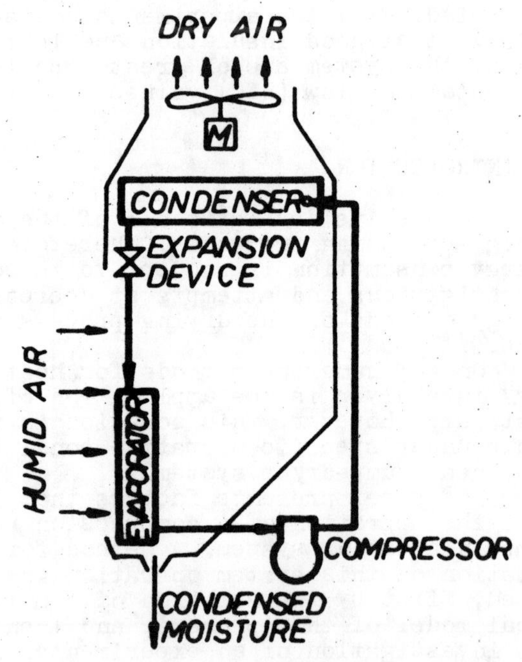

Fig. 2 Commercial high-temperature dehumidifier

In the system presented in Fig. 2 an increased gas flow through the condenser in order to limit the maximum temperature of the refrigerant was applied. On the basis of experience we can say that this causes an increase of $(COP)_A$ and allows standard refrigeration components to be used. In this system $(COP)_A = 3.7$ is claimed. In large systems for drying plaster blocks several heat pumps operating in various temperature ranges were applied concurrently [14,15,3] and thus for each heat pump advantageous values of $(COP)_A > 5$ were obtained.

Heat pump dehumidification is also applied in drying of corn [16], malt [17,3] and tea leaves. Analysis of these cases shows that the method is employed mainly in the constant drying rate period, i.e. when air parameters are approximately constant despite that the process is of a batch type. In the vast literature dealing with heat pump drying systems there are no design instructions concerning optimization or selection of heat pumps for the drying process into which heat pumps are to be introduced. Usually, heat pumps are adjusted using a trial-and-error method and therefore a systematic method useful in other cases cannot be formulated. Among many authors only Flikke et al. [16] stated that in grain drying the lowest total power consumption by heat pump and air blower was attained at air flow rate equal 0.15 m^3/s for an evaporator of cooling capacity equal 2.5 kW, at $0.76 \cdot 10^{-3}$ kg water/s being condensed. In the system (SEC) = 1100 kJ/kg water was obtained. The quoted figures refer only to the system investigated by the authors at air temperature at the dryer inlet limited to 43°C. Heat pump temperature range $(T_{co}-T_{ev})$ was at the optimum point equal to 33°C.

5. REMARKS ON HEAT PUMP DEHUMIDIFIER DESIGN

Numerous studies discussing the application of heat pumps in drying rarely provide information on the methods of designing the systems which would operate efficiently. The data contained in air-conditioning handbooks are not very useful as in air-conditioning dehumidification is to be avoided. Nevertheless, from the so far published literature data some design indications can be taken. They may be presented briefly as follows.

- At high values of air humidity low energy consumption is attained.
- Heat pump temperature range $(T_{co}-T_{ev})$ should not exceed 40°C in order to maintain a high value of $(COP)_A$.
- The surface of condenser tubes and fins is usually designed twice as large as the evaporator surface.
- The evaporator should have large face area and short air path.
- The distance between fins must be large enough to prevent condensed water from being retained between the fins.
- Gas velocity should be smaller than 2 m/s so that gas does not carry away droplets.
- The mass flow rate of gas through the condenser should be much higher than in the evaporator of heat pump to decrease T_{co} and increase $(COP)_A$.
- In a chamber dryer with a heat pump higher gas velocities are usually applied.
- Additional heaters for heating the dried material should be used.

The above mentioned remarks are not sufficient, however, to design an optimum heat pump-dryer system. Lack of experimental data in the literature does not allow to verify Geeraert's theoretical considerations on the optimum, from the (SEC) point of view, choice of air enthalpy change in the heat pump evaporator at higher temperatures. There are not sufficient data to choose properly heat pump dehumidifier (HPD) capacity depending on the quantity and type of dried material. There are no experimental data concerning the choice of flow rate of evaporator air stream and refrigerant evaporation temperature either. Thus, it is not possible to adjust these parameters to given drying process and type of material being dried. It is not known either, whether due to an increase of process temperature by supplying additional heat, i.e. by intensifying the drying process, specific energy consumption (SEC) is minimized.

Properly chosen refrigerant evaporation temperature and dried air flow rate are of basic importance for the optimum operation of the system discussed.

The paper presents results of model and experimental investigations carried out for simulation of the heat pump-dryer system. The dependence of dehumidification capacity on the difference between dew point temperature of air being dried and evaporator coil temperature will be discussed in detail. The dependence of dehumidification capacity on the difference between air dew point temperature and refrigerant evaporation temperature as well as the dependence of dehumidification capacity on dried air flow rate will be determined experimentally.

6. OPTIMIZATION OF HEAT PUMP-DRYER SYSTEM WITH CLOSED AIR CYCLE

The model of air drying process, apparatus taking part in the process, heat and mass balance equations, and kinetic heat and mass transfer equations are the basis for the process optimization.

In the present paper a simplified mathematical description of the process in a dryer and heat pump is applied. Only the model of air drying on the evaporator surface is more complex and is based on a correlation equation derived from the experimental data.

Two mathematical models of the system denoted as model A and model B will be presented below. The system described by the model A is characterized by low heat losses, while the model B system is of the high heat losses.

6.1. The Mathematical Model of Heat Pump--Dryer System of Low Heat Losses. Model A

Figure 3 presents a schematic diagram of the process in a model A system of adiabatic dryer with a heat pump with closed gas cycle.

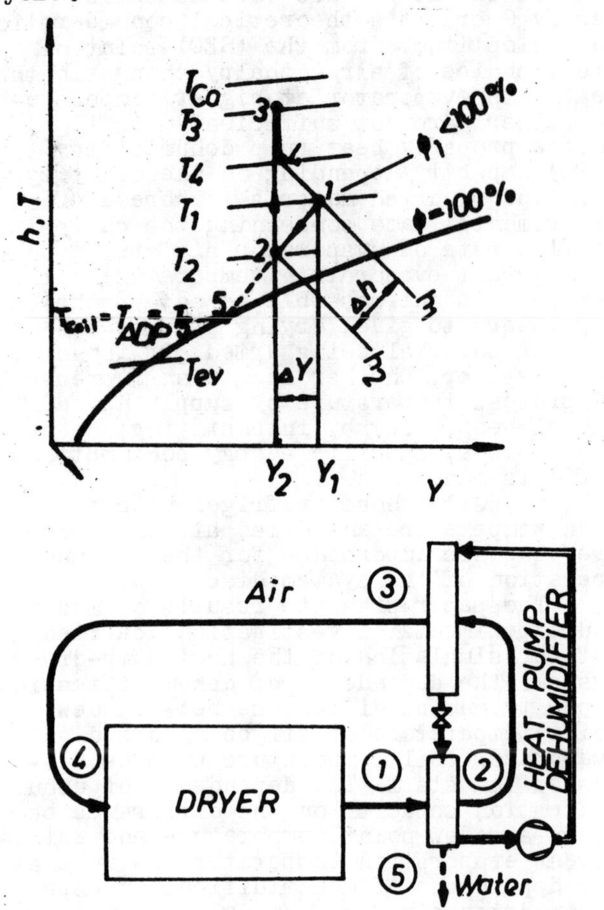

Fig. 3 Air dehumidification cycle

Point 1 represents air humidity and temperature at the dryer outlet. Air flowing through the evaporator is cooled down to an average temperature T_2. Simultaneously moisture is condensed. It leaves the system at the mean coil temperature T_{ADP}. The air flowing next through the heat pump condenser at the temperature T_{co} is heated up to the temperature T_3. Enthalpy at point 4 is lower than that at point 3 due to heat losses and/or to material heating within the dryer. It was assumed that the process between points 4 and 1 is the process of an adiabatic air humidification during the constant drying rate period. The amount of heat delivered by the condenser $Q_{co} = G_A \cdot c_p \cdot (T_3 - T_2)$ is larger than the amount of heat absorbed in the evaporator $Q_{ev} = G_A \cdot (h_1 - h_2)$ due to power input of the compressor which balances heat losses of the system.

The process of moisture condensation denoted by points 1, 2 and 5 in Fig. 3 is a process of a simultaneous heat and mass transfer in a quite complex geometry of heat exchanger finned coil. In a simplified model of heat transfer it is assumed that in the evaporator the refrigerant is boiling under a constant pressure and at a constant temperature. It was taken after Geeraert [6] that the difference between evaporation temperature T_{ev} and coil temperature T_{coil} is 5 K, in practice, however, it may range from 3 to 15 K [9].

If coil temperature is lower than dew point of the inlet air, then on the heat exchanger surface moisture condensation occurs. Heat and mass transfer between air and coil can be described by eq.(3) obtained experimentally by Nagaraja and Krishna Murthy [18] for air conditioner heat exchanger.

$$Nu = 0.0627\, Re^{0.502} \cdot Pr^{1/3} \cdot Syl^{-0.02} \quad (3)$$

where

$$Nu = \frac{h_E \cdot G_A \cdot c_p \cdot d_{equ}}{A_t \cdot \eta A}$$

$$Re = \frac{G_A \cdot d_{equ}}{A_f \cdot \eta A}$$

$$Syl = \frac{c_p \cdot \Delta T}{\Delta h_v \cdot \Delta Y}$$ sensitive heat to latent heat ratio

$$h_E = \frac{h_1 - h_2}{h_1 - h^*_{ADP}}$$ enthalpy exchange efficiency

This equation allows to determine the enthalpy exchange efficiency h_E. In optimization calculations dimensions of the heat exchanger being considered in the paper [18] and constant air velocity 3.7 m/s were taken.

As an optimization parameter specific energy consumption (SEC) was taken (eq.(4))

$$(SEC) = \frac{W}{G_M} = \frac{\frac{Q_{ev}}{(COP)_A - 1}}{G_A \cdot (Y_1 - Y_2)} = \frac{h_1 - h_2}{((COP)_A - 1) \cdot (Y_1 - Y_2)} \quad \frac{kJ}{kg\, H_2O} \quad (4)$$

For the process assumed air in the condenser will be heated up by $Q_{co} = Q_{ev} + W$ to the temperature $T_3 = T_2 + Q_{co}/(G_A \cdot c_{p2})$. For simplicity it is assumed that in the condenser freon condensation temperature is higher by the constant temperature difference $\Delta T = 5$ K than the final gas temperature and equals $T_{co} = T_3 + 5$, which in practice can be obtained at heat transfer surface of the condenser twice as large as that of the evaporator. The Actual Coefficient of Performance $(COP)_A$ was assumed to be 50% of Carnot cycle (COP)

$$(COP)_A = 0.5 \frac{T_{co}}{T_{co} - T_{ev}} \qquad (5)$$

On the basis of the above presented assumptions the optimization calculations were made. A target function was specific energy consumption, and a control parameter was the difference between dew point temperature of dryer outlet air and evaporator coil temperature. Figure 4 presents relevant results for various temperatures and humidity of air at the dryer outlet.

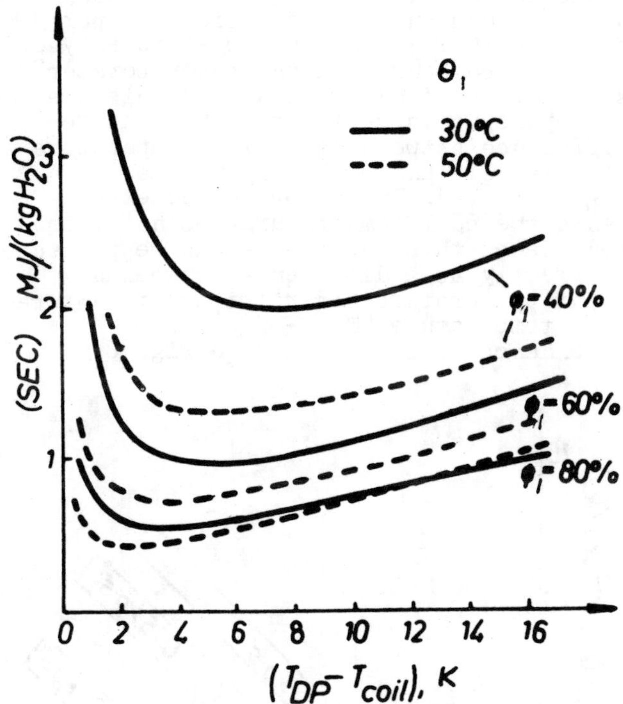

Fig. 4 Dehumidifier specific energy consumption. Model A

The minima occurring on the curves point that there are optimal parameters of heat pump-dryer system operation. It is worth noting that for low values of relative humidity, the minima are shifted towards higher temperature differences $(T_{DP} - T_{coil})$. Simplifications and assumptions concerning heat losses made in the calculations according to the above model, cause that each point of the curve in Fig.4 represents dryer model of different share of heat losses in the system heat balance and of various conditions of mass transfer in the dryer. Thus, the (SEC) resulting from these calculations can be treated as a minimum which can be attained theoretically in various drying systems for the same heat pump and at relatively small heat losses.

6.2. Simulation Model for the Dryer-Heat Pump System with High Heat Losses. Model B

In the model B the same equations describing evaporator characteristics, i.e. moisture condensation and heat transfer are applied as in model A. $(COP)_A$ of the heat pump obtained for the water-water heat pump was assumed after Gutierrez et al. [19]

$$(COP)_A = 0.57 \frac{T_{co}}{T_{co} - T_{ev}} + 0.448 \qquad (6)$$

It is assumed additionally that thermal power input Q_{AD} is supplied to the system and the heat loss value is defined by the equation

$$Q_{loss} = \alpha \cdot A \cdot (T_4 - T_{Amb}) \qquad (7)$$

where $(\alpha \cdot A)$ is taken as 100 W/K and $T_{Amb} = 293$ K = 20°C.

In this model it is also assumed that material is dried in the constant **drying** rate period according to the equation given by Perry [15],

$$\dot{m}_e = 0.013 \cdot \left(\frac{G_A}{A_D}\right)^{0.8} \cdot (Y_{WB} - Y)_m \qquad (8)$$

where: $\dot{m}_e$ - drying rate, $kg/(m^2 s)$;

and the dried material surface area is $A_W = 20\ m^2$, mass flow rate is $G_A = 0.3$kg/s, and the dryer cross section area is $A_D = 0.5\ m^2$. In the calculation of the (SEC) the energy consumption in the heat pump and heater was summed.

Due to such assumptions the values of temperature and humidity in the dryer are self adjusting at such a level that there is an equilibrium of heat input and losses, moisture evaporation and condensation. Thus, both the gas temperature and humidity in the dryer will depend on heater power Q_{AD}. Figure 5 presents results of simulation calculations for the system according to model B at various values of the control parameter $(T_{DP} - T_{coil})$. On the curves outlet air parameters are marked for the optimum point. It is worth noting that the change in parameter $(T_{DP} - T_{coil})$ apart from changing (SEC) causes a change in air parameters, i.e. T and Y at a constant value of heat added Q_{AD}. In this case, contrary to the model A, with an increase of outlet air temperature, the optimum temperature difference $(T_{DP} - T_{coil})$ increases, and for the temperature $\theta_1 = 37$°C ($\varphi_1 \approx 60\%$) corresponding to the additional power $Q_{AD} = 2$ kW the optimum is obtained for the value of control parameter $(T_{DP} - T_{coil}) \approx 20$K

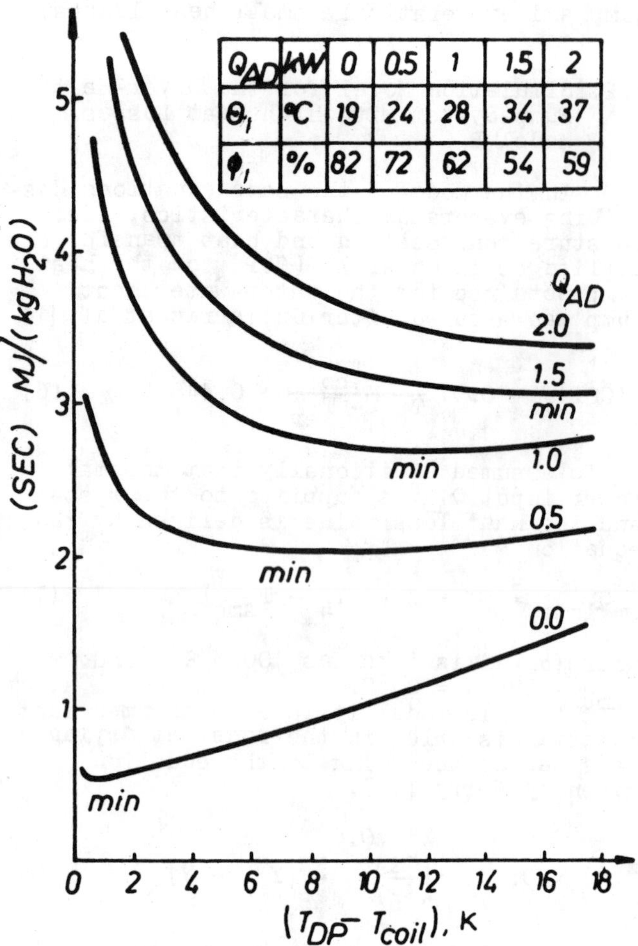

Fig. 5 Dehumidification specific energy consumption. Model B system simulation

Since the model B represents more closely the investigated heat pump-dryer system, it should be expected that for real system the dependence considered would be similar.

Mathematical modelling of the heat pump-dryer system taking into account most of the system parameters such as, for instance, the type of working fluid, compressor characteristic, requires a very detailed mathematical model of the evaporator (cf. the study of Hodgett and Lincoln [21]) as well as of the dryer and other components to be built.

Another way for optimizing the system operation is an experimental determination of controllable parameter values so as to obtain the minimum power consumption per unit of mass of condensed moisture.

7. EXPERIMENTAL OPTIMIZATION

The experimental system was designed aiming at investigating the effect of heat pump control on specific energy consumption and also a possibility of testing the interdependences between controlled variables and other parameters in the heat pump-dryer system.

The heat pump was designed to be used in the investigations on air drying of temperature 30-70°C and relative humidity 30-70%, the system being intended to operate in the middle of that range. On the account of high temperatures expected a refrigerant Arcton 114 was used. In order to simulate the adiabatic dryer operation a humidification chamber was built. In that chamber fabric of total wet surface 17.8m^2 was employed. The spraying water flow rate was controlled in the range from 0 to 0.02 kg/s. A detailed description of the experimental system was presented in the previous paper [22].

According to the earlier considerations the specific energy consumption (SEC) depends on the coefficient of performance $(COP)_A$ and on the efficiency of moisture condensation on the evaporator surface. The flow of condensed moisture G_M depends on air flow rate and on the difference between dew point temperature of dried air and coil temperature $(T_{DP} - T_{coil})$. In the presented experiments the coil temperature was not measured. Hence this dependence can be presented indirectly as a difference between dew point temperature and refrigerant evaporation temperature $(T_{DP} - T_{ev})$. The above dependence is illustrated in Fig. 6.

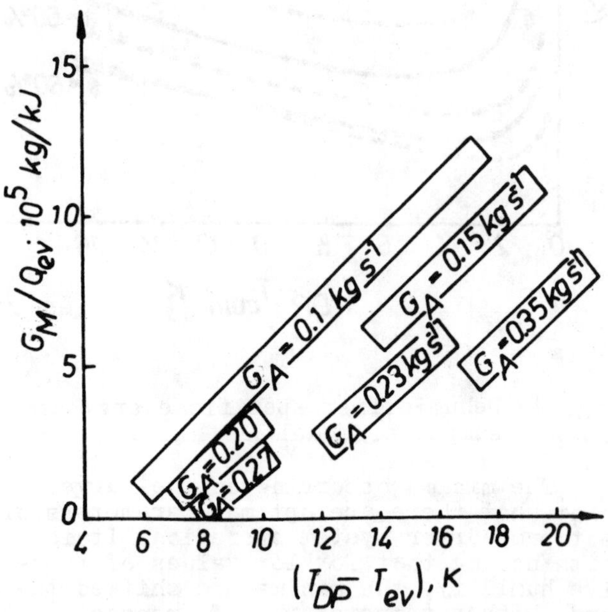

Fig. 6 Dehumidification capacity vs. control parameter $(D_{DP} - T_{ev})$

The points were collected in over 110 experiments carried out in the range of air temperature being 30 to 50°C and at humidities Y = 0.02 ~ 0.035 kg H_2O/kg dry air.

From the data given in Fig. 6 it follows that dehumidification ratio defined as $G_M/Q_{ev} \simeq \Delta Y/\Delta h$ increases with an increase of the difference in controlled parameter $(T_{DP} - T_{ev})$. When the gas stream flowing through the evaporator increases, the dehumidification ratio decreases significantly. For the temperature difference $(T_{DP} - T_{ev}) < 4K$ no moisture condensation occurs. When the air stream exceeds 0.25 kg/s condensation does not take place even for the temperature difference $(T_{DP} - T_{ev}) \simeq 7K$. In the evaporator refrigerant vapours are usually superheated. This is advantageous for the compressor since ensures its proper performance, but it is disadvantageous for the moisture condensation process, as on the evaporator outlet the coil temperature can exceed the air dew point temperature - then on part of the surface moisture condensation does not occur. A parameter that well characterizes average condensation conditions is the difference between the dew point temperature and average temperature of refrigerant in the evaporator $(T_{DP} - T_R)$. This dependence is presented in Fig. 7.

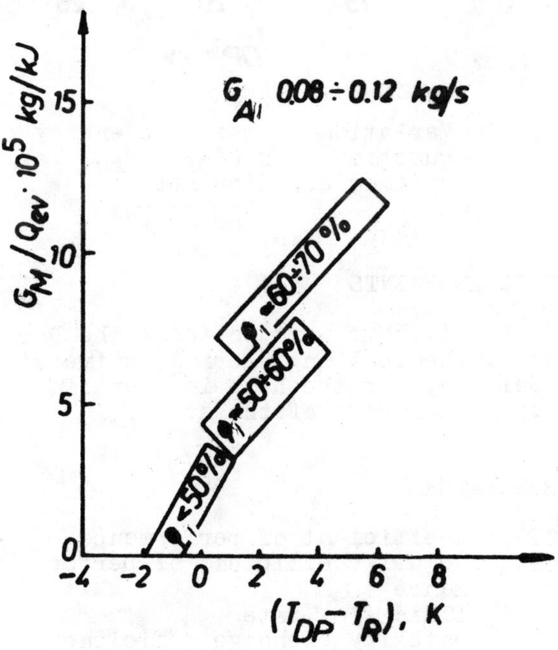

Fig. 7 Dehumidification capacity vs. temperature difference $(T_{DP} - T_R)$

A maximum value of the dehumidification ratio is obtained for $(T_{DP} - T_R) \simeq 5K$.

It follows from the above graphs that when the lowest rate of air flow through the evaporator, i.e. $G_A \simeq 0.1$ kg/s, is applied, the highest dehumidification ratio is obtained. Since this is an element constituting the formula of specific energy consumption (SEC), it can be expected that the application of low gas flow rates makes the system performance closer to the optimum. Figure 8 shows the dependence of heat pump specific energy consumption (SEC) on the expansion valve closing (additional heater energy consumption not included) for the gas flow range being considered.

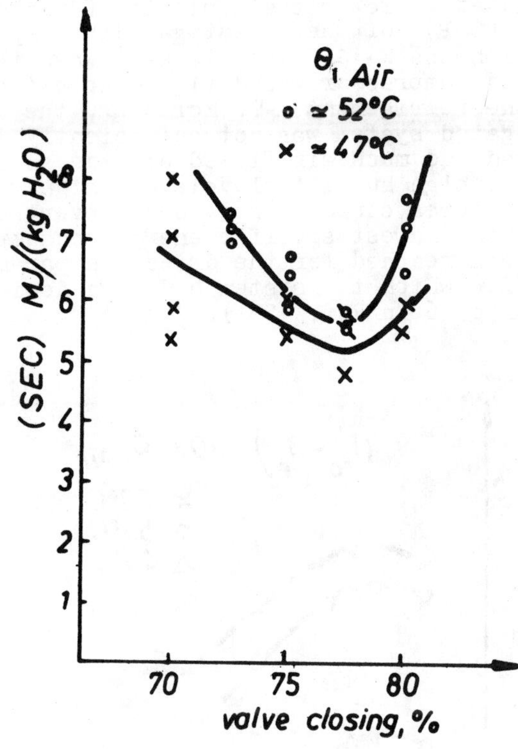

Fig. 8 Dehumidification specific energy consumption vs. valve closing

When the valve is closed too tightly the dehumidification ratio increases which is caused by a decrease in refrigerant evaporation temperature. Such a decrease in the temperature T_{ev} has, however, a disadvantageous effect on the coefficient of performance $(COP)_A$ and thus on the specific energy consumption (SEC). Figure 9 extracted from the previous paper [22] presents how $(COP)_A$ depends on the stream of gas flowing to the evaporator with heat pump temperature range $(T_{co} - T_{ev})$ kept constant. An increase of $(COP)_A$ with an increase of air temperature is a result of improved compressor efficiency at higher refrigerant vapour density. From the above graphs it follows that the maximum $(COP)_A$ is within the range of gas streams $G_A = 0.08 \div 0.12$ kg/s. This range overlaps with the one determined previously for the maximum dehumidification ratio. Thus, in the case of heat pump under consideration the range maxima of $(COP)_A$ and $\Delta Y/\Delta h$ overlap. For so selected range of gas flow rate through the evaporator the dependence of specific energy consumption (SEC) on control parameter $(T_{DP} - T_{ev})$ was investigated. The calculations of (SEC) included compressor power, additional heater and fans power input. Figure 10 presents results of these investigations. As can be observed, the minimum on curves is within the range $(T_{DP} - T_{ev})$ = 18÷20 K, irrespectively of air temperature. However, the value of this minimum is much above the theoretical value determined by models A and B. This is caused by lower coefficients of performance $(COP)_A$ of the investigated heat pump in which the R-114 fluid is used when the applied compressor was designed specifically for the refrigerant R-12. Moreover, the investigated system was not sufficiently insulated and much air flowed outside so that neither high humidity levels nor high gas temperatures could be attained. Nevertheless, the lowest specific energy consumption was reached for the dried gas parameters for which the system had been designed (cf. Fig. 6 in Ref. [23]).

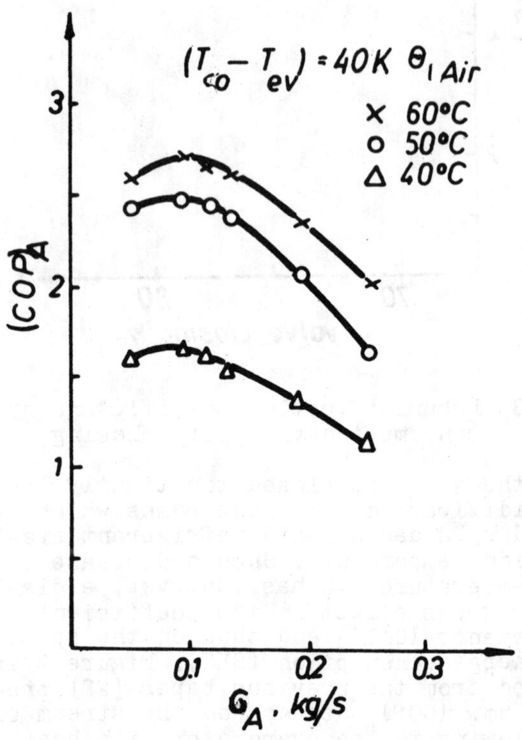

Fig. 9 Heat pump coefficient of performance vs. air flow rate

8. CONCLUSIONS

Two mathematical models of heat pump-dryer systems were proposed. The optima of specific energy consumption predicted on the basis of optimization performed for the model B which takes into account heat losses have been confirmed experimentally.

The obtained results allow for a wider application of the simulation model in predicting the dependence of specific energy consumption (SEC) on control parameters, i.e. the temperature difference $(T_{DP} - T_{ev})$ and gas flow rates through the evaporator and condenser.

From the experimental data it follows that good insulation and leak proofing of the system are of great importance for obtaining low (SEC) values. For this reason it seems that such systems operate far better in the dryers in which leakage can be avoided, e.g. in batch chamber dryers.

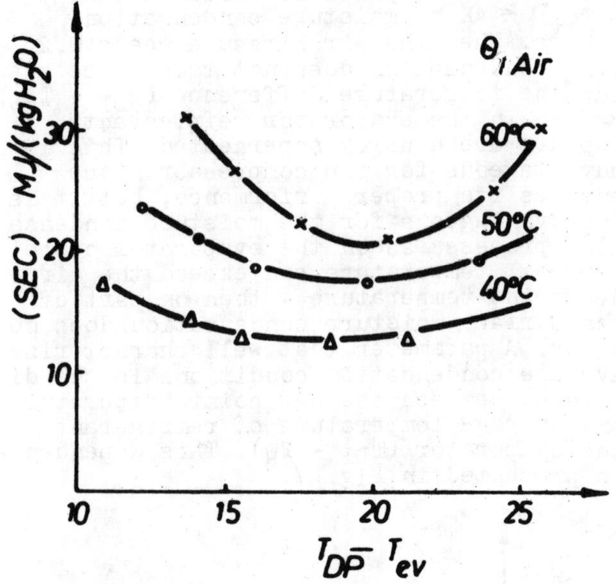

Fig. 10 Variation of specific energy consumption with $(T_{DP} - T_{ev})$ for optimum air flow rate

ACKNOWLEDGEMENTS

The Authors want to thank the Department of Chemical Engineering, University of Salford, for the help in preparing the experimental part of the study.

NOMENCLATURE

(COP)	coefficient of performance	-
$(COP)_A$	actual coefficient of performance	-
G	fluid flow rate	kg/s
h_E	enthalpy exchange efficiency	-
$\dot{m}_e$	water evaporation flux density	kg/(m²s)
Q_2	power delivered by the heat pump	kW
Q_{AD}	power of the additional heater	kW
(SEC)	specific energy consumption for air dehumidification	kJ/kg moisture
Syl	sensible heat to latent heat ratio	-
W	shaft power input to heat pump	kW
Y	air absolute humidity	kg/kg

Subscripts

A	air
ADP	apparatus dew point

co condensation
DP dew point
ev evaporation
f face area
M moisture
R refrigerant
t total
WB wet bulb

REFERENCES

1. Duminil, M., in Heat Pumps and Their Contribution to Energy Conservation, ed. E. Camatini and T. Kester, NATO Advanced Study Institute Series..., Noordhoff, pp.97-154 (1976)
2. Lascelles, D.R., and Jebson, R.S., Bull. IIF/IIR. Annexe, vol. 1 (1976)
3. Curis, O., and Laine, J.D., Int. Symp. Industrial Application of Heat Pumps, Coventry, UK, Paper C4, 99-116 (1982)
4. Kolbusz, P., Industrial Application of Heat Pumps, (see ref. 1 , pp.201-217)
5. Solignac, M., Union Internationale d'Electrothérmie VIIIth Congress, Liège Section III, Ref. no. 5 (1976)
6. Geeraert, B., Air Drying by Heat Pumps with Special Reference to Timber Drying see Ref. 1 pp.219-246
7. Żyłła, R., Abbas, S.P., Tai, K.W., Devotta, S., Watson, F.A., and Holland, F.A., Int. J. Energy Research, vol. 6, 305 (1982)
8. Goodman, W., Heating Piping Air Conditioning, vol. 10,11 (1938-1939)
9. Gogolin, A.A., Air Drying with Refrigerators, Gosiztorg, Moscow (1962),(in Russian)
10. Shaw, D.R., private communication
11. Hodgett, D.L. and Friedel, W., Commission of the European Communities Report EUR 8077 EN (1982) after J.Heat Recovery Systems, vol. 3, 91 (1983)
12. Lawton, J., Heat Pumps-Energy Savers for Process Industries, Salford, UK, 7-8 Apr. (1981)
13. Oliver, T.N., Int. Symp. Industrial Application of Heat Pumps, Coventry, UK Paper C2, pp.73-88 (1982)
14. Teculescu, N., PAC-Industrie, No. 5, 25-31, (1977)
15. Perry, E.J., IEE Conf. Publ. 192, Int. Conf. Future Energy Concepts, pp.246-254, (1981)
16. Flikke, A.M., Cloud, H.A. and Hustrulid, A., Agricultural Engineering, vol. 38, 592 (1957)
17. Malkin, L.S., Proc. 3rd Int. Drying Symp. Birmingham UK, vol.1 256 (1982)
18. Nagaraja, S. and Krishna Murthy, M.V., Int. J. Heat Mass Transfer, vol. 21, 87 (1978)
19. Gutierrez, A,G., El-Meniawy, S.A.K., Watson, F.A. and Holland, F.A., Ind. Chem. Eng., vol. XXI, 3 (1979)
20. Perry, I., Chemical Engineering Handbook, Chapter 15, 4th ed., McGraw-Hill (1963)
21. Hodgett, D.L. and Lincoln, P., Electricity Council Research Centre, Report M1147 (1978)
22. Tai, K.W., Żyłła, R., Devotta, S., Diggory, P.J., Watson, F.A. and Holland, F.A., Int. J. Energy Research, vol. 6, 323 (1982)
23. Tai, K.W., Devotta, S., Watson, F.A. and Holland, F.A., Int. J. Energy Research, vol. 6, 333 (1982)

LIMITS TO THE RATE OF DRYING OF MATERIALS

R.B. Keey

Department of Chemical and Process Engineering
University of Canterbury, Christchurch, New Zealand

ABSTRACT

The literature on process drying is dominated by the concern for drying kinetics as a basis for sizing process plants. Yet not always are these kinetics the sole or principal factor in limiting drying. The extent of drying is better described in terms of dimensionless parameters such as the number of transfer units (NTU) or loading ratio. In practice, there may be situations where the extent of drying appears dependent more on the gas dynamics or the particle-size distribution in disperse systems than on the relative drying rates as determined from batch tests. Thus, whenever the material is pieceform particulate or loose, the distribution of that material becomes important. In other cases, whenever the mass-transfer coefficient is very small or very large, the drying will be equilibrium-limited. Examples include the slow aeration of grains and possibly some cases of fluid-bed drying.

1. INTRODUCTION

The conventional approach to dryer design is well stated in the Heat Exchanger Design Handbook (Schlünder, 1983). "The prediction of the size of the dryer is based on the so-called drying rate function, which correlates the drying rate with the product moisture content .... Usually the drying rate function must be determined through laboratory-scale experiments with a sample of the particular product .... Knowing the drying rate function, the remaining question is the residence time and the residence-time distribution in the large-scale dryer".

This drying rate function is called by others, the characteristic drying curve. It is a postulate, based on limited experimental evidence, that this characteristic drying curve is independent of the external drying conditions, such as temperature and air-rate. Besides these parameters, in the through-circulation of woven woollen fabrics, characteristic drying curves have been observed which are also independent of the number of transfer units, that is the extensiveness of the drying. Recent tests (Keey et al., 1985) have demonstrated that, when the wetted-surface model is appropriate to the drying of a material, then a characteristic drying curve is found. When evaporation takes place within the body itself, then a characteristic curve is found only in the limiting cases of low and high-intensity drying.

Reay (1983), in evaluating this approach to dryer design, notes that there is no difficulty when the product's dwell time in the dryer can be determined with certainty, as in the case of band dryers where neither back-mixing of material nor a significant deviation from plug flow in the air is found. With equipment such as rotary and airlift dryers, there is considerable uncertainty regarding the motion of the solids and the time they spend in the equipment. However, it is implied that continuing research may develop more certain prediction of rate coefficients and better models of product movement in disperse-systems drying so that, eventually, the conventional approach, or some variant of it, can be applied rigorously.

Nevertheless, this approach may not be entirely appropriate in all cases, assuming that the problems of evaluating the rate coefficients and the residence-time distributions have been overcome. There may be instances when non-kinetic factors become so predominant that the conventional approach is either unnecessarily pedantic or can be misleading if inappropriate rate-averaging is done.

The point can be demonstrated in a trivial way. Consider pieceform goods being dried on a conveying band. If the band were extensive enough, or the band speed slow enough, the goods would reach moisture contents close to equilibrium by the time the goods are discharged from the band. Increasing the band length will not bring about any further significant amount of moisture loss, and the mean evaporation rate per unit band area would fall. Therefore, within limits, the mean evaporation rate per unit band area will depend upon the loading of the goods on the band, that is the spacing between individual pieces. The dryer's performance is no longer kinetically controlled.

This situation is illustrated in Fig. 1; the mean evaporation rate is halved by halving the number of items on the band.

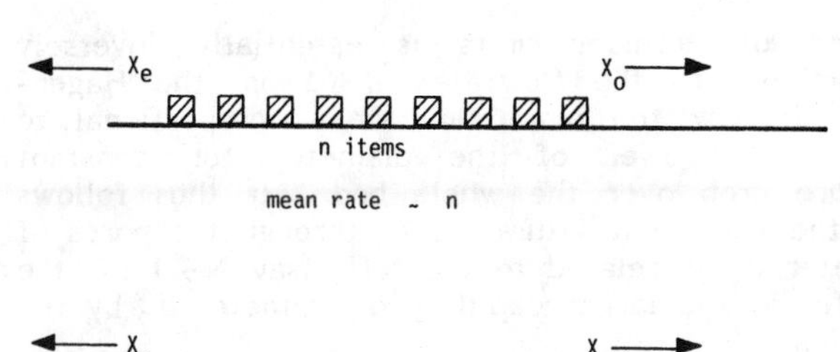

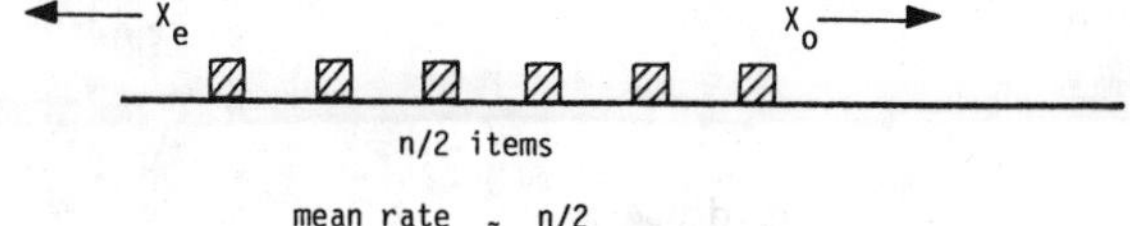

Figure 1. Drying of pieceform goods on an extensive band.

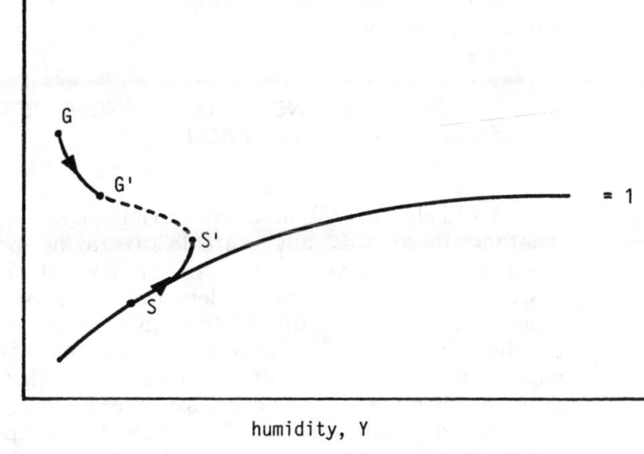

Figure 2. State condition changes in cocurrent drying.

2. EXTENSIVENESS OF DRYING

To characterise the extent or extensiveness of drying, the rate coefficients or the residence time is not the prime concern, although material constraints may place certain bounds on these variables. The removal of moisture and the interchange of enthalpy between the material and its environment represent changes in extensive or capacity-related parameters, rather than intensive factors. It is thus important to consider drying processes primarily in terms of their extensiveness, as opposed to the intensive parameters that determine rate of change. Rates of change are only predominant at conditions far from equilibrium. Particularly with material that is disperse or divided, local pockets of equilibrium can exist, and a view of drying concerned only with kinetic considerations can be misleading.

Figure 2 illustrates state paths for a co-current, convective drying process. The driving force causing the change in moisture content is $(Y_S - Y_G)$ over an infinitesmally small surface dS the moisture pick-up into the air will be dY_G . Conversely, a surface area dS will release a quantity of moisture per unit mass of dry gas, dY_G , if the driving force is $(Y_S - Y_G)$. The quantity $dY_G/(Y_S - Y_G)$ is thus the measure of the surface needed per unit of driving force. Since this ratio is also the number of transfer units over that surface, the total number will give us a measure of the total extensiveness of the drying process.

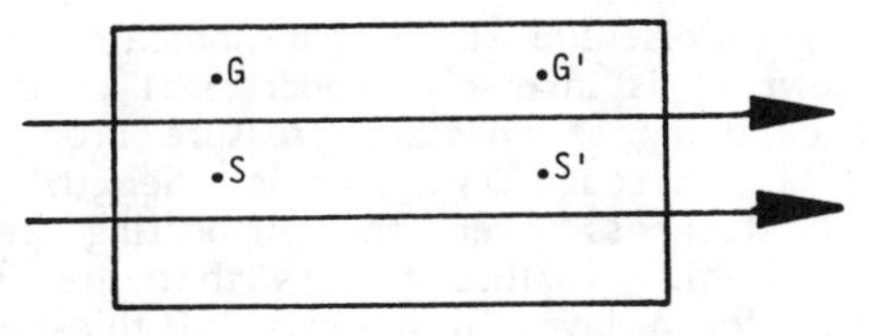

$(Y_G - Y_S)$ $(Y_{G'} - Y_{S'})$ driving forces

$$dN_t = \frac{dY_G}{(Y_G - Y_S)} \qquad dN_t = \frac{dY_{G'}}{(Y_{G'} - Y_{S'})}$$

The importance of considering the extensiveness rather than the drying kinetics in certain circumstances can be illustrated by considering two extreme examples.

Evaporation of a bidisperse spray cloud. Consider a cloud of sprayed slurry particles evaporating to dryness in a hot-air chamber. Suppose this spray cloud is bidisperse; that is, composed only of two particle sizes, one very large and one very small. The evaporation rate is inversely proportional to the square of the particle diameter. Thus the group of very small particles would quickly dry out and reach equilibrium with the surrounding air. The very large particles would be slow to dry out, and may have barely lost any moisture by the time they emerge from the dryer. In the extreme, the small-sized particles emerge at equilibrium and the large-sized particles are virtually as wet as when they were first formed at the atomizer. The lumped (mean) moisture content of the outlet material, then, does not depend upon the drying kinetics, but on the relative amounts of small and large-sized material.

Through-drying of a binary porous solid. Consider the through-circulation drying of a porous bed of solids which is composed of two zones. In one zone, the air passes through a highly open bed of solids; in the other, the air is forced through a much more densely-packed material. These zones can be conceived as being side by side. Clearly, the air will tend to channel through the more open part of the bed and the more densely-packed part will get correspondingly less air. The maldistribution of air will cause a consequential uneveness in drying. In the very porous zone considerable amounts of moisture can be picked up; in the confined zone very little evaporation will take place.

Thus the bed-averaged drying rate depends primarily on the airflow distribution between the two zones and not on the drying kinetics.

3. THROUGH-DRYING ON NONUNIFORMLY-SIZED POROUS MATERIAL

Schlünder (1977) in a pioneering paper examined whether there was any suitable averaging procedure whereby the mass-transfer efficiency of a whole system composed of various elements could be deduced from the mass-transfer "efficiency" or effectiveness of the elements. To answer the question, Schlünder examined a simple bundle of parallel capillaries with uniform diameter d_1 except for one of much larger diameter d_2, as sketched in Figure 3. Further, it is supposed that these capillaries form part of a wet, capillary-porous body in which hygroscopic effects are negligible.

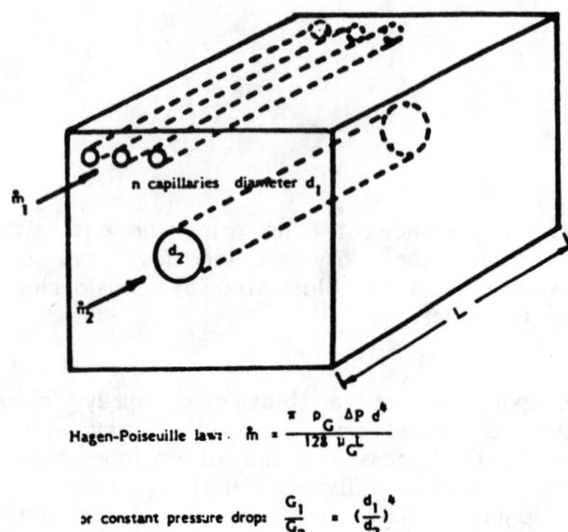

Figure 3. Schlunder's model of nonuniformly-porous body.

Under these conditions, the change in the bulk-air humidity from the inlet Y_{in} to that at the outlet Y_{out}, under constant wet-bulb conditions, is given by adiabatic-saturation expression,

$$\frac{(Y_{out} - Y_W)}{(Y_{in} - Y_W)} = \exp(-\beta\phi A/\dot{m}S) = \exp(-NTU),$$

in which β is the mass-transfer coefficient and ϕ is the humidity-potential coefficient. The area A is the total exposed area of the drying body, and S is the cross-sectional area open to the airstream. The left-hand side of the expression represents the fractional saturation of the air. If the through-circulation is gentle and the capillaries relatively extensive, then the flow through the pores may be assumed to be fully developed and laminar. Under these circumstances the mass-transfer coefficient is almost independent of the flowrate, and thus the number of transfer units is essentially inversely proportional to the flowrate $\dot{m}$. From the Hagen-Poiseuille law the mass flowrate is proportional to the fourth power of the diameter for constant pressure drop over the whole bed. It thus follows that the fractional saturation f_1 through the pores of diameter d_1 is related to the NTU (say N_2) for the transfer in the larger capillary of diameter d_2 by the equation:

$$f_1 = \exp\{-N_2 (\frac{d_2}{d_1})^4\}$$

Suppose $N_2 = 1$ and $d_2/d_1 = 2$, then $f_1 \sim 10^{-7}$ - virtually no transfer is taking place in the smaller pores. The transfer is governed solely by the transfer possible in the larger capillary. Indeed, with a binary pore-size distribution, with diameters that differ by about 2, it is possible to get saturation of the airstream in the larger pores whilst essentially no moisture uptake occurs in the smaller pores. The drying is no longer kinetically-limited, but controlled by the airflow distribution between the larger and smaller pores. This effect is likely to be highly significant in the drying of loose materials when irregularities in packing or openness can easily occur. The extreme sensitivity of the fractional saturation to the pore-size ratio is illustrated in Figure 4.

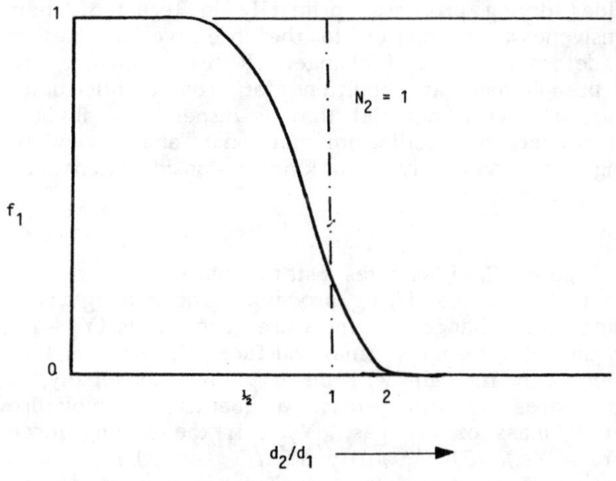

Figure 4. Fractional saturation in pores compared with reference capillary ($N_2 = 1$, $f_1 = 0.368$) as a function of pore-size ratio.

The Hagen-Poiseuille law also indicates that the mass flowrate is inversely proportional to the length of the capillary at constant pressure drop. In the context of through-drying, this "length" is the material thickness over the supporting perforated band or drum. It thus follows that the fractional saturation for a layer of material of thickness b_1 is related to the NTU (say N_2) for transfer in a zone with a different layer thickness b_2 by the relationship,

$$f_1 = \exp\{-N_2(\frac{\dot{m}_2}{\dot{m}_1})\} = \exp\{-N_2(\frac{b_1}{b_2})\} .$$

A fivefold variation in thickness is not uncommon in the drying of loose material such as scoured fleece wool. For $b_1/b_2 = 1/5$ and $N_2 = 1$, $f_1 = 0.819$. This compares with the fractional saturation for an NTU of 1, which is 0.368. Thus, over the thin-layered material wherein the fractional saturation of the airstream is greater, one would expect to see, as the material moves through the dryer, increasingly larger absolute differences in moisture content normal to the band (or drum) in the airstream direction, compared with the differences observable in the thick-pile portions. Since the material is also thinner when the absolute moisture-content differences are larger, then the moisture-content gradients are also larger too. The gradient ratio is given by:

$$\frac{(\Delta X/\Delta b)_1}{(\Delta X/\Delta b)_2} = \frac{f_1/b_1}{f_2/b_2} .$$

In the example discussed, the ratio of gradients, $(\Delta X/\Delta b)_1 : (\Delta X/\Delta b)_2$, is over elevenfold.

Effects of this magnitude ought to be observable. Before the experimental evidence is considered, it is useful to consider these ideas in the context of the loading diagram.

Loading diagram. For a given set of process conditions, a particular dryer will have a fixed capacity for abstracting moisture from the material to be dried. Therefore, one can define the moisture-transfer effectiveness of a dryer in terms of the degree to which the drying medium is "loaded" with moisture compared with the maximum possible loading. This quantity is called the loading ratio (Keey, 1978b).

The following expressions may be derived for this loading ratio ε:

In co-current movement of solids and gas,

$$\varepsilon_{co} = 1 - \exp[-P]/\{1 - (1 - \nu)\exp[-P]\} ;$$

And for countercurrent movement,

$$\varepsilon_{ctr} = 1 - \exp[-P].$$

In both these expressions,

$$P = (N_1 - (1 - \nu)N_2/\ln\nu) ;$$

$$\nu = \dot{m}_v/\dot{m}_v|_{max} \text{ (the relative drying rate);}$$

$$N_1 = \text{NTU in unhindered-drying zone;}$$

and

$$N_2 = \text{NTU in hindered-drying zone.}$$

For moisture loss wholly within the first period of unhindered drying, $P = N_1$ and $\varepsilon_{co} = \varepsilon_{ctr} = f$.

These expressions can be conveniently displayed on a loading diagram, in which the loading ratio is plotted as ordinate, the NTU as abscissa and ν_0, the relative drying rate at the solids-discharge end of the dryer, as parameter.

The loading diagram also has one interesting feature. The overall rate of moisture removal may be related to some "mean" driving force ΔY_m by

$$\dot{m}S(Y_{out} - Y_{in}) = \beta\,\phi\,A\,\Delta Y_m .$$

The number of transfer units (NTU) is given by

$$NTU = \frac{\beta\,\phi A}{\dot{m}\,S} = \frac{\beta\,\phi\,aZ}{\dot{m}} .$$

Further, for adiabatic drying, the loading ratio becomes

$$\varepsilon = \frac{Y_{out} - Y_{in}}{Y^*_{out} - Y_{in}} = \frac{Y_{out} - Y_{in}}{Y_W - Y_{in}} .$$

It follows that

$$\varepsilon = \frac{\Delta Y_m}{(Y_W - Y_{in})} . NTU .$$

In particular, when the NTU is 1, the loading ratio takes the meaning of ratio of the mean driving force to the inlet driving force. In the so-called "constant-rate" period of drying, the mean driving force is independent of the relative directions of solids and air movement, whether counter- or co-current. Should the solids be in the falling-rate region of drying and the relative drying rate ν less than 1, the value of the mean driving force is not independent of the direction of solids movement with respect to the air.

In through-circulation drying, it is the co-current profiles that describe conditions in the airstream direction. Suppose a material is composed of two pore sizes, with a diameter ratio $d_1/d_2 = 0.8$. Further, let $N_1 = 1$. Then the loading ratios, when the material is wholly within the constant-rate period, become,

$$\varepsilon_2 = 1 - \exp(-1) = 0.632$$

$$\text{and } \varepsilon_1 = 1 - \exp(-0.8^4) = 0.336$$

respectively. Suppose, on the other hand, the material is wholly within the falling-rate period with $\nu = ½$. The loading ratios now become

$$\varepsilon_2 = 1 - \frac{½\exp[+½/\ln ½]}{1 - ½\exp[+½/\ln½]} = 0.678$$

$$\text{and } \varepsilon_1 = 1 - \frac{½\exp[+½.(0.8)^4/\ln½]}{1 - ½\exp[+½.(0.8)^4/\ln½]} = 0.407 .$$

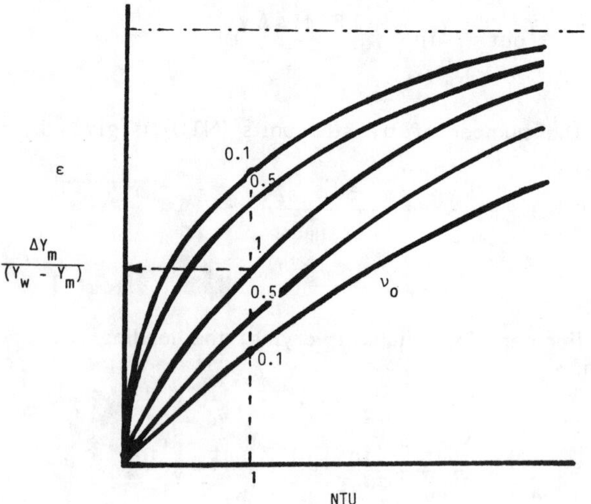

Ordinate value ε_1 is the loading ratio and also the humidity-potential ratio when NTU = 1

Figure 5. Loading diagram.

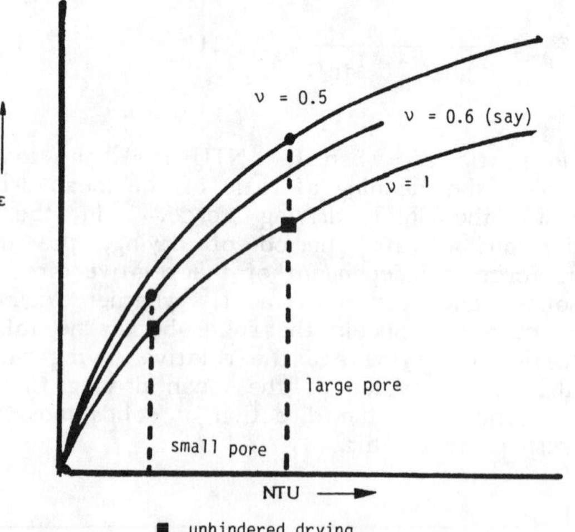

■ unhindered drying

● hindered drying (uniform moisture content at inlet-air face).

Figure 6. Loading conditions for nonuniformly-porous material

The effect of the difference in capillary sizes, as measured by the loading of the airstream, becomes somewhat less as the material enters the falling-rate period. As the material dries out, so the air can pick up less moisture and the loading limit is more closely approached at a given number of transfer units. In the example taken, $\varepsilon_1 / \varepsilon_2$ is 1.88 for drying within the first period and 1.66 for drying in the falling-rate period. The comparison taken, for ease of calculation, is for the unusual circumstance of uniform moisture contents at the face where the air **leaves** the material. In the case where the moisture contents are uniform at the air-inlet face, the moisture uptake in the smaller-sized capillary would be somewhat larger due to the somewhat greater local moisture contents, with a corresponding small fall in the loading ratio. The overall effect would be to enhance the moisture-loss difference between the differently-sized pores.

Some experimental evidence will now be reviewed.

4. EXPERIMENTS

4.1 Through-drying of Textile Webs

Gummel (1977) in his doctoral work at the University of Karlsruhe, examined the drying behaviour of textile and paper webs when through-circulated. The materials included polyester, wool and wool-acrylic fabrics, with the number of threads ranging between 1.28 and 2.80 per mm, yielding rectangular holes of sides between 49 and 449 mm. Air at velocities in the range of 0.06 to 1.5 $m\ s^{-1}$ were used, with the inlet-air temperatures of 20°C, 40°C and 70°C for the textiles and 20°C, 60°C and 90°C for the paper webs. The webs were presoaked in water, and superficial moisture was driven off (as indicated by the air pressure at the upstream face of the through-circulated web). Tests began when the mechanical entrainment was judged to have finished. The drying was monitored by measuring the outlet-air humidity by means of an infra-red hygrometer of rapid response. From the variation of humidity with time, the instantaneous mean moisture content of the fabric could be calculated, and thence the mass-transfer coefficient.

For moisture loss in the first period of drying, the mass-transfer coefficients, expressed in terms of the Sherwood numbers $\beta\ d/D$, were compared with values estimated by assuming laminar flow in a capillary. In all cases, the mass-transfer coefficients (or Sherwood numbers) fell well below the "theoretical" values for laminar flow. In other instances, when air at 90°C was passed through a paper web, the experimental value was an order of magnitude less.

A number of possibilities were examined. Molecular diffusion was ruled out because that too gave an overestimate. The assumption of perfect mixing in the gas phase yielded estimates of the right order, but physically it was an unrealistic picture. On the other hand, the results could be fitted to an arbitrary two-pore model of the webs. It was assumed that the fabric was composed of two sets of pores spaces, the smaller spaces being 10 times more numerous than the larger ones, with a pore-diameter ratio of either 1.5 or 2. It was also assumed that the larger pores would empty first. For a given pore-number and pore-size ratio, it was possible to draw up a family of curves for the Sherwood number as a function of the Peclet number, for various degress of drying out of the larger pores. The experimental data scatter about these curves. While the agreement is not conclusive evidence, and choice of pore-distribution parameters speculative, the explanation is nonetheless plausible.

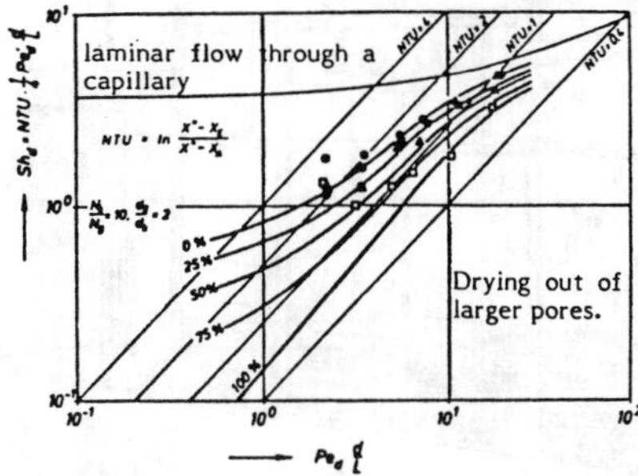

(a) Polyester fabric, 324 mm dia. threads, 2.68 x 1.97 per mm; d_1/d_2 = 2; n_1/n_2 = 1/10.

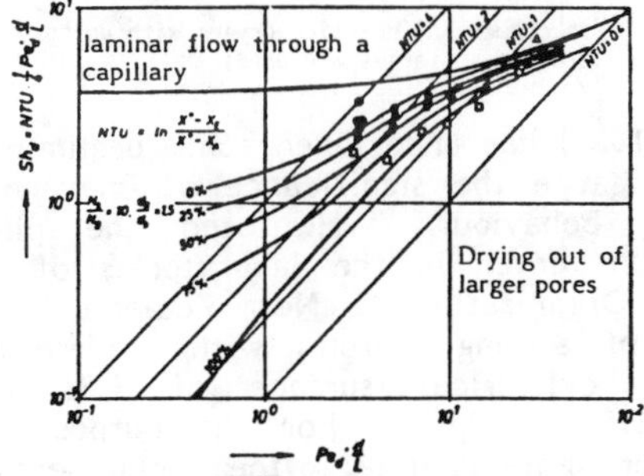

(b) Wool-acrylic fabric, 323 x 449 mm dia. threads, 2.11 x 1.28 per mm; d_1/d_2 = 1.5; n_1/n_2 = 1/10.

Figure 7. Through-circulation drying of textile webs. (Gummel, 1977).

4.2 Fluid-bed Drying of Capillary-porous Solids

In another doctoral study, Zabeschek (1977) looked at the fluid-bed drying of free-flowing, capillary-porous solids. The experimental technique was essentially that for the experiments with webs: the mean drying rate, and thus the instantaneous moisture content, was estimated from measurement of the outlet-air humidity as determined by an infra-red hygrometer. The inlet-air temperatures, however, covered a wider range: from about 35°C to 200°C. The materials tested included aluminium silicate , type-13X molecular sieve and clay particles of diameters between 0.92 and 4.12 mm. The standard deviation of the particle diameters is given as ranging between 5.1 and 10 percent.

The experimental data and those of Mossberger (1964) were compared with values to be expected from transfer from an isolated sphere. The relationships used were:

$$Sh_{tot} = [Sh^2_{lam} + Sh^2_{turb}]^{\frac{1}{2}} [1 + 1.5(1 - \Psi)]$$

$$Sh_{lam} = 0.664 \, Re^{\frac{1}{2}} Sc^{1/3}$$

$$Sh_{turb} = \frac{0.037 Re^{0.8} Sc}{[1 + 2.44 (Sc^{2/3} - 1) Re^{0.1}]}$$

in which Sh_{tot}, Sh_{lam} and Sh_{turb} are respectively the Sherwood numbers for the total particle-gas transfer, that due to laminar causes and that due to turbulence effects. There is also a porosity "correction function" to allow for differences in transport behaviour as the bed expands.

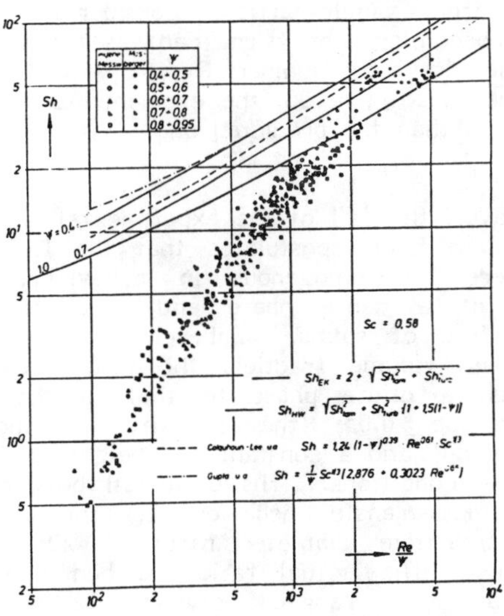

Figure 8. Mass transfer coefficients in a fluidised bed.

Data of Mossberger (1964) and Zabeschek (1977)

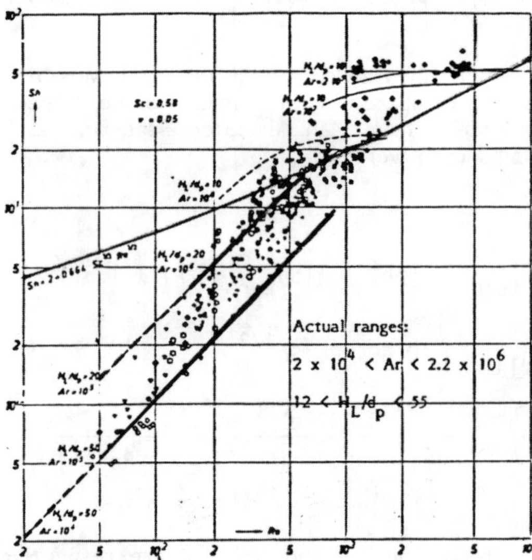

Figure 9. Mass transfer in a fluid bed compared with that for an assumed bypass volume fraction of 0.05 and $Ar = 10^5$.

The experimental data points for nearly 200 separate runs generally fell below the values predicted from single-particle considerations, the deviations becoming more significant as the value of the modified Reynolds number, Re/Ψ, becomes less. At a value of $Re/\psi \simeq 100$, the experimental value is an order less than the predicted one.

However, the fall of the experimental data may be explicable if one postulates that the fluid beds tested were non-homogenous in behaviour, being composed of a dense phase and a dilute phase relatively free of solids. Zabeschek took the so-called bypass volume fraction, the volume ratio of air through a dilute phase to that in the dense phase, to be 0.05; the porosity at incipient fluidisation 0.45 and a constant bed height of 50 mm under these conditions. His data fell between the estimated mass-transfer behaviour for the limits of bed height/particle diameter ratio of 20 and 50 respectively. The actual ratios of bed height to particle diameter ranged between 12 and 55.

Again the proof is not conclusive, (it is not explicit, for instance whether two-phase behaviour was actually observed); but the explanation is plausible. The fall in mass-transfer coefficients could not be attributable to feasible macromixing models such as ideal mixing of the solids and plug flow with axial dispersion of the air.

In this context, it is interesting to compare this kind of behaviour with that found in uniform undistended beds of particles when channelling of the continuous phase is unlikely. Under these conditions Wakao and Kaguei (1982) find that the Sherwood number for the bed is closely given by the correlation for transfer is an isolated particle, provided due account is taken of dispersion in the continuous phase.

4.3 Suction-drum Drying of Loose Wool

Greasy loose wool is scoured in a series of wash-bowls each followed by squeeze rollers. The scoured wool is then opened out and conveyed directly to a suction-drum dryer. It is extremely difficult in practice to feed the dryer uniformly, and variations in fleece loading and openness occur in the dryer.

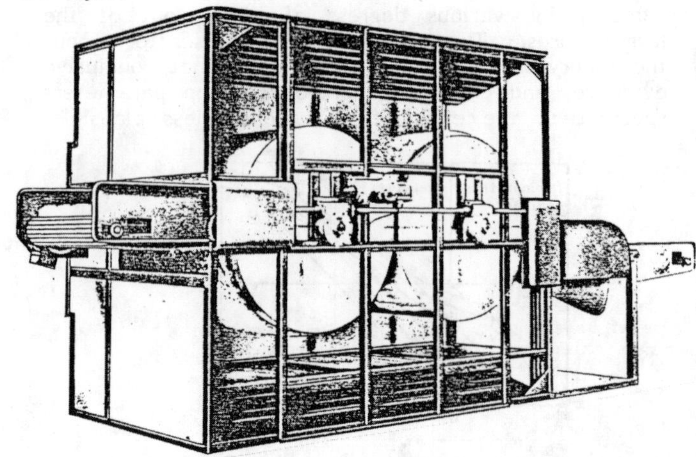

Figure 10. A cut-away view of a Suction Drum Dryer.

(In commercial practice dryers with up to 10 drums in series are used).

Marshall (1983) has undertaken some preliminary experiments to gauge the significance of unevenness on the drying behaviour. He used the pilot-plant, single-drum dryer in the laboratories of the Wool Research Organisation of New Zealand. The dryer consists of a single drum, with perforations over half the cylindrical surface of 1.35 mm diameter and 1.5 m width. For the purposes of obtaining experimental data, the commercial-sized unit gave difficulties with feeding, and a blanking strip was sewn around the drum leaving an active area of 600 mm width. The wool used was presoaked in a tub of water, allowed to drain and then spun in a hydroextractor to reach a fairly uniform moisture content of 52 percent regain on a wet basis. All experiments were performed with a scoured dagwool blend, DB, of average bulkiness (bulk = 25.2 arbitrary units) and a fibre diameter of 33.1 µm. Each sample of wool was soaked, spun and dried twice, and then discarded, before significant felting was caused by repeated handling of the same wool. Each run was replicated three or four times, and the results averaged. The data obtained are summarized in the following table.

TABLE 1

Suction-drum drying of scoured dagwool blend.

Inlet-air humidity: 0.46 kg kg^{-1}

Inlet-air temperature: 70°C

Through-circulation air-rate: 1.5 ~ 3.0 m s^{-1}.

(Marshall, 1983).

Runs	Condition	Bed depth/ mm	Wool feed/ kg s^{-1}
A	full cover	25	0.022
B	full cover	30	0.018
C	full cover	55	0.022
D	full cover	55	0.018
E	half cover	30	0.018
F	half cover	30	0.0087
G	half cover	55	0.017

Runs	Residence time/s	Outlet moisture/%	Evaporation/ g m^{-2} s^{-1}
A	62	12.4	7.81
B	90	8.5	7.33
C	115	9.4	8.53
D	140	7.3	7.33
E	48	25.0	4.22
F	90	19.3	2.69
G	65	24.9	4.22

Runs A to C show the effect of doubling the bed depth at essentially constant wool-feed rate. The air-rate is effectively halved (because of the increased back-pressure), but the increased residence time compensates, so that the moisture loss changes little. In practice, feed-depth irregularities may also be matched by local feed-rate variations, and such compensation may not be observed. More significant, in the context of this paper , are the runs (E to G) in which the wool was deliberately fed over only half the drum area. Specifically, compare runs B and F, in which the residence time was the same, the bed depth the same, and the wool-feed rate per unit width the same. However, in run F, much of the air would bypass the bed and issue through the open areas of the drum; that air passing through the bed would emerge more saturated with moisture and its moisture-exchange capacity less. The mean outlet moisture content of the wool in run F is much higher than that in run B, demonstrating this loss in capacity.

Ma (1984) has examined the moisture-content profiles over a suction-drum dryer for the simplifed case when the wool mat is presented to the dryer in two streams, each of uniform but different thickness. Although the difference in the thickness-averaged local moisture contents are small between those for a wholly-uniform mat and those for a mat with two thicknesses, significant local vairations in moisture content between the thin and thick zone can be found. For a 3:1 thickness variation, at the end of the first drum from the wool-feed end, the ratio of maximum to minimum local moisture content is double that for uniform feeding. The overall unevenness, measured as $(X_{max}-X_{min})/X_{mean}$, can reach a value of 0.9 at the end of the third drum for a 3:1 thickness ratio, and thus the drum-averaged drying rates can give a misleading indication of the extent of drying in terms of the existance of "wet spots".

5. EQUILIBRIUM LIMITS

Ultimately any thermal drying process is equilibrium-limited, when the material being dried has reached a hygrothermal state only slightly different from that of its environment. This usually is viewed as the end state of drying, but it may not be. Let us look at some examples.

5.1 Slow Aeration of Fixed Beds

If the dryer is extensive, or the air-rate very small, the number of transfer units (NTU) $\beta\phi aZ/m$ becomes large. The loading ratio,

$$\varepsilon = 1 - \nu \exp[-P] / \{1 - (1 - \nu) \exp[-P]\},$$

approaches 1 as P, being dependent on the NTU, is also very large. The air leaves the dryer almost saturated, and a drying zone slowly progresses downstream in the dryer. At each place in the drying zone, local equilibrium can be assumed.

Under these conditions, the separate potentials for heat and mass transfer can be combined into a new potential which is a linear combination of humidity and temperature when the state-change range is small:

$$F_i = T + \alpha_i Y \quad .$$

The equations for mass and energy change can be consolidated into the single expression (Nordon and Banks, 1973),

$$\frac{\partial F_i}{\partial \tau} + \left[\frac{u_G}{1 + \mu\gamma_i}\right] \frac{\partial F_i}{\partial z} = 0,$$

with $\mu = \rho_S (1 - \Psi) / \rho_G \Psi$

and $\gamma_i = \left[\frac{\partial X}{\partial Y}\right]_{F_j} ; i, j = 1, 2; \quad i \neq j.$

This is a kinematic wave equation. Two waves F_i propagate through the bed with a velocity $u_G / [1 + \mu\Psi_i]$, where u_G is the cup-mixing velocity of the air. The heat and mass-transfer process, although coupled, essentially manifest themselves in two distinct waves. For a cereal grain, $\mu \simeq 1000$, $\gamma_1 \simeq 0.5$ and $\gamma_2 \simeq 20$. Thus the first (F_1) wave comes through at a velocity of about $u_G/500$ and second (F_2) wave at about $u_G/20\,000$. The state path changes accompanying this more sluggish front virtually follow an adiabatic saturation path. This kind of behaviour has been observed in the aeration cooling of grain in northern Australia by Sutherland and colleagues (1983). Air velocities are so low that the fast wave travels through the grain bed during aeration. The grain moisture content (and thus air relative humidity) appears almost constant during the passage of the fast wave, while the wet-bulb temperature remains almost constant during the slow wave as the grains cool and loose moisture.

5.2 Fluid-bed drying

Because of very high gas-to-particle heat-transfer rates in a fluid bed, particularly in the zone close to the grid, it might be expected that moisture contents approaching equilibrium with the fluidising gas might be obtained in fluid-bed drying. If this view is correct, increasing the bed height over some value corresponding to the drying zone should have no influence on the course of drying.

Reay and Allen (1982) describe some experiments with a pilot-plant fluid-bed dryer. The drying chamber was 300 mm in diameter and incorporated a sintered-mesh distributor with an average pore size of 50 μm. Three materials were used in the experiments: one was an ion-exchange resin with a minimum fluidisation velocity of 0.15 m s^{-1}, with most of the material lying in a size band from 200 to 1000 μm; the second was crushed iron ore with a minimum fluidisation velocity of 0.33 m s^{-1}, with somewhat more fines (down to 100 μm); the third was wheat with a fluidisation velocity of 0.9 m s^{-1} and with axial dimensions typically 6 x 3 mm.

Bed weight was chosen as the index of bed height since the latter is difficult to estimate with precision. Should an equilbrium-limited zone appear close to the grid, then the effect of increasing bed depth or bed weight is to reduce the frequency at which particles enter this zone. The time taken to reach a given moisture content should thus be directly proportional to the bed weight. Indeed, Reay and Allen found this to be the case for two of the materials examined, ion-exchange beads and iron ore particles. For the ion-exchange material with wet charge weights of 1.5, 2.0 and 2.5 kg, corresponding very roughly to bed heights of 60, 80 and 100 mm respectively, the drying time was proportional to the bed height. Shallower beds needed a relatively longer time to reach a given moisture content. These tests imply that the equilibrium-limited zone extended for about 40 to 60 mm above the grid. Similar results were found in the fluid-bed drying of the iron ore. Wheat grains, however, behaved differently. With wheat, the high internal resistance to moisture movement reduced the drying rates, and no equilibrium-limited zone was discerned over the range of bed heights explored.

Although the tests with the fast-drying materials suggested strongly that there was equilbrium between the air leaving the bed and the particles therein, the actual humidity was found to be very much less than the equilibrium value corresponding to the solids moisture content. The tests were undertaken with air velocities well above the minimum needed to induce fluidisation. (With ion-exchange beds, the tests to determine the effect of bed weight were done uniformly with an air velocity 3.3 times the incipient fluidisation value). Under these circumstances, it is possible that some air is bypassing the bed, in the way that Zabeschek suggested had taken place in his experiments. Air in the dilute phase would not reach equilibrium, whereas the smaller volume of air in the dense phase would. The mean humidity of the leaving air, when mixed together, would thus be much less than the equilibrium value based on the average moisture content.

There is also the possibility that the particles reach equilibrium at the surface very quickly, while their interiors slowly dry out. This is a point picked up by Ashworth (1979) in another context.

5.3 Kiln-drying of Timber Boards

Ashworth postulated that, for the drying of hygroscopic microporous solids such as wood, the evaporation of moisture takes place exclusively at the surface under normal convective conditions. The rate of drying is assumed to be small enough so that the surface layers remain in hygrothermal equilibrium with the surrounding airstream. From the known moisture-sorption behaviour, Ashworth was able to draw up a reduced driving-force function which varied with surface moisture content and, to a lesser degree, with wet-bulb depression. The curves, virtually characteristic drying curves, take the same form as those obtained experimentally on drying pieces of timber boards and were also similar to other curves drawn from other experimental data (Ogura, 1951). Consistent with Ashworth's postulate is the finding that the average moisture content of timber can be monitored by following the surface temperature of wood during kiln-seasoning (Kotak et al., 1969).

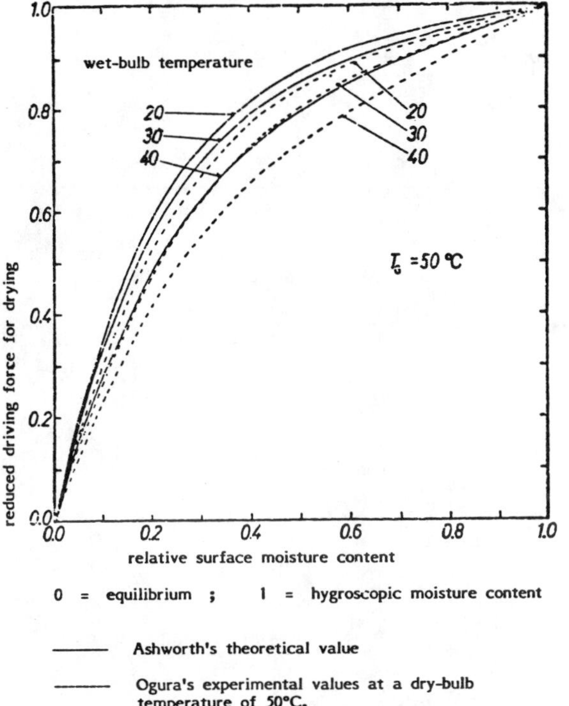

Figure 11: Reduced driving-force function for the drying of pinewood.

6. CONCLUSION

Tentatively, certain conclusions may be drawn from this analysis. Drying may be regarded as an equilibrium-limited, mass-transfer operation. The fundamental quantity to characterise a drying process is thus is extensiveness or the extent to which equilibrium is approached rather than its rate-determined drying intensity. In particular, non-kinetic limits will have significant influence on the course of drying whenever the moist material is unevenly distributed in the drying medium. Should the material be pieceform, particulate or loose, such unevenness is likely, and the course of drying thereby influenced. In these cases, considerable variations in extensiveness may occur, and the dryer's moisture-exchange performance becomes more dependent upon the way the material is spread in the dryer than on the drying kinetics themselves. In other cases, where the mass-transfer coefficient is either very small or very large, equilibrium limits appear directly.

NOMENCLATURE

a	interfacial area per unit volume
A	surface area
b	layer thickness
d	diameter or linear dimension
d_p	particle diameter
D	diffusion coefficient
f	fractional saturation
F	$[T + \alpha Y]$
H_L	bed height
$\dot{m}$	specific mass gas rate
$\dot{m}_v$	evaporation flux
N	number of transfer units (NTU)
P	$[N_1 - (1-\nu) N_2 \ln\nu]$
S	cross-sectional area
T	temperature
u_G	gas velocity (cup-mixing mean)
X	moisture content
Y	humidity
Y_G	bulk-gas humidity
Y_S	surface humidity
Y_W	wet-bulb humidity
z	distance
Z	total distance
α	coefficient
β	mass-transfer coefficient
γ	$[\partial X / \partial Y]_F$
ε	loading ratio
μ	$[\rho_s (1-\Psi)/\rho_G \Psi]$
ν	relative drying rate, $[\dot{m}_v / \dot{m}_{v\|max}]$
ρ_G	gas density
ρ_S	solids density
τ	time
ϕ	humidity-potential coefficient
Ψ	porosity

REFERENCES

Ashworth, J.C. (1977), "The Mathematical Simulation of Batch-Drying of Softwood Timber", Ph.D. thesis, University of Canterbury.
Gummel, P. (1977), "Through-circulation Drying - Experimental Determination and Analysis of the Drying Rate and Pressure Drop of Air Through-circulated Textiles and Paper", (in German), Dissert. University of Karlsruhe.
Gummel, P. and Schlünder, E.U. (1980), in Mujumdar, A.S., "Developments in Drying", **1**, pp 357-366, Hemisphere, Washington.
Keey, R.B. (1978a), "Introduction to Industrial Drying Operations", pp 154-162, Pergamon Press, Oxford.
Keey, R.B. (1978b), ibid, pp 245-255.
Keey, R.B., Langrish, T.A.G., and Reay, D. (1985), "The Application of the Characteristic Drying Curve to the Drying of Porous Particulate Material", Proc. 3 Australasian Conf. Heat Mass Transfer, Melbourne, Vic.
Ma, K. (1984), unpublished work, University of Canterbury.
Marshall, A.D. (1983), "Investigations Into Loose Wool Drying", WRONZ Internal Report, Christchurch.

Mossberger, E. (1964), "Concerning the Heat and Mass Transfer between Particles and Air in Fluid Beds as well as their Expansion Behaviour", (in German) Dissert. TH Darmstadt.

Nordon, P. and Banks, P.J. (1973), "Interacting Heat and Mass Transfer - An Australian View", Proc. 1 Australasian Conf. Heat Mass Transfer, pp R41-56, Monash, Vic.

Ogura, T. (1951), "Studies on the Mechanism of drying of wood. 3. On the evaporation rate of moisture in wood", Bull. Govt Forest Expt Station, No.51, Tokyo.

Reay, D. (1983), in "Heat Exchanger Design Handbook", Sec. 3.13.7, VDI-Verlag/Hemisphere, Washington.

Reay, D. and Allen, R.W.K. (1982), in Ashworth, J.C., Proc. 3 Int. Drying Symposium, Vol. 2, pp 130-140, Birmingham.

Schlünder, E.U. (1977), Chem. Eng. Sci., **32**, pp 845-851.

Schlünder, E.U. (1983), in "Heat Exchanger Design Handbook", Sec. 3.13.1, ibid.

Sutherland, J.W., Banks, P.J. and Elder, W.B. (1983), "Interaction between Successive Temperature or Moisture Fronts during Aeration of Deep Grain Beds", J. Agric. Res. **28**, pp 1-19.

Wakao, N. and Kaguei, S. (1982), "Heat and Mass Transfer in Packed Beds", Gordon and Breach, New York, p 158.

Editors' Note :

This paper is the revised text of a Keynote Lecture presented by the author at the Third International Drying Symposium, Kyoto, Japan.

CONTRIBUTORS INDEX

SUBJECT INDEX